T0221439

CREATIVE SYSTEMS IN STRUCTURAL AND CONSTRUCTION ENGINEERING

PROCEEDINGS OF THE FIRST INTERNATIONAL STRUCTURAL ENGINEERING AND CONSTRUCTION CONFERENCE/HONOLULU/HAWAII/24-27 JANUARY 2001

Creative Systems in Structural and Construction Engineering

Edited by

Amarjit Singh
University of Hawaii at Manoa, Honolulu, Hawaii, USA

A.A.BALKEMA / ROTTERDAM / BROOKFIELD / 2001

The texts of the various papers in this volume were set individually by typists under the supervision of each of the authors concerned.

Authorization to photocopy items for internal or personal use, or the internal or personal use of specific clients, is granted by A.A. Balkema, Rotterdam, provided that the base fee of US$ 1.50 per copy, plus US$ 0.10 per page is paid directly to Copyright Clearance Center, 222 Rosewood Drive, Danvers, MA 01923, USA. For those organizations that have been granted a photocopy license by CCC, a separate system of payment has been arranged. The fee code for users of the Transactional Reporting Service is: 90 5809 161 9/01 US$ 1.50 + US$ 0.10.

Published by
A.A. Balkema, P.O. Box 1675, 3000 BR Rotterdam, Netherlands
Fax: +31.10.413.5947; E-mail: balkema@balkema.nl; Internet site: www.balkema.nl
A.A. Balkema Publishers, 2252 Ridge Road, Brookfield, VT 05036-9704, USA
Fax: 802.276.3837; E-mail: info@ashgate.com

ISBN 90 5809 161 9
© 2001 A.A. Balkema, Rotterdam
Printed in the Netherlands

Creative Systems in Structural and Construction Engineering, Singh (ed.) © 2001 Balkema, Rotterdam, ISBN 90 5809 161 9

Table of contents

4 *Constructability and design-construction*

5 *Construction methods*

Creative Systems in Structural and Construction Engineering, Singh (ed.) © 2001 Balkema, Rotterdam, ISBN 90 5809 161 9

Preface

The importance and value of constructed structures to the development of nations and civilization must be specially underscored and emphasized. The super technology of the 21st century available today is built on the efforts, structures, and infrastructure put in place by our heroic construction and structural engineers.

The cost of civil engineering construction projects is so large in some cases that small and developing nations have construction costs amounting to 50% of their annual national budget. Clearly, the attention given to construction projects by the people of the world and elected representatives needs to be of high priority.

Our construction and structural engineers are in a unique position to make immense potential contributions to civilization, in general. While their great landmark contributions are evident for all to see, it must be said that there is considerable wastage of resources, information, and material on construction projects; structural designs do not consciously and adequately design for cost or constructability; drawings have errors, while value engineering is used quite infrequently or incompletely at design stage; and excellence is a perpetual concern. There is a lot of house cleaning to do for the structural and construction engineers themselves.

While we pursue creative and innovative techniques in materials, structural analysis, and construction management, we must keep in perspective the broader picture linking structural and construction engineering. Often structural and construction engineers have different ways of approaching a problem. This international conference brings together researchers from 25 countries with an aim for them to understand each other and each others' research work. It is the purpose of ISEC to provide an international forum for the discussion of topics important to developing new knowledge in construction and structural engineering.

We have a wide variety of topics for this set of proceedings. Approximately half the total papers are in the different specialty areas of structural engineering, while the other half of the papers are in the different specialty areas of construction management, housing and architecture, materials engineering, education and ethics, and soils and pavements.

The editor expresses gratitude to all authors, members of the scientific and review committee, local organizing committee, and conference attendees in bringing about a successful conference.

Dr Amarjit Singh
Editor
University of Hawaii at Manoa

Creative Systems in Structural and Construction Engineering, Singh (ed.) © 2001 Balkema, Rotterdam, ISBN 90 5809 161 9

Acknowledgements

The editor gratefully acknowledges the sponsorship, promotion, and endorsement of the following organizations:

Co-sponsors:
– American Concrete Institute
– International Council for Research and Innovation in Building and Construction
– CIB Working Commission 99 for Construction Safety and Health

Promoters:
– The Association for the Advancement of Cost Engineering International

Endorsers:
– University of Hawaii at Manoa
– Hawaii Council of Engineering Societies
– Hawaii Department of Transportation (Highways)

The editor extends his sincere appreciation and acknowledgement to Professor Emeritus Harold S. Hamada, University of Hawaii at Manoa, for helping with the review of multiple abstracts submitted in the area of structural engineering, and for his subsequent help in the allocation of chapters belonging to structural engineering. Thanks are due to Ms Erinn Orimoto for data management of all abstracts and papers throughout the review process, and to Mr Ross Anderson for review of paper formats.

Creative Systems in Structural and Construction Engineering, Singh (ed.) © 2001 Balkema, Rotterdam, ISBN 90 5809 161 9

Scientific and Technical Committee

Amarjit Singh, Conference Chair, University of Hawaii at Manoa, Honolulu, USA

John Abel, Cornell University, USA
Hojjat Adeli, Ohio State University, USA
Franco Bontempi, University of Rome 'La Sapienza', Italy
Mark Bradford, University of New South Wales, Australia
Albert Chan, Hong Kong Polytechnic University, China
Richard Coble, University of Florida, USA
Zbigniew Cywinski, Technical University of Gdansk, Poland
Dejan Dinevski, University of Maribor, Slovenia
Shizhao Ding, Tongji University, China
Brian Eksteen, University of Port Elizabeth, South Africa
Roger Flanagan, University of Reading, United Kingdom
Arie Gottfried, Polytechnic of Milan, Italy
Daniel Halpin, Purdue University, USA
Harold Hamada, University of Hawaii, USA
Takashi Hara, Tokuyama University of Technology, Japan
Gary Holt, University of Wolverhampton, United Kingdom
Rong-yau Huang, National Central University, Taiwan
Nabil Kartam, Kuwait University, Kuwait
Kincho Law, Stanford University, USA
In-Won Lee, Korea Advanced Institute of Science and Technology, Korea
Ahmed Morgan, Nihon Hitek, Inc., Japan
Anura Nanayakkara, University of Morutawa, Sri Lanka
Indubhushan Patnaikuni, RMIT University, Australia
M.Azadur Rahman, Bangladesh University of Engineering and Technology, Bangladesh
Swapan Saha, University of Western Sydney, Australia
Tamim Samman, King Abdulaziz University, Saudi Arabia
Asko Sarja, VTT, Finland
Takahiro Tamura, Nagaoka University of Technology, Japan
Ali Touran, Northeastern University, USA
Koshy Varghese, IIT Madras, India

1 Keynote papers

Creative Systems in Structural and Construction Engineering, Singh (ed.) © 2001 Balkema, Rotterdam, ISBN 90 5809 161 9

Better constructed facilities through improved education

R. N. White
School of Civil and Environmental Engineering, Cornell University, Ithaca, N.Y., USA

ABSTRACT: This paper addresses a host of educational activities needed in the 21st century to ensure that we provide the best possible constructed facilities for the public and private sectors. Educational needs discussed here include specific improvements in how we teach structures to civil engineering undergraduates and graduate students; new emphases on materials engineering and selection; sustainability and life cycle issues; professional master's programs; cultivating the "constructed facilities mentality"; continuing education at all levels and how the internet will make a huge difference; achieving a new level of mutual respect within the membership of the project team; and educating the owner. Each of these activities is aimed at providing better-constructed facilities with proper attention to actual life cycle costs, a healthy respect for the environment, and a renewed sensitivity about the aesthetic impact of everything we design and build.

1 INTRODUCTION AND SCOPE OF COVERAGE

This paper explores how education can be better utilized to achieve substantial improvements in designing and building all types of structures. "Improvements" should be assessed from the multiple perspectives of safety, serviceability, durability, constructability, esthetics, and lifetime cost. Discussion is aimed at new facilities as well as the rehabilitation of existing facilities.

This paper is based on the premise that improved and expanded education is more critical than ever before as we cope with construction needs in our crowded urban areas, utilizing a mix of new construction and rehabilitated existing infrastructure, with both types of construction utilizing improved "conventional" materials and a host of new materials. Owners face increasingly complex alternative scenarios with optimal solutions dependent upon a complex set of technical, technological, financial, social, environmental, and aesthetic issues. Implications for the detailed technical side of education are rather obvious, but we also must give expanded attention to our responsibilities in educating owners and clients.

As professionals, we collectively have the obligation to design and build structures and facilities to achieve an optimal blend of about a dozen parameters, including overall quality, safety, acceptable initial cost, economy of materials usage, durability, low maintenance, definable life cycle cost, appearance and aesthetics, impact on the external environment, internal environment of the facility occupants, and meeting all other expectations and needs of the owners. Some of these parameters complement each other and some are in direct contradiction. Hence the challenges to the project team -- engineer, architect, materials supplier, contractor, sub-contractor, quality control and quality assurance personnel, maintenance organization, and owner -- are daunting.

Meeting these challenges during the coming decades will provide almost unlimited opportunities for our young people, and we bear the responsibility of informing them about the excitement and challenges in the planning, engineering, and construction of tomorrow's infrastructure. Unfortunately, our ability to attract the best and brightest young people into the civil engineering business seems to have suffered from the perception that we are an old-fashioned, overly mature industry. Let's make some changes!

2 BASIC INSTRUCTION IN DESIGN

2.1 *Definition of issues*

This discussion in this section is directed primarily at four-year B.S. degree programs in civil engineering, but this does not at all dilute the importance of our many excellent technology programs leading to Associate degrees or to the Bachelor of Technology. And given my background and experience, there is a natural bias towards coverage from the American perspective. But much of what I have to say applies internationally.

The scope, depth, and quality of structural design education varies considerably across our undergraduate and graduate programs in civil engineering. Civil engineers tend to be quite conservative, and as a profession we are slow to abandon old approaches and emphases developed over generations of teaching civil engineering students. Much of the constructed world around us shows that we have done quite well in the past, but many rather poor facilities also exist. The challenge is to devise strategies to do better! I believe there is urgent need for improvements in a number of key areas, including (a) preliminary design, (b) simple methods for estimating loads in members, (c) approximate design, (d) variability in loads and the influence of such factors as shrinkage and creep, (e) design of the entire structure, (f) the proper role of analysis in design, (g) effects of modeling assumptions on analysis results, (h) identification and control of errors encountered in design and construction, and (i) integration of materials selection and behavior into structural design and analysis processes.

2.2 *Analysis and design of members and systems*

We simply must teach more preliminary design, including simple methods for quickly estimating loads in members (slabs, beams, columns, etc.) and for sizing these primary components. Even modest coverage of approximate preliminary design can have a profound effect on a student, helping to develop an intuitive common sense about structures and giving a new sense of confidence about dealing with real structural design.

I recently taught undergraduate reinforced concrete structures in a reinforced concrete classroom building (one-way slabs supported on beam-and-column framing), and the very first assignment (given in the first lecture) was to have the students determine the amount of reinforcement needed at the center of the multi-span reinforced concrete beam spanning over the classroom. I provided the students with an absolute minimum of information: design live load, load factors for dead and live load, floor slab thickness, and reinforcement yield strength. They measured the actual structure for other dimensions. Many students struggled over this assignment, having to determine the load transfer paths, make meaningful estimates on bending moment values, and then reduce the flexural action to an internal couple approach with estimated location of the compressive resultant in the T-beam at midspan. But completion of the exercise was at a very high level of learning. It made later lecture coverage much more meaningful, particularly that on

flexural strength of a section and calculation of bending moments and shears in frames.

The ability to quickly calculate approximate forces in any structure is also an absolutely necessary skill in checking the results of computer-based analysis and design. I advocate allocating at least 1/4 of the total time spent in teaching analysis to determining approximate forces. The sketching of approximate deflected shapes is an integral part of the approximate analysis.

Students deserve to get better explanations of the approximate nature of design, including the great variation in actual live loads, the real-life variability of material properties, and how poorly-defined effects such as shrinkage, support motions, unexpected restraining forces, and the like can often "overwhelm" the calculated effects of assumed loadings.

An obvious starting point is to use no more than two significant digits in expressing any final design calculation result. I cannot begin to count the times I've seen (in books, journals, and other publications) such quantities as M = 2, 248.9 m-kN or beam deflection = 4.187 mm. Students should be seeing these results as M = 2.3 m-mN and deflection = 4 mm. Without this first step, how can we possibly expect students to appreciate the fact that real-world design is not a precise science?

We must provide more emphasis on design of the entire structure, including the critical process of selecting the basic structural type to be used, and how the structural elements are connected. Perhaps the best way to accomplish this is to have a substantial portion of the total problem set assignments in the form of open-ended problems -- rather simple structural situations at the undergraduate level and much more comprehensive and realistic projects at the graduate level (see Section 5). An example from the author's own undergrad class at Cornell follows:

Working in teams of 3 engineers, we will be designing selected portions of a three-story general-purpose academic building with a footprint of 5000 square feet and a total gross floor area of 15,000 square feet, to be located in the current parking lot behind Thurston Hall. The building is to have a full basement, a freight elevator, and two sets of stairs. In this first assignment, you are to propose a typical floor framing plan (location of columns, definition of slab spans and supporting beams and girders). Clear height between top of floor and lowest part of underside of floor framing above is to be 10 feet. Live load is to be 125 psf (conservatively selected to permit changing use of the building space by different groups). Material strengths are to be selected by each team." In subsequent assignments the building design was refined, with final design being completed for all typical elements of the building, including foundations.

The proper role of analysis in the design process needs better definition. We need to continually stress that the prime role of analysis is to improve design and hence improve the overall quality of the completed project. Student attitudes on any subject correlate well with faculty attitudes. There is far too much separation of analysis and design in academia; often we see one group of faculty teaching analysis and another group teaching design. A solution is pretty obvious! However, if tradition or some other reason still dictates that separate courses be given, then it is essential to have some modest coverage of the behavior of steel and concrete structures prior to teaching analysis, in order to have realistic and meaningful discussions on effective stiffness of members and behavior at loads beyond the proportional limits.

Effects of assumptions on types of supports, boundary conditions, foundation stiffness, and the like need to be integrated into the lecture material and in the problem assignments. Students too often begin to think that supports are either hinged or fixed, and that building frames stop at ground level -- a rather ridiculous state of affairs considering real-world conditions. Fundamental concepts on foundations should be integrated into the structures classes, with continual linkages between courses in structures and in soil mechanics and foundations.

At the upper-class level, and in graduate education in structural engineering, we need to convey the sensitivity of results (forces and displacements) to the modeling assumptions involved in the analysis, particularly the more complex problems that utilize finite element approaches. Behavior beyond the elastic limit is particularly sensitive to material properties, mesh fineness, and load increment size. An effective approach is to have students (or teams of students) model and analyze the same structure, using the finite element approach, and then compare results and critically discuss the effects of different modeling assumptions.

Students should have at least a minimal understanding of errors in design and construction -- the most typical errors and mistakes, potential impact on safety and serviceability, checking processes, and error control in the office and in the field. A single lecture on this topic, supplemented by readings from Nowak 1986, will suffice as long as the topic is pursued further at appropriate times in other courses. My experience indicates that students become highly motivated in these discussions -- the topic is perfect for problem assignments given to teams of three or four students, followed by class discussions.

2.3 *Integration of materials into structural analysis and design*

The specification of proper materials and the integration of the material properties into all phases of the analysis and design process may well be the most critical step in arriving at a durable, top-quality constructed facility. Some brief comments on how materials might be best integrated into the teaching of structural design will be given here, with the major coverage on materials to follow in Section 3 below.

In contrast to modern design practice, where the individual project needs dictate the choice of materials, academia has tended to have separate courses on the design of concrete, steel, timber, and masonry structures. There are certain advantages of immersing students in, say, steel structures design exclusively for several months, and then follow up with a similar immersion in concrete structures design. But at some point some merging of materials needs to be done. I suggest that the first course in structural engineering have a limited coverage of the primary issues met in designing and building with a variety of materials, with later courses having more concentrated coverage, and perhaps the final course going back to "general design" where the students have to make the decisions on which material(s) to use. And certainly graduate level design should have substantial components of "multi-material" design, a topic to be discussed more in Section 5 below.

As an example of how most designs incorporate two or more primary construction materials, most modern high-rise buildings now utilize some type of composite (hybrid) framing to carry gravity loads and lateral forces (wind and seismic). Composite construction utilizes concrete and steel, each to its best advantage, providing a fire-resistant, economical structural system that is fast to construct and provides substantial stiffness to resist lateral drift. The extensive activity in composite construction (Xiao & Mahin 2000) is expected to continue, with an ever-increasing usage of high strength concrete and high-performance structural steel, and with extensions to buildings of medium and moderate height. One of the greatest challenge in designing and building a high-rise composite framed building is developing a strategy that will guarantee reasonably flat floors after time-dependent deformations in the columns have essentially stabilized. This requires a level of knowledge about the concrete in the columns that far exceeds what we typically know about concrete in the usual structure.

2.4 *How do we fit everything into the curriculum?*

This is an age-old problem because there is always more to teach than time permits, particularly when an ever-expanding list of important topics needs coverage (e.g., computer science, materials science, probability and statistics, systems engineering, environmental issues, costs and economics, etc.). The apparent solution is to decrease existing coverage at the B.S. level and then rely on students to continue their study at the Master's level, either full-time or part-time while working, or in on-the-job training. The undergraduate courses then must strive to teach the fundamentals in an integrated fashion, and to teach the students how to think critically and how to learn.

Economies in coverage can begin early, say during the teaching of mechanics of materials, where problems should be designed to illustrate practical situations, with later courses building directly on the mechanics coverage. We probably spend more time on flexure than is really needed; it is a "comfortable" topic with results conforming very well with basic assumptions. Teaching flexure from the standpoint of two internal forces (C and T) separated by an internal moment arm provides a fundamental basis that can be then readily adapted to any material and to both elastic and inelastic behavior. It also helps in teaching shear, with horizontal shear stress resulting from differential values of either C or T along the beam axis.

Hard decisions must be made on what displacement-calculation methods to teach. Virtual work is the most general; if a student understands virtual work concepts and how to apply them, then he or she can solve any displacement problem. I also like students to know the moment-area method for quick and relatively error-free calculation of displacements of simple structures

Structural analysis procedures can be some combination of stiffness-based computer methods and a thorough coverage of approximate analysis utilizing assumed points of inflection. The teaching of moment distribution has "fallen by the wayside" in many curricula. I still like to have students know how to do beams and frames (without sway) by moment distribution, both because it is a very good method to get answers and also because it illustrates so convincingly the dependence of bending moment on flexural stiffness.

3 MATERIALS

3.1 *How did the situation get so critical?*

As evidenced by the generally deteriorated state of a sizable portion of our infrastructure, we have an urgent need for the development and employment of better materials for constructing new facilities, as well as improved materials for use in repair and retrofitting. Each of the traditional construction materials (concrete, steel, masonry, wood, and of course, soil) is a promising candidate for improvement, and at the same time, there are numerous new materials being developed with great potential for successful application.

At the same time when we need more coverage of materials, the typical U.S. civil engineering department has reduced the time given to traditional construction materials, particularly to understanding the complexities of concrete, soils, and rock. Several issues are involved -- the pressure to reduce required hours for graduation to ensure that most students finish in four years, the desire to cover new materials such as plastics, and the gradual reduction in laboratory sessions to reduce costs.

The difficult problem we face in providing the proper coverage on construction materials is well illustrated by considering the Civil Engineering Research Foundation (CERF 1993) statements on high-performance construction materials and systems, expressed in terms of the long list of desirable characteristics of infrastructure performance. These characteristics include (1) superior strength, toughness, and ductility, (2) enhanced durability/service life, (3) increased resistance to abrasion, corrosion, chemicals, and fatigue, (4) initial and life-cycle cost efficiencies, (5) improved response in natural disasters and fires, (6) ease of manufacture and application or installation, (7) aesthetics and environmental compatibility, and (8) ability for self-diagnosis, self-healing, and structural control.

Selecting the one material that I know best, high performance concrete (HPC), the list of attributes (Carino & Clifton, 1991) includes: "adhesion to hardened concrete, abrasion resistance, corrosion protection, chemical resistance, ductility*, durability, energy absorption (toughness)*, early strength, high elastic modulus, high compressive strength, high modulus of rupture, high tensile strength, high strength-to-density ratio**, high workability and cohesiveness, low permeability, resistance to washout, and volume stability", where * denotes "fiber-reinforced concrete" and ** denotes "especially with high-strength, lightweight concrete". Carino & Clifton then state that there is an urgent need to replace our traditional reliance on empiricism with a technology firmly rooted in materials science and structural mechanics.

There have been rather massive recent additions to the literature on new materials and methods of repair for infrastructure applications, such as the 1250 page proceedings of the Third Materials Engineering Conference (Basham 1994).

3.2 *What can be done to improve "materials engineering" education?*

How do we possibly do justice to this complex set of material characteristics that goes so far beyond the typical mechanical properties we've always covered in undergraduate courses? How do we get young practicing engineers educated to the point where they can make informed decisions on choices of materials for each new project they work on? It seems rather obvious that all of this cannot be done in the typical B.S. degree program, but at the same time we must educate our students to be sensitive to these issues (particularly durability) and to be motivated to continue learning about materials as an integral part of improving their design skills.

A number of formidable educational tasks must be addressed, with programs aimed at several quite different user groups: undergraduate and graduate engineering students, engineers in practice, materials suppliers, construction workers, quality control personnel, etc. We shall return to these topics in Sections 5 and 7 on graduate programs and continuing education.

3.3 *Advanced composites*

Advanced composite materials consisting of strong fibers (glass, Kevlar, carbon) embedded in a resin matrix have already found substantial use in construction and in retrofitting projects, and they have excellent potential for becoming an important construction material. But these materials have properties much different than traditional civil engineering materials, and design approaches may well be quite different than for, say, steel and concrete. We must provide young engineers with the opportunity to become proficient in designing and building with advanced composites.

4 SUSTAINABILITY & SERVICE LIFE ISSUES

Space limitations do not permit any degree of coverage of these far-ranging issues here, but a brief mention is made to emphasize the importance of including these topics in modern civil engineering curricula. Again, I draw on my own experience in the field of concrete to point out some of the key issues.

For too many decades we've regarded concrete structures as "permanent", while at the same time spending excessively on repairs, rehabilitation, and premature demolition. This is not a proper use of our natural and financial resources. We now know enough about materials and construction practices to eliminate most of these problems, but it will take new levels of technical effort along with a continuous commitment to achieving a defined service life. Our current students will be the generation most directly charged with these responsibilities, so it is essential that they be exposed to the ideas and concepts and that they are in the proper position to build up their expertise in the many issues of sustainability and service life definition and extension.

A suggested format for one or two lectures is provided by (Mehta 1999) in a paper entitled concrete technology for sustainable development. He defines and discusses three crucial elements needed for an environment-friendly concrete construction technology: (a) conservation of concrete-making materials, (b) enhancement of durability of concrete structures, and (c) a paradigm shift from reductionistic to holistic approach in concrete technology research and education. The reader is urged to study this paper, as well as the volume of 33 papers on concrete technology for a sustainable development in the 21st century (Gjorv & Sakai 2000). These issues would make a great topic for a continuing seminar series for all civil engineering students.

5 PROFESSIONAL DEGREES BEYOND THE B.S.

We need to supplement our many fine B.S. programs with more professional education programs to produce engineers with the type of technical background that just cannot be achieved in a four year B.S. program. This degree, typically called the Master of Engineering, is inherently different from Master of Science programs, with heavy emphasis on design, no research components, and strong involvement of practicing engineers in helping teach design. At Cornell, about 80% of student effort during the intensive 9 month program is spent in taking graduate level courses in civil engineering topics (including advanced materials courses) and in supporting disciplines such as management, labor relations, architecture, and the like.

Our experience with the Master of Engineering degree at Cornell over the past 30 years has shown that a comprehensive design project, accounting for other 20% of the total effort, is the cohesive element that most sharply defines the program. It provides students with a rich experience in open-ended preliminary design and in formulating and modeling real structural systems. It provides them an opportunity to work in guided team situations with students from several different concentrations (typically structures, geotech, and management), with constant stress on constructability, practicality, and economics, and in close contact with eminent practicing professional engineers who bring the design projects to campus. Conceptual and preliminary design is done in the fall semester, followed by a full-time three-week final design period in January, and a formal, public presentation in February covering financial feasibility, design results, and construction planning and costs.

The response to this program by practicing engineers has been truly phenomenal -- they are exceptionally cooperative in donating their time and talents in bringing projects to campus and returning repeatedly to work with students and faculty. They also like to hire our M.Eng. graduates. And added benefits include informal talks to undergraduates by the visiting engineers, along with undergraduate attendance at presentations given by the M.Eng. students.

ASCE's decision that the professional Master's degree should be the first designated civil engineering degree, and thus serve as the entry point into the civil engineering profession, has received considerable discussion, both pro and con. The basic concept is fine, I think, because it is very difficult to argue against the considerable additional knowledge gained during this intensive advanced study. But I have trouble with requiring everyone to follow this path. Some students may be better off pursuing additional graduate studies on a part-time basis while working, either at a local campus or by internet-provided programs. Some need the maturity and direction provided by getting into the full-time job market after getting their B.S. degree; those who follow this path and then come back for the M.Eng. degree always do exceptionally well in their graduate studies. Others may choose to supplement the B.S. in civil engineering with a degree in business or management. So my personal position is to enthusiastically recommend the 5th year, but not require it.

6 CULTIVATING THE CONSTRUCTED FACILITIES "MENTALITY"

Construction engineering and construction management are disciplines by themselves; hence a regular civil engineering program cannot begin to do justice to the full scope of these critically important topics. We are not talking simply of teaching "content" here; we must also find ways to transmitting the philosophy and way of thinking involved in the successful execution of any construction project. Construction projects, particularly the larger ones, are wonderful examples of applied systems engineering, and we need to take advantage of this fact in attracting students into civil engineering.

Courses in construction engineering, with substantial systems engineering content, should be available -- perhaps a simplified, motivational version at the freshman level, followed by an upperclass course or two.

I believe that it is absolutely essential that constructability and pertinent construction-related issues be integrated into every course in the civil engineering curriculum.

7 CONTINUING EDUCATION; THE INTERNET

The audience for continuing education spans a broad spectrum, ranging from construction personnel to highly experienced designers. The offering of short courses and seminars to civil engineers has been a "big business" for some time. Most typically, teams of professors and practicing engineers prepare and present these materials to an assembled audience at locations around the country. But the delivery mechanism is changing! The rapidly evolving distance learning opportunities afforded by the Internet will have a major impact on education at all levels. For example, the popular masonry design seminar put on by American Concrete Institute (ACI) in various cities has been converted into a high-quality format for distribution on the internet and is now (July 2000) under review by its sponsoring societies (ACI, TMS, and ASCE). It should be available on the Internet in late 2000. This offering will be followed by a host of other educational offerings on the web; literally a revolution in how the individual can learn new technology and techniques.

More publications directed specifically at materials technicians and construction workers are needed. These must be highly specific, brief, and in the right language. A specific example might be a half-hour explanation, along with a single sheet of text, on handling and placement of high-strength concrete containing silica fume and other admixtures. ACI is producing such a "tool-box series" on a number of construction techniques, involving both regular and high-performance materials.

Designers, material specifiers, and contractors must have ready access to material properties and related data on the performance of each new material. As stated so well in (CERF 1994), "To transmit the benefits of high-performance construction materials and systems to all potential users, we must make reliable data about composition, properties, and performance available to owners, designers, contractors, and the construction community."

As an example of what might be done, one of the information transfer mechanisms envisioned by the CONMAT concrete group is the establishment of a computer bulletin board with detailed data on properties and behavior of HPC, directed to designers, architects, ready-mix suppliers, contractors, facility owners, and researchers. Given the international nature of the Internet and its 24 hour-a-day availability to anyone, we have a powerful new tool with tremendous potential for enhancing the use of high-performance materials. However, it is crucial that common formats for data be agreed on, and that the on-line electronic data base system be well designed, user friendly, and properly integrated. International agreements on formats, methods of updating, etc., are needed.

8 ENGINEER-ARCHITECT-BUILDER RELATIONSHIPS

It is a sad fact that we often hear engineers taking "potshots" at architects, and vice versa; in fact, a recent national news release had a prominent architect complaining that he had been treated "like an engineer", rather than "like a creative person". Conflicts between contractors and the design team are too frequent, with attitudes that almost guarantee discord. I contend that our educational process could do a better job in reducing these conflicts. More faculty need to gain practical experience in design and project management, to spend sabbatical time in industry, and to really understand what construction is all about. We need to expose our students to a balanced mix of outside speakers. Faculty in civil engineering and architecture programs need to cooperate and offer lectures and courses for students from each other's discipline

9 EDUCATING THE OWNER AND THE PUBLIC

9.1 *Why should we educate the owner? How do we do it?*

Members of the construction project team (architect, engineer, materials supplier, contractor, and sub-contractor) must become more proactive in educating owners, both private and public, about the realities of design and construction. With public facilities, the owners arc you and I and our neighbors and friends. Our political leaders usually control spending for new public infrastructure; they naturally want to build as many new facilities as possible to "get credit" and to maximize their chances (or their party's chances) of winning the next election. Thus the emphasis is usually on the lowest possible initial cost, with very little attention given to maintenance programs and costs. But it is more important to stress total lifetime project costs if we are to avoid producing too many structures requiring extensive maintenance and repair.

The owner must be taught enough about the physical side of construction to fully appreciate what can be done with a given sum of money. We need to be completely frank with owners, showing them examples of how things have gone bad on earlier

projects, and describing in no uncertain terms precisely what it takes to avoid these problems on the current project, including expected costs and approximate future savings in maintenance and repair costs when highly durable construction is used.

No one is better equipped for this owner education role than ourselves, the designers and project managers and constructors -- professionals who have dedicated our lives to planning and providing constructed facilities. I believe that if the public really understood the issues, they would vote for high quality construction even if it required that new facilities were completed at a slower rate. They might even be willing to pay higher taxes for top-quality, durable construction. The old adage "You can pay me now or you can pay me later" rings very true on all construction projects. The saying is even better if it is modified to "You can pay me now or you can pay me a lot more later."

Getting back to specific issues, the pros and cons of alternative designs must be explained fully. For example, if reinforced concrete is being compared with prestressed concrete, we need to clearly explain the two processes to the owner, perhaps using simple models to illustrate just what prestressing does to a structure in terms of adding residual compression and thus delaying the onset of cracking and all the problems that can come with excessive cracking. The essentials of durable concrete, including what happens to concrete cover when reinforcing bars corrode, need to be clearly explained in layman terms. With a little imagination, this type of education can be done using simple physical models supplemented with visual aid presentations on the basics of material performance and structural behavior.

The same educational approach also needs to be used in working with owners who are rehabilitating, repairing, retrofitting, and remodeling their facilities. Many important issues need to be quantified and factored into the decision process -- not only the value of the facility, but the cost of lost business opportunities during the down time needed for rehabilitation, loss of image if a structure fails, and potential lack of trust on the part of their clients and employees.

9.2 *An example of owner education*

An excellent example of how an engineering team educated the owners is provided by (Gates et al 1992). The project was the seismic rehabilitation of the Rockwell Building in California, the world headquarters for the Rockwell Corporation. The philosophy followed by the project team in doing this job is applicable to almost any modification project on an existing structure. It also has strong lessons for new construction.

9

The 8-story reinforced concrete frame Rockwell Building was built in Seal Beach, California, in the mid-1960s under the provisions of the UBC-1964, which produced a structure with reinforcing quantities and details not adequate by today's standards. The basic design criteria for the rehabilitation were defined very carefully and made completely clear to the owners: (a) to assure life safety, and (b) to minimize operational downtime after a damaging earthquake. After extensive studies, four alternatives were identified as candidates for modifying the building to meet the seismic design criteria: (1) base isolation plus exterior diagonal bracing, (2) conventional braced frames added to the exterior of the building, (3) exterior shearwalls in the perimeter frames, and (4) jacketing of non-ductile reinforced concrete beams and columns.

The owner's management team was educated by the consulting engineers on several key earthquake engineering concepts, including the probabilistic approach to hazard prediction, definition of those conditions which typically result in closing of facilities by building officials after earthquakes, and areas of engineering uncertainty (ground motion, design and analysis, and constructability) in predicting seismic structural behavior of the retrofitted building. These are difficult topics, both conceptually and technically.

The four retrofit designs were carried forward and analyzed for relative risk, cost, etc., in comparison to the "do nothing" alternative. The building owner established a ranked listing of factors to be considered in reaching a decision on the best scheme, as follows: (1) life safety and minimization of bodily injury, (2) interruption of vital administrative operations, (3) technical uncertainties associated with each retrofitting concept, (4) disruption of vital building functions during construction, (5) comparative dollar losses resulting from the design earthquake, (6) total project construction costs, (7) economic payback in terms of protection provided vs. total cost, and (8) esthetic impact of the adopted scheme.

The final decision process also took into account the non-quantifiable issues, including potential for loss of life and bodily injury costs, future lost business opportunities, impact on the image and reputation of the company, and loss of confidence in building safety by its occupants. Most of these additional issues were non-technical, showing that what's important to an engineer is often not that important to the owner, and vice versa.

The selected retrofit method was base isolation combined with strengthening of weak perimeter frames to ensure that they could safely resist expected seismic inelastic deformations which might occur even with base isolators in place. This scheme had retrofitting costs nearly twice that of the other three, but by far the lowest expected loss in a major earthquake. The overriding message of this example is that detailed, constant communication between engineer and owner is absolutely essential to help the owner make the most rational decision and ensure that the engineer does the best job in meeting the owner's needs.

10 CONCLUSIONS

We all must be involved in the very serious obligation to help provide the technical bases needed to meet the rising expectations and demands of the public -- to provide better structures and facilities, integrating the complex issues of durability, serviceability, and total lifetime cost issues with safety and other parameters.

What can we, the members of professions of structural and construction engineering, do to help improve the teaching and learning processes? Specific suggested activities include:

a. become involved in teaching design at the undergrad and grad levels, bringing in actual project situations, with real constraints and conditions.

b. volunteer to meet with student groups to discuss recent interesting projects, comment on life in the design profession, etc.

c. materials suppliers can likewise come to campus to give talks on new products and new ways of producing better concretes.

d. construction companies, fabricators, and precast suppliers can host field trips to show students real world issues that cannot be taught effectively in the classroom or lab.

e. become members of college civil engineering department advisory councils.

f. support your young employees in continuing education efforts.

g. volunteer to work with educational activities committees of national technical societies, such as the American Concrete Institute, and get involved in putting on seminars (both live and on the internet).

h. assume central responsibility for educational efforts (including substantial use of Internet capabilities) to inform owners and the public of their critically important roles.

REFERENCES

Basham, K. (ed) 1994. *Infrastructure: New Materials and Methods of Repair*. Proceedings of the Third Materials Engineering Conference. San Diego: ASCE.

Carino, N.J. & J.R. Clifton 1991. *High-Performance Concrete: Research Needs to Enhance its Uses*. Concrete International. 13(9): 70-76. Detroit: ACI.

CERF 1993. *High-Performance Construction Materials and Systems -- An Essential   Program for America and its Infrastructure*, CERF Executive Report 93-5011.E, April, Civil Engineering Research Foundation, Washington DC: ASCE.

CERF 1994. *Materials for Tomorrow's Infrastructure: A Ten-Year Plan for Deploying High-Performance Construction Materials and Systems,* CERF Executive Report 94-5011.E, Dec. 27,   Civil Engineering Research Foundation, Washington DC: ASCE.

Gates, W.E., M.R. Nester, & T.R. Whitby 1992. *Managing Seismic Risk: A Case History of Seismic Retrofit for a Non-Ductile Reinforced Concrete Frame High Rise Office Building.* Proc. Tenth World Conference on Earthquake Engineering. 9: 5261-5266.

Gjorv, O.E. & K. Sakai   (eds) 2000.   *Concrete for a Sustainable Development in the 21st Century,.* Great Britain: E&FN Spon.

Mehta, P.K. 1999.   *Concrete Technology for Sustainable Development.* Concrete International . 21(11): 47-53. Detroit: ACI.

Nowak, A.S. (ed) 1986. *Modeling Human Error in Structural Design and Construction.* ASCE. Xiao, Y.. & S.A. Mahin 2000. *Composite and Hybrid Structures*, Proc. 6th ASCCS Conf. Los Angeles. March 2000, 1220 pp. ASCCS Secretariat: Univ. of Southern Cal.

Creative Systems in Structural and Construction Engineering, Singh (ed.) © 2001 Balkema, Rotterdam, ISBN 90 5809 161 9

New computing paradigms for innovative construction and structural engineering

Hojjat Adeli
Department of Civil and Environmental Engineering and Geodetic Science, Ohio State University, Columbus, Ohio, USA

ABSTRACT: Several new computing paradigms for innovative construction and structural engineering are discussed. They include neurocomputing, evolutionary computing/genetic algorithms, object-oriented programming, parallel processing, Internet computing, and collaborative computing. Areas of applications include CAD/CAE, large-scale design automation and optimization, active control of structures, construction scheduling, resource scheduling, and construction cost estimation. Examples of research performed by the author and his associates in recent years are described briefly.

1 NEUROCOMPUTING

Adeli and Yeh (1989) published the first archival journal article on structural engineering application of neural networks nearly a dozen years ago. Since then a large number of articles have been published on applications of neural networks in structural engineering, construction engineering and management, and other civil engineering disciplines. Recent examples of the research performed by the author and his associates are described briefly in this section.

1.1 Design Automation and Optimization

Automation of design of large one-of-a-kind civil engineering systems is a challenging problem due partly to the open-ended nature of the problem and partly to the highly nonlinear constraints that can baffle optimization algorithms (Adeli, 1994). Optimization of large and complex engineering systems is particularly challenging in terms of convergence, stability, and efficiency. Recently, Adeli and Park (1998) developed a neural dynamics model for automating the complex process of engineering design through adroit integration of a novel neurocomputing model (Adeli and Hung, 1995), mathematical optimization (Adeli, 1994), and massively parallel computer architecture (Adeli, 1992a&b, Adeli and Soegiarso, 1999). The computational models have been applied to fully automated minimum weight design of high-rise and superhighrise building structures of arbitrary size and configuration, including a very large 144-story superhighrise building structure with 20,096

members. The structure is subjected to dead, live, and multiple wind loading according to the Uniform Building Code (UBC, 1997) and according to the AISC LRFD code (AISC, 1998) where nonlinear second order effects have to be taken into account. The patented neural dynamics model of Adeli and Park finds the minimum weight design for this very large structure subjected to multiple dead, live, and wind loadings in different directions automatically.

Optimization of space structures made of cold-formed steel is complicated because an effective reduced area must be calculated for members in compression to take into account the non-uniform distribution of stresses in thin cold-formed members due to torsional/flexural buckling. The effective area varies not only with the level of the applied compressive stress but also with its width-to-thickness ratio. For statically indeterminate structures a new effective area has to be calculated for each member in every iteration of the optimization process. As such, the constraints are implicit, non-smooth, and discontinuous functions of design variables. Tashakori and Adeli (2001) adapt the patented neural dynamics model of Adeli and Park for optimum (minimum weight) design of space trusses made of commercially available cold-formed shapes in accordance with the AISI specifications (AISI, 1996, 1997). A CPN network (Hecht-Neilsen, 1987, 1988) is developed to learn the relationship between the cross-sectional area and dimensions of cold-formed channels. The model has been used to find the minimum weight design for several space structures commonly used as roof structures in long-span commercial buildings and canopies, including a large structure with 1548 members with excellent convergence results.

1.2 Construction Scheduling and Management

Adeli and Karim (1997) present a general mathematical formulation for scheduling of construction projects. An optimization formulation is presented for the construction project scheduling problem with the goal of minimizing the direct construction cost. The nonlinear optimization problems is then solved by the patented neural dynamics model of Adeli and Park (1998). For any given construction duration, the model yields the optimum construction schedule for minimum construction cost automatically. By varying the construction duration, one can solve the cost-duration trade-off problem and obtain the global optimum schedule and the corresponding minimum construction cost. The new construction scheduling model is particularly suitable for studying the effects of change order on the construction cost.

1.3 Resource Scheduling

Senouci and Adeli (2001) present a mathematical model for resource scheduling considering project scheduling characteristics generally ignored in prior research, including precedence relationships, multiple crew-strategies, and time-cost trade-off. Previous resource scheduling formulations have traditionally focused on project duration minimization. The new model considers the total project cost minimization. Furthermore, resource leveling and resource-constrained scheduling have traditionally been solved independently. In the new formulation, resource leveling and resource-constrained scheduling are performed simultaneously. The resource scheduling model is solved using the neural dynamics model of Adeli and Park.

1.4 Construction Cost Estimation

Quality of the construction management depends on the accurate estimation of the construction cost. Automating the process of construction cost estimation based on objective data is highly desirable not only for improving the efficiency but also for removing the subjective questionable human factors as much as possible. The problem is not amenable to traditional problem solving approaches. The costs of construction materials, equipments, and labor depend on numerous factors with no explicit mathematical model or rule for price prediction. Highway construction costs are very noisy and the noise is the result of many unpredictable factors. Adeli and Wu (1998) present a regularization neural network model and architecture for estimating the

cost of construction projects. The new computational model is based on a solid mathematical foundation making the cost estimation consistently more reliable and predictable. Moreover, the problem of noise in the data is taken into account in a rational manner.

2 EVOLUTIONARY COMPUTING/ GENETIC ALGORITHMS

Evolutionary computing and genetic algorithm (GA), based on the Darwinian evolution principle of the survival of the fittest, has received considerable attention in the optimization and artificial intelligence communities (Adeli and Hung, 1995). An early formulation of GA for structural optimization was presented by Adeli and Cheng (1993). Three most recent examples of the work of the author and his associates are reviewed in this presentation.

In the traditional optimization algorithms, constraints are satisfied within a tolerance defined by a crisp number. In actual engineering practice constraint evaluation involves many sources of imprecision and approximation. When an optimization algorithm is forced to satisfy the design constraints exactly it can potentially miss the global optimum solution within the confine of commonly acceptable approximations. Extending the augmented Lagrangian genetic algorithm of Adeli and Chang (1994), Sarma and Adeli (2000b) present a fuzzy augmented Lagrangian GA for optimization (minimum weight design) of steel structures subjected to the constraints of the AISC ASD specifications taking into account the fuzziness in the constraints. The membership function for the fuzzy domain is found by the intersection of the fuzzy membership functions for objective function and the constraints using the max-min procedure of Bellman and Zadeh (1970). The advantages the new fuzzy GA algorithm include increased likelihood of obtaining the global optimum solution, improved convergence, and reduced total processing time.

Only a small fraction of hundreds of papers published on optimization of steel structures deals with cost optimization; the great majority deal only with minimization of the weight of the structure (Sarma and Adeli, 1998, 2000a). Those few that are concerned with cost optimization deal with small two-dimensional or academic examples. Sarma and Adeli (2000c) present a fuzzy discrete multi-criteria cost optimization model for design of space steel structures subjected to the actual constraints of commonly-used design codes such as the AISC ASD code (AISC, 1995) by considering three design criteria: a) minimum material cost, b) minimum weight, and c) minimum number of different section types. The computational model starts with a continuous-variable minimum weight solution with a preemptive constraint violation strategy as lower bound followed by a fuzzy discrete multi-criteria optimization.

Genetic algorithms are usually implemented by binary representation of parameters which is simple to create and manipulate. The binary representation, however, requires excessive computational resources when the number of variables is large or a high level of precision is required. An alternative to the popular binary genetic algorithm is floating-point parameter GA where each variable is represented by a floating point number and linear interpolation is used to combine variable values to crossover from parents to an offspring. Kim and Adeli (2001) present cost optimization of composite floors using floating point genetic algorithm. The total cost function includes the costs of a) concrete, c) steel beam, and c) shear studs. The design is based on the AISC Load and Resistance Factor Design (LRFD) specifications (AISC, 1998) and the plastic design concepts.

3 OBJECT-ORIENTED PROGRAMMING

In the 1990's the object technology gained increasing popularity for development of flexible, maintainable, and reusable CAD/CAE software systems (Yu and Adeli, 1991, 1993, Adeli and Kao, 1996, Adeli and Kumar, 1999). In addition to its desirable characteristics of abstraction, encapsulation, and inheritance and its reusability advantage, the object-oriented programming (OOP) paradigm facilitates the message passing constructs needed for collaborative computing on the Internet (to be discussed later).

Karim and Adeli (1999a) present an object-oriented (OO) information model for construction scheduling, cost optimization, and change order management based on the new construction scheduling model discussed in section 1.2, with the objective of laying the foundation for a new generation of flexible, powerful, maintainable, and reusable software systems for the construction scheduling problem. The model is presented as a domain-specific development *framework* using the Microsoft Foundation Class (MFC) library and utilizing the software reuse feature of the *framework*. The information and computational models have been implemented into a new generation prototype software system called CONSCOM (for CONstruction Scheduling, Cost Optimization, and Management) (Karim and Adeli, 1999b&c).

CONSCOM includes a superset of all currently available models such as CPM plus new features such as (Karim and Adeli, 1999c):
- Integrated construction scheduling and minimum cost model based on the patented robust and powerful neural dynamics optimization model of

Adeli and Park that provides reliable cost minimization of the construction plan, time-cost trade-off analyses, and change order management.
- Support for a hierarchical work breakdown structure with tasks, crews, and segments of work.
- Capability to handle multiple-crew strategies.
- Support for location (distance) modeling of work breakdown structures (very useful for modeling linear projects such as highway construction).
- A mechanism to handle varying job conditions.
- Nonlinear and piecewise linear cost modeling capability for work crews.
- Capability to handle time and distance buffer constraints in addition to all the standard precedence relationships.
- Ability to provide construction plan milestone tracking.

4 PARALLEL PROCESSING

Recently, Adeli (2000) presented a review of the articles on high-performance computing in the areas of analysis, optimization, and control published in archival journals since 1994. The review is divided into three main sections: a) parallel processing on dedicated shared memory and distributed memory parallel machines, b) distributed computing on a cluster of networked workstations, and c) parallel computing and object-oriented programming. It is noted that the great majority of the journal articles published since 1987 deal with fundamental issues and academic problems. The trend in the future, at least in part, should be more toward solution of large-scale and complicated real-life engineering problems, the kind of problems that cannot be solved readily by conventional uniprocessor computers. A few examples are presented in this keynote lecture.

In the area of optimization, Park and Adeli (1997) present distributed nonlinear neural dynamics algorithms for discrete optimization of large steel structures. The algorithms are implemented on a distributed memory machine, CRAY T3D (CRI, 1993a&b). For the solution of resulting linear simultaneous equations a distributed PCG algorithm is developed employing the worksharing programming paradigm. Parallelism is exploited at both node and member levels by allocating each member of the structure and nodal degree of freedom to a processing element (PE). Other examples of parallel algorithms for large-scale structural optimization can be found in the recent books by Adeli and Kumar (1999) and Adeli and Soegiarso (1999).

A major bottleneck in optimal active control of large structures is the solution of the complex eigenvalue problem encountered in the solution of the resulting Riccati equation as well as the solution of both open loop and closed loop systems of equations (Adeli and Saleh, 1999). The methods reported in the literature yield satisfactory results for small problems but often become unstable for large problems (Gardiner, 1997). Saleh and Adeli (1996) present robust and efficient parallel algorithms for solution of the eigenvalue problem of an unsymmetric real matrix using the general approach of matrix iterations (Rutishauser, 1990) and exploiting the architecture of shared memory supercomputers through judicious combination of vectorization, microtasking, and macrotasking. They apply the algorithms to large matrices including one resulting from a 21-story space truss structure.

The solution of the Riccati equation is the most time-consuming part of any optimal control problem. It requires an inordinate amount of processing time when applied to large problems. Saleh and Adeli (1997) present robust and efficient parallel-vector algorithms for solution of the Riccati equations encountered in the structural control problems using the eigenvector approach (Meirovich, 1990). The algorithms have been implemented on the shared-memory multiprocessor Cray YMP 8/8128 and applied to three large problems resulting from a continuous bridge structure, a 21-story space truss structure, and a 12-story space moment-resisting building structure.

5 INTERNET AND COLLABORATIVE COMPUTING

In the second half of the 90's decade the Java programming language was developed to provide the high-level of user-machine interactivity needed on the Internet. Java is an OOP language with all its aforementioned desirable features. Java also provides multi-threading, that is a Java program can execute different processes simultaneously, for example loading an image while performing numerical processing (Adeli and Kim, 2000).

An attractive feature of Java is that it can be used in a heterogeneous environment using various operating systems such as Unix, Windows, and Linux without any need to change the source code. Java applets can be implemented on Java-enabled browsers such as Netscape and Internet Explorer. Further, Java's multithread function provides multiple and secure access to servers by many clients simultaneously without any need to write specific multitasking programs.

A Web-collaborative remote computing *framework* architecture is being developed for CAD/CAE applications (Qi and Adeli, 2001) using an advanced Java function, the Remote Method Invocation (RMI) (Sun Microsystems, Inc.,1997). The framework is a skeleton software that can be used to develop Web-collaborative CAD/CAE applications. RMI provides a remote procedure call mechanism through which objects at separate locations can communicate with each other. The proposed framework consists of two main components: clients and server. The server creates the remote objects (for transferring data to the client side, for example) and combines them with the *registry*. The client invokes the remote objects' methods through the registry located on the server side. RMI supplies a message passing mechanism between the client and the server including the registry to facilitate their communication.

ACKNOWLEDGMENT

This Keynote Lecture is based upon work sponsored by the U.S. *National Science Foundation* under Grant No. MSS-9222114, *American Iron and Steel Institute, American Institute of Steel Construction, Ohio Department of Transportation, Federal Highway Administration*, and Cray Research, Inc. Supercomputing time was provided by the *Ohio Supercomputer Center* and *National Center for Supercomputing Applications* at the University of Illinois at Urbana-Champaign. Part of the work (the neural dynamics model described briefly in section 1.1) resulted in a United States Patent entitled *Method and apparatus for efficient design automation and optimization, and structure produced thereby*. The patent was issued by the *U.S. Patent and Trademark Office* on September 29, 1998 (Patent 5,815,394). The inventors are Hojjat Adeli and H.S. Park.

REFERENCES

Adeli, H., Ed. (1992a), *Supercomputing in Engineering Analysis*, Marcel Dekker, New York.

Adeli, H., Ed. (1992b), *Parallel Processing in Computational Mechanics*, Marcel Dekker, New York.

Adeli, H., Ed. (1994), *Advances in Design Optimization*, Chapman and Hall, London.

Adeli, H. (2000), "High-performance computing for large-scale analysis, optimization, and control", *Journal of Aerospace Engineering*, Vol. 13, No. 1, pp. 1-10.

Adeli, H. and Cheng, N.T. (1993), "Integrated Genetic Algorithm for Optimization of Space Structures", *Journal of Aerospace Engineering*, ASCE, Vol. 6, No. 4, pp. 315-328.

Adeli, H. and Cheng, N.T. (1994), "Augmented Lagranjian Genetic Algorithm for Structural Optimization", *Journal of Aerospace Engineering*, ASCE, Vol. 7, No. 1, pp. 104-118.

Adeli, H. and Hung, S.-L. (1995), *Machine Learning – Neural Networks, Genetic Algorithms, and Fuzzy Systems*, John Wiley & Sons, New York.

Adeli, H. and Kao, W.-M. (1996), "Object-Oriented Blackboard Models for Integrated Design of Steel Structures", *Computers and Structures*, Vol. 61, No. 3, pp. 545-561.

Adeli, H. and Karim, A. (1997), "Scheduling/Cost Optimization and Neural Dynamics Model for Construction", *Journal of Construction Engineering and Management*, Vol. 123, Np. 4, pp. 450-458

Adeli, H. and Kim, H. (2000), "Web-Based Interactive Courseware for Structural Steel Design Using Java", *Computer-Aided Civil and Infrastructure Engineering*, Vol. 15, No. 2, pp. 158-166.

Adeli, H. and Kumar, S. (1999), *Distributed Computer-Aided Engineering for Analysis, Design, and Visualization*, CRC Press, Boca Raton, Florida.

Adeli, H. and Park, H.S. (1998), *Neurocomputing for Design Automation*, CRC Press, Boca Raton, Florida.

Adeli, H. and Saleh, A (1999), *Control, Optimization, and Smart Structures - High-Performance Bridges and Buildings of the Future*, John Wiley & Sons, New York.

Adeli, H. and Soegiarso, R. (1999), *High-Performance Computing in Structural Engineering*, CRC Press, Boca Raton, Florida.

Adeli, H. and Wu, M. (1998), "Regularization Neural Network for Construction Cost Estimation", *Journal of Construction Engineering and Management*", Vol. 124, No. 1, pp. 18-24.

Adeli, H. and Yeh, C. (1989), "Perceptron learning in engineering design", *Microcomputers in Civil Engineering*, Vol. 4, No. 4, pp. 247-256.

AISC (1995), *Manual of Steel Construction - Allowable Stress Design*, American Institute of Steel Construction, 9th edition, 2nd revision, Chicago, Illinois.

AISC (1998), *Manual of Steel Construction – Load and Resistance Factor Design – Volume I Structural Members, Specifications, & Codes*, American Institute of Steel Construction, Chicago, IL

AISI (1996), *Specification for The Design of Cold-Formed Steel Structural Members*, American Iron and Steel Institute, Washington, D.C.

AISI (1997), *Cold-Formed Steel Design Manual*, American Iron and Steel Institute, Washington, D.C.

Bellman, R. E. and Zadeh, L. A. (1970), "Decision-making in a fuzzy environment," *Management Science*, 17, B141-164.

CRI (1993a), *CRAY T3D System Architecture Osverview Manual*, Cray Research, Inc., Eagen, MN.

CRI (1993b), *MPP Fortran Programming Model*, Cray Research, Inc., Eagen, MN.

Gardiner, J.D. (1997), "A stabilized matrix sign function algorithm for solving algebraic Riccati equations", *SIAM Journal on Scientific Computing*, 18:5., pp. 1393-1411.

Hecht-Neilsen, R. (1987), "Counter Propagation Networks", *Applied Optics*, Vol. 26, No. 23, pp. 4979-4985.

Hecht-Neilsen, R. (1988), "Application of Counterpropagation Networks", *Neural Networks*, Vol. 1, No. 2, pp. 131-139.

Hegazy, T., Fazio, P., and Moselhi, O. (1994), "Developing Practical Neural Network Applications using Backpropagaton", *Microcomputers in Civil Engineering*, Vol. 9, No. 2, pp. 145-159.

Karim, A. and Adeli, H. (1999a), "OO Information Model for Construction Project Management", *Journal of Construction Engineering and Management*, ASCE, Vol. 125, No. 5, pp. 361-367.

Karim, A. and Adeli, H. (1999b), "CONSCOM: An OO Construction Scheduling and Change Management System", *Journal of Construction Engineering and Management*, ASCE, Vol. 125, No. 5, pp. 368-376.

Karim, A. and Adeli, H. (1999c), "A New Generation Software for Construction Scheduling and Management", *Engineering, Construction, and Architectural Management*, Vol. 6, No. 4, pp. 380-390.

Kim, H. and Adeli, H. (2001), "Discrete cost optimization of composite floors using floating-point genetic algorithm", *Engineering Optimization*, Vol. 33, No. 4, to be published.

Meirovich, L. (1990), *Introduction to Dynamics and Control*, John Wiley & Sons, New York.

Qi, C. and Adeli, H. (2001), "A Web-Collaborative Remote Computing Framework Architecture for Computer-Aided Design and Engineering", in preparation.

Park, H.S. and Adeli, H. (1997), "Distributed neural dynamics algorithms for optimization of large steel structures", *Journal of Structural Engineering*, ASCE, 123:7, pp. 880-888.

Rutishauser, H. (1990), *Lectures on Numerical Mathematics*, Birkhauser, Boston, Massachusetts.

Saleh, A. and Adeli, H. (1996), "Parallel eigenvalue algorithms for large-scale control-optimization problems", *Journal of Aerospace Engineering*, ASCE, Vol. 9, No. 3, pp. 70-79.

Saleh, A. and Adeli, H. (1997), "Robust parallel algorithms for solution of Riccati equation", *Journal of Aerospace Engineering*, ASCE, Vol. 10, No. 3, pp. 126-133.

Sarma, K. C. and Adeli, H. (1998), "Cost optimization of concrete structures," *Journal of Structural Engineering, ASCE*, 124(5), 570-578.

Sarma, K.C. and Adeli, H. (2000a), "Cost optimization of steel structures," *Engineering Optimization*, 32(6), 777-802.

Sarma, K.C. and Adeli, H. (2000b), "Fuzzy genetic algorithm for optimization of steel structures," *Journal of Structural Engineering, ASCE*, 126(5), 596-604.

Sarma, K.C. and Adeli, H. (2000c), "Fuzzy discrete multicriteria cost optimization of steel structures," *Journal of Structural Engineering, ASCE*, 126(11), accepted for publication.

Senouci, A. and Adeli, H. (2001), "Resource scheduling using neural dynamics model of Adeli and Park", *Journal of Construction Engineering and Management*, accepted for publication.

Sun Microsystems, Inc.(1997), Remote Method Invocation Specification, http://java.sun.com/products/jdk/1.1/docs/guide/rmi/spec/rmiTOC.doc.html

Tashakori, A. and Adeli, H. (2001), "Optimum Design of Cold-Formed Steel Space Structures using Neural Dynamics Model", to be published.

UBC (1997), *Uniform Building Code - Volume 2 - Structural Engineering Design Provisions*, International Conference of Building Officials, Whittier, California.

Yu, G. and Adeli, H. (1991), "Computer-Aided Design using Object-Oriented Programming Paradigm and Blackboard Architecture", *Microcomputers in Civil Engineering*, Vol. 6, No. 3, pp. 177-189.

Yu, G. and Adeli, H. (1993), "Object-Oriented Finite Element Analysis using an EER Model", Journal of Structural Engineering, ASCE, Vol. 119, pp. 2763-2781.

2 Ethics and education

Creative Systems in Structural and Construction Engineering, Singh (ed.) © 2001 Balkema, Rotterdam, ISBN 90 5809 161 9

Ethics in engineering today

J. Belis & R. Van Impe

Laboratory for Research on Structural Models, Department of Civil Engineering, Ghent University, Belgium

ABSTRACT: Every human action, including the scientific, has an ethical side. Due to a growing number of negative aspects of science and technology, ethics have become a widely discussed issue. Because of their knowledge and important position in society, engineers should have lines of conduct to answer ethical questions. The question is considered whether or not it is advisable or even possible to limit scientific freedom. Our fundamental attitude is that no limits can be tolerated, as long as the respect for life, in its largest meaning, is assured. It is made clear to what extent engineers are responsible for negative consequences of science and technology. Within certain conditions, this responsibility can shift to other levels. Engineers do have, however, ethical duties. Three universal baselines are dealt with. Finally these theoretical baselines are made clear according to the practical lines of conduct developed by the Royal Flemish Society of Academic Engineers.

1 INTRODUCTION

This paper is the written result of a further elaboration of ideas about ethical engineering as they were developed at the Laboratory for Research on Structural Models of the Ghent University (Vandepitte 1989). The reader should be aware that the presented is a rather personal view on the subject, even if the authors claim to be familiar with recent literature in this important and interesting tangent plane of engineering sciences. In an analogy with architectural critique –which starts from an individual point of view also- we believe that this approach does have a value in opening a window on the subject for other engineers.

In the history of mankind the changes of human society have never succeeded each other as rapidly as they do now. The exponential growth of social and cultural changes is inseparably connected with a parallel explosive –we choose our words carefully-development of science and technology. The public opinion recognizes and welcomes the enormous advantages offered by new technologies but at the same time also becomes more and more aware of their negative aspects. What is meant by "negative aspects"? Sometimes "negative" is understood to be related to the "public interest". This is a wrong interpretation. First, we find it hard to believe that there is such thing as "public interest" in dictatorial or totalitarian states. But even in a well-established

democracy, there are innumerable social groups, some with and others without a political face, but all with different interests and varying power. The confrontation of those powers will lead up to a social equilibrium that is acceptable for the major players, but there will always be certain groups that want exactly the inverse of the final outcome. "Negative" is not related to public interest, but to public health or public safety, two conceptions that are more exact and objective to work with (McFarland 1986).

2 NEGATIVE OUTCOMES OF SCIENCE AND TECHNOLOGY – SOME EXAMPLES

In order to make clear what we mean with negative aspects and consequences of science and technology, we give a few examples.

Nuclear physics led to the development of atomic bombs, two of which exploded on Hiroshima and Nagasaki in 1945 and killed more than a hundred thousand people.

Experiments with hydrogen bombs brought a haze of radioactive isotopes into the earth's stratosphere. Years after the explosions strontium and cesium particles still reach the surface of the earth. As a consequence, they get in the food chain and eventually in the human body, where they can cause genetic mutations, the effects of which are poorly known.

Certain chemical industrial processes produce enormous quantities of polluted air –by turning it to "acid rain"- and water, two vital elements for life on this planet, with considerable environmental and even ecological impacts (Vandepitte 1989).

Intensive artificial manuring makes phosphates and nitrates land in brooks and ditches. This is one of the reasons why some rivers are biologically almost or totally dead.

Disasters like that of the nuclear power plant of Chernobyl or accidents with large oil tanker ships together with environmental impacts on the ozone layer and quality of water and air are still clear in everybodies mind.

3 ESTABLISHING ETHICAL GUIDELINES: A PARADOX?

Being among engineers, our important role and responsibility in the development and control of technological know-how should not need explanation. In practice however, this responsibility can put the engineer in very serious moral problems and conflicts, which he cannot possibly solve without a well-thought basis and guidelines to answer ethical questions.

The idea of ethical guidelines is a paradox in itself. On the one hand we have the guidelines, which are, as a whole, meant as a universal instrument. In order to be applicable to different situations and to different persons, all with their own beliefs and values, guidelines of any kind should be of general value. This universality is a necessary condition to allow a group –engineers in this case- to accept guidelines as being theirs, and to identify their own faith with it. Hence, a driven generalization brings along the danger of meaninglessness. In an attempt to address all, guidelines might miss a basic profoundness, resulting in hollow phrases and overshooting the goal. If so, guidelines become redundant and useless.

On the other hand, we have the ethical aspect. When we look at ethics as a science, we notice very different tendencies, going from consequentialism over utilitarianism to situationism and universalism (Mortier & Raes 1997). The differing emphasis in ethical thinking is a source of discussions –which can be a moral enrichment *in se*- and can lead to completely contradictory, yet ethically perfectly tenable conclusions. Ethics are based on moral values, which are personal. The moral consciousness and sensibility are depending on education, knowledge of life and culture. It is a personal matter.

Combining those apparently irreconcilable elements of generality and personality at the same time, the concept of "ethical guidelines" appears like a *contradictio in terminis*. If this were the case, our argument would stop here, but, as mentioned before, we are dealing with a paradox.

4 LIMITATIONS

Before figuring out the answers to these difficulties, it should be examined whether or not it is advisable or even possible to limit scientific freedom.

Ingenuity is what made the difference between the *homo sapiens* and other living species. Throughout the biological evolution it became part of the human nature to be subjected to a continuous hunger for knowledge. Practically, trying to impose boundaries on this genetic characteristic of humanity would be an illusion.

Even facing this practical impossibility, it would be interesting from a philosophical and ethical point of view to examine the desirability of circumscribing scientific freedom. Can negative consequences of technological developments be a valuable reason for curtailing further scientific research? We believe that scientific research itself is morally neutral, elusive for moral values by definition. It escapes the level of "good" and "bad", judgements that belong to the way in which –and for what purposes- science is used.

Microwaves were originally developed in the 1920's to serve as a weapon, but their wavelengths made attack distances much too short to be used. Several decades later this technology was rediscovered and applied in magnetron ovens that are present in every modern kitchen today. Developing a new technology for human destruction is seldom seen as a noble and ethically sound purpose. That same knowledge however, applied this time in a housekeeping-apparatus, would not probably challenge any one's morality.

If new scientific progress is applied with a philantropic purpose, but seems to cause unforeseen disastrous consequences, even then our reasoning is still valid. When the new pharmaceutical product called Softenon was released in Europe in the 1960's to prevent women from pregnancy inconveniences, they had no idea that that same product would cause serious physical handicaps to their children. The scientific research to answer the need of such products can ethically perfectly be justified. Fatal moral and human errors, however, have been committed when Softenon had been released to the market without a thorough investigation to possible medical side effects. We come back on this issue later.

5 RESPONSIBILITIES

Following the reasoning above, ethics should be situated at the level of using (and deciding to use) new technologies and not at the technical level itself. When we look at employment, it is clear that most engineers are working in a company owned by others. Usually they start as a junior engineer and throughout their career they can climb to the top: the upper management or staff. Normally it will be the manager who decides about the use of new technological developments together with the client, the user or the government. When we talk about ethics in engineering, people in management functions are not our prime goal. When the engineers who developed the whole technology have finished their tasks, they usually won't have a dictate in the final applications. It is clear that any further responsibility will be on the behalf of the management, users or government involved.

This won't, however, ethically discharge the engineer! An engineer should always follow his own conscience. He should have space to hang back or even refuse certain projects if they are not morally sound to him. Once he accepts a new project, he has to keep in mind three baselines to maintain his ethical equilibrium at all times.

First we refer to the Softenon case. Developing new technologies or scientific offspring means: examining all aspects of the problem, including the negative ones. It should not need mentioning that costs or deadlines are no arguments to skip the "dark side" of the examination.

We began our argumentation starting from the negative aspects of science and technology to develop the guidelines as they are presented here. They do have, however, a more general applicability in ethical engineering. We will illustrate this shortly in the following example.

The Laboratory for Research on Structural Models was asked for assistance by an engineering assessment agency in a case of thin-walled steel shell silo failure. We presented several possible failure scenarios to be examined, one of which was refused by the assessor because it would not be favourable to his client. This is a violation of the first guideline, resulting in a behaviour that is ethically inadmissible.

Secondly, acquiring a global picture is not enough. Being a specialist in complicated high-tech problems, the engineer has the duty to share his knowledge with managers, clients or government according as relevant. Being wrongly or incompletely informed will drastically reduce the latter's responsibility for negative consequences; not only ethically but most of the time even legally. Only fulfilling his information duty profoundly, exhaustively and in time will liberate the engineer from further responsibilities. These are the limits of what can be generally imposed about the engineer's responsibilities. Occasionally it happens that the engineer has fulfilled this duty, but that management neglects or overrules his advice. Having observed his responsibilities and therefore being objectively spoken free of charge, he may feel that the decisions of his superiors are inconsistent with his own conscience. At this point the engineer should weigh his concerns against each other, including the importance of the subject and possible consequences on his personal career and job opportunities, and could decide the matter to be important enough to bring his grievances, knowledge and information to the levels where they will be heard. This can be to a higher level of management, but *in extremis* even to the government, to the press, or in court. In literature this is usually referred to as "whistle blowing" (Johnson 1991, Unger 1994). We feel that this is a real ethical problem, where the solution will depend on the personal moral values of the engineer concerned. It is not possible to make general statements about this kind of problem. Those who want more profound information about this phenomenon should review the references.

Thirdly, the engineer is best placed to solve eventual problems. With his knowledge and understanding of both technological renewals and their occurring negative side effects, he has the moral duty to try to solve or counter everything that endangers the public health or the public safety. The question of "who is bearing the final responsibility" should not influence the engineer's assiduity for attempting to find a solution to the problems he is facing.

As the attentive reader will have noticed, the difficulty in practical suitability of the guidelines increases towards the third one. In a healthy employment situation the company staff normally will want to know all possible side-effects of a new product. With this information they normally will take precautions to avoid disasters. The tendency of building up a career by working for a rather short period for the same employer and changing jobs –and company- frequently could make managers decide in favour of short-term advantages instead of durable solutions. By the time negative consequences evolve, the manager will probably have left the company. Such an attitude will make it hard for the engineer to ensure that his information will reach the level where it will be heard, according to the second ethical guideline.

The third guideline -finding a solution- will be impossible most of the time when taken literally. The engineer won't usually get time to start such research during or besides his own full-time job. The third guideline is meant as follows: the engineer has the ethical duty to search for solutions or, if this is impossible, to stimulate and defend to his superiors any research to find solutions to solve negative consequences.

The idea behind those three baselines always goes back to the same fundamental principle: respect for human life and welfare. In the preceding argumentation we have called this public safety or public health.

New technological developments are inseparably woven into economical progress. Prominent philosophers have expressed their fear about the growing power of the marketplace (Kruithof 2000). Politics, the traditional ultimate power in our civilisation, sees its power restricted today because of geographical, ethnical, linguistic and historical boundaries. These kinds of boundaries do not have, on the contrary, serious obstructional effects on today's real powers: the power of the free market economy. Huge multinationals are present across the whole world, working in an atmosphere that is quasi immune to national fluctuations. The scale at which today's multinationals are operating exceeds the scale of every political activity, placing the latter's objective power in a very relative perspective, to the advantage of the first.

Let us take a look now at the relation between today's new power, the economy, and the engineering working field of technology and high-tech sciences. Technology is the motor of our economy. Until the first part of this year, technological and telecommunication shares were the best a stockholder could imagine, yielding him profits he could hardly have dreamed of before.

Regarding the ethical position of the engineer, this is a blessing and a curse at the same time. The ideal as mentioned, ethically guiding today's engineers, will bring them from time to time into positions that are straight opposite to the objectives of the marketplace. Remember the manager's decision countering the engineer's advice, and putting him into a situation of obeying or damaging his career. There is a very serious economic dependency of the engineer on his employer. The blessing is that this is the case the other way around, too. Without engineers, there will be no technological progress and no spectacular economical growth anymore. Using this establishment is not self-evident. Forming a block as a profession and all pulling the same engineering rope is a far-away ideal. The guidelines here presented should form a firm basis, general enough to reach most engineers, but respecting their own morality at the same time, to grow closer to that ideal that would provide engineers the space to work in a responsible and morally healthy way. Ways in which this can be realised are education, professional organisations, publications and open colloquia like this. The idea of "union is strength" is not new, of course: hence, numerous professional engineering organisations, often with their own ethical codes, have been founded in the past

As an example, we will take a closer look in the following at the KVIV, the Royal Flemish Engineering Society of Belgium.

6 KVIV

The "Koninklijke Vlaamse Ingenieurs Vereniging" or KVIV in short was founded in 1928 and groups over 12,000 members in the only official academic engineering society in Flanders, Belgium. KVIV is an open and flexible society. Its continuing advanced education programs and publications aim at staff, entrepreneurs, teachers and officials, regardless of their KVIV membership. Due to its structure and many working groups, it can continuously offer efficient and accurate services. Thanks to this, a systematic contact with both a various public and academic and industrial circles is possible. In a European context, KVIV is a member of CLAIU (Comité de Liaison des Associations d' Ingénieurs Universitaires de la Communauté Européenne) and FEANI (Fédération Européenne des Associations Nationales d' Ingénieurs).

What makes the Flemish situation interesting is that KVIV is a relatively small engineering association that does not suffer from heavy administrative structures, and that it represents almost all academic engineers of Flanders (over 85 percent of all engineers graduated from Flemish universities).

6.1 Guidelines for the engineer: Moral values

The ethical codes of the KVIV, representing most of the ideas mentioned in this paper, will be given in the following as an example (KVIV 1999).

6.1.1 Responsibility
In exercising their profession, engineers are responsible for their personal decisions, as individuals, as members of a group and of their profession.

6.1.2 Commissions
Engineers will only accept mandates they think they are capable of, individually or in collaboration with others.

Engineers will mention conflicts of interest to all parties involved.

Engineers will guard the confidentiality of information as far as this won't cause injustice to others.

Engineers will inform their employer's customers correctly and will accomplish their agreements faithfully.

6.1.3 Civic spirit
Engineers will qualify themselves the best they can in technological sciences to contribute efficiently to the broad debate where all scientific disciplines and social tendencies are important. This debate will lead to choices of technological developments where the certainty of existing insights and the uncertainties of future developments are carefully balanced in an acceptable way. Engineers will pursue their chosen goals loyally.

Engineers will obey the law and respect the culture of the country they are working in. They will actively make propositions to enhance the legislation quality.

6.1.4 *Good-fellowship*

Engineers will handle the power that results from the execution of their function with circumspection. They will postulate the spiritual and physical integrity of everyone concerned as an unconditional priority. Engineers will support their colleagues and co-workers in their professional development and in their ethical behaviour. They will respect the inventions of others as their intellectual property and they will recognise their essential contributions.

Engineers are aware that they belong to a professional group with a specific vocation. It is from there on that their loyalty and good-fellowship will grow. They will not see each other as competitors, but as colleagues who support and learn from one another. They are fully aware of bearing personal responsibility for the image of their professional group.

7 CONCLUSIONS

Knowing the huge powers of technological developments, engineers should be very aware of their enormous responsibilities towards others. Possible negative side effects will often endanger the public health and the public safety, which represent the idea of respect for human life and environment and which should both be the major concerns of whoever is working in the field of engineering. Holding back this as a main moral value, three basic lines of conduct have been presented, solving the paradoxical requirements of combining generaltity and personality in ethical guidelines. Taking into account today's overwhelming influence of the free market economy, it is clear that the presented engineer's way of avoiding, handling and solving negative technological consequences will come into collision with main economic principles of quick profit. We have to remark that this is not to be seen as a totally pessimistic view. Often, even most of the time, managers will listen to their engineer's advice, accepting their professional know-how with respect. Furthermore, under pressure of the public opinion, a phenomenon that is important even to huge multinationals, a lot of companies will pay high prices just to avoid any negative association with their products. The effects on pollution, animal tests, general product reliability and safety, general employment conditions and security are legion. This automatically leads to a healthier working atmosphere for the ethical engineer.

Despite this positive evolution, probably every engineer will have to face decisions that are morally not compatible with his own ethical insights. The authors of this paper, being convinced that every ethical problem has to be solved individually by the person involved, share the opinion that the engineer should look for a solution according to the baselines here presented, and reckoning with his personal ethical beliefs. This is due to the personal character of ethical problems by definition. It is clear that different individuals can come to different but ethically justified solutions to the same problem.

In order to extend a healthy ethical working atmosphere, it is important that engineers become aware of these problems in time –i.e. before they become their own problems- and form a pressure group to achieve this goal. In addition to education and publications, professional organisations can play an important role for that. Most of these organisations have formulated their own version of an ethical code. These statements are under no condition to be understood as rules or as a "cook book", but as guidelines. The KVIV guidelines are presented as a good example. Throughout the world there is too much variation in cultural, political and social factors to strive for a unified, world-wide engineering code of ethics. We believe that the three baselines here presented are at the same time general and meaningful enough to serve as a basis for every ethical engineer or ethical engineering code.

REFERENCES

Johnson, D. G. 1991. *Ethical issues in engineering.* New Jersey: Prentice Hall inc.

Koninklijke Vlaamse Ingenieursvereniging (KVIV) 1999. *Member proceedings.*

Kruithof, J. 2000. *Het neoliberalisme*. Berchem: Epo.

McFarland, M. C. S. J. 1986. The public health, safety and welfare: an analysis of the social responsibilities of engineers. *IEEE Technology and society magazine.* 5(4):18-26.

Mortier, F. & Raes, K. 1997. *Een kwestie van behoren. Stromingen in de hedendaagse ethiek.* Gent: Mys & Breesch.

Unger, S. H. 1994. *Controlling technology. Ethics and the responsible engineer.* New York: John Wiley & Sons.

Vandepitte, D. 1989. Ethiek in wetenschap en techniek. Speech at the annual meeting of the Algemene vereniging der ingenieurs uit de Universtiteit Gent (AIG).

Creative Systems in Structural and Construction Engineering, Singh (ed.) © 2001 Balkema, Rotterdam, ISBN 90 5809 161 9

Ethics – Essential quality of the civil engineering profession

Z. Cywinski
Technical University of Gdansk, Poland

ABSTRACT: In the paper various issues of the engineering ethics are discussed. Referring to some foundations of the normative ethics, the particular problems of the applied ethics are considered. The opinion is expressed that at present the relevant philosophy and its practical applications must recognize that "true development" should reflect the dignity of man - basing upon the priority of spirit versus matter, of person versus object, of ethics versus technology.

1 INTRODUCTION

On the verge of the 21st century, our human gender faces many new and extraordinary challenges. The globe is presently stamped by vigorous changes, to mention only the strong rise of world population, the enormous move in the development of science and technology, and the acceleration of the industry production. Their relevant main determinants are the industrial civilisation and the overwhelming consumption - on the expense of the decay of the traditional society and the damage of our natural environment. It is apparent that a new ideology for the man's way of life is required, in order to secure his survival (Glomb 1999). A sustainable development of mankind is needed, the feature of which should be not only material but - first of all - spiritual; its practical implementation must be preceded by a relevant philosophical vision (Glomb 1999, Cywinski 2000).

The mentioned vision should be addressed to the human being as its central subject. Therefore, the present technology has to be introduced by ethics. This refers, fundamentally, to the profession of the civil engineer who must strive for a harmonious habitat of ordinary people. This paper is aimed on the accentuation of ethics in such performance of the civil engineer. Hereby, more stress will be put rather on the philosophical basis of that performance, than on its practical connotations. The development of a proper intellectual potential of the civil engineer is here of top importance (Cywinski 1997). It is the

"conditio sine qua non" to connect the purely technical activities of people with the strategy of sustainability - preserving the eco-system of mankind in its local and global dimension. A rational and creative thinking and a sound moral feeling of the civil engineer appears here to be the necessary and unquestionable precondition.

2 BACKGROUND

Within the profession of civil engineering, certain problems of ethics are being observed since several years; various codes of ethics - that of the American Society of Civil Engineers (ASCE) in particular - are here the representative examples (ASCE 2000, Cywinski 1993). Codes of ethics have been developed mainly for practical, professional use - supplementing properly the building codes and other acts of law. Accordingly, while the ethical aspects were considered also as far as their fundamental theory and concepts were concerned (normative ethics) - especially those based upon man's natural morals (Drabarek & Symotiuk 1999, Whitbeck 1998) - the problems of engineering ethics were considered, typically, more in terms of their practical implications in form of applied ethics (Schaub & Pavlovic 1986, Whitbeck 1998).

The relevant need appeared mainly due to the extreme development of the profession and of its relations with the public. Specific reciprocal position of employers, workmanship, peers, and clients; rise of characteristic moral dilemmas; deepening of the understanding of professionalism; appearance of "whistleblowing", i.e. - awareness of ethical conflicts; as well as the relationship between the standards of ethics and the constitution of the professional authorization and the maintenance of competence - were some topics of interest (Schaub & Pavlovic 1986).

Ethical questions of engineering design became another issue of the professional ethics. The basic and central scopes of the adequate responsibility, models of professional behaviour, workplace rights and responsibilities, ethics in the theoretical and experimental research, special responsibility for the environment (ecological thinking), and the credit and intellectual properly in engineering practice, were other matter of concern (Whitbeck 1998).

The standards of professional ethics in civil engineering have been widely promoted by ASCE - not only through the mentioned Code of Ethics. By installing a special award (Daniel W. Mead prize), granted annually to younger members and students - in recognition of their relevant original input (Barkdoll 1998, Mason 1998) - ASCE acknowledged the values of ethics in the performance, competence and professional development of civil engineers; their direct involvement into the design, construction and maintenance of various forms of habitat and public infrastructure became here the basic reason. The problem of the engineers' responsibility for the safety of their structures was specially raised and the importance of the related morals, underlined (Robinson 2000). Particular attention was paid to some real ethic cases in professional practice - by introducing, in the Journal of Professional Issues in Engineering Education and Practice, a regular series of columns on engineering ethics. Similarly, structural failure cases have been chosen to develop appropriate determination of students and practitioners (Delatte 1997). Ethical maturity has been found as the mandatory attribute in generating the engineer's social leadership (Duffield & McCuen 2000).

Naturally, corresponding efforts are presently conducted by many other professional bodies, as well - to mention only the International Association for Bridge and Structural Engineering (IABSE) that recently formed an Ethics Committee, in order to encourage the observation of high ethical standards in the practice of structural engineers, members of IABSE (Hanson 1999).

3 HEART OF MATTER

In this paper, normative ethics are dominant points of attention. As it was shown earlier (Cywinski 1997), ethics are largely influenced by different items of the cultural landscape, such as society, culture, economy, environment, aesthetics, heritage, sustainability, information technology, etc., whereby the "cultural landscape" was understood as a "place that has been created, shaped and maintained by the links and interactions between people and their environment". A particular junction combines the environment and sustainability - producing the term of "environmental sustainability". It should by added that "sustainability" results from the notion of "sustainable development" that, originally, has been identified in the following form: "sustainable development meets the needs of the present without compromising the ability of future generations to meet their own needs" (Cywinski 1997).

According to ASCE, "sustainable development is the challenge of meeting human needs for natural resources, industrial products, energy, food, transportation, shelter, and effective waste management while conserving and protecting environmental quality and the natural resource base essential for future development" (ASCE 2000).

There is a particular relation between ethics and sustainability. As it follows from the 1st Fundamental Canon of the ASCE Code of Ethics, "engineers shall hold paramount the safety, health and welfare of the public and shall strive to comply with the principles of sustainable development in the performance of their professional duties" (ASCE 2000); protection and improvement of the environment through the primary practice of sustainable development became here a top recommendation of the relevant guidelines.

This particular situation was studied by several scholars; their main question was: "What are the ethical responsibilities (of the civil engineer) towards sustainable development?" (Brennan 1998, Johnson 1998, McWhorter 1998). The preservation of our global supplies and the development of renewable resources have been acknowledged to be dependent on the common foundation of human ethics. The opinion was expressed that engineers, as professionals and regular humans - inspired by ethical motivation, can become serious "facilitators of sustainable development" (Brennan 1998). Civil engineers have been found "ethically obligated to practice the profession in a manner that encourages the development of eco-sensitive technologies and planning practices"; "the use of sustainable technologies, the practice of cost-effective risk management, and the integration of environment and economics in engineering decisions" have been recognized to be "ethical methodologies" (Johnson 1998). Finally, it was stressed that the actual development should be governed by the "use of recycled materials, source reduction and choosing rehabilitation over replacement" whereby engineers have been addressed "to do all they can to incorporate those practices into their designs"; this was declared as another ethical duty of the engineer (McWhorter 1998).

The evolution of the engineers' applicable disposition requires high standards of their thinking capabilities - their proper spiritual formation. Unfortunately, today those attributes decline and a "mechanistic", purely materialistic interpretation of the human nature becomes more popular. However, there are also numerous examples opposing such interpretation.

Reference to the past achievements of engineering - its heritage - becomes presently an important help in the strive to form thinking and creative engineers. "By understanding ... the humanness of those who once dreamed of what we now so often take for granted, we not only engage ourselves in the technosocial endeavour that involves engineers at its core, but we also understand how their human natures and their dreams affect the way we experience our cities and towns, our borders, and our open spaces" (Petroski 1996); it is evident that our forefathers were convinced that technology relied on "the full human dimensions of engineers". Simultaneously, they were aware of limitations in our resources, to quote only the bridge engineer G. Lindenthal (1850-1935), who stated: "The large sources of energy in nature, coal, water power, wind, tides, heat of the sun, etc., can none of them be utilized without large masses of iron ..." and "the colossal consumption of iron will have come to an end" (Petroski 1996).

History and heritage of engineering have been strongly referred to also elsewhere. By introducing some famous civil engineers and their projects, several of the latter being now accepted landmarks, the problem of ethics and professionalism was approached - resulting in the birth of the "civilized engineer" idea. The view was expressed that the civil engineering profession should develop, above all, "an appropriate level of interest and concern for the humanistic dimension of the profession" (Frederich 1989).

There is a group of scholars who specially emphasize the importance of the spiritual foundation of the environmental ethics for engineers (Vesilind & Gunn); author of this paper is one of them. Both former authors believe that a spiritually based ethics could largely effect the performance of engineers. Hereby "ethics may be defined as the systematic analysis of morality". Classical ethics addressed mainly the mutual human-human relations but had difficulties when considering nonhuman nature. There can be a help when the notion of anthropocentric environmental ethic is considered, in term of its influence on people. It is worthwhile to mention that the environmental policy is largely effected also by aesthetic considerations. Spiritually based environmental ethic can be realized by two approaches:

• acceptance of an organized religion or faith,
• development of a personal spirituality.

The opinion was advocated that both of them may help to form the proper moral behaviour and the professional performance of engineers.

Author of this paper represented the view that "true development" should not be limited to pure technical problems only; "human ecology" must be taken into account. The spiritual motivation can be often a decisive argument in the professional undertaking of engineers. This means that human development should be referred, all of first, to the dignity of man - basing on the priority of spirit versus matter, of person versus object, of ethics versus technology. Accordingly, author suggested the following definition of sustainable development "Sustainable development meets the holistic - spiritual and material - needs of the present without compromising the ability of future generations to meet, correspondingly, their own needs" (Cywinski 2000); ethics were here the decisive determinants.

4 CONCLUSION

The performed analysis has shown that ethics are presently going to be a very important issue of the engineering profession. Basing upon the principles of the normative ethics, its applied shape becomes one of the foundations of any engineers' endeavour. Such standing should be paramount in the education and practice of the civil engineers, in particular.

REFERENCES

American Society of Civil Engineers (ASCE) 2000. *ASCE Official Register 2000.*

Barkdoll, B.D. 1998. Standing firm: when ethics are challenged. ASCE, *Civil Engineering* 68(11): 65-68.

Brennan, R.A. 1998. What are the ethical responsibilities towards sustainable development? ASCE, *J. Profl. Issues in Engrg. Educ. and Practice* 124(2): 32-35.

Cywinski, Z. 1993. American code of ethics for engineers (in Polish). *Inzynieria i Budownictwo* 50(8): 319-321.

Cywinski, Z. 1997. Humanities & arts – essential agents of contemporary engineering education. *Proc. SEFI Annual Conf., Cracow (Poland), 8-10 Sept 1997*: 22-35. Cracow: Oficyna Cracovia.

Cywinski, Z. 2000. Current philosophy of sustainability in civil engineering. ASCE, *J. Profl. Issues in Engrg. Educ. and Practice* 126: in print.

Delatte, N.J. 1997. Failure case studies and ethics in engineering · Mechanics Courses. *J. Profl. Issues in Engrg. Educ. and Practice* 123(3): 111-116.

Drabarek, A. & S. Symotiuk (eds) 1999. *Hominem quero · studies of ethics, aesthetics, history of science and philosophy, history of social ideas* (in Polish). Lublin (Poland): Wydawnictwo UMCS.

Duffield, J.F. & R.H. McCuen 2000. Ethical maturity and successful leadership. ASCE, *J. Profl. Issues in Engrg. Educ. and Practice* 126(2): 79-82.

Fredrich, A.J. (ed) 1989. *Sons of Martha*. New York, NY: ASCE.

Glomb, J. 1999. *On certain determinants of the present time* (in Polish). Gliwice: Wydawnictwo Politechniki Slaskiej.

Hanson, J.M. 1999. Should IABSE have a statement of ethical principles for the practice of structural engineering? IABSE, *Structural Engineering International* 9(2): 156.

Johnson, D.B. 1998. What are the ethical responsibilities of the civil engineer to provide sustainable development. ASCE, *J. Profl. Issues in Engrg. Educ. and Practice* 124(2): 35-37.

Mason, R.R. 1998. Ethics: a professional concern. ASCE, *Civil Engineering*, 68(12): 63-64.

McWhorter, R. 1998. What are the ethical responsibilities towards sustainable development. ASCE, *J. Profl. Issues in Engrg. Educ. and Practice* 124(2): 37-39.

Petroski, H. 1996. *Engineers of dreams*. New York NY: Vintage Books.

Robinson, Ch. 2000. Ethics: a design responsibility. ASCE, *Civil Engineering*, 70(1): 66-67.

Schaub, J.H. & K. Pavlovic 1986. *Engineering professionalism and ethics*. Malabar, Fl: Robert E. Krieger Publishing Company.

Vesilind, P.A. & A.S. Gunn 1999. Spiritual dimensions of environmental ethics for engineers. ASCE, *J. Profl. Issues in Engrg. Educ. and Practice* 125(3): 83-87.

Whitbeck, C. 1998. *Ethics in engineering practice and research*. Cambridge (UK): University Press.

Creative Systems in Structural and Construction Engineering, Singh (ed.) © 2001 Balkema, Rotterdam, ISBN 90 5809 161 9

Mentorship: An empowerment tool for emerging contractors in a developing country

A.C. Hauptfleisch
Department of Quantity Surveying and Construction Management, University of Pretoria, South Africa

ABSTRACT: In developing countries one of the problems encountered is the empowerment (ability to manage a sustainable construction enterprise) of construction contractors who have not progressed through institutional education and training, but through unstructured experience on site. These emerging contractors experience a multitude of problems due to a limited knowledge base, resulting in low performance, low quality and uncompetitiveness. To contribute to the solution of this problem, mentorship was researched as a possible solution for this deficiency in the construction industry. The conclusions regarding which elements need to be addressed in a mentorship programme highlighted the following:
1. An affirmative procurement policy has to be adopted by government to support the process.
2. A mentorship assistance programme is needed to empower emerging contractors in as short a time span as possible.
3. An institutional education and training programme for emerging contractors' development has to be available.
This paper describes the research and findings, and how the three elements above were instituted, and are now being applied in practice in South Africa. The principles which were established is transferable to other developing countries.

1 INTRODUCTION

South Africa is a developing country, albeit a unique one. After the arrival of the first white settlers in 1652 an amazing sequence of events lead to the development of a nation, and an economy, which now displays all the development categories of a first world, third world and developing economy, rolled into one. The following specific features are observed:

- A population of approximately 40 million with the number of illegal aliens unknown, but possibly in excess of 7 million.
- Population growth is presently in the region of 3%.
- The first democratically elected government came into power in 1994, following a remarkable negotiated transition.
- With sweeping policy changes taking place in government and civil service, the learning curve is very steep regarding the process of capacity building and delivery.
- With one of the highest incidences (1 in 5) of the disease in the world, the impact of AIDS on the population is now starting to manifest itself socially, with important economic implications.
- Being in large part a developing country, demands on limited resources are severe.
- Speaking 11 main languages and having even more minorities, the country consists of minorities by virtually any norm except than on the basis of colour with approximately 70% black, 15% white, 5% Asian and 10% of mixed descent.
- An estimated 8 million households in South Africa have a combined income below the minimum subsistence level of R1,400.00 per month ($200 per month).

An obvious outcome from the above is that the affluent minority (mainly whites) and the poor majority (mainly blacks) have to find ways whereby a strong middle class, which includes all sections of the population, can be developed by empowering people to create a better life for themselves. In the construction industry this translates into one option (there are others), being the empowerment of emerging contractors. An emerging contractor is a previously disadvantaged (marginalized) person (mainly black) who strives towards the establishment of an economically sustainable enterprise - an enterprise that can compete with the well-established first world construction contractors. The latter having grown out of generations of experience and access to first world, internationally accredited, tertiary education institutions.

During a national workshop on emerging contractors in 2000, the following Construction Industry Development Board (CIDB) draft legislation definition of an emerging contractor was adopted: "*An emerging contractor (enterprise) means an enterprise which is owned, managed and controlled by previously disadvantaged persons and which is overcoming business impediments arising from the legacy of apartheid.*" (Department of Public Works & South African National Roads Agency Ltd 2000)

2 POSITION PRECIS

Since the establishment of the 1994 democratic government many role-players have worked, researched and offered "recipes" (solutions) to expedite the empowerment process relating to emerging contractors. These included every thinkable possibility, ranging from economic and social engineering to strict labour legislation to ensure equal opportunities.

The view and interest at the University of Pretoria are twofold, being firstly to provide tertiary education and research to the community and industry at large, and secondly to provide community service by assisting with the empowerment of emerging contractors. The first route is taken mainly by young people following the sound, though be it the longer process of tertiary education. The second being those people who cannot afford the first route for reasons of age, financial resources and family commitments - they simply have to be empowered in the "run".

The topic of this paper is the second group; providing an outline of the remedies offered by government and the support structures which have been identified through our ongoing research and work in this field. This has been endorsed by government and has been introduced as a national policy.

3 ELEMENTS OF AN EMPOWERMENT PROGRAMME SUPPORTED BY STRUCTURED MENTORSHIP

Through a process which involved a wide range of stakeholders from all sectors of industry, the following present structure has evolved:

3.1 *Affirmative Procurement Policy*

During its first term in office (starting in 1994) government embarked on a comprehensive affirmative procurement policy to empower people who have been disadvantaged economically under the previous government. In essence, a policy which enables people to enter the market by offering protection against established businesses, in this case, construction contractors.

In the construction industry the cornerstone of the policy is to provide tenderers with a tender advantage against competitors, effectively taking the latter out of the targeted sections of the market. This approach however showed immediate flaws as the emerging contractors are lacking knowledge, skills, resources and education and training, which the policy alone (awarding of contracts) could not address.

3.2 *Mentorship Programme*

It soon became clear from the type of practical problems arising from the above affirmative procurement policy, supported by field observations, that the following two very important elements were missing from the equation for success:

3.2.1 *Mentorship*
The first element that was insufficiently promoted was the provision of mentors to assist in knowledge transfer and to provide support to emerging contractors on a day-to-day basis. In this context mentorship is defined as:

"*The provision of assistance in construction project management and the purposeful transfer of skills.*"

In practice a missing ingredient in the empowerment of emerging contractors was thus the availability of knowledgeable mentors who could supply mentorship services.

"Training the trainer (mentor)" was rejected as it would render limited results in the short term. With affirmative procurement as a government policy, it was clear that the nett result would be abundant opportunities for emerging contractors, with shrinking prospects for the previously advantaged sector of the industry. Two core possibilities was thus evident early on, being firstly joint ventures (between existing well-established contractors and emerging contractors) and secondly the utilization of capacity becoming available in the established market as a result of the affirmative procurement policy of government and a sluggish economy. The first outcome (joint ventures) met with some success, but was not the subject of the research. The second, however, presented the unique opportunity to utilize first world skills to enhance the emerging contractor market, in the process supporting affirmative procurement by empowering emerging contractors and utilizing a growing oversupply of highly competent persons.

In South Africa no progress can be made with projects of this nature if all stakeholders are not involved. The modus operandi was thus to involve as many stakeholders as possible in our research in order to develop a mentorship programme. Part of the outcome was the development of the following guidelines:

- Guidelines for utilization of mentors by emerging contractors
- Mentor's guidelines (This guideline describes the requirements to become a Certified Accredited Mentor - by the University of Pretoria).
- Contractor-mentor agreement (incorporating a Mentor's Code of Professional Conduct).
- Emerging contractor evaluation.
- Bank manager's guidelines.

From the above it was soon clear that whichever way mentorship was approached, some core issues remained, namely who will pay the mentor and how does the payee ensure that the mentorship service is of a high standard. The other problem which was addressed to a certain extent (mainly in theory) by government, was to establish an Emerging Contractor Development Programme (ECDP), which was further addressed in the research discussed below.

3.2.2 Institutional Education and Training

To ensure that emerging contractors do not remain dependant on a mentor indefinitely an Emerging Construction Contractor Entrepreneurial Development Programme (ECCEDP) was designed for institutional education and training. This programme is delivered on a night school-basis and is mainly funded through sponsorships. Elements of the government's ECDP (mostly undelivered) was introduced into our programme, which is presented on an ongoing basis to groups of approximately 20 emerging contractors at a time.

3.2.3 Overview

The outcomes of the above were the creation of a practical institutional education and training programme (as stated above), and the establishment of private sector initiatives which translated into organizations that present mentorship assistance and on-site training. The most successful of the latter is established on campus at the University of Pretoria, which company operates in close association with the ECCEDP. Figure 1 provides an overview of this venture.

Figure 1.   Emerging Construction Contractor Mentorship and Entrepreneurial Development Programme

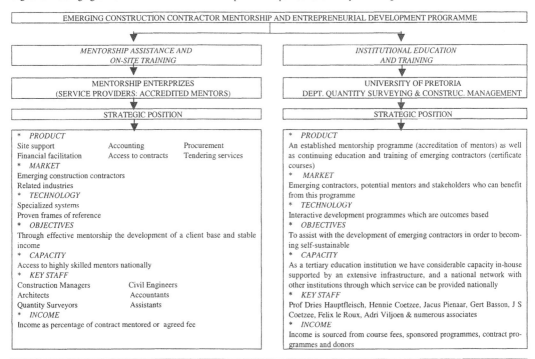

The net result is practical assistance and on-site training, supported by our ongoing institutional night school education and training programme.

4 NATIONAL RECOGNITION

The research that was done and the outcomes have attained national recognition as sighted in the following:

4.1 Budget Report (speech) of the Minister of Public Works, May 2000 (Department of Public Works 2000)

The following abstract is derived from the above speech of the minister, recognizing the outcomes of the work which has been done:

"*To ensure the success of this new programme, the Department has developed a mentorship programme in collaboration with the University of Pretoria and each project will incorporate mentorship as well as special measures to promote access to finance and sureties. We are determined to create an environment that enables the banks and financial institutions to shift their current stance in support of the emerging construction sector.*

In the challenges we face as a developing country, I want to emphasize that the construction industry is a national asset, central to the delivery of infrastructure and to the process of renewal in Southern Africa. Its full potential has yet to be realized in the changing environment that is impelled by our transition to democracy and, indeed, by increasingly rapid global change. A number of structural problems and constraints retard the industry's progress and require focussed action."

4.2 Registration and accreditation for the Department of Public Works Mentorship Programme

The following abstract has been made from a letter which was sent out by the Department of Public Works (Department of Public Works June 2000) nationally to all persons and enterprises (numerous of which are professional built environment practitioners) which have previously indicated that they are interested in providing mentorship services:

"*The Mentorship programme is to be included in specific projects as part of the Department of Public Works' Emerging Contractor Development Programme within the Strategic Project Initiative. You have previously forwarded information with regard to yourself to this office for inclusion onto the database of applicant mentors. Based partly on the response that we received from persons and organizations such as yourselves, we have continued the*

development of the Mentorship Programme and are ready to commence its implementation.

The process of accreditation of mentors is as follows:

(1) Accreditation with the University of Pretoria/Khula Enterprise mentorship programme;

(2) Completion of a Department of Public Works questionnaire that is aimed at evaluating the applicant's practical knowledge; and

(3) Interview with officials from Department of Public Works or their representatives.

The accreditation with the University of Pretoria involves a written examination that, depending on demand, can take place in Pretoria, Cape Town, Durban, Pietersburg, Bloemfontein and Port Elizabeth and a psychological evaluation."

The ECCEDP is now supported by government, also having agreed to include a financial provision in projects in order to finance mentor assistance.

5 CONCLUSION

The programme as developed has now advanced beyond the implementation stage and has the ability to contribute substantially to the empowerment of emerging contractors in South Africa. Important future steps to be taken include the establishment of a mentor association in order to create a professional home for mentors; whilst it will be of great benefit to the emerging contractors to become fully entrenched in the organized industry by becoming members of one of the member organizations of the Construction Industries Confederation (national umbrella organization).

The final question remaining is whether this South African model is transferable to other developing countries. The answer lies in a positive response to the following questions:

- Is it possible to obtain national consensus by all stakeholders that this is an acceptable way forward?
- Is it possible to establish an affirmative procurement strategy where necessary to protect emerging contractors during the initial stages?
- Are potential mentors available (or can they be imported) for accreditation to act as mentors in a professional capacity?

Although this approach is primarily based on a skewed free market situation (affirmative procurement) it seems inevitable that society needs to make that contribution to empower emerging contractors whilst the following generation develops through normal educational and training processes to ensure a free and competitive future market.

The time required to normalize and create a sustainable competitive society will be determined by the rate at which society itself reaches the stage when they insist on open free market competition. In South Africa's case, this is estimated to take between 10 and 20 years.

REFERENCES

Department of Public Works & South African National Roads Agency Ltd. 2000. *Towards the implementation of a national Emerging Contractor Development Programme: A framework for implementation.* Public sector workshop (March 2000). Pretoria: Department of Public Works.

Department of Public Works. 2000. *Budget Report 1999/2000* (May 2000). Cape Town: Department of Public Works.

Department of Public Works. 2000. *Registration and accreditation for the DPW mentorship programme.* Letter to all prospective mentors (June 2000). Pretoria: Department of Public Works.

Creative Systems in Structural and Construction Engineering, Singh (ed.) © 2001 Balkema, Rotterdam, ISBN 90 5809 161 9

Development and delivery of an engineering project management course for Indonesian nationals

W.A.Young, C.F.Duffield & B.Trigunarsyah
Department of Civil and Environmental Engineering, University of Melbourne, Vic., Australia

ABSTRACT: This paper outlines the development and delivery of a bilingual course in engineering project management for Indonesian nationals comprising of two-days of face to face lectures linked to a distant education program. The course was pilot tested on 28 Indonesian employees on a large mining operation in East Kalimantan Indonesia. If the course proved successful in matching in-country and organisational needs then a sustainable training program would be established and modelled using it as a base. The paper provides the background to the development and delivery of the course and its innovative approach to engineering management education in Indonesia. It outlines the course's development philosophy, the stakeholders involved and provides a summary of development surveys, including student performances and their response to training sessions. The practical aspects of adapting course materials into an Indonesian context are discussed, along with lessons learnt throughout the actual development.

1 INTRODUCTION

Engineering Project Management is a generic set of management skills used to facilitate the development of both civic infrastructure and the manufacturing industry. It is recognized now by the Indonesian Government and industry broadly, that for projects to achieve success, in terms of meeting budgets, schedules, and quality parameters, they must be led by skilled project managers.

The training program outlined in this paper was developed by the University of Melbourne, as part of an Australian aid (AusAID) private sector linkage program. The program partners comprised of PT Prakarsa Insinyur Indonesia (PT PII - the Institution of Engineers Indonesia, education provider), Engineering Education Australia (EEA – the Institution of Engineers Australia, education provider), Universitas Indonesia, and the University of Melbourne.

The consortium of EEA, PT PII, the University of Melbourne, and Universitas Indonesia have understood the national demand and priority of developing project management training with respect to the benefit it will have on supporting the management of Indonesia's physical development. The Competency Standards for Professional Engineers recently adopted by Persatuan Insinyur Indonesia (the Institution of Engineers Indonesia), have also stressed the need for skilled engineering project managers, who are able to foster sustainable project success. There

has consequently been a growing demand for training in engineering project management from many Indonesian organizations.

This paper concentrates on the development of a project management course resulting from a Training Needs Analysis (TNA) conducted by PT PII. An existing distance education course, 'Project Management: Conception to Completion' (Duffield 1997), was identified as suitable for adaptation for this purpose. Much of the body of project management knowledge has lent itself to adaptation contextually due to its generic nature. Thus, most of the modules were considered suitable for presenting internationally.

The new course was pilot tested on 28 Indonesian employees on a large coal mine in East Kalimantan Indonesia. If suitable, the course would be used for an integrated continuing education program for Indonesian engineers. The course comprised of a two-day on-site lecture program, followed by a one-semester distance education program.

2 EDUCATIONAL PHILOSOPHY

The TNA, along with specific client interviews (in this case mine management staff), identified the requisite subjects that had to be incorporated into the program. It was determined that the materials of an existing high quality distance education course

'Project Management: Conception to Completion' was suitable to be adapted to match the specific needs, conditions, and requirements of the Indonesian program. (An overview of the open learning mode of delivery has been presented previously (Duffield 1993))

Adapting the existing materials included:

- Developing a range of suitable project management case studies selected in conjunction with Indonesian engineering companies on the basis of their relevance to key sectors (mining, infrastructure, and energy).
- Learning and assessment activities which required application to the individual company's workplace specifications.
- Translation of key components into Bahasa Indonesia (Indonesian language).

Quality education programs in Indonesia have historically been delivered through face to face lectures and not via distance education. It was necessary for the consortium to demonstrate to the client the strengths and benefits of such programs. Apart from the cost benefits, one of the salient features of the combined short course / distance education program was that it created minimal disruption to normal business flow, particularly when the program was conducted within the company environment.

The educational philosophy of the distance education course is to deliver quality, industry relevant education, regardless of location. Such a program is reliant on access to swift forms of communication, such as the 'Internet', and appropriate student support throughout the course. Checks and balances were required to ensure constant participation by students. The strategy adopted to overcome perceived difficulties with distance education was to develop an alternate approach combining a balance of face to face teaching with distance education materials. This approach is detailed more fully in Section 3.

2.1 Specific Requirements for the Indonesian Program

- It was important to recognize that a course developed in Australia was based upon western theories of thinking and philosophy and it was necessary to be relevant to the cultural context, organizational priorities and the personnel needs of the Indonesian participants. This understanding was achieved by having educators who had sound work experience in project management in Indonesia.
- Before embarking on the program, it was also essential to understand the needs and expectations of the client. This was achieved through questionnaire surveys and communications via the Internet.

- A local partner, in the Universitas Indonesia, was selected who was knowledgeable on the topic area and able to provide background advice on local and cultural issues.
- International presenters prepared material on a range of topics that may have been of interest to the students and were able to present other team members material. This was important in case of illness of a presenter or difficulties with spoken accents (Somers 1996).
- With the face to face education program, maximum provision for audience participation was recognized as a key success factor (Somers 1996). The program incorporated group discussions, mini presentations, and simulation exercises.
- The course was developed on the principle that the training should make the most effective and efficient contribution possible to achieving the successful implementation of the client's organizational goals. There was a concern for doing the 'right' training, rather than doing training 'right' (Brinkerhoff 1989).
- The training program aimed to (Brinkerhoff 1989):
  - Emphasize client service - needs, expectations (sometimes these were different). Including, surveys of client satisfaction.
  - Create close ties with management (to ensure ongoing relevance).
  - Focus evaluation on specific training 'targets'.
  - Foster an environment that promoted professional thinking and practices.
  - Recognize individual patterns of learning and be sufficiently flexible to be able to adapt alternative styles.
- It was appropriate to use the course as a pilot program because:
  - No training program in its first run, regardless of how well designed would be error free.
  - Thorough evaluation would produce useful information for revision.
  - Training goals required constant revision to match the organization's evolving needs.
  - Despite the generic nature of course materials, various organizations / environments have localized needs that have to be determined.
- Training and performance outcomes need to be considered in context with other work elements. Training can change skills, knowledge, and attitudes / beliefs (SKA). Evaluation of SKA's is complex as they are only a part of the many factors make up the human performance equation. Other factors include job design, work amenities, and compensation / reward structures (Brinkerhoff 1989).

38

- Feedback questionnaires were administered to all participants with a view to collecting information to further improve the program. Virtually all formal training in today's organizations is concluded with such end of program surveys (Brinkerhoff 1989).

3 COURSE STRUCTURE

The approach of combining face to face lectures with the distance education program comprised:
- A two day course on site to introduce the participants to the material and the distance education trainers. 28 students participated in this program.
- A semester length (4 months) distance education program involving regular electronic and telephone communications between students and educators. 10 students were selected from the initial two day course to participate in the distance education program.
- In the distance education program, face to face tutorial support was provided on a weekly basis by an expert on location, under the guidance and supervision of the distance educators.
- Formal submissions and feedback were made on a regular basis by way of activities, assignments and a supervised examination.
- In the case of this pilot program, the level of ongoing support for students was made possible because of the well resourced training facilities and experienced personnel at the Mine site.

3.1 Course Development

The course provided the fundamental concepts of project management along with an overview of current thinking in the discipline. Emphasis was on principles of management, project relationships, communication, contract and risk management. Topical case studies were used to introduce key principles and opportunities were provided for students to consider applying these in their own work places.

A range of learning modes was employed including; lectures, class discussion, syndicate groups, videos, and an interactive project management computer simulation.

3.2 Evaluation of the Program

A critique of the program was conducted using the time-scale of Pre-Course, During Training and Post-Training. These time-scales were termed Phases 1 to 3 respectively. The specifics of these Phases are briefly summarized in Table 1.

Table 1. Summary of Development Phases

Phase 1: Pre-Training (development)	Phase 2: During Training	Phase 3: Post-Training (evaluation)
Conduct a Training Needs Analysis (TNA)	Conduct the training	Trainees attempt to use what they have learnt
Gain management support	Teach trainees	Identify and overcome barriers to training
Identify appropriate course content / materials	Lead training	Identify emerging training needs
Identify barriers to performance & transfer of training	Measure learning	Feed-back information for further program development
Develop a training program plan		

3.2.1 Critique of Implementation Strategy

The greatest returns on the training investment come from making activities effective in Phases 1 and 3 (Brinkerhoff 1989). Despite the importance of the actual training itself, what will determine the success of the program is the effectiveness of the pre and post training evaluation. For training to be effective: a) trainees need to be appropriately selected so that they are able and willing to apply what they learn. b) it is important to follow up to determine if there are any institutional barriers preventing the application of what has been learnt. If Phases 1 and 3 are not adequately addressed then many inappropriately selected students may be thoroughly trained in Phase 2, only to find they are either not willing or able to apply their learning within the organization. This can lead to a significant waste of organizational resources.

3.3 Stakeholders

Surveys and individual questionnaires were issued throughout the development of the program to support decision-making. Stakeholders in the program were involved in all aspects of the development via their input on surveys / questionnaires. Table 2 summarized the stakeholders that participated in the program.

Table 2. Summary of Stakeholders

Client:	The party that represented the Mines, management's views and interests.
Manager:	The person responsible for coordinating course development.
Developer:	The party developing the training program.
Deliverer:	The lead course trainer in terms of face to face lectures.
Students:	In this case 28 Mine employees undertook the face to face lectures, and 10 continued on to complete the full program.

3.4 Evaluation Plan

A series of 13 questionnaire / surveys were issued. Ten were designed for the Client, Manager, Developer, Deliverer, for course development purposes, whilst three were used to survey Students for feedback purposes. Questionnaires were written in English, except those issued to students were in both English and Bahasa Indonesia. This systematic method of surveying was employed to gain feedback throughout the process of course development and actual delivery.

Emphasis was placed on evaluation throughout the full training program for the reasons outlined in section 3.2.1 and the knowledge of, "what gets measured, gets improve"(Levitt 1976).

3.5 Survey Details

Example summaries of survey results appear in Tables 3, 4, and 5. Table 3 shows the results of a knowledge test that was given to students in the initial face to face course, prior to the course commencement. This was to gain a basic understanding of the student's knowledge of the topic. On completing the two-day face to face course, the students were issued the same test to ascertain some dimensions of learning.

Table 3. Summary of 28 Student Performances & Survey

Student Surveys for the face to face program	Topics Questioned	Score
Pre - Lecture Test (Knowledge Questions)	Project Management?(out of 25%)	15
	Option Evaluation? " " "	13
	Risk Management? " " "	12
	Contract Management? " " "	15
	Score out of 100% (average)	**55**
Post - Lecture Test (Knowledge Questions)	Project Management?(out of 25%)	21
	Option Evaluation? " " "	18
	Risk Management? " " "	19
	Contract Management? " " "	20
	Score out of 100% (average)	**79**
Post - Lecture Survey (Questions on course structure & evaluation) **(1 = very poor, 5 = excellent)**	Sufficient Content? (ratings)	4
	Sufficient Time?	3
	Presentation Elements (videos, etc)?	4
	Case Studies?	4
	Relevance to Work?	4
	Preferred Bahasa Indonesian?	3
	Expectations Satisfied?	3
	Overall Evaluation?	**4**

Table 4 provides a survey summary of the distance education students, six months after completion of their course. The survey time lag was pre-set in order to give students time to determine what they had learnt was able to be assimilated into their work environment.

Table 4 Summary of Distance Education Participants

Survey Questions (very much, rate 1 - not at all, rate 5)	Rating (Ave.)
Were the study materials clear and specific?	2.6
Doing the assignments and other activities has been a valuable learning experience?	2.3
Was the assignment written work load manageable?	2.5
Was working in groups effective for you?	2.2
Was the overall workload of the Course manageable?	2.8
Were Case Studies relevant and useful?	2.0
The feed back on my work included clear suggestions for further improvements?	2.8
The Course demonstrated innovative practical / professional skills?	2.5
Was the arrangement of lectures and distant education material an effective combination?	2.2
I have learnt to apply principles from the Course in new situations?	2.5
Have you found applying the principles from the Course straight forward?	2.3

Table 5 provides a summary of surveys throughout the development and evaluation stages of the training.

Each questionnaire / survey had specific objectives which provided key information towards the development of the course. Combined, they allowed continual refinement of the program as it developed.

4 CONCLUSION

The aim of the pilot program was twofold; to improve the project management competencies of employees on the Mine site, and test the suitability of the training program to develop participants' professional skills across the broad scope of project management. The training aimed to impart new practical skills, increase knowledge of the subject of project management, and develop professional attitudes. The realization of this aim would enable participants to take leading roles on projects in their departments within the organization. The program achieved both objectives.

In addition, Universitas Indonesia is reviewing the dual-language distance education course with a view to incorporating it into their own education programs. This will be a notable achievement in a prestigious university that has traditionally been reluctant to recognize such delivery methods. The development may well have a long-term impact on the way in which the University offers professional development services to engineers' (Hartnell 1999), and on other educational institutions that follow their lead.

Table 5. Summary of Key Survey Responses

```
┌─────────────────────────────────────────────────────────┐
```

Phase 1: Pre-Lecture Survey

For the Client:
What were your objectives in seeking this training?

- To improve the Project Management Competencies of a number of our Indonesian staff. Concerns are; time / cost / quality / deliverable's. "Getting it right the first time"

For the Manager:
What were your Clients objectives?

- Additional to above, was to facilitate the 'Indonesianisation' of their workforce

For the Developer:
What was most difficult about developing the course?

- Distance to the Client. Was unable to discuss issues face to face during development
- Bilingual course took longer to develop than expected
- Case studies difficult to obtain. This is partially due to a lack of direction / knowledge of data required. Again a distance problem

For the Students: (Refer to Table 3)

Phase 3: Post-Lecture Survey

For the Client:
How could the course be improved?

- Make it 4 days...not 2 days. Use a lot more local case studies
- General Comment:
- Handouts in dual language, and the standard of course material was excellent

For the Manager:
Did the course achieve your Client's objectives?

- The course appears to have achieved at least their short-term objective of increasing the knowledge and skill levels of the participants
- It will be difficult to tell, without a follow up survey, whether it achieves their (assumed) longer term objectives of achieving desirable behavioural change which ultimately has a favourable impact on the 'bottom line' delivery of their projects

For the Developer:
How could the course be improved?

- Provide additional time for students to present case study material from their workplace
- Have notes replicate examples a little more closely

For the Deliverer:
Did you have sufficient time to present the course?

- Not really sufficient, however, it was enough to cover the main points to be presented

What could be changed to improve the delivery?

- More time for workshops (exercises and discussion)

For the 28 Students in the face to face course:
Course Strengths:

- Course material was relevant to participants work
- Presenters were very knowledgable of the material
- Case Studies were good
- Working through examples was a good way to learn (including computer simulation)

Course Weaknesses:

- The most commonly noted comment was "not enough time" to cover all materials
- Presenter's explanation did not directly follow notes. Notes could be structured better
- Some language difficulties. First language should be Indonesian and then English

For the 10 Students who also completed the Distance Education course: (most common comments)
What barriers have you experienced implementing what you learnt?

- Co-operation and understanding across levels of the organisation by those involved in a project. Often difficult to achieve
- Time to put into practice what has been learnt

The vision of the consortium was for this and other similar programs to become widely adopted across industry and thus contribute to raising skill levels of Indonesia's engineers to enable the effective management of the nation's physical development. The benefits of adequately skilled engineers in project management are now well recognized by the Government, other national institutions, and industry broadly. If programs, such as the one outlined in this paper, are embraced by Indonesia's engineering profession they will eventually deliver many economic and social benefits for Indonesia.

REFERENCES

Brinkerhoff, R. O. (1989). "Using Evaluation to Transform Training." Evaluating Training Programs in Business and Industry: 5 to 20.

Duffield, C. (1993). Communicating Project Management Through Open Access Learning. 2nd Australian Conference for Engineering Management Educators, The University of Melbourne, ACEME.

Duffield, C. F., Young, D.M., Wilson, J.L. (1997). Project Management - Course Notes. Project Management: Conception to Completion. E. E. Australia. Melbourne, Engineering Education Australia Pty Ltd: 151.

Hartnell, R. (1999). Activity Completion Report to AusAID - Private Sector Linkage Program. Melbourne, Engineering Education Australia / AusAID.

Somers, M. J., Kumar, A,. (1996). Internationalisation of Short Courses in Continuing Education. Higher Education in the 21st Century, Mission and Challenge in Developing Countries, Melbourne, RMIT.

Levitt, R. E., Parker, H.W. (1976). "Reducing Construction Accidents - Top Management Role." Journal of Construction Division, ASCE 102(3): 465 - 478.

Creative Systems in Structural and Construction Engineering, Singh (ed.) © 2001 Balkema, Rotterdam, ISBN 90 5809 161 9

Generic and specific IT training: Capability alignment using process protocol

J.S.Goulding & M.Alshawi
School of Construction and Property Management, University of Salford, UK

ABSTRACT: Construction organisations should alter their recruitment strategies, performance appraisal systems, and education and training policies to benefit from information technology (IT) capabilities. Flexible and coherent business strategies (BS) can also help adapt to the changing economic environment, the details of which should envelop the organisation's skill base. Trained operatives can make organisations more customer responsive, and readily adaptable to change initiatives. Furthermore, training is an investment; being part of a capacity-building infrastructure geared to meet current and future business requirements. In this context, the IT training strategy should make provision for generic and specific IT training needs. This research introduces an IT training model using Process Protocol (PP). Key sequential stages (gates) and links are identified, and a method of applying this technique to close the 'performance gap' between the BS and ITTS is demonstrated.

1 INTRODUCTION

The fragmented nature of the construction industry (Emmerson, 1962; Banwell 1964; Latham, 1994) has been cited as a primary factor that can often adversely affect performance and productivity. Contemporary 'change' initiatives have tried to improve performance by focussing on time, quality or cost elements, but Kagioglou et al (1998) noted that the majority of problems in the construction industry were process related, and not product related. The use of IT in strategic planning can help deliver strategic objectives (Rockart et al, 1996), but managers need to assess the capability of their current services and determine the IT and business maxims needed to either clarify gaps between what exists and what is required, or find a reasonable match between actual and desired capabilities. IT can assist in the strategic planning process (Hinks et al, 1997), but this requires mangers to align their IT strategy to the company's IS strategy; more fundamentally, it is equally important to ensure that the IT training strategy is aligned to the IT strategy and BS.

Issues surrounding the impact and value of organisational learning are increasingly becoming more important (Gratton et al, 1996; Joia, 2000; Raghuram, 1994; Barlow and Jashapara, 1998; Kim, 1993), and organisations should therefore try to match opportunity with corporate capability (Andrews, 1987). In this context, the development of skills through training is particularly important (Raghuram, 1994; Ahmad et al, 1995; Cooper and

Markus, 1995; Rockart et al 1996). Furthermore, the remit of process management is continuing to provide enhanced business opportunities (Davenport, 1993; Al-Mashari and Zairi, 2000; Aouad et al, 1999; Chan and Land, 1999), the benefits of which should not be overlooked.

This research demonstrates the use and application of Process Protocol applied to a generic IT training model, and assesses its contribution to performance gap analysis.

2 THE BUSINESS STRATEGY AND IT TRAINING STRATEGY

The BS can be described as a pattern of decisions made to determine organisational goals and objectives, the nuances of which have been explored by many authors, not least Ansof (1968); Porter (1985); Mintzberg and Quinn (1991); Robson (1997); Ward and Griffiths (1997). The BS should focus on how a company competes and positions itself in the marketplace, in particular, how it channels resources to convert competence into strategic advantage. The current vogue in BS development centres on understanding the concepts and principles that govern competition, and issues using 'systems thinking', 'chaos theory' and the 'learning organisation' are increasingly being progressed (Stacey, 1993; Huber, 1991). These improvements are increasingly using IT as a core tool and enabler of strategy (Ward and Griffiths, 1997; Robson, 1997; Rockart et al, 1996) and issues surrounding the impact and value of organisational learning on strategy is also being given more importance (Gratton et al, 1996; Joia, 2000; Raghuram, 1994; Barlow and Jashapara, 1998; Kim, 1993).

Any IT needs (or skills) needed to deliver the BS can be categorised into three main groups for ease of use – specifically: operational, managerial and strategic (Goulding and Alshawi, 1997). Any deficiencies in skills are determined through a skills assessment/priorities process, and grouped into generic (common to all) and specific (specialist) needs. These training needs can then be addressed using the IT training strategy, the relationship of which is shown in Figure 1.

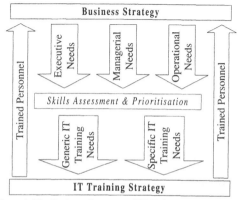

Figure 1. The Business Strategy and IT Training Strategy Relationship

3 PERFORMANCE (GAP) ANALYSIS

Performance analysis is a process used for measuring performance levels against predetermined targets, the purpose of which is to analyse any gaps in performance. The difference between the current level of achievement and the desired target is known as the performance gap (or opportunity gap) – the details of which are indicated in Figure 2. This procedure is known as gap analysis or variance analysis. In this representation, three areas for assessment are identified, specifically: Area 'X', Area 'Y' and Area 'Z'. Performance target levels for each of these areas are represented by the dashed line, and the greatest area for improvement can be seen for Area 'Z'. Therefore, if this activity was classed as critical to the delivery of the BS objectives, then resources should be prioritised accordingly to address these deficiencies and close this performance gap.

Performance analysis has been widely used in the construction, manufacturing, and financial sectors, being a particular valuable tool for controlling re-

source driven activities for predicting trends over time. In this context, it could therefore be extended to include skills development (Van Daal et al, 1998); the development of intellectual capital (Joia, 2000); or gaps in information system (IS) and IT needs (Ward and Griffiths, 1997). However, from strictly an IT training perspective, performance analysis can be used to ascertain the precise contribution IT training can have on the BS.

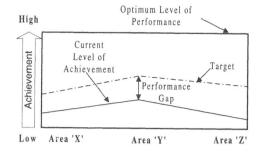

Figure 2. Performance (Gap) Analysis

4 PROCESS IN CONSTRUCTION

Process has developmental links with systems thinking, industrial engineering and the quality movement. Historically however, the origins of process as a discipline could be attributed to Taylor (1911), as this work focussed on process improvements using the science of work-study. Process related improvements are also evident in the works by Mayo (1933), and Fayol (1949). More recently however, and particularly with the advent of Business Process Reengineering (BPR), business improvements have been cited as achieving many benefits using process as the focus of attention (Hammer, 1990; Davenport, 1993; Valiris and Glykas, 1999). Organisations are therefore increasingly using process to improve business performance (Soares and Anderson, 1997; Chan and Land, 1999), the concepts of which focus on activities (how things are done), rather than 'what' is produced (Hammer and Champy, 1993; Davenport, 1993). Various process tools and initiatives can be used and applied in this context, from BPR (Hammer, 1990; Davenport and Short, 1990; Hammer and Champy, 1993) through to capability maturity models (Humphrey, 1989; Paulk et al, 1993) and process protocol (Cooper et al, 1998; Kagioglou et al, 1998; Aouad et al, 1999).

The construction industry is continually striving to improve process, and many large construction organisations in the UK are now using maps and protocols to deliver business benefits (Hinks et al, 1997; Aouad et al, 1999). In this context, using a process

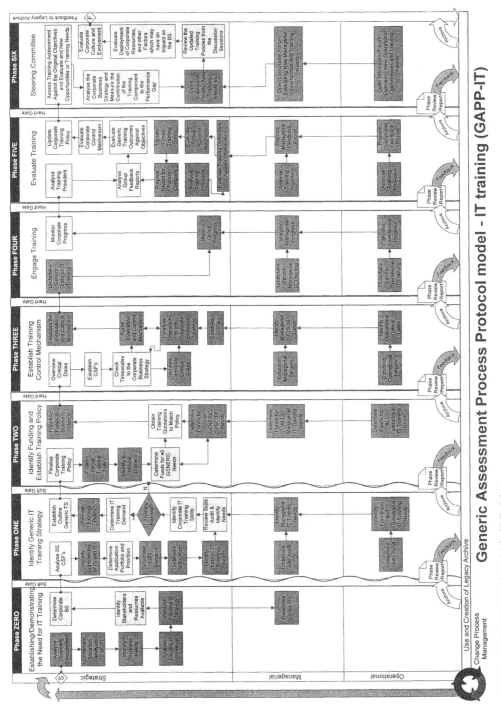

Figure 3. GAPP-IT Protocol Model for IT Training

45

approach to assess the impact of IT training on the BS could offer many benefits.

5 PROCESS PROTOCOL

Process Protocol (PP) is a modelling tool that is capable of representing all diverse parties interested in a process; the flexibility and clarity of which allows generic activities to be represented in a framework that encompasses standardisation. PP was developed by a research team from the University of Salford UK, together with seven collaborating companies, specifically: Alfred McAlpine Construction; Advanced Visual Technology Ltd; British Airports Authority Plc; British Telecom; Capita; Waterman Partnership; and Boulton and Paul Ltd. The Engineering and Physical Sciences Research Council (EPSRC) funded this research project under the Innovative Manufacturing Initiative (Kagioglou et al, 1998).

The PP framework encourages users to appreciate process more easily; affording improvements in communication and co-ordination, the control and management of resources, and the adoption of 'shared vision'. The key attributes of this framework encompasses the following concepts:

Activity Zones: A structured set of sub-processes designed to support the solution.

Deliverables: Outputs from project and process information, used to create the Phase Review Report.

Phase Review & Stage Gate Processes: Generic processes within the stages, separated by decision gates (Phase Review Meetings) needed to fix and approve the information prior to progression.

Gates & Phase Reviews: Project and process review points, used to examine progression, dependant upon predefined criteria. Gates are either Hard (prevent progression) or Soft (accept conditions and allow concurrency).

Legacy Archive: Mechanism for storing, recording and retrieving project and process information.

Phase Review Report: Document of deliverables presented at the phase review gates, the information of which is subsequently stored in the legacy archive.

Processes must therefore be managed and controlled in a consistent and predictable manner, and the generic nature of PP can be adapted to suit many disparate and diverse project environments (Cooper et al, 1998). In this context, PP seems to offer a framework that could be used for assessing the impact of IT training on the BS.

6 DEVELOPMENT OF A PROCESS PROTOCOL IT TRAINING MODEL

A Generic Assessment Process Protocol model for IT training (GAPP-IT) was developed with two industrial partners to analyse the key sequential stages (processes) required for evaluating the impact of IT training on the BS. In this context, IT training issues and processes were 'mapped' into the GAPP-IT framework for discussion and evaluation. This model went through various iterations during the prototyping and testing stage to secure consensus, especially on the validation of processes and taxonomic understanding.

This research identified three main target groups for analysis, specifically, operational, managerial and executive levels. Seven key phases are offered for discussion, the representation of which is shown in Figure 3. Each of these phases can be broken down to show lower level information, and analysis of this lower level detail will be discussed in later works. The GAPP-IT model is divided into three horizontal levels (covering the operational, managerial and executive requirements), and the model is entered at Phase Zero, and exited in Phase Six. A brief summary of each of these phases follows:

Phase ZERO: This allows users to establish the need for IT training by evaluating the existing business need (contemplating the current IS and IT strategy) and assessing all potential resource requirements. The BS is established at this phase, reflecting the prevailing market forces and stakeholder involvement. If the BS indicates an outline need for IT training, users pass through a soft gate into Phase One.

Phase ONE: This identifies the processes involved in formulating a generic IT training strategy. It uses information from Phase Zero, and identifies Critical Success Factors (CSFs) derived from the BS. A skills audit is used to ascertain the type and level of IT training needed. If IT training is not required, users exit this model through a decision icon, and the information is stored in the Legacy Archive. If IT training is needed, an outline generic training strategy is formed, and users pass through a soft gate (as no financial commitment has been made) into Phase Two.

Phase TWO: This phase enables financial resources to be allocated to the training needs identified in Phase One. The training policy is finalised at this point, and funding is determined for the generic operational, managerial and strategic training needs. Specific IT training needs (deemed non-generic) are incorporated at group/subsidiary level, where alternative funding may need to be sought. A Phase Review board meeting is required before progression to the next phase is allowed, as a hard gate (requiring agreement on financial commitment) exists.

Phase THREE: This phase establishes the training and control mechanism needed to ensure critical needs and deadlines can be met. The CSFs generated by the BS, are agreed, and the type of training evaluation and control mechanism is chosen (cognisant of resource implications and appropriateness to the task). This information is confirmed and agreed at the Phase Review board meeting, and users are then able to pass through the hard gate into Phase Four.

Phase FOUR: This phase is used to monitor and control the IT training in accordance with the control mechanism agreed in Phase Three. A Phase Review board meeting is used to ratify and record progress – users then pass through a hard gate into Phase Five for evaluation.

Phase FIVE: This phase evaluates the overall training experience from the training providers. Training outcomes are measured against the original objectives, and the effectiveness of the control mechanism is evaluated. The existing training policy is updated, and information is stored in the Legacy Archive. A Phase Review report records this information – users then pass through the final hard gate into the feedback phase (Steering Committee).

Phase SIX: This is the final and most important phase in the GAPP-IT model. It uses a steering committee to overview the whole process of IT training, and evaluates achievements against original objectives. Training is assessed and measured against the performance gap, and open discussion forums are used to foster new ideas and stimulate discussion for new training initiatives. This information is then evaluated in context, noting the company's current deployment of resources, culture, and level of evolvement. The Legacy Archive is used to feed back information into the process management wheel (incorporating any revisions, changes in policy, or change management issues), and the whole process is able to start again.

7 SUMMARY AND CONCLUSIONS

Process models can be useful tools for 'mapping' ideas and concepts into tangible frameworks for reflection and evaluation. Process Protocol takes this one step further by absorbing unprecedented levels of generic detail into its architecture, thereby allowing users to more readily conceptualise and appreciate the links and relationships between processes and sub-processes. In this context, the relationship between IT training and the BS can be better understood using the PP approach. The GAPP-IT model

can therefore be used to assess the impact of IT training initiatives on the BS. Moreover, investment decisions can be evaluated against the performance gap using targets that measure IT training's contribution to the delivery of BS CSFs. However, it is important to note that the effectiveness of any IT training initiative can be influenced by many factors, not least, the prevailing organizational culture and overall commitment to training.

REFERENCES

Ahmad, I, Russell, J & Abou-Zeid, A (1995), Information Technology (IT) and Integration in the Construction Industry, *Construction Management and Economics Journal*, Vol. 13, Part 2, pp 163-171

Al-Mashari, M & Zairi, M (2000), Revisiting BPR: A Holistic Review of Practice and Development, Business Process Management Journal, Vol. 6, No. 1, pp 10-42

Andrews, K.R (1987), The Concept of Corporate Strategy, Irwin Inc, Illinois, USA

Ansoff, H.I (1968), Corporate Strategy: An Analytic Approach to Business Policy for Growth and Expansion, Harmondsworth, Penguin Publishers

Aouad, G, Cooper, R, Kagioglou, M & Sexton, M (1999), The Development of a Process Map for the Construction Sector, Proceedings of the CIB W55 & W65 Joint Triennial Symposium, Cape Town, South Africa

Banwell, H (1964), Report of the Committee on the Placing and Management of Contracts for Building and Civil Engineering Works, HMSO, UK

Barlow, J & Jashapara, A (1998), Organisational Learning and Inter-Firm "Partnering" in the UK Construction Industry, The Learning Organization Journal, Vol. 5, Part 2, pp 86-98

Chan, P.S & Land, C (1999), Implementing Reengineering Using Information Technology, Business Process Management Journal, Vol. 5, No. 4, pp 311-324

Chang, W.P & Cox, R.P (1995), A Balance in Construction Education, CIB W89 Proceedings of Conference on Construction/Building Education and Research Beyond 2000, Orlando, USA, pp 235-242

Cooper, R & Markus, M.L (1995), Human Reengineering, Sloan Management Review, Summer, pp 39-49

Cooper, R, Kagioglou, M, Aouad, G, Hinks, J, Sexton, M, Sheath, D, (1998), The Development of a Generic Design and Construction Process, Proceedings from the European Conference on Product Data Technology, 25-26 March 1998, Building Research Establishment, Watford, UK

Davenport, T.H (1993), Process Innovation: Reengineering Work Through Information Technology, Harvard Business School Press, Boston, Massachusetts, USA

Davenport, T.H & Short, J.E (1990), The New Industrial Engineering: Information Technology and Business Process Redesign, Sloan Management Review, Vol. 31, No.4 pp 11-27

Emmerson, H (1962), Survey of Problems Before the Construction Industries, HMSO, UK

Fayol, H (1949), General and Industrial Management, Pitman Publishing, London, UK

Goulding, J, & Alshawi, M (1997) Construction Business Strategies: A Synergetic Alliance of Corporate Vision, I.T and Training Strategies, First International Conference on Construction Industry Development, University of Singapore, 9-11 December 1997, Singapore

Gratton, L, Hope-Hailey, V, Stiles, P & Truss, C (1996), Linking Individual Performance to Business Strategy: The People Process Model, Human Resource Management (USA), Vol. 38, No. 1, pp 17-31

Hammer, M (1990), Reengineering Work: Don't Automate, Obliterate, Harvard Business Review, July/August, pp 104-112

Hammer, M & Champy J (1993), Reengineering the Corporation: A Manifesto for Business Revolution, Harper Business, New York, USA

Hinks, J, Aouad, G, Cooper, R, Sheath, D, Kagioglou, M, Sexton, M, (1997), IT and the Design and Construction Process: A Conceptual Model of Co-Maturation, International Journal of Construction Information Technology, Vol. 5, No.1, pp. 1-25

Huber, G.P (1991), Organizational Learning: The Contributing Processes and the Literatures, Journal of Organization Science, Vol. 2, No. 1, pp 88-115

Humphrey, W.S (1989), Managing the Software Process, Addison-Wesley, Massachusetts, USA

Joia, L.A (2000), Measuring Intangible Corporate Assets – Linking Business Strategy with Intellectual Capital, Journal of Intellectual Capital, Vol. 1, No.1, pp 68-84

Kagioglou, M, Cooper, R, Aouad, G, Hinks, J, Sexton, M, Sheath, D, (1998), Final Report: Process Protocol, University of Salford, UK, ISBN 090-289-619-9

Kessels, J & Harrison R (1998), External Consistency: The Key to Success in Management Development Programmes?, Journal of Management Learning, Vol. 29, Part 1, pp 39-68

Kim, D.H (1993), The Link Between Individual and Organizational Learning, Sloan Management Review, Vol. 35, Spring, pp 37-50

Krogt, F & Warmerdam, J (1997), Training in Different Types of Organisations: Differences and Dynamics in the Organisation of Learning at Work, International Journal of Human Resource Management, Vol. 8, Part 1, pp 87-105

Latham, M (1994), Constructing the Team, HMSO, UK

Mayo, E (1933), The Human Problems of an Industrial Civilization, Macmillan, New York, USA

Mintzberg, H & Quinn, J.B (1991), The Strategy Process: Concepts, Contexts, Cases, Prentice Hall International (UK) Ltd, London, UK.

Paulk, M.C, Weber, C.V, Garcia, S.M, Chrissis, M.B & Bush, M (1993), Key Practices of the Capability Maturity Model Version 1.1, Software Engineering Institute Technical Report CMU/SEI-93-TR-25 and ESC-TR-93-178, Carnegie Mellon University, Pennsylvania, USA

Porter, M.E (1985), Competitive Advantage: Creating and Sustaining Superior Performance, Free Press, New York, USA

Raghuram, S (1994), Linking Staffing and Training Practices with Business Strategy: A Theoretical Perspective, Human Resource Development Quarterly, Vol. 5, Part 3, pp 237-251

Robson, W (1997), Strategic Management & Information Systems, Pitman Publishing, London, UK

Rockart, J.F, Earl, M.J & Ross, J.W (1996), Eight Imperatives for the New IT Organization, Sloan Management Review, Vol. 38, Fall, pp 43-55

Soares, J & Anderson, S (1997), Modelling Process Management in Construction, Journal of Management in Engineering, Vol. 13, Part 5, pp 45-53

Stacey, R.D (1993), Strategic Management and Organisational Dynamics, Pitman Publishing, London, UK

Taylor, F.W (1911), The Principles of Scientific Management, Harper, New York, USA

Valiris, G & Glykas M (1999), Critical Review of Existing BPR Methodologies – The Need for a Holistic Approach, Business Process Management Journal, Vol. 5, No.1, pp 65-86

Van Daal, B, De Haas M, & Weggeman M (1998), The Knowledge Matrix: A Participatory Method for Individual Knowledge Gap Determination, Knowledge and Process Management Journal, Vol. 5, No.4, pp 255-263

Ward, J & Griffiths P (1997), Strategic Planning for Information Systems, John Wiley and Sons, Chichester, England

Creative Systems in Structural and Construction Engineering, Singh (ed.) © 2001 Balkema, Rotterdam, ISBN 90 5809 161 9

Use of Mathcad as a teaching and learning tool for project scheduling

Ahmed B. Senouci
Civil Engineering Department, University of Qatar, Doha, Qatar

ABSTRACT: This paper demonstrates the use of Mathcad to supplement and enhance teaching and learning methods for project scheduling courses at the University of Qatar. A Mathcad program for the scheduling of linear construction projects is presented to show the attractive computational environment of Mathcad and to illustrate its importance in teaching project scheduling. Successful use of Mathcad programming as a teaching and learning tool in project scheduling courses resulted in an increased students' understanding of the topic.

1 INTRODUCTION

Commercial programs, such as Primavera and Microsoft-Project, are usually used by students and instructors in project scheduling courses. However, these programs present a number of shortcomings as a teaching aid. First, the programs are inadequate for the scheduling of linear construction projects (i.e., projects with repeating activities/tasks). Second, these programs are not able to teach project scheduling procedures to the students. Third, these programs are usually time-consuming and not flexible enough to suit the project scheduling requirements of the students.

Mathcad (1995) is an efficient learning environment for technical topics such as project scheduling. Mathcad contains powerful presentation capabilities such as the use of charts and graphic objects and offers significant learning enhancements to students of technical subjects. Mathcad makes possible new learning strategies for students and teachers such as what-if discussions, trend analyses, trial and error analyses, and optimization. Taking advantage of the computational power and speed of Mathcad, instructors and students can quickly cycle through project scheduling scenarios.

This paper describes the use of Mathcad program as a teaching and learning tool for project scheduling courses. A Mathcad program for the scheduling of linear construction projects is discussed and demonstrated to show the attractive computational environment of Mathcad and to illustrate its importance as a teaching and learning tool for construction courses.

2 COURSE DESCRIPTION

CVE 482, Construction Planning and Scheduling, is an elective course for civil engineering students. The course is offered in the fall semester of the fourth year in engineering. There are three lecture hours per week and one weekly two-hour tutorial session. Seven distinct topics are covered within the course including work breakdown structure, linear scheduling, critical path method, PERT, resource leveling, least-cost scheduling, and project cost control. During the semester, the students must complete a detailed scheduling of a construction project.

3 LINEAR PROJECT SCHEDULING OVERVIEW

Scheduling of construction projects has typically been accomplished using the critical path method (CPM). However, CPM is inadequate for the scheduling of linear construction projects (Vortser et al. 1992) because of its inability to model the repetitive nature of linear construction projects, to provide work continuity for crews or resources, to plan the large number of activities necessary to represent linear construction projects, and to accommodate changes in the sequence of work between units/locations.

Linear projects are defined as the projects in which the activities are performed repetitively at different work locations. Examples of such projects include high-rise buildings, pipelines, tunnels, and highways. A number of linear scheduling methods were developed for the management of such projects (Russell and Caselton 1988, Al-Serraj 1990, Reda 1990, Eldin and Senouci 1994, Harmelink and Rowings 1998, Harris and Ioannou 1988).

3.1 Terms and Definitions

In a typical linear project, activities are performed sequentially from one location to another. A "location" references a level in a high-rise building, a section in a pipeline, etc. Each activity has only one predecessor. NA and NL represents the number of activities and locations, respectively. The array AD represents the activities' durations at all locations. The component AD(n,i) represents the duration of activity (n) at location (i). Lag times between an activity and its predecessor are associated with the typical precedence relationships (finish-start, start-start, finish-finish, and start-finish). The lag times between each activity and its predecessor are described by a two-dimensional array (LT). The component LT(n,i) represents the lag time between activity (n) at location (i) and its predecessor at the same location. The array APR describes the precedence relationships between each activity and its predecessor. The component APR(n,i) represents the precedence relationship between activity (n) at location (i) and its predecessor at the same location.

3.2 Time Computations

The component ST(i), which represents the start time of activity (n) at location (i), is determined by one of the following equations (Figure 1):

$$ST(1) = AST \qquad (1)$$

$$ST(i) = ST(i-1) + AD(n,i-1) \qquad (2)$$

where AST = start time of activity (n) at the first location.

The component FT(i), which represents the finish time of activity (n) at location (i), is determined by the following equation:

$$FT(i) = ST(i) + AD(n,i) \qquad (3)$$

where ST(i) = start time of activity (n) at location (i).

The start time (AST) of activity (n) is computed using the following equation :

$$AST = \max\{CST(i) - (i-1) * AD(n,i)\} \qquad (4)$$

where CST = one-dimensional array describing the constraining start time of activity (n) at each location (i). CST(i) is calculated by either of the following equations based on the precedence relationship

between the activity in question and its predecessor:

For finish-start (FS):

$$CST(i) = FT(i) + LT(n,i) \qquad (5)$$

For start-finish (SF):

$$CST(i) = ST(i) + LT(n,i) - AD(n,i) \qquad (6)$$

For finish-finish (FF):

$$CST(i) = FT(i) + LT(n,i) - AD(n,i) \qquad (7)$$

For start-start (SS):

$$CST(i) = ST(i) + LT(n,i) \qquad (8)$$

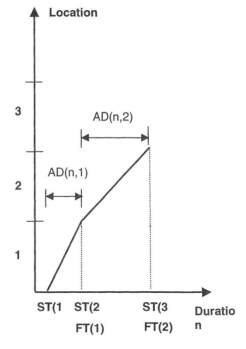

Figure 1. Activity Start and Finish Times

4 MATHCAD PROGRAM FOR LINEAR SCHEDULING

A Mathcad program has been written to automate the manual procedure of the linear scheduling method. The program consists of the following computational steps:
1. Read the following input data: 1) the number of activities (NA), 2) the number of activity locations (NL), 3) the activity duration array

(AD), 4) the activity precedence relationships array (APR), and 5) the activity lag time array (LT).

2. Compute the start time AST for each activity. Based on the activity precedence relationships, the constraining start time array CST is computed using either Eq. 5, 6, 7, or 8. The activity start time AST is then computed using Eq. 4.

3. Compute the activities' start time at all locations.

4. Print the activities start time at all locations and draw the project schedule.

The second step of the Mathcad program is shown in Figure 2.

5 ADVANTAGES OF MATHCAD ENHANCED INSTRUCTION

A Mathcad enhanced teaching method can be successfully integrated into a project scheduling course. The Mathcad program is projected directly from the instructor's computer onto a large screen in an appropriately equipped classroom. Different formatting, including various fonts, colors, patterns, and borders are used. The readability of the text exceeds what instructors can produce by hand on the classroom board. The equations look the same as they are written on a blackboard or in a reference book. By using different drawing entities and varying their color, pattern, and line weight attributes, highly readable drawings are produced to illustrate the computations. To free student attention from transcription, students are given a hard copy for taking additional notes. An electronic copy of the Mathcad program is also made available for the student to review and practice later.

Mathcad is relatively easy to learn and straightforward and offers powerful tools to create sophisticated programs. The equations look the same as they are written in a reference book. Once the equations are entered into the program, it is easy to check the validity of the logic because the calculations are immediate.

Mathcad program provides outstanding graphics capabilities. The way in which Mathcad tend to relate numbers to graphics is important. Since Mathcad can generate graphs from a range of numerical values, it is easy to generate a graphical depiction of a solution. Furthermore, it is possible to directly alter the graphical output by changing the desired parameters. Like spreadsheets, as soon as a change is made in the input data, the results are updated and the plots are redrawn. Other types of charts, such as pie and histogram charts, can also be easily generated. The Mathcad program allows for the determination of an optimum project schedule simply by changing the input data and observing the changes in the schedule.

START TIME FUNCTION

Compute Start Times of Activity (n)

$$ST(n, AST) := \begin{vmatrix} s_1 \leftarrow AST \\ \text{for } i \in 2 .. NL \\ \quad s_i \leftarrow s_{i-1} + AD_{n,i} \\ s \end{vmatrix}$$

Compute Finish Times of Activity (n)

$$FT(n, AST) := \begin{vmatrix} \text{for } i \in 1 .. NL \\ \quad s_i \leftarrow ST(n, AST)_i + AD_{n,i} \\ s \end{vmatrix}$$

Compute start time (AST) of all activities

$$AST := \begin{vmatrix} s_1 \leftarrow 0 \\ \text{for } i \in 2 .. NA \\ \quad k \leftarrow i - 1 \\ \quad time \leftarrow s_k \\ \quad \text{for } m \in 1 .. NL \\ \qquad t1_m \leftarrow FT(k, time)_m + LT_{i,m} \\ \qquad t2_m \leftarrow ST(k, time)_m + LT_{i,m} \\ \qquad t3_m \leftarrow FT(k, time)_m + LT_{i,m} - AD_{i,m} \\ \qquad t4_m \leftarrow ST(k, time)_m + LT_{i,m} - AD_{i,m} \\ \quad \text{for } m \in 1 .. NL \\ \qquad CST_m \leftarrow t1_m \text{ if } APR_{i,m} = 1 \\ \qquad CST_m \leftarrow t2_m \text{ if } APR_{i,m} = 2 \\ \qquad CST_m \leftarrow t3_m \text{ if } APR_{i,m} = 3 \\ \qquad CST_m \leftarrow t4_m \text{ if } APR_{i,m} = 4 \\ \quad \text{for } m \in 1 .. NL \\ \qquad Delta_m \leftarrow CST_m - (m-1) \cdot AD_{i,m} \\ \quad s_i \leftarrow \max(Delta) \\ s \end{vmatrix}$$

Figure 2. Mathcad Program

There are several benefits of a Mathcad enhanced approach to teaching. The time saved from tedious transcription frees student and teacher for the discussion of concepts, and exploration of alternate problem scenarios, observation of trends, and expansion of the discussion to related topics. Outside the classroom, the instructor uses the same program to quickly generate test questions and solution keys. Trial and error solutions are cycled through rapidly. The student can review the classroom material by changing input variables and observing results.

OUTPUT RESULTS

$$Out = \begin{bmatrix} 0 & 0 & 3 & 5 & 12 & 17 \\ 1 & 2 & 6 & 10 & 16 & 23 \\ 2 & 5 & 10 & 13 & 20 & 28 \\ 3 & 7 & 13 & 17 & 23 & 32 \\ 4 & 11 & 18 & 23 & 25 & 37 \end{bmatrix}$$

Column 1 = Activity locations
Column 2 = Start times of activity 1
Column 3 = Start times of activity 2
Column 4 = Start times of activity 3
Column 5 = Start times of activity 4
Column 6 = Start times of activity 5

Figure 3. Mathcad Output Results

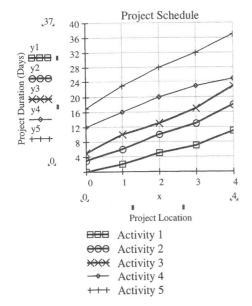

Figure 4. Mathcad Project Schedule

Homework assignments can be developed to encourage students to use the program. Making the program available to students encourages them to learn by exploring on their own. The time spent using the program to explore problem scenarios posed by the instructor, can lead students to a better understanding of the concepts involved in the problems. Students can learn to write Mathcad programs using their own way of problem solving.

6 ILLUSTRATIVE EXAMPLE

The project example is presented to illustrate the capabilities of the Matchad program. The project consists of five activities repeated at four locations. The activity durations, precedence relationships, and lag times are summarized in Table 1. Figure 3 shows the Mathcad output results of the project, which consists of the start and finish times of the project activities. The last row of the array represents the finish times of the actvities at the fourth location. Thus The finish time of activity (5) at the fourth location, which represents the project duration, is equal to 37 days. Figure 4 show the project schedule, which was automatically drawn by Mathcad program.

Table 1. Project Example Input Data

Activity	Duration Array	Preced. Relation.	Lag Array
1	2, 3, 2, 4	----	-----
2	3, 4, 3, 5	FS	1
3	5, 3, 4, 6	SS	2
4	4, 4, 3, 2	SS	1
5	6, 5, 4, 5	FS	1

7 CONCLUSIONS

Mathcad contains tools, which can enhance and supplement traditional methods of teaching and learning. The versatility, accessibility, and ease of use make Mathcad a platform for creating learning modules for technically based courses. Mathcad contains the capabilities for traditional classroom computation, but at a greater degree of accuracy, reliability, and presentation quality. In addition, its speed at repetitive tasks, and its programmability, make new learning strategies possible. Mathcad programs take time for an instructor to develop, but with many benefits in return. By freeing the instructor and student from tedious computation and transcription, Mathcad programs create opportunities for meaningful understanding of technical material. A well-designed Mathcad program can engage both student and teacher, inviting their exploration and discovery of the subject, drawing them deeper into the secrets that it holds.

REFERENCES

Al-Serraj, Z. M. 1990. Formal development of line-of-balance technique. *Journal of Construction Engineering and Management.* 116(4): 689-704.
Eldin, N. N. & A. Senouci 1994. Scheduling and control of linear projects. *Canadian Journal of Civil Engineering.* 21(2): 219-230.

Harmelink, D. J. & J. Rowings 1998. Linear scheduling model: development of controlling activity path. *Journal of Construction Engineering and Management.* 124(4), 263-268.

Harris, R. B. & P. Ioannou 1998. Scheduling projects with repeating activities. *Journal of Construction Engineering and Management.* 124(4): 269-278.

Reda, R. M. 1990. RPM: repetitive project modeling. *Journal of the Construction Division.* 116(2): 316-330.

Russell, A. D. & W. Caselton 1988. Extensions to linear scheduling optimization. *Journal of the Construction Division.* 114(1): 36-52.

Vorster, M. C. and T. Bafna 1992. Discussion of 'Formal development of line-of-balance technique' by Z. M. Al-Serraj. *Journal of Construction Engineering and Management.* 118(1): 210-211

3 Housing and architecture

Creative Systems in Structural and Construction Engineering, Singh (ed.) © 2001 Balkema, Rotterdam, ISBN 90 5809 161 9

Prefabricated housing in developing countries: India

E. Koehn & M. Soni
Lamar University, Beaumont, Tex., USA

ABSTRACT: In many developing countries there tends to be a housing shortage. This presentation describes an investigation conducted to determine the feasibility of using pre-cast elements (reinforced concrete or pre-stressed units) for construction of various facilities in the developing countries. It was found that pre-cast housing is approximately 25% less expensive compared to conventional construction. In addition, factory built components tend to be of higher quality.

Apart from usual advantages of utilizing pre-cast concrete sections, the objective of this study was to develop a system to minimize or eliminate the use of water and heavy equipment at the work site. Another criteria was to design a structure that may be dismantled and reused elsewhere with minimum waste or damage to the individual components. This could possibly be accomplished with pre-cast elements.

1 INTRODUCTION

Today, as developing countries such as India are looking forward to moving into the 21st century, it is expected that the basic needs of the citizenry should be fulfilled. Specifically, where the demand and supply ratio of living units is 3:1, the supply of apartments and houses is much less then the demand (IA&B 1996). This is due, in part, because the increase in the cost of homes is approximately 50% greater then the overall inflation rate. Here, the role of the construction industry could be questioned. Nevertheless, to assist in meeting the demand, the government of India has enacted legislation to increase the resources to the building industry. These include low interest loans and various tax incentives for developers and builders to construct additional housing units. Authorities are also recommending that the private sector work with semi-governmental and governmental agencies to establish a Building Center or Institute. The task for this group would be to develop efficient construction systems so that the cost of housing could be reduced (Murashev, Singlove & Baikov 1998).

There is also an urgent need to initiate environmental friendly or green building construction, which will consume less material, labor and other resources. This may be accomplished by maximum use of pre-fabricated industrial components manufactured from locally available materials which tend to consume less resources and should be of consistent quality (All India Nov.1998).

2 USE OF PRECAST COMPONENTS

In India, and other developing countries pre-cast and pre-fabricated components are not used to the same extent compared to that in the developed countries. This is illustrated in detail in Table 1 (Nirman Bharti 1987).

Table 1. - Worldwide Utilization of Pre-Fabricated Elements.

Country	Use of pre-fabricated elements (%)
U.S.S.R	80%
Sweden	60%
U.K.	40%
France	22%
Switzerland	20%
India	Negligible

As shown, the use of pre-fabricated building elements is above 50% in some countries. In India, however, the difficulty in transportation and erection of pre-fabricated components tends to inhibit their utilization. In addition, the availability of laborers trained in this field is at a minimal level.

3 TOTAL COST

It is known that the cost of construction varies for different facilities. However, as shown in Table 2, the cost of materials in building construction amounts to an average of approximately 66% of the total expenses (IA&B 1990). Labor may be estimated to be approximately 30.5% of the total cost as illustrated in Table 3 (IA&B 1999). Supervision, design and other fees vary between 3-4%.

Table 2. Material Cost in Building Construction

Material	Cost, as a Percentage		
	Maximum	Minimum	Average
Cement	19	14	16.5
Steel	14	12	13
Bricks	13	10	11.5
Sand	06	04	05
Aggregates	06	04	05
Timber	13	05	09
Miscellaneous	08	04	06
Total	79	53	66

Table 3. Labor Cost in Building Construction

Material	Cost, as a Percentage		
	Maximum	Minimum	Average
Masons	12	10	11
Carpenter	05	03	04
Painter	1.5	0.75	01
Unskilled	15	14	14.5
Total	33.5	27.5	30.5

It is shown in Table 2 that the cost of cement varies between 14 to 19% of the total cost. On a typical construction site in developing countries the amount of cement utilized is, unfortunately, often left to the discretion of masons or laborers, who generally have very little knowledge concerning the behavior and characteristics of this material. In order to save money, less or an inferior type of cement and additional water may be used than that required in the specifications. This can reduce the quality and strength of the concrete mix, which may cause unsafe conditions to exist during construction and in the finished product.

In many cases, it is difficult to control quality on the work site in developing countries especially if there are no engineering inspectors assigned to the project. In addition, the findings of previous studies have shown that the quality and productivity of labor-intensive construction tends to be low in developing countries (IA&B 1987,1999).

4 ROLE OF PRE-CAST IN COST EFFECTIVE HOUSING

Pre-cast elements are generally constructed under factory conditions. Here, the quality of the work is easier to control compared to that on an unregulated construction site. In addition, due to the educational learning curve effect, productivity tends to be higher with repetitive work. The second writer has considerable experience in this field and recommends that the components to be manufactured should be limited to those that can be easily erected on the work site without the assistance of heavy equipment. Table 4 lists the components that may be produced efficiently (Bljuger 1997, Klein 1998, Murashev et al. 1996). Representative samples are illustrated in Figures 1-4.

Figure 1. Pre-Cast Door Frame

Figure 2. Pre-Cast Stairs (Outside)

Figure 3. Pre-Cast Stairs (Inside)

Figure 4. Pre-Cast Barrel Section

It is known that through out the world, the use of prefabricated systems has become a cost effective construction technique. In India and other developing countries, however, this approach has been difficult to introduce due, in part, to a shortage of equipment needed for casting, curing, transportation and lifting of modules. Nevertheless, a scheme for partial prefabrication of roofing systems has evolved that does not require large-scale mechanical equipment. This has been achieved by using lightweight pre-cast elements weighing less than 100 kg. (Koehn & Ganapathiraju 1996).

Many other designs have also evolved to satisfy specific construction conditions. They have, generally, contributed to the reduction of the duration and cost of various construction projects. In addition, these systems are capable of providing work for laborers through out year. Another example is a stone block/masonry pre-cast unit for walls, which has been economically adopted in both urban and rural areas where stone is a readily available resource. Pre-cast waffle units, T-sections and semicircular rectangular slabs shown in Figure 4 are also manufactured. These generally weigh less than 160 kg. Miscellaneous items such as stairs, lintels, door and window frames, and shutters are also available in pre-cast units. Many of these are listed in Table 4, and shown figures 1-4. (IA&B 1996)

Various industrial systems have also been developed for the use of indigenous available raw materials. In addition, many of these schemes can be utilized on a small scale for village or rural level operations. Nevertheless, problems have developed involving the application of new systems. These difficulties are generally due to ignorance of the specific requirements needed to operate the systems, and the usual resistance of workers to the adoption of new techniques. In addition, in a labor-intensive industry, workers are concerned about their jobs if labor saving devices are utilized. (Economic 1996).

Table 4. Pre-Cast Versus Conventional Construction

Pre-cast	Conventional
Blocks for walls	Bricks for walls
Pre-cast Lintels (3 cm to 7cm thick)	Cast in place 10 cm thick. Lintels.
Ferro-cement overhangs with Lintels	Cast in place Lintel and overhangs.
Pre-cast cantilever stairs	Cast in place stairs
Joists and plank system	Conventional beam & slab system
RCC Pre-cast door and window frames	Teak /Non-teak wood door and window frames
Ferro-cement shutters for door & windows	Teak / Non-Teak shutters
Pre-cast Ferro-cement wall panels	Brick walls with stucco
Architectural blocks	Brick walls with stucco
Structural columns	Cast-in-place columns.
Use of concrete spacer below reinforcement	Use of stone, wood or steel chairs below reinforcement
Ferro-cement water tank.	PVC, brick or RCC water tank

5 CASE STUDY

A comparison of the cost of housing utilizing cast-in- place or pre-cast construction methods can be calculated. Here, the amenities provided are similar except that pre-cast components instead of cast-in-place building systems are utilized. Specifically, the cost of a 45m^2 home is estimated to be Rs 4958/- (US$ 115.00) per m^2, whereas the cost of a pre-cast unit is roughly Rs.3729/- (US$90.00) per m^2. This calculates roughly to be @ 25% less then conventional systems. Even though the cost is less, pre-cast building construction is scarcely utilized in India. This is probably due to the fact that few professionals, and governmental agencies promote its use. One reason may be related to past practices whereby a number of contractors became wealthy by building pre-cast facilities. They insisted on a high profit margin, which increased the price of pre-cast construction approximately three times that of conventional systems. This practice, unfortunately, inhibited the growth of the industry. Nevertheless, the actual quality and cost of pre-cast housing remains excellent (Alternative Building Materials 1998). As an example, Figures 5-8 illustrate typical buildings utilizing this system.

Figure 6. Load Bearing Building With Pre-Cast and Conventional Construction Systems

Figure 7. R.C.C Framed Building. Pre-Cast and Conventional R.C.C Framed Construction

Figure 5. Complete Pre-Cast Building (Multi Story)

Figure 8. Small Pre-Cast Building

6 SUMMARY & CONCLUSION

It is time for developing countries such as India to begin using pre-fabrication in building construction. This will assist in the effort of building houses that are more economical for and affordable to the common man.

It has been shown that cost effective and quality construction can be achieved by using pre-cast technology. In fact, the cost of construction utilizing pre-cast components is roughly 25% less than that of using conventional cast-in-place methods. It appears, therefore, that a concentrated effort should be initiated to encourage the growth of the pre-fabrication industry in India and other developing countries.

REFERENCES

All India Conference on *Housing challenges & Solutions.* Nov. 1998. Indian Institute of Engineers Mumbai, India

Alternative building materials. 1998. MTPC Publishers, Mumbai, India

Architecture + Design. 1987. Vol 3. No.4. Bljuger. 1997. *Design of Pre-cast Concrete Structures.* Building Research Station, Mumbai, India

Design of Reinforced Concrete Structures. Translated from Russian by G.Leib.

Economic Technique of building low cost houses 1996. Indian Institute of Engineers, Mumbai, India.

Indian Architect & Builder (IA&B). Oct 1996. Mumbai, India.

Indian Architect & Builder (IA&B). January 1990. Mumbai, India.

Indian Architect & Builder . (IA&B). January 1999. Mumbai, India.

Klein, S. 1998. *Production of Pre-cast Concrete.* Translated from the Russian.

Koehn, E. & V.R Ganapathiraju. 1996. *Industrialization of construction in developing countries:*

India. Proceedings, (CIB) International Symposium on shaping theory and practice. Glasgow, Scotland

Murashev, V.I., E.E. Singlove, & V. N. Baikov. *National housing & habitat policy.* 1998.

Published by : Govt. of India, Ministry of urban affairs and employment. New Delhi, India.

Nirman Bharati. Oct. 98. New Delhi, India.

Creative Systems in Structural and Construction Engineering, Singh (ed.) © 2001 Balkema, Rotterdam, ISBN 90 5809 161 9

Study on the housing business operation by Consumers' Life Cooperative Association in Japan

Kazuya Ohnishi
Aichi Konan College, Japan

ABSTRACT: Defective problems by the harmful building material in the new construction of the residence and remodeling, the corner-cutting construction increase rapidly in Japan recently, and a consumer raises concern against the safety and the health of the residence. It is a connection post with the construction dealer at first in such conditions the development of the home business that the consumers' cooperative society which had piled up a movement around the safety of food took a consumer's position seriously. It can think with the thing that this creates the new form of home-making in future Japan. The present condition of the home business of this consumers' cooperative society is investigated, and this research aims at explaining those characteristics.

1 INTRODUCTION

In recent years, many serious housing troubles have arisen in Japan, such as a rush of complaints on defective houses and incomplete renovation work, and the "sick-house syndrome" caused by new construction materials. (The sick-house syndrome is a series of sicknesses -e.g. pains in the mouth and throat- caused by the air polluted with volatile organic compounds in newly-built houses.) More and more owners of defective houses have started to consult to lawyers and / or take the cases into court.

The government amended the Building Standards Act after 1999, were introducing several systems to enhance the quality of housing. The principal amendments were made on the following three points. The first point is the expansion of a system on building confirmation and building inspection. Under the system, mid-term inspections must be made even for small wooden housing, which has high rate of defects. The second is the expansion of a warranty system against defects concerning the contracts for newly-built housing. The system obliges constructors to bear the liability for defects for ten years in regards to the principal parts necessary for structural strength and flashing parts. The third is the establishment of a housing performance indication system, which enables consumers to compare housing performance objectively before making contracts. Practically speaking, these

systems include several controversial points. The legal circumstances around defective housing, however, surely took a step forward.

One of the reasons why defective housing and breaches of contracts arise so often is that no contracts or rules have been socially established between consumers (house-owners) and suppliers (designers / constructors). Another factor is that architects, who are expected to conduct design and management of housing for consumers, seldom participate in renovation work on small scales. Recently the Consumers' Life Cooperative Associations (co-ops hereafter), which are engaged in activities and movement mainly for "safety of foods," have just started to develop a housing business, stressing consumers' stances. This attempt is expected to be a new style of the future of the housing business in Japan.

Many kinds of studies and research have been conducted so far on the housing business by co-ops in the U. S. A. and in Europe, and by nonprofit organizations, or on the history of the business operated by housing co-ops after the World War II. The housing business lately started by co-ops in Japan, however, has not been fully studied. This study aims to clarify its features through the investigation of the present situation of the housing business by co-ops in Japan. It will help to think about how an alternative system of supplying houses should be.

2 MATERIALS AND METHODS

Statistic materials, research papers, and publications by co-ops were collected and examined to clarify the present situation of the housing business by co-ops. Some preliminary surveys were also made for several co-ops, which have developed the housing business aggressively. Next, interviews were held in February 1999 with the persons in charge of the housing business at three co-ops: Consumer's Co-operative Kobe, Nagoya Citizen Coop, and *Seikatsu Club* Consumer's Cooperative Tokyo.

Thirdly, a questionnaire on their housing business were sent out to 178 co-ops throughout Japan, and they were requested to send back the response sheets by mail or by fax. As a result, 71 out of 178 co-ops (the collecting rate - 39.9%) did as requested. As for the rest 107 co-ops, telephone calls were made to them after the deadline to get the response sheets. The principal questions of the questionnaire are:

1) whether or not the co-op is engaged in a housing business;
2) the time when the co-op started the business;
3) the total income received from deals with the business;
4) the business organizations and systems (e.g. the number of the employees in charge of the business);
5) the contents (details) of the business;
6) the contents of activities for the business;
7) the standards of design and construction; and
8) the themes and issues to be solved for the future of the business.

Finally, further interviews were held about the present situation of their housing business and the themes and issues to be solved with the three co-ops previously mentioned since they are engaged in the business aggressively.

3 RESULTS OF THE RESEARCH

3.1. *The Outline of the Housing Business by Co-ops - drawn from the questionnaire-*

3.1.1. *Whether or not the co-op is engaged in a housing business (Fig.1)*

53 out of 178 co-ops (almost 30%) to which the questionnaire was sent out were already engaged in a housing business. Closely 70%, or 122 out of 178 did not manage the business at the time of the investigation. 27 out of those 122 co-ops (approximately 20%), however, had intentions to start the business in future. The rest 3 co-ops did not make a clear response.

The following analysis is based on the responses from 33 co-ops as the rest 20 out of 53 co-ops which are engaged in a housing business did not send back the response sheets.

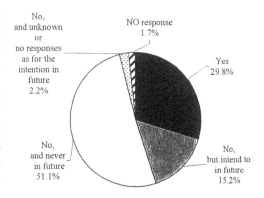

Figure 1. Question: Is this co-op engaged in a housing business?

3.1.2. *The time when they started the housing business*

Though one co-op started the business in as early as 1973, 7 did in 1980's and 24 in 1990's. The result shows that it is quite a new field of business to co-ops.

3.1.3. *The total income received from deals with the housing business*

7 co-ops replied that their total amount was less than 100 million yen (about $0.94 mil.), while 13 co-ops 100 to 500 million yen (about $0.94 - 4.7mil.). These two occupied approximately 80% of all responses to this question. To any co-op, the total income with this business occupied not more than two percent of the total amount with all fields of the business conducted by them. It means that the housing business plays only a small role in their whole business.

3.1.4. *The number of the employees dedicated to the housing business*

The majority (22 co-ops - approximately 81.5% of the co-ops which responded to this question) had one to three dedicated employees. As for the approved architects outside, six co-ops had none, while 11 had one to three persons. 17 co-ops also had 10 approved constructors or less, while eight had more than 20. The responses also varied on the number of outside cooperators. The replies, however, can be divided into two groups: one group with 10 or less and the other with 20 or more. It is supposed that the number of the approved constructors of each co-op is in proportion to its dealing amount. The more the amount is, the more constructors the co-op has. On the other hand, there is a tendency that the co-ops which hold seminars so often and conduct the design and construction management aggressively, have many architects and outside cooperators.

3.1.5. *The contents of the housing business (Fig.2)*

The activities most frequently cited were "measures against termite (white ants)" and "agencies of constructors for renovation, "both of which were conducted by 81.8% of the co-ops, followed by "consulting services"(69.7%), "agencies of constructors for new construction" (63.6%), and "consulting sessions"(54.5%). Most co-ops considered to attach importance to "agencies of renovation work" and "consulting sessions" in future. The tendency is that the co-ops which manage new construction intend to lay emphasis on "consulting sessions," while those which are devoted to renovation work have intentions to stress "design and execution of renovation work" and / or "checkups of existing houses by specialists."

3.1.6. *The activities for the housing business as a consumers' movement (Fig.3)*

"Seminars for the co-op members" was the response most frequently cited (63.6%), followed by "training of architects and constructors"(54.5%) and "training of the co-op staff" (51.5%). This result implies that all of co-op members, architects / constructors, and co-op staff try to study from their own angles to enhance their knowledge and skills. In addition, many co-ops had held "liaison with the other co-ops" (42.4%) for the business as a part of their movement. The majority of the co-ops mentioned "seminars for the co-op members" as an activity that they intended to stress in future, whereas some co-ops pointed out " development of secure construction materials" and / or "reciprocation of information among the members," both of which are rarely conducted so far.

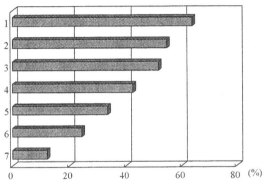

Figure 3. Question: What kinds of activities does this co-op do for the housing business?

1.	Seminars for the members	63.6%
2.	Training of architects and constructors	54.5%
3.	Training of the co-op staff	51.5%
4.	Liaison with the other co-ops	42.4%
5.	Development of safe construction materials	33.3%
6.	Reciprocation of information	24.2%
7.	Others	12.1%

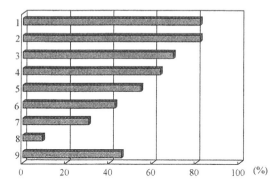

Figure 2. Question: What kinds of work does this co-op conduct for the housing business?

1.	Agencies of constructors for renovation	81.8%
2.	Measures against termite (white Ants)	81.8%
3.	Consulting services	69.7%
4.	Agencies of constructors for new construction	63.6%
5.	Periodical consulting sessions	54.5%
6.	Execution of renovation work	42.4%
7.	Execution of new construction work	30.3%
8.	Checkups of existing houses	9.1%
9.	Others	45.5%

3.1.7. *Whether or not the co-op has standards on design and construction (Fig.4)*

This questioned was asked to 15 co-ops which have conducted design and construction management. 10 co-ops (66.7%) had the standards for "simple (easy to understand for the members) estimates" and on "safety of construction materials." 9 co-ops had "timing and standards of mid-term inspections on site," while "construction manuals for assurance of safety" were made by 8 co-ops (53.3%).

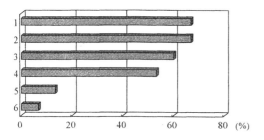

Figure 4. Question: What kinds of standards does this co-op have for design and construction?

1.	Simple estimates	66.7%
2.	Safety of construction materials	66.7%
3.	Timing and standards of mid-term inspections on site	60.0%
4.	Construction manuals	53.3%
5.	Standard of the final inspections (after the building completes)	13.3%
6.	Others	6.7%

3.1.8. *The present themes and issues to be solved for the future of the housing business*

The co-ops were also asked about issues to be solved and keynote for their housing business. They mainly cited following 7 points as key topics:

1) safe design and construction of houses with emphasis on environment and welfare;
2) shift from passive to active attitude on renovation (i.e. not only implementing what they are requested but also making proposals for consumers from their side);
3) clear pricing / charging and / or competition among constructors;
4) training of the co-op staff to make specialists in the field of the housing business;
5) significance of the housing business in relation to the co-op movement (from foods to living);
6) styles of their PR activities; and
7) ways of complying with requests by the co-op members (in the case of agency business).

3.2. *Research of advanced cases*

3.2.1. *The Consumer's Cooperative Kobe*
This is the first co-op in Japan which started a housing business. It has been engaged in the business since 1973. It mainly conducts two categories of the business: one is the sales of interior goods at D.I.Y. shops; and the other is the design and construction for renovation. It recently has exerted itself for proposals on renovation in consideration of welfare, health, and environment. It also offers such services as "consulting sessions on renovation" and "checkups of housing conditions" to the co-op members. Many members entrust it with construction work repeatedly as it undertakes any renovation work, even of small scales. This co-op asks the members to reply to a questionnaire after construction work is completed. According to the responses, approximately 90% of the members are satisfied with the work. It is supposed that the members' high estimation partly comes from its thorough instructions on the constructors' manners on site (e.g. greetings and cleaning).

3.2.2. *The Nagoya Citizen Co-op*
This co-op started a housing business in 1996. It periodically holds "consulting sessions on housing." It transfers construction work, especially renovation work, to reliable approved companies (design offices / building constructors). As for the details of construction (e.g. the way of selecting construction materials), it has guidelines on health and safety to comply with the members' requests. It also makes mid-term and final inspections of buildings with manuals from the members' standpoint. It holds "seminars on living and housing" for the members to learn health and safety of housing. This attempt helps the members to enhance their consciousness about housing. It has also promoted the ideas of establishing an area network in connection with surrounding co-ops. On the basis of the concept of the network, it shares information and develops construction materials with the other co-ops.

3.2.3. *Seikatsu Club Consumer's Cooperative Tokyo*
This co-op, which started a housing business in 1991, has transferred new construction and renovation work to approved companies / offices. As it mainly consists of the members with high consciousness about health and safety, it has purchased safe construction materials and introduced reliable constructors / designers to its members for the purpose of creating better living environment. It holds a "housing school for comfortable living" at which it offers technical knowledge and legal consultation on house-building (e.g. how to make construction plans including the concept of barrier-free and safe materials and how to make fund raising plans) to the members who hope to build houses conforming to their styles of living. This attempt lays emphasis on enlightening the members rather than conducting the housing business. It also offers "checkups of housing conditions," which examine the safety and health of houses and suggest in detail the ways of house maintenance to the members.

4 CONSIDERATION

The following are the major six causes of troubles with renovation work by general (commercial) constructors:

1) explanations for their profit when consulted;
2) obscure estimates;
3) no written contracts but verbal ones made in the case of small scale construction work;
4) construction materials inexpensive but of poor quality instead of secure ones (no care about health and safety);
5) careless and incomplete work by carpenters and craftsmen on site due to a multi-tiered subcontracting structure; and
6) no faithful reactions to the troubles and problems found after construction work completes.

These are, of course, bad cases by some constructors. It is true, however, that possibility of such troubles makes consumers anxious about making requests to general (commercial) constructors.

Judging from the results of the research, co-ops conduct the housing business from the standpoint of the members. They are mainly engaged in renovation work of very small scales, which general (commercial) constructors are reluctant to do. In fact, their policy is:

1) not only making consultation / explanations for the standpoint of the members, but also giving information on or hold seminars on house-building for the members;
2) not only making clear and easy-to-understand estimates but also explaining them in detail;
3) taking the responsibility for contracts even though the construction work is done by approved companies;
4) supplying the members with safe and healthy construction materials for health inexpensively due to cooperative purchasing;
5) introducing reliable constructors and making mid-term and final inspections securely; and
6) providing periodical maintenance.

The housing business by co-ops has a function as a coordinator to enhance the quality of house-building through seminars for the members and constructors.

There are still several problems to be solved for co-ops to spread their housing business throughout Japan. According to the responses to the questionnaire, some co-ops which did not conduct a housing business at the time when the questionnaire was sent had intentions to start agency services of renovation and new construction work in future. Still, they expressed such difficulties and anxieties about starting the business as: "they are not big enough to conduct the business;" "they have to compete with general (commercial) constructors;" "they do not have enough funds, managerial and technical know-how, or employees;" and "few members make a request for the housing business as they are concentrating on "foods." To solve these problems, it is necessary for co-ops to cooperate with the other co-ops and to establish network with experts.

5 CONCLUSION

The demand for renovation work (i.e. refurbishing or restoring the existing houses) will increase in Japan. The arrangement of the living environment with renovation work is expected to increase more and more since "nursing and caring for the aged at home" is getting more important in a society where the elderly are in the majority.

The public organizations as housing suppliers (e.g. local governments and urban development corporations) restrict their business range to the sale of newly-constructed housing, the supply of housing for rent, and the renovation of existing public apartment houses. On the other hand, general (commercial) constructors are not eager for small scale renovation work, which does not produce high profits. Even though the work is on a small scale, the expenses are high and the contents of the work are too complicated for consumers. These factors make it very difficult for consumers to find reliable constructors. Nowadays many kinds of problems have arisen from the housing as merchandise supplied by general (commercial) companies. These are partly caused by the separation of production from consumption. For building safe and healthy houses, it is important to re-connect production with consumption and to enhance the quality of both. In this sense, the housing supply in a cooperative style is promising.

The present methods of the housing supply in a cooperative style can be classified into four types as follows:

1) cooperative construction (mutual aid, D.I.Y.);
2) financing building funds (by the housing cooperatives in Japan);
3) cooperative housing (owned by lots); and
4) renting houses (in consideration of residential rights, especially for people with low income).

Any of these four types, however, are not able to cope with small scale (renovation) work of individually, which requires safe design and construction. Then, the housing business by co-ops, that is, the business with specialist network with co-ops as coordinators (for studying and for design and management), could be the fifth type of the housing supply in a cooperative style. As mentioned before, the demand for renovation work has increased recently. Under such circumstances, the housing supply and renovation work by co-ops as the third arm supplier following after public organizations and commercial companies could be one of the options for consumer-centered supply, solving such problems as defective houses and breaches of contracts.

ACKNOWLEDGMENT

Acknowledgment is made to the following people and organizations for their assistance with this study: Consumers' Cooperative Associations, Shinji Isobe, Ken Ohbayashi and Ikuyo Suzuki. Gratitude is gratefully expressed to Professor Katsuyo Ueno for her advice for this study.

REFERENCES

Hirayama, Yosuke. 1993. Community-based Housing. Japan Domes Shuppan.

Housing and Community Foundation. 1997. Non Profit Organization. Japan. Fudosha.

Kamiya, Koji et al. 1988. Co-operative Housing. pp.146– 165. Japan. Kashima Shuppan.

Nohara, Toshio. 1996. Contemporary Co-operative Society Discussion., pp.172–213. Japan. Nagoya University Shuppan

Ohmoto, Keino. 1991. Housing Policy of Japan. pp.373– 397. Japan. Nihonhyoronsha.

Sakaue, Kaori & Ueno, Katsuyo. 1998. A Study on Co-operative Housing in Norway. Urban Housing Sciences, no. 23, pp. 35 – 40. Japan.

Creative Systems in Structural and Construction Engineering, Singh (ed.) © 2001 Balkema, Rotterdam, ISBN 90 5809 161 9

Product orientation in housing construction – More competitive edge with innovative products and methods

J.T. Pekkanen
National Technology Agency, Finland

P.S. Pernu
YIT-Rakennus Oy, Finland

ABSTRACT: Private-sector building developers and contractors have dominated the Finnish housing construction market over the last few years. Almost all these housing projects have been carried out using traditional procurement methods and all the contractors have utilized the same construction technology. Competitive tendering has been based on labour and material costs. Contractors have neither been encouraged to develop their own concepts for production nor to abandon the price competition approach. The result has been an end product lacking individuality. Competitiveness has not been based on product development, because innovation has not traditionally been an intrinsic part of the business.

The need for innovative development is increasing, however, as clients in the construction market are increasingly demanding new housing concepts that offer more flexibility and variety. According to recent research, home buyers nowadays assume that their new homes will be capable of being adapted to their evolving life circumstances. The key issues regarding the fulfilment of clients' dwelling needs are the functionality and adaptability of the spaces within ecologically sustainable solutions, together with the possibilities afforded by information technology for various new fixtures. Today's contractors are being forced to change their strategies in order to secure a viable future. The most advanced contractors differ from their competitors by integrating their technical and service strategies. As a result, these contractors have taken advantage of their own innovations, such as the lean building process in residential construction.

This paper will discuss experimental housing projects in which product orientation has been implemented successfully. The aim of these research projects has been to improve competitiveness by utilizing technological innovation and building processes with proven success. A concrete example of sophisticated product orientation is the YIT Media Home launched by YIT-Rakennus Oy, one of Finland's biggest contractors in the housing sector. A major aim of the YIT Media Home brand is to incorporate and integrate the latest information technology innovations into its development and implementation. These include fixed broadband Internet access with a local area network, the introduction of modern security services and HVAC controlling systems.

1 HOUSING CONSTRUCTION IN FINLAND

In the last few years, construction growth has been driven by new housing construction. The volume of housing construction grew 14 per cent in 1999. Confidence in the future, low interest rates and growth in purchasing power suggest that this trend will continue.

Finland currently has 5 million inhabitants, but this figure is expected to start falling after 2015. The aging of the population and, especially, the retirement of the baby boom generation will be reflected in the population structure after 2010. At the end of 1999, the housing stock totalled 2.4 million units: 1.1 million in apartment blocks, 1 million one-family houses and 0.3 million row houses. Forty-three per cent of dwellings in Finland are in apartment blocks. In 1999, a total of 260,000 people moved to a new municipality. As a result of these migration patterns, the outlook for housing construction varies by region. Demand remains strong in the growth centres, while in municipalities with net out-migration, vacant dwellings are an increasing problem.

A significant share of housing production is financed through loans from the State-owned Housing Fund of Finland. This State-subsidized housing construction typically involves strict government control with regard to prices, amenity level and quality. This has resulted in a very limited range of plan solutions, where any differences between plans have usually been limited to architectural detail. These housing projects have usually been carried out in accordance with the 'design-bid-build' approach.

Because the design-bid-build approach and seeking the lowest production prices have long been the main criteria in bidding competitions, the main interest of contractors in developing their own operations has been to minimize costs and their main strategy has been to keep prices low. Inevitably, this has meant that there has been little product differentiation in quality and detail.

By contrast, non-subsidized housing has provided a far greater variety of dwellings for clients, who in this sector are typically more demanding. Competition amongst contractors for these clients is also tougher. Contractors seek to gain a competitive edge by complying with potential clients' requirements and by differentiating their end products.

In the future, construction sector clients will increasingly focus on the need to improve the building process and its end products, while also emphasizing the building's properties over its entire life cycle. Technological developments in other fields will force construction companies to invest in better quality, productivity and service. (ProBuild, 1997)

2 NEED FOR NEW INNOVATIVE CONSTRUCTION METHODS AND PRODUCTS

Along with the transition from an industrial society to an information and service society, the new-building market has now reached a saturation point in terms of volume, resulting in a transition from a seller's market to a buyer's market. In addition, as a result of more general trends towards democratization and individualization, the individuality and influence of residents and clients will increase, particularly with respect to the end products of construction, while the market itself will also need more modifiable space.

Problems in housing construction have resulted in high construction costs and expensive buildings, as well as quality problems and features that do not match the client's needs. If they are to be more attractive to the client, the products of the construction industry should be of a higher quality and should be produced more efficiently and economically. Otherwise, people will increasingly invest in consumer goods and other alternatives instead.

As a result of the problems and the compulsive drive for competitiveness, the parties involved in the building process have introduced new technology and sought to develop the process. The advances in information technology, automation and new

materials open up new potential for construction, and we have to be able to take full advantage of them. The increasing speed of change, demand for sustainability and customer-orientation, and the increasing individuality also create pressures to improve housing construction methods. (Lahdenperä, 1998)

A key element in developing Finnish housing production will be a product-oriented approach, which will mean thinking in terms of modules. A building is a product, as is the space inside it. Building systems and components are also products, and the building process itself must be regarded as a product, too – specifically, a service product. The interfaces within this production chain must respond very quickly to new trends in order to add know-how and service to products. Adding know-how and service will make a product more attractive to the market and help it to satisfy the client's true needs. (Technology and the Future, 1998)

From the contractor's point of view, this has required a change of operating strategy. Because the traditional price strategy alone cannot guarantee success, the more forward-looking contractors endorse a quality and service strategy, adopting a product-oriented approach for their housing construction business.(Grönroos, 1990)

3 CUSTOMER EXPECTATIONS CONCERNING NEW APPROACHES TO HOUSING CONSTRUCTION

The objective of increasing the resident's influence on the built environment has gained wide acceptance. The policy of building for an average dweller has led housing producers to incorporate features which are not always needed, while they remain unable to respond to the specific needs of individual groups of users. The differentiation of values and lifestyles has created housing requirements and wishes that are becoming both more diverse and more difficult to foresee. Notions concerning the degree of spatial differentiation and the fixtures needed in the dwelling vary more and more according to the choices and special needs of the individual dweller or household. (Tiuri, 1998)

There are considerable pressures to change the traditional approach to the house as a product and to housing production. Home buyers are gradually becoming as demanding as the other clients of the building industry and want individual, high-technology homes. Flexibility, adaptability and low overall life-cycle costs have a high priority. In addition, providers of rental housing are interested in construction innovations, concerning for example the properties and potential of materials and systems.

Home buyers can be divided into three groups according to their life situations: first-time buyers, those moving house, and those looking for a retirement home. The fact that each group has different needs and expectations regarding the properties of a dwelling must be taken into account.

The main targets of the product-oriented approach and of innovation in housing are:

- Properties of the end product, e.g. easy access for disabled people and flexibility of plan
- Materials, components and systems used in the building: ready broadband Internet connections, intelligent, resident-controlled HVAC systems, etc.
- Construction process and procedures for user participation in the planning process
- Supplementary services available to the building or apartment, e.g. home care and security services for senior citizens.

Actual product orientation has, nevertheless, so far been implemented on a fairly small scale. The main features applied are:

- Flexibility of dwellings on the 'open building' principle
- Installation of information technology and data networks as part of the basic structure of the dwelling
- Use of new system solutions based on intelligent systems and functions

The following is an example of a construction company's product-oriented concept for housing production: here, houses and flats are regarded as quality branded products with value-added properties for the benefit of clients.

4 YIT MEDIA HOME

The YIT Corporation provides a wide range of services for the construction sector and in industrial and infrastructure investment and property maintenance. The YIT-ARK Group is one of the biggest contractors in market-financed housing construction in Finland, with net sales of USD 100 million in 2000. The strength of the YIT Home brand is reflected in the increased interest in YIT Media Homes, with more than 1200 such homes under construction in 2000 (YIT Corporation 2000).

The aim of the YIT Media Home concept is to offer additional value to home buyers. With this in mind, YIT and Europe's leading telecommunications company, Sonera, have established a joint enterprise to develop and market home networking services. The purpose is to offer technological solutions that improve the usability and resale value of the homes. The YIT Media Home provides the opportunity to take advantage of future Internet services that will benefit all types of home buyers from first-time buyers to those looking for retirement homes. Due to the rapid evolution of telecommunications technology, new kinds of services for households are not anymore possible using standard telephone and TV cables.

Initially only broadband data networks were included in the housing projects, but later a fixed broadband Internet access was also included. The capacity of the local network is 1000 megabytes per second, provided via copper cables of the latest standard. The capacity of a single home is between 256 and 2000 kilobytes per second, depending on the number of homes that share the Internet access. Every room has a fixed data socket ready to use with a PC fitted with an ethernet-card. The residents themselves enter into an agreement for the desired services with a network operator or with a service broker.

Figure 1 shows the variety of future services via the Internet and the different levels of communication (Massachusetts Institute of Technology 2000). Via the Internet it is possible to automate functions such as heating or ventilation in the home or the building. The occupants can manage and control electronic devices and lights even from outside their homes. Fire and burglar alarms could also be connected to the network for safety reasons. With access to different on-line health care services it is possible to create a transgenerational environment and maintain the residents' independence as they age. The Internet also give access to entertainment such as network games or video-on-demand, and e-commerce offers, for example, easy delivery of goods and banking services. Increasingly, work and learning will take place directly in the home. Keeping in touch will also be possible via video-conferencing techniques, allowing, for example, grandparents and grandchildren of the future to communicate with ease. Local news and public transport information will be available too. All these activities are designed to satisfy the needs of households in the future.

5 CONCLUSIONS

The YIT Media Home is a prime example of an innovative, product-oriented approach in the building market. The concept produces additional value for the client and at the same time provides a competitive edge both in market-financed and state-subsidized housing construction. The latest information technology innovations are fully integrated into the end product.

By making building procurement more product-oriented it is possible to ensure that the construction industry utilizes the available technology and expertise to the full, in the drive to satisfy the changing needs of clients.

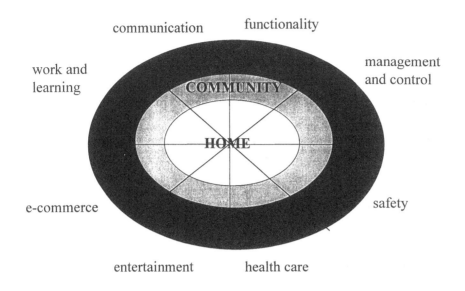

communication   functionality

work and   management
learning   and control

COMMUNITY

HOME

e-commerce   safety

entertainment   health care

Figure 1. Fixed broadband Internet access can respond to the changing needs of future homes.

REFERENCES

Grönroos C. 1990. *Nyt kilpaillaan palveluilla.* ISBN 951-35-5011-7. Finland: Helsinki.

Lahdenperä P. 1998. *The Inevitable Change*, ISBN 951-682-504-4, Finland: Tampere.

Massachusetts Institute of Technology 2000. *House_n The MIT Home of the Future.*USA.

Tekes (National Technology Agency) 1998. *Teknologia ja tulevaisuus (Technology and the Future).* ISBN 951-53-1411-9. Finland: Helsinki

Tekes (National Technology Agency) 1997. Technology Program *ProBuild – Progressive Building Process.* Finland: Helsinki.

Tiuri U. & Hedman M. 1998. *Developments Towards Open Building in Finland*, ISBN 951-22-4127-7, Finland: Otaniemi.

VTT Building Technology 2000. *Well-Being Through Construction in Finland 2000.* ISBN 952-5004-31-7. Finland: Tampere

YIT Corporation 2000. *Interim Report January 1 - April 30.* Finland: Helsinki.

Creative Systems in Structural and Construction Engineering, Singh (ed.) © 2001 Balkema, Rotterdam, ISBN 90 5809 161 9

Architectural design of structural masonry buildings

S. L. Machado & H. R. Roman
Federal University of Santa Catarina, Florianopolis, Brazil

ABSTRACT: The aim of the work was to verify the knowledge of architects to design buildings on structural masonry in the south of Brazil. Preliminarily, a literature review was used to systematize basic principles associated with the development and elaboration of architectonics projects using this type of structure. Then, a questionnaire was applied to obtain data from the south-Brazilian architects. It was verified how the architects use the structural masonry concepts. At the same time, the project management was analyzed in these architectural offices. The work is qualitative, treating organizational features, and know-how transference. Our study revealed that the methodology employed by most of the architects is inappropriate. The South Brazilian architects are not organized to project on structural masonry and demonstrated insufficient technological knowledge on this type of structure.

1 INTRODUCTION

In Brazil the Construction Industry suffered dramatic advance after the inflation's fell from a rate of 80 % monthly to less than 10 % a year. This new economic reality imposed large seek for competitiveness to this business sector, that now struggles in order to improve the construction processes. This situation seems appropriate to structural masonry, which appears as a construction process plentiful of conditions to supply the lack of quality, and the competitive requirements of the market. Since this type of building construction allows the decrease of the execution period and cost reduction of construction when compared with the traditional masonry. However to expand the construction market is important that the project's stage to structural masonry must to be completely evaluated and implemented. To achieve satisfactory performance of these projects must to contemplate actions that seek than knowledge technologic as management actions.

1.1 Structural Masonry in Brazil

Roman (1996) asserts that this construction system showed good performance before the needs and the existing conditions in most part of the country. Structural Masonry was used, in the very first time, in modesty homes and buildings 4 or 5 story high, after the 1960's. Franco (1992) points out that this system was not used technically correct and many cases of pathologies occurred as a consequence. This created a negative view of structural masonry with a consequent retraction of the construction market.

The application of this system had grown after 1980's, when one construction company used the system in wide scale in the country, to get more productivity. The level of utilization of structural masonry is not the same that in Unites States and Europe.

2 THE PROJECTS MANAGEMENT

According to Gus (1996), the aim of the project management is to integrate the several agents, to define responsibilities and the way of reviewing the projects.

One efficient management might add various benefits to the project stage. The systemization, documentation of proceedings and projects coordination leads to the rationalization of project and execution too.

2.1 Performance of projects

Nascimento & Formoso (1998) argue that performance of project during of service's execution is associated with the way the project is interpreted by the workmanship and by its level of communication.

The establishment of project criteria helps to lead the project process and to measure the project performance. The performance expected from projects might be enhanced by the control of this process.

2.2 Project Stages

The project stages are defined by Gus (1996) as the several phases of development that one project might be divided, as function of amount of data collected and the relationship of these data with the other phases of the process; clients, legislation organs, cash flow, etc. The considered fundamental project stages are

2.2.1 Briefing
The briefing contains information that will lead the development of projects. These informations are the requirements of clients, that can be obtain through formal mechanisms, like questionnaires and market research.

Gus (1996) asserts that the briefing possess subsidy to evaluate the quality design, documentation and general performance of the project. If the project has one incomplete briefing, it will generate a deficient solution and the evaluation of the project also tends to be imperfect. The briefing reflects the involvement level of client in the process of conception.

2.2.2 The compatibility of projects
The objective of compatibility stage of projects is to agree the physical, geometric and technical components, which relate mutually to the horizontals and verticals constructive elements of buildings. The verification of sob-position and identification of interference between them will result into compatibility. The computer might be utilized to dispose the information in real time.

2.2.3 The conception of details
The conception of architectural projects to structural masonry demand more attention to detail. According to Curtin (et al. 1984) these details must be carefully thought, simple, clean, and informative enough to reduce failure, inadequacies, misunderstandings and misinterpretation.

2.2.4 Documentation
The amount of documents issued and sent to the construction site must be descriptive enough to allow ample understanding of the project as a process as representing of the final product. Franco (1992) states this project coordinates the wall's elevation - that is the critical stage of execution- without interference and resultants problems. The scales of presentation must be in 1/50 to plants and 1/25 to elevations.

Gus (1996) suggests that during the elaboration of the architectural project, some documents should be issued to aid the organization of the conception, as flow sheets, checklists and the schedules of stages. And the computer might help to issue these documents.

3 THE ARCHITETURAL PROJECT OF STRUCTURAL MASONRY BUILDINGS

3.1 Technology transfer

The structural masonry is still considered as an innovating technology in Brazil. To implement this system is necessary to be familiar with its process. Sabbatini (1984) states that all individuals involved should know how each service is executed and how the projects are conceived, in order to optimize the production.

The practice of architectural projects in structural masonry does not introduce radical changes in the conception process; it only presumes some organizational changes. The elaboration of architectural projects in structural masonry is one process, which is essentially interactive.

3.2 Basic technologic principles of architectural design of Structural Masonry Buildings

The architectural project of masonry requires enormous qualification from the designers for its creation. It must to be based in a profound knowledge of geometric, mechanic and aesthetic properties of the components of structural masonry to create safe and economic projects. To confer transparency, avoid waste and augment quality of the of the execution process services, these projects must to have high level of accuracy, and high integration between the structure and close functions.

This project's features can be achieved by the application of the following concepts: constructibility, rationalization and modulation.

3.2.1 Constructibility
The constructibility is the ability pertinent to the project, which allows an efficient utilization of construction resources. The understanding of the structural properties of the walls is an essential tool to provide constructibility for the architectural projects of the structural masonry.

The work of the architects Dieste and Louis Kahn demonstrate that the knowledge of structural properties may lead in expressive works.

The use of exposed blocks also confers constructibility for the project. However Brazilian architects have been disdained this cultural values due to the large diffusion of reinforced concrete.

3.2.1.1 Fundamental concepts of structural properties of the wall
The masonry has good compressive strength. On the other hand, the tensile, shear and torsion strengths are weak, unless reinforced.

The walls have two functions: load bearing and space definition. According to Curtin (et al. 1984), the wall designed to resist loads are called structural walls, and can not be removed or to be cut out in their surface.

In structural masonry the way in which the intercession's walls are create will determine the stability of whole. Moreover the geometry determines the resistance of the building. For example, symmetric constructions are more efficient in terms of construction costs and less susceptible to damages.

3.2.2 Rationalization

A rationalized project shows perfect compatibility and integration between several subsystems.

One action of rationalization is to specify components from the same "family" of blocks and to select uncomplicated elements to the others components. These components must follow the design according to a pattern and coordinate its dimensions to allow large amount of repetitive operations.

3.2.2.1 Principal details and basic unity: the block

Each kind of block composes one "family of components" with different dimensions and chemical, mechanical and physical properties.

The principal details to be note in the drawing are:

➤ intersection of the walls, integration between slabs and the last layer brick, type of slab, stairs, elevator's walls,

➤ spans, cells with grout, pillars,

➤ veneers (arches), windows,

➤ layout of wall (cellar system, transversal system, complex system), installation's layout and shafts

3.2.3 Modulation

The modular coordination is a reference system which has as it foundation the block's dimensions It is possible to create an entire system from these modular dimensions. The measurements are defined and coordinated by the dimensions of the blocks in the horizontal and in the vertical plans. The structural elements, (beams, slabs, veneers, stairs, etc.) must to be coordinated with the grid used for masonry. It is important to observe the tolerance of each situation.

4 METHODOLOGY

In order to verify the knowledge of the South-Brazilian architects regarding how to make structural masonry projects and also to analyze the way that the information were treated in process of project, we used the following techniques: direct observation, interview and questionnaires.

The interview aimed to analyze the transference of technology, with subjective questions about acceptance of this system, the contentment to elaborate these projects and to point out difficulties or easiness to make these projects.

The second part of the study was an objective analysis with the application of a questionnaire asking specifics knowledge of structural masonry and management of projects.

The research was accomplished in two cities of South of Brazil: Florianopolis and Porto Alegre. Ten architects were interviewed five in each one of cities. The blocks referred in this study are of concrete and ceramic blocks, because structural masonry applied in South of Brazil is most of them.

5 RESULTS

5.1 Project process management

All architects interviewed implanted management of project process system in their offices, to seek refine the quality in their projects. The management concepts adopted by these architects were basically driven to graphic representation. They were preoccupied to produce drawing of easy comprehension, and they practiced sparse actions of systemization and documentation of internal and construction proceedings.

This management of project process system implantation was endemic, in other words, made by internal employees, which became the process without critical analysis.

5.1.1 Project's Performance

The South-Brazilian architects interviewed thought that project ought to pay attention the final customer's requirements. Few architects considered important that the project acquire the requirements of execution's process.

The control of the process of project is made by external consultants and has been well accepted by the architects.

5.1.2 Stages of Projects

All architects interviewed formalize the stages of projects. In the briefing's elaboration they make use of checklists.

The architects agree that compatibility stage is extremely important to structural masonry constructions. However the most architects made it erroneous, they revise information in gatherings with another designers. No one draw in similar time with the partners, and just one those architects incorporates computational resources in his methods to intensify compatibility. The CAD would help to get efficient this stage, beyond to create basis to information's systematization.

5.1.3 Documentation

The architects fulfilled the literature's recommendations when they employed the right scale, 1/50 in plants, and 1/25 in elevations.

Some of these architects systematize the process the project, using schedule's tasks, and checklists with the requirement's client.

Nonetheless these architects did not prepare manuals of internals proceeding's register and manuals of proceeding's construction. As the result the transmission of information were prejudiced.

5.2 Analyze of technical knowledge

5.2.1 The technology transfer

The most of architects questioned conceived their first project of structural masonry buildings through adaptation from a pre-existent project of traditional masonry. This adaptation is characterized by a change of the measurements based in the size of the block.

In order to obtain knowledge how to project structural masonry the architects took courses and they searched for information in brochures of block manufactories. The architects visit the construction site to verify if the building was been executed according to the project. They could use this visit also to improve their knowledge about construction.

The architects did not demonstrate reluctance to perform structural masonry projects, nevertheless the preferred to project traditional system, because they have more experience on this area.

5.2.2 Concepts of structural walls properties

The south Brazilian architects interviewed were confident how to arrange the walls to get stability for the building. They project their buildings with symmetry.

5.2.3 Concepts

The south Brazilian architects applied adequately modulation and they applied constructibility during shafts project, which contain the electric and hydraulic pipeline. They did not use exposed block in their projects due to conceptual reasons.

The solely principle of rationalization which they applied was to have the bathroom, kitchen and/or other houserooms which needed water pipeline in close physical distance.

The scarce application of these recommended concepts emphasized that South-Brazilian architects do not consider the construction process part of the project requirement.

5.2.4 Drafting of the details and the block selection

The South-Brazilian architects produced details in large amount, but the details recommended by literature are not part of final project, with exception of the intersections of the walls, which are detailed by the most of them. Only one of the South-Brazilian architects interviewed, drawn details in 3D.

The selection of a specific block was generally made by the client, that choose it by economic reasons, instead technical reasons.

6 CONCLUSIONS

The depreciation of the project process occurred every time that these architects were contracted to adapt projects and did not participate effectively in the initial planning of the product.

The transference of technological data in the project process of structural masonry depends on the right systematization and proper record of all the information.

The query revealed that for the most of South-Brazilian architects interviewed, the project management did not provide optimization of resources. It also did no provide use of adequate methods for designing of architectural masonry projects. South-Brazilian architects demonstrated limited technological knowledge on this type of structure through lack of the fundamental details in their projects.

6.1 Recommendations

In general it is recommended the application of project management to obtain ample knowledge about structural masonry. The project feedback must to be formally executed, for instance, using manuals of constructive proceedings, which also helps in the elaboration of essential details.

The compatibility must to be implemented in towards to promote acquisition of broad knowledge and to avoid mistakes to process the project and its execution.

7 REFERENCES

Curtin W. G., Shaw, G. Beck, J. K., Parkinson, G. I. 1984. *Structural Masonry Detailing.* London. Granada Technical Books. 254p.

Franco, L. S. 1992. *Application of rationalization constructive guidelines to technological evolution of constructive process in structural masonry.* Thesis. São Paulo, EPUSP, 319p

Gus, M. 1996. *Method to generating of management system of project stage of Civil Construction; One case study.* Dissertation. CPGEC –UFRGS. Porto Alegre, 150 p.

Nascimento, C. E. & Formoso, C. T. 1998. Method to evaluete the project from production point of view In: Entac – *Quality in the Constructiv Process* 27-30 de abril, 1998, *Anais...* Florianópolis, UFSC,. vol. II, 801p. p. 151-158.

Roman, H. Structural masonry. 1996. Index technics: How build. *Téchne.* Ed. Pini. São Paulo, nº. 24, October.

Sabbatini, F. H. 1984. *The constructive process of buildings of structural masonry in Sílico-Calcária.* Dissertation. EPUSP/USP, São Paulo, 298 p.

Creative Systems in Structural and Construction Engineering, Singh (ed.) © 2001 Balkema, Rotterdam, ISBN 90 5809 161 9

An educational computer design tool for the passive thermal design of housing

I. L. Steenkamp & A. C. Malherbe
Department of Construction Management, University of Port Elizabeth, South Africa

ABSTRACT: An uncomplicated educational computer design tool for the passive thermal design of housing is presented. The analytical basis of the program is illustrated and explained. The educational features of the program are emphasized. The simplicity and ease of use of the program makes it perfectly suited for developers and designers to evaluate the thermal performance of proposed housing designs. The results of passive thermal design improvements to existing buildings are also easily assessed. The database of the program (climatic data, building materials, thermal properties, etc.) is readily adjusted and varied by a user, to suit local conditions.

1 INTRODUCTION

TERMO-2000 is an uncomplicated computer design tool, incorporating prominent educational features, specifically for use in the passive thermal design of housing. It is aimed essentially at non-expert users such as architects, local housing authorities, home builders and students in the building disciplines. The program is meant mainly to compare the relative thermal performance of different passive thermal design strategies, and to test innovative building methods against the known performance of conventional building methods. Executing the above assignments provides a valuable learning opportunity for those inexperienced in the technologies of passive thermal design.

The analytical procedure of the program is empirically derived (Wentzel, Page-Shipp & Venter 1981) from thermal measurements of many different types of buildings in various climatic regions of South Africa. Reasonably well designed buildings were studied where, for example, windows had been shaded against excessive solar heat gains, and adequate cross-ventilation had been provided in summer. The program's use should, therefore, be confined to such buildings and would not, for instance, accurately model a building that has been orientated with large west-facing windows.

User inputs of complex variables, such as interior convective, radiative and latent heat generation due to occupants, lights and equipment, as well as variable ventilation rates for natural venti-

lation, etc., are not required. As a result, the simplicity of use of the program is enhanced.

Due to its many limitations the program is not ideally suitable for research purposes, or able to compete with sophisticated building energy and thermal simulation computer programs. However, the program provides a satisfactory degree of precision when its results are compared with that of more powerful programs.

2 ANALYTICAL BASIS

The thermal performance of a building can be expressed in various quantitative ways, for example, in terms of the expected summer and winter indoor temperatures, or by means of the amplitude ratio (the ratio of the daily maximum to minimum indoor temperature swing, to the corresponding outdoor temperature swing).

The thermal performance of a building is controlled by its physical properties, such as the heat storing capacity (thermal mass) of the various structural components, the thermal resistance (insulation) of the building enclosure, solar heat gain through windows, solar heat absorption of building surfaces, ventilation rate, etc.

The thermal analytical procedure of the program is based on correlations derived from thermal measurements of a variety of buildings, under both summer and winter conditions, in different parts of South Africa (Wentzel et al. 1981). A correlation was established between the amplitude ratio, the total active heat storing capacity of a building and the equivalent thermal resistance of the building shell. A similar empirically verified correlation was obtained between the difference in the mean daily indoor and outdoor air temperature, and the total daily solar heat gain through the windows, the thermal resistance of the building shell and the ratio of window area to floor area.

The reliability of the program was considerably enhanced by incorporating features of a semi-empirical procedure (Mathews 1985) which is founded on electric analogue theory, as well as on experimental results.

Graphs depicting typical variations in summer outdoor and indoor daily air temperatures in South Africa are shown in figure 1.

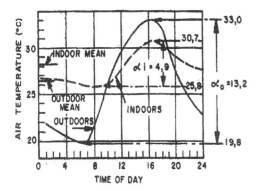

Figure 1. Typical indoor and outdoor air temperature graphs and amplitude ratios for a relatively well designed building in the hot inland region of South Africa (Wentzel et al. 1981).

The shape of the indoor air temperature curve is very similar to that of the outdoor air temperature curve, but the temperature amplitude is usually smaller, and a time-lag is evident between the occurrence of maximum and minimum indoor, and maximum and minimum outdoor air temperatures. This so-called 'cushioning effect' is typically manifested in buildings where mass and insulation are judiciously applied in the building structure.

3 CALCULATION PROCEDURE

The general calculation procedure, illustrated in figure 3, procures the following quantities:
1. The total active heat storing capacity of the building, utilizing the dimensions of the various building components, and the mass densities and specific heat values of the different materials.
2. The equivalent thermal resistance of the building shell, employing the U-values of

the various components of the building envelope.
3. The amplitude ratio, from the above two quantities, by means of the experimentally established correlation.
4. The amplitude of the daily indoor air temperature fluctuation, given the amplitude of the daily outdoor air temperature fluctuation and the amplitude ratio, calculated above.
5. The difference between the mean daily indoor air temperature and the mean daily outdoor temperature (for both summer and winter) from the previously described empirical relationships.
6. The mean daily indoor air temperature (for both summer and winter), given the known mean daily outdoor air temperature.
7. The summer and winter indoor daily minimum and maximum air temperatures from the indoor mean daily temperature and the indoor temperature amplitude, calculated above.

The precision with which temperatures in different buildings are predicted by the program, is illustrated in figure 2.

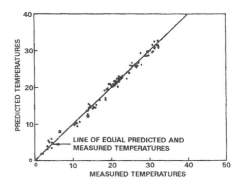

Figure 2. Comparison of predicted and measured temperatures in different buildings (Wentzel et al. 1981).

4 DATABASE

A pre-constructed climate and building database is provided in the program.

The climate database contains summer and winter climatic data for seven locations in South Africa, consisting of the daily air temperature amplitude, the mean daily air temperature and daily solar radiation on vertical and horizontal surfaces. These values portray the so-called 90 per cent design day weather data (Wentzel, Page-Shipp, Kruger & Meyer 1985), i.e. they represent the mean

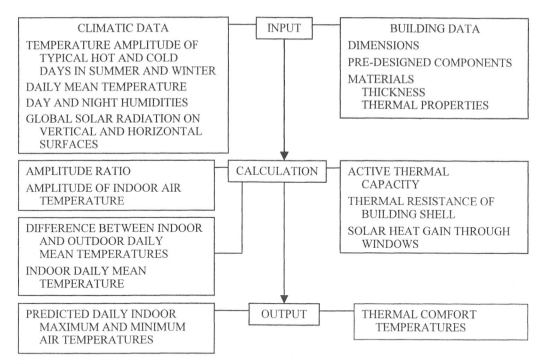

| CLIMATIC DATA
TEMPERATURE AMPLITUDE OF
TYPICAL HOT AND COLD
DAYS IN SUMMER AND WINTER
DAILY MEAN TEMPERATURE
DAY AND NIGHT HUMIDITIES
GLOBAL SOLAR RADIATION ON
VERTICAL AND HORIZONTAL
SURFACES | **INPUT** | BUILDING DATA
DIMENSIONS
PRE-DESIGNED COMPONENTS
MATERIALS
THICKNESS
THERMAL PROPERTIES |

Figure 3. Block diagram showing general calculation procedure.

of the 10 per cent warmest days in summer (or coldest days in winter). Generally, this implies that the design temperature limits will be exceeded on only about 14 days in an average year.

The building database is comprised of building materials and their physical attributes, thermal properties, as well as pre-designed building components, such as different wall, roof and floor types. Information on the properties of most building materials is available in technical literature.

A user can easily extend or edit the database.

5 BUILDING INPUT DATA

The bulk of the input data (dimensions, composition of building components and materials) is derived from the working drawings.

A series of input screens (drop-down menus) are used, for climate and for each of the main components of the building (external walls, internal walls, roof, floor, doors and windows).

6 OUTPUT

Maximum, minimum and mean indoor air temperatures are predicted for typically hot and cold design days in summer and winter.

Graphs display the predicted indoor air temperatures in relation to outdoor air temperatures.

The calculated inside temperatures are compared to thermal comfort temperatures to ensure thermally efficient design.

The results are also available in tabulated format.

7 EDUCATIONAL FEATURES

In addition to the usual program <help> features, unique educational menus are provided to instruct the user in the fundamentals of passive thermal design of housing.

REFERENCES

Wentzel, J.D., R.J. Page-Shipp & J.A. Venter 1981. The prediction of the thermal performance of buildings by the CR-method. CSIR research report BRR 396, Pretoria, RSA.

Mathews, E.H. 1985. The prediction of natural ventilation in buildings. D.Eng dissertation, Potchefstroom University for Christian Higher Education. Potchefstroom, RSA.

Wentzel, J.D., R.J. Page-Shipp, H. Kruger & P.H. Meyer 1985. Climatic design data: typical summer and winter design days for 23 stations and daily mean temperatures and temperature amplitudes for 113 stations in South Africa. Special report Bou 73, CSIR. Pretoria, RSA.

Creative Systems in Structural and Construction Engineering, Singh (ed.) © 2001 Balkema, Rotterdam, ISBN 90 5809 161 9

Ecological and quality oriented planning and building

V. Premzl

Faculty of Civil Engineering, University of Maribor, Slovenia

ABSTRACT: The building of residences and residential settlements is in all societies one of the primary social aims, as well as being one of the targets in the spatial development of cities and settlements. The realisation of these aims is most often connected with conflicts, which derive from the different interests of the participants in the process of planning, and in the realisation of the socially determined aim. With the purpose of obtaining the optimal efficiency in the elaboration of the planning aims of residential building, encompassing the economic, social, and space ecological aspect, we have developed a model to assure an adjustment of conflicts among particular spatial results- or for physical and urban management.

1 INTRODUCTION

The pluralism of interests is a fact in the construction of residences, residential settlements, it is not owned by one or another social system. One the one hand we can speak about elementary human needs, and one the othere of the priority targets of the long-term development of settlements. At the same time there appears a long line of interests and values, in developing aims of the residential building, for example economics, rationality, social and spatial ecological balance of building. Very often conflict situations appear which have the following sources:

- In the argumentation of the aims of the spatial development or the building of residences in the phase of the realisation of these aims .
. In the dispersion of the argumentation in the way of obtaining of these aims.
- Because of the obsolete or non-actual goals which lead to the formalisation of aims, especially in the realisation of ancient building plans.

All the described sources and typesof conflicts in the planning and building of residential settlements are to be analytically described, if we want to form and use a suitable model to find out the optimal success of the particular aims, or various solutions based on the target determination, and to harmonise conflicts among them in such a way.

2 OUTLINE OF THE MODEL IN THE COMPATIBILITY OF CONFLICTS AND AIMS

An approach to define a model in the compatibility of conflicts and aims in the spatial development and building of settlements is based on an optimistic approach of the social development from the optimistic evolution, and does not cause an essential change of the developing programmes, mainly the most important planed aims in the construction of residences and residential settlements.

Within the presented aspect to the formulation of the model to solve the conflict in the spatial planning one has to make some basic assumptions:
- conflicts in the explanation of the aims of the spatial planning are an unavoidable aspect of the planed targets,
- solving the conflict of the aims of space planning is usually possible by a process of the resolution of conflicts,
- resolution of the conflict aims is not possible without participation of the affected, as well as the participation of experts,
- it is not permissible to discuss only one of the conflict aims, it is necessary to judge the purpose of all the alternative conflict aims,
- because the intensity of the conflict aims of space managing may be small, they might be reckoned by a simple procedure of resolution on the basis of the clear evidence of causes for the clash of aims,
- we have to professionally estimate the decision about the optimality of the success of the particular conflict aim of the spatial development, by using quantitative methods and expert opinion for the economic, social, and space-ecological aspect.

3 FORMING THE CRITERIA TO EVALUATE THE MODEL

In forming the criteria to determine the efficiency of the model we put the primary thesis that in the solution of conflicts and aims, which are solved by results, we resolve for the question of optimal effects of one or another aim, and mainly for the problem of the possibility of optimal achieving of the determined aim, to satisfy its goal. In the case of the conflict of two or more sub-aims we have to search for an optimal effect according to the criterion of a superior aim, etc.

Another thesis, obligatorily satisfied by the criteria, claims that the compatible conflict of aims should satisfy the criterion of the principle aim of the space managing.

3.1 The evaluation criteria

The criteria to evaluate the efficiency of the module formation are as follows:
- The criteria of objectiveness of the defined aims conflict. The model should be based on the optimal possible objectiveness of searching for the conflict aims of all, or particular aspects of space managing;
- The criteria of the wholesome discussion of the conflict aims that enable the views of all three aspects of space managing: space-ecological, social, and economic aspect;
- The criteria which offer hierarchy and priority of aims;
- The criteria of step-by-step solution.
According to a different intensity and degree of the space managing of the conflict planning aims the model for its solution should enable resolution of conflicts in the first phase. The model should also enable the solution of the particular conflict aims of the same aspect, as well as all aspects of the space managing (economic, social, space-ecological).

3.2 The criteria of the optimal efficiency of the planed aims.

With the model of solving of conflict aims we follow achieving of the optimal aim effects of overordinate aims, and a common effect of all the partial aims.

The criteria of possibilities to use quantitative and qualitative measures should enable us to use quantitative as well as qualitative measures, typical for the particular aspects of the space managing.

3.3 Limitations

In forming the model some limits will be considered:
- Recognition of interests, planning of aims depend from temporal changing of values, needs and fromthe long-term orientation of the spatial management. It confirm that time essentially influence the degree of conflicts, or on their adjustment. Therefore it is possible to solve planed aims conflicts in advance, but only simultaneously when they appear.
- The model does not solve the possibility of the simultaneous realisation of the controversial aims. We suppose that it is not possible to anticipate, to plan or even to realise the conflict aim solutions at the same physical location.

4 FORMATION OF THE MODEL TO ADJUST CONFLICT AIMS

We derive from an empirical acknowledgement that in managing urban planning and building of residential settlements there exists an interaction between particular planed solutions or their aims in space, which are denoted as T_1, T_2.....T_n between the particular aspects of the space managing: space-ecological (O), economic (E) and social (S) aspect.
The interaction between particular aims of the space managing and the building of residential settlements T_1, T_2.....T_n has a line of consequences (P) on the particular aspects of the spatial management and the building of residential settlements. This interaction may be clearly presented by the following register.
Where in this register we find out a negative interaction among the aims of the particular planed solutions and the particular aspects of the space managing (O,E,S) we register their conflict.

Table 1. Interaction of Aims of the Particular Space Solutions of the Aims in the Particular Aspects of the Space Managing.

Consequences on particular aspects of managing	Aims of managing	T_1	T_2	T_n
P_O	P_{O1} P_{O2} . P_Om	P_{OT1}	P_{OT2}.......$P_{OT}n$	
P_E	P_{E1} P_{E2} . P_Em	P_{ET1}	P_{ET2}.......$P_{ET}n$	
P_S	P_{S1} P_{S2} . P_Sm	P_{ST1}	P_{ST2}........$P_{ST}n$	

In Table 1 presents the conflicts of aims in the particular solutions we try to determine the essential, we mark them with an X, and the non-essential, are marked with O.

On this basis of the common number of essential consequences is possible to realise the first assessment of effects of the particular aims of the expressed variant solutions. The objectivity of such an assessment is questionable in this degree without the proper criteria to estimate the suitability of the aims. Criteria that are important for the suggested space solution we determine for all three aspects of the spatial management, but in equal number.

Table 2. Criteria to Evaluate Variant Solutions

Space-ecological	(O)	O_1,	O_2, ... On
Economic	(E)	E_1,	E_2, ... En
Social	(S)	S_1,	S_2 ...... Sn

Table 3. Estimation of the Importance of the Determined Criteria to Achieve the Aims of the Particular Solutions

Criteria (K)		Aims T_1, T_2, ... Tn
(O)	K_{O1} .	
	K_Om	
(E)	K_{E1} .	
	K_Em	
(S)	.	
	K_Sm	

Table 4. Evaluation of the Priority of Criteria

Criteria	K_{O1}.....K_Om	K_{E1}...K_Em	K_{S1}....K_Sm
Priority	+,++	+,++	+,++

Since criteria may be quantitative or qualitative we will estimate them according to their importance for the efficiency of the particular aim solutions by help og graphic marks in the span from 0 to 000, where there are the following items:
000 Very Important Criterion
00 Criterion of medium importance
0 Necessary Criterion
All the criteria to estimate the propriety of aims for the spatial solution are not equally important. The estimation of the importance of the criterion of the particular aspects of the spatial management or settlement we realise in the following table.

Table 5: Table of the Expected Efficiency of Aims or Solution according to the Priority and Importance of Criteria

Aims	++			+		
	000	00	0	000	00	0
T_1						
T_2						
-						
-						
-						
Tm						

Table 6. Numerical evaluation of Efficiency of the Particular Aims

Aspects of spatial management	Aims T_1, T_2, T_3, ...Tn
K_{O1} -	
K_Om K_{E1} -	
-	
K_Em K_{S1} -	
K_Sm	• T_1, • T_2, • T_3, • T_n

Not only the importance of criteria but also their priorities are important for the objectiveness of the as assessment of the effects of the particular aims of the solution. To assess the priority we introduce graphic (the way of expression of quantitative and qualitative criteria is taken from Nijkamp, P. Environmental Policy Analysis. John Willey and Sons,1980) evaluation + for lower and ++ for higher priority of the criteria.

In the evaluation comparison of the particular alternative aims of solutions we may remove the less efficient aims in the comparison with those which are closer to the optimal planed aim. According to the whole assessment and the survey of success of the particular alternative solutions in achieving the planned aims, in table 5 we change the graphic values with a numerical evaluation of the efficiency of the particular aims according to the planed criteria in the particular aspects of the space management and building of settlements.

With this final evaluation of the efficiency of the particular aims of various solutions we should have bases enough to decide about an optimal variation of the space management, or else the construction of a residential settlement on the basis of the presented model. Although a large, previously presented control, there might be equal results, which do not enable clear evaluation about the optimal achievement

about the previously performed aim, we may accomplish previous findings with an additional control of impacts on the efficiency of aims.

The influence on the optimal efficiency of aims is by hierarchy, priority of aims, possibility of their gradual realisation, and the contribution of a particular aim to a common effect on the mutual realisation of other aims.

The evaluation of these additional effects of aims is realised by a weight. The common evaluation of the weights of all the previously quoted four additional impacts is 1.0.

Weight values to evaluate additional particular impacts on the efficiency of aims have a common value 1.0, and are determined according to their influence.

The calculation of the assessment of the efficiency of particular aims according to the aspects of the space management from the table 6 we multiply by weight values from table 7.

The evaluation of the final success of the aims according to the particular aspects of the space management may also be presented graphically, if the obtained values of the product (for aims 0, E and S) are set into the axis of the three-dimensional vector space, and we find out the achieving of the value of vector $r = \sqrt{x^2 + y^2 + z^2}$ (where $x = O$, $y = E$, $z = S$). Vector r has a starting-point in the middle of the co-ordinate system and the length $r = \bullet \; x2+y2+z2$ which in our case means the value of the optimal efficiency in point T.

Therefore, in the choice of the most efficient aims and solutions we shall choose those which are closest to the ideal point T.

Procedure of finding out and adjusting the aim conflicts has the following 7 steps:
- Social, consenzual accepted aims of new settlement building and spatial development .
- Finding out of plan aims conflicts.
- Identification of the particular concurence or conflict aims.
- Forming of criteria to find out the efficiency of the fifth aims.
- Determination of weight values for evaluation of the additional impacts on the aims efficiency.
- Finding out the achievement of optimal common effect of the particular aims.
- Coordination and decision about the most convinient aim.
(All these items are connected other, and the last one is through the eventual change of goals, aims and criterias connected with second item).

Table 7. Evaluation of the Additional Efficiency of Aims

Additional efficiency of aims	Aim T_1 Weight U_1....T_2 U_2 ....T_n U_n
Degree of hierarchy (H)	
$\quad H_1$	
$\quad H_2$	
$\quad -$	
$\quad -$	
$\quad H_m$	
Priority of aims (C)	
$\quad C_1$	
$\quad C_2$	
$\quad -$	
$\quad -$	
$\quad C_m$	
Possibility of gradual achieving of aims M)	
$\quad M_1$	
$\quad M_2$	
$\quad -$	
$\quad -$	
$\quad M_m$	
Contribution to common effect of other aims (D)	
$\quad D_1$	
$\quad D_2$	
$\quad -$	
$\quad -$	
$\quad D_m$	
$\bullet \; U_1 \text{-} U_n = 1$	$\bullet \; U_1 \qquad \bullet \; U_2\text{--} \qquad \bullet \; U_n$

5 CONCLUSION

The presented model was checked in the laboratory through a number of exemples, which are in connection with the evaluation of the alternative solutions of the spatial development of settlements.

It was shown to be a valuable help in deciding the optimal concepts of the settlement development. Naturally,. the model itself cannot replace the professional ca-pability of people, who use it as a tool.

The heterogenity of interests of the particular subjects of planning and space management, which are, in the final phase, presented as conflict aims, particular aspects, or even as conflict among particular aspects of the space management, becomes more and more obvious. The acknowledgement about the natural limits contributed a lot, not just in material like minerals and agricultural land, but likewise in water and air.

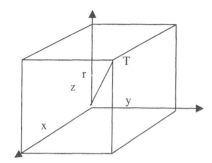

Figure 1: Space Presentation of the Optimal Efficiency Achievement

The economic aspect of the space management discussed space as a basis to dispose settlements and objects, which exploit natural sources in the space. An outcome to such an approach to the space is seen in a catastrofal form in all countries. This is only one of the sources of conflict aims.

The consequences are not only expressed verbally, but also materially, since there are expenses of the sanation of the state, or because the preventive from the particular intruding into the space management, not talking about the false economic development.

6 REFERENCES

Buckley, M. 1988. Multicriteria Evaluation: Measures, Manipulation, and Meaning, Environment and Planning. Planning and Design. Vol. 15 Nijkamp. P. 1980 Environmental Policy Analysis. John Wiley and Sons, Am Amsterdam.

Piha, B. 1987. Tehnološka revolucija i prostorno planiranje, Založba Centar. Beograd.

Premzl.V. l989. Konfliktnost družbenih ciljev v planiranju prostorskega razvoja. Fakulteta za arhitekturo, gradbeništvo in geodzijo. Ljubljana.

Premzl, V. l992. Die Umweltverträglichkeitsprüfung als technisches und administratives Instrument, Fallstudie Slowenien. Umweltverträglichkeitsprüfung. 236-248. München. Bayerisches Staatsministerium fur Landesentwicklung und Umweltfragen.

4 Constructability and design-construction

Creative Systems in Structural and Construction Engineering, Singh (ed.) © 2001 Balkema, Rotterdam, ISBN 90 5809 161 9

Constructability issues in Indonesian construction

B.Trigunarsyah & D.M.Young
Department of Civil and Environmental Engineering, University of Melbourne, Vic., Australia

ABSTRACT: Many construction projects in Indonesia are awarded by competitive bid where professional designers and constructors are engaged in separate contracts. The contractors usually would not be involved until the designs have been completed. This division has been suggested as being responsible for the lack of constructability in construction projects that leads to projects exceeding budgets and schedule deadline. The studies in the US and Australia have demonstrated that improved constructability has lead to significant savings in both cost and time required for completing construction projects. However, in implementing constructability improvement, it is important to consider the uniqueness of the construction industry in a specific country. This paper will present the study performed on the construction contractors in Indonesia in regard to their current constructability practices and its impact on the project performance.

1 INTRODUCTIONS

1.1 *Construction industry in Indonesia*

Construction industry is an essential contributor to the process of development in Indonesia. It influences most, if not all, sectors of economy. Roads, dams, irrigation, houses, schools and others construction works are the physical foundations on which development efforts and improved living standards are established.

The construction industry in Indonesia has grown significantly since the early 1970s. Its contribution to the GDP has increased from 3.86% in 1973 to just above 8% in 1997. It constitutes about 60% of the gross fixed capital formation. The annual growth of the manpower employed by the construction industry is about 13.1%.

More than 50% of the construction works in Indonesia have been in heavy engineering construction, which provides the infrastructure, i.e. roads, ports, irrigation, power generations and distribution, gas and water distribution, and telecommunication. Building construction constitutes about 35%. The remaining works are shared by residential and industrial construction.

Most public work projects, including any construction projects under government authority or under state owned companies, are awarded on a competitive basis using the traditional approach. Professional designers and constructors are engaged in separate contracts. The contractors usually would not be involved until the designs have been completed. Many private sector projects have also been using the similar approach.

The separation of design from production in the construction process has led to a certain amount of isolation of the professionals from technical development in construction industry (Wells 1986). This division has also been suggested as being responsible for the lack of constructability of the construction projects (Griffith 1984), which was cited as a reason for projects exceeding budgets and schedule deadlines (Construction Industry Institute Australia 1992). By separating construction from design function the project stakeholders are ignoring opportunities of significant savings in both project cost and completion time resulting from the careful interaction of planning, design, and engineering with construction (Tatum et al. 1986).

1.2 *Constructability defined*

The concept of constructability in the US or buildability in the UK emerged in the late 1970s, which evolved from studies into how improvement can be achieved to increase cost efficiency and quality in the construction industry. It is an approach that links the design and construction processes. In this paper the constructability is defined as "the optimum use of construction knowledge and experience in planning, design, procurement, and field operations to achieve the overall project objectives" (Construction Industry Institute 1986).

1.3 Prior research

Early research work in the UK has focused attention upon only one genuine aspect, that of design rationalization. The primary aim of the research was to identify those factors in the design of a construction project that have an impact on site construction techniques (Griffith 1984). As a result of the research work done in 1979, the Construction Industry Research Information Association (CIRIA) (CIRIA 1983) provided seven 'Guidelines for Buildability', which was later expanded into sixteen 'Design Principles' for practical buildability (Adams 1989).

In contrast to the UK approach, constructability researchers in the US have placed emphasis on the management systems and involvement of owners and contractors. Constructability improvement has been considered as an integrated part of the whole project life cycle. The Construction Industry Institute (CII) (Construction Industry Institute 1987) has fourteen developed Constructability Concepts, which were later appended with three additional concepts (Russell et al. 1992).

In Australia, the Construction Industry Institute Australia (CIIA) has also developed 12 principles of constructability based on the CII constructability concepts, which were tailored to the Australian construction industry.

2 CONSTRUCTABILITY IMPROVEMENT IN INDONESIA

There has not been a particular study in Indonesia regarding the impact of the separation of responsibility between design and construction. However, the study by Wells (Wells 1986) in several developing countries indicated that this separation has led to a certain amount of isolation of the professional from technical developments, which is reinforced by the rigid compartmentalization of training. Furthermore, the separation also means that the designer (architects/engineers) is isolated from knowledge of actual construction costs and the costs of construction based upon alternative designs. Wells argued that as long as design is divorced from the responsibility for building, the builder has practically no say in the opportunities or occasions for introducing, or making use of, innovations except to some extent in the technical organization of the actual building process.

As suggested by the research findings in Australia, implementation of constructability improvement has to consider the uniqueness of the construction industry in a specific country. Therefore, in order to improve constructability in Indonesian construction projects, it is important to assess the existing constructability practices.

2.1 Constructability survey

This paper will assess the current constructability practices among the construction project stakeholders in Indonesia. The scope of study was limited to the current constructability practices among the construction contractors particularly in their involvement in the early stages of the project life cycle and its influence in the project performance. The study was done using questionnaire surveys, which have been developed based on previous surveys and research done in the US. As the concept of constructability has been developed in the US, it is assumed that this concept was not known by most of the respondents. A brief explanation of the terminology and the basic concept of constructability has been included in the questionnaire survey.

A total of 88 questionnaires were received from 200 that were sent to various construction projects. Only 77 of the 88 responses were used in analysis due to the incomplete information.

The type of respondents can be classified into three different categories, i.e. state owned contractors (BUMN), private (national) contractors, and multinational contractors. The type of the project that they performed can be grouped into four categories, i.e. heavy engineering, industrial, buildings, and residential. Table 1 summarizes the type of the respondents and the type of projects that they performed.

Table 1. Type of respondents & types of projects

Type of projects	Type of respondents			Total
	BUMN	Private	Multinational	
Heavy Engineering	12	9	3	24
Industrial	5	12	3	20
Buildings	6	12	12	30
Residential	1	1	1	3
Total	24	34	19	37

The project values range as follow:
- Up to US$ 10 million   62.3%
- Between US$ 10 - 25 million   19.5%
- Between US$ 25 - 50 million   10.4%
- More than US$ 50 millions   5.2%

Some of respondents (2.6%) did not state the project value.

51.6% of the projects had a duration of less than 12 months; 29.9% between 12 and 24 months; 13% between 24 and 36 months; 2.6% between 36 and 48 months; and another 2.6% more than 48 months.

About 40% of the projects have been delivered using the traditional approach; 35% using the design-construction approach; 21% using the construction management approach; and the remaining projects have been delivered using the design-manage and owner-builder approach.

From the contingency table between the type of the project and the project delivery selected by the owner it was found that heavy engineering (infrastructure) and residential projects are more likely to be delivered using the traditional approach. Industrial construction projects tend to use the design-construct approach. Delivery of building construction projects varies from the traditional to the design-construct and the construction management approaches. However, in the construction management (CM) approach the CM function was performed by the general contractor and the general contractor did not get involved in the earlier phases of the projects.

2.2 Early contractor involvement

The role of contractor prior the construction stage of the project will be limited by the contractual approach implemented by the Owner. Only 27 (35.1%) of the respondents were engaged in the project as early as in the conceptual planning stage, and 31 (40.2%) started to be involved in the project during design-procurement phase.

In the conceptual planning phase the most common activities that the contractors have been involved with were providing advise/suggestions to the project owner regarding structural systems, selection of major construction methods & material, and preparation of schedule, estimate and budget. As their involvement was limited by the scope of the contract, the survey results suggested that it is because of their expertise and experience in those areas that the owner required the contractors' services. The least common activities that the contractor was involved in were conducting the feasibility studies and the selection of sites, and developing the basic designs. The reason for this can be attributed to the preference of the owner to perform the studies by themselves or by engaging a consultant firm. Table 2 presents the list of activities that the contractors were involved in during this project phase.

Table 2. Contractor involvement during conceptual planing phase

The activities in which the contractor was involved	*
a4. Suggest structural systems	67%
a5. Selection of major construction methods & materials	67%
a6. Preparation of schedule, estimates & budget	67%
a3. Advise owner in the contracting strategy	48%
a1. Advise owner in setting the project goals & objectives	41%
a2. Feasibility studies & advise in site selection	15%
a7. Others (development of basic design)	11%

*as % of the number that were involved in this phase

During the design-procurement stage the most common activities that the contractors have been involved with included: reviewing & providing advice regarding accessibility for personnel, materials & equipment; and analyzing & revising specification to allow easy construction. As in their involvement in the conceptual planning phase, it was their expertise and experience in conducting these services that was required by the owner and/or the designer in order to improve the constructability of the project. The least common activities included the analysis and promotion of design that facilitates construction under adverse weather. It is not surprising as Indonesian weather is relatively stable throughout the year and does not vary much in different area except for some areas with high precipitation. Table 3 summarized the involvement of the contractor (respondents) during the design-procurement phase of the project. Other activities included the preparation of safety regulations and procedures.

Table 3. Contractor involvement during design-procurement phase

The activities in which the contractor was involved	*
b2. Review & advice re. accessibility of personnel, material & equipment	77%
b4. Analyzed/revised specification to allow easy construction	71%
b6. Preparation of schedule, estimates & budget	61%
b1. Analyzed design to enable efficient construction	61%
b3. Advice on source of materials & engineered equipment	58%
b5. Analyzed/promoted design to facilitate construction under adverse weather condition	26%
b7. Others	6%

*as % of the number that were involved in this phase

Table 4. Early contractor involvement

Activities	No	Yes (%)				
	(%)	1	2	3	4	5
c8. Provide timely input	6.5	-	3.2	38.7	32.3	19.4
c4. Use pre-construction plan for input to design	6.5	-	19.4	22.6	25.8	25.8
c5. Study construction method	9.7	-	6.5	12.9	38.7	32.3
c3. Proactively involved in developing project plan	9.7	-	3.2	29.0	29.0	29.0
c6. Review & select constructability issues	12.9	6.1	6.5	35.5	22.6	6.5
c7. Provide means to monitor constructability	16.1	3.2	22.6	35.5	19.4	3.2
c1. Assign appropriate personnel	19.4	3.2	-	22.6	29.0	25.8
c2. Locate personnel close to design team	25.8	9.7	19.4	25.8	12.9	6.5

The respondents have also been asked if they were involved in those phases in regard to improving the constructability of the project. If the response is 'Yes', the respondents were asked to rank the degree of their involvement from 1 (very low) to 5 (very high). Their responses are summarized in Table 4.

Most of the respondents stated they have provided a timely input to the designer using a pre-construction plan as the base for their input. These results suggested that for the contractor input to have a positive impact on constructability of the project, the input should be given in time. And a pre-construction plan is the best reference on when is the best time to incorporate the input. Attaching the construction personnel (representatives) to or locating them in close physical proximity to the design team is the least common method that the respondents did during those stages. One possible reason could be the use of regular project meetings using the pre-construction plan as the basis to facilitate construction input to design.

2.3 Constructability improvement during construction phase

The involvement of the contractors in the early stages of the project is very dependent upon the owner and owner selection of contractual approach. Therefore, it is then important to assess the role of the contractor in improving constructability during construction phase. Decisions related to field operations constructability tend to be relatively low leverage decisions. However, CII (Construction Industry Institute 1987) suggested that collectively they offer substantial benefits. The concept that was recommended by CII to improve constructability during field operation is a development and utilization of innovative construction methods, which can simplify construction effort and reduce project cost.

Table 5. Constructability improvement during construction

Activities	No (%)	Yes (%) 1	2	3	4	5
d2. Plan the sequence of field tasks to improve productivity	-	-	1.3	10.4	24.7	63.6
d1. Analyzed layout, access & temporary facilities	2.6	1.3	-	9.2	35.5	1.3
d3. Innovative uses of hand tools	11.7	-	-	22.1	31.2	63.6
d5. Use innovative construction equipment	11.7	5.2	7.8	46.8	10.4	18.2
d4. Customize or upgrade construction equipment	14.3	2.6	2.6	27.3	29.9	23.4
d6. Use of constructor-optional pre-assembly	19.5	3.9	6.5	32.5	22.1	15.6

Table 5 summarized the contractor activities during construction phases related to constructability improvement. The most common activities performed by the contractor (respondents) are preparing an effective sequence of field tasks and carefully analyzing site layout, access and temporary facilities to improve productivity. Good construction planning is the basis for effective and efficient construction activities that could be monitored and controlled to achieve project objectives (ASCE 1990). Hence, it would be a logical priority for the contractor to achieve optimum project results. In addition to that an efficient layout, access and temporary facilities can have the effect of reducing labor intensity, reducing the likelihood of delays in providing for utilities, or improving the work environment (O'Connor and Davis 1988).

2.4 Typical constructability problems

In order to assess how early contractors' involvement improve the constructability of the project, the respondents have been asked about typical problems which they encountered during the construction period. For each problem they encountered, the respondents were asked to rank the degree of the problem from 1 (very low) to 5 (very high). The most common problems encountered by the respondents were related to construction tolerances and specifications. Table 6 summarizes the common constructability problems faced by the contractors during construction period.

Table 6. Typical constructability problems

Problems encountered	No (%)	Yes (%) 1	2	3	4	5
e2. Tolerance problems	16.9	11.7	22.1	36.4	7.8	5.2
e1. Specification problems	18.2	6.5	36.4	23.4	10.4	5.2
e5. Unrealistic schedule	29.9	10.4	19.5	24.7	7.8	7.8
e4. Weather related problems	30.3	17.1	13.2	25.0	9.2	5.3
e3. Problems with physical interference	31.6	18.4	19.7	21.1	5.3	3.9

2.5 The impact of early contractor involvement on the project performances

Contingency tables and correlation analysis were used to assess the impact of early contractor involvement on the project performances. The project performances in this case are the typical problems that the contractors encountered during field operations as presented on Table 6. Table 7 and 8 present the contingency tables between the involvement of contractors in pre-construction phases and the project performances, and the related chi-square results.

Table 9 summarizes the non-parametric correlation coefficient between how the contractors involved in the early stages of the project and the project performances.

In preparing the contingency tables, the degree of the problems was divided into two groups, i.e. low and high. The problems were considered low when the responses were no or yes with the score of 1 and two. The yes answers with the score of 3 to 5 were considered high degree problems. The chi-square results from the contingency tables show that there is a relationship between the contractor involvement in conceptual planning as well as design-procurement phases of the project and the project performance. As it is very difficult to totally avoid these problems occurring during field operations, the contingency tables indicated that contractor involvement in the early stages of the project could reduce the degree of problems. It can be seen from Table 7 that when the contractors were involved during the conceptual planning and the design-procurement phases, it resulted in lower degrees of problems in construction tolerances (e1), specifications (e2) and weather related problems (e3) that can be avoided during the design phase.

Table 7. Contingency table between contractors' early involvement and typical problems

| Degree of problems | | Contractor involvement during | | | |
| | | Conceptual planning | | Design-procurement | |
		No (%)	Yes (%)	No (%)	Yes (%)
e1	Low *	48.0	85.2	47.8	80.6
	High**	52.0	14.8	52.2	19.4
e2	Low	38.0	74.1	34.8	74.2
	High	62.0	25.9	62.2	25.8
e3	Low	71.4	66.7	73.3	64.5
	High	28.6	33.3	26.7	35.5
e4	Low	46.9	85.2	48.9	77.4
	High	53.1	14.8	51.1	22.6
e5	Low	54.0	70.4	54.3	67.7
	High	46.0	29.6	45.7	32.3

* Low degree means from no problem to score 2 (re. Table 6)
** High degree from score 3 to 5 (re. Table 6)

Table 8. Chi-square results

| Typical problems | Contractor involvement during | | | |
| | Conceptual planning | | Design-procurement | |
	c-s*	p**	c-s	p
e1	10.194	0.001	8.388	0.004
e2	9.128	0.003	11.508	0.001
e3	0.187	0.665	0.676	0.411
e4	10.658	0.001	6.254	0.012
e5	1.954	0.162	1.381	0.240

* Chi square
** p-value

Negative correlation on Table 9 suggests that early contractor involvement can reduce the problems during the construction period. The two problems that had been influenced the most by the contractor involvement in the pre-construction phases were construction tolerances and specifications problem. It is not a coincidence that the highest influence occurs against the two most common problems that contractors encountered during this period. As a logical consequence of having the knowledge and experience of facing these problems the contractors would concentrate their early involvement in avoiding or reducing the occurrence of these problems during the field operation.

Table 9. Non-parametric correlation

| Activities | | Typical problems faced during construction | | | | |
		e1	e2	e3	e4	e5
c1	*)	-.262	-.335	0.011	-.197	-.137
	**)	0.022	0.003	0.928	0.089	0.236
c2		-.194	-.259	0.013	-.179	-.093
		0.091	0.023	0.910	0.122	0.422
c3		-.286	-.357	-.012	-.240	-.204
		0.012	0.002	0.919	0.038	0.078
c4		-.295	-.309	0.088	-.146	-.208
		0.009	0.006	0.448	0.207	0.069
c5		-.311	-.379	-.033	-.248	-.198
		0.006	0.001	0.779	0.030	0.084
c6		-.220	-.264	0.039	-.152	0.121
		0.055	0.020	0.741	0.191	0.101
c7		-.233	-.321	-.014	-.189	-.216
		0.042	0.004	0.903	0.102	0.059
c8		-.268	-.300	0.013	-.171	-.207
		0.018	0.008	0.910	0.139	0.071

*) correlation coefficient
**) p-value

Most of the respondents (97%) agreed that their involvement in the early phases of the project could avoid these problems and produce better constructable projects. They also agreed that during the early phases of the project construction should be included as another specialty like architectural, structural, mechanical, electrical etc.

3 CONCLUSIONS

3.1 Current constructability practices among construction contractors in Indonesia

The concept of constructability was developed in the US. However, a constructability survey conducted among construction contractors suggested that, with limitations, many contractors in Indonesia have been implementing part of the concept in their projects. The concepts that usually applied during the construction stage as part of the overall construction plan were planning the sequence of field tasks and analyzing layout, access, and temporary facilities.

Contractor involvement in pre-construction phases would depend on the owner and owner selections of project delivery systems. When engaged early in the project life cycle, contractors in Indonesia have also shown some implementation of the constructability concept. In the conceptual phase of the project, the main activities performed by contractors were providing advice/suggestions to the project owner regarding structural systems, selection of major construction methods & material, and preparation of schedule, estimate and budget. In the design-procurement stage, the most common activities contractors involved with were reviewing & providing advice regarding accessibility for personnel, materials & equipment; and analyzing & revising specification to allow easy construction. The most common activities or services that contractors were involved with were in line with their expertise and experience as constructors of projects.

Contractor early involvement in the project life cycle could lead to a better project performance in term of less problem occurrences. The survey results point to the fact that contractor involvement in pre-construction phases could reduce the problems during field operation. The survey results also suggested that it is important for the contractors to provide a timely input to design using a pre-construction plan as the basis of their input.

3.2 *Further studies*

Constructability improvement required contribution from all project stakeholders. It is then important to conduct a similar constructability survey against the other main construction project stakeholders, i.e. owner and designer.

The next step would be case studies for in-depth investigation of the constructability practices in the Indonesian construction industry in implementing the CII's constructability concepts. As suggested by the Australian CII (Francis and Sidwell 1996) it is important to consider the uniqueness of the construction industry in a specific country in implementing constructability improvement.

REFERENCES

Adams, S. 1989. *Practical Buildability*. London, The Construction Industry Research and Information Association.

ASCE 1990. *Quality in the Constructed Project: a Guide for Owners, Designers and Constructors*. New York, American Society of Civil Engineers.

CIRIA 1983. *Buildability: an Assessment*. London, Construction Industry Research and Information Association.

Construction Industry Institute 1986. *Constructability: a Primer*. Austin, Construction Industry Institute.

Construction Industry Institute 1987. *Constructability Concepts File*. Austin, Construction Industry Institute.

Construction Industry Institute Australia 1992. *Constructability Principles File*. Adelaide, Construction Industry Institute, Australia.

Francis, V. E. and A. C. Sidwell 1996. *Development of the Constructability Principles*. Adelaide, Construction Industry Institute, Australia.

Griffith, A 1984. *A Critical Investigation of Factors Influencing Buildability and Productivity*. Department of Building, Heriot-Watt University.

O'Connor, J. T. and V. S. Davis 1988. "Constructability Improvement During Field Operations." *ASCE Journal of Construction Engineering and Management* 114(4): 548-563.

Russell, J. S. et al. 1992. *Project-level Model and Approaches to Implement Constructability*. Austin, The Construction Industry Institute.

Tatum, C. B. et al. 1986. *Constructability Improvement During Conceptual Planning*. Austin, The Construction Industry Institute.

Wells, J. 1986 *The Construction Industry in Developing Countries: Alternative Strategies for Development*. London, Croom Helm Ltd.

Creative Systems in Structural and Construction Engineering, Singh (ed.) © 2001 Balkema, Rotterdam, ISBN 90 5809 161 9

Barriers to the implementation of constructability in project procurement

Tony Y. F. Ma
School of Geoinformatics, Planning and Building, University of South Australia, S.A., Australia

Patrick T. I. Lam
Department of Building and Construction, City University of Hong Kong, People's Republic of China

Albert P.C. Chan
Department of Building and Real Estate, Hong Kong Polytechnic University, People's Republic of China

ABSTRACT: The emergence of "Constructability" is partly due to the long existing separation between the processes of design and construction, which resulted in many problems in the construction industry today. Although constructability is being used in the United States, this concept has yet to be widely accepted and adopted by the industry in other countries. This paper analyses the potential barriers to constructability implementation and their relation to the building procurement process through a recent survey conducted in Australia.

Whilst a measure of constructability has yet to be fully developed, Singapore has gone some way in implementing a scoring system for quantifying "buildability", which is recognized generally as the ease of construction. The modification of such framework to evaluate constructability may be useful in sorting out the quantification problem, which can lead to higher objectivity and hence overcome some of the existing barriers..

1 INTRODUCTION

The traditional contracting system has remained predominant in the construction industry for a long time because it still has the perceived merits in providing a reasonable degree of certainty in design & documentation through which quality and functional standards can be met. However, this system does not invite any design contribution from the contractor thus causing the responsibility for design to be far removed from the responsibility for production. As such the project time is usually longer and buildability of the design may suffer. Earlier reports such as the Simon Report (1944), the Emmerson Report (1962) and the Banwell Report (1964) had highlighted the problems then existing in the construction industry. However, the recommendations brought up in those reports were not widely implemented (Francis et al 1996a). This could be attributable to the characteristics of the construction industry as follows:-

- Unique nature of the project, basically no two projects are same;
- Project location is fixed but with wide geographical spread of sites and conditions are often not similar;
- Organisational fragmentation and majority of works are done by sub-contractors;
- Adversarial contractual relationships and lack of co-ordination and communication between many parties.
- The lowest bid usually gets the job with quality being sacrificed;
- Poor working conditions and continuity of work is uncertain, etc.

As a result of these deficiencies, fast track procurement methods such as Design & Construct, Construction Management, etc. emerged in the industry. They feature overlapping of design and construction through early involvement of contractor in the design process. These methods rely to a large extent on the contractor's expertise in advising the client on aspects of buildability, which may have time and cost implications thereby improving the overall viability of the project to the client's advantage. On the other hand, the concept of 'Partnering' is also introduced to overcome the adversarial relationship between parties. Each of these strategies will of course tackle one or two inefficiencies in the procurement process, but none takes a holistic approach from inception till project completion. Up to this time the Australian construction industry has applied piecemeal solutions to improve the process, seeking efficiencies within each phase such as Value Engineering being introduced in times of need.

Therefore, the main aim of this paper is to look at the '*constructability*' principles as developed by the Construction Industry Institute and analyse the potential barriers to implementing constructability and its relation to the procurement process.

2 EVOLUTION OF "BUILDABILITY" AND "CONSTRUCTABILITY"

The concept 'Buildability' only emerged in 1979 when a research program was undertaken to identify the major problems in the construction industry by the Construction Industry Research and Information Association (CIRIA) of United Kingdom. CIRIA defined buildability as "*the extent to which the design of building facilitates ease of construction, subject to the overall requirements for the completed building*".

About the same time, the term 'Constructability' emerged in the United States, as the result of the Business Roundtable. A report was produced in 1983 after a four-year study by the Construction Industry Cost Effectiveness Project task group into how to promote quality, efficiency, productivity and cost effectiveness in the construction industry (as cited in Francis et al, 1996). Most of the research studies into constructability in the United States were carried out by the Construction Industry Institute (CII), which was established in 1983. The CII defines constructability as "*a system for achieving optimum integration of construction knowledge and experience in planning, engineering, procurement and field operations in the building process and balancing the various project and environmental constraints to achieve overall project objectives*"

'Constructability' was a new term to the Australian construction industry in the early 1990s. It was only in 1992 when the Construction Industry Institute, Australia (CIIA) was established. The mission of the CIIA was to tailor and develop a constructability process that would be suitable for the Australian condition based upon the Constructability Concept File developed by the CII in the United States. CIIA defines constructability as "*the integration of construction knowledge in the project delivery process and balancing the various project and environment constraints to achieve the project goals and building performance at an optimal level*" (Sidwell, Francis & Chen 1996).

CIRIA's definition focused only on the link between design and construction and implies that factors which have a significant impact on the ease of construction of a project are solely within the influence or control of the design team . Its focus is limited by concentrating on design. Through the development, it can be seen that current definitions of constructability provide a holistic perspective of the procurement process focusing on the consideration of all stages in the whole project life cycle.

However, ever since the term 'Constructability' emerged in Australia, the response appears not to have been as enthusiastic as was expected by the researchers. Moreover, the potential benefits on cost and time saving as well as the improvement on the quality of constructed product have not been the major focus in the industry, which has instead remained on mitigating the extra time and cost incurred on the project.

Some professionals in the industry claim that the present procurement system in Australia does not facilitate the implementation of constructability; others claim that constructability is just another cost cutting measure. The reasons for the lack of acceptance to implement constructability in Australia is still under investigation.

3 PRINCIPLES OF CONSTRUCTABILITY

Constructability is an attempt to solve many problems of performance technology and management that are encountered in the practical construction process. It is not a concept that should be witnessed as an encumbrance and should not have any separation and boundary between the contractual parties. It should become an implied and accepted characteristic in the construction process where contribution comes from all the various construction professionals (Sidwell et al, 1995).

The Constructability System developed by the CIIA provides a framework with which the project team can employ to exploit fully construction knowledge in time and structure manner. It is said that the system provides companies with the comprehensive approach which will increase the level of constructability that can be achieved on their projects. This will lead to improvements in cost efficiency and quality in the construction industry. The CIIA advocates a structured approach which identifies the following five stages in the procurement process:

- feasibility
- concept design
- detailed design
- construction
- post-construction

The principles of constructability are the core of the system. There are 12 principles which are then mapped on to the procurement process. However, not all principles are applicable to each stage of the project life cycle, or to every project. These principles are listed in Table 1, whereas Table 2 shows the distribution of the 12 principles within five project stages.

Table 1: Constructability Principles

1	*Integration*	(Constructability) must be made an integral part of the project plan.
2	*Construction Knowledge*	Project planning must actively involve construction knowledge and experience.
3	*Team Skills*	The experience, skills and composition of the project team must be appropriate for the project.
4	*Corporate Objectives*	(Constructability) is enhanced when the project teams gain an understanding of the clients' corporate and project objectives.
5	*Available Resources*	The technology of the design solution must be matched with the skills and resources available.
6	*External Factors*	External factors can affect the cost and/or programme of the project.
7	*Programme*	The overall programme for the project must be realistic, construction sensitive and have the commitment of the project team.
8	*Construction Methodology*	Project design must consider construction methodology.
9	*Accessibility*	(Constructability) will be enhanced if construction accessibility is considered in the design and construction stages of the project.
10	*Specifications*	Project (constructability) is enhanced when construction efficiency is considered in the specification of the development.
11	*Construction Innovation*	The use of innovative techniques during construction will enhance (constructability).
12	*Feedback*	(Constructability) can be enhanced on similar future projects if a post-construction analysis is undertaken by the project team.

(Source: CIIA 1996)

4 IMPLEMENTATION OF CONSTRUCTABILITY PRINCIPLES

McGeorge (1997) claimed that the roles and responsibilities of the participants varied in term of the 12 principles and at the different stages of the project's lifecycle. He further proposed a Constructability Implementation Planning Framework which identified and coordinated the decision roles and responsibilities of individual project participants throughout a project's lifecycle.

This framework provides awareness to individual participants with better understanding of the tasks carried out by other participants throughout the project's lifecycle, thus allowing better integration and co-ordination.

Even though the CIIA Constructability Principles File (1992) does not address a specific implementation program procedure, a number of key issues were listed and discussed. The following issues should be incorporated into the procedures established by each individual organisation:

- a constructability policy.
- project objectives which can be enhanced by constructability.
- reference to the constructability principles files.
- organisational chart with responsibilities clearly defined.
- training.

Table 2: The distribution of the 12 principles over the procurement process.

Feasibility	Conceptual design	Detailed design	Construction	Post construction
[1]	[1]	1	1	1
2	[2]	2	2	4
[3]	[3]	[3]	(7)	[12]
[4]	[4]	(5)	(9)	
[5]	(5)	6	[11]	
[6]	(6)	(7)		
7	[7]	[8]		
8	[8]	[9]		
	[9]	[10]		

[6] - very high importance; (6) - highly important; 6 - lesser importance

(Source: Construction Management - new directions, McGeorge & Palmer 1997, p60)

5 BENEFITS OF IMPLEMENTATION OF CONSTRUCTABILITY

Clients in the industry today demand a timely project hand-over, minimum cost over-run and good quality for completed facilities. After all, what they are asking for is a better value for money and high quality performance in the services provided by the industry. All these expectations are achievable in the completed facilities if clients encourage the implementation of constructability (Sidwell et al, 1995).

Similarly, constructability not only works in favour of the clients but also benefits other parties involved in the project. It provides the designers with opportunities to gain practical construction advice to simplify design, reduce overhead costs and perform an easier management role on site. For contractors, they will enjoy a more systematic, better-coordinated and efficient working environment. Constructability also enhances the reputation for both designers and contractors. It further reduces wastage and provides better quality products with fewer defects.

In addition, Hon (citing from Francis et al, 1996) claims that buildability not only benefits quality assurance but also client's satisfaction and maintenance of good industrial relations. It was claimed that the benefits of constructability not only occur in the conceptual planning and design stage, but also at all stages of the procurement process where constructability thinking is incorporated.

A case study was conducted by CIIA (1996) for the construction of Australis Media Limited - National Customer Service Centre in Technology Park, South Australia, which included the implementation of Constructability. As a result, the project was completed with the benefits of:-

- Completion in less than 9 months compared with 12-15 months for an equivalent project delivered under normal design approval and construction procedures
- Cost with savings on the project in excess of A$1.1 million for a contract value of A$12 million.

6 BARRIERS OF IMPLEMENTATION OF CONSTRUCTABILITY

The implementation of constructability has the potential to make significant savings on constructed facilities, in terms of time and cost without compromising design and quality. However, the perception from industry people does not always reflect the same attitude. Many still think that constructability is similar to cost cutting exercises as they perceive what Value Engineering does for projects. Or they just simply perceive constructability as only the ease of construction (i.e. buildability). Even so, Moore (1996) identifies a number of reasons why buildability in the United Kingdom has not been accepted by the industry as a whole. One reason given is the inconsistent approach to both defining and applying buildability. Without a clear definition, the problem of developing a relevant and effective buildability strategy will persist.

In the United States, O'Connor and Miller (1994a) and the CII set out to identify the prevalent barriers to constructability by conducting detailed studies. The severity of the barriers so identified varies widely from company to company, particularly with respect to constructability program ranking, organisation type, project type and annual turnover. There were 8 common barriers to the effective implementation of constructability as identified by O'Connor (1994b), whilst 5 common barriers were identified by CIIA (1996) in their workshop survey.

They are listed as follows.

O'Conner (1994b)
- Complacency with status quo
- Reluctance to invest additional money & effort in early project stage
- Limitations of lump-sum competitive contracting
- Lack of construction experience in design organizations
- Designer's perception that "we do it"
- Lack of mutual respect between designers & constructors
- Construction input is required too late to be of value
- Belief that there are no proven benefits of constructability

CIIA (1996)
- There is a lack of awareness of the benefits of the constructability system
- Construction input is requested too late to be of value
- The traditional lump-sum project delivery system limits the inclusion of construction input
- Contractors are unwilling to divulge too much prior to the award of the contract
- Owners are reluctant to invest additional money on construction input early in the project

7 RECENT SURVEY IN AUSTRALIA

In a recent attempt to gain a further insight into the barriers, Eu (1998) conducted a questionnaire survey. Of the 50 sets of questionnaires sent out through electronic mails, 20 were returned in July & August 1998. In summary, the five barriers being identified are not much different from the previous studies and are reported as follows:-

- Traditional form of contract limits the inclusion of construction output.
- Reluctance to invest additional money and effort in early project stages.
- Unwillingness to disclose too much information prior to the award of contract by the contractor.
- Lack of construction experience in design organisation.
- Lack of qualified personnel to implement constructability.

The majority of the respondents (80%) claim that they are familiar with the term 'constructability' and unanimously agree that constructability is useful in their current fields of work. However, many opine that this concept has yet to be widely accepted and implemented into their projects. There are also others who claim that these concepts indeed exist in many past projects but are not given the term

'Constructability'. There is still a lack of understanding and use of formal constructability programs in today's industry as it was in the past, despite CIIA's effort of promoting constructability over the past 5 years. Most of the respondents also claim that constructability is often being incorporated into value management or other cost cutting measure exercises instead of being seen to be independent on its own. This demonstrates their confusion over the use of terminology and the precise definition of constructability.

8 EFFORTS TO IMPROVE "BUILDABILITY" IN SINGAPORE

Due to labour shortage and the increasing expectation for better quality buildings, the Singapore Government took the lead to foster the concept of "Buildability" in the construction industry since the early 90s. Modelling from the in-house system used by Takenaka Corporation in Japan, the then Construction Industry Development Board (now renamed as Building Construction Authority – or "BCA" in short) developed the Buildable Design Appraisal System for use in its public sector projects as a pilot study.

The appraisal system focuses on the 3Ss: standardization, simplicity and integration of elements.

Standardization refers to the repetition of grids, sizes of components and connection details. A repeated grid layout, for example, would facilitate speedier construction and result in less cutting waste for formwork and infill panel materials. Columns and external claddings of repeated sizes would reduce the number of mould changes whether on-site or in the factory.

Simplicity means uncomplicated building systems and installation details. A flat slab system, for example, eases formwork and rebar fixing work considerably. Use of precast components reduces wet trade operations on site and improves productivity and quality.

Integration of elements refers to the combination of related components into single element which may be prefabricated under factory condition and installed on site. Pre-finished and precast external walls and prefabricated toilet cubicles are good examples.

A set of labour saving indices have been devised for many types of building systems based on site productivity studies. A high index indicates that the design is more buildable and fewer site workers are required.

The formula for the Buildability Score (BS) is as follows:

$$BS = 50[\sum(A_S \times S_S)] + 30[\sum(A_W \times S_W)] + N \quad (1)$$

Where:

A_S	$= Asa / Ast$
A_W	$= A_{wa} / A_{wt}$
A_S	$=$ Percentage of total floor area using a particular structural design.
Ast	$=$ Total floor area which includes roof & basement areas
Asa	$=$ Floor area using the particular structural design.
Aw	$=$ Percentage of total external & internal wall areas using particular wall design
Awt	$=$ Total wall area, including basement wall
Awa	$=$ External & internal wall areas using particular wall design
S_S	$=$ Labour saving index for structural design
S_w	$=$ Labour saving index for external & internal wall design
N	$=$ Buildability Score for other buildable design features, e.g., grid layouts

Minimum Buildability Score have been stipulated for the purpose of building plan approval for different types of building categories (Table 3).

Table 3: Minimum Buildability Score

Category of Building	Minimum Buildability Score	
	5,000m² =< GFA < 25,000m²	GFA >= 25,000m²
Residential (1)	52	55
Residential (2)	58	61
Commercial	65	68
Industrial	67	70
Institutional	64	67

(Source: Code of Practice, BCA, April 2000)
(1): landed category, (2): non-landed category

With effective from 1 January, 2001, all projects due for Planning Approval Submission have to comply with the above stipulations. The declared data as submitted by consultants and developers at the planning stage would be subject to re-checking from as-built drawings before issuance of the statutory occupation permit.

Although the overall success of this legislated system has yet to be seen, design consultants have been steadily moving towards the goal of improving buildability of their designs in order to meet these impending requirements.

9 CONCLUSION

Although a lot of barriers still remain to be overcome to enhance constructability, sustained efforts have to be made in our quest to achieve economy (in terms of time and cost) and quality for our construction. Whilst the measures taken by the Singapore government are focused only on "buildability", the quantification framework is nevertheless worth extending to the wider scope of "constructability". Some may consider legislation inappropriate or restrictive from the angle of free-hand design. Hence, under the present climate, the best way to break the ice is through continuous education and provides proper training workshops to industry personnel such as project managers and design consultants. At the same time, clients have to be demonstrated on the benefits of constructability through ongoing project case studies.

Based on the evaluation in the earlier part of this paper, it can be seen that the traditional procurement system has little room for the implementation of constructability. The recent move towards Design & Construct and Construction Management might offer better opportunities for early contractor input. In view of the cited developments in Australia and Singapore, it seems apt to end this paper with a new question: Who should best be the constructability facilitator- Design Consultants, Project Manager, Client, Construction Manager or the Government?

REFERENCES

Building & Construction Authority 2000, *Code of Practice on Buildable Design*, Singapore

CIRIA United Kingdom 1983, *Buildability : an assessment*, CIRIA London.

Construction Industry Institute Australia (CIIA) 1992, *Constructability Principles Files*. CIIA University of South Australia.

Eu, B.L. 1998, The Analysis of the barriers to the Constructability Implementation and their relations to the Building Procurement Process, Unpublished Research Project, University of South Australia.

Francis VE and Sidwell A.C. 1996, *The development of constructability principles for the Australian construction industry*, CII Australia.

Griffith A & Sidwell A.C. 1995, *Constructability in building and engineering project*, Macmillan Press Ltd.

McGeorge D & Palmer A 1997, *Construction management new directions*, Blackwell Science Ltd.

McGeorge WD & Sidwell A.C. 1996, "*Current Management Concepts in the Construction Industry - Where to From Here*", Australian Institute of Building Paper 7.

Moore D 1996, *Buildability assessment and the development of an automated design aid for managing the transfer of construction process knowledge*, Engineering, Construction & Architectural Management, Vol 3, no.1, p29-46.

O'Connor JT & Miller S.J. 1994a, *Constructability programs: Method for Assessment and Benchmarking*, Journal of Performance of Constructed Facilities, Vol. 8, no.1 Feb, p.46-64.

O'Connor JT, Miller S.J. 1994b, *Barriers to constructability implementation*, Journal of Perforamance of Constructed Facilities, Vol. 8, no.2, May, p.110-128.

Sidwell A.C., Francis V.E. & Chen S.E. 1996, *Constructability manual/prepared for CII, Australia(2nd edition)*, CII Australia.

Sidwell A.C.& Mehrtens V.M. 1996, *Case studies in constructability implementation*, CII Australia.

Creative Systems in Structural and Construction Engineering, Singh (ed.) © 2001 Balkema, Rotterdam, ISBN 90 5809 161 9

Integrating services design to simplify the use of manufacturing

M.J. Mawdesley, A. Brankovic & G. Long
School of Civil Engineering, University of Nottingham, UK

G. Connolly
Crown House Engineering, M&E Services Division, UK

ABSTRACT: This paper describes a research project investigating the use of off-site manufacturing for building services provision, in particular for services distribution, within the UK construction industry. The low uptake of off-site manufacturing is investigated using the construction process as a basis for hypothesis. Three models of the construction process are given and explained in detail, showing how the late involvement of the services designer or engineer can affect the decisions taken regarding building services provision. It is found that a lack of experience and knowledge regarding the use of off-site manufacturing is the reason for its limited use in the UK industry. The development of a system for the dissemination of knowledge and evaluation of building projects as to their suitability for use of off-site manufactured services is then discussed. The system is intended to be used by all parties involved in the UK construction industry.

1 INTRODUCTION

The work described here is a research project that examines the use of prefabrication for building services distribution. It is the result of a UK company recognising the need and putting together a team of academics and industrialists to provide an objective assessment of the situation.

The research was divided into two main parts: An investigation into what factors affected the possibility of using off-site manufacture and the development of a computer-based advisor to evaluate the suitability of projects for off-site manufacture of building services. Initially the research has focussed on a specific building type, i.e. large office buildings. This may be expanded as the research continues.

The first part of this was carried out by means of questionnaires and interviews with a wide range of industrial parties. It resulted in a report (Brankovic, 2000) showing the major factors. It became apparent that the procurement methods employed in the UK have a significant effect and these are discussed in section 2 of this paper. The second part, the advisor, is detailed in section 3 of the paper. It became evident that each party to a project recognized different aspects as important and therefore the advisor has to be *intelligent* in the manner in which it assists the user. The difficulties involved in this and the solution approaches adopted are discussed.

This research is funded by the UK Engineering and Physical Sciences Research Council (EPSRC); Carillion Construction plc; Crown House Engineering; Ove Arup; Laing Technology Group; TPS-Consult Ltd and the Building Services Research and Information Association (BSRIA). Members of all of these organisations contributed to the work in this paper.

2 CONSTRUCTION PROCESS

The possibility of using off-site manufacturing for building services is closely linked with the procurement method for any project. Gibb (Gibb, 1999) discusses off-site manufacturing and gives a number of examples of its usage in the UK construction industry. Whilst it is recognized that all projects are different, several generic procurement methods are commonly used in the UK. This section outlines three such methods and in particular considers the timing of the involvement of the participants and their interaction with each other.

The models of the methods presented are not intended to be full descriptions of them. Rather, they are intended to highlight how some of the features of the construction process can affect the usage of off-site manufacturing as a method of building services provision. The models are similar in format to a *Gantt* or activity diagram, charting the involvement of project members over time. Bi-directional arrows have been added to indicate interactions between project members. The interactions are simplified in these diagrams and in reality will progress over time. For example, the architect will interact continuously with the client to determine to form of the structure. In this diagram, for clarity, only the initial and final interactions are marked.

The object of each interaction (indicated by its label) is the result of the tasks undertaken by one of the roles involved in the interaction or, in some cases, is a request for alterations to an existing result. For example, the interactions labelled *structure* in Figure 1 represent the output of the structural design phase. The interactions labelled *feedback* in Figure 1 state an interaction whose object is a request for alterations to the result of a previous interaction.

2.1 A "Traditional" Construction Process

This model shows the standard way in which many construction projects are organised and processed. The important features of this model are that the various roles involved in the construction project each act sequentially and pass on the output of their respective process to the next role in the process as well as feeding back to the client or their representative. Each role has little or no involvement in the process prior to the commencement of their individual tasks. This means the building services are not designed until after the structural design is complete. This seems reasonable since the structural detail is a necessary input for the design of the services; however, it means that there is less flexibility to make large changes to the structural design. Changes to the structural design to accommodate services may occur and some are shown in Figures 1 and 2. However, in most cases these changes will be relatively small due to the cost, in both time and money, of revising a large amount of the structural design.

Each party within the process brings with it its own experience (and prejudices) in terms of construction. If the structural engineer has not been involved in any projects involving off-site manufacture of building services then it is unlikely that he will design the structure with this in mind (unless it was a requirement of the client or architect). By the time the building services engineer is involved there is little scope for changing the building form or structure.

The sequential nature of the process can lead to difficulties in cooperation between the parties that in turn leads to alterations being required late in the process. Indeed, some of the alterations will not be realized until construction work has begun, as indicated by the *structural changes* and *services changes* interactions shown in Figure 1. A set of tolerances set by one party can sometimes leave little scope for

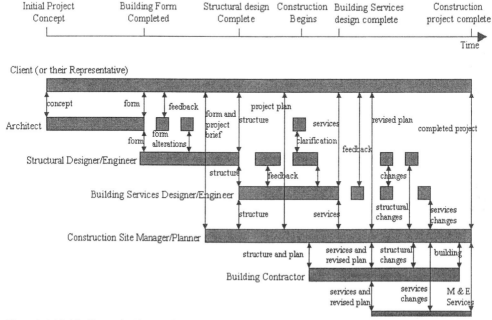

Figure 1: A Model of interactions in a *Traditional* Construction Project

error or variation on site. Such variations, which occur after the notional completion of a phase, are here called post-phase variations. They usually have a greater effect than might be anticipated because of their lateness and the involvement of several parties. This often means that the construction contractors on site undertake changes to the structure or services.

In this model, if the services have been designed and created using off-site manufactured modules there is very little scope for variations to be made to their dimensions or structure. In addition, if many on-site alterations are required then the major benefits of employing off-site manufacturing are removed.

2.2 Construction Process employing Design and Build Provider

Figure 2 shows a construction project where a design and build provider (referred to as the provider from this point) is hired to handle the project. This is a common occurrence, as many clients have neither the time nor the expertise to undertake the management of, and the time-intensive participation in, a construction project. This would be especially true for a more complex project. The provider offers their services to manage the project and liaise regularly with the client to keep them informed of progress.

The majority of the interactions are between the provider and each specific party involved in the construction process. Some interactions between the other roles in the process occur when a specific need to clarification or alteration is urgently required by another party. For example, in Figure two the structural engineer interacts with the building services engineer to pass on revisions to the structure that was requested by the services engineer. In general, most interactions occur via the provider which maintains overall management over the project.

From Figure 2, it can be seen that the number of post-phase revisions is less than was the case in Figure 1. The increased experience and project management skills of the provider account for some of the reduction in the post-phase revisions. Another factor, for building services, is the earlier involvement of the building services engineer, who starts the services design before the structural design is completed. This is achieved because the experience of the provider enables them to produce a preliminary brief of the services requirements with some partial structural details.

The earlier commencement of the services engineer should enable some potential problems between the services and structural design to be highlighted earlier in the process therefore reducing the cost of revising these designs. Even with the reduction in alterations, some will still occur as shown in Figure 2.

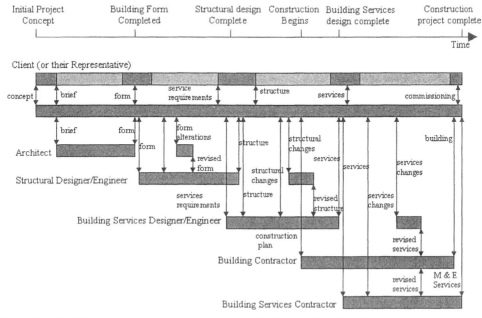

Figure 2: Model of interactions in a construction process managed by Design and Build Provider

The main difference between the model shown in Figures 1and 2 is the integrating presence of the provider in Figure 2. However, the benefits that this brings will be greatly dependent upon the individual expertise and experience of the provider involved. Even though the services engineer is involved earlier in the process, the structural design is still well advanced by this time and it will still be difficult for the engineer to promote the use of off-site manufacture if the structural engineer is resistant to the idea unless the provider can be convinced. This will be influenced by their experience of the use of off-site manufacturing of services.

2.3 *"Team" based Construction Process*

Figure 3 shows this version of the construction process. It describes a more inclusive method of undertaking the construction process. The idea of using a team to manage and oversee the project including involvement from all those parties involved in the construction process is not new and indeed forms the basis of much of the Egan report (Egan, 1998) on improving the UK Construction industry. However, it is uncommon for the services engineer to be involved in a construction project from its inception.

This model represents the type of process that is most suited to the use of off-site manufacture of building services. All parties in the process, except

for the services contractor in this example, are involved in the project form its commencement and the initial concepts for the building. The early involvement of all parties enables the experience and expertise of each project member to be fully utilised throughout the process. This introduces the knowledge and experience of the building services designer into the process much earlier on allowing them to discuss the idea of using off-site manufacturing before any critical building design work has been completed.

With this model, a reduction in the number of *post-phase* variations is shown compared to the previous two models. The multi-skilled experience of the project team means that any conflicting requirements for the buildings form, structure and services will be identified at an earlier point in the process enabling these conflicts to be resolved before the process moves on to the next phase. This may mean that some phases of the process, particularly the reviewing of each completed phase, will be longer but this is countered by the reduction in *post-phase* - variations that, as discussed in section 2.1, are expensive to the project in terms of both time and cost. These *post-phase* variations, especially those that occur once the actual building work has commenced, are one of the major obstacles to the use of off-site manufacture of services. It is proposed that by adopting this more integrated approach to the de

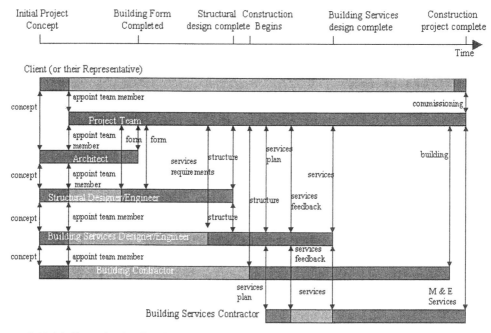

Figure 3: Model of interactions in a *Team* based construction process

104

sign of the building's structure and services even greater benefits could be gained from the use of off-site manufacturing.

A limitation of this approach is that it requires that all those parties to work on the project have been decided at the very start of the process. It is sometimes impractical to choose the services designer until some details regarding structure and form have been decided. This is especially true if the contract for design and build of services is to be put out for tender.

2.4 *Summary*

From the investigation of the UK Construction industry, it is evident that in a large number of instances that the provision of building services is a peripheral consideration until very late in the construction process. Once the building services designer or engineer is involved in the project, a large number of decisions regarding the building's structure will have been determined and any alterations to these will be more expensive than if they were taken earlier in the process. This tendency to treat the provision of building services as an addendum to the building project is a likely factor in the slow uptake of off-site manufactured services.

A method for rectifying this problem is a change to the overall construction process to include the services as a key element of the project from its inception, as shown in Figure 3. This sort of change is occurring in the UK industry but it will time and refinement before it can reap the benefits it is ascribed.

A major influence on the decisions taken throughout the construction process is the experience and knowledge of the individuals and organizations involved. This is the other major factor affecting the use of off-site manufactured building services. Even amongst those who design building services, there is sometimes a lack of specific knowledge relating to the use of off-site manufacturing. It is therefore not surprising that the other parties involved in construction may not have a great deal of knowledge in the use of off-site manufacture. Whilst there remains little experience or knowledge relating to their use present in those involved in UK Construction it is unlikely that they will be considered as an option for a specific project.

The knowledge gained as a part of the MEDIC project will be used to inform those involved in UK Construction about the use of off-site manufacturing. This will be primarily achieved by the development and dissemination of a computer system to include training and advice on the use of off-site manufacturing as a method of building services provision.

3 THE MEDIC SYSTEM

A major output of the MEDIC project is the development of a computer-based system to transmit knowledge regarding the use of off-site manufacturing as a method of building services provision. The system produced should inform those interested parties, i.e. those involved in the construction industry, of potential benefits and methods for using off-site manufacturing. Figure 4 shows a map of the proposed system, indicating the various elements that it consists of. The function of the system is split into two main areas:

- A tool for *training* users about the benefits, limitations, possibilities and methods for employing the use of off-site manufactured building services.

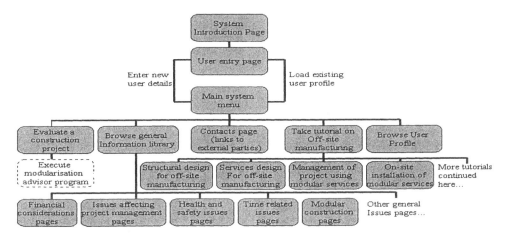

Figure 4: Overview map of the MEDIC system

- A computer based *advisor* to evaluate a given project and return a response as to its suitability for the use of off-site manufactured services.

To avoid problems with ease-of-use the system is delivered via the user's Internet browser and uses a web-based format for displaying information. The system will be published on CD-ROM and will be disseminated to those involved in the UK Construction industry. The system will record details of the user and their specific area(s) of interest (user profile) in order to deliver content appropriate to the specific user. For example, if the user is a building services engineer then a greater level of technical detail may be required when discussing the construction and installation of modular services. This profile will also be used to determine initial areas of exploration for the user that can be determined from our investigative work with those in the UK Industry (Brankovic et al, 2000).

It has been found that different parties to a construction project have different opinions as to what is important when deciding how to procure parts of a building. For example, designers consider the three most important aspects to be overall project time, quality and installation time for services whilst clients consider them to be money, quality and building efficiency. These differences not only indicate possible sources of conflict in the project but also suggest benefits from being able to tailor the systems presentation to the user.

The *training* component of the system will provide the user with a series of tutorials on the use of off-site manufactured services. For example, videos illustrating all stages of construction using both manufacturing and traditional techniques have been collected and are linked to relevant pages. The user will also be able to browse through information in an unstructured manner or to specifically search for a piece of knowledge if they wish. Information will be held on both general issues relating to the use of off-site manufactured services, comparisons with traditionally procured services and a set of example *case studies* of individual projects where off-site manufacturing has been employed.

The computer-based *advisor* element of the system will act as a *decision support system* to accept data from the user regarding their project and produce a response as to the suitability. The system will start by asking a series of general questions regarding the building's shape, size and service requirements. Depending upon the response to this a series of other questions will be asked until the system is ready to give an evaluation as to the suitability for modularisation of building services. The response can range from the project being unsuitable for the use of modularised services to the project being suitable for large-scale modularisation i.e. both vertical and horizontal services distribution being practical for the project

The series of questions asked will also be affected by the user profile and the current state of the project. For example, if the user is a structural engineer then the questions will be more detailed in respect of the structural aspects of the project.

4 CONCLUSIONS

The key points that can be drawn from this paper are as follows:

- A lack of knowledge and experience is the main reason for the low uptake of off-site manufacturing in the UK construction industry.
- The likely promoter of the use of off-site manufacturing for building services are those who have experience of its use or are involved in building services provision.
- The provision of building services is not considered until very late in the construction process.
- The traditional procurement system employed in the UK militates against the use of prefabrication of building services.
- A less serial, more team-based approach to procurement increases the possibilities for manufacturing.
- Different parties require different aspects of knowledge at different times and the advisor must therefore be intelligent to the users and their requirements.

REFERENCES

Brankovic, A., Mawdesley, M. J. (2000). Factors affecting the decision to use off-site manufacture of building services. MEDIC Project Report. Dissemination via project web site. www.civeng.nottingham.ac.uk/medic/publications/report01.doc

Egan, J. (1998) *Rethinking Construction* DETR, London

Gibb, A. G. F. (1999) Off-site Fabrication: Prefabrication, Pre-assembly and Modularisation Whittles Publishing, Caithness, Scotland

Creative Systems in Structural and Construction Engineering, Singh (ed.) © 2001 Balkema, Rotterdam, ISBN 90 5809 161 9

Integrated design, production and safety in reinforced concrete structures: Operating tools and real experiences

Arie Gottfried & Barbara Maria Marino
Building Engineering and Territorial Systems (ISET) Department, Polytechnic of Milan, Italy

ABSTRACT: The new Health and Safety standards and regulations in the Italian Construction Industry have led to a re-examination of the building process. This has generated a greater awareness regarding the need to consider Health and Safety matters during design stage. It is therefore suggested that the new approach to design identifies those different construction methods taking into account process optimisation, human and material resources and Health, Safety and Environmental management. In order to achieve Quality in design and Safety in construction, professionals involved in project activities must consider construction factors during the design phase and this has been tested through experiences on sites focusing on specific working activities concerning the execution of reinforced in-situ concrete structures. It has been learnt through these experiences how project must be integrated with a set of variables strictly connected to site's characteristics, building context and works' scheduling in order to minimise or reduce all risks for site based workers.

1 INTRODUCTION

The introduction in Italy of new Safety standards and regulations has acted as a starting point for considerable re-examination of the building process. This has generated a greater awareness regarding the need to consider Health and Safety design factors at the pre-construction phase, these factors being specific to site operational activities. The new standards have introduced new professional roles which include; the project supervisor and the co-ordinator during project preparation stage. Their tasks involve the development of specific Health and Safety documentation which consists of design based risk assessment. This results in the development of a project schedule that relates specific safety procedures to working activities as well as site logistic matters.

The possibility of monitoring a sample of building sites involving the production of reinforced concrete structures by means of modern formwork systems has provided an opportunity to focus attention on a range of specific safety problems. From the collection of project planning data through to the management of on-site operations, it has been possible to identify the project planning and construction problems arising within the specific framework of the work being performed, which inevitably has consequences for Safety Co-ordination. For the purposes of the present article, the most recent changes in Italian legislation in relation to building safety, and the effects that these have had on the co-ordination of planning and construction activities within the building process, will be first illustrated.

The safety problems involved in the construction of reinforced concrete structures, and how these problems can be solved through the activities of the Safety Co-ordinators, will then be identified. The problems encountered are sub-divided by safety and operational categories and the analysis of them is based on detailed reports of actual experiences on building sites.

2 ITALIAN NEW STANDARDS: SAFETY DURING THE PLANNING PHASE

The legislative provision which has most significantly modified the approach to the management of Safety on building sites is Legislative Decree n. 494/96. This decree, revised in November 1999, is the Italian interpretation of European Union Directive 92/57/EEC on "Health and Safety on Temporary and Mobile Building Sites".

As a result of this legislative change, significant modifications have been undertaken in the building process, involving the introduction to the planning and construction phases of both new professional personnel and new operating procedures.

On the basis of the entire contents of the decree, as well as the underlying principle of the European Directive, there is a belief, supported by research and statistics, that a large percentage of the accidents on site is attributable to incorrect project planning decisions. The actual prevention of accidents must therefore begin in the course of the design stage and responsibility for safety must be assigned to the project promoter (the Client). The Client now finds himself directly involved in the project which he himself is promoting, from the start of the initial project development phase through to the planning and management of safety. In order to fulfil his legal responsibilities, the Client must, therefore, nominate two professional persons to be responsible for the management of safety problems: The Planning Phase Co-ordinator and the Execution Phase Co-ordinator. The former, during the project planning stage, prepares the Safety and Co-ordination Plan, which should identify, analyse and assess safety risks and contain details of procedures and actions required to ensure that, for the entire duration of the project, all legal norms relating to the protection of workers' safety and health are respected.

The Execution Phase Co-ordinator is required to perform his duties while the construction work and operations are being performed on site, and must ensure, by means of appropriate control and co-ordination measures, that the provisions contained in the Safety Plan are implemented. The duties of the Execution Phase Co-ordinator also include keeping the contents of the Safety Plan up-to-date in relation to the performance of the work and any possible alterations undertaken.

The Leg. decree n. 494/96 has therefore laid down that safety planning must start during the design phase of the project and so becomes an assessment parameter for architectural, technical and technology decisions. This means that the Client is responsible, through the persons appointed by him, for analysing the methods used in performing the construction work, in the form of the specific project planning decisions, and for assessing the risks in terms of possible alternative technology and construction solutions.

The Client is, therefore, obliged to involve himself in the planning of the project and to be familiar with production technology and construction methods (or at least envisage a scenario if precise details of the building contractor are not known) so as to be capable of assessing the risks and taking steps to develop procedures designed for the protection and safety of workers on site.

3 INTEGRATED DESIGN FOR SAFETY

The requirement of the legislators that the effects of architectural and technology decisions on the safety of workers on site have to be assessed during the project planning phase has inevitably led to a more thorough evaluation of the construction methods employed in the individual phases of the project.

The Planning Phase Co-ordinator, ideally working alongside the project planning team, must be capable of developing a series of hypotheses concerning these construction methods, taking into account parameters such as the organisation of the site; the time, manpower and materials required for the completion of each individual phase; and the sequencing scheduling of each of the stages of the process.

Should the Safety Co-ordinator have insufficient resources available to assess the characteristics of the building contractor handling the project (operating procedures, machinery, equipment, experience of the workforce), a situation which often arises in the case of public tenders, he will be obliged to perform a "simulation" of the construction phases. At this stage the Co-ordinator will have to envisage the parameters set out above and then select the building methods, which should take account of:
– the optimisation of the construction process;
– the optimisation of human and material resources;
– the management of Health, Safety and Environment

This procedure will necessarily involve an analysis of the project planning decisions in terms of the assessment of risks during the construction phase, and can lead to:
– a process of feedback between the project designer and planner and the Safety Co-ordinator which can then lead to modifications to the project plan in order to reduce or eliminate risks for the workers on site;
– the development of operating procedures (contained in the Safety and Co-ordination Plan) designed to permit individual phases of the work to be completed in safety.

This integrated approach to safety leads to a much greater awareness of the construction and organisational aspects of the building site. In fact, such aspects, if overlooked at the project planning phase, could provoke unforeseen consequences or "last minute" alterations which, if not managed with the required professionalism and experience, are among the primary causes of accidents on site.

In relation to the use of modern formworks' systems for the production of reinforced concrete structures, it seems clear that the involvement of the Co-ordinator will most likely lead to the development of operating procedures taking account of the safety implications. At this point, therefore, the factors which the Co-ordinator should take into account for the development of such procedures in relation to a specific formwork technology envisaged by him will be illustrated. This begins with an analysis of the normal risks which could be encountered during the construction of reinforced concrete structures and then envisaging them in the context of either the particular formworks' system used or the building site.

4 IN-SITU CONCRETE STRUCTURES: SAFETY PROBLEMS AND PROCEDURES

In the course of work involved in the completion of a building project, the construction of the structural elements is a particularly critical moment from the Safety point of view, for a number of reasons:
1. complexity of the operations to be performed (erection of the formwork system, positioning of the reinforcements, pouring of concrete, disassembly of the formwork system);
2. the large number of workers and the limited co-ordination between teams of workers involved;
3. the extent and size of machinery and equipment used (cement mixers and/or cement production units, pump units, cranes with steel buckets);
4. the duration of the work phases;
5. the particular characteristics of the workplaces: working at height, restricted work space, gaps and openings, holes, trenches etc…

In relation to the work involved in assembling the formwork system and pouring the concrete it is to some extent possible to identify phases in which a significant level of risk for workers is to be noted, and during which it is necessary to have available appropriate safety operating procedures:
– raising and moving the formworks: the workers run the risk of being struck by the equipment in motion or by the casing if it should unexpectedly become detached from the machinery raising it; therefore the operating range of the lifting machinery, hooks and stabilisation systems as contained in the construction plans must be indicated.
– erection and removal of the formworks: also in this case the workers run the risk of being struck by parts of the casing if an appropriate system for stabilising the individual parts is not used; the principal external factor most likely to create problems of stability is wind. Wind can produce swinging movements which prevent safe manoeuvres. For this reason, once the parts of the formwork have been positioned on the work plane or the storage area, they may be only detached from the lifting machinery, even when clearly static, when the supporting props or other stabilisation devices are seen to be functioning.

– pouring and vibration of the concrete: if workers are working at a height above walking level which is in excess of the minimum distance prescribed in the safety regulations and who run the risk of falling, additional protective scaffolding with appropriate parapets must be erected. In any case, general protection equipment must be present (such as safety nets) which are capable of breaking a person's fall from a height of more than 3 metres.

What has been discussed so far deals with the risks and the operational safety regulations in the construction of the structural elements but ignores the organisation of the building site. The production of a reinforced concrete structure is on the other hand one of the most complex phases from the point of view of the logistics of the building site. Starting from the provisional construction plans and executive project for the formworks and the equipment (documentation which is specifically required by Italian technical legislation concerning the prevention of accidents on building sites) the following characteristics must be capable of being expressed for the purpose of planning the stages of the construction work and the management of safety:
– identification of the construction sequence involved in the erection of the casing system;
– the type, layout and dimensions of the provisional formwork's equipment (service gangways, protective scaffolding etc…);
– the erection of general protective equipment when there is a risk of falling from a height;
– the convenient storage areas arranged in such a way as not to interfere with other work taking place, perhaps at the same time;
– the arrangements for the movement and parking of vehicles for pouring the concrete.

In order to organise the planning of safety on site prior to the actual beginning of the construction work, it is necessary during the planning phase to purchase the equipment for manufacturing the products (such as casing systems, traditional or modern, the arrangements for supplying concrete etc..), thereby bringing the planners and the Co-ordinator into contact with the world of the building contractor. Therefore the planner and Co-ordinator are required to make the utmost effort to familiarise themselves with the production technology in order to increase their operational effectiveness to the maximum extent possible, both in terms of quality and the scheduling of construction as well as of the safety of the workers.

Gaining this knowledge should also be possible by means of collaboration with the building contractors, who should be prepared to:
– provide the technical documentation requested;
– provide appropriate information relating to the storage and transport of goods on site;

– assist the project planner in drawing up a complete and easily understandable construction plan for assemblies on site.

During this phase of purchasing the manufacturing technology and the planning of safety measures it is necessary to make provision for the proper training of the workers in the operation of the technology. Indeed, inappropriate work practices on the part of the workforce, or the non-adherence to established safety operating procedures, frequently leads to the safety plans being jeopardised and to accidents occurring on site.

5 REAL EXEPERIENCES

Based on the recommendations developed earlier, and prompted by the new legislative norms, it has been possible to carry out investigations on site and to check the theories developed in the university.

The collaboration between the Building Engineering and Territorial Systems Department of the Polytechnic of Milan and one of the leading global producers of industrial (or modern) casings has enabled a significant number of building sites to be monitored during the phase of constructing reinforced concrete structures. In the course of the monitoring work it was possible to identify some common problems and group them into three categories:
1. Quality of the project planning information;
2. Planning and organisation of the building site;
3. Constructing vertical structures;
4. Constructing horizontal structures;
5. Training of the workforce.

These categories have been found useful for systematically classifying project planning and construction problems which, although in this case restricted to one single context, are widespread in the building industry. These problems will be discussed below, by briefly describing the evidence found on building sites and what proposals have been advanced to remove these operational shortcomings.

5.1 Quality of the project planning information

In relation to project planning information it was found that, to a varying extent, depending on the specific project, the project planning documentation (relating to the structural elements) was frequently missing or vague, forcing both the contractors and the planners to continuously make modifications to work already in progress. Specifically, the quality of the level of detail found in the construction plans was frequently very limited. This meant that workers had difficulties understanding the plans and had to make special efforts to attempt to understand them. As if this was not enough, even the designers of the formworks' systems, who are members of the technical personnel, often found themselves in difficulty as a result of incomplete architectural and construction plans. The low level of definition and of detail inevitably has consequences for the planning of the formworks' system itself, and as such creates a sort of "definition" tolerance which is only decided at the moment of production on site.

Another shortcoming in planning which could only be solved during the construction phase was integrating the formworks' assembly with the preparation for plants' parts or equipment such as electric cables or pipes. Openings, holes, trenches and various installations must be already prepared when the casing for the concrete is assembled, in order to avoid having to subsequently break the concrete structure. Such awareness at the planning stage is absolutely essential if one wants to avoid having to later perform dangerous operations such as perforating a reinforced concrete structure.

Unfortunately it must be stated that the inadequacies identified in project planning occurred regardless of the complexity or the size of the project, indicating that this a very real and serious problem in the project planning discipline, a fact which is confirmed by the statistical data relating to the causes of accidents on site.

5.2 Planning and organisation of the building site

The use of modern formworks' systems presupposes that, given the dimensions of these elements, suitable storage areas of sufficient size appropriately arranged around the building site are available. This must obviously be considered both by the building company tendering for the contract when developing the site plan (as well as the Operational Safety Plan designed for the specific building site), and by the Co-ordinator during the planning phase when the minimum site safety standards are being prepared for the Safety and Co-ordination Plan.

What was found during the monitoring of building sites has shown that such areas, in the majority of cases, are carefully studied and planned, perhaps due to the fact that the sheer size and awkwardness of the elements to be stored is such that they can not be ignored. The accessories supplied by the manufacturer of the formworks for moving the panels (small and medium-sized), support props, stabilisation hooks, vies, grips and stabilisation elements, contributed significantly to the orderly management of the storage areas. This facilitated a smoother organisation during erection, disassembly and cleaning of the formworks' panels, thus ensuring a high level of safety for the workers involved in these operations. In addition, the possibility of using suitable buckets for moving support props, vices, grips and hooks significantly reduced the risks of being struck by objects while they were being moved.

5.3 *Constructing vertical structures*

The phases involving the production of vertical reinforced concrete elements, such as columns, reinforced concrete supporting walls and dividing walls, can be differentiated from the point of view of the analysis of risk on the basis of the following operating characteristics:
- formworks' system used (traditional or modern);
- method used to pour concrete (automatic pump or steel bucket);
- activities taking place at the same time the operations are being performed.

In relation to the casing system used, the majority of the building sites monitored were found to use a system of pre-fabricated formworks, occasionally combined with traditional elements. The use of such elements presupposes an accurate assessment of the risks involved while they are being moved. The elements being moved can be very large and contain a significant risk for the worker on the ground, who must guide the assembled formwork system during the positioning phase, and can be struck while the lifting machinery is in use. In this specific situation the climatic conditions should be carefully assessed in order to avoid the element being moved becoming uncontrollable.

The presence of holes in the panels of the formwork, used for inserting hooks and transporting the structure, enabled the structure to be moved with greater ease and speed and reduced the risks to workers involved in positioning the element.

During the assembly the vertical formwork's elements, the most dangerous phase was found to be the moment when the structure was being stabilised with support props or similar, mounted on a horizontal structure below.

Awareness of the operating procedures involving mounting and co-ordinating with the lifting machinery (it should be noted that it is preferable to have a worker at ground level using a remote control rather than a crane operator working in a cabin at height) has led, in the majority of cases analysed, to a reduction in the number of dangerous situations. The choice of operating procedures for pouring the concrete significantly influenced the layout of suitable temporary structures on which the workers controlling the equipment used to pour the concrete (pump or bucket) were positioned. From the results obtained from monitoring, the automatic pump is the equipment which caused the least problems, but its use should be assessed on the basis of other factors, such as the actual quantity of concrete required and the suitability of the area in which the automatic pump itself is positioned. The same recommendations apply to the concrete vibration phase and should make sure that suitable temporary structures and appropriate protective metal scaffolding and support props are in position. The steps involved in erecting formworks, pouring concrete, and baring vertical reinforced concrete columns can create a very critical moment for the workers from the point of view of safety, especially if other work is being performed on the site at the same time. On the building sites examined, however, it was found that there was a desire to "isolate" from each other the individual operations involved in completing a single stage of the total project. Such choices have undoubtedly led to improved management of the construction work and thus reduced the risks of accidents which could have resulted from other activities being carried on at the same time and interfering with the work itself.

5.4 *Constructing horizontal structures*

The problems associated with the production of horizontal structural elements, especially those of providing suitable temporary structures, are accentuated compared with vertical structures. Erecting formworks designed to hold the concrete subjects the workers involved to an increased risk of falling.

During this operation, the workers have to work close to the opening which has not yet been covered by a gangway, which is still being erected, and this often creates significant differences in height levels between the work level and the level below. The sites which were monitored used casing systems for beams and floors which could be assembled "from below", with the worker working on the scaffolding below, thus reducing the risks from operations performed at height. However, it was observed that this system led to a significant increase in the time required to perform the work, as a result of which the workers opted for the traditional working method, working on the horizontal elements already erected.

As it is not possible, in this case, to erect protection parapets (with the exception of the top of the slab) it has been suggested that workers should be equipped with a safety belt connected to safety ropes, and a safety footbridge should be erected where possible to reduce the risks of falling. A final risk factor for workers involved in assembling the formwork and the subsequent pouring of concrete (the methods of which are governed by the same recommendations prepared for vertical structural elements) is in the form of the gaps and openings used to create a framework structure and to join the units to the vertical structures. Such openings represent not only the risk of falling in the event that the worker should inadvertently trip. There is also the risk of workers being struck by materials or equipment falling through the openings. The systems analysed had the advantage of possessing prefabricated safety elements which could be put in place once the scaffolding around the pillar had been completed, thereby avoiding waiting times during which the openings are left unguarded and unprotected.

5.5 *Training the workforce*

The sites monitored on which particular forms of casing are used has enabled the work practices of workers using non-traditional construction technology according to precise operating instructions to be observed.

The speed of the construction processes and the awareness of safety in carrying out the work, characteristic of these systems, are reduced when workers have not received sufficient training and practice, and therefore the initial assembly processes are used for the purposes of training the workforce. Only after the completion of this training is it possible for the system to be fully functional and for its full potential to be exploited.

From the Safety point of view it was found that correct work practices on the part of the workforce are absolutely essential to enable the safety regulations already planned for in the technology of the system to provide the expected prevention and protection while the system itself is being operated. It was therefore possible to identify, on the different sites monitored, how the efficiency of the construction methods used and the safety of the operating procedures were directly correlated to the level of practice (in using the formworks) and the training of the workforce. Indeed, the rare situations which led to accidents, or which at least indicated work practices with a high level of risk, have incorrect work practices on the part of the workers as their principal causes (and not adhering to the safety operating instructions for the casings system). Specifically, the following improper work practices were noted:

– not using appropriate PPE;
– not using the temporary structures prescribed for working at height;
– lack of co-ordination during the movement of components;
– not putting the safety elements in place;
– not using the correct containers for moving components.

On the basis of the data discussed so far, it has been possible to develop a complete picture which has enabled us to produce a series of conclusive recommendations, listed below, which have highlighted several points where it is necessary to make changes both in the areas of technology and training in order to ensure the safety of workers on site.

6 CONCLUSIONS

The present article has illustrated that, beginning with explicit legal provisions, it is necessary to consider right at the project planning phase the operating procedures to be used on site in order to achieve and maintain safe working conditions. Experience has shown that, if the operating procedures are con-

sidered and assessed according to their risk profile for the workers, it is possible to identify in advance, before the commencement of the project, those aspects which are critical to safety, and to make provision for operating procedures which provide protection. This has been shown for sequential operations involved in the construction project as well as for individual tasks such as the production of reinforced concrete structures. In the specific case mentioned, the technical documentation accompanying the casings used, (the producer of which permitted not only unlimited consultation but also the monitoring of some sites on which such systems are employed), was of great help to the Co-ordinator during the project planning phase when planning the performance of the operations in a safe manner. In the cases studied it was additionally shown that it is the collaboration between project planners, Co-ordinators and building contractors which allow the exchange of information necessary to envisage in advance all the likely problems during the construction phase and to develop the pre-planning of the construction work and the procedures for safe operation, adherence to which is necessary to ensure safe working conditions for the workers. Such an exchange of information, even in the case in which the building company and/or the contractors have not been chosen, nevertheless helps the Co-ordinator during the project planning phase, in that it provides the conditions for a sort "operational simulation" of the construction work and serves as a guide for the selection of the most likely construction solutions, on the basis of which the documentation for the planning and management of safety can be produced.

Finally, it should be emphasised again that, in order to achieve conditions of Quality and Safety, it is necessary to include at the project planning phase recommendations concerning the construction phase on site, all of which is designed to create a more complete and integrated performance of the construction work. In any case, some members of the workforce, depending on their training and experience, should be included in the safety planning process, both at the request of the Client and as an expression of the attitude of the building contractor, so as not to jeopardise what has been agreed upon and organised during the project planning phase.

REFERENCES

Gottfried, A. & Trani, M.L. 1999. *Il Coordinatore per la sicurezza nelle costruzioni in fase di progettazione e di esecuzione.* Milano: Maggioli.
Illingworth, J.R. 2000. *Construction Methods and Planning.* London: E & FN Spon.
Nunnally, S.W. 1998. *Construction Methods and Management.* Upper Saddle River (NJ): Prentice Hall

Creative Systems in Structural and Construction Engineering, Singh (ed.) © 2001 Balkema, Rotterdam, ISBN 90 5809 161 9

Avoiding structural design failures as a contractor's engineer

D.A.Cuoco
LZA Technology, Thornton-Tomasetti Group, New York, N.Y., USA

ABSTRACT: A contractor's engineer is frequently required to provide structural design services. In these cases, it is imperative that the contract documents clearly define the responsibilities of the contractor's engineer and the design team. It is neither recommended nor wise for the contractor's engineer to take responsibility for the work of the design engineer. However, the contractor's engineer can sometimes avoid failures by identifying structural design deficiencies, even though this is beyond its scope of responsibility and despite the fact that only limited design information is available.

1 INTRODUCTION

There are many situations in which contractors are required to provide structural design services, e.g., design/build projects, performance-based specifications, etc. In some cases, the contractor has the in-house capability to provide these services; in other cases, he retains the services of an independent professional engineer.

In the case of products specified on a performance basis, the contact documents should be very specific as to the scope of design work for which the contractor is to be responsible and for which the design team is to be responsible. Thus, there should be a clear "dividing line" that separates the design responsibility of the design team and that of the contractor's engineer.

All too often the design team and the contractor's engineer "put on blinders" with regard to the other's work. Of course, one does not want to (nor should he) assume liability for work that is someone else's responsibility, and for which he is not getting paid to design or review. In some cases, however, major blunders can be discovered simply by "looking" at what the other person has designed, possibly by making a quick calculation or just by exercising some common sense. If a design discrepancy is evident, a telephone call can resolve whether the problem is nonexistent or needs to be addressed.

Several case studies are presented in which the writer's firm, retained as a contractor's engineer, was able to identify major structural design problems, even though this was beyond its scope of

responsibility and despite the fact that only limited design information was available. Although all design problems certainly cannot be detected this way, the ones that can will eliminate the subsequent finger pointing and lawsuits when something goes wrong and, most importantly, will eliminate potentially unsafe conditions.

2 CASE STUDY NO. 1

A glass-covered atrium roof structure for an office building was specified on a performance basis by the design team, requiring that the contractor perform the design, fabrication, and erection. The selected contractor was the supplier of a proprietary structural system, and retained the writer's firm to serve as the contractor's engineer and perform the structural analysis and design.

The atrium roof structure was to be supported from a series of brackets connected to the main building structure. The contract documents were clear in defining the responsibilities of the design team and the contractor: the brackets and their connections to the building were the responsibility of the design team, and the roof structure and its supports sitting on top of the brackets were the responsibility of the contractor.

The contract documents showed a relatively simple bracket detail (Fig. 1). It consisted of a steel plate attached to a post-tensioned concrete beam using expansion bolts, and a steel WT bracket that was to be field-welded to the steel plate. The post-tensioned concrete beam was a major transfer girder supporting columns from many stories above, and thus was highly congested with post-tensioning tendons as well as mild reinforcement. As a result, the expansion bolts were called out on the design drawings as having the minimum embedment length into the concrete transfer beam, i.e. approximately 5 inches (125 mm), in accordance with the manufacturer's catalog.

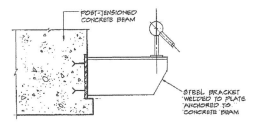

Figure 1. Bracket detail shown in contract documents.

Due to the fast-track nature of the project and the fact that the atrium was at the lower level of a high-rise building that was still in construction, the steel plates and expansion bolts had already been installed by the time the atrium roof structure contract was awarded. After the contract award, the analysis and design of the roof structure proceeded, and the final design of the roof structure and supports, including reactions applied to the brackets, was submitted to the design team for review. This submittal was signed and sealed by the contractor's engineer.

Although beyond its scope of responsibility, the contractor's engineer decided to check the WT bracket and expansion bolts that were specified on the design drawings. It was found that the WT bracket would be severely overstressed when subjected to the atrium roof structure reactions. It was also found that the expansion bolts would be marginally adequate, and then only if they could develop the full tension capacity advertised in the manufacturer's catalog for the minimum embedment length used. The contractor's engineer became concerned that the ability of the brackets to support the roof structure depended upon the reliability of expansion bolts subjected to very high tension forces, whose pull-out capacity can be significantly affected by field installation conditions, e.g. hole size, drilling method, etc.

As a result, the contractor's engineer called the design engineer to advise of this concern. The contractor's engineer further noted that it would not

be unusual to require some modification of the bracket design since that design took place prior to the availability of the final roof structure reactions. Surprisingly, the design engineer elected not to avail himself of this reasoning, and instead adopted a stance of insisting that the design shown on the contract documents was correct, and that there was no need for any concern regarding the expansion bolts. After several more telephone calls, the two opposing engineering positions became engraved in stone.

This left the contractor's engineer with the choice of either silencing his concern or putting it in writing. The latter option was selected, also explaining that the final reactions were not available at the time of the bracket design, in order to encourage the design engineer to modify the expansion bolt design without incurring the wrath of the owner, who would have to pay for it. In response, the design engineer reiterated his opinion that the concern regarding the expansion bolts was unfounded, but he also recommended the reasonable approach of in-place load testing of each expansion bolt in order to confirm its adequacy.

The load testing was performed. Each expansion bolt was tested in tension, and each of them failed to achieve the required capacity. Some of the expansion bolts pulled out at only 10% of the required test load! It was now clear that a modification to the bracket design was in order.

The modified bracket design is shown in Figure 2. The WT shown in the original design was replaced with a built-up section comprised of channels for the top and bottom flanges, and a web plate. The existing steel plate and expansion bolt

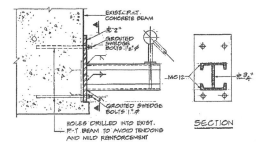

Figure 2. Modified bracket detail.

arrangement was replaced with a new steel anchor plate, 2 inches (50 mm) in thickness, and swedge bolts drilled and grouted 18 inches (450 mm) into the existing post-tensioned concrete beam. The upper swedge bolts were 1.5 inches (38 mm) in diameter in order to resist the high tension forces, and the lower swedge bolts were one inch (25 mm) in diameter. Prior to drilling, tendons and mild reinforcement were located using a pachometer, thus dictating the location of the swedge bolts and the required thickness of the new steel anchor plate.

3 CASE STUDY NO. 2

A large space frame roof structure, containing approximately 20,000 members and covering an area of 160,000 ft^2 (15,000 m^2), was specified to be only fabricated and erected on a performance basis, with the design of all members and connections contained in the contract documents prepared by the design team. Although the design of the space frame was specifically excluded from the performance specification, the contractor was required to have the shop drawings prepared under the direction of a licensed professional engineer. The contractor was also required to perform the detailed design of the 130 space frame supports using the reactions supplied by the design team. As a result, the space frame supplier retained the writer's firm to serve as the contractor's engineer in order to perform these functions.

The space frame reactions for all of the various design loading combinations were furnished to the contractor's engineer so that he could proceed with the design of the supports. In reviewing the computer output supplied by the design team, however, several inconsistencies became apparent. First, the lateral stiffness of many of the supporting columns beneath the space frame supports seemed to be taken equal in all directions, e.g. the strong-axis and weak-axis stiffnesses of wide-flange columns of different size all had the same values (Fig. 3). Second, the reactions for load combinations that included thermal loads were identical to the values that excluded thermal loads. Rather than proceed blindly with the design of the supports, questions were raised to the design team in order to obtain an explanation for these apparent discrepancies. Obviously, the concern went far beyond the design of the supports, as these issues could also have a significant effect on the design of the space frame structure itself.

In the months that followed, the design team issued repeated statements that the structural design of the space frame was proper, and accused the contractor's engineer of attempting to conduct an unsolicited peer review of the design. Since the design team provided no quantitative evidence to verify the space frame design – only verbal assurances – the contractor authorized his engineer to conduct a preliminary analysis of the space frame in order to attempt to confirm its adequacy.

The preliminary analysis performed by the contractor's engineer indicated that hundreds of members in the space frame would be severely overstressed, many by several hundred percent. This was reported to the design team, who again maintained that the design was adequate. Finally, the owner demanded that the design team satisfy all doubts by providing an engineering report that demonstrated the structural adequacy of the space frame, to which the design team agreed. One month later, the design team issued an engineering report that concluded that the space frame design was adequate – provided that approximately 2,000 member sizes were changed! This, of course, aggravated the already delayed schedule of the project, and resulted in substantial claims by the contractor. The alternative of having an unsafe structure, however, would have been far worse.

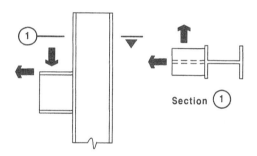

Figure 3. Inconsistent stiffnesses.

4 CONCLUSION

The contractor's engineer is usually privy to a very limited amount of information, if any, relating to the design of a structure. However, in many cases he can identify potential design deficiencies just by "unofficially" looking slightly beyond his scope of responsibility and exercising some engineering judgment. It is also important for the design engineer to be receptive to questions raised by the contractor's engineer. An informal telephone conversation between the contractor's engineer and the design engineer will determine whether a problem may actually exist – and ultimately may avoid a failure.

Creative Systems in Structural and Construction Engineering, Singh (ed.)© 2001 Balkema, Rotterdam, ISBN 90 5809 161 9

Design and construction of the Intensively Rammed Composite Foundation (IRCF)

Z.Q.Xie
Quanan Construction Supervision Company, Shenzhen, People's Republic of China

Q.M.Zhang & M.D.Li
Baoan Urban Construction Investment and Development Company, Shenzhen, People's Republic of China

ABSTRACT: In recent years, the Intensively Rammed Composite Foundation (IRCF) has been developed into an integrated construction technique of the composite foundation and a quality guaranteed system. This paper describes the design and strengthening principle, parameter choices and calculating and construction procedures. It raises the methodology and standards of construction quality tests. As an engineering case, the project was at a tide silt area. The design parameters of the IRCF were used in one municipal project located at Shenzhen City, Southern China.

1 INTRODUCTION

The Chuangye Road Municipal Project was located at a main trunk road at Baoan District, Shenzhen City of China. This road consists of 6 lane motor ways, which are 80m wide. The chainage from 0+060 to 0+660 lies the tide silt strip of the mouth of the Jewelry River. Before this project, there are many fishing pools with thousand square meters in this area. The geology investigation showed that the top stratum was gray and black marine silt layer, which had high water content, high porosity, high compressibility, low permeability and low solid coefficient. Their heights varied from 4.6m to 9.1m. The second stratum was the flooding silt which located at 0+040 ~ 0+240 or the coarse sand layer at 0+240~0+660. They were designed as the IRCF to bear loadings. The main mechanical indices are shown on Table 1 below.

Table 1. The physical mechanical index of main strata (average)

Stratum name	Stratification	Flooding silt layer
Nature water content w%	75.7	29.5
Porosity c	2.019	0.886
Compressibility modulus Es [Mpa]	1.814	4.17
Internal cohesive force • °	2.675	30.14
Cohesive force C [kPa]	8.04	32.5
	4.42×10^{-4}	$n \times 10^{-2} \sim n \times 10^{-3}$
Main solidity coefficient (100 kPa) Cv [cm^2/s]		
Standard values of bearing capacities Fk, [kPa]	50	180

2 THE DESIGN OF THE IRCF

2.1 Technical Index of the Design

The strength of the IRCF under the surface of contacting layers ≥100kPa. The settlement of the composite foundation: residual settlement not exceeding 15cm; The longitudinal and transversal difference of settlement not exceeding 10cm within 100m.

2.2 The Choices of Design Parameters

The depth of filled stones: The depth must not be less then 3m. It was changed up to 4.5m within the scopes besides two ditch strips (14m wide). The backfill materials consist of middle weathering and micro-weathering stones. The maximum size is not great then 80cm.

The Rammer and dropping height: The weight of the Ramer was 15 tons, diameter 1.0~1.2m and height 2.0~2.3m. The dropping height was 20m. The maximum singly impacting energy was 300 T·M.

The arrangement of ramming: 2.5x2.5m. The distance between the centers of two neighboring piers is 2.5m. The displacement ratio m=18%.

Every ramming impacted 3 times. 7 groups were divided. The total ramming numbers were great than 21 times. The interval ramming were applied.

The standard of ramming: The total ramming reached 21 times. The last two ramming settlements were less then 30cm.

The requirement of the bottom: Every pile must be through the silt layer and up to the bearing layer.

The ditch strip areas: In order to dig further, a 15 T flat hammer was applied. Ramming energy was 200~250T·M. The backfills with 2.6m high were rammed to 0.8~1.0m.

Ramming surface: Using the flat rammers with 2.0m diameter, the energy was controlled with 100~150T·M. After Ramming, the connecting areas were focused to re-ram. The ends of the connecting parts were strengthened. Their overlapping length was 20cm. The repeating impacts were limited within 4 times.

Survey: During the construction, the level of the backfilling must be controlled well. The 10x10m level surveying grids were used. One bench mark in this area was established in order to monitor the final level of the surfaces and amounts of the settlement.

3 THE CALCULATED PRINCIPLE AND RESULTS

3.1 *The bearing capacities of the composite foundation*

The bearing capacity Fap of the IRCF may be calculated by the below formula (1).

$$Fap = m \cdot Fp + (1-m) \cdot \cdot Fs \qquad (1)$$

Where m = displacement ratio (as the space between piers is 2.5m, m=0.18); Fp = the bearing capacity of the block stone pier (the value may be chosen as 354-442kN/m^2); • = coefficient (the value equals 1); Fs = the bearing capacity of the soil between piers (the value is 50kN/m^2 here).

Using the above formula, the bearing capacities of the composite foundation in the project were calculated. Their values varied between 104.72kPa and 120.74kPa. All of them were greater then 100kPa, which met the needs of the requirement by the client.

3.2 *The final settlement*

The final settlement includes the residual settlement and the difference settlement. The formula of the calculation consists two parts, i.e.:

$$S = S_1 + S_2 = \cdot P \cdot H_1 / Es_1 + (\cdot P + \cdot Po) \cdot H_2 / Es_2 \qquad (2)$$

Where •P = the additional stress by live loads from the upper road structure; H_1 = the height of the composite foundation; •Po = the additional stress by block stone piers in the composite foundation; H_2 = the bottom bearing layer; Es_1 = the compressibility modulus of the composite foundation; Es_2 = the compressibility modulus of the bottom layer.

Es_1 can be got by Weighted Average Method (WAM) from the following formula:

$$Es_1 = m \cdot Eps + (1-m)Ess \qquad (3)$$

Where Eps = the compressibility modulus of block

stone piers; Ess = the compressibility modulus of soil between piers; m = the displacement ratio.

Considering silt depth, bottom layer and level factors, the Chuangye Road was divided into three stages to be calculated. The calculating results are listed as follows:

Stage A: 0+060~0+240. The length was 180m and the final settlement 20.88cm;
Stage B: 0+240~0+600. The length was 360m and the final settlement 20.49cm;
Stage C: 0+600~0+660. The length was 60m and the final settlement 20.93cm.

From the above results we got that the final settlements of the three stages were very close.

3.3 *The Residual Settlement After Construction (RSAC)*

The RSAC was defined as the final settlement minus the settlement during construction. The composite foundation was subjected to the additional stress during loading. Its form was a multi-step loading procedure. The solidity settlement was also a step and transient procedure. It was calculated as a transient and equal velocity procedure. Ut' is the transient and average solidity with one time. For the multi-step solidity with equal velocity, the revised solidity can be obtained in (5).

$$Ut' = q_1[T_1 + 2(e^{-at} - e^{-\cdot (t-T_1)})/\cdot]/\cdot \cdot P + q_2[T_3 - T_2]$$
$$/\cdot \cdot P + a[e^{-\cdot (t-T_2)} - e^{-\cdot (t-T_3)}]/\cdot + \cdots \cdots \qquad (4)$$

$$Ut = \cdot Ut' \cdot \cdot [1-(T_n + T_{n-1})/2] \cdot P_n / \cdot \cdot P \qquad (5)$$

Where T_n, T_{n-1} = the stating time and ending time (day); • = the revised coefficient between 0.5 and 0.7; P_n = the increase of loading (kPa); • • P = the accumulated increase of loading (kPa); q_n = the average velocity ratio of n step(kPa/day); • = the revised coefficient related to the diameter of piles, piers, etc.

Through calculations, the RSACs of Stage A, Stage B and Stage C were 3.99cm, 4.84cm and 3.41cm respectively. They were much greater then 15cm, which the client required.

3.4 *The Different Settlement*

Due to smooth changes of the silt basement, the depth of silt along longitudinal direction kept almost flat. The RSCA presented small differences (3.44~4.84cm) which well satisfied the client's standard (10cm per 100m).

118

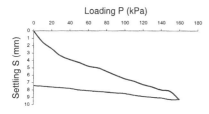

Figure 1: Static loading test P-S curve 1

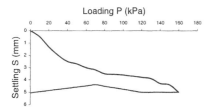

Figure 2: Static loading test P-S curve 2

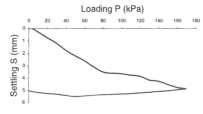

Figure 3: Static loading test P-S curve 3

4 CONSTRUCTION PROCEDURES

In the municipal project at the Chuangye Road, the construction procedure of the IRCF is described as follows:

a). To set up the survey control points along the whole project.

b). To drive water in ditch away and keep the bottom of pools and ditches dry.

c). To remove silts and flat fields.

d). To fill mixed stones with 3.0m high. A 10x10m level grid was used.

e). To ram intensively both sides of ditches twice. The filling stones were accumulated to 4.5m high.

f). To arrange 2.5x2.5m square grids as ramming points.

g). To choose a 50 T rammer.

h). To apply a radar with inclined drilling method to test the bottoms of the piers.

i). To flat fields with 10x10m grid levels.

j). To ram on the surfaces of block stones.

k). To measure the settlement by a gauge meter.

l). To collect all information and summarize the project.

5 THE TESTS AND EVALUATIONS OF THE RESULTS

The Chuangye Road Project by the IRCF started from Oct. 1995 and ended April 1996. The engineering supervisors of the project tested and traced the whole project.

The depth of filling stones: A pulse EKKOIV radar and PC IBM-486 were applied to collect data in site. In every 150~200m, supervisors would tested once to guarantee the filling depth over 3m.

The conditions of block stone piers and their bottoms: A geology SIR-10A radar was used to inspect their longitudinal forms.

The inclined displacement of piers: 1736 piers were tested. 1612 piers among them reached the bottoms. 124 piers were inclined or moved horizontally. They only shared 7.145% of the total amount.

The shapes of block stone piers: The block stone piers stand as a cylinder form. The bottoms perform a round shape. There were backfilling stones between piers. The bottoms touched to bearing layers.

The assessment of the condition of the bottoms of piers: The state standard and a book in the references of the paper allows the maximum 5% as the ratio of piers without touching bottoms. In this project, 46 piers shared 2.65% of the total 1736 piers.

The loading tests on the plate: A 2.5x2.5m loading plate was adopted. The design bearing capacity was 100kPa. The maximum loading was 1.5 times of the design loads. 9 stage loads were applied. The first stage was 20 tons, then 10 tons in 7 times. The last one was 4 tons. According to the regulation by the supervision station, 3 sites were chosen to do loading tests within 600m. During tests, the records were taken before loading and after loading about a half hour. If the increase volume was less than 0.25mm within 1 hour, the next stage was proceeded.

6 CONCLUSIONS

The project in Chuangye Road was very successful to apply the IRCF. The following points can be obtained from our experiences.

1). The calculating methods of the bearing capacities, the last settlement and residual settlement after construction in this paper can satisfy the requirements of the engineering designs.

2). Because the construction speed of the IRCF was fast, the composite foundation can more quickly solidify in order to expedite the process of the soft foundation. The whole construction period was shorter than the planned period.

3). Since the intensively rammed block piers have the bigger diameter and the certain depth, the composite foundation can better control the settlements and transformation and improve the solidity of the foundation.

4). The IRCF is a potential technique to strengthen foundations. There are still some issues to be studied, such as its working principle, calculating theory, construction procedure and testing means. The construction equipment needs improvements.

7 REFERENCES

The State Standard of People's Republic of China, 1991. *The Technical Specifications of Building Foundation Construction (JGJ79-91)*. China Building Industry Publish House.

Ye S.L., Han J. & G.B.Ye 1994. *Treatment and Transfer Techniques of Foundations*, China Building Industry Publish House.

5 Construction methods

Creative Systems in Structural and Construction Engineering, Singh (ed.) © 2001 Balkema, Rotterdam, ISBN 90 5809 161 9

Modeling construction strategies for large concrete highrise projects

Asad Udaipurwala & Alan D. Russell

Department of Civil Engineering, University of British Columbia, Vancouver, B.C., Canada

ABSTRACT: Effective scheduling of large concrete highrise buildings such as commercial office buildings and hotels, requires efficient modeling of the interplay between a number of distinct operations and trades that compete with each other for use of the workspace as well as shared resources such as cranage, while at the same time depending on each other's output to maintain a continuous and orderly workflow. Current project planning tools, with their activity-centric paradigm, lack the ability to provide the scheduler with metrics that can aid in this modeling. In this paper, we present aspects of a planning tool that uses the concept of multiple project views and a family of repetitive planning structures that enhance the user's ability to express more fully the intent of the schedule, that address the realities of high-rise construction, and which facilitate the speedy formulation and evaluation of alternative construction strategies.

1 INTRODUCTION

Commercial concrete highrises are characterized by the presence of a large number of similar horizontal components (or typical floors) consisting of a number of vertical subcomponents (or columns and walls) and a single horizontal subcomponent (or slab). These are served by one or more elevators and stairways generally arranged in vertical cores running centrally through the slabs. This configuration lends itself to a constant cycle construction strategy. In constant cycle construction, the intent is to optimize the workflow so as to have the same duration for each subcomponent from one floor to the next. To achieve this optimized workflow, it is imperative in the planning stages to choose the right blend of construction methods so as to provide continuity for work crews, allow for cycling of complex forming systems without having to transport them to ground level after each use, and allow optimum use of shared resources such as cranage, given the project's physical context such as typical floor dimensions, available work space, site location, storage areas available, expected environmental conditions and the regulatory climate.

In the next section, we use a hypothetical example to illustrate some of the decision-making requirements for the choice of a construction strategy. This example also serves as a backdrop against which we introduce key elements of a project-planning tool that enables efficient modeling of repetitive construction activities. Of special interest is the use of project Views. Views provide a mechanism for isolating the project information into related categories. This allows for ease of maintenance, and the development of new value-added functions by allowing data exchange between the Views. In addition, Views help in the reuse of planning information from one project to another, thereby providing construction companies with the ability to bank the expertise of their key personnel and use it for training new personnel. It is a departure from the current activity-centric approach, where all information is associated with the activities, thus causing duplication and inconsistency, and difficulties in filtering out relevant information. We have developed six Views that we believe cover the entire gamut of information generated during the project planning and construction phases. These are the Physical, Process, Cost, As-Built, Quality and Change Views. Two of these views have been highlighted in the ensuing discussion. These are the Physical view – what is to be built and the site context, and the Process View – how, when, where and by whom.

In the final section, we use the insights gained from this example to present some extensions that we believe would provide valuable aids in making construction strategy related decisions in the planning stages of a project.

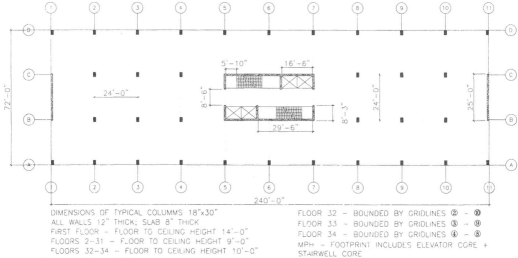

DIMENSIONS OF TYPICAL COLUMNS 18"x30"
ALL WALLS 12" THICK; SLAB 8" THICK
FIRST FLOOR - FLOOR TO CEILING HEIGHT 14'-0"
FLOORS 2-31 - FLOOR TO CEILING HEIGHT 9'-0"
FLOORS 32-34 - FLOOR TO CEILING HEIGHT 10'-0"

FLOOR 32 - BOUNDED BY GRIDLINES ② - ⑩
FLOOR 33 - BOUNDED BY GRIDLINES ③ - ⑨
FLOOR 34 - BOUNDED BY GRIDLINES ④ - ⑧
MPH - FOOTPRINT INCLUDES ELEVATOR CORE +
STAIRWELL CORE

Figure 1. Floor plan for typical floor

2 HIGHRISE BUILDING EXAMPLE

2.1 Developing a construction plan

As shown in figure 1, the example project consists of thirty-four floors, with a main floor, thirty-one typical floors, three penthouse floors, and a mechanical room on top. A central core containing four elevators and stairs services all the floors. A typical floor consists of an eight inch thick slab supported on thirty four columns spaced equidistant both along the length and width, and two end shear walls.

Before diving into planning the actual construction, we first populate the project's Physical View with data about the work locations, the various constituent components of the designed structure, along with their attributes in terms of installed quantities as well as temporary material requirements. The Physical View consists of a hierarchical breakdown structure organized as a semantically predefined tree. This is known as the Physical Component Breakdown Structure (PCBS). The PCBS has two main functions. First, it serves as a reference for the available work locations, which are used by the repetitive planning structures in developing the project schedule. Second, it acts as a central store for all information related to the project's physical attributes, as well as any known environmental and regulatory requirements. A PCBS for the example project is shown in figure 2(a). Each of the PCBS components can be described in terms of a set of attributes, as shown in figure 2(b). The attributes can take either Quantitative, Linguistic or Boolean values. For this example, we have chosen to describe them using only quantitative values. A component can inherit the attribute set of its parent much like inheritance in an object-oriented framework. These attributes can

be assigned values corresponding to the work locations on which they occur. A higher-level component can derive its attributes' values by aggregating the values of its children. An example of this is shown in figure 2(c). Each component can also be associated with a set of activities that act upon it. Higher-level components can aggregate the activity associations of their children to derive their association set, as shown in figure 2(d). We will return to this topic of activity association later.

Having modeled the project's physical context, we can now use this information to advantage in developing our plan. The slab of a typical floor has a gross area of 17280 square feet. A quick calculation shows that it would be difficult to do the entire slab in one continuous pour. Hence, we have to divide the floor plate into zones. A number of factors govern the division. Primary among these is that the core has to be placed in one integral pour, thus dictating that it be contained entirely in one zone. Another important consideration is the selection of the number and capacities of the crane(s). The geometry of the building suggests that if we were to use a single crane to serve the entire floor, it would swing out about a hundred feet over the short edge of the building, as it would have to be placed off-center along the long edge of the building on one side of the core. Placement in the core itself would necessitate the use of a 300+ foot high tower increasing the rental costs. Placing it off-center would afford the advantages of using a climbing crane. Thus, use of two smaller cranes seems a better alternative. We have decided to use gang forms for the columns and walls, and flyforming for the slab. Considering the number of column and wall forms required, the number of flyforms required, the desire to minimize

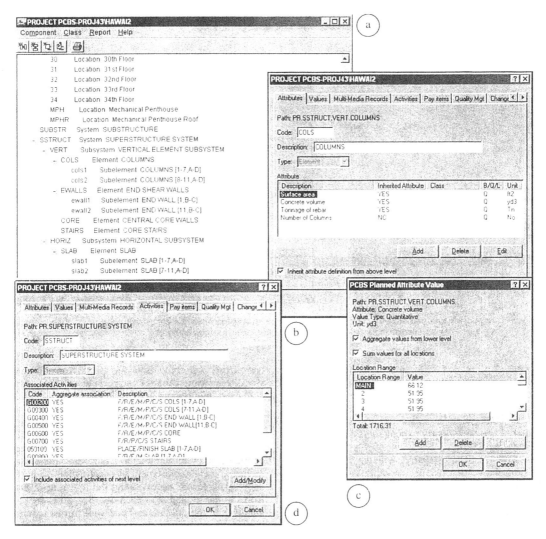

Figure 2. Physical Component Breakdown Structure (PCBS)
    (a) Hierarchical breakdown showing various components of the building.
    (b) Declaring attributes of a physical component. Note the use of inheritance.
    (c) Determining attribute values of higher level components by aggregating them from lower levels.
    (d) Activity association and its aggregation up the hierarchy.

lowering of forms to the ground, and the amount of re-handling of forms required, dividing the floor at gridline 7 seemed to be the best alternative. Consequently, we would need one complete set of fly-forms for the entire floor and half a set of column forms that we would cycle horizontally once per floor. The core would be cast as a part of the first zone, and the core wall forms would be lowered to the ground after each use. A change was necessitated to the PCBS to include column, wall and slab subelements as shown in figure 2(a) because of the zoning strategy adopted. This construction strategy

was translated into a schedule, with an emphasis on achieving a constant cycle as quickly as possible. A listing of the activities used is shown in figure 3(a). Note the use of aggregated activities for constructing each of the lower level components in the PCBS tree. This was a conscious decision, which has a significant effect on the resource modeling as discussed shortly. Figure 3(b) shows one of the recurring themes central to the tool's design philosophy. Note the use of location ranges for specifying the duration per location, as well as the ability to specify a pattern of working the locations in that range using a

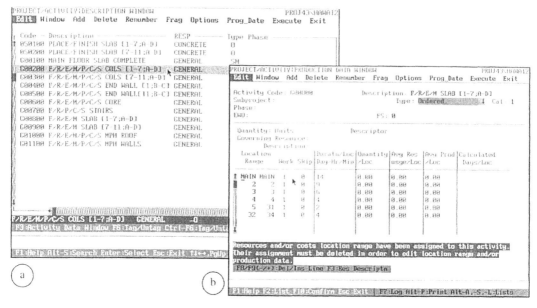

Figure 3. (a) Activity listing showing the use of aggregated activities to construct a physical component.
(b) Specifying the duration per location for an activity, using location ranges for brevity.

work-skip function. Also, notice, that the current activity is specified to be Ordered. This combination provides the system with a concise directive for computing the final schedule. Logic relationships between the various work locations for different activities can be similarly expressed using typical relationships.

Cranage being a shared resource can easily become the bottleneck. Its modeling is therefore described here in more detail. Since we used aggregated activities, we had to derive a profile for the usage of the crane across the activity on a daily basis. This therefore necessitated a disaggregation and subsequent reaggregation exercise. Considering the expressiveness of the activity breakdown in representing a construction cycle, we thought this to be a worthwhile tradeoff. To determine the time the crane was used by each activity on a daily basis, we counted the number of lifts required, considering the item being lifted, the carrying capacity of the crane, and the activity duration; and multiplied this by the time per lift calculated considering the site geometry, crane location and manufacturer's specifications regarding travel speeds. A similar exercise was done for labour of each type, i.e., formwork, reinforcement, mechanical, electrical and slab placement.

2.2 Results and Analysis

A six day constant construction cycle was achieved for all typical floors starting from the fifth floor. The cycle took three days for the vertical components in each zone, and three days for the slab. Dedicated work crews were used for each component, with the core crew doubling up as the stair crew.

Figure 4(a) shows the resource usage profiles for the tower cranes. As can be seen, the planned resource usage for the cranes exceeds the available crane hours per day. There are two bottlenecks in the construction cycle that lead to this result. First is the intensive crane requirement of constructing the core using gang forms, necessitating the lowering of the forms to the ground after each use. Second is the re-handling of the column and end wall forms, requiring significant time on the first day of slab construction. The results suggest that a rethink of the construction strategy is required. Alternatives include studying the possibility of building intermediate platforms in the core to store the core forms, jump forming the core instead of the current gang forming technique, reassessing the advantages of using a single crane so as to reduce re-handling of forms, and reassessing the zoning strategy. Figure 4(b) shows the usage profile for labour as a function of time and location. As can be seen from the chart the variation in the labour usage is quite large. However, this by itself does not provide any metric by which one can assess the workability of this resource loading. Figure 5(a), shows this same resource load from the perspective of the work area available to each labour on a typical floor. Using a threshold of 300 ft^2/worker (Riley & Sanvido 1995), this chart shows that there is no congestion. This is just one example of the benefit of using Project Views. The

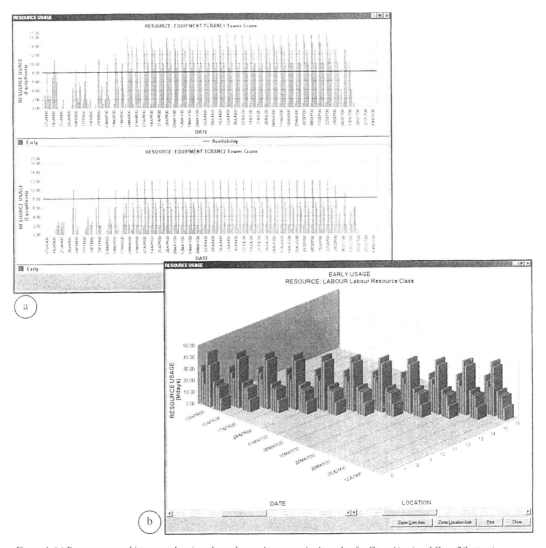

Figure 4. (a) Resource usage histogram showing planned crane hours required per day for Crane1(top) and Crane2(bottom).
(b) Daily manpower required on a location. Note the variation in manpower levels and the cycling every six days.

data for this chart has been amalgamated from the Physical and Process Views. Another example of a useful project control metric is shown in Figure 5(b) (Russell & Udaipurwala 1999). This figure shows the planned productivity values required of the forming labour to achieve the time and cost targets. To create this chart use has been made of the activity associations previously defined in Figure 2(d).

3 CONCLUSIONS AND FUTURE WORK

We have demonstrated that representing a project in terms of multiple Views leads to the ability to sup-

port value-added analysis and knowledge-management functions. In turn, these functions help in assessing of the workability of a construction strategy.

The example presented points to further avenues of investigation. First is the need for a Methods modeling and reasoning environment that can aid in the quick and easy formulation and assessment of alternate construction strategies. We have already taken the first steps in this direction with the development of a framework for representing knowledge of construction methods and resources within the tool (Udaipurwala & Russell 2000).

Second is the need for a hierarchical scheduling

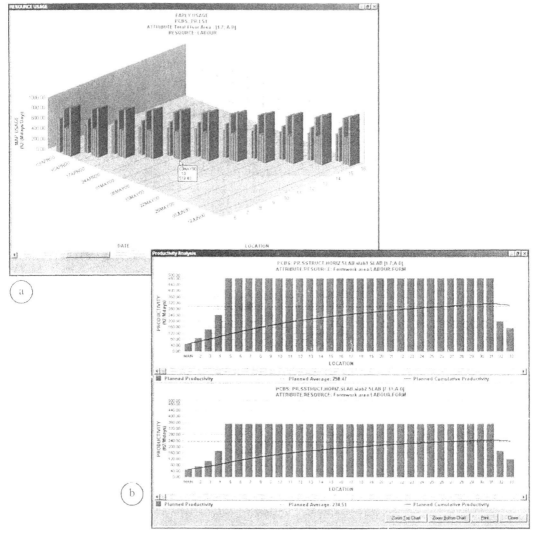

Figure 5. (a) Chart showing working available per man on a daily basis. This can be used to assess the congestion levels on site.
(b) Planned productivity levels required to meet targeted project duration and cost.

framework. Such a framework will for example let the user specify a breakdown of the activity F/R/E/M/P/C/S as a network of constituent tasks, but still retain the ability to specify logic at the aggregate level, thus eliminating the need for the tradeoff we made in working this example.

Third is the development of reasoning mechanisms that can suggest feasible construction methods based on the attribute values in the PCBS.

REFERENCES

Riley, D. R. & Sanvido, V. E. 1995. Patterns of Construction-Space use in Multistory Buildings. Journal of Construction Engineering and Management, Vol. 121, No. 4, December 1995: 464-473

Russell, A. D. & Udaipurwala, A. 1999. Modeling Construction Productivity. Annual Conference of CSCE, Regina, Saskatchewan, 2-5 June 1999: 249-258

Udaipurwala, A. & Russell, A. D. 2000. Reasoning about Construction Methods. Proc of ASCE Construction Congress VI, Orlando, FL, 20-22 Feb 2000: 386-395

Creative Systems in Structural and Construction Engineering, Singh (ed.) © 2001 Balkema, Rotterdam, ISBN 90 5809 161 9

Wharf embankment and strengthening program at Port of Oakland

F. Lobedan
Port of Oakland, Calif., USA

ABSTRACT: The Wharf and Embankment Strengthening Program (WESP) is a structural modification project involving approximately 12,000 linear feet of pile-supported, marginal wharf structures. WESP is necessary because the Port of Oakland plans to deepen its berths from –42' Mean Lower Low Water (MLLW) to –52' MLLW, in conjunction with a Federal Government-sponsored channel dredging project. Unless they are structurally reinforced prior to the dredging, the waterfront components (i.e. wharves and embankments) will be weakened by the berth deepening project. WESP is a three-phase program that establishes the existing structural and seismic capacities of waterfront components, develops designs for improvements necessary to maintain these capacities after the berth deepening, and constructs the improvements. WESP also includes consideration of seismic upgrade improvements. The Port is currently completing the first phase of the WESP program. This paper will describe the design criteria, project phasing, construction type of waterfront components, project organization, and results to date for WESP.

1 INTRODUCTION

The Port of Oakland occupies 19 miles (30.6 kilometers) on the eastern shore of San Francisco Bay reaching from the Oakland-San Francisco bay bridge to the Metropolitan Oakland International airport (see Fig. 1). The maritime area of the Port consists of 1110 acres (449 hectares) of terminal and support areas, including 27 deepwater berths and 30 gantry container cranes. The Port handles 98% of all containerized cargo that passes through northern California ports and is the fourth largest port in the United States.

Figure 1. Vicinity map.

The Port is in the process of deepening selected berths in conjunction with a Federal Government-sponsored channel dredging project covering both the Inner and Outer Harbors (see Fig. 3). The dredging project will deepen the Port's channels and berths from –42' mean lower low water (MLLW) to –52' MLLW. The Port has carried out preliminary investigations to determine the structural capacities of the waterside crane rail girders and check the static slope stability of the embankments beneath the wharves for the proposed –52' dredge depth. The investigations found that modifications to the waterfront components (i.e. wharf structures and embankments) are required prior to the berth deepening. These modifications fall into two categories:

1. Embankment Stabilization: Installation of sheet pile walls at the toe of the embankments or some other form of ground stabilization improvement.
2. Structural Modifications to the Wharves: The pile supported crane girders will require strengthening. Also, many existing batter piles will intrude into the berth area and will have to be replaced.

Figure 2 depicts a transverse cross section of a wharf and embankment together with the proposed modifications.

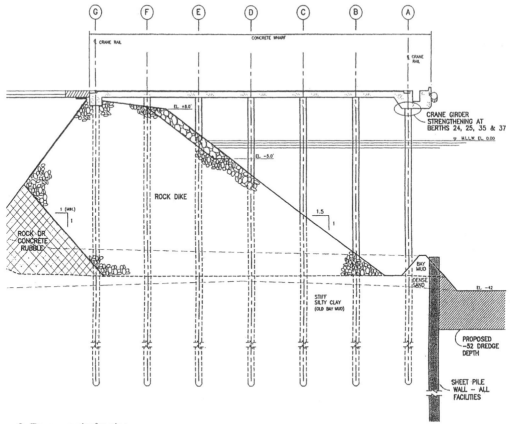

Figure 2. Transverse wharf section.

The design of the modifications is required, at a minimum, to maintain similar structural capacity as currently exists prior to berth deepening. The objective of the Port's Wharf and Embankment Strengthening Program (WESP) is to design and construct the structural modifications necessary to mitigate the effects of the proposed –52' MLLW deepening project.

In addition, WESP includes a seismic risk evaluation of its waterfront components in order to consider potential seismic upgrade improvements in conjunction with the design of embankment stabilization and structural modifications. Conceptual seismic upgrade designs will be developed for increasing levels of seismically induced forces. Economic and risk analyses will then be utilized to determine the most cost-effective seismic upgrade alternative for the specific waterfront components.

2 DESIGN CRITERIA

The analysis and design of WESP will be based on a revised Port of Oakland wharf design code. The Port is in the process of revising its wharf design code, which was originally adopted on May 1, 1990. The revised code will be based on a multi-level, performance based design approach. The performance levels being proposed are as follows:

- Level I- Establish ground motions having a 50% probability of exceedance in 50 years. The wharf and embankment system shall be designed so that under this level of shaking, only minor, repairable damage is anticipated and that operations will not be interrupted.
- Level 2- Establish ground motions having a 10% probability of exceedance in 50 years. The wharf and embankment system shall be designed so that under this level of shaking, controlled, economically repairable damage is anticipated and that operations may be limited and/or interrupted. In addition, collapse shall be prevented and damage shall be observable and accessible for repair.

- Post- Level 2-The wharf and embankment system shall be designed so that under this level of shaking, significant uneconomically repairable damage may occur, but will not be sufficient to endanger the life-safety of the users of the structure. Collapse shall be prevented.

3 PROJECT PHASING

WESP is comprised of three phases; a seismic analysis phase, a seismic upgrade planning phase, and a design/bid/construction phase. These phases are further described below.

Phase I – Seismic analysis to determine the current seismic capacities of waterfront components.
A. Conduct analysis to determine the peak horizontal ground accelerations associated with the levels 1 and 2 performance criteria. Determine the associated probabilities of exceedance in 50 years.
B. Conduct analysis to determine the peak horizontal ground accelerations associated with the Post-Level 2 performance criteria. Determine the associated probability of exceedance in 50 years.
C. Prepare a Wharf and Embankment Seismic Risk Evaluation Report summarizing the results of the above analysis and describing the vulnerability and specific failure mechanisms of waterfront component.

Phase II – Preliminary design of seismic upgrade improvements together with economic and risk analyses for consideration of seismic upgrade improvements.
1. Develop preliminary design and associated cost estimates for improvements to waterfront components necessary to maintain an equivalent seismic risk for the future deepened berths (i.e. e structural modifications necessary to mitigate the effects of the proposed –52' MLLW deepening project).
2. Conduct seismic risk reduction planning at selected facilities. Develop conceptual strengthening concepts and associated construction cost estimates for up to 4 increasing levels of design associated with increasing levels of seismically induced forces. Evaluate seismic performance and expected damage that would result from a range of plausible earthquake events for the region. Prepare corresponding repair costs and business interruption costs (i.e. operational costs) for the range of plausible earthquake events. Compute the mean value and standard deviation of total cost (strengthening costs, repair costs and operational costs) for the distribution of plausible earthquakes for each level of design.
3. Develop Port-wide strategy for implementation of WESP based on business goals and the above analysis.

Phase III – Implementation of WESP. Prepare Plans and Specifications to construct the structural modifications necessary to mitigate the effects of the proposed –52' MLLW deepening project together with the selected seismic upgrade improvements. Phase I was being concluded at the time of the preparation of this paper.

4 GENERAL DESCRIPTION OF WESP BERTHS

As stated previously, only selected berths are being considered for deepening in conjunction with the Federal Government-sponsored channel dredging project. The berths selected are described in Table 1 and Figure 3 below.

Table 1. Berth descriptions.

Berth identification #	Year	Notes
Berth 23*	1978	Original construction
Berth 24,25*	1975	Original construction
150' Ext. @ Berth 24	1971	Original construction
Berth 26/30	1986	Original construction
Berth 32,33	1966	Original construction
Berth 35,37	1967	Original construction
Berth 35,37	1990	Earthquake Damage Repair
Berth 60-63	1971	Original construction
Berth 67,68	1979	Original construction
Berth 69	1994	Wharf Extension at Berth 68

*Berths 23 was formerly designated as Berths 5
*Berths 24, 25 were formerly designated as Berths 2, 3 and 4

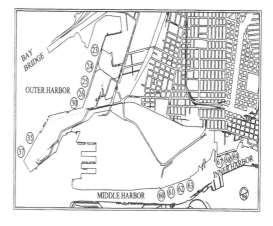

Figure 3. Site plan.

Berths 23, 24 and 25 were constructed between 1975 and 1978 and were previously designated as Berths 2,3,4 & 5. These berths consist of three separate structures built at separate times. Berths 4 & 5 were constructed on each side of a 150-foot wharf structure of yet another structural design. Each structure is separated from the others with a transverse, unkeyed expansion joint. With the exception of the 150-foot extension, these wharf structures are quite similar consisting of an approximate 2-foot thick reinforced concrete deck and an asphalt concrete pavement supported on 18" square, pre-cast, pre-stressed concrete piles. The lateral force resisting system consists of two sets of battered (4H:12V) piles. The 150-foot wharf section consists of a 1-1/2-foot thick reinforced concrete deck ballasted with a 3-foot layer of fill and an asphalt concrete pavement supported on 16" square, pre-cast, pre-stressed concrete piles. The lateral force resisting system for the 150-foot extension consists of a set of battered (4H:12V) piles. The land-side crane rail is not structurally connected to the deck and is supported on a grade beam. The structures are located on land reclaimed from the Bay and are underlain by fill materials. The embankment beneath the wharf consists of sand fill, and young bay muds which slopes at a 1-1/2H:1V and is armored by rubble.

Although berths 26 and 30 were constructed at different times, they are all of similar design. The 109' wide, reinforced concrete deck is supported by seven rows of vertical, 24" octagonal pre-cast, pre-stressed concrete piles. The concrete deck includes both crane rails. The embankment beneath the wharves consists of a rock dike, which slopes at a 1-1/2H: 1V. The lateral force resistance is supplied by the back rows of vertical piles which embed into the rock dike.

Berths 35-37 were originally constructed between 1967 and 1971. The wharves as originally built are 1-1/2-foot thick reinforced concrete deck ballasted with 2-1/2 foot of fill and an asphalt concrete pavement. The deck was supported on 16" square, pre-cast, pre-stressed concrete piles with lateral force resistance provided by a row of alternating battered piles at the landward edge of the wharf. The land-side crane rail is not structurally connected to the deck and is supported on a grade beam. The structures are located on land reclaimed from the Bay and are underlain by fill materials. The embankment beneath the wharves consists of sand fill, and young bay muds which slopes at a 1-1/2H:1V and is armored by rubble.

The Loma Prieta earthquake of 1989 caused extensive damage to the battered piles at berths 35-37. The Port constructed repairs consisting of two rows of vertical, 24" octagonal pre-cast, pre-stressed concrete piles, one at the landward row of original battered piles, and the other beneath the new land-side crane rail. An extension the concrete deck was constructed to connect the two row of new piles, and the battered piles were cut off so that the new vertical piles provide the lateral force resistance. In addition, portions of the sand fill embankment were densified by installation of stone columns.

Berths 60-63 were originally built in 1971. The wharves are approximately 1-1/2-foot thick reinforced concrete deck ballasted with 2-1/2 foot of fill and an asphalt concrete pavement. The deck is supported on 16" and 18" square, pre-cast, pre-stressed concrete piles with lateral force resistance provided by a row of alternating battered piles at the waterside edge of the wharf. The land-side crane rail is not structurally connected to the deck and is supported by battered piles. The embankment beneath the wharves consists of sand fill, and young bay muds which slopes at a 1-1/2H:1V and is armored by rubble.

Berths 67-68 was constructed in 1980. The 80' wide reinforced concrete deck is supported by five rows of vertical, 24" octagonal pre-cast, pre-stressed concrete piles. The landside crane rail is separated from the wharf structure and supported by a grade beam on battered piles. Berth 69 is actually a 300' extension to berth 68 constructed in 1995. It consists of a 109' wide reinforced concrete deck supported by seven rows of vertical, 24" octagonal pre-cast, pre-stressed concrete piles. The concrete deck includes both crane rails. The embankment beneath the wharves consists of a rock dike, which slopes at a 1-1/2H: 1V. The lateral force resistance is supplied by the back rows of vertical piles, which embed into the rock dike.

5 PROJECT ORGANIZATION

The Port issued a request for proposals (RFP) for consultant services for the WESP. The RFP organized the work of WESP into 7 contracts; 2 geotechnical contracts, 4 structural contracts and a seismic risk reduction-planning contract. Each contract reports directly to the Port. The delegation of work is depicted in Table 2. The facilities were grouped in accordance with their structural similarities.

Table 2. Delegation of work.

Consultant contract	Berth identification #'s
Structural #1	23 - 25
Structural #2	60 - 63
Structural #3	30, 67 - 69
Structural #4	35 & 37
Geotechnical #1	23 – 25 & 60 - 63
Geotechnical #2	30, 67 – 69, 35 & 37
Seismic risk reduction planning	all berths

Table 3. Relative ranking of seismic vulnerability.

Berths identification #	Probability of Exceddance in 50 Years (%)		
	Level I	Level II	Post Level II
Berth 23	60	40	30
Berth 24 & 25	60	35	25
150' Ext. @ Berth 24	95	95	45 to 35
Berth 26	20	5	5
Berth 30	26	26	13
Berth 35 & 37	60	20	10
Berth 62	65	62	62
Berth 63	70	67	67
Berth 67 & 68	50	20	15
Berth 69	30	30	15

6 RESULTS TO DATE

As stated previously, at the time of the preparation of this paper, Phase I, determination of current seismic risk for existing conditions of waterfront components, was concluding.

The geotechnical analysis methodologies used by the two geotechnical consultant teams were consistent (URS Greiner Woodward Clyde June 26, 2000 and Harding Lawson Associates April 24, 2000). Static slope stability was analyzed using the limit-equilibrium method. Seismic slope stability analysis was performed using a pseudostatic approach. Slope deformations were estimated using the Newmark approach. In addition, computer code FLAC version 3.4 (2-dimensional, finite difference) was also used to evaluate seismic deformations of the embankment. The FLAC analysis was performed with and without the inclusion of piles in the embankment. Liquefaction susceptibility evaluations were also performed.

The structural analysis methodologies used to determine the seismic capacity of the wharf structures varied. Three of the structural consultant teams utilized a simplified, 2-dimensional pushover/ response spectrum analysis (CH2MHill & B.C. Gerwick, Inc. May 2000, TranSystems Corporation & Han-Padron Associates May 1, 2000 and F. E. Jordan Associates & B. C. Gerwick, Inc. May 1, 2000). In all three cases, the computer program SAP2000 was used for the analysis. The fourth structural team was assigned berths 60-63 and utilized a 3-dimensional time history analysis (Parsons Brinkerhoff et. al. May 25, 2000). The computer program ADINA was used for the 3D analysis. Soil-structure interaction effects were incorporated in both the 2D and 3D models by providing "soil-spring" supports along the pile. Both the 2D and 3D modeling procedures decoupled the inertial seismic response from the embankment response. However, in one case, the two load cases (i.e. forces due to the structure's inertial seismic

response and those due to the embankment response) were combined to a certain extent (CH2MHill & B.C. Gerwick, Inc. May 2000). A California Department of Transportation (CALTRANS) approach was adopted entitled " CALTRANS Interim Seismic Design Criteria for Pile Survivability".

Phase I was intended to culminate in a relative ranking of seismic vulnerability of the selected WESP berths. Table 3 presents a preliminary version of this relative ranking.

7 CONCLUSION

The engineering analysis performed on WESP to date has identified, in general, three failure mechanisms for the wharf structure/embankment systems when they are subjected to seismically induced forces. These behavioral effects are summarized as follows:
1. Wharf Structure's inertial Response: Response of the structure under inertial seismic forces.
2. Below-grade Grade Pile Response: Response of the buried portions of the structure to relative ground movements at soil boundaries.
3. Embankment Response: Response of buried portions of the structure to embankment deformations.

The 3D ADINA structural modeling captures the behavioral effects 1 and 2. Presently, additional analysis is being considered to further calibrate the model. By using ground motions from the 1989 Loma Prieta earthquake as input to the model, the wharf structure/embankment behavior can be compared against the observed/recorded damage from the earthquake. The model can then be calibrated as appropriate.

The 2D SAP2000 structural modeling captures behavioral effect 1 and behavioral effect 3 was incorporated to a certain extent. The geotechnical consultants provided lateral soil pressures intended to represent the limiting upper bound of soil deformations around the existing piles. Efforts are currently underway to determine if the FLAC results can be utilized to estimate behavioral effect 2 so that it can be accounted for in the analysis.

Additional analysis may also be required to better understand the performance of the berth 35-37 wharf and embankment under seismic loading conditions. As stated previously, the structure was upgraded following the 1989 Loma Prieta earthquake. The Phase I geotechnical analysis predicts significant ground spreading motion of the non-densified sand fill embankment, up to several feet of soil mass slide, at earthquakes of the same magnitude that achieves the Level I limit state of the structure. In the event of such slides, the 16" square piles of the original construction can be expected to fail catastrophically due to the drag forces exerted on them. The failure of several piles in a bent can be expected to result in an actual collapse of the deck above. This conclusion is of the greatest concern to the Port and does not coincide with the intent of the 1989 upgrade. Further verification and validation of this conclusion is necessary before proceeding with Phase 2 of WESP at this facility.

REFERENCES

CH2MHill & B.C. Gerwick, Inc. May 2000. *Wharf and Embankment Seismic Risk Evaluation Report, Berths 23, 24 & 25.* (Technical report).

F. E. Jordan Associates & B.C. Gerwick, Inc. May 1, 2000. *Phase I Seismic Evaluation Report Berths 26, 30, 67-68 ext.* (Technical report).

Harding Lawson Associates April 24, 2000. *Ground Motion Report Berths 26-30, 35-37, 67-68 ext. Wharf and Embankment Strengthening Program.* (Technical report).

Parsons Brinkerhoff et. al. May 25, 2000. *Seismic Risk Evaluation Report Port of Oakland Wharf and Embankment Strengthening Program Berths 60-63.* (Technical report).

TranSystems Corporation & Han-Padron Associates May 1, 2000. *Seismic Risk Evaluation Report WESP Phase I (Berths 35-37).* (Technical report).

URS Greiner Woodward Clyde June 26, 2000. *Port of Oakland Wharf and Embankment Strengthening Program Berths 60-63 Geotechnical Report of Ground Motions.* (Technical report).

Creative Systems in Structural and Construction Engineering, Singh (ed.) © 2001 Balkema, Rotterdam, ISBN 90 5809 161 9

Microtunnel – Recouping from pitfalls

H. Hessing
STV Incorporated, New York, N.Y., USA

ABSTRACT: This paper describes a microtunnel project that was in planning for five years and whose actual construction time was three months. The paper describes the study report, plans and specifications, subsurface investigations, manufacture and inspection of jack pipe, description of concurrent projects on the same site, the microtunnel project including construction procedures, what actually happened, the contractor's claim and the determination of the claim's merits. On the whole, the constructed project achieved the goal of the plans and specification.

1 INTRODUCTION

STV Incorporated (STV) has been involved with a $200,000,000 utility restoration program being performed at Co-op City, Bronx, New York for River Bay Corporation and New York State Housing Finance Administration. The work spanned a period of twelve years. STV provided comprehensive mechanical, electrical, structural, and civil engineering services for in-depth investigations and testing of all underground utilities, mechanical and electrical systems, and subsequent design and construction support services for the upgrade of systems and complete rehabilitation at the site.

Detailed investigations of all building systems and utilities on the 350-acre site were performed to determine existing conditions and the extent of damage due to the settlement of structures of up to three feet. The complex consists of 35 hi-rise buildings ranging from 24 to 33 stories with 15,372 apartments and seven groups of 472 townhouses presently housing 60,000 people.

The "Co-op City Housing Project Repair Program" consisted of replacing the dual temperature and high temperature hot water distribution system; electric ductbank distribution system; reconstruction and repairs to underground utilities and appurtenant structures including gas, water, sewer; site settlement repairs including grading, drainage, paving and landscaping; and miscellaneous repairs to building interior mechanical and electrical systems, structure, and interior finishes, and replacement or repairs to building envelops and roofs.

The Hutchinson River and the Hutchinson River Parkway separate the southern section of the site from the power plant. STV Incorporated prepared a study to bring the new dual temperature system and electrical system to the south site.

Soil borings were taken and STV prepared plans and specifications entitled "TES-221" for installing five tunnels consisting of 54 " reinforced concrete casing pipe under the Hutchinson River Parkway in the vicinity of the intersection of Bartow Avenue. The work included excavating and backfilling of jacking and receiving pits and temporary surface restoration in the work areas. The Contractor was not limited as to the method to accomplish the installation other than the fact that open cut and cover was not permitted because traffic could not be interrupted on the Parkway. The size and type of equipment was the Contractor's choice.

1.1 Study Report

The heating, cooling hot water and electrical power for Co-op City's housing complex are supplied to the buildings through three central underground systems that emanate from the central Power Plant located at the northeast quadrant of the intersection of Bartow Avenue and Co-op City Boulevard. The power plant services both the North and South sites of Co-op City.

Heating or cooling, depending on the season is supplied through a two-conduit (supply and return) Dual Temperature (DT) syst em. Similarly, hot water is supplied through a two-conduit High

Temperature (HT) system. Electrical power is supplied through cables in multiple conduits, encased in concrete forming an electrical duct bank (EDB) system.

For over a decade, severe problems had been experienced with all three systems resulting in frequent service disruptions, efficiency degradation, excessive maintenance repairs costs and general inconvenience to the tenants and the administration. The decision was made to reconstruct all three systems.

STV prepared a report entitled "The Routing of the Dual Temperature, High Temperature and 15 kV Duct Bank Systems from MH #1 to the South Site" in April 1993. The purpose was to identify alternate routes and determine the most suitable and acceptable alternative for reconstruction of a portion of the systems, namely, from MH # 1 to the beginning of the South site east of the Hutchinson River Parkway (HRP).

The need for the study was precipitated by the lack of a clear unimpeded corridor to construct the new systems while the existing systems remained in service. The roadway bed of Bartow Avenue/HRP east is highly congested with various underground utilities including the existing DT/HT/EDB systems. The northerly shoulder area of Bartow Avenue/ HRP East (area north of the northerly curbline) is limited in width and is immediately followed by the bank of Hutchinson River. The southerly shoulder area of Bartow Avenue/HRP east is extremely narrow and is constricted at the Hutchinson River Parkway's bridge abutment and embankments. The physical restraints dictated the need to consider various alternative routes and to perform careful research, study, assessment and evaluation before a selection was made.

Three corridors were identified as suitable for the crossings of the HT/DT/EDB systems. Within the corridors, four possible schemes were possible. Based on the findings of the study an above ground scheme along the shoulder area requiring steel support structures was recommended.

1.2 Plans & Specifications

The study recommendation was not approved. There were many meetings and discussions with several agencies including but not limited to: the New York State Parks Department, New York State Department of Transportation, New York State Department of Environmental Protection, New York City Department of Transportation, New York City Department of Environmental Protection, New York City Department of Parks and recreation and the U.S. Coast Guard. It was decided that microtunneling would be permitted to be constructed through the embankment of the Hutchinson River Parkway.

STV prepared plans and specifications for the microtunnel project. The drawings indicated the layout of piping, size, dimensions, grades and other details. Coordinates of the center tunnel were given with horizontal offsets shown for the remaining tunnels. Preliminary specifications were prepared in February 1994 and final specifications completed in April 1996. Research entailed obtaining literature from diverse sources including New York City Department of Environmental Protection- Bureau of Sewers, New York State Department of Transportation Standard Specifications, ASTM C76 – 85a Standard Specification for Reinforced Concrete Culvert, Storm Drain, and Sewer Pipe. Design requirements determined that Class IV Reinforced Concrete Pipe would be specified.

1.3 Subsurface Investigations at the Hutchinson River Parkway

A subsurface investigation program was performed by Warren George, Inc. for Riverbay Corporation between September 18, 1995 and October 18, 1995. Ten soil borings were taken. Nine were directly on top of the embankment of the Hutchinson River Parkway Bridge and one on Bartow Avenue near the southwest abutment of the bridge. The depth of borings into bedrock ranged from 53 feet at Boring B-40 on Bartow Avenue to 70 feet at Boring B-43 on top of the embankment. Ground water readings were not recorded. The ground water table fluctuates with rainfall. It is higher or lower based on a dry or wet season.

There were five basic layers of soil strata encountered. They are as follows: A) Embankment Fills, overlaid by B) Gray, c-f Sand, little to some Silt, with Cobbles, overlaid by D) Gray, weathered rock (Mica), overlaid by E) Gray Schist with Quartz Lens.

1.4 Embankment Fills

The embankment fills ranged from a depth of 27 feet on the southbound side to 19 feet on the northbound side of the HRP. The soils consisted of a brown c-f Sand with little to some silt, containing pieces of cinder ash, brick chips, glass, uniformly distributed. The material may have been placed in concurrent lifts from the same stockpile. The fill is essentially a medium dense soil as per the blow counts. Some blow counts were higher because the split spoon sampler may have hit pieces of cobble causing a higher reading. These cobbles could be penetrated using a roller bit. In one instance a large cobble had to be cored. This occurred at a depth of 10.0 feet at Boring B-50. The toe of the embankment on the southbound side is at approximate elevation 8.0 feet and on the northbound side it is at approximately 12.0 feet.

1.4.1 Gray, c-f Sand, little to some Silt, Cobbles

These soils underlay the embankment fills and are nine feet in thickness on the southbound side. On the Northbound side of the HRP, this soil was approximately ten feet thick. The soil may be considered Medium density. Cobbles of Mica were encountered; however, the cobbles could be penetrated using a roller bit.

1.4.2 Gray Clayey Silt

This soil underlies the Gray c-f Sand, little to some Silt with Cobbles. Based on the N value, (number of blow counts per foot) the Clayey Silt is very stiff. The soil was essentially Silt turning to clay.

1.4.3 Gray Weathered Rock (Mica)

This stratum is under the Gray Clayey Silt. The soil consisted of medium density soil of weathered mica and cobbles.

1.4.4 Gray Schist with Quartz Lens

Bedrock was encountered at a depth of 48 feet below Bartow Avenue, Boring B-40. Core recovery ranged between 49 inches and 60 inches. Due to the high percentage of Clear Quartz found in the rock, it may be considered a soft through hard rock.

1.4.5 Observations

Although clusters of cobble were found during observations, it seemed feasible that concrete pipe could be installed in the Gray c-f Sand with little silt, underlying the embankment fills. No borings had to be moved due to an inability to penetrate cobble. One boring had to be moved due to equipment failure.

2 54" RCP JACKING PIPE (3)

STV performed a plant inspection at the pipe manufacturer's facility on September 16, 1997. Vianini Pipe, Inc. is located in Somerville, New Jersey. STV witnessed the manufacturing of 54-inch diameter reinforced concrete jacking pipe required for the Microtunneling project at Co-op City. On the day of inspection weather conditions were sunny and 80° F.

STV met the plant production manager who showed us the areas used for manufacturing the 54" RCP. He introduced us to the inspection staff and Mr. Kevin Brown who was a great source of information.

It was explained that Vianini Pipe, Inc. has production and engineering divisions. Quality control is part of engineering.

Quality control of the manufactured pipe follows the procedures submitted for approval on 7/21/97. The inspector uses the form entitled "Quality Assurance Jacking Pipe Report." A form is completed for each pipe and cross-referenced by production number and date cast. The inspector marks the pipe with the appropriate designation. The form is filled out as the

pipe goes through the various stages of construction. It cannot be completed at one time or in one day.

Quality control (QC) has a testing laboratory on site. They have the capability to test aggregate, run sieve analysis, take concrete slump and concrete cylinder samples. They have curing boxes. They can break the cylinders to determine compressive strength. They have separate steel tapes of each diameter pipe to measure circumference.

Aggregates are stockpiled. Certain clients such as New York State Department of Transportation (NYSDOT) require separate stockpiles labeled for their use only. The stockpile next to it may be the same material but they are separated just the same.

Woven wire fabric (WWF) reinforcement arrives at the site in rolls. In one plant it is unrolled through a machine and set in the shape and required diameter. The Class D round pipe required for the TES 221 Project had two layers of reinforcement which were two different diameters.

The inner reinforcement is placed around an inner steel form. The form and WWF are set vertically on a round, level plate in an outdoor, open area by a Manitowoc crane. The outer cage reinforcement is lifted and lowered into place. Spacers are used to maintain the required cover. The outer form is lifted and lowered into place. Vibrators are permanently attached to the outside jacket. The entire assembly is checked for alignment before the pour.

A preset computer controls the aggregate, cement and water. An operator can adjust the mix based on his experience and his observations of the water content of the aggregate as it enters the hopper. The material goes into a mixer and then onto a conveyor belt. The concrete mixture is placed into a crane bucket, placed in the forms from the top and vibrated. Two men climb on top to verify the height of the pour. Two pipes are poured in the morning and two in the afternoon after which the crane places a tent over them. On site steam generation allows for curing. The allowable ambient temperature range is 90 to 130 degrees. The setting was 110 degrees.

After the initial set the forms are removed. The pipe is transferred to another area. Excess concrete is removed from both ends of the pipe. The inspector takes preliminary measurements of the wall thickness and inside radius. These are not recorded as it is performed just to eliminate any pipe which is out of round or otherwise does not meet the specification.

Tremendous effort is made to ensure that the bell and spigot ends are flush, round and manufactured to the required depth and tolerances of the acceptance criteria. Each man has a square (tool) to measure the depth of the bell and spigot ends. A hand held jackhammer is used to get the bell and spigot to the correct depth. A smooth section is roughened in the same manner. Sika 123 is applied by hand trowel to prepare a smooth surface. This can only be done in 1/8 of an inch layers. Excess material is removed. Half the pipe is done and then the pipe is rolled so the other half can be done. Air holes are patched. Exterior "gate" refers to the area where the forms come together. If it is greater then 1/8 of an inch it is ground smooth. The inspector uses a caliper, a square, and a tape to verify acceptance of all of the above and then places a check mark on the form for each item.

When the results from the concrete cylinder breaks come back they are recorded on the inspector's form.

The ends of the pipe are sounded with a Swiss hammer. This is done to check for voids, which are not observable. A weak section is cause for pipe rejection. Rejected pipe is removed from the production line and stored in a separate location. I did not see this test performed on the day of my inspection but I was shown a pipe rejected on that basis.

External load crushing test is performed by the three-edge-bearing method in accordance with ASTM C76-85a, 11.3.1. The three-edge bearing test machine is located in another area. This is a destructive test in which a load is applied on the top of the pipe. A crack usually develops on the bottom inside radius. A feeler gauge is used until 1/100th of an inch can be measured. The reading on the pressure gauge indicates the strength of the pipe. If the ultimate strength is required the pressure is continued until the pipe destructs. This test was not performed the day STV visited the production site.

Gaskets are sent with the pipe for the Contractor to install.

STV asked Kevin Brown if there was an independent quality assurance person or group who implements the quality control program and sees to it that QC is being performed correctly. He said that this is not done for reinforced concrete pipe in the industry. It does exist for pressure pipe and Vianini follows the same quality control procedures with RCP as they do for pressure pipe with the only difference being the material. He said that they are the only manufacturers in the northeast who has an independent agency perform quality assurance of their quality control program. The independent QC auditor is Lloyd's Register Quality Assurance. STV asked for and received copies of this certification.

STV was told that NYSDOT, NYCDOT, NYCDCC as well as NJDOT have approved the plant.

In conclusion, the fabrication facility and the QC exercised in the plant met all the requirements of the TES 221 Specifications.

3 CONCURRENT PROJECTS

Although not all inclusive, the following is a listing of the major work whose time frame spanned the same time the Housing Project Repair Program was underway at Co-op City. The Contractor for the microtunnel project was expected to coordinate his work with these and other work contracts.

3.1 RIC/EDB-250

Under this contract, the thermal systems and electric distribution system servicing the entire north site was replaced. This included Buildings 1 through 25 as well as townhouses and shopping centers in the area. A limited amount of surface restoration was performed under this contract.

3.2 SIT-207

Under this contract, the domestic water, fire, gas, storm and sanitary sewer lines were replaced in the vicinity of Buildings 12, 13 and 14 of the north site. Surface restoration was performed in the area following this work.

3.3 SIT-205

Under this contract, the domestic water, fire, gas, storm and sanitary sewer lines were replaced on the eastside of the north site, except for Buildings 12, 13 & 14. Surface restoration was performed in the area following this work.

3.4 TES-220

Under this contract, the thermal and electric distribution system servicing the entire south site was replaced, except for work performed under contract TES-222. The domestic water, fire, gas, storm and sanitary sewer lines were replaced. This included Buildings 26 through 35, except Building 29, as well as townhouses in the area and Shopping Center 3. Surface restoration in the work area was performed under this contract.

3.5 TES-222

Under this contract, thermal and electric lines were run from the proximity of the Power Plant through the casing pipes installed under contract TES-221 and into the south site and Building 29. Surface restoration in the work areas will be performed under this contract.

3.6 ASB-252

Under this contract, asbestos abatement to support all construction work within the north was performed, except for asbestos abatement within Buildings 12, 13 & 14.

3.7 ASB-SOU

Under this contract, asbestos abatement to support all construction work within the south site was performed.

3.8 Garage Rehabilitation

Contracts were to be issued for the repair, renovation and/or replacement of garages. The work scope was unknown at the time. A garage consultant was engaged to do all investigation and engineering work to accomplish this work.

3.9 Emergency/Priority Repairs

Repair work to the existing piping and electrical systems servicing Co-op City. This work was ongoing throughout the life of the Housing Project Repair Program.

3.10 Power Plant Work

The Power Plant required various contracts to rehabilitate the facility. Some of the work entailed roof replacement and exterior brickwork to the existing facility.

4 MICROTUNNELING PROJECT

Microtunneling is a term used to describe methods of horizontal earth boring. It can be subdivided into two groups: slurry method and auger method. The tunnel boring machine is laser guided, and remotely controlled to permit precise line and grade installation to an accuracy of +/- 1-inch.

The low bidder was disqualified as he submitted an improper bid. He had tried to qualify his bid by tying it into his anticipated production rate. This was unacceptable. The second low bidder for the TES 221 Contract was Northeast Construction Inc. (NCI) of New Jersey. Their bid was accepted. They chose the Microtunneling method to install the casing pipe. The boring span of each tunnel was 230 feet. Work was started in January 1998 and completed in March 1998. One bidder chose the jacking method rather then microtunneling but his bid was not competitive.

Based on his interpretation of the boring data NCI chose the Harrenknecht AVN 1200 tunnel-boring machine (TBM) which utilizes the slurry method. As it was known that there were boulders in the fill, NCI was questioned about the capability of the TBM to bore through the anticipated soil layers. NCI gave verbal assurances of the same. The slurry method involves jacking pipe and simultaneous cutting of soil at the face of the machine by a cutting head. The tunnel was supported at the face by pressurized slurry. The spoil was removed hydraulically in the form of slurry. The conveying fluid is simultaneously used to counteract hydrostatic forces created by ground water pressure at the face of the as well as spoil removal.

4.1 Construction Procedure

1. A reinforced concrete jacking pit was constructed on the southwest side of HRP. The back wall was thicker then the other walls, as it had to absorb and distribute the thrust of the jacking reaction force into the surrounding ground. For the 54 inch RCP with a factor of safety of 3.2 the jacking force required was 583 tons.
2. The Contractor used a Grove RT635C cherry picker to place the jacking equipment, discharge pump in the pit and pipe into the pit.
3. An entrance ring with rubber seal was placed on the wall around each bore location to form a seal against groundwater and slurry penetration into the shaft.
4. Slurry settling tanks were placed at grade near the shaft. Piping between the tanks and the shafts make a circuit for the slurry. A charge pump is set near the slurry tanks.
5. A control cabin was set up at grade. The electrical equipment and operation board was set up with power and control cables. The main power supply was a diesel-operated generator.
6. Hydraulic hoses between the power pack and the jacking equipment were connected.
7. The Harrenknecht shield was lowered into the jacking pit and set on the guide rails.
8. Flexible hoses for slurry lines were connected to the TBM shield from the pit by-pass unit, the power and control cables were connected to the machine and the slurry tanks were filled with water. Bentonite was added via a separate pump.
9. The system was tested, adjusted and retested.
10. The hydraulic jacks were engaged to push the TBM close to the work face through the entrance rubber seal.
11. The pit by-pass unit and the slurry pumps were operated to circulate the slurry between the TBM shield and the slurry tank.
12. The cutter head of the TBM shield is rotated, and the jacks extended to push it forward and start the excavation.
13. During driving, the operator controls the jacking speed, the torque of the cutter head, the slurry flow rate, the slurry pressure at the work face, the earth pressure and the inclination of the TBM.
14. After the TBM shield is driven into the earth, the operation of the machine and slurry pumps are stopped and the jacks are retracted. The electric cables and slurry lines are disconnected in the jacking pit, in order to allow placing of the first 54 " RCP onto the rails.
15. The hydraulic jacks are extended to push the pipe forward until it fits to the tail of the TBM shield.
16. The pipe joint with the shield tail was checked for fit. The electric cables and slurry lines were reconnected to the TBM shield. The laser and target were checked.

17. The microtunneling operation was resumed with the restart of the slurry pumps, the pit by-pass unit operated, the cutter head rotated, and the hydraulic jacks extended.
18. This microtunneling process is repeated to jack each pipe, one after another.
19. While the pipe jacking operation is carried out, a lubricant is pumped continuously to the periphery of the concrete pipe to reduce jacking friction.
20. The receiving pit is constructed at the same time the jacking pit is. It is used to recover the TBM shield on completion of the drive.
21. When the shield was within one foot of the receiving pit the TBM was stopped and the timber sheeting removed as the jacks pushed the pipe ahead. The TBM was recovered and brought back to the jacking pit. The equipment was aligned to begin the next tunnel.
22. The TBM was checked and the shield cleaned. It was lowered into place and the electric cables and slurry hoses reattached.

4.2 Actual Experience

Microtunneling worked well on the first two tunnels but encountered difficulty on the third. The head could not advance beyond approximately the halfway point. This meant the head of the TBM was directly under the median of the Parkway. NCI stated that the TBM was not advancing and was obstructed by solid sheeting, oil tank, or excessive timber bulkhead.

To recover the TBM, NCI constructed a jacking pit on the opposite side of the Hutchinson River Parkway. They installed 1" thick, 84" diameter steel casing pipe. Excavation was done by hand from inside the steel casing pipe. Rail was placed and a cart was utilized so that the excavated material could be removed. The steel pipe was advanced using hydraulic jacks. No obstruction was found when the head of the TBM was reached.

Ears were welded on the TBM and cable attached. The TBM was hauled out as the 54 " RCCP was pushed in place from the opposite side. The TBM was dismantled and cleaned. The intake port was clogged. Small steel bolts nails and debris was found in the bottom of the cone section. It was hypothesized that the steel disturbed the turbulent flow required to make the operation work. Thereafter, hose, pipe, and pumps were cleaned and flushed more often. The next two tunnels were completed without any problems.

Monitoring points indicated no settlement to the Parkway. The next contract required that the Contractor install the carrier pipe from the power plant to the south site through the five, 54" RCCP tunnels under the Hutchinson River Parkway.

5 THE CLAIM

NCI filed a claim with the construction manager based on differing site conditions. Throughout the claim NCI stated they relied on the test boring information and they were part of the Contract documents. In fact, the borings were part of an appendix and marked "For information only, not part of contract documents".

As noted earlier, NCI stated at the pre-award meeting that they had no concern about any material other than encountering bedrock. Several interested parties attended the pre-award meeting. Everyone understood those words. The claim attempted to put a different twist on what was said. The owner, construction manager and engineer were unwilling to accept new interpretations.

It was pointed out to NCI that they were not the low bidders. The low bidder put forth-unacceptable conditions with the regard to extras in the event that certain materials were encountered. The owner paid a premium based on NCI's assurances. The award was recommended based on these assurances.

The claim expressed the contractor's view as to the responsibilities of the engineer. He ignored to mention his responsibility to read the entire contract documents and abide by the pre-award understanding.

The remainder of the claim dealt with the title of the specification, selection of the equipment, definition of microtunneling and their interpretation of the "intent" of the Contract. These were put forth to justify the claim.

During the bidding stage questions were asked about jacking, were responded to in an addenda which, by definition, were made an integral part of the contract. Language in the specifications permitted jacking and/or boring. At least one bidder intended to do jacking.

6 CONCLUSION

It was interesting that the contractor sought 100 % compensation, when retrospectively "solid sheeting, oil tank, or excessive timber" as anticipated by him did not obstruct the TBM when the TBM was not advancing. There was no mention at that time of bolts, nails or debris. Their opinion was that it had to be a total blockage until it was discovered there was none. Then it became a question of the validity of borings, miscellaneous small hardware, title of the specification, intent of the specification, etc. Accordingly, the claim was rejected.

REFERENCES

Urbahn/Seelye, 1993. Report on: The Routing of the Dual Temperature, High Temperature and 15KV Duct Bank Systems from MH #1 to the South Site
SSV&K, 1995. Draft Report of Subsurface Explorations
Hessing, H. 1997. Inspection report prepared for SSV&K.

Creative Systems in Structural and Construction Engineering, Singh (ed.) © 2001 Balkema, Rotterdam, ISBN 90 5809 161 9

The leakage-proof way of concrete

M.C.Chen & Y.L.Zhan
Department of Civil Engineering, Ching Yun Institute of Technology, Taiwan

W.C.Shih
Department of Chemical Engineering, National Chin Yi Institute of Technology, Taiwan

ABSTRACT: The article is going to make use the acrylic –emulsion filling-in principle. It is to aim at the crack prevention before the construction of the concrete roof and crack processing after construction. Based on tens of construction cases, we can prove that spraying warm acrylic-emulsion into the roof crack by means of gravity makes the crack of self-permeation concrete slab have immediate, evident leakage-proof effect to the roof. In addition, about the construction of slab, we can mix acrylic-emulsion with concrete in advance in order to replace part of mixed water. It will not influence work ability and reduce W/C, and reduce capillary. Based on the experiment, we know added to acrylic-emulsion, the concrete will not deteriorate compression force. So it can become watertight concrete of the roof. There is no need to use waterproof material.

1 INTRODUCTION

Concrete has the advantages of good compression, and cheapness. But by contrast, it has the disadvantages of bad tension (liable to brittle). Therefore, it is liable to leak. If it encounters great deformation (such as earthquake, or settlement), the leakage will be greater, and it is not easy to renovate. So, ordinary, roofs have the pavement of waterproof layer.

The crack can be concluded as below: (1). The bond force between aggregate and cement is not enough (insufficient). (2). Concrete is of not homogeneous material with weakness. (3). During hydration, cement and aggregate direr in thermal coefficient; therefore, there is tiny crack, in concrete. (4). The surfaces of aggregate and cement have negative charges, which repel each other to have crack. If the crack connect with capillary that will leakage. (5).If there is a crack in the concrete, the water and the CO_2 will combine into H_2CO_3 accelerates neutralization.

Therefore we can say concrete is the structure of sucking water as sponge.

According to the tightness theory of high-performance concrete (HPC) if decreases these tiny holes, and should be able to increase compression force and tension force, and anti-leakage effect; however, in doing so, the concrete will increase stiffness relatively, and will decrease the ability to resist deformation.

In order to attain the quality of HPC, the materials must be under strict control these conditions adverse to the construction of site. If the adverse conditions can be improved, the leakage of roof will be less. Therefore, the treatment of waterproof layer can be saved.

The idea of the research is to process the crack of concrete with softer water-base emulsion, without increasing the stiffness of concrete, and without strict construction.

2 PAPER-REVIEW

2.1 Same View

There is several construction ways to choose waterproof on construction's roof. But because of the aging of waterproof material, on concrete .the leakage problem of construction has been a problem, which cannot be effectively solved.

The correct concept is making the concrete to be tightness. And therefore, there is a good possibility of long-effect waterproof.

There are many papers in filling in polymer emulsion in concrete; the papers have the same conclusion filling in polymer emulsion may reduce sucking-water-rate, and permeation rate. Such as Zheng,W.H (1993), Zuo,X.J (1990), Song,M.S (1988), Qiu,E.T (1998), Cai,E.J (1995), Rao, V.V.et al (1998); Bloomfield, Thomas D. (1997);. Rebeiz, K.S & Lafayette Coll. The above can prove no matter which polymer emulsion material it is, it has the effect of filling in the crack of concrete because its particles are very small.

Under these conditions, the polymer concrete (PC) has durability, acid-alkaline poof, and environment change resistant. And the polymer emulsion material of long-chained structure can reduce the elastic coefficient of concrete, which has no controversial conclusion. Like Zheng,W.H (1993), Qiu,E.T (1998), Mantrala, Syam K.& Vipulanandan, C.; (1995).

In addition, there are some reports about better bond-force of reinforced steel because of the adhesion force of emulsion. Like Zheng,W.H (1993), Rebeiz, K.S. said.

2.2 Different View

As for tension force, bending force, and impact strength, there are positive. Such as Zuo,X.J (1990), Song,M.S (1988), Qiu,E.T (1998), Mantrala, Syam K.; Vipulanandan,C.; (1995). Some are negative views. Such as Zheng,W.H (1993), Cai,E.J (1995) Aniskevich, K.; Hristova, J.;(1999). These different conclusions make us believe choosing different makers of emulsion, different kinds of emulsion different water-cement ratio, different cements, or different pH values of emulsion.

Under so many conditions, it is not easy to directly compare real variations of strength.

3 FOUNDATION OF THEORY

3.1 Concept

Theoretically when the tiny hole of concrete decreases, the concrete can increase compression strength. But water-base polymer emulsion has many kinds; there are different particles in size under the same constituents and different producing process.

Generally, big particle emulsions are easy to mix with cement; small particle emulsions can increase adhesives and have better effect of permeability-proof. If there are no sufficient experiments, which particle is better is not easy to conclude. In addition, different makers have different commercial formulae in additives, such as antibacterial, antioxidants, and surfactants.

Based on the above reasons, different statements of papers have different results; especially emulsions usually include about 50% water. If the mixed water of concrete is not reduced, adding emulsion to concrete directly is equal to increasing mixed water, which decreases concrete strength contrastive.

According to the above results, to conjecture emulsion can decrease concrete strength is improper. Moreover, emulsion can make the workability of concrete better. Under controlling the workability of concrete, adding emulsion to concrete is equal to decreasing mixed water, which increases concrete strength naturally. According to the above results to conjecture emulsion can in crease concrete strength is improper, too.

Furthermore, the uses of emulsion are different, so there are cationic dispersions, anionic dispersions, and nonionic dispersions in emulsion forms. And acid and alkaline (pH values) can influence setting time of concrete, but in some papers, to infer the effect of emulsion with 28-day strength also seems improper.

On the whole, water-base polymer emulsion is a kind of long-chained structure; the diameter of every particle is very small, and we can control the size of its particle in the producing process. As said above, the particle of emulsion can fill in the size of small holes in concrete. It is believed that the particle of emulsion can enter get into the crack of hardening concrete more probably.

3.2 Fill In Principle

Therefore, in the first step of the article, making use of the characteristic of small-size particle of emulsion and easy-thermal deformation pours warm acrylic emulsion into the crack of concrete by gravity-infiltration. We can test whether to attain the waterproof effect of roof.

In the second step, we fix mixed water content, and add acrylic emulsion to concrete (the water include inside water of acrylic); then we extend the observation time to 90-day compression strength. The purpose is not to consider chain structure, and varieties of emulsion. Whether or not concrete is slowly retarding, we can confirm the variation of strength. Then we can provide the reference quantity of the addition of emulsion.

4 PROCEDURES FOR EXPERIMENTS

4.1 Repair for Roof Cracking

4.1.1 Testing for Optimal Temperature for Repairing
The creamy acrylic –emulsion at temperature 50° C to 70° C was found easily to penetrate to crack. From M.C.Chen & Y.L.Zhan, (1999), therefore the warm water at 60°C without and with a spoon of soap were used to test the permeability of roof. The results (Figure 1) indicated that the case of warm water with soap were able to penetrate the cracks.

Figure 1. Acrylic Resin Permeate to the Bottom(left)

Figure 2. Spray Warm Acrylic Resin on the Roof

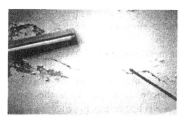

Figure 3. The Roof Moisture and Have Moss

Figure 4. Contrast With the After Construction

Figure 5. Acrylic Resin Include 50%Water

Figure 6. Mix Acrylic Resin

Figure 7. Comp. Stress Test

Figure 8. Outlook of Result

Table1. Case1 Sand and cement adding Acrylic
           Case2 Concrete adding Acrylic

Acrylic volume (%) not include inside water

A	B	C	D	E	F
0.0	1.3	2.6	3.9	5.2	6.5

case1.   comp. stress (kg/cm^2)

14d	180	175	165	160	155	150
28d	223	183	180	217	170	167
56d	227	238	230	203	202	195
90d	227	240	235	208	205	200

case2.   comp. stress(kg/cm^2)

28d	220	218	220	204	183	175
56d	228	230	280	250	225	201
90d	230	235	306	255	230	208

4.1.2. Field Testing

For a roof leaking with moisture and mold on it, without waterproof Cover on its surface, crack at its corners (Figure 2), with (moss) inside the surfaces, or, due to a bad quality of cement mixture, was improved by spraying once or twice of warm acrylic resin (Figure 3). And the roof leaking was stopping for one year after many heavy rains and typhoon (Figure 4).

143

4.2 Tests of Cement Mixed with Acrylic Resin Cream

4.2.1. In Order to Understand the Water Contain of the Creamy Acrylic

A creamy acrylic weighted 1 kg was spread over a plate and put into an over with 120°C for 24 hours (to avoid the vaporization and under the vaporized temperature of acrylic). The result showed that the creamy acrylic consist of 50% of water (Figure 5).

4.2.2. Way of Experiments

According to CNS 1010, R3032 a mixture of ordinary Portland cement and standard sand with weight ratio at 1 to 2.75 was prepared:

Case (a) W/C was kept in constant with increasing in content of acrylic (Figure 6) (total water contain is fixed). There were six groups at A to F of 18 specimens at 5cm x 5 cm x 5cm. The specimens were separated from their frames after 24 hours of setting. A half of the 18 specimens were moisture curing and the rest were dry curing, for a period of 14,28,56 and 90 days. Then the compression tests (Figure 7) were conducted for each group of specimens.

Case (b) W/C was kept in constant but in addition to the standard sand during mixture, the thick aggregate were used during adding acrylic there were six groups at A to F of 9 cylindrical specimens at 5cm x 10 cm. The specimens were separated from their frames after 24 hours of setting and in dry curing for a period of 28,56 and 90 days. Then the compression tests were conducted for each group of specimens.

All the test results were shown in Table 1 and (Figure 8). It indicated that in both cases of (a) and (b) the content of acrylic at certain level would increase compressive strength of specimens although different length of chain in acrylic may have different results, its effort could not be neglected. The variance in strengths in less acrylic might cause by poor mixing among their contents and too small amount of acrylic.

The Strength dropped with high content of acrylic was due to the molecular connected segment were replaced by acrylic. It is worth to mention that the specimens with moist curing and dry curing result in same strength in the test of (a) and (b). But it showed lower strength for specimens were water curing. It might cause due to the separation of frame before acrylic was dried, or, the acrylic could not get dry in water. However, it agreed to the results in other paper like Qiu, E.T. (1998). And we concluded that for the cement mixture with acrylic as additives do not needed moisture curing and

since it has moisture retention. Thus, it is suggested to use it in dry locations.

Although, all the acrylic may not have the same Contents and their length of molecular chain are not known. Our conclusion is that with adequate acrylic additives will increased the compressive strength of cement mixture but delay the setting time due to its pH value is less then concrete. (The concrete pH value is 12,but acrylic only 8)

5 CONCLUSION AND DISCUSSION

1. The field tests indicated that to improved for a roof leaking caused by small cracks, it needed to clean up its surface by even remove the waterproof cover. Then, spraying with acrylic in warm water (50° C to 60° C) to let acrylic penetrated to crack surfaces by its own weight during its cool down period. It stays inside the crack surfaces to improve roof leaking. Thus, the procedure is an easy, cheap and without the needy of professionalism. It causes less waste need less material.

2. The compressive strength of cement mixture may increase by the additive of acrylic. But the setting time for cement mixture will increase also.

3. Other kind of emulsion (PU) or deferent condition of electric charge may have the same conclusion but it needs test results to prove it.

4. Acrylic needs no additional water for its setting therefore it is different to ordinary chemical additives. Thus it should not relate to aging and it needs test results to prove too.

5. From the outlook result of compress test. We can judge the polymer concrete have good ductility than general concrete.

REFERENCES

Zheng,W.H (1993) Study on strength characteristics of Polymer Concrete ---National Taiwan Ocean University, The Institute of Harbor and Ocean Engineering, thesis.

Zuo,X.J (1990) Study on polymer later modified cement mortar ---National Taiwan Chiao Tung University.

Song,M.S (1988) The Effect of Addition SBR polymer Latex on the property of the ordinary Portland cement Motor—Report of National Science Council (Taiwan) (NSC76-0410- E151,01)

Qiu,E.T (1998) Study on Basic Properties of Polymer Concrete with oxide Anion National Chia-Tung University, The Institute of Applied Chemistry, thesis.

Cai,E.J (1995) Charatenstics of sea-sand-blended concrete Using Blast-furnace Slag and polymers --- National Cheng Kung University, The Institute of Civil Engineering, thesis.

Aniskevich, K & Hristova, J.; (Mar 21 1999); Prediction of creep of polymer concrete; Journal of Applied Polymer Science; v71; n12; p1949-1952

Rao, V.V.Lakshmi Kanta & Krishnamoorthy, S.; (Mar-Apr 1998) ; Influence of resin and microfiller proportions on strength, density, and setting shrinkage of polyester ; ACI Structural Journal ;   v95 ; n2 ; p153-162.

Gunasekaran, Muthian; (1997); Polymer concrete:A viable low-cost material for innovative power systems; proceedings of the IEEE International Conference on Properties and Applications of Dielectric Materials; v2 ; p 770-   773

Bloomfield, Thomas D.; (1997); Sewers and manholes with; Practical Applications Proceedings of the conference on Trenchless - Pipeline Projects; p 466-472

Palmese, G.R & Chawalwala, A.J.; (1996); Environmental durability of polymer concrete; International SAMPE Symposium and Exhibition (Proceedings); v41; n2; p 1642-1654

Rebeiz, K.S.[Corporate Source]Lafayette Coll; Strength properties of non reinforced and reinforced polymer concrete using recycled PET ; polymer Recycling ; v2 ; n2 ; p 133-139

Mantrala, Syam K.& Vipulanandan, C.; (Nov-Dec 1995); nondestructive valuation of   polyester polymer concrete; ACI Materials Journal; v92; n6; p 660-668

M.C.Chen & Y.L.Zhan, (1999) waterproof on roof be used by acrylic resin ---The Seventh East Asia-Pacific conference (p1538~p1544).

Related CNS Specifications of premixed Concrete-- National Central University

Quality Control of Waterproof Engineering of public Castrations (1999) ---Taiwan Construction Research Institute.

Creative Systems in Structural and Construction Engineering, Singh (ed.) © 2001 Balkema, Rotterdam, ISBN 90 5809 161 9

Differences between steel vs concrete bridge and transitway construction management

Kenneth H. Dunne

The Louis Berger Group Incorporated, Naples, Fla., USA

ABSTRACT: This paper addresses the Construction Engineering Management issues faced on four bridge and transitway structure projects in Miami and Naples, Florida, USA with respect to the necessary differences in building steel vs. post-tensioned concrete structures. Significant increases in construction productivity, and decreases in construction claims, can be achieved by more closely interfacing Construction Management into the Design Phase through an Inter Disciplinary Engineering Review. Planned implementation of these construction engineering issues into plans and specifications, can be achieved through the lessons learned in the four case studies presented herein. Quality Assurance, Constructability, and Alternative Contracting Methods are presented.

1 INTRODUCTION

This paper presents actual case histories of four primary structural projects, in two main categories: Bridges and Transitways for the Florida Department of Transportation (FDOT) and the Miami-Dade Transit Agency (MDTA). AASHTO Fracture Critical Steel Material (FCM) was specified for Transitway Construction at MDTA, and Pre-Stressed/Post-Tensioned (PS/PT) Florida Type VI Bulb Tee Beams for Concrete Bridge construction for the FDOT. (FDOT 1991)

Delivering the projects to the Clients on-time and at a fixed cost on major Bridge and Transitway infrastructure projects is the primary focus of the paper, and represents the growing trend in the industry to consolidate lines of responsibility on multi-million dollar infrastructure projects around the world. Design/Build project delivery systems are a growing trend around the world, and many incorporate the issues presented in the case studies. (FTBA 1999)

2 QUALITY ASSURANCE

In January 1991, MDTA started a successful transitway project for the $30 Million Omni Metromover 1.5 Mile Extension to the Downtown Miami Peoplemover, finishing within time and budget considerations with structural engineering and construction in June 1993. Quality Assurance during the AASHTO FCM fabrication process was critical in maintaining tolerances and specifications in accordance with the FDOT Specifications for Road and Bridge Construction, 1991 Edition, the American Welding Society Bridge Welding Code, and AEG/Westinghouse System Supplier Specifications. (Dunne 1993)

2.1 *Constructability and QA Review*

In July 1999 the MDTA approved an $8 Million Sub-Contract for Transitway Steel Fabrication on the $80 Million Palmetto Extension Project to Metrorail in Miami for 0.6 miles of elevated Transitway fabricated to AASHTO FCM requirements, AWS welding Code, and FDOT Specifications. Under a very tight time frame for erection of structural steel tub girders, manpower resources were directly affecting the ability to prepare detailed shop drawings before fabricating the FCM tub girders could begin.

A delay in the Owner's property acquisition for a portion of the project caused a stop work order to be issued after the contract had been awarded (MDTA

1998). This resulted in a six month delay in starting steel fabrication, and created a heavy backlog of steel fabrication work at the shop when the Notice to Proceed was finally issued. The fabrication sub-contractor requested an alternate detail for continuous welding and flange stiffener attachments to the webs, and intermediate diaphragms, which is the focus of the Constructability issue, to speed up fabrication and deliver the critical steel box girders on-time.

In June 1993 the FDOT began the construction of the $38 Million MacArthur Causeway Bridges in Miami, Florida over Biscayne Bay, adjacent to the Port of Miami. The bridges' foundation and superstructures were designed and constructed to meet the requirements of additional lateral loads from Ship Impact, and long-term scour effects on the drilled shaft bridge foundations within the channel from Biscayne Bay to the Atlantic Ocean, adjacent to the Port's turning basin for returning vessels.

The MacArthur Causeway Bridges are also designed to accept future MDTA Metrorail Commuter Rail Cars to pass over the bridge in the lateral separation between individual, ½ mile long, post-tensioned concrete bridges between Miami and Miami Beach on I-395. The project was successfully completed in December 1996.

2.2 Alternative Contracting Methods and QA

In October 1998 the FDOT initiated Alternative Contracting methods in their bridge and road construction projects when time is of the essence for construction completion due to heavy congestion along some of its primary arterials, in an effort to reduce congestion and improve levels of service.

On the $7 Million State Road 951 Project in Naples, the primary method utilized was the Incentive/Disincentive Contract Special Provision. This provision rewards the Contractor with $2,500/calendar day for completing early, and imposes the same amount per day as a Disincentive Cost for completing the project late. Four bridges utilizing pre-stressed concrete Type IV beams were used to span McIlvane Bay and McIlvane Creek in very sensitive, environmental wetlands.

Issues addressed here focus on the Contractor/Subcontractor relationship and problems that can develop when the Incentive bonus is not shared with subcontractor performing the bridge construction. The project was successfully completed

in March 2000, four months ahead of schedule, and several construction claims amounting to approximately $1 Million have been filed. The General Contractor completed the project 84 days ahead of schedule, but did not request that the $210,000 Incentive Bonus be paid since the contract documents required releases of all claims in order for the bonus to be awarded.

3 QA/QC REVIEW

In the steel fabrication processes, between Fracture Critical Material purchase, cutting, fit-up and tack welding and final weld-out, Requests for Information (RFI) and coordinate geometry must be addressed upstream (Karash 1992) of the detailing process needed for material orders for thousands of tons of specialized rolling patterns. This is specifically due to the spiral, vertical and horizontal curves of the guideway structure, signaling, and to coordinate locations of Automatic Train Operations (ATO) and Power Distribution System (PDS) (AEG/Westinghouse 1990) inserts and penetration openings, at the earliest point after the Notice to Proceed is issued. These elements are crucial for the stopping distances at Stations, and communications installation for the Automated Peoplemover Vehicles traversing the 1.5 mile alignment.

The Interdiscipilinary Review, between QA and Surveying to verify coordinates was very effective during this project, and is very significant in consideration of the long lead time for FCM Material. Over 6,000 Tons of Guideway Steel were fabricated, delivered and erected on the project site in Downtown Miami. Locating ATO and PDS inserts in the guideway webs and attaching to Vehicle Guidebeams mounted on the Guideway bridge steel can be adjusted by cutting and re-welding in areas of the Guidebeam, and re-attaching certain components within the permissible areas of the webs and flanges pursuant to AASHTO guidelines.

For concrete bridge construction, the pre-stress and casting operations QA/QC program implemented at the yard is critical in attaining required wire stresses and concrete bonding for the beams, to prevent spalling and long-term deterioration of reinforcement due to seawater exposure in an extremely aggressive coastal environment. These concrete beams are less forgiving in "cut-to-suit" field conditions as with steel, and erection and fabrication tolerances as they accumulate along normal bridge vertical and horizontal curves, must be closely addressed for proper P/T duct alignment at the diaphragms.

As the use of PS/PT concrete tangent beams can be utilized in relatively long stretches of Peoplemover Systems in airports, the QA/QC of ATO/PDS locations, and rail or guidebeam attachments to structural elements, must continue to be "upstream" of detailing and casting operations.

Pre-stress forces and concrete cover must be inspected before each casting process. End-bearing stresses at the jacking points above the bearing plate, necessary camber calculations considering LL and construction deflections and DL storage deflections, P/T duct alignment between continuous beams considering all deflections, must be addressed by QA/QC, Survey and the Resident Engineer during the Interdisciplinary Review process.

3.1 Constructability Review

Within the technical specifications for the MDTA Omni Metromover transitway, the Blocking and Bolt-Up of finished sections in the fabricator's yard required support at every 5' on-center under the bottom flanges (MDTA 1991) This was intended to help account for the spiral and vertical curves to assure that no problems would be encountered with erection in the field. This, however, is not the representative deflected shape of the transitway structure under construction loads and live transit loads.

Modifications were required at the bolted field splices, and increasing crane resources for a multiple crane pick, reduced the deflection-displacement of end conditions. Typical span lengths were between 90-108' (295-354M) in length.

Of primary interest in the constructability of steel structures, is the welding of fracture critical members and the many components that are attached (AASHTO/AWS D1.5-88) Continuous submerged-arc welding is the most common and efficient method of producing thousands of tons of FCM material, with the proper consideration to the modification and scheduling of attaching web stiffener plates during the fabrication process.

In delivering the project to the Client at the best cost on the fastest schedule, the project management team must consider the detailing of a stiffener plate on design plans that will prohibit the continuous submerged-arc welding process described above for approximately 4,500 tons of FCM material on the Palmetto Extension of Metrorail.

An alternate stiffener design, previously approved and endorsed by FDOT on bridge structures and taught to various Consultants and FDOT Engineers at the annual Design Conference in Orlando, Florida on many occasions, was questioned by the Owner during the initial meeting with the steel fabricator. (FDOT 1998)

The Owner accelerated the construction schedule by four calendar months, and it did finally consider the productivity factor increase in reduced construction costs, and accelerated construction schedule to finish on-time, despite the four-month delay in acquiring property to construct the project.

This issue is one of many which, after due consideration, leads to the need for more Interdisciplinary Reviews with Construction, and Design/Build infrastructure projects where the whole process is consolidated into one responsible entity.

PS/PT concrete girder production and erection operations for the FDOT MacArthur Causeway Bridges presented similar, but different problems to be solved in the field during construction. The first 7 beams delivered to the jobsite were found to be 9.5" (24cm) too long upon inspection before erection. These beams were detailed 9.5" (24 cm) too long in the plans, from bearing plate to bearing plate, and could not be "cut-to-suit" in the field due to the enormous build-up of end-stresses, and the actual physical location of the steel bearing plates at each end of the girders, beginning at only 3.5" (9 cm) from the end of the concrete girder. These girders were rejected at the jobsite by the Resident Engineer. The Contractor was paid over $110,000 for additional work and time to cast new girders.

Some of the most difficult and time-consuming operations occurred during the post-tensioning operations of shooting multiple tendons through narrow steel duct in the classic drape pattern at mis-aligned ducts between adjacent beams which are 145' (475M) in length. Specifications permit a 0.25" (0.63cm)(+ or -) vertically and horizontally, at each duct opening at both ends of each beam.

P/T operations were typically 4-span units, 580' (1900M) with 6 intermediate duct interfaces. This leads to an accumulation of construction tolerances in any given direction, and presents problems with shooting the remaining several tendons in a 9-strand configuration, due to the limited room available and random placement of tenons over the entire length.

Technical specifications prohibit the re-use of steel tendon if the first insertion is unsuccessful for any reason (FDOT 1991). This is a very expensive item and the sub-contractors are keenly aware of the need for proper P/T duct alignment. There are always disagreements between the General Contractor and the P/T sub-contractor, on not only this issue, but also the need for additional QA/QC efforts in the tendon jacking stresses applied to the tendons, and the subsequent grouting of the P/T ducts to eliminate the water-intrusion that can have serious long-term negative effects on the post-tensioning stresses in the tendons.

3.2 *Alternative Contracting Methods Review*

Many old ideas in the construction industry have emerged as "new ideas" with Bonuses, Incentive/Disincentive, A+B Bidding, and a host of others. In days past, the Engineer and Contractor would discuss the work to be performed and the time and material it would take to accomplish the task at hand. When problems arose, as they always do, a fair price and time were determined without the presence of litigation, generally.

Alternative contracting methods encourage the quick completion of major transportation projects to limit motorist delays and inconvenience, and also, to mitigate the potential business loss from diverting traffic, and major traffic delays encountered when replacing bridges when additional Right-of-Way land acquisition is not possible.

When a potential Incentive bonus of $2,500/day can be earned for the early completion of a project, the Contractor/Sub-contractor relationship plays an increasingly important role in Completion vs. Bonus $ Earned.

If the G.C. will not share the incentive money with the sub-contractor, then what incentive does the bridge sub-contractor have to complete the bridge phase early?

This is exactly the case with the FDOT SR-951 from Naples to Marco Island, when the roadway contractor was the prime and the bridge sub-contractor encountered underwater utility line crossing interferences with piling. The G.C. was also the Utility Contractor for a private water utility within the FDOT Right-of-Way, called "By Others" in the FDOT plans,

under a separate contract. (Florida Water Services 1998) This water utility was constructing over 10 miles of various size water lines adjacent to the roadway construction. The FDOT roadway contractor was now also the Water Utility contractor which experienced delays with installing new vs. old utility lines.

This case is used to illustrate the Critical Path Method (CPM) schedule differences between the Prime Contractor and the various sub-contractors, in efficiently completing their respective work tasks within the overall FDOT project schedule, but which are driven by the different sub-CPM's required the Prime of all sub-contractors. Even though the G.C.'s CPM schedule submitted and approved by the Owner indicated an on-time completion of the pile driving activity interfering with the waterline, the G.C. has litigation pending against the bridge sub-contractor for loss of incentive money.

The Bonus or Incentive is tied to the CPM schedule and the date of Final Acceptance by FDOT, and can take the form of a drop-dead bonus, either finish all work by a certain date or no bonus at all, or the incentive per day which allows for some amount of bonus money for each day completed ahead of schedule.

When the "loss of bonus" can be tied to the sub-contractors' work, the Prime Contractor can, and usually will, pursue the lost bonus and lost profit from the sub-contractors and the Owner.

The bonus and incentive clauses in Florida are written with a specific requirement that "only in the event of a declared state of emergency by the Governor, or a Hurricane" can lead to an increase in bonus and incentive time calculations. It specifically states that "all changes, conflicts and interferences are anticipated by the Contractor and are not compensable." (FDOT 1998)

4 UPSTREAM QUALITY ASSURANCE

As we look at the cases presented above, we must begin to develop a strategy to implement the lessons learned, and upstream quality assurance (Karash 1992). This simply means that the construction engineering management resources required for the project, must be applied much earlier, and in deeper concentration into the planning and design phases. This serves to avoid quality control problems in casting and steel fabrication, and minimize construction field problems.

Shop drawings' detailing, in placing both post-tensioning ducts in concrete bridge girders, and ATO/PDS inserts in FCM transitway steel structures, must be closely coordinated in the interdisciplinary engineering review process with surveying and construction management personnel early in the project planning and development phases.

These early phases are critical in the appropriation of millions dollars of funding for construction projects from Federal and State Transportation Agencies, the World Bank, International Development Bank, and other world-wide funding agencies and lending institutions.

The Engineer's Estimate of construction costs is among one of the most crucial elements in providing these funding entities with complete, and accurate capital cost expenditure estimates for bonding and insurance.

QA details, such as the 5" on-center blocking requirements in the technical special provisions for the Omni Metromover Extension, and the tub girder stiffener detail for Metrorail governing continuous arc-welding, can have a tremendous financial impact on the funding institution's future obligations for the project. This additional obligation comes in the form of construction claims and extended project durations to solve problems in the field.

4.1 Constructability Review

The accumulation of construction tolerances, as discussed with the P/T duct alignment above, also becomes critical in assessing construction loads when determining deflections to a higher degree than a given uniform load per unit. This is very clear in the 5' blocking of Metromover steel girders discussed above.

As part of the Interdisciplinary Review process, constructability and biddability reviews are becoming increasingly valuable to Owner Agencies and lending institutions in reducing costs and time to complete construction.

Experienced construction engineers must be involved early and deeply in the planning and design phases to address the optimum methods that a reasonable contractor would use to accomplish a given task within the CPM schedule. This includes heavy equipment access, material delivery lead times, and dewatering foundation caissons, as just a few examples.

Value Engineering Cost Proposals (VECP) which identify alternative construction methods at a reduced cost to the Owner Agencies and increased monetary reward to the innovative contractor are frequently used in FDOT construction projects for the past ten years. This has led to a steadily increasing trend towards

Design/Build project delivery systems in the last several years, recognizing the fact that constructability and biddability are valuable assets for the Owner to explore.

4.2 Alternative Contracting Methods Review

It is extremely important to determine the optimum alternative contracting method for each specific project during the initial design phase. If the time to complete estimate is unrealistic, then a bonus or incentive clause will not be effective because it will lead to claims regarding lost bonus money.

If the time to complete is too long, then obviously incentive money will have been wasted. The constructability review is very important in this determination, and the total time to complete may have to be driven by incentives to account for emergency repairs, evacuation routes from Hurricanes and Tropical Storms, Earthquakes, and other natural disasters such as floods and tornados.

This is a significant consideration in major, urban areas of the world where transportation gridlock necessitates the need for new infrastructure. However, it is also a major concern in the rural areas where economic benefit is gained for the region's population in developing countries where new trade infrastructure is placed.

The ability to effectively address the speed with which transportation infrastructure is constructed, whether steel or concrete materials are used, must be addressed in the Interdisciplinary Review process by selecting the best alternative contracting method to be utilized on a given project.

5 CONCLUSION

All proven, documented design and construction techniques within the bridge-transitway structures industry must be pursued towards the goal of providing the Owner with the fastest, most responsible and economically feasible project that can be constructed within a given time frame. All of these costs are measured against required results, and the best approach should be pursued with all vigorous energy by the team.

REFERENCES

Florida Department of Transportation (FDOT) 1991, *Standard Specifications for Road and Bridge Construction.*

Florida Transportation Builders Association, 1999 Annual Meeting.

Dunne, K. H. 1993, *Quality Control: Bridge Steel Fabrication and Erection,* Fourth International Conference on Automated Peoplemovers, Las Colinas, Texas.

Miami-Dade Transit Agency, Miami, Florida 1998

Karash, K. H. 1992, *USDOT Federal Transit Administration QA/QC Guidelines.*

MDTA/AEG/Westinghouse 1990 Technical Specifications.

MDTA 1991 Technical Specifications, 663172-05120.

AASHTO/AWS 1988 Bridge Welding Code D 1.5.

FDOT Annual Design Conference 1998.

FDOT Supplemental Technical Specifications

FDOT Special Provisions 03030-3509, 1998

Creative Systems in Structural and Construction Engineering, Singh (ed.) © 2001 Balkema, Rotterdam, ISBN 90 5809 161 9

A review on common technology employed for the construction of buildings in Hong Kong

Raymond W. M. Wong

Division of Building Science and Technology, City University of Hong Kong, People's Republic of China

ABSTRACT: When the hilly and extremely congested environment is taken into account, Hong Kong may be regarded as one of the most difficult places in the world for construction profession. Unimaginable complex site environment such as to construct a 40-storey building with a 4-level basement close to a busy mass transit railway tunnel in the center of the city; or to form a vertical cut along a steep slope in order to construct a semi-basement type podium with a series of building blocks above it, are just some typical examples. Beside all these complications, most jobs are done in super-fast track manner to minimize the tying up of huge capital due to extremely high land cost. All these are nightmare situations for builders. Yet, construction professionals in Hong Kong still manage to accomplish such difficult task that satisfy the requirements as set by most stake-holders.

1 INTRODUCTION

As an international megacity, Hong Kong is world-famous for her crowded metro environment with out-portioned hilly relief. Within the 1050 sq km territory, there are about 240 outlining islands occupying one-third of her total area, the remaining less than 700 sq km of land has to accommodate her 6.8 million populations. This figure includes a series of mountain ranges that stretch all over the entire territory of Hong Kong.

In order to supply the required land for accommodating the huge population, and to provide ample infrastructure and community facilities to substantiate an acceptable standard of living, commercial operations and other necessary developments, many critical locations which may be unsuitable for development to most international yardsticks, are built with very large sized and high-rise buildings. The following situations are some of these examples.
a) To build in close proximity to very steep, or sometimes quite unstable slopes.
b) To build very tall building, often with deep basement, in extremely congested urban environment.
c) To build in close proximity, or sometimes even within, very sensitive and congested underground facilities like the Mass Transit Railway subways, surcharged areas of building foundations, or layers of large-sized drains, gas and water pipes, and culverts etc.

d) To build in close proximity of very large and complex traffic interchanges or busy transportation facilities.
e) To build in newly reclaimed or previous dump-filled areas.

Owing to the long period of experience working with such kind of harsh environment, practitioners in the construction field in Hong Kong have adopted their own practice to build, making full utilization of the technology and resources available locally. This paper tries to summary the local practices, common methods and choices employed in Hong Kong to construct high-rise buildings.

2 COMMON STRUCTURAL FORMS FOR HIGH-RISE BUILDINGS IN HONG KONG

Structural forms adopted to construct high-rise buildings in Hong Kong are in fact quite limited owing to some special local reasons, such as:
a) Regulations governing the land control, or planning and design of various kinds of buildings,
b) scale of development,
c) design and construction practices due to long years of tradition and custom,
d) efficient use of local labour and contractor expertise,
e) marketing trends to fulfil the needs of building consumers and the maximization of profit of the developers etc.,

f) design fashions and sales strategies.

As a summary, the popular structural forms employed for the construction of high-rise buildings in Hong Kong can be highlighted as follow:

a) In-situ Reinforced Concrete Frame – this is the most popularly used system in Hong Kong. Usual spans range from 4m to 10m depending on design and use of buildings. Recently, spans above 20m can also be seen but majority of which are tensioned in order to minimize the size of beams. Floor slab supported by beams is a major part of the horizontal stiffening member in framed structure. However, flat slab structure, often post-tensioned, is growing in popularity especially in commercial buildings for the benefit of providing a clear ceiling void to accommodate services.

b) Load Bearing or Shear Wall Structure – load bearing walls in this case are used to replace columns. The use of beams is often reduced due to the avoidance of complicated formwork detail between the junction of beams and wall. Usual spans range from 4m to 8m. Due to the limitation in the layout arrangement confined by the load bearing walls, buildings using this structural form are commonly limited to residential or apartment buildings. Panel type or large-sized shutter forms for walls and table forms for slab are often used for the highly repetitive nature of the building. However, detailing arrangement between walls and slabs, beams or staircases still imposes certain complexity to the design and erection of formwork.

c) In-situ RC Core Wall with RC External Frame – this structural form is used mainly in commercial buildings. The core wall which accommodates the lift shafts, staircases, toilet facilities and other building services provisions, is usually square or rectangular in section in order to make the forming process more efficient. Thickness of the wall ranges from 0.6m to 2.0m depending on height of building and loading requirements. Due to the lack of complicated architectural feature and highly repetitive nature, large-sized panel shutter forms, sometimes mechanically self-lifting, are used in the forming of the core structure. The external frame, in majority, is formed by the use of more traditional manually operated panel-type formwork, constructed in the same phase or separated phases with the core wall structure. Span ranging from 10m to 12m is most common.

d) In-situ RC Core Wall with Structural Steel External Frame – similar to the above form but with the external frame constructed in structural steel. Almost without exception, the core wall in this case is constructed using self-lifting formwork system in an advanced phase, with the connection of the steel frame follows. Record of 3 days per floor cycle can be achieved with floor area up to 2000 sq m. Effective spans range from 12m to 15m. However, the incorporation of an anchor frame in the core wall, especially in floors where bracing members or the out-rigger frames are located, often complicates the construction process and retards the overall progress.

e) Mega-Structure using pure Structural Steel Frame – this kind of structure is not too common for it is not rigid enough to take up strong wind load under typhoon situation which occurs in Hong Kong during summer period. To strengthen the structure, often very complicated stiffening members in the forms of transfer trusses, sectional floor plates, out-riggers or heavy-sectioned bracing members, are required to add into the steel frame. This makes the design, fabrication, handling and erection become very difficult. However, certain forms of composite design such as the use of composite floor with reinforced concrete topping or concrete filled columns, are often used to increase the efficacy of the building. Recent high-rise examples in Hong Kong are the 70-storey China Bank Building and the 80-storey "The Center", which were completed in 1990 and 1998 respectively.

f) Semi-fabricated Structure using Precast Concrete Components – it is not too practical for high-rise buildings to be constructed using totally precast methods for the relative flexible in nature especially under typhoon situations in Hong Kong. However, as a means of industrialization to minimize the intensive use of expensive labours, semi-fabricated structure using a certain number of precast concrete components are growing its popularity. In this case, the main structural members, such as for the core walls, columns and main beams, are cast insitu, often making use of some kinds of patented metal forms. While the secondary members like the flight of stairs, secondary beams, short span tie beams, slabs (or semi-slabs) and external façades etc., are constructed using precast methods. In order to improve the rigidity of the joints, most precast elements are placed in position with build-in link bars and cast at the same time together with the main elements. Post-tensioning is sometimes employed to increase the overall performance of the structure.

3 FOUNDATION SYSTEMS AND METHODS

In addition to the usual high-rise and heavy nature, buildings in Hong Kong are facing very severe typhoon environment which imposes very complication loading effect to the foundations of buildings. Wind speed above 200km/hr is not uncommon during the period of tropical typhoon. Furthermore, some other unique features such as the existence of shallow-laying hard rock over the territory, subsoil with large amount of boulders of volcanicous nature, or work sites are very close to developed areas with sensitive and congested underground or above-ground structure, or requiring to work under extremely tight schedule with complicated phasing arrangements, are quite usual under Hong Kong's experience.

The following foundation methods are thus adopted after long years of practice and proven to be quite effective under local situations.

a) Steel H-pile – standard universal sections are used as pile with load taking up by end-bearing and skin friction. Driving operation and equipment requirement is relatively simple except that the noise and vibration so created is highly restricted especially under urban environment. In case of boulders exist, pre-drilling can be used before the insertion of the pile. This method is economical and effective for use in buildings sometimes up to 30 storeys or above.

b) Precast concrete pile – precast pile can be of square or circular sections. A prestressed hollow-section circular pile, modulated to 10m or 12m in length, is becoming more common in Hong Kong due to reasons of cost, convenience and reliability. However, problems such as occasion failures of concrete during the driving process, smoothness of pile surface reducing skin friction, as well as noise and vibration, are the major drawbacks.

c) Mini-pile or Pipe-pile – by the use of compact drilling machines, steel pipes of size 150mm to 250mm are inserted into ground and grouted as pile. Various loading requirements can be obtained by controlling the numbers of piles used, or by additional reinforcing bars inserted into the pipe section before grouting. Due to the small diameter of pile, drilling can be done fairly easily and cause limited disturbance to neighborhood. This kind of foundation is suitable for use in congested areas with restricted working space or headroom. Besides, the pile can be tensioned and provide very good resistance to overturning due to wind load.

d) In-situ Concrete Pile, medium sized – generally refers to pile sizes ranging from 300mm to 900mm. The use of drilling rig of appropriate capacity does the forming of the bore conveniently. Drilling process can be facilitated by the use of steel casing or bentonite drilling fluid. Due to the rapid development in a wide range of highly effective mechanical drilling equipment, this kind of foundation choices is becoming quite popular in the construction of medium to high-rise buildings in Hong Kong.

e) Large Diameter Concrete Bore Pile – boring process can be done manually or mechanically. In general, pile sizes ranging from 1m to 3m can be formed by the use of mechanical-dug methods. While pile of sizes 3m and above, manual-dug methods are to be employed. However, starting from the beginning of 1998, manual-dug method has been banned due to high accident rate in work, unless special approval is obtained satisfying certain safety requirements. Mechanical means for the forming of the bore can be achieved by grab-and-chisel methods or reverse circulation drilling methods. Both methods require the use of steel casing as a means to stabilize the bore during excavation. Sometimes, super large-sized pile up to 6m to 8m can be constructed. In this case a cofferdam formed by sheet piles, soldier piles or in-situ concrete piles is to be provided for soil retaining purpose.

4 BASEMENT AND SUBSTRUCTURE

In addition to the extra building area obtained by the provision of a deep basement, substructure of this type can provide very good buoyancy effect to relieve the dead load of superstructure and to counter-balance the uplifting effect due to wind load acting on the building surfaces.

In Hong Kong, large basements can cover a site of more than 20,000 sq m in size, and 20m down into ground. A recent project case, with features briefly summarized below, can best illustrate the extremely complex construction environments that such a job may face.

Name of building – Festival Walk
Area of site – 210,000 sq m (about 90m x 240m)
Use of building – shopping mall, entertainment center and office spaces
Average depth of basement structure – 20m for full area of site
Neighborhood facilities/environment – a 3-lanes 2 ways motorway on one side, single track 2 ways railway line on the other, 2 single track tunnel tubes for mass transit railway cutting across in the middle of site, one 18m-deep pedestrian and ventilation shaft touching one side of the building boundary, inhabited residential and institutional buildings located on both ends, very limited access provisions, sub-soil obtaining about 30% of slightly decomposed volcanic hard rock.

One major element to facilitate the excavation and construction of basement is the provision of a cut-off wall. There are a wide variety of choices for cut-off walling design depending on the scale and depth of basement, period of work, neighborhood environment, mechanical equipment available, methods to construct the basement or cost planning requirements etc. Below is some common cut-off walling systems being used in Hong Kong.

a) steel sheet pile wall – most efficient for excavation up to 8m to 10m deep. However, complicated horizontal support in the form of strut or bracing frame may be required which restrict onward excavating operation. It is not suitable for use in areas with large amount of scattered boulders.

b) Soldier pile – similar to the principle of sheet pile wall but H-piles are inserted into ground at intervals with lagging structures to seal up soil surfaces between the piles. This is particularly suitable in areas with boulders for pre-drilling can be carried out in a fairly convenient manner. The H-piles can also be inserted into concrete bored pile to produce a 2-stage retaining design.

c) Insitu concrete pile wall – concrete bored piles ranging from 0.9m to 1.5m are often used. The piles can be arranged in secant, contiguous or at spaced interval. Effective retaining depth can be up to 12m or above and is suitable for use in more sensitive ground due to basically vibration-free drilling operations.

d) Pipe-pile wall – similar to in-situ concrete pile wall but smaller pipe piles are used. This system is most effective for use in very delicate environment where disturbance can be kept to minimum, or for site which is small in size where large-sized machines are inconvenient to operate.

e) Diaphragm wall – panels of trench wall are first formed by grab and chisel or by semi-automatic trench cutting machines. Wall thickness ranges form 0.9m to 1.2m usually. The forming of the diaphragm wall unavoidably requires significant amount of plant facilities and may not be economical for job of smaller scale. However, for larger site and deeper retaining requirement, the cost effectiveness increases due to the elimination of too complicated shoring supports during the onward excavation process. In addition, the diaphragm wall can be used as the permanent wall to save up the fixing of formwork within confined space inside the excavated pit.

In addition to the use of an appropriate cut-off wall to facilitate the excavation, grouting is often introduced as a means of subsoil strengthening and to improve water-tightness of ground. Besides, as the excavation proceeds, horizontal support using strut frame or other shoring and bracing systems are to be installed in order to counteract the lateral pressure due to the newly exposed cut. Sometimes when situation allows, ground anchors can be used as lateral support to eliminate the strut system in order to gain more working space inside the work pit. Finally, appropriate dewatering provisions to suit specific geological or neighbouring environment should be incorporated to keep the pit safe and free from the entering of ground water.

Construction of basement can be done by traditional bottom-up method, or on the contrary, using the top-down method. The sequence of works for the two methods can be summarized as follow:

Using Bottom-up method

a) Construct the cut-off wall.
b) Start excavation within the basement parameter.
c) Erect lateral support, layer by layer, as excavation proceeds until the required depth is reached.
d) When reaching the formation level, construct the foundation rafts, pile caps or ground beams.
e) Construct the basement slab and other internal structure starting from the lowest basement level. The works usually done in carefully scheduled sections to avoid the disturbance of the strut members.
f) Repeat the basement works until it reaches the ground level.
g) Release and dismantle the strut members at

suitable stages as the basement structure is completed.

Using Top-down method

a) Construct the appropriate cut-off walling system, usually, diaphragm wall is used for this method is more effective in handling project of larger scale.

b) Erect temporary columns, usually at the same position of the permanent columns and in the form of steel stanchions, as support to the basement structure that is construct from the top level downward.

c) Construct the first basement floor slab, usually starts from the ground level, which will be used also as the horizontal support to the cut-off wall as excavation proceeds.

d) Excavate downward and construct the second level of basement slab similarly. Erect intermediate temporary shoring support where required.

e) Repeat the excavation and basement slab construction until the required depth is reached.

f) Construct the foundation caps, rafts, ground beams or sub-soil drains as required.

g) Construct other internal structure where required.

h) Encase the temporary columns in concrete to transfer them into permanent columns.

The employing of bottom-up method has the benefit of requiring fewer plant facilities to operate and thus it is more suitable for use in smaller site where the depth of cut is relatively shallow. However, disadvantages will appear as the size of site getting bigger, basement deeper, or project under tighter schedule. Under this situation, very complicated horizontal support is to be erect. This will increase the working time and cost of the project, and at the same time limited the working space within the basement pit. Besides, the superstructure can only be commenced until the completion of the basement structure, unlike employing the top-down method that the superstructure and the basement can be carried out at the same time making use of the first basement slab as a separating plate. Hence, basement projects of larger scale at recent years are, almost without exception, all constructed using top-down method.

However, no matter which method is used to construct the basement, some very fundamental considerations such as the detail arrangement for the erection or later dismantling of the temporary support; sub-division of the basement structure to workable sections and the provision of construction joints; the co-relationship of the phased sections to cope with the access of labour, plant and materials; arrangement to remove spoil; or how basement work is to be merged to match with the overall progress of the superstructure, are some very critical factors in the planning and execution of a successful basement project.

5   CONSTRUCTION OF SUPERSTRUCTURE

The construction of the superstructure involved 3 major activities, that is, the provision and erection of a suitable formwork system for work, steel fixing and the placing of concrete. And of course, an efficient site layout with the required plant facilities properly set-up is a "must" condition.

There are a wide variety of formwork choices suitable for use in high-rise construction. However, under Hong Kong practice, the more traditional manually operated timber form is still the most frequently used system due to its flexibility to meet difficult shapes, non-standard layout requirement and less demand on other plant facilities or logistic supports. Though obvious drawbacks can easily be observed, such as more labour intensive, slower speed in work, environmental unfriendly or unsatisfactory concrete quality may be resulted, it is still readily received in many projects of smaller scale for its simplicity and independence in the planning and coordination of works.

For projects of larger scale with tighter schedule and higher performance requirements, the following formwork systems are often employed:

a) large-sized panel shutter, often in mild steel or coated timber/metal combination.

b) Manual operable alumimium panel forms, various options to use as wall or slab formwork.

c) Table form and flying form for slab construction.

d) Self-lifting formwork system such as the slip-form, jump-form and climb-form, often under special patented design.

e) Other patented formwork systems such as the SGB, RMD or PERI systems etc.

The practice in fabricating and fixing of steel reinforcement in Hong Kong is still very traditional. The current way, in majority, is to cut and bend the steel bars on site, transport the bars to the work spots by the use of tower crane, then fix the bars into position. The use of prefabricated components in workshop and transport to spot for erection is uncommon. This can be explained by the usual non-standardization in design, using of slim structural elements and congested reinforcement condition in order to minimize the size of structure.

Concrete with strength ranging from 20 to 35 N/mm^2 (Grade 20 to 35) is used most commonly in construction due to easier in performance control under the built environment of Hong Kong where supervision and quality of labour standard is not consistent. However, in order to produce higher and slimmer structure, as well as to reduce dead weight of building, grade 40 or even up to Grade 60 concrete is occasionally used in some projects. Typical examples of use are for core wall, transfer structure, and other large-span or tensioned structures etc. Most of the concrete used are supplied by specialist suppliers and delivered to site using concrete mixer trucks.

The placing of concrete for high-rise building is mainly by the help of mechanical equipment such as using booster pumps, placing boom; or using skip, hopper or bucket lifted by tower crane or hoist rack etc. The placing process may still be quite labour intensive for horizontal movement of concrete on the floor slab level, as well as to render satisfactory placing and compaction result within the congested placing environment, greater amount of human labours is still unavoidable.

In term of speed of work, fast track schedule is almost a "must" under most contract requirements. In general, for a building of about 800 to 1000 m^2 in size and with the arrangement and set-up as mentioned, it should be of no difficult to complete a typical floor cycle within 5 to 6 days' time. For building of larger size and with more difficult layout, progress may be up to 10 days or above for a typical floor. Other planning options can also be made such as to sub-divide the floor area into two convenient phases, or to cast the wall first and the floor/beam follows, or even work with staggered floor arrangement, so as to obtain the best result in scheduling and efficient use of resources to suit the layout and other design constraints.

6 CONCLUSION

The construction of high-rise building is a very broad subject by itself. The author tries to make use of a few pages to discuss on this subject matter can only highlight a few of the local features. Besides the application of mere technology in construction, the overall structure and common practices in the local industry, in fact, form the most powerful drive to shape the actual outcomes.

In recent years, due to the drift from sellers' market to buyers' market after the economical crisis occurred after 1997, there is a very great tendency for developers to produce really high quality buildings to satisfy the genuine need of the public. Hence, some forward-looking developers begin to invest more in investigating possibility and options to employ advanced technology in order to make building more reliable and better perform. At the same time, thanks to the change in attitude by the government to become more answerable and accountable to public, tighter control in the building development process is adopted. As a result, smaller firms which operate under traditional managerial ideology with limited capital and resource can hardly survive. Construction firms of the new generation are more ready to establish a work organization with stronger professional capability, more proactive to input more resources in research and development and to restructure themselves by setting up more efficient administration system in the running of projects. The gradual change in this direction is obvious and it is expected that a modernization process will eventually lift the overall professional standard of the Hong Kong construction industry from the backbone within this decade or so.

REFERENCE

Andrew Fan, Building Construction in Hong Kong 1998. Construction of High-rise Concrete Buildings in Urban Areas. *Buildings Department, Hong Kong SAR Government.*

R Grosvenor, Building Construction in Hong Kong 1998. Comparison of Hong Kong & Overseas Construction Practice & Management. *Buildings Department, HK SAR Government.*

C. Gibbons, G. Ho, J. MacArthur, 1999. Developments in High Rise Construction in Hong Kong. *Proceedings of Symposium on Tall Building Design and Construction Technology.*

W.M. Wong, 1999. Fifteen Most Outstanding Projects in Hong Kong. *China Trend Building Press, Hong Kong.*

W.M. Wong, 2000. Construction of Residential Buildings – Developments and Trends in Methods and Technology. *China Trend Building Press, Hong Kon*

6 Construction contracting and practice

Creative Systems in Structural and Construction Engineering, Singh (ed.) © 2001 Balkema, Rotterdam, ISBN 90 5809 161 9

Intelligent agent applications in construction engineering

C.J. Anumba, O.O. Ugwu & Z. Ren
Centre for Innovative Construction Engineering (CICE), Loughborough University, UK

L. Newnham
Centre for Construction IT, Building Research Establishment (BRE), Watford, UK

ABSTRACT: This paper explores the use of intelligent agents within the construction industry, starting with a general introduction of the fundamental concepts of agent technology. It draws on two research projects – ADLIB, which addresses the collaborative design of light industrial buildings, and MASCOT, which focuses on the development of a multi-agent system for claims negotiation. Key issues in the development of intelligent agent applications for construction engineering are discussed – these include agent ontology, negotiation protocols and strategies, knowledge acquisition, and usability. The paper concludes with an outline of the benefits of intelligent agents in construction, and key research issues that need to be resolved.

1 INTRODUCTION

Construction engineering often requires collaborative working between members of a multi-disciplinary project team. A typical project involves a wide range of disparate professionals - clients, architects, structural engineers, building services engineers, quantity surveyors, contractors, materials suppliers, etc. - working together for a relatively short period on the design and construction of a facility. In many cases, the participants are geographically distributed making the need for effective information and communication technologies acute.

For effective collaborative working between the parties in a construction project team, it is essential that enabling information and communications technologies are available. The additional problems posed by the use of heterogeneous (legacy) software tools are well known and need to be overcome by the adoption of new approaches. One such approach, which has significant potential for use in the construction industry, involves the use of distributed artificial intelligence, which is commonly implemented in the form of intelligent agents. Intelligent agents consist of self-contained knowledge-based systems that are able to tackle specialist problems, and which can interact with one another (and/or with humans) within a collaborative framework.

This paper discusses the use of intelligent agents in construction engineering. The early part of the paper reviews the basic concept of intelligent agents, highlighting related work in the construction do

main. Two research projects that are developing intelligent agent applications for construction engineering are then presented. The benefits of agent applications and the research issues that need to be resolved are discussed in the later sections of the paper.

2 INTELLIGENT AGENTS

2.1 Definitions

There is much discussion about whether some particular system is an agent, an intelligent agent or merely a program. This is the manifestation of a general problem in AI of defining 'intelligence' that had led to much fruitless discussion. The result is that there are as many definitions as there are researchers, leading to the term being substantially overused. However, there are several broad qualities that have some measure of general agreement (Wooldridge & Jennings, 1995). The key feature would appear to be 'autonomy', the ability of the agent to formulate its own goals and to act in order to satisfy them.

Nwana (1996) defines an agent in terms of three behavioural attributes, any two of which must be possessed by a software agent. Quoting from Nwana, these are:

- *"Autonomy:* This refers to the principle that agents can operate on their own without the need for human guidance, even though this would sometimes be invaluable. Hence agents have individual internal states and goals, and [an agent] acts in such a manner as to meet its goals on behalf of its user. An important element of their autonomy is their pro-activeness, i.e. their ability to 'take the initiative' rather than acting simply in response to their environment.

- *Co-operation:* Co-operation with other agents is paramount; it is the *raison d'être* for having multiple agents in the first place in contrast to having just one. In order to co-operate, agents need to possess a social ability, i.e. the ability to interact with other agents and possibly humans via some communication language. Having said this, it is possible for agents to co-ordinate their actions without co-operation.

- *Learning:* For agent systems to be truly 'smart', they would have to learn as they react and/or interact with their external environment. In our view, agents are (or should be) disembodied bits of 'intelligence'. Though we will not attempt to define what intelligence is, we maintain that an important attribute of any intelligent being is its ability to learn. The learning may also take the form of increased performance over time."

Nwana's requirements for agent-hood may be neatly shown as a Venn diagram (Figure 1):

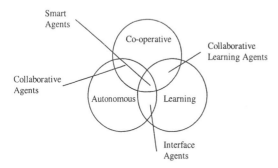

Figure 1: Taxonomy for Agents

The inclusion of learning as, at least, an aspiration recognises a core quality of intelligent human behaviour. In addition, it provides a framework by which software agents, as well as biologically based agents from worker ants up to higher mammals, can be visualised.

2.2 *Agent applications*

Intelligent agent research grew out of Distributed Artificial Intelligence (DAI). The aim of this area was to gain the benefits of modularity when tackling large, real-world problems. These benefits are essentially speed and reliability. One of the problems researchers faced was the sheer scale of the task; in order to get anything working the complete system and its environment needed to be modelled. With the advent of the wide usage of the Internet, new and somewhat more manageable research areas started to open up, as the Internet is essentially an ideal electronic environment for agents.

In the last few years, there has been an explosion in the amount of information available on a daily basis and, as it has increased, so has dependence on it. This information may be stored as passive stored databases and files or it may be information we need to actively request in order to make a decision, for example, in the case of scheduling a meeting. Much of this information is stored remotely in a variety of formats and sources; much of it badly labelled, if at all, and much of it time-consuming to locate. This has led to a state of affairs where traditional IT systems are increasingly hard-pressed to meet many information-gathering challenges. Whereas, in the past, humans would take on the role of sifting and co-ordinating gathered information in order to take decisions, agent-based software technology is rapidly evolving to perform all of these functions.

Agents are considered particularly useful for tackling large-scale, real world problems involving multi-disciplinary perspectives. They are currently applied to a variety of application domains including workflow management, telecommunications network management, air traffic control, business process re-engineering, information retrieval and management, electronic commerce, personal digital assistants, e-mail filtering, command and control, smart databases, and scheduling/diary management (Ndumu & Nwana, 1997).

2.3 *Collaboration models and the role of agents*

There are essentially four modes of collaboration depending on the nature of separation and pattern of communication, between the participants in a project. The classification in the space-time communication matrix shown in Figure 2 is useful but further characteristics of such group collaboration (such as group sizes) can be identified and used to generate a more elaborate matrix. The types of collaboration in the generated matrix are briefly summarised below (Ugwu et al 1999):

	Same Time	Different Times
Same Place	Face-To-Face Collaboration	Asynchronous Collaboration
Different Places	Synchronous Distributed Collaboration	Asynchronous Distributed Collaboration

Figure 2: Collaboration Models.

- Face-to-Face Collaboration: This involves meeting in a common venue such as a meeting room, and participants engaging in face-to-face discussions;

- Asynchronous Collaboration: this mode of communication can be conducted using such medium as notice/bulletin boards within an organisation;

- Synchronous Distributed Collaboration: this involves real time communication using any of the current technologies such as telephony, computer mediated conferencing, video conferencing, electronic group discussion or editing facilities;

- Asynchronous Distributed Collaboration: this includes communication via the post (e.g. periodic letters, news bulletins), fax machines, telephone messages or voice mail, pagers, electronic mail transmissions, etc.

Agents are particularly appropriate for distributed collaboration. While there are tools, such as video-conferencing (Baldwin et al, 1999) that support distributed synchronous collaboration by enabling 'virtual co-location' of project team members, there are very few design tools that adequately support distributed asynchronous collaboration.

3 RELATED WORK

There have been several projects that have sought to apply intelligent agents to problems in construction engineering. For example, some work has been done at the Centre for Integrated Facility Engineering (CIFE), Stanford University on the use of agents in a federated collaborative framework (Khedro, 1996). Pena-Mora at MIT has also investigated the use of intelligent agents in change negotiation (Pena-Mora & Hussein, 1996). However, specific design applications are very few, with no commercial systems available for use by practising engineers.

Much of the work in the area of automated design has been done on building design support. Systems such as the Lawrence Berkeley Laboratories' Building Design Advisor (Papamichael & LaPorta, 1996) and the Building Design Support Environment (Papamichael & Selkowitz, 1996) rely on expert knowledge residing on one machine, being used solely to advise the user of the consequences of some design choices.

The US Army Corps of Engineers' Construction Engineering Research Laboratory (CERL) has done work on collaborative engineering design (Chiou & Logcher, 1996; McGraw et al, 1999), taking the above decision support ideas a step further by implementing an agent-based system. Here, agents have areas of specialist knowledge and perform design and checking tasks. They interact directly with a user who is responsible for design changes. However, there is no provision for direct negotiation between agents during the design process.

The DESSYS Project (URL1) is part of a wider research project - Virtual Reality Design Information Systems (VR-DIS) that is investigating the deployment of multiple software agents to improve collaborative decision-making in a multidisciplinary architectural design environment. The DESSYS research covers knowledge modelling for a decision support system in geotechnical design.

A common limitation in the above projects and applications is that the agents do not negotiate amongst their peers to converge on a solution. Since there is no provision for direct negotiation between agents during the design process, the number of designs evaluated remains small. These projects fall short of fully automated collaborative design as most of them rely on human input to suggest design changes and they do not make use of intelligent agents. Hence, convergence to a near-optimal design is dependent on the user and does not make full use of the available computational power, which can allow for the evaluation of many slightly differing designs automatically. There is therefore the need to build on the achievements of these projects and take the research investigations further.

The two projects described in the next two sections of the paper address the limitations of the above projects and focus on the use of agents at the design and construction stages.

4 THE ADLIB PROJECT

The aim of the ADLIB (Agent-based Support for the Collaborative Design of Light Industrial Buildings) project is to investigate the issues involved in collaborative design using intelligent agents, within the context of light industrial buildings. It builds on the work done at CERL by developing a multi-agent system for distributed collaborative design, within a more powerful agent environment (Ugwu et al., 1999).

The ADLIB domain agents include both an Interface/Architectural Agent (IAA), and Specialist Agents - Structural Design Agent (SDA), Building Services Agent (BSA), Costing and Constructability Agent (CCA), and Safety Advisory Agent (SAA). These agents participate in a collaborative design and negotiate for an optimum design based on the constraints outlined in the project specifications (Anumba & Newnham, 1998). These agents are shown in Figure 3.

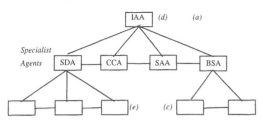

Note:
- Vertical arrows denote delegation and reporting back
- Horizontal arrows denote peer-to-peer negotiation

Figure 3: ADLIB Domain Agents

A pilot system has been developed based on the above conceptual model. Provision is made, within the pilot system, for peer-to-peer negotiation between the agents, within a collaborative design environment. This facilitates the automation of some basic design tasks in the problem domain. The use of agent-based systems in this context is expected to result in increased competitiveness of the construction industry, as the decentralisation of complex, large-scale problems and the collaborative input to their resolution, will lead to better quality, more economic, safer and more optimal designs.

A detailed discussion of the approach adopted in developing the ADLIB pilot is provided elsewhere (Ugwu et al, 1999).

5 THE MASCOT PROJECT

The MASCOT (Multi-Agent System for Construction Claims NegoTiation) project is concerned with the applicability of intelligent agents to the negotiation of claims. This is an important application area, as most project managers consider claims negotiation the most time and energy-consuming activity in claims management (Hu, 1997). This is because of the following:

- Some negotiation items are ambiguous and sensitive;

- Engineers and clients typically respond in a tough and unyielding manner to contractors' claims, which often result in cost overruns;

- Claims management procedures are inefficient;

- Negotiation process is often lengthy.

- Negotiation requires skill and experience for a successful outcome.

The MASCOT system provides a novel approach for the improved management of claims negotiation and involves three main agents (see Fig. 4) – a client agent (CLA), a contractor agent (COA) and an engineer agent (ENA).

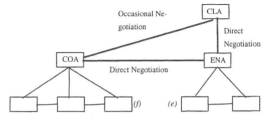

Figure 4: Agents in MASCOT

The agents in MASCOT, on behalf of their owners, are designed to directly negotiate with each other about negotiation items to reach a solution within a specified time frame. The main negotiation takes place between the contractor and engineer agents, both of which could have sub-agents (e.g. representing sub-contractors or resident engineers). A learning mechanism based on the Bayesian approach has also been incorporated in the multi-agent system so that agents can adapt their negotiation strategies based on previous interactions. Thus, both the efficiency and effectiveness of claims negotiation can be improved.

6 ISSUES IN AGENT SYSTEM DEVELOPMENT

There are several important issues in the development of multi-agent applications for construction engineering. These are discussed below with the approach adopted in the ADLIB and MASCOT projects highlighted, as appropriate.

6.1 Development Environment

The choice of a development environment is of utmost importance, as it often dictates the functionality of the system. Both the ADLIB and MASCOT prototypes are based on Zeus, a proprietary tool developed by BT that uses TCP/IP messaging built on Java to achieve interoperability.

6.2 Agent Knowledge

Agents need to have adequate knowledge for their intended function, and conventional knowledge acquisition techniques can be applied here. Domain experts and published documents were the main sources of knowledge for the two prototypes here.

6.3 Agent Ontology

In order for the domain agents to communicate effectively, they need to have a common understanding and shared knowledge of the concepts associated with the domain. The ten industrial partners on both the ADLIB and MASCOT projects contributed to developing this for each prototype.

6.4 Agent Interaction and Negotiation

In order to achieve effective interaction, each agent must have some mechanism to ensure that it reaches an agreement with the other agents. There are two main elements in agent interaction – a negotiation protocol and negotiation strategies.

6.4.1 Negotiation Protocol

The negotiation protocol is the set of rules of interaction that the agents will follow to converge on a solution in design space. A common method, which has been adopted in the prototype, is the Monotonic Concession Protocol (MCP) (Rosenchein & Zlotkin, 1994, Jennings et al, 1996). For two agents, the standard version is as follows. Agents start by simultaneously proposing a deal (in this case a design). Agreement is reached if one agent matches, or exceeds what the other has asked for (in terms of utility). If agreement is not reached negotiation proceeds to another round. An agent can only propose deals that have a greater or equal utility for the other agent, that is, concede or do nothing. If neither agent concedes then negotiation ends as conflict is reached, otherwise negotiation continues.

6.4.2 Negotiation Strategy

These are the strategies that an individual agent follows to propose alternate deals within the negotiation protocol above. Given that the final system is designed so that any agent that adheres to the negotiation protocol above can work within the system, then this will be completely up to the designers of these agents.

6.4.3 Approach Adopted

Within ADLIB, a simple gradient descent algorithm is used. In this case, an agent makes an offer that will decrease its utility by the least amount with the minimum step size being governed by the time cost; the more deals it has to offer before one is accepted the greater will be the time cost, so the more it will have to concede on the design.

The approach adopted in MASCOT is somewhat different in that the agents are regarded as being involved in a bounded self-interested negotiation (i.e. each party is only concerned about his own benefit but is bounded by the contract not to break the negotiation). The Zeuthen model is used and integrated with a Bayesian learning mechanism that facilitates convergence (Ren et al, 2000).

6.5 Usability

It is important in developing multi-agent systems for construction engineering that the system provides adequate support to the key decision makers. In this regard, agents should have an appropriate user-interface such that each agent's owner can modify its negotiation strategy and/or update the agent's knowledge base. Where possible, the agents should be able to link into legacy IT systems being used by the project team members. These usability issues are being addressed to some extent in the two projects.

7 OTHER POTENTIAL AGENT APPLICATIONS

There are several other application areas for agents in construction engineering. Some of these are listed below and include one or two (such as information retrieval) that are already well established:

- Information filtering, customisation and retrieval to meet the needs of various users in distributed decision-making environments;
- Automation of basic design and project management tasks including negotiation to reach an optimum solution;
- E-Commerce: automatically linking customers and suppliers, and negotiating to complete a transaction;
- Managing the construction supply chain to ensure accurate and just-in-time delivery of components and services;
- Management of the corporate knowledge within an organization or virtual project team to facilitate retrieval;
- Project co-ordination to ensure that the collaborative input by the disparate disciplines involved is timely and efficient.

8 SUMMARY AND CONCLUSIONS

This paper has discussed the use of intelligent agents in construction engineering. The concept of agenthood has been introduced and previous work reviewed. It was shown that previous attempts at the development of multi-agent system are limited by the lack of a provision for peer-to-peer negotiation between the agents. Two applications, ADLIB and MASCOT, which involve multi-agent negotiation at the design and post-design phases, have been presented. Issues in the development of multi-agent systems have also been discussed.

There are numerous benefits in the use of multi-agent systems for construction engineering: distributed collaboration, reduction of information overload, automation of mundane tasks, improved coordination/communication, and better integration. However, there are also a number of issues that need to be tackled. These include:

- Development of a domain ontology to facilitate communication between multiple agents;
- Coordination of the collaborative input to decision making;
- Development of appropriate negotiation protocols and strategies;
- The development of Web-based systems for effective distributed collaboration;
- How to effectively interface with users for efficient distributed decision making; and
- Integration of agent-based systems with legacy software systems used by project team members.

The construction industry stands to benefit from agent technology and should make the necessary investment to take advantage of the opportunities.

ACKNOWLEDGEMENTS

The work described in this paper is funded by the Engineering and Physical Sciences Research Council, UK. For further details see:
http://helios.bre.co.uk/adlib.

REFERENCES

Anumba C. J. & L. Newnham 1998. Towards the Use of Distributed Artificial Intelligence in Collaborative Building Design, *Proceedings, 1ˢᵗ International Conference on New Information Technologies for Decision Making in Civil Engineering,* Miresco E. T. (Ed.), Sheraton Hotel, Montreal, Canada, 11-13 October, pp 413-424.

Baldwin A. N., A. Thorpe, C. Carter & G. Taylor 1995. Conferencing Systems Within Construction Organizations, *Civil Engineering*, Proceedings of the Institution of Civil Engineers, Vol. 132, No. 4, pp 174-180.

Chiou J. D. & R. D. Logcher 1996. Testing a Federation Architecture in Collaborative Design Process, - Final Report, *Report No. R96-01*, US Army Construction Engineering Research Laboratory, Urbana-Champaign, USA.

Hu, J. X. 1997. *Chinese Oversea Construction Contractors' Claims Management (Chinese Version)*. Beijing: China Construction Publisher.

Jennings, N. R., P. Faratin, & M. J. Johnson, 1996. Using Intelligent Agents to Manage Business Processes, *Proceedings of the First International Conference on the Practical Application of Intelligent Agents and Multi-Agent Technology* (PAAM 96), London.

Khedro T. 1996. AgentCAD: A Distributed Cooperative CAD Environment, *Information Representation and Delivery in Civil and Structural Engineering Design*, Kumar B. and Retik A. (Eds.), Civil-Comp Press, Edinburgh, pp 15-19.

McGraw K D, P. W. Lawrence, J. D. Morton & J. Heckel 1999. The Agent Collaboration Environment an Assistant for Architects and Engineers (For details, see URL: http://www.cecer.army.mil/pl/ace.htm)

Ndumu D. T. & H. S. Nwana 1997. Research and Development Challenges for Agent-Based Systems, *IEE/BCS Software Engineering Journal*, Special Issue on Agents Technology, Vol. 144, No. 1, February, pp 2-10.

Nwana, H.S. 1996. Software Agents: An Overview, *Knowledge Engineering Review*, 11, (3), pp 205-244.

Papamichael K. & J. LaPorta 1996. The Building Design Advisor, *ACADIA96*, Tucson, Arizona.

Papamichael K. & S. Selkowitz 1996. Building Design Support Environment, *Internal Report*, Lawrence Berkeley Laboratory.

Pena-Mora F. & K. Hussein 1996. Change Negotiation Meetings in a Distributed Collaborative Engineering Environment, *Information Representation and Delivery in Civil and Structural Engineering Design*, Kumar B. and Retik A. (Eds.), Civil-Comp Press, Edinburgh, pp 29-37.

Ren Z., C. J. Anumba & O. O. Ugwu 2000. Towards a Multi-Agent System for Construction Claims Negotiation, *Proceedings ARCOM 2000 Conference* (in press).

Rosenschein, J. S. & G. Zlotkin 1994, *Rules of Encounter*, MIT Press.

Ugwu O. O, C. J. Anumba, L. Newnham & A. Thorpe 1999. Agent-Based Decision Support for Collaborative Design and Project Management", *International Journal of Construction Information Technology*, 7(2), pp. 1-18.

URL1: DESSYS Project:
http://www.ds.arch.tue.nl/Research/Agents/DessysIntro.stm

Wooldridge, M. & N. R. Jennings 1995. Intelligent Agents: Theory and Practice, *The Knowledge Engineering Review* 10 (2).

Creative Systems in Structural and Construction Engineering, Singh (ed.) © *2001 Balkema, Rotterdam, ISBN 90 5809 161 9*

Predicament of developing and sustaining local technical carriers in the construction industry of Saudi Arabia

Ibrahim Al-Hammad
Faculty of Civil Engineering, College of Engineering, King Saud University, Riyadh, Saudi Arabia

ABSTRACT: There exists a manpower imbalance between the number of expatriates versus the number of natives in the labor market of Saudi Arabia. This led the public policy planners, to direct the research towards Saudisation studies; i.e. the process of substituting foreign technical and skilled labor with counterpart Saudi labor in general. In this paper, the challenges of Saudisation focused on the construction industry, is illustrated through a case study of a national corporate company, operating on a commercial basis. A decision making tree is proposed to represent the various alternate scenarios of Saudisation. Gradual Saudisation based on education, qualification, and experience, is envisaged. Finally, this study serves as a model for similar challenges that face a representative sample of private sector firms operating in the construction industry and/or industrial industry in Saudi Arabia and other Arabian Gulf countries.

1 INTRODUCTION

Arabian Gulf countries, as middle income oil producing un-industrialized states, have invested tremendously in infrastructure, education, agriculture, and other economic sectors. The demand for services and goods in those countries substantially surpasses the existing capacity of its own. The magnitude and complexity of development in the last three decades, required importing all sorts of labor, technicians, and professionals in almost every field of specialty. Presently in Saudi Arabia alone, the largest Arab nation in the Gulf, there are more than 120 countries identified as origins of laborer's recruitment (Al-Aqtsadia 1999). It is also reported that Asian laborers are the dominant foreign labor forces in the Construction Industry. More recently, in the year 2000, it is expected to have 7.1 millions in the labor force whereby 65.3% are of expatriate origin, who constitute 73% of the labor force in the private sector, and 20% in of the labor force in the public sector respectively (SAMA 2000).

A case study, in one construction supervision department of a large private corporate company, is selected to examine the challenges of Saudisation in the construction industry profession.

2 STATEMENT OF THE PROBLEM

Saudi Aramco, one of the largest Oil Company in the world, is selected as a case study for Saudisation.

This company employees more than 60,000 persons with annual budget of more than 6 US $ Billion. While Saudi manpower makes up 70% of the total number of employees, there is a shortage or lack of interest the by natives to fill specific positions in the company due to various reasons.

The company management, under public pressure to employ more Saudis, has succeeded in its drive for Saudisation in many sectors of work; however, it recognized that it faced a problem of filling positions in other sectors. The company management commissioned the author to study this problem.

For the scope of this paper, emphasis is placed on examining the reluctance of Saudi professionals to work full time at the construction sites at the level of residential engineers and site superintendents.

3 BACKGROUND ON THE CONSTRUCTION SUPERVISION DIVISION

In the organization chart, of Saudi Aramco; there are many divisions to carry out its mission as stated in its Establishing Charter. Under the hierarchy of Project Management Department, comes the Central Area Projects Design and Construction Division, shorthanded for CAPD&C Division, which we will focus on. This division is supportive in nature, and consists of the following sectors: utilities, schools, medical and industrial, community, and office and support. Each sector is headed by senior project en-

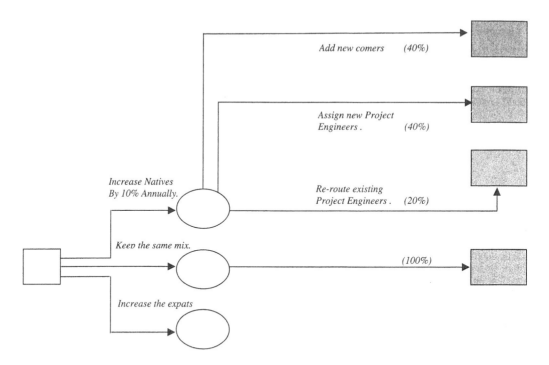

Figure 1. Decision Tree for Possible Alternatives.

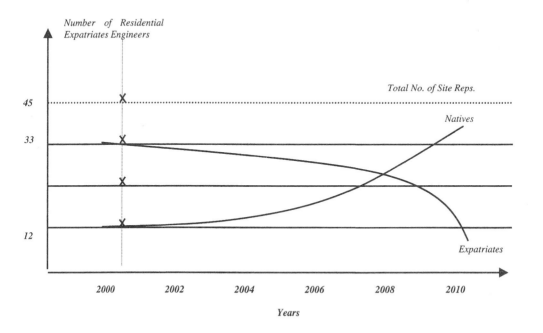

Figure 2. Gradual Pace of Saudisation

gineer assisted by project engineers. Those project engineers supervise many projects under construction and simulatounsly follow up on other projects in its conceptual, preliminary design, and new projects budgetary approval stages. Each project engineer, direct construction projects in various geographically disparate locations, with the assistance at the project site, by residential engineers (out soured) and by a technical assistant (company employee).

The problem in this division, as exemplary of this research problem of slow Saudisation, lies on the fact that there is a reluctance of qualified locals to take the positions of residential engineers and technical assistants.

4 METHODOLOGY FOR RESEARCH

This problem demanded extensive data and information collection including structured interviews with senior management down to the site superintendents, reviewing the existing operating procedures, reviewing resumes of the personnel at the division and their performance evaluation report, discussing the problem with the staff of the company management, and visiting more than 10 construction project sites.

The aforementioned stage helped in expressing the present situation as it is. Second stage in this research, is to explore research tools to address such problem. Since there are several scenarios of Saudisation planning, at the preliminary phase, it is found that decision-making tree, as prescriptive mode, is convenient to use.

Table 1. Comparisons of Selected Features for the Construction Supervision Team

Feature	Project Engineer	Residential Engineer	Technical Assistants
Age (in years)	45	55	27
Nature of Work	Coordination and Follow up	Field Supervisor	Field Ass. Supervisor
Field Experience	Good	Excellent	Average
Suitability of Work	Suitable	Suitable	Suitable
Future Carrier Path	Clear and Promising	Does Not Apply	Not Clear
Employ. Contract	Permanent	Temporary	Permanent
Satisfaction	High	Med. To High.	Medium
Qualification	B.Sc.	B.Sc.	Diploma
Cost	High	High	High
Type of Nationality	Maj. Saudi	West. Expat.	All Saudi

5 PROFILE COMPARISONS OF PROJECT ENGINEERS VS. RESI   DENTIAL ENGINEERS VS. TECHNICAL ASSISTANTS

The first step in this research is to draw some distinguishing features for the role and profile of the personnel involved in Saudisation process. Table 1 presents the profiles for the three categories, i.e., project engineers, residential engineers, and technical asistants.

This table reveals diversity of supervision team in its make up. As mentioned previously, the services of the residential engineer is provided by local A/E firms. This setup, as practiced by Saudi Aramco, is intended to have an external auditor for its staff as well as promoting local A/E firms. Also, such firms provide other services such as design, tenders preparation, review of bids, and construction management services to name a few.

6 TOOLS FOR SAUDISATION

6.1 Decision Making Tree

Possible scenarios for Saudisation can be simply represented by a decision tree shown below, in Figure 1.

This figure displays three possible alternatives, to deal with Saudisation in the company. The first alternative proposes a 10% annual increase in Saudi employees. The second alternative decision, is to keep the balance as it is. The third option, to increase the expatriates ratio, is deemed infeasible. Although, the third option is highly unlikely; however, it is observed by the author that there are few private sector companies, adopt the third alternative covertly.

The first alternative decision, is the most desirable for the company. Although, the 10% increase is proposed, it is up to the management to determine the proper ratio that they can accept. When we dwell further in the first alternative, we see that to achieve a specified increase in the local labourers; there is another problem arises. That is to identify the source of the new entry of the local laborers/employees. The figure shows that there are three possible alternatives to replace the expatriate residential engineers by counterpart Saudis as follows: add newly hired Saudis from outside the company, assign company new project engineers as residents in the construction sites, or re-route the existing project engineers into residential engineers assignment. The recruitment and personnel policy by the company, combined with the economics of each previous alternative; affect the selection either of the previous choices or combination of those alternatives.

Again, the percentages attached to each alternative, are exemplary; and thus, the management has to decide its make up.

6.2 *Gradual Saudisation Plan*

To farther implement the Saudisation alternative as a desirable option in figure 1, it is envisaged to phase out expatriates in a timely manner in the construction sites of the CAPD&C Division. Figure 2 portrays two curves reflecting the two types of resident engineers; i.e., that is Saudis vis-à-vis Non-Saudis. The first curve represents the phase out pace of expatriates residential engineers which is offset by the increase of the qualified Saudi counterparts. The figure shows that there are about 33 expatriates residential engineers at present. Those can be substituted with counterpart Saudis over a span of 10 years from year 2000, at a rate of 3 to 4 annually. The numbers, as a guideline, may be adjusted as conditions change with respect to the business and economic cycle. The time span allowance of 10 years, is presumed to allow for a smooth and gradual substitution of Saudi engineers and technical assistants, after gaining fair experience and farther to be induced to take in more laborious jobs with time.

7 CONCLUSIONS

This study highlighted the problem of developing and promoting local technical employees to take on jobs in the construction industry. This is a typical predicament facing all the Arabian Gulf countries; and mimics indirectly, other oil rich countries.

Further research in Saudisation, is to explore treating major constraints of 1) maintaining the quality of work, and 2) minimizing the labor costs.

REFERENCES

Al-Aqtsadia Daily Newspaper (October 10th, 1999), Report On Laborers Recruitment from Indonesia, pp.1.

SAMA (Saudi Arabian Monterey Agency), (2000), Annual Report, Department of Economic Research and Statistics.

Creative Systems in Structural and Construction Engineering, Singh (ed.) © 2001 Balkema, Rotterdam, ISBN 90 5809 161 9

Risk and remuneration of authorized persons in Hong Kong

Sai-On Cheung, Ka-Chi Lam & Shek-Sau Wong
Department of Building and Construction, City University of Hong Kong, People's Republic of China

Sik-Hung Wong
Morrison Hill Technical Institute, Vocational Training Council, Hong Kong, People's Republic of China

ABSTRACT: Authorized Persons (AP) play a critical role in the design and construction of building works in Hong Kong. Under the Building Ordinance, design and construction of building and street works need to be coordinated and monitored by an authorized person appointed by the owner. Professionally qualified Architects, Engineers and Surveyors are eligible to register as authorized Person after completing the prescribed procedures. This paper reports a study on the correlation between remuneration levels of Authorized Persons with their risk exposures.

The data measuring the risk exposures and the corresponding remuneration level were obtained through interviews and questionnaire survey. A multiple regression model was developed for the research objective. It was found that liabilities to obtain design approval and civil obligations towards the client showed the highest correlation with the level of remuneration. The result is consistent with the key role to be played by Authorized Persons in building projects.

1 INTRODUCTION

Despite the recent downturn of the property market, construction activities in Hong Kong remain relatively high as compared with other countries in the region. As basic necessity, buildings should be fit for human habitation. This calls for adequate design and construction. Hence building control is heavily regulated. In Hong Kong, the primary legislation for building control is the Building Ordinance (BO). Supporting regulations include Building (Administration) Regulations (B(A)R), Building (Construction) Regulations (B(C)R) and Building (Planning) Regulations (B(P)R).

The system of Authorized Persons (AP) is also implemented by the Building Authority (BA). Under this system, ensuring the design and construction of building works meet with the legislative requirements rest with the Authorized Persons. This study seeks to examine the relationship between risks exposures and remuneration level of Authorized Persons.

2 ROLE OF AUTHORIZED PERSON

2.1 *Under the Building Ordinance*

In order to relieve the burden on the Government in performing building control but at the same time ensure that the statutory standard is maintained, the

Hong Kong Government introduced a professional licensing system of "Authorized Architect" in 1903 and re-titled as "Authorized Person" in 1974. Authorized Persons are being recognized as statutory agents on building control. There are three lists of AP for which the Building Authority keeps a separate register for those persons who are qualified to perform their respective statutory duties under the Building Ordinance. List I consists of Architects, List II is for civil or structural engineers and surveyors will be included in List III. Applicants for inclusion in the register must possess the relevant qualifications, experience and competence and are required to attend an interview before the respective registration committee constituted under the Building Ordinance. Under s.4(1) of the BO, whoever wants to carry out building works, except for those exempted works stipulated in s.41, an AP has to be appointed to coordinate the building works. Therefore, an AP is the coordinator of the building works.

If structural elements are involved, a Registered Structural Engineer (RSE) would also be appointed as a consultant to the AP. In other words, the AP is recognized by the BA as a competent person and held responsible for all statutory matters under the Building Ordinance.

Section 4(3) of the Building Ordinance summarizes the duties of an AP. An AP appointed shall (a) supervise the carrying out of the building works; (b) notify the BA of any contravention of the regulations which would result from the carrying out of any work shown in any approved plan; and (c) comply generally with the provisions of the Ordinance. This is important that as s.37 of the BO, Government Authorities and building officials will not attract any liability for their action in carrying out their duties under the Ordinance. Therefore, an AP is primarily responsible for assuring that all building works are designed and constructed in compliance with the statutory requirement under the BO.

2.2 *The significance of AP in Hong Kong development projects*

APs play a significant role throughout the development cycle i.e. from design through to completion. Any person intending to carry out building works is required to obtaining the approval of plan and consent from the Buildings Department (BD), which is a condition precedent for the commencement of the building works (s.14(1) of the BO). Under s.4(1) and 14(1) of the BO, it is necessary for the appointment of an AP to submit building proposals to the BD for approval and consent. An AP may seek assistance from other consultants such as fire services etc. for design preparation. Nevertheless, it is the AP who has to certify that the submitted plans comply with the provision of BO and allied regulations (B(A)R, s.18A). This is formalized by the submission of prescribed form BA 5. For Addition and Alterations works involving structural alterations, under B(A)R s.18 the AP and RSE concerned are required to certify that the structural stability of the existing building in a prescribed Form BA 6.

Furthermore, an AP will also coordinate and monitor the works of other consultants and contractors involved in the projects. B(A)R s.37 requires an AP to provide periodic supervision, and make inspection such that works are carried out in accordance with the approved plan and the requirements under the BO. In this regard, under s.17(3) of the Building (Amendment) Ordinance 1996, a supervision plan prepared jointly among the AP, RSE, Registered General Contractor (RGC) and Registered Specialist Contractor (RSC) need to be lodged with the BA by an AP prior to or at the time of application for consent for commence building works. The supervision plan defines the supervision role of each party involved in the project and specifies the degree of supervision required. The BA may (s.23(2) of the BO) order the cessation of building works if there has been a material deviation from the supervision plan submitted.

Once the project is completed, it is necessary to obtain an Occupation Permit (OP) before a building or part of the building can be occupied. Under B(A)R s.25, within 14 days after completion of building works, Form BA 13 is to be submitted by the AP with signatures of the AP, RSE and RGC jointly. It serves to endorse that the works have been carried out in accordance with the approved plans and are structurally safe and in full compliance with requirements under the BO. It is also the duty for an AP to liaise with government departments. For example, an AP needs to submit relevant documents to the Fire Services Department to obtain the relevant certificates, which are also condition precedent for the issuance of an occupation permit

An AP could also be liable in tort to a third party for personal injury or damage to property. Normally, he owes a duty of care to the adjoining building owners to ensure their interests are not affected by the construction work. Also he owes a duty of care to the general public for the safe execution of the proposed works, save for the negligence on the contractor's part. Whether a breach of such a duty is judged by what a professional of equal experience and knowledge would do in similar situation.

Apart from the role as project team's leader, an AP is usually named in a building contract as the contract administrator. Although the AP is not a party of the building contract, he is authorized to issue instructions and certificates as stipulated in the contract. Therefore, he may be challenged by the contractor against his decisions.

The responsibilities of an AP towards his client are mainly governed by the contract of engagement. This often imposes duties and standards beyond the statutory requirements and may include obligations such as e.g. maximizing the development potentials. Standard engagement contract is often published by the relevant professional institutions.

Apart from the preparation and submission of the building proposal, it is not uncommon for an AP treat as the project manager and lead the project team. In this connection, it was held that the AP has a duty to warn the client of any failing of the project team members (*Chesham Properties v Bucknall Austin Project Management Services and others (1996) CILL 1189*).

3 RESEARCH DESIGN

3.1 *Purpose of study*

In discharging his duties as above mentioned, an AP may expose to risks due to:

1. Liabilities arising from duties of an AP under the Building Ordinance at the design stage.
2. Liabilities arising under the Building Ordinance at the construction stage.
3. Liabilities arising from duties under the Building Ordinance at the post-construction stage.
4. Liabilities towards the client.
5. Liabilities towards the contractor.
6. Liabilities towards third parties.

7. Liabilities towards his/her fellow project team members.

This study seeks to investigate the relationship between the remuneration of APs and the risk exposures arising from the seven types of liabilities as listed.

3.2 *Methodology*

A regression model is used to investigate the relationship between remuneration level and risk exposures of APs. In this regard, data on fee and risk were sought.

The data obtained was analyzed by a multiple regression analysis, a multivariate statistical technique used to examine the relationship between a single dependent variable and a number of independent variables. In this research, the dependent variable is the remuneration level of APs. The independent variables are the risk exposures of APs. In practice, the measurement of the remuneration of the AP is usually on project basis. The fee can be a lump sum or more commonly as a percentage of the contract sum. Usually the condition of engagement issued by the professional institutions will set up the fee scale. The usual remuneration level is approximately 3% of the contract sum, although there may be a mandatory fee scale for the minimum charges, and rules not allowing the members to complete on fees. As the information obtained was private and commercially confidential, the questionnaire was designed to obtain the remuneration level in terms of the percentage of the contract sum. It was not within the scope of this research to determine whether it was reasonable to support a mandatory fee scale or whether it was unprofessional for the fee cutting.

Table 1. Summary of the risk exposure surveyed

Risk Score		Risk Exposures
X_1	1.	Statutory role as an AP;
	2.	Submission of building proposals & endorsement of building proposals;
	3.	Endorsement of structural stability;
	4.	Endorsement of building plans prepared by sub-ordinates;
	5.	Endorsement of demolition plans;
	6.	Endorsement of site formation plans;
	7.	Endorsement of statutory test;
	8.	Temporary absent of AP without notification to the Building Authority;
	9.	Acting as a temporary AP; and
	10.	Explanation to the client on the grounds for disapproval of the building proposal submitted.
X_2	1.	Carrying out of amendment work without a valid consent for commencement of works;
	2.	Submission of further particular to the Building Authority;
	3.	Periodic Supervision;

	4.	Inform contractor on works which do not comply to approved plans and corresponding building regulations;
	5.	Fail to advice the client on the need to appoint additional specialist consultant; and
	6.	Ensure timely amendment plan submission and consent application for all changes.
X_3	1.	Certification of completion of building works in accordance with the statutory requirements;
	2.	Certification based on the advises from other consultants; and
	3.	Liaison with other government departments for the relevant certificates which affect the issuance of the occupation permit.
X_4	1.	Fail to achieve full development potential;
	2.	The brief requirements cannot be achieved;
	3.	Fail to give proper advice on the implications on innovative design approach;
	4.	Delay in giving necessary information to the contractor;
	5.	Delay due to variation order;
	6.	Fail to give proper advice on the specialist items;
	7.	Fail to control the site progress and construction cost due to inadequate supervision given;
	8.	Improper issue of the Certificate of Practical Completion;
	9.	Improper certification on the saleable floor area in sale brochures;
	10.	Defects due to design fault;
	11.	Certification of the substandard work;
	12.	Fail to give proper advice on the effect of variation works; and
	13.	Acting ultra vires to the contract provisions.
X_5	1.	Late issue of drawings;
	2.	Delay caused by client's approval;
	3.	Late response in request for information or approval of shop drawings;
	4.	Biased contract administration;
	5.	Feasibility of the construction details; and
	6.	Fail to provide proper technical support in related to the building control matters.
X_6	1.	Duties of care towards the adjoining building owners; and
	2.	Duties of care towards the general public safety.
X_7	1.	Fail to lead consultant & control of the programme to achieve design brief; and
	2.	Fail to give proper advice on his/her design in related to the requirements of the BO.

3.3 *Data Collection*

A questionnaire was devised for data collection. The questionnaire consists of two sections, section 1 solicits the background information such as the type of appointment. The corresponding professional fee expressed in terms of the percentage of the contract sum is also solicited.

Section 2 seeks to measure the seven types of risk exposures as identified as aforesaid. Each risk category is separately measured by a number of questions. Table 1 provides the summary of the questions asked.

The risk exposure is measured by a Likert scale of 1 (very low risk) to 7 (very high risk). The measurement of each risk category is calculated by the following equation:

$$X_i = \frac{\sum_{j=1}^{m} Q_{ij}}{m}$$

where X_i = Risk score for liability; Q_{ij} = Risk measurement questions under liability type i and j = number of questions under X_i.

A total of 150 questionnaires were sent out to the practicing APs. A total of 42 responses were obtained with 9 of them had missing information. As a result, 33 data sets were used for establishing the regression equation.

The multiple regression was performed by the SPSS programme (SPSS, 1993).

The multiple regression model is of the form:

$R = a_0 + a_1 X_1 + ... + a_7 X_7$

where R = remuneration level; X_1= Risk score for liabilities arising from duties of an AP under the BO at the design stage; X_2= Risk score for liabilities arising under the BO at the construction stage; X_3= Risk score for liabilities arising from duties under the BO at the post-construction stage; X_4= Risk score for liabilities towards the client; X_5= Risk score for liabilities towards the contractor; X_6= Risk score for liabilities towards the third parties; X_7= Risk score for liabilities towards his/her fellow project team members; and a_0 = constant.

Table 2: Correlation Coefficient of Variables

	R	X_1	X_2	X_3	X_4	X_5	X_6	X_7
R	1.00							
X_1	0.76	1.00						
X_2	0.59	0.50	1.00					
X_3	-0.35	-0.26	-0.31	1.00				
X_4	0.73	0.40	0.48	-0.31	1.00			
X_5	0.47	0.27	0.20	-0.23	0.41	1.00		
X_6	0.23	0.30	0.19	-0.15	0.10	0.12	1.00	
X_7	-0.33	-0.39	-0.18	0.14	-0.29	-0.24	-0.17	1.00

The correlation coefficient of the independent variables were first examined. Table 2 gives the coefficients. It can be observed from Table 2 that the correlation coefficients between the dependent variable (R) and the independent variables. The correlations existed among the independent variables suggest that it might cause multi-collinearity if all independent variables are included in the regression analysis. In this regard, Step-wise Multiple Regression approach was adopted. Selection of independent variables into

the equation was by way of examining the correlation coefficient of the independent variables (Table 2). As X_1 (i.e. Risk score for liabilities arising in the submission) exhibited the highest correlation coefficient with remuneration score (0.76). X_1 was chosen as the first independent variable for inclusion. Table 3 illustrates the statistical results after step one.

Table 3. Stepwise Multiple Regression: Step One

Variable(s) entered on step number 1 Included variable: X_1				
Multiple R	0.764			
R Square	0.583			
Adjusted Square	0.570			
Standard Error	0.3047			

Analysis of Variance

Setting	DF	Sum of squares		Mean Square
Regression	1	4.030		4.030
Residual	31	2.879		9.286E-02
F = 43.401				

Variable in the Equation

Variable	B	SE B	Beta	Tolerance
X_1	0.498	0.076	0.764	1.000
Constant	-0.886	0.318		

Variable	VIF	T	Sig T	
X_1	1.000	6.588	0.000	
Constant		-2.784	0.009	

Variables not in the Equation

Variable	Beta In	Partial	Tolerance	VIF
X_2	0.279	0.374	0.748	1.336
X_3	-0.165	-0.246	0.932	1.073
X_4	0.508	0.723	0.842	1.118
X_5	0.289	0.431	0.930	1.076
X_6	0.006	0.010	0.911	1.098
X_7	-0.034	-0.049	0.847	1.180

Variable	Min Toler	T	Sig T	
X_2	0.748	2.209	0.035	
X_3	0.932	-1.390	0.175	
X_4	0.842	0.052	0.000	
X_5	0.930	2.619	0.014	
X_6	0.911	5.726	0.959	
X_7	0.847	-0.270	0.789	

The R square (R^2) indicates the percentage of total variance of remuneration explained by the X_1 liabilities. The total sum of squares (i.e. 4.030 + 2.879 = 6.909) is the squared error that would occur if only the mean remuneration level was used to predict the dependent variable. Using the values of X_1 reduces this error by 58.33% (4.030 ÷ 6.909 = 58.33%).

The regression coefficient and beta value of X_1 are 0.498 and 0.764 respectively. The t-value was calculated by dividing the regression coefficient by the standard error (0.498 ÷ 0.076 = 6.553). The t-value of the variable in the regression equation measures the significance of the partial correlation of the variable reflected in the regression coefficient.

The next step is the selection of the second variable for inclusion in the regression analysis. The partial correlations of remaining variables were examined. The partial correlation is a measure of the variation of the remuneration not being accounted for by X_1, and could be accounted for by other variables. The next independent variable with the highest t-value is X_4 with a partial correlation of 0.723. This means that 21.80% $[(1-0.583) \times (0.723)^2]$ of the total variance could be explained by adding Client score into the regression model. This would then increase the value of R^2 from 0.583 to 0.801.

Another selection criterion is to examine the t-value as it measures the significance of the partial correlations for variables not in the equation. For a

thirty-one degrees of freedom and a significance level of 0.05, the threshold value is 2.040. Accordingly, X_4, with a t-value of 5.726 (>2.040), was selected for inclusion in step two. The statistics of the regression analysis after step two are presents in Table 4.

The R square of the resultant regression is 0.801. The final regression equation is:

$R = -2.348 + 0.366\, X_1 + 0.487 X_4$

where R = remuneration level; X_1 = Risk score for liabilities arising from the submission stage; and X_4 = Risk score for liabilities towards the client.

3.4 Assessing the overall fit of the regression model

The overall fit of the regression model was assessed by checking the linearity and homoscedasticity of the residual. These were found to be satisfactory. The degree if multicollinearity can also be measured by observing the tolerance and Variance Inflation Factor (VIF) values. The tolerance is one minus the proportion of the variable's variance explained by the other independent variables. Thus, a high tolerance value indicates little collinearity. The VIF is the reciprocal of the tolerance value. In the present case, the tolerance values all exceed 0.8 and the VIF values are 1.188 suggesting that the level of collinearity is not excessive. The results indicate that the interpretation of the regression variate coefficients should not be affected by multicollinearity.

4 DISCUSSION AND CONCLUDING REMARKS

The final regression equation includes two risk categories; risk exposures arising from the submission of design and towards the client. This suggests that the remuneration for an AP is most highly correlated to these two risks. The result matches well with the actual situation as the primary role of an AP, as envisaged by the Building Ordinance, is to ensure the proposed development meet with the statutory requirements.

In Hong Kong, under a centralized building plan processing system, the BD is responsible for the checking of submissions whether the requirement under the BO and allied regulations are complied. However, because of the enormous work involved, only curtail checks will be performed by the BD. Under s.37 of the BO, the BD and his officers shall not be held liable for any act in executing the BO as part of their duty, although internal reprimand for the serious and obvious case of negligence is possible.

In general, the AP of a project shall ensure that the building proposal is in compliance with the BO, and notify the BA of any contravention of the regu-

Table 4. Stepwise Multiple Regression: Step Two

Variable(s) entered on step number 2
Included variable: X_1, X_4

Multiple R	0.895
R Square	0.801
Adjusted Square	0.788
Standard Error	0.2141

Analysis of Variance

Setting	DF	Sum of squares	Mean Square
Regression	2	5.534	2.767
Residual	30	1.375	4.585E-02
F = 60.350			

Variables in the Equation

Variable	B	SE B	Beta	Tolerance
X_1	0.366	0.058	0.562	0.842
X_4	0.487	0.085	0.508	0.842
Constant	-2.348	0.339		

Variable	VIF	T	Sig T
X_1	1.188	6.325	0.000
X_4	1.188	5.726	0.000
Constant		-6.918	0.000

Variables not in the Equation

Variable	Beta In	Partial	Tolerance	VIF
X_2	0.098	0.178	0.652	1.534
X_3	-0.057	-0.119	0.882	1.133
X_5	0.139	0.282	0.820	1.099
X_6	0.019	0.040	0.910	1.220
X_7	0.049	0.099	0.825	1.212

Variable	Min Toler	T	Sig T
X_2	0.652	0.973	0.339
X_3	0.797	-0.645	0.524
X_5	0.742	0.218	0.829
X_6	0.773	1.581	0.125
X_7	0.759	0.536	0.596

lations resulting from the carrying out of any work shown in the approved plan that is known to him, (s.4(3) of the BO). Correspondingly, B(A)R s.18A requires an AP to endorse the submitted plans comply with the BO. As an AP assumes also the role as a coordinator for the building works, the endorsement of the building plans does not confine to those parts that are designed by him/her.

There will be a penalty imposed on an AP for breach of these statutory duties imposed. Under s.5 of the BO, the BA shall appoint a disciplinary board to conduct disciplinary proceedings when an AP appears to have contravened the provisions of the BO. The disciplinary proceeding is governed by s.7 of the BO. The board has the power under s.7(3), to decide whether to remove an AP from the registrar permanently or for a period of time and or any other reprimand. The most significant penalties against an AP liabilities arising from violating the BO are stipulated under s.40(2A). For example, an AP who authorizes or permit incorporation of defective materials into building works shall be liable on conviction to a fine and to imprisonment.

Another critical factor that affects the remuneration level of an AP is the claim from his client. The responsibilities of an AP towards the client arising from the contract of engagement, the requirement thereof may be much higher than those imposed on an AP under the statute. For example, due to the high land costs in Hong Kong, the client will be highly concerned with the full development potential of the site in order to maximize their profit from the development. An AP should work towards the benefit of his client by ensuring the building proposal prepared by him/her meets with the client's objective. It will be likely to be claim by the client if the AP's design fail to achieve the permitted plot ratio and site coverage, under s.21 and s.20 of the B(P)R respectively. On the other hand, if the proposed design exceeds the permitted plot ratio or site coverage, it will inevitably be disapproved by the BD (s.16(3) of the BO).

An AP also acts as contract administrator in the development process. The building contract provisions authorized him to issue instructions and certifications to enable the works to be carried out as intended on changed. If the changes are initiated due the fault of the AP, the contractor can seek extension to the contract period as well as loss and expenses resulting therefrom. The client may turn to the AP to claim for the loss suffered.

Normally an AP should provides periodic site supervision to give control to the site progress and construction cost. In *Wharf Properties Ltd. & Another v Eric Cumine Associates Architects Engineers and Surveyors (1991) 52 BLR 1*, the developer sued the architect for breach of contract arising out of a building contract. It was alleged the architect had failed to properly supervise the contractor and the subcontractors. The works were delayed. The developer was held liable to damages sustained by the contractors. He also claimed for loss of rental income against the architect. Site supervision is now a statutory duty, particular, for the site safety matters upon the introduction of site supervision plan system.

In summary, the results of the empirical work are well supported by actual practice, in particular, many liabilities are imposed under the Building Ordinance.

REFERENCES

Buildings Department, *Building and Development Control in Hong Kong*, Hong Kong Government Printer, 1999.

Buildings Department, *Practice Notes to Authorized Persons, and Registered Structural Engineers*.

Chan, H.W. & Leung, W.Y., "The Uncertain Liabilities of Authorized Persons in Hong Kong" in *Asia Pacific Building and Construction Management Journal*, Vol 3 No.1 1997: pp.13-19.

Chan, H.W., "Indemnity Insurance for Architects/ Authorized Persons in Hong Kong", in *Professional Practice for Architects in Hong Kong*, Hong Kong: Pace Publishing Limited, 1997.

Hair, J.E. & Anderson R. et. al., *Multivariate Data Analysis with Readings*, Prentice Hall, 1993.

Hong Kong Government, *Buildings Ordinance*, Hong Kong Laws Chapter 123, Hong Kong Government Printer, 1996.

SPSS, SPSS Base System Syntax Reference Guide Release 6.0, SPSS Inc. 1993.

Creative Systems in Structural and Construction Engineering, Singh (ed.) © 2001 Balkema, Rotterdam, ISBN 90 5809 161 9

A hierarchy fuzzy MCDM method for studying Taipei Rapid Transit System contracting model

How-Ming Shieh & Hung-Kun Ku
National Central University, Taiwan

ABSTRACT: Civil & Architecture, Water & Electricity and Electrical & Mechanical are the three main engineering categories in the planning of Rapid Transit System (RTS). The appropriately use of the contracting models is the key to the project success. The decisions made in this procedure are concerned with the complexity and coordination of the construction interfaces as well as the compatibility and the integration of the system. The determinations should be considered seriously because of the impacts on the schedule, quality and cost of the construction phase. Therefore the contracting models of each Taipei Rapid Transit System (TRTS) Lines were organized into six models for the study. Firstly, the criteria were established for the further experts' evaluations of their weights based on the AHP method. Then the performances of these six contracting models were evaluated by utilizing the Triangular Fuzzy Number. Consequently, the "F model" is found to be the better one through the performances ranking.

1 THE REVIEW OF THE PAST CONTRACTING STRATEGIES OF TRTS

TRTS project had started on since 1986. Tamshui, Mucha, Chungho, Hsintien, Nankang and Panchiao Line had been accomplished to date with the total distance of 86.8 km and the total cost of 14.3 billion dollars. The follow-up constructed lines include Hsinchuang Line, Luchou Extension Line, Hsinyi Line, Sungshan Line and an eastward extension from the Nankang Line. In the case of other cities, the BOT contracting operation of the Kaohsiung Rapid Transit System had finished, whereas the system of Taoyuan, Hsingchu, Taichung and Tainan is in the progress of planning. Confronting with the enormous quantities of the above projects, the contracting operation becomes the focus of this study.

In the past, there are two types of contracting operations in Taiwan:

i. After the designer was authorized to finish the design work, the contractors fill in the total prices based on the bidding documents (the design drawing and the quantity of each pricing item) for the further regular open bidding procedure. Then the one who has the lowest price will win the bid.

ii. The owner delivers the bidding documents to the assigned contractor (public contractor) for the construction. The contractor will achieve the construction contract provided that his price is lower than the upset price of the owner.

In order to become the member of WTO, Taiwan government has begun implementing "Government Procurement Law" since 27th of May in 1999. In the regulations, the negotiation, the open tendering or the design-build method is approved to be selected for the contracting operation.

Regardless of any contracting method, the preparation before the contracting will influence the following progress of the project. In America, either the public or the private sector tends to adopt design-build method for the contract (Songer and Molenaar 1996; Molenaar, etc. 1999), whereas there are fewer design-build cases in Taiwan due to the incompleteness of the related laws.

Owing to the special laws of managing the construction industry in Taiwan, the scale of the early construction firms is hard to be compared with the international construction firms except for the public firms. Driven by gradually raising of the public construction scale in Taiwan, a lot of the projects will choose the Joint Venture method for the international open tendering procedure or be divided into several parts for the further contracting procedure.

The contracting model in this study is defined as the segmenting method of a project. Either the whole project is taken full charge by one contractor or shared out between different contractors respectively in terms of their specialty, for instance, the contracting model of each route in TRTS project. After synthesizing and arranging the past experiences and the future possible models, six contracting models are yielded and the best model is obtained through a series of literatures review, experts interview and performance evaluation.

2 THE EVALUATION OF THE CONTRACTING MODELS

Formerly the researchers in Taiwan couldn't study deeper into the issue of evaluating the contracting models because of the difficulty in acquiring the related information (Wang, 1997). A Fuzzy Multiple Criteria Decision Method is used in this study. The author intends to achieve the evaluation criteria weights and determine the ranking of various contracting models through questionnaires of experienced experts in rapid transit projects.

2.1 Building a Hierarchy Evaluation Criteria

A typical multiple criteria evaluation is defined as the process of ranking the performance of all feasible models considering more than one evaluation criterion. Five fundamental principles that suggested by Keeney & Raiffa (1976) for building the criteria are listed below:

i. Completeness : The evaluation criteria have to cover all important characteristics of decision problems.
ii. Operational : The evaluation criteria have to be significant and applied effectively by the decision maker.
iii. Decomposable : The evaluation criteria can be decomposed from the Higher Hierarchy into the Low Hierarchy.
iv. Non-Redundancy : The evaluation criteria have to avoid repeated measuring the same type of performance value.
v. Minimum Size : The evaluation criteria items should be lessen as much as possible in order to lower the use of manpower, time and cost.

The hierarchy evaluation criteria of the contracting model for the rapid transit projects is built according to the above principles, the related literature review (Aditi & Messiha 1999, Lin 1999) and the expert brainstorming. The most influent elements affected on the assemblies in releasing the contracts are divided into three portions, i.e., engineering management, site construction, and revenue service stage. We expect to determine the principal weight by analytic hierarchy process (Saaty 1980). The principal structural diagram is as shown in figure 1. and the details for every elements are as followed :

First Layer: Successful Assemblies on Releasing Contracts.
To evaluate the priority from different assemblies on releasing contracting model is the goal of this research.
Second Layer
To give further explorations to the influent factors (i.e., engineering management, site construction, and revenue service stage) on the successful contracting model.
Third Layer
Further details or explorations on Second Layer.
Contract Price:
Different releasing compositions influence the total cost.
Interface:
The numbers of interface contracts effect the engineering management.
Total Completion Dates:
Different releasing compositions arisen different compromising factors will impact final opening dates.
Successful Rating on Bidding Procedure:
Different releasing compositions impact the successful rating on bidding procedure.
Numbers for Change Order:
The numbers for change order happened during the contract period.
Contract Failure Risk:
Different risks occur on different releasing compositions.
Site Construction:
Expertise of Sub-Contractor:
The expertise of Sub-Contractor will affect the wills on bidding and the achievement on completion the contract of main contractor.
Construction Management of Main Contractor:
Different releasing assemblies derive different methods in management. It also shows the main weight to their projecting ability.
To Interflow Resources:
To dispatch machines, tools, and personnel to match different releasing assemblies.
Integration:
Different releasing policy will influence their flexibility in integration.
Obtainable Profits:
Different profits compare with different releasing assemblies.
To Master the numbers of Expertise Sub-Contractor:
To compare main contractor's abilities in controlling and coordinating sub-contractors on different releasing compositions.
Revenue Service Stage:
Completion Quality:
Unwholesome releasing composition will impact the quality of the whole contract.
Interface Between Construction Period and Revenue Service Stage:

After the project is completed and transferred to the operator, the various interface problems arise from the different contracting model.

Maintainability in Revenue Service Stage :
Different contracting model has varied spares. The increasing cost of managing the spares and the interface problems that may derived from it will affect the maintenance situation.

Reliability in Revenue Service Stage :
The influences of a lot of coordination factors yielded by varied contracting models on the operation period

Operation Maintenance :
Different contracting model will affect the follow-up maintenance condition.

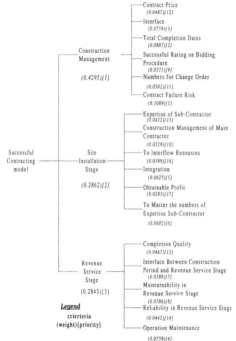

Figure 1. Relevance System of Hierarchy contracting model

2.2 *TRTS Contracting Model*

Six routes of the TRTS project had been contracted recently. There are 13 work items from arranging all project items of each route (TRTS Annual Report, 1998)(Table 1). Table 2 lists all possible contracting model of TRTS from the past to the future. Each model is represented by one of the seven alphabet letters (A, B, C, D, E and F). In each model, if each item code is listed alone, the item isn't combined with the other items for contracting. If two or more than two of the items are linked by the symbol of "+", then these items are contracted together.

Table 1. Classification of project items

Functional Classification	Work Breakdown Items
Civil & Architecture (C & A)	Lane and Station Landscape Work Water & Environmental control Trackwork Permanent Locking & Master Key Depot construction
Electrical & Mechanical System (E & M)	Escalator Power Supply Train Manufacture Signal Communication Automatic Fare Collection (AFC) Depot facilities

Table 2. Contracting Models

Models	Contents	Remarks
A	a,b,c,d,e,f,g, h,i,j,k,l,m	All items are contracted independently.
B	a+b+c+d+e+ f+g+h+k+m, l,i,j	The train, Signal and AFC system are contracted independently, whereas the remaining items are contracted by a single contractor.
C	a+b+c+e+f+ g, h+i+j+k+l+ m+d	The trackwork in C & A is exchanged with the escalator system in E & M. After the exchanging, the two parts of C & A and E & M are contracted by a single contractor respectively.
D	a+b+c+d+e+ f+g, h+i+j+k+l+ m	The escalator system is drew out from E & M and combined into C & A. After that, the two parts of C & A and E & M are contracted by a single contractor respectively.
E	a+b+c+d+e +g, h+i+j+k+l, f+m	Three divided parts : (1)C & A including escalator system but Depot construction; (2)Depot construction is combined with Depot facilities system. Each part is contracted by a single contractor.
F	a+b+c+d+e+ f+g+h+i+j+k +l+m	All items are contracted by a single contractor.

2.3 *The Process of Evaluating the Hierarchy module*

The process of evaluating the hierarchy modules includes three steps:

2.3.1 *Evaluating the weights for the hierarchy relevance system*

The AHP weighting (Saaty,1997,1980)is mainly determined by the evaluators who conduct pairwise comparisons, so as to reveal the comparative importance of two criteria. If there are evaluation criteria/objectives (items or nodes), then the decision-makers have to conduct pairwise comparisons. Furthermore, the relative importance derived from these pairwise comparisons allows a certain degree of inconsistency within a domain. Saaty used the principal eigenvector of the pairwise comparison matrix derived from the scaling ratio to find the comparative weight among the criteria (as aspects and criteria/objective, see Fig.1) of the hierarchy systems for contracting model.

Suppose that we wish to compare a set of n criteria/objectives in pairs according to their relative importance (weights). Denote the criteria/objective by C_1, $C2$, ..., C_n and their weights by w_1, w_2, ..., w_n. If $w=(w_1, w_2, ..., w_n)^t$ is given, the pairwise comparisons may be represented by a matrix A of the following formulation:

$$(A-\lambda_{max}I)w=0$$

Equation(1)denotes that A is the matrix of pairwise comparison values derived from intuitive judgment (perception) for ranking order. In order to find the priority eigenvector, we must find the eigenvector w with respective λ_{max} which satisfies $Aw=\lambda_{max}w$. Observing from intuitive judgment for ranking order to pairwise comparisons to test the consistency of the intuitive judgment that since small changes in elements of matrix A imply a small change in $\lambda j, (\sum_{i=1}^{n} \lambda i$

$=tr(A)=$sum of the diagonal elements$-n$, therefore only one of λ_j , we call it λ_{max}, equals n, and if $\lambda_j=0$, the $\lambda_j \neq I_{max}$), the deviation of the latter from n is a measure of consistency,i.e., C.I.$=(\lambda_{max}-n)/(n-1)$, the consistency index (C.I.), as our indicator of "closeness to consistency."In general, if this number is less than 0.1, we may be satisfied with our judgment (Saaty 1977, 1980).

In this problem, the group decision–makers should at least need to include three groups:a) engineering management, b) site construction, c) revenue service stage.

2.3.2 *Getting the performance value*

The evaluators (experts in contracting model) choose a score (performance value) for each contracting model based on their subjective (intuitive) judgment. This way the methodology for estimating the achieving level of each criterion/objective in each contracting model can use the methods of fuzzy theory for treating the fuzzy environment. Since Zadeh introduced fuzzy set theory (Zadeh, 1965), and Bellman and Zadeh(1970) described the decision–making method in fuzzy environments, an increasing number of studies have dealt with uncertain fuzzy problems by applying fuzzy set theory. The application of fuzzy theory to get the performance values can be described as follows:

Linguistic variable: According to Zadeh (1975), it is very difficult for conventional quantification to ex-press reasonably those situations that are overtly complex or hard to define; thus the notion of a linguistic variable is necessary in such situations. A linguistic variable is a variable whose values are words or sentences in a natural or artificial language. For example, the expressions of criteria/objectives as "Contract Failure Risk," "To Master the numbers of Expertise, or maintainability," "Interface Between Construction Period and Revenue Service Stage," "Completion Quality, " and so on all represent a linguistic variable in the context in these problems (see Fig. 2). Linguistic variables may take on effect–values such as "very high (very good), " "high (good), " "fair, " "low (bad), " "very low (very bad), "The use of linguistic variables is rather widespread at present and the linguistic effect values of contracting modules found in this study are primarily used to assess the linguistic ratings given by the evaluators. Furthermore, linguistic variables are used as a way to measure the achievement of the performance value for each criterion/objective.

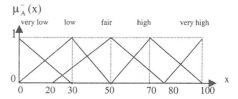

Figure 2. The Membership Function of the Five Levels of Linguistic Variables (hypothetical example)

2.3.3 *Evaluating contracting model*

Bellman and Zadeh (1970) were the first to probe the decision-making problem in a fuzzy environment, and they heralded the initiation of fuzzy multiple criteria decision–making (Fuzzy MCDM). Our study uses this method to evaluate the electronic marketing model and ranks it for each strategy. The method and procedures of Fuzzy MCDM theory are as follows:

a. *Measuring criteria/objectives*: Using the measurement of linguistic variables to demonstrate the criteria/objectives performance (effect–values) by expression such as "very high," "high," "fair," "low," "very low," the evaluators are asked to make their subjective judgments. Each linguistic variable can be indicated by a triangular fuzzy number (TFN) within a range of 0–100. Also the evaluators can subjectively assign their personal weights to the linguistics variables.

Let E_{ij}^k indicate the fuzzy performance value of evaluator k toward strategy i under criterion/objective j, and let the performance of the criteria/objectives be indicated by the set S, then,

$$E_{ij}^k = (LE_{ij}^k, ME_{ij}^k, UE_{ij}^k), j \in S \qquad (1)$$

Since the perception of each evaluator varies according to the evaluator's experience and knowledge, and the definitions of the linguistic variables very as well, the study uses the notion of average value so as to integrate the fuzzy judgment values of m evaluators, that is,

$$E_{ij} = (\frac{1}{m}) \otimes (E_{ij}^1 \oplus E_{ij}^2 \ldots \oplus E_{ij}^m) \qquad (2)$$

The sign $\otimes$ Denotes fuzzy multiplication, the sign $\oplus$ denotes fuzzy addition. E_{ij} is the average fuzzy number of the judgment of the decision–maker, and it can be displayed by a triangular fuzzy number as follows:

$$E_{if} = (LE_{ij}, ME_{ij}, UE_{ij}) \qquad (3)$$

The preceding end–point values $LE_{ij} = \frac{1}{m}(\sum_{k=1}^{m} LE_{ij})$, $ME_{ij} = \frac{1}{m}(\sum_{k=1}^{m} ME_{ij}^k)$, and $UE_{ij} = \frac{1}{m}(\sum_{k=1}^{m} UE_{ij}^k)$ can be solved by the method introduced by Buckley (1985).

b. *Fuzzy synthetic decision* : The weights of the criteria/objectives of electronic marketing as well as the fuzzy performance values (effect–values) have to be integrated by the operation of fuzzy numbers so as to be located at the fuzzy performance value (effect–value) of the integral evaluation. According to the weight w_j derived by AHP, the weight vector can be obtained, and the fuzzy performance matrix E of each of the model can also be obtained from the fuzzy performance value of each strategy under n criteria/objectives, that is,

$$w = (w_1, \ldots, w_j, \ldots, w_n)^t \qquad (4)$$
$$E = (E_{ij}) \qquad (5)$$
$$R = E \circ W \qquad (6)$$

and the sign "o" indicates the operation of the fuzzy numbers, including addition and multiplication. Since the operation of fuzzy multiplication is rather complex, it is usually denoted by the approximate multiplied result of the fuzzy multiplication and the approximate fuzzy number R, of the fuzzy synthetic decision of each strategy. The expression then becomes,

$$R_i = (LR_i, MR_i, UR_i), \forall \qquad (7)$$

Where $LRi = \sum_{j=1}^{n} LE_{if}* wj \qquad (8)$

$$MRi = \sum_{j=1}^{n} ME_{if}* wj \qquad (9)$$

$$URi = \sum_{j=1}^{n} ME_{if}* wj \qquad (10)$$

c. *Ranking the model (fuzzy number)* : The result of the fuzzy synthetic decision reached by each strategy is a fuzzy number. Therefore, it is necessary that the nonfuzzy ranging method for fuzzy numbers be employed during the comparison of the model. In other works the procedure of defuzzification is to locate the Best Nonfuzzy Performance value (BNP). Methods of such defuzzified fuzzy ranking generally include mean of maximal (MON), center of area (COA), and α–cut, three kinds of method (ZHAO & Govind, 1991, Teng & Tzeng, 1996). To utilize the COA method to find out the BNP is a simple and practical method and there is no need to bring in the preferences of any evaluators. For those reasons, the COA method is used in this study. The BNP value of the fuzzy number R_i can be found by the following equation:

$$BNP_i = [(UR_i - LR_i) + (MR_i - LR_i)]/3 + LR_i, \forall i \qquad (11)$$

According to the value of the derived BNP, the evaluation of each electronic marketing strategy can then proceed.

3 EMPIRICAL STUDY AND DISCUSSIONS

There are 13 experts participating in this questionnaire study. 7 of them are representative of Track, Construction and Mechanical & Electrical firms, whereas 4 and the rest of 2 experts are on behalf of Rapid Transit Department and Rapid Transit Corporation respectively. However all of them had taken part equally in the rapid transit projects for more than 8 years. Owing to the difficulty of acquiring the practical contracting data, the questionnaire method is used for analyzing the related data. The questionnaire is divided into 2 sections. In the first part, each expert is invited to conduct pairwise comparison for the evaluation criteria of the contracting models in order to determine the weight of each criterion. In the second part, the experts have to give their individual subjective assessments on the 5 linguistic variables (Very low, Low, Fair, High, Very High) with a value range of 0 to 100. The higher point indicates the better model, and vice versa. The experts must define the assessment range of the 5 linguistic variables before proceeding the evaluation of the contracting models.

3.1 Computing the weights

With the frame diagram of the contracting models selection criteria, the evaluators can be acquainted more clearly with the relationships and the weight scale of each criterion (Figure 1). The weights are computed by the AHP method and the "Expert Choice" software. The criteria weights that filled by the 7th expert couldn't pass through the consistency test (Satty, 1980), so the computation of weights is based on the results of the remaining 12 experts. From the aspects in Figure 1, the weight of construction management is the highest while the revenue service stage is the lowest. Inside the aspects, according to the rankings of all criteria significance, the contract failure risk has the highest weights while the total completion dates has the second priority and the following items are the interface of different contracts in construction stage, the operation maintenance, the integration and so on.

The contract failure risk which ranks as the highest criterion obviously indicates the scruple of Taipei Rapid Transit Department in awarding the whole route contract to a single contractor. Additionally, the total completion dates and the interface of different contracts are 2 significant issues in construction management, so they are in the second and third ranks. Formerly, the completion is the only factor that considered in the contracting procedure but the results of the experts questionnaires still rank the operation maintenance in the fourth position.

3.2 The rankings of the evaluated contracting models

According to the weight of each criterion through the AHP method and the fuzzy performances of all criteria given by the experts, the fuzzy performances of the contracting models are evaluated by formulas (6), (8), (9) and (10). Finally, the defuzzified performance values are figured out by formula (11) as shown in Table 3. The resulting rankings of each model illustrated by Table 3 are listed in sequence as below :

F<D<C<B<E<A

4 CONCLUSIONS

Presently, the contracting models of large-scale public projects in Taiwan can be roughly categorized into design-build, independent contract, combined contract of large-section, B.T. and B.O.T. However, its final objective is not beyond the scope of reducing the construction interfaces, producing the products that meet the quality demand and implementing the project successfully according to schedule.

The department of TRTS has completed some of the initial network continually and began revenue service. Driven by the increasingly growing up of the project scale of each TRTS route in the future, there are few local contractors be able to contract the project with design-build model according to their financial condition, construction capability and design ability. Moreover, with the "Construction Industry Management Regulations", the government hasn't permitted the contractors involving in design work. Nevertheless, as a result of this study, the model that awarding the contracts of the whole routes to a same contractor is the best one in views of the construction management, the field construction and the revenue service stage.

Table 3. The performances of each model

MODEL	Ri	BNPi
Plan_A	(39.7,49.4,61.1)	50.042 (6)
Plan_B	(56.3,66.4,75.9)	66.205 (4)
Plan_C	(56.3,66.4,77.0)	66.558 (3)
Plan_D	(56.4,66.6,77.4)	66.797 (2)
Plan_E	(54.8,65.0,75.9)	65.204 (5)
Plan_F	(62.0,72.3,82.1)	72.114 (1)

REFERENCES

Aditi, D.,and Messiha, H.M. (1999). Life cycle analysis (LCCA) in municipal organizations, Journal of infrastructure systems, 5(1), 1-10.

Bellman, R.E., and Zadeh, L.A. (1970). Decision-making in a fuzzy environment. Management Science, 141-164.

Buckley, J.J (1985). Ranking alternatives using fuzzy numbers. Fuzzy Sets and Systems, 15(1), 21-31.

Drew, D.S., and Skitmore, R.M.(1997). The effect of contract type and size on competitiveness in bidding, Construction Management and Economics,15,469-489.

Keeney, R.L, and Raiffa, H (1976). Decision with multiple objectives preferences and value tradoffs, John Wiley and Sons, New York.

Lin, C.P. (1999). A decision support system for determining the optimal contract size in a construction, PhD thesis Tokyo University , Japan.

Saaty, T.L. (1977). A scaling method for priorities in hierarchical structures. Journal of Mathematical Psychology, 15(2), 234-281.

Saaty, T.L. (1980). The analytic hierarchy process. New York: McGraw-Hill.

Songer, A.D., and Molenoar, K.R. (1997). Project Characteristics for Successful Public-Sector Design-Build, Journal of Construction Engineer and Management, ASCE 123(1)34-40

Tsaur, S.H., Tzeng, G.H., and Wang, K.C. (1997). Evaluating tourist risks from fuzzy perspectives. Annals of Tourism Research, 24(4), 796-812.

Zadeh, L.A. (1965), Fuzzy sets. Information and Control, 8, 338-353.

Zadeh, L.A. (1975), The concept of a linguistic variable and its application to approximate reasoning , Parts 1,2 and 3. Informance Sciences, 8, 199-249, 301-357; 9, 43-80.

Wang,M.D.,etc, (1996). A Study of Tendering Strategy and Construction Management Performance in TRTS,NSC Report.

TRTS Annual Report, TRTS, 1998.

Creative Systems in Structural and Construction Engineering, Singh (ed.) © 2001 Balkema, Rotterdam, ISBN 90 5809 161 9

Demystifying 'best value' – A contractor selection perspective

E. Palaneeswaran, X.Q. Zhang & M. Kumaraswamy
Department of Civil Engineering, University of Hong Kong, People's Republic of China

ABSTRACT: The crucial exercise of contractor selection is a complex decision-making process. Tender price is a dominant decision criterion in many contractor selection exercises. Generally, the performance potential of contractors differ, some of which may be reflected in the quality of the output. The 'price based' contractor selection practices ignore many significant value elements and potential performance levels of contractors. Although such approaches are preferred on various grounds such as simplicity and public accountability, they may potentially result in some 'false economy'. Furthermore, the relative significance of price and the risk transference patterns in procurement routes such as Design-Build and Build-Operate-Transfer (BOT) type arrangements render the purely low bid approaches less effective. In such complex procurement routes, the value of the end product depends on various factors including the technical quality as provided for and the performance of the contractor in delivering the promised quality levels. Therefore, in such demanding situations, a structured multi-criteria approach accounting for all the client goals and objectives is needed in a best value framework. This paper portrays an array of some interesting practices in some contractor selection approaches by various clients using the Design-Build and BOT routes. A summary of collective perceptions on some best value factors derived from a recent survey on Design-Builder selection is also provided. The discussions in this paper aim at facilitating benchmarking of better practices by establishing a knowledge base of good practices and innovative approaches.

1 BACKGROUND

Selecting appropriate contractors is one of the most crucial tasks facing a client's team. Multitudes of procurement options are available to today's construction clients and different types of contracts are fundamental in all those options. For instance, in Design-Bid-Build (DBB) routes, the clients contract sequentially with consultants and contractors; in a Design-Build project the client contracts with a single source, commonly known as Design-Builders; and in a Build-Operate-Transfer (BOT) project, the contract is signed between clients and the franchisee (who is responsible for designing, constructing, operating during the franchise period and then transferring the contracted facilities to the clients). Traditionally, 'low bid' selection is common in both public and private contracts. In general, the lowest bid may not necessarily provide the most economical end results. This is because: (a) there are many attributes other than price that influence the procured value of a construction contract and (b) contractors may seek other means to compensate for unrealistically low bids during the construction itself.

Furthermore, client requirements and project attributes also govern the best value definition. Therefore, the contractor selection exercises should also aim for achieving best value appropriate to a particular procurement scenario, matching specific client requirements. Thus, accomplishing best value goals in any contractor selection exercises is not just about economy. A discussion on contractor selection in a best value framework is presented in this paper. Some comparisons and lessons drawn from best practices and innovative initiatives (in contractor selections) of several clients from Asia, Europe, Australia and North America are also provided.

2 BEST VALUE CONTRACTOR SELECTION FRAMEWORKS

The following are some of the definitions provided by the Oxford English Dictionary for the term 'value': (a) quality of being useful or desirable; (b) worth of something in terms of money or other goods for which it can be exchanged; (c) what something is considered to be worth. All those definitions could be attributed to any contractor selection tasks in a 'best value' framework. Several research papers such as by Gransberg (1997) discussed the apparent cost savings illusions of purely price based contractor selections. Singh and Shoura (1998) suggested considering an 'overall profitability index' of bidders as one of the contractor selection criteria. United States Army Material Command (1998) recommended a 'price realism analysis' and 'cost versus quality' trade-off analysis for best value contractor selection.

Project attributes and client parameters such as time, cost, image, aesthetics/appearance, operation and maintenance, safety, security and environment aspects normally influence 'best value' in the construction procurement context. Specifics such as emergency situations (in general) and 'benefits to the local economy' (in public clients) also further sway the definition of 'best value' in construction procurement. Furthermore, patterns of risk management, flexibility requirements and resource constraints such as schedule (time) limitations, cost (budget) constraints and space limitations also govern the 'best value'.

2.1 *Value functions*

Best value procurements focus on selecting the contractor with the offer most advantageous to the client, when price and other factors are considered (Gransberg and Ellicott, 1997). Thus, a best value contractor selection should establish relevant evaluation criteria and their corresponding value functions in an appropriate 'best value' framework. The value function (of an evaluation criterion) may not be always linear and/ or uniform. For example, more floor area may be useful up to a certain limit and beyond that it may not be cost effective (with increased maintenance and operational expenses). Similarly, earlier completion/ product delivery while often beneficial may not be desired in another context. Figure 1 has been developed by the authors to portray probable relationships between actual criteria values and corresponding realizable best values. As portrayed in Figure1, the value function may not be always linear and/ or uniform. For example (Curve IV in Figure 1), more floor area may be useful up to a certain limit and beyond that it may not be cost effective (with increased maintenance and operational expenses). Similarly, earlier completion/ product delivery while often beneficial may not be desired in another context. Examples such as operating costs (in Design-Build-Operate projects) and tolls/ tariffs (in Build-Operate-Transfer) projects could be leading examples for decreasing realizable

best values with increasing criterion values (Curve V in Figure 1). While some evaluation criteria may have linear relationships with realizable best value (e.g. Curves I and II in Figure 1), others may have non-linear relationships. Some criteria may reach saturation value levels after certain limits and beyond which there may not any further increase in realizable value (Curve III in Figure 1).

In an idealistic source selection, the value functions of client's goals/objectives should be clearly spelled out at the initial stages. These should be translated into the source evaluation criteria that should similarly be clearly defined and made transparent. Transparency in source selection procedures has the potential to improve 'value' by enabling a level playing field, in which contractors can submit optimized bid proposals including reasonable profit margins, while clients can select a best value-yielding source. Furthermore, the 'best value' focuses in different procurement routes are also different. Figure 2 has been developed to portray some best value focuses in different procurement routes.

3 BEST PRACTICES TARGETING BEST VALUE IN DESIGN-BID-BUILD

3.1 *Background*

In Design-Bid-Build (DBB) procurements, design consultants are initially contracted separately through various means such as design competition and then suitable constructors are selected to construct the already completed designs. Hence, the scope for innovations are limited in this procurement route. Within such limited scope, client's focus on 'best value' selection approaches normally relies on constructor's track record information such as past performance, past experience, safety records and claims history. Various best practices such as - prequalification/ registration/ short-listing, considering non-price attributes in constructor selections, incentive/disincentive arrangements, performance specifications and performance based contracting are targeting best value in DBB route. Some such best practice approaches are listed in this section.

3.2 *Past performance*

In general, past performance is one of the significant forecasting parameters of quality performance that contribute to potential best value. Several clients have well established monitoring systems for contractor performance. While those performance management systems are useful in controlling and improving contractor performance in ongoing projects, some clients also use the past performance information for contractor prequalification and tender evaluation. For instance, many US Departments of Transportation use such past performance data in computing contractor prequalification ratings (such as 'maximum capacity rating'). Contractor performance scores could be advantageously blended into the contractor selection, if there is a well established performance measurement system. For example, the Hong Kong Housing Authority has established a 'PASS' system for contractor performance evaluation, on the basis of which they recently formulated a 'Preferential Tender Score' system. Furthermore, some clients such as the Works Departments under the Hong Kong Works Bureau (WB) allocate an '80:20' breakup for price and non-price attributes

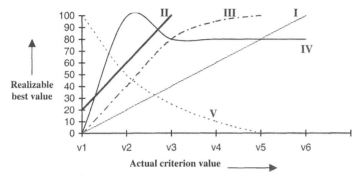

Figure 1. Some patterns of probable relationships between realizable value and actual criterion value.

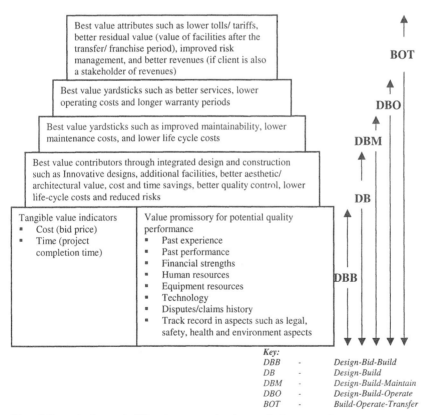

Figure 2. Best value focuses in different procurement routes – a selection perspective.

185

(such as past performance) in their tender evaluation.

3.3 *Incentive/ Disincentive arrangements*

In general, 'bonuses and penalties' are used in contracts to motivate and regulate contractors' performance pertaining to various criteria. In addition, some incentive/disincentive arrangements are also used to augment the ultimate value in DBB procurements. Several researchers such as Schexnayder and Ohrn (1997) and Arditi and Yasamis (1998) have discussed various aspects, including the significance of incentive/ disincentive schemes.

3.4 *Performance based contracting*

Under the philosophy of performance-based contracting, the client pays for the work it intends to accomplish for a given purpose and the client will not tell the contractor 'how to get the job done'. It says 'what needs to be achieved'. Performance specifying means specifying the required result rather than prescribing the means of achieving it. Key elements of performance based contracting are (i) outcome based contracts; (ii) performance specifications/ standards; (iii) compensation coupled with incentive/ disincentives; and (iv) advanced monitoring and measurement techniques. Palaneeswaran and Kumaraswamy (1999) described the use of performance based contracting in different sectors. Such performance specifications and performance based contracting approach could be advantageously used in a best value contractor selection framework.

4 BEST VALUE FRAMEWORK FOR DESIGN-BUILDER SELECTION

The single point of responsibility and potential gains in Design-Build increases the incentives for innovations dramatically. Design-Builder selections should aim in bringing 'best value' in an amicable win-win environment, balancing both quality and price in technical and price proposals. Any selection process in which proposals contain both price and qualitative components, and award is based upon a combination of price and qualitative considerations is called a 'Best Value' (also known as greatest value) selection (Design-Build Institute of America, 1999). Despite the growing popularity of Design-Build, only few detailed studies of specific selection methodologies such as Potter and Sanvido (1994 and 1995), Gransberg and Senadheera (2000) and Palaneeswaran and Kumaraswamy (2000) could be traced.

4.1 *Best practices and innovative approaches*

Some re-engineered 'low bid' based Design-Builder selection approaches in a best value framework are: 'Equivalent Design – Low Bid'; 'Complying Design – Low Bid'; 'Lowest Price – Technically Acceptable'; and 'Low Bid Design-Build with bid price adjustment factors for non-price attributes' (e.g. time). Furthermore, some best practices targeting best value Design-Builder selection are:

- Some clients prefer to adjust bid prices for certain higher quality elements or for other client preferences (e.g., the Florida Department of Transportation, USA has a bid price adjustment scheme for early completion proposals).
- Some clients include some adjustments for other client preferences in the evaluation of Design-Build proposals (e.g., the Asian Development Bank recommend discounting bid prices of 'domestic' contractors by 7.5%, in order to increase their work opportunities - Asian Development Bank, 1996).
- Some public clients use a 'Best and Final Offer' (BAFO) approach to obtain best value for money when choosing the best offer (e.g., a BAFO approach was used in the I 15 corridor reconstruction project by the Utah Department of Transportation, USA).
- Some clients adopt an 'Best Design-Best Price' approach in their prestigious projects (e.g., the U.S. Department of Agriculture earmarked a bottom-line bid price and called for proposals to 'bid up for quality' in their Beltsville Headquarters Office Complex project by setting a cost target price of US$ 37.7 million - Wright, 1998).

4.2 *Best value contributors in Design-Builder selections*

In general, in Design-Build projects the 'best value' could be realized from various aspects of a Design-Build proposal, such as project completion time, product life, maintainability, operating costs, life cycle costs, aesthetic value, environmental impact, standardization and flexibility. Table 1 displays a consolidated summary of opinions on aspects contributing to 'best value' from a recent questionnaire survey (on Design-Builder selection - conducted by the first author). Furthermore, in the proposed Design-Build proposal evaluation criteria, 'value engineering potential' is introduced for potentially eliminating items in bid proposals that do not add any value.

4.3 *Best value approaches in Design-Build variants*

Design-Build-Maintain (DBM) and Design-Build-Operate (DBO) are some innovative detours from the general Design-Build procurement in which, the clients transfer risks and responsibilities involved in 'maintenance' and 'operations' of constructed facilities to the contractors. In those procurement paths, the aspects of operational aspects such as quality of services, as well as cost of services and maintenance also contribute to the best value. For example, in the Tolt water treatment plant of in Seattle, USA the key procurement objectives set by the clients were as follows: (a) optimization of water treatment processes; (b) minimization of design & construction costs; (c) acceptable project schedule; (d) integrated operations; (e) quality services to the public; (f) optimized maintenance & operation costs.

4.4 *DBO case study*

In the proposal evaluation of that Tolt water treatment plant DBO project, financial criteria (including cost effectiveness and financial qualifications) made up 40 percent and technical criteria (including project implementability, technical reliability, technical viability, environmental aspects, past performance and 'women' and 'minority business enterprise' utilization) constituted the remaining 60 percent. In addition to the normal bonus/ penalty arrangements and liquidated damages for the design and construction phase, comprehensive liquidated damages were framed for the operation phase. For example, if the fluoride content in the treated water is beyond the specified range of (1.0 ± 0.05) units, based upon continuous monitoring, the liquidated damages will be $10 per million gallons after 60 minutes of non-compliance.

Table 1. Best value contributors consolidated from responses of a recent survey*.

Best value contributors	Rank
Improved maintainability and lesser operating costs	1
Earlier project completion time	2
Increased product life	3
Lower life-cycle costs	4
Additional benefits	5
Better aesthetic/ architectural value	6
Enhanced benefits for local economy	7
More environment-friendly	8
Fitness for multiple/ flexible use	9
Modular and repeatable design/ construction	10

* From a preliminary summary of 66 responses from various countries (in the second phase of two-phase survey on contractor selection for Design-Build projects)

5 BEST VALUE PERSPECTIVE IN BOT FRANCHISEE SELECTION

Rapid economic development, increasing population, and quick urbanization have resulted in an enormous demand for improved and new infrastructure facilities. However, the shrinking governmental financial resources cannot meet such demand.

Among the new procurement routes in public infrastructure development, the BOT approach is more popular. In addition to the utilization of money from the private sector, this popularity is also due to the recognition of the governments that some project development processes can be efficiently carried out by the private sector.

5.1 *Best selection values in BOT*

Besides additional funds from the private sector, the public client aims to achieve two more main objectives in choosing the BOT procurement route. One is the transfer of risks (which may be not quantifiable) to the private sector. Risks that can be transferred to the private sector include (a) design and construction risks; (b) operating risks (such as agreed quality and availability of the services); (c) revenue stream risks; and (d) technological obsolescence.

The other is to obtain a better value for money (e.g. in terms of reduced costs and enhanced services at lower prices) than a traditional procurement route would do. This judgement can be made by comparing the net present value of costs of the BOT scheme with that of a similar reference project procured by the traditional route, assuming that the full range of services of the BOT arrangement would be provided by the reference project. Other things being equal, the one with lower total net present cost is better. This comparison should be a fair one. The reference project should incorporate capital investment, operation, maintenance and ancillary service. It should realistically reflect all costs for all services provided by the BOT solution and if possible, value the risks which are not retained to the client. The reference project should also be affordable to the public client (Construction Industry Council 1998).

5.2 *Franchisee selection in BOT projects*

The significant paradigm shift from traditional route to the BOT scheme necessitates a corresponding change from traditional contractor selection techniques. The tenderers are usually consortiums that comprise several companies and may have no track records. Tenders should be evaluated against various package criteria, including compliance to the concession, deliverability, finance, operational issues, quality, technical factors, environment impacts and price.

There are two commonly used techniques for BOT tender evaluation, the multi-attribute utility analysis (MAUA) and the Kepner-Tregoe decision-making technique (Kepner & Tregoe 1981). In both techniques, the BOT tender is divided into several packages (e.g. legal, financial, technical, operational and environmental), which are assigned varying weights according to their relative importance.

Maximum available points are given to each criterion within a specific package. Each proposal is then evaluated and scores are given against each subcriterion. The proposal with the highest final total weighted score will be chosen for the concession. The Kepner-Tregoe technique differs from the MAUA in that the former classifies the various criteria into MUSTs and WANTs. The MUST criteria are mandatory, functioning as a screen to eliminate nonconforming alternatives. After screening through the MUST criteria, the remaining alternatives will be judged on their relative performance against WANT criteria, which is conducted in a way similar to the MAUA method. The critical success factors (CSFs) in BOT tendering have been discussed, for example, by Tiong and Alum (1997), identified the distinctive winning elements (DWEs) in the three main CSFs: (1) technical solution advantage, (2) financial package differentiation and (3) differentiation in guarantees.

6 CONCLUSIONS

Best value should be the goal in every contractor selection exercise. In best value contractor selection frameworks appropriate balance between 'most economically advantageous' perspectives and the 'best value' prospects should be structured. The best value selections should examine some key questions such as 'what is expected', 'what is delivered' and 'how can things improve'. Furthermore, Palaneeswaran and Kumaraswamy (2000) mentioned that the clients may consider questions such as 'how much more should be paid in order to achieve benefits from added value /best value?' and 'does the added value justify the additional price? Thus, in any best value contractor selection approaches, the prime focus for clients lies in establishing a structured framework of what 'value' they wish to achieve, where and how those value contributors exist, and how those value contributors could be evaluated in terms of both objective and subjective indicators.

In general, direct selections, competitive negotiations, cost/ design competitions, cost competitions and purely technical quality based competitions are all used in different contractor selection exercises. Some clients use a 'Best and Final Offer' approach to obtain best value in their procurements. Some good contractor selection practices that target best value are briefly enumerated in this paper with twin objectives of dissemination and future benchmarking. Although, best value selections may expectedly yield economical and other strategic benefits in the long run, some clients opt for 'desired value' instead of 'best value' due to some of their other constraints such as budget limitations.

REFERENCES

Arditi, D. and Yasamis, F. 1997. Incentive/disincentive contracts: perceptions of owners and contractors. *Journal of Construction Engineering and Management*, 124(5): 361-373.

Construction Industry Council 1998. Constructor's key guide to PFI. Thomas Telford Services Ltd., London, UK.

Design-Build Institute of America (1999). *Design-Build Manual of Practice*

Gransberg, D.D. 1997. Evaluating best value contract proposals. *1997 AACE Transactions*, C&C.04.1 to C&C.04.5.

Gransberg, D.D. and Ellicott, M.A. 1997. Best value contracting criteria. *Cost Engineering*, 39(6): 31-34.

Gransberg, D.D. and Senadheera, S.P. 2000. Design-Build contract award methods for transportation projects. *Journal of Transportation Engineering*, 125(6): 565-567.

Kepner, C.H. & Tregoe, B.B. 1981. The new rational manager. New Jersey: Princeton Research Press, USA.

Palaneeswaran, E. and Kumaraswamy, M.M. 2000. Contractor selection for Design-Build projects. *Journal of Construction Engineering and Management*, publication pending.

Potter, K. and Sanvido, V. 1994. Design-build prequalification system. *Journal of Management in Engineering*, ASCE, 10 (2): 48-56.

Potter, K. and Sanvido, V. 1995. Implementing a design-build prequalification system. *Journal of Management in Engineering*, ASCE, 11 (3): 30-34.

Schexnayder, C. and Ohrn, L.G. 1997. Highway specifications – quality versus pay. *Journal of Construction Engineering and Management*, 123(4): 437-443.

Singh, A. and Shoura, M. 1998. "Optimization for bidder profitability and contractor selection", *Cost Engineering*, Morgan Town, USA, 40(6): 31-41.

Tiong, R.L.K. & Alum, J. 1992. Distinctive winning elements in BOT tender. *Engineering, Construction and Architectural Management*, 4(2), 83-94.

United States Army Material Command (1998) "Contracting for best value – A best practice guide to source selection", *AMC Pamphlet 715-3*, United States Army Material Command, USA.

Wright, G. 1998. Tapping the team synergy. *Building Design & Construction*, Cahners Publishing Co., Chicago, USA, 39(8): 52-56.

7 Performance improvement

Creative Systems in Structural and Construction Engineering, Singh (ed.) © 2001 Balkema, Rotterdam, ISBN 90 5809 161 9

A framework for evaluating design project performance

A. Robinson Fayek & Z. Sun
Department of Civil and Environmental Engineering, University of Alberta, Edmonton, Alb., Canada

ABSTRACT: This paper describes a framework of factors for use in predicting and evaluating design project performance, and for measuring performance in terms of three criteria: cost, time, and quality. Because design is a relatively subjective process, many of these factors are linguistic in nature. A method is presented of quantifying each of these factors and relating the multiple inputs to the output. This method is implemented in the form of a fuzzy expert system for use in predicting and evaluating design project performance. An overview of this fuzzy expert system is presented. The main conclusion of this paper is that fuzzy logic provides an effective method of modeling the factors that impact design project performance. The fuzzy expert system provides a tool for design performance evaluation and prediction, which may be useful to project management personnel.

1 INTRODUCTION

Managing engineering design performance is a significant issue, although it has traditionally received less attention than managing project performance during the construction phase. One reason may be that the cost associated with the engineering phase is only 3-10% of the total project cost (Eldin, 1991). This relatively small percentage of total cost, however, can amount to a significant dollar value on a large project. Since most construction costs are fixed by the design features of the project, management of the engineering phase warrants more attention in order to reduce overall project costs.

This paper presents a detailed framework of factors for use in evaluating design project performance. An overview is presented of a fuzzy expert system for use in predicting the performance of a design project in terms of three criteria: cost, time, and quality. The use of fuzzy logic in the development of this expert system enables the assessment of the project and its environment to be made in linguistic terms, which would naturally be used by project personnel. A complete description of the development and validation of the fuzzy expert system is provided in Fayek and Sun (2000).

2 A MODEL FOR USE IN EVALUATING DESIGN PROJECT PERFORMANCE

Design is such a complex process that no single nor static factor can be used to predict nor evaluate its performance. Its evaluation requires a complete, dynamic, and comprehensive set of factors that influence performance, and a complete set of criteria to measure performance. The factors considered in the model can be classified according to three groups, based on their functions: context variables, input factors (and sub-factors), and output factors (and sub-factors). The structure of the model, which forms the basis of the fuzzy expert system, is as follows:

- Input sub-factors are used to determine each input factor.
- Input factors are used to predict output sub-factors (i.e., the performance of the design project).
- Output sub-factors, once predicted (using the model) or determined (after completion of construction) are used to determine each output factor.

2.1 Context Variables

Context variables are used to classify design projects into similar groups; these variables are qualitative in nature. Examples of context variables are: type of project, type and scope of design contract, type and scope of construction contract, method of tendering for the construction contract, and project priorities.

2.2 Input Factors

Input factors describe the project, its environment, and its participants. Each of these factors is variable, i.e., can vary from one project to another in a fixed context. Some of these factors become known only through the course of the design process. Each of these factors is subjective in nature and can be described using a linguistic term (e.g., small, average, large) or on a numerical scale (e.g., from one to ten). Each input factor can be further broken down into a number of sub-factors. These sub-factors are easier to quantify and serve to reduce the subjectivity associated with assessing the input factors. Table 1 lists examples of the input factors and their associated sub-factors.

Table 1. Examples of Input Factors and Sub-factors Impacting Design Performance

Factor No.	Factor
1	Overall size of design firm:
1.1 - 1.3	Number of employees, Annual volume of work, Number of projects held
2	Level of competition in the market:
2.1, 2.2	Number of similar design firms, Number of projects available in the market
5	Continuity of manhour commitment for project:
5.1, 5.2	Number of manhours per week per designer on the project, Total manhours per week on the project
10	Quality of owner's profile:
10.1 - 10.5	Time taken by owner to make a decision, Number of times owner changes mind or interferes, Number of changes in owner's personnel, Number of years of experience of owner's representative, Owner's attitude toward risk

2.3 Output Factors

Output factors are used to measure the performance of the design project in terms of three criteria: cost, time, and quality (i.e., accuracy of the design). Design performance can not be fully evaluated until construction is complete, at which point all design outcome information becomes known. The model described in this paper can be used to predict the level of each of these output factors prior to construction being complete. Each output factor is detailed by a number of sub-factors, as shown in Table 2.

3 DATA COLLECTION

The data required to develop, calibrate, and test the fuzzy expert system was collected from a survey of industrial design projects in Alberta and British Columbia in 1999 (Gue & Fayek, 1999). Eighteen case studies were obtained. For each input and output factor and sub-factor, the respondents were asked to provide a linguistic description (e.g., small, average, large) and a corresponding numerical value (e.g., number of employees). The linguistic terms and the corresponding numerical values were used to develop the fuzzy set membership functions describing each factor. The relationships between the factors were used to develop the expert rules.

4 GENERATING MEMBERSHIP FUNCTIONS

Membership functions are used in fuzzy set theory to represent linguistic concepts. A membership function indicates the degree to which an element of a set fits the linguistic concept, ranging from 0.0 (no membership) to 1.0 (full membership). One of the weaknesses of fuzzy set theory is in the inherent subjectivity involved in defining membership functions. A method of systematically generating membership functions on the basis of objective data was developed. Its basis is that the frequency of a given numerical response is proportional to the membership value of that response in the fuzzy membership function (Li & Yen, 1995). Each factor and sub-factor in the model has three linguistic descriptors, each of which can be represented by a fuzzy membership function. The frequency with which each numerical value is associated with each linguistic term describing a factor was plotted. The distribution of the responses was used to determine the shape and range of each membership function. Two common shapes of membership functions were used: triangular and trapezoidal (Lorterapong & Moselhi, 1996). These two shapes reflect the trends found in the frequency plots of the collected data: triangular shapes are used to model frequency plots with a distinct peak, and trapezoidal shapes are used to model frequency plots with a range of peak values.

The data collected from the survey were separated into a training data set (for developing the membership functions) and a testing data set (for validating the accuracy of the membership functions). The procedure for developing and validating the membership functions was conducted twice, using two different training and testing data sets (13 projects for training and 5 for testing in the first case, and 11 for training and 7 for testing in the second case, with no common projects in the two testing data sets). Fifteen membership functions out of 89 failed in both trials, yielding an 83% accuracy rate. These results are satisfactory, indicating the feasibility of the proposed technique for generating membership functions. Much of the inaccuracy lies in the lack of sufficient data with which to generate and test the membership functions. Nevertheless, the proposed technique is significant in that it illustrates how membership functions can be developed on the basis of objective (i.e., numerical) data.

Table 2. Output Factors and Sub-factors Used to Measure Design Performance

Factor No.	Factor
1	Level of performance against cost of design:
1.1 - 1.6	Percent change in design manhours (actual/budgeted), Percent design manhours due to change orders, Percent design manhours due to rework, Percent change in cost of design (actual/budgeted), Percent design cost due to change orders, Percent design cost to total construction cost
2	Level of performance against schedule for design:
2.1, 2.2	Percent change in duration of design (actual/scheduled), Percent of design document release deadlines missed
3	Level of accuracy of design documents:
3.1 - 3.5	Number of approved changes during construction, Total cost of approved changes, Percent value of construction due to changes, Total number of design rework manhours during construction, Number of problems during construction due to design errors/omissions

Table 3. Examples of Expert Rules for a Sub-model

If	Number of Employees (Input Sub-factor 1.1)	and	Annual Volume of Work (Input Sub-factor 1.2)	then	Overall Size of Design Firm (Input Factor 1)
	Small		Small		Small
	Small		Average-large		Average
	Average-large		Small		Average
	Average-large		Average-large		Average

If sufficient data can be collected for a given context, these membership functions can be developed to reflect a widely held concept of a subjective (i.e., linguistic) term. These membership functions can be calibrated to suit different contexts if the effect of the context variables on the shape and range of membership functions is identified.

5 MODEL SIMPLIFICATION USING CORRELATION ANALYSIS

The proposed model for design performance evaluation is complex, consisting of 14 higher-level input factors and 3 higher-level output factors, 16 of which have from 2 to 12 sub-factors. Given that each factor and sub-factor can take on any one of three linguistic terms, this complex structure leads to an unmanageably large number of expert rules. Furthermore, its complexity may make it difficult for the decision-maker to pinpoint the main factors influencing positive or negative design performance. Consequently, to yield a more realistic and manageable model, a correlation analysis was

conducted on each sub-model to eliminate input factors that are not significantly correlated to the output. Some sub-models were eliminated completely because there exists no strong relationship between any of the input factors to the output factor. No significant correlation exists between any of the output sub-factors and the higher-level output factors; consequently, all 3 of these sub-models were eliminated. As a result of the correlation analysis, 21 simplified sub-models remained.

6 GENERATING FUZZY EXPERT RULES

A technique for generating fuzzy expert (i.e., if-then) rules was developed. Rules are established based on existing patterns in the survey data, relating different levels of input factors to different levels of output factors in each sub-model. In building the rulebase, the direction of correlation is used to determine the logical relationship of the rule premise and the rule consequent. For each sub-model, the frequency of the rule indicates its relative likelihood of occurrence. Inconsistent rules are eliminated by choosing rules with the highest frequency and that are consistent with the correlation results. Rules that are not represented in the actual data are derived to cover all other possible combinations of input, thus ensuring completeness of the rulebase. Examples of the expert rules developed for one sub-model are shown in Table 3.

7 MODEL VALIDATION AND SENSITIVITY ANALYSIS

The fuzzy expert system was implemented in MATLAB Fuzzy Logic Toolbox (Mathworks, Inc., 1998). Model validation was performed using the 7 test cases not used for training. Base case testing was done using the "min" (minimum) operator in the premise of the rules, the min-max (minimum-maximum) method for rule implication-aggregation, and the centroid method for defuzzification. The output of each testing case was a crisp number after defuzzification. Whether or not a given sub-model is successful depends on two criteria: (1) how close the defuzzified crisp number is to the actual numerical value given in the test case (i.e., a numerical match); (2) to what extent the defuzzified value represents the linguistic term given in the test case (i.e., a linguistic match). Of the 21 sub-models, only 38% produced good numerical matches. Linguistic matching, on the other hand, produced good results, with a success rate of 71%. Because of the roughness of the membership functions, the fuzzy expert system can not conclude an accurate enough crisp number. In the design context, however, linguistic terms are more commonly used than crisp numbers to describe the dynamics of design project performance. Consequently, the performance of the model is acceptable on the basis of linguistic term prediction.

A sensitivity analysis was performed on the fuzzy expert system to determine which of its components affect its accuracy to the greatest degree. The sensitivity analysis was conducted by changing the defuzzification method, the implication-aggregation method, and the operator method, each in turn. The largest of maximum (LOM) defuzzification method is the best method for increasing the accuracy of the model. Changing the implication-aggregation method or the operator does not change the model's accuracy. Regardless of the method used, the results for numerical data prediction are not satisfactory. On the other hand, all linguistic term prediction results are good, especially for the LOM defuzzification method.

8 USES OF THE FUZZY EXPERT SYSTEM

The constructed fuzzy expert system can be used in several ways. First, given detailed information on the project, as represented by the input sub-factors, the expert system can determine a crisp value and a linguistic term to describe each input factor. This information provides the design manager with a good description of the project and its environment. Second, given all pre-construction information (i.e., values for the input factors), the performance of the design project can be predicted before construction is complete. Project management can use this information to alert project participants of any potential problems that may arise, or to modify the input factors to improve design output to meet the targeted time, cost, and quality objectives. The system can be used to optimize one or more project objectives and to assess the tradeoff between objectives.

9 CONCLUSIONS AND FUTURE RESEARCH

This paper describes a comprehensive framework of factors that affect design project performance and factors used to measure performance. It illustrates how fuzzy logic can be used to model design performance evaluation and prediction, both of which involve linguistic and subjective assessments. The use of natural language for reasoning in the model is a realistic and desirable feature for decision making in project management.

Future research is being conducted to: simplify and refine the proposed model through case studies of design projects; develop improved data collection techniques to increase the quality and quantity of data collected; and, explore how the context variables affect design performance and how to use this information to define and calibrate the membership functions.

ACKNOWLEDGEMENTS

The authors would like to express their sincere gratitude to the numerous companies that participated in this study, to Mr. K. Gue for administering the survey, and to Dr. W. Pedrycz and Dr. M. Allouche for their guidance. The financial support of the Natural Sciences and Engineering Research Council of Canada is gratefully acknowledged.

REFERENCES

Eldin, N.N. 1991. Management of engineering/design phase. *Journal of Construction Engineering and Management* 117(1): 163-175.

Fayek, A. Robinson & Z. Sun 2000. A fuzzy expert system for design performance prediction and evaluation. Submitted to the *Canadian Journal of Civil Engineering*, February.

Gue, K. & A. Fayek 1999. *Report on the University of Alberta's study of the factors affecting design performance.* CEM technical report CC-99/01. Edmonton: University of Alberta.

Li, H.X. & V.C. Yen 1995. *Fuzzy sets and fuzzy decision-making.* Boca Raton: CRC Press.

Lorterapong, P. & O. Moselhi 1996. Project-network analysis using fuzzy sets theory. *Journal of Construction Engineering and Management* 122(4): 308-318.

Mathworks, Inc. 1998. *Fuzzy Logic Toolbox user's guide for use with MATLAB.* Natick: Mathworks, Inc.

Creative Systems in Structural and Construction Engineering, Singh (ed.) © 2001 Balkema, Rotterdam, ISBN 90 5809 161 9

Determinants of construction planning effectiveness

O.O. Faniran
School of Civil and Environmental Engineering, University of New South Wales, Sydney, N.S.W., Australia

ABSTRACT: Previous research studies have investigated the impact of influence factors on construction planning effectiveness. However, these studies have focused mainly on assessing the individual effects of identified influence factors on construction planning effectiveness, and have not addressed the interrelationships that exist between the influence factors themselves. The aim of the study reported in this paper is to examine how individual influence factors work together to determine construction planning effectiveness. 52 building projects in Australia were examined and the interrelationships within predefined variable sets were examined using Principal Component Factor Analysis (PCFA). The paper concludes with a discussion of strategies for improving construction planning effectiveness.

1 INTRODUCTION

Construction planning can be defined as the development of appropriate strategies for the achievement of predefined construction project objectives. Previous research studies investigating the effectiveness of construction planning efforts have focused mainly on measuring directly-observable variables, and assessing the individual effects of these variables on the construction planning process (Cohenca et al 1989; Laufer and Cohenca 1990; Laufer 1991; Faniran *et al* 1994; Faniran *et al* 1998). However, there is also a need to examine the interrelationships existing within variables affecting project planning in order to understand how individual variables work together in influencing planning effectiveness; and identify variables that measure similar factors. This would allow conceptual not-directly-measurable determinants of project planning effectiveness to be identified. It would also be particularly useful for developing a classification scheme or taxonomy which can be incorporated into a model for assisting in organizing project planning activities and predicting the effectiveness of project planning efforts in different situations.

The objectives of this paper are: (i) to identify determinants of construction planning effectiveness that are characterized by related groups of variables; and (ii) to evaluate the relative significance of determinants of construction planning effectiveness.

2 LIMITATIONS OF PREVIOUS STUDIES

Previous research studies have examined the impact of a variety of influence factors on the effectiveness of construction planning efforts. Cohenca et al (1989), Laufer and Cohenca (1990) and Laufer (1991) examined the effect of eight situational variables (classified into three project dimensions - project complexity, project uncertainty and attitudes towards construction planning) on the efforts invested in construction project planning, and on the outcomes of construction project planning. Findings from the studies showed how these situational variables individually influence project planning.

Faniran *et al* (1994) and Faniran *et al* (1998) evaluated the influence of situational factors in project environments and organizational characteristics of performing organizations on project planning efforts and project planning effectiveness. As in the studies of Cohenca et al (1989), Laufer and Cohenca (1990) and Laufer (1991), significant relationships were also found between the variable sets representing the individual situational factors and the variable sets representing construction planning efforts and effectiveness in the studies of Faniran *et al* (1994) and Faniran *et al* (1998).

Although the findings from the studies highlighted above (and other related studies) have contributed to understanding how the construction planning process interacts with its environment, the potential applications of the findings to the development of strategies for improving construction planning effectiveness has certain inherent limitations. The major limitation of the findings arises from the fact that the methodologies used in the previous studies to evaluate the interactions between the construction planning process and influence factors have focused mainly on the interrelationships *between* the variable sets of the influence factors and the construction planning process. However, the methodologies used in the studies have not considered the interrelations that exist *within* the variable sets of the influence factors. The significance of the relationships developed in the previous studies is therefore valid only for the situation of the current relationships that existed within the variable sets at the time of analysis.

3 CONCEPTUAL FRAMEWORK OF CURRENT STUDY

The construction planning process can be conceptualized as consisting of two major components: (i) development of construction project plans; and (ii) execution of construction project plans. Both of these components are in a state of constant interaction with their respective environments. This is illustrated in Figure 1. The development of construction project plans is undertaken by project planners within the organizational environment of the construction firm. The planners are therefore bound to be constantly changing and adapting to influences from the organizational environment. Similarly, the execution of the plans (after they have been developed) is undertaken within the project environment and is bound to be affected by situational factors in the project environment. Therefore, if appropriate strategies for improving project planning effectiveness are to be developed, there is a need to understand how the construction planning process interacts with its respective environments. As discussed in the previous section of this paper, previous studies have examined the interactions that occur between situational factors in the project and organizational environments and the construction planning process.

The study reported in this paper examined the interactions that occur within the variable sets that constitute the project and organizational environments and the variable sets that constitutes the project planning process, and examined how the variable sets, as individual entities, interact with each other. Specifically, the study examined how construction planning effectiveness is related to the organizational environment of construction firms and to the project environment. Figure 2 illustrates the conceptual framework of the study.

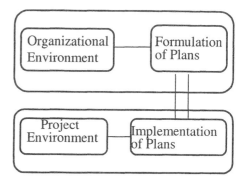

Figure 1. Interactions between the project planning process and the project planning environments

4 METHODOLOGY

4.1 Data Collection

The variables used in the study (see Figure 3) were identified from an extensive literature review of factors influencing project planning. Data was collected using a structured questionnaire. The questionnaires were distributed to 85 planning/contract management personnel in a sample of construction firms. Respondents were required to select a building project completed within the last five years and complete the questionnaire with respect to the selected project. Fifty-two of the firms responded, giving a 61% response rate. Tables 1 and 2 show the profile of the sample.

Table 1. Profile of Sample – Project Sizes

5 Project Size (Australian Dollars)	% of respondents
5.1 5.2 Under $100,000	-
$0.1 million- $1 million	14
$1 million- $10 million	39
$10 million- $15 million	9
$15 million- $20 million	11
Above $20 million	27

Table 2 Profile of Sample - Size of Construction Firms

Size of Construction Firms (average annual volume of work – Australian Dollars)	% of respondents
Less than $1 million	0
$1 million- $10 million	29
$10 million- $30 million	35
$30 million- $50 million	12
above $50 million	24

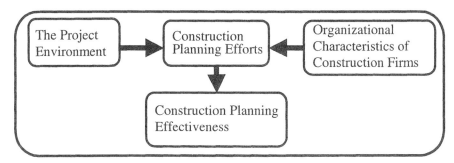

Figure 2. Conceptual Framework of Study

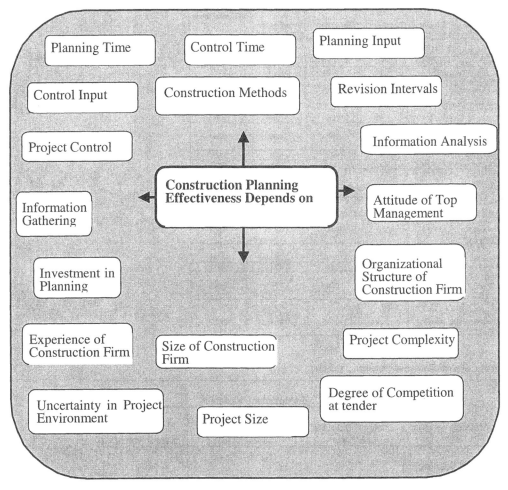

Figure 3. Summary of Variables Influencing Construction Planning Effectiveness

Table 3. Initial Statistics of Principal Component Factor Analysis

5.1 5.2 Factors	5.3 5.4 Eigenvalues	5.5 5.6 Percentage of variance	5.7 5.8 Cumulative percentage of variance
1	9.786	40.5	40.5
2	6.257	31.4	71.9
3	4.868	19.6	91.5
4	4.029	8.50	100

Table 4. Component Loadings – correlations between factors and loadings

Variables	FACTOR 1 Planning Efforts	FACTOR 2 Planning Effectiveness	FACTOR 3 The Project Environment	FACTOR 4 Organizational Characteristics	Communalities
Planning Time	0.75	0.15	0.06	0.11	0.950
Control Time	0.72	0.08	0.26	0.11	0.886
Planning Input	0.67	0.16	0.00	0.25	0.956
Control Input	0.62	0.02	0.24	-0.10	0.989
Revision Interval	0.61	0.08	0.07	0.24	0.972
Information Analysis	0.56	0.16	0.12	0.30	0.983
Focus on Construction Methods	0.54	0.14	0.06	0.43	0.841
Cost Variance	0.07	0.91	0.05	0.04	0.970
Time Variance	0.10	0.85	0.04	0.15	0.956
Man-Hour Variance	0.12	0.81	0.01	0.06	0.917
Workmanship	0.05	0.78	0.08	0.08	0.951
State of Design Completion	0.03	0.04	0.88	0.01	0.957
Difficulty of Performance Objectives	0.14	0.04	0.84	0.07	0.967
Past Experience	0.27	0.12	0.61	0.22	0.981
Age of Firm	0.06	0.10	0.09	0.85	0.927
Average Annual Volume ot Work	0.27	0.14	0.06	0.77	0.949
Level of Centralization in Organizational Structure	0.28	0.24	0.06	0.31	0.980

Table 5. Regression Analysis (standard β values)

Dependent Variable	Independent Variables		
	Project Planning Efforts	Organizational Characteristics	The Project Environment
Project Planning Effectiveness	0.54	0.06	0.78

4.2 Data Analysis

A two-stage data analysis approach was adopted. In the first stage, Pearson product-moment correlations were computed in order to examine the strength of the associations that occurred between the variables *within* the respective variable sets. To check for curvilinear relationships, scatter diagrams of data points were plotted. There was no indication of curvilinearity in any of the cases examined. Associations with a 95% confidence interval ($p<0.05$) were examined further in the second stage.

The second stage of the analysis involved examining the interactions within the variable sets using the principal component factor analysis (PCFA) technique. The use of factor analysis in this study enabled an in-depth understanding of the variable groupings underpinning the effectiveness of construction planning efforts.

5 RESULTS

The determinant of the correlation matrix was found to be 0.0083, which is greater than 0.00001, indicating that the data matrix is not affected by multicollinearity or singularity (Kinnear and Gray 1993). The Kaiser-Meyer-Olkin measure of sampling adequacy was found to be 0.514, which is greater than 0.5, indicating that the sampling adequacy is acceptable (Norusis 1993).

Table 3 shows all factors with their eigenvalues, percentage of variance, and cumulative percentage of variance. The 35 factors represented four basic factors. The first factor was interpreted as 'Project Planning Efforts' and accounted for 32.1% of the total variation (the variation was derived from the original principal component before rotation). The second factor was interpreted as 'Project Planning Effectiveness' and accounted for 16.4% of the total variation. The third factor, 'The Project Environment' accounted for 10.9% of the total variation and the fourth factor, 'Organizational Characteristics of Construction Firms' accounted for 9.5% of the total variation.

Table 4 shows the factor loadings of all the variables on the four factors except for loadings with coefficients less than 0.05. The table also contains the communalities that show how much of the variance in the variables have been accounted for by the three factors that have been extracted. A close analysis of the communalities revealed that the four factors in the analysis account for over 60% of the variance in all the variables, suggesting that the factor analysis has been quite effective. Like the percentage of variance the eigenvalues indicate the relative importance of the various factors in accounting for the total variance in the data set. Note that factors with an eigenvalue of less than one are not se-

lected because an eigenvalue is a measure of standardized variance with a mean of zero and standard deviation of one. The variance that each standard variable contributes to the principal component extraction is one. A component with an eigenvalue of less than one is less important than an observed variable and can therefore be ignored.

Table 5 shows the results of the regression analyses of the extracted factors. The dependent variable in the regression analyses was the extracted factor representing "Construction Planning Effectiveness". The independent variables were the extracted factors representing "Construction Planning Efforts", "The Project Environment", and "Organizational Characteristics of Construction Firms". The relationships found between the extracted factors show clearly that The Project Environment has the highest influence on Construction Planning Effectiveness ($\beta=0.78$). Project Planning Efforts also has a high influence on Construction Planning Effectiveness ($\beta=0.54$), though not as significant as The Project Environment. Organizational Characteristics of the Construction Firms had a relatively insignificant influence on Project Planning Effectiveness ($\beta=0.06$).

A major implication of these results is that any strategy that is developed for the purpose of improving project planning effectiveness should not only focus on improving Project Planning Efforts, but should also focus on identifying and eliminating (or at least minimizing) any potentially adverse impacts which The Project Environment might have on Project Plans. This could be achieved by designing appropriate and effective control systems that are capable of detecting potential problems early, so that corrective measures could be taken before the problem manifests itself. The results also indicate that the organizational environment in which the project planning process is undertaken has no significant influence on the outcome of the process.

6 CONCLUSION

Previous studies examining factors that determine construction planning effectiveness have focused on assessing the individual influences of situational variables on construction planning effectiveness. However, these studies have failed to consider how the different variables work together in influencing project planning effectiveness. Therefore, the significance of the relationships developed in those studies is valid only for the situation of the current relationships that existed within the variable sets at the time of analysis. In the study reported in this paper, PCFA was employed to determine the inherent structure of the variable sets. Four factor groupings were extracted, representing The Project Environment, Organizational Characteristics of Performing Organizations, Project Planning Effectiveness, and Project Planning Efforts.

The results of the study showed that the factor group Project Environment exerted the highest influence on Construction Planning Effectiveness, followed by Construction Planning Efforts. The influence of Organizational Characteristics was found to be relatively insignificant. Strategies developed to improve construction planning effectiveness therefore need to recognize the influence which the project environment has on project performance and design appropriate measures to address this potential influence. One way of reducing potentially adverse impacts of the project environment is through the use of effective control systems that are capable of detecting potential problems and rectifying them at an early stage.

REFERENCES

Cohenca, D., Laufer, A. and Ledbetter, W.B. (1989). Factors Affecting Construction Planning Efforts. *Journal of Construction Engineering and Management*, ASCE, Vol. 115(1), 70-89.

Faniran O.O., Oluwoye, J.O. and Lenard, D. (1994). Effective Construction Planning. *Construction Management and Economics*, Vol. 12, 485-499.

Faniran O.O., Oluwoye, J.O. and Lenard, D. (1998). Interactions between construction planning and in-fluence factors. *Journal of Construction Engineering and Management*, ASCE, Vol. 124(4), 245-256.

Kinnear, R.P. and Gray, D.C. (1993). *SPSSPC+ made simple*. Lawrence Erlbaum Associates, Publishers, Hove, East Sussex, U.K.

Laufer, A. (1990). Decision-Making Roles in Project Planning. *Journal of Management in Engineering*, ASCE, Vol. 6(4), 416-430.

Laufer, A. and Cohenca, D. (1990). Factors Affecting Construction Planning Outcomes. *Journal of Construction Engineering and Management*, ASCE, Vol. 116(1), 135-156.

Norusis, M.J. (1993). *SPSS for Windows, base system user's guide*. SPSS Inc., Chicago, Ill.

Creative Systems in Structural and Construction Engineering, Singh (ed.) © 2001 Balkema, Rotterdam, ISBN 90 5809 161 9

Performance-reason-process model for managing design

A.S.Chang

Department of Civil Engineering, Cheng Kung University, Taiwan

ABSTRACT: Ineffective control systems are one major reason for inadequate design performance. Traditional project control systems focus on detecting cost and schedule results, but seldom further analyze causes and provide process feedback. In addition to result feedback, an effective control system should be able to help analyze causes and provide process feedback. This paper presents a performance-reason-process (PRP) model, to complement traditional systems in controlling and improving design performance. This model links design performance, reasons for inadequacy, and design processes. In a situation of inadequate design performance, this model assists clients and designers in finding out the reasons, responsibility, and improving processes.

1 INTRODUCTION

Design affects the life cycle cost of a construction project to a great extent. However, design performance is usually not satisfactory to the owner. A survey reveals that about one third of design projects miss cost and schedule targets (Anderson & Tucker 1994). It seems that inadequate design management has become the norm for many design projects (Barlow 1985).

Design process includes many complex variables. According to Sternbach (1988), the process variables causing design project delays are extra work beyond the original scope of agreement, failure of the owner or its other consultants to provide information, and objections by environmental or community groups. Many construction problems are due to design defects and can be traced back to the design process (Bramble & Cipollini 1995). For these problems, design complexity can be a major cause (Glavan & Tucker 1991). Furthermore, few effective measures exist for evaluating design performance (Tucker & Scarlett 1986).

Managing design projects needs information to control design performance. This management will rely on control systems, especially for more complex and larger projects. Good control systems will more likely ensure successful projects (Ashley et al. 1987).

Control systems should provide process feedback in addition to result feedback because quality comes from process improvements (Deming 1982), and organizations that have quantified process variables have significantly outperformed those that have not (Juran 1992). Process performance can be evaluated earlier and more frequently, which will allow the project to avoid undesired results, and modify its goals to fit changing conditions.

This paper presents a process model for managing design projects. This model works more than a control system. It links process and result information as well as reasons for inadequacy to diagnose, control, and improve design performance.

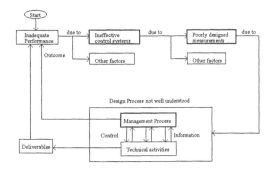

Figure 1. Logic of inadequate design performance

2 INADEQUATE DESIGN PERFORMANCE

The logic of inadequate design performance can be shown in Figure 1. Inadequate performance has many causes. Ineffective control systems are a major cause occurring during the project execution. Ineffective control systems are interlocked with other causes. Control systems can be ineffective because measurements are poorly designed so useful data are not generated. Measurements are poorly designed because the design management process is not well understood.

2.1 Poorly designed measurement

Measurement is an essential step in any control process (Koontz & Weihrich 1990). Project performance is reflected by measures, upon which corrective action is suggested and taken. However, Sink & Tuttle (1989) claim that measurement systems in organizations are poorly designed. Drucker (1993) states that the measurements most needed to allow for proper business control are lacking. Oglesby et al. (1989) also indicate that the performance measures for construction are often unsatisfactory.

2.2 Not-well-understood design management process

While efforts have been aimed at improving the construction process management, design process management is less studied. Edlin (1991) says that design management depends almost solely on a *drawing control log* although it is not an ideal control tool. Also, labor-hours have been a common measure for many tasks although they aren't always the most effective measure for the work of professionals (Sink & Tuttle 1989).

As seen in Figure 1, the management process directly influences performance outcome. The influence is greater than that of the deliverables mainly produced from technical activities. But project management is only classified as overhead or "level-of-effort" work in many situations (McConnell 1985). Managers and engineers tend to focus on technical activities rather than management process (Tarricone 1993).

3 STEPS TOWARDS ADEQUATE PERFORMANCE

To address the above problems, the following studies have been taken (Chang 1997):

3.1 Identifying the reasons for inadequate performance.

Identifying reasons is always the first step when addressing a problem. Categorizing the reasons helps spot problems and trace responsibility. The reasons for inadequate design performance (cost and schedule overrun) were investigated by analyzing four design projects with cost and schedule increase. They are categorized and listed in Table 1.

Table 1. Reasons for Inadequate Design Performance

A. Mainly Within the Owner's Control

R1 Owner's request
  a. Additional work
  b. Convenience
  c. Optimistic schedule
  d. Omissions

R2 Owner's failure
  a. Failure to provide information
  b. Incomplete or incorrect information
  c. Other consultants

B. Mainly Within the Consultant's Control

R3 Consultant's failure
  a. Consultant inability
  b. Underestimates or omissions

C. Beyond Either the Owner's or Consultant's Control
R4 Growing needs

R5 Stakeholders
  a. Agencies
  b. Public

R6 Others

Among these reasons, growing needs (R4) have to be explained. Since design is complex, at the scoping stage extra work or additional level of effort is often not anticipated, and is found needed after more studies, engineering, or design has been done. The work grows "naturally" without requests from the owner. This is an excusable reason.

3.2 Developing Measurement

Measurement helps clarify the definition of objectives (Keeney 1992). Under conditions of imperfect process knowledge, measurement can also be used to create visibility (Kurstedt 1985). In other words, measurement helps increase the understanding of measured objects.

Table 2. Deliverable Measures

A. Accuracy
1 Consistency/change
2 Errors and omissions
3 Rework hours
4 Impact of change/error
B. Usability
5 Completeness
6 Clarity
7 Conformance
8 Format
C. Constructability
9 Construction time
10 Construction knowledge
11 Construction safety
12 Maintainability
13 Adaptability
D. Design Economy
14 Construction cost estimate
15 Resources
16 Value engineering
17 Field engineering
E. Milestone Schedule
18 Timely delivery
19 Milestone progress
20 Milestone forecast
F. Milestone Cost
21 Milestone hours
22 Milestone forecast

As shown in Figure 1, the management process needs further understanding. In order to achieve the necessary understanding, deliverable measures were developed to help understand the deliverable standards, process measures developed to understand the management process, and cost and schedule measures developed to understand performance outcome.

Contracts always specify an owner's requirements for a project's products. The requirements from a typical design contract were studied and translated into deliverable measures (Table 2). Project management practices that create satisfactory performance, i.e. the best practices, can be captured by analyzing management processes of successful projects. From these practices process measures were developed (Table 3). The performance outcome, in terms of cost and schedule results, reflected in the four projects' documents

were investigated to develop cost/schedule measures. Milestone and periodic cost and schedule results provide result and process feedback, respectively (in Tables 2 and 3).

Table 3. Process Measures

G. Communication
1 Coordination
2 Stakeholder needs
3 Meeting effectiveness
4 Responsiveness
5 Schedule effectiveness
6 Administrative soundness
H. Consultant Team
7 Team turnover
8 Teamwork
9 Knowledge & experience
10 QA/QC functioning
11 Project documentation
12 Project management methods
I. Owner Satisfaction
E. Project Schedule
14 Monthly progress
15 Project forecast
F. Project Cost
16 Project monthly cost
17 Project forecast

These deliverable and process measures and the reasons identified above constitute the performance-reason-process model.

4 THE PERFORMANCE-REASON-PROCESS MODEL

The three groups listed in the three tables can be linked to explain the relationships among inadequate deliverable performance, the reasons for the inadequacy, and causal processes (see Figure 2). Since deliverables are the products of design input and process, unexpected deliverable performance can be traced back to its cause: poor input or design process. Since many of the developed process measures have been either theoretically or empirically proven to be associated with better performance, they will help analyze and identify possible cause-and-effect relationships between processes and deliverable performance. Since the reasons for inadequate performance are the lessons learned from real project operations, these reasons will also help clarify the cause-and-effect relationships between processes and deliverable performance.

Table 4. Deliverable performance analysis

Project: Eng. & Road Design Deliverables: Due Dates:
Milestone: 11 A. PSSR Draft: 4/4/99 Rec'd: 4/4/99
Amount:
Duration: 3/14/99 ~ 5/31/99 Reviewer: K, J Date: 6/10/99

Measures (1)	Weight (2)	Score (3)	Reasons * R1 - R6 (4)	Improvement ** needs on G, H, I (5)	Adjusted Score # (6)
A. Accuracy	10	8			
1 Inconsistency/change	0	N/A			
2 Errors and omissions	35	7			
3 Rework hours	30	10			
4 Impact of change/error	35	8			
Subtotal	100	83			
B. Usability	30	17			19
5 Completeness	35	4	R3, R6	G2	6
6 Clarity	30	6			
7 Conformance	25	7			
8 Format	10	7			
Subtotal	100	56			63
C. Constructability	20	14			
9 Construction time	50	6			
10 Construction knowledge	30	9			
11 Construction safety	0	N/A			
12 Maintainability	0	N/A			
13 Adaptability	20	6			
Subtotal	100	69			
D. Design economy	0	0			
14 Construction cost estimate	0	N/A			
15 Resources	0	N/A			
16 Value engineering	0	N/A			
17 Field engineering effort	0	N/A			
Subtotal	0	0			
E. Milestone schedule	20	13			
18 Timeliness	34	10			
19 Milestone progress	33	5	R3	G5, H12	
20 Completion forecast	33	5	R3, R2	G5, H12	
Subtotal	100	67			
F. Milestone cost	20	10			
21 Milestone hours	50	5	R3	G5, H12	
22 Completion forecast	50	5	R3, R2	G5, H12	
Subtotal	100	50			
Total	100	62			64

* Unexpected performance is due to reasons R1 - R6, listed in Table 1.

** The processes in G, H, and/or I (listed in Table 3) that should be improved.

For reasons beyond the consultant's control, lower-rated scores can be adjusted.

4.1 *Implementation Procedures*

To use this model to gain desired control and improvement, an implementation procedure was established as shown in Figure 3. The steps included in this procedure are: assign weights to measures, use measures to evaluate deliverable and process performance, and calculate performance scores.

Measures do not have equal influence on overall performance. This prompts the need to assign weights to measures. Evaluation of a deliverable can be made after it has been submitted. Process evaluation can be made regularly, monthly or bimonthly. Instruments were designed for both deliverable and process evaluations, and scores can be calculated from the ratings.

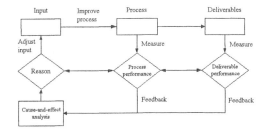

Figure 2. The performance-reason-process model

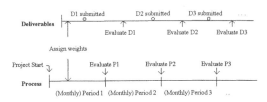

Figure 3. The performance evaluation procedure

4.2 *Performance Analysis Tool*

Unexpected performance can be analyzed after the evaluation of a deliverable has been completed. Traditionally, comments are sent to the consultant after his/her deliverables have been reviewed by the owner. These comments generally identify errors or inadequacy about the deliverables, and request the consultant to correct or clarify. But the reasons for inadequacy and methods for improving work processes are usually not addressed. This can be improved by linking deliverable performance, processes, and reasons for inadequate performance.

Table 4 is a tool for analyzing the performance. For measures rated low, e.g. less than 6, one must identify what the reasons are, which work processes need improvement, and if there is a need to adjust the rated score. Then entries are made in the corresponding blanks.

Identifying reasons for inadequacy helps reduce the chances of occurrence and helps track responsibility. Constant responsibility tracking will make project participants responsible for their own actions, and reduce future argument and disputes. Identifying improvement needs in the work process is essential to improving performance.

5 CONCLUSION

This paper presents a systematic performance-reason-process model for managing design. The complete and flexible measures can disclose areas of excellent and poor performance for both deliverables and the design process.

Measurement instruments were also prepared.

This model can generate data for cause-effect analysis. The linkage of deliverable performance, reasons for inadequate performance, and processes is a tool to aid this analysis. This linkage allows the owner or consultant to identify what the reasons are for inadequate performance and within whose control, and what processes need improvement. Then corrective action can be taken accordingly.

Substantial productivity increases can be directly attributed to the use of productivity measurement systems (White and Austin 1989). The developed model being equipped with measurement function can be used to improve project performance.

ACKNOWLEDGEMENT

The author would like to thank the National Science Council of Taiwan, Republic of China, for the grant of NSC89-2211-E006-090.

REFERENCES

Anderson, S. D. & Tucker, R. L. 1994. Improving Project Management of Design. *J. of Management in Engineering*, 10(4): 35-44.

Ashley, D. B., Lurie, C. S., & Jaselskis, E. J. 1987. Determinants of Construction Project Success. Project Management J. June, pp. 69-79.

Barlow, K. J. 1985. Effective Management of Engineering Design. *J. of Management in Engineering*, ASCE, 1(2): 51-66.

Bramble, B. B. & Cipollini, M. D. 1995. *NCHRP Synthesis of Highway Practice 214: Resolution of Disputes to Avoid Construction Claims*. Transportation Research Board, National Research Council, Washington D. C.

Chang, A. S. 1997. Consultant Performance Measurement and Evaluation for On-Call Projects. Ph.D. dissertation, Univ. of California at Berkeley, Spring.

Deming, W. E. 1982. *Quality, Productivity and Competitive Position*. Massachusetts Institute of Technology Center for Advanced Engineering Study.

Drucker, P. F. 1993. We Need to Measure, Not Count. *The Wall Street Journal*, April 13.

Eldin, N. N. 1991. Management of Engineering/Design Phase. *J. of Construction Engineering and Management*. ASCE, 117(1): 163-175.

Glavan, J. R. & Tucker, R. L. 1991. Forecasting Design-Related Problems--Case Study. *J. of Construction Engineering and Management*, ASCE, 117(1): 47-65.

Juran, J. M. 1992. *Juran on Quality by Design*. The Free Press, New York.

Keeney, R. L. 1992. *Value-Focused Thinking*. Harvard University Press, Boston, Mass.

Koontz, H. & Weihrich, H. 1990. *Essentials of Management*. 5th edition, McGraw-Hill, NY.

Kurstedt, H. A. 1985. The Industrial Engineer's Systematic Approach to Management. MSM Working Draft Articles and Responsive Systems Articles, Management System Lab. Virginia Tech.

McConnell, D. R. 1985. Earned Value Technique for Performance Measurement. *J. of Management in Engineering*, ASCE, 1(2): 79-94.

Oglesby, C. H., Parker, H. W., & Howell, G. A. 1989. *Productivity Improvement in Construction*. McGraw-Hill, Inc., New York.

Sink, D. S. & Tuttle, T. C. 1989. *Planning and Measurement in Your Organization of the Future*. Industrial Engineering and Management Press, Institute of Industrial Engineers.

Sternbach, J. 1988. *NCHRP Synthesis of Highway Practice 137: Negotiating and Contracting for Professional Services*, Transportation Research Board, National Research Council, Washington D. C., July.

Tarricone, P. 1993. What Do You Mean By That? *Civil Engineering*, April, 60-62.

Tucker, R. L. & Scarlett, B. R. 1986. *Evaluation of Design Effectiveness*. A Report to the CII, The University of Texas, Austin.

White, C. R. & Austin, J. S. 1989. Productivity Measurement: Untangling White-Collar Web. *J. of Management in Engineering*, ASCE, 5(4): 371-378.

Creative Systems in Structural and Construction Engineering, Singh (ed.) © 2001 Balkema, Rotterdam, ISBN 90 5809 161 9

Project management improvement – A case study of a large contractor in Hong Kong

H.Y.Sam, K.W.Wong & P.C.Chan
Department of Building and Real Estate, Hong Kong Polytechnic University, People's Republic of China

ABSTRACT: The prosperity of Hong Kong is highly associated with the level of construction activities. The nature of construction work is heterogeneous and enormously complex. Construction projects are intricate and time-consuming. Construction companies have to provide effective management skills in order to maintain competitive advantage in the market. Company A, being one of the leading contractors in Hong Kong operates actively in the local construction market. In order to sustain its influence in the industry by occupying a relatively high market shares, continuous improvement in its management system is necessary. The objective of this paper is to carry out an investigation on Company A and to explore its existing management problems. Firstly, the paper describes different school of management theories and their evolution. It is followed by a case study. Company A's past history and its existing project organization structure are being reviewed. In the concluding section, possible solutions are recommended to resolve those identified management problems.

1 MANAGEMENT THEORIES

Business entity is essentially a continuing and internal activity, which involves a company's human, finance, and other resources. 'Management' relates to the planning, organizing, directing, and controlling of a business entity. Four management theories evolved from the beginning of this century:

1.1 *Classical theory*

Fayol (1841-1925), Urwich and Taylor (1865-1915) developed the classical theory in the middle of 1930's. Its characteristics are authoritarian leadership style, specialization of works and centralization. This management theory mainly focused on the tasks of the business. Labour was treated as a resource, to be worked as hard as possible.

1.2 *Human relation theory*

Elton Mayo (1880-1949) is the founder of the human relation theory. He found that employees must be understood as human beings if organizations are to be run efficiently. This theory focused on communication, worker's participation, leadership, stress, labour turnover, performance and motivation. The result revealed that workers considered job security, good working environment, and career advancement path are much more important than payroll. The human relation theory is more appropriate in dynamic situations.

1.3 *System theory*

The system theory was a new way of thinking during the 1960's and 1970's. Under this management theory, employees and tasks are considered to have a close relationship and managers should manipulate and formulate their approach to suit the dynamic situations.

The system theory views an organization as an open system, which interchanges information with external environment through its permeable boundary. That is to say, the organization itself is part of the input-output model in which resources are brought from the environment and then transferred within the organization system into desirable output in order to fulfill the objectives of the system.

1.4 *Contingency theory*

This theory was developed in the 1960's as the result of a number of research studies (Lawrence and Lorsch's, 1967). This theory seeks to understand the interrelationships within and among subsystems, as well as between an organization and its environment. It is different from the traditional management approach in which it does not suggest type of organization to suit all situations. In fact, contingency theory provides guidelines for management's considerations in designing the organizational structure. Burns and Stalker (1996) found that there were two kinds of management structure, which called 'Mechanistic' and 'Organic'. For the former, the more rigid and mechanistic stable conditions, where technology and market conditions are changing slowly. For the latter, the more flexible and organic approach works well for firms which are operating in unstable conditions, where market conditions are unpredictable or technology is changing rapidly. Mintzberg (1993) stated that structuring of organization requires a close fit between contingency factors and the designing parameters- its structure to match its situation. The concept that no one best structure will suit all organizations has prevailed (Nahaphiet & Nahaphiet, 1985).

For a typical construction project, the site organization structure will need to reflect the uniqueness of the project and the dynamics of the industry to be fully effective. A careful examination of the contingency factors reveals that the best design should be a mix of organic and mechanistic systems although the ideal structure leans itself more towards the organic system (Chan & Tam, 1994; Chan, 1995a; Chan 1995b).

There are various definitions for management, including 'getting things done through people' and 'a process of planning, organising and controlling activities' (Kast & Rosenzweig, 1985). However, nowadays the 'open system' approach and 'self-discipline' based management concept advocated by Karl Weich and James March seems to be widely accepted by managers (Calvert, Bailey and Coles (1995)):

1. firm control and strong direction,
2. human nature of employees,
3. external factors, and
4. self-discipline, humanity, outside world of customers and society.

In essence, organization and management concepts follow a logical evolution. There has not been a radical transformation that eliminated the old and substituted the new. Rather, the resulting contingency theory has been a changing of the old and retaining enduring concepts.

2 RESEARCH METHODOLOGY – CASE STUDY APPROACH

Case study method based on a structural survey was used to gather data for this study. Yin (1989) defines a case study as an empirical inquiry that investigates a contemporary phenomenon within its real-life context; is appropriate when the boundaries between phenomenon and context are not clearly evident; and in which multiple sources of evidences are used.

It is the intention of this paper to research on the structure of Company A by applying different school of management theories, identify the major management problems and suggests solution to resolve the problems in order to achieve continuous improvement.

3 BACKGROUND OF COMPANY 'A'

Company A is a large state-owned construction enterprise in the People's Republic of China. In the late 1970's, it expanded its business to the construction market in Hong Kong.

At present, Company A is one of the Hong Kong Special Administrative Region (SAR) Government's Public Works contractors. Furthermore, it is also registered in the Hong Kong Housing Authority's List for Contractor Group NW2 (i.e. New Works 2) which enables Company A to submit tenders of unlimited amount. In 1997, the company has a total asset value of HK$6.6 billion and a staff of over 2,000. The annual business turnover exceeds 1 billion since 1993. Table 1 illustrates the annual turnover of Company A.

Table 1. Annual Turnover of Company A (1994-1998).

Year	1994	1995	1996	1997	1998
Business Turnover (Hong Kong $)	3,700M	3,200M	7,560M	3,032M	5,427M

Source: Annual Reports of Company A (Year 1994 to Year 1998)

4 IDENTIFICATION OF EXISTING MANAGEMENT PROBLEMS

4.1 *Typical organization structure*

Basically, Company A adopts a team approach in organizing its project works, i.e. the project organization, which is common in large construction projects. For each construction project, the site manager is the project leader. All key members of the site team including project manager, site agent, quality manager, quantity surveyor, engineer, safety officer etc. are employed on site and are responsible for one particular project only.

Under this typical organization structure in Company A, resources may not be fully utilized since the functional expert only serves for one particular project. Besides, it does not provide job security to staff of the site team as redundancy is likely when existing project has completed and there is no new project to follow.

4.2 *Authoritarian style of leadership*

According to the policy of Company A, the senior management at the Head Office nominates all site managers. The site managers are mostly from the Chinese Mainland. Although they may possess the technical knowledge required for construction and have been working in Hong Kong for a few years, they do not fully participate in day-to-day operations. They concentrate mainly on internal administrative work. They are reluctant to interact with outsiders, especially with architects or clients. Moreover, they do not normally delegate authorities to their subordinates. The limited knowledge of local construction practices is a hindrance to their judgement, decision making and issuing directions during the progress of work.

Besides, most site managers apply authoritarian style of leadership. Most team members are being regarded as passive, which require tight control. Motivation and satisfaction of the subordinates is not a priority. The effect of such management style leads to low-morale, high labour turnover, disputes, low productivity and poor sense of belonging to the company.

4.3 *Top-down Communication System*

Company A is made up of people from different educational background and levels. Their contributions vary and a lot of project information has to be communicated and shared. This requires a sound and well-organized network of communication system. Even a network exists, communication still breaks down because management staff either fail to keep simple messages, or they pass on too little information. Similar to other construction companies, the senior management of Company A have different formal and informal communication channels to pass on their instruction or information to their subordinates. On the other hand, there is no proper bottom-up communication channel for the subordinates to communicate with their superiors.

4.4 *Recognition of achievement to an individual rather than to a team*

Each year the senior management of Company A will nominate a number of staff who have demonstrated excellent performance to the award of 'Best Staff of the Year'. However, such award has its own deficiencies. Construction is basically teamwork. Should target be achieved, the recognition from the company should be shared by the whole team instead of an individual. According to past records, only the senior management staff has the opportunity to get the award. Under this circumstance junior staff consider such award as ear marked for senior staff only, thus creating a de-motivated working atmosphere within the organization.

4.5 *Inadequate training*

It is undoubtedly that provision of on-the-job training to staff will enable them to accomplish their assigned tasks more productively. Knowing what to do allows them to work more quickly and make fewer errors than they would without the training. Output is also enhanced because of the motivational benefit of the training experience itself. The employees are pleased with themselves upon finishing job-training courses because not only they have learned new skills, but also the skills they have mastered also increase their earning power. Furthermore the training experience brings further satisfaction, which are the concept of challenge and reward.

Unfortunately, there is no training department in Company A. The senior management have overlooked the desire of staff and the benefits in relation to training. It is the perception of the senior management that staff should be designated to work on their assigned responsibilities instead of attending 'non-productive' training activities. Another reason is that training is a long-term investment on staff and financial support is required. It is disappointed that the senior management does not encourage staff to attend training courses organized by outside professional bodies. They do not sponsor their staff in monetary term nor to grant paid study leave to the attendants.

4.6 Lack of a standardised project cost control system

Company A does not have a standard project cost control system at company level. Every construction site has its own method to control the cost. There is a cost plan prepared by the quantity surveyor during the construction stage of the project. All works are itemized and priced in the cost plan, which leads to an estimated final contract sum. However, the cost plan is usually furnished to the senior management for approval at a late stage of the contract period. It is not uncommon to see site managers to manipulate the estimated final sum of the cost plan in order to meet the expectation of the senior management. As a consequence, the estimate does not represent the true cost.

5 RECOMMENDATIONS

5.1 Matrix organization

To overcome the deficiencies of the existing project organization structure of Company A, a matrix organization is recommended. In the matrix organization all staff are responsible for the various trades and specialists report vertically to their 'line' managers in the head office and laterally to the site manager (Chan, 1991).

The functional managers will allocate their subordinates to a number of projects. The matrix organization structure can retain the benefits of a project organization and at the same time achieving good resource utilization and giving the staff a high job security as they can revert back to their functional department once the project is completed.

5.2 Good leadership

In order to choose a suitable leader to supervise a project team, the senior management should appoint a competent person regardless of the cultural background. There should not be a standard mode of leadership style. A site manager must be flexible and be able to alter his management style to suit the circumstances. The Hersey & Blanchard's (1982) model of leadership would provide managers with general guidelines on appropriate leadership style.

5.3 Effective communication system

Apart from the existing communication channel for Company A, graphical and numerical communication tools should be introduced. In a construction project, there is heavy reliance on graphical and numerical communication. They are in the form of drawings, diagrams, schedules and charts. A single drawing often conveys a great deal of information in a much clearer way than would be possible by using text alone. Bar charts and network diagrams are popular means of presenting information, which is partly numerical and partly written. Those are valuable management tools. Another consideration to improve communication is by the use of computing network. Network computers can acquire, process, store, re-arrange and disseminate large amount of information almost instantly. It is also an efficient tool to communicate over long distances with various concerned parties. By the use of network computers, it can help to improve the speed and accuracy for the transmission of project data.

5.4 Appropriate recognition of achievement

Company A should set up a new award for the 'Best Team of the Year' rather than for the 'Best Staff of the Year'. In order to enhance the 'motivation effect' of such award, a few inexpensive arrangement and preparation are all that required. For instance, to organize a special ceremony to award the certificates. The sense of achievement for the staff will be higher than it would be if certificates were simply handed out in plain envelopes. Under this situation, the 'Best Team' will derive not only a sense of 'team-spirit' but also a sense of 'belonging' to the company.

5.5 Provided on-the-job training

On-the-job training to the employees is being regarded as a long-term investment to the company. A full training programme gives the employees an opportunity to learn and to advance from low to high management level. The programme provides basic entry-level skills training, upgraded skills training, lateral skills training, supervisory training, as well as specialized training in areas such as site safety and first aid. These training should be offered in addition to the existing training schemes provided by Company A.

5.6 Project cost control system

Project cost control begins with the preparation of the original cost estimate and the subsequent construction budget. Keeping within the cost budget and knowing when and where job costs are deviating are the two major factors that constitute the key to profitable operation. As the work proceeds on site, cost accounting methods are applied to determine the actual costs of production. The costs, as they actually occur, are continuously compared with the budget. In addition to monitor current expenses, periodic reports should be prepared to forecast final project costs and to compare these predicted costs with the established budget. Cost reports should be prepared at regular time intervals. These reports are designed to determine the cost status of the project and to pinpoint those work classifications where expenses are excessive. In this way, the senior management's attention can be focused on these job areas instantly. Timely information is crucial for effective action against cost overruns.

6 CONCLUSION

Company A is such a large organization that site managers must be equipped with effective skills for managing people. These roles have collectively become known as personnel management. To carry out the personnel function effectively, site managers have to take a closer look at their own leadership style and the effect they have on others. At the same time, the value of productivity management techniques and personal organization skills are becoming more widely appreciated.

In this paper, management problems are being identified in Company A. Due to the continuous changing working environment, the existing management system is found to be inappropriate. Matrix organization structure should be adopted in managing construction projects.

Furthermore, alternative management tools are proposed to cope with the communication problems in Company A. Further study is recommended to develop effective project management procedures, structures, and reward systems in order to create a working climate that is favorable to improve safety, timeliness and quality on construction sites. The identified problems and the recommendations are also useful references for other Chinese Mainland based construction companies in Hong Kong.

REFERENCES

Bent, A. J. & A. Thumann 1989. *Project Management for Engineering and Construction*. The Fairmont press, Inc.

Calvert, R.E., G. Bailey, & D. Coles 1995. *Introduction to Building Management*, Butterworth-Heinemann Ltd., 6th Ed., p. 28.

Chan, A.P.C. 1991. *Matrix Organisational Structure, UNIBEAM Journal*, National University of Singapore, Vol. x1x, 60-63.

Chan, A.P.C. 1994. *Leadership and Project Performance*. International Conference on Engineering Management. The Institution of Engineers, Australia, 247-255.

Chan, A.P.C., & C.M. Tam 1994. Mechanistic or organic? - Effective organisational structure for a medium sized building project, *NZIOB Papers, New Zealand, Paper 1*.

Chan, A.P.C. 1995a. Contingency approach to organisational design, Campus Construction Papers, *The Chartered Institute of Building, March, Vol. 3, Issue 1, Spring, 13-15*.

Chan, A.P.C. 1995b. Engineer the structure to fit the organisation, *The Chartered Institute of Building 1996 Handbook, September, C14-C18*.

Clough, R.H. & G.A. Sears 1991. *Construction Project Management, Third Edition*. A John Wiley Publication.

Hersey, P. & K.H. Blanchard 1982. Management of Organisation Behaviour: Utilizing Human Resources, 4th Edition, 262.

Kast, F.E. & J.E. Rosenzweig 1985. *Organisation and Management – A System and Contingency Approach*. McGraw Hill, 4th Edition.

Langford, D. et al. 1990. *Construction Management Vol. 1 and 2*. First published Mitchell in association with the CIOB.

Lawrence, P.C. & J.W. Lorsch 1967. *Organization and Environment - Managing Differentiation and Integration*. Harvard, Harvard Business.

Mintzberg, H 1993. *Structure in fives: designing effective organisations*, Englewood Cliffs, N.J.: Prentice Hall.

Nahaphiet. H & J. Nahaphiet 1985. A comparison of contractual management for building projects, *Construction Management and Economics, 3, 217-231*.

Pilcher, R. 1985. *Project Cost Control in Construction*. First published in Great Britain by Collins Professional and Technical Books.

Walker, A. & S.M. Rowlinson 1990. *The Building of Hong Kong - Constructing Hong Kong Through the Ages*: Published for the Hong Kong Construction Association, Hong Kong University Press.

Yin, R.K. 1989. *Case Study Research – Design and Methods*. Applied Social Methods Series. Vol. 5, SAGE Publication.

Creative Systems in Structural and Construction Engineering, Singh (ed.) © 2001 Balkema, Rotterdam, ISBN 90 5809 161 9

A decision framework for construction technology selection

M. Hastak & V. Thakkallapalli
Department of Civil and Environmental Engineering, University of Cincinnati, Ohio, USA

S. Gokhale & E. Sener
Department of Construction Technology, School of Engineering and Technology, IUPUI, Purdue University, Indianapolis, Ind., USA

ABSTRACT: The decision to replace a conventional construction process by an automated system requires careful analysis of tangible and intangible factors such as need based criteria, economic criteria, technological criteria, project specific criteria, and safety/risk criteria. As every construction project is unique, it is necessary to evaluate the feasibility of an automated system on a project-to-project basis. This paper presents AUTOCOP, a decision support system, for *AUT*omation *O*ption evaluation for *CO*nstruction *P*rocesses. AUTOCOP has been designed to assist construction managers in systematically evaluating the two options: conventional construction process versus automated system. The concept developed through this research has been applied to several case studies in trenchless technology selection. The development, selection, and utilization of trenchless technology (TT) has expanded rapidly over the past ten years and includes a wide range of methods such as *Pipebursting and Cured-in-place Lining* to rehabilitate existing underground pipelines with minimal excavation of the ground. This paper will discuss a case study and application of AUTOCOP.

1 INTRODUCTION

Open-trench method is currently the most widely used method for installation of underground pipelines and conduits of all sizes. Traditional construction methods of underground conduits have included plowing and trenching. However, open-cut construction has several shortcomings, chief amongst which are: safety concerns of workers, surface disturbance, disruption to vehicular/pedestrian traffic and reduction of pavement life. Trenchless technology (TT) offers a wide range of methods, materials, and equipment to install new or to rehabilitate existing underground pipeline and utility systems to overcome some of the drawbacks mentioned above. However, it is necessary to evaluate the suitability of an automated system on a project-to-project basis by considering all the criteria and subcriteria important for the decision task.

This paper presents AUTOCOP, a decision support system, for *AUT*omation *O*ption evaluation for *CO*nstruction *P*rocesses. AUTOCOP has been designed to assist construction managers in systematically evaluating the two options: conventional construction process versus automated system. The concept developed through this research has been applied to several case studies in trenchless technology selection. The development, selection, and utilization of trenchless technology (TT) has expanded rapidly over the past ten years and includes a wide

range of methods such as *Pipebursting and Cured-in-place* to rehabilitate existing underground pipelines with minimal excavation of the ground. This paper will discuss one of the case studies and application of AUTOCOP.

2 AUTOCOP

The factors that need to be considered in a decision problem of this nature have been organized in a hierarchy of five criteria and associated subcriteria in the proposed system (Figure 1). The five criteria include: need based criteria, technological criteria, economic criteria, project specific criteria, and safety/risk criteria. The criteria and subcriteria are arranged in a hierarchy to establish their interdependencies and facilitate their analysis through the Analytical Hierarchy Process (AHP).

2.1 Decision Framework

AUTOCOP includes an analytical model and a group decision model (GDM). The GDM assists the primary decision maker (PDM) in (i) evaluating the group members and (ii) in collecting and evaluating the opinion of other team members and experts who are involved in the decision making process. This model synthesizes the information obtained from other team members into a group decision that is

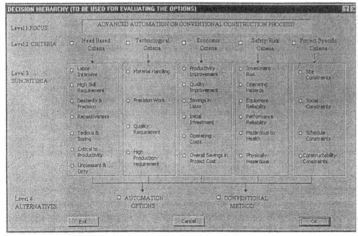

Figure 1. Sample Hierarchy of decision Factors

then used by the analytical model to process the various criteria and subcriteria.

Subjective assessment is often used in comparing options involving new technology. Therefore, it is important to have a team consensus in order to perform a more responsive analysis of the decision problem. Both models in this decision framework utilize Analytical Hierarchy Process (AHP) to determine the preference among various criteria, subcriteria, and alternatives. For more information on AHP refer to Saaty (1982) and Hastak and Gokhale (2000).

2.2 Group Decision Model

In the group decision model (GDM), each team member is evaluated by the primary decision maker (PDM) in two stages with respect to four criteria (i) their technical knowledge, (ii) experience, (iii) current project knowledge, and (iv) knowledge about the technology. In the first stage, the PDM evaluates the four criteria to determine his or her value based preference among the four criteria (Figure 2). The evaluation is based on the AHP. Each criterion has four associated intensities or subcriteria namely: extensive, significant, moderate, and low. AHP analysis of the subcriteria with respect to each criterion establishes the distinction between the four intensities. This analysis has been illustrated in Figure 2 for a hypothetical project situation.

In the second stage of the group member evaluation, the PDM evaluates (or grades) each group member's technical knowledge, experience, current project knowledge, and knowledge about the technology as extensive, significant, moderate, or low. The score obtained by each group member is the sum of the weighted score obtained under each cate

gory (Figure 2). The scores obtained by each group member are normalized to determine the weight for each group member's input for the decision problem. AUTOCOP allows a maximum of five team members to evaluate the decision problem (Figure 3). Also five alternative technologies can be evaluated simultaneously.

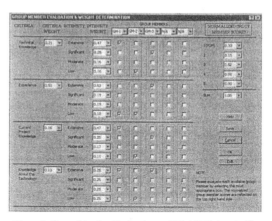

Figure 2. Group Member Evaluation and Weight Determination

Each group member evaluates the decision hierarchy (Figure 1) to establish relative importance of the various criteria and subcriteria in reference to the project situation and their knowledge based preference (Figure 3). The input provided by each group member is weighted according to the group member evaluation performed earlier (Figure 2). The sum of the weighted input from team members provides the group decision. The group decision thus obtained is used in the analytical model for further analysis.

Once the group priorities have been established for the subcriteria, alternative technologies are graded by the group members based on a 0-5 relative scale where 0 indicates least desirable and 5 indicates most desirable option with respect to a specific subcriteria and the project under consideration. The input provided by each group member is again weighted according to the group member evaluation to determine the group decision with respect to the technologies under consideration (Figure 3).

2.3 Analytical Model

The objective of the analytical model is to evaluate the input provided by the group members and to establish the group's preference among the various criteria, subcriteria, and alternatives (Figure 1). The Analytical Hierarchy process is used for establishing the comparison (or weight) matrix for each level of the hierarchy and for computing the priority vectors (Figure 3).

In the absence of significant historical performance data it is important that all the important aspects are carefully analyzed and sensitivity analysis performed before selecting a construction process, particularly advanced automation. An easy-to-use user interface has been designed to assist the user in this decision problem and for performing sensitivity analysis.

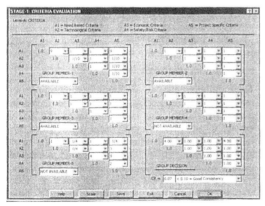

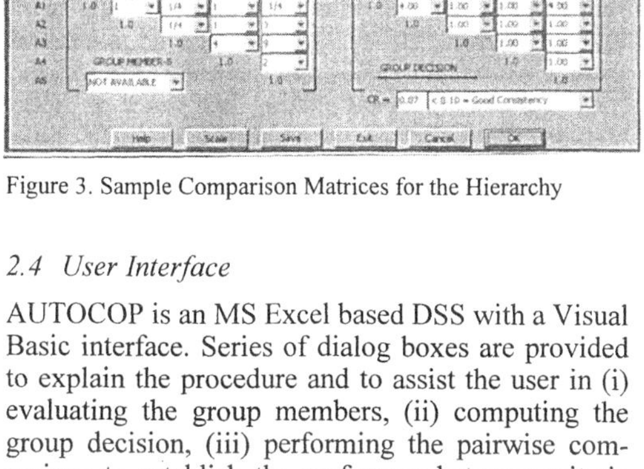

Figure 3. Sample Comparison Matrices for the Hierarchy

2.4 User Interface

AUTOCOP is an MS Excel based DSS with a Visual Basic interface. Series of dialog boxes are provided to explain the procedure and to assist the user in (i) evaluating the group members, (ii) computing the group decision, (iii) performing the pairwise comparison to establish the preference between criteria and subcriteria, and (iv) the evaluation of alternatives.

The user interface allows easy interaction with the system and facilitates sensitivity analysis of the inputs and the corresponding results. The user can provide the input with respect to the pairwise comparison by utilizing the drop-down combo boxes. The values in the combo-boxes represent a predetermined comparison scale. A button labeled "scale" has been included on the dialog box to provide easy access to the comparison scale and it's explanation. The user interface not only facilitates data input but also provides real time feedback in terms of results (final priority vector, intermediate priority vectors, group decision, etc.) and also sensitivity analysis. The user interface also provides a tabulation of results to include the final decision and priority vectors established at each level of the hierarchy. A printout of the results allows the user to critically assess the evaluation of the criteria, subcriteria, and alternatives and perform modifications and sensitivity analysis as desired. A sample summary of results has been shown in Figure 4. Additionally, data collection sheets and help file has been included to assist the user in the data collection and analysis phase.

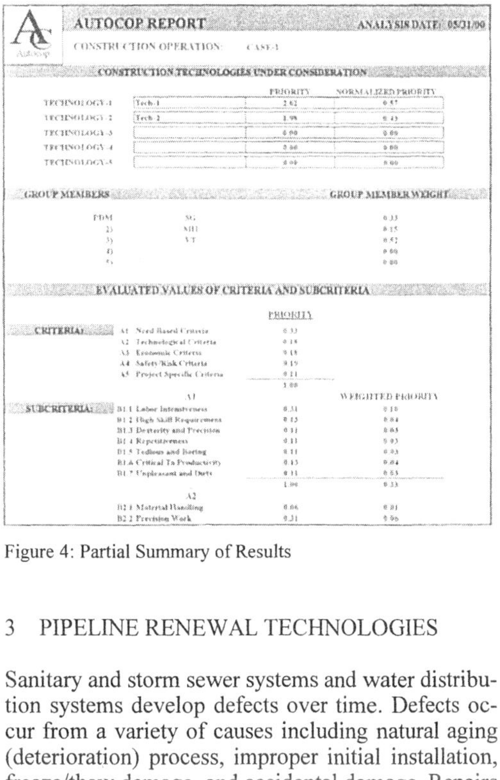

Figure 4: Partial Summary of Results

3 PIPELINE RENEWAL TECHNOLOGIES

Sanitary and storm sewer systems and water distribution systems develop defects over time. Defects occur from a variety of causes including natural aging (deterioration) process, improper initial installation, freeze/thaw damage, and accidental damage. Repairs or renewal of a failing pipe system can be done by replacing the entire pipe or by replacing only the defective portion. Construction requirements for repair and renewal are similar to original construction requirements. The following discussion highlights the different methods of pipeline renewal.

3.1 Pipeline Renewal-Open Cut

When it is determined that an underground pipe has deteriorated to a point where its structural or functional integrity is compromised, replacement is typically the prudent alternative. To replace an entire pipe segment (manhole to manhole for sewer pipe), the old pipe is excavated. Paved surfaces are saw cut prior to excavation to prevent damage beyond the construction limits. The original backfill is removed, if it is acceptable material by today's standards, it can be reused, or else it must be removed and disposed. When the pipe is exposed, it is removed along with the bedding material. Lateral connections are disconnected before the old pipe is removed to prevent fracturing of the laterals. New segments of pipe are then installed following the procedures for new pipe installation. During new pipe construction, external or by-pass pumping may be required to allow the flow in the pipeline to continue uninterrupted. External pumping comes at an additional cost.

3.2 Pipeline Renewal-Trenchless

Trenchless pipeline renewal methods offer several advantages over conventional dig-up and repair/replace methods. The trenchless pipeline-system renewal methods can be divided into three categories (Table 1):

- Sliplining
- Cured-In-Place Pipe (CIPP)
- Pipebursting

Table 1: Trenchless Pipeline Renewal Methods

Method	Diameter Range (Inches)	Maximum Installation Lengths (ft.)	Liner Material	Application
Sliplining:				
Segmental	12 - 150	5000	PE, PP, PVC, GRP	Gravity & Pressure
Continuous	4 - 60	1000	PE, PP, PVC, GRP	Gravity & Pressure
Spiral Wound	4 - 100	1000	PE, PP, PVC, PVDF	Gravity
CIPP:				
Inverted in Place	4 - 108	3000	Thermoset Resin	Gravity & Pressure
Winched in Place	4 - 54	500	Thermoset Resin	Gravity & Pressure
Pipebursting:				
	4 - 36	1000	HDPE	Gravity & Pressure

Definition of Acronyms:

PE - Polyethylene; PP - Polypropylene; PVC - Poly Vinyl Chloride; HDPE - High Density Polyethylene; PVDF - Poly Vinylidene Chloride; GRP - Glassfiber Reinforced Polyester

3.3 Sliplining

Sliplining, also called as pipe-in-pipe, is one of the earliest forms of trenchless pipeline rehabilitation. There are three main types of sliplining: Continuous, segmental, and spiral wound. A new pipe of smaller diameter is inserted into the host pipe by pulling, pushing, or spiral winding, and the annulus between the existing pipe and the new pipe is grouted. Small liners may be pulled in manually, but most require winching. The winch applies a steady, progressive pull to place the liner inside the host pipe. The liner pipe is generally butt fused to its design length. Numerous designs of pipe pushing machines, both manual and hydraulic, are available. To reduce the size of the insertion pit, segmental liners are generally used (Fig. 5). Pipe joints are generally of mechanical type with either a snap-fit or a screw-on mechanism (Fig. 5).

This sliplining method has the merit of simplicity and is relatively inexpensive. One of the chief drawbacks of sliplining method is the resulting decreased cross-sectional area. In some instances however, despite the reduced cross-section, the hydraulic capacity of the pipeline may actually increase due to the superior flow characteristics of the new pipe.

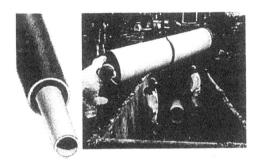

Figure 5: Segmental Sliplining

3.4 Cured-in-place Pipe (CIPP)

The main alternative to sliplining in the non-person-entry pipeline renovation market is the Cured-in-place Pipe (CIPP). The common feature of CIPP is the use of polyester or epoxy resin impregnated fabric tube (Fig. 6). The tube is inserted into the existing host pipe and inflated against the wall of the host using a hydrostatic head or air pressure. The inflated liner is cured by re-circulating hot water or steam (Fig. 6). The CIPP process creates a "close-fit" pipe, which has quantifiable structural strength and can be designed for specific loading conditions.

Figure 6: Cured-in-place Pipe (CIPP)

The chief advantages of CIPP method are that it minimizes the reduction in cross-section, and the ability of the liner pipe to conform to non-circular cross-sections. The laterals can be reopened remotely after lining, by using a remote controlled robotic cutter. The chief disadvantage of this method is the need to take out the host pipe from service during the installation and curing. Diversion or by-pass pumping adds to the installation cost. CIPP is not cost effective for large diameters.

3.5 Pipebursting

Pipe busting is an on-line replacement technique that offers a means for replacing existing deteriorated pipelines. Pipe bursting, also referred to as "pipe cracking", is a method where the host pipe is broken up and pushed aside, while a new pipe of equal or larger diameter is pulled or jacked into place (Fig. 7). Pipe bursting can be utilized for a "size-to-size" replacement or for "upsizing" of the existing pipe. While greatly influenced by site conditions, upsizing of 25% to 100% of original pipe diameter is not uncommon.

Due to the outward expansion of the old pipe fragments, it is necessary to disconnect laterals and service pipes prior to using pipe bursting. While remote disconnection is made possible through recent developments, the most common method is by means of excavation pits at the lateral/service pipe location. The number and frequency of lateral or service connections is a determining factor when assessing the economic viability of pipe bursting.

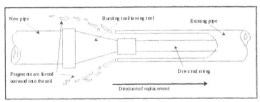

Figure 7. Pipebursting Process

4 CASE STUDY-City of Ft. Wayne, Indiana

As with other metropolitan areas in the United States, the City of Fort Wayne, Indiana, is struggling with sewer overflows associated with wet weather events. The homeowners in the Sunny Meadows Subdivision, located in the northern part of the City of Fort Wayne, were often on the receiving end of problems with sewer backups into homes and basements. This was caused in part by the hydraulically limited sanitary sewer pipe that served this area. The original 12-inch, asbestos cement (transit) pipe was constructed in the 1970s. The system would get backed up in the event of rain, generally exceeding 1-inch. During these times, sewage would overflow into the adjacent Schoppman Drain, which is a tribu-

tary of the St. Joe River. Hence the Water Pollution Control Maintenance (WPCM) department was forced to pump flow out of a manhole to avoid sewage backups. The annual cost to the City of Fort Wayne to send pump crew to the site exceeded $5,400 ($137/hour x 8 hours per visit x 5 visits per year).

Therefore, in 1999 the City of Fort Wayne commissioned an engineering study of the problem. A CCTV inspection of the existing sewer line revealed the following trouble spots:

- Of the 2,634 lft of sewer line, 2,170 lft (82%) has a flat or reverse slope
- 14 "bellies" in the sewer pipe, making up 722 lft (27%) of pipe length
- Significant I/I from leaking service taps, pipe joints and manholes
- Inside lining of the pipe severely deteriorated
- Bellies and flat slopes (lack of cleaning action) cause hydrogen sulfide (sewer gas) that is further corroding the exposed pipe walls

The main issues to be addressed by the rehabilitation were: increasing flow capacity, removing restrictions to flow such as flat slopes and bellies, reducing I/I, and enhancing the structural integrity of the pipeline at a reasonable cost. Sliplining was not an alternative as it results in a decreased cross-section and thus a decreased hydraulic capacity. The rehabilitation alternatives considered include: lining with cured-in-place pipe (CIPP); pipe bursting with the installation of a larger diameter pipe; traditional open-cut sewer replacement; and a combination of CIPP and pipe bursting. A review was conducted to investigate the three alternatives and to recommend to the City of Fort Wayne the best possible option. Ideally, such option would provide additional hydraulic capacity for conveying peak wet weather wastewater flows and reduce sanitary sewer backups.

- *Alternative 1* – Rehabilitation of existing 12-inch sewer with a CIPP liner and construction of a new 12-inch polyvinyl chloride (PVC) relief sewer to convey peak wet weather wastewater flows;
- *Alternative 2* – Excavation and replacement of the existing 12-inch sewer with a new 15-inch PVC sewer; and
- *Alternative 3* – On-line replacement (pipe bursting) of the existing 12-inch sewer with a new 16-inch high-density polyethylene (HDPE) (14-inch I.D.) pipe.

4.1 Interference With Existing Utilities

Existing buried electrical lines for streetlights and water lines were located and determined to conflict with excavation. Mature trees and privacy fences in the residential subdivion were also in conflict with the excavation along the pipe path.

4.2 Easement

In Alternative 1, no additional easement would be required for the CIPP liner installation. However, acquisition of significant permanent and temporary easement would be necessary for the construction of a new PVC relief sewer. In Alternative 2, no additional permanent easement would be necessary, however significant temporary easement would be necessary to dig down to the existing pipe and replace it with a new PVC pipe. In Alternative 3, no permanent easement would be necessary. Some temporary easement would be required for excavation at the entry and exit pits during the pipe bursting process.

Alternative 1 would require open-cut construction of the 12-inch relief sewer along Knollcrest Road and Meadowbrook Drive and would cause major disruptions to the backyards of property owners. Construction of the 12-inch relief sewer would require deep excavations and clearing of mature trees and flowering shrubs. Alternative 1 would provide protection from wet weather sanitary sewer overflows that occur along North Brookwood Road but would be inadequate for conveying peak wet weather flows during rain events greater than 1 inch.

Alternative 2 would require open-cut construction, as in the previous case, causing major disruption to property owners and general public. Construction would require bypass pumping, potential re-paving of the roads, reinstatement of service laterals and disposal of the asbestos cement pipe. Implementation would provide the greatest protection against wet weather sanitary overflows but would be the most expensive and the most disruptive alternative.

Alternative 3 would require insertion pits approximately 2 ft x 30 ft in size and pull pits approximately 13ft x 25 ft in size. This alternative would require deep excavations at each service tap. However, overall this alternative had the potential of being the least disruptive to the homeowners and the public. Additionally, as the existing pipe would be replaced in-line, it would not create any hazardous waste issues. On the basis of this analysis pipe bursting was recommended for the installation of a new 16-inch HDPE sewer.

4.3 AUTOCOP Analysis

An analysis was undertaken using AUTOCOP to establish the preference between the three alternatives for the pipeline rehabilitation case study for the City of Fort Wayne, Indiana. The results of the analysis are shown in Fig. 8. The analysis validated the utilization of pipebursting as the preferred method of pipeline renewal. The primary reasons were:

- Site congestion and safety concerns
- Speed of construction

- High groundwater levels would require extensive dewatering for open-cut construction
- Settlement concerns in the existing pipeline
- Depth of existing pipe

CASE and ALTERNATIVES		ACTUAL VALUES	NORMALIZED VALUES	COMPARISON % PRIORITY	
CASE: Storm Sewer Rehabilitation, City of Ft. Wayne, Indiana					
Alternative 1 - CIPP	C1	1.89	0.36	C1/C2 =	1.40
Alternative 2 - Dig & Replace	C2	1.35	0.26	C3/C2 =	1.47
Alternative 3 - Pipebursting	C3	1.98	0.38	C1/C3 =	0.95

Fig. 8. Results of AUTOCOP Analysis

The existing pipe was almost 30 feet deep at certain locations, requiring extensive shoring and dewatering operation, which would make any open-cut construction disruptive to the neighborhood and cost prohibitive. Of the three methods pipebursting would be the fastest however CIPP requires little or no excavation as the liner can be inserted through existing manholes and the lateral reconnections can be achieved through the pipe by means of an automated robotic cutter. Pipebursting requires some excavation, both for the machine pits as well as the exit pits and at location of each lateral service connection (20 in all in this case). In this particular application it was expected that open cutting would have to take place over as much as 25% of the pipe length for pipe bursting due to the numerous lateral connections.

In the end, CIPP and pipebursting were fairly competitive alternates, but pipebursting edged out CIPP by virtue of being able to provide the greatest hydraulic capacity of all three alternates, an overwhelming concern in this particular case study.

5 CONCLUSIONS

The benefits offered by AUTOCOP include (i) systematic analysis of various criteria involved in the decision process, (ii) capacity to analyze both quantifiable and intangible decision criteria, and (iii) identification of the most suitable option on the basis of priority. AUTOCOP was able to predict the optimum rehabilitation application. Although the evaluators ahead of time knew the final outcome the results from the analysis are a strong validation of the system.

REFERENCES

Saaty, T. L. (1982). *Decision Making for Leaders*. Belmont, CA: Lifetime Learning Publications.

Hastak, M. and S. Gokhale. (2000). "AUTOCOP: A System for Evaluating Underground Pipeline Renewal Options." Paper accepted for publication in the *Journal of Infrastructure Systems*. Tentatively scheduled for September 2000 issue.

8 Procurement and project management

Creative Systems in Structural and Construction Engineering, Singh (ed.) © 2001 Balkema, Rotterdam, ISBN 90 5809 161 9

An overview of procurement systems in Hong Kong

Albert P.C. Chan & Esther H. K. Yung
Department of Building and Real Estate, Hong Kong Polytechnic University, People's Republic of China

Tony Y. F. Ma
School of Geoinformatics, Planning and Building, University of South Australia, S.A., Australia

Patrick T. I. Lam
Department of Building and Construction, City University of Hong Kong, People's Republic of China

ABSTRACT: The objective of the paper is to analyse the prevailing procurement systems in Hong Kong. Three major systems are identified, namely, the separated approach, the integrated approach, and the management-oriented approach. The characteristics and the process of each system will be examined. Variations to these main forms of procurement systems will be analysed.

1 INTRODUCTION

The construction process in Hong Kong has always been regarded as complicated. The complexity of construction is compounded by the following characteristics (Rowlinson, 1997). Firstly, the difficulty of sites and ground conditions in Hong Kong. Secondly, in times of high demand, there is invariably a labour shortage and a restriction on importing labour. Thirdly, both public and private clients demand high speed of construction. Fourthly, most sites are congested due to high plot ratios. The logistics of site transport both horizontally and vertically are very difficult. Fifthly, the heavy dependence on imported materials creates uncertainty in delivery.

Despite the complexity, the construction industry in Hong Kong is rather conservative. The majority of the building projects in Hong Kong are carried out by the traditional procurement system where a client appoints consultants to act on his behalf to produce the design and supervise the construction phase (CIOB, 1998). In current dollars, expenditure on capital works in Hong Kong over the last twenty years has exceeded HK$1600 billion. Many clients, however, are becoming increasingly dissatisfied with the traditional form's operational characteristics and seek other methods of procurement, organisation and management to meet their more exacting needs.

The objective of the paper is to analyse the prevailing procurement systems in Hong Kong. Three major systems are identified, namely, the separated approach, the integrated approach, and the management-oriented approach. The characteristics and the process of each system will be examined. Variations to these main forms of procurement systems will be analysed.

2 SEPARATED PROCUREMENT SYSTEM

2.1 Traditional system

Building projects in Hong Kong have been mainly delivered in a traditional contract system. The client appoints consultants to act on his behalf to produce design and supervise the construction phase. The architect traditionally acts as team leader and coordinates the other consultants of the design team. The quantity surveyor provides preliminary cost advice to the client in the early stage of the project. After the architect prepares alternative proposals for the client to select, the quantity surveyor estimates the cost of the alternatives.

When the client accepts a proposal, the architect develops the design. At this stage, consultations with specialist engineers and negotiation with specialist contractors come in. When drawings and specifications are prepared, the quantity surveyor provides regular monitoring of the alternative designs and ensures that the cost implications of the design decisions are known to all concerned. The quantity surveyor then prepares the bills of quantities (Frank, 1998).

Together with the bills of quantities, tender drawings and forms of tender that are prepared by the architect are sent to selected contractors. The contractors estimate the costs of the operations involved in the project and submit tenders for the work. The duration of the project is assessed from the pre-tender plan prepared by each contractor's production planners and managers. Management decisions determine the margin to be added to the tender for profit. When the tender submitted by a particular contractor is accepted by the client, the contractor sets up his site management system, plans and organizes the works, schedules material deliveries. Concurrently, he also places orders with his own subcontractors and those nominated by the architect (Sanvido, 1997).

The common variations of traditional procurement system are:

i) sequential

In a sequential traditional lump-sum building contract system, a substantially completed design is prepared together with the cost documentation for the contractors to bid the job in competition. In general, construction starts when the design is finished.

ii) accelerated

A contractor is appointed earlier in the sequence of design on the basis of partial information, either by negotiation or in competition. Negotiation, from the basis of the initial, partial information, takes place once the final design information becomes available. Construction starts when the design is developed to the final stage. Overlapping of design and construction works occurred (Chan, 1995).

The major advantage of the accelerated traditional system is the overlapping of design and construction phase, so that the project duration can be reduced. On the other hand, construction commences when the design work is not yet finalised would generate a lot of changes at the construction stage. Consequently, these variations mean extra cost and time.

3 INTEGRATED PROCUREMENT SYSTEM

The potential need for greater cost control and the need for co-ordination between detailed design and site construction have promoted integrated organisational forms such as project management, design and build contracts (Ganesan et al, 1996).

With the various methods of D&B contracting, the traditional form of D&B, where the contractor undertakes the design and construction of the project with the architect being employed by the contractor, is the most commonly adopted method.

Over 40 % of the design and build projects undertaken by the firms are public works. ASD accounted for about 1/3 of all these 40% of projects (HKIA, 1998). The percentage of the nature and the type of projects are stated as followed.

Preliminary research indicated that most experienced user-client using design and build were government-funded bodies. These include the Architectural Services Department, the Hospital Authority, the Housing Authority, City University of

Hong Kong and the Hong Kong University of Science and technology. Private user-clients tend to use D&B for specific projects only, for instance, - the Hong Kong Jockey Club at Happy Valley Racecourse Redevelopment, - South China Morning Post at their Tai Po Printing Plant.

The design and build contractors that have worked on D&B projects are most often appointed by the user-client to assist in preparing the user-client's brief, to prepare sketch design and to see that the user-client's requirement are met during construction. The second most common arrangement is where the D&B contractor undertakes design and cost advice. There are also cases where the architectural firm is a member of a D&B consortium and where the firm is appointed by the contractor to undertake checking role (HKIA, 1998).

Design and build is the procurement system where one organisation is responsible to the client for both design and construction. The process begins with establishing the need to build and the client's requirement. In order to prepare these, professional advice in preparing employer's requirement and in assessing contractor's proposals needs to be significant in order to select the right tender. Then the client starts selecting and inviting tenders to bid. After several contractors prepare their proposals for design, time and cost, the client evaluates the proposal. The winning contractor then enters into contract with the client and the design and construction of the project commence (Turner, 1994).

The essence of this system is that the contractor is in control of both design and construction aspect of the project. The contractor will commission a firm of architects for the design work. In some instances, the employer may nominate the firm. Essentially, design and build contract is only appropriate if the conceptual design is firm at the time of contracting (Ndekugri & Turner, 1994).

The common variations of design and build are:

3.1 *Direct*

In direct design and build, only one tender is obtained and therefore no competition is achieved in tenders. Contractor is usually obtained from negotiation and peer relationship may possibly exist. This form of design and build has been rarely used in the public sector of Hong Kong where public accountability is the prime concern (Chan, 1995).

3.2 *Competitive*

Competitive design and build is the most usual procedure with tenders being obtained from documents defining the project prepared to enable several contractors to offer competition in designs and in prices. A consultant or contractor for a fee may prepare this Employer's requirement. The winning contractor is responsible for the design and construction of the building to the completed stage (Chan, 1995).

3.3 Enhanced

The term "enhanced" signified that the client would develop the design using their own team of consultants, to a point where the significant planning issues and inter-department relationships were all determined, and require tenders to submit a conforming bid based on this design. Client either employs a consultant team or engages its in-house staff to develop the design of the project to the level of 1:200 scale plans. The winning contractor employs his own team of design consultants to assist him to work on the design development of the project. The contractor as in traditional design and build contract is also responsible to select sub-contractors and carry out the construction work of the project (Skues, 1999).

The enhanced design and build procurement system has only been adopted for two Hospital projects in Hong Kong. The specialised requirements of the hospital required an in-house consultant team to prepare a detailed employer's requirement to ensure the contractor would be able to perform the role of both design and build.

3.4 Novation

Novation can be seen as a variation of the design and build system. It can also be considered as a hybrid of traditional procurement system and the design and build system. In principle novation is the process whereby a contract between A and B is transformed into one between B and C in such a way that A no longer has any rights or obligations under the contract (Ndekugri and Church, 1994).

Consultants design the building required to a partial stage, then obtain competitive tenders from contractor whom develop and complete the design and then construct the building. The extent of consultant design varies from giving broad indications of external appearance of materials required or, at the other end, with a consultant preparing a scheme design that specifies all the major components, materials, elevations and fenestration outline. Thus the contractor only has to prepare the working details for construction.

Novation is usually employed by a client to produce quite detailed drawings which competitive tenders are then obtained. Then the client novates that design team to the contractor by asking the

contractor to take on the employment of the design team with the responsibility of all its previous design.

3.4.1 Pre-novation stage
At the pre-novation stage, client initiates the project by commissioning design consultants to develop brief and commence design work in a manner very similar to the traditional system. The role of the consultants is to complete the designs to a stage where all the requirements of the client are adequately described to a level of legal clarity. The employer requirements are normally defined, drawn and specified in the range of 30-80% of the overall design. Then the client is ready to call for tenders from contractors to undertake the completion of the design and construction process (Waldon, 1993).

3.4.2 Post-novation stage
Once the contractor has been novated, the contractor has a direct contractual link with the client and the design consultants. The contractor has become the designer and is responsible for all the design work as well as the construction. The contractor instead of the client pays the consultants. The clear distinction drawn between novation contract and design and construct contract is that the contractor must employ the designated designer who has carried out the preliminary design for the client (Chan, 1995).

The involvement of the contractor in the detailed design allows the contractor to implement changes to the design that suit his particular construction practices and equipment. By including the appropriate clauses in a novation contract, the client can retain the right to monitor and comment on the design.

4 MANAGEMENT ORIENTED SYSTEM

4.1 Management contracting system

Management contracting system is a procurement system whereby a management contractor is appointed to the professional team during the initial stages of a project to provide construction management expertise. The management contractor employs and manages works contractors who carry out the actual construction of the project and he is reimbursed by means of a fee for his management services and payment of the actual prime cost of the construction. The management contracting system has not been widely used to deliver buildings in the Hong Kong industry. Several clients who have used management contracting were the HK Jockey Club, KCRC, HK Bank and Swire properties.

The management contract arrangement does not fit neatly into the conventional pattern of pre-contract and post-contract stages. Dearle and Henderson (1988) have identified three distinct periods: the period before the appointment of management contractor; the preconstruction period; and the construction period.

The pre-appointment of the management contractor will encompass the carrying out of a feasibility study and formulation of the brief. During the early stage, the employer will appoint an architect, a quantity surveyor and any other professional advisers who are considered necessary. These consultants form the design team, which initially assists in the preparation of the brief and prepares drawings and a specification, which describe the scope of the project. Once this documentation has been prepared, the employer, with the advice from appropriate members of the professional team, invites tenders from management contractors using one of the methods of selection and appointment that have been devised for this purpose.

Under the management contract form of agreement, the client employs the management contractor who assumes all the duties of a traditional main contractor in aspects such as the legal and statutory responsibilities (Dearle & Henderson, 1988).

Generally, during the pre-construction period, the management contractor's roles and duties include the following:

1. to prepare of the overall project program;
2. to prepare material and component delivery schedules;
3. to advise on the buildability and practical implications of the proposed design and specifications;
4. to establish the construction methods;
5. to prepare a detailed construction program;
6. to advice on the provision and planning of common services and site facilities;
7. to advice on the breakdown of the project into suitable packages for trade and works contractors and suppliers;
8. to prepare a potential tenderers;
9. to assist in the preparation of tender documents and to obtain tenders from approved trade;
10. to prepare, in consultation with the client's consultants, of the documentation necessary to ensure the efficient placing of the proposed trade contracts.

The management contractor will endeavor to obtain more than one quotation for each works package on an open book basis. The consultant QS, relevant consultant and management contractor will make recommendations for the employer's acceptance prior to the award of contract. A budget and cost plan will be established as early as possible after the management contract is awarded (Asiabuild, 1998).

The management contractor does not carry out any of the construction work him. The actual work is divided into a series of separate packages. The duties of the management contractor during the construction period can be summarised as the setting out, management, organizing and supervising of the implementation and completion of the project using services of his trade/ work contractors. Payment is made to the management contractor during the construction period by means of interim certificates based on the prime cost of the work so far completed. At the same time directions are given for the amounts that are to be paid to the individual works contractors. The certificates will also include the reimbursement of an installment of the construction period management fee and the cost of the management contractor's directly employed resources. Retention is deductible on all elements of the prime cost including the management contractor's fee.

In management contracting system, the design consultants will maintain independence in terms of design and statutory obligations. The management contractor will plan and control the schedule of information release and the construction activities, co-ordinate the parties concerned and provide early practical input on the buildability of the design (Naoum, 1994).

The system is most useful for large and complex contracts when considerable co-ordination of specialist required and early completion is vital. Flexibility of the management contracting system enables variations on the original design and specifications throughout the course of construction (Frank, 1997). For example, the KCRC Kowloon Station Renovation & Extension Project had a very tight program. There were also many design changes even though the project had gone into its construction phase. Therefore, management contracting was adopted because the management contractor had reduced potentially negative impacts on the projects through the flexibility options available to them. If this project were let on a traditional form of contract, it would be difficult to imagine the magnitude of the claims submitted by the main contractors and the time delays to the project as a whole (Ho, 1995).

Although MC has been adopted in Hong Kong since the 1980's, there were only a few projects, which have adopted management contracting system. The project uncertainty is regarded as the main constraint due to the absence of a tendered lump sum price prior to construction (Naoum and Mustapha, 1996). Clients are subject to a greater risk in respect to costs because to the staggering and phasing of orders for specific work over a long period time (Naoum, 1991).

Moreover, the form of management contracting contract seems to hinder its use in Hong Kong.

Although the JCT has published a standard form in 1987, there seem to be difficulties in applying it in the Hong Kong industry. This is because of the different organizational structure and also partly due to culture practice and customs (Naoum, 1994).

High final project cost is also the constraint for the use of the system. The managing contractor has no incentive to keep costs down and preliminaries are often greater than the traditional system because a managing contractor is paid on a fee basis (Naoum, et al 1994).

Other criticism of management contracting is the possible conflict, which may occur between the management contractor and the architect. There may be some jealousy by the other professional consultants since the management contractor acts as a team leader in the project.

5 CONCLUSION

The prevailing procurement systems in the Hong Kong construction industry were investigated in this paper. They included the traditional system, design and build, novation and management contracting system. Despite the variability of the prevailing procurement systems, the construction industry in Hong Kong is rather conservative. The majority of the building projects in Hong Kong are carried out by the traditional procurement system where a client appoints consultants to act on his behalf to produce the design and supervise the construction phase.

ACKNOWLEDGEMENT

The authors gratefully acknowledge the Hong Kong Polytechnic University for providing funding to support this research effort.

REFERENCES

AsiaBuild, (1998) Management contracting and Asiabuild, Asiabuild Construction Ltd. Publication, 1998.

Chan, A.P.C. (1995) Towards an expert systems on project procurement, Journal of construction procurement, Volume 1, no.2, November 1995, 124-149.

CIOB (1998) Code of estimating Practice, Longman Ltd. 1998.

Dearle, & Henderson (Firm), (1988) Management Contracting, A Practice Manual, E & F.N. Spon, London. 1988.

Franks, J. (1998) Building Procurement systems, A client's guide, 3rd edition Addison Wesley Longman Limited and the charted Institute of Building, 1998.

Ganesan, S. Hall, G. & Chiang Y.H. (1996) Construction in Hong Kong, Athenaeum Press, Ltd., 1996.

Ho, T.O.S. (1995) Managing for the future management contracting-KCRC Kowloon Station Renovation & Extension Project, Asia Pacific Building and Construction Management Journal, Volume 1, 1995. 94-101.

Naoum, S.G. (1994) Performance of management contracts, ASCE-Journal of construction engineering and management, Volume 120, no.4, 687-705, 1994.

Naoum, S.G. Chan, H. & Mustapha, F.H. (1994) The potential use of management contracting in the Hong Kong construction industry attitudinal survey, CIB W92 East Meets West, Procurement Systems Symposium,1994. Hong Kong. 169-177.

Ndekugri, I. & Church, R. (1997) Construction Procurement by the design and build approach: a survey of problems, CIB W92, volume 3, May, 1997, 452-4.

Ndekugri, I. & Turner, A. (1994) Building procurement by design and build approach, ASCE-Journal–of-construction-engineering-and- management. Volume 120. No.2, 1994, 243-256.

Rowlinson, S. (1997) Procurement systems: The view from Hong Kong, Proceedings of the CIB W92 Symposium on Procurement, May 1997. Montreal, 665-672.

Sanvido, V.(1997) A comparison of project delivery systems, ASCE Construction Congress V. Management Engineered Construction in expanding Global Markets, 1997, 573-581.

Turner, A. (1997) Building Procurement, 2nd Edition. Macmillan Press Ltd. 1997.

Waldron, B.D. (1993) Design and Construction through novation, Construction Project Law (International) Seminar Paper presented on 27 April.

Creative Systems in Structural and Construction Engineering, Singh (ed.) © 2001 Balkema, Rotterdam, ISBN 90 5809 161 9

An investigation into the application of total quality management within small and medium sized constructional organisations

N.Chileshe & P.Watson
Sheffield Hallam University, UK

ABSTRACT: The main of this paper is to investigate the application of Total Quality Management (TQM) within Small and Medium Sized (SMEs) Construction Enterprises. The advocated rationale for the implementation of TQM is the attainment of a sustainable competitive advantage and this study seeks to understand whether organisational performance is directly linked to the adoption of TQM. Empirically identified sources of competitive advantage are presented along with the necessary conditions for attainment. This paper concentrates on the methodological aspects such as construct validity , measurement errors, sampling errors, choice of respondent etc which have previously affected studies in the field of Production and Operations Management (POM). The literature review draws extensively from the *Journals of Operational Management International Journal of Quality & Reliability Management.*

1 INTRODUCTION

The focal point of the research project is on the opportunities and benefits associated with TQM, and in particular its effectiveness for Small and Medium-sized Enterprises in enabling a competitive advantage. The rationale for investigating SME's is based on the fact that over 95% of construction companies with the UK employ fewer than 10 people, and over 50% of the labour force is self-employed. Whereas previous research has concentrated on large organisations, it is evident that excluding such a group from any research would be folly, as they perform an important role in the UK economy. Though the above figures cited relate only to the construction industry, on a national scale, SME's account for approximately 99.8% of total UK business (as illustrated in table 1) (DTI, 1998) and support approximately 67.2% of total UK employment. (DoE 1997). The importance of TQM for SME's is widely acknowledged by various authors (Quiz & Padibjo, 1998, Barrier, 1992, and Ghobadian & Gallear, 1996) who state that SME's are often suppliers of goods and services to larger organizations and therefore a lack of product quality from SMEs would adversely affect the competitive performance of larger organizations. Another proposition is that a relatively high level of quality is a key factor in the attainment of a superior sustainable competitive position. A literature search identified those successful businesses in both manufacturing and construction which are engaged in making and taking competitive opportunities. Hardy advocates that the development of a competitive advantage automatically creates an opportunity and so the reasoning may be modified to: "Successful businesses are engaged in the creation and exploitation of competitive advantage."

Table 1: Employment in the construction industry by size of firm

Number of employees in firm	Approximate number of firms	% of total
1-13	78 000	85.04
14-59	11 000	12.00
60-114	1 400	1.53
115-599	1 100	1.20
600-1199	125	0.14
1200 and over	80	0.09

2 LITERATURE REVIEW AND THEORY DEVELOPMENT

The purpose of the literature review is to compare various insights from different authorities on TQM and competitive advantage, and the development of a contingency model for the quality dimension of competitive advantage. The literature also identified methodological problems, which are associated with previous studies on the relationship between TQM and performance/competitive advantage. Some of these are stated under the limitations of previous research.

2.1 Limitations of Previous Research

• Not controlling for industry related factors.

According to Handfield & Melnyk (1998), this has to be considered when setting boundary assumptions on observations as it leads to a biased sample, which affects the way observations, are interpreted, and therefore affects parameter estimation.

• Exclusion of non-TQM firms in the research.

This inevitably leads to sampling error, as one of the most critical elements of the sampling procedure is the sample frame used to represent the population of interest. (Malhotra & Grover 1998). Another limitation is that of studies conducted by parties with vested interest in their outcomes, that did not conform with generally accepted standards of methodological rigor. This is summarized by Samson & Terziovski (1999) who stated that rigorous statistical analysis is required in order to meet professional standards of reliability and validity.

• Exclusion of medium and small sized firms.

Where research conducted included SMEs it found that the implementation of BS EN ISO 9000 series improved the management operations of organisations, (Rayner, 1991). The limitation was that there was no evidence to suggest that BS EN ISO 9000 series could be used as a vehicle for the achievement of TQM.

• Not tracking the performance of comparable non-TQM firms over the same time period.

• Failure to assess construct validity.

One major problem with the research process is that of ensuring the measurement of constructs is free of error. According to O'Leary-Kelly & Vokurka (1998), this omission leads to ignoring the many corrupting elements embedded in measures such as measurement error and informant bias, which could affect the conclusions drawn. Construct validity pertains to the degree to which the measure of a construct sufficiently measures the intended concept. The reliability assessment to be adopted in this research will be Cronbach's (1951) α Coefficient. This is generally accepted as one of the most popular methods for assessing reliability, (O'Leary-Kelly & Vokurka 1998). This study considers the different components of construct validity - unidimensionality, reliability, and convergent and discriminate validity. Construct validation can be described as a multifaceted process that is comprised of three basic steps as illustrated fig 1, (O'Leary-Kelly & Vokurka, 1998).

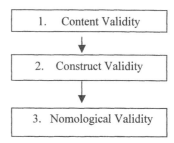

Fig 1 : Construct Validation Process

The first step requires the identification of a group of measurement items (empirical indicators) which are thought to measure the construct.

3 TQM & COMPETITIVE ADVANTAGE

Competitive advantage has been defined as:

"an advantage your competitors do not have ... once they have access to the special formulation, the new process, high speed machinery, or whatever the advantage is, then it is no longer competitive" (Hardy, 1983.29).

Powell (1996) shows that under the resources model, success derives from utilizing economically valuable resources that other firms cannot imitate, and for which no equivalent substitute exists. According to Moreno-Luzon (1993), quality management can improve a small firm's competitiveness through co-operation. Cherkasky (1992) opines that when quality concepts are applied to every decision, transaction, and business process, quality becomes a competitive weapon. However, processes which have the greatest impact on customer satisfaction would have to be targeted for improvement and only market research would identify the "key customer drivers" or those products and service attributes of greatest concern to customers.

Chapman et al. (1997) argues that although there is a perception that a quality driven strategic advantage has a direct link with increased business performance, the latter has been difficult to achieve without the development and implementation of a TQM philosophy. Chien et al (1999) highlights the factors related to competitive advantage under the four sub-headings of Manufacturing, Marketing, R&D and Engineering and Management. This research draws heavily upon this model and applies the factors adopted under management as follows;

♦ management competence;
♦ planning and control system;
♦ nature of organization culture and values;
♦ quality of corporate staff;

◆ capability for negotiation with the environment; and financial strength.

Whereas the above mentioned factors do affect competitive advantage, it is outside the scope of this study to examine them in detail as each factor warrants a separate investigation.

3.1 Conditions Necessary For Competitive Advantage

Fahy (1996) contends that competitive advantage for service firms lies in the unique resources and capabilities possessed by the firm. Not all resources or capabilities are a source of competitive advantage. Only those that meet the stringent conditions of value, rareness, immobility, and barriers to imitation are true sources. The actual sources of competitive advantage are likely to vary depending on the nature of the service, the particular traits of the firm, the nature of the industry, and the country of origin.

Literature review highlights the fact that superior skills, superior resources and superior controls are the major sources of competitive advantage, therefore Fahy (1996) concludes that service firms must seek to identify the skills and resources they possess and they must satisfy the above criteria in order to realize a sustainable, competitive advantage.

Similar studies conducted by Piercy & Morgan (1996) showed that the linkage between a quality strategy and competitive advantage, though pursued, were never understood within the organisations involved. Improving organizational competitiveness is one of the primary goals of quality management (Rao *et al*, 1997). Similarly Small and Medium sized construction organizations need to identify their sources of competitive advantage in order to fully satisfy their clientele.

4 REVIEW OF METHODOLOGY

The matching of research strategy with theory building activities draws heavily from Handfield & Melnyk (1998) who have outlined the following stages:

◆ purpose;
◆ research question;
◆ research structure;
◆ data collection techniques;
◆ data analysis procedure.

Due to the limitations and scope of the subject area, this paper will only address the purpose and research question, taking into account the methodological aspects which have previously affected the studies.

4.1 *Purpose*

The purpose of this study is the development of an Implementation Model focused on the provision for achieving a competitive advantage via the application of Total Quality Management for construction related organizations.

The importance of different research methodological approaches available with their inherent strengths and weaknesses have been fully considered. Particular attention is drawn to comparing and contrasting the qualitative and quantitative approaches. The choice of research tools is guided by Wing *et al* (1998) who advocates that many research issues in construction management are practical problems which involve generalization of experience and the formulation of hypothesis that can generate empirically testable implications. For problems of this nature, the testability of hypothesis and reproducibility of results are important, and the naturalist approach of discovering casual relationship is more likely to produce practical solutions.

Handfield & Melnyk (1998) provide the basis for major research objectives as being to create knowledge. Knowledge is created primarily by building new theories, extending old theories and discarding either those theories or those specific elements in current theories that are not able to withstand the scrutiny of empirical research.

4.2 *The Research Question*

This project's research question is based on the assumption that SME's adopt TQM with the intention of achieving a sustainable competitive advantage. A series of research questions have been developed. These questions will be explored in the main using empirical evidence drawn from the UK construction industry.

Q_{1-1} : That the Implementation of a total quality management system can provide a competitive advantage for construction related organizations as it does for manufacturing organizations.

Q_{1-2} TQM construction related organizations outperform non-TQM construction organizations.

Q_{1-3} : Manufacturing TQM organisations outperform TQM construction related organizations.

Q_{1-4} : TQM performance is correlated with the organisation's structure and culture.

This project adopts a triangulation methodology involving both a qualitative and quantitative approach.

As noted from the research methodological model, the following methods will be used for data collection. The choice of data collection is governed by Malhotra & Grover (1998) who purport that survey designs with questionnaires are the most commonly used methodology in empirical Production and Operations Management (POM) research.

5 DEFINITION OF TQM

In Construction terms, the digest of data for the construction industry have the following definition ; TQM is a management philosophy which aims to produce a better performance from a whole project team and to result in better quality products & services, delivery & administration, which ultimately satisfy the client's functional & aesthetic requirements to a defined cost and completion time. For this to work, the client himself has to accept the responsibility as being part of that project team.

6 APPLICATION OF TQM WITHIN CONSTRUCTION

Oakland & Aldridge (1995) found that the construction industry is associated with a patchy reputation, with many projects that are not completed on time. Similarly Chileshe (1996) showed that most organisations in the construction industry are quite happy with accreditation to ISO 9000 series than pursuing TQM programs, among the reasons given for non-implementation being that ISO 9000 was enough of a " culture shock ", secondly due to the current industrial climate, particular in the construction industry, most directors had more 'pressing' matters to consider, those of financial survival .

6.1 Application Problems

The Problems facing the Implementation of TQM in the Construction Industry are well researched and documented. Earlier studies indicate that the nature of the industry in itself creates problems for the development of an effective quality management systems. Grover (1987) notes that when the construction industry is stripped to its basic elements, its one that designs and assembles structures made up of other industries, a task which involves formidable problems of organisation. His sentiments are shared by Pheng (1994) who states : " The nature of the construction industry is, however, unique as most building projects encompass the participation of numerous parties, including design consultants, con-

tractors, subcontractors, building materials , manufacturers and suppliers ." Here, Pheng is advocating for the integration of all parties involved in a building project for quality to be achieved.

Finally , the constructional industry has been reluctant to embrace TQM. Schriener *et al.* (1995) point out that Construction historically is an industry reluctant to change, but it is now trying to catch up with the Total Quality Management revolution that has transformed many businesses.

However, the road to TQM is not without its dangers. Creating a culture of continuous improvement offers many opportunities to go astray. (Laza and Wheaton, 1990)

7 THE IMPLEMENTATION IMPACT OF TQM IN SMES

One of the objectives is concerned with the establishment of the business forces necessary for the attainment of sustainable competitive advantage for construction related organisations. This involves the application of Porter's (1990) competitive strategic model to the construction industry using it to identify generic categories of opportunities for competitive advantage. In this framework, the research examines the structural implications of TQM for the construction industry and its effect on rivalry within the industry, its impact upon the industry's relations with its customers and suppliers and its implications for prospective entrants and substitute products.

Porter's (1990) framework for the analysis of competition in specific industries shows that an industry has a high level of competitive rivalry when

- it is easy to enter
- both buyers and suppliers have a bargaining power;
- and there is a threat of substitute product/services.

Although Porter's analysis of competitive forces does not specifically address TQM, it can provide a framework for establishing the role TQM can play in an organisation's competitive strategy.

7.1 Can TQM be utilised to build barriers against new entrants to the industry ?

The barriers of entry are largely dependent on the size of the organisation . Small and medium sized organisation may gain entry into the construction market, they are however likely to face competition from other smaller firms wishing to become suppliers to large organisations. This is due to the increasing demand for a higher quality of service from large organisations. (Ghobadian and Gallear, 1995). TQM could provide a barrier if clients insisted that it be a pre-requisite for entry onto tender lists.

7.2 Can TQM change the basis of competition ?

Competition in the construction industry is no longer between firms from different sectors. Hasegawa (1988) noted that with the interface between the construction and non-construction industries growing increasingly wider, it exposes contractors to competition from a wider range of outside companies. Mohrmam *et al* (1995) established a correlation between various market condition and the application of TQM practices. The practices included organisational approaches such as quality improvement teams, quality councils, cross-functional planning, etc. One encouraging fact is that these studies showed that companies experiencing foreign competition and extreme performance pressure were more likely to use most of the TQM practices, tools and systems. This is suggested, provided evidence that competitive pressures lead to the adoption of TQM. Constructional related SME's within the UK are bound to come under pressure due to globisation and harmonization of the Single market.

Betts and Ofori (1992) opines that as trade barriers come down, construction enterprises will face real competition from firms from other countries, even for small construction projects.

7.3 Can TQM change the balance of power in supplier relationships ?

Many companies in the manufacturing industry ensure the quality of their component delivery by requiring suppliers to adopt TQM programs. (Powell, 1995). Similarly in construction , some owners & contractors have been requiring their suppliers (vendors) to implement TQM if they wish to be considered for future work (Mathews and Burati 1989)

Ghobadian & Gallear (1996) identified that Small and medium sized enterprises (SMEs) were often suppliers of goods and services to larger organisations and in order to remain competitive, they would have to consider the application of TQM due to the increasing demand for higher quality from the larger organizations .

7.4 Can TQM change the balance of power in supplier relationships ?

The thoughts of Ghobadian and Gaellear are corroborated by Moreno-Luzon (1993) who comment that if a small firm wants to become a supplier to a large company, the increasing demand for quality by the latter creates a strong influence on the former to consider the application of TQM. Organizations should *create supplier partnerships* by choosing collaborative ventures on the basis of quality, rather than solely on price (Gummer 1996). Companies today are not only placing demands on their own organization to become world class suppliers they place equally heavy demands on their own suppliers to become world class. (Steingraber 1990). Moreno-Luzon identifies other factors influencing the spread of TQM between small and medium-sized firms as, the pressure of costs, increasing competition, and more demanding customers requiring small firms to implement TQM. However, aware of the importance of quality in improving the competitiveness of the local economy, some public institutions promote and facilitate the efforts of small firms to take on this innovation .

Bricknell (1996) emphasizes upon communication and relationships extending beyond the organization. A good long term relationship with a particular supplier allows you to have a strong influence on the quality of products and service that you receive far beyond that of a conventional supplier customer relationship. Powell (1996) concluded that process improvement and supplier certification improves performance, but the performance impacts of the remaining TQM features vary depending on the firm's stage of TQM advancement. Bergstrom (1996) states that it should go without saying that suppliers play a critical role in any TQM process: If what is coming down stream to you is laden with waste, in form of poor quality or erratic delivery schedules, your TQM efforts, regardless how aggressive, can only suffer.

8 CONCLUSIONS

The paper has highlighted, through an extensive literature review, the methodological issues associated with construction management research and how they can be overcome. The model to be developed will be of great value to construction related SMEs as it will be based upon the existing best practice of all industrial and service sectors. This model will address the implementation issues of applying a Total Quality Management philosophy within the operational environment of construction related SMEs. A further contribution will be the testing of the proposition by the group 11 (WG 11) that the implementation of TQM by construction organizations provides a means for achieving the 30% increase in productivity. Although Quality Management has been advocated there is no research to date to underpin the pursuit of this strategy. If such a strategy did not lead to the attainment of the set objectives then construction related enterprises would be wasting valuable organizational energy. This would in fact detract from obtaining an increase in productivity. Therefore the results of this research project will be of interest to practitioners as the positive outcome of the verification of the strategy would vindicate the working group 11 and add impetus to the implementation of TQM in particular to construction related SMEs.

REFERENCES

Betts, M. and Ofori, G., (1992). Strategic planning for competitive advantage in construction. *construction Management and Economics,* 10: 511-532.

Chapman, RL; Murray, PC and Mellor, P. 1997. Strategic quality management & financial performance indicators. *international Journal of Quality & Reliability Management* 14(4): 432-448.

Cherkasky, SM. 1992. Total Quality for a sustainable competitive advantage. *Quality* August: 4-7.

Chien, T.W, Lin,C Tan, B, and Lee, W,C. 1999 A neural networks-based approach for strategic planning. *Information and Management* 35: 357-364.

Chileshe, N., 1996. *An Investigation into the problematic issues associated with the implementation of Total Quality Management (TQM) within a constructional operational environment and the advocacy of their solutions.* Unpublished MSc dissertation, Sheffield Hallam University.

Cronbach, L.J 1951 Coefficient Alpha and the internal structure of tests. *Psychometrika,* 16(3): 185-202.

DTI 1998 Statistical press release p/98/597, small and medium enterprises (SME) statistics for the UK, 1997. Available from World Wide Web :http://dti.gov.uk/sme4

Fahy, J. 1996, Competitive Advantage in International Services : A Resource-Based View, *International Studies of Management & Organisation,* , (2) 24-37

Ghobadian, A., and Gallear, D.N., 1996. Total Quality Management in SMEs. *Omega International Journal of Management Science* 24(1). 83-106

Gummer, B. (1996), " Total Quality Management : Organizational transformation or passing fancy ", *Administration in Social Work,* Vol. 20 pp. 75-95

Hardy, L., 1983. Successful Business Strategy-How to win the market place, Kogan Page.

Handfield, R.B., and Melnyk, S.A., 1998. The scientific theory-building process : a primer using the case of TQM *Journal of Operations Management ,* 16 321-339

Hasegawa, F. (1988) , Built by Japan : Competitive strategies of the Japanese construction industry, John Wiley & Sons

Malhotra, M.K., and Grover, V., (1998). An assessment of survey research in POM: from constructs to theory. *Journal of Operations Management ,* 16 . 407-425.

Mohram,S.A., Tenkasi, R.V, Lawler III E.E and Ledford Jr. G.E., (1995). Total quality management : practices and outcomes in the largest US firms. *TEmployee Relations,* 17(3) 26-41.

Moreno-Luzon, M.D., (1993). Can total quality management make small firms competitive ? *Total Quality Management* 4(2) 165-181.

Morgan, N.A., and Piercy, N.F., 1996. Competitive advantage, quality strategy and the role of marketing *British Journal of Management ,* Vol. 7, pp. 231-245.

Oakland, S.J., (1993). *Total Quality Management ,* 2nd Ed. Dartmouth.

O'Leary-Kelly, S.W., and Vokurka, R.J., 1998. The empirical assessment of construct validity. *Journal of Operations Management ,* 16 387-405.

Laza, R.W. and Wheaton, P.L. (1990), " Recognizing the pitfalls of Total Quality Management ", *Public Utilities Fortnightly,* Vol 125, pp17-21

Pheng, L.S., and Hwa, G.K (1994), " Construction Quality Assurnace: Problems of Implementation at Infancy Stage in Singapore ", *International Journal of Quality & Reliability Management,* Vol.11, No.1, pp. 22-37

Porter, M.E., 1990 The Competitive advantage of nations. *Harvard Business Review.* March-April

Powell, T.C., 1995. Total Quality Management as Competitive advantage: a review and empirical study. *Strategic Management Journal.* 16, 16-37.

Rayner, P., 1991. The experience of small firms with BS 5750 registration. *International Journal of Quality & Reliability* Management.

Rao, S.S., Ragu-Nathan, T.S., & Slis, L.E., 1997. Does ISO 9000 have an effect on quality management practices ? An international empirical study. *Total Quality Management* 8(6). 335-346

Steinngrader, F.G 1990. Total Quality Management: A new look at a basic issue. *TVital Speeches* 57(13). 415-416

Quazi, H.A., and Padibjo, S.R., 1998. A journey towards total quality management through ISO 9000 verification - a study on small-and medium-sized enterprises in Singapore *Journal of Quality & Reliability Management* 15(5) 489-508.

Wing, C.K., Raftery. J., anf Walker, A., 1998. The baby and the bathwater: research methods in construction management. *Construction Management &Economics* 16(1) 99-104.

Creative Systems in Structural and Construction Engineering, Singh (ed.) © 2001 Balkema, Rotterdam, ISBN 90 5809 161 9

Motivational factors in Australian construction industry workforce

Swapan Saha
Construction and Building Sciences, University of Western Sydney, N.S.W., Australia

Wade Cruickshanks
Richard Crookes Constructions Pty Limited, Australia

Ron R. Wakefield
Jamerson Professor of Building Construction, Virginia Polytechnic Institute and State University, Va., USA

ABSTRACT: With the decline in Australian construction worker productivity over the past twenty years, it could be suggested that construction workers are not motivated to perform their work in the same manner as workers of the previous generation. The predominant aim of this paper is to discover variables of motivation that are most important to construction workers in Australia's construction industry. Many motivation theories have been postulated to explain the complex relationship between variations in human personality and motivation, with resultant conflicting interpretations of worker motivation and the effects of different motives. This paper presents results obtained after interviewing workers from a cross section of the industry and performing a statistical analysis of the results. Results of this study of motivation of construction workers in Australia are compared with results of a similar study conducted with construction workers in the UK and Nigeria. The study demonstrates that motivating variables are very interdependent, with a high correlation between motivating and de-motivating factors. The results show that construction workers motivation involves a complex interplay of different variables. The results show that the most important motivational variable is a person's safety needs, including job security and provision of safety at work.

1 INTRODUCTION

Construction Companies in recent years are focusing on shifting their attentions and efforts to provide better service products to satisfy the customer needs. This emphasis is a result of increased competition at the national level as well as the global scale. The challenge of meeting client desires, coupled with the fierce construction market, has forced contractors to look for methods of improving their overall performance in terms of quality, productivity, customer satisfaction, and safety records and cost effectiveness. Construction company efforts to make such improvements are being hindered by what has been described as the declining level of worker satisfaction, morale, motivation and performance. This is apparent from the fact that the construction industry has reported a decline in worker productivity for more than twenty years (Construction Industry Institute 1992). It has been questioned whether or not management has accurately identified the motivational needs of their workers in order to concentrate their improvement efforts (Cox & Beliveau 1994).

It is widely recognised that some attributes of the workers affect their productive capacity in a particular trade or craft. These attributes can be broadly defined as:

i) the skills, qualifications, training and experience of the worker;

ii) the worker's innate physical ability and

iii) The intensity of the application of both skill and ability to the production process.

Motivation studies focus on the third attribute. This depends on an integral behavioral generator in individuals, which in turn is a product of the environment. This study seeks to resolve the confusion about the motivational needs of construction workers in Australian construction companies. Workers at various commercial and industrial building sites have been surveyed to formulate the results for this study. The predominant aim of this study was to discover variables of motivation that are most important to construction workers in Australia's construction industry.

2 BACKGROUND

2.1 Classical motivation theories

"Motivation refers to a person's desire to perform particular behaviors and willingness to expend effort on them" (Collins 1996). Since the human relations movement in the 1930's motivation has become a focus of managers as a way of increasing the productivity of their workers.

During the early 1920's and early 1930's Elton Mayo and Fritz Roethlisberger of Harvard, conducted research into how physical working conditions affected the productivity and efficiency of factory workers. They changed the lighting conditions of the workplace thinking that if they increased the lighting, productivity would increase accordingly. However, they found that productivity increased irrespective of whether the lighting was decreased or increased. Their conclusion was that 'the attention' they were giving to the workers that motivated them to increase productivity. This became known as the "Hawthorn Effect" (Collins 1996).

Abraham Maslow a psychologist who was very focused on workers needs and motivation, formed his own theory of motivation, commonly known as "Maslow's hierarchy of needs", which is still very prevalent today. Maslow believed that there are 5 levels of a person's needs whereby for a higher level of need to be reached the lower level must first be fulfilled. The needs begin with survival needs and ascend to self-fulfilment needs. Maslow's Hierarchy of Needs (Collins 1996):

1. Physiological: food, water, shelter, sex, sufficient wages

2. Safety: security, stability, freedom, from fear

3. Social: friendship, interpersonal interaction, group membership

4. Self Esteem: achievement, recognition, personal worth

5. Self-actualisation: realisation of one's full potential, personal growth, accomplishment, self expression

Frederick Herzburg developed a complementary way of thinking to Maslow's theory, by categorising factors required by a worker for them to perform their work adequately and furthermore be motivated in their given position at work. Herzburg put these factors into two groups (Collins 1996):

- *Hygiene factors*
  1. Salary
  2. Physical working conditions
  3. Job security, growth and learning
  4. Policies and procedures
  5. Good relationship with supervisor and peer
- *Motivating factors*
  1. Interesting work
  2. Opportunities for personal growth
  3. Sense of achievement
  4. Responsibility

The hygiene factors are concerned with a persons working conditions, if these factors are satisfactory then the worker will be content in their position but not necessarily motivated to make an extra effort in performing their tasks at work.

For a person to be motivated in their position the motivating factors must be present, meaning the work must be interesting, having possibilities for personal growth through learning and developing new skills. It is accepted that we as humans like to feel that we are achieving something and not wasting our time regardless of what we are achieving. Responsibility motivates people, because through responsibility people know that what they achieve will be recognised as their achievement.

2.2 Motivation vs. Demotivation

Frederick Herzberg drew his motivation hygiene theory from an examination of events in the lives of engineers and accountants, along with other investigations involving many different populations.

The results suggested that the factors that people felt satisfied & motivated by are separate and distinct from the factors that caused dissatisfaction and demotivation. This meant that when looking at motivation or demotivation, different factors have to be considered, meaning the two feelings are not opposites of each other. The opposite of satisfaction is not dissatisfaction and similarly the opposite of dissatisfaction is not satisfaction (Herzberg 1991).

This is a proposition which confuses many people because we usually consider satisfaction and dissatisfaction as opposites of each other, what is not satisfying is dissatisfying and vice versa.

Dissatisfiers are made up essentially of such matters as pay, supplementary benefits, company policy and administration, behaviour of supervision, working conditions and several other factors associated with the task.

Motivators, for the most part are the factors of achievement, recognition, responsibility, growth, advancement, and other matters associated with self-actualisation of the individual on the job.

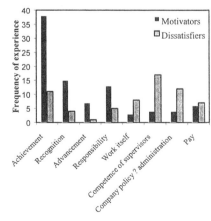

Figure1. Factors affecting hourly rate technician (Myers 1991)

Olomolaiye et al. (1998) became a supporter of Herzburg's theory, when he found similar results whilst conducting a survey of UK and Nigerian workers. The workers ranked the motivating factors and demotivating factors in order of importance and the results disclosed that the motivating and demotivating factors were not opposite each other. Olomolaiye et al. (1998) in their study showed that in both the UK and Nigeria the construction workers considered supervision was of great importance and would be a major dissatisfier if not of a competent level. More about this study is discussed in the results section of this paper.

Myers (1991) based his study on Herzberg's original study and further focused his study on five (5) occupational groups within the company, Texas Instruments. He looked at engineers, scientists, manufacturing supervisors, hourly technicians and female assemblers. The overall results where very similar to those of Herzberg although when each profession was looked at individually there where varying results shown in the Figure 1.

3 DATA COLLECTION

The survey was distributed to seven (7) sites of Richard Crookes Constructions, Australia. Response was received from sixty (60) workers representing 60% of the questionnaires distributed. This is an adequate response rate considering the lukewarm and suspicious attitude of construction workers towards productivity studies in general. According to Babbie (1991) it is a relatively good result. Achieving a relatively high response rate reduces the chance of significant response bias.

Data from this survey was analysed using a spreadsheet. Data were analysed based on all workers responses, their age group, payment methods, and employers. Detail about these results is presented below.

4 RESULTS

4.1 Ranking methods

A survey questionnaire included 15 motivators and 12 de-motivators as shown in the tables presented below were distributed to respondents for evaluation based on a four (4) points scale. In response to how respondents felt each item is important for 'motivation' and 'demotivation', they were asked to tick a box containing three choices, namely, somewhat important (scaled 1), important (2) and very important (3). For the same item they were asked to place an importance of the item for motivating the respondent at workplace on a scale between one (1) to four (4), where, one (1) being least motivating and four (4) being most motivating.

For example if a respondent saw "good relationship with mates' was very important (3) and provided a importance score of four (4) (most motivating), then he would have scored 12 points towards the total score. An index for an item can be obtained by adding the scores from all respondents divided by common denominators. Based on Olomolaiye (1988) PhD study, common denominator was the number of respondents times ten (10).

4.2 Ranking for all workers

Results shown in table 1 indicate 'job security' is number one motivating variable for all the workers. Good safety program ranked as the 2nd highest motivating variable overall with an index of 0.88.

Table 1. Ranking based on all workers

All workers	Score	Index	Rank
Motivators			
Good relationship with work mates	456	0.76	4
Good safety programme	528	0.88	2
Your work itself	442	0.74	6
The possibility of overtime	212	0.35	15
Recognition on the job	312	0.52	9
Level of Pay	514	0.86	3
Accurate work description	434	0.72	7
Participation in making decisions	318	0.53	12
Good supervision from your foreman	426	0.71	8
The possibility for promotion	314	0.52	9
More responsibility	294	0.49	13
Having a challenging task	448	0.75	5
Job security	570	0.95	1
Being able to choose your work mates	222	0.37	14
Bonuses and fringe benefits	310	0.51	11
Demotivators			
Disrespect from your supervisors	380	0.63	2
Little accomplishment in your work	318	0.53	8
Repetitiveness of work	288	0.48	10
Recognition for your work	330	0.55	7
Under utilization of your skills	286	0.48	12
Incompetent work mates	370	0.62	3
Mates who don't co-operate	348	0.58	6
Poor inspection	360	0.60	5
Unsafe conditions	434	0.72	1
Hot weather00	302	0.50	9
Cold weather	254	0.42	10
Too much work	370	0.62	3

A very surprising result considering most of the respondent's site managers reported that they were required to remind the workers about keeping up their safety practices. The only category who did not consider safety to be a high factor was the workers between the 18-25 years of age bracket who ranked this item 9th as shown in table 2. This would indicate that young workers provide less importance to their own safety compared to the 'relationships with mates at work'.

4.3 Ranking by age category

It is well known that as you get older your priorities in life change. When you are young you are interested in having fun and living life to the full then you move into your thirties and work and family start to become a priority. After the thirties your look at you personal development (Collins 1996). Because of these factors it appeared interesting to check for any independence between different age groups.

Table 2. Ranking based on age group

Motivators	Age group		
	18-25	25-35	35-50
Good relationship with work mates	1	5	7
Good safety programme	9	2	1
Your work itself	5	9	6
The possibility of overtime	14	14	15
Recognition on the job	11	13	9
Level of Pay	3	3	3
Accurate work description	7	10	4
Participation in making decisions	12	11	10
Good supervision from your foreman	6	7	7
The possibility for promotion	10	6	13
More responsibility	13	12	11
Having a challenging task	4	7	5
Job security	1	1	2
Being able to choose your work mates	15	15	12
Bonuses and fringe benefits	8	4	14
Demotivators			
Disrespect from your supervisors	2	4	3
Little accomplishment in your work	3	11	2
Repetitiveness of work	8	9	7
Recognition for your work	8	9	4
Under utilization of your skills	12	5	10
Incompetent work mates	6	2	6
Mates who don't co-operate	11	5	5
Poor inspection	9	1	8
Unsafe conditions	6	3	1
Hot weather	3	8	11
Cold weather	3	11	12
Too much work	1	7	9

Of the sixty (60) respondents there were 17% in the age group of 18-25 years. There is a considerable difference in response compare to their older counterparts. The major distinction is their low consideration of a 'good safety program' as shown in table 2. This reflects the values of people at this age as discussed earlier. Whilst there were some similarities in ranking there was a considerable variance in their indexes.

37% of the respondents were in the age group of 25-35 years. The index scores overall were higher than for all respondents, showing that they consider the motivating variables to be more important than most. The main difference was *'the possibility for promotion'* with an index of 0.71, which is 0.19 points higher than for all respondents. This substantiates the theory that at this age people tend to concentrate on their careers as an important part of their life (Collins 1996). 43% of the respondents were in the 35-50 years age group. This group considers safety is the top priority with an index value of 0.98 compared to overall workers index value, 0.82. It is evident that at this age the workers will generally have a family to support, which influences them to rank safety as a number one motivating variable. The result in table 2 indicates that 'job security' and 'level of pay' were ranked highly. Then consideration was given to the factors that helped in performing their work such as 'accurate work description', ' challenging tasks', and 'good relationship with mates' which were all ranked highly.

4.4 Ranking by type of contract

Head contractors in the building industry engage a number of subcontractors for different trades. A subcontractor for a smaller trade can be a one man company. In this study anybody working for a head contractor was called an 'employee'. Responses from 'employee' and subcontractors were analysed separately and presented in the table 3. This was attempted to see whether there is any difference in opinion between subcontractors and employees. 80% of the respondents were employees therefore, the overall workers ranking generally reflects the employees thoughts with minor variations. Subcontractors represented 20% of the respondents with some interesting results. This group ranked 'job security' as 1st with a very high relative index of 1.10, this is quite interesting considering the nature of work they perform as a subcontractor. This establishes that many subcontractors like to work regularly for well-established builders to reduce their risk of being out of work. Subcontractors compared to the employee interestingly ranked 'Good supervision from your Foreman' lower. From this result it is reflected that subcontractors are not under direct control of a foreman, while, employees work under a foreman and they thought a good foreman can make the job considerable easier through efficient coordination of trades.

Table 3. Ranking based on contract type

Motivators	Contract type	
	Employee	Subcon.
Good relationship with work mates	6	5
Good safety programme	2	3
Your work itself	5	8
The possibility of overtime	14	15
Recognition on the job	10	10
Level of Pay	3	3
Accurate work description	7	6
Participation in making decisions	11	9
Good supervision from your foreman	4	12
The possibility for promotion	9	13
More responsibility	13	11
Having a challenging task	8	2
Job security	1	1
Being able to choose your work mates	15	14
Bonuses and fringe benefits	11	7
Demotivators		
Disrespect from your supervisors	2	7
Little accomplishment in your work	9	1
Repetitiveness of work	7	11
Recognition for your work	7	4
Under utilization of your skills	10	6
Incompetent work mates	4	10
Mates who don't co-operate	5	8
Poor inspection	2	12
Unsafe conditions	1	1
Hot weather	10	5
Cold weather	12	9
Too much work	5	1

Table 4. Ranking based on payment

Motivators	Payment type	
	Salary	Wage
Good relationship with work mates	3	4
Good safety programme	1	3
Your work itself	5	7
The possibility of overtime	12	15
Recognition on the job	3	14
Level of Pay	7	2
Accurate work description	2	8
Participation in making decisions	5	11
Good supervision from your foreman	11	4
The possibility for promotion	13	10
More responsibility	9	12
Having a challenging task	9	6
Job security	7	1
Being able to choose your work mates	15	13
Bonuses and fringe benefits	14	9
Demotivators		
Disrespect from your supervisors	10	2
Little accomplishment in your work	3	11
Repetitiveness of work	11	8
Recognition for your work	2	9
Under utilization of your skills	11	9
Incompetent work mates	9	3
Mates who don't co-operate	8	5
Poor inspection	4	4
Unsafe conditions	1	1
Hot weather	6	7
Cold weather	7	11
Too much work	5	6

4.5 Ranking by method of payment

In this study an assumption was made that wage earners are generally paid for the number of hours they work per week for a task based on an hourly rate in the construction industry. Salary earners are paid based on a lump sum amount per annum.

Results from this study presented in the table 4 shows that the method of a workers payment would greatly influence their motivation to work.

Wage earners represented 70% of the respondents. They consider job security as their top priority for motivation. Regulation by the Australian government regarding the termination of their employment may offer some explanation to this result. Wage earners are paid weekly and the employer normally requires only 1 (one) weeks' notice before termination of this type of worker.

Salary earners accounted for 30% of the respondents with some notable differences of opinion. All the respondents who where on a salary were employed in a supervisory role. Safety was by far the utmost of their concerns with a motivating index of 1.13 and a demotivating index of 0.83. This was followed by 'accurate work description' and 'recognition of the job'. Other motivation variables including 'participation in making decisions', 'more responsibilities', 'challenging tasks' and 'possibility for promotion' were scored much more highly by salary workers than the workers who received a wage. In the demotivators category 'little accomplishment in work' and 'recognition for your work' were the only other demotivators to score well.

The results clearly shows that salary workers were motivated by their work description, position, and clear direction from supervisors. These also inspired them to become more career-minded than the average construction worker. It is evident from the results that 'level of pay' was not very important to them and was ranked 7th compare to 3rd by all respondents.

Salary workers did not see 'job security' as a number one priority and ranked this item 7th, compared to wage earners ranking of 1st. This suggests that a salary earner normally has higher job security than wage earners. On the other hand 'safe working conditions' and 'recognition of their work' were more important to salary earner as reflected in their ranking in both motivation and demotivation categories.

Table 5. Comparison of Motivation Ranking on Maslow's scale

Theoretical ranking (after Maslow)	Ranking		
	AU	UK	NGR
Physiological needs			
Earnings related :			
Level of pay	3rd	1st	1st
Bonus fringe benefits	11th	5th	2nd
Overtime	15th	15th	7th
Safety needs			
Good safety program	2nd	8th	13th
Job security	1st	6th	3rd
Belonging needs			
Good relationship with mates	4th	1st	9th
Good supervision	7th	3rd	12th
Need for esteem			
Recognition on the job	9th	13th	8th
Promotion	9th	14th	5th
Need for self actualisation			
Challenging tasks	5th	11th	6th
Participation in decision making	12th	10th	9th
More responsibility	13th	12th	4th
Choosing workmates	14th	9th	-
The work itself	6th	3rd	14th

4.6 Comparison of Australia to UK and Nigeria

Because of the recognition accorded to Maslow's motivation theory in literature it was chosen as base of comparison for comparing the motivation ranking of Australia with UK and Nigerian workers (Collins 1996). The 15 motivation variables were regrouped to reflect the 5 broad groupings on the Maslow scale.

The results from table 5 suggest that Australian workers considered 'job security' and 'safety programmes' as their number one and two motivating items respectively. The UK counterparts ranked 'level of pay' and 'good relationship with mates' as their number one priority. Safety was not ranked highly by UK and Nigerian workers.

The results presented in table 5 indicate that there is a difference in importance placed on different factors by construction workers from three nations. This clearly shows workers from each country have different priorities in their life. It did not matter whether the workers were from a developed country or a developing one. The workers from any countries have their own culture and work ethic which was reflected from the variation in opinion.

5 CONCLUSION

The findings from this study regarding factors influencing construction workers motivation provides some guidance to the construction industry management in Australia.

Firstly the study has demonstrated that motivating variables are very interdependent, as shown from the correlation's found between motivating and demotivating factors. Construction workers' motivation involves a complex interplay of different variables, the most paramount of which has been seen to be of a person's safety needs, including job security and provision of safety at the work place.

Secondly, workers motivation is dependent on their age and method of payment. Therefore, employers have to consider methods of motivation carefully based on the employee's age and the payment scheme they fall into.

Thirdly, Maslow's theory of basic human motivation suggests that once a motivating factor is acquired it no longer remains motivating, seems to apply to the construction industry. This was demonstrated from the result that salary workers did not consider 'job security' as the most important factor in motivation because their form of employment provides a certain level of job security compared to other types of workers. In the UK construction workers did not regard job security as a number one priority compared to their Australian counterparts.

Fourthly, according to Olomolaiye et al. (1998) study, comparison between UK and Nigerian workers revealed that motivating variables are dependent on the construction management system and the economy. This could not be confirmed due to significant variations found in regards to the ranking of motivating factors by construction workers between Australia and the UK where both countries are recognised as western and developed countries.

Finally, this study has only identified the key areas that needed more attention in the construction industry in order to satisfy workers from different payment system and age group. Further investigation has to be carried out to determine an effective method of motivation for implementation.

REFERENCES

Babbie E. 1991. *Survey Research Methods*. Second Edition, Wadsworth publishing Company.

Construction Industry Institute, 1992. *The Challenge of Workforce Retention*. CII Task Force Product, CII Conference.

Collins, R. 1996. *Effective Management*. CCH Australia, North Ryde, Australia.

Cox R & Y. Beliveau 1994. Cost Management through Employee Participation Programs. *Transaction of the American Association of Cost Engineers*.

Herzberg F. 1991. One More Time: How Do You Motivate Employees. *Harvard Business Review paperback*. No. 90010, USA.

Myers, S. 1991. Who are your Motivated Workers? *Harvard Business Review Paperback* No. 90010. USA

Olomolaiye P. O., A. K. W. Jayawardane & F. C. Harris, 1998. *Construction Productivity Management*. The Chartered Institute of Building, UK, Longman.

Olomolaiye P. O. 1988. *Construction Operative Motivation in U.K. and Nigeria - A Comparison*, A dissertation submitted to fulfil PhD at Loughbrough University, UK

Creative Systems in Structural and Construction Engineering, Singh (ed.) © 2001 Balkema, Rotterdam, ISBN 90 5809 161 9

Proposed support framework for the procurement of public-private partnered infrastructure projects

X.Q. Zhang & M.M. Kumaraswamy
Department of Civil Engineering, University of Hong Kong, People's Republic of China

ABSTRACT: Build-Operate-Transfer (BOT) type procurements are popular vehicles for infrastructure development. Such schemes bear the advantages of additional funds availability, managerial efficiency, technology transfers and multi-skills development. However, they are much more complex than traditional procurement routes. Many countries are still inexperienced in the complexities and implications of such scenarios. To achieve win-win results, concerted and coordinated efforts from both the public and private sectors are essential, in addition to favorable political, legal and economic environments. Based on the experiences and lessons from evolving cross-country practices in public/ private partnership (P/P P) projects, a preliminary support framework is formulated, which after further improvement is expected to assist public sector clients to improve their BOT-type procurement processes in particular and P/P P initiatives in general. Key development issues throughout the project life cycle are taken into consideration in this framework, ranging from the feasibility study to the post-transfer management and operation.

1 INTRODUCTION

The increasing gaps between dwindling public funds and escalating demands for improved infrastructure facilities make public/private partnerships (P/P P) a popular option in public infrastructure procurement. Furthermore, low productivity and high construction dispute levels have raised arguments against the adversarial scenarios inherent in most traditional procurement paths (Egan 1998), which often position the constructor against the architect/ engineer/ client, rather than encouraging teamwork towards common targets. Such shortcomings are attributed to an inappropriately structured construction industry and the consequent procurement arrangements (Cox & Townsend 1997).

To counter the above-mentioned shortcomings, various BOT-type P/P Ps have been explored world over in public infrastructure development, where managerial expertise, technical innovations, operational efficiencies as well as enormous funds from the private sector are incorporated. Diverse results have been reported, with both successes and failures. Furthermore, many countries are inexperienced in infrastructure P/P Ps. There is a need to formulate a workable and efficient support framework to facilitate the procurement of public infrastructure projects through P/P Ps. The authors have thus attempted to develop such a framework. As a first step to do this, evolving governmental practices

in procuring BOT-type infrastructure projects are compared across countries to identify strengths of successful approaches and to draw lessons from less successful or abortive projects. Particular examples are drawn from experiences in Hong Kong and Mainland China, UK, USA, and Thailand.

2 SUGGESTED PROCUREMENT FRAMEWORK

The preliminary procurement support framework is simply expressed in a flowchart (Figure 1). The BOT spells out a significant shift in the philosophy of project procurement. By integrating financing, design, construction and operation, it transcends the traditional adversarial procurement environment by reversing the over-fragmentation of functions that has previously led to divergent or even confrontational agendas of the multiple participants.

The framework provides an overview on the development process through the BOT-type schemes, including BBO (Buy-Build-Operate), BLT (Build-Lease-Transfer), BOO (Build-Own-Operate), BOOM (Build-Own-Operate-Maintain), BOOT (Build-Own-Operate-Transfer), BT (Build-Transfer), BTO (Build-Transfer-Operate), DBFO (Design-Build-Finance-Operate), DBOM (Design-Build-Operate-Maintain), DOT (Develop-Operate-

Transfer), LDO (Lease-Develop-Operate), ROO (Rehabilitate-Own-Operate) and ROT (Rehabilitate-Operate-Transfer). The whole project development process is mainly divided into five stages: feasibility study, concessionaire selection, design and construction, operation and services provision, and transfer/ post-transfer management. The framework indicates what appraisal needs to be done and what decisions have to be taken in each stage. The following sections provide detailed discussion of this preliminary framework.

3 FEASIBILITY STUDY

3.1 *Evaluating the viability of BOT-type projects*

Not all projects are suitable for BOT-type procurements. A commercially viable BOT project requires the following conditions: (1) a stable political system; (2) a predictable and reasonable legal system, including investment codes and dispute resolution mechanisms; (3) a long-term promising economy; (4) adequate local capital markets; (5) predictable currency exchange risks; (6) a long term demand for the service provided by the project, and the service is affordable to the public; (7) clear governmental objectives and commitments; (8) limited competition from other projects; and (9) revenue streams from the project are attractive to both the investors and financiers (Ogunlana 1997). Many projects designated to develop through such schemes never materialize due to a lack of such conditions such as those projects in Turkey (Birgonul & Ozdogan 1998), and some even get trapped such as the Second Stage Expressway System and the Don Muang Tollway in Thailand (Ogunlana 1997).

Ashley et al. (1998) have developed the Project Scoring Table (PST), a tool that can identify the key considerations and decisions in a P/P P project and show how these factors affect the values of the project to the public client and the private developer. The PST framework evaluates the viability of a potential P/P P project through nine 'decision clusters': (1) political clearance; (2) partnership structure; (3) project scope; (4) environmental clearance; (5) construction risk; (6) operational risk; (7) financing package; (8) economic viability and (9) developer financial involvement.

It is advisable for the public client to seek the opinions of commercial organizations (e. g. project developers, contractors, and financiers) on whether there is a BOT market for the project under consideration, the suitability for a particular BOT-type procurement and the appropriate scope of the project. This can be achieved through a consulting process. Then, the client can make appropriate adjustments to facilitate the perceived BOT scheme, taking into consideration the suggestions of commercial organizations.

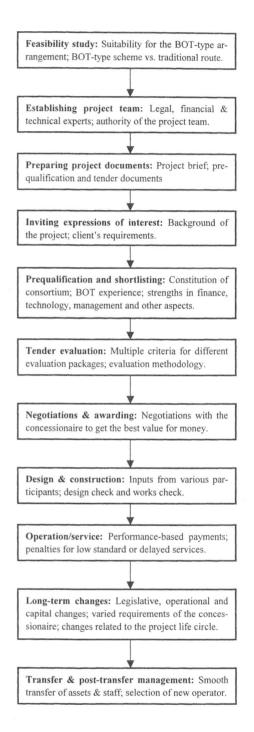

Figure 1. Proposed support framework for BOT-type project procurement

3.2 *Conventional procurement vs. BOT*

For an infrastructure project, the public client should initially consider the need for the project (its social and economic benefits) and its viability. Then, adequate consideration should be given to the suitability of using the BOT strategy for its procurement. Even if the project is suitable for a particular BOT-type arrangement, still there is a need to decide whether this arrangement is better than a traditional procurement route for this specific project. For example, in the PFI project in UK, this evaluation is done by the generation of a reference project, the public sector comparator (PSC). A PFI scheme must demonstrate value for money when compared with the PSC, substantial risk transfer to the private sector and being affordable to the client. A PFI approach should not cost more than a conventionally procured comparator (CPC) would have done. Such comparison should be exercised over the whole contract life and reflect all the constituents of the contract. This is achieved by comparing the net present value of the cash flow of costs from the PFI option and that of a CPC. Other things being equal, the option with the lower net present cost should be the preferred one. All the costs and benefits of the CPC to the client should be taken into calculation, assuming that the whole range of PFI services were provided by a conventional means. This would include facilities costs, full life circle operating costs and any external income. The CPC would also include the costs of the risks being retained by the client. These costs would be discounted at the client's own real long-term cost of capital (Construction Industry Council 1998 and Higher Education Funding Council for England 1997).

4 SETTING UP THE PROJECT TEAM

The value of a BOT contract should be based on the delivery of services over a long period. It is likely to be several times the cost of the total assets involved to deliver the services, and it is this enormous overall cost that justifies the relatively higher expense of the BOT-type procurement process.

The BOT scheme is complicated and many aspects may be new territories to the client, demanding substantial clerical and administrative support. To promote the whole procurement process, it is a normal practice for the client to establish an ad hoc full time project team, incorporating legal, commercial, financial, technical and environmental advisers. Reputable advisers can enhance the credibility and standing of a BOT project. This team is responsible for the feasibility study, detailing the client's requirements in the project, preparing prequalification and tender documents, prequalifing and evaluating tenders, conducting negotiations with the tenderers to select the most suitable concessionaire, assisting the concessionaire to obtain various permits, and facilitating the design and construction process.

For example, in Seattle's Tolt Treatment Facilities (a DBO scheme), a project team was established. This team comprised of water quality, engineering, policy and administrative leadership from Seattle Public Utilities and specialty skills from representatives of the City Council Staff, Mayor's office, and other independent technical, legal, and financial advisors. The project team continuously monitored the project progress, maintained timely and productive team communications and discussions of quality control and quality assurance measures (Kelly et al. 1998).

5 CONCESSIONAIRE SELECTION

5.1 *A fresh approach needed*

The BOT shifts from a traditional reactive *modus operandi* to a proactive scenario. The concessionaire is involved in more project development aspects (legal, financial, technical, operational and environmental) and more long-term risks and uncertainties. Furthermore, since the proposed consortiums usually have no track records their prequalification for a BOT-type project is particular to that project. This is different from conventional contractor prequalification, in which the client may consider contractors for a number of projects within certain design, service and/ or works categories.

In addition, tender costs for BOT projects are normally very high (in some cases between 5% and 10% of the project costs), as the resources involved in preparing a tender package to finance, design, build and operate a facility are greater than those required for a conventional form of procurement (Merna & Smith 1996). The total tendering costs for the preferred tenderer have sometimes been upwards of 10% of the total capital cost for smaller projects. It was found that on smaller PFI projects (with a capital cost of around £10 million) the cost of tendering to a short-listed consortium is in the order of 1% (Higher Education Funding Council for England 1997).

The significant difference of the BOT-type procurement from the traditional route, the many issues that need to be included in the tendering package and the very high tendering costs necessitate a fresh approach to the selection of the BOT concessionaire. The tender evaluation methodology and criteria should be quite different from those for traditional projects. For example, in traditional projects the lowest tender price may be the only selection criterion, but this is not suitable for a BOT-type project.

5.2 Prequalification and tender documents

Hatush and Skitmore (1997) have discussed five criteria packages that need to be considered in contractor prequalification. These criteria are still useful for concessionaire prequalification after proper adjustments. The prequalification criteria should include the constitution of the consortium and each constituent company's qualification level, the consortium's business, financial, technical, and commercial strengths, and experience in BOT-type projects. The tender documents should include criteria against which the client can evaluate how well its objectives and required standards are met. Issues need to be included in the tender documents mainly include: (1) the structured contract (comprising of general, specific, common terms and the project conditions regarding the construction, operation and maintenance, finance and revenue generation packages); (2) output specifications (including the project operating environment and risks transferred to the concessionaire); (3) evaluation criteria and evaluation methodology; (4) Obligations of the client and the concessionaire (e.g. risk allocations and supports/ guarantees from the client).

5.3 Prequalification/ shortlisting

Because costs and resources expended in tendering a BOT-type project would be very high, the number of tenders invited should be kept to a maximum of four to six. This can be achieved in the prequalification stage, in which the interested companies or consortiums are firstly prequalified and reduced to a relatively small number (e.g. 6 to 8). Then, the prequalified ones are shortlisted to an appropriate number (e.g. 3 to 4) according to their prequalification results, subsequent presentations and/ or their outline project proposals required in case of complex projects.

5.4 Tender evaluation

In BOT-type tender evaluation, various packages (legal, technical, financial, commercial, operational, safety/ health and environmental) should be assessed. To facilitate evaluation and comparison, the tenderers in their tender preparation should follow a standard format. For example, in the financial aspect, a standard cost calculation formula may be used for the tenderers to input their costs and discount them to the net present value.

Zhang and Kumaraswamy (in print) have discussed the current evaluation practices and formulated a basic model for concessionaire selection. Simple scoring system, multi-attribute analysis and the Kepner-Tregoe decision analysis technique have been used in BOT-type tender evaluation. Sensitivity analysis and value engineering are two useful tools that can be used to facilitate financial and technical evaluation respectively.

Low-priced finance, low service charges (tolls/ tariffs) and innovative technical solution that can both enhance the service and reduce life circle cost of the project are critical features of the tender and are usually assigned much higher weights than other aspects. For example, in the Laibin B BOT power plant in China, electricity tariff, financial proposal and technical proposal are assigned 60%, 24% and 8% of the total weight respectively.

During the tender evaluation stage, there are may be discussions between the client and the tenderers for clarification and possibly for some adjustments of the tenders. For example, in the PFI project in UK, there are separate and confidential discussions between the client and the tenderers during tender evaluation, after which the tenderers may be requested to submit a best and final offer (BAFO), which is a well defined last and formal tender that incorporates all critical points that have been made during the course of discussions.

5.5 Contract negotiation

The negotiation stage comes after the client announces the preferred or the winning tender. The negotiations deal with specific terms rather than general terms discussed during the tender evaluation stage. A final contract will be entered into between the client and the winning tenderer. This contract reflects the client's transfer of risks and value for money compared with a traditional procurement route, as against the concessionaire's money-making opportunities in project construction and operation, and possibly in related property development.

6 DESIGN AND CONSTRUCTION

The design, construction and commissioning of the project can be handled by a construction company that is part of the concessionaire. These activities can also be assigned by the concessionaire on a Design and Build contract to a contractor or contractors. Early successful completion of the construction works is the common objective of the client, the concessionaire and the financiers. Large amounts of the project cost would be spent in the construction stage. The construction progress might be delayed or even be put in jeopardy if there were substantial changes after construction commencement in the project design either requested by the client or the operator. It is advisable to incorporate the efforts of the client, the operator, the construction project manager, the main contractor and/ or subcontractors to make sure integrated quality design. In addition, an appropriate change management procedure should be followed. Value management and buildability review should be exercised to keep the design to the core objectives and to ensure that the construction works will be carried out smoothly.

The client should monitor the construction progress, checking time and quality issues against the agreed technical specifications and construction milestones. For example, in the procurement of its BOT tunnel projects, the Hong Kong government requires the concessionaire to employ an independent third party checking of design and construction works (Kumaraswamy and Zhang in print).

7 OPERATION AND SERVICE QUALITY

7.1 Contract monitoring

Safe and smooth operation of the project and provision of quality services are a long-term issue for a BOT-type project. Key operation conditions should be included in the contract: performance and equipment specifications; testing procedures; methods of measuring offtake; warranties; accounts and records; service quality; monitoring and inspection procedures; payment procedures and penalties for delayed or substandard services. The client should monitor the concessionaire's performance to ensure the availability and quality of the agreed services. Any deterioration in quality or availability would be informed timely of the concessionaire for proper corrections within a specific time.

7.2 Performance based payments

Service payments from the client to the concessionaire should not be guaranteed, which would otherwise make the project 'on balance sheet' to the client. If the service provided meets the agreed quantity and standard then the client will pay full service charges. However, if the concessionaire failed to offer services by an agreed date or the services were of low standard, penalties would be applied, e.g., in the form of liquidated damages and/ or reduced payments or non-payments. The concessionaire's serious and persistent failure to commence service delivery or to provide acceptable services in the agreed quantity could even result in contract termination without client compensation.

8 CHANGES OVER THE CONCESSION PERIOD

Some major changes may happen over the long-term concession period. These changes can gener-

ally be classified into three types: (1) legal changes comprising of legislative changes and changes to the original concession agreement; (2) changes due to the varied requirements of the concessionaire; and (3) changes relating to the life circle of the BOT-type project (Construction Industry Council 1998).

Provisions on the principles and procedures to deal with long term changes should be established to facilitate both the client and concessionaire in managing such changes should they appear. For example, legislative changes may be either general that have wide influence or specific that only affect certain public sector. It is a normal practice that the concessionaire would be responsible for the effect of the general legislative changes on its operations, while the public client would compensate the concessionaire's losses resulted from sector-specific legislative changes. A BOT-type project should aim at achieving a win-win result. The concessionaire should obtain a 'reasonable but not excessive' return commensurate to its performance, while the client achieves value for money. For this purpose, the client and the concessionaire may agree in the concession agreement on a range of percentage of the concessionaire's return, and later in the operation stage they conduct periodic reviews of the concessionaire's actual profit level and make adjustments to bring the level within the agreed range. This practice has been adopted in Hong Kong's BOT tunnel projects. The public client and the concessionaire agreed on in advance the maximum and minimum levels of 'estimated net revenue' (ENR) and a defined number and level of 'anticipated toll increases' (ATIs). The concessionaire may implement an ATI on a designated date provided that the 'actual net revenue' (ANR) is below the maximum ENR. The concessionaire may also advance an ATI should the ANR fall below the minimum ENR. If the ANR exceeds the maximum ENR, 'excess revenues' are siphoned into a 'toll stability fund' that the government has right to use for the deferment of specified ATIs by subsidizing the tolls.

9 TRANSFER AND POST-TRANSFER MANAGEMENT

At the end of a BOT-type contract, all the assets of the project should be transferred to the public client at no cost. The transfer process should be carried out smoothly so that the service provision will not be interrupted. As the contract approaches its termination date (may be two to three years before this date for a large long concession period project) the client should begin to prepare the transfer issues. For example, in the Cross Harbor Tunnel (CHT) project in Hong Kong, a BOT project transferred to the government in September 1999 after a 30-year concession period, the following critical issues were addressed to ensure a smooth transfer: (1) Necessary legislation for future management of the CHT; (2) Preparation of tender documents for a MOM (Management, Operation and Maintenance) contract for the post-transfer running of the CHT; (3) Agreement on the list of assets to be transferred by the concessionaire; (4) Following up on the outstanding maintenance works with the concessionaire and (5) Smooth transition of the staff of the concessionaire (Zhang and Kumaraswamy await print).

One important issue is to check and ensure the proper maintenance of the project by the concessionaire, who may lack interest in maintenance as the contract is drawing to a close, especially when the concessionaire is losing money in running the project or when the maintenance cost becomes much greater than the sum total service charge penalties. It was a normal practice that any remedial works for dilapidation would be carried out at the concessionaire's cost.

After the transfer, the public client has three options to run the project: by the client itself, by employing the former operator or by a new operator selected through a tendering process. For example, in the CHT project, the Hong Kong government selected a new operator through public tendering based on a MOM contract.

10 CONCLUSION

The relatively short history, inexperience and increasing popularity of BOT-type public private partnerships in public infrastructure development and the complexities of such schemes call for a support framework for the whole procurement process. Based on experiences and lessons from cross-country P/P P practices an indicative preliminary procurement support framework is developed to assist the potential public client contemplating the P/P P in their particular sector. First, this framework suggests the client to conduct a feasibility study to evaluate the viability of a potential P/P P project and to decide which particular partnership structure to be adopted. Second, an appropriate tender evaluation technique should be established to select the most suitable concessionaire. Third, inputs from various participants to a project should be involved in the design stage for proper management of value, risk and buildability, and independent design and works checking activities should be conducted to secure quality design and construction. Fourth, a monitoring team should be set up to monitor the performance of the concessionaire, based on which

the client makes the service payments. Fifth, proper measures should be taken to realize a smooth transfer of the project to the client and the continuing successful management and operation. Sixth, principles and procedures for dealing with long-term changes should be incorporated in the concession agreement, and periodic reviews conducted to safeguard the interests of both the client and the concessionaire.

REFERENCES

Ashley, D., Bauman, R., Carroll, J., Diekmann, J. & Finlayson, F. 1998. Evaluating viability of privatized transportation projects. *Journal of Infrastructure Systems*, Vol. 4 (3), 102-110.

Birgonul, M.T. & Ozdogan, I.1998. A proposed framework for governmental organization in the implementation of build-operate-transfer (BOT) model. *ARCOM 1998 Conference Proceedings*, 517-526.

Construction Industry Council 1998. Constructor's key guide to PFI. Thomas Telford Services Ltd, London, UK.

Cox, A. & Townsend, M. 1997. Latham has a half-way house: a relational competence approach to better practice in construction procurement. *Journal of Engineering, Construction & Architectural Management*. 4(2), 143-158.

Egan, J. 1998. Re-thinking construction. Dept. of Environment Transport and Regions. July 1998. London, UK.

Higher Education Funding Council for England 1997. Practical guide to PFI for higher education institutions. Bristol, UK.

Hatush, Z. & Skitmore, M.1997. Criteria for contractor selection. *Construction Management and Economics*. 15(1), 15-38.

Kelly, E.S., Haskins, S., & Reiter, P.D. 1998. Implementing a DBO project – the process of implementing Seattle's Tolt Design-Build-Operate Project provides a road map for other utilities interested in alternative contracting approaches. *Journal of American Water Works Association*, June 1998, 34-46.

Kumaraswamy, M.M. & Zhang, X.Q. in print. Governmental role in BOT-led infrastructure development. *International Journal of Project Management*.

Merna, A. & Smith, N.J. 1996. Guide to the preparation and evaluation of build-own -operate-transfer (BOOT) project tenders. Asia Law & Practice Ltd, Hong Kong, China.

Ogunlana, S.O.1997. Build operate transfer procurement traps: examples from transportation projects in Thailand. *CIB Proceedings: A key to innovation*. Rotterdam, The Netherlands.

Wang, S.Q., Tiong, R.L.K., Ting, S.K., Chew, D. & Ashley, D. 1998. Evaluation and Competitive Tendering of BOT Power Plant Project in China. *Journal of Construction Engineering and Management*, 124(4), 333-341.

Zhang, X.Q. & Kumaraswamy, M.M. await print. Hong Kong experience in managing BOT projects. *Journal of Construction Engineering and Management*.

Zhang, X.Q. & Kumaraswamy, M.M. in print. Choosing concessionaires in BOT-type projects. *Proceedings of the 16th ARCOM Annual Conference*, Glasgow, UK.

Creative Systems in Structural and Construction Engineering, Singh (ed.) © 2001 Balkema, Rotterdam, ISBN 90 5809 161 9

Transit project success criteria

J.A. Kuprenas
University of Southern California and Vanir Construction Management, Los Angeles, Calif., USA

A. Nowroozi
University of Southern California, Los Angeles, Calif., USA

E.B. Nasr
California Department of Transportation, Los Angeles, Calif., USA

ABSTRACT: This research identifies both positive and negative factors in the management of local transit projects. These factors influence the local agency satisfaction with the project delivery process and affect project budget performance and schedule performance. The research is based upon a one-page survey that was created and distributed to local agencies for data collection on completed projects. Eighteen completed surveys were used in this study. The data contained in these surveys is summarized in this report and analyzed with respect to project characteristics, performance, and key project success / hindrance factors.

1 INTRODUCTION

1.1 Background

Within the United States, state and federal transportation agencies have responsibility to effectively manage the delivery of thousands of design and construction projects worth billions of dollars. One state in the western United States annually makes $250 million available for local transit projects to improve transit services to all communities within the highly decentralized cities of the state. Unfortunately, recent delivery success of these local transit projects has been only 40%, meaning only 40% of the funded transit design and construction projects are completed in the year that funding is available.

The objective of this paper is to identify management factors that lead to improved delivery of transit services for this state. The goal of this research will be to identify both positive and negative factors that affect the local agency satisfaction with the project delivery process, the project budget performance, and the project schedule performance.

1.2 Literature Review

Success factors are well established in the construction industry. Jaselskis and Ashley (1991) found that key success factors affect project outcomes differently. Sanvido et al (1992) also found that when certain success factors related to the project owner, engineer, contractor, or operator are completed, the likelihood for project success is increased. Recent researchers have begun to apply these factors to specific types and subsets of the construction industry.

Success factors are also well used on the design/procurement stages. Based on a survey of over 450 respondents, Anderson and Tucker (1994) identified 52 specific best practices in 5 project management categories for project management of the design process.

Chua et al. (1999) came up with a simple and comprehensive hierarchical model that categorizes all these factors into four groups
- External Project Characteristics
- Contractual Arrangements
- Project Participants
- Monitoring & Control

Chua then conducted a survey and collected information about influence of these factors on three performance criteria (cost, schedule, and quality), using a pairwise comparison technique.

Chua's performance criteria can be viewed as either positive or negative enhancements to a project delivery model process. If defined as an action or attribute that enhances the likelihood of process success, then the criterion is denoted a success factor. If defined as an action or attribute that decreases the likelihood of process success, then the criterion is denoted a hindrance factor.

Otto and Ariaratnam (1999) and Poister (1999) have attempted to define success factors for the transit project process, but little emphasis was placed on the factors relating to design and construction. This study seeks to define key success factors and key hindrance factors that influence Local Transit Agency Project Delivery (LTPD) design and construction performance.

2 DATA COLLECTION

2.1 *Process*

The first step in the study process was collection of local transit agency project performance data. The data collection process consisted of four steps

- Identification of project characteristics
- Identification of types of funding
- Identification of types of projects
- Creation and distribution of survey form to local agencies

The research team from the University of Southern California worked with the state transportation department staff to identify all descriptive elements of any local transit agency project. Dozens of possible data elements of a typical local transit agency project were identified.

Based upon the literature review of typical key data elements, and the fact that such detailed data is not maintained by the state or the local agencies, several key project characteristics were identified. The characteristics were of two types – descriptive (i.e. where the project was located) and performance (i.e. how the project performed with respect to planned funding, budget, and schedule)

2.2 *Survey*

In order to provide accurate and reliable information, a standard data collection survey was created to gather descriptive and performance data. A first draft survey was completed in February 2000. After a first review of the draft survey, the project team decided to add two questions to the survey to gather information related to a third characteristic – key success / hindrance factors (as identified as critical within the literature review). During the months of March, April, and May 2000, over 150 surveys were distributed to local transit agency staffs throughout the state. At the time of drafting of this study (May 2000), 18 were returned. One (1) cost record and six (6) schedule records had incomplete information and were not useful. The remaining survey data was entered as it was received into a database that was used to conduct the initial data analysis.

3 DATA ANALYSIS

3.1 *Process*

The analysis began when the first completed survey forms were received in late April 2000. As additional data was obtained, it was added to the analysis database. The final data set was analyzed across the three characteristics/elements of the survey:

- descriptive data
- performance data
- key success / hindrance factors

The sections below detail the results of the analysis through a series of tables.

3.2 *Descriptive Characteristics*

In an effort to check that the projects of the data sample were representative of all Local Transit Agency projects, the data was analyzed with respect to project location, type, and funding.

Analysis showed that for our limited sample size, the projects were spread throughout the twelve (12) state districts (no district had more that 23% of the projects). However, three (3) districts were not represented at all and two (2) respondents did not identify the district number. Analysis also shows that the projects were somewhat representative of the dozen typical transit project types identified in the drafting of the survey form, with the actual data showing six (6) types of projects identified. "Transit Operations" projects were most common (38.9%) and two (2) projects were not assigned any project types.

With more data, the research team expects all one dozen project types to be represented. An examination of funding shows large number of funds (31 funds) was used on the eighteen (18) projects of the database. Note that several projects (12 projects, 67% of the sample) used more than one fund, and some projects used more than two funds (8 projects, 44% of the sample). The analysis shows that state approved transit improvement funding was most frequent used (6 occurrences but only $2.2 million per occurrence), but the highest funding amounts (in dollars) came from federal rail construction sources and local sources ($133.0 million for the one occurrence and $59.2 million average for two occurrences, respectively).

Additional analysis was conducted with respect to the project funding. Table 1 shows a summary of funding variance by type of variance (i.e. positive, negative, or none). The table shows how well local agencies were able to satisfy their anticipated project funding requirements since a lack of funding could certainly be one reason why a project would not be completed as planned. The table shows only two thirds of the projects received funding at their planned level and that over 25% of the projects experienced a negative funding variance, meaning the projects did not receive funding up to the amount estimated to be needed by the local agency. In addition, the magnitude of the under-funding deviation was found to significant – nearly 40% of the project expected funding. Additional analysis to test whether any particular funding source is particular susceptible to contributing to a funding deficit was conducted; however, a lack of data at this point does not allow this information to be determined.

Table 1. Funding Variance

Project Subgroup (1)	Number of Occurr. (2)	Percentage (3)	Funding Deviation	
			Average Amount (6)	% of Budget (7)
No Deviation	26	66.67%	$0	0.00%
Postive Deviation	3	7.69%	$376,867	126.70%
Negative Deviation	10	25.64%	($320,285)	-39.84%
TOTAL / AVERAGE	39	100.00%	($462)	1.98%

Table 2. Budget Variance

Project Subgroup (1)	Number of Occurr. (2)	Percent. (3)	Average Budget (4)	Budget Deviation	
				Average Amount (6)	% of Budget (7)
No Deviation	6	35.29%	$13,826,448	$0	0.00%
Postive Deviation	5	29.41%	$1,829,673	$51,930	24.23%
Negative Deviation	6	35.29%	$3,690,956	($62,498)	-69.45%
TOTAL	46	100.00%	$11,118,080	($4,897)	-8.58%

3.3 Performance characteristics

The second characteristic of the data collected was performance information. Two levels of performance were studied. One level of performance of the project is with respect to its budget. Once a project was funded, how close did the actual expenditures for the completed project come to the available funding (initial budget)? As defined in this report, a positive budget variance would be considered bad; meaning the project ran over its expected budget. The second level of performance relates to schedule. How do a project's start and finish dates compare to the original plan? In addition, once a project's expected start and completion dates were established, how close did the actual project duration (defined as difference between completion date and start date) compare to the planned duration? As defined in this study, a positive duration variance would be considered bad; meaning the project ran over its expected duration.

Two tables show the analysis. Table 2 shows a summary of budget variance by type of variance. The table shows that over three quarters of the projects had no budget variance from the funded amount; the remaining one quarter of the projects were almost equally divided between performing over and under budget. Note, however, that the projects that exceeded their budgets did so by a large amount – 24%.

An examination of schedule performance shows that that half of the projects started later than planned and three quarters of the projects surveyed were completed later than planned. Table 3 shows a summary of project duration deviations by type of variance. The table shows, with respect to project duration, two thirds of the projects took longer to complete than originally planned and the magnitude of the deviation was very significant – 231 days or 34% of the original duration.

3.4 Key Hindrances / Successes

The third characteristic of the data collected was key factors. These factors were items identified by the local agencies that were deemed to have been keys to success or key hindrances for a specific project.

Table 3. Duration Variance

Project Subgroup (1)	Number of Occurr. (2)	Percent. (3)	Average Project Duration (4)	Schedule Deviation	
				Average No. of Days (6)	% of Total Duration (7)
No Deviation	2	16.67%	364 days	0 days	0.00%
Postive Deviation	8	66.67%	1005 days	231 days	34.13%
Negative Deviation	2	16.67%	197 days	-160 days	-12.30%
TOTAL / AVERAGE	12	100.00%	764 days	127 days	16.65%

Two tables are used to show these results. Most surveys listed several key factors (both success and hindrance) for any individual project. Table 4 shows a summary of keys to success and the eight key factors identified through the surveys. Table 5 shows a summary of key hindrances and the eleven key factors identified through the surveys. As shown in the tables, the two primary keys to project success were identified as "State Staff Assistance" and "Established Funding Procedures". The primary key hindrances were "Bureaucracy", and "Poor Local Staff Assistance".

Table 4. Keys to Success

Success Criteria (1)	Number of Occurrences (2)	Percentage (3)
State Staff Assistance	9	26.47%
Cooperation among entities	4	11.76%
Established Funding Procedures	9	26.47%
Local Staff Assistance	4	11.76%
Ongoing Operations	4	11.76%
Program Flexibility	1	2.94%
Suppliers	1	2.94%
Training Programs	1	2.94%
No Comments	1	2.94%
Total	34	100.00%

Table 5. Key Hindrances

Hindrance Criteria (1)	Number of Occurrences (2)	Percentage (3)
Bureaucracy	4	16.00%
State Process & Procedures	2	8.00%
State Staff Assistance	1	4.00%
Contractors	3	12.00%
Engineering	1	4.00%
Environmental Issues	1	4.00%
Established Funding Procedures	1	4.00%
Local Staff Assistance	3	12.00%
State Process & Procedures	1	4.00%
Suppliers	2	8.00%
Unexpected Issues	1	4.00%
No Comments	5	20.00%
Total	20	100.00%

3.5 Differentiation Analysis

Despite a lack of data, the research team next attempted to identify whether any project characteristics were more common to projects which performed better (with respect to budget and schedule) than in projects which did not perform as well. In order to make this differentiation; the team categorized the results into "Successful Projects" and "Special Projects" as defined below:

- Successful Project: Neither cost variance (expended – allocated) nor schedule duration variance (actual duration – planned duration) should be greater than zero (i.e. no cost overrun AND no duration / schedule slippage)
- Special Project: At least one of the two variances (cost and/or schedule) performed poorly (i.e. either cost overrun or schedule slippage, or both)

Table 6 shows the data sample breakdown between successful and special projects. As seen in the table, nearly two thirds of the projects were categorized as special, with most of the special projects resulting from schedule problems (almost 65%). The table shows the average cost deviation for special projects to be $32,000 (over budget) or about 10% of the expected budget. The average schedule duration deviation for special projects was 264 days (delayed) or 36% of the total duration. Note that only 11 projects are shown in the table since records with incomplete data were not included.

The one characteristic of the differentiated data that is immediate value, despite the lack of surveys, is key factors. These factors were items identified by the local transit agencies that were judged to have been keys to success or key hindrances for the spe-

cific project, but, in this analysis, the factors are divided based upon the successful/special project differentiation explained above.

Two tables are used to show these results. Table 7 shows a summary of keys to success for successful and special projects. The table shows that "Established Funding Procedures" and "Ongoing Operations" were the two key success factors for projects that performed well. In other words, the success of the projects that were truly successful was believed to be a result of appropriate funding and development and implementation of a sound and well-structured procedure. Table 7 also shows that "Established Funding Procedures" and "State Staff Assistance" were the two key success factors for projects that did not perform well. That means even special projects were perceived successful due to the above two factors.

The summary of key hindrance factors divided based upon the successful/special differentiation is not as clear. Table 8 shows a summary of key hindrances for successful and special projects. The table shows a large number of keys for projects that performed well and for projects that did not perform well. The keys are diverse and mostly common to both the successful and special projects types. Additional data is needed to reach conclusions, but it appears that "Contractors", "Bureaucracy", and "Local Staff Assistance" may be critical factors.

4 CONCLUSIONS

The study has accomplished several major milestones in the analysis and improvement of the local transit agency project delivery process. The study achieved the following:

- Formalization of the data collection process – identified list of data items to be collected (survey form), identified list of types of funding, identified list of types of projects
- Collection of data on 18 completed local agency transit projects
- Development of a data analysis methodology and presentation formats, using databases and spreadsheets. Capability to perform automated statistical analysis upon compilation of additional information through a user-friendly database form. A diskette version of the Microsoft Access Database file (containing data to date, input screens, and queries) has been supplied to the state transportation agency. The program can be enhanced to incorporate additional analysis tools, as needed.
- Completion of data analysis for 11 completed projects. Specific findings to date are:

Table 6. Characteristics of Successful and Special Projects

Project Classification	Number of Occur.	Percent	COST				SCHEDULE		
			Average No. of Funds	Average Amount	Average Deviation		Average Duration	Average Deviation	
					Amount	Percent		Amount	Percent
(1)	(2)	(3)	(4)	(5)	(6)	(7)	(8)	(9)	(10)
Successful Projects	4	36.36%	3	$1,670,206	($10,632)	-3.51%	281 days	-80 days	-31.11%
Schedule Slippage	7	63.64%	1.71	$13,643,452	$32,138	9.85%	1097 days	264 days	36.31%
Funding Sleepage	2	18.18%	2	$300,333	$136,494	69.66%	758 days	271 days	40.93%
Sch. & Fund Sleepage	2	18.18%	2	$300,333	$136,494	69.66%	758 days	271 days	40.93%
Total Special Projects	7	63.64%	2	$13,643,452	$32,138	9.85%	1097 days	264 days	36.31%
TOTAL / AVERAGE	11	100.00%	2	$9,289,545	$16,585	4.99%	800 days	139 days	11.79%

Table 7. Keys to Success - Successful/Special Differentiation

Success Category	Successful Projects		Special Projects	
	No. of Occurrences	Percentage	No. of Occurrences	Percentage
(1)	(2)	(3)	(4)	(5)
State Staff Assistance	1	10.00%	4	28.57%
Cooperation among entities	0	0.00%	2	14.29%
Established Funding Procedures	3	30.00%	5	35.71%
Local Staff Assistance	2	20.00%	1	7.14%
Ongoing Operations	3	30.00%	1	7.14%
Suppliers	0	0.00%	1	7.14%
Training Programs	1	10.00%	0	0.00%
Total	10	100.00%	14	100.00%

Table 8. Key Hindrances - Successful/Special Differentiation

Hindrance Category	Successful Projects		Special Projects	
	No. of Occurrences	Percentage	No. of Occurrences	Percentage
(1)	(2)	(3)	(4)	(5)
Bureaucracy	2	28.57%	2	20.00%
State Process & Procedures	1	14.29%	0	0.00%
State Staff Assistance	1	14.29%	0	0.00%
Contractors	1	14.29%	2	20.00%
Engineering	0	0.00%	1	10.00%
Established Funding Procedures	1	14.29%	0	0.00%
Local Staff Assistance	0	0.00%	2	20.00%
State Process & Procedures	0	0.00%	1	10.00%
Suppliers	0	0.00%	1	10.00%
No Comments	1	14.29%	1	10.00%
Total	7	100.00%	10	100.00%

1. Nearly two thirds of the projects experienced a positive funding variance, meaning the projects did not receive funding up to the amount estimated to be needed by the local agency.
2. Three quarters of the projects had no budget variance from the funded amount;
3. Half of the projects start later than planned and three quarters of the projects surveyed are completed later than planned.
4. With respect to project duration, two thirds of the projects took longer to complete than originally planned.
5. The average cost deviation for special projects was $32,000 (over budget) or about 10% of the expected budget. The average schedule duration deviation for special projects was 264 days (delayed) or 36% of the total duration.

- Creation of a list of key success/hindrance factors based on initial data set. Findings to date are:
  1. The two primary keys to project success were identified as presence of "State Staff Assistance" and having an "Established Funding Procedure".
  2. The primary key hindrances were excessive "Bureaucracy", and poor "Local Staff Assistance.

Some work remains to be researched by future research teams and/or the state. Specifically with respect to the local transit agency process, the following items are needed:

- Collection of additional data (to an amount so as to allow statistical justification of conclusions)
- Development of a framework to facilitate the data collection process (web-based/email)
- Development of automated project performance analysis methods through standard software packages in order to facilitate the state and local agency management and reporting efforts.

As additional data becomes available, the power of the differentiation analysis can be truly recognized. Practically all of the tables of this report can be re-run based upon the two categories or even upon the subdivisions within the special project category. Analyses of particular interest would be

- Examination of successful/special projects verses type of project
- Examination of successful/special projects verses size of project
- Examination of successful/special projects verses type of funding
- Examination of successful/special projects verses number of funds per project
- Examination of successful/special projects verses funding variation

Again, as was the case for the analysis that has already been done, once the database queries for these examinations has been done, monitoring and reporting of the results can take place as data comes available and/or as the analysis is needed

REFERENCES

Anderson, Stuart D. and Tucker, Richard L. (1994). Improving Project Management if Design." *Journal of Management in Engineering*, Vol. 10, No.4.

Chua, D. K. H., Kog, Y. C., and Loh, P. K. (1999). "Critical Success Factors for Different Project Objectives." *Journal of Construction Engineering and Management*, Vol.125, No.3.

Jaselskis, Edward J. and Ashley, David B. (1991). "Optimal Allocation of Project Management Resources for Achieving Project Success." Journal of Construction Engineering and Management, Vol. 117, No. 2, p. 321-340.

Otto, Steve and Ariaratnam, Samuel T. (1999). "Guidelines for Developing Performance Measures in Highway Maintenance Operations" Journal of Transportation Engineering, Vol. 125, No. 1, p. 46-54.

Poister, Theodore H. (1999). "Performance Measurement in State Departments of Transportation", National Cooperative Highway Research Program Synthesis of Highway Practice 238, Transportation Research Board, Washington D. C.

Sanvido, Victor, Grobler, Francois, Parfitt, Kevin, Guvenis, Moris, and Coyle, Michael (1992). "Critical Success Factors for Construction Projects." *Journal of Construction Engineering and Management*, Vol. 118, No. 1, p. 94-111.

9 Cost Systems – I

Creative Systems in Structural and Construction Engineering, Singh (ed.) © 2001 Balkema, Rotterdam, ISBN 90 5809 161 9

The effect of a new prompt-pay measure on the contractor's cash flow

A. Touran & I. Bhurisith
Northeastern University, Boston, Mass., USA

ABSTRACT: In this paper, we have examined the issue of contractor's cash flow and interim financing and the impact of the new D.O.T. rule on prompt payment to subcontractors. The new rules require that the general contractor pay subcontractor's retainage after the sub has completed his work satisfactorily, even if the owner is withholding retainage from general contractor. Using a financial model, and data collected from contractors in the Boston area, we have quantified the effect of major financial parameters on the contractor's profit margin. We have also shown that the effect of new rules on contractor's profit is small and in many cases may be negligible.

1 INTRODUCTION

Cash flow problems are responsible for many contractors' financial failures. On many occasions the contractor needs to finance his negative cash flow on a short-term basis until the owner pays him. It is common for the contractor to apply owner's payment provisions to his subcontractors and to not pay them until he is paid. The new U.S. Department of Transportation (D.O.T.) regulations (Federal Register 1999) is designed to reduce subcontractors' hardship by requiring that general contractors pay all their subs in a certain number of days after satisfactory completion of their work. The payments must be made regardless of whether the state D.O.T. is withholding the general contractor's retainage. This rule increases the financial risk for the general contractor by widening the gap between his revenues and expenditures during the course of a D.O.T. sponsored contract.

We have developed a financial model for quantifying the effect of this new rule on the contractor's profit margin. The model considers important parameters that affect contractor's project cash flow, including project planned S-curve, typical owner payment and retainage terms, percent of job done by general's own forces, general's payment and retainage terms with regard to subs and material suppliers, and typical financing cost and requirements. In order to assure the validity of assumptions, a questionnaire covering all of these parameters was created and sent to a select group of contractors. Based on their responses, the model was adjusted and the effect of the

new contractual requirements on the contractor's cost was quantified. This was done by running the model with and without implementing the new "prompt-pay" measure and comparing the results.

2 BACKGROUND

Cash flow management is an indispensable part of overall project management (Peer & Rosental 1982; Singh & Lakanathan 1992). In a typical construction project, there is a time lag between the time that contractor incurs expenditures and the time that he gets paid. During this time lag the contractor needs to finance the project costs. The cost of financing this gap, sometimes known as *interim financing*, cuts into contractor's profit and if left unmanaged, can turn a successful project into a financial loss for the contractor. Financing a construction business is difficult. In a survey of nearly 600 contractors reported by Chang (1989), about 40% of respondents indicated that they had difficulty in obtaining working capital. Many contractors manage this financing need aggressively and try to minimize associated costs. Actions taken include negotiating with the owner to obtain front money and/or reduce retainage amount, delaying payments to suppliers and subcontractors, and negotiating more favorable rates with banks. It is common for the contractor to apply the owner's payment provisions to his subcontractors and to not pay them until he is paid. Furthermore, often contractors use harsher procedures for subcontractor payments, the most common being not to pay the most common being not to pay the subcontractors long after they receive payments themselves.

The same policy applies to retainage; the contractor usually keeps the subcontractors' retainage sums until after the owner pays his retainage. The retainage rate held from subcontractors is usually the same but sometimes higher than that the owner uses in the main contract (Hinze & Tracey 1994). Even after the owner pays the retainage, the general contractor may delay payment to subcontractors.

These payment practices create difficulties for subcontractors, especially small and DBE (Disadvantaged Business Enterprise) firms. Small firms usually have limited working capital and they also have to cope with late payments. In the survey cited earlier (Chang 1989), 56% of DBE contractors surveyed indicated that they had difficulty in securing working capital for their business. In a study conducted by Beliveau *etal* (1991) on the state of DBE program in construction, several DBE contractors indicated that withholding of retainage was disastrous to them as these contractors usually had a weak financial position to begin with.

3 THE NEW D.O.T. PROMPT-PAY MEASURE

The D.O.T.'s new DBE Program makes significant changes that will affect the recipients (*e.g.,* state and local transportation agencies), DBE, and non-DBE contractors that participate in the program. In 1999 the Congress reauthorized and the President signed legislation for this new program. This new regulation attempts to create a level playing field for all parties that benefit from D.O.T. projects. As part of this new regulation, a prompt payment provision will be required of all subcontractors, DBEs and non-DBEs alike. All recipients (such as highway agencies) must include a provision in their contracts requiring prime contractors to make prompt payment to their subcontractors. It would obligate the prime contractor to pay subcontractors within a given number of days from the receipt of each payment. The clause also applies to the return of retainage from the prime to the subcontractor. Retainage would have to be returned within a given number of days (to be determined by the agency) from the time that the subcontractor's work has been satisfactorily completed, *even if the prime contract has not yet been completed and the prime contractor has not received his retainage.*

On October 11, 1999, Engineering News Record reported the AGC's response to the new D.O.T. rule. AGC is concerned because the new rule prohibits prime contractors on any federally funded highway or transit project to withhold the retainage from any subcontractor (not just the DBE firms) after the completion of its portion of the work, even when the state is withholding retainage from the prime. The requirement also removes any leverage the prime has

over the subcontractor should any problems later be found with the subcontractor's work.

4 CASH FLOW MODEL

We developed a cash flow model for contractor's interim financing using Excel™ spreadsheet. The general approach is similar to what is reported in many references (Halpin & Woodhead 1998). The model, which is based on a weekly time unit, is sufficiently detailed and flexible so that the effect of each important parameter on the financing requirements can be readily calculated. The following assumptions and data are used in the model.

4.1 *Planned S-Curve*

The shape of the planned S-curve impacts the cash flow requirements. Several researchers have suggested equations that provide percent project completion or expenditures as a function of time (Perry 1970; Drake 1978; Peer 1983; Miskawi 1989). These equations assume that the percent completion is either a polynomial or a trigonometric function of duration. Some of these equations are derived for specific project types such as hospitals. We examined these equations and chose to use a beta distribution function for modeling the planned S-curve. Beta function is quite flexible and one can fit the function to many different patterns by changing the shape factors. We verified the appropriateness of the function by comparing the function with the point estimates of S-curve that were provided by the contractors responding to our questionnaire. The beta function used for modeling data in the three projects given in Table 1 had the following shape parameters: $\alpha=1.70$ and $\beta=1.85$.

4.2 *Inputs to the Model*

The cash flow model was developed using the *typical* input values. These values were obtained from textbooks, papers, and industry surveys. For example, an industry-wide CFMA survey reported that on average, the number of days in accounts payable for contractors is a little less than 45 days. In most of Massachusetts's state contracts, typical progress payment duration from owner to contractor is 30 days. After the model was developed for these input values, a limited questionnaire survey was conducted to verify the assumed values.

We contacted several contractors in the Boston metropolitan area to collect information on the cash flow problem. Five questionnaires were completed (three heavy and two building projects), and follow-up interviews were conducted to clarify ambiguities that usually arise in most detailed questionnaires. The shape of the S-curve was adjusted based on input from contractors. Table 1 summarizes general project information for the three heavy contractors contacted. We considered only heavy projects because most of D.O.T. sponsored projects tend to fall under heavy category. In these projects because the general's share of the work is relatively large, the effect of the new prompt-pay measure would not be as profound compared to building projects. It should be noted that some of these are hypothetical projects; we were more interested in obtaining typical values rather than specific data on any one project. We believe that these responses realistically portray these contractors' practices.

Where a range is given rather than a single value, the contractor intended to provide possible values in a range of projects rather than the specific project considered. As for profit, this is usually confidential information. In the model we used 4.03%, the value suggested by *ENR* in their 1999 Top 400 Contractors issue. Table 2 summarizes responses that relate to subcontractors and material suppliers. These are based on information supplied by general contractors.

For progress payment to subcontractors, the first contractor's policy was to pay 10% of sub's bill less retainage immediately, 65% when paid by the owner, and 25% two weeks after being paid by the owner.

The second contractor paid 10% of the sub's bill less retainage immediately and 90% two weeks after

TABLE 1. General Project Information

	Project No.1	Project No.2	Project No.3
Project type	Heavy	Heavy	Heavy
Contract price($)	15m	20m	50m
Duration (months)	18	24	60
Markup (%)*	5	10	19
Interest on loan (%)	-	8	7.5
Interest on income (%)**	2.5	2.5	2.5
Time to get paid (weeks)	3-12	4	4
Retainage (%)	5-10	5	5
Retainage duration (% completed)	50	50	100
Front money (%)	2-3	5	5
Final payment paid (months after compl)	6-24	12	6
Interest on retainage?	No	No	Yes

* The first gave % profit rather than the markup, which includes office overhead.
** Assumed value based on current rates.

TABLE 2. Subcontracted Work

	Project No. 1	Project No. 2	Project No.3
Portion of work done by general (%)	40-60	60	85
Retainage withheld from subs (%)	5-10	5	5
Material suppliers payment delay	6-8 wks with retain.	4 wks	6 wks

being paid by the owner. The third contractor paid 100% of the bill less retainage 10 days after being paid by the owner. For retainage, the general contractors paid the subcontractors' retainage after the owner paid the general's retainage.

5 RESULTS

5.1 *The effect of financing on profit*

We ran the model using data from the three projects described above. For each case, the total cost of interim financing was calculated. An interesting finding was that contractors have managed to minimize the impact of interim financing using the procedures already described. A large portion of the work is subcontracted and they do not pay subs and in many cases material suppliers until they have been paid. So there is no need for financing the gap between expenditures and revenues. The only exception was the third project where the general contractor performed 85% of the work and its indirect costs were high.

The effect of interim financing for these projects was to reduce the profit by 4%, 5%, and 13%, respectively. A sensitivity analysis was performed on Project No. 1 to see the effect of various parameters on profit. This sensitivity analysis showed that the parameters affecting general's profit the most, in order of importance, were the percent of work done by the general contractor (the higher the percentage, the lower the profit), the owner's delay in paying the contractor, the rate of interest on interim financing, and the amount of front money received. The summary of results is given in Figure 1. The base case scenario (profit adjusted to 1.0) is for a case where original profit rate was 5%, there was a 4-week delay for payment from owner to contractor, 2% front money, and general performed 40% of work with own forces. Changing each of these parameters using the ratios given on the horizontal axis, will result in the profit as reflected in ratios on the vertical axis. For example, if the amount of work done by general contractor is increased by 25% (from 40% base value to 50%), his profit will decline by 2%.

5.2 *The effect of the prompt-pay measure*

In order to compare the traditional procedure with the new prompt-pay regulation, we noticed that the effect of progress payments is minimal because the regulation allows the contractor to delay payment to subs until he is paid by the owner. The contractors that we interviewed contended that they usually pay the subcontractors promptly after they have been paid. There is much more sensitivity with respect to retainage because the new rule requires that the contractor pay the subs after they complete their work regardless of the fact that the owner has paid the general contractor. Even in this case, the negative effect on the general's bottom line seems small. We considered two cases. In the first case, we assumed that the general would not pay subs until six months after he is paid by the owner. This value was used because in a subcontractor's survey, 50% of respondents mentioned that it took them more than six months after project completion date to get their retainage (Hinze & Tracey 1994). In the second case, we assumed that the general contractor does not withhold any retainage from subcontractors. Although this is unrealistic, it serves as a worst case scenario for the general contractor. In both cases the model calculated profits and the results are given in Table 3. In this table, general contractor's profit using subcontractors' retainage is presented. As can be

seen in the worst case, the effect is about 6.5% of the profit margin; this value is an overestimate because the new rules allow for retainage and also Project No. 3 has a relatively high overhead rate. In most cases the range of impact would be less than the 3-4% of the profit calculated for the first two projects.

Table 3. Comparison of Contractor's Profit with and without subcontractor retainage

	With Ret.	Without Ret.	Diff.
Project No.1	$695,499	$673,172	-3.2%
Project No. 2	$746,893	$715,202	-4.2%
Project No. 3	$1,718,356	$1,605,967	-6.5%

It should be noted that data in Table 3 is based on the feedback that we obtained from general contractors in regard to their dealings with subcontractors. With this information, the effect of new prompt pay measure seems trivial. The problem can be more profound in building projects where the contractor subs out almost all of the work, but that case usually does not apply to D.O.T. sponsored projects.

REFERENCES

Beliveau, Y.J., D. A. Snyder & M. C. Vorster. 1991. DBE programmes new model. *Journal of Constr. Eng. & Management.* 117(1):176-192. ASCE.

Chang L. 1989. Method to deal with DBE issues. *Journal of Professional Issues in Eng.* 115(3):305-319. ASCE.

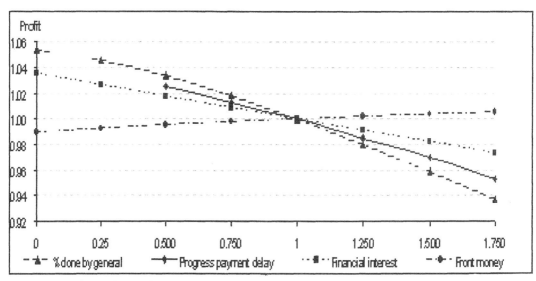

Figure 1. Results of sensitivity analysis

Drake, B.E. 1978. A mathematical model for expenditure fore-casting post contract. *Proc. CIB W-65 Symposium on Organization and Management of Construction,* Haifa, Israel, II-163-II-183.

Federal Register. 1999. Participation by Disadvantaged Business Enterprises in Dept. of Transp. Programs. *49 CFR Parts 23 & 26.* 64(21).

Halpin, D.W. & R.W. Woodhead. 1998. *Construction Management.* 2nd Ed., John Wiley & Sons. New York, New York.

Hinze J. & A. Tracey. 1994. The contractor-subcontractor relationship: the subcontractor's view. *Jornal of Construction Eng. and Management.* 120(2):274-287. ASCE.

Miskawi, Z. 1989. An S-Curve equation for project control. *Construction Management & Economics.* Vol.7:115-124.

Peer, S. 1982. Application of cash flow forecasting models. *Journal of the Construction Div.* 108(2):226:232. ASCE.

Peer, S. & H. Rosental. 1982. Development of cost flow model for industrialized housing. *Nat. Bldg. Res. Station.* Technion, Israel.

Perry, W.W. 1970. Automation in estimation contractor earnings. *The Military Engineer.* No. 410:393-395.

Singh, S. & G. Lakanathan. 1992. Computer-based cash flow model. *Proc. 36th Annual Transactions of AACE.* Morgantown, W. Va., R5.1-R5.14.

Creative Systems in Structural and Construction Engineering, Singh (ed.) © 2001 Balkema, Rotterdam, ISBN 90 5809 161 9

Cost saving using principles of value engineering in the construction industry

N. Bokaie & C. S. Putcha
Department of Civil and Environmental Engineering, California State University, Fullerton, Calif., USA

M. F. Samara
Brown and Root Energy Services, Tupman, Calif., USA

ABSTRACT: Value engineering (VE) is a disciplined effort to analyze the functional requirement of a project for the purpose of achieving the essential functions at the lowest total costs (U.S. Environmental Protection agency, 1976). The total cost as referred here encompasses capital, operating and maintenance cost. This implies that Value Engineering in any construction program is closely connected to Cost Reduction Program (CRP). It should be noted that while VE is a system to develop the most effective way to achieve a function, CRP is a system for reporting the benefits from individually identifiable cost-saving actions (Schuman, 1969). The purpose of VE is mainly to eliminate costs (associated with any project) related to non-essential functions and minimize the cost of the project to provide the essential functions.

1 INTRODUCTION

The construction industry is the largest industry in the United States with approximately $80 billion in annual expenditures. Historically, it has been shown to be extremely competitive, with low profit margins compared with the risks involved and with other areas of the economy. Building products continually rise in cost and wages are consistently higher than those of other industries.

Looking at the present construction industry methods, the private sector and government methods differ somewhat. A reappraisal of the current methods has become necessary due to the advent of the systems concept, the rising inflation and the increasing importance of time. More turnkey projects are being tried and a short-cut method of design and build (called "fast-track" or project delivery system/ scheduling) is becoming more prevalent (Bokaie, 2000).

These new trends are reducing the 3–5 year period between concept and occupancy that is required by the current construction methods. The result is an overall savings to the owner. There is substantial difference in the time for designing and constructing a large office building between the federal govern-ment and the private sector. Major discrepancies between the two are caused by the approvals required, approval agents involved, government laws enforced, and general insensitivity to time requirements in federal government procedures. Controls are needed to force the construction cost index to become more aligned with the industrial segment of the economy.

2 ELEMENTS OF A VE PROGRAM

The following are the main elements:

2.1 *Support of organization management*

It is necessary that the intermediate managers and top management understand and support the VE program.

2.2 *Policy directive development*

An effective VE program should be based on a policy of where, when, how and to which specific areas of work the VE effort must be directed.

2.3 *Administrator/coordinator of VE*

It is the role of the VE administrator/coordinator to ensure company's value engineering program functions.

3 VE GENERAL PRINCIPLES AND GUIDELINES

3.1 *Project selection and other VE considerations*

Paretos curve or the law of distribution (80–20 principal) is a generally acceptable approach for selecting both projects and items to be studied within a project. Based on this curve, not all projects will be candidates for a successful VE study. The timing of VE study is important. If VE study is done right after the basic elements are available, the design recommendations can be incorporated into the project. The report by Bokaie (2000) discussed, through the use of life-cycle phases diagram, how the potential for savings is greater when VE is applied earlier since decisions made in the beginning phases of project development have considerably more influence on life-cycle costs than those made later in the PS&E (plan, specification and estimates), construction and maintenance phases.

3.2 *Team structure in VE study*

The following aspects should be considered in forming a VE team so as to ensure the success of the project.

1. Structure of team
2. Leader of the team
3. Location and site of team
4. Timing
5. Basic design
6. Duration of study
7. Presentation of VE team

3.3 *Review and implementation*

There are five common barriers to implementation of VE study recommendations:

1. Poor documentation or insufficient supporting data
2. Imbalance of priorities
3. Inadequate appreciation, understanding and acceptance of VE potential
4. Resistance to change
5. Study completed too late

4 VE GUIDELINES TO CONSTRUCTION AND BENEFITS

4.1 *Value engineering change proposals*

Value Engineering Change Proposal (VECP) programs where the construction contractor decides whether or not to participate in developing VE recom-mendations. The program allows the contractor to show ingenuity and construction excellence and to receive financial incentive. It is essential that the VECP does not compromise any necessary design criteria or any preliminary engineering commitments.

4.2 *Benefits*

This program allows a low-cost opportunity for the contractor to use their experience and creative talents. The VECP program offers the following benefits.

1. Reduce the cost and enhances the design
2. Increase the net saving over the contract cost
3. Early project completion date

4.3 *Other benefits of a VE program*

Some of these benefits include:

1. Continuous reviewing of design, construction and maintenance standards through VE team activities
2. VE provide functional approach to any problem
3. Improve communication
4. Teamwork skills and team dynamics enhances through design process
5. Improves and develops the designer's skills by preparing and delivering organized presentations to management
6. Simple application of VE principles in the design process
7. Improve relations through cooperative processing of change proposals utilizing a VECP program
8. When proven VE designs or techniques and VECP accepted changes are utilized continuing savings and other benefits are achieved.

5 APPLICATION EXAMPLE

5.1 *Project description*

The project proposes to construct direct high occupancy vehicle connectors between I-405 HOV lanes and the North leg of the SR-55 HOV lanes, HOV access ramps at SR-55 Alton Avenue, and local access ramps to I-405 between SR-55 and Bristol Street. The local access improvements include:

1. A grade separation of the southbound SR-55/northbound I-405 branch connection and the northbound I-405 off ramp to Crystal Street
2. In ramp connection from the northbound I-405/Bristol Street off ramp to the Avenue of arts.
3. A new on-ramp from Anton Boulevard to the SR-55/I-405 northbound branch connection
4. Improvements to the existing Bristol Street on-ramp to the northbound I-405

The project is divided into three minimum operating segments (MOS) described as to follow:

- MOS-1: HOV direct connector from NB-I-405 to North Leg of SR-55 (South transitway), widen I-405 and SR-55, and reconstruct Red Hill over crossing.
- MOS-2: construct branch connection structure from southbound SR-55, add Anton Boulevard on-ramp and Avenue of Arts off-ramp
- MOS-3: HOV direct connector from SB SR-55 to NB I-405 (North transit way), modify Main Street and MacArthur Boulevard under crossing and widen I-405 and SR-55.

Costs of the project, including the local access improvement described, are estimated at $110,600,000. Funding is to be provided by Federal Transit Administration, the Orange County Transportation Authority (Measure M), and the City of Costa Mesa.

5.2 The value analysis process

General

Following procedures were used during the value analysis study:
1. Value analysis study agenda
2. Value analysis daily study participants
3. Cost model
4. Life-cycle cost/benefits analysis
5. Function analysis/fast diagram
6. Creative idea listing

A systematic approach was used in their value analysis study and the key procedures followed where organized into three distinct parts:
1. Pre-study preparation
2. Valued analysis study
3. Post study procedures

5.3 Pre-study preparation

In preparation for the value analysis study, the facilitator reviewed the project record dated July 1993 and prepared a costs summary, based on that report, for distribution to the value analysis team at the start of the value analysis study. The cost data was revised during the first day of the study to correlate to the supplemental project report distributed at the start of the study.

5.4 Value analysis study

This value analysis study was a 3-1/2 day effort. The VA Job Plan was followed by guide the team in the search for areas of opportunity in the design, scheduling and/or process, and in developing alternative solutions for consideration. The Job Plan phases are:
1. Investigation (Information and Function Analysis) Phase
2. Speculative Phase
3. Evaluation Phase
4. Development Phase
5. Presentation Phase
6. Implementation Phase

5.4.1 Investigation phase
Key to the VA process is the function analysis techniques used during the Investigation Phase. A Function Analysis System Technique (FAST) diagram, or function model of the project, was developed to better understand the relationships of the key functional requirements.

5.4.2 Speculative phase
This VA study phase involved identifying and listing creative ideas. During the phase, the VA team participated in a brainstorming session to identify as many means as possible to provide the necessary functions within the project. Judgment of the ideas was not permitted at this point.

5.4.3 Evaluation phase
This phase systematically reduced the large number of ideas generated during the creative phase to a number of concepts that appear promising in meeting the project objectives. The VA team, as a group, judged the ideas relative to performance of the functions required. Ideas were rated on a six-point system with a maximum possible ranking of 5 points to a mini-mum of 0 points. The rating system is as follows:

5 Significant Improvement
4. Some Improvement
3 No Significant Change
2 Slight Degradation
1 Significant Degradation
0 Fatal Flaw

Based upon the rating, ideas rated 4 or 5 were developed further and documented on the Value Analysis Alternative forms.

5.4.4 Development phase
During this phase, each idea was expanded into a workable solution. The development consisted of the recommended design and a descriptive evaluation of the advantages and disadvantages of the proposed alternatives.

5.4.5 Presentation phase
The VA study concluded with a preliminary presentation of the VA alternatives, which were developed. This provides others impacted by the results of the study and opportunity to review the proposals and develop an understanding of the rationale behind them.

6 POST-STUDY PROCEDURES

This study includes the preparation of a Preliminary Value Analysis Study Report incorporating a description of the VA study and the alternatives developed for consideration.

6.1 Function analysis

Based on the functions identified by the team, a Function Analysis System Technique (FAST) diagram was developed, and appears on "Functional Analysis System Technique (FAST) Diagram. This diagram illustrates the functions arranged in a logical sequence, such that when reading from left to right each function answers the question "how do I". Conversely, when reading from right to left, each function answers the question "why do I". Functions beneath under functions, connected with only a single line, represent functions that occur at the same time as the required function on the critical path of the diagram. Certain functions were identified as "All the Time" functions and are shown in a separate box beneath the diagram.

6.2 Cost model

The VA Team Leader prepared a cost summary from the designer's cost estimate. The summary is organized to segregate the costs between the three MOSs of the project, as well as mark-ups included in total costs. The cost summary was then converted to a Bar Chart, for a graphic illustration of project costs. This clearly indicated the cost drivers for the project as was used to help the tem quickly focus on where the dollars are allocated within each MOS (Bokaie, 2000).

6.3 Life-cycle benefit/cost analysis

The Caltrans Transportation Economics life Cycle Benefit/Cost Analysis (LCB/CA) program was used to evaluate the overall project for benefit/cost. The LCB/CA is structured to include only MOSs -1, -2, -3, and is based on the following assumptions:

1. Ninety percent (90%) of total ADT for the project was used to approximate the ADT for MOSs -1, -2, -3.

2. The average vehicle operating speed for the existing facility is conservatively estimated at 50 MHP for the base year and 45 MPH for the year 2015, based on the current merging and weaving required, as well as the shorter radius curves, on the facility.

3. 3-year accident data was obtained from the Project Report dated August 1993

4. Construction of MOS-1 is scheduled to begin in the year 2001. MOS-2 is not currently scheduled, but is anticipated to begin shortly after MOS-1 based on current funding expectations. MOS-3 is not currently scheduled, nor is it known when funding for this segment may become available. Thus, the project cost data was entered in two-year increments for the purpose of this analysis. Based on the assumptions used, the project results in a Rate of Return on Investment of 16.9 percent.

7 VA STUDY RESULTS

The VA team generated 56 ideas for change during the function analysis and creative idea phases of the VA job plan. The evaluation of these ideas were based on each idea's impact on project cost, construct-ibility, construction schedule, facility operation, safety, traffic management and whether the idea adds value to the project. This evaluation resulted in 11 VA alternatives to pursue.

The VA proposal summaries show the impact of the alternatives on the total project. All of the alternatives are presented in detail in the report (Bokaie, 2000). The savings identified are first cost savings, except for VAP 9 which shows savings realized after allowing for life-cycle costs related to Caltrans' long term maintenance.

8 CONCLUSION

Value Engineering (VE) has shown to be a proven tool for both product improvement and design enhancement. To improve design excellence and achieve both efficient cost and quality control, it is the industries' position that:

1. Companies should establish an ongoing VE program

2. Through the use of VE methods in project development, construction, operation, maintenance and other areas, the challenges of inflation and decreasing resources may be addressed

3. In order to assist organizations in accepting and using VE, guidelines should be provided with the provision of flexibility to adapt to individual project needs.

REFERENCES

Bokaie, N. 2000. The application of value engineering in the construction industry. Thesis submitted in partial fulfillment of the requirements of the degree of M.S. in Civil Engineering, California State University, Fullerton, CA.

Schuman, R 1969. Cost reduction and value engineering. Value Engineering in Federal Construc-tion Agencies, Symposium/workshop Report No. 4, National Academy of Sciences, Washington, D.C.

U.S. Environmental Protection Agency 1976. Value engineering workbook for construction grant projects. Office of Water Program Operations, Municipal Construction Division, Washington, D.C., July.

Creative Systems in Structural and Construction Engineering, Singh (ed.)© 2001 Balkema, Rotterdam, ISBN 90 5809 161 9

Decision analysis of competitive bidding in construction

Hing-Po Lo & Ma-Li Lam
Department of Management Sciences, City University of Hong Kong, People's Republic of China

ABSTRACT: The construction industry is a major business and there are more than ten thousand registered construction companies in Hong Kong. The competition for projects is very keen. The most common method of distributing the projects among the construction companies is competitive sealed bidding. Development of successful bidding strategies is therefore a key factor to the survival of construction companies. Two major criticisms of most of the bidding strategies are the use of a single distribution with fixed parameters to model the bidding patterns and the assumptions of independence of bids among competitors. The present paper intends to answer these criticisms by developing realistic multivariate distributions to represent different bidding patterns and dependence among bidders. Data on more than 3,500 bids from 172 contractors for 267 contracts awarded between 1990 and 1996 from the Architectural Service Department of the Hong Kong SAR Government were collected. Statistical techniques are first used to detect non-serious bids before bidding models are developed for the data. Truncated multivariate log-normal distributions are found to be most suitable to represent the bidding patterns and associations among bidders. Decision rules for competitive bidding are formulated once the probabilities of winning the projects are found by numerical integration.

1 INTRODUCTION

The construction industry is a major business in Hong Kong. Output of building and construction industry increased almost threefold from $3.2 billion (HK$24.9 billion) in 1989 to $8.2 billion (HK$64.3 billion) in 1996. In terms of GDP contribution of the industry, it increased from 5.1% in 1992 to 6.1% in 1998. In 1999, the gross value of construction work performed by contractors amounted to $16.2 billion (HK$126.4 billion).

There are more than ten thousand registered construction companies in Hong Kong and the competition for projects is very keen. The most common method of distributing the projects among the construction companies is competitive sealed bidding. Development of successful bidding strategies is therefor a key factor to the survival of construction companies.

The first important contribution to contract bidding was published in 1956 by Friedman who modelled the distribution of mark-ups of all the competitors by a Gamma distribution with fixed parameters. The probability of winning the project, P(M), could then be found by using the Gamma distribution. Friedman then suggested that the best bidding level for the collaborating contractor was the value that maximises the profit. Since then, a series of paper which extended

or modified the Friedman's model were published. For example, in an attempt to simplify the method of finding P(M) and to reduce the data requirements, Hanssman and Rivett (1956) suggested using the distribution of lowest competitor's bids. Gates (1967) considered the results of 381 contracts and discovered a relationship between the spread of bids and the value of the contract. Gates also suggested a heuristic approach to obtaining the probability of winning the contract. Sugrue (1982) described a method of optimising the expected profit quickly without the assistance of a computer. Carr (1982) developed a general bidding model by using standardized bid to cost ratios. All the above approaches assumed a fixed bidding pattern for each competitor, which has a significant implication. This assumption means that there is a single distribution for the competitors' mark-ups and that this distribution not only has a fixed shape but also has constant parameters over all contractors. Thus this assumption implies that different contractors perform equally well and their mark-ups are randomly selected from one static bidding pattern, irrespective of any variations in characteristics of the contractors. This may sometimes be a realistic assumption but often it is not.

The first attempt to model different distributions of mark-ups was suggested by Mercer and Russell (1969) but only until recently that different statistical techniques such as multiple regression, analysis of variance or likelihood function maximisation were used to analyse or model the distributions of bid/cost estimate ratios and bid/price estimate ratios (King and Mercer (1991), Skitmore (1991, 1999).

Another major criticism of most of the bidding models is the assumption of independence of bids among competitors. Most models assumes that the performance of the bidders are independent of each other and that the probability of winning a project by a particular contractor is simply the nth power of the chance of having a lower bid against a single opponent, where n is the number of competitors. This approach ignores the interaction effect among the contractors.

This paper intends to answer these two criticisms by using a large set of bidding data to develop a multivariate distribution that takes into account different bidding patterns and dependence among bidders. Data on more than 3,500 bids from 172 contractors for 267 contracts awarded between 1990 and 1996 were collected from the Architectural Service Department of the Hong Kong SAR Government. This represented general construction contracts (including new work and alteration work) for 53 types of building according to CI/SfB classification. For comparison, all bid prices were updated to a common base date based on the Hong Kong Government tender price index for building work.

The paper is set out as follows. In the next section, an index of competitiveness similar to the one used by Carr (1982) is introduced. In section 3, a truncated multivariate log normal distribution is developed. Its application in the determination of optimal mark-up is demonstrated in section 4. The findings of the paper and conclusions are given in the last section.

2 A MEASURE OF COMPETITIVENESS

In order to build a decision model for competitive bidding, predictive information concerning the competitiveness of potential competing contractors is essential. One approach is to obtain the competitiveness information by analysing past bidding performance of the contractors as suggested by Flanagan and Norman (1982) who examined the tendering pattern of building contractors. As competitive tendering appears to be a valid means of attracting a realistic price for a project – the most *efficiently* organised contractor for a given project may be regarded as the one who produces the lowest bid - a measure of the competitiveness (efficiency) of a contractor for a project can be defined as the ratio of the bid submitted by the contractor to the lowest bid of the project, i.e.

$$\text{Index of competitiveness} = \frac{\text{contractor's bid}}{\text{lowest bid}} \quad (1)$$

This index has values ranging from 1 to infinity. Small values indicate high competitiveness and the value of 1 means the contractor's bid is the lowest.

It should be pointed out that Carr (1982) use a similar definition in his bidding model. He defines a variable "Bid to Cost ratio" as a measure of the performance of a contractor. However, it is in general difficult to collect the information of cost estimates from all the participating contractors and the lowest bid could be used as a close estimate of the cost of the project.

3 THE BUILDING OF A MULTIVARIATE DISTRIBUTION

The ideal approach in building bidding models is to develop separate distributions for the bidding patterns of different contractors. The problem of this approach is the lack of sufficient data for individual contractors even six years of data have been accumulated. We therefore combine contractors according to the grouping system used by the Hong Kong Government. Contractors are classified into 6 groups based on their sizes and experiences. Assuming the bidding patterns among the contractors within each group are the same, separate distributions can be built for each group. It is obvious that the distribution thus obtained can only serve as an approximation of the bidding pattern of an individual contractor. Our purpose here is to illustrate the potential application of the multivariate distribution in the forming of bidding strategy. More characteristics can be used to further subdivide the groups if more accurate results are needed and that more data are available.

The indices of competitiveness "bid to lowest bid ratios" within each group are used to build the distribution. It is observed from the histograms of these ratios that all exhibit a long right hand tail and hence a logarithmic transformation is used.

Another data treatment that is necessary before the calculation of the estimates of the parameters is the removal of high outliers. It is well known that there are non-competitive bids in sealed-bid auctions for a variety of reasons (Skitmore and Lo, 2000). For example, full order books, low projected profit levels and short period allowed for bid preparations are some of the reasons why contractors who are invited (pre-selected) but prefer not to bid for a particular contract. One means of achieving this is through what is known as 'cover' pricing, by which a competitor's bona fide bid is used with the addtion of a few percent to ensure non-competitiveness (Skitmore and Lo, 2000). The presence of these kind of non-competitive bids certainly would affect the development of the statistical model and should be removed. In this paper, we follow a method similar to the one used by the Hong Kong Government. Bids that are 1.8 times higher than the lowest bid are treated as outliers and they are removed from the data sets.

The following figure shows the histogram of log(bid/lowest bid) for group 1.

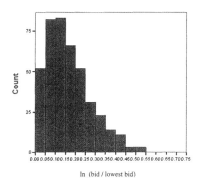

In (bid / lowest bid)

Figure 1. The histogram of ln(bid/lowest bid) for group 1

It is clear from the histogram and from the definition of the index that the distribution is left truncated at zero. Otherwise, a normal distribution could be used to fit the data. We therefore propose a truncated log-normal distribution for the bidding patterns.

$$f(x) = \frac{1}{\sqrt{2\pi}\sigma} e^{-\frac{1}{2\sigma^2}(x-\mu)^2} \left[\frac{1}{\sqrt{2\pi}\sigma} \int_A^B e^{-\frac{1}{2\sigma^2}(t-\mu)^2} dt \right]^{-1} \quad (2)$$

where:

$$x = \ln\left(\frac{Bid}{LowestBid}\right) \quad 0 < x < \infty$$

The following table shows the estimates of the parameters of the truncated log normal distribution and the Chi-squared values for the goodness of fit test.

Table 1. The estimates of means and standard deviations for the truncated normal distibution and the Chi-squared values of the goodness of fit test:

Group	1	2	3	4	5	6
$\bar{x}^*$	0.168	0.189	0.161	0.155	0.149	0.127
s^*	0.111	0.118	0.108	0.108	0.092	0.082
χ^2	8.338	7.750	7.815*	13.722	4.858	3.285
p-value	0.304	0.355	0.175*	0.089	0.302	0.511
No. of data	420	332	999	676	120	107

The number of classes used in the goodness of fit is 15 except in the case of group 3. The test for group 3

turns out to be significant (p value<0.01) unless fewer classes are used. For example when the observations are grouped into 8 classes, a p-value of 0.175 is obtained. Hence, we may conclude that in general, a truncated log-normal distribution can be used to fit the data for the bid to lowest bid ratios.

In order to estimate the correlation coefficients between pairs of contractors and to find the probability of winning, the parameters of the full normal distributions have to be obtained. The following formulae relate the parameters μ and σ^2 of a normal distribution with the parameters μ^* and σ^{2*} of the corresponding truncated normal distribution where the portion to the left of zero is deleted:

$$\mu^* = \mu + \frac{z\left(\frac{-\mu}{\sigma}\right)}{1 - \Phi\left(\frac{-\mu}{\sigma}\right)}\sigma \quad (3)$$

$$\sigma^{2*} = \left[1 + \frac{\left(\frac{-\mu}{\sigma}\right)z\left(\frac{-\mu}{\sigma}\right)}{1 - \Phi\left(\frac{-\mu}{\sigma}\right)} - \left\{\frac{z\left(\frac{-\mu}{\sigma}\right)}{1 - \Phi\left(\frac{-\mu}{\sigma}\right)}\right\}^2 \right]\sigma^2 \quad (4)$$

where:

$$\Phi\left(\frac{-\mu}{\sigma}\right) = \frac{1}{\sqrt{2\pi}\sigma} \int_{-\infty}^0 e^{-\frac{1}{2\sigma^2}(t-\mu)^2} dt$$

The following table shows the values of the sample mean $\bar{x}^*$ and sample variances s^2 obtained from the truncated distribution and the estimates of the population parameters of the full normal distibution by using formulae (3) and (4).

Table 2. The estimates of means and standard deviations for the truncated and full normal distibution

Group	1	2	3	4	5	6
$\bar{x}^*$	0.168	0.189	0.161	0.155	0.149	0.127
s^*	0.111	0.118	0.108	0.108	0.092	0.082
$\hat{\mu}$	0.107	0.143	0.096	0.074	0.126	0.089
$\hat{\sigma}$	0.150	0.151	0.148	0.155	0.104	0.108

It is clear that the correlation coefficient obtained from the truncated bivariate log-normal distibution would over estimate the correlation coefficient of the full bivariate log-normal distibution. However, the estimation of the correlation coefficient between contractors is not as straight forward as the mean and variances. Lo (2000) discusses a method to obtain the estimates of the correlation coefficients from the sample moments of the truncated bivariate normal distribution. Using Lo's method, the covariances of X_i and X_j, where i and j could be contractors from any groups are given in the following table:

Table 3. Estimate of covariance, $\hat{\sigma}_{x1x2}$, between and within groups (*Diagonal entries are variances)

Groups	1	2	3	4
1	*0.0224	0.0022	0.0070	0.0044
2	0.0022	*0.0227	0.0096	0.0047
3	0.0070	0.0096	*0.0220	0.0022
4	0.0044	0.0047	0.0022	*0.0241
Within gp	0.0023	0.0068	0.0002	0.0048

With the findings of the covariance matrix, we may now propose to make use of a multivariate truncated log-nomal distribution to develop bidding strategy. That is, if we let Y=bid/lowest bid, then X=ln(Y) follows a multivartiate normal distibution truncated to the left at ln(x)=0. The parameters for the full distribtiuon can be estimated from the truncated sample as described in this section and the results are given in tables 2 and 3. This distibution can be used to estimate the probability of winning as explained in the next section.

4 PROBABILITY OF WINNING

Assuming that the lowest bid is a good estimate of the cost of the project, the percentage makr-up, i, is related to X in the following way: $X = \ln(Y) = \ln(1+i)$. The relationship between the probability of winning w(i) and its markup i can be found as follows:

4.1 *Probability of winning*

$$w(i) = P\left[\min(X_i) > \ln(1+i)\right]$$
$$= \int_{\ln(1+i)}^{\infty} \cdots \int_{\ln(1+i)}^{\infty} f(x_1,\ldots, x_n) dx_1 \ldots dx_n \quad (5)$$

where $f(x_1,\ldots,x_n)$ is the multivariate truncated normal distribution of the ln(Bid/Lowest bid) ratios of the competitors which is given as:

$$f\left(\underset{\sim}{X}\right) = \frac{1}{C} \frac{1}{(2\pi)^{p/2}|\Sigma|^{1/2}} \exp\left\{-\frac{1}{2}\left(\underset{\sim}{X}-\mu\right)^T \Sigma^{-1}\left(\underset{\sim}{X}-\mu\right)\right\}$$

where $E\left(\underset{\sim}{X}\right) = \mu$, $Cov\left(\underset{\sim}{X}\right) = \Sigma$ and C is a constant.

Mathematical software MATHCAB version6.0 can be used to evaluate the multiple integration to find the probability of winning.

4.2 *Example 1: All competitors are in different groups*

Assume that we are in group 1 and there are three competitors from groups 2, 3 and 4 respectively. We want to set up the markup % to be 38%. The probability of winning is found to be equal to

$$\int_{\ln(1+i)}^{\infty}\int_{\ln(1+i)}^{\infty}\int_{\ln(1+i)}^{\infty} f(x1,x2,x3)dx1dx2dx3 = 0.00511$$

where i=0.38 and

$$\Sigma = \begin{bmatrix} 0.0227 & 0.0096 & 0.0047 \\ 0.0096 & 0.0220 & 0.0022 \\ 0.0047 & 0.0022 & 0.0241 \end{bmatrix} \quad \underset{\sim}{\mu} = \begin{bmatrix} 0.143 \\ 0.096 \\ 0.074 \end{bmatrix}$$

The following figure shows the relationship between the probability of winning *w(i)* and markup *i* when competitions are from groups 2, 3 and 4.

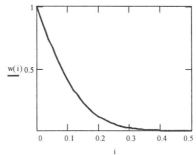

Figure 2. Probability of winning as a function of mark-up in Example 1

4.3 *Example 2: Some competitors are in the same groups*

Assume that we are in group 1 and there are four competitors from groups 1, 1, 3 and 4 respectively. We want to set up the markup % to be 38%. The probability of winning is found to be equal to

$$\int_{\ln(1+i)}^{\infty}\int_{\ln(1+i)}^{\infty}\int_{\ln(1+i)}^{\infty}\int_{\ln(1+i)}^{\infty} f(x1,x2,x3,x4)dx1dx2dx3dx4 = 0.00117$$

where i=0.38

$$\Sigma = \begin{bmatrix} 0.0224 & 0.0023 & 0.0070 & 0.0044 \\ 0.0023 & 0.0224 & 0.0070 & 0.0044 \\ 0.0070 & 0.0070 & 0.0220 & 0.0022 \\ 0.0044 & 0.0044 & 0.0023 & 0.0241 \end{bmatrix} \quad \underset{\sim}{\mu} = \begin{bmatrix} 0.107 \\ 0.107 \\ 0.096 \\ 0.074 \end{bmatrix}$$

The following figure shows the relationship between the probability of winning w(i) and markup i when competitions are from groups 1, 1, 3 and 4.

4.4 *Expected profit and optimal markup*

Using the probability of winning and the markup values, we can evaluate the expected profit p(i).

$$p(i) = i \times w(i) \qquad (6)$$

4.5 Example 1 (continued)

The following figure shows the relationship between the expected profit and its markup precentage using the results in example 1 with all competitors coming from different groups. It is found that there is a maximum value of the expected profit which occurs at the markup percentage around 0.09. That means, profit can be maximized using 9% of markup when there are three competitors from groups 2, 3, 4.

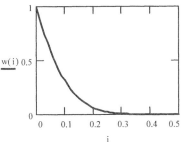

Figure 3. Probability of winning as a function of mark-up in example 2

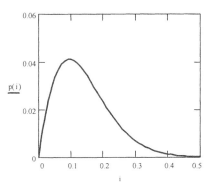

Figure 4. Expected profit as a function of mark-up in example 1

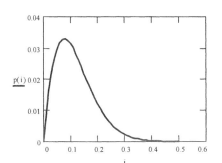

Figure 5. Expected profit as a function of mark-up in example 2

4.6 Example 2 (Continued)

The following figure shows the relationship between the expected profit and its markup precentage using the results in example 2 with some competitors are in the same group. It is found that there is a maximum value which occurs at the markup percentage around 0.08. That means, profit can be maximized using 8% of markup when there are four competitors from groups 1, 1, 3, 4.

5 CONCLUSION

Many bidding models used in construction industry are being criticized as unrealistic. Two major criticisms are the use of a single distribution with fixed parameters to model the bidding patterns and the assumption of independence of bids among competitors. This paper describes an approach to build a truncated multivariate log-normal distribution based on a large data set of bidding records. Methods using formulae and graphs to estimate the parameters of the distributions are illustrated. This distribution can handle both variations in bidding patterns of different bidders and associations among tenderers at the same time. This statistical model is then applied to obtain the probability of winning and expected profit. Bidding strategy can be formulated accordingly.

ACKNOWLEDGEMENT

This research is supported by the Research Grants Council, University Grants Committee of Hong Kong through Grant CPHK 737/95H.

REFERENCES

Carr, R.I., 1982, General bidding model, *Journal of the Construction Division, Proceedings of the American Society of Civil Engineers*, 108(C04), 639-50.

Flanagan R. and Norman G. 1982, An examination of the tendering pattern of individual building contractors, *Building Technology and Management*, 28: 25-28.

Friedman L. 1956, A competitive bidding strategy, *Operations Research*, 1(4): 104-12.

Gates M. 1967, Bidding Strategies and probabilities, *Journal of the Construction Division, Proceedings of the American Society of Civil Engineers*, 93(C01): 75-107.

Hanssman, F. and Rivett, B.H.P. 1956, Competitive bidding, *Operational Research Quarterly*, 10: 49-55.

King M. and Mercer A. 1991, Distributions in Competitive Bidding, *Journal of Operational Research Society*, 42:151-155.

Lo H.P. 2000, Bidding Strategy in Construction, paper under preparation.

Mercer A and Russell J.I.T. 1969, Recurrent competitive bidding, *Operational Research Quarterly*, 20: 209-221.

Skitmore R.M. 1991, The construction contract bidder homogeneity assumption: An empirical test, *Construction Management and Economics*, 9: 403-429.

Skitmore R. M. 1999, The probability of tendering the lowest bid in sealed bid auctions: an empirical analysis of construction contract data, submitted to *Operational Research*.

Skitmore R.M. and Lo H.P. 2000, A method for identifying higher outliers in construction contract auctions, submitted to *Journal of Construction Engineering and Management (ASCE)*.

Sugrue P. 1982, Optimum bid estimation on fixed cost contracts, *Journal of Operational Research*, 33: 949-956.

Creative Systems in Structural and Construction Engineering, Singh (ed.) © 2001 Balkema, Rotterdam, ISBN 90 5809 161 9

An investigation of industry practices on change order markup allowances

A. Robinson Fayek & M.Y. Nkuah
Department of Civil and Environmental Engineering, University of Alberta, Edmonton, Alb., Canada

ABSTRACT: Allowable markups and the manner in which they are applied to the total cost of change orders vary greatly amongst owners. The purpose of this paper is to explore existing practices employed in reimbursing contractors' profit, overhead, and labour burden cost on changed work, and to evaluate the adequacy of such change order markup allowances. The results of a questionnaire survey on change order markup allowances on lump sum building contracts are compared to the allowances contained in industry standards and actual contracts. This comparison indicates that existing predefined allowances for profit, overhead, and labor burden reimbursement on change order work inadequately compensate the general contractor. The results of this study may assist researchers and industry practitioners in developing a fair and equitable change order pricing strategy for building contracts.

1 INTRODUCTION

On a project of any size, change orders have become an integral part of the construction process. Most construction contracts contain a "change" clause that allows the owner to make changes to the original contract work, usually within the scope of the original contract. A change made while work is in progress may significantly impact the construction costs and schedule.

In practice, on lump sum contracts, the contract documents contain clauses that stipulate the items to be considered in assessing the cost associated with performing a change in the work. The prime components of these costs (labor, material, equipment, and subcontracts) are usually well defined. Most lump sum contracts also contain predefined allowances (i.e., markup percentages) for the reimbursement of labor burdens, overhead, and profit on change order work. These allowances and the manner in which they are applied to the total cost of the changed work vary greatly amongst owners. The purpose of this paper is to examine industry guidelines and contract conditions for allowances for change order reimbursement, and to compare these allowances with the actual costs incurred by contractors in administering and performing change order work.

2 CANADIAN INDUSTRY STANDARDS ON CHANGE ORDER MARKUP ALLOWANCES

A review of Canadian construction industry standards on change orders revealed a lack of uniformity in allowable markups. The CCDC (Canadian Construction Documents Committee) Standard Construction Documents (CCDC, 1994), which are widely used throughout Canada, do not contain any statement on the payment of markup allowances on change orders. The Canadian Construction Association (CCA, 1992) provides guidelines for the reimbursement of markup for overhead and profit on change orders; they recommend a combined percentage of 20% for work done by a contractor's own forces and 15% for work done by subcontractors. Neither organization provides guidelines for allowable payroll burden on change order work.

Four actual contracts were reviewed to examine their conditions for the reimbursement of profit and overhead on change order work. Allowable markups on a general contractor's own work ranged from 12% to 20% of the total cost of the change (inclusive of both profit and overhead). A general contractor's allowable markup on change order work performed by a subcontractor ranged from 5% to 10%. These markups are inclusive of contractor's bond, head and field office staff, banking charges, and small tools.

Three of the contracts surveyed contained predefined percentage allowances for payroll burden cost: 25% in two cases, and 35% in one case. The findings of previous research further indicate that allowable markups on change order work do not adequately compensate the contractor performing the work (Nkuah, 1999). A survey of Canadian contractors was therefore conducted to examine the adequacy and uniformity of existing procedures employed in reimbursing contractors' profit, overhead, and labour burden on change order work on lump sum building contracts.

3 MARKUP ALLOWANCES ON CHANGE ORDERS ON LUMP SUM BUILDING CONTRACTS

A questionnaire survey of markup allowances on change orders on lump sum building contracts was conducted amongst contractors in Alberta in 1999 (Nkuah, 1999). Six case studies were obtained from a total of 11 questionnaires sent (a 55% response rate). This paper presents a summary of the results only as they pertain to the adequacy of change order markup allowances; a more extensive analysis of the findings is found in Fayek and Nkuah (2000). Contractors were asked to provide actual allowances for labor burden components as a percentage of their base wage rates (including vacation pay). Table 1 shows the average percentage allowances for labor burden. All respondents include worker's compensation, unemployment insurance, Canada Pension Plan, and medical-dental coverage in labor burden. Sixty-seven percent of respondents include insurance and payroll tax as labor burden. Fifty percent include their company pension plan as labor burden. Other items of labor burden include a safety allowance, a wage protection fund, time keeping, union benefits, and small tools.

Table 1. Average allowances for labor burden components (as a percentage of the base wage rate including vacation pay)

Labour burden component	Average percentage allowance
Vacation pay and statutory holidays	9.44
Worker's compensation	2.76
Unemployment insurance	3.43
Canada Pension Plan	3.50
Company pension plan	3.25
Payroll tax	0.73
Safety	2.70
Medical-dental coverage	2.80

Employing the percentage allowances in Table 1 and the average labor base rates reported in the survey, the actual total burdens carried by the general contractor are calculated for three different wage rates, as shown in Table 2.

Table 2. Example labour burden calculations (in Canadian dollars)

Component of labour burden	Labourer	Journeyman	Foreman
Base wage rate	14.25	19.00	21.67
Vacation pay and statutory holidays	1.35	1.79	2.05
Subtotal	15.60	20.79	23.72
Worker's compensation	0.43	0.57	0.65
Unemployment insurance	0.54	0.71	0.81
Canada Pension Plan	0.55	0.73	0.83
Company pension plan	0.51	0.68	0.77
Payroll tax	0.11	0.15	0.17
Safety	0.38	0.56	0.64
Medical-dental coverage	0.44	0.58	0.66
Total	18.56	24.77	28.25
Total percentage burdens	30.25%	30.37%	30.36%

Each company was asked to report on an actual recently completed commercial building project with a lump sum contract. All projects obtained had a stipulated fixed percentage markup on change orders. The average allowable change order markup for overheads on contractors' own work was 7.0%. Average allowable markup for profit on contractors' own work was 5.4%. An average of 6.2% was provided as allowable markup for overhead and profit on subcontractors' work.

On all projects, labor wage rate, material, and equipment were priced as direct costs on change orders. Eighty-three percent of respondents priced subcontracts as direct costs on change orders. Overhead costs on change orders on all contracts included bonding and insurance, off-site administration costs, and planning, estimating, and scheduling of work. The next highest group of change order overhead (83%) was for on-site administrative costs, small tools and consumables, and permit, legal and accounting fees. Sixty-seven percent had clean up as part of overhead costs on change orders, and 33% included surveying. The ratio of total overhead costs to direct costs of all change orders ranged from 11% to 50%. Payroll burden cost for change orders was fixed in most of the contracts surveyed (83%). In the majority of the contracts (50%), the percentage allowable for payroll burden cost was in the range of 20% to 25%.

Table 3 lists the number of change orders contemplated and priced, and the number of change orders approved for 5 of the projects surveyed. The percentage of priced change orders that were approved ranges from 56% to 100%. In 40% of the projects, each change order was priced an average of once; in 40% the average number was twice; and, in 20% the average was three times. Forty percent of the projects had an average of 2 trades impacted by each change order; 40% had 3 trades impacted; and, 20% had 5 or more trades impacted by each change order.

Table 3. Number of change orders contemplated and priced versus number approved

Project number	Number of change orders priced	Number of change orders approved	Percentage of change orders approved
1	4	4	100
2	18	15	83
3	38	37	97
4	104	99	95
5	126	70	56

4 ADMINISTRATIVE COST OF CHANGE ORDERS

For the majority of the projects surveyed (50%), the approximate time spent by site supervisory, project management, accounting, and secretarial staff in processing and pricing all change orders was between 101 and 499 hours. On 25% of the projects the time spent was less than 100 hours, and on 25% the time spent was more than 500 hours.

A breakdown of the average time spent at each stage of change order administration is shown in Table 4. This breakdown of time can be used in evaluating the appropriateness of markup provisions for change orders. For example, assume a request for a change is received that is 100% trade content with a value of $5000. According to Table 4, the general contractor spends an average of 3.03 hours (182 minutes) of direct supervision administering the change. At an in-house charge of $75 per hour, the administration cost alone would amount to $227.25, which is 4.5% of a $5000 change order. Given that the average allowable markup allowance for subcontractors' work is 6.2%, this leaves very little allowance for the remaining overhead costs and profit. This scenario assumes that every single trade quote is received on time and is accurate, and that the owner does not cancel the change. The results of the survey, however, indicate that 60% of all change orders are priced more than once. Furthermore, all contractors surveyed indicated that on some occasions change orders contemplated and priced were cancelled by the owner, and that no compensation was provided.

Table 4. Average time spent administering a change order

Change order administration	Average time spent (minutes)
Design review/verification	17
Site inspection	12
Preparation of cover letter and faxing of request for change to trades	12
Entering of request for change into change order log	5
Clarification/field questions and quotations from trades	16
Preparation of quotations for consultants	18
Receipt of approved change order, submission of approval to trades	30
Revision of schedules and work sequence	13
Cross-checking of change with drawings	16
Posting of changes in specifications and on all drawings	16
Layout and ensuring that trade forces are complying with change	19
Accounting processing of change in monthly billings	8
Total time spent	182

5 CONCLUSIONS AND RECOMMENDATIONS

The results of this preliminary investigation of industry practices on change order markup allowances suggest that there is no accepted standard set of markups for change order reimbursement in the Canadian building construction industry. The percentage allowable for payroll burden in the majority of contracts surveyed was inadequate, ranging from 20% to 25%, with actual labor burdens being a minimum of 30%. On average, 3 hours are spent in administering a change before work even commences on the change, thereby substantially reducing the markup allowance to cover additional overhead costs and profit. Furthermore, although contractors spend a significant amount of time processing change orders, they are not compensated for the time spent on cancelled change orders, which add to the overhead cost of the project.

Preliminary conclusions indicate that existing pre-defined markup allowances for overhead, profit, and labor burden costs on change orders do not fully compensate the general contractor. Further research is warranted to develop a more equitable change order pricing strategy that suits the actual practices of the construction industry. Such research should focus on a number of issues, including:

- Development of a standard definition of items comprising home office overhead, project overhead, and labor burden.
- Establishment of a set of realistic percentages for allowable markup to account for the complexity of the change, the time the change order is issued during the work, the value of the change, the number of trades impacted by the change, the time spent in pricing the change, and the number

of times the change is re-priced.

- Assessment of an appropriate allowance to compensate the general contractor for the time spent in pricing change orders that are subsequently cancelled.

ACKNOWLEDGEMENTS

The authors would like to express their gratitude to the companies that participated in this survey for their valuable time and information. A special thanks goes to Mr. Chuck Burnett, who helped formulate the survey. This research was funded by the Construction Research Institute of Canada (CRIC).

REFERENCES

CCA 1992. *Guidelines for determining the costs associated with performing changes in the work.* Document No. 16. Ottawa: Canadian Construction Association (CCA).

CCDC 1994. *Stipulated price contract.* Document No. 2. Ottawa: Canadian Construction Documents Committee (CCDC).

Fayek, A. Robinson & M.Y. Nkuah 2000. Analysis of change order markup allowances on lump sum building contracts. Submitted to the *Project Management Journal*, February.

Nkuah, M.Y. 1999. *Analysis of change order markup allowance in lump-sum building contracts.* M.Eng. Project Report. Edmonton: University

Creative Systems in Structural and Construction Engineering, Singh (ed.) © 2001 Balkema, Rotterdam, ISBN 90 5809 161 9

Why do fabricators loose money on tubular trusses?

R.H. Keays

Keays Engineering, Melbourne, Vic., Australia

ABSTRACT: It seems that every new tubular truss erected in Victoria costs the fabricator far more than anticipated at tender time. This paper aims to explain the reasons for success or otherwise in the competitive environment of building structures. Matters addressed include design details, design documentation, material supply, fabrication processes, and erection techniques. Most aspects of tubular truss design are covered by CIDECT publications. Following their rules leads to an economical structure in most instances. However there are problems with implementation, especially in choosing amongst the design options available for tube/tube joints. Joints with overlapping members create the most problems for the fabricator. Field joints are another problem area. Practical solutions such as flanged joints are often considered far too ugly and agricultural, and welded joints have problems associated with temporary support. The joints used on a number of projects are compared, and indicative costing included

1 INTRODUCTION

Tubular trusses are fashionable for the structural framing of sports stadiums in Australia. Trusses have the advantage of high structural efficiency, allowing large spans at minimum weight, and painted tubular members have clean lines that appeal to the architect. Many of the major sporting venues and public buildings constructed in recent years in Sydney and Melbourne feature exposed trusses with Circular Hollow Section chords.

Fabrication and erection of these structures has not been a particularly pleasurable experience for those involved, and rarely has the recompense matched the expenditure. There appear to be problems at every step of the process from drafting through to touch-up painting. The purpose of this paper is to document those problems, and point the next generation of monumental masons along a path which might be just a little less rocky.

In Australia "fast-tracking" is the norm, and tenders are called on incomplete drawings. Connections are normally designed by the structural engineer (not the fabricator), with details becoming available after award of the fabrication contract. Detail drafting is carried out in parallel with procurement. Fabrication, painting, erection and touch-up follow in order.

The paper is structured to follow the sequence of events from concept to completion.

2 DESIGN DOCUMENTATION

Design documentation consists of Specification, General Notes, and Engineering and Architectural Drawings. As tender time, the builders will be intent on "building a book", with bets on the risky elements laid off amongst the trade sub-contractors. Occasionally mention will be made of a "Head Contract, a copy of which is available for perusal in the builder's head-office". The Clause that the builder invariably forgets to mention is that the trade contractor assumes the risk for all the essential bits missing from the tender documentation.

The drawings may be incomplete. Perhaps one or two connection details will be included, or there might just be a note "All welds to be 6mm continuous fillet welds unless noted otherwise". This paucity of information is an accepted part of competitive quotation, and the fabricator must draw on his experience to fill in the missing information. Where the structural form is familiar, the fabricator can still produce a reliable estimate. With unusually structures such as tubular trusses, the fabricator is moving into unfamiliar territory, and may underestimate the complexity of the task. The risk is further compounded by the potential for an inexperienced engineer to unwittingly add to the fabricator's task, as will be discussed below.

The builder rarely provides the fabricator with the architect's drawings and details. As a consequence, artistic requirements such as "welds ground flush" and "concealed bolted splices" are frequently missed at tender time.

Inadequacy of design documentation has been a topic of concern to the Australian fabrication industry for some time. Research by CSIRO (Tilley, 1998) has identified a significant reduction in the quality of design documentation over the past ten years. Tilley reports that competition has resulted in a reduction in fee levels for engineers. As a consequence, engineers have reduced the level of service they provide, resulting in inadequate/impractical /unworkable details on the construction drawings. This leads to the detailer and fabricator taking responsibility for development of practical details (instead of focusing on doing their designated task efficiently).

The writer's experience has been that poor initial documentation is the most reliable predictor of future losses. If the designer hasn't the time or ability to prepare a proper brief before asking prices, then it is unlikely that an easily built structure will result.

3 MATERIAL SELECTION

One problem that is perhaps unique to the Australian market is the limited range of tubular sections readily available. There are just two manufacturers of pipe and RHS in the Australia. They produce only thin-walled sections (up to 9mm only).

To differentiate their product from competing imported tube, the manufacturers have run a campaign warning of potential problems with unidentified steel, or steel not made to the local standard. This has resulted in designers trying to work within the wall thickness limits of the local product. Instead of inserting a short length of thicker wall pipe at truss nodes, the designer attempts to make the thin-wall pipe work – often with grossly inefficient results. (Fortunately for those in the know, there are merchants who stock imported heavy wall pipe that has been checked for compliance with the local Standard.)

There is also a tendency for designers to use every section in the catalogue. The common structural design packages include a list of all the available sections. Because tubular sections are a by-product of the piping industry, there is a wide range of sizes available, with axial strength increments averaging 6% (compared to 14% for the bending strength of Universal Beam sections). This can result in a proliferation of sizes of truss members. Trusses with six different web member sizes are not uncommon, and buildings with twenty-five different pipe sizes have resulted.

Using too large a range of sizes adds to the cost of procurement and detailing, and decreases the efficiency of fabrication through excessive offcuts and additional handling. One large building truss used five different wall thicknesses in one pipe size for the chords. Not surprisingly, this is one of the jobs where the fabricator lost money.

4 BASIC DESIGN DECISIONS

Truss design has always been more complicated than conventional framing, and becomes even more so when the members are tubular. With a simple beam, the designer can select the size without giving any thought to the end connections. Only after he has checked the beam's strength and stiffness, does he start to worry about the end connections, and rarely will he need to reconsider the initial decision on member size.

Trusses are different. The designer must consider the connection details early in the design process – even before deciding on the section types to be used for the chords and webs. Given that this paper is about tubular trusses, there is little to be gained in discussing whether the truss members should be anything but tubes.

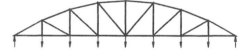

Figure 1. Pratt Truss

The form of the truss is the first critical decision. A Pratt Truss (such as the arch shown in Figure 1) will give the lightest practical top chord. Before committing to this form, the designer should consider other possibilities such as the modified Warren Truss in Figure 2. The spans for the top chord are longer, and so will be heavier, but there are three fewer web members, and that will lead to a saving in connection costs. Whilst addressing the truss form, it is wise to consider the number of bays – an even number produces a simpler detail at mid-span.

Figure 2. Modified Warren Truss

Having decided the form, the dimensions of the chords and webs follow. Optimizing the weight without consideration of the connections will give members with large diameters and thin walls. Adding in the cost and weight of connections changes the optimum. The chords will have a reduced diameter and thicker walls. The web members will generally be the thinnest wall available, and have an outside diameter higher than needed for strength. This might result in a structure 10% heavier, but the saving will be made in fabrication and possibly painting if the surface area is reduced.

It has been the writer's experience that structural engineers have left no room in their budget for consideration of design changes to save the fabricator time and money. All too often the fabricator takes the easy way out, wears the cost, and "just builds it".

5 JOINT DETAILS

Tube web/chord splices are covered in detail by CIDECT publications such as Wardenier (1991). This gives formulae for the tension/compression capacity of the web member for a reasonable range of member geometries. For tubes there are two critical modes of failure – chord plastification and punching shear, with the former normally more critical.

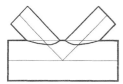

Figure 3. Tubular Joint – K Joint with gap.

Where the (D/t) of the chord is less than 35, and where column buckling dictates the web member size, the simple end connection shown in Figure 3 has sufficient capacity without reinforcement. Outside this range, or where the designer is aiming to make the connections stronger than the members, CIDECT offers some options.

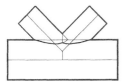

Figure 4. Tubular Joint – K Joint with Overlap.

The first option is to change the inter-points for the web members so they overlap (Figure 2). This gives a 25-50% increase in joint capacity, apparently for very little effort. The designer simply adds a note "All tube/tube joints have been designed as overlap joints – fabricator to provide details for approval". But this creates a minefield for the fabricator -
- All the tubes must have double preparations.
- The inter-points move from the chord centerline, making checking of assembly more difficult.

- The truss must be assembled and welded member by member, making the division of activities in the workshop more complicated.
- Errors in set-out accumulate, because every member must mate accurately at both ends.

The next option is even more unpalatable. Gussets plates can be slotted through the web and chord members to provide an alternative load path for the truss forces. Each of these plates must be profile cut to the right size, and the tubes must be carefully slotted to provide a good fit-up. Then there is the time to weld this plate in position, invariably with "full strength butt welds" because the designer hasn't the time or understanding needed to calculate the weld size really required.

Rarely is the practical option of cutting in a short length of heavier wall pipe considered. In one instance, the writer was able to intercept a particularly nasty detail before the fabricator was committed. A square metre of 20mm plate with an estimated fabrication time of 20 hours was replaced by a short length of heavier wall pipe, with a the fabrication time of just one hour.

Design strategies involving the fabricator's connection designer at an early stage of structural design are a rarity. Work by Tizani et al (1993) has shown that benefits will result. In the Australian context, the designer is on a fixed fee and the fabricator is not appointed until after commitment to the structural form. As a consequence, expensive forms of construction result.

6 WELDING

Inexperienced designers are renowned for overwelding. When there is any doubt about a joint, they simply specify "full strength butt weld", without realizing the consequences. Tube/tube joints are best done as fillet/butts, as shown on Figure 5. The weld that does all the work is the natural fillet at the shoulders and toe. The fillet weld at the heel is really just there for appearances.

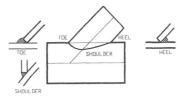

Figure 5. Fillet/butt weld preparation.

The profiling of the tube for welding the fillet/butt is a simple square cut. The shape is simply defined by the shorter of the inside and outside generators of the tube. Occasionally, the welder will need to grind off the edge of the toe, but otherwise fit-up time is easily predicted at tender time.

Profiling for a full butt preparation is far more complex, with careful calculation of the bevels. The correct profile is only achieved by using three torches, instead of the single torch at right angles to the tube axis. Fit-up and pre-weld inspection takes longer, and there is more weld metal to deposit. The experienced estimator, aware of the additional hours involved, looses the contract to the amateur competitor. The costs are real, and the poor fabricator becomes poorer.

7 FIELD JOINTS

Different possible field joints for one particular tube size are illustrated on the following Figures. Obviously, the simple butt joint is easiest to define and represents the least fabrication cost, but there can be severe cost penalties associated with holding the joint in position during welding, and the access requirements for welding and painting.

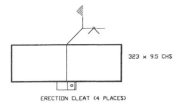

Figure 6. Field-welded Joint

The flanged joint is disliked by architects, but can be used where less obvious. The flange thickness is as recommended by CIDECT. The bolt size is set 25% high to allow for prying effects and the limited ductility of bolts in tension.

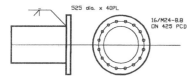

Figure 7. Flanged Joint

The third joint is one used in a number of projects associated with the Olympic Games facilities and at Melbourne's Colonial Stadium. The details are shown are styled upon joints used in the latter. If the architect's opinion prevails, this joint can be covered with sheet metal to hide the detail, and avoid rainwater collecting in the pockets.

The fourth joint is one of the writer's preferred options where the architect requires a concealed joint. It has the ductility associated with bolts in shear, and all welds are fillet welds or single-pass partial penetration butts.

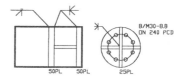

Figure 8. Hidden Tension Joint

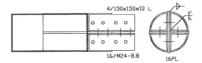

Figure 9. Hidden Shear Joint

Cost Centre	Field welded	Flanged	Hidden Tension	Hidden Shear
Design & Drafting	2	4	9	9
Material Supply	7	101	104	72
Fabrication				
Cutting	6	14	31	58
Assembly	27	42	114	72
Welding	14	33	232	75
Holes	4	115	72	35
Painting additional area		8	8	17
Bolt supply	4	59	53	40
Erection (all the same)				
Field welding	145			
Remove er. cleats	20			
Final Tighten Bolts		17	17	17
Paint welded area	44			
Cover Plate (supply & fit)			24	24
Total cost	271	394	664	418
	100%	145%	245%	154%
Additional costs for extra trade on critical path.				
Field welding	145	none	none	none
Adjusted total cost	416	394	664	418
	100%	95%	160%	100%

Table 1. Field Joint Costs (in US$)

Estimated costs of these four joints are tabulated in Table 1. Rates have been taken from Watson (1996), using an all-inclusive hourly rate of $US24 per hour for labor, and $US600/t for steel plate. At first glance, the cost of the filed welded joint is less than any of the bolted alternatives, even after allowing for erection cleats and painting afterwards. However, welding critical joints adds another trade to the field force, and exposes the construction manager to the industrial muscle of another group of workers. Allowing for this by doubling the welding cost, shows that the flanged joint and "hidden shear" joint are cost-competitive. The "hidden tension" joint remains an expensive last option.

8 CONCLUSIONS

The paper illustrates means of keeping the cost of fabrication and erection of tubular trusses to a minimum.

Designers have a duty to produce efficient designs. When the documentation is lacking in depth and quality, the members are sized with the wrong purpose, or the details are overly complex, the job will suffer.

Fabricators need to be aware of their true costs when quoting for unusual jobs. Fabrication of tubular trusses will always be more expensive than the normal beam and column structural framing. The experienced fabricators need to document their knowledge so that an inexperienced competitor does not undercut them on the next job.

Builders and owners need to allow time at the start of the job for proper planning and design. Based on the Australian experience, fast-track is neither fast nor on the right track.

We all need to be aware of our duty as an industry to produce the best answer at the right price, and to distribute the costs in an equitable manner.

REFERENCES

Tilley, P.A., 1998, "Design and Documentation Deficiency and its Impact on Steel Construction", Steel Construction, Vol.32 No.1, Australian Institute of Steel Construction.

Watson, K.B., Dallas, S., Van der Kreek, N., 1993, "Costing of Steelwork from Feasibility through to Completion", Steel Construction, Vol.30, No.2, Australian Institute of Steel Construction.

Wardenier, J. et al, 1991, "Design Guide for Circular Hollow Section (CHS) Joints under predominantly Static Loading", Comite International pour le Developpement et l'Etude de la Construction Tubulaire (CIDECT).

Dutta, D. et al, 1998, "Design Guide for Fabrication, Assembly and Erection of Hollow Section Structures", CIDECT.

Tizani, W.M.K., Davies, G., McCarthy, T.J., Nethercot, D.A., and Smith, N.J., "Economic appraisal of tubular truss design", Fifth International Symposium on Tubular Structures, Nottingham, United Kingdom, August 1993, Proceedings pp.290-297.

APPENDIX 1. LESSONS FROM CRACKS

Our team of crack engineers has been at work again!

Cracks have been found in the chords of the tubular trusses at the Colonial Stadium in Melbourne. The cracks were through the thickness of the chord wall, and some were over 100mm long. Cracks were generally parallel to the chord axis.

The cracks were adjacent to the weld connecting the vertical and inclined web members to the bottom and top chords in a Pratt truss. Cracks initiated at the shoulder of the joint (as defined in Figure 2).

The chord members were 508x8 (bottom) and 508x12 (top), in 450MPa yield / 500MPa ultimate steel. The vertical member was 168x4.8 and 12.4m long. The inclined member was 168x6.4 and 14.5m long. These were 350/450 MPa material.

Soon after these members had been assembled into the truss (prior to lifting from the ground), the site engineer observed significant vibration in the afternoon sea breeze (15-20 knots). Vortex shedding from the web members (or adjacent chords) was the obvious cause.

Notice was taken and measurements made. Dampers designed to reduce the vibration to minimal levels, but not installed immediately. The cracks were noticed after 12 months of exposure. Soon after, the cracks were welded closed, and dampers installed.

There is a lesson here for all of us. Small amounts of vortex-induced vibration can be tolerated. Bad connection details can be tolerated. But when the two are combined, there is potential for a disaster.

10 Cost Systems – II

Creative Systems in Structural and Construction Engineering, Singh (ed.)© 2001 Balkema, Rotterdam, ISBN 90 5809 161 9

A fuzzy logic decision-support system for competitive tendering

M.Wanous, A.H.Boussabaine & J.Lewis
School of Architecture and Building Engineering, University of Liverpool, UK

ABSTRACT: For a construction contractor, deciding whether or not to submit a bid for a new project is a highly complex process. Contractors usually relay on their experience and intuitively make this decision. An innovative fuzzy logic system is presented in this paper to help contractors in making their "bid/ on bid" decisions. The rule base of this system was extracted automatically from one hundred and sixty two real life bidding situations using the neurofuzzy technique. The model was tested on another twenty real projects and proved to be 90% accurate in simulating their actual decisions. Although the proposed model is based on data from the Syrian construction industry, it could be easily modified to suit other countries.

1 INTRODUCTION

The tendering process involves two crucial decisions. The first is whether or not to bid and the second is associated with selection of a suitable mark up. Both decisions are of great importance as the success or failure of a company lies in their outcomes. This importance has been attracting many researchers in the last fifty years. Fuzziness in information on a new construction project, the client, the potential competitors, and the overall construction market make it a very complex process to decide whether to bid or not to bid on this project. Usually, this decision is derived from intuition and subjective judgement based on past experience (Ahmad 1990). However, such practice does not guarantee consistent decisions due to the lack of a binding mechanism that relates present cases to past patterns. Thus, a structured framework for making the bidding decisions can be of great help to construction contractors especially new contractors who do not have considerable experience in dealing with different bidding situations. The following section provides a brief review of the bidding literature.

2 LITERATURE REVIEW

The literature contains a great number of theoretical bidding models based on the works of Friedman (1956) and Gates (1967) and concerned with estimating the probability of winning a contract with a certain mark up. All these mathematical models proved to be suitable for academia but not for prac-

titioners (Gates 1983). The mathematical complexity of these models made them unpopular in the construction industry (Ahmad and Minkarah 1988, Dawood 1995). Very few qualitative approaches, which study how the bidding decisions are made in practice, have been carried out.

Recently, the bidding problem has been approached practically rather than mathematically using artificial intelligence techniques such as expert systems (ES) and artificial neural networks (ANN). It has been argued that the ANN techniques are more suitable for mark up estimation because it is an unstructured problem and affected by many factors the influence of which is not easy to quantify (Moselhi et al 1993).

However, research on the "bid/ no bid" decision has received less attention compared to the mark up part of the bidding process despite the reliance of the later on the decision to bid (Dawood 1995). Very few publications that address the "bid/ no bid" process can be found in the construction literature.

Ahmad and Minkarah (1988) conducted a questionnaire survey to uncover the factors that characterise the bidding decision-making process in the United States. Subsequently, Ahmad (1990) proposed a bidding methodology based on the decision analyses technique for dealing with the "bid/ no bid" problem. This model demands many inputs some, of which the bidder, especially those with limited experience, might not be able to provide. Also, it assumes that all factors contribute positively to the "bid" decision.

No distinction was made between some factors that count for the "bid" decision, such as profitability, and others that count against it, such as "degree of hazard". However, this approach is the most promising step on the road of modelling the "bid/ no bid" decision.

Shash (1993) identified, through a modified version of the same questionnaire used by Ahmad and Minkarah, fifty five factors that characterise the bidding decisions in the UK. The need for work, number of competitors tendering and experience in similar projects were identified as the top three factors that affect the "bid/ no bid" decision.

AbouRizk *et al* (1993) proposed an expert system called BidExpert. This model was integrated with a database management program, call BidTrak, that retrieves historical information from past bids submitted by the company and its competitors. BidExpert processes the outcomes using its knowledge base and provides the user with a "bid/ no bid" recommendation. The necessity for historical information limits the applicability of this model. BidExpert has other drawbacks. For instance, the company capacity is evaluated by the number of projects the company has handled in the last five years and the number of the current projects, without any consideration of the projects' sizes.

Abdelrazig (1995) carried out a literature review and identified thirty seven factors that affect the "bid/ no bid" decision in Saudi Arabia. The analytic hierarchy process (AHP) was utilised and computer software, named Expert Choice, was developed to help contractors in making their "bid/ no bid" decisions.

Wanous *et al* (1998) conducted a questionnaire survey among Syrian general contractors to uncover the factors that characterise their "bid/ no bid" decision-making process. Thirty eight factors were ranked according to their relative importance in making the "bid/ no bid" decision in Syria.

Subsequently, Wanous et al (1999) considered the most important factors and developed a parametric profile for each one. All a contractor needs to use this parametric model is his/ her subjective assessments of the considered bidding situation in terms of certain criteria. The contractor's assessment of a certain factor is compared with its parameters to quantify the contribution of this factor in the final recommendation. If the collective contribution of all factors is positive, a "bid" recommendation will be made with a certain degree of confidence. This model was tested on twenty real bidding situations and succeeded to simulate the actual decisions of 85% of them.

Dawood 1995 rejected the suitability of the ANN technique to the bidding process due to its "blackbox" nature and used expert systems to help in making the "bid / no bid" decision in the "make-to-order" precast concrete industry. The explicit knowledge representation and the explanation facility are the main advantages of the ES. However, the practicality of applying this technique can be questioned because it is extremely difficult to explain the process of making the "bid/ no bid" decision through if-then rules. Even experienced contractors might not be able to articulate their way of thinking when dealing with new bidding situations. These properties call for hybrid decision-support systems that can learn from real examples and can justify the rational behind their recommendations.

Combining the neural networks systems with fuzzy logic models could be a very suitable answer to this problem. Moreover, fuzzy and neural network hybrid decision support systems are able to mimic the ability of the human mind to effectively employ modes of reasoning that are approximate rather than exact (Zadeh 1994).

The present study investigates the application of a new technique called the Neurofuzzy technology to develop a fuzzy expert system for the "bid/ no bid" process. The neurofuzzy module of a fuzzy logic development software called "*Fuzzy*TECH 5.10b for Business Professional" (*Fuzzy*TECH 1997) was employed in this study. The theory and principles of this technology are dealt with in the following section.

3 NEUROFUZZY MODELLING

A combination of the explicit knowledge representation of fuzzy logic with the learning power of neural networks (ANN) could provide a very powerful tool for modelling the bidding process. This combination is called "Neurofuzzy" and it reaps the advantages of both technologies. The basic idea of the composition method of fuzzy logic and ANN is to achieve fuzzy reasoning by a neural network whose connection weights represent the parameters associated with a set of fuzzy rules. These parameters are called degrees of support (DoS). Neurofuzzy methods are purposely developed to automatically identify fuzzy rules and tune both the shape of the membership functions and their position in the universe of discourse. Neurofuzzy modelling involves the extraction of rules from a typical data set and the training of these rules to identify the strength of any pattern within the data set. The system creates membership functions from which linguistic rules can be derived as opposed to real values. Many alternative methods of integrating neural networks and fuzzy logic have been proposed in the literature. Amongst these is the Fuzzy Associate Memories (FAM) method, which is the most common approach. FAM is a fuzzy logic rule with an associated weight. This method is based on a mathematical function that maps FAMs to neurons in the neural network to enable the use of a modified back-propagation learning algorithm with fuzzy logic. This is possible by modifying the weights of the connections of a suitably defined feed-forward ANN with a learning procedure based on the back propagation algorithm.

The mathematical background behind this methodology is explained in details in (Kosko 1992). The neurofuzzy module used works as an intelligent assistant during the development process. It helps to generate and adjust membership functions and rule bases from sample data. This makes it unnecessary to worry about mathematical details of the underlining mapping algorithm. Fig. 1 shows the general framework of a neurofuzzy bidding model.

1. There is much experience in neural networks as extensive research has gone on for more than fifty years. Neurofuzzy in contrast is still a young technology; and,
2. Neurofuzzy training features fewer degrees of freedom for the learning algorithm compared with a neural network.

Therefore, in applications where massive amounts of data are available but there is little knowledge of the system's structure, Neurofuzzy may not be the best tool to be used.

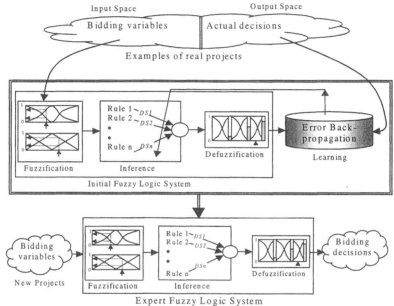

Figure 1. Integrating neural networks with fuzzy logic

Neurofuzzy models are fundamentally different from neural networks and expert systems as they have the following characteristics (Boussabaine and Elhag 1999 and Altrock 1997):
1. Automatically extract a set of fuzzy rules from input/output data sets;
2. Automatically train and change the shape of member functions according to data patterns;
3. Allow the inclusion of knowledge and expertise in choosing system topology;
4. Allow the interpretation of results since they contains self-explained fuzzy logic rules and linguistic variables; and,
5. Lead to a model the performance of which can be directly modified using all the available engineering know-how.

However, there are a few disadvantages of neurofuzzy compared with the ANN techniques.

These include the following (Altrock 1997):

4 DEVELOPMENT OF A NEUROFUZZY "BID/NO BID" SYSTEM

The development of neurofuzzy models involves a series of interactive processes. Figure 2 shows the sequence used in the development of the neurofuzzy bidding model.

4.1 Data elicitation and analysis

The developement of fuzzy logic models requires one or more of the following types of data:
• Rules of thumb the collection of which is highly difficult;
• Preprocessed data, which are rarely available; and,
• Raw data, which have to be processed prior to implementation.

283

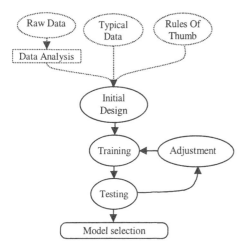

Figure 2. Framework of the development process

Table 1. The most influential "bid/ no bid" factors

Factors	r	|r|
1. Fulfilling the to-tender conditions	+0.69	0.69
2. Site accessibility	+0.64	0.64
3. Site clearance of obstructions	+0.57	0.57
4. Availability of capital required	+0.52	0.52
5. Availability of materials required	+0.51	0.51
6. Proportions that could be constructed mechanically	+0.49	0.49
7. Confidence in the cost estimate	+0.46	0.46
8. Financial capability of the client	+0.44	0.44
9. Public objection	-0.43	0.43
10. Current workload	-0.42	0.42
11. Relation with/ reputation of the client	+0.42	0.42
12. Favourability of the cash flow	+0.41	0.41

In the absence of typical data and adequate rules, new data was collected.

First, the most important bidding factors identified by Wanous et al (1998) were selected. Factors whose importance indices less than 0.40 were omitted. The remaining factors were used in preparing a simple form to elicit situation-outcome data about recent bidding situations in Syria. Three hundred copies of this form were sent to thirty general contractors. Respondents were requested to record their subjective assessments of twenty two criteria listed in the form and to provide the actual "bid/ no bid" decision made in each bidding situation. The contractors assessment of a certain factor is made simply ticking a box on a scale between 0 and 6, where 0 is extremely low and 6 is extremely high. One hundred and eighty two forms were filled in and returned. Twenty cases were randomly selected and reserved for the validation process (validation sample). A detailed statistical analysis was made on the remaining cases (modelling sample) to check their validity and to study the cause-effect relationships between the considered criteria and the actual "bid/ no bid" decisions.

Factors whose absolute correlation with the actual bidding decision is less than (0.40) were omitted. Table 1 shows the remaining twelve factors with their Pearson correlation coefficients.

In addition to indicating the most influential factors, the modelling sample provides a set of inputs (subjective assessments between 0 and 6) and the associated output (o for "no bid" and 1 for "bid") for training.

4.2 Initial design

In this step, an empty fuzzy logic system was developed and made ready for training. Figure 3 shows the main components of the fuzzy logic system.

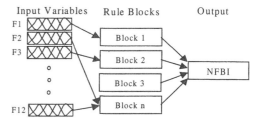

Figure 3. Components of a fuzzy logic system

Developing such a system involves the following tasks:

1. Definition of input and output variables. The most influential factors identified in section 4.1 and shown in Table 1 were considered as input variables. Only one output called the Neurofuzyy bidding Index (NFBI) is expected from the fuzzy bidding model. The closer the value of NFBI is to one, the more confidence in the "bid" recommendation and the closer it is to zero, the more confidence in the "no bid" recommendation.

2. Setting the linguistic variables for the considered inputs and output. The number of linguistic terms for each variable need to be determined. As a start, the number of terms in all input variables was set to three and the number of terms in the output variable was set to five (Altrock 1997). Also, types of membership functions (MBFs) of each input variable should be selected. For all input variables, the cubic interpolative S-shaped MBF was used because it has proved to be more accurate models of human concepts for complex decision-support applications. For the output variable, the • -type, i.e. linear (L), was used because most applications use this type of MBF for output variables (Altrock 1997). All the selected membership functions are standard, i.e. maximum is always (• =1) and minimum is (• =0). Figure 4 shows the " Accessibility" linguistic variable as an ex-

ample of the input variables. The output linguistic variable is shown in Figure 5.

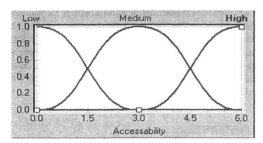

Figure 4. Linguistic variable of the "Site Accessability" input factor

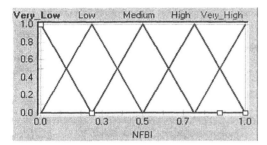

Figure 5. Linguistic variable for the output variable

3. Setting the inference rule base. This involves the following tasks:
   - Definition of the fuzzy rules:
   The collection of these rules from the domain experts is a highly difficult process. Therefore, the neurofuzzy module of the *fuzzy*TECH development software was used to generate the fuzzy rule base from the real projects included in the modelling sample. The rule base was arranged in four rule blocks similar to figure 3. The degrees of support (DoS) of the generated rules were set to zero. DoSs will be modified during the training phase.
   - Selection of the fuzzy aggregation operators:
   The "MinMax" premise aggregation operator with no compensation parameter and the "Max" result aggregation operator was adopted. These will be optimised later in the modification phase.
4. Selection of the output inference method, i.e. defuzzification method. The center of maximum (CoM) was used in this stage.

For information about fuzzy aggregation operators and inference methods, see Altrock 1997. The resultant model was called model 1. The structure of this model identifies the fuzzy logic inference flow from the input space to the output space. The input values are fuzzified and translated into linguistic expressions such as "Low", "Medium", or "High". Fuzzy inference takes place in the rule blocks, which

contain the linguistic control rules. In the output interface, the linguistic bidding index (Very Low, Low, Medium, High, Very High) is defuzzified and translated back into a numerical value. Model 1 does not have any knowledge at this stage because the degrees of support, weights, of all its rules are zeros. The following section explains how these DSs are adjusted by training the mode on real life bidding examples (the modelling sample).

4.3 *Training*

When training a fuzzy logic model, important rules will have high degrees of support, i.e. close to one. Unimportant rules will have low DoSs, i.e. close to zero. These rules can be deleted, as they do not have significant influence on the model's behaviour.

The average deviation between the actual output values of the training examples and the predicted values for these examples is produced automatically by *fuzzy*TECH. The generated average deviation is a measurement of the training performance. Model 1 was able to map the input space of the training samples to the output space with an average error 22.55% after 5 iterations. This parameter is recorded and many other models with different characteristics were trained for the same number of iterations. This is to enable a fair comparison between different development parameters as explained in section 4.5.

4.4 *Testing*

After completing five training iterations, model 1 has been tested using twenty real bidding situations (validation sample) reserved for the testing process. The average testing error produced by model 1 (27.55%) was recorded. The average deviations of training and testing are useful to judge the performance of model 1 and to compare it to others.

4.5 *Adjustment*

The adjustment phase was basically an iterative trial and error process, which involved the following actions:
1. Revision of the aggregation operators/ parameters;
2. Revision of the linguistic variables, i.e. terms, membership functions;
3. Revision/ extension of the rule base;
4. Revision of the output inference, i.e. defuzzification method; and,
5. More learning iterations

In each experiment, the resultant model was trained for five iterations and, then, tested. The average deviation parameters were recorded for all the examined models.

4.6 Model selection

Twenty five models were experimented with during the adjustment phase. The best model was selected using the following criteria:

- Low training average deviation; and,
- Low testing average deviation.

The training and testing average deviations of the selected model were 6.9% and 9.15% respectively.

Table 2 summarises the main characteristics of this model. Table 3 shows a sample of the rule base.

Table 2. Characteristics of the final model

Number of terms		MBFs		Number of rules
Inputs	Output	Inputs	Output	
5	5	S-shaped	Linear	59

Table 3. Examples of the model's fuzzy rules

IF			THEN	
Financial (F8)	Fulfillment (F1)	Relations (F11)	DoS	NFBI
Very low	Very low	Very low	1.00	Very low
	Very low		1.00	Very low
Very low	Very low	Low	0.03	Medium
Medium	Very high	Medium	0.34	Very high

The selected fuzzy bidding model requires subjective assessments of the bidding situation under consideration in terms of the factors listed in Table 1 to produce a bidding index (NFBI). Only where the NFBI is greater than 0.5, will the model makes the "bid" recommendation for a situation.

5 VALIDATION

The final model was used to predict the actual "bid/ no bid" decisions of the twenty projects in the validation sample. The actual decisions of eighteen cases were successfully predicted. This means that the model is 90% accurate in simulating the actual practice of making the "bid/ no bid" decisions in Syria. The parametric model developed in previous work (Wanous et al 1999) predicted the actual decisions of seventeen cases of the same sample (85%). Therefore, it can be concluded that the developed fuzzy model is a reliable tool and can used with high confidence to assist contractors in making their "bid/ no bid" decisions.

6 CONCLUSION

This paper has demonstrated that the neurofuzzy technology is a very suitable technique for modelling the "bid/ no bid" decision-making process. A novel fuzzy bidding model was developed using this technique. Testing the developed model on real life bidding situations suggested that it could be used with high confidence to help contractors in making their bidding decisions. Although this model is based on data collected form the Syrian construction industry; it can be modified very easily to suit other practices. This can be done be retraining on new bidding situations and/ or by direct adjustment to the rule base and the membership functions.

REFERENCES

Ahmad, I., and Minkarah, I. A. 1988. Questionnaire survey on bidding on construction. *Journal of Management in Engineering*, ASCE, 4 (3), pp. 229-243.

Ahmad, I. 1990. Decision- support system for modelling the bid/ no bid decision problem. *Journal of construction engineering and management*. Vol. 116, No. 4, pp 595-607.

AbouRizk, S.M., Dozzi, S.P. and Sawhney, A. 1993. BidExpert- An Expert System for Strategic Bidding. *Annual conference of the Canadian Society of Civil Engineering, Fredericton*, NB, Canada. Pp. 39-48.

Abdelrazig, A. A. 1995. MSc dissertation. *King Fahd University of Petroleum and Minerals/ Saudi Arabia*.

Altrock, C. 1997. *Fuzzy logic & neurofuzzy applications in business and finance*. Prentice Hall PTR.

Boussabaine, A.H. and Elhag, T.M.S. 1999. Applying fuzzy techniques to cash flow analysis. *Construction Management and Economics*, Vol. 17, p. 745-755.

Dawood, N. N. 1995. An integrated bidding management expert system for the make-to-order precast industry. *Construction Mgmt. and Economics*, Vol.13, No. 2, pp. 115-125.

Friedman, L. 1956. A Competitive Bidding Strategy. *Operational Research*, 4, pp. 104-112.

*Fuzzy*TECH 1997. User's manual. Inform Software Corp

Gates, M. 1967. Bidding strategies and probabilities. *J. Construction Division*, ASCE 93(1), pp. 75-103.

Gates, M. 1983. A bidding strategy based on ESPE. *Cost Engineering*, Vol. 25, No. 6, pp. 27-35.

Kosko, B. 1992. *Neural Networks and Fuzzy Systems*. Englewood Cliffs, New Jersey: Prentice Hall.

Moselhi, O., Hegazy, T. and Fazio, P. 1993. DBID: Analogy Based DSS for Bidding in Construction. *ASCE Journal of Engrg. and Management*. 119, 3, pp. 466-470.

Shash, A. 1993. Factors considered in tendering decisions by top UK contractors. *Construction Management and Economics*, Vol.11, No.2, pp.111-118.

Wanous, M., Boussabaine, A.H. and Lewis, J. 1998. Tendering factors considered by Syrian contractors. *ARCOM, 14th annual conf. Proc.*, Vol. 2, (535-534), Oxford, England.

Wanous, M., Boussabaine, A.H. and Lewis, J. 2000. Bid/ no bid: a parametric solution. *Construction Management and Economics*. Vol. 18, No. 4, pp 457-466.

Zadeh, L.A. 1994. Fuzzy logic, Neural networks and soft computing. *Communication. of the ACM*, Vol. 37 (3), pp 77-84.

Creative Systems in Structural and Construction Engineering, Singh (ed.) © 2001 Balkema, Rotterdam, ISBN 90 5809 161 9

A cost model for predicting earthquake damage

A. Touran
Northeastern University, Boston, Mass., USA

U. Kadakal & N. Kishi
Applied Insurance Research, Boston, Mass., USA

ABSTRACT: The objective of this project is to develop a comprehensive cost model for estimating repair/replacement costs for various levels of earthquake damage in buildings. The approach is based on the Advanced Component Method (ACM^{TM}) and provides objective estimates rather than subjective opinions for quantifying damage costs caused by an earthquake. For any specific building the cost of repairs for various structural and nonstructural components, for various damage levels is estimated. Using the building's period and damping, and given the earthquake characteristics, one can calculate the *Spectral Displacement (S_D)*. Using S_D, *Percent Physical Damage (D)* can be estimated for a specific component such as column or slab. Using the percent damage values, the cost of repair and replacement can be applied for the members to arrive at an estimate of the total monetary damage. Eventually, the objective is to use these estimates to assess the overall cost of earthquake on any selected building inventory (portfolio) in a given region. This will be a significant step in creating an objective relationship between the physical damage and monetary damage caused by an earthquake.

1 BACKGROUND

Earthquake vulnerability assessment is the science of estimating the probability and extent of earthquake damage before an earthquake occurs. Most recent attempts to assess building vulnerability to earthquakes have been based on observations made in the aftermath of earthquakes and on the opinion of experts as to the potential impact of an earthquake. These approaches use peak ground acceleration (*PGA*) as the measure of intensity, but *PGA* is a measure of ground motion and not a measure of how buildings respond to ground motion. In other words, using *PGA* as a measure of earthquake's intensity, assumes that during an earthquake, the top of a building moves exactly with the bottom of the building, *i.e.*, like the ground itself. This ignores any effect of a building's natural period (which is a function of its flexibility and mass), or of the resonance between the building's motion and ground motion.

The Advanced Component Method^{TM} (ACM) developed by Applied Insurance Research (The ACM Method, 2000), uses spectral displacement, S_d, as an objective measure of earthquake's intensity. S_d is the maximum horizontal displacement experienced by

building during an earthquake. When displacement occurs, the building and its component parts are deformed; deformation causes damage. In ACM, damage depends on the deformation response of individual structural and nonstructural components to ground motion, rather on how the ground is shaking. Because each building has different mechanical characteristics and a different natural period, each will be subjected to a different seismic intensity (spectral displacement) and, hence, a different damage state.

Using a nonlinear seismic analysis procedure, the building deformations can be calculated. Using deformation values, damage percentages are calculated for various structural components such as beams and slabs. Separately, a damage function for nonstructural components, such as glazing and partitions is estimated. Damage functions are developed separately for structural and nonstructural components because different occupancy types will show different proportions of structural and nonstructural elements. For example, a hospital and a parking garage may have very similar structural systems and structural components. Their nonstructural components, however, differ dramatically.

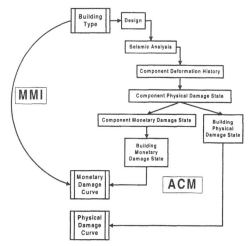

Figure 1. Overview of the ACM System

Traditionally, monetary damage is estimated either by using ATC-13 curves (ATC-13 1985), or by multiplying HAZUS (1997) physical damage ratios by replacement cost of the building. These methods more or less rely on expert opinion to derive monetary damage curves as a function of building type. ACM on the other hand, estimates monetary damage at component level, and the component damages are subsequently aggregated to produce building monetary damage curves. Figure 1 gives an overview of the ACM methodology.

2 COST MODEL

This paper describes the cost models developed during the first phase of the ACM's costing project. This phase considered low and mid-rise office buildings, with four different structural systems. Structural systems considered included moment resisting steel frame, braced steel frame, moment resisting concrete frame, and concrete frame with shear walls. The objective was to estimate repair and replacement cost estimates for various levels of earthquake damage. Each building type was broken down into three structural and five nonstructural components. Structural components were the following: (1) beams, (2) columns, and (3) floors; nonstructural components were the following: (1) partitions, (2) cladding, (3) glazing, (4) suspended ceiling, and (5) mechanical, electrical, and plumbing. These eight components were chosen because sufficient historical damage reports on these

elements were available. Naturally, the building consists of other components such as roofing system, or stairs. Each of these components were "attached" to one of the eight components listed above, so that the total building cost could be estimated accurately.

2.1 *Specifications for buildings*

For each building model, a structural design was done by design firms. Local design firms were enlisted to ensure conformity with local and regional construction practices. Design documents included physical dimensions of each component; each building type is to be a representative of the average in the modeled geographical region, such as Southern California. For each building, a set of specifications (type of cladding, finishes, *etc.*) was developed. The main source of cost data consisted of various R.S. Means (1999) publications. Use of R.S. Means costs allows us to update cost models on a regular basis as current costs become available. Because the main emphasis was to consider *typical* structures, we chose the typical R.S. Means models (Square Foot Cost Data, 1999) to define specifications. These specifications do not vary from building to building because they all represent office buildings in this phase. Using these specifications and structural designs, a replacement cost was estimated for each building model.

2.2 *Damage states and repair strategy*

Five damage states are considered in this work: *negligible*, *slight*, *moderate*, *extensive*, and *total*. In order to calculate repair costs for each of the damage states, we first defined these damage states. These definitions are based on literature search of previous earthquakes and group discussions. Repair cost depends on repair strategy, which depends, in turn, on the physical damage state of each component. In the case of a steel column, for example, if damage is negligible or slight, with minor deformations in connections or hairline cracks in a few welds, the recommended action may be a minimal repair or even to do nothing at all. Hence, the repair cost, in this instance, is negligible or zero. There may be alternative methods of repair, each associated with a different cost; the appropriate method may depend on the degree of damage and accessibility. At high levels of damage, replacing the affected member may be considered.

3 COST ESTIMATES

For each building, R.S. Means *Square Foot Costs (1999)* was used to arrive at a total cost estimate by incorporating building specifications and modifying the R.S. Means base estimate. The total cost would be the cost of constructing the building as a new project. It should be noted that the cost estimate is broken down into cost of the eight components that define the building behavior in case of an earthquake, and also into cost of components in various stories. *Total replacement cost* was then calculated by adding the cost of demolishing the old building to the cost of a new building. For each building model, the repair costs for each damage state was calculated using several of R.S. Means publications. In many cases the cost of repair for *extensive* or *total* damage states were larger than the replacement cost of the component. This was to be expected because in repairing an isolated component, many factors add to the cost of construction such as cost of protection of existing structure, cutting and patching to match existing structure, and shoring and bracing other components.

3.1 *Cost ratios*

For each building type, a table of cost ratios was developed that expressed the cost of repairing each component as a percentage of the total replacement cost for that component. These cost ratios may be represented by the following equation:

$$f_{ijk} = c_{ijk}/C_{ik} \tag{1}$$

where f_{ijk} is cost ratio for component i, for damage state j, on floor k of a specific building type. c_{ijk} is the cost of repair for component i, for damage state j, on floor k, and C_{ik} is the total replacement cost for component i in floor k. The total replacement cost for the building would be equal to $\sum C_{ik}$.

After the building is subjected to the desired seismic load and damage states for each component, i, in various floors, k, is established, one can use f_{ijk} values to calculate the total repair cost. This can be accomplished by multiplying appropriate pairs of f_{ijk} and C_{ik}s and summing them up.

3.2 *Probabilistic aspects of the cost model*

It is recognized that a single building type cannot adequately represent all buildings of the same population. Because of this, the cost estimates can at best be considered parameter estimates of an underlying distribution. Previous research (Touran & Wiser 1992; Touran 1993) had shown that lognormal distribution fitted well to the component costs of office buildings. For this phase of the project, component repair and replacement costs were treated as random variables with lognormal distributions. Further, it was assumed that structural component costs (beams, columns, floors) were correlated and also nonstructural component costs (partitions, cladding, glazing, suspended ceiling, HVAC) were correlated. Using these assumptions, the distribution of the total repair cost for each case can be simulated. These distributions can in turn be used for evaluating the monetary exposure in case of an earthquake with specific characteristics.

4 FUTURE WORK

The next step is to expand the cost estimation effort to other types of structures. Currently we are working on a large number of structures such as high rise, light industrial, low and mid-rise masonry, and residential buildings. Also, more work is needed in the statistical aspects of the project. We are collecting and analyzing data on the distributions of various building types and their costs. These will serve in estimating the total cost exposure in case of various intensity earthquakes in a specific region.

ACKNOWLEDGMENT

The work reported here was funded by Applied Insurance Research. This support is gratefully acknowledged.

REFERENCES

Advanced Component Method 2000. ACM™: a breakthrough in vulnerability assessment technology. Applied Insurance Research. Boston. Mass.

ATC-13 1985. Earthquake damage evaluation data for California. *Report no. ATC-13.* Applied Technology Council.

HAZUS 1997. Earthquake loss estimation methodology. Vol. 1. National Institute of Building Sciences.

R.S. Means 1999. Square Foot Cost Data. R.S. Means, Inc. Kingston. Mass.

Touran, A. & E. P. Wiser 1992. Monte Carlo technique with correlated random variables. *J. of Construction Eng. and Management.* 118(2): 258-272.

Touran, A. 1993. Probabilistic cost estimating with subjective correlations. . *J. of Construction Eng. and Management.* 119(1): 58-71.

Creative Systems in Structural and Construction Engineering, Singh (ed.) © 2001 Balkema, Rotterdam, ISBN 90 5809 161 9

Modeling of construction cost estimating for shared AEC models

T. M. Froese

Department of Civil Engineering, University of British Columbia, Vancouver, B.C., Canada

K. Q. Yu

Timberline Software Corporation, Beaverton, Oreg., USA & Department of Civil Engineering, University of British Columbia, Vancouver, B.C., Canada

ABSTRACT: A cornerstone of information technology for the Architecture, Engineering, Construction, and Facilities Management Industry (AEC/FM) is the ability to exchange information and integrate the industry's wide variety of computer applications. This integration requires standard data models of construction project information to provide the language needed to exchange information. The International Alliance for Interoperability (IAI) is an organization dedicated to developing such data standards, the Industry Foundation Classes (IFCs) for AEC/FM. Although much of the IFC's focus is on the representation of the physical components that make up constructed facilities, the IFC scope includes other project information such as costs, schedules, etc.. Furthermore, the IAI has combined with another industry-wide effort, aecXML, which is aimed at producing XML-based standards for AEC/FM. This paper discusses the general role of these standards relative to the area of construction cost estimating, for both traditional and new estimating tools. The paper presents details of how the cost estimating domain is modeled within the IAI, and provides examples of how these emerging data standards have been applied to date in support of AEC/FM data integration.

1 INRODUCTION

Integration and data exchange are major themes within the field of information technology for the Architecture, Engineering, Construction and Facilities Management Industries (AEC/FM). Recent trends in Web-based services for AEC/FM (construction ".COM's") have only increased the need for data exchange solutions. This integration requires standard data models of construction project information to provide the common language needed to exchange information. The International Alliance for Interoperability (IAI) is an organization dedicated to developing such data standards in the form of the Industry Foundation Classes (IFCs) for AEC/FM and aecXML. These standards represent not only the physical components that make up constructed facilities, but also costs and other project information.

This paper discusses the status of AEC/FM data standards relative to the area of construction cost estimating, and the role that such standards could play for both traditional and new estimating tools and processes. The paper also presents details of how the cost estimating domain is modeled within the IAI, and provides examples of how these emerging data standards have been applied to date in support of cost-related AEC/FM data integration.

2 DATA STANDARDS FOR AEC/FM

2.1 *Industry Data Standards Efforts*

The Industry Alliance for Interoperability (IAI) is a global, industry-based consortium for the AEC/FM industry. Their mission is to enable interoperability among industry processes of all different professional domains in AEC/FM projects by allowing the computer applications used by all project participants to share and exchange project information. The IAI's scope is the entire lifecycle of building projects including strategic planning, design and engineering, construction, and building operation. Since its creation, the IAI's goals have been to define, publish and promote a specification--called the Industry Foundation Classes (IFCs)--for sharing data throughout the project lifecycle, globally, across disciplines and technical applications. The IFCs are used to assemble a project model in a neutral computer language that describes building project objects and represents information requirements common throughout all industry processes. Recently, another industry-based data standards effort, aecXML (aecXML 2000), has merged with the IAI organization. The goals for the aecXML standards are similar to those for the IFCs, but the approach is based on XML technology, the focus is on a range of project documents and information *other than* the basic product model (i.e., other than the "facility" itself), and the emphasis is on rapid "time to market" for the resulting standards (aecXML 1999).

2.2 Development methodologies within the IAI

To developing the IFCs, IAI Domain Committees carry out projects to analyze the data exchange requirements for specific application areas and design initial data models to satisfy these requirements. IAI data modeling experts then integrate these results into the overall IFC standard. Several commercial AEC/FM software products now support the import and export of the IFCs' current release 2.0 or its predecessor release 1.5.1 (notably CAD packages such as AutoCAD Architectural Desktop) and a new release is under development (currently called 2.x, though this name will change prior to release). (See Froese et al 1998) for more detail of the IAI development methodology)

At this time, the aecXML effort does not have a well-defined development methodology.

2.3 Status of cost-related data within the IAI

Much of the IFCs' focus is on representing the facilities that are being designed and constructed (i.e., the *product model*), but the IFCs' scope also includes project management information such as costs, schedules, work tasks, resources, etc. The Project Management (PM) Domain Group of the IAI North American Chapter (IAI NA PM) has developing portions of the IFCs to support estimating and scheduling processes (Froese at al. 1999). Much of these processes have now been incorporated into the IFCs, and the cost-related portions of these models will be described later in this paper.

No significant cost-related development has yet occurred within aecXML. However, Timberline Software (Timberline 2000), a leading developer of estimating and accounting software for the AEC/FM industries, has published some preliminary estimating XML schemas to the Biztalk.org XML repository (Biztalk 2000) and is expected to submit additional estimating-related XML schemas to aecXML in the near future.

Timberline Software has publicly demonstrated prototype software that imports IFC files to generate construction estimates. To date, no commercial implementations of the cost-related portions of the IFCs have been released.

3 THE ROLE OF DATA STANDARDS FOR SUPPORTING COST ESTIMATING

The development methodology for the IFCs calls for the IAI Domain Committees to analyze specific industry processes and to develop detailed use cases (typical scenarios) of the way in which IFC's could be used to support AEC/FM work practices. This was done, for example, as part of the cost-estimating project. However, these use cases are developed in an ad hoc manner and they are used to support model development only, they do not appear in any form in the actual IFC standards released by the IAI.

In contrast, some data standards being developed for other industries include standardized specifications of specific partner-to-partner business transactions. (e.g., the Partner Interface Processes defined by the RosettaNet effort for the electronics industry, RosettaNete 2000). These "transaction standards" are developed in addition to the data model standards, and they set out the context and criteria under which the data exchange occurs between partners. We suggest that the IAI consider defining AEC/FM transaction standards to supplement the IFC and aecXML data model standards.

Some of the cost-estimating-related transactions involving data exchange are as follows:

– *Estimating unit-cost categories:* If a quantity takeoff task is performed using software outside of an estimating application (e.g., in a CAD package), it may be necessary to first send a list of estimating unit-cost categories from the estimating system to the takeoff system. This allows the takeoff system to associate takeoff quantities with specific assemblies or cost items in the estimating system.

– *Estimating takeoff:* One application capable of generating takeoff quantities, such as a CAD system, can send a quantity takeoff to an estimating system for pricing.

– *Estimate reporting:* Once a cost estimate has been created, the resulting estimated cost can be submitted to various parties or deposited into a project repository. Different situations will call for different levels of detail in the estimate report. Here, there's no requirement that the estimate report contain all of the information necessary to justify or recreate the estimate.

– *Estimate exchange:* Under certain circumstances, an estimator will wish to exchange a complete estimate with another party, archiving estimates, or use it for some of the purpose. This type of estimate exchange should include all of the information associated with creating the estimate.

– *Estimate changes:* A mechanism is required for submitting changes to existing estimates.

– *Estimates as input to other activities:* Information generated during estimating is useful for purposes other than determining final costs. Estimating cost information may feed forward into the budgeting, buyout, cost control, accounting, and other activities. Estimate information may also be used as input to scheduling or resource planning tasks, for example.

4 MODELING COSTS IN STANDARD AEC/FM MODELS

This section explains how costs and related information are represented as object-oriented classes in the current IFC data model. A complete example is extremely helpful in understanding these representation structures, but this requires more space than is available in this paper. An example has been prepared and made available on the Web at: http://www.civil.ubc.ca/~tfroese/pubs/fro01a_ifc_est/

4.1 *Products and kernel IFC objects*

Buildings, and all of their various systems and components, are represented as types of products—that is, as sub-classes of the IfcProduct class. IfcProduct is one of the basic or kernel IFC classes—along with IfcProject, IfcProcess, IfcResource, IfcActor, IfcControl, and several others—which are all sub-classes of the abstract IfcObject and IfcRoot classes.

4.2 *Cost as an attribute of product and other objects*

A simple way of representing costs is to associate a basic cost attribute with any building product or other kernel object. The IFCs release 1.5.1 implemented this representation, which is sufficient for many data exchange purposes. However, any cost value is only meaningful if the context of the cost value is known. For example, the fact that a door in a construction project costs $100 is only useful information if we know to whom this cost applies (is this the cost to subcontractor, to general contractor, to owner, etc.?), what is included in the cost (does it include delivery, installation, taxes, etc.?), whether the cost is a rough estimate or a firm quote, etc.

When cost information is sent from one project participant to another, the context of the cost information is usually implicit in the context of the data exchange (i.e., the information was sent for a specific purpose and both parties know that purpose). However, if cost information is stored in an IFC-based project database, there is no implicit context to that cost information: the cost context *must* be explicitly represented. The approach of associating a simple cost attribute with IFC objects does not capture the context of the cost information, leaving a high potential for misinterpretation of cost information, and this approach is no longer used in the current IFC model.

4.3 *Basic Cost Model*

The current IFC model represents costs using four basic classes (see Figure 1). IfcCost represents a specific cost associated with the project. Each IfcCost must be associated with an IfcCostSchedule

object, which describes the specific context for a collection of IfcCosts. Each IfcCost is typically associated with an IfcObject, which is the representation of the thing in the project that is being costed. Finally, each IfcCost is associated with one or more IfcCostValue objects, which capture the actual monetary value of cost. Each of these classes is described in detail in the following sections.

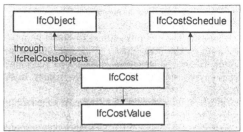

Figure 1. Basic Cost model

4.4 *IfcCostSchedule*

The IfcCostSchedule entity represents an overall cost assessment of something for some purpose. The cost assessment is represented by a collection of IfcCosts and the purpose is represented by attributes that describe the context for the cost schedule. IfcCostSchedule is used to represent estimates, budgets, price lists, etc. Some of its attributes are as follows:
– *Title:*
– *SubmittedBy, ApprovedBy, PreparedBy:* People or organizations.
– *SubmittedOn, UpdateDate:* The dates the schedule was submitted and updated.
– *TotalCost:* The total cost on the schedule.
– CostElements: The collection of IfcCosts.
– *Status:* The current status of a cost schedule.
– *IntendedUse:* Intended use for the cost schedule.
– *Comments:* Assumptions, qualifications, conditions, and other context information.
– *TargetUsers:* The actors for whom the schedule was prepared; can also control access permission.
– *ValidFromDate, ValidToDate:* Time period during which the cost schedule is valid.
– *Controls:* A cost schedule can be associated with any individual objects in a Project through the *IfcReControls* relationship entity.
– *Decomposes / IsDecomposedBy:* Allows IfcCost objects to be nested into hierarchical groupings of costs through the *IfcRelNestsCostSchedules* relationship entity.

4.5 *IfcCost*

The IfcCost entity represents the cost of goods, services, or the execution of work, under certain condi-

tions and context. Some of the attributes used to describe these costs are as follows:
- *Description:* General description.
- *ContextDescription:* Contextual information.
- *ElementCost:* The unit or item cost of a single item of each 'Quantity' (an IfcCostValue).
- *ExtensionCost:* The total cost (an IfcCostValue).
- *CostUse:* Indicates how element calculations are to be carried out (e.g., does this cost provide an element cost only, an extension cost only, an extension of a nested set of element costs, etc.)
- *PreparedOn:* The date the cost is provided.
- *Quantities:* The quantity of the associated items.
- *CostType:* The cost type. Allowable values include: basic cost, overhead cost, profit cost, purchase cost, subcontract cost, contingency cost, etc.
- *CostSchedule:* A reference to the cost schedule to which the cost element belongs.
- *Controls:* Associates this cost with a one or more IfcObjects through the *IfcRelCostsObjects* relationship entity.
- *Decomposes / IsDecomposedBy:* Allows IfcCost objects to be nested into hierarchical groupings of costs through the *IfcRelNestsCosts* relationship entity.

4.6 *IfcCostQuantity*

Items costed through IfcCosts are quantified using IfcCostQuantity, a complex measure made up of the following attributes:
- *BaseQuantity:* The "as measured" quantity resulting from a quantity takeoff or similar operation.
- *FinalQuantity:* Value used for the actual price calculations after modifications are applied.
- *WasteFactor:* Percentage added to account for wasteage.
- *RoundOffBasis, RoundOffIncrement:* The basis and increment size for rounding off quantities, e.g., when materials that must be ordered in specific size lots.

4.7 *IfcRelNestsCosts and IfcRelNestsCostSchedules*

IfcCosts and IfcCostSchedules can be assembled into nested hierarchies (i.e., breakdown structures or trees) through the IfcRelNestsCosts and IfcRelNestsCostSchedules relationship entities. These relationships also add attributes to represent the criteria used for nesting the costs or cost schedules and a conversion rate for correlating nested unit costs or quantities to their parents if different units are used or other conversions are required.

4.8 *IfcCostValue*

The actual monetary amounts associated with IfcCosts are represented by IfcCostValue, which can

be a simple currency value or a more complex assembly of cost values and modifiers. Figure 2 shows the relationships between IfcCostValue and the entities relating to cost modifiers. IfcCostValue includes the following attributes:
- *BaseCostValue:* The cost amount before the application of cost modifiers.
- *UnitCostBasis:* If the cost value is a unit cost, this provides the quantity and unit of measure (e.g., cost per 100 board feet of lumber).
- *CostDate*: The date at which the cost is assessed.
- *Currency:* Currency for this cost that can override the global currency setting for the project.
- *CostType:* The cost type (as described above).
- *FinalCostValue:* Cost amount after modifiers.
- *ModifierAssignments:* Modifiers for this cost (*IfcAssignsCostModifier* relationship entities)..

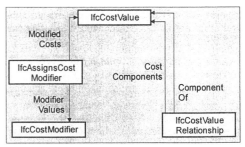

Figure 2. IfcCostValue

4.9 *IfcCostValueRelationship*

An IfcCostValue can represent the aggregated value of several different cost types. For example, a labor unit cost may reflect a base wage, benefits, taxes, etc. IfcCostValueRelationship relates one IfcCostValue object with a group of additional IfcCostValue objects representing cost components. The properties of IfcCostValueRelationship include:
- *CostComponents:* Costs that are components of another cost and from which that cost may be deduced.
- *ComponentOf:* The cost value of which the value being considered is a component.
- *Description:* Description of relationship.

4.10 *IfcAssignsCostModifier*

IfcAssignsCostModifier assigns a defined set of IfcCostModifier objects to a set of IfcCostValue objects. A collection of IfcCostModifier objects can be defined and assigned to many IfcCostValue objects to represent changes in cost of many objects due to a single set of circumstances. Properties include:
- *ModifierValues:* Modifiers (IfcCostModifiers) which may be applied to a cost to change its value.

– *ModifiedCosts:* Costs (IfcCostValues) to which a set of modifiers are applied.

4.11 *IfcCostModifier*

IfcCostModifier represents a modifier that influences a cost. Cost modifiers can potentially be applied on a complex bases (e.g., a percentage markup on hours of overtime labor). Cost modifiers can accommodate a limited set of calculation alternatives (static/running and add/subtract/multiply). Modifiers applied on a more complex basis can be converted within cost applications to these simple bases (e.g., the more complex modifier can be resolved into a fixed currency value modifier). Properties include the following:

– *Purpose:* The purpose for which a cost modifier is applied (e.g., trade discount, bulk purchase rebate, small quantity surcharge, delivery charge, etc.)
– *ModifierValue:* The cost modifier value.
– *ModifierDate:* The date at which a modifier is applicable.
– *CostOperator:* A mathematical operator that determines how the cost modifier is applied to the cost (add, subtract, or multiply)
– *ModifierBasis:* The manner in which cost modifiers are applied to a cost (running or static).

4.12 *Attributes inherited from IfcObject*

Because several of the cost-related IFC objects are subclasses of the IfcObject entity, they inherit many general-purpose attributes, such as the following:

– *DocumentedIn*: Reference to documentation applied to the object.
– *ClassifiedBy:* Reference to a classification, applied to the object.
– *OperatedInProcesses:* Set of Relationships to processes that operated on the object.
– *IsDefinedBy:* Set of Relationships to properties (statically or dynamically defined) that further defines the object.
– *IsActedUpon:* Set of Relationships to actors that act upon the object.
– *IsControlledBy:* Set of Relationships to controls that apply a control to the object.

5 IMPLEMENTATIONS OF COST-RELATED IFC MODELS

An overall use scenario for an IFC implementation of cost estimating is to import project information such as building products or construction activities from IFC data into a cost estimating system to perform cost analysis based on databases of costing assemblies and unit costs. The result of the cost estimates (represented by the IFC classes discussed earlier) can then be exported back to the IFC data source. A software design of such a system needs to consider the following key issues:

Cost Estimating Object Views: for cost estimating, IFC data, once imported, needs to be represented through cost estimating views. The IFC objects are structured in a generic form intended to support all AEC/FM functions. For example, the dimensions of a wall in the IFCs can be represented by a bounding box or a surface extruded along a linear path. This type of data representation is essential for CAD engines, but it doesn't necessarily reflect the typical views for cost estimating. The Cost Estimating Object Views are collections of objects that present the underlying IFC data in a form that is more suited to cost estimators. They also filter out the IFC details that are not needed for estimating functions. It is also desirable that the views can be configured at runtime for different projects and users.

Data Mapping: a cost takeoff engine such as Timberline's Precision (Timberline 2000) works with cost databases. To perform cost takeoff, therefore, the IFC data must be mapped to the assemblies or items in the cost databases. However, the cost databases are typically created at runtime with different schemas, object types and properties for different project types. Hence, object data mappings are required at runtime to map the IFC objects and their attributes to database assemblies, items, and their variables accordingly. When the project IFC data are imported, the IFC instances and their attribute values can then be converted to the variable values of the corresponding assembly or item.

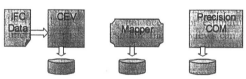

Figure 3. IFC Implementation Architecture

Figure 3 depicts a high level conceptual architecture of the Precision IFC Integrator developed by Timberline Software. The application imports IFC files, generates cost estimates, and exports IFC data files with the cost information. The CEV (Cost Estimating View) is a software component that exposes the functionality for defining object views through software APIs. The object views configured at runtime are stored in a database. The Precision COM is a component that exposes Precision's cost database assemblies and items. It also supports Precision's takeoff functions and generates cost estimates. The Mapper is a software engine that maps the CEV objects to cost database objects (i.e. assemblies and items). The mapping specifications must also be persistent to support data conversions. On top of this component layer, user interfaces can be developed that perform object view configurations, browse project object information, and specify the object mappings.

6 CONCLUSIONS

We have outlined the status and treatment of cost estimating data within standard AEC data models. These data structures are emerging as industry standards. We have briefly introduced a prototype system developed by a leading cost estimating software vendor that uses these data standards for input and output of the estimating process.

We believe that these are positive steps towards the development of industry-wide data exchange capabilities in the area of cost estimating and its relationships with other AEC functions. However, a great deal more experimentation and experience is required with the application and use of these data standards, and we encourage others to become involved in this process.

REFERENCES

aecXML (2000). http://www.aecXML.org/
aecXML (1999). aecXML, A Framework for Electronic Communications for the AEC Industries. White paper available at http://www.aecXML.org/
Froese, T. et al. (1999). "Industry Foundation Classes For Project Management—A Trial Implementation", Electronic Journal of Information Technology in Construction, Vol.4, 1999, pp.17-36. Available online at http://itcon.org/1999/2/
Froese, T. Grobler, F. and Yu, K (1998). "Development of Data Standards for Construction--An IAI Perspective," Proceedings of the CIB W78 conference, Stockholm, Sweden, June 3-5, 1998. KTH, Stockholm, pp.233-244.
Biztalk (2000). http://www.biztalk.org/
RosettaNet (2000). http://www.rosettanet.org/
Timberline (2000). http://www.timberline.com/

Creative Systems in Structural and Construction Engineering, Singh (ed.) © 2001 Balkema, Rotterdam, ISBN 90 5809 161 9

The necessity for accuracy in building investigations: A case study

E.C. Stovner & J.M. Coil
LZA Technology, Tustin, Calif., USA

G.G. Thater & E.E. Velivasakis
LZA Technology, New York, N.Y., USA

ABSTRACT: A large 12-story building in Guam suffered damage during extreme wind and seismic events. Several engineering firms investigated the type and extent of damage and reported grave concern about the building's strength. Through focused structural and building code analysis as presented in the paper, LZA Technology determined that the building had, in fact, primarily been designed in accordance with the building code.

1 INTRODUCTION

Tumon Bay, Guam, is a tourist destination for travelers from many nations. In the Bay area are several mid-rise hotels. A Richter magnitude 8.1 earthquake hit the area in 1993 (Comartin 1995). In December 1996, Super-Typhoon Paka pounded the buildings in the Bay with measured peak wind velocities up to 260 km/hr (161 mph), and high winds lasting as long as 12 hours (Hagemeyer 1997, Naval).

A large 12-story concrete frame hotel suffered damage during both extreme events. Some damage to the structural frame, subsequently repaired, was caused by the earthquake while the building was still under construction. During the Super-Typhoon, the building's non-structural cladding suffered some damage, which has also since been repaired.

Several structural engineering firms investigated the type and extent of damage and its significance to the building's remaining years of service. Many of the investigators recommended major costly retrofits to the building. Through diligent and focused structural engineering and building code analysis, LZA Technology determined that with only relatively minor amounts of supplemental structure and cladding remediation, the building would provide many years of further use.

2 BUILDING DESCRIPTION AND HISTORY

2.1 *Building description*

The building is approximately 43 m (140 feet) in height with a high-rise plan area of about 15 m (50 feet) by 105 m (350 feet). The low-rise portion of the building, extending 2 stories in height, has a plan area of approximately 30 m (100 feet) by 150 m (500 feet).

The building structure consists of cast-in-place concrete beams, columns, and shear walls. The floor slabs consist of 75 mm (3-inch) thick cast-in-place concrete topping slabs and 100 mm (4-inch) deep precast concrete floor planks spanning in the north to south direction. The high-rise columns carry loads down to a mat foundation; the low-rise columns are supported on spread footings. The lateral load-resisting system consists of concrete Special Moment-Resisting Frames (SMRF) from the 3rd floor and above, and a concrete shear wall system below the 3rd floor.

The building façade is primarily constructed of a combination of EIFS (Exterior Insulation and Finish System) cladding, concrete (in the form of exposed structural columns and walls), and a glass curtain wall system.

2.2 *Building history*

The building structure was designed under the requirements of the 1985 Uniform Building Code (UBC). The structural drawings for this particular building were filed in 1989 with the local building department and a building permit was granted subject to the requirements of the 1985 UBC. Construction began soon afterward in late 1989, but was halted after three levels had been completed. Shortly thereafter, the local building department adopted the 1988 UBC for the design and construction of subsequently permitted buildings. In 1996-97, construction of the building recommenced and was completed.

The capacity of the building's structural frame and cladding system to resist the code-required wind and seismic forces had been questioned in engineering reports by other investigators. LZA Technology addressed these questions as follows.

3 CODE AND ENGINEERING ISSUES

3.1 Governing code

During the investigation of this particular building, several engineering firms evaluated the building's structural design conformance with an incorrect or non-governing building code.

When evaluating the duty of care of a building's design and construction team, it is necessary to determine the time period of the building's design and at which date the building design was submitted to the local building department for permit applications. Generally, such permit applications require that the building be designed to the latest adopted building code. Building departments generally adopt a new code edition one to two years after publication. When a building department adopts a new code, projects currently in the design or construction phase are generally not required to conform to the newly adopted code provisions. Moreover, completed structures, rarely, if ever, are required to upgrade to the provisions of a newly adopted building code.

This standard of care was crucial to the standard to which to hold this particular building's design as the code had changed significantly over a relatively short timeframe. The 1988 UBC had revisions in the wind exposure category definitions and additions to certain seismic ductility requirements. In 1995, the Legislature of Guam enacted the requirement that future permitted buildings be designed for Seismic Zone 4 instead of the previous Zone 3 (Twenty-Third 1995).

3.2 Wind exposure classifications

During the engineering investigations of this building, there were disagreements as to which wind Exposure Category was appropriate.

At the time of the original design of the building, the basic design wind speed for Guam was 250 km/hr (155 miles per hour). In determining wind pressures for the design of building frames and cladding, the building code requires consideration of the characteristics of ground surface irregularities near the building site.

Exposure Categories are site-specific in that they are dependent upon the local topography and the extent and height of surrounding structures, trees, or other obstructions to the flow of wind. The general philosophy of the UBC, and wind design in general, is that structures surrounded by a topography having surface irregularities, buildings or trees need not be designed for wind loads as high as those required for structures constructed on flat land with few nearby structures or trees.

The 1985 UBC definitions for Exposure Categories are, as follows:

Exposure B: has terrain which has buildings, forest or surface irregularities 6.10 m (20 feet) or more in height covering at least 20 percent of the area extending 1.609 km (one mile) or more from the site.
Exposure C: represents the most severe exposure and has terrain which is flat and generally open, extending 0.80 km (one half mile) or more from the site.

Upon first impression, an engineer might assume that a tall hotel structure near the shore would be classified as Exposure C, with a generally open and flat terrain at sea. However, upon review of the site-specific conditions at this particular building, Exposure B was the more appropriate Exposure Category, as the building was mainly surrounded by large forest trees and two-, three- and multi-story buildings.

3.3 Seismic ductility

The building's conformance to ductile detailing requirements was called into question during the investigation.

For the design of buildings or other structures, Building Code design seismic loads are typically reduced from the actual earthquake demands that the building may be subjected to, whereas the code design wind loads are the full maximum wind loads applied to the building. The reduction in the seismic demand is allowed for several reasons. The primary reason is the ability of a structure to deform beyond the elastic range and still remain standing after the event, with repairable damage. Designing the structure to remain elastic would be economically unfeasible.

By permitting the design of buildings with reduced seismic loads, the Code requires that the lateral-load resisting system be ductile. Ductility is the ability of a structural member to deform beyond its elastic limit repeatedly before fracturing. The UBC has specific detailing requirements to provide ductility in structures. For example, the close spacing of column ties and beam stirrups provides confinement to the column and beam longitudinal bars and prevents them from buckling out of the member, and failing in a brittle fashion.

A thorough review of this building's Construction Document Addenda and shop drawings revealed conformance to ductile detailing requirements that was not evident from a review of the original design drawings alone.

There were concerns by an engineering firm that the 3rd floor beams and columns were not built in a ductile manner; however, as the shear wall lateral-resisting system was built up to the 3rd floor, the seismic demands on the 3rd floor beams and columns were negligible.

3.4 *Cladding repairs*

Portions of this particular building are comprised of an Exterior Insulation and Finish System (EIFS) cladding attached to a light-gage cold-formed steel stud framing system. Approximately 2,500 m^2 (27,000 square feet) of EIFS was installed on the building in 1996.

On December 17, 1996, Super-Typhoon Paka hit the Island of Guam. It was reported that this particular building experienced localized exterior wall damage that included the detachment of a small number of EIFS cladding panels from the building. Approximately 18.5 m^2 (200 square feet) of panels, less than 1% of the total installed area, reportedly detached from the corners of the building. Measured peak wind velocities ranging between 240 km/hr (150 mph) and 260 km/hr (161 mph) and gusting to 380 km/hr (235 mph) were reported during the Super-Typhoon. In addition, it was reported that the high winds caused by Super-Typhoon Paka lasted for as long as 12 hours, making it an extremely long-duration storm.

The failure mode of the EIFS cladding panels appeared to be a detachment of the EIFS cladding and the exterior sheathing board from the steel studs due to excessive loads caused by high suction or negative pressures generated by the Super-Typhoon. Photos taken after the storm indicated that the interior drywall was still intact, which is consistent with a failure due to negative pressure forces. The photos also indicated that the screws attaching the exterior sheathing board to the studs were still attached to the studs and that the sheathing board pulled away from the screws. To accurately determine the sheathing-screw pull-through capacity, LZA Technology designed mechanical load tests that were performed in a laboratory.

Although an engineering firm had recommended wholesale repair of nearly the entire cladding system, upon a careful review of the actual failure mechanism, continued performance of the cladding system after the storm, load tests and analysis, the cladding system only required selected supplementary anchorage at the corners of the building.

3.5 *Building performance*

During the investigation, an engineering firm recommended that "the building not be occupied until the retrofitting has been completed." However, the building structure had performed admirably under the most severe of environmental tests: Super-Typhoon Paka.

Various meteorological reports indicated maximum wind speeds of 260 km/hr (161 mph) with gusts up to 380 km/hr (235 mph). It was estimated that a billion gallons of water were dumped on the island.

The reported wind velocities tend to indicate that structures on the island of Guam likely experienced winds reasonably close to or perhaps exceeding their design wind loads. In a sense, Super-Typhoon Paka essentially "load tested" the entire structure at wind velocities in the general vicinity of its design requirements. Despite the extremely high winds, there was no reported structural damage to the building in question and only relatively minor non-structural damage to the cladding and glazing. This serves as rather strong evidence that the building, as designed and constructed, could safely withstand the wind design loads it was required to withstand by the governing building code.

4 CONCLUSION

This case study of an investigation of the design and performance of an existing building illustrates the necessity for the investigating engineer to objectively evaluate the evidence at hand. Only through such objective and careful analyses, can a realistic and rational value be placed on the costs of retrofit, if needed.

REFERENCES

Comartin, C.D. (ed) 1995. Guam Earthquake of August 8, 1993 Reconnaissance Report. *Earthquake Spectra*. 11B.
Hagemeyer, R.H. 1997. U.S. Department of Commerce, National Oceanic and Atmospheric Administration, National Weather Service. *Service Assessment Super Typhoon Paka*. Honolulu, Hawaii.
Naval Pacific Meteorology and Oceanography Center West – Joint Typhoon Warning. *A Synopsis of Super Typhoon Paka*.
Twenty-Third Guam Legislature 1995. *Bill No. 417 (LS)*.
Uniform Building Code 1985. Whittier, California: International Conference of Building Officials.
Uniform Building Code 1988. Whittier, California: International Conference of Building Officials.

Creative Systems in Structural and Construction Engineering, Singh (ed.) © 2001 Balkema, Rotterdam, ISBN 90 5809 161 9

A new project cost forecasting model in China

Guangbin Wang
Research Institute of Project Administration and Management, Tongji University, Shanghai, People's Republic of China

ABSTRACT: The effect and process of the project cost control system at the construction process mainly relies on the application of cost forecast modelling techniques and the environment of the project construction. Research of the cost forecast models has concentrated on the time-series analysis techniques with there being less understanding of the environment of the models application and the role of the judgement. In China, as there are often great changes of policy and market during a project implementation period which greatly affect the project cost and lack of history project cost data during the last decade , many cost professionals are not satisfied with the many cost forecast models based on the time-series analysis. A new simplified model is developed in this paper to meet the numerous requests in China for forecasting of the total project cost. On base of current practice of cost estimate and control in China, using "2-8 principle(the Pareto's Law)"of cost management practice, the model systematically takes account of Cost Key Items(CKI) changes and All Cost Items(ACI) changes(the influence of policy and market). A computer program of the model is developed using Oracle RDBMS and proved to be very valuable and effective when coming into application. The model is also can be used in some countries or areas which development of economy is not smooth during some period.

1 INTRODUCTION

Forecasting techniques can be classified into three categories: qualitative forecasting technique, causal forecasting technique and time series analysis forecasting techniques(Nick T. Thomopoulos 1980). Although many methods have been probed in the cost forecast of a construction project, they tend to be only focused on one aspect, for example, time factor, change, or personal judgement. There is no cost forecast model that takes all influential factors into account and can also be applied in the computer. Time series analysis forecasting technique is widely used overseas, but is not most suitable to China due to the lack of scientific and accurate data of construction costs. Another requirement from time series analysis technique is the continuity (stability) of the construction cost during the whole process, because 'there will be no use using forecasting techniques based on statistical data, if outside environment changes greatly' (Makridakis, 1983). Under the existing immature market economy in China, it is impossible not to have the unexpected events occurring during the project process, which will have severe effect on the cost. Among all the factors causing these unexpected events, policy change and unstable economic environment are most influential to construction projects. On the other hand, qualitative forecasting techniques have stringent requirements of the forecaster in terms of technical experience, methods and procedures, which makes its application in the construction practice very difficult due to the limitation of labor resources and time. Another drawback associated with qualitative forecasting techniques is that it can not be applied to computer. So far there is no mathematical model existing for cost forecasting of the total project cost in construction due to many influential factors and their complexity.

It has far reaching importance for construction industry in China to establish a sound cost forecasting model. Cost forecasting should take account of the various factors which have influence on cost in order to achieve the accuracy and practicality, and at the mean time to build up mathematical model for computerized information system. This paper has proposed a cost-forecasting model that is part of the research project ' cost control information system', which has solved the above problems to a great extent.

2 THE PRINCIPLE FOR ESTABLISHING THE COST FORECASTING MODEL

With reference to the existing cost forecasting techniques and combining the functional requirements of a cost control system and the practice in China, establishing the cost-forecasting model shall follow the principles as listed below:

2.1 Simple and easy to use

During the process of construction, the cost forecast is not a recalculation of unfinished work based on norms, drawings, and changes, but an analysis and adjustment of original calculation according to actual changes to the project. If recalculating the cost of unfinished work, it will involve too many labor resources and take too much time so as to render the cost forecasting useless. Therefore, the cost-forecasting model shall be simple and easy to use.

Many researchers point out that the focus of the cost forecasting shall be on items, where there are cost changes compared with original cost plan. The key to cost forecasting is to adjust the original cost plan according to changes (John K. Nabors, 1982) (Neil N.Eldin, 1989). For the item where there is no change, the planned figure shall not be adjusted. All these suggest that cost forecasting is not the recalculation or reestimate of project cost, therefore, it shall be simpler and easier to use when compared with cost estimate of project cost.

2.2 "2-8" Principle (Pareto's Law)

"2-8" Principle (Pareto's Law) states that the large amount of statistic data shows that among the total project cost, the sum of 20% cost items makes up around 80% of the total cost. The other 80% cost items only takes about 20% of the total cost.

This principle has considerable importance for project cost control and is similar to ABC analysis in the management theory. It reminds that one shall always analyze and emphasize on the factors that have rather substantial impact on the research subject. Construction projects are of high degree of complexity and involve many related aspects. Therefore, it is unnecessary and impossible to forecast all the changes to the project in order to achieve 100% accuracy of project cost forecasting. In the process of establishment of the cost model, we give the name 'Key Cost Item' to the items that have significant influence on the project cost - the 20% cost items in "2-8" Principle. And our cost forecasting is mainly focused on the Key Cost Items.

Key Cost Items vary from project to project. The cost-forecasting model allows the user to analyze and define Key Cost Items according to the project and their experience. During the development process of the project, it is likely that Key Cost Items, which have important effect on unfinished work, can change. As a result the computer program based cost model allows the user to adjust the Key Cost Items according to the reality.

The analysis and determination of Key Cost Items is a very important task in the cost forecasting. In practice, statistics can be used combined with personal experience. First, every single cost item will be calculated its percentage of the total project cost, then they will be sorted and accumulated according to the magnitude of their respective percentage. The items will be selected as key cost item when their accumulated sum approximates 80% of the total project cost. In general, the number of key cost items selected through this method is about 20% of the total cost items. Besides the mathematical method by utilizing computer program, personal judgement of the cost forecaster based on their experience shall also play an important role in the selection of key cost items. Although some cost items do not involve large sum of money and are normally not within 20% according to mathematical method, they might vary tremendously during the whole process due to uncertain probability so as to cause change to the project cost. Such cost items shall be considered as key cost items at the definition stage. For example Shanghai is characteristic of its soft soil condition, for a large size building with high standard, the excavation shoring cost item does not occupy a relative high percentage of the total project cost, but its cost can vary a lot during the construction thus causing changes to total project cost. Therefore, it normally is defined as key cost item. Along the path of a project, the actual costs of certain key cost items become available, the computerized system allows the user to reanalyze and redefine the key cost items.

2.3 Suitable for China's current practice

The establishment of the cost-forecasting model shall be based on the current practice in construction industry in China. China is still at the stage of improving its socialism market economy step by step to achieve its perfection, the establishment of the cost-forecasting model shall take full account of the two main characteristics of the current economical development in China.

(1)Imperfection of market economy

Investment management in construction is controlled by central government policy and is very much influenced by the macro economy. The development of the whole economy is unbalanced and fluctuating, sometimes there maybe economy slump. At the period when economy is speeding, the government needs to enforce policy and regulations to enhance the control and adjustment of the macro economy. Every singly policy from the government can influence the investment and result in the changes in project cost (sometimes tremendous changes). When building up the cost-forecasting model, all these influences and factors shall be taken into account, especially their influences on the key cost items.

(2) Lack of substantial reliable cost data

It had been a long time that china was under the central planning economy and did not accumulate enough cost data in the past. On the other hand, due to the special influence from government's administrative body and preferences laid down by the government policy, the then cost data for the construction projects do not reflect the reality. For example, a high rise office building completed in 1995 in Pudong, Shanghai, is of high standard (all imported marble façade, high standard finishes with part of them equal to 4 star hotel standard, all main equipment are from oversea) and the unit price for design and construction is about RMB5600/ m^2. The main reason for this relatively low unit price is because all the foreign currency used on this project was bought according to government's standard exchange rate (1 US$ = 5.8 RMB) due to the special administrative power associated with this project. The cost data obtained from this project does not reflect the true investment, when using such kind of cost data to analyze and forecast the new project, it is necessary to readjust the data before their application. Establishment of the project cost-forecasting model shall fully recognize such current limitation in China by basing the system on eliminating unnecessary analysis of statistical data from the past and at the same time controlling the key factors of the forecasting in order to achieve the accuracy of cost forecasting.

3 ESTABLISHING THE COST FORECASTING MODEL

3.1 The factors for consideration

Following the above principles, the main factors to be considered when establishing the cost forecasting models include following items:
(1) accumulate cost of actually finished work.
(2) change of quantities of key cost items, which is caused by change order and quantities in the construction different from the contract document.
(3) change of standard and specification of key cost items, which is caused by changes to the specification, standard initiated by client or designer. These changes lead to higher unit cost of key cost items.
(4) Changes of market condition. There are many market changes that can cause change on investment, such as inflation, fees requested from government administrative institution or ad-

justment of labor cost and material cost from government, change of exchange rate in the international market. All the changes of market condition have effects on all the unfinished work.

The above (2) (3) only takes care of changes that affect part of the unfinished work, while the changes of market condition is panoramic.

3.2 The cost forecasting model

The cost forecasting of the total project cost is not the recalculation of the total project cost, instead it shall consider the two types of cost additions based on the original cost plan: the first being the changes on the project cost caused by factors that affect part of the project, the second being the changes on the project cost caused by factors that affect the whole project. In another word, the forecasted project cost consists of three components: the latest total project cost and two cost additions. The model is expressed as below:

$$C_t = C_0 + \Delta KC + (C_0 - AC_t) \times \sum_{j=1}^{3} \alpha_j\% + \sum_{j=1}^{3} A_j + B$$

(1)

Where:

$$\Delta KC = \sum_{i=1}^{k} [(FQ_i \times FP_i) - (PQ_i \times PP_i)] \qquad (2)$$

Where:

C_t: the forecasted total project cost at "t" time point

AC_t: the accumulated actual cost before "t" time point

ΔKC: cost additions caused by changes of key cost items (changes of quantity or unit price), the cost additions can be positive or negative, the value can be calculated through formula (2)

FQ_i: the forecasted quantity of number i key cost item at "t" time point

FP_i: the forecasted unit price of number i key cost item at "t" time point

PQ_i: the planned quantity of number i key cost item

PP_i: the planned unit price of number i key cost item

K: the total number of key cost items to be adjusted

C_0: the latest total project cost

$\alpha_1\%$: the inflation rate of market price (affecting all unfinished work after "t" time point)

A_1: the other adjustment caused by change of market price (lump sum figure)

$\alpha_2\%$: the change rate of labor cost (affecting all unfinished work after "t" time point)

A_2: the other adjustment caused by change of labor cost (lump sum figure)

α_3%: the change of fees (affecting all unfinished work after "t" time point)

A_3: the other adjustment caused by change of fees (lump sum figure)

B: the adjustment to the total project cost caused by other factors (lump sum figure)

In the above forecasting model, Formula (2) is the adjustment to part of the key cost items during forecasting of the total project cost, which reflects the cost additions (it can be positive or negative) to the total project cost caused by key cost items due to the changes of their quantities or specifications (unit price). It is imperative to define the key cost items before using the cost-forecasting model for the first time.

In the formula (1),

$$(C_0 - A C_t) \times \sum_{j=1}^{3} \alpha_j\% + \sum_{j=1}^{3} A_j$$

is the cost addition to the total project cost caused by changes of inflation rate, labor cost and fees, which is the adjustment to all the unfinished work. They are referred to as All Cost Items in this paper.

In formula (1), B stands for the influence to the total project cost caused by all the factors other than those mentioned above. It is a lump sum figure.

In formula (1), one important task is to decide C_0, which is the base figure in the cost forecasting model. During the whole project life, there are many versions of estimate of the total project cost, such as budget cost estimate, conceptual cost estimate, cost plan, bids, contract price and actual cost. In the cost forecasting model, C_0 shall be decided as the latest and most detailed version of the estimated total project cost to forecast the future project cost.

4 CONCLUSIONS

(1)The cost forecasting model introduced above has relatively satisfied the project cost forecasting principle. The application of the cost forecasting model does not require the use of large amount of historical cost data, instead it is based on the control and adjustment of the changes to the project during the project life and focus on the key factors (key cost items) and market conditions. The cost model meets the needs of the current practice in China and bears the characteristics such as being simple and easy to use. The cost forecasting has the standard formulas to follow and can be applied to computer program. According to the programming software developed by using Oracle database based on Windows NT or Unix operation system, the results are reliable and promising.

(2)In terms of forecasting techniques, the cost forecasting model introduced above is the combination of qualitative forecasting technique and causal forecasting technique, and the combination of qualitative and quantitative technique as well. Like all the other qualitative forecasting techniques, the cost-forecasting model requires tremendous work from the forecaster. The definitions of every single key cost item, adjustment of their quantities (KIQ_1), changes of their unit prices (KIP_1), the figure of every single α_j% and a_j, all requires the analysis based on personal judgement before they are put into the computer program to calculate the forecasted project cost. Here, every parameter directly affects the accuracy of the forecasting.

The process of deciding these parameters is also a process of analyzing and exercising judgment, where some qualitative methods can be utilized, for example *expert score method*. It is necessary to avoid the limitation and bias of personal judgement through team working. An axiom existing in the forecasting area: the collective judgements is more accurate and reliable than individual judgment (Lcwrlic E. McMullan, 1996).

(3)The cost forecasting model introduced above avoids the limitations associated with time series forecasting technique and qualitative technique respectively. Time series forecasting technique allows for the time factor, but not take into account of changes during the whole project life, while qualitative technique is not suitable for computer analysis and has difficulty in its application. The establishment of the aforesaid cost-forecasting model creates a new way of thinking in terms of methods and means. The cost forecasting model has been put into the application in several projects, such as the Dongnan Garden, a high rise development in Pudong, Shanghai (two connected 30- story high standard residential apartments with total investment of 100 million US$) and Shanghai Yangzi Estuary Project(Total invest is about 1.9 billion US$). Through the discussion with the cost control practitioners, the feedback for the cost-forecasting model is that it is easy to use and data are reliable and it possess the relative good application value. With the development and perfection of market economy, when combining with other forecasting techniques this cost-forecasting model will undoubtedly have great value.

Until now there is no forecasting model for the total project cost that is mature and can be applied everywhere. The forecasting itself can not be 100% accurate. Based on the detailed analysis of each forecasting technique, current practice of construction and the development of forecasting techniques, it is extremely important in terms of both theory and practice to continually probe and develop new techniques and new method of cost forecasting suitable for construction.

REFERENCES

Nick T. Thomopoulos, Applied forecasting methods, Prentice-Hall Inc., Englewood Cliffs, N.J, 1980

Shizhao Ding, Project Management, Shanghai Kuaibida Software Publishing Inc., Shanghai, 1991

Lewslie E. McMullan, Cost forecasting ---- Beyond the crystal ball, 1996 TRANSACTIONS of AACE INTERNATIONAL, 1996

John K. Nabors, Forecasting final project cost with econometric techniques, AACE TRANSACTIONS, 1982

Neil N. Eldin, Cost control system for PMT use, 1989 AACE TRANSACTIONS, 1989

Ayman H. AL-Momani, Construction cost predication for public school building in Jordan, Construction Management and Economics, Vol. 14, Num. 4, July 1996

Leslie E. McMullan, Cost control – the tricks and traps, AACE Transactions, 1991

Hashem Al-tabai and James E. Diekmann, Judgemental forecasting in construction projects, Construction Management and Economics, (1992) 10, 19-30, 1992

Malik Ranasinghe, Total project cost: a simplified model for decision makers, Construction Management and Economics, (1996) 14, 497-505, 1996

11 Planning and knowledge management

Creative Systems in Structural and Construction Engineering, Singh (ed.) © 2001 Balkema, Rotterdam, ISBN 90 5809 161 9

The impact of out of sequence work on linear construction projects

Alan D. Russell, Asad Udaipurwala, Marina Dimitrijevic & William Wong
University of British Columbia, Vancouver, B.C., Canada

ABSTRACT: Examined in this paper is the modeling of linear construction projects and features desired of a planning and control tool for such projects. An abstracted example derived from actual projects is used to highlight the challenges associated with projects that should be executed in an ordered work location sequence because of constraints imposed by the contract or methods adopted and/or the need to capture efficiencies. The usefulness of various visual representations of as-planned and as-built information within an integrated environment is demonstrated.

1 INTRODUCTION

Rapid transit projects of various forms (elevated, at grade and subsurface) and significant length are being pursued in various countries as a way of coping with ever increasing congestion, capacity problems, and environmental concerns. Typically, these projects are constructed in very busy urban corridors. The need to capture economies of scale in order to control costs coupled with the need to localize the impact of the construction process on traffic dictate that such projects be built in an orderly location sequence. Direct benefits of doing so include minimizing the cost of equipment spreads in part through the sharing of key resources, minimizing construction travel time, minimizing the length of the construction corridor that is active, maximizing the potential for learning and maximizing the cycling of key construction resources. However, substantial difficulties can be encountered in the form of delayed property acquisitions, multiple regulatory jurisdictions, changing geotechnical conditions, overhead and underground utilities, work day and work week restrictions, and traffic control, to name a few. These difficulties make an orderly location sequence hard to maintain. Their impacts on a construction plan and schedule can be disastrous, and can lead to significant delays and substantially increased costs. These difficulties are often compounded by the fast-track nature of such projects. Increasingly, they are being procured using some sort of public-private partnership arrangement, such as BOT or design-build.

This paper examines the modeling of such projects along with the representation of their as-built job history within a single integrated environment. The importance of conveying the intent of the original construction plan and the significance of not being able to follow it, in visual terms that are readily understood by the client and other authorities, is discussed. An example that is abstracted from lessons learned on past projects is used to illustrate features of such projects as well as a modeling environment that is well attuned to them. It should be noted that actual projects are some 10 to 50 times larger than the example provided, which reinforces the need for the kind of environment described.

2 PROJECT SCENARIO

The project consists of the construction of a 16 pier, 15 span segment of an elevated rapid transit guide way under the terms of a design build contract. Contract award is as of 03 January 2000 and completion must be achieved by 31 August 2000. This time window was set by the client, and is seen as being very aggressive by the contractor. The client organization is responsible for ensuring access to all aspects of the site, and as per the contract, the whole construction corridor is to be delivered effective 03 January 2000. The construction corridor traverses a number of properties, some of which are publicly owned and others that are privately held. The contract has been awarded on the basis of a preliminary design submitted by the contractor. Features of the design include drilled caissons for the foundation, cast in place piers, and a segmentally constructed guideway using a launching truss. Final design of the foundations will depend on timely feedback from the site. The launching truss is to be acquired from another project, and modified to suit the project at hand. Immediately following this project, the truss is to be used on another project. The contractor will incur significant penalties, both for late delivery of the project, as well as for late delivery of the truss. The contractor intends to construct their own precast yard, and has determined that a total of 5 casting lines will be sufficient. Each span consists of 12 precast segments. It is noted that the construction strategy adopted by the contractor is very unforgiving, in that an ordered sequence is essential for the precast operation to be in sync with field work, and because the launching truss must start at one end and move to the other end. The contractor is concerned that the client may not fully comprehend the significance of the methods adopted, and wants to convey as clearly as possible the construction strategy along with the consequences of having to work out of sequence.

3 THE AS-PLANNED SCHEDULE

Figure 1(a) provides a list of activities used to model the project. As used herein, an activity corresponds to a planning structure that allows an activity to have multiple locations assigned to it, an order of execution, variable production rates, internal logic, variable crewing assignments, continuity constraints, and so forth. The set of structures used represents an extension of those originally described by Russell and Wong (1993). In drafting the schedule, consideration has been given to learning curve effects, extra time needed to set up at the first work location, work continuity requirements where appropriate, and design and procurement activities that are reflective of a design-build project. Use has also been made of lag relationships to reflect curing times, and overlapping of work of the same activity at different locations. Dedicated crews have been assumed for major operations (foundations, piers, guideway), and continuity constraints have been imposed on activities G02000 – Pier construction, G02200 – Set-up launching truss, and G02600 – Erect guideway. Figure 1(b) shows a partial bar chart for the repetitive work for locations Pier 1 and Span 1. Note that precast work is critical and that the hammock activity G00700 spans the foundation work. One difficulty with a bar chart representation is that it does not readily convey how well production rates have been matched, learning curve effects, opportunities for work continuity, or how best to modify the construction strategy.

Figure 2(a) is an early start linear planning chart representation of the same project, and clearly shows the intent of the contractor. The vertical axis corresponds to locations (of interest are the pier locations P01 through P16 and span locations S01 through to S15). Some of the activities in Figure 1(a) have been filtered out to provide clarity. Use has been made of the hammock activity for representing foundation work. Important observations are as follows: access to all pier locations is shown effective 03 January 2000; the project completes on 30 August 2000; the critical path flows through precast work; learning curve effects are shown by way of the changing durations at each location; the use of 5 casting bays is clearly illustrated; the duration for the precasting operation for each span includes the production of 12 segments and the curing time essential before they can be erected; the distance between the end of curing and erection shows the additional time required for storage of the segments; and, work continuity is shown for both the pier construction and guideway erection. Other observations are that only one set of forms is required for the piers, field work is not critical, the possibility exists of reducing the number of precasting forms while still not lengthening the project, the storage yard capacity need not exceed the number of segments for 10 spans, and, the contractor has done a reasonable job of matching the production rates. Finally, the requirement for both on site and off site work to be carried out in an ordered location sequence is obvious.

Figure 2(b) depicts the late start linear planning chart for the project. This figure demonstrates two things. First, the integrity of the methods statement is maintained under the late date scenario, providing some confidence in the quality of the schedule (see Russell and Udaipurwala (2000)). Second, a fair amount of contingency time (approximately one month) has been built into the schedule to deal with the uncertainties surrounding the foundation work. Note the latest time for delivery of the site. An interesting question is – who owns the float?

4 THE AS-BUILT SCHEDULE

Figure 2(c) depicts actual performance as of 03 March 2000 along with results from the first step in the updating process, which results in a new completion date of 29 September 2000. The client has been unsuccessful in delivering access to the pier locations in both a timely manner and in the sequence required. The first step in the updating process involves using daily status data recorded in the as-built view (described later) to derive actual dates through a batch updating procedure, followed by re-execution of the schedule. When re-computing the schedule, an attempt is made to adhere to the original location sequence assigned to each activity structure. In fact, given that work has started on locations P05 to P07, this work is likely to be completed before relocating the drilling rig to location P01, because of the high mobilization costs involved. As a consequence, foundation work at location P01 could be delayed further. Modification of location sequences would involve a second step in the updating process. Although not shown, for the scenario considered, the precast yard work and precasting is still on schedule.

What is not possible to show in a schedule representation is the uncertainty faced by the contractor by the seemingly random delivery of access. While this process is underway, it is not clear whether one should delay start of the work in the hope that work can start at the most important locations, or commence work out of location sequence, and then relocate forces and equipment

spreads later. The difficulty in real life is that often the client cannot provide date certain for each location, because property acquisitions can be beyond their control. Nevertheless, they must recognize the impact on the contractor, rather than take the position that work at one location can be substituted for another with no consequences for the methods adopted or the project delivery date.

Figure 3 is a comparison linear planning chart showing the original plan (figure 2(a)) versus the updated schedule (figure 2(c)) for the geotechnical investigation (G00800), pier construction (G02000), set up launching truss (G02200) and erect guide way (G02600) activities. In terms of reading the figure,

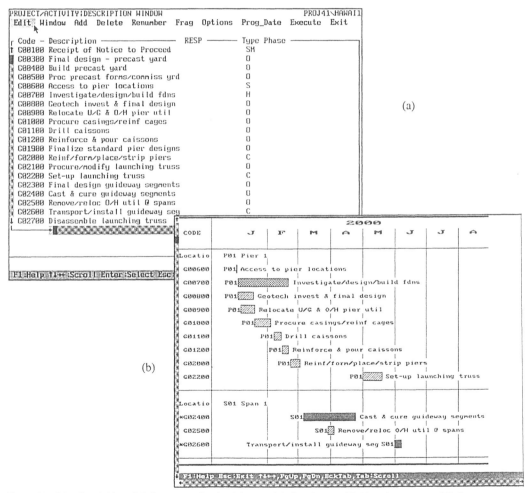

Figure 1(a). List of Activities – O indicates an ordered activity –work is done in a specified location sequence; C indicates a continuous activity – once work starts, it proceeds continuously as it moves from location to location in an ordered way; S indicates a shadow activity – as soon as predecessor work is done at a location, the activity can be carried out; and, H indicates a hammock activity.

1(b). Bar chart of activities, location by location. The hammock activity (G00700) is used to span the foundation work. The hammock is shown on the linear planning charts in order to simplify the representation.

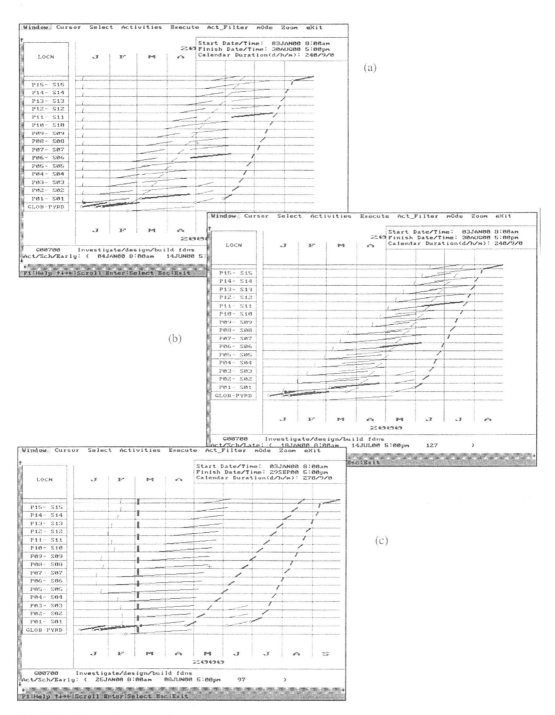

Figure 2(a). EST linear planning chart. Only selected activities are shown for clarity. The critical path goes through the precast work.

    2(b). LST linear planning chart. When compared with the EST linear planning chart, it shows that a fair amount of contingency time has been built into the schedule to deal with the uncertainties surrounding the foundation work.

    2(c). The schedule as updated effective 03 March 2000. Note the out of sequence work for the hammock activity.

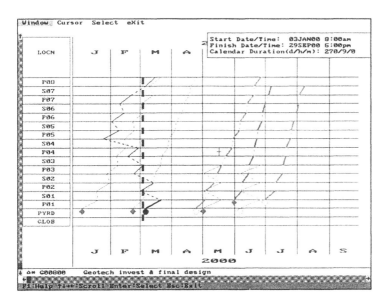

Figure 3. Comparison linear planning chart showing the original plan versus the updated schedule for activities G00800, G02000, G02200 and G02600. Figure key: diamonds correspond to the original schedule; circles correspond to the updated schedule.

diamonds correspond to the original schedule and circles correspond to the updated schedule. Note the significant shifts involved for activities G02000, G02200 and G02600. Left as is, significant additional costs are involved for storage for the precasting operation, for rental of the launching truss, and for penalties associated with late delivery of both the project and the truss (the latter assumes an unsympathetic hearing from the client and other third parties). If the capacity of the storage yard is reached and additional storage is required, then even more costs would be incurred through rental charges for space and for the double handling of the precast segments. Further, the more handling that has to be done, the more potential there is for damage to the precast sections. Of even greater concern, if no adjustment to the construction strategy is made, is the likelihood of further delays. This is because the entire contingency has been used up, prior to carrying out the work that involves the most uncertainty – i.e. utility relocation and caisson drilling. Strategies that could be pursued to speed up construction are suggested through a close examination of Figure 2(c) (another benefit of using a linear planning representation). For example, two sets of forms along with two crews could be used for pier work, thereby allowing an earlier start of the guideway erection while still ensuring work continuity.

To assist in tracking project performance, it is very useful to integrate an as-built view of a project within the scheduling environment. Figures 4(a) and 4(b) depict the recording of daily status for each activity, along with the assignment of problems en-

countered and their consequences in terms of time lost on the activity and/or manhours lost. Lower case letters in 4(a) correspond to the planned status as derived from the schedule; upper case letters correspond to actual status (P – postponed; S – started; O- ongoing; I – idle; F – finished; D- started and finished the same day). An exclamation mark signifies that one or more problems were recorded against the activity's status on that day. Having recorded the status of each activity, the first step in schedule updating involves reading the daily site files, requesting estimates of time to complete for activities in progress, batch entering them into the activity files and re-computing the schedule.

Figure 4(c) shows a three-dimensional depiction of an activity's daily status across its location set. It is one aspect of data mining supported by the system. For this example, a problem code profile consisting of only No access has been used to explain the reason for delays in the activity, Access to pier locations. The vertical axis depicts the activity's status on a day-by-day, location-by-location basis. Note the significant postponement of work because the client did not provide access. While only one problem code is shown, multiple problem codes can be through colour coding. Such graphs assist in detecting patterns of problems, which helps in identifying remedial actions or explaining performance.

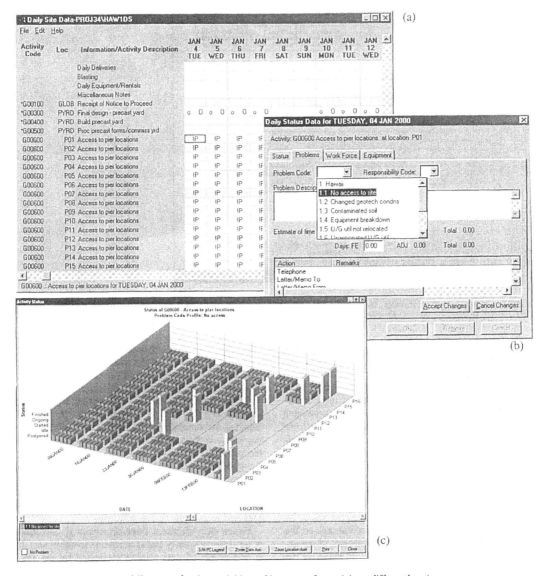

Figure 4(a). As-built View - daily status of various activities and instances of an activity at different locations.
    4(b). As-built View - assigning a problem code to an activity's status on a given day.
    4(c). Three-dimensional depiction of an activity's daily status across its location set.

5 REFERENCES

Russell, A. D. & Wong, W. C. M. 1993. A New Generation of Planning Structures. Journal of Construction Engineering and Management, ASCE, 119:2; 196-214.

Russell, A. D. & Udaipurwala, A. 2000. Assessing the Quality of a Construction Schedule. ASCE Construction Congress VI, Orlando Florida, 17-21 February, 2000; 928-937.

Creative Systems in Structural and Construction Engineering, Singh (ed.) © 2001 Balkema, Rotterdam, ISBN 90 5809 161 9

Decision rules for site layout planning

A.W.T. Leung
Division of Building Science and Technology, City University of Hong Kong, People's Republic of China

C.M. Tam & T.K.L. Tong
Department of Building and Construction, City University of Hong Kong, People's Republic of China

ABSTRACT: Site Layout Planning (SLP) is one of the critical planning processes in site production planning. However, systematic research in this area is difficult due to the unique site conditions of each building project. Hence, construction managers usually apply their experience in formulating site layouts. The interrelation between allocations of temporary site facilities, the construction schedule and construction methods throughout the construction process further augments the difficulty and complexity in developing any SLP models in a scientific approach. As there is no systematic approach for the overall SLP, construction managers learn SLP through years of practical experience or trials-and-errors. It leads to a gap between the industrial practices and theoretical approaches. The objective of this paper is to build up a heuristic model on SLP by extracting knowledge from well-experienced construction managers in Hong Kong and the quantitative and qualitative results obtained from interviews with construction managers form a database for analyses. A set of decision rules for public housing projects in optimizing positioning of site facilities has been developed. The rules are classified into two levels. The first deals with the criteria in determining the positions of site temporary facilities. The second determines the priority in planning the temporary facilities. This pilot study provides a rational base for further investigation on the development of computer-based decision models for SLP.

1 INTRODUCTION

Site Layout Planning (SLP) is one of the critical planning processes in construction planning. However, to conduct systematic analysis in the area is difficult due to the fact that SLP is governed by the complicated multi-interrelationship between planning constraints. The site conditions including the topography layout, building layout and adjacent environment of construction sites are unique. Consequently, it results in great variations in formulating strategies and approaches in SLP. The allocation of temporary facilities is interrelated to the construction processes that vary throughout the construction period. Optimization of SLP, a non-linear problem in nature, in a scientific approach is difficult. Apparently, there are little trace of systematic approach and logical quantitative frameworks for SLP. Construction managers learn SLP through years of practical experience. It leads to a gap between the industrial practices and construction education.

The objective of this research is to build up a series of decision rules of SLP based on the results of interviews with experienced construction managers. In this study, public housing projects, which share about 50% of the residential housing market in Hong Kong, are focused for the ease of comparison.

2 DEFINITION OF SITE LAYOUT PLANNING

Cheng and O'Connor (1996) stated that many temporary facilities were included in a construction site to support construction operations. Thus, an efficient temporary site layout is to locate facilities such as job offices, warehouses, workshops, etc. to optimize workforce traveling, material and equipment handling, travel distance, traffic interference, and cater for the need of expansion and relocation.

Similar definition is made by Calvert, Bailey and Coles (1995) that site layout plan is a plan that should be drawn up showing the relative location of facilities, accommodation and plant, with the overall intention of providing the best conditions for optimum economy, continuity and safety during building operations.

In view of the above definitions, SLP is

basically defined as a site space allocation for materials storage, working areas, site accommodations, plant positions and general circulation areas within a site. These facilities are integrated to form a temporary production line for the construction works, which usually remain on-site within the project duration. As a result, optimal resources, time, costs and safety can be achieved through this planning exercise.

In short, site layout entails three main points as defined by Tommelein, et al. (1991). They are:

(1) identifying facilities that are temporarily needed to support construction operations in a project, but that do not form a part of the finished structure;
(2) determining the size and shape of these facilities and;
(3) positioning them within the boundaries of available on-site or remote areas.

3   AUTOMATIC SLP SYSTEMS

Cheng and O'Connor (1996) developed an automatic site layout system called "ArcSite" to assist planners in assessing the site layout design. ArcSite is a geographic information system integrated with database management systems. This system is a new tool assisting planners in identifying suitable areas to locate temporary facilities on site. As a result, construction conflicts can be minimized whereas the production efficiency can be improved.

Although ArcSite can be applied to most construction sites, some drawbacks need to be overcome. Since ArcSite highly relies on a database system, numerous data are required for expanding the system to cover all types of sites. This is a clear limitation of its application. On the other hand, the system does not consider the location of construction equipment and materials on sites. However, the integration of site equipment and materials transport routing is one of the important considerations in SLP.

Besides the automatic site layout system, Tommelein, et al. (1991) developed a model called "SightPlan". This model provides an expert strategy and modeling systems for planning layout of construction sites by using artificial intelligence programming techniques. In addition, it provides planners with suggestions on how to make improvements upon their layout strategies. Under SightPlan, an artificial intelligence system is used to provide modeling solutions to site layout. That means a symbiotic human-machine system is utilized in planning. It aims to overcome human cognitive limitations in affecting solution interpretation.

In the application of SightPlan, it can create a partial arrangement, including the site boundaries as well as the objects with defined positions on site. The system can then identify the remaining available space, identify the sub-areas of the arrangement and include all areas that need to be positioned in the arrangement. SightPlan is efficient in the arrangement of available site space. This provides an alternative tool to design site layout. However, as regards the technological issue of the system, the proposed plan is not easy to be adjusted and revised when site conditions have been changed.

4   IMPLICATIONS OF THE AUTOMATIC SYSTEMS

Having reviewed the recent development in SLP systems, it is recognized that computer technology has not yet fully substituted human brains. Firstly, it is difficult to include all the data in precise details to form a knowledge-based system. Secondly, a computer system cannot provide a flexible tool for SLP unless it is updated regularly regardless of the cost implications.

The strength of human perception is its ability to recognize the various dimensions and geometry, for which human brains are more flexible than a computer while that of a computer system is its objectivity and ability to compute a large quantity of data at fast speed. This study aims at developing a set of decision rules for guiding construction practitioners in SLP while allowing flexibility to consider various unique site situations.

5   RESEARCH METHODOLOGY

In this study, qualitative approach was adopted to investigate site layout planners' strategies. With reference to Naoum's model (1998), we used a qualitative approach in collecting opinions with emphasis on meanings, experiences, description and others. The data collected could be quantified to some extent (Coolican 1993) for subsequent easy analysis.

An abstract of the questionnaire adopted in this attitudinal research is shown in Table 1. The example shows that attitudinal research allows respondents to rank the importance of different factors.

Table 1: Abstracts of the questionnaire on SLP

Equipment	Ranking
Tower Crane	□ access
	□ site condition
	□ suit working radius
	□ within operator's view
	□ overlapping between cranes
	□ others:

Table 2 Evaluation scheme for allocating the positions of tower cranes

Project	Access		Site condition		Suit working radius		Within operator's view		Overlapping between two cranes		Others	
	Priority	Mark	Priority	Mark	Priority	Mark	Priority	Mark	Priority	Mark	Priority	Mark
1	2	5	0	0	1	6	0	0	3	4	0	0
2	1	6	0	0	2	5	0	0	3	4	0	0
3	1	6	0	0	2	5	0	0	0	0	0	0
4	3	4	1	6	2	5	5	2	4	3	6	1

6 QUESTIONNAIRE SURVEY AND SITE INTERVIEWS

Twelve face-to-face interviews have been carried out with a structured questionnaire to solicit experienced construction managers' views on projects that they have come across. In order to obtain, build up and compare data for deducing the general rules, standardized public housing projects have been focused for the ease of comparison.

The main part of the survey is an evaluation scheme, which examines the priority of the factors in allocating the temporary facilities.

Table 2 demonstrates an example of the evaluation scheme. The data shows the priorities of the considerations for allocating the position of tower cranes. The first column shows the code number of the sample projects. The subsequent columns represent the six considerations. Construction managers were requested to rank the considerations. The highest score to a consideration is assigned with the highest priority. For a facility with six considerations, the scoring scale is represented as: (priority, score) - (0,0), (1,6), (2,5), (3,4), (4,3), (5,2) and (6,1).

The objective of this paper is to build up a heuristic model on SLP from knowledge of the construction managers. It is presumed that the attitude survey reflects what the construction managers have considered during the SLP process. The planning process for SLP is complicated with the interrelationship between the temporary facilities. The key facilities listed in Table 3 are all interrelated, which indicates that a change in certain facility would lead to a change in other key facilities.

With a large number of these interrelated facilities, human brains will have great constraints in dealing with such large number of considerations at the same time. However, it is presumed that one can solve a certain question, one at a time, as that set in the questionnaire.

7 DECISION RULES FOR SLP

Referring to the considerations for allocating tower cranes, the factor "suiting working radius" and "close to the access road" are the most important factors considered by construction managers.

Checking the working radius of a tower crane is the first priority in planning the position of a tower crane in relation to facilities such as storage areas for precast units and reinforcement, bending yard and formwork yard. On the other hand, accessibility is closely related to the other key facilities namely, material hoist, concrete pump, loading point for concreting, bending yard, car park and refuse chute. The survey and interviews deduce that "position and size of the tower crane" and "space available for access road" would be the most important parameters to control the relative positions of other facilities.

SLP is a non-linear planning problem for which planners cannot deduce a solution from a polynomial function or functions. There exists a higher degree of relationships between different key elements. Hence, the proposed rule-based approach can help solve this complicated problem.

The next step is to integrate the relationships of the key elements and to build up a model based upon the logical framework developed from the survey of construction managers.

Following the concept of decisions with multiple objectives, Richard A. Epstein (1977) proposed a process of strategic selection by weighted statistical logic. There are four steps to generate a probable best decision:

(1) Determination and evaluation of the individual factors pertinent to the final decision.
(2) Expression of the alternate courses of action in terms of the weighting assigned to each of the individual pertinent factors.
(3) Derivation of a composite ranking, representing the overall relative merit of each alternate course of action.
(4) Recognition and designation of the highest-ranking alternative as the optimum course of action.

Equation 1 is the general equation describing a decision based on a weighted score derivation. The results of the survey identify the weighting factors for setting up an evaluation system. This is to help the planners calculate the relative score of each alternative of SLP.

Table 3: Score Summary of allocating the Temporary Facilities

Key Facility	Prority Score
Tower Crane	4.96
Material Hoist	4.06
Passenger Hoist	2.48
Concrete Pump	3.08
Batching Plant	2.82
Refuse Chut	2.97
Site Office	4.23
Plant Workshop	3.32
Toilet	2.32
Loading Point for concreting	3.95
Bending Yard	4.53
Formwork Yard	3.74
Site Laboratory	3.12
Car Parks	2.00
Canteen	1.56
Fencing and Hoarding	2.65
Entrance	3.67
Washing Wheel Bay	2.79
Scaffolding	1.98
Working Area for sub-contractor	2.26

8 APPLICATION OF THE EVALUATION SYSTEM

The following are the steps for applying the SLP model for a project:

(1) Define the constraints of the site conditions, which include the positions and dimensions of the buildings, and other permanent facilities.

(2) Start with the most critical facility, i.e. the tower crane, as shown in Table 3 as revealed from the survey. Determine the number, positions and capacity of the tower crane's coverage. Table 4 shows the parameter of the evaluation system for first stage of the SLP model: tower crane.

(3) The second stage would be the allocation of key facilities associated with that of access road. The same evaluation will be used to verify the site layout plan.

(4) Go back to step 2 to review and re-plan the site layout so that the overall score on both stages increases at the end of each cycle of calculation. Table 5 shows the calculation of score for the SLP.

(5) The final stage would be the remaining key elements such as the site offices and laboratory which positions are usually governed by contractual requirements. The overall objective of the evaluation is to optimize the SLP score.

$$R_i = \sum_{j=a}^{n} W_{ij}Q_j = W_{ia}Q_a + W_{ib}Q_b + \cdots + W_{ij}Q_j + \cdots + W_{in}Q_n \dots\dots\dots\dots\dots\dots\dots\dots\dots(1)$$

where:

R is the composite score representing the relative merit of the ith course of action;

Q is defined as the constant numerical value (quality level) of individual strategic elements;

W is defined as the weighting factor that modifies the jth strategic element in accordance with its relative contribution to the ith course of action.

Table 4 Calculation of R-value for Tower Crane

Factors	Priority Score	Weighting	Rank*	Weighted Score
Suiting working radius	8	0.30 (e.g. 8/27)	60	17.78 (e.g. 0.30x60)
Overlapping between cranes	5.83	0.22	50	10.80
Site Condition	4.67	0.17	60	10.38
Access	4.33	0.16	41	6.58
Others	2.67	0.10	75	7.42
Within operator's view	1.5	0.06	53	2.94
Sum	27		Score	**55.89**

*Rank is the score of the factor assessed by a planner when a tower crane is positioned into a particular defined point.

Table 5 Calculation of Score for SLP

Stage I of SLP				
Factors	Priority Score	Weighting	Rank	Weighted Score
Tower Crane	27	0.406 (27/66.5) **55.89**		22.69 (0.406x55.89)
Systems formwork yark	3	0.045	61.23	2.76
Formwork yark	3.5	0.053	45.28	2.38
Storage Area for Precast Units	12	0.180	70.15	12.66
Storage Area for Fabric-reinforcement	8	0.120	80.21	9.65
Bending Yark	13	0.195	53.16	10.39
Sum of PS	66.5		Score	*60.54*

9 CONCLUSION

This paper has identified the difficulties of SLP and proposed an evaluation system to assist site planners in optimizing positioning of site facilities. The decision-making process of SLP is very complex and composed of a lot of constraints, which is not easy to manage by any mathematical models. This study tries to integrate the shared experience of construction managers and organize them into a logical framework and a set of decision rules. This is aimed to provide a model to quantify SLP and identify the priority order of site facility positioning. The study has built up heuristic decision rules, which can simply the evaluation process of SLP. A part of an application example is illustrated that demonstrates the robustness of the model. The model can be further enhanced by computerizing the calculation process and combining human intelligence input and computing capability of the computing machine.

ACKNOWLEDGEMENTS

The authors are grateful to the Quality Enhancement Fund and the Strategic Research Grant of City University of Hong Kong for the funding provided.

REFERENCES

Calvert R.E, Bailey G. and Coles D. (1995), Introduction to Building Management, Butterworth Heinemann, 6th Ed., UK.

Cheng. M.Y. and O'Connor J.T. (1996), ArcSite: Enhanced GIS for Construction Site Layout, *Journal of Construction Engineering and Management., ASCE, Vol.122, No. 4, December, pp. 329-336.*

Coolican H., (1993), Research Methods and Statistics in Psychology, Hodder and Stonghtion.

Epstein Richard A (1977), The Theory of Gambling and Statistical Logic, San Diego: Academic Press, c1977, Rev. Ed.

Naoum, S.G. 1998. *Dissertation Research and Writing for Construction Students.* Oxford: Butterworth-Heinemann.

Tommelein, I.D., et al. 1991. SightPlan Experiments: Alternate Strategies for Site Layout Design, *Journal of Computing in Civil Engineering, ASCE, Vol. 5. No. 1, January, pp. 42-61.*

Creative Systems in Structural and Construction Engineering, Singh (ed.) © 2001 Balkema, Rotterdam, ISBN 90 5809 161 9

Knowledge management in a multi-project environment in construction

J. M. Kamara, C. J. Anumba & P. M. Carrillo
Department of Civil and Building Engineering, Loughborough University, UK

ABSTRACT: The management of project and organisational knowledge is now recognised as a vital ingredient for competitive business performance in the construction industry. This can involve the management of knowledge within, and across projects, within individual firms, and across the supply chain involved in project delivery. This paper presents the initial findings of a research project, which is aimed at developing a viable framework for knowledge management in a multi-project environment, within a supply chain context. The background and methodology being adopted in the research are described, and case ('as-is') study results of current knowledge management processes within selected firms are discussed. The paper concludes with suggestions on the development of a framework for knowledge management.

1 INTRODUCTION

The implementation of a construction project is carried out within a temporary multi-disciplinary organisation through which the resources and the people involved in the process are managed. Until recently, the management of projects and other construction business processes have focused on the interrelationships between resources, and the integration, monitoring and control of the contributors to a project and their output. However, it is now recognised that the knowledge (or intellectual assets) of project participants and construction firms also need to be managed if construction businesses are to remain competitive, and adequately respond to the needs of their clients

This paper describes an ongoing research project (Cross-sectoral LEarning in the Virtual entERprise - CLEVER) which is aimed at developing a framework for knowledge management in a multi-project environment, within a supply chain context. Following a review of the context of knowledge management in construction, the aim and methodology of the CLEVER project are described. Preliminary results from pilot studies are presented and the findings used to discuss the management of knowledge within a multi-project environment.

2 KNOWLEDGE MANAGEMENT

Knowledge Management (KM) can be defined as the identification, optimisation and active management of intellectual assets to create value, increase productivity, and gain and sustain competitive advantage (Webb, 1998). It can involve the capture, consolidation, dissemination and reuse of knowledge within an organisation (Kazi et al. 1999). The need to manage intellectual assets arises from the realisation that both the content and value of knowledge deteriorate over time. The emergence of the knowledge economy also suggests that what organisations know is becoming more important than the traditional sources of economic power (i.e. capital, land, etc.) which they command (Scarbrough and Swan, 1999). It is not surprising therefore, that the management of knowledge is now recognised as a core business concern (Drucker, 1993).

Within the project-based construction industry, the need for innovation and improved business performance requires the effective deployment and utilisation of the intellectual assets of project team members (Egan, 1998; Egbu et al. 1999; Kamara et al. 2000). Knowledge therefore needs to be managed at different, interrelated levels including:

- Knowledge management (KM) within projects (i.e. across different stages of a project). This is from the perspective of the temporary 'virtual' project organisation, and the supply chain (i.e. network of organisations) that is responsible for delivering a project.
- KM within construction firms (e.g. consultants, contractors, etc.) to enhance their ability to adequately respond to client requirements (in association with other firms involved in a

project), and the changing business environment in which they operate. This involves the transfer of knowledge/learning across different projects, as well as harnessing per capita knowledge into a corporate knowledge base.

There is no doubt that some form of knowledge management is being practised within the construction industry, whether in the documentation of best practice (e.g. codes of practice for certain operations), or in the use of information technology (IT) applications (Anumba et al. 2000; Kazi et al. 1999). However, there is no known framework for the management of knowledge in the context of project organisations (Egbu et al. 1999). Even in other sectors (e.g. manufacturing, IT, etc.) where various firms have developed 'best practice' methods for improvements to project knowledge management (Scarbrough and Swan, 1999), there is the lack of a formal framework underpinning it. What is needed is a framework whereby different methods and philosophies may be compared, and other companies may 'mix and match' these 'best practice' approaches, with some degree of confidence that the choice will suit their own circumstances and operational contexts. This framework must address the different processes necessary for formal and tacit knowledge management. It must also address contextual issues, and should be useful to firms in the construction industry who may want to implement knowledge management programmes in their organisations. The CLEVER project, which is described below, seeks to address these issues.

3 THE CLEVER PROJECT

The main focus of the CLEVER project is on the organisational and cultural dimensions of knowledge management (KM) within a project context. This acknowledges the growing understanding that good knowledge management does not result from the implementation of information systems alone (Grudin 1995; Davenport 1997; Stewart 1997). However, consideration will also be given to the applicability of IT tools within the organisational and operational contexts of projects.

3.1 *Project aims*

The specific aims of the CLEVER project are as follows:
- To generate 'as-is' representations of knowledge management practices in project environments both within and across enterprises in the manufacturing and construction sectors;
- To derive generic structures for these practices by cross-sectoral comparisons;

- To develop a viable framework for knowledge management in a multi-project environment, within a supply chain context, together with requirements for support; and
- To evaluate the framework using real-life projects and scenarios supplied by the participating companies.

3.2 *Project strategy and methodology*

The basic strategy being adopted is to investigate KM practices in the manufacturing and construction sectors with a view to facilitating cross-sectoral learning to the mutual benefit of both sectors. The methodology includes the following:
- continuous review of literature to maintain awareness of current developments in the field;
- the development of a common process model for project research within the academic team;
- the adoption of a user-centred approach to the classification of current practice and the identification of use cases.

3.3 *Framework for project research*

Figure 1 shows a theoretical framework that is being used in the development of 'as-is' models of knowledge management practices within various companies. It is underpinned by the understanding of knowledge as a component of a task-performing system. That is, a state of that system which warrants task completion, and the future repetition of this task. The lack of this component implies a failure when completing a task. If this lack is sustained over time, it means that this system ceases to exist. (Blumentritt and Johnston, 1999).

The interrelationships between the four elements in Figure 1 illustrate the fact that knowledge management is usually not a linear process of creation, capture, storage, transfer, reuse etc. that available literature on the subject might suggest (e.g. Laudon and Laudon, 1998; Webb, 1998). The 'knowledge base' (used in a wider sense) refers to the kind of information, data or project knowledge that is to be managed. 'Knowledge management processes' refer to the tasks and activities that are implemented to manage knowledge, within the context of the project and/or organisation ('process shaping factors'). 'Performance measurement' deals with the assessment of the real-time usefulness of knowledge management efforts, since KM is not an end in itself, but a means to add value and increase competitive advantage. The next section describes the research that was carried out to determine how knowledge is managed within a project context.

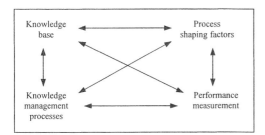

Figure 1: Framework for project research.

Table 1: List of companies involved in CLEVER

Company	Industry	Number of Interviews
1	Construction/Services	7
2	Airport Management/ Construction client	3
3	Construction/Consultants	2
4	Construction /Contractor	4
5	Manufacturing (French)	7
6	Water Industry/Utilities	1
7	IT Developer	1
	Total Interviews	25

4 STUDIES OF KM PROCESSES

Case studies of knowledge management practices within the industrial partners of the CLEVER project were conducted. Table 1 shows the list of companies involved in the research project and the number of interviews conducted to date. In addition to these official partners, the KM practices of 8 other companies (mainly in the manufacturing/aerospace sector) were also investigated. The research investigation involved a series of semi-structured interviews, which lasted for about 2 hours. Questions asked by researchers revolved around the following themes:

- What is the organisational context for the management of project knowledge?
- What knowledge is currently transferred between projects?
- How is that knowledge transferred?
- What are the challenges and opportunities for cross-project knowledge management?

The preliminary findings from the research are described below, and are discussed with respect to the organisational context for KM, organisational/ project knowledge, and the mechanisms/constraints for knowledge acquisition, sharing and transfer.

4.1 Organisational context for KM

The organisational context for knowledge management refers to those issues that have influenced (or are influencing) the development of KM strategies within these companies. They include:

- The need to cope with organisational changes with respect to high staff turnovers and changing business practices (e.g. from a hierarchical organisational structure to 'virtual' teams).
- The need to minimise waste, prevent the duplication of efforts and the repetition of similar mistakes from past projects, and for improved efficiency.
- The need to cope with growth and the diversification of a firm's business activities (e.g. from traditional main contractor to design and build operator/facilities manager).
- The effective management of the supply chain in project delivery (need for knowledge of suppliers and their capabilities, etc.).

These factors (process shaping factors) influence the way companies deliver projects, and the kind of knowledge that needs to be managed within that context.

4.2 Organisational/project knowledge

Organisational/project knowledge can be described as the 'knowledge base' (Fig 1) that needs to be captured, stored and transferred. Preliminary findings from the research suggest that the following categories of knowledge are important for KM in a project context:

- Knowledge of organisational processes, and procedures. This includes knowledge of the construction process, statutory regulations and standards, and the management of the interfaces between different stages/components of a project. In-house procedures and best practice guides would also come under this category.
- Technical/domain knowledge of construction design, materials, specifications, and technologies. It also includes knowledge of the environment in which the construction industry operates.
- Know-Who knowledge of people with the skills for a specific task, knowledge of the abilities of suppliers and subcontractors. Knowing who to contact when there is a problem was considered to be a key aspect of any KM strategy.

Another categorisation of knowledge used was the distinction between current account and deposit account knowledge. Current account knowledge includes forms of knowledge required to complete a specific task within a specific project. This form of knowledge is unlikely to be used/needed in another project. Deposit account knowledge includes forms of knowledge that have a longer time-span (e.g. a database of suppliers/subcontractors).

The identification/categorisation of types of knowledge that need to be managed is related to its capture, storage and reuse, and therefore have a direct bearing on the infrastructure and processes for its management.

4.3 *Knowledge acquisition and storage*

The acquisition of project knowledge within participating companies is mainly through experience gained while a person is directly involved in an activity. Knowledge is also acquired through published material (both internal and external), internal workshops and seminars, and through formal training programmes. Informal networks are also vital in the acquisition of knowledge. Thus it involves both the reuse of existing knowledge and the creation of new knowledge.

Mechanisms for the storage (or capture) of knowledge include people, project drawings and specifications, databases (computer or otherwise), files, and organisational procedures, standards and best practice guides.

4.4 *Knowledge sharing/transfer*

The transfer/sharing of knowledge derives from its acquisition and storage. Project/organisational knowledge is usually shared through the following:

People. This is brought about through the involvement of people in various projects, with the assumption that the learning acquired from a previous project will be used in subsequent projects. Interactions among people can either be through the formal organisational structure, structured workshops and seminars, or through informal relationships.

Organisational procedures/practices, standards and best practice guides are also used to capture learning from previous project/activities, and are therefore used in the transfer of project knowledge. Organisational practices used in the transfer of knowledge include job rotation and mentoring, and the use of virtual teams that cut across departmental barriers.

Published documents within the organisation. This includes procedures, newsletters, journals, memos, financial statements, project drawings and specifications. For example, one of the firms involved in the CLEVER project encourages the sharing of knowledge through an internal newsletter, which is both paper-based and available on the company's Intranet.

Contractual arrangements, while not directly used in the transfer of knowledge, can facilitate this to happen. For example, the adoption of framework agreements with a number of suppliers to deliver a service over a certain period of time (say 5 years) can facilitate the transfer of project knowledge. The use of design and build contracts can also allow key people from the contractor to be involved in every stage of a project and thereby facilitate the transfer of knowledge across the stages of that project.

Information technology in the form of Intranets, local area networks, document management systems, etc. was used by various companies to transfer knowledge.

4.5 *Constraints in the transfer of Knowledge*

Some of the constraints in the transfer of knowledge derive from the mechanisms used to facilitate this. For example, the use of virtual teams can inhibit the sharing of knowledge if there is inadequate support that will minimise or discourage the rivalries and competition between departments. The reliance on informal relationships for the transfer of knowledge can be less effective if staff are not co-located. In one of the projects studied, staff who were located in satellite offices did not benefit from the informal sharing at the project head office, and therefore did not perform as well as the others.

There can also be constraints in the sharing of knowledge through framework agreements. This is because, members within the framework may be in competition elsewhere (e.g. on other projects) and may not always be willing to share their knowledge with other members.

5 DEVELOPMENT OF A KM FRAMEWORK

The research into the KM practices of participating companies reveals that some form of knowledge management is being practised. It is observed that there is heavy reliance on people and the use of procedures to capture and transfer the learning from previous projects/activities to subsequent tasks. However, for these strategies to be effective, a number of issues have to be addressed.

5.1 *Issues for concern*

Post-project reviews need to be more structured, consistent and should be focused on capturing and reusing the learning from a project. The impression from the research suggest there is usually insufficient time for project reviews, and even when they are carried out, the scope of the review is often too narrow to capture the learning from that project.

The *process of documenting best practice* needs to be well defined to ensure that there is consistency in what is documented. The findings from the research suggest that even though some companies have a comprehensive set of in-house standards, the process for documenting these standards is not documented.

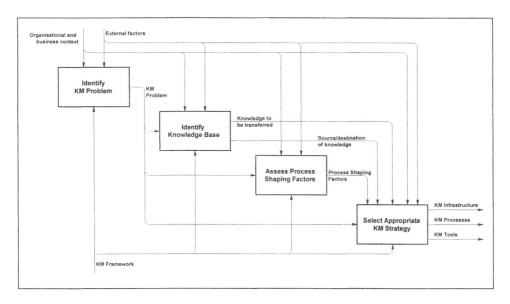

Figure 2: Framework for knowledge management

The *database of peoples' skills*, which usually contains information on the activities they have been involved in, needs to be more comprehensive and should not only be dependent on what people know (in their heads) about others.

5.2 *Need for a structured framework for KM*

The findings from the research suggest that there is need for a structured framework for knowledge management. Figure 2 is an IDEF-0 representation of a framework for knowledge management within a project context. It consists of four major activities: "identify KM problem", "identify knowledge base", "assess process shaping factors" and "select KM strategy".

Identify KM problem refers to the knowledge management problem that an organisation intends to solve. A typical problem can be, "how do we transfer knowledge between two projects that are simultaneously being implemented by the organisation?" *Identify knowledge base* refers to the kind of knowledge that needs to be managed/transferred, and the source and destination (i.e. people, software, paper, product) of this knowledge. *Assess process shaping factors* deals with the identification of the enabling/constraining factors for knowledge management. These can both be internal and external to the organisation. *Select KM strategy* involves the mix of processes, tools and infrastructure that is needed to solve the KM problem identified earlier. This strategy can be in the form of "If-then" rules. For example, 'if the knowledge that needs to be managed undergoes a rapid rate of change, but the source of this knowledge is external to the organisation, then the organisation can either buy-in this knowledge, or merge with other firms to survive'.

6 CONCLUSION

This paper has presented and discussed the preliminary findings from a research project, which is aimed at developing a knowledge management framework within a multi-project environment. While there is evidence that construction firms do practice some form of knowledge management, it is clear that there is need for a structured process to facilitate this. The framework that is presented in this paper is intended to achieve this. Further research will focus on identifying the strategies for effective knowledge management in specific project environments in the construction industry.

REFERENCES

Anumba C. J., Bloomfield D., Faraj, I. and Jarvis, P. (2000). *Managing and Exploiting Your Knowledge Assets: Knowledge Based Decision Support Techniques for the Construction Industry*, BR 382, Construction Research Communications Ltd, UK (ISBN 1 86081 346 1).

Blumentritt, R. and Johnston R. (1999). "Towards a Strategy for Knowledge Management," *Technology Analysis and Strategic Management*, 11 (3): 287-300.

Davenport, T. (1997). "Secrets of Successful Knowledge Management," *Knowledge Inc.* 2, February.

Drucker, P. (1993). *Post-Capitalist Society*, Oxford, Butterworth-Heinemann.

Egan, J. (1998). "Rethinking Construction," *Report of the Construction Task Force on the Scope for Improving the Quality and Efficiency of UK Construction*, Department of the Environment, Transport and the Regions, London, UK.

Egbu, C., Sturgesand, J. and Bates, B. (1999). "Learning from Knowledge Management and Trans-Organisational Innovations in Diverse Project Management Environments," W. P. Hughes (ed.), *Proc. 15 Annual Conference of the Association of Researchers in Construction Management*, Liverpool John Moores University, Liverpool, 15-17 September, pp. 95-103.

Grundin, J. (1995). "Why Groupware Applications Fail: Problems in Design and Evaluation," in R. Baecker, et. al. (eds.), *Readings in Human-Computer Interaction: Toward the Year 2000,* Morgan Kauffman.

Kamara, J. M., Anumba, C. J. and Carrillo, P. M. (2000). "Integration of Knowledge Management within Construction Business Processes" in Faraj & Amor, (eds.), *Proc. UK National Conference on Objects & Integration for AEC, 13-14 March,* Building Research Establishment Ltd. 95-105.

Kazi, A. S., Hannus, M. and Charoenngam, C. (1999). "An Exploration of Knowledge Management for Construction," in M. Hannus et al. (eds.), *Proc. 2nd International Conference on CE in Construction* (CIB Publication 236), Espoo, Finland, 25-27 August, pp. 247-.256.

Laudon, K. C. and Laudon, P. L. (1998). *Management Information Systems*, 4th edition, Prentice-Hall, New Jersey, USA.

Scarbrough, H. and Swan, J. (1999). *Case Studies in Knowledge Management*, Institute of Personnel Development, London,UK.

Stewart, T. A. (1997). *Intellectual Capital: The New Wealth of Organisations*, New York, Doubleday.

Webb, S. P. (1998). *Knowledge Management: Linchpin of Change*, The Association for Information Management (ASLIB), London, UK.

Creative Systems in Structural and Construction Engineering, Singh (ed.)© 2001 Balkema, Rotterdam, ISBN 90 5809 161 9

Stochastic simulation and schedule optimization

N.M.Cachadinha
RWTH Aachen, Germany

ABSTRACT: This paper starts by reviewing the concept of uncertainty and the most common problems associated to it in construction scheduling. It then looks at traditional and recent methods used to overcome them in the construction industry , as well as their limitations.

A new paradigm is then set by proposing quantifiable parameters for assessing quality in construction scheduling and putting this matter within the range of the State of the Art quality assurance concepts, principles and methods used in the manufacturing industry. A process quality optimization method, known from the telecommunications industry, is then chosen and a methodology for its appliance to this field described. Within this context, two new ratios for quantifying the uncertainty of activities and their potential impact on the project delivery date are proposed .

1 INTRODUCTION

1.1 *The importance of meeting the delivery date in present day construction industry*

The present trend in the construction industry follows the general economic trends of this end of century: globalization and mergers. Construction projects become bigger, more complex, more interdisciplinary. The owner wants maximal revenue in shortest time. This is due not only to the large financial amounts invested in the larger and larger projects being awarded, but also to the fact that clients are becoming increasingly demanding on what concerns the performance and range of services of the project management team (PM team).

A survey mentioned by Mulholland & Christian (1999) on 40 U.S. construction managers and owners, covering projects of an average total cost of US$ 5,000,000 reveals that, from a scope and design objectives perspective, the overwhelming majority of the construction projects have medium to very high uncertainty at the beginning of the construction. 65% of the projects considered had medium to very high uncertainty. This figure was risen to 80% on a more recent report by Laufer & Howell (1993), confirming the trend indicated above.

Thus, all stakeholders loose when the project delivery date is not met. These primarily comes in form of:

– Financial penalties

Globalization, operational integration, increased speed of process and the financial amounts involved force the stakeholders to protect their positions by spreading the financial consequences of the risk of delays through the inclusion of heavy financial penalties in the contracts.

– Prestige loss

Increasingly demanding clients force the PM team to respond with improved technical means and solutions, as well as increased competence of their staff. Only so can one company survive in the highly competitive market of large scale Project Management. For this, prestige is vital. It is the business card of a company when bidding a project. Therefore, is has to be maintained at all costs.

The major task of a PM company is to complete the project within the cost, quality and time foreseen. Thus, not meeting the project end date is one of the severest setbacks it can face.

2 UNCERTAINTY

Increased complexity necessarily brings about increased uncertainty. A schedule is always a forecast, therefore itself subject to uncertainty. Besides, there aren't two construction projects which are exactly the same, which leaves previous experience in a weak position to be the major guarantee for a work well done.

2.1 Where does uncertainty manifest itself?

Uncertainty is linked with every step in life, and construction is not an exception. It is much more one area where it has to be dealt with on a daily basis.

Schedules define the work sequence and duration for a specific, generally unique project. They are completed during the work preparation, and involve decisions and choices on:
- Duration of activities
- Ties between them

The definition of these two points is typically where the scheduler is confronted with uncertainty, and where most improvements to reduce it can be achieved.

2.2 How is it presently dealt with?

Construction scheduling has been using quantitative, network based modeling processes ever since the 60s . These have produced good results and grown to become classical standard methods. However, traditional scheduling processes have treated uncertainty and risk as if they wouldn't exist.

Network based planning processes, such as CPM, by giving deterministic durations to the activities, originate seemingly precise predictions and frequently produced unrealistic project performance times in the past , as mentioned by Laufer & Howell (1993). Knowing this limitations, schedulers have been using down to earth precautions to counteract this problem. The traditional approach to reduce uncertainty when preparing a schedule consists mainly on adopting the following measures:
- Involvement of experienced schedulers
- Storing productivity rates and other data from previous works to improve the deterministic duration assessment in the future
- Updating the construction progress and reorganizing the schedule at regular intervals during the construction phase

These procedures have proven to be adequate but insufficient on several occasions in the past.

2.3 Limitations of the traditional methods

The major enemy of these methods is the non stationary, ever changing character of the construction business. A new project involves new crews, equipment, different owners and local authorities, posing different problems of communication, relationship, performance levels and expectations. The fringe conditions, such as soil, accesses, design and others will also change, making all the previous experience and the data gathered insufficient to assure the reduction of uncertainty to the minimum.

2.4 Consequences

Uncertainty frequently leads to situations that harm all the stakeholders in the construction process. Some of the consequences are as follows:
- To the project manager

Loss of credibility due to innumerous schedule changes, visible is uncountable schedule updates and recovery plans
- To the PM and owner

High fines and prestige loss due to delays
- To the contractor

Constant changes to the schedule cause extra costs due to reorganizing resource assignments and work sequence.

2.5 Recent solutions

Other methods have recently been used to improve the response to uncertainty. Their basic principle is trying to gather as much information as possible, both historical and virtual. Some of the most important ones are:
- Monte Carlo simulation
- Expert systems

The first generates virtual statistical data about one project through simulation based on more or less loose parameters known to describe the situation dealt with.

The second gathers and organizes the experience of several experts into a body of knowledge, enabling decisions based on close approximations of the problem analyzed.

2.6 Objective

The methods already mentioned and the approach and method proposed in this paper share a common objective:
- Minimal uncertainty in the project delivery date

The approach described in this paper further aims at a
- Minimum number of schedule changes during the construction phase

To achieve these objectives, a baseline study was carried out by Cachadinha (1999), aiming at determining the influence of the schedule structure in the input of inaccurate activity durations.

3 BASELINE STUDY

3.1 Fundamental questions

Having in mind the objective proposed, the study by Cachadinha (1999) set up a body of qualitative, theoretical knowledge to backup schedule optimization efforts. It answered the following questions:

- Does shortening of duration of some activities make up for the increase of others (i.e. does the end date remain unchanged)?
- What are the variables that characterize a schedule? Which of those can be controlled by the scheduler?
- What is the influence of those variables on the intensity of the impact of activity duration changes (increases and reductions) on the total duration of the project?

To answer these questions, the author created the software application ProSim.

3.2 *ProSim*

This software application applies Monte Carlo simulation to network scheduling.

3.2.1 *Functioning principle*

Taking a given schedule structure and percent duration value, the simulator randomly reduces or increases the duration of the activities by the value entered. This procedure is repeated the number of runs defined by the user. In this study, that number was previously set by determining the threshold of data consistency for the worst case schedule structure simulated.

The schedule is then calculated according to the traditional Network Scheduling Theory (CPM/PERT). Finally, the average of the total project durations obtained is calculated. For a comprehensive description of the study and software utilized, please refer to Cachadinha (1999).

3.2.2 *Variables that characterize the schedule*

The variables isolated by the study as significant were separated in two classes:

1 Within the range of influence of the scheduler
2 Partially or completely outside his range of influence

The first class was divided into two items:
- Structure
1 Number of activities in one row (vertical development), where row is defined as a set of activities connected by simple "finish to start" ties, corresponding generally to a constructive process (e.g: form – cast – strip).
2 Number of parallel rows (horizontal development)
- Ties between rows
1 Density
2 Orientation

The second class is composed by:
- Balance of the size of the activities' durations
- Percent value of the changes in the activities' durations

3.2.3 *Results*

More than 190 different structures and schedules were simulated at 1000 simulation runs each. The results were then validated on a real project of a public facility near Aachen, Germany, with more than 60 activities and a total duration of 117 days.

The results obtained identified clear tendencies in the relationship between the characteristics of the structure and the impact of single activity duration uncertainty on the total project duration. They are summarized in the two tables below. It is to be noted that the expected value of the project total duration obtained through simulation was always higher than the deterministic project duration. Therefore, when a fall of the percent variation is mentioned in the second column of the tables, a convergence to zero from the positive, delay side is meant.

Three characteristics were observed:

1 Percent variation of the total project duration expected value (EV) to the deterministic total project duration (DPD) obtained through the classical Critical Path Method (CPM), hence considering defined and constant activity durations for the structure entered
2 Left dispersion of the simulation values, considering the lowest total project duration obtained
3 Right dispersion of the simulation values, considering the highest total project duration obtained

Table 1. Schedule structure – summary of tendencies.

Characteristics	1	2	3
Structure change			
Positive horizontal development	Falls moderately	Falls	Falls moderately
Positive vertical development	Rises logarithmically	Falls	Rises
Ties' density increase	Falls moderately	Falls significantly	Always maximum

Table 2. Activities' durations – summary of tendencies.

Characteristics	1	2	3
Structure change			
Increase of uniformity	Falls moderately	Falls	Always maximum
Increase of variation percentage	Rises proportionally	Always maximum	Always maximum

3.2.4 *Further observations and conclusions*

1 The simulations carried out show that in one situation of random increase and increase of the activities' durations (all having the same deterministic duration), the project delay caused by the activities with increased duration is not compensated by the activities with decreased durations. On the contrary, there is one clear tendency to an increase of the total project duration.

2 The schedule structure has direct and quantifiable effects in the total project duration. Thus, it is possible to define qualitative tendencies for the impact of changes in the activities' durations as function of the above mentioned schedule characteristics. This allows the scheduler to optimize his schedules on what concerns the reduction of the impact on the project total duration due to changes in the activities' durations.

3 After analyzing the correlation between effects of changes in the various variables studied, it was concluded that, at a practical level of detail, their effects can be considered not correlated. Even in large and complex schedules, it is possible to identify and isolate the effect of each variable. This was observed in the simulation and analysis of the validation project mentioned in section 3.2.3.

4 ACTIVITIES' DURATIONS AND THEIR UNCERTAINTY

4.1 Concept definition

In the context of this research uncertainty reflects itself as the difficulty of determining an exact duration and sequence for the schedule.

4.2 Proposals for its quantification

4.2.1 Structure

The common and widely used deterministic methods consider just one critical path, focusing all the attention and efforts to assure the delivery date in the activities of the critical path. However, due to the inherent uncertainty of the activities´ duration and even of the work sequence, these methods are likely mislead the PM into overseeing activities which have an important role to play in the assurance of the delivery date.

Stochastic methods generate different activity durations for each simulation according to a known or assumed statistic distribution. This makes near reality schedules possible, in which the critical path is not static.

Han (1997) proposes one probability approach to this matter, assigning each activity one degree of criticality K, according to the frequency of its presence in the critical path during the simulation. The higher the K, the more likely the activity will impact the total duration of the schedule.

4.2.2 Duration

Uncertainty in the duration plays a major role in the overall uncertainty of the schedule. The traditional PERT method considers an optimistic duration (a or OD), a pessimistic duration (b or PD) and a most

likely, realistic duration (m) for the definition of average activity duration by the formula

$$\mu = \frac{a + 4m + b}{6} \tag{1}$$

This method has been subject to a lot of criticism due to the inaccuracy it may cause. The value with the highest contribution to the definition of the average duration is precisely to one which is harder to estimate and hence most likely to be inaccurate.

In his work, Han proposes the substitution of m by a tendency c, with ranges from very optimistic (1) to very pessimistic (5). The in depth description of the approach, method and its justification can be found in Han (1997). The value of the tendency is easily and more accurately estimated by an experienced scheduler.

Thus, uncertainty in the duration of the activities can be quantified by the evaluation of the range of duration between a and b and by the tendency assigned.

1 Range

Uncertainty is directly proportional to the span of the range between OD and PD

2 Tendency

Assigning a tendency to the duration actually reduces uncertainty, since it means focusing the generation of durations by the simulator on a given part of the range. However, confronted with high uncertainty in determining the duration of an activity (due, for instance, to lack of information), the scheduler will, as a defensive measure, typically assign a rather pessimistic tendency. Hence, the author proposes two concepts for two different approaches to the tendency assignment:

– Effective

focuses on a part of the range of durations (optimistic – beginning; medium – middle; pessimistic – end), hence decreasing the uncertainty

– Defensive

corresponds to a pessimistic tendency assignment due to caution and/or lack of information, thus depicting an increased level of uncertainty

4.2.3 Ratios for uncertainty quantification

Based on the considerations made in the previous sections, we propose two ratios for the quantification of activity duration uncertainty and for the level of seriousness of its impact on the project duration.

Uncertainty ratio (UR)

Quantifies the intrinsic uncertainty of an activity

$$UR = \frac{PD}{OD} \times C \tag{2}$$

Where PD = pessimistic duration; OD = optimistic duration; and C = tendency, ranging from 1 to 5 (very optimistic to very pessimistic).

Seriousness ratio (SR)

Takes into consideration the duration and structural uncertainty and its seriousness in the context of the whole schedule

$$SR = \frac{PD}{OD} \times C \times BD \times K^3 \qquad (3)$$

where BD = average duration of the activity, obtained through simulation; and K = critical degree of one activity, obtained through simulation.

5 QUALITY IN SCHDULING

5.1 *Proposal of concept*

The quality of a schedule is defined by the level of accuracy, with which it correctly foresees the construction process. That includes the sequence of activities, individual activity duration and, last but not least, the project delivery date.

5.2 *Assessment*

The author proposes the assessment of this quality concept by considering the number of updates and changes of the master schedule during the construction phase, excluding rescheduling due to major changes in the scope of works.

A low number of schedule changes and updates implies a high level of schedule quality.

5.3 *How can this be achieved?*

1 By guaranteeing the relevance of the activity sequence and of the project delivery date by making the schedule robust to its intrinsic uncertainty, defined by the parameters proposed in previous sections.
2 By changing the way a schedule is looked at, from a mere work preparation auxiliary or contractual obligation to an integral part of the production of major importance, thus a product itself. Its quality should be considered in the evaluation of performance, just as much as the quality of the concrete.
3 This paradigm change has the purpose of leading to the appliance to scheduling of the quality concepts, trends and optimization methods common to the manufacturing industry. Specifically, the author proposes the appliance of manufacturing optimization methods to the morphology of the schedule, in order to make it robust to its intrinsic uncertainty.

6 FRACTIONAL FACTORIAL EXPERIMENTS APPLIED TO SCHEDULING

A wide range of methods for reducing the variance in general are known. But given the nature of construction industry and previous good experiences in adopting techniques from the manufacturing industry, Dr. Genichi Taguchi's approach to Fractional Factorial Experiments, termed Taguchi Method, has been selected to tackle this problem. This method fits perfectly in the project management perspective, for the reasons described further on in this article.

6.1 *What is the Taguchi method?*

This method was developed by Genichi Taguchi in the late 1950s to improve the quality of the telecommunication system in Japan. It is based on the design of experiments to provide near optimal quality characteristics for the specified objective. It employs orthogonal arrays, the matrices based on Latin squares, which dictate the experimental setup for the defined factors. Taguchi method belongs to a set of approaches that attempt to ensure quality through design, in this case through identification and control of critical variables that cause deviations to occur in the process quality.

Ross (1996) puts it this way:" The purpose of process development is to improve the performance characteristics relative to the customer needs and expectations. The purpose of experimentation should be to understand how to reduce and control variation of a product or a process; subsequently, decisions must be made concerning which parameters affect the performance of a product or a process".

6.2 *Features*

This method is very versatile in application and produces good, objective results, which have considerable importance in modifying the process. Important features listed by the ASI (1998) are:

– It is a system of quality engineering driven by cost that emphasizes on the effective application of engineering strategies rather than complex statistical techniques
– This method provides a strategy for dealing with multiple and interrelated problems and gives guidelines for better understanding of the process
– Aids rational decision making
– Provides a method for designing processes that are minimally affected by the negative external forces
– Identifies the ideal function of a process as against the traditional method whose focus is on analyzing the symptoms as a basis for improvement

– Serves as an effective tool to optimize the number of experiments for obtaining significant results

This method highlights on the fact that the quality is best achieved by minimizing the deviation from the target. The process should be made insensitive to uncontrollable factors. In other words, it should be robust.

One of the main reasons for integrating this methodology into the project management perspective is the concept of Loss Function. The Loss Function takes into the consideration the consequences of uncertainties mentioned earlier. The method's basic principle is that the quality is defined as the total loss caused to the stakeholders from the time the product is shipped to the reception by the customer (compare with Taguchi & Clausing (1990)). This loss should be measured in monetary terms and includes all the costs in excess of the cost of a perfect product. It quantifies the consequences of variability associated with the process.

According to the "Goal Post Syndrome", (Ross 1996), any process or product is acceptable if the value of the specified parameter is within a specific range of tolerance, as shown in Figure 1. No loss is assumed to occur in this range. However, outside this range, 100% functional deterioration occurs. This view does not reflect reality on an effective manner. What Taguchi stated is that such limits do not exist and performance, hence the customer's satisfaction, gradually deteriorates as it deviates from the target. Instead of fixed limits, there is a continuous function. Thus, contrary to traditional methods which are determined by producer's specifications, these methods are fully driven by customer satisfaction.

The Taguchi Loss function shown in Figure 1 portrays quite accurately the cost-time relationship on construction: costs increase both when efforts are made to compress the schedule and when the project overruns the contracted delivery date.

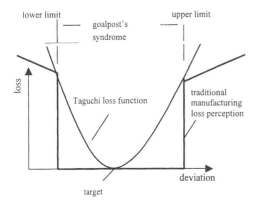

Figure 1. Goal post syndrome vs. Taguchi loss function (Compare with Ross (1996) p.4 ff)

6.3 Validation and future perspectives

This methodology was adopted not only for validating the results described at Cachadinha (1999), but also to test its applicability to Scheduling Optimization. It has served both purposes with very positive results.

The validation results obtained so far in 4 large scale projects carried out in Germany are very encouraging and anticipate further papers on the subject with the latest and final conclusions.

7 CONCLUSIONS

The construction industry time and cost overruns are generally considered genetic disorders in projects, thus part of their essence. This method proposes a tool to overcome this situation and bring about the paradigm shift.

The new concept of quality of scheduling proposed in this paper profiles as a framework to guarantee delivery dates to the owners, a major and often forgotten point of the "Quality in Construction" concept as a whole.

The new revision of ISO 9000, about to be released, will surely take into consideration the recommendations issued in May 97 by the Members of ISO/TC 176/SC 2 (1997), which state as "Principle 4" of their "Quality Management Principles and Guidelines on their Application" the Process Approach: "A desired result is achieved more efficiently when related resources and activities are managed as a process." This is precisely the core of the methodology proposed.

REFERENCES

American Supplier Institute ASI 1998. Taguchi Methods. http://www.amsup.com

Cachadinha, N. 1999. Estudo da Influência da Estrutura do Planeamento no Impacto de Alterações à Duração de Actividades Individuais na Duração Total do Projecto. *Gestão de Projectos 1999 Conference Proceedings*. Lisbon: APDIO & FUNDEC.

Han, G. 1997. *Computergestützte Ablaufplanung zur Optimalen Baustellensteuerung mit Hilfe Stochastischer Simulation*. Cottbus: Lehrstuhl Baubetrieb und Bauwirtschaft, Brandenburgische Technische Universität Cottbus

ISO/TC176/SC2/N133 1997. Quality Management Principles and Guidelines on their Application. http://www.iso.ch

Laufer, A. & Howell, G.A. 1993. Construction planning: revising the paradigm. *Project Management Journal* XXIV(3): 23-33

Mulholland, B. & Christian, J. 1999. Risk Assessment in Construction Schedules. *Journal of Construction Engineering and Management*, Jan./Feb. 99. ASCE.

Ross, P.J. 1996. *Taguchi Techniques for Quality Engineering*. New York: Mc-Graw Hill.

Taguchi, G. & Clausing, D. 1990. Robust Quality. *Harvard Business Review*, (Jan-Feb).

Creative Systems in Structural and Construction Engineering, Singh (ed.) © 2001 Balkema, Rotterdam, ISBN 90 5809 161 9

Database approach for construction data management

A. Chandran
Parsons Brinckerhoff, Cincinnati, Ohio, USA

M. Hastak
Department of Civil and Environmental Engineering, University of Cincinnati, Ohio, USA

ABSTRACT: Data management is a common problem on many construction projects across the world. The amount of raw data generated on projects is bound to increase but the time available to collect, summarize, and analyze the data is decreasing. Thus the traditional practice of processing information using a paper-based system may be inefficient and cumbersome. There is a need to trim down the paper mountain and lay a more convenient and accessible path for data management. Databases are an effective tool that can be used to model construction applications. The development and utilization of database technology has increased rapidly over the past few years. This exceptional growth can be attributed to the desire to use faster and convenient tools with associated cost benefits. This paper will compare the different database technologies available for use in construction and highlight their benefits. The paper will present a model developed using one of the database technologies and illustrate the effectiveness of these automated techniques for construction data management.

1 INTRODUCTION

1.1 Current status of data management in construction

To some degree, each construction project is unique, and no two are quite alike. In its specifics, each structure is tailored to suit its environment, arranged to perform its own particular function, and designed to reflect personal tastes and preferences. The vagaries of the construction site and the infinite possibilities for creative and utilitarian variation of even the most standardized construction product combine to make each construction project a new and a different experience. This complex nature of the construction life cycle has led to the development of different approaches to standardize the tracking system. The truth that we cannot always standardize the process remains, and the need for new models to accurately represent the tracking system is strengthened.

1.2 Importance and need for data integration

The volume of data on a construction project cannot be handled efficiently by a paper-based system. The need for the design of an integral system that will help the construction process function better in today's fast changing world becomes apparent. For years, this process has been very labor intensive and highly redundant methodology was being used.

Projects were the responsibility of a single contractor before, but the current reports do not provide that indication. Nowadays projects are fragmented and broken down into many parts. Different entities undertake different parts of the project, both for design and construction. For an average sized construction project, there are several firms or units of the same firm involved in the project. The transmittal and exchange of information becomes an integral part of any such set up. Hence data integration is a key step to develop relationships among the various elements and reduce the storage for the existing volume of data. Data integration is essential to unifying existing data and the data sources into a single framework. The difficulties that exist in the industry at present due to poor documentation, obscure semantics, heterogeneous systems, and duplication can be addressed effectively with data integration. The models developed can be best put to use by developing a relational tree and designing appropriate forms that can be used on a construction project. Numerous database methods can be applied to solve the data integration problem. There is a need to study and analyze the various database methodologies and then use the most appropriate one. The first part of the paper discusses the various database techniques. The advantages and the disadvantages of these methodologies are highlighted and the most efficient methodology is chosen to develop the data integration model. Integrated Construction Data

Management System (ICDMS) is a prototype database system developed using the relational database techniques. ICDMS is developed using a relational tree for the various data elements associated with a construction project. The parameters that are associated with the forms are then identified and channeled as inputs to the main elements of construction like time, cost and scope. The data in the forms is normalized using the relational database techniques and a generic list is created. The generic set is based on subsections where the field and the office personnel can differentiate and identify the type of forms they will be using on the project. This will serve as a stepping-stone in creating a contract management system with associated relations and will enable the integration of a construction project to a large extent.

2 DATABASES

2.1 Types of databases

Data can be described as known facts that can be recorded and have implicit meaning. A database is a collection of related data. A database management system (DBMS) is a collection of programs that enables users to create and maintain a database. It is a general-purpose system that facilitates the processes of defining, constructing and manipulating databases for various applications. There are various types of database systems. File systems are the first generation data management tool. The application program uses the programming language feature to interact with the operating system. The programmer is responsible for describing the layout of data records on the file. The file system only provides sequential access to a file and the records can be read only in their physical sequence. It does not support multiple sequential accessing. For example, project records are sometimes needed by contract number, sometimes by work breakdown structure and so on. The file system will only provide sequential access and the records can be read only in one order. To accommodate multiple needs, multiple copies of the same data need to be made and stored. Network and Hierarchical systems are the second generation of the data management systems. They solve the basic problem of multiple access seen in the file system. The records maintained in this kind of system need not be sequentially accessed. They support sharing and recovery features. They typically do not provide an ad hoc query facility. These systems are characterized by having their own data language. They provide index structures, tables and inter-record pointers. Their structures are relatively difficult to change and programming for new applications can

be a time consuming proposition. The third type of data management system is the relational system. The relational model views the data in a table format. The various columns in the table represent the table properties and its fields. The values of these properties are represented by the rows in the table. Connections between the various tables are established by forming a connection between the columns of the table. The relational model also separates the logical and the physical design of the database. However, the user need not know anything about the physical structure or the design of the database to use the database. The representation of tables is available to the user as forms. These forms may be built using information from one or more tables. Hence, the representation seen by the user on the exterior does not mean that the data is internally stored in the same format. The relational system provides for the ad hoc query facility, which is not available in the network and the hierarchical system. Relational database design is a process of determining the tables that are needed for data storage and then optimizing those tables for efficient performance. Typically a technique called normalization is used which minimizes the replication of data across the tables. In addition to the functions of the relational system, many new functions are a part of the next-generation database systems. Classical database systems do not manage well the modeling and handling of composite objects. New types of data, such as image or audio, need to be maintained. Search and query capabilities for these new data types have to be added. New organizational structures like work groups have to be supported, and the handling of different versions of the same data is also an important requirement. To meet these requirements for new database function current relational systems can be extended. This system deals with objects and supports object oriented programming languages. The main appeal of the object-oriented system is that the data model closely matches the real world entities. The objects can be stored and manipulated directly and there is no need to transform application to tables. The user rather than being constrained to pre-defined types can define the data types.

2.2 Databases in construction

The architecture-engineering-construction (AEC) industry is heavily dependent upon both inter and intra- organization communication. Hence, integrating data throughout the AEC process and across all disciplines is viewed as a very broad problem (Fischer 1996). Small parts of this AEC construct were the initial undertakings of the researchers

334

(Fischer 1989). These small parts consist of the design and the construction cycle. The design process has always been a challenging area to work on and early research efforts were concentrated on developing a good automated design system. To a large extent, the current practice in the construction industry continues to involve sifting through numerous boxes of printed documents, determining which items are relevant, and then copying, compiling, collating and organizing the paper into individual folders related to particular issues (Dorris 1989). Many of the construction applications can be modeled using the classical (relational) database function, such as the ability to offer concurrent access to large amounts of data that are maintained in a reliable and consistent fashion.

3 INTEGRATED CONSTRUCTION DATA MANAGEMENT SYSTEM (ICDMS)

3.1 Introduction

Construction project management deals with a broad range of functions to be performed. So far the literature shows various researchers working on different aspects of data integration (Abudayyeh 1993, Fischer 1994, Froese and Paulson 1994). The area, which is still not fully researched, is the development of a small-scale construction database model, which will track the required project information and will be project oriented. Work in the area of product modeling suggests that localized models are now perceived as the solution to the difficulties of representing different components of the integrated system and the relationships between them (Ford et al 1994). There is already sufficient evidence in the use of relational database management techniques and object-oriented techniques in construction (Abudayyeh 1993, Froese and Paulson 1994). But there is little literature in the development of a system, which concentrates on information flow and tracking on a project. A prototype database model (ICDMS) was developed to facilitate the need for efficient construction management (Chandran 2000). ICDMS provides a conceptual framework for a comprehensive integrated and coordinated information system to optimize the benefits inherent in available technology. The main concentration of ICDMS is on data management and documentation. It seems logical that the genesis of an intelligent approach to develop a consistent framework for comprehensive integrated information systems lies within the domain of databases.

3.2 Design

Once the problem has been defined (Phase I) and the requirements have been collected and analyzed (Phase II), the next step is to create a conceptual schema for the database, using a high-level conceptual data model (refer to Figure 1). This step is called conceptual database design. The conceptual schema is a concise description of the data requirements of the users and includes detailed descriptions of the data types, relationships, and constraints; these are expressed using the concepts provided by the high-level data model.

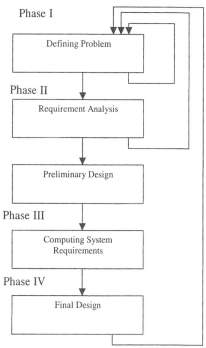

Figure 1: Application Design Steps

The non-technical user group benefits from conceptual schema, as they do not include any implementation details. The high-level conceptual schema can also be used as a reference to ensure that all user data requirements are met and that the requirements do not include any conflicts. This approach enables the database designers to concentrate on specifying the properties of the data, without being concerned with storage details. Consequently, it is easier for them to come up with a good conceptual database design. After the conceptual schema has been designed, the basic data model operations can be used to specify high-level transactions corresponding to the user-defined operations identified during functional analysis. This also serves to confirm that the conceptual schema meets all the identified functional requirements. Modifications to the conceptual schema can be introduced if some functional requirements cannot be specified in the initial schema. Phase III in database design for ICDMS is the actual implementation of the database, using a commercial DBMS such as MS Access 97. Most currently available commercial DBMS's use an implementation data model, so the conceptual schema is transformed from the high-level data model into the implementation data model. This step is called logical database design or data model mapping, and it results into a database schema in the implementation data model of the DBMS. Finally in Phase IV the physical database design is developed in which the internal storage structures and file organizations for the database are specified.

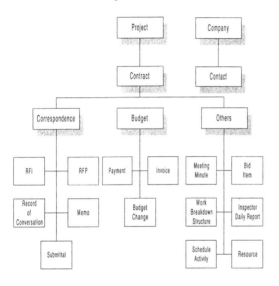

Figure 2: ICDMS Framework

3.3 Framework for ICDMS

Figure 2 represents the framework for ICDMS. Most of the elements in the list have sub-elements, which are already accounted for under their respective tree. For example, the invoice has an invoice number, date, and amount associated with it. These are already considered a part of the invoice details. The basic breakdown of the entire table involves three sub categories namely, cost details, correspondence details, and other. All the necessary elements that are typically tracked on construction projects are incorporated in the ICDMS framework.

3.4 Entity relationship (ER) model and constraints

The conceptual schema and the ER model are developed from these entities identified in the data collection phase and the framework. The contact and the personnel information are separated from the main entities. The hierarchy is set up to encompass all the entities under the contract. Figure 3 consists of all the entities identified at the various phases and

represents the relational schema between the entities. The Min/Max notation is used to draw this representation. For example, the project and contract are two different entities. The project consists of a minimum of one contract and a maximum of M contracts. A contract can be associated with only one project. The notation (1,M) between the project and the contract entity represents the relationship from

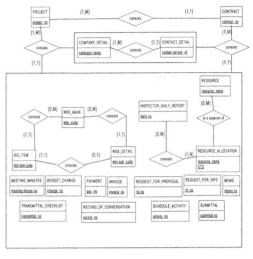

Figure 3: Entity Relationship Diagram

the project to the contract. The notation (1,1) between contract and project represents the relationship from contract to project, where a contract can be associated with a project only. Considering another example, a company can have many employees but an employee can be associated with only one company. The primary key or the element that is unique in that entity group is underlined in the entity group schema. The tree structure enables assigning multiple primary keys. For example consider a Request for Information (RFI). The RFI has a unique number but has to be associated with the contract and the project. So the combination of the project number, the contract number and the RFI number form the primary key in this case. The project number and the contract number get assigned because of the relationships established and the hierarchy built into the model. The primary key is established for all the entities in a similar fashion.

The ICDMS application keeps track of a project, the contracts under the project, the bid items and other contract details under the contract. It also consists of a company and a contact directory listing, which can be used to track the personnel and the company details. The requirements analysis study

336

was performed and the following descriptions were identified for the ICDMS to be represented in the database.

1) The construction project is organized into many contracts. Each project has a unique number, a project name, a project start date and a project end date. The project information is used at the top of the hierarchy (refer to Fig. 3) and the other details are encompassed under it. The project original budget and current budget are also tracked but they can only be viewed after the detailed cost information is entered at the bid item level.

2) A contract is the primary element, which is directly related to all the other sub information. It consists of the correspondence information, resource information and the cost information. Each contract has a unique number, a name and a start and an end date associated with it. A contract can be associated with only one project but a project can be associated with many contracts. The contract original budget and current budget work on similar lines as the project budget. Information at the bid item level is rolled up to the contract level and the contract level information is rolled up and extended to the project level. Consider a project P1 with two contracts C1 and C2 under it. Both the contracts have their respective bid items assigned to them. The sum of all the bid items under C1 summarizes contract C1's cost. Similarly, the sum of all the bid items under C2 summarizes contract C2's cost. Project P1's cost is summarized by adding C1 and C2's cost.

3) The company and the personnel details are related to the project and also the contracts. These details need to be globally maintained. The company and the contact details pertain to the whole group of projects. For example, the same company may be associated with two projects. By assigning them globally, they can be referenced for any project in the database. A company is assigned a company ID based on which the company information is tracked. The personnel information is identified based on the company the person works for. The contact person's ID, name, title, phone, email and the cell phone number are stored in this module. This information is useful, when entering the information in the document modules. An employee of a company can be assigned to one or many projects, which means that he can be assigned to one or many contracts in a project also.

4) The bid item is identified by a bid item code and consists of the bid item description, unit of measurement, bid quantity, quantity change, unit cost and the unit cost change. The quantity change is used to track any changes in the original bid quantity. The unit cost change is used to track any changes in the unit price due to inflation or material shortage. The bid item amount is then calculated based on this information. The original bid item amount is the product of the quantity and the unit price and the current bid amount is the product of unit cost + unit cost change and bid quantity + quantity change. This keeps track of all costs on the project and helps to monitor the fluctuations in the cost due to the any price or the quantity variations.

5) The work breakdown structure (WBS) is identified by a main WBS code and a sub code. The main WBS code is predefined in the database as design (DES), construction (CON), and management (MGT). WBS budgets can then be developed for each element and summed up to yield a budget for the entire project.

6) The elements invoice, payment, and budget change are identified with a unique identifier along with the project and contract identifier. This means that each project is associated with numerous invoices, payments and budget changes and so is each contract. The primary key for this group of elements consists of the project identifier, contract identifier and the invoice/payment/budget change identifier. The individual identifiers for these elements are the invoice number, payment number and the budget change number respectively.

7) The correspondence section follows the same rules as the budget section. The correspondence folders are request for proposal, request for information, record of conversation, memo, submittal and meeting minutes. A unique identifier identifies each of these elements. The identifiers for these folders are as follows: RFI number for request of information, RFP number for request of proposal, submittal number for submittal, memo number for memo, record of conversation number for record of conversation, and meeting minute number for meeting minute. The primary key is established by the uniqueness of the combination of the project number, contract number and the individual identifier number. The schedule activity is also identified in a similar way.

8) The inspector daily report element is associated with the resource allocation table. The inspector daily report is identified by the combination of the project, contract, and the inspector daily report number. Each inspector daily report is assigned resources. These resources are a subpart of the inspector daily report. The resources hence branch out to a fourth level. The resources are identified by a combination of the project number, contract number, IDR number and the resource name. The fourth normalization rule is used to develop the resource and the resource allocation table. Resource allocation table can be associated with zero to many resources from the

337

resource table. The resource name comes from the resource table where the resources for the project are maintained. All the elements of the resource allocation table are present in the resource table but all resource tables' elements need not be present in the resource allocation's table.

The ER model describes data as entities, relationships, and attributes. The entity identification and the attribute assignment were the first step performed before establishing the relationships. The relationships between the various elements have been described above. The most important constraint (on the entities) the uniqueness constraint or the primary key has also been assigned. Figure 3 shows the relationships between the entities and establishes the primary key for each entity. The next step is to establish the entity group. The entity groups can be defined in various ways. The most suited grouping for this case would be, grouping the correspondence, budget and the resource information together as shown in figure 3. The other groups consist of the project group, the contract group and the details group (company and contract elements). The ER model now represents all the entities with their relationships and attributes. The ICDMS application was developed using Access 97 and is windows based. It consists of a customized menu and a set of navigation bars, which help the user in navigating across the various forms. ICDMS was tested using the data from the projects the author was involved in. The ER model was first tested based on the relationship flow. This was achieved by studying the relationships that tied the different entities together and validating each relationship using a sample example from a construction project. For example, the inspector daily report may or may not have resources allocated to it. The relationship between the inspector daily report and the resource allocation table takes care of this condition. The prototype model was then tested based on actual project data and users.

3.5 Conclusions

The work undertaken in this research concentrates on developing an integrated information management system for the construction industry. The relational database approach was used to develop an entity relationship model. The entity relationship model was developed based on the information available from construction projects the author was involved in. This ER model was further developed into a prototype model, where the relational features of the database were put to use. The prototype integrated system (ICDMS) could be used for informa-

tion management on a construction project. The main feature of the relational model developed is that, in ICDMS, not only the underlying data structures and relationships are represented, but also the required structural details of how to use this model in the industry is elaborated. This paper emphasizes the importance of establishing an information framework, which can be developed using different techniques.

REFERENCES

Abudayyeh, O. Y. (1993). "Electronic Data Acquisition Forms in Construction Management." *Advances in Engineering Software*, Vol. 16, pp. 187-193.

Chandran, A. (2000). *Data Integration for Construction Management using Relational Database Approach*, Thesis in partial fulfillment of the requirement for the degree of MS in the Department of Civil & Environmental Engineering, University of Cincinnati.

Dorris V.K. (1989). "Treading the Paperwork Trail." Engineering News Record, pp. 36-40.

Fischer, M. (1989) "Design Construction Integration Through Construction Design Rules for the Preliminary Design of Reinforced Concrete Structures." *Proc., CSCE/CPCA Struct. Concrete Conf.*, Montreal, Canada, pp. 333-346.

Fischer, M. A., Betts, M., Hannus, M., Yamazaki, Y. and Laitinen, J. (1993). "Goals Dimensions and Approaches for Computer Integrated Construction." *Management of Information Technology for Construction*, World Scientific Publishing Co., Singapore.

Fischer, M., and Froese, T. (1996). "Examples and Characteristics of Shared Project Models." *Journal of Computing in Civil Engineering*, Vol. 10, No.3, pp. 174-182.

Ford, S., Aouad, G.F., Kirkham, J.A., Cooper, G. S., Brandon, P.S., and Child, T. (1994). "An Object-Oriented Approach to Integrating Design Information." *Microcomputers in Civil Engineering*, Vol. 9, pp. 413-423.

Froese, T. M. and Paulson, B. C., (1994). "OPIS: An Object Model-Based Project Information System." *Microcomputers in Civil Engineering*, Vol. 9, pp.13-28.

12 Information technology

Creative Systems in Structural and Construction Engineering, Singh (ed.) © 2001 Balkema, Rotterdam, ISBN 90 5809 161 9

Development and utilization of information technology systems in construction

John Christian
Construction Engineering and Management, University of New Brunswick, N.B., Canada

ABSTRACT: This paper reviews the creation and development of systems utilizing information technology in construction engineering and management which have been developed within the Construction Engineering and Management Group at the University of New Brunswick. Systems in the areas of risk, equipment selection, change orders, facilities management and life cycle costs, and claims are briefly described.

The effectiveness of the creation of innovative systems in the construction industry using high technological tools is briefly discussed. The ease of the introduction and integration of software and other tools such as spreadsheets, data base software, expert systems, shell programs, artificial intelligence into an industry which is fundamentally conservative and historically slow to change is examined.

1 INTRODUCTION

Analytical techniques and computer technology have played an important part in the creation and development of systems in construction engineering and management in the past two decades. The construction industry, a naturally conservative industry, has taken time to accept emerging technologies, but, in some instances, once the initial barrier has been overcome, it embraces the new technology fervently. In what appeared to be in a period of five years, the more technical segments of the industry went from regarding expert systems as too academic to readily adapting expert system technology in their systems. Indeed, sometimes it took only two years from renunciation to practical utilization.

Many analytical methods have recently been used to develop systems, such as regression analyses, modelling, neural networks, random deviation detection, and many more. Combined analytical and computer methods have also been used in decision support systems, expert systems, and modelling. Computer software has been utilized in systems such as shell programs, spreadsheets, hypertext, databases, and project management systems.

Computer programming was essential in the early days of the information technology era in construction and structural engineering but now is used mainly in systems which utilize software where some programming is required such as Visual Basic.

The scope of this paper is limited to an overview of some systems which have been developed in one group using emerging information technologies in construction engineering and management. Systems in the areas of claims, risk, change orders, equipment selection, and life cycle costs are briefly described to show which technologies have been used.

The objective of the paper is to show how the introduction and integration of innovative systems can utilize emerging technologies and thus benefit management in the construction industry.

2 COMPUTERS AND THE CONSTRUCTION INDUSTRY

In the last two decades an important decision for construction companies has been the selection of the most suitable type of computer software. Generally the cost of developing extensive in-house computer software within a small or medium sized company is prohibitively expensive. Construction companies therefore have usually resorted to purchasing application package programs to cater to their needs.

It has therefore often been left to research groups in universities or government agencies to suggest and point the way in the use of innovative technologies. Skeletal systems have often been developed as a proof of concept. Often these systems were not very practical but at least they showed what could be achieved.

Precise software requirements must be determined when computer software is to be chosen. There may also be secondary and peripheral requirements which should be considered. After this stage, a decision can then be made on the most compatible type of hardware for the software requirements. Many mistakes in the past have been made by reversing this decision making process.

Many general software packages are used in computer aided design and drafting, estimating, project management and scheduling. These software packages replace older methods in basic drawing to solid modelling visualization and animation. Software packages also assist in other construction management functions involving spreadsheets, word processing, project management, time reporting, work requests, scheduling and resource allocations, charts, accounting, cost control, and many other functions.

3 EXAMPLES

In this section systems are described which have been developed by the Construction Engineering and Management Group at the University of New Brunswick. These descriptions are just an overview of a limited number of systems which have been developed to show how emerging technologies and tools can be used, with more traditional analytical methods, to improve procedures in construction engineering and management. In many ways they are in the initial stages of development in the overall scheme of things and should be regarded more as proofs of concept.

3.1 *Claims*

A system was developed which used hypertext in the processing of construction claims (Christian & Bubbers, 1992). As is well known today hypertext is a method of text retrieval.

Rather than using conventional text which is designed to be read linearly from start to finish, hypertext consists of a collection of consequential text items or nodes. In hypertext, access to information is provided by links from one node to another. Jumps from one node to another are accomplished by activating the desired link, usually by selecting it with a mouse. Links are indicated within nodes usually by highlighting a word or icon which is relevant to the node to which it leads.

Before the advent of Windows and the many common uses of hypertext, this research helped pave the way in the early 1990's in the practical use of hypertext in construction management. It showed that a hypertext system could inform the user of the rights and obligations contained in the contract conditions. The hypertext nodes guide the user to relevant information contained in various documents. The system can inform less experienced users what constitutes reasonable costs and is capable of presenting the user with a contrary opinion thus making the user aware that claim decisions in construction are not always straight forward.

3.2 *Equipment Selection*

A computer based integrated interactive decision support expert system was developed for earthmoving equipment selection and estimating using a shell program (Christian & Xie, 1998).

The system contains a knowledge base, into which the knowledge obtained from experts in the industry and from other sources is stored, and three databases. It has the capability of selecting appropriate fleets of machines from sixty machines considered in the system, and estimates their outputs and costs, based on the given working conditions. Through a relatively easy question-answer routine of consultation, recommendations, with relevant results, can be rapidly made and presented by the system.

In the planning for equipment in an earthmoving operation, a decision should be made on what machines should be employed in the operation. In making such a decision, many interactions between engineering and economic considerations must be taken into account. By searching different sources and consulting with experts, knowledge on various common types of earthmoving operations was acquired and stored in the knowledge base of the expert system developed. To acquire knowledge regarding the type of machine versus the type of operation, and to implement the knowledge in the expert system, a rating system for selecting machines that are suitable to each operation was developed. Machines were weighted according to their appropriateness to each type of operation. Many experts contributed to this knowledge base.

Haul distance, haul road condition, rolling resistance, grade resistance, traction, availability of machines, machine costs such as depreciation, investment, fuel, service, repairs, operator's wages and many other factors were built into the system. This expert system used human expert knowledge, in a specific domain, to reach a level of performance achievable only by skilled human experts.

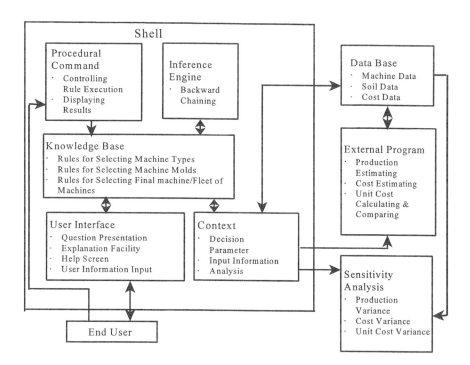

Figure 1: Structure of the System

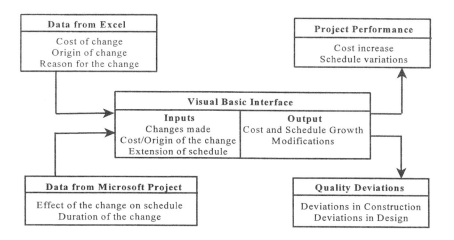

Figure 2: Interaction between the Various Components of the Software

The research showed that as a decision support tool, an expert system in earthmoving can assist an earthmoving planner, especially one who is less experienced. Even a skilled planner may benefit from such an expert system as it can quickly reach a solution that could be used as a reference or a datum. Refer to Figure 1 for the structure of the system.

3.3 *Life Cycle Costs*

A computer based decision support system was developed to provide facility managers assistance with life cycle costs (Christian & Pandeya, 1997). Regression analyses, neural networks, random deviation detection methods, and artificial intelligence were used to develop an expert system to act as the equivalent of an advisor or analyst and to provide supplementary information.

Quantitative data on the historical operating and maintenance costs of twenty-two facilities, along with knowledge of the factors affecting the costs were elicited through various sources. A costs prediction decision-support system was created using the analytical results and the knowledge acquired. The system created may be used to assist and advise on certain aspects of facility management, such as the estimation of operating and maintenance costs, and the development of preventive and general maintenance plans for facilities similar to those investigated. Linear and non-linear regression analyses, neural networks and random deviation detection methods were used in models to predict costs in future years.

The decision support system consists of five databases that contain information such as long-term space growth projections, overall cost information, overall maintenance information, the organization's difficulties and policies, and energy conservation opportunities.

The decision support system which was developed using the knowledge acquired showed that the application of an expert system is both feasible and appropriate in life cycle cost analysis because it takes into consideration the importance of human expertise in facility management.

3.4 Change Orders

Forty different projects were examined in detail to determine the effects of change orders on performance for various types of contract (Christian & Cariappa, 1999).

The number and the cost of changes were identified and changes were tracked to show the effect on performance and quality. The tracking of changes was made easier with the use of computers.

One of the aims of the research was to create an interface between suitable software, with the objective of facilitating the monitoring of the performance of projects, and to measure deviations in design and construction quality. It was necessary to select suitable software to achieve this objective. The decision was made that a spreadsheet and a scheduling package would be best suited for recording and analyzing the data generated during the projects. A search and analysis of the available software packages that could be customized to meet the required objectives was made. The software was required to keep track of the changes occurring during the project and show the effect the changes had on the schedule and the cost of the project.

As a result of the search, it was decided that Visual Basic from Microsoft would be used to customize certain Microsoft applications. Visual Basic programming language for Windows is the modern dialect of BASIC (Beginners All-purpose Symbolic Instruction Code) programming language. The two packages that were selected, in this case, to interface with Visual Basic were Microsoft Project and Excel as they could be modified using the macros and the Visual Basic editor. The requirements were that two software packages could be used to share the data and a single programming language could be used to modify the packages. The individual features of Excel, such as using formulae to calculate and derive output data, were significant. For example, the individual cells in a spreadsheet can be programmed to produce an output as soon as the inputs have been entered into the spreadsheet. Microsoft Project features include the ability to allocate resources to individual activities and to monitor the completion of an event and the time extensions to that event. Data is easy to record and obtain using these features. A Visual Basic interface was created for this purpose. Refer to Figure 2.

An advantage of the customized software is that it enables the user to input data for the individual activity simultaneously in both packages.

3.5 Risk and Schedules

This research developed a systematic way to consider and quantify uncertainty in construction schedules (Mulholland & Christian, 1999). The system incorporates knowledge and experience acquired from many experts, project-specific information, decision analysis techniques, and a mathematical model to estimate the amount of risk in a construction schedule at the initiation of a project. The model provides the means for sensitivity analyses for different outcomes wherein the effect of critical and significant risk factors can be evaluated.

The system includes the following three key features:

1. A hypertext information system for schedule risk identification
2. A spreadsheet to describe and evaluate project uncertainty
3. Direct pictorial information to assist the decision makers in selecting a realistic yet acceptable project completion time.

Part of the development of the system utilized a Macintosh PC and commercially available application programs called HyperCard and Excel.

In the system, the HyperCard application program provides an information module that can be used in identifying schedule risks. The Excel spreadsheet is the tool used for modelling the effects of the risks on the project performance time.

The hypertext system is used to store and give access to information concerning previously experienced schedule risks. The main system is composed of schedule risk information (facts, data, and heuristics) linked together using hypertext tools. The information can be in the form of text, graphics, or pictures. Thereby, typical project risks can be documented and made available to assist new project teams in becoming aware of general risk information and the possible inferences for a specific project.

The system's hypertext links provide the means to access the documents within the database. The links, which guide the user through the database, are divided into two types: organizational and navigational. Organizational links connect the structure of the system. In the database the basic elements of information are contained in objects. Once an object has been defined, it is possible to define navigational links that lead to other documents in the database. These navigational links were used to provide features such as importance factor criteria and confidence level definitions. A spreadsheet database was created to model uncertainty in the engineering design phase. The spreadsheet is also used to model the total project schedule risk. The information can then be displayed pictorially on the computer screen using the graphical functions of the spreadsheet program.

4 EMERGING TECHNOLOGIES

There have been rapid technological advances in computer hardware and software recently. In the future it is likely that a technological plateau will be reached.

Recently several researchers in the construction area have developed intelligent computer aided design and drafting (CADD) systems and decision support systems using artificial intelligent techniques, which included expert systems and neural networks. The real beginning of the computer age in construction had arrived.

There are now digital techniques for the storage, transmission, and presentation of much construction information by electronic means. This has enhanced integration, networking and connectivity processes. The electronic information highway is already further enhancing the distribution of construction information.

Current virtual reality technologies are not yet readily usable, but there is great promise. These virtual reality displays are interactive with head/hand referenced computer displays that can give the user an illusion of displacement to a site location. The illusory virtual reality display can be created by head set sensors to detect the user's movements, effectors to stimulate the user's senses, and hardware to link the sensors and effectors. Virtual reality could be extremely beneficial on construction sites because it is often difficult and expensive to correct problems which occur during the actual construction of a complex project. Problems can even be anticipated by spatial displays with a time dimension.

Research described in this paper has played a small role in the developments with emerging technologies in construction. A sample of notable developments have included the following:

- CAD and virtual reality in house building
- New generation of software for construction scheduling and management
- Information and communication technology in large scale engineering construction
- Construction planning knowledge based system for information exchange
- Software for case based reasoning
- Information technology decision support and business process change
- Opportunities for CAD in construction
- Case based reasoning and computers
- A computer system for assessing house renovation
- Use of computers in quantity surveying
- Application of expert systems to quantity surveying
- IT requirements for European construction industry
- Information modelling and sharing
- Modelling design process and knowledge
- Construction planning and monitoring technologies integrated with vision system
- Case-based reasoning model for contractor prequalification
- Construction information technology: where information is located
- Case based reasoning for materials selection in design
- Exchange of 3D CAD data and building models
- IT and the design and construction process
- Project management application models and computer assisted planning in total project system
- A control-oriented approach to management information systems

- A proposed information technology framework for road maintenance management systems
- Artificial neural network system for predicting project duration
- An expert system for construction planning and productivity analysis
- Using internet as a dissemination channel for construction research
- Computer representation of design standards and building codes - Canada - USA
- Architectural foundations of a construction information network

The list is not intended to be an exhaustive list of all the developments of information technology in construction but attempts to show the range in developments.

5 CONCLUSIONS

The construction industry is generally a traditional and conservative industry. It has taken time to accept emerging information technologies, but once the initial barrier is overcome, it often embraces the new technologies fervently. Emerging technologies have become simpler to use. For example, shell programs made expert systems much easier to develop.

The rapid development of information technology contributes to effective ways in reaching solutions which are sometimes difficult to obtain manually using conventional mathematical methods. Traditional methods employed in construction management are mostly quantitative. Construction management decisions, however, involve more qualitative information. Experience, judgement and intuition are often needed to make decisions against multiple decision criteria. Because of this, traditional approaches are often inadequate. The development of emerging information technologies provide a way to overcome the limitation of traditional methods, to a certain degree, by incorporating the experience, rules of thumb, and judgement of many knowledgeable experts.

Some of the systems described in this paper have been remarkably simple to create, but they have been time consuming to develop.

Systems which are developed for commercial use must be appropriate for the procedure. In the past, researchers have not been fully aware of the practicalities of the procedures, and the practitioners have not been fully aware of the capabilities of emerging information technologies. However, enormous strides have been made as we enter the new millennium. We have entered an era of solid modelling, computer based documentation and financial management, web and CD-ROM based continuous upgrading and updating, with virtual reality and animation on the horizon.

ACKNOWLEDGEMENT

Financial support from the Natural Sciences and Engineering Research Council of Canada for some of the research projects described is gratefully acknowledged.

REFERENCES

Christian, J. & Cariappa, A. 1999. "Using Software to Monitor Performance of Construction Projects", p.229-234, *Civil-Comp. Press*, Oxford.

Christian, J. & Mulholland, B. 1999. "Risk Assessment in Construction Schedules", *Journal of Construction Engineering and Management*, Vol. 125, No. 1, pp.8-15, American Society of Civil Engineers, New York, January/February.

Christian, J. & Xie, T.X. 1998. "A Computer-Based Integrated System for Earthmoving Equipment Selection and Estimates" (Invited paper). Vol. 5, No. 2, pp.1-18, *International Journal of Construction Information Technology*, England, Winter, 1998.

Christian, J. & Pandeya, A. 1997. "Cost Prediction of Facilities", *Journal of Management in Engineering*, Vol. 13, No. 1, pp.52-61, American Society of Civil Engineers, New York, February.

Christian, J. & Bubbers, G. 1992. "Hypertext and Claim Analysis", Journal of Construction Engineering and Management, Vol. 118, No. 4, pp.716-730, *American Society of Civil Engineers*, December.

Creative Systems in Structural and Construction Engineering, Singh (ed.) © 2001 Balkema, Rotterdam, ISBN 90 5809 161 9

A prototype portal to web based collaborative engineering

Z. Turk & T. Cerovsek
Faculty of Civil and Geodetic Engineering, University of Ljubljana, Slovenia

ABSTRACT: Internet is causing the fourth big wave of introduction information technology to construction. The first Internet services for engineers provided information. Current services attempt to provide a horizontal integration platform, that brings together the virtual construction company. The next generation should use the Internet for both horizontal and vertical integration of the construction industry thus achieving computer integrated construction (CIC) which also includes the providers of software, information and communication technologies and services. The topic is researched European Union's project "Intelligent Services and Tools for Concurrent Engineering" (ISTforCE). A prototype system is under development. In the paper we present the conceptual foundation, the six layered architecture and the prototype itself. In conclusions we analyze strengths, weaknesses, opportunities and threats of the approach.

1 INTRODUCTION

In 1993, at University of Ljubljana, we were among the firsts to study the use of the Web and related technologies and set up quite a few services aimed at the international scientific community, as well as practice. To date, the Internet has matured from being an academic toy to being the driving force of the current economic growth in all sectors of industry.

1.1 *Computers in construction*

Since the first uses of computers in the construction industry in the 1960s (Grierson, 1996), the industry has embraced new technology in three waves:

1. The first was powered by the number numerical analysis programs that were able to crunch larger and larger FEM models and have changed the way in which structural engineers do load analysis and proportioning.
2. The second was enabled by computer graphics. Drawing boards were replaced first by 2D drafting and more recently by 3D modeling packages.
3. In the mid 1980s microcomputers became ubiquitous and general office software, such as Microsoft Word, Excel and Project, is used to get most of engineer's work done or at least documented.

These three types of "killer applications" in construction businesses were all supporting just one type of engineers activity, also considered a core activity - the processing of input requirements into output designs. The other important aspect of engineers work (see Section 2.2) - interaction with others - have not been supported by technology. However, since the mid 1990s, the Internet and other communication technologies are changing that.

1.2 *Communication technologies in construction*

Communication technologies are the technologies dealing with the transmission of information. They are supporting the process by which information is exchanged. Figure 1 shows the evolution of some information technologies (mainly communication technologies) in construction. They are positioned according to their complexity (vertical axis) and construction specifics (horizontal axis). Note that the more specific services tend to use or rely upon generic ones and that more complex services tend to integrate simpler ones.

The first services were generic and included networked file archives, email communication and text based group conferencing. The Web provided a much friendlier navigation and presentation of the files on remote machines. It was as that time when first construction related content appeared. In the first 1000 websites there were only a couple construction related and one was by the author at the University of Ljubljana.

The first construction specific services used the Web to publish information, such as scientific papers, building codes, product specifications etc. Content was not project related. The next generation services were staring to use the Internet as a collaboration platform for the companies involved in a construction project. One such infrastructure, for example, was developed in an EU project ToCEE (Scherer at al, 1998). Just a few years ago it was proposed that a large design, consulting or investment company would host a collaboration platform for the projects it is involved in.

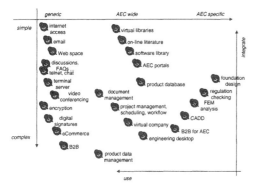

Figure 1: Evolution of Internet related communication technologies.

Since 1998, however, construction too has been following trends of general service providing on the Internet. A business models of the providers of various kind of Internet related software (e.g. for managing mailing lists, discussion forums, help desks, photo albums etc.) has evolved from selling software which the users would install on their servers, to providing a service on their Website, that offers the same functionality to the end user. In a construction context there are now dozens of companies providing collaboration tools such as document managing, project coordination and scheduling on the (Cerovsek and Turk, 2000).

1.3 Problem statement

Through the Internet based collaboration tools, engineers and architects can exchange data and communicate. But can they get their work done? Collaboration services cover only one aspect of engineering work. We claim, that an engineering desktop should be conceptualized in such a way that it offers the support for all kinds of human work. Most of the pins in Figure 1 show the services that support the unavoidable, but essentially overhead, non-value-adding activities. Support for activities where engineering work, such as design, actually gets done, is in only in the bottom right of the figure.

Secondly, Internet should provide a platform where not only horizontal, but also a vertical integration of construction businesses can occur. In ad-

dition to the engineers and architects, the platform should also integrate providers of services, software, consulting and knowledge.

For construction companies, such a platform should be accessible with current infrastructure, be portable and easy to use, without much training. To engineers it should provide a customizable desktop where they would go to get their work done. To providers of software and services it should provide a platform where they can safely sell their software, on-line services or consulting. Safety, security and functionality to charge for services is important.

1.4 Paper structure

In Section 2 we present the conceptual analysis of the system, based on an ontological framework of construction. Section 3 presents an architecture of the platform, illustrated in Section 4. Conclusions in Section 5 include a SWAT analysis of the proposed approach.

2 CONCEPTUAL FOUNDATION

The conceptual foundation for a Web based collaborative engineering consists of (1) the people, (2) the work they do, (3) the information they need and/or create, and (4) the tools and services they use or provide.

2.1 The people

Three main groups of people that will be meeting on the platform are:
1. Users or the traditional project participants that would typically be involved in the designing or planning within a project or several projects at a time. This would include engineers, architects, but also the project manager and the chief internet officer of the project.
2. The client or the owner of the construction project.
3. Providers of knowledge, services and tools which are using the platform to sell their expertise. Typically, their involvement would not be project specific, for example a company that writes a FEM analysis package would integrate it into a platform and charge for its use, but would not offer extra services in the context of a specific project. However, some services may be project specific, for example bringing a consultant into a project.

2.2 Human work

In a proposed ontology of construction IT and based on the philosophical theories about work and intelligence, two different kinds of understanding of what humans essentially do were identified (Turk and Lundgren, 1999):

- In the 1920s Taylor defined work as processing. In the assembly lines, workers got inputs, such as parts, performed some operations, and produced some outputs. In this paradigm construction is seen as a flow or transformation process in which raw materials and components are transformed into construction products. Similarly, the design and planning processes are seen as processes where, based on input data (e.g. architectural design) outputs (e.g. structural design) are generated.
- Based on the philosophical tradition of Heidegger, Winograd and Flores (1980) claimed that work is asking for, negotiating, accepting and fulfilling commitments. In this view, construction can be understood as a social network of people sharing a common goal of building. Similarly Schön (1983), in his observation of architects and managers, found that design and management are research processes with strong social components.

Each of these paradigms focuses on a different perspective of construction. Based on the first one, numerous research and development efforts were hoping to integrate the fragmented construction business by (1) standardizing the inputs and outputs of the various processes, so that the information flow is smoother. Standardized product models have been developed. Additionally, (2) the processes were analyzed in an attempt to complement the product models with the process models.

These approaches, however, neglected how construction has been functionally integrated, in spite of its objective fragmentation. It has been held together by the people sharing common background knowledge, common goals and who were communicating with each other.

We believe that the two paradigms complement each other and that a synthesis is required. The proposed taxonomy of human activities is shown in Figure 2. Humans communicate (with each other) or are involved in processing. The two communication types are commitment negotiation, where people give or accept commitments, for example a client talks to a designer, wants her to do a design and they agree what the design should be like so that the client would be satisfied in the end. In more traditional views they would be understood as command, monitoring and control of workers by the superiors.

Another type of communication is not explicitly related to design or planning work work, however, it is very important for a group of people working together. They are supposed to create a social network through which they learn about each other and establish a wide shared background. The speech-act communication theory (Searle, 1969) claims that the

motive of communication is not information exchange but an "act" which is supposed to trigger some kind of a response by the receiver of the message. By knowing the receiver the sender can be more confident, that the message would have the desired effect. One of a more important attributes of the partners in a project is the mutual trust and feeling of belonging to a group.

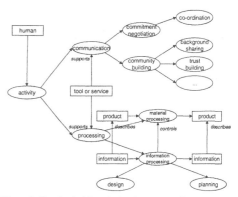

Figure 2: Ontological foundation for construction.

The processing works can be further decomposed according to their dominant output (Bjork, 1997). Construction processing works can be split into material processes that take raw materials, components and energy as input. Their results are material items such as components, materials or energy. They take similar material items as input. The other type of processes are Information processes - those of which major input and output is information. The main information processes in construction are design and planning.

The processing types are interrelated and the split is not entirely clean. For example, information is always stored on some material media that needs to be moved around - be it on paper, or as electrons. All processes create data that can be observed and used by information processes. Material processes are controlled by information processes – e.g. design information specifies how much reinforcement should be placed into a concrete slab; production plan defines when and who will install the reinforcement; payment scheme calculates when and how much should be paid.

2.3 Information

Information can be of one of the following types:
- project related - created in a context of a current project.
- company related - company policies, templates, knowledge bases, previous projects.

- wide area specific -building codes and regulations, legislation, best practices, product databases etc.
- global - in the globalization age, any of the last two information items can in fact have a global scope.

2.4 *Tools and services*

Tools and services support or enable the communication and the processing activities. They include:

- Software, either (a) such that has to be installed on client's machine, (b) dynamically loaded through the Internet or (c) rented in such a way that all or some parts of it do not run on a workstation, but on the author's computer. Examples of processing support are CADD packages or proportioning software. Examples of communication support are email, voice and video conferencing and collaboration tools, such as NetMeeting (installable), or Centra (for rent).
- Infrastructure services that don't generate or influence the design and planning information of a project, but enable the processes with a e-commerce support, security, digital signatures, authentication, backup, archiving, accounting.
- Databases that provide access to the information types listed in the previous subsection. Document management systems and product model databases, for example, handle project related data. Other types of databases, such as manufactured product catalogs provide access to the other types of data.
- Help desks, discussion forums, FAQs related to the above.
- Consulting and assistance services by a human.

3 ARCHITECTURE

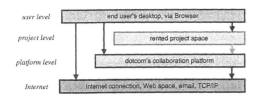

Figure 3: Architecture of current Web based engineering platforms.

The previous section has outlined the concepts that are integrated in a Web based collaboration portal. This section shows its architecture. The model of some of the existing Web based services is shown in Figure 3. Through the browser, the end user is accessing the rented project space. It was set up based on the collaboration platform provided by the Inter-

net company. The user can access any other resource on the Internet as well. Typically both the user and the project manager who rented the project space, can to an extent, parameterize it. Such platforms try to integrate the project participants but do not integrate the providers of services.

Figure 4 shows a proposed collaboration platform that also integrates bottom up. It provides means for the providers of construction related services and tools to offer and charge for their services through the platform. The platform has a layered structure, with project specific layers on top towards more and more generic services towards the bottom.

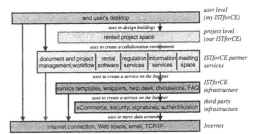

Figure 4: Architecture of the proposed collaboration platform.

The user and project levels remain essentially the same, but the functionality and the customization that can be brought onto the engineers desktop is not static, as offered by the platform provider, but a dynamic selection of anything that any provider of online construction services can provide.

The rented project space is not only a parametrically customized solution from one provider, today typically integrated around a document management system, but a dynamically assembled set of services, based on service template. A service template is a ready to run application based on which highly customized applications can be built. For example a document management system could have attributes, controls, and logic added to suit local needs. Process support could be created in such a way that local process models could be supported.

The partner services level contains the services that the providers are offering through the platform. For example document management solutions, workflow support, software that can be rented on the Web, on-line regulations, conferencing tools, FAQs, help desks and other real time collaboration support.

The fourth and fifth layers contain support for the providers of services in the previous layer. For example a consulting company would create a help desk related to the area of her expertise. To build it, it could use a template for making help desks easily, so that they concentrate not on the Web technology, but on the content. Services would not be free and to be able to charge for them, the providers might rent a service that handles on line payment. Similarly, they might track their users using some user profiling solution. These 5th level services are generally not construction specific.

Figure 5: CIO's interface while creating collaboration platform on the Web.

Figure 6: Registration of a new user.

Figure 7: Project data assess through mobile phone.

4 IMPLEMENTATION

The prototype is being implemented on the Internet using CGI interface and Perl programming language as the main technologies. All the user needs is a Web browser. All pages are delivered by a program, so that the specifics of the user's environment can be preserved, or pages for small mobile device's screen generated.

4.1 *End user session*

A typical session with the system would look as follows.

On request of the project manager or project owner, the chief Internet officer of the project would create project collaboration support. He would define the key project properties, select and customize the core collaboration functions that will be used in the project, for example a data management system, calendar, project planning, etc (Figure 5). He would then suggest other tools that might be useful in the project, for example some on-line analysis programs, a products database or an estimation tool. This would effectively create a home page for a project.

The manager would then manage human resources and invite various people from different companies to work in the project, assign their roles and associated right to the project data.

Each of the participants would see this new project on his desktop, as one of the projects on which he is supposed to work on.

Thins that he has to do, tools that he needs to use and people he has to work with, all will be available through a single platform.

The platform would be available using different styles - also from a WAP enabled mobile phone (Figure 7).

5 CONCLUSIONS

An architecture and some prototype work on a portal for collaborative engineering on the Web has been described. It is not only concerned with the horizontal integration of the architects and engineers, but a vertical integration of all professions providing construction related services on-line. The key component of the portal is a platform for the collaboration on construction projects as well as a market place for selling construction related services, tools and knowledge. The key components of the infrastructure for collaborative work are information exchange and communication tools. The key components of the infrastructure for the providers are service templates, security and e-commerce tools, so that they could concentrate on their core knowledge and not on Internet technology. In the remaining subsection we perform a brief SWOT analysis of the proposed approach.

5.1 *Strengths*

The presented approach, at least as presented conceptually, integrates the entire profession in which small and medium companies are in a large majority. It provides them a new model of doing business and all the necessary infrastructure. As such it can integrate the fragmented construction profession.

5.2 *Weaknesses*

The prototype is created using tools that allow for rapid prototyping but lack the robustness of the tools with which a professional platform would be built. We did not address issues like security and privacy. The system is not based on a monolithic product model and static project model, which is both a weakness but also an advantage. In the future we should expect that only some of the product information will be structured in product models. Coexistence of product models and unstructured, document based data remains an issues to be resolved.

5.3 *Opportunities*

Central management of project information should result in a digital archive of previous project. This could enable better reuse of old project data, analysis of the processes as well as synthesizing new knowledge about construction. The data could be used to support full life cycle of the structure. To service providers, a common point of entry of all users and a centralized user tracking could lead to better understanding of the users and their needs.

5.4 *Threats*

Companies providing core collaboration services could be tempted into using project data, either discretely or synthetically, to learn about the participants, about the ways construction work is done and therefore exploiting the implicit knowledge of the construction companies using the service and benefiting from them. An open collaboration platform where many small providers of services and tools can offer thme to construction are also a threat to established players in the field, who are interested in exploiting collaboration platforms to extend their monopoly in one segment of the market (e.g. CADD or project planning) over the whole industry. Such portals are also threatened by the general lack of economic soundness on the internet. In order to establish market shares, the dot coms are offering services nearly for free. Ingineering consultants and software authors cannot operate at a similar price.

ACKNOWLEGEMENTS

The presented research has been done in the context of an EU funded 5th Framework project called IST-forCE. The contribution of the funding agency as well as the project partners is acknowledged.

REFERENCES

Björk, B.-C. (1997). INFOMATE: A framework for discussing information technology applications in construction, In CIB Working Commission W78 workshop "Information Technology Support for Construction Process Re-engineering",Cairns, Australia, July 9-11, 1997.

Cerovsek T. and Turk Z., 1999-2000. Evaluation of enterprise collaboration solutions for the AEC Industry, IKPIR-ITC Report 2000/1, University of Ljubljana, FGG.

Grierson, D.E., 1996, Information technology in civil and structural engineering: Taking stock to 1996, in B. Kumar, editor, Information processing in civil and structural engineering design, Civil Comp Press, Edinburgh, UK.

Scherer, 1998. ToCEE.

Schön, D.A. (1983). The Reflective Practitioner - How Professionals Think in Action, Basic Books, UK.

Searle, J. R., 1969. Speech Acts. An Essay in the Philosophy of Language, Cambridge University Press.

Turk Z. and B. Lundgren, 1999. Communication Workflow Perspective on Engineering Work, CIB Publication 236, VTT, Finland, pg. 347-356.

Winograd T and R Flores (1987). Understanding Computers and Cognition, Addison-Wesley.

Creative Systems in Structural and Construction Engineering, Singh (ed.) © 2001 Balkema, Rotterdam, ISBN 90 5809 161 9

The impact and potential of the Internet for civil and structural engineers

D. M. Lilley
Department of Civil Engineering, University of Newcastle, Newcastle upon Tyne, UK

ABSTRACT: Modern electronic communication facilitates rapid access to vast quantities of information in ways which were inconceivable a few years ago. New technology brings new challenges, opportunities and threats for civil engineers. New possibilities exist for promoting engineering activities, enhancing contacts with clients, and improving recruitment into the engineering profession. Ease of communication can produce information overload, and recent media reports suggest that this can lead to stress amongst staff in the workplace. This paper reviews some of the facilities that are already available on existing Web sites, including registration of construction companies in the UK and developments that are likely to improve provision for distance learning. The potential effect of these and other developments in the not-too-distant future is considered. References are made to Web sites that may be useful to practicing engineers and academics.

1 INTRODUCTION

The rapid growth of the Internet has created the possibility of major changes in the management and availability of information, and offers exciting and imaginative opportunities for civil and structural engineers. Most of the effort in the early days of the Internet was aimed at providing basic information about companies and products, but this has rapidly developed under commercial pressure towards a more advertising-orientated culture.

Little information is available from existing literature about the impact of the Internet on professional activities of civil and structural engineers. Conflicts between existing business practices and culture and changes to requirements when new technology is introduced have been examined (Cowperthwaite et al. 1999).

Poor public perception of the work of civil and structural engineers over many years has led to lower financial rewards for most people within the industry when compared with other professions, such as medicine or law. This is particularly true for professional engineers working in the UK, and has been a recognized long-term problem. The situation has been made substantially worse by the highly competitive, almost cut-throat, approach to work taken by many consultants and contractors, and must be completely self-defeating when carried to extremes. A change in this competitive ethos to one of mutual collaboration and recognition would have a dramatic beneficial effect on rewards for work, self-esteem, and recruitment into the profession.

Electronic data interchange (EDI) has been steadily growing between clients, engineers, architects and other construction personnel and is now established as a reliable means of communication. Drawings and other documents can now be sent in electronic form to speed the design process. It also has the effect that drawings are usually printed at the place of use rather than the place of origin and this can generate a significant cost to the recipient.

Rapid retrieval of information is a basic aim of those civil and structural engineers who use the Web, but the vast quantity of information available often prevents retrieval of a particular item which may be most useful. Accurate searching techniques are essential in such cases, but existing intelligent systems for directing information to those with particular interests as soon as it becomes available are becoming more widespread.

Evidence of increasing use of the Web by engineers is provided from statistics available through the Internet (Institution of Civil Engineers 2000). In March 2000 there were a total of 274,074 page-requests to the Institution of Civil Engineers (ICE) Web site, which was substantially more than during the previous month, and an increase of more than 80% on the number of requests in March 1999. Most (17500) page-requests were made on Wednesday 1 March, and the busiest time of day was 1.30 pm. The most popular pages were those that referred to membership and professional examination matters, although this may have been influenced by forthcoming plans to change the ICE rules regarding some grades of membership.

Information provided by software automatically monitoring the use of the ICE Web pages indicated international interest from many countries including the United States, Sweden, Hong Kong, Brunei Darussalam, Australia, Canada, Singapore, Malaysia, and France. In common with previous months, most requests were made during week-days and Saturday was the most popular day at weekends. Similar trends in access and usage have been observed for Web pages provided for structural engineers (Institution of Structural Engineers 2000).

Access to up-to-date information from recent research or changes to regulatory codes of practice is highly desirable to engineers seeking advice or guidance on particular issues. Most of the major engineering journals (such as those of the Institution of Civil Engineers or the American Society of Civil Engineers) are available in electronic format, and can usually be accessed remotely through an Internet link to a university library or directly through the institution (American Society of Civil Engineers 2000) at modest cost. However, it is interesting to note that one of the UK engineering institutions recently decided not to allow electronic access to its journal for the time being as it thought this would jeopardize its marketing strategy for the existing paper copy. A development to notify those interested in a particular topic whenever new information on that subject was published would be very welcome, especially if this could include changes to codes of practice and standards.

2 SOFTWARE DISTRIBUTION

Design software is available over the Internet, and in some cases can be obtained free of charge. An example of this is SCALE (Fitzroy 2000), a large ensemble of programs for structural design and detailing. The software is supplied free to any Corporate Member of the Institution of Structural Engineers or the Institution of Civil Engineers on the understanding that it is for non-commercial use and subject to a number of other conditions. Programs in the form of text files are available for structural calculations, detailing, drawing and scheduling of reinforced concrete, steel, steelwork, masonry, and timber. Freely available software of this nature is a useful marketing strategy in that it allows a professional audience to be targeted to try a commercial product. Success can be measured in some way by the number of sales of commercial software made to those who first tried the free version.

Facilities commonly available now from software suppliers include the ability to update software and obtain "software patches" through the Internet. Some modern commercial software has become so complex that completely thorough testing of every routine and condition is virtually impossible, and problems can remain hidden for long periods after software has been released for public sale. Sometimes the effect of inevitable changes in working data or knowledge were not predicted or considered important at the time software was produced. This was a major concern towards the end of 1999 when problems arose with date information stored in particular formats within commercial software (the "Millennium bug"). Many identifiable problems were often readily resolved by reference to regularly updated information provided through the Web sites of software suppliers. The speed and extent of distribution of revised software that occurred prior to the end of 1999 could not have been achieved without the Internet.

3 EDUCATIONAL ISSUES

3.1 *Distance learning*

The Internet now provides access to facilities for distance learning, the benefits of which have yet to be fully realized by civil and structural engineers within the UK. A notable recent innovation (Blackboard Inc. 2000) has introduced on-line teaching and learning environments in a large number of colleges, universities, schools in the USA and overseas. The facility for participating students to search courses through a catalogue of listings offers a potential means of remote access for engineers wishing to study material for professional development. Flexibility in time and place of access for participants should reduce financial costs when compared with those normally associated with sending staff to courses away from the workplace.

Course developers can now build and manage their own course Web sites for any subject using software downloaded through the Internet. Institutions can create online campus "portal" environments that aggregate information from their courses and provide focus points and on-line campus communities. At the time of writing, the potential advantages of this facility are being evaluated within the University of Newcastle in relation to providing courses for post-graduate continuing professional development for civil engineers.

Some elements are provided as a free service to allow individual course leaders to add on-line components to their courses or deliver entire courses through the Web. If courses are to be offered on a commercial basis, facilities are available to allow course providers to register their courses at an annual cost of £65 ($100) and then charge students for access to their courses through an e-commerce system.

Numerous useful features are available to broaden the learning experience; these include threaded discussions, real-time informal discussion and whiteboard activity, assessment tools, messaging system and on-line exchange of files between course leaders and students. The system providers claim that the system has grown rapidly to include more than 27,000 course sites with 240,000 registered users and approximately 12 million page-views each month.

3.2 *Recruitment of students*

Stimulating interest in the work of civil and structural engineers amongst school-children and young people has been a long-standing problem in the UK, and lack of progress is thought to be a contributing factor to the year-on-year reduction in numbers of students applying to study engineering at university.

A common approach to recruitment in the past has involved a small number of engineers attending "careers" evenings or similar events and trying to attract young people using one or more display stands. This rarely seems to have been very successful, given the broad nature of civil engineering and usually limited availability of resources available for display. The Internet now provides the means to communicate with every secondary school in the UK, and its imaginative use could produce a major revival in interest amongst young people. Information about "good" Web sites (when available) needs to be communicated to teachers so that they can encourage their students to explore the Web sites. This encouragement could be made more effective if reinforced by a suitable advertising campaign aimed at young people.

All universities in the UK now have information published on the Web. This includes a wide range of material intended to promote activities relevant to prospective students, and often provides reference information such as University regulations, policies, staff lists, access to library, sport, and welfare facilities for current students, etc., etc.. A useful guide (University of Wolverhampton 2000) now provides a simple means of linking to the Web sites of all universities and higher education institutions in the UK.

Teaching provision within higher education in the UK is now subject to periodic scrutiny and inspection by a Government-funded body. Civil engineering and related subjects were assessed during 1996-1998 and reports produced following the assessment of each institution are available through the Internet (Quality Assurance Agency 2000). This has enabled students and employers to understand the strengths, interests, and performance of different university departments.

Academics are now very conscious of the need to improve or maintain academic standards and resources, and to have due regard for student support and welfare services. It is not at all clear how public availability of information about the quality of educational provision has affected recruitment to individual departments, mainly because national numbers of students applying to study civil engineering at university are reducing.

In the future more use is likely to be made of Web cameras showing major events that occur in construction (such as the erection in London of the "Millennium Eye") and it will be interesting to see how the industry reacts to such publicity. Sometimes there is resistance to providing information publicly from those who feel that this provides opportunities for critics to obtain information that may harm the business or interests of those concerned.

4 RECRUITMENT OF ENGINEERING PERSONNEL

Engineers seeking new or different employment can find information about appropriate opportunities through the Internet. In the academic field, many UK universities and higher education institutions advertise employment opportunities (Universities Advertising Group 2000) through the Web. Persons who are seeking employment in one or more academic fields can electronically register their interest and will then be automatically notified by e-mail when new posts in that area are advertised through the Web. The facility to send information to those with direct interest will undoubtedly benefit all concerned, and may well have long-term implications for employment, particularly at times of shortages of skilled and talented people.

An enhancement of this scheme has been suggested to allow prospective academic employers direct access to curriculum vitae of those individuals willing to provide them. This would mean that they could be automatically considered for suitable posts as and when these become available. If implemented, the new scheme could have major effects on academic recruitment as "head-hunting" would be easier for employers, and fixed-term contracts might become increasingly more common.

A similar scheme for sending important up-to-date information to those requesting it is provided by the UK Government (Foreign and Commonwealth Office 2000). Travel advice is issued for UK citizens travelling abroad and can provide warnings of problems in distant overseas places. Information is provided about every country of the world; and it is updated regularly, sometimes two or three times a day.

5 ARCHIVE OF INFORMATION

A common problem for many engineers undertaking refurbishment of existing structures is lack of available information relating to the final form of construction. It is possible to conceive the development of an electronic archive of such information that could be provided through the Web. In some cases access to sensitive information relating to strategically important structures would need to be controlled for reasons of security to reduce the potential risk of attack by terrorists or other malicious persons. There may also be strong commercial reasons why some details of a building or structure should not enter the public domain, and a balance may need to be found so that potential major benefits from such a scheme are not lost.

6 THREATS

6.1 Security and well-being of staff

Free access to personal information can give rise to potential problems, particularly if it provides data to those seeking to harm those to whom the information relates. In the UK, care needs to be taken to ensure that the Data Protection Act is not breached.

Disclosure of personal information can infringe the privacy of those involved, and may be deemed as undesirable by those individuals. An example of this in a university context is the publication through the Internet of photographs of members of staff, which is meant to assist students and others in identifying their tutors. Some staff may feel that the photographs could be used for malicious purposes and may attract the unwanted attention of "stalkers". At Newcastle, the decision of whether to publish a personal photograph on the Web has been left to individual members of staff.

As more people and organizations become connected to the Internet, the volume of e-mail appears to be continually increasing. This volume can itself create a problem of management for recipients who need to find time to read incoming messages and prioritize subsequent actions. Recent reports in the UK media have cited the increasing volume of e-mail messages received by office staff as a significant contributing factor to personal stress in the workplace.

6.2 Commercial security

Transmission of important data for commercial reasons is still regarded with suspicion by many because of a perception of lack of security and potential fraud. Encryption of data transferred via the Internet is meant to provide high levels of security, but few believe that the coding systems are completely reliable.

International law is unclear in relation to fraudulent dealing through the Internet, and this should be borne in mind when considering payment for goods or services offered for sale through the Web.

Uncertainty arises with regard as to when contracts made through the Internet become legally binding. The "old rules" stated that acceptance was confirmed when a letter or paper document was placed in a post-box, not when it arrived at its destination. The UK Government is currently working on its electronic commerce Bill, which may become law by the end of 2000, but until then it is advisable to seek confirmation of the arrival of important e-mail messages from the intended recipient.

Potential legal problems are sometimes identified by companies providing professional indemnity in that electronic records of information can be changed after technical problems have arisen. If clear evidence of original design is required then this is only possible with a paper copy, which is usually agreed to be kept by the consultant. There are considerable costs in producing, storing, and retrieving paper copies of documents for projects. The author is aware the documents of one project valued at £100M ($150M) required 2500 lever arch files, and an estimated saving of £250,000 made by one company on one major contract by using electronic transmission and archiving of documents.

6.3 Loss of expertise

An ad-hoc survey of local civil and structural engineers in Newcastle revealed their Web pages were written either by a willing member of their IT staff or were contracted out to professional Web page providers. In both cases the company or organization is vulnerable either to changes in personnel, i.e. if the member of staff decides to move to another company, or to changes in the pricing policy of the Web page authors.

At present it is rare for practising engineers to have the necessary skills to be able to develop in-house Web pages, and the situation is very similar to that which existed some 15 years ago when word-processing software and personal computers were becoming established. At that time, few engineers had the skills to compose and electronically process a report or other document directly, and most would have relied on the skills of a typist. Most students entering UK universities already have a reasonable working knowledge of word-processing. There is little need to teach basic skills in this respect at undergraduate level, but advanced topics such as use of graphics or equations are often taught as useful supplements to existing knowledge. A question remains as to whether or not skills in authoring Web pages should be taught as a "transferable skill" to undergraduate students.

6.4 *Viruses*

Viruses within files transmitted through the Internet can pose major threats to organizations, and much effort is expended by Internet Service Providers (ISPs) to prevent viruses entering or being transmitted around computer systems. Recent media publicity surrounding the arrival of the "Love Bug" virus in May 2000 and its ensuing problems highlighted rapid changes which were made to impenetrable electronic barriers or "fire walls". These filtering systems were modified to examine all mail entering or leaving major sites and any items containing identifiable viruses were automatically destroyed.

Electronic communication systems are and probably always will be vulnerable to threats from malicious persons, or those who simply wish to try to beat security systems. Every strategy or guard developed to prevent unauthorized access to information on the Internet poses a challenge to would-be "hackers", and keeping one step ahead will always be difficult for companies providing Internet services.

7 LOCAL INFORMATION FOR ENGINEERS

Recent reductions in the number of practicing engineers attending evening meetings and seminars organized by the Institution of Structural Engineers in North East England prompted the introduction of a system of information transmission and reminders sent by e-mail and the Internet. Over a period of about six months, e-mail addresses of many civil and structural engineers were collected into an electronic distribution list. In June 2000, the system allowed a single message to be sent to almost 200 engineers within a few minutes, with more names being added to the list every week. Reminders are sent about two or three days prior to meetings and attendance at meetings has significantly increased, in some cases by a factor of four. Local engineers have welcomed this development, and similar systems are being developed in other regions.

The advantage of sending information such as that outlined above by e-mail is that it does not require the target audience to be pro-active in searching the Internet. As there is no guarantee that information made available through the Web will be seen by its target audience, systems which ensure delivery of important messages are much to be preferred.

Information about local meetings will also be provided on Web pages in addition to contact details of the current Branch officials, and those relating to Continuing Professional Development (CPD) and training activities. Pages describing local student/ graduate activities, and those which might appeal to local teachers and school-children of different ages are being considered.

Risk management is a very important factor for engineers supervising site construction, and bad weather can cause expensive delays and damage to partially completed structures. Knowledge of the likelihood of sudden changes in weather allows those responsible to make decisions about the timing of site operations with greater confidence. Internet mail systems can easily be used to give up-to-date weather information and widespread advance warning of adverse conditions. It is unlikely that weather forecasting will ever be an exact science, and individual engineers will still need to make decisions based on the probability of weather events and its consequences.

8 REGISTRATION OF ENGINEERS AND COMPANIES

Imaginative and appropriate use of the Internet within the construction industry has allowed the development of Web pages to provide useful information for prospective clients. In the UK a register of civil and structural engineering companies (Capita 2000) has recently been developed to which prospective public sector clients have free access via the Internet. The register is one of the first UK Government's new Public Private Partnerships, with key aims to cut costs, improve the quality of public sector procurement and reduce the risk of fraud.

Contractors and consultants seeking public sector construction work in the UK can apply to be placed on the register regardless of size of organization. Checks are made against financial, managerial and technical criteria to pre-determined standards agreed with the UK Department of the Environment, Transport, and the Regions (DETR). These include an evaluation of professional qualifications and competency of key staff and a check of three references from previously completed work.

Firms fulfilling the criteria for registration benefit from automatic qualification for public sector projects, avoiding the need to complete the qualification forms otherwise required by individual public sector organizations. It has been estimated to cost a firm approximately £300 (US$450) in time and money for every qualification to a public sector approved list. This cost may not be recovered as there is no guarantee that any work will be awarded. Centralized collection and assessment of information eliminates the need for clients to make separate checks and considerably reduces the cost to clients in both time and money.

The overall annual cost of repetitive qualification procedures in the UK has been estimated at about £130M (US$200M), and it is envisaged that the register has the potential to save 75% of this cost. Annual public sector spending on construction in the UK is approximately £24,000M (US$36,000M); in 1998 contracts totaling £9,000M (US$14,000M) were placed with firms on the register. Over 60% of contracts last year were valued at less than £100,000 (US$150,000), and numerous clients requested information about more small and regionally-based firms.

The author is, however, aware of reports in civil engineering media indicating that some users are less than happy with the service provided by the register. The main problem appears to be poor response time from computer systems providing electronic access. Such problems are not unusual, particularly when new services become available and future demand cannot be accurately forecast. Improved technology is planned to resolve this problem and it is understandable that those responsible would wish to wait for reliable estimates of use before committing expenditure to more responsive network systems.

Another initiative is the Construction Industry Trading Electronically (CITE) set up in 1995 by some major contractors to develop and encourage the use of electronic trading in the UK construction industry. Member companies, who now include contractors, major suppliers of building products and quantity surveyors, can transfer data in common electronic interchange formats to any global location through Internet e-mail systems.

There is a need in the UK to provide information at local level for potential clients who may need structural engineering advice, or possibly a structural investigation or design. Details of suitably qualified engineers and organizations in local areas could be provided through the Internet by a "responsible body" such as the Institution of Civil Engineers or the Institution of Structural Engineers at modest cost. Provision of hypertext links would enable easy access to Web pages provided by individual companies or organizations. Public facilities for checking professional qualifications of individual engineers could be provided through the Internet by the engineering institutions (ICE, IStructE, ASCE, etc.).

9 CONCLUSION

Future developments in Internet technology will provide civil and structural engineers with important information as soon as it becomes available. It seems highly likely that, in cases of public safety, mandatory requirements will be enforced to ensure appropriate action is taken on receipt of information through electronic means. Ignorance of public information is no defence in British law, and modern information systems not only send messages instantly, but can record the arrival time at destination and when messages were read. It thus becomes possible to identify the exact time information was received and places

an increased obligation on the recipient to react appropriately.

In the near future, contact with Internet systems by mobile telephone will become more common, although it seems unlikely that this will have the same dramatic effect as that already provided by mobile telephones for personal communication.

Recent history of computer applications such as word-processing and spreadsheets suggests systems that will be used in, say, fifteen years' time will be considerably different to those used now both in appearance and in method of operation. The only certainty about the future is its uncertainty.

Doubt and uncertainty will always exist in civil and structural engineering projects, and no information system, however good, can ever remove the element of risk from construction in the real world.

Reliable systems for filtering and prioritizing Internet mail will be required so that important messages are clearly identifiable.

Limits to the use that can be made of Internet are those that exist in our imagination. Future success of the industry and the engineering institutions can be greatly influenced by innovation today and enhancement tomorrow. The Internet offers enormous opportunities for development of the engineering professions, and unparalleled opportunities to influence public perception of and access to civil and structural engineers. This will undoubtedly challenge the role of the professional engineering institutions (such as the Institution of Civil Engineers) in terms of promoting their profession and their members.

REFERENCES

American Society of Civil Engineers 2000. Civil engineering database. http://www.pubs.asce.org/cedbsrch.html.
Blackboard Inc. 2000. Blackboard – Bringing education online. http://www.blackboard.com/.
Capita 2000. Home page for the Constructionline register for clients and consultants. http://www.constructionline .com/
Cowperthwaite, S., Raven, G. & Richards, M. 1999. Integrating culture and technology. *The Structural Engineer*, London, January, 77, 1, 16-20.
Foreign and Commonwealth Office 2000. Home page for the UK Foreign and Commonwealth Office, London. http://www.fco.gov.uk/.
Fitzroy 2000. Home page for SCALE suite of programs for structural design. http://www.fitzroy.com/.
Institution of Civil Engineers 2000. Home page of the Institution of Civil Engineers. http://www.ice.org.uk/.
Institution of Structural Engineers 2000. Home page of the Institution of Structural Engineers based in London http://www.istructe.org.uk/.
Quality Assurance Agency 2000. Quality Assessment Reports - Civil Eng. Index. http://www.niss.ac.uk/education/hefce/ qar/civil_engineering.html.
Universities Advertising Group 2000. Jobs and studentships in the academic community. http://www.jobs.ac.uk/.
University of Wolverhampton 2000. University of Wolverhampton UK Sensitive Maps - Universities and HE Colleges. http://www.scit.wlv.ac.uk/ukinfo/uk.map.html.

Creative Systems in Structural and Construction Engineering, Singh (ed.) © 2001 Balkema, Rotterdam, ISBN 90 5809 161 9

An integrated web-based virtual model to support water treatment projects

G. Aouad, M. Sun, N. Bakis & W. Swan
School of Construction and Property Management, University of Salford, UK

ABSTRACT: This paper describes an integrated web-based virtual model that is being developed at the University of Salford in the UK to support water treatment projects. The model supports the integration of design, cost estimating and time planning information using the concept of an integrated project database that is interfaced to a web-based virtual reality environment to enhance the visualisation and sharing of information within water treatment projects. The model has been developed in response to users requirements and a phase of requirements capture has been undertaken. The data has been collected through a series of interactive workshops whereby the industrial organisations involved in the project have clearly articulated the system's specifications for a better water treatment market. In the developed system, the user can perform what-if analysis which will allow the optimal design of a certain site to accommodate as many water tanks as possible. The system would then generate all the cost estimates and time plans which can be visualised in a web-based virtual reality environment. This will give clients the opportunity to remotely view any water treatment development and look at the cost associated with it and in some instances perform some changes to specifications to look at the time and cost implications. The system is now being validated by a water treatment firm in the UK and some feedback on this will be included in the paper. In addition, this paper will describe the conceptual modelling phase as well as the implementation issues associated with the integrated system. Finally, this paper presents some of the findings in benchmarking the types of benefits that could be attained from using such a system when compared with the conventional approach to water treatment developments.

1 INTRODUCTION

Water treatment projects are generally very expensive to design, construct and maintain. The industry can undoubtedly benefit from simulation techniques that would allow what if analysis to be performed at a very early stage of a project in order to produce designs that are constructable, maintainable and cost effective. This industry needs to develop and use innovative and creative technologies, being construction or IT to overcome some of these problems. This paper addresses some of the state of the art IT technologies that could be used to enhance the performance of the water industry. This paper presents a web-based virtual model and its potential uses within the water industry. The model is centred around a suite of integrated applications that communicate within a central model. The prototypes have been developed in C++ on the PC. The integrated database supports the functions of design, estimating and planning by allowing these phases to effectively share information dynamically and intelligently. The system revolves around a central object-oriented information model. The information model contains a core which captures knowledge about the means of transferring information across domains involved in a water project. All the models in the system are fully independent of specific applications, and each domain model provides support for general classes of a given application.

Many studies have been conducted in recent years recommending the use of integration and VR/3D technologies to simulate and model construction processes. Fischer (1997) proposed the use of 4D technologies (3 dimensional technologies plus a time factor) to model and simulate construction processes. Recent work in the application of VR in construction includes that of Jones and Webb (1997) and Ha (1997). Jones and Webb developed an open VR system that allows for better communication across the various participants of a construction project. Three demonstrators have been developed as part of this work which support activities such as contractor briefing, environmental impact and acoustic modelling. VRML (Virtual Reality Modelling Language) standard was used to implement these demonstrators.

This has resulted in a relatively cheap web-based open system which can allow many participants in the construction sector to benefit from such work. However, it has to be said that this development is still at the prototyping stage. Ha (1997) has developed a VR-based design co-ordinator that can be used by designers at the early stage. The system has been implemented by integrating VR, databases and video conferencing technologies. The main limitation of this system is the lack in its web-based capabilities. Other work includes that of Stone (1995) and Griffin (1995) on how VR can be used as design tools. Retik (1995) and Lorch(1995) highlight the importance of VR as animation and simulation tools. Alshawi & Faraj (1999) suggest that VR should be used as the interface for 3D models and databases. Most of the aforementioned work has impacted on the types of technologies proposed in this paper which can be used for water specific projects.

2 THE PROPOSED SYSTEM

Figure 1 illustrates the system architecture for the proposed model that supports better integration of information within the water market. The targeted software applications include AutoCAD, a project management tool, an estimating tool, and a virtual reality package. Each package stores and retrieves information from the database independent of the other packages. For each package, an interface has been built that translates data from the package's internal representation to the data model. The synchronization of these packages is controlled by a separate module, called the Process Controller. This is in-line with the process-driven integration approach adopted for the system. This Process Controller performs the role similar to a project manager. It allocates database access permission to the included software packages, their operation sequences, information flow, etc.

2.1 The Design Interface

A data model describing the design of water tanks was developed in AutoCAD. This model has been implemented as an object oriented database (objectstore). Each concept when implemented can be referred to as a class. The classes supported by the object-oriented database have their counterparts in the AutoCAD drawing package.

An interface between AutoCAD and the Objectstore proprietary database was implemented. The AutoCAD user interacts with its drawing editor using familiar commands. Graphical representations of water project elements elements can be produced within the AutoCAD package. However, the information regarding these elements will be stored

in the object-oriented database rather than in the AutoCAD drawing database. This approach allows information about the geometry, shape, cost, material, of an element object to be stored within the same database.

AutoCAD was chosen because of its popularity in the construction industry. However this package is used as a front end to display images and graphical representations, but the information related to these objects is stored in the object oriented database rather than the AutoCAD drawing files to enable the proper modelling of design information.

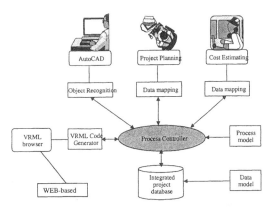

Figure 1. the system architecture

2.2 The estimating interface

Once a design model is created through AutoCAD, the estimating application can be used to examine the costs of the project. A resource-based estimating application was developed. It includes a library of standard work items and standard resources, which can be company specific, to identify work items and resources for a project. The user can start a new project and copy this library into the new project and use the most appropriate work items. Changes can be made to the quantity, rate, etc. and see the cost implications. Alternatively, the user can generate the work items and quantity take-offs from the design model. The work items will then pull the appropriate resources associated with them. The user can experiment with the design model and see the cost changes within the estimating package. The cost data was taken from recently completed work and the individual work items from analysis of the construction process and pre-defined cost/item libraries used by the UK industry. The application used to represent the cost-estimating package was Microsoft Excel. The choice of application is significant and a decision is based on connectivity (technical) and rate of use (economic).

2.3 *The project planning interface*

The project planning interface is based on the Microsoft Project application. The choice of this application is based on the popularity of the software among the potential client base, as well as the ease with which it can be readily integrated with the database. Principally, the project planning interface adds a timeline to the cost/item equation, and is used to sequence and time individual work items identified in the analysis of the construction process. In some instances, aggregated sequences are already pre-defined by the industry, and these are identified and incorporated into the design of the database. The user is free to make adjustments to the time scale for each work item or sequence, and see how this impacts on the overall cost.

2.4 *The virtual reality interface*

The potential of using Virtual Reality (VR) as an interface for an integrated project database using the World Wide Web is of crucial importance. VRML (Virtual Reality Modelling Language) is one of the newest open technologies on the web. It allows the creation of 3D views that can be explored in real time. With the rapid development of computing and communications technology, it is now possible to access information residing on a computer remotely. This capability can be exploited in an integrated construction environment to maximise the information sharing between different professionals. For instance, site engineers can query the database from their sites if they have access to a modem and the Internet.

Virtual Reality has usually been regarded as a visualisation tool. However, the system described in this paper has used VR as a user interface. This will allow the construction practitioners better access to information and motivate them to use integrated databases.

The Virtual Reality (VR) application reads information about the design produced in AutoCAD from the database and displays it in a virtual reality environment. This provides better visualisation using the web-based VRML. This utility is used as a means of interrogating the integrated database remotely over the Internet.

The user can interact with individual components of a structure represented in VRML by simply clicking with the mouse. Information relating to the component is then retrieved from the database and displayed in an information window inside the browser.

2.5 The process management interface

Projects in the construction industry are characterised by large numbers of actors working concur-

rently at different locations and using heterogeneous technologies. The system is designed to support such kind of collaboration by developing an integrated construction database that can be accessed by any of the actors. To ensure consistency and integrity of the shared project database, there need to be some constraints on how each actor can and cannot interact with the database.

The Project Manager utility developed aims to provide process management functions usually performed by a project manager. These functions include give permission to communications between the different actors at a given time and to monitor the progress of each task of the project. An example of use of this utility is to limit access and manipulation of a given project design to only participants which are involved in this process. It can also be used to freeze or unfreeze a design on the request of the client who might be happy with a version of design and therefore needs not to be altered. This can be also applied to project costing, planning and virtual reality interface.

3 THE POTENTIAL USES OF THE SYSTEM

The potential benefit of using the system is in the automation of quantities take off, cost estimating, time planning and the remote access of such information. The various parties will be able to quickly produce a prototype that can be evaluated in terms of design, cost and time implications and the identification of buildability problems. The application of the system in the water industry will undoubtedly change the way projects are conducted within this industry. Through a pilot study implementation, the project seeks to demonstrate how integrated project database can bring immediate business benefit in water projects. The open architecture of the system will ensure that this will be achievable. The functionality, however, will be enhanced by developing a large number of objects that allow the system to be used on real life projects and tested.

4 COLLABORATION SUPPORT

The developed system intends to support the collaboration of a multi-disciplinary project team located in different parts of the country; some may be on-site, others located at an administrative office. What each has in common is access to a computer and the Internet. Computer based communication and collaboration tools will then be used to connect project participants and support the information flow and processes of the design and construction activities.

Table 1. Performance Framework

Process - Process Title		
Description of Process	Description of process prior to the implementation of System	Description of process after the implementation of the system Database, indicating how the process has changed.
Stakeholder/s	Identify those involved in system	Identify those affected by change to the system

Benefit Type	Indicator	Benefits
Efficiency (£)	These are quantitative indicators that will generally be measured in cost or time.	These will be the value of the change in the quantitative indicators between the pre and post implementation stages
Effectiveness (Q)	These are qualitative benefits that the system may bring. They may not have a direct monetary value, but improve the quality of the way a process is carried out	The qualitative benefits will be described. This could include customer satisfaction or an improvement in quality.
Business Performance (S)	These are benefits that reflect wider improvements in the process that may reflect long term strategic goals.	These are broader qualitative issues that will be recognised at the strategic level such as development of IT strategy or improved long term competitiveness.

Collaborative working using computers is the theme of two research areas; computer mediated communication (CMC) and Computer Supported Co-operative Work (CSCW). In practice the two areas often overlap in producing actual technical solutions. CMC is concerned with both synchronous and asynchronous communication using computer networks as a medium. The communication media include audio, visual or a mixed format. CSCW applies CMC technology to solutions for collaborative working practices by providing a centralised work store, version control, concurrent work processes, etc.

The advent of the Internet has greatly enhanced the operational scope of both CMC and CSCW. There is now a wide range of ready-made tools aimed at supporting projects where participants are potentially widespread. The developed system will explore the full potentials of these tools and design a communication infrastructure to ensure that the integrated project database prototype is used effectively by the participants.

To facilitate flexible communication, as well as file sharing, between participants in the design process, a computer supported collaborative work tool known as BSCW, was chosen. BSCW has been developed by a German Institute for Applied Information Technology and its principal advantage is that it is completely web based, without the need for a separate client application or plug-in. BSCW provides a secure environment to provide access to user specified interest groups, exchange messages and arrange meetings.

5 PERFORMANCE INDICATORS

The following framework has been developed to assess the performance of the developed system. It describes the types of processes and stakeholders involved

The performance model is being tested within a contracting firm to evaluate the benefits that can be obtained from using the model. Savings in terms of time and cost and improvements to quality have been identified by the company evaluating the system. Also, better innovation and learning have been considered as long term goals that the company will be able to realised from using the developed system.

The performance model is being tested within a contracting firm to evaluate the benefits that can be obtained from using the model. Savings in terms of time and cost and improvements to quality have been identified by the company evaluating the system. Also, better innovation and learning have been considered as long term goals that the company will be able to realised from using the developed system.

6 CONCLUSIONS

This paper presented a web-based virtual model that supports the better integration and visualisation of water projects information. The model is centred around a project integrated database that supports the function of design and cost and time planning information. The system offers potential benefits to the water market through the proper integration of the various participants involved in a water project. The whole supply chain within the water market including clients, consultants, and constructors will benefit from the development of an integrated model. This paper has demonstrated the types of benefits that can be obtained from using the developed system.

7 REFERENCES

Alshawi, M & Faraj, I. Integrating CAD and VR in construction. Proceedings of the Information Technology Awareness Workshop. January 1995, University of Salford.

Fischer, M. (1997). 4D technologies. Proceedings of Global Construction IT Futures Lake District, UK, April 1997, pp 86-90.

Griffin, M (1995). Applications of VR in architecture and design. Proceedings of the Information Technology Awareness Workshop. January 1995, University of Salford.

Ha, R (1997). VR-based design co-ordination in distributed environments. PhD thesis, Strathclyde University.

Hubbold, R and Stone, R (1995). Virtual reality as a design tool in Rolls Royce. Proceedings of the Information Technology Awareness Workshop. January 1995, University of Salford.

Jones, B and Webb, I (1997). Interactive visualisation through applied virtual reality in construction. Proceedings of Visualisation in Construction, London, August 1997, pp166-172.

Lorch, R (1995). Animation in communication. Proceedings of the Information Technology Awareness Workshop. January 1995, University of Salford

Marir.F, Aouad.G, Cooper, G. (1999). OSCONCAD: a model based CAD system integrated with computer-related construction applications, Accepted by ITcon the electronic journal.

Penn, A et al (1995). Intelligent architecture: rapid prototyping for architecture and planning. Proceedings of the Information Technology Awareness Workshop. January 1995, University of Salford.

Retik, A & Hay, R (1994). Visual simulation using VR. ARCOM 10th conference, Vol 12, 1994. pp 537-546.

Creative Systems in Structural and Construction Engineering, Singh (ed.) © 2001 Balkema, Rotterdam, ISBN 90 5809 161 9

A study acquiring for 3-dimension data by measurement method using digital camera and GPS

Hirokazu Muraki, Shigenori Tanaka, Hitoshi Furuta & Etuji Kitagawa
Faculty of Informatics, Kansai University, Japan

ABSTRACT: In this part of study we have got efficiency 3 dimensions data by using GPS and digital camera and we devised the measurement method. After this according to measurement method of one reference point we constructed one system and verify it. As for the condition of observation we took photograph from at least three low places and we established one reference point in the side of the survey object, also we requested and learned this coordinates by using GPS. The process of the analysis was, observation of the place from where we took photograph and set up the reference point, measured the image of the digital photograph, occurred the one by one digital photogrammetry, we requested the relation between survey object and digital photograph, and then by using method of space inquire area correlation we acquired automatic three dimension data.

1 INTRODUCTION

In our country we are suffering from damages which occur frequently by natural disasters, especially many lose from disasters of earthquakes, heavy rains etc. Also in town structure and mountain, which has been collapsed by earthquake, by cause of some small external force or rains etc it is easy to occur secondary disasters. For prevent this disasters it is necessary to have detailed three dimensions data of the places which is on dangerous to occur disasters. In this part of study we made as purpose to establish one technique in which by using GPS and digital camera we be able to get efficiency and in details three dimension data.

2 METHOD OF MEASUREMENT

In the former photogrammetry, we have already studied about combination analysis of GPS and terrestrial survey data on reference literature 1, and on reference literature 2 we have studied the systematized of three dimensions data handle by using GPS. In the former photogrammetry there was a method to get 3 dimensions data of survey object without set up any reference point but it was hard to get adjustment of the existing geographical features data. Because of this reason it was necessary for the survey object to set up many reference point. However in the collapsed such as dangerous place the establishment of ref-

erence points is substantial but also and impossible. In this part of study, with basic the way of thinking of reference literature 1 and 2 we established a theory of measurement method with one reference point. In this method on the side of survey object we set up one reference point, and with a digital camera that its position has been observed by using of GPS took a picture and with use the method of differential positioning we calculated the position.

Differential positioning method has the advantage to be able to make sure of position error which is±0.5m~1.5m.This error become because there is a compact of machine and material, and also because we need real time for the point positioning. We made as condition that we took photograph from at least three different viewpoints (see figure1). Then the M*N mesh which has been occurred on survey object by the acquisition of the three dimensions data, by using the law of space inquire area correlation on the digital photo we made automatic analyze of intersection depth of the mesh and we examined the price which Z got for the fix X, Y mesh. The characteristic of this method is that we can measure with out use the image that was impossible to made up by stereo matching method; measurement width becomes equable so we can get data, and we can choose the place from where we take photo without restriction.

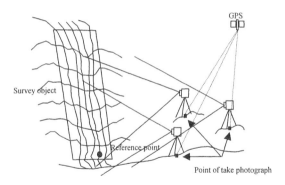

Figure 1 Measuring image

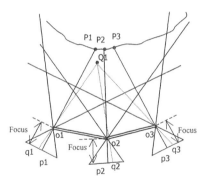

Figure 2 Establish of reference point

3 OBSERVATION OF THE PLACE FROM WHERE TAKE PHOTOGRAPH AND THE REFERENCE POINT ESTABLISHMENT

In this part of study we establish the three dimensions measure theory with one reference point. As figure1 shows, on the side of measurement object or left or right of it, in the less dangerous place we set up the reference point, and from the relation of the place from where we take the photograph (at least three) and the reference point, by using the triangulation method we look for the coordinates of the new reference point position. The new reference point position become when from the coordinates (a) of the place from where we took photograph and coordinates (b) of the reference point which we took from the digital image of the photograph we look for the position of the P1, P2, P3 by using the theorem of sine. The angle that we use for the theorem of sine is $\angle p1o1q1, \angle p2o2q2, \angle p3o3q3$ and calculate according to the coordinates of the observation (b)

(a) already established point coordinates(point of take photograph and reference point)

$$O1 = (X_{01}, Y_{01}, Z_{01})$$
$$O2 = (X_{02}, Y_{02}, Z_{02})$$
$$O3 = (X_{03}, Y_{03}, Z_{03})$$
$$Ol = (Xl, Yl, Zl)$$
$$\text{(1)}$$

Unit (m)

(b) coordinates of the photograph observation(digital picture's reference point)

$$q1 = (x1, y1)$$
$$q2 = (x2, y2)$$
$$q3 = (x3, y3)$$
$$\text{(2)}$$

Unit (mm)

(c) New reference point coordinates

$$P1 = (X1, Y1, Z1)$$
$$P2 = (X2, Y2, Z2)$$
$$P3 = (X3, Y3, Z3)$$
$$\text{(3)}$$

Unit (m)

On the digital picture which we take from many place, because take each other project and also take the intersection point $(p1, p2, p3)$ of measurement object, we thoughts that this point is the new reference point, and we used the digital photogrammetry.

4 DIGITAL PHOTOGRAMMETRY

From the line that is in the same condition in method (4), (5) on the basic method of every photograph as for orientation technique we consider the small relation of camera position, (X_0, Y_0, Z_0) survey object (X, Y, Z) and camera slope.

$$Fxij = f(x_{ij}, X_0, Y_0, Z_0, \omega, \phi, \kappa, X, Y, Z) =$$
$$(-CA1 - x_{ij}C1)/C1 \qquad (4)$$

$$Fyij = f(y_{ij}, X_0, Y_0, Z_0, \omega, \phi, \kappa, X, Y, Z) =$$
$$(-CB1 - y_{ij}C1)/C1 \qquad (5)$$

At this point

$$A1 = a_{11}(X - X_0) + a_{12}(Y - Y_0) + a_{13}(Z - Z_0)$$
$$B1 = a_{21}(X - X_0) + a_{22}(Y - Y_0) + a_{23}(Z - Z_0)$$
$$C1 = a_{31}(X - X_0) + a_{32}(Y - Y_0) + a_{33}(Z - Z_0)$$

X, Y, Z : The ground coordinate s of survey object

X_0, Y_0, Z_0 : The ground coordinate s of camera lens

C : Focal distance

x_{ij}, y_{ij} : Coordinate s oh photograph of the survey object

a_{ij} : Rotation matrix

The (4) and (5) is co linearity condition of every one photograph. In case in which we use the photographs, which we have got from many different place for analysis them, it is necessary to find the slope of every photograph, and the camera's position by using bundle adjustment. In this part of study we set up the point which we took photograph, and the coordinates of the one reference point and after we made the adjustment calculate.

5 BUNDLE ADJUSTMENT METHOD

If we compare the bundle adjustment with the polynomials method and independent method, in the bundle adjustment method is not necessary stereo photograph and also the place from where we take the photograph is free. With the height adjustment ability, which this method has, and by using survey point we can decide the terrestrial coordinate value and exterior orientation parameter of every photograph, when the remains different of square sum at the measurement values of reference point that has been measurement come to the smallest price. The next is Bundle adjustment development theory. In general we can express the bundle adjustment co linearity condition as in the expression (4) and (5)

$$Fx_{ij} = F(x_{ij}, X_0, Y_0, Z_0, \omega, \phi, \kappa, X, Y, Z) = 0 \quad (6)$$

$$Fy_{ij} = F(y_{ij}, X_0, Y_0, Z_0, \omega, \phi, \kappa, X, Y, Z) = 0 \quad (7)$$

(x_{ij}, y_{ij}) : i is the point of photograph,

   j photograph ic coordinate

$(X_0, Y_0, Z_0, \omega, \phi, \kappa)$: i the point of exterior

   orientation parameters

(X, Y, Z) : terrestrial coordinate of point j

The co linearity condition express which is concerning with photographic coordinate X and Y we can express it with total 27 unknown functions of exterior orientation parameters and terrestrial coordinate. In the express (6) and (7) because there is not line which is concerning the unknown co linearity condition and terrestrial coordinate we are using the tear develop with the around approximate value and make liberalization, and with square smallest method we look for the correction and the answer of converge. From expression (6) and (7) when we make the liberalization of tear development we get the next express.

$$Fx_{ij} + vx_{ij} - \partial Fx \bullet \Delta X_0 / \partial X_0 - \partial Fx \bullet \Delta Y_0 / \partial Y_0$$
$$- \partial Fx \bullet \Delta Z_0 / \partial Z_0 - \partial Fx \bullet \Delta \omega / \partial \omega - \partial Fx \bullet \Delta \phi / \partial \phi$$
$$- \partial Fx \bullet \Delta \kappa / \partial \kappa - \partial Fx \bullet \Delta X / \partial X - \partial Fx \bullet \Delta Y / \partial Y$$
$$- \partial Fx \bullet \Delta Z / \partial Z = 0 \quad (8)$$

$$Fy_{ij} + vy_{ij} - \partial Fy \bullet \Delta X_0 / \partial X_0 - \partial Fy \bullet \Delta Y_0 / \partial Y_0$$
$$- \partial Fy \bullet \Delta Z_0 / \partial Z_0 - \partial Fy \bullet \Delta \omega / \partial \omega - \partial Fy \bullet \Delta \phi / \partial \phi$$
$$- \partial Fy \bullet \Delta \kappa / \partial \kappa - \partial Fy \bullet \Delta X / \partial X - \partial Fy \bullet \Delta Y / \partial Y$$
$$- \partial Fy \bullet \Delta Z / \partial Z = 0 \quad (9)$$

In this place

$$a1 = \partial Fx / \partial X_0 \qquad a6 = \partial Fx / \partial \kappa$$
$$a2 = \partial Fx / \partial Y_0 \qquad a7 = \partial Fx / \partial X$$
$$a3 = \partial Fx / \partial Z_0 \qquad a8 = \partial Fx / \partial Y$$
$$a4 = \partial Fx / \partial \omega \qquad a9 = \partial Fx / \partial Z$$
$$a5 = \partial Fx / \partial \phi \qquad b1 = \partial Fy / \partial X_0$$
$$b6 = \partial Fy / \partial \kappa \qquad b2 = \partial Fy / \partial Y_0$$
$$b7 = \partial Fy / \partial X \qquad b3 = \partial Fy / \partial Z_0$$
$$b8 = \partial Fy / \partial Y \qquad b4 = \partial Fy / \partial \omega$$
$$b9 = \partial Fy / \partial Z \qquad b5 = \partial Fy / \partial \phi$$

To show the express (8) and (10) by a simple purity we use matrix and will get the next:

$$\begin{pmatrix} Vx_{ij} \\ Vy_{ij} \end{pmatrix} + \begin{pmatrix} -a1 - a2 - a3 - a4 - a5 - a6 \\ -b1 - b2 - b3 - b4 - b5 - b6 \end{pmatrix} \begin{pmatrix} \delta X_0 \\ \delta Y_0 \\ \delta Z_0 \\ \delta \omega \\ \delta \phi \\ \delta \kappa \end{pmatrix} +$$

$$\begin{pmatrix} -a7 - a8 - a9 \\ -b7 - b8 - b9 \end{pmatrix} \begin{pmatrix} \Delta X \\ \Delta Y \\ \Delta Z \end{pmatrix} = \begin{pmatrix} Fx_{ij} \\ Fy_{ij} \end{pmatrix}$$

$$(10)$$

And also we can show as follow:

$$V_{ij} + \alpha_{ij} \bullet \delta_i + \beta_{ij} \bullet \Delta j = F_{ij} \quad (11)$$

The express (10) and (11) is the equation of liberalization which is concerning the photographic coordinate which has been measurement. In case where there is many photograph point we use the next express.

$$V + \alpha \bullet \delta + \beta \bullet \Delta = F \quad (12)$$

Now we will write clear the size of matrix (12).

$V:(2mn,1), A:(2mn,6m), \delta:(6m,1),$

$\beta:(2mn,3n), \Delta:(3n,1), F:(2mn,1)$

M:number of the row (number of photograph)

N:number of and also

V:The difference remains vector of the photographic coordinate

A:The partial derivatives matrix which is concerning to the exterior orientation parameter

δ:The correction vector of the exterior orientation parameter

β:The partial derivatives matrix which is concerning to terrestrial coordinate.

Δ:The correction vector of the terrestrial coordinate

F:The difference between the measurement values and the revision values of the photographic coordinate

Therefore using the bundle adjustment method we can find the slope of every camera (κ, ϕ, ω) the camera's position (X_0, Y_0, Z_0), and the small relation of the survey object.

6 METHOD OF SPACE INQUIRE AND AREA CORRELATION

We use the exterior orientation parameters of every photograph which we have found by bundle adjustment and we use two photograph of every mesh intersection depth and with the expression (14) of space inquire area correlation method occur the accurate verification of the measure method

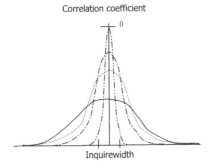

Figure 3 Inquire width and correlation coefficient

$r = \sigma_{12}/\sigma_1 \bullet \sigma_2$

$\sigma_{12} = \sum (e1 - \bar{e}1)(e2 - \bar{e}2),$

$\sigma_1 = \sum (e1 - \bar{e}1), \sigma_2 = \sum (e2 - \bar{e}2)$

r : correlation coefficient value

e1, e2 : The element values of the image into the correlative windows

$\bar{e}1, \bar{e}2$: The average element values of the image into the correlative windows.

The figure 3 shows the relation between the correlation coefficient values and space inquire width. This figure in dictates the many step of correlation coefficient change on the left and right photograph. The solid line is the outline position request on the coarse first step. The graph becomes by decide the last position through dot line, 1dot line, 2dot line.

Table1. Relativity position error

Point name	Change calculation value (H)	Height above sea level (H)	Difference remain (H)
101000	9.34	7.971	1.369
101400	8.76	7.766	0.994
100900	6.85	7.966	-1.116
101300	7.1	7.765	-0.665
100800	8.95	7.939	1.011
100500	9.48	7.945	1.535

7 VERIFICATION OF MEASURE RESULT

In this part of study has been founded the point in where take photo according to the differential method and also occurred the accurate verification of the new reference point. Next we got vertical photographic with digital camera, the traverse survey has been occurred by established the nation triangulation point, on the slop we set up 5 reference point, we got by camera terrestrial photogrammetry, and by using analytical picture instrument we got form data. On the result of differential measure method by compare the verification position, few relativity position error (Table1, and Katsumi, 1994) take place but we understood that from the side of form we be able to measure. About space inquire area correlation method the verification occur by using slope image, if we compare it with the matching method (Chuji, Susumu, & Osamu, 1985) looked like comparative and we reach to got automatic measure result.

CONCLUSION

We have studied how early we can get efficient emergency information. On this article we structure a method to get easy three dimension data by combine of digital camera which in recent years has spread and GPS. But for space inquires area correlation method it is necessary adjust of. Inquire width in every survey object. After this is necessary to study a method of proper emergency measure by using former information which we have get and make automatic adjustment of inquire width of the method of space inquire and area correlation.

REFERENCE

Japan photogrammetry learned society analytical photogrammetry committee: analytical photogrammetry (revise edition), 1997.

Katsumi, Nakane: three dimension handle of the smallest 2 multiplication survey data in the GPS period, Toyo bookshop, 1994.

Chuji, Mori, Susumu, Hattori, & Osamu, Uchida: improvement of photograph stereo matching according to the method of area correlation, essay of learned society of civil engineering No.356/VI-3, pp. 61-70, 1985

Creative Systems in Structural and Construction Engineering, Singh (ed.) © 2001 Balkema, Rotterdam, ISBN 90 5809 161 9

Computers in construction: Strategy or technology?

V. Ahmed

Department of Civil and Building Engineering, Loughborough University, UK

ABSTRACT: This paper reviews a strategic approach adopted to effectively develop and implement Computer Assisted Learning tools to aid the learning process in construction, highlighting important measures of effective learning through the used of CAL, derived from learning theory. A strategic framework is introduced to aid effective implementation of CAL tools in construction. To show the validity of this framework, two case studies are reviewed. The first case study describes the methodology adopted to evaluate the MERIT2 construction management simulation game, introducing quantitative and qualitative measures of its effectiveness. The second case study, shows a strategic approach adopted to develop a multimedia CAL tool to meet the needs and learning strategies required for novice quantity surveyors. This paper highlights the importance of well thought strategies for the use of computers in construction, rather than focusing on the technology to lead vague strategies.

1 INTRODUCTION

Over the last few years, the use of computers within the construction field has suffered many criticisms addressing the need for the technology to be implemented to meet the demands of a highly competitive construction industry. Although the growth in applying the new technology has been slower within the construction domain than others, the lack of its use is no longer a major issue. This is because the use of computers progressed from merely being productivity tools, to their use for communication, teaching and learning, solving problems and for information exchange.

Within higher education, the implementation and use of these tools in construction has widely spread, and is catching up with other fields of engineering. The environment within which it operates has also become more flexible. However, the learning outcomes from the use of this technology are often assessed through qualitative rather than quantitative measures. Therefore, this causes a failure in identifying whether the objectives for introducing the new technology have been fulfilled, rather than the failure of imbedding the technology itself.

This paper introduces a generic approach to the development and implementation of Computer Assisted Learning tools, reviewing two case studies that illustrate the importance of strategic approaches to the development and implementation of these tools.

2 A FRAMEWORK STRATEEGY

To help develop a strategy for implementing CAL tools effectively within the construction domain, several approaches were reviewed. None of these addressed the needs and demands of the construction domain, nor did they address measures of effective learning through the use of CAL. This conclusion led to a study of learning theory to derive measures of effective learning and relate them to learning with CAL. As a result of this study, a framework a strategic framework was developed to promote effective implementation of CAL tools and aid learning in construction (see Figure 1). The framework is defined by three stages (planning, implementing and evaluating) and the role of the educator, developer and user of CAL tools at each stage. The planning stage is considered to be the task of the educator, and requires the educational needs for CAL, its targeted users, motivational factors and the targeted learning strategies to be identified. It is proposed that effective planning, leads to effective structuring of CAL tools. The second stage of the framework is the implementation CAL, which is considered the task of the educator and the developer. The implementation of CAL tools requires effective structuring (i.e. the content of CAL) and effective operations (i.e., the environment of CAL). To assist in monitoring effective implementation of CAL, formative (efficiency of CAL) and summative (effectiveness of CAL) evaluations are required, using various methods of testing and evaluating the effectiveness of CAL. This stage of the framework requires the involvement of the educator, developer and the users of CAL.

The need for CAL to target the learning strategies associated with a particular domain, is also addressed. This resulted in identifying the learning strategies associated with novice Civil Engineers and novice Quantity Surveyors, and to select a CAL tool that is capable of promoting these strategies, as described in section 3.

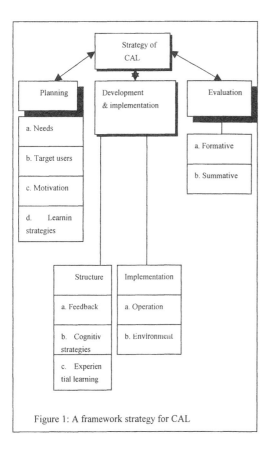

Figure 1: A framework strategy for CAL

3 CASE STUDIES

This section describes two case studies carried out adopting the frame-worked strategy introduced in the previous section. According to Kolb (1991), novice quantity surveyors are convergent in their learning styles, whereas novice quantity surveyors are divergent in their learning styles (Lowe, 1992). Such findings, led to study CAL tools, and their role in facilitating different learning strategies. It was therefore found that simulation games are suitable tools

to facilitate the convergent learning strategies and multimedia CAL tools are suitable for facilitating the divergent learning studies. A review of this study is described below

3.1 MERIT2 simulation game

3.1.1 Background.
MERIT2 is a construction management simulation game that demonstrates the interacting nature of variables that need to be considered in the tendering and production phases of a construction project. The MERIT 2 simulation game has been running for the last 10 years, in industry and in higher education. It was developed in Loughborough University for Balfour Beatty and been played by at least 15,000 participants (Ahmed et al, 1996).

MERIT2 allows up to 1000 teams (each team has 3-4 participants) referred to as companies to operate a construction company for up to 16 periods or quarters, representing 4 trading years. The participants are required to control and manage the direction of their company through inter-related marketing, tendering, overhead allocation, labor staffing and general financial decisions. The companies operate in a computer-simulated market based on current UK statistics.

This simulation game illustrates an example, of a computer based simulation game, which has been successfully running over the last ten years, but no measures of its effectiveness have been undertaken. This resulted in a case study developing quantitative measures of the effectiveness of the MERIT2 Construction Management simulation game, guided by the evaluation criteria proposed by the framework earlier.

3.1.2 Surveys.
A number of pilot surveys were carried out, investigating the participants' opinion of the simulation game for promoting CPD while in industry, and as a CAL tool complimenting students' learning on the postgraduate level. Accordingly, 600 previous MERIT2 participants were contacted by posting a questionnaire survey, of which 236 participants responded. Similar survey was carried out by direct interviews. A number of useful samples of participants were those of postgraduate students who took part in MERIT2 at Loughborough University. This provided an opportunity to carry out a number of knowledge improvement tests, before taking part in the MERIT2 and afterwards. These surveys generated some interesting results.

3.1.3 Results.
The results of the pilot surveys were coded and analyses using various computer based statistical

techniques. The interpretation of these results showed that:

a. *MERIT2 can be more effective should its operation be improved, such as:*
- the introductory stage of the game,
- the quality of information in the manual is improved to suit participants' to match participants' background and previously gained experience,
- enough time must given to participants to make decisions after each round of the game, and
- enough feedback must be generated, clarifying the consequences of the decisions made for filling in the bidding forms.

b. *Measures of the success of MERT2 were reflected upon motivational factors such as:*
- participants' response to their enjoyment of the simulation game;
- the value of the simulation game as a learning medium,
- their interest to take part again.
- its relevance to the participants' practical experience in the construction industry.

c. *MERIT2 met the participants' educational needs in:*
- demonstrating the nature of decisions made within a construction company;
- demonstrating the interlocking nature and complexity of construction management decision made within a construction company;
- enhancing their knowledge of company decisions and improved awareness of team work, communication and leadership skills, particularly by team leaders.

However, MERIT2 did not meet the participants' needs in enhancing their knowledge of decisions made on overhead allocation and cash distribution and borrowings, resulting in poor knowledge gained in other areas of construction management promoted by MERIT2.

3.1.4 Summary.
MERIT2 is a valid CAL tool for promoting Continuous Professional Development and for complementing the teaching of construction management at postgraduate level. However, by adopting a strategy to investigate the effectiveness of MERIT2, it is proposed that a better mode of delivery should be applied to cut down on costs and time for delivering postal information to the participants. It is also proposed that clearer instruction should be given to the participants to develop a better understanding of the main principals of the simulation game. An in depth study of the effectiveness of MERIT2 showed some gaps within the simulation to successfully promote participants' understanding of some construction management aspects.

A new version of MERIT is currently up and running successfully on line, taking into consideration the findings of this study.

4.1 *QSMM package*

4.1.1 *Background.*
One of the main stumbling blocks in teaching quantity surveying measurement is that, students need to have an understanding of the construction technology process before they can begin to understand the measurement rules that govern the quantification process. This leaves the student facing a problem of not being able to understand the construction process, due to their lack of construction technology knowledge (i.e. of the elements that the measurement rules are based upon). This also presents major problems with undergraduate quantity surveying measurement module, has been evidenced by educators in the Department of Civil and Building Engineering at Loughborough University. This led to the development of QSMM (Quantity Surveying Measurement in Multimedia) to aid students' learning and to meet their educational needs in this subject area, by providing visual modes of information such as, images, animations, video clips and text.

4.1.2 *Strategies.*
For QSMM to be effectively and efficiently developed, the role of the developer was not confined to the development and implementation stage of QSMM, but also included the planning and evaluation stage. For effective planning of QSMM, clear communications between the quantity surveying educators (lecturers and tutors), the developer of QSMM and the students were needed, by surveying individuals' opinions to identify the existing problems. Therefore a number of steps undertaken at the pre-implementation stage showed that:
- the conventional method of teaching promoted a surface approach to learning, and the recognition of students' needs is essential;
- although the QS tutors had defined their own educational needs and goals, and showed interest and motivation for implementing new technology, they were not fully aware of the students' educational needs. There was a mismatch between what the educators thought the students' difficulties were, and what the actual students' difficulties were,
- Learning style analysis showed that, the Quantity Surveying students did not have a defined learning style during their first year in the undergraduate domain, although the required one is that of an imagery style (divergent style),
- Educators' learning styles differed from the students' learning styles (as expected due to their professional experience), Kolb (1986). QS tutors preferred students with learning styles that matched their own, and were aware of what style of learning they wanted the future graduate quantity surveyors to have.

To aid the effective development and implementation of QSMM, the followings steps were undertaken:

- Recognising the students' background and their previous experience, assisted the structuring of QSMM in such a way, that suits students' individual levels and backgrounds.
- Defining aims and objectives, learning styles and transferable skills initially assisted with effective structuring of information in presented by QSMM.
- Defining the relationship between time spent in learning and the integrated media assisted in planning for QSMM to be both time and cost effective.

4.1.3 *Surveys.*

Following the development of QSMM a number of post-task surveys were carried out, eliciting the following:

- the formative evaluation, showed that QSMM was run efficiently on the departmental server and was easily accessed by the students at all times. Also, QSMM was found user friendly with more than satisfactory levels of mapping and usability;
- the summative evaluation showed that QSMM was more effective than the conventional way of teaching. Its effectiveness was mainly concentrated on its functionality to assist students with their imagery learning difficulties. It enhanced students' learning skills and reduced the time taken to understand complex aspects. However, there was no evidence that QSMM was more effective for a particular style of learners;
- more than two thirds of the population found all aspects of QSMM an interesting way of presenting information and very helpful. Non of the students found QSMM confusing.

4.1.4 *Summary.*

QSMM proved an effective and useful tool for teaching quantity surveying measurements from first principles. Its values is in its ability to divide elements of construction into its constituents and show how these fit together in animated sequences. This, in turn, improves students' wholistic abilities of thinking, their imagery thinking style and also helps improve their analytic style of learning. By doing so, QSMM assists in overcoming the difficulties faced in a traditional lecture-led class room environment.

5 CONCLUSIONS

The main conclusions drawn from this study is that, the effective use of computers in construction must be led by strategies rather than the technology. This conclusion was drawn from the following finding:

- The framework developed is an effective guide for the strategic approach to planning, developing and evaluating QSM.
- The direction of learning must target the required learning strategies associated with the future graduates, and focus on enhancing the related skills and abilities.
- A CAL tool can only be described as effective, if it meets the educational needs of educators and learners.
- Educators' lack of unawareness of the difference in students' learning styles, may be a contributing factor to students' learning difficulties.
- Educators must become aware of the functionality of different computer based applications and the role they can play to promote learning strategies.

REFERENCES

Ahmed V., Thorpe A. and McCaffer R. (1996). The effectiveness of MERIT2 as a training tool in construction management: In Thorpe, A.ed., *Proceedings of the ARCOM conference,* Sheffield Hallam University, Sheffield, 1, p277-285.

Lowe, D. (1992). Experiential learning: A factor in development of an expert pre-tender estimator. A thesis submitted for the degree of master of science, Department of surveying, University of Salford.

Kolb, D. (1991). *Experiential learning: Experience as a source of learning and development.* Prentice-Hall, New Jersey.

13 Safety and quality

Creative Systems in Structural and Construction Engineering, Singh (ed.) © 2001 Balkema, Rotterdam, ISBN 90 5809 161 9

Education of design professionals on construction safety

M. A. Usmen, S. Baradan & T. Dikec
Department of Civil and Environmental Engineering, Wayne State University, Detroit, Mich., USA

ABSTRACT: The need for educating design professional (engineers and architects) on construction safety is discussed in this paper in view of the modesty of the current efforts on addressing this issue by the academic community. A course developed and implemented at two U.S. universities over the past thirteen years is described in detail, and presented it as a model for similar efforts elsewhere.

1 BACKGROUND

Construction industry has had a poor record of site safety over the years, leading virtually all industries in related statistics in the United States and abroad. A major factor contributing to this situation has been the lack of sufficient knowledge and understanding of the relevant safety (and health) principles, and in many cases, a lack of full appreciation of the importance and implications of safety in design and construction by the project participants; i.e. owner, A/E, general and subcontractors, CM, and others.

Although engineers and architects often find themselves in responsible positions as employees or consultants of one or more of the construction project participants, their knowledge of this important field may often be less than adequate for providing the appropriate level of professional service in design, management, or supervision. Traditionally, the emphasis in the technical curricula has been on the safety of a facility as designed and constructed and little attention has been given to public and occupational safety during the construction process. Most of the knowledge related to work site safety is gained through on-the-job experience.

A Center for Excellence on Construction Safety (CECS) was established in the Civil Engineering Department of West Virginia University in 1986 with support provided by the National Institute for Occupational Safety and Health (NIOSH). The principal author served as the Technical Director of this center for three years. The mission of CECS was to conduct education, research and information transfer activities primarily addressing the civil engineering community (Usmen 1988). As part of the education

the development of a three credit hour elective course on construction safety, suitable for seniors and graduate students (Usmen 1992). This course was taught at West Virginia University twice, in 1987 and 1989, receiving excellent reviews from the students. The same course has been taught at Wayne State University every other year in Winter semesters since 1990 after the principal author's move to this institution in 1989. In its current form, the course is offered for 4 credit hours, and is taken mostly by students who have degrees in engineering and architecture.

Many updates and improvements have been incorporated in this course over its thirteen years of development and implementation. During this period, extensive course notes and related handout materials have been developed and transformed into an instructional module by the principal author (Usmen 1994). This document was distributed in 1994 by NIOSH to all the Civil Engineering and Civil Engineering Technology Departments in the United States to disseminate its contents and to promote its utilization by the academic community. The contents is intended to be used as a supplement to textbooks and course notes in existing courses and can be used as resource materials in courses covering design, construction engineering and management, and professional practice. The course content and the associated materials have been continuously updated to ensure that the students are exposed to current information on construction safety and health. The student interest in and appreciation of the course continues to be very high as detected from the very high scores the course has been receiving in student evaluations. The contents have been added to enable the students to access the course materials and related safety information from remote sites.

Designer involvement in construction safety has recently been gaining increasing attention from the standpoint of constructability and the need to address jobsite hazards while designing the project (Gambatese 2000 & Hinze 2000). It is agreed that designers can impact worker safety through careful consideration of the appropriate scenarios when making decisions about site-layout, structural details, construction materials selection, and other relevant aspects. The American Society of Civil Engineers has recently revised its policy on Construction Site Safety (ASCE Policy A – 350) to better define the role of the engineer in regard to site safety. The policy states that contract terms and industry standards establish the engineer's duties and obligations, and the engineer is responsible to review and check the plans for their design adequacy, constructability and safety. It further emphasizes that project safety can be enhanced with construction friendly designs. This policy is still under review, and comments are being provided by various constituencies.

In order for the designers to accomplish (occupationally) safe designs, and perform their roles effectively in project implementation oversight, they need to be knowledgeable about the principles and practices associated with jobsite safety. This requires their exposure to this body of knowledge as part of their college education, which can be reinforced through continuing professional development, including training and on-site experience. The mentioned course is one of the numbered efforts in the engineering and architecture programs in the U.S., which attempts to cover this subject in an in-depth fashion. We feel that more such courses should be included in the curricula and made available to the students, if we expect to realize further improvements in safety on construction projects. The following information on our course at Wayne State University is presented with the idea that it will serve as a model for these interested in developing and teaching a similar course on this topic.

2 COURSE OVERVIEW

The Construction Safety course introduced in this paper consists of essentially five parts:
1 Justification for safety
2 Accident causation, investigation, and prevention
3 Safety standards and practices
4 Engineering for safe design and construction
5 Safety and health management with TQM integration.

These are briefly described in the following paragraphs:

2.1 Justification for Safety

This introductory part of the course starts by defining occupational safety and health as the area of science and technology that deals with the identification and control of environmental and personal hazards in the workplace. Unique aspects of the construction industry, i.e., fragmentation, transient workforce, multi-employer projects, and work under variable conditions, are emphasized to draw the students' attention to the need for consideration of safety. Following the coverage of national and worldwide statistics on injuries and fatalities which show the highly dangerous nature of construction work, an economic analysis of construction accidents is presented. It is shown through comparisons of costs and benefits that safety saves money, in addition to saving lives. Direct versus indirect costs of injuries, workers compensation and Experience Modification Rating (EMR) concepts are introduced to provide the necessary tools for an in-depth understanding of the benefits of safety programs in the industry.

2.2 Accident Causation, Investigation and Prevention

An accident is defined as an unexpected, unforeseen, or unintended event, which has a probability of causing personal injury and/or property damage. Causes of accidents are discussed as unsafe acts and/or unsafe conditions (Table 1). Several theories (Usmen 1994 & Hinze 1997) on accident causation are introduced, using examples from real construction projects.

Table 1. Causes of Accidents

Unsafe Acts
Unsafe use of tools and equipment
Unauthorized operation of equipment
Use of defective equipment
Failure to use personal protective equipment
Unsafe material handling
Failure to follow safe procedures
Poor housekeeping
Attitude problems (horseplay, inattentive behaviors, etc.)
Improper guarding
Unsafe Conditions
Improper illumination, ventilation
Hazardous substances (chemicals, explosives, etc.)
Poor site layout, housekeeping
Defective tools and equipment, lock-out tag-out practices
Unsanitary conditions
Unsafe design and construction

Common types of accidents (Table 2) and injuries/illnesses (Table 3) are described, along with the principles of accident investigation (Table 4) and the Occupational Safety and Health Administration (OSHA) requirements for reporting and record keeping. This part of the course is concluded by covering the hierarchy of controls principle applicable to accident prevention. This four-step approach consists of the following:

1 *Eliminate* the hazard from machine, method, and material system.
2 *Control* by isolation, enclosure, guarding.
3 *Train* personnel for hazard recognition and safe job procedures.
4 Prescribe *Personnel Protective Equipment* to shield personnel from hazard.

Table 2. Common types of accidents

Falls from elevations
Struck by or against objects
Electrocution
Industrial vehicles, equipment and tools
Exposure to temperature extremes
Overexertion
Explosions
Inhalation of fumes; contact with chemicals

Table 3. Types of injury / illness

Cuts, amputations
Bruises, contusions, abrasions
Foreign body
Puncture wounds
Fracture
Sprain and strain
Hernia
Dermatitis
Burns, hypothermia
Heart attacks
Latent illnesses (cancer, AIDS, hepatitis)

Table 4. Accident Investigation

Understand need
Prepare
Gather Facts
Analyze Facts
Report conclusions / recommendations
Correct situation
Double check corrective action

2.3 *Safety Standards and Practices*

In this part of the course, various specific hazard recognition and prevention/control techniques are taught as related to OSHA standards (29 CFR 1910 and 29 CFR 1926). The coverage includes the following:

– Material Handling (Hazardous materials/Hazard Communication Standard; manual lifting; housekeeping; rigging; personal protective equipment-PPE).
– Welding and cutting (Fire prevention; fumes and UV).
– Steel erection (Advanced planning; shop drawings; erection sequence).
– Concrete construction (Cement; silica; rebars; formwork and shoring; design and construction aspects).
– Fall prevention and protection (Leading edges; scaffolding; roofs; floor openings; ladders and stairways; fall arrest systems; safety nets).
– Excavation and trenching (Soil classification; cave –ins; buried utilities; sloping and shoring; trench shields/boxes).
– Confined space entry (Types of confined spaces; advance planning/permitting; PPE; rescue operations).
– Cranes and Heavy Equipment (Operation and maintenance; site lay-out; signals)
– Hand and power tools (Manual; power; powder-actuated; proper storage and up-keep; guards).
– Electrical Safety (Electric shock; proper grounding; overhead lines; signing and barricading; fire prevention).
– Lockout/Tagout (Energy control program for electrical, mechanical, air, fluid, thermal and gravitational energy sources).
– Hazardous waste handling (OSHA part of RCRA/CRCLA)
– Ergonomics (Pending standard on cumulative/repetitive work related trauma).

Training aspects are emphasized throughout the presentation of these topics, and the hazards and their prevention are illustrated by videotapes covering the entire spectrum. A guest lecture by a safety professional or a compliance agency (OSHA) representative is also included.

2.4 *Engineering for Safe Design and Construction*

This part focuses on the causes and prevention of construction (and structural) failures by covering well known case studies, such as Kansas City Hyatt Regency Walkway, Willow Island Cooling Tower, and L'Ambiance Plaza (Usmen 1994, Feld & Carper 1997). Emphasis is placed on knowledge (training and education), competence (experience), and care (control). Designer's role in construction safety is discussed in detail alongside risk management and liability considerations. A guest lecture by a design professional with expertise on structural safety considerations during construction nicely complements this topic.

2.5 Safety Management with TQM Integration

All of the principles covered and technical knowledge acquired in the previous parts of the course are placed into an implementation framework in this final part of the course. A complete treatise on how to set up and run an effective safety and health (S&H) program is attempted with the coverage of the following elements:

- Objectives of S&H management (Minimize losses due to injuries; avoid injuries, OSHA fines and litigation)
- Top management commitment (Top priority, written policy statement)
- Goal setting (Clear goals, communicated to all employers)
- Organization and administration (Understanding roles and responsibilities; communication network; safety meetings; safety personnel)
- Planning (On-going activity, project S&H plans)
- Company / Project safety rules (Written and provided to each employee)
- Worker orientation & training (New worker orientation, general & specific training)
- Accident investigation and reporting (Every accident & incident should be investigated and recorded)
- Safety budget (Line item on the project budget)
- Safety in contracts (Contracts should include safety requirements)
- Safety audits (Evaluate S&H program; site inspections)
- Incorporating Total Quality Management (TQM) principles in S&H programs (Deming's principles).

Considerable emphasis is placed on the TQM framework in program implementation with focus on customer service, continuous process improvement and documentation, employee involvement and leadership, and the option of ISO 9000 certification.

Although the material is presented primarily from the perspective of the contractor/subcontractor, attention is also directed to the owners' role in S&H management and the elements of owners' safety programs.

3 WEB SUPPORT AND OTHER DETAILS

The effectiveness and reach of the course is significantly enhanced by a web-site. Grading is based on two exams (mid-term and final) and a term paper prepared in written from and presented in class by the students, using PowerPoint. The web-site (www.ce.eng.wayne.edu/ce7020) includes:

- Course outline and schedule
- Messages from the Instructor
- Lectures notes (including materials posted by guest lecturers)
- Listing of term-paper topics
- PowerPoint presentation files
- Study guides for exams
- Exam questions
- OSHA standards
- Listing of videotapes and summaries of contents
- Links to other web-sites (OSHA, NIOSHA, BLS, ASSE, NSC, etc.)

It is mandated in the course that all submissions (assignments, term papers, exams, etc.) are made both electronically (by e-mail) and in hard copy. E-mail is encouraged as a communication tool with both fellow students and the course instructor. Students have expressed very favorable feelings about the benefits of the course web-site.

4 SUMMARY

The need for inclusion of construction safety aspects in engineering and architecture curriculum is emphasized in this paper, because designers are generally not very well educated on this important topic. The development and delivery of a course on construction safety (and health), primarily focused on the education of design professionals is described in detail. It is desired that increased attention will be given to this field by the academic community, which in turn can improve safety on construction project sites. The course presented here can serve as a model in the development of similar courses around the world.

REFERENCES

Feld, J. & Carper, K.L. 1997. Construction Failure. John Wiley & Sons, Inc., 2nd Edition.

Gambatese, J.A. 2000. Safety Constructability: Designer Involvement in Construction Site Safety. Proceedings, K.D. Walsh Editor, Construction Congress VI, ASCE. Orlando, Florida.

Hinze, J. 1997. Construction Safety. Prentice Hall.

Hinze, J. 2000. The Need for Academia to Address Construction Site Safety through Design. Proceedings, K.D. Walsh Editor, Construction Congress VI, ASCE. Orlando, Florida.

Usmen, M.A. 1988. A Center for Excellence in Construction Safety. Proceedings, ASEE Zone II Meeting. Louisville, Kentucky.

Usmen, M.A. 1992. Development of Educational Modules on Construction Safety for Civil Engineering Students. Proceedings, NIOSH SHAPE Project Workshop. Corvallis, OR.

Usmen, M.A. 1994. Construction Safety and Health for Civil Engineers. Instructional Module. NIOSH/ASCE. New York, NY.

Creative Systems in Structural and Construction Engineering, Singh (ed.) © 2001 Balkema, Rotterdam, ISBN 90 5809 161 9

A performance approach to construction worker safety and health – A survey of international legislative trends

Theo C. Haupt
Department of Construction Management and Quantity Surveying, Peninsula Technikon, South Africa &
M.E. Rinker, Sr, School of Building Construction, University of Florida, Gainesville, Fla., USA

Richard J. Coble
M.E. Rinker, Sr, School of Building Construction, University of Florida, Gainesville, Fla., USA

ABSTRACT: This paper describes the legislative efforts in several countries and regions to facilitate and promote the improvement of construction worker safety and health. In particular, the legislative frameworks in Australia, New Zealand, United Kingdom, Europe and the United States are examined. The paper concludes from the examination that the value of legislation outside of the United States lies in the requirements of all participants in the construction process to make safety and health a mandatory priority in a structured way.

1 INTRODUCTION

Both legislators and safety professionals in the construction industry have held that responsibility for safety and health should be placed on those indirectly involved in construction as well as the contractors who actually carry out the works. Designers, architects and, particularly, clients influence the construction process. A great contribution to the avoidance of accidents would be made if that influence were used with accident prevention in mind - from project inception through project execution and then throughout the life of the facility until its final demise through demolition (Joyce 1995; Berger, 1999).

Given the unique nature of the construction industry and the interdependence of the large number of stakeholders, the team building approach to construction safety and health has been advanced as being pivotal to achieving safety and health on construction projects (Smallwood and Haupt, 2000). The monumental task facing the construction industry is to encourage every person involved in the design, management, and execution of construction projects to give priority to safety and health issues which have until now failed to attract the necessary attention, especially from clients and designers (Joyce 1995).

The exclusion of health and safety from specifications, and health and safety being the sole responsibility of the contractor have been identified as primary causes of accidents in construction (Ngowi and Rwelamila 1997). Further, in a study conducted in South Africa, planning was identified as the primary preventative action that could have been taken in 40% of the cited cases (Szana and Smallwood 1998).

Additionally, in a study into scaffolding accidents in the United States, South Africa, and Turkey, designing for safety and enforcement of regulations and standards were suggested as reasonably practicable preventative precautions (Müngen, Kuruoglu, Turkoglu *et al* 1998).

The poor safety and health performance record of the construction industry has resulted in safety and health regulations around the world being subjected to major revisions during the last three decades. In some cases, new legislative and regulatory approaches have replaced existing regulations and legislation. The emphasis of these new pieces of legislation has been on individuals and their duties. Additionally, they represent significant departures from previous prescriptive approaches (Coble and Haupt, 1999). They have been based on principles designed specifically to increase awareness of the potential hazards associated with safety and health issues. They demonstrate a new approach and commitment to the management of construction projects.

In this paper, we examine the approach advocated by the Council Directive 92/57/EEC that forms the basis for construction worker safety and health legislation in Europe, The Construction (Design and Management) Regulations (CDMR) 1994 in the United Kingdom, The National Model Regulations, and the National Code of Practice for the Control of Workplace Hazardous Substances 1994 in Australia, and the Health and Safety in Employment Act 1992 and Regulations 1995 in New Zealand. These examples of safety and health legislation are performance based and have as their main thrust the redistribution of responsibility for health and safety on construction sites away from the contractor to include clients and planning professionals (ILO, 1992; Lorent, 1999; Caldwell, 1999). The performance approach identifies broadly-defined goals, ends or targets that must result from applying a safety standard, regulation or rule without setting out the specific technical requirements or methods for doing so.

Additionally, the Occupational Safety and Health Act of 1970 (OSHA) in the United States is also examined, as legislation that is largely prescriptive in nature, but is slowly moving towards a performance approach. The prescriptive approach requires strict, rigid, and enforced conformity to a safety standard, regulation or rule, and specifies in exacting terms the means or methods of how employers must address given conditions on construction sites.

2 THE CONSTRUCTION (DESIGN AND MANAGEMENT) REGULATIONS (CDMR) OF 1994

The CDMR were introduced in the United Kingdom (UK) in March 1995 in compliance with the European Union Council Directive 92/57/EEC in 1992, in terms of which all European Union member states were to implement the terms of the directive into national legislation by 1994. The directive was, however, not implemented in its entirety by the CDMR. Rather the CDMR implemented the organizational and management aspects (Caldwell, 1999). The regulations were, additionally, a response to the study conducted by the Health and Safety Executive (HSE) which recorded that during the period 1981 through 1985, 739 people were killed in the construction sector (Munro, 1996). An analysis of the main causes of accidents in UK construction revealed the following:
- a lack of supervision by line managers in the industry;
- inadequate equipping of workers to identify dangers and take steps to protect themselves from these; and
- a lack of co-ordination between the members of the professional team at the pre-construction phase (Joyce, 1995).

They were consequently designed to provide a legislative framework aimed at achieving co-operation and co-ordination in the drive to improve construction safety and health on construction sites.

The regulations promote the teamwork approach during the design and construction life of construction projects, which has been advocated by Sir Michael Latham in his 1994 report, *Constructing the Team*. They place new responsibilities and duties on clients, designers, and contractors (Caldwell, 1999). The CDMR carry a criminal sanction of up to 2 years imprisonment and unlimited fines for non-compliance with their provisions. The primary objective of the CDMR is to ensure proper consideration of safety and health issues throughout

each phase of the construction process from project inception through to the eventual demise of the building by demolition (Tyler and Pope, 1999). The CDMR have been described as a management solution. They involve co-ordination in a notoriously fragmented industry as well as the integration of the major participants in the construction process.

Major distinguishing characteristics of this legislation include:
- a departure from the traditionally prescriptive or "deemed-to-comply" or "command-and-control" approaches to a performance based approach in terms of which no standards for compliance are set;
- the compelling of safety and health management as an obligation into the planning and design of virtually all but the smallest of construction projects;
- emphasis on the identification of construction hazards and the assessment of risks in order to eliminate, avoid or at the very least reduce perceived risks;
- consideration of safety and health issues not just during the construction life of the project, but from project inception through to the final demise of the facility by demolition, including the operation, utilization and maintenance periods;
- the redistribution of responsibility for construction worker safety away from the contractor, who was previously solely responsible, to include all participants in the construction process from the client through to the end-user;
- the introduction of a new participant to the construction process, the planning supervisor, with responsibility to co-ordinate the other participants and documents to facilitate better management of safety and health on construction projects;
- mandatory safety and health plans as instruments facilitating exchange and communication of safety and health issues between all participants in the construction process, on all "notifiable" projects where the construction phase is longer than 30 days or will involve more than 500 person days, and where there are more than 5 persons carrying out construction work at any one time; and
- mandatory compilation of a safety and health file by the planning supervisor to be handed over to the client upon completion of the facility.

The CDMR acknowledge the roles of each participant in construction. For example, whereas designers were not previously extensively involved in giving advice about systematic consideration of health and safety issues, they are now required to avoid foreseeable risks as a duty for all construction projects.

The establishment cost to the industry in the UK was calculated to be in the region of $825 million with the cost of compliance by designers an additional annual amount of about $435 million.

3 THE COUNCIL DIRECTIVE 92/57/EEC OF 24 JUNE 1992

The Council of European Communities committed itself to ensuring greater protection of the safety and health of construction workers through the adoption of minimum requirements for encouraging improvements in working environments on construction sites to ensure a better level of protection. In particular, increased responsibility was placed on employers accompanied by new obligations for workers and greater involvement by all participants in the construction process – owners to workers – in the management of risks (Loren, 1999). The imposition of additional administrative, financial, and legal constraints that would impact negatively on small and medium-sized undertakings was not intended. Rather the Council Directive 92/57/EEC of 24 June 1992 was designed to guarantee the safety and health of workers on construction sites in the European Community wherever building or civil engineering works were carried out. The Directive was transposed into national law in most member countries of the European Union with minor changes in the management or personnel structure and/or the safety measures advanced by the original Directive. In some countries the adoption of the Directive was necessitated by the need for organizational change due to developments to improve the cohesion of the construction process and communication, as well as the structural changes caused by the cluster of subcontracting arrangements characterizing their construction industries (Lorent, 1999).

The Commission recognized that more than 50% of occupational accidents on construction sites were attributable to unsatisfactory architectural and/or organizational options, or poor planning of the works at the project preparation stage (Lorent, 1999). Moreover, the Commission recognized that large numbers of accidents resulted from inadequate coordination especially where various undertakings worked simultaneously or in succession at the same construction site. This recognition represented a major paradigm shift. Previously all responsibility for safety and health on construction sites was attributed solely to contractors. The provisions of the Directive were directed to bring about a cultural change to improve the poor safety culture prevalent within the industry (Schaefer and De Munck, 1999).

The main distinguishing features of the Directive include:

– the performance based nature of the provisions of the Directive;
– ensuring that safety and health issues are taken into account through all phases of the construction process, extending to the operation, utilization, and maintenance periods, and the final demise of the facility through demolition;
– the redistribution of responsibility for construction worker safety away from the contractor, who was previously solely responsible, to include all participants in the construction process from the client through to the end-user;
– the introduction of the project supervisor who is responsible, while acting for the client, for all applicable general safety and health requirements during the stages of design and project preparation, including ensuring that the safety and health plans and files are accordingly adjusted;
– the appointment of one or more safety and health co-ordinators by the client or the project supervisor, for either or both the project preparations and project execution stages, their duties in terms of each stage being different;
– the compilation of mandatory safety and health plans by the client or project supervisor before actual work commences on site;
– the giving of a prior notice, which must be updated periodically and displayed on the construction site, submitted to the authorities responsible for safety and health at work on all construction sites where the work is scheduled to last longer than 30 working days, and on which more than 20 workers are employed at the same time, or on which the amount of work to be carried out is scheduled to be more than 500 person-days;
– the mandatory preparation of a file appropriate to the characteristics of the project containing relevant safety and health information to be taken into account during any subsequent works; and
– the fact that the entire Directive, together with all annexures, is contained in a total of 17 pages.

Examples of performance-based standards taken from the Council Directive 92/57/EEC (1992) are contained in Figure 1.

These sections are the equivalent of OSHA 29 CFR 1926 Subparts L (1926.450-453) and T (1926.850-860).

Architects, in particular, across Europe felt very uncomfortable with this change in responsibility from the contractor to the client, who was required to take appropriate steps with respect to safety and health in the planning and execution of a construction project. Further, the client was responsible for organizing the work on the construction site in such a way that risks to life and health were avoided as far as is possible, and where not possible, to maintain residual risk at the lowest level possible (Berger 1998).

However, concerns remain among many of the member countries of the EU about the cost to implement the revised structure embodied in the provisions of the Directive. This cost has been estimated to range between 0.2 and 2% of the total project cost distributed on the basis of 35% for co-ordination during the project preparation phase and 65% during the project execution phase (Lorent, 1999; Berger, 1999). Further there is concern about the lack of a standard and simplified system of reporting construction-related accidents, injuries, fatalities and diseases which might have been embodied in the Directive (Papaioannou, 1999; McCabe, 1999; Casals, Exteberria and Salgado, 1999; Onsten and Patay, 1999). This lack makes it difficult to conduct comparative analyses of the effectiveness and impact of the introduction and implementation of the Directive in member countries on the safety performance of the industry on a country-by-country basis. Additionally, there is confusion in some countries about the need for and content of the project-specific safety and health plan (Onsten and Patay, 1999; Caldwell, 1999). A final concern revolves around the poorly defined competence and qualification requirements of project supervisors and safety coordinators with mutual recognition of training and development programs and qualifications (McCabe, 1999; Dias, 1999; Gottfried, 1999; Caldwell, 1999).

---

6. *Scaffolding and ladders*

6.1 All scaffolding must be properly designed, constructed and maintained to ensure that it does not collapse or move accidentally.

6.2 Work platforms, gangways and scaffolding stairways must be constructed, dimensioned, protected and used in such a way as to prevent people from falling or exposed to falling objects.

11. *Demolition work*

Where the demolition of a building or construction may present a danger:

(a) appropriate precautions, methods and procedures must be adopted;

(b) the work must be planned and undertaken only under the supervision of a competent person.

Figure 1. Examples of performance based standards

4 THE NATIONAL MODEL REGULATIONS AND THE NATIONAL CODE OF PRACTICE FOR THE CONTROL OF WORKPLACE HAZARDOUS SUBSTANCES OF 1994

It was realized in Australia that it would be impossible to draft appropriate standards to cover each of the between 21,000 and 37,000 chemicals individually that are used in Australian workplaces. Further, it was recognized that specific substance controls were insufficient to deal with the wide range of workplace situations where large numbers of hazardous substances were used.

The National Model Regulations and the National Code of Practice for the Control of Workplace Hazardous Substances of 1994 are consequently generic rather than substance-specific. They provide cover for all hazardous substances used in workplaces throughout Australia. The model regulations apply to all workplaces where hazardous substances are used or produced, and to all persons with potential exposure to hazardous substances in those workplaces (Lawson, 1996).

The regulatory package is an example of performance based regulations. The health and safety outcomes are specified in the regulation, but not the means to achieve them, as has been the case for previous prescriptive Australian safety and health regulations and legislation of the past. The regulations provide a comprehensive approach to the control of health risks from exposure to hazardous substances by setting the outcomes to be achieved and by setting the processes to be followed. They do not prescribe how risks must be controlled. The regulations give industry the flexibility to select the most appropriate control measures for different workplace conditions, based on the identification and assessment of risk (Lawson, 1996).

A risk management process is incorporated in the National Model Regulations for the Control of Workplace Hazardous Substances. Features of this process include:

– Establishment of the context with respect to scope and objective.
– Identification of hazards or risks.
– Risk assessment.
– Risk control.

In attempting to evaluate the effectiveness of the new performance risk management style regulations when compared with the former prescriptive, rules-based approach, Gun (1994) referred to the report of the Health and Safety Executive in the UK, where it was established that there had been significant improvements in the assessment and control of risks arising from hazardous substances in the workplace since the introduction of the new regulations. About 49% of the survey respondents reported more efficient use of chemicals, and a similar percentage reported a range of other benefits including better management of plant.

The regulations had enabled companies to focus on the individual realities of their own workplaces and develop appropriate and effective action.

5 THE HEALTH AND SAFETY IN EMPLOYMENT ACT 1992 AND REGULATIONS 1995

The New Zealand Building Code (NZBC) is an integrated performance based code, divided into clauses, that sets out descriptions of objectives, general functional requirements, and specific mandatory performances that must be achieved to comply with the law (Table 1).

Methods for compliance are not prescribed. The NZBC originated from building industry requests for reform dating back to 1980. The national building code had to be performance oriented, consistent with public interest, and within a suitable economic framework with respect to efficiency and accountability underlying the restructuring of the New Zealand economy. The code, and its performance base, is regarded as the best building control tool to encourage innovation, remove barriers to international trade, and to minimize the guessing game of why regulators insist upon particular prescriptive requirements (Hunt and Killip, 1998). These benefits are being gained through a custom-made administrative legislative framework uniquely designed for New Zealand.

The Health and Safety in Employment Act 1992 (HSE Act) demonstrates the confidence which the New Zealand government has in the performance approach. It extends the application of the performance approach to worker safety and health. The HSE Act has reformed the law and many separate regulations and altered their nature from a prescriptive base to a performance-based platform of legislation. In this way, it provides, for the first time, comprehensive coverage and a consistency of approach to the management of safety and health in all workplaces. The HSE Act provides comprehensive coverage for all work situations, clearly defines responsibilities, promotes systems for identifying hazards and dealing with them, enforces involvement of employees in health and safety issues along with requirements for health and safety training and education.

The guidelines to the HSE Act with respect to the construction industry include checklists to aid in identification of risks, and the assessment and control of those risks.

Table 1 Performance code from New Zealand Building Code

Objective	F4.1
	The objective of this provision is to safeguard people from injury caused by falling
Functional	F4.2
Requirement	Buildings shall be constructed to reduce the likelihood of accidental fall
Performance	F4.3.1
	Where people could fall 1 meter or more from an opening in the external envelope or floor of a building, or from a sudden change of level within or associated with a building, a barrier shall be provided

6 THE OCCUPATIONAL SAFETY AND HEALTH ACT (OSHA) OF 1970

OSHA in the United States applies specifically to employers, which in the case of construction are contractors. The OSHA standards have historically been formulated on the basis of traditional prescriptive and "deemed-to-comply" approaches. The prescriptive approach describes means, as opposed to ends, and is primarily concerned with type and quality of materials, method of construction, and workmanship (CIB, 1982).

It attempts to standardize the work process using prescriptive rules and procedures usually backed by the monitoring of compliance and by sanctions for non-compliance (Reason, 1998). The approach has been described as being conservative in that it is difficult to take account of variations in workmanship and materials (Walsh and Blair, 1996). It is problematic to refine the approach to keep pace with innovation, better construction techniques, and new materials..

The effort to change the culture of the current regulatory system needs to be based on sound science and good information. To this end, OSHA has been pilot testing a system which will give both construction managers and workers the primary responsibility for ensuring safety and health at their individual work sites.

– For its part, OSHA, in a May, 1995 report, entitled "The New OSHA," has committed itself to promoting common sense regulations, encouraging partnerships, and eliminating red tape, while at the same time ensuring greater safety and healthier working conditions for American workers (Office of Management and Budget 1996).

The August 1996 revision of the OSHA standard protecting approximately 2.3 million workers on scaffolds in the construction industry is an example of a performance based approach. The standard establishes performance-based criteria, where possible, to protect employees from scaffold-related hazards such as falls, falling objects, structural stability, electrocution, and overloading (Office of Management and Budget 1996). Employers are allowed greater flexibility in the use of fall protection systems to protect workers on scaffolds. This flexibility extends to workers erecting and dismantling scaffolds. The training of workers using scaffolds is also strengthened. Further, the standard specifies when retraining is required.

7 FINAL COMMENT

While the benefits of the adoption of the Council Directive 92/57/EEC in Europe, the CDMR in the UK, National Model Regulations and the National Code of Practice for the Control of Workplace Hazardous Substances in Australia, and HSE Act 1992 and Regulations 1995 in New Zealand have not been extensively measured and evaluated yet, it is anticipated that the paradigm shift promoted by this type of regulatory framework will have positive results for the construction industry and contribute to the common vision of accident free construction sites. Further, for the fully successful introduction of a performance-based code an effective and efficient administrative and legal underpinning be supported.

The value of the CDMR and Council Directive 92/57/EEC, in particular, lies in the requirements of all participants in the construction process to make safety and health a mandatory priority. They are performance based, permitting flexibility in dealing with safety and health issues and the relationships, which are common for construction projects. Additionally, they provide a framework within which all the activities of all participants in the construction process, are co-ordinated and managed in an effort to ensure the safety of those involved with, or affected by, construction. While OSHA is still largely prescriptive in nature, there are signs of increasing acceptance of a paradigm shift towards a performance-based approach. There is a steadily growing recognition that new approaches are necessary to reduce accidents and fatalities on construction sites throughout the United States.

REFERENCES

Approved Code of Practice (ACOP) (1995) : *Managing Construction for Health and safety : Construction (Design and Management) Regulations 1994 :* Health and Safety Exec.

Berger, J. (1999): "Construction Safety Coordination in Germany," In Gottfried,A., Trani, L., and Dias, L.A. (eds), *Safety Coordination and Quality in Construction*, Proceedings of International Conference of CIB W99 and Task Group 36, Milan, Italy, 22-23 June, pp. 51-60

Caldwell, S. (1999): "Construction Safety Coordination in the United Kingdom," In Gottfried,A., Trani, L., and Dias, L.A. (eds), *Safety Coordination and Quality in Construction*, Proceedings of International Conference of CIB W99 and Task Group 36, Milan, Italy, 22-23 June, pp. 141-148

CDMR (1994) : *Construction (Design and Management) Regulations,* SI 1994/3140 HMSO

CIB Publication 64 (1982): *Working with the Performance Approach in Building,* Rotterdam, CIB

Coble, R. and Haupt, T.C. (1999): "Safety and Health Legislation in Europe and United States: A Comparison," In Gottfried,A., Trani, L., and Dias, L.A. (eds), *Safety Coordination and Quality in Construction*, Proceedings of International Conference of CIB W99 and Task Group 36, Milan, Italy, 22-23 June, pp. 159-164

Council Directive 92/57/EEC (1992) : "Council Directive 92/57/EEC of 24 Jn 1992 on the implementation of minimum safety and health requirements at temporary or mobile construction sites (eighth individual Directive within the meaning of Article 16 (1) of Directive 89/391/EEC)" *Official Journal of the European Communities* no. L 245/6

Gottfried, A. (1999): "Construction Safety Coordination in Italy," In Gottfried,A., Trani, L., and Dias, L.A. (eds), *Safety Coordination and Quality in Construction*, Proceedings of International Conference of CIB W99 and Task Group 36, Milan, Italy, 22-23 June, pp. 141-148

Gun, R. (1994): "The Worksafe Model Regulations for Chemical Safety: How much Benefit?" *J. of Occ. Health and Safety- Australia & Auckland*, V10, no.6, pp.523-527

Joyce, Raymond (1995) : *The Construction (Design and Management) Regulations 1994 Explained :* Thomas Telford Publications : London

Latham, M. (1994): *Constructing the Team,* HMSO

Lawson, J. (1996): "Workplace Hazardous Substances Regulations - A Performance-based and Risk management Approach" *Risk Engineering,* Univ. of New South Wales, The Munro Centre for Civil and Environmental Engineering

Lorent, P. (1999): "Construction Safety Coordination in Belgium and Luxembourg," In Gottfried,A., Trani, L., and Dias, L.A. (eds), *Safety Coordination and Quality in Construction*, Proceedings of International Conference of CIB W99 and TG 36, Milan, Italy, 22-23 June, pp. 7-26

McCabe P.J. (1999): "Construction Safety Coordination in Ireland," In Gottfried,A., Trani, L., and Dias, L.A. (eds), *Safety Coordination and Quality in Construction*, Proceedings of International Conference of CIB W99 and Task Group 36, Milan, Italy, 22-23 June, pp. 69-82

Müngen, U, Kuruoglu, M, Türkoglu, K, Haupt, T.C, and Smallwood, J. (1998): "Scaffolding accidents in Construction : An International Examination" Haupt, Theo C., Ebohon, Obas John and Coble, Richard J. (ed) : *Health and Safety in Construction : Accident Free Construction :* Peninsula Technikon : Cape Town pgs 43-54

Munro, W.D. (1996) : "The Implementation of the Construction (Design and management) Regulations 1994 on UK Construction Sites" In Alves Dias, L.M. and Coble, R.J. (eds), *Implementation of Safety and Health on Construction Sites,* Proceedings of the First International Conference of CIB W99, Lisbon, Portugal, 4-7 September, Rotterdam, A.A. Balkema, pgs.53-66

Ngowi, A.B. and Rwelamila, P.D. (1997) : "Holistic approach to occupational health and safety and environmental impacts" In Haupt, Theo C. and Rwelamila, Pantaleo D (ed) : *Health and Safety in Construction : Current and Future Challenges :* Pentech : Cape Town pgs 151-161

Office of Management and Budget (1996) : "Report to the President on the Third Anniversary of Effective Order 12866 : More Benefits Fewer Burdens : Creating a Regulatory System which Works for the American People," *Office of Management and Budget and Office of Information and Regulatory Affairs*

Onsten, G. and Patay, A. (1999): "Construction Safety Coordination in Sweden," In Gottfried,A., Trani, L., and Dias, L.A. (eds), *Safety Coordination and Quality in Construction*, Proceedings of International Conference of CIB W99 and Task Group 36, Milan, Italy, 22-23 June, pp. 135-140

Papaioannu, K. (1999): "Construction Safety Coordination in Greece," In Gottfried,A., Trani, L., and Dias, L.A. (eds), *Safety Coordination and Quality in Construction*, Proceedings of International Conference of CIB W99 and Task Group 36, Milan, Italy, 22-23 June, pp. 61-68

Reason, J. (1998): "Organizational controls and safety: the varieties of rule-related behavior," *Journal of Occupational and Organizational Psychology,* vol. 71, no. 4, pp. 289-301

Rogers, B. (1999): "Developing Pro-Active and Effective Strategies to Manage Health and Safety Risk on Site - A Practitioners Perspective," Presentation at 2nd Annual Building and Construction Series, Auckland

Smallwood, J. and Haupt, T. (2000): "Safety and Health Team Building," In Hinze, J., Coble, R., and Haupt, T. (eds), *Construction Safety and Health Management,* New Jersey, Prentice-Hall, pp. 115-144

Szana, Tibor and Smallwood, John (1998) : "The Role of Culture and management systems in the Prevention of Construction Accidents" In Haupt, Theo C., Ebohon, Obas John and Coble, Richard J. (ed) : *Health and Safety in Construction : Accident Free Construction :* Peninsula Technikon : Cape Town pgs 21-35

Tyler, A.H. and Pope A. (1999) : "The Integration of Construction (Design and management) Regulations into Small and medium Companies" *Implementation of Safety and Health on Construction Sites,* Proceedings of the Second International Conference of CIB W99, Honolulu, HI, 24-27 March, Rotterdam, A.A. Balkema, pgs.447-452.

Creative Systems in Structural and Construction Engineering, Singh (ed.) © 2001 Balkema, Rotterdam, ISBN 90 5809 161 9

Safety deficiency during construction and their effects on life span of buildings

J.A. Igba, X.L. Liu & D.P. Fang
Department of Civil Engineering, Tsinghua University, Beijing, People's Republic of China

ABSTRACT: The reasons for buildings' collapse during construction are usually traced to the deficiencies or inadequacy in safety provisions during design and construction. It has been confirmed that, more often than not construction failure occurs because of underestimation of the magnitude of prevailing live load on the young concrete or inaccurate structural analysis. Therefore, in trying to understand and finally solve the construction safety question, many researchers have accomplished tremendous and recommendable evaluation of safety in the field of construction. However, these achieved records deal better with the problems of structures in their service life stage (the normal use stage). In this paper however, efforts are made to highlight the most likely reasons for construction disasters. Considering the potential seriousness of the consequences in the wake of construction disaster, the authors of this paper made some important recommendations at the end of this write-up.

1 INTRODUCTION

As a matter of fact, much has been accomplished so far in the field of construction safety, but lots of work is still desired to be done. In terms of structural reliability during the process of construction of high rise buildings, what has been achieved is still short of clear definition of a universal structural safety index (β) for construction. Meanwhile, there are accumulated records of building collapse occurring during construction in the absence of construction process code. It is in the light of these events that, this area of structural engineering started attracting the attention of researchers, including the present authors. The pioneering researchers have all made their enormous contributions in developing structural reliability theories. Names like Shitailman, P. Grundy and Kabaila, (1963) Agarwal R.K., Lasisi and Ng (1979) and the Raymond C. Reese research award winners Liu et al. (1985) and a host of others are always mentioned in this area. All these researchers took giant steps in the right direction to solving the construction safety question. For example, the devastating consequences of the progressive collapse that rocked an entire structure at Bailey Crossroads, near Alexandria, Virginia (Chen and Mossalam, 1991) is still flesh on our minds. In 1987, the L'Ambiance Plaza in Bridgeport, Connecticut collapsed killing 28 workers (Liu et al., 1985). On July 2nd 1997, a building under construction collapsed in Karachi and 24 deaths were recorded. The persistence occurrence of these accidents can justify the new momentum to the dimension of research activities in this area of engineering.

Even though all these efforts are aimed at achieving the desired safety in construction, they apply most effectively to structures at the service stage. The implication is that, The codes have not gone deep enough to cover for buildings under construction, and those problems that are normally encountered during construction are as a matter of fact not envisaged at the design stage and hence not provided for. In essence what we have as a result of this, is a persistent increase of construction failure on the world scenario. Some statistics put their occurrence at every other day, with 90% of all collapse happening at the construction stage (ISODIS 2394, 1996). This phenomenon has called for more attention to research in safety control measures during construction in recent times. The question is then, "how can we achieve the much needed safety during construction?" How realistic are those safety measures we have been adapting during the design & construction of high rise buildings?

2 RESEARCH SIGNIFICANCE

First and foremost, it is generally noticed that most

building structures collapse during the construction stage. Secondly, to construction crews (both skilled and unskilled) as a matter of profession, spend their active working life on construction sites. Therefore, their likelihood of being injured due to construction failure should not be greater than those for whom these structures are designed. Unfortunately though, currently there are no codes or standards for specific safety margin during construction. The third aspect is, in an event of construction collapse, the whole process is viewed as gross economic loss which must be averted. That, for all the monetary input, man-hour, labor, time and miscellaneous activities sum up to a huge economic loss. Worse of all is that, more often than not these construction accidents involve human lives. It is particularly in these regards that, various researchers including the authors of this paper have made several efforts, to investigate the safety of structures during construction. The investigation includes the assessment of loads, strength available and their influences due to decision on construction management (construction cycle). In the wake of present speedy construction, the problem is even compounded the more; especially with the concrete strength being far lower than their 28-day design values. It then implies that, the load acting on the partially stiffened R.C. element might be more than their available capacities. Based on the current work, it is hoped that, a recommendation will be made as to what aspect structural designers should watch out for in the course of their designs. It is also believed that, a professional advice will be given in relation to what aspect of safety should be carefully looked into during construction. With all these at back of our minds during this research program, we believe in its relevancy.

3 CONSTRUCTION LOAD MEASUREMENTS

During the current study, there were several investigations conducted in respect of construction loads. The first investigation was mainly concerned with aspect of construction live loads on the newly poured slabs. This measurement was conducted in the same manner as Haitham Ayoub's "Live Loads on Newly Poured Slabs" (1994) report, published in May 1994. On the other hand, during the subsequent measurements a close attention was given to dead and live load condition particularly during casting of a concrete slab. The surveyed information on the influence of loads on forms and shores or steel props was of paramount importance during this part of investigation.

In the first load investigation process was on for several months in 1997. Within the said time, various projects in several cities within the Peoples Republic of China were carried out. The structures

investigated also cut across a number of buildings with vast functional assignments. Some of the structures investigated were designed as apartment buildings, office blocks and scientific research complexes. The building sites were geographically distributed randomly within the Chinese territorial integrity, though most were in Beijing, Shanghai, Linxi as well as in Shangdong provinces respectively. In terms of their level of technology application and management rating, all these contractors rated first class in the Chinese construction Industry.

4 SITE SURVEY REPORT

The setup: A data-assembling task was arranged, and the aim of this task was to collecting data in respect to construction loads on the slab and shores from building under construction. One of the load surveys was carried out during construction of the new Law Department situated towards the east entrance of the University. It is hoped that, at the end of this investigation a conclusion could be drowned that might assist professionals understand better the influence of shores on general reliability during construction process. The knowledge expected here is important and will serve as good contribution during decision-making and during construction.

The contractor's scheduled a 14-day construction cycle on this project. Our main concern here was to observe the nature of loading and the reaction of temporary elements during and after loading of the slab under construction. This survey setup also has helped shade more light on how realistic or otherwise is the assumption of infinite and unyielding axial stiffness of steel shores. We hope to use the same analogy to prove or disprove the assumption that, all the interconnected slabs deflect equally to carry their own weight when a new load is introduced. As a matter of fact, this test enable us describe closely what takes place when shores are performing their desired function of load transmission from floor to floor. Although these results are mostly from multistory structures, logically the same phenomenon is applicable to high-rise structures.

After the whole test operation, it was noted that each shore had to some extent a degree of stiffness but not entirely infinite. So the current work has also contributed in removing the illusion that shores axial stiffness is infinite. However, the shores axial deformation observed during this investigation was quite small and insignificant, such that could (usually) be ignored. Even at that, the point is still that axial stiffness of shores is finite. The current work recognized that, the insignificant shore axial deformation is due as a result of a degree of deflection in the interconnected slabs. These deformations normally take place when new loads are added or when shores are removed (Liu, 1985).

5 TEST ACCEPTABILITY

On the whole, the shores general dimensions were determined i.e. their cross sections, their lengths, their distribution and so on. However, there exist variation tendencies in coupling of joints, thermal loading, and variation in the individual stability of the entire shoring network. The question of acceptability level of workmanship during their erection is another factor that could be investigated when conducting a sensitivity analysis in future. At any rate, these factors are mild and therefore it's expected that, the conclusion that will be drawn from this report does not fall short of the reality any way. At this level the mentioned shortcomings are regarded as falling within acceptable range. Another form of upgrading the result of this study is, by extending the work to encompass sensitivity analysis of the shores as already mentioned above. Unfortunately, there is no enough space in this paper to include such detail analysis. This can be achieved by first outlining all the possible characteristic of the individual shores used in the test. The random variables in the whole test operation can then be considered one after another during sensitivity studies. The second aspect could be achieved, by obtaining manufacturer's technical data about the shores used in the projects. These data may include specifications like designed axial stiffness of shores, their various dimensions, deflection-calibration limits and any other relevant information.

6 ANALYSIS OF SAFETY DURING CONSTRUCTION

The failure of a concrete slab may be defined by exceeding the flexural bending capacity, the punching shear capacity or prescribed deflection limits. Though in the current work, analyses were made based on three selected failure modes, a brief mention is made of how other factors also affect the value of safety index β. The basic reality is that, not all failures can be treated alike because there may be wide range of consequences, depending on the mode of failure. As mentioned above, it is generally recognized that discussions relating to safety of structures involve not only the engineering analyses but also the social, human factors and sometimes even cultural/ethical values of the society within which the project is located (ISODIS 2394, 1994).

During the analysis both dead and live loads were considered as random variables, in fact all the parameters involved in load distribution themselves were treated as random variables during computation. The dead load, which is the self-weight of the structure, was assumed to be normally distributed with a mean value and standard deviation worked out. On the other hand, the construction load that was being observed was randomly spread on the partially cured slabs. At first the survey considered the non-zero loading areas and the zero load distributed areas of the construction loading. However, it was more reasonable to apply the Equivalent Uniformly Distributed Load (EUDL) theory. The EUDL option was good for further analysis because the slab is an element of a structural system bounded together by concrete materials and reinforcement and so can afford redistribution of loads. The resistance at each stage was determine using the relative (partial strength) development of concrete strength as compared with the design strength of concrete assumed to be developed at the age of 28 days. Coefficients for development of concrete strength depending on the age and type of concrete were also used (Chen and Mosallam 1991).

The first failure mode considered in the present work is deflection, it is advised that in general case where the failure condition is non-linear in nature, an iteration method must be used to find the convergence point of the iterative equation. In the present work an iterative method is used to find the convergence point 'A' on the assumed failure surface. In the analysis the so-called failure surface is defined by equation:

$$f(Z_1, Z_2, ...Z_n) = 0 \qquad (1)$$

The analysis proceeds first with deflection failure mode criterion given by limit state provision, that the following condition is not exceeded:

$$U\max = \frac{1}{9.6}\frac{l^2 M}{EI} \leq \frac{1}{250}l \qquad (2)$$

where U_{max} is the limiting maximum deflection, E is the modulus of elasticity, I is the relevant moment of inertia, l is the slab span so that U_{max}, Q, l, E, and I, are realization of the non-correlated random variables U_{max}, Q, L, E, & I respectively. The coefficients of 0.79 and 0.90 were used in the analysis to determine the actual strength available at 7 and 14 days of construction cycle period respectively. The final limit state function used for iterations in the analysis was formulated thus:

$$(E,I) - D - L = 0 \qquad (3)$$

where E, I are modulus of elasticity and moment of inertia respectively; D, L in (3) are construction dead and live loads respectively.

Table 1. Calculated β values by iteration

PARAM-ETERS	START VALUE	ITERATION NUMBERS				
		1	2	3	4	5
β	3.0	4.01	3.58	3.24	3.14	3.14
α_1	-0.58	-0.37	+0.02	+0.03	+0.04	+0.04
α_2	-0.58	-0.95	-0.99	-0.99	-0.99	-0.99
α_3	+0.58	+0.19	+0.15	+0.11	+0.11	+0.11

From the reliability analysis, using the same method as was used by Thoft-Christensen, the safety index β = 3.14 and the design iteration converged at point (α_1=+0.04, α_2=-0.99, α_3=+0.11). The 'αs' are used in solving the basic variables equations: α_i, i = 1,2,3, and their corresponding signs were chosen for the starting values. Its usual to chose a negative sign when a variable is strength or of geometric characteristic like modulus of elasticity (E) and moment of inertia (I) for instance. A positive sign is recommended for a positive variable (e.g. loading variable Q or P as the case may be). The total sum of the squared values of α_i, i = 1...n from the start values and all subsequent iteration numbers at each point should total up to 1 (one)

($\sum_{i=1}^{n} \alpha_i^2 = 1$) as in equation (4) below.

$$\sum_{i=1}^{n} \alpha_i^2 = \alpha_1^2 + \alpha_2^2 + \alpha_3^2 = 1 \qquad (4)$$

The second failure mode considered was based on concrete compressive strength in relation to the construction cycle. This form of analysis is conducted with a view to determining the resistance available at the time of one construction cycle, the relevant level of reliability and its consequence. Generally, the resistance available during one construction cycle is described by the ratio of development of concrete strength within the stipulated time designated as the so-called construction cycle. In this work, the cycles are 14 and 7 days as already as already mentioned, with 0.90 or 90% and 0.79 or 79% maturity respectively. Chen and Mosallam (1991) also gave these values for development of concrete strength. The limit state equation in this mode is given as given in Chinese Code for Design of Concrete Structures (1994).

$$\mu_R - \mu_S \geq 0 \qquad (5)$$

where

$$\mu_R = A_s f_y (h_o - x/2) \qquad (6)$$

$$x = f_y . A_s / f_{cm} . \qquad (7)$$

μ_R and μ_S are the mean values of resistance and load respectively
A_S and f_y are area of cross sectional

reinforcement and tensile strength respectively
h_o is the effective depth of slab
f_{cm} is flexural compressive strength of concrete

The construction process considered during the analysis is the most usual practice of 2-levels of shores and a level of reshore. With all the condition and parameters as stated above, the results of the reliability index values for the two construction cycles were β = 3.18 and β = 2.42 for 14 and 7 days respectively. A change in construction technique from 2-levels of shores and one reshore to 2-levels of shores with no reshore indicated a significant change in the result of the same failure mode bringing down remarkably the values of reliability index β. This phenomenon shows that the choice of construction process also plays a significant role in the safety of structures during their construction.

The third mode considered during work is the analysis using the punching shear failure limit state model. In the flat slab construction it is unquestionably clear that punching capacity forms one of the most delicate issues as safety of this form of structure is concerned. Especially when the dimensions (span) of the structure in question is large. In light of the above the current work did a careful analysis of column sizes avoiding flare head columns to achieve a relatively reasonable result. Using the same loading case as in the failure mode above, since the analyses was done on common loading and resistance quantities a failure probability was defined by the expression

$$P_f = P(\phi V_n - V_u \leq 0) \qquad (8)$$

where V_n, V_u are the nominal and ultimate punching shear strength
ϕ is the strength reduction factor = 0.85 for shear

The limit state formula for punching shear load bearing for a slab provided by the Chinese Code for Design of Concrete Structures, (1994) was then calibrated with the actual strength available at the two different construction cycles and results were obtained. The punching shear load bearing capacity is defined by the formula:

$$F_l \leq 0.6 f_t \mu_m h_o \qquad (9)$$

where F_l is design value of local loads
f_t is design concrete strength under tension = 1.5
μ_m is perimeter with a distance of $h_o/2$ from the column periphery

h_o is effective depth of slab

From the result of the above analysis, the values for reliability indexes for the two construction cycles 14 days and 7 days were $\beta = 2.75$ and $\beta = 2.18$ respectively. The calculated values obtained from this mode of failure were discovered to be the most sensitive of all the three modes considered during this work. The numerical values relative to safety are often described in relation to their failure probability (P_f) defined as:

$$\beta = \Phi^{-1} (P_f) \tag{10}$$

The data collected from most of the load experiments were analyzed using a computer program TDSA-2, the same was used to compute the results. To check for the reliability index β, the distribution of both modulus of elasticity (E) and moment of inertia (I) were considered normally distributed. Their coefficient of variation COV were 0.25 and 0.2 respectively. The distribution of dead load (Q_1) was considered lognormal with mean value $\mu_{Q1}=2.65kN/m^2$ and coefficient of variation $cov_{Q1} =0.074$. The live load on the other hand was considered extreme distribution with the mean value $\mu_{Q2}=0.5662kN/m^2$ with coefficient of variation $cov_{Q2}=1.2$. The final result of this analysis shows that at the 7th and 14th day the reliability indexes were as high as 3.2115 and 3.6951 respectively. These high reliability values could be attributed to the satisfactory quality control of concrete as well as the grade of concrete that was being use on this particular project.

7 CONCLUSION

A practical field test of construction loads is undertaken, and the results of their effects are presented in this paper. Prior to that a brief background of prominent works done by various researchers in the area of structural reliability is given. This paper then, uses iteration method to calculate the safety index for which the iteration converged with the value of β at 3.14, considered as reasonable by the authors. With acquisition of more data, a computer program TDSA-2 was employed for further computation and with the normal construction cycle of 7 days the safety index β was 3.2118, for 14days construction cycle β was as high as 3.6951.

A comparison of the calculated values of safety index during construction seems similar to the values in the literature for reliability index during construction and safety index during service life of a building. The result of this investigation also prompted the authors to recommend that, the safety index during construction be kept within a range between 2.8 and 3.2 given that the construction cycle is more than 3 days. Igba, Liu, Fang (2000).

ACKNOWLEDGEMENT

This study would not have been possible without the cooperation of the companies that allowed access to their construction sites for data collection. The authors also wish to thank Mr. Zhu Hongyi for allowing us make adequate use of some of the data that was collected by him.

REFERENCES

Chen, W.F., and Mosallam K.H., "Concrete Buildings: Analysis for Safe Construction, *CRC P-Boca Raton, PP. 221, 1991.*

H. Ayoub S. Karshenas "Construction Live Loads on Slab Formworks Before Concrete Placement" *Structural Safety Journal 14 (1994) 155-172*

ISO/DIS 2394 "General Principles on Reliability for Structures" *International Organization Journal of Standardization*, July 1996.

Lasisi M.Y & Ng. S.F. "Construction Loads Imposed on High-rise Floor Slab", *Concrete International*. February, 1979

Liu X.L., "Analysis of Reinforced Concrete Buildings during Construction". *Doctoral Dissertation, Purdue University*; 1985.

Liu X.L., Chen W.F., & Bowman M.D., "Construction Loads on Supporting Floors", *Concrete International vol. 7*, No.12 Dec. 1985.

National Standard of the People's Republic of China; *"Code for Design of Concrete Structures" GBJ 10-89, New World Press*, Beijing 1994.

Paul Grundy & A. Kabaila, "Construction Loads on Slab with Shored Formwork in Multistory Buildings" *ACI Proceedings, v.60. no.12. Dec.1963, pp.1729-1738.*

Creative Systems in Structural and Construction Engineering, Singh (ed.)© 2001 Balkema, Rotterdam, ISBN 90 5809 161 9

XML based database system for accident cases in construction of prestressed concrete bridges

M. Hirokane – *Department of Informatics, University of Kansai, Japan*

A. Miyamoto – *Department of Computer and Systems Engineering, University of Yamaguchi, Japan*

H. Konishi – *Department of Design, Japan Bridge Corporation, Japan*

Y. Fujioka – *Department of Civil Engineering, P.S. Corporation, Japan*

ABSTRACT: In the present study, the authors constructed a system for workers and site foremen engaging in construction work of PC (Prestressed Concrete) bridges to retrieve and input information on accidents. This system was constructed by using the XML (eXtensible Markup Language) so that workers and site foremen can retrieve information from any construction sites through the WWW (World Wide Web) and use such information for their safety planning and management. Besides, the system offers visualized information on the situations in which accidents occurred so that they can have pseudo experience of accidents in their offices. Furthermore, by using one and the same format, information can be shared among all concerned and a large number of accident cases can be accumulated easily.

1 INTRODUCTION

In Japan, newly constructed PC bridges are becoming larger and longer with the advance of technology. Precision required of their construction work is getting higher, and their construction processes are becoming more and more complex. Accidents occurring in the circumstances can be categorized into three types; i.e. fall, accident associated with construction machinery and cranes, and accident by collapse of members and structures (Japan Ministry of Labor 1992, Japan Ministry of Labor 1994).

Among these accident types, falls due to human factors are occurring most frequently. Major accidents due to poor understanding of the situations of erection sites, inadequate routine inspection, insufficiency in making erection schedules known to all persons concerned, insufficiency in giving signals to other workers during work, and so on are reported (The Association of PC Construction 1993). Some of such accidents can be prevented if workers and site foremen have always in mind what accidents can happen next or what measures to take if the situation threatens any accident. In other words, it is important that workers and site foremen have safety consciousness and have in mind what accidents are likely to occur in the present work. To help them prevent such accidents, it would be useful to accumulate information on accident cases, visualize the information as far as possible, analyze the information by work type, date and time, age, years of experience, accident scale, etc., and offer the information to them. At present, however, individual companies are accumulating such information for their in-house use, and their data sources are insufficient for the above purpose. Industrial safety and health regulations and other literature are their sources of information on preventive measures of accidents. In such literature, however, matters to be observed are prescribed in letters, and it would be difficult to get concrete pictures in relation to actual erection from such literature.

In the present study, the authors constructed a system for workers and site foremen engaging in erection of PC bridges to retrieve and input information on accidents using a microcomputer. The system was constructed by using the XML (Kobayashi, S., Takagi, Y. & Miyamoto, Y. 1999, Shimizu, W., Satoie, M. & Ogawa, T. 1998) so that workers and site foremen can retrieve information from any erection sites through the WWW and use such information for their safety planning and management. Besides, the system offers visualized information on the situations in which accidents occurred so that they can have pseudo experience of accidents in their offices. For example, after an erection method was decided, workers and their site foreman will retrieve information on accidents associated with the erection method and, using of the retrieved information, make safety planning. Moreover, as the XML is used in the system as mentioned above, information can, by adding tags to specific items of information, be analyzed by nature of work, date and time, age, years of experience, accident scale, and so on. Furthermore, by using one and the same format, information can be shared among all concerned and a large number of accident cases can be accumulated easily.

2 SAFETY MANAGEMENT SYSTEM

2.1 *Contents of accident cases*

To build a database of accident cases by using the XML, data on accidents were accumulated. Among various items in such data, the following seven items were chosen for the database.

1. *Accident Type*: This item enables to derive statistics of accidents by accident type from the database. Accidents were classified into types such as fall, overturn, and being caught between objects. Such classification and statistics help us grasp trends in accidents.

2. *Scene of Accident*: Scenes of accidents were classified into factory and erection sites. The erection sites were sub-classified into various types.

3. *Nature of Work*: Stages in erection process where accidents occurred. This system is designed to enable users to retrieve accident cases by stage in erection process and grasp stages of frequent occurrence of accidents.

4. *Degree of Injury*: This and the immediately preceding item enable to derive the statistics of what degrees of injuries have happened at what stages in erection process.

5. *Age, Sex, and Years of Experience*: Workers of different ages and careers are working in erection sites, and human factors such as physical strength and experience play important parts in the occurrence of accidents. These data are useful for making placement plans of workers.

6. *Occurrence Time*: The physical strength and attentiveness of workers change with the passage of working time, which is an important factor of accidents. By identifying time zones of frequent occurrence of accidents, intermissions can effectively be set in working hours.

7. *Causes and Measures*: These information are very useful for the development of appropriate measures to prevent accidents.

2.2 *Description of accident cases*

The accumulated accident cases were described by using XML. Figure 1 is an example of actual accident cases that are described by this language. The contents of each data item can be understood easily because each data item is marked by individual tag. For example, the data item for the accident type was marked by tag named the accident type, and the data

```
Sample1 - メモ帳
ファイル(F) 編集(E) 検索(S) ヘルプ(H)
<?xml version="1.0" encoding="Shift_JIS" ?>
<?xml:stylesheet type="text/xsl" href="sample.xsl"?>
<Accident xmlns:HTML="http://www.w3.org/Profilse/XHTML-transitional">
        <Division>
            <Accident_type>Crash or Fall</Accident_type>
            <Nature_of_accident>Fall in dismantling staging</Nature_of_accident>
        </Division>
        <Accident_data>
            <Scene_of_accident>Hanging scaffold above city road</Scene_of_accident>
            <Nature_of_work>Dismantling of staging</Nature_of_work>
            <Injurer>
                    <Sex>Male</Sex>
                    <Age>40</Age>
                    <Job_type>Construction worker</Job_type>
                    <Years_of_experience>18 years</Years_of_experience>
                    <Injury_degree>Death was caused by rib bone fracture and abdomen internal hemorrhage
                    </Injury_degree>
            </Injurer>
        </Accident_data>
        <Situation>
                <Time_1>00:45</Time_1>
                <Detail_1>The workers A, B, C and overseer D finished to dismantle some protection nets,
                        flared steel pipes and protection floors from the staging.</Detail_1>
                <Time_2>02:30</Time_2>
                <Detail_2>For the arrangements of the next erection process, the worker C was walking toward
                        the ladder that led from the staging (on the side of the bridge in construction) to the
                        ground, but lost his balance near the ladder and fell.
                </Detail_2>
        </Situation>
        <Cause>No main pile equiptments were set up</Cause>
        <Cause>The worker C was new on the erection sites</Cause>
        <Measure></Measure>
</Accident>
```

Figure 1. Description example of an accident case by XML.

item for the scene of accident was marked by tag named the scene of accident. As stated above, each data item was marked by tag that described the contents of its data. Moreover, the accident type and the nature of work are marked together by tag named the division because they are used in retrieving information on accident cases. In case of describing the accident cases by using XML, it is not necessary to describe the output format on screen and paper for all cases. Therefore, their data structure is very simple. People other than the database manager can also understand easily the contents of all data items by only looking at these descriptions, because the letters that described precisely the contents of each data item was used as its tag.

2.3 Input of accident cases

In this research, the past actual accident cases are described by using XML, but they are still insufficient for diffusing safety consciousness. It is important to grasp enough the causes and the measures of actual accidents for preventing the similar accidents. For this purpose, it is necessary to accumulate the past accident cases as much as possible. At the same time, it is necessary to prepare the friendly interface so that workers and their site foreman can input easily the accident that will occur in the near future.

So, the system with the above interface was constructed by using JAVA language (Kawanishi, A. 1996). Figure 2 is an example of interface for inputting the accident case that had occurred in the erection by erection girder. It was considered for the simplification of input to realize that by the check box or the selection box as much as possible. For example, if an erection method is selected, the flowchart of its erection method is shown, and a nature of work is selected in this flowchart. If a nature of work is selected, the careers of workers who seem to work in the selected nature of work are shown, and a career of sufferer is selected in these careers. By realizing the above input method, information can be standardized and shared among all concerned, and a large number of accident cases can be accumulated easily.

2.4 Operation of safety management system

Files made in the XML can be browsed through a browser. Typical browsers are Netscape Navigator and Internet Explorer. Figure 3 and 4 shows the browsing screen with Internet Explorer.

When the user makes access to this system, the initial screen appears on the display, which is the first step of retrieval of relevant information. On this screen, the user chooses an erection method. If the user chooses the erection with staging, the flowchart of the erection with staging appears on the dis-

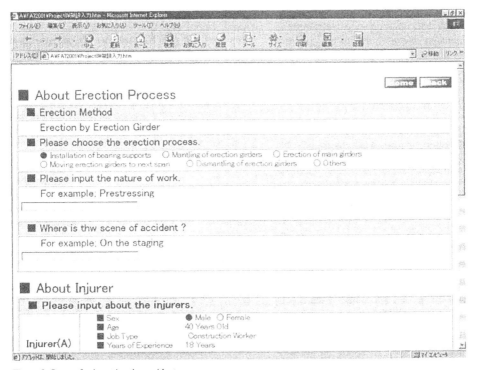

Figure 2. Screen for inputting the accident cases.

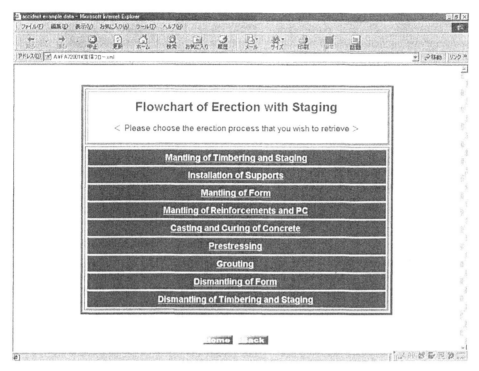

Figure 3. Screen showing the flowchart of erection with staging.

play as shown in Figure 3. When the user chooses a stage in the flowchart, a list of relevant accident cases is displayed. This list shows the situations and types of accidents to give users the general pictures of the accident cases. Then, the user can browse the details case by case as shown in Figure 4.

3 ANALYSIS OF ACCIDENT CASES

3.1 *Analysis by occurrence time of accidents*

The graph of Figure 5 shows the relation between time zone and number of accidents. The time zone of "Others" includes midnight and early morning. According to this graph, accidents are frequently occurring in the time zone of 9:00-11:00 and 15:00-17:00. The reason for frequent occurrence of accidents in the time zone of 9:00-11:00, soon after the opening hour, would be that workers have not yet physically adapted themselves to their work. The reason for the time zone of 15:00-17:00, near the closing hour, would be that their physical fatigue has built up.

At around 9:40 a.m. in a site where internal timbering were being assembled, while a worker was guiding a cargo (five clamp bags on a pallet) lifted by a crane, the pallet touched a pipe and the cargo fell on and broke a scaffold. A worker on the scaffold lost his balance and fell about nine meters to injure his head. The main causes of this accident

were that 1) the slinging was inappropriate, 2) no signaler was set, 3) there was a worker below the hoisted cargo, and 4) the worker did not wear a safety belt. It can be said that the accident was caused by carelessness of the workers concerned. It also seems that the discipline for abidance by the safety rules and the working procedure had been inadequate.

At around 4:20 p.m., a worker was engaging in dismantling work of a wooden form on a 2.25 meters high T-beam. When he pushed a panel forward with a crowbar, the bar slipped off the panel, and he lost his balance and fell to break his pelvis. The main causes of the accident were that 1) no safety guard was provided, 2) his working posture was unstable, 3) he was wearing no safety belt, and 4) no main rope was provided. The second and third factors suggest that his physical fatigue damped down his concentration.

3.2 *Analysis by years of experience*

The graph of Figure 6 shows the relation between years of experience and number of accidents. According to Figure 6, accidents are occurring most frequently among workers with experience less than five years. As the years of experience build up, the number of accidents decreases. The high frequency of occurrence of accidents among workers with experience less than five years seems to imply that

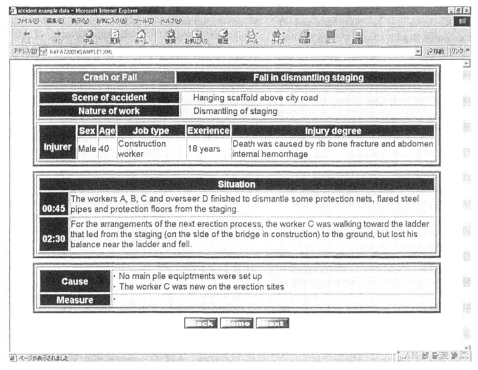

Figure 4. Screen showing details of an accident case.

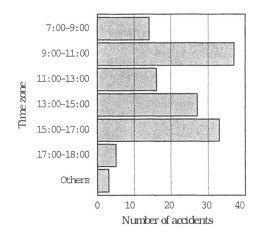

Figure 5. Analysis by time zone.

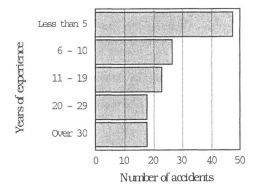

Figure 6. Analysis by years of experience.

there is a high risk of accidents when they get used to work.

In the driving work of landslide-protection piles, a worker with three years experience was, on the working platform, guiding the wire with his left hand to prevent disorderly winding of the wire on the drum and operating, with his right hand, the hand-held switch box of the winch of a boring machine to move a sheet pile. The third and fourth fingers of his left hand were caught between the drum and wire, and mutilated. The main causes of the accident were that 1) the working method was wrong, and 2) he put his hand directly on the wire. The basic factor seems to be inexperience.

On the other hand, a worker with 25 years experience was, on floor slabs, engaging in the work of dismantling timbering and temporarily placing H-steel bars (588mm high and 12m long) on floor slabs. A H-steel bar was lifted and placed on floor slabs by a movable crane, and the clamp removed off the H-steel bar caught an edge of one of H-steel bars, which fell and broke his left hand. The main causes of the accident were that 1) the handling of the clamp was inappropriate, 2) signaling was inadequate, and 3) no consideration was given to signaler. In other words, the accident was caused by getting used to erection work and playing less attention to the working procedure and safety ascertainment.

4 CONCLUSIONS

The authors constructed a system for workers and site foremen to retrieve and input information on accidents occurred in the erection work of PC bridges. The features of the system are summarized below.

1. *Easiness of Operation*: Because this system was constructed by using the XML, workers and site foremen can, from anywhere that a microcomputer and a telephone line are available, retrieve relevant information from the database system through the WWW. Besides, by making good use of the system, workers and site foremen at various erection sites can improve their safety programs in accordance with the natures of their work, and safety education and training can also be conducted effectively.

2. *Visualization of Accident Cases*: This system offers visualized information on the situations in which accidents occurred so that workers and site foremen can have pseudo experience of accidents in their offices. Besides, after the workers and their foreman decide on an erection method, they can retrieve information on accidents associated with the erection method and, using the retrieved information, make safety planning. Moreover, as the XML is used in the system, information can, by adding tags to specific items of information, be analyzed by nature of work, date and time, age, years of experience, accident scale, and so on.

3. *Information Sharing*: By using one and the same format, information can be standardized and shared among all concerned, and a large number of accident cases can be accumulated easily, which enable us to construct a more practical and effective system.

The matters to be improved in near future are as follows:

1. *Visualization of Illustration Data*: This system was constructed by XML. This language is suitable for describing the only documents, but it is difficult to describe and visualize the illustration data by this language. It is necessary to simplify the input and the visualization of illustration data so that workers and their site foreman will make the effective safety planning through they have pseudo experience of accidents in their office on the basis of the visualized information.

2. *Linkage with Regulations*: This system has to be developed further so as to be able to retrieve and analyze information from the point of view of relevant regulations such as the Industrial Safety and Health Law. Such laws and regulations include guidelines for proper working procedure. If a system is made available to offer such guidelines in visual forms through microcomputers to workers and site foremen, it will contribute to the enhancement of their safety education and training.

3. *Use in Actual Sites*: It is necessary to make the accident cases full through the workers and their foreman use positively this system in the actual erection sites. Through their positive use, the deficiencies and the improvements of this system will be shown clearly. By modifying these deficiencies and improvements, this system can be changed into more practical one.

REFERENCES

Japan Ministry of Labor 1992. The work indication of concrete bridges erection. *The association of labor disaster prevention in construction*: 63-77.

Japan Ministry of Labor 1994. The guideline of safety and health supervisor, *The association of labor disaster prevention in construction*: 187-201

The Association of PC Construction 1993. The indication of safety management for the prestressed concrete construction.

Kobayashi, S., Takagi, Y. & Miyamoto, Y. 1999. A practical use of XML on WWW. *Tokyo electro-technological college press.*

Shimizu, W., Satoie, M. & Ogawa, T. 1998. Development of spot information network system. *Report of kawasaki construction co. ltd.*: 146-151

Kawanishi, A. 1996. Internet language 3 • An introduction to JAVA. *Technique criticism co. ltd.*

Creative Systems in Structural and Construction Engineering, Singh (ed.) © 2001 Balkema, Rotterdam, ISBN 90 5809 161 9

Quality management tools vs ISO 9000 in the construction industry

Swapan Saha
University of Western Sydney, N.S.W., Australia

Mark Glading
Lendlease Pty Limited, Australia

Ron R. Wakefield
Jamerson Professor of Building Construction, Virginia Polytechnic Institute and State University, Va., USA

ABSTRACT: There has been growing concern about the performance and competitiveness of the Australian Construction Industry. It is largely believed that the current quality measures are neither specifically tailored to the needs of the construction industry nor adequately implemented on construction sites. The main tool, up until now, to combat this problem has been the introduction of the ISO 9000 series Quality Assurance System. According to a number of studies this system has many limitations and does not adequately service the construction industry in its fight for quality. There are many Quality Management methods including, Quality Circles, BPR, QFD, TQM etc., emerging, which it is believed could better service the industry than the current systems. A quantitative method of analysis is used in this study through the distribution of a survey to a sample population of top level management in large construction firms operating within the Australian (NSW) Construction market. It is found that ISO 9000 has some limitations. Through the research, it is proven that the industry is already indirectly using the above mentioned quality management principles, and that they perceive these tools to be extremely beneficial in terms of quality management for the industry. Further, the industry clearly indicates the formal use of quality management tools in conjunction with ISO 9000 to overcome the limitations of the current ISO 9000 system.

1 INTRODUCTION

The area of Quality Assurance and Quality Management systems in the construction industry is extremely large. As such the main aim of this paper is to look at the inherent limitations of ISO 9000 and then examine the Quality Management tools emerging in the industry such as Quality Circles, Quality Function Deployment (QFD), Total Quality Management (TQM) and Business Process Reengineering (BPR).

The use of the ISO 9000 Quality Assurance System in the Australian Construction Industry has escalated over recent years. Increased acceptance has occurred due to a number of reasons, such as the building boom of the mid 80's which resulted in many development projects being built rapidly, the consequence of which was a reduction in quality of both design and construction. At the same time there was a desire on the part of the developer to complete projects in as short a period of time as possible, so as to minimize financial holding costs and maximize returns (Mohyla 1996).

Unfortunately, this has resulted in the quick adoption of the ISO 9000 Quality Assurance System, which by and large, is not suited to the Construction Industry and has many inherent limitations.

Adding to the problem, the trend towards non-traditional forms of project delivery has not helped in achieving a satisfactory level of quality. This is possibly due to the effect such systems have on the standard of design and documentation (Mohyla 1996).

An ever increasing cost of poor quality has been investigated by the Quality Management Taskforce of the Construction Industry Institute (CII), their findings concluding that, the average cost of rework on projects exceeded 12 per cent of the total project cost (Abdel-Razek 1998).

The current ISO 9000 Quality Assurance System could be largely accountable for much of the problem by concentrating on the quality of the processes taken to achieve the end product, rather than the quality of the end product itself. The system gives no consideration to the motivation of employees, which plays a big role in the quality of a final product in construction. Therefore, there is an increasing need for Quality Management tools to be introduced into the industry.

Over the years there have been numerous studies undertaken to determine whether the application of ISO 9000 has brought any real benefits to the construction industry, although most of these have been specifically targeted on the overseas arena.

There have been both many advantages and disadvantages highlighted for the ISO 9000 system with the disadvantages now significantly outweighing the advantages. Many of the problems experienced with the system overseas can be largely compared with the problems being experienced in the Australia Construction Industry.

The findings from Thailand (Ogunlana et al. 1998) were that the implementation of ISO 9000 has resulted in some minimal improvements in the quality of the finished product. It is generally perceived that the image of companies and indeed that of the construction industry have been enhanced through its implementation. The major benefits identified were clarification in roles and responsibilities and improved information flow. This is in line with the work of Tam & Tong (1996) in Hong Kong that confirms most companies think that getting ISO 9000 certification is crucial for their businesses survival, therefore, if this is the underlying reason for its adoption, then it won't produce the expected results that the industry urgently requires.

Another study was conducted regarding the adoption of ISO 9000 in UK (Keivani et al. 1999). In this study the majority of the firms that responded to the survey indicated that they had incurred significant increases in paperwork and administration cost. None of these firms thought that the negative effects were so great that they would offset the beneficial effects of the certification system. However, one of the respondents claimed that the biggest drawback was that it was "not really geared to the construction industry. It can be a bit too rigid since construction is not a production line activity". Similar results were recorded by a study conducted in Australia (Wiele & Brown 1997). The ISO 9000 system does not focus enough on the needs of the workers. On the other hand Quality Management tools such as quality circles aim to achieve quality through motivating and involving employees. In effect, achieving quality is not the prime concern of quality circles, increasing motivation and productivity are, but it just happens that there is a positive end result, which works very effectively (Rosenfield et al. 1992).

Research is also starting to emerge regarding the use of Quality Management tools in the Construction Industry, which up until now were largely confined to the manufacturing industry. Wang & Lu, (1998) defined Quality Function Deployment (QFD) as a systematic approach that deploys a series of relationships among customers needs and provide a satisfactory product in the end. According to this study QFD can begin at the design stage, to understand the demand of the customer by firstly identifying customers needs, structuring these needs

into quality attributes, prioritizing these attributes, comparing customer perception and extracting quality elements from each attribute. Mallon & Mulligan (1993) have stated that QFD can be a facilitator in achieving the improvement of quality and productivity with an attendant reduction in cost. Before QFD can be fully implemented into an organization "an expanding quality orientated organization must already be in place".

This study finds that these Quality Management Tools could do a better job of controlling the quality of projects in the Construction Industry. According to findings from literature it can be viewed that the industry urgently needs to supplement the ISO 9000 system with the QFD system in the design stage, and Quality Circles in the construction stage. It is also evident from the study that both TQM and BPR also have a future in the Australian Construction Industry (Mohamed & Tucker 1996 and Rounds & Chi 1985).

2 METHODOLOGY

The aim of this study is to determine top level management's perceptions and attitudes to the performance and drawbacks of the ISO 9000 system. It will also look at their current knowledge on quality management tools, their usage and the respondent's receptiveness to trying these tools, such as, QFD, Quality Circles, TQM and BPR. In order to achieve the research objectives a field survey is conducted using quantitative methods.

The sample for the study will consist of top level management of large construction firms operating within the NSW Construction market. This will include Project Managers, Construction Managers and Company Directors. The survey, designed by the authors, will be of a quantitative nature, utilising closed-ended questions with a small comment area at the end of the survey for respondents to make further comments. The survey questionnaire consisted of four parts described below.

3 DATA COLLECTION

In this study eighty (80) survey questionnaires were distributed to the Project Managers & Site Superintendents. Forty nine (49) respondents returned the completed questionnaire. Response rate was 61 %. According to Babbie (1991) response rate is satisfactory. Achieving a high response rate results in less chance of significant response bias than achieving a low rate. The results outlined below, part A, B, C & D correspond to the four sections of the survey. Each part of the questionnaire is analysed and interpreted below.

4 RESULTS

4.1 Part A: respondent's detail

The purpose of part A of the questionnaire was to develop an understanding of the respondents, without them having to reveal their identity. This included their position within the company, the company size and the expertise of the company.

The results indicate that of the total number of respondents, 43% were involved in Site Management, 43% were Top level managers and 14% were Company Directors. The survey was purposely directed at such respondents as they have the most influence over the quality procedures of their companies, and hence, the strongest opinions on the topic. Quality Managers were purposely left out of the survey, as it was believed that their opinions might contain some bias on the subject.

There was quite a large spread of industry sectors in which the respondents operated. As such, 24% operated within all categories, 8% operated in Residential only, 4% operated in Civil only, 14% in Commercial Industrial, 33% in Commercial/ Industrial/ Residential, 6% in Commercial/ Industrial/ Civil, 6% in Commercial/ Residential and 4% in Industrial/ Civil. This would indicate that the use of ISO9000 by head contractors is by no means sector specific, rather it is industry wide.

Results indicate that of the total number of respondents, 59% of the companies were involved in only construction, while 41% were categorised as Design and Construction companies.

4.2 Part B: limitations of ISO 9000

The purpose of part B of the questionnaire was to analyse the usage and drawbacks of the current ISO 9000 system. This included how long the respondents had held accreditation to the system, what they thought were the advantages and disadvantages of the system and whether they conducted training on the system.

The result indicates that a large majority of the respondents have ISO 9000 system accreditation. The results indicate that 88% are accredited, while only 12% are not. Of the 12% who are not accredited, most of these respondents indicated that they were in the process of achieving accreditation.

Of the respondents, who were accredited to the system, large percentages have been accredited for a period longer than 3 years. The results indicate that 41% have been accredited for between 3-5 years, while 22% indicated 5-7 years, 9% over 7 years, 21% for 1-3 years and 7% for under 1 year. Hence, as the majority of the respondents have been accredited for over 3 years, it would be reasonable to say that any problems they are experiencing with the system, would be due to the system drawbacks, and

not due to teething problems during its implementation.

Respondents from ISO 9000 accredited companies were asked in the survey to provide an estimate of initial set up cost. The result form the survey shows that 49% of respondents estimated the cost to be in the order of $500,000 - $1 million, while 40% stated $0 - $500,000, 9% claimed $1-$2 million and a small minority, 2% estimated the cost to be in excess of $3 million. The results also show 76% of respondents stated, that estimated turnover of their company is more than $20 million. This indicates that expenditure on current quality system is fairly reasonable.

The table 1 explains the main advantages of ISO 9000 indicated by the respondents from forty three (43) ISO 9000 accredited companies. The table shows that the respondents thought that increased marketability was the main advantage of ISO 9000. The 'keeping of records and files' item closely followed this. The majority of the items in this question have quite low percentages of agreement. This would indicate that the respondents do not feel strongly about any of the advantages of ISO 9000 except its marketability.

Table 2 indicates that the most highly agreed disadvantage of ISO 9000 is the 'large increase in paperwork'. This was closely followed by 'hard to enforce all subcontractors to adopt the system' and 'not being implemented by all parties to a project'. Respondents did not agree that 'top down management ignoring workers need' and 'high cost of implementation' was major disadvantages to the ISO 9000 system.

Table 1. Advantages of ISO 9000

Statement	Agree	
	Yes	No
Increased Marketability for your company	65%	35%
Keeping of records & files	51%	49%
Improved client satisfaction	33%	67%
Formalised delineation of responsibilities	28%	72%
Reduction in variation & rework	26%	74%
Improved communication between parties involved	23%	77%

Table 2. Disadvantages of ISO 9000

Statement	Agree	
	Yes	No
Large increase in paperwork	58%	42%
Hard to enforce all sub-contractors to adopt the system	56%	44%
Not being implemented by all parties to a project	51%	49%
Tedious audit process	47%	53%
Top down management ignoring workers needs	35%	65%
High cost of implementation	28%	72%

4.3 Part C: status of QM tools

The purpose of part C of the questionnaire was to analyse the usage and current knowledge of quality management tools. This included whether the respondents knew about or used, either, Quality Function Deployment, Total Quality Management, Quality Circles and Business Process Reengineering.

Table 3 shows that out of the total number of respondents, 96% indicated that they had heard of Total Quality Management, while only 4% indicated that they had not. Further, result indicates that only 29% of respondents formally use the tool, while 71% do not. Results from Table 3 indicate that TQM is more extensively used than QFD, however, its use in the construction industry is not as prevalent as first thought. Like QFD, in order to implement this tool as a supplementary system to ISO 9000, much more promotion of its benefits will need to take place.

The table 3 also indicates that out of the total number of respondents, 69% indicated that they had heard of BPR, while only 31% indicated that they had not. Further from the table it shows that 33% of respondents formally use the tool, while 67% do not. This shows that BPR is used more frequently than QFD, TQM and Quality Circles.

Table 4 illustrates the use of the main principles of TQM by the respondents. The table shows that the majority of the respondents extensively use all the items except Venn Diagrams. This is represented in the high index values of items. Therefore, it is evident that the majority of respondents indirectly use TQM, as they make extensive use of the tools main principles. A sample calculation regarding index values related to tables 4,5 & 6 is presented in Appendix 2.

Table 5 illustrates the use of the main principles of Quality Circles by the respondents. This table again shows that the majority of the respondents extensively use all items except 'Histogram and Pareto analysis'.

Table 3. Status of QM tools

Statement	Yes	No
Heard of Total Quality Management	96%	4%
Your company formally use TQM	29%	71%
Heard of Quality Function Deployment	24%	76%
Your company formally use QFD	4%	96%
Your company concentrate on identifying the clients needs when forming the design brief for a project	85%	15%
Heard of Quality Circles	61%	39%
Your company formally use Quality Circles	14%	86%
Heard of Business Process Reengineering	69%	31%
Your company formally use BPR	33%	67%

Table 4. Use of TQM Principles

Item	Total Score	Index	Rank
a) Flow Charting	196	0.8	1
b) Client Feedback Surveys	194	0.79	2
f) Peer Reviews	187	0.76	3
c) Staff Feedback Surveys	182	0.74	4
d) Brainstorming	182	0.74	4
e) Venn Diagrams	82	0.33	6

Table 5. Use of Quality Circle Principles

Item	Total Score	Index	Rank
c) Management presentations	196	0.8	1
e) Implementing subordinates solution to a problem	188	0.77	2
d) Regular meetings involving all levels of employees	187	0.76	3
a) Cause & Effect diagrams	158	0.64	4
b) Histogram & Pareto Analysis	91	0.37	5

Table 6. Use of Business process reengineering

Item	Total Score	Index	Rank
b) Regular client progress meetings	204	0.83	1
d) Just in time material delivery	194	0.79	2
e) Electronic Data Interchange between all parties	191	0.78	3
f) Selective tendering	176	0.72	4
c) Reorganising work packages into larger modules	148	0.6	5
a) Business Restructuring	144	0.59	6

In the table 5 the all the items have high index values except 'Histogram and Pareto Analysis'. Therefore, it could be fair to say that the majority of respondents indirectly use some of the principles of Quality Circles system. It can be recommended here that the formal implementation of Quality Circles in construction phase could be more beneficial to the industry and can overcome some limitations of ISO 9000.

Table 6 shows the use of the main principles of Business Process Reengineering by the respondents. This table highlights that the majority of the respondents extensively use all items. This is represented in the high index values of items. Therefore, it can be concluded that that the majority of respondents indirectly use the main principles of BPR as indicated in table 6 and that most of the time index values exceeded by 0.5.

4.4 Part D: QM tools vs ISO9000

The purpose of part D of the questionnaire was to analyse the respondents overall perceptions of ISO 9000 and whether they agreed that ISO 9000 should be either replaced by a new quality management tool or supplemented by one.

Table 7. QM tools vs ISO900

Statement	Agree		UI (95%)	LI (95%)
	Yes	No		
ISO900 system has brought any benefits to your company	60%	40%	77%	44%
Clients have received any benefits from use of the ISO 9000 system, in terms of higher levels of quality or better value for money	60%	40%	77%	44%
Company conducts training on meeting the ISO 9000 requirements	51%	49%	67%	35%
ISO 9000 is an adequate system in achieving high levels of quality	21%	79%	34%	8%
ISO 9000 should be replaced with a new quality management tool.	35%	65%	49%	21%
A new quality management system should be implemented to supplement the current ISO 9000 system	70%	30%	84%	56%
The main principles of the Quality Management tools beneficial in achieving high quality	88%	12%	98%	78%

Table 7 explains that by applying the confidence interval method with a 95% degree of accuracy, that between 56% and 84% of respondents agree with the statement, that a new quality management system should be implemented to supplement ISO 9000. Again, this serves as a strong test of the hypothesis, as it reflects that the top level managers of large construction firms, have recognised the need for an additional quality management tool to be urgently implemented to supplement the failing ISO 9000 system. Results from table 7 conform that the principles of quality management tools are beneficial in achieving high quality. A sample calculation for confidence interval is shown in Appendix 1.

5 CONCLUSION

Through the statistical analysis of the results, of the sample population of top level managers from large construction firms operating within the New South Wales, Australia, construction market, it can be concluded that:

The industry's attitude towards the ISO 9000 system was mixed. They saw that some changes are required to make to overcome the limitations of the system. The general perception of the system, which became evident through the survey, that the system generates more paper work. It does not improve communication between all of parties involved in the construction projects. Respondents agreed that ISO9000 has failed to reduce variation and rework in construction.

Apart from this negative viewpoint on the system, the industry still believes that it is beneficial to keep the system, due to its underlying objectives and the few advantages that it may bring to their company. Respondents to the survey agreed that due its international recognition, companies with ISO 900 accreditation had better market position. There was a general agreement among the participants that there is no need to replace the ISO9000 systems with a new quality management system. But they indicated that quality management tools such as Quality Circles, QFD, BPR and TQM should be implemented to supplement the current ISO 9000 system.

The current knowledge that the industry has on quality management tools is somewhat disappointing for an industry that must be aware of all the new developments in quality management tools. However, in general a large number of the respondents indirectly use a number of these tools as discussed in the result section. This is an encouraging finding that many companies are using the tools indirectly considering the majority of the sample population stated that they either have not heard of these Quality Management tools or do not formally use them.

It is evident from this study that the industry strongly believes that their use of these quality management tools, whether directly or indirectly has provided enormous benefits to the industry in terms of quality management.

Furthermore, the industry has almost universally indicated the need for inclusion of ideas of quality management tools to supplement some limitations of the current ISO 9000 system. Although the research does not conclusively provide evidence as to which systems should be implemented. Based on the literature reviewed and findings from this study, it is clear that the use of Quality Management Tools, in conjunction with ISO 9000, may bring about better facilitation of quality in the Australian construction industry.

6 RECOMMENDATIONS

This research study specifically focused on the perceptions of top level managers of large construction firms. A study with a similar hypothesis, but focusing on lower level employees in large construction, such as, tradesmen, would indicate which system works best at the place where it really counts, that is, where the actual construction takes place, rather than in the board room.

Techniques such as Quality Circles are specifically designed to operate in such an environment and such a study may indicate that it is the best system to implement.

A study with a similar objective, but specifically targeting a quality management based company and an ISO 9000 based company through the use of a case study system would pinpoint the real advantages and disadvantages of each system. This type of study is however limited by the ability to find such companies that are willing to participate.

The current knowledge that the industry has on these quality management tools is quite disappointing for an industry that needs to be at the forefront of quality management. As such, it is recommended that the quality management tools discussed herein be heavily promoted towards the construction industry, as they are in the manufacturing industry.

REFERENCES

Abdel-Razek, R H. 1998. Quality Improvement in Egypt: Methodology & Implementation. *Journal of Construction Engineering and Management*, Vol. 124 No. 5 pp 354-360.

Babbie, E. 1991. *Survey Research Methods*. 2nd Edition, Wadsworth Publishing Company

Gilly, B A, A. Touran, & T. Asai, 1987. Quality Control Circles in Construction. *Journal Of Construction Engineering & Management*. Vol, 113, No. 3 pp 427-439.

Holt, G. 1998. *A guide to successful dissertation study for students of the built Environment*. Second Edition, The Built Environment Research Unit, UK.

Keivani, R. M., A. R. Ghanbari-Parsa, & S. Kagaya 1999. ISO 9000 Standards: PerceptionsAnd Experiences in the UK Construction Industry. *Construction Management And Economics*. Vol. 17, No. 1 pp 107-119

Kirchner, D. & S. A. Wood, 1992. Guide to Quality Management", *The Building Economist*. pp11-15

Leedy, P. 1997, *Practical Research Planning & Design*. 6th Edition, Prentice Hall, New Jersey.

Mallon, J. C. & D. E. Mulligan 1993. Quality Function Deployment, A System for meeting Customers needs. *Journal of Construction Engineering & Management*. Vol. 119, No. 3 pp 516-531

Mohamed, S. & S. Tucker 1996. Options for Applying BPR in the Australian Construction Industry. *International Journal of Project Management*,. Vol. 14, No. 6, pg 379-385.

Mohyla, L V. 1996. *Construction in Australia: Law and project delivery*. Law Book Company, Sydney

Ogunlana, S, A. Sutandi & P. Phasukyud 1998. Experiences with International Quality Standards (ISO 9000) in Thailand. *The Sixth East Asia-Pacific Conference on Structural Engineering and Construction*, Jan, 14-16th, pp 1179-1184.

Rosenfeld, Y., A. Warzawski, & A. Laufer 1992. Using Quality Circles to Raise Productivity & Quality of Work Life. *Journal of Construction Engineering & Management*. Vol. 118, No. 1, pp 17-33

Rounds, J. L. & N. Y. Chi, 1985. Total Quality management for Construction. *Journal of Construction Engineering and Management*. Vol, 111, No. 2, pp 117-128

Siegel, A., 1988. "Statistics and data Analysis- An Introduction", John Wiley Sons.

Tam, C. M. & T. K. Tong 1996. A Quality Management System in Hong Kong: A lesson for The Building Industry Worldwide. *Australian Institute of Building papers*. Vol. 7 pp 121-131

Wang, M. T. & C. C. Lu 1998. Quality Function Deployment for Public Facilities. *The Sixth East Asia-Pacific Conference on Structural Engineering and Construction*. Taiwan, Jan 14-16th, pp 1191-1196

Wiele, T. V. & A. Brown 1997. ISO 9000 series experiences in small and medium sized Enterprises. *Total Quality Management*. Vol. 8, No. 2&3, pp 300-304.

APPENDIX 1. SAMPLE CALCULATION FOR CONFIDENCE INTERVAL

Part D, Question 3. A new quality management system should be implemented to supplement the current ISO 9000 system?

Disagreed = 13/43 x 100 = 30.23%
Agreed = 30/43 x 100 = 69.77%
Average = 0.697 = proportion of agreed responses
Standard Error
= Sq. Rt. of [proportion of positive responses x (1 – proportion)]
                              Sample size - 1
Standard Error = Sq. Rt. Of {[0.697 x (1 – 0.697)]/(43 – 1)}
                = 0.07
Degree of Freedom
= Sample size – 1
∴ 43 – 1 = 42, yielding a t-value of approximately 2.
∴ Multiplying t-value by the standard error
∴ = 2.021 x 0.07
   = 0.14147 = 14%
Subtracting and adding this to the average, we find the 95% confidence interval from:-
Lower 95% confidence limit = 0.697 – 0.14147 = 0.56 (56%)
Upper 95% confidence limit = 0.697 + 0.14147 = 0.84 (84%)

APPENDIX 2. CALCULATION FOR INDEX FOR TABLES 4,5, & 6

1   The selections from never to very often are scaled from 1 – 5. Scales used here are; 1=never, 2=almost never, 3= sometimes, 4=fairly often and 5=very often

2   All the scores obtained under each sub category are tabulated to give the total score for each tool. This is calculated by multiplying the frequency of occurrence for each item by the scale number assigned to that selection.

3   The index number for each item is then calculated by:
[(total score)/(no of respondents x no. of scales)]
total score = sum(frequency *scale number)

4   The items are then ranked from highest to lowest index.

5   Example: total score for "Reorganising work packages"
= (2 x 1) + (11 x 2) + (24 x 3) + (8 x 4) + (4 x 5)
= 148,
Therefore, Index = {148/(49*5)}= 0.60 (see table 6)

Creative Systems in Structural and Construction Engineering, Singh (ed.) © 2001 Balkema, Rotterdam, ISBN 90 5809 161 9

Recent technological systems and safety in cultural heritage building restoration

R.G. Laganà
DASTEC, University of Reggio Calabria, Italy

ABSTRACT: Recent experiences have shown that in the maintenance and restoration works carried out in building of cultural importance, there is the need of implement a safety code of safety which takes into account the application of new technological systems. Cultural heritage restoration often is place of accidents, due to various causes and conditions. The organisation of the site concerned with any particular interventions needs therefore specific attention. It is particularly important to define the typology of the risks which these buildings are subject to, in order to be able to impose an effective preventative safety program for workers.
In Italy and in other European countries, the increasing need for interventions on these building, either for maintenance and restoration has often meant a bad organisation of the concerned site.

1 INTRODUCTION

Building trade, along with new buildings erection, is more and more involved with ancient buildings maintenance and restoration. Even in this field, traditional technological systems are supported by new systems adopting procedural innovations together with the application of contemporary technologies.

In this field building sites safety culture finds difficult to cope with strong innovations trends concerning above all interventions techniques.

Yet under other circumstances we stressed how it is important to foresee safety measures not only considering building static restoration. When we plan these works or during building site organisation, we have to think carefully about safety requirements implementation: they must be implemented during working operations and must be defined in works performances specifications drawn up for each work.

Every week we read news about work accidents occurred in restoration or maintenance building sites. Negligence, poor predictability in risks evaluation, technicians lacking professionalism and workers' poor technical knowledge are among the causes which have highly marked accidents in recent years.

We believe that an important feature in risk occurrence is the use and the implementation of new technologies in this kind of works.

2 THE RENEWAL OF BUILDING TRADITION

Last century technological evolution introduced productive improvement of traditional materials and new building procedures. The result of this evolution brought the implementation of new technologies (steel, concrete, etc.) which are widespread in new buildings, and which have to perform particular functions in the static balance of ancient buildings.

The contribution of new technologies has also been outstanding for monuments preventive study. Besides lab studies which consent to recognise materials employed centuries ago, it is possible nowadays to determine mortars consistence, their mechanical features and even their places of origins.

Technologies transfer from other scientific fields allows the implementation of non destructive systems in order to write out precise diagnosis and to collect as many data as it is possible even in extremely difficult evaluation situations.

Few examples that can be quoted are: resonance techniques implemented on walls in order to detect the presence of vacuums, studies performed by the use of fibre optic viewers inside microdrillings.

So we see the development of a new planning culture for ancient buildings maintenance and restoration where the study phase becomes very, very important.

This, obviously, needs high initial investments costs that have to anticipate working plan. We can detect, in doing so, a previous risks evaluation phase; in the past, when operational choices were taken during execution, it was passed on site engineer's decisions.

3 PRACTICE CODES AND NEW TECHNOLOGICAL SYSTEMS

In a recent memorandum about building maintenance safety and quality aspects, I stressed my opinion that, among risks causes in maintenance and restoration building sites, bad evaluation of structural stability ranks the first.

Then follows the poor knowledge of building systems that sometimes are very complex because they have been the object of different interventions in various historical periods.

We believe that, instead of relying on broad technical specification items, it is necessary to rely on building custom which depict practice codes.

It is important to detect in them, besides manufacturing descriptions, all fixed aspects concerning safety.

The panel in which I take part in my university copes with the problems linked to practice codes implementations in restoration works of ancient towns in southern Italy. In this framework, even on the strength of implemented planning experiences, we are going into the subject thoroughly in order to link quality, safety and sustainability aspects. Safety aspects are led towards implementations of necessary data for working phases.

In walls stabilisation works the introduction of new kinds of manufacturing often can require the use of machinery's; so we must evaluate not only their risks factors on workers' safety and health, but even all eventual implications that could occur from their misleading use (vibrations, weight, impact, etc.)

In order to run and best check all eventual risk situations it is better to implement, before each manufacturing process, specific interim works that have to be managed through an improved operational planning.

Planning activities development within restoration works has allowed, in these last years, to obtain many references about different intervention techniques that can be compared with specific technological choices.

In Europe cultural heritage recovery has lately seen the implementation of a progressive building sites modernisation. As it occurred in painting and in other artistic fields, even in architecture we have achieved new results, and technical evolution offers new kinds of manufacturing that can fit to different building sites needs.

• Preventive building research	STATIC SAFETY **SAFETY IN CONSTRUCTION SITE**
• Evaluation of risks	STRUCTURAL RISKS **WORKING RISKS**
• Application of codes of practice	EXECUTIVE PROJECT **DEVELOPMENT BY RECENT TECHNOLOGICAL SYSTEMS**

Figure 1. Safety role in building restoration project.

Figure 2. The Aragonese Castle before the collapse

Figure 3. The collapsing north-west wall.

Nowadays the implementation of EU norms about workers' health and safety allows us, after a decade, to collect a selection of positive and negative data coming out from different work experiences.

The need to have an exchange of experiences about methods and technological systems adopted for risks prevention's in building sites is fundamental in order to give the scientific community and professionals all the tools which are necessary to improve health and safety performances and to suggest systems for new technological approaches in this field.

Figure 4. Compression text.

Figure 5. Sonic surveys.

4 A WORKING EXPERIENCE

In the town of Reggio Calabria, located in South of Italy, an working experience in restoration has allowed to draw interesting hints about the application of safety rules both in the disposition of building sites and in the practical phases of the work.

It handles about the restoration of an ancient monument, the Aragonese Castle, built at the ending of the ending of the XIV century and partly survived

to the destruction by some earthquakes, but early in the last century changed because of town-planning events.

Works started about ten years ago, but they were stopped because of a wall collapse on May 7, 1986 due to contractor's inability.

The following static restoration project has been drawn up by new designers and has made the monument object of a particular study in order to know the physical state of the structures.

Figure 6. Endoscopic test.

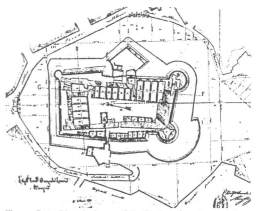

Figure 7. An historical map of the Castle before the partial destruction by 1908' Earthquake.

4.1 *Photogrammetric survey*

Besides monument façades survey and particularly collapsed wall survey, we performed microsurveys by close shootings of all external walls presenting cracks.

4.2 *Not destructive investigations on the materials.*

After the study of different masonry and of all building techniques related to different historical periods (all this referred to a preventive survey on structure evolution), we have then collected samples of all mortars in order to detect physical and chemical properties of inerts and cements adopted (by tracing section test, X rays diffraction, thermogravimetric analysis, porousness level, calcimetry).

We also performed lab tests on bricks and stone ashlars used in building process.

4.3 Continuous core sampling and rotary mechanical drillings

Alterations and tampering suffered by the area where the monument is located have modified its morphological technical aspect. Since there has been a different loads distribution on the soil with an eventual alteration of response coefficients related to static and dynamic stresses, it was momentous to get a specific knowledge of geological formations which are present under building area and particularly nature determination, physical state, lithotypes succession and their spatial variability.

4.4 Seismic refraction and geoelectrical prospectings

Since the monument is located on a highly seismic area, refraction prospectings allow the detection of seismic propagation velocity connecting it with soils mechanical features.

Collected data allowed the delimitation of discontinuity surfaces geometry and the discovery of buried structures.

Thanks to the use of geological resistance gauges we performed a geoelectrical prospecting in order to detect vacuums in masonry on the whole monument area.

4.5 In situ masonry tests

In different areas of masonry we performed compression tests by a flat jack, obtaining a check of pressure on triangular samples by a precision manometer inserted into oleodinaqmic circuit.

Strains have been measured by a removable comparator.

By unthreading tests we evaluated masonry breaking stresses through the examination of strains in relationship with intervals at strength increase, kept constant up to unthreading, supported by photographic and metric survey of the surface of collected sample.

We performed sonic surveys using an impulsive transmitter and a piezoelectric receiver placed in a symmetric position. We noticed transparency measurements in order to know longitudinal waves propagation velocity through walls so that to perform comparisons of degradation of different masonry.

Figure 8. The temporary scaffolds for the first safety phase.

Figure 9. Section by central vault and lesion's sewing.

Figure 10. Microposts application in the west wall.

Permeability tests performed by having water falling up to the middle of the wall thickness allowed us to measure the amount of liquid coming out from sloping holes done in the lower part.

4.6 *Masonry endoscopic test*

We performed about 130 direct visual observation inside masonry. We prepared inspection holes, with a diameter about 3 or 4 cm, putting inside them stiff endoscopes for front vision, side vision endoscopes and fibroscopes for detailed tests of cracks and holes which were in the masonry.

The results of these surveys allowed to detect morphological changes inside walls thickness and the discontinuity between masonry and faces; the mortar thickness between stone ashlars; mortars repair, the presence and condition of cracks and internal holes.

The correct knowledge of constructive work system and of the building area's traditions had supported the safety implementation during final plan drawn and working plan. It was important to get the knowledge of the history of the building. There are original technical papers acquired by a careful historical research done in the historical archives; they allow us to perform an attentive analysis between project prescriptions and actual performance.

The collection of acquired data through different kinds of surveys allowed the detection of some intervention areas and at the same time to mark off most at risk areas inside the building site.

Planning intervention has been set on static restoration plan according to this organisation chart:
1. Building site disposition and precarious elements stabilisation;
2. Collapsed walls fragmentation and recovery of lithoid and ceramic materials;
3. Front walls stabilisation by the injection of new legant material and safety measures restoration;
4. Cracked masonry stabilisation and sewing;
5. Rebuilding of collapsed masonry linked to previous structure by sewing with microposts;
6. Rebuilding of collapsed ceilings and static realignment by walls precompression obtained with the use of "dividag" bars;
7. Recovery of towers interiors;
8. Outside areas disposition.

The need not to influence structures with tremors required the use of particular tools (rotary drills, hydraulic hammers, etc.).

An original "Report of Work" that led planning choices was drawn up not only in order to know how to do practical choices but even to define safety measures to implement during preparation and management of the building site.

Figure 11. Precompression wall's apparatus and rebuilding of barrel vault.

N.	Working plan	Executive plan Measures	Digital photo	Safety
.1a.				
.2a.				

Figure 12. File scheme of "Report of Work".

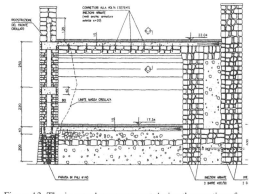

Figure 13. The image show a moment during the creation of courts in order to reinforce elements. After a comparison the site engineers and the contractor have preferred a technique using drills ordered at a fixed distance at interim while the other using the demolishing hammer had high effect of vibration.

Figure 14. The rebuilding walls.

Even if restoration projects are not included in government decree # 494/1996 (application of EEC Directory # 57/1992: Safety in construction sites) the designers carried out an attentive evaluation of safety rules.

The contractors drawn up an attentive and scrupulous Safety Plan for prevention of accidents into processes involved in the plan.

The site engineers, during the works, agreed with the contractor a series of safety and health implementations to improve standards measures.

First of all a series of tight interventions on damaged works during the organisation of the building site in order to reinforce workplaces which were more exposed to risks of collapsing or cracks, always keeping in mind even high seismic risk of this area.

Besides we performed a continuous watch even through a suitable signalling apparatus of walls with serious structural lesions.

The workers were made responsible for the particular conditions of works and for technical work conditions and technical expected doings.

We at last improved technical operations by using alternative contractor's equipment in some processes to implement workers' safety and health.

Even if it was not compulsory to make a Building Maintenance Safety File, the site engineers wrote a special paper documenting all operations that have been carried out in the building site using an electronic newspaper supported by a rich range of digital images, put into an electronic archive.

5 CONCLUSIONS

The creation of Systems for future technologies cannot exclude the building restoration sector.

The improvements in cultural heritage building restoration offers the opportunity of refined technics especially in diagnostic phases.

The applied methodologies now use traditional systems implemented by recent technologies. We should reach the balance between the two systems.

The shown example help us to become aware of a working methodology in relation to a new safety culture.

ACKNOWLEDGEMENT

The municipal administration of Reggio Calabria had charged Prof. Eng. Calzona Remo, Prof. Eng. Arena Giuseppe and Prof. Arch. Laganà Renato to draw the working plan. The works were financed by government decree July 5,1986 n.246 carried out by contractors Ortega & Emma C. Assoc. Works Superintendente was Eng. Casile Antonino.

REFERENCES

Laganà, R.G. 2000. Quality and Safety Systems in Building Maintenance. In *Implementation of Construction Quality and Related Systems: A Global Update*.(Dias A.L. et al.) CIB, Lisbon.

Nesi A. (resp. Et alt.) 1998. Analisi tecnologica e codici di pratica per gli interventi nei centri storici minori della Calabria, Reggio Calabria: Falzea.

14 Concrete mixes

Creative Systems in Structural and Construction Engineering, Singh (ed.) © 2001 Balkema, Rotterdam, ISBN 90 5809 161 9

Effectiveness of admixtures on the reduction of drying shrinkage of concrete

C.Videla & C.Aguilar
Pontificia Universidad Católica de Chile, Santiago, Chile

ABSTRACT: Drying shrinkage has been largely recognized as one of the main causes of cracking in concrete structures. Continuous research has been performed in order to develop procedures to control this phenomenon and to produce high performance concrete with regard to drying shrinkage. However results reported on specialized literature are not conclusive and sometimes are contradictory. Therefore an extensive experimental program has been conducted in order to evaluate the effectiveness of admixtures claiming to decrease drying shrinkage strain. The analysis of the effect of admixtures of different brands and amounts on drying shrinkage were considered. The study involved the measurement of drying shrinkage strains up to 112 days of 52 different concrete mixtures. The different admixtures were prioritized according to their effect on the reduction of drying shrinkage strains. It is concluded that shrinkage reducing admixtures presents the best behavior and can reduce up to 50 % of the free shrinkage strains of normal concrete without admixtures. However these admixtures also reduce around 10% the concrete compressive strength.

1 INTRODUCTION

Volumetric changes of concrete are of concern in all construction projects. Of particular importance are those associated with changes in moisture content, i.e. drying shrinkage (ACI Committee 224, 1980). A significant proportion of the cracking problems generally experienced can be attributed to shrinkage effects, with cracks occurring between 1 to 6 months after casting. These cracks produce important technical, economic and quality impacts over a project because they impair the durability, serviceability and aesthetic of a structure. Therefore cracking of concrete should be prevented or controlled and the measures taken depend upon the use and exposure of the structure and on the economics of the situation.

Due to the implications of concrete cracking continuous research to develop procedures to prevent concrete cracking due to drying shrinkage is carried out worldwide. However, there are no widely research efforts involving concrete made with Blended Portland Cements (Videla & Aguilar, 1999). Particularly there is some evidence in Chile showing that concrete made with Portland Pozzolan Cement shrinks more than concrete made with Ordinary Portland Cement probably due to the intrinsic characteristics of the materials used. Furthermore some commonly used procedures in practice to reduce shrinkage do not appear to be really effective. For example, there is experimental evidence showing

that the use of water reducing admixtures allowing to reduce the water content of the concrete for a given slump and concrete strength, does not reduce shrinkage (Videla, 1996).

Therefore research is needed on the effectiveness of worldwide recommended and new procedures to reduce drying shrinkage based on the use of admixtures claiming to modify the main concrete parameters affecting shrinkage when they are applied to concrete made with Blended Portland Cements (Brooks, 1996; Brooks, 1989).

2 RESEARCH SIGNIFICANCE

This research project is intended to provide engineers with valuable quantitative information about the consequences of using new and common technologies in order to reduce drying shrinkage strains. It is expected that the obtained results will allow minimizing shrinkage cracking of concrete structures and to avoid their technical and economical consequences. National and oversees studies clearly show that a significant proportion of the cracking problems generally experienced by concrete structures between 1 and 6 months after casting can be attributed to shrinkage strains. According to these studies shrinkage cracks represents of the order of 35% to 40% of the total number of cracks produced (Videla, 1994; Campbell Allen, 1974).

3 OBJECTIVE AND SCOPE OF THE STUDY

The aim of the research project was to verify the feasibility of reducing drying shrinkage strains and cracking of concrete structures. Specifically, the use of different types and amounts of admixtures to reduce drying shrinkage cracking was analyzed. Therefore alternative types of concrete to those typically used in building construction practice were fabricated considering the most relevant variables affecting shrinkage in order to obtain less free drying shrinkage. The admixtures considered were water reducers, high range water reducers, expansive agents and shrinkage reducers. Both commercially and experimental admixtures were studied.

4 EXPERIMENTAL PROGRAMME

An extensive experimental program was carried out, including 52 trial mixes, in which the type and proportion of admixtures were varied to analyze the principal factors directly affecting shrinkage properties. The characteristics of the different concretes were as follows.

– Concrete without Admixture (Reference Concrete):
  Specified Compressive Strength: 35 MPa.
  % defectives: 10%.
  Slump: 60 mm y 120 mm.
  Cement Type: Portland Pozzolan Cement
  Aggregate Maximum Size: 40 mm.
  Aggregate Grading: Between Curves N°2 and N°3 of Road Note N°4 - Transport and Road Research Laboratory.
– Concrete with Water Reducing Agents.
– Concrete with Superplasticizer Agents (High Range Water Reducers).
– Concrete with Shrinkage Reducing Agents.
– Concrete with Expansive Agents.

It must be noted that the admixture dosage used for the fabrication of the concrete batches was obtained considering the limits at third points of the range recommended by the producers of the particular admixture considered.

5 CHARACTERISTICS OF THE MATERIALS

5.1 *Cement*

A mixture of equal proportions of two brands of Portland Pozzolan cement was used in the trial mixes in order to avoid the effect of cement brand. Table N°1 shows the physical and chemical properties of the cement mixture employed.

Table 1. Cement properties

Property	Unit	Type of Cement		
		Cement A	B	Blended Portland Cement[*]
Vicat Initial Set	HH:MM	04:10	04:10	04:10
Vicat Final Set	HH:MM	05:10	05:10	05:10
Blaine Fineness	cm²/g	4249	4363	4300
Compressive Strength				
7 days	MPa	29.2	29.8	29
28 days	MPa	35.4	38.5	37
28 days	MPa	39.2	38.4	39
Flexural Strength				
7 days	MPa	6	6.3	6
28 days	MPa	7.8	8	8
28 days	MPa	7.7	8.1	8
Specific Gravity	---	2.87	2.84	2.86
Consistency	%	33.6	34.8	34.14
Insoluble Residue	%	25.55	26.54	26
Loss on Ignition	%	2.86	3.24	3.03
SiO_3	%	2.16	3.04	2.56

[*] Portland Pozzolan Cement

5.2 *Aggregates*

Standard tests were performed to characterize the properties of the aggregates. They included unit weight, specific gravity, absorption and grading. The test results are shown in Table N°2.

5.3 *Admixtures*

The types of admixtures used and their characteristics are summarized in Table N°3.

6 MIX PROPORTIONS

For each theoretical concrete mix design four preliminary trial mixes were cast in order to adjust the water content of the mix to satisfy the slump and strength requirements of the concrete. The concrete was mixed in a 0.25 m³ (0.327 yd³) capacity-vertical rotating mixer. One 0.18 m³ (0.235 yd³) batch was required to carry out the freshly mixed tests and to cast the eighteen cylindrical samples, two cubic samples and six prisms to be tested. The concrete was mixed according to the procedure specified in ASTM C 192.

7 TESTING PROCEDURE

Several standard tests were carried out on every mix.

Table 2. Aggregates properties

Property		Coarse Aggregate	Coarse Aggregate	Coarse Sand	Fine Sand
		A	B	C	D
Volumetric Coefficient	---		0.23	0.27	---	---
Crushed Effective	%	89	80	---	---
Rounded Effective	%	11	20	---	---
Oven dry unit weight	kg/m³	1651	1623	1780	1657
Absorption	%	0.9	1.01	0.95	1.59
Void Content	%	39.4	40.3	33.5	37
Fines Content	%	0.2	0.2	0.5	1.8
Maximun Size	mm	40	20	10	5
MixProportion	%	37	28	23	12

Table 3. Admixtures properties

Type and Brand of Admixture		Amount of Admixture		Density (kg/m³)
		D1	D2	
WR1	Adiplast 11 - Polchem	3,83 cc/kg cement	4,17 cc/kg cement	1170
WR2	Platiment HE - Sika	0,367 rc	0,433 rc	1150
WR3	Plastiment HER - Sika	0,367 rc	0,433 rc	1150
HWR1	Adiplast 101 - Polchem	1,23 rc	1,87 rc	1160
HWR2	Sikament FF 86 - Sika	0,83 rc	1,17 rc	1220
HWR3	Sikament 10 - Sika	0,367 rc	0,433 rc	1100
SR1	Adipol SI 2 - Polchem	1,33% rc	1,66rc	1025
SR2	Sika Control - Sika	1,33% rc	1,66rc	1020
SR3	Tetraguard - MBT	1,67 rc	1,83 rc	1000
SR4	Adipol SI 1 - Polchem	1,33% rc	1,66rc	952
SR5	Adipol SI 3 - Polchem	1,33% rc	1,66rc	962
E1	Meyco - MBT	1,67 rc	1,83 rc	3037
E2	Stabilmac - MBT	23,33 kg/m³	26,66 kg/m³	1892
E3	Flowcable - MBT	4,33 rc	4,66 rc	1834

WR: Water Reducing Admixture
HWR: Superplasticizer
SR: Shrinkage Reducing Admixture
E: ,   Expansive Agent
rc   With respect unit weight of cement

For fresh concrete the slump (ASTM C 143), freshly mixed density (ASTM C 138), and ambient and concrete temperatures were measured.

In hardened condition compressive strength tests were performed using a TONIPAC 3000 testing machine (ASTM C 39): 2 cylindrical specimens at 7, 28 and 90 days and 2 cubic specimens for testing at 28 days. Also splitting strength (ASTM C 496) at 7, 28 and 90 days were performed. Modulus of elasticity tests (ASTM C 469) at 7, 28 and 90 days were also carried out.

Finally, the drying shrinkage tests were performed according to ASTM C157 and C490. For each concrete mix 3 prisms specimens of 75*75*285 mm and 3 prisms of 100*100*500 mm were tested for shrinkage strains up to 112 days of drying. The samples were dried in an special ambient conditioning room at 23 ± 2 °C and 50 ± 4 % relative humidity after been cured for 1 day in a standard curing room and 6 days in water saturated with lime at 20 °C.

8 RESULTS AND DISCUSSION

In the following paragraphs the results of the test program are presented grouped by type of admixture and are analyzed with respect to the reference concrete. Then a comparison is made between the best solutions for each admixture family according to their effectiveness to reduce drying shrinkage strains.

8.1 Water Reducing Admixtures

Figure N°1 presents the shrinkage results for concrete made with water reducing admixtures with respect to concrete without admixture. These results suggests that at long term (112 days of drying) the majority of concrete made with admixtures that reduce the water content have a slightly smaller shrinkage than the reference concrete. The percentage of reduction varies between 4% and 10%, depending on the brand and dosage of admixture used. However it can be observed that up to 56 days of drying the concrete with WR admixtures show larger shrinkage than concrete without admixture. Therefore there is a higher risk of cracking at short term. It can be concluded that although the smaller value of long-term shrinkage this does not assure to avoid concrete cracking. The best behavior was obtained with WR 3 – D1 admixture (Plastiment HER) with a 15% shrinkage reduction.

8.2 Superplasticizer Admixtures

Figure N°2 illustrates the effect of brand and amount of Superplasticizer on the drying shrinkage of concrete compared with the results of concrete without admixture. From the analysis of this figure it can be concluded that the use of High Water Reducing Admixtures is beneficial because it is possible to reduce drying shrinkage strains for up to 38%, depending on the brand and amount of admixture. It is believed that this result is due to the high reduction of water content of the concrete (approximately 20 to 30%) obtained with this type of admixture allowing to have a larger aggregate content on the mix and therefore a smaller shrinkage (Whiting 1992; Whiting 1979). The results indicated that admixture HWR 3 – D2 (Sikament 10) has the best behavior with 38% shrinkage reduction.

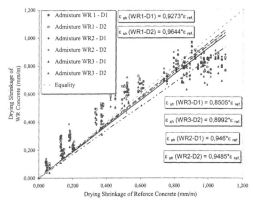

Figure Nº1. Drying Shrinkage of Concrete with Water Reducing Admixtures (WR) versus Reference Concrete.

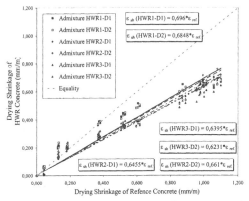

Figure Nº2. Drying Shrinkage of Concrete with High Water Reducing Admixtures (HWR) versus Reference Concrete.

8.3 *Shrinkage Reducing Admixtures*

Figure Nº3 clearly shows the high impact of using Shrinkage Reducing Admixtures over drying shrinkage of concrete. The results show that a reduction of up to 50% can be expected with respect to concrete without admixtures. The latter can be explained by the effect of this admixtures over the surface tension

of the water held in the capillary voids of the cement paste (Perenchio, 1997; ACI 209, 1997). The decrease of this surface tension reduces the volumetric change of the cement paste and therefore a smaller drying shrinkage should be expected. However these admixtures also reduce approximately 10% the concrete compressive strength. The measured loss of strength is smaller than values reported on published research (Shah, 1992; Holland 1999, Nmai, 1998). The results highlight the large effect of SR 3 admixture (Tetraguard) with 56% shrinkage reduction.

8.4 *Expansive Agents*

Figure Nº4 presents the results of drying shrinkage of concrete fabricated with different expansive agents with respect to the results of concrete without it (Reference Concrete). From the analysis of this figure it is observed that used of this type of admixtures implies a decrease of the value of drying shrinkage of the order of 28%. The initial expansions induced by these expansive agents that compensate the later shrinkage are probably the cause of these results.

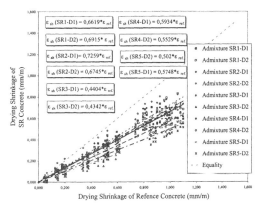

Figure Nº3.- Drying Shrinkage of Concrete with Shrinkage Reducing Admixtures (SR) versus Reference Concrete.

However, the same as the case of shrinkage reducing admixtures, the use of expansive agents implies a strength loss of near 12% at 28 days, except for the case of admixture E3 (Flowcable), which also possesses water reducer characteristic. The results indicated that admixture E2 (Stabilmac), has the best behavior with 28% shrinkage reduction.

8.5 *Relationship between admixture type and drying shrinkage of concrete*

Figure Nº5 compares the evolution of drying shrinkage strains for concrete made with admixtures that show the best behavior with regard to shrinkage and also of the reference concrete. It is apparent from this figure that all the analyzed alternatives exhibit a much smaller drying shrinkage that the reference concrete.

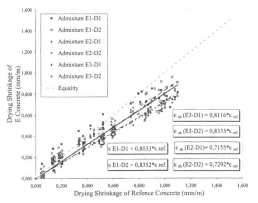

Figure Nº4. Drying Shrinkage of Concrete with Expansive Agent (E) versus Reference Concrete.

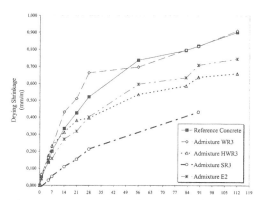

Figure Nº5. Comparison of the Effect of Admixture Type on Drying Shrinkage of Concrete for V/S = 25 mm.

Therefore, the use of water reducing, superplasticizer, shrinkage reducing or expansive admixtures improve the shrinkage behavior of concrete, but in different amounts depending on the type and dosage of the admixture.

Table Nº4 presents a relative comparison of the concretes presenting the best behavior regarding shrinkage cracking risk.

A total of 8 concretes were selected according to the value of drying shrinkage that their drying shrinkage values at 90 days of drying and were also compared with the reference concrete.

>From the analysis of this table it is observed that concrete made with shrinkage reducing admixtures presents 50% less risk of cracking than concrete

Table Nº4. Top eight solutions to reduce drying shrinkage strains

Position	Admixture Family	Dosage	ε_{sh} (90 days)	Ratio ($\varepsilon_{sh} / \varepsilon_{sh}$ plain) with respect to Reference Concrete
1	SR3	D2	0,431	0,46
2	SR3	D1	0,457	0,49
3	HWR3	D2	0,619	0,66
4	HWR3	D1	0,676	0,72
5	E2	D1	0,720	0,77
6	WR3	D1	0,758	0,81
7	E2	D2	0,774	0,83
8	WR3	D2	0,851	0,91
9	None	---	0,938	1,00

without admixture (reference concrete), assuming equal tensile strength and modulus of elasticity for all tested concrete. It should be noted that this assumption can be made because although concrete made with shrinkage reducing admixture showed a reduction of compressive strength, this effect was not noticed for the splitting strength.

9 CONCLUSIONS

The objective of this study was to determine the effectiveness of admixtures to reduce the drying shrinkage of concrete. As it was shown in the previous sections, practically all the analyzed alternatives satisfied this objective.

Nevertheless, to select a concrete type for a particular project it should also be considered in the analysis the strength and modulus of elasticity of the concrete, the restraining conditions and the cost of the solution.

From this study it can conclude that shrinkage reducing admixtures present the best behavior with regard to preventing shrinkage cracking of concrete.

On the other hand, the use of superplasticizers, admixture used to achieve high slump concrete, show to be the second best alternative to control the cracking of concrete.

With respect to concrete made with water reducing admixtures it can be concluded that they present a good behavior at long term. However, at early ages they generate larger shrinkage values that could lead to early cracking of concrete.

Lastly, it should be notice that expansive agents diminishes the values of free shrinkage and present at 90 days a similar behavior than concrete made with water reducing admixtures.

10 ACKNOWLEDGEMENTS

The authors would like to express their gratitude for the grant made by the Council for Technological and Scientific Research (Project Nº1980943 – 1998) to carry out the reported work.

REFERENCIAS

American Concrete Institute - 209, (1997), Factors Affecting Shrinkage, Creep and Thermal Expansion of Concrete and Simplified Models to Predict Strains.

American Concrete Institute - 224, (1986), Cracking of Concrete Members in Direct Tension, Journal of American Concrete Institute, Vol.83, N°1, pp 3-13, and Discussion, Vol. 83, N°5, pp 878.

Brooks, J.J., Neville, A.M., (1992), Creep and Shrinkage of Concrete as Affected by Admixtures and Cement Replacement Materials, Creep and Shrinkage of Concrete: Effect of Materials and Environment, SP 135, American Concrete Institute, Detroit, pp. 19-36.

Brooks, J.J., (1989), Influence of Mix Proportions, Plasticizers and Superplasticizers on Creep and Drying Shrinkage of Concrete, Magazine Concrete Research., Vol.41, N° 148, pp 145-154.

Campell, A., (1979), The Reduction of Cracking in Concrete, The University of Sydney.

Charif, H., et al, (1990), Reductiond of Deformations with the Use of Concrete Admixtures, Proceding of the International Symposium held by Rilem, pp. 402-429.

Holland, T., (1999), Using Shrinkage Reducing Admixtures, Concrete Construction.

Nmai, C., (1998), Shrinkage - Reducing Admixtures, Concrete International, April, pp. 31-37.

Perenchio, W., (1997), The Drying Shrinkage Dilemma, Concrete Construction.

Shah et al, (1992), Effects of Shrinkage Reducing Admixtures on Restrained Shrinkage Cracking of Concrete, Journal of American Concrete Institute, Vol. 89, N°3, pp. 289-295.

Videla, C., (1994), Desarrollo Tecnológico de Aditivos, X Jornadas Chilenas del Hormigón, Santiago, Chile.

Videla, C., (1996), Retracción Hidráulica y Propiedades Mecánicas y Elásticas de Hormigones, Instituto Chileno del Cemento y del Hormigón .

Videla, C., Aguilar, C., (1999), Evaluación de la Efectividad de Procedimientos para la Reducción de la Retracción Hidráulica en Hormigones fabricados con Cementos Portland Puzolánicos, Revista de Ingeniería de Construcción, N°20, Pontificia Universidad Católica de Chile.

Whiting, D., (1979), Effects of High Range Water Reducers on Some Properties of Fresh and Hardened Concretes, Portland Cement Association.

Whiting, D., Dziedzic, W., (1992), Effects of Conventional and High Range Water Reducers on Concrete Properties, Portland Cement Association.

Creative Systems in Structural and Construction Engineering, Singh (ed.) © 2001 Balkema, Rotterdam, ISBN 90 5809 161 9

Performance of high alumina concrete in sulphate environment

Abu Bakar Mohamad Diah, Nor Azazi Zakaria, Nordin Mohd Adlan,
Mohamad Ibrahim Alla Pitchay & Kamarul Badlishah Kamarulzaman
School of Civil Engineering, Universiti Sains Malaysia, Tronoh, Malaysia

ABSTRACT: This paper presents the results of an investigation on the effect of sulphate environment on mortar prism specimens made of ordinary portland cement (OPC), high alumina cement (HAC) and sulphate resisting portland cement (SRPC). Prism specimens (25mm*25mm*100mm) with ratio 1:3 were prepared using water cement ratio 0.5. Specimens were demoulded after 24hours and air cured in room temperature (28° ± 2° C) with high humidity environment for 7 days before immersion in sulphate solution (0.3M $Na_2 SO_4$). Results show that high alumina mortar is very vulnerable in sulphate environment hence using such cement is not advisable despite being highly recommended unless extra precautions are taken during preparation and strict control of curing temperature. Result also implies using SRPC is still the best choice over HAC

1 INTRODUCTION

The deterioration of mortars and concrete associated with sulphate salts presents in sulphate bearing soils and ground waters has been known for years. To make concrete more durable in such environment high alumina cement was developed and used with successful results until problem crops up (Grutzeck and Sarkar, 1994). Later sulphate resisting portland cement was introduced to replace high alumina cement and proved successful (Lawrence, 1990). Today besides sulphate resisting portland cement, combinations of SRPC and cement replacement materials are recommended to deal with such aggressive environments (BRE 363). Sulphate attack on concrete proved to be a complex process due to several factors that combine to aggravate the situation (Murdock et al.1991).

Sulphate attack on concrete or mortars results from a chemical reaction between the sulfate ion and hydrated calcium aluminate and/or the calcium hydroxide components of hardened cement paste in the presents of water (Mehta, 1986). The products of the reactions are calcium sulphoaluminate hydrate or ettringite and calcium sulfate hydrate or gypsum. These solids have a very much higher volume than the solids reactants and as a result stresses are produced that may result in expansion, cracks and eventually breakdown of mortar or concrete (Osborne, 1992).

The needs for more reliable standard tests and models to predicts performance and service life of a long-term structure has attracted several agencies to study these areas more seriously (Cohen and Mather, 1991). This paper presents the results of experiment done in high humidity environment couple with high average room temperature (28° ± 2°C). Hopefully the results presents here can contribute towards developing a systematic research effort on sulphate attack and developing appropriate standard and predictive models that is applicable in the temperate, middle east as well as tropical region.

2 MATERIALS

2.1 Cement

Locally produced equivalent ASTM C150 types I, V and high alumina cement was used in preparing mortar specimens. Table 1 shows the chemical composition of the cements used in this investigation.

2.2 Fine aggregate

Locally available river sand conforms to BS 812 were used.

2.3 *Sulphate salt*

Sodium sulphate anhydrous salt with molecular formula $Na_2 SO_4 = 142.04$ g was used.

Table 1. Chemical composition of cements

Constituent: wt%	Type I cement	Type V cement	HA cement
Silicon dioxide	20.6	22.0	3.9
Aluminium oxide	4.2	4.1	38.6
Ferric oxide	3.2	4.2	12.3
Calcium oxide	62.3	64.1	38.6
Magnesium oxide	3.4	2.2	1.5
Sulphur trioxide	2.2	2.0	1.5
Potassium oxide	-	0.3	-
Titanium oxide	-	-	1.9
Iron oxide	-	-	4.0
C_3S	56.7	54.6	-
C_2S	16.1	21.9	-
C_3A	8.5	3.5	-
C_4AF	11.6	12.9	-

3 EXPERIMENTAL PROCEDURES

3.1 *Casting of specimens*

The mortar with ratio 1:3 and water cement ratio 0.5 were thoroughly mix in a mixer before casting into a mould size 25mm*100mm*500mm. The mould was vibrated for 15 seconds before the top were strike to smooth the surface. After casting, the specimens were covered with plastic sheet for 24 hours prior to demoulding. The mortar prism was air cured in laboratory temperature (28° ± 2°C) with high humidity environment (tropical climate) for 6 days. There after the prism were cut using concrete cutter to required sizes (25mm*25mm*100mm). They were then air cured for another day. The specimens were labeled, their weight and length were recorded before immersing them in the sulphate solution.

3.2 *Preparation of test solution*

The test solution were prepared by mixing sodium sulphate anhydrous salt with tap water (average pH 9.5) to produce the required sulphate solution ($0.3M\ Na_2 SO_4$). The solution was kept in a plastic container in the laboratory and covered with a polythene sheet to minimize evaporation. The volume of the solution and its level was adjusted from time to time so that the specimens were always immersed in the solution. The solution's pH was monitored and adjusted on a

weekly basis by adding sulphuric acid so that the pH variation was within ± 10% of the original solution.

3.3 *Weight change*

Physical determination due to sulphate attack on the hardened mortar was evaluated in term of change in weight, change in length and visual appearance after every 10 days of immersion. At the scheduled time the specimens were retrived, air-dried for 1 day in laboratory environment and weigh. The percentage weight change was determined using the following relationship:

$$W_c = \frac{(W_t - W_i)}{W_i} \times 100 \qquad (1)$$

W_i = average initial weight of 5 specimens, g
W_t = average weight of 5 specimens after exposure period of t days, g.

3.4 *Length change*

The above specimens were also measured to record any change in length that occurred. The percentage length change was determined using the following relationship:

$$L_c = \frac{(L_t - Li)}{L_i} \times 100 \qquad (2)$$

L_c = average initial length of 5 specimens, mm
L_i = average length of 5 specimens after exposure period of t days, mm

3.5 *Visual inspection*

The specimens above were also observed to detect any visual changes that occur. The following guide were used to record visual changes:

Stages	Visual changes
0	No visible change
1	Partly whitish
2	Fully whitish
3	Small crack/cracks < 0.5mm
4	Cracks > 0.5mm
5	Break and/or crumble

After all the required test were done, the specimens were placed back in the test solution for another 10 days before taken out for subsequent measurement.

4. RESULT AND DISCUSSIONS

4.1 *Weight Change*

The data on weight change of mortar made from OPC, HAC and SRPC specimens were plotted against the exposure period as in Figure 1. Each data point in the figures is an average of 5 reading taken from 5 different samples. Figure 1 indicates that the weight gain of OPC prism varied from 0%, when the specimens were initially exposed to the solution up to 4.08% after an exposure period of 240 days. For HAC prism varied form 0% to 4.35% and for SRPC prism from 0% to 3.36%. Up to 240 days exposure period the maximum change in weight was observed in HAC prism (4.35%).

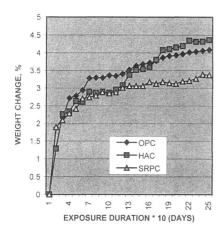

Figure 1. OPC, HAC and SRPC mortar prism in 0.3M sodium suphate solution

4.2 *Length change*

The data on length change for the prisms are shown in Figure 2. Each data point in the figures is an average of 5 reading taken from 5 different samples. These data indicates higher length change in prisms made of high alumina cement after exposure of 240 days compared to OPC and SRPC.

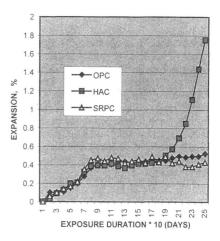

Figure 2. OPC, HAC and SRPC in 0.3M sodium sulphate solution

The length change after 240 days of exposure was 0.43% for SRPC mortar prisms, 0.53% for OPC mortar prisms and 1.75% for HAC mortar prisms. Up to 240 days of exposure we can see that HAC mortar prisms expanded the most 1.75%.

4.3 *Visual Inspection*

The recorded visual changes that have taken place are shown in Table 2. For OPC mortar prisms all 5 were at stage 2, only colour changes occurred. For HAC mortar prisms one at stage 2, one at stage 3, one at stage 4 and 2 at stage 5. For SRPC mortar prisms all 5 were at stage 2, only colour changes occurred. Clearly we can from Table 2 that 4 out of 5 HAC mortar prisms were at very damaging stage.

Table 2. Visual Inspection of 5 samples after 240 days of exposure.

Stage	OPC prism	HAC prisms	SRPC prisms
0	-	-	-
1	-	-	-
2	5	1	5
3	-	1	-
4	-	1	-
5	-	2	-

5 CONCLUSION

1. Mortar made of high alumina cement gains weight considerably, suffer greatest expansion and visually shows sign of deterioration compared to ordinary portland cement and sulphate resisting portland cement.
2. Mortar made of high alumina cement is not suitable in sulphate environment associated with high temperature and high humidity.
3. Mortar made of sulphate resisting portland cement proved to be very durable in sulphate environment.

ACKNOWLEDGMENT

The author acknowledgment the research grant provided by Universiti Sains Malaysia, Penang that has resulted in this article.

REFERENCES

Grutzeck M.W. & S.L Sarkar 1994. Advances in cement and concrete. Proceedings of an Engineering Foundation Conference, Durham. ASCE. New York.

Lawrence C.D.1990. Sulphate attack on concrete, Mag. Of Con. Research, 42.no 153, Dec 1990. Pp 249-264.

Building Research Establishment. Sulphate and acid resistance of concrete in the ground. Digest 363: 1991 edition. Garston, Watford, WD2 7JR.

Murdock et al. 1991. Concrete materials & practice.London: Edward Arnold.

Mehta P.K. 1986. Concrete structure, properties and materials. New Jersey: Prentice Hall.

Osborne G.J. Proceedings of the National Seminar , University of Dundee 1992. Concrete Technology Unit, Dept. of Civil Engineering , Dundee, Scotland.

Cohen M.D. & Mather B., Sulphate attack on concrete: Research needs, ACI Materials Journal, Jan-Feb. 1991. pp 62-69.

Creative Systems in Structural and Construction Engineering, Singh (ed.) © 2001 Balkema, Rotterdam, ISBN 90 5809 161 9

Voids in coarse aggregates: An aspect overlooked in the ACI method of concrete mix design

Z. Wadud, A. F. M. S. Amin & S. Ahmad
Department of Civil Engineering, Bangladesh University of Engineering and Technology, Dhaka, Bangladesh

ABSTRACT: The effect of different ACI normal concrete mix design parameters in predicting mix proportions and strength attainment has been studied. The parametric study reveals that the inter-particle voids, a function of gradation, in the coarse aggregate plays a significant role in mix proportion prediction. However, the ACI design method has failed to properly address this aspect. This gives unrealistic mix proportions when coarse aggregates of high void ratio is used. In this case, the amount of fine aggregate becomes very high with an increased surface area to be covered by the constant amount of cement. The strength attainment of such mixes should be very poor. The hardened strengths of trial mixes have substantiated this fact.

1 INTRODUCTION

Concrete is a ubiquitous and versatile construction material. Different desirable properties of concrete make it a widely used material for civil engineering constructions throughout the world, specially in the developing countries. Also, the flexibility of using locally available ingredients as the aggregates to produce concrete of required properties makes concrete unique among other construction materials. Although plant mixed concrete is rapidly gaining popularity in the developed countries, concrete in the developing countries is still produced and laid at the field. This calls for proper selection of concrete ingredients and their proportions in the mix design.

Among the different methods available to design a normal concrete mix for a given strength under various weather and workability conditions, American Concrete Institute method (ACI 1996) is one of the most popular ones. The method is based on the following principles:

1. Water content determines the workability for a given maximum size of coarse aggregate.

2. The water-cement ratio (w/c ratio) is solely dependent upon the design strength with a restriction from the durability point of view. The w/c ratio is inversely proportional to the design strength.

3. The bulk volume of coarse aggregate per unit volume of concrete depends on the maximum size of the coarse aggregate and the grading of the fine aggregate, expressed as the fineness modulus.

The design starts with the selection of a water content for a given maximum size of coarse aggregate. Cement content is then found out simply from this water content and the w/c ratio, which depends on the design strength. The volume of coarse aggregate is then determined as per 3, and fine aggregate content is found out by subtracting the volume (or weight) of other ingredients from the total volume (or weight) of concrete.

However, some recent experiences and subsequent comprehensive studies made at the Bangladesh University of Engineering and Technology (BUET) have revealed that there are cases where the ACI mix design philosophy fails in proportioning the relative ratio of coarse and fine aggregates for a particular amount of cement content. In such cases, the designed mix fails to attain the desired strength. In this context, a careful observation shows that in the ACI method, cement content determination process is not directly related with aggregate gradation. But in reality, the binding action of the hydrated cement paste takes place mostly on the surface of the aggregate particles. Again, so far as the aggregate surface area is concerned, fine aggregate is the major contributor. Therefore, the quantity of fine aggregate is essential to the determination of the cement content. In this course, the earlier communication by the authors (Amin et al. 1999) reported that ACI method suggests for higher proportion of fine aggregate for the cases where coarse aggregates of lower unit weights are to be used. However, unit weight

Table 1. ACI mix design parameters, variation ranges and assigned values

Sl. No.	Mix design parameters	Unit	Variation range	Assigned value
1	Void ratio of coarse aggregate	-	0.1 – 0.5	-
2	Unit weight (SSD) of coarse aggregate	kg/m³	800 – 1900	1200
3	Design strength	Mpa	13.8 – 34.5	27.6
4	Specific gravity (SSD) of fine aggregate	-	2.25 - 3.00	2.65
5	Fineness modulus of fine aggregate	-	1.75 - 3.00	2.4
6	Maximum size of coarse aggregate	Mm	10 – 75	40
7	Slump	Mm	25 – 150	50

of the coarse aggregate is closely related to the inter-particle voids, which depends on the gradation of the coarse aggregate particles. Hence a further but closer look at the initial findings revealed some more interesting conclusion. The inter-particle void has been found to have a governing role on mix proportion prediction in the ACI method, but it has not been duly addressed in the method. The present paper deals with such conclusion together with the experimental substantiation.

2 PARAMETRIC STUDY

2.1 Study methodology

The ACI method of mix design requires in total seven parameters to design a non-air entrained normal concrete mix. These are: coarse aggregate unit weight, design compressive strength, fine aggregate specific gravity, coarse aggregate specific gravity, fine aggregate fineness modulus, coarse aggregate maximum size and slump.

Specific gravity has been defined as the ratio of mass (or weight in air) of a unit volume of material to the mass (or weight) of same volume of water at a specified temperature. However, as the aggregate contains pores, both permeable and impermeable, specific gravity term may have different meanings. The *absolute* specific gravity refers to the volume of the solid material excluding all pores, while the *apparent* specific gravity (ρ) refers to the volume of solid material including the impermeable pores, but not the capillary ones. It is the *apparent* specific gravity that is normally used in concrete technology. It is defined as the ratio of the weight of the aggregate particle (oven-dried at 100°c to 110°c for 24 hours) to the weight of water occupying the volume equal to that of the solid including the impermeable pores. This specific gravity has to be multiplied by the unit weight of water (γ_w, approximately 1000 kg/m³) in order to be converted into absolute density. However it must be carefully noted that this absolute density refers to the volume of individual particles only, and it is not physically possible to pack these particles such that there are no voids between them. This is where the unit weight (or bulk density, γ) comes into action. It is defined as the weight of the aggregate as a whole per unit volume, the volume including all void spaces between the

aggregate particles. The relation between *apparent* specific gravity, unit weight and void ratio can be expressed by the following:

$$\text{Void ratio} = 1 - \gamma/(\rho\gamma_w) \tag{1}$$

Here it should be noted that the total void content in a mass of unit volume of coarse aggregate can be equated as follows:

$$\text{Total void content} = \text{Permeable pores} + \text{Impermeable pores} + \text{Inter-particle voids} \tag{2}$$

However the proportion of permeable and impermeable pores is much lower than the inter-particle voids. In this paper the authors focus on the inter-particle voids. Therefore, the void ratio, as in Equation 1, refers to the inter-particle voids of the coarse aggregate. It is evident from Equation 1 that the unit weight and specific gravity of the coarse aggregate can be replaced by the void ratio of the coarse aggregate and either of unit weight or specific gravity of the same. The authors opt for void ratio and unit weight of the coarse aggregate. The void ratio has been varied within a selected range, all others being assigned a constant value (Table 1). The effect of this variation on the ratios of fine aggregate to coarse aggregate and cement to aggregate has been investigated. Figures 1, 3, 5, 7, 9 and 11 graphically present the effect of these variations on fine aggregate/coarse aggregate ratio whereas, Figures 2, 4, 6, 8, 10 and 12 illustrate those effects on the cement/aggregate ratio. All these quantities are on weight basis.

2.2 Study Findings

An inspection of the parametric study curves reveals that, with the increase of void ratio in the coarse aggregates, the proportion of the fine aggregate increases in comparison to the coarse aggregate. This increase in the fine aggregate greatly increases the total surface area of the aggregates. As the binding action of cement takes place on the aggregate surface, this increase in total surface area would have required a larger amount of cement, i.e. an increased cement aggregate ratio is expected in a rational design. However, the cement content remains the same in the ACI method of mix design irrespective of the increase in surface area, which leads to a lower cement/aggregate ratio, as depicted in all

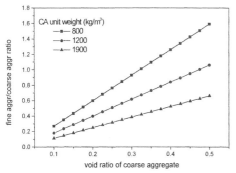

Figure 1. Effect of variation of coarse aggregate void ratio and coarse aggregate unit weight on fine aggr./coarse aggr. ratio

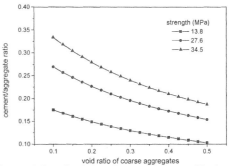

Figure 4. Effect of variation of coarse aggregate void ratio and design strength on cement/aggregate ratio

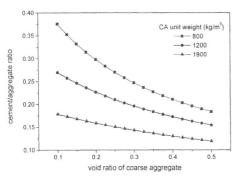

Figure 2. Effect of variation of coarse aggregate void ratio and coarse aggregate unit weight on cement/aggregate ratio

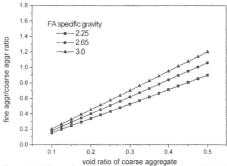

Figure 5. Effect of variation of coarse aggregate void ratio and fine aggregate specific gravity on fine aggr./coarse aggr. ratio

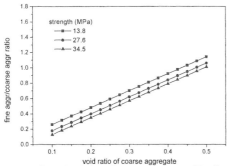

Figure 3. Effect of variation of coarse aggregate void ratio and design strength on fine aggr./coarse aggr. ratio

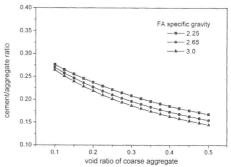

Figure 6. Effect of variation of coarse aggregate void ratio and fine aggregate specific gravity on cement/aggregate ratio

cases of right hand side figures. The situation worsens further when the fineness modulus or specific gravity of the fine aggregates varies, which is explained by the sharper slope of the curves.

However, at a given void ratio of the coarse aggregate, the ACI method suggests a rational design. For example, a higher fineness modulus means a smaller total surface area, requiring a smaller cement/aggregate ratio. This is clearly depicted in Figures 7 and 8, where the cement/aggregate ratio

curve for a higher fineness modulus always lies below that for a lower fineness modulus. Similarly at a constant void ratio and fine aggregate/coarse aggregate ratio, it is expected that the cement/aggregate ratio will be higher to achieve a high slump mix, evident from Figures 11 and 12. Therefore the study reveals that the ACI method fails to explain the increase of fine aggregate/coarse aggregate ratio with increase in voids in the coarse aggregate, with consequent decrease in the cement/aggregate ratio.

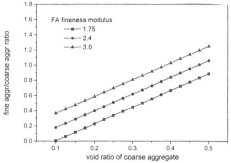

Figure 7. Effect of variation of coarse aggregate void ratio and fine aggregate fineness modulus on fine aggr./coarse aggr. ratio

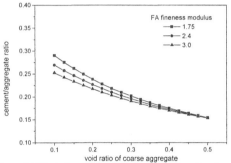

Figure 8. Effect of variation of coarse aggregate void ratio and fine aggregate fineness modulus on cement/aggregate ratio

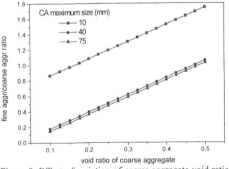

Figure 9. Effect of variation of coarse aggregate void ratio and coarse aggregate maximum size on fine aggr./coarse aggr. ratio

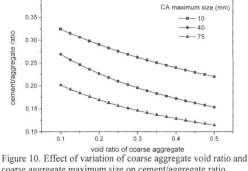

Figure 10. Effect of variation of coarse aggregate void ratio and coarse aggregate maximum size on cement/aggregate ratio

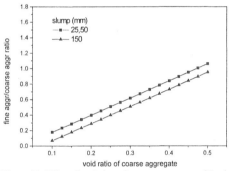

Figure 11. Effect of variation of coarse aggregate void ratio and slump on fine aggr./coarse aggr. ratio

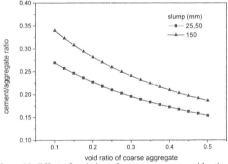

Figure 12. Effect of variation of coarse aggregate void ratio and slump on cement/aggregate ratio

3 EXPERIMENTAL DETAILS

Eight different mixes were designed following the ACI method with a view to substantiate the findings of the parametric study. With these designs at hand, trial mixes were cast in the laboratory following standard ASTM procedures. Apart from these regular trial mixes, two mixes (Mix 3 and Mix 6) were also cast with some readjustment in the proportioning of the mix. The fine aggregate content was arbitrarily reduced by 50% in these cases, as it was found that higher voids of coarse aggregate led to

higher fine aggregate content and failed to produce the design strength. The mix design, casting, curing, and testing procedure of cylindrical concrete specimens are presented in the following subsections.

3.1 Materials

In Bangladesh, crushed brick, an indigenous material, is widely used as coarse aggregates because of scarcity of natural stone aggregates. Locally produced brick can attain a compressive strength as high as 35 MPa, with the most commonly found one ranging from 17 to 24 MPa. Earlier studies on brick aggregate concrete (Akhtaruzzaman & Hasnat 1983) revealed that the modulus of elasticity of brick aggregate concrete is 30% lower and the tensile strength 11% higher than the same grade of stone aggregate concrete. The unit weight of brick aggregate concrete is also lower because of the lower unit weight of the brick aggregate and is around 1900 kg/m³. Because of the high absorption capacity of brick aggregate, the concrete prepared from this has also a high absorption capacity (greater than 10%).

To have a comparative idea involving different types of aggregates, in the present study, both crushed brick and crushed stone were used as coarse aggregates for different mixes. To ensure strength, brick aggregates were produced from well-burnt clay bricks (locally known as *picked jhama* bricks). Local riverbed sand of different gradations were used as fine aggregates. Ordinary Portland Cement (ASTM Type I) was used as the binder. Normal potable water was used for mixing.

Material properties required to design the concrete mix, such as specific gravity, unit weight, fineness modulus and absorption capacity of both fine and coarse aggregates were determined following standard ASTM procedures (1988). These properties are reproduced in Table 2.

3.2 Mix design

Mixes were designed according to ACI method for normal non-air entrained concrete for different strengths with a slump of 50 mm for all of those. The proportions of various ingredients thus obtained are presented in Table 3.

3.3 Preparation of cylinder specimen

On the basis of the design, required quantities of the materials were weighed on SSD weight basis. The coarse aggregate, fine aggregate and cement were mixed thoroughly and continuously in the mixing machine with the required amount of water added gradually until a uniform concrete mix was produced. The slump was checked following standard cone method.

After the mixing was complete, the fresh concrete was placed in reusable cylindrical moulds. Concrete compaction was done by a mechanical vibrator in two layers.

3.4 Curing

After the casting was complete, the cylinders were stored in moulds for 24 hours in moist condition at room temperature. The moulds were then removed and concrete cylinders immersed in saturated lime-water at room temperature for curing. Curing was done continuously until the specimens were removed for strength tests at different durations.

3.5 Testing

Specimens of all the batches were tested for compressive strength at the age of 7 days and 28 days. All of the cylinders were tested in a moist condition. The top surface of the cylinders were capped with sulfur mortar in accordance with standard specification, in order to ensure a uniform stress distribution. Dimensions of the specimens were recorded. The cylinders were then crushed in a Universal Testing

Table 2. Properties of aggregates

Mixes	Aggregates	Unit weight (SSD), kg/m³	Specific gravity (SSD)	Specific gravity (OD)	Void ratio, %	Absorption capacity, %	Fineness modulus
Mix 1	Brick chips	1145	2.08	1.83	44.91	13.66	6.88
	Sand	1522	2.68	2.66	-	0.75	2.74
Mixes 2 & 3	Brick chips	1185	1.95	1.70	39.18	21.95	6.90
	Sand	1522	2.64	2.60	-	1.54	2.30
Mix 4	Brick chips	1009	1.92	1.69	47.42	13.61	7.13
	Sand	1466	2.82	2.80	-	0.71	2.54
Mix 5	Stone chips	1778	2.27	2.22	21.64	2.25	6.97
	Sand	1458	2.79	2.75	-	1.45	2.40
Mix 6	Stone chips	1470	2.30	2.25	36.07	2.22	6.93
	Sand	1466	2.82	2.80	-	0.71	2.54
Mix 7	Brick chips	1214	2.19	2.03	44.53	7.88	6.77
	Sand	1493	2.84	2.81	-	1.07	2.77
Mix 8	Stone chips	1634	2.67	2.66	38.8	0.38	7.46
	Sand	1493	2.84	2.81	-	1.07	2.77

Table 3. Proportions of ingredients in the ACI mix

	Unit	Mix 1	Mix 2	Mix 3	Mix 4	Mix 5	Mix 6	Mix 7	Mix 8
Design strength	MPa	27.6	27.6	20.7	20.7	20.7	20.7	20.7	20.7
Cement	kg/m³	312	312	262	262	262	262	262	262
Fine Aggr.(FA)	kg/m³	900	713	756	1009	467	786	994	887
Coarse Aggr. (CA)	kg/m³	774	854	854	702	1263	1023	819	1098
Water	kg/m³	180	180	180	180	180	180	180	180
Cement : FA : CA	By weight	1:2.9:2.5	1:2.3:2.7	1:2.9:3.3	1:3.9:2.7	1:1.8:4.8	1:3.0:3.0	1:3.8:3.1	1:3.4:4.2

Table 4. Performance of the ACI trial mixes

Mixes	% of design strength attained at 7 days	% of design strength attained at 28 days
Mix 1	44.50	67.00
Mix 2	51.50	73.25
Mix 3	33.83	62.10
Mix 4	39.33	64.00
Mix 5	75.00	105.33
Mix 6	50.33	64.67
Mix 7	47.67	72.80
Mix 8	43.23	66.37

Table 5. Performance of the readjusted trial mixes

Mixes	Cement: FA : CA	% of design strength attained at 7 days	% of design strength attained at 28 days
Mix 3a	1:1.5:3.3	72.67	97.33
Mix 6a	1:1.5:3.9	73.33	97.67

Machine. The strength attainment features of all eight mixes are summarized in Table 4. Strength results of the readjusted mixes (Mix 3a and Mix 6a) are presented in Table 5.

3.6 Experimental findings

The material properties (Table 2) and subsequent mix design computations (Table 3) of different sets of materials indicate the limitations of the ACI method of mix design in proportioning the fine aggregate content in respective mixes. The prediction of high fine aggregate content is more pronounced for those mixes, where the coarse aggregate void ratio is higher. This is similar to the theoretical predictions of the parametric study. As anticipated, all the mixes except Mix 5, which has a considerably lower void ratio in the coarse aggregate, have failed to attain the design 28-day strength (Table 4).

The two separate mixes (Mix 3a and Mix 6a), in which the fine aggregate content was reduced by 50% from the original mix (Mix 3 and Mix 6), showed much better performance than their parent mixes, very nearly attaining the design strength (Table 5). This is most likely due to the subsequent reduction of fine aggregate surface area, which could be sufficiently covered up by the constant amount of cement, determined earlier.

A graphical presentation (Figure 13), showing the relation between void ratio and percent strength attainment is produced on the basis of the experimental results. This plot clearly shows that the percent design strength attainment varies inversely with the void ratio of the coarse aggregate.

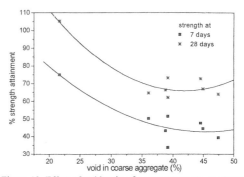

Figure 13. Effect of void ratio of coarse aggregate on strength attainment of cylinder specimens

4 CONCLUSION

The ACI method of mix design for normal concrete mixes fails to rationally design the mixes where coarse aggregates of higher inter-particle voids are used. The occurrence of higher inter-particle voids is related to the gradation of the coarse aggregate. Unfortunately, the ACI method has no adequate parameter to take this aspect into account. In such cases, the design provides higher fine aggregate content, which increases the total surface area of the aggregates. This, accompanied by the fact that the cement content is determined even before the consideration of any aggregate type, implies that the cement/aggregate ratio becomes lower. This is why mixes designed by the ACI method fail to gain the required strength, if coarse aggregates of higher voids are used. The present study indicates the need for further research with a view to incorporate some modifications in the ACI method regarding the voids or gradation of the coarse aggregates.

REFERENCES

American Concrete Institute 1996. Standard practice for selecting proportions for normal, heavyweight, and mass concrete. *ACI Manual of Concrete Practice, Part 1-1996.* Detroit.

Akhtaruzzaman, A. A. & A. Hasnat 1983. Properties of concrete using crushed brick as aggregate. *Concrete International: Design and Construction* 5(2): 58-63.

American Society for Testing and Materials 1988. ASTM standard test methods: C 127-84, C 128-84, C 29-87, C 136-84, C 143-78, C 470-87, C 31-88, C 617-87, C 39-86. *1988 Annual Book of ASTM Standards, Volume 04.02.* Philadelphia.

Amin, A. F. M. S., S. Ahmad, & Z. Wadud 1999. Effect of ACI concrete mix design parameters on mix proportion and strength attainment. *Proceedings of the Civil and Environmental Engineering Conference-New Frontiers & Challenges, 8-12 Nov. 1999, Bangkok, Thailand* III: 97-106.

Creative Systems in Structural and Construction Engineering, Singh (ed.) © 2001 Balkema, Rotterdam, ISBN 90 5809 161 9

High performance grouts for durable post-tensioning

A. J. Schokker
Department of Civil and Environmental Engineering, Pennsylvania State University, Pa., USA

J. E. Breen & M. E. Kreger
Department of Civil Engineering, University of Texas at Austin, Tex., USA

ABSTRACT: The use of post-tensioning in bridges can provide durability and structural benefits while expediting the construction process. Corrosion protection of the post-tensioning system is vital to the integrity of the structure because loss of post-tensioning can result in catastrophic failure. In bonded post-tensioned structures, the most common cause of corroded tendons is poorly grouted ducts. An optimum grout combines a high level of corrosion protection with necessary fresh properties such as workability and adequate bleed resistance for the intended use. Bleed resistance is crucial in applications with tall vertical rises such as bridge piers. A series of fresh property tests and accelerated corrosion tests were used to develop grouts that combine properties such as fluidity and bleed resistance with good corrosion protection. The grouts were then pumped into a clear duct with multiple drapes to observe behavior under simulated field conditions. Two high performance grouts for post-tensioning applications are recommended from this testing.

1 INTRODUCTION

Portland cement grout is often used in post-tensioned structures to provide bond between internal tendons and the surrounding concrete, to discretely bond external tendons at diaphragms and deviators, and also as corrosion protection for the tendons. Grout for post-tensioning is usually a combination of Portland cement and water, along with any admixtures necessary to obtain required properties such as fluidity, bleed resistance, and reduced permeability.

Voids can be formed in the post-tensioning duct from incomplete grouting, trapped air pockets, or from the evaporation of bleed water pockets. Bleed lenses form as a result of the separation of water from the cement that is accentuated by the addition of seven-wire strand. Ducts with vertical rises will typically experience more bleed due to the increased pressure within the grout column. Grouts containing anti-bleed admixture, or thixotropic grouts, can be bleed resistant even when used in ducts with large vertical rises (Schupack 1974).

After three publicized collapses of European post-tensioned structures, the United Kingdom placed a moratorium on the use of grouted post-tensioned structures in 1992. In 1996, the United Kingdom's Working Party completed a document entitled *Durable Bonded Post-Tensioned Bridges* and the moratorium on bonded post-tensioning in the UK was lifted based on adoption of these recommendations. Corrosion of post-tensioned bridges in the United States does not appear to be a major problem at this point (Perenchio et al. 1989), although many post-tensioned structures are relatively new. By far the most common problem found in corroded post-tensioning tendons is poorly grouted or completely ungrouted tendons. Recently, substantial corrosion was reported in a few poorly grouted external tendons in the Florida Keys. No collapse occurred. In response to the need for proper grouting, the Post-Tensioning Institute has recently completed the *Guide Specification for Grouting of Post-Tensioned Structures* (in press).

2 TESTING PROGRAM

Numerous grout designs were evaluated through three phases of testing: fresh property tests, an accelerated corrosion test, and a large-scale clear duct test that simulated field conditions. The objective was to develop a grout with suitable workability and bleed resistance that also provided good corrosion protection.

3 FLUIDITY AND BLEED RESISTANCE

The first phase involved testing of over 30 grout designs for adequate fluidity and bleed resistance. This testing built on the information gained from numerous fresh property tests by Hamilton (1995). The testing focused on low water-cement ratio mixes with different combinations of admixtures and pozzolans. Often admixtures may improve one property while degrading another, so this type of performance based testing is beneficial for selection of appropriate dosages and combinations of admixtures.

3.1 *Mix Design*

Many types of admixtures were tested to evaluate their effect on the fresh properties and corrosion protection properties of the grout. Products from several different manufacturers were included, and a list of the products chosen is shown in Table 1. Other brands of admixtures may be used to achieve the desired results, but the data presented in this paper is specific to the brands chosen and may differ for other brands.

Table 1. Admixtures included in testing.

Admixture Type	Brand Name	Manufacturer
Superplasticizer (HRWR)	Rheobuild 1000	Master Builders
Expansive	Intraplast-N	Sika Corporation
Anti-bleed	Sikament 300SC	Sika Corporation
Fly Ash	Class C	
Silica Fume	Sikacrete 950DP	Sika Corporation

3.2 *Test Methods*

Table 2 summarizes the test methods used to investigate fluidity, standard bleed, and pressurized bleed. At the time of testing, the Post-Tensioning Institute (PTI) was working on the *Guide Specification for Grouting of Post-Tensioned Structures* (1997). Criteria were taken from this document for fluidity and standard bleed testing. At the time of testing, no established standard existed for pressurized bleed, so performance was judged using recommendations of Hamilton (1995) who used as his criteria a maximum bleed of 2% of the grout sample volume at 345 kPa (approximately equivalent to the pressure at the base of a 38 m tall grout column).

Table 2: Fluidity and bleed tests

Property	Test Method	Description	Criteria
Fluidity	ASTM C939	Flow cone*	20-30 seconds
Standard Bleed	ASTM C940**	Standard bleed	1% maximum bleed
Pressure Bleed	Gelman test	Pressurize in steps to 552 kPa	2% maximum bleed at 345 kPa

*Modified for thixotropic grouts: fill cone to top and time evacuation of 1 liter of grout (PTI, 1997)
**Modified to include 3 strand bundle

3.3 *Fluidity and Bleed Results*

The results from the fluidity and bleed testing phase are summarized below. Further details of all testing can be found in Schokker (1999).

- The addition of superplasticizer increases bleed.
- The addition of fly ash (Class C) or silica fume decreases bleed.
- Most grouts containing anti-bleed admixture exhibited little or no bleed in the standard bleed test.
- Only grouts containing anti-bleed admixture met the bleed resistance criteria for the Gelman pressure test. A large gap in pressurized bleed performance exists between grouts that include an anti-bleed admixture and grouts that do not. The grouts that did not contain an anti-bleed admixture typically began to bleed instantly in the Gelman filter.

4 ACCELERATED CORROSION TESTING

Three grouts were chosen to proceed into the accelerated corrosion testing phase based on their performance in the fresh property test phase: a 0.33 water-cement ratio grout with anti-bleed admixture, a 0.35 water-cement ratio grout with 30% cement replacement class C fly ash, and a 0.35 water-cement ratio grout with 15% cement replacement silica fume. Additionally, the Texas DOT standard grout (0.44 w/c with 1% expansive admixture) was included as a comparison. A plain grout with 0.40 water-cement ratio was used as a baseline to check the repeatability of the present tests with previous tests.

4.1 *The Accelerated Corrosion Testing Method (ACTM)*

An accelerated corrosion testing method was developed by Thompson et al. (1992) in a FHWA sponsored study and was refined at The University of Texas at Austin by Hamilton (1995) and further refined by Koester (1995). This corrosion test uses anodic polarization and applies a potential to the specimen to speed the onset of corrosion. The applied potential (or voltage) is more positive than the free corrosion potential of the specimen, and the potential gradient developed tends to drive the negatively charged chloride ions through the grout to the steel surface. The applied potential in this study is +200 V_{SCE}. The test was intended to provide a fairly simple and quick method for the relative comparison of grout corrosion protection under extremely harsh corrosive conditions.

4.2 Test Setup

The specimen, or working electrode, consists of a short length of prestressing strand in a grouted clear PVC mold casing as shown in Figure 3. The strand is cleaned with acetone to remove surface contamination prior to filling the duct with grout. A portion of the PVC casing is removed to expose the grout to the 5% NaCl solution. The corrosion current (i_{corr}) is calculated from the measured potential (E_{meas}). A spike in corrosion current indicates the onset of corrosion. The average time to corrosion for each grout design is found from the average value for six stations.

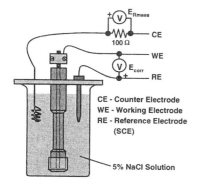

Figure 1: Accelerated Corrosion Test station

4.3 Accelerated Corrosion Testing Results

Previous ACTM studies at The University of Texas at Austin were carried out by both Hamilton (1995) and Koester (1995). Koester's studies used the same test setup, applied potential and materials (cement from the same batch and strand from the same reel) as the present study and are included with some of the results presented.

Figure 2 shows the average time to corrosion for each grout series tested with values from the same mix designs averaged together. Since these are the results of an accelerated corrosion test, the actual times shown are not intended to directly relate to actual service life. The test is used as a relative comparison between different grout types. The longer bars in the figure indicate longer times to corrosion and therefore better corrosion protection. The addition of fly ash to the mix gave the longest time to corrosion with over 40% improvement in corrosion protection over the next best grout. The mix containing anti-bleed admixture also tested well. Although anti-bleed admixture has been known to reduce time to corrosion, the very low water-cement ratio of 0.33 allowed this mix to perform favorably. This mix was the only grout design included in the

ACTM testing that passed all of the fresh property tests including the pressurized bleed test.

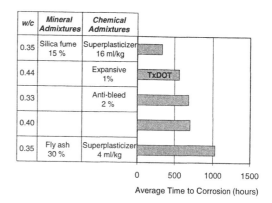

Figure 2: Accelerated Corrosion Test results

Most of the grouts containing chemical admixtures did not perform well in the accelerated corrosion testing. These included the Texas DOT grout that contains an expansive admixture / superplasticizer and a silica fume grout with superplasticizer. The poor performance of the silica fume grout is likely due to the large amount of superplasticizer that was necessary for this grout to maintain fluidity at a water-cement ratio of 0.35. Silica fume is likely to be more beneficial in smaller percentages of replacement (Whiting & Detwiler, 1998), but the replacement percentage of 15% of the cement weight was chosen to correspond with the suggestions of 15-25% replacement silica fume in the draft of the PTI grouting specifications available at the time of testing. The current PTI specifications (in press) recommend 0-15%.

A grout that combines low permeability and minimal chemical admixtures gives the best corrosion performance in the ACTM tests.

5 SIMULATED FIELD TESTING

Grouts that performed favorably in the fresh property test phase and the accelerated corrosion test phase were tested in a large-scale duct with multiple drapes to simulate field conditions. This test used a clear duct to investigate grout flow along the duct and strand, along with the formation of bleed water lenses due to changes in duct elevation. The workability of the grout during pumping was also observed.

5.1 Test Setup and Procedure

A clear vinyl flexible tube (38 mm inside diameter) with a bundle of three untensioned 13 mm, 7 wire prestressing strands was used to simulate conditions in a post-tensioning duct. The three-strand bundle provides a realistic steel area to duct area ratio and promotes water movement along the duct. The duct was connected in a parabolic shape to a wooden frame as shown in Figure 3. The length of frame was 9.8 m and the height of the vertical rise in the duct was 0.9 m.

Grouting procedures followed those recommended by the Post Tensioning Institute (1997). The hardened grouted duct was autopsied by removing approximately 50 mm long slices from critical locations to investigate voids.

Figure 3: Draped duct with vent locations

5.2 Results

Table 3 gives a summary of the performance of the grouts tested in all three tests phase. Three grout mix designs were tested in the draped duct. All grouts were very workable during pumping.

The first grout tested was a grout with a 0.33 water-cement ratio and 2% anti-bleed admixture. No voids or bleed water were noticed in the duct at any time, and autopsy slices revealed no voids along the entire length of the duct.

The next grout tested was a 30% fly ash (Class C) grout with a 0.35 water-cement ratio. No voids were observed during pumping. Approximately 10 minutes after pumping, a long, very thin void was observed near the intermediate crest. Within 24 hours the water in the void had reabsorbed and the void was no longer noticeable. Upon autopsy, no voids were found in any of the duct slices.

The standard Texas DOT grout was also tested in the parabolic duct as a comparison to the other grouts. Immediately after pumping, bleed water was visible in the duct. Bubbles traveled to the intermediate crest where a large void pocket was formed. After 24 hours, the bleed water had reabsorbed and a large void remained at the intermediate crest.

A comparison of autopsy slices at the intermediate crest for each of the three grouts is shown in Figure 7. The anti-bleed grout and fly ash grout have no noticeable voids, while the TxDOT grout has a large void that exposes the tendon.

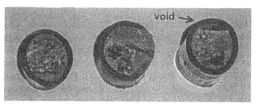

| anti-bleed grout | fly ash grout | TxDOT grout |

Figure 4: Comparison of slices at intermediate crest

6 RECOMMENDED GROUTS

The two grouts discussed below showed the best performance in this testing program. Even slight variance in materials (such as the fineness of the cement or variations in fly ash) can change the workability and bleed properties, so standard recommendations should be performance-based.

- *Fly ash grout (0.35 w/c, 30% Class C fly ash, 4 ml/kg superplasticizer)*

This grout is recommended for situations requiring a high resistance to corrosion without extreme bleed conditions (vertical rise of less than 1 meter). This grout may also be appropriate for larger vertical rises (1-5 m), but simulated field-testing should be performed on a case-by-case basis. The superplasticizer dosage may need to be adjusted depending on conditions (and brand used), but the dosage should be kept at a minimum (while attaining the necessary workability) to ensure maximum corrosion protection. Class F fly ash may be substituted for Class C fly ash in applications where sulfate resistance is a concern. The properties of the grout will change with different fly ashes, so mix proportions should be adjusted accordingly based on testing. If Class F fly is used, the percentage of fly ash used should be 25% or less as recommended by the Post-Tensioning Institute *Guide Specification for Grouting of Post-Tensioned Structures* (in press).

- *Anti-bleed grout (0.33 w/c, 2% anti-bleed admixture)*

This grout is recommended for situations requiring a high resistance to bleed along with good corrosion protection. The anti-bleed admixture dosage may need to be adjusted slightly depending on conditions (and brand used), but the dosage should be kept at a minimum (while attaining the necessary workability and bleed resistance) to ensure maximum corrosion protection.

7 SUMMARY AND CONCLUSIONS

The cement grout injected into the tendons in post-tensioned bridge structures has the important dual role of providing bond between the strands or bars

432

and the concrete, as well as providing corrosion protection to the prestressing steel. An optimum grout combines good corrosion protection with desirable fresh properties so that the ducts can be completely filled with ordinary grouting techniques. Numerous grouts were tested in three phases of testing to develop a high performance grout for corrosion protection. The testing phases included fluidity and bleed testing, accelerated corrosion tests, and a large-scale clear multiple drape duct test that allowed observation of the grout under simulated field conditions.

Two grouts showed promising results in the testing program. For situations requiring a high resistance to corrosion without extreme bleed conditions (vertical rise of less than 1 meter), a fly ash grout (0.35 water-cement ratio, 30% cement weight replacement fly ash, 4 ml/kg cement weight superplasticizer) showed the best performance. This grout may also be appropriate for larger vertical rises (1-5 m), but field-testing should be performed on a case-by-case basis. For situations requiring a high resistance to bleed along with good corrosion protection, an anti-bleed grout (0.33 water-cement ratio, 2% cement weight anti-bleed) showed the best performance.

ACKNOWLEDGEMENTS

The authors would like to thank the Texas Department of Transportation for sponsoring this research through Project 0-1405, "Durability Design of Post-Tensioned Bridge Substructures." The conclusions drawn in this paper are the opinions of the authors and do not necessarily reflect the opinions of the sponsors.

The authors also would like to thank Brad Koester and Trey Hamilton for sharing their knowledge of the Accelerated Corrosion Test Method. Jeff West's continuous help and support were extremely valuable in the development of the grout research in Project 0-1405.

REFERENCES

American Society for Testing and Materials 1987. *Standard Test Method for Expansion and Bleeding of Freshly Mixed Grouts for Preplaced-Aggregate Concrete in the Laboratory, ASTM C940-87*, Philadelphia: ASTM.

American Society for Testing and Materials 1994. *Standard Test Method for Flow of Grout for Preplaced-Aggregate Concrete (Flow Cone Method), ASTM C939-94a*, Philadelphia: ASTM.

Hamilton, H.R., III 1995. *Investigation of Corrosion Protection Systems for Bridge Stay Cables*, Ph.D. Dissertation, The University of Texas at Austin.

Koester, B.D. 1995. *Evaluation of Cement Grouts for Strand Protection Using Accelerated Corrosion Tests*, MS Thesis, The University of Texas at Austin.

Perenchio, W.F., Fraczek, J. & Pfeifer, D.W. 1989. *Corrosion Protection of Prestressing Systems in Concrete Bridges*, NCHRP Report 313, Washington D.C.: Transportation Research Board.

Post-Tensioning Institute 1997. *Guide Specification for Grouting of Post-Tensioned Structures, 5th Draft*, PTI Committee on Grouting Specifications.

Post-Tensioning Institute 2000 (in press). *Guide Specification for Grouting of Post-Tensioned Structures*, PTI Committee on Grouting Specifications.

Schokker, A.J. 1999. *Improving Corrosion Resistance of Post-Tensioned Substructures Emphasizing High Performance Grouts*, Ph.D. Dissertation, The University of Texas at Austin.

Schupack, M. 1974. Admixture for Controlling Bleed in Cement Grout Used in Post-Tensioning, *Journal of the Prestressed Concrete Institute*, November-December.

Thompson, N.G., Lankard, D. & Sprinkel, M. 1992. *Improved Grouts for Bonded Tendons in Post-Tensioned Bridge Structures*, FWHA Report No. RD-91-092, Cortest Columbus Technologies.

Whiting, D. & Detwiler, R. 1998. *Silica Fume Concrete for Bridge Decks*, NCHRP Report 410, Washington D.C.: National Academy Press.

Working Party of the Concrete Society 1996. *Durable Bonded Post-Tensioned Concrete Bridges*, Technical Report No. 47.

Creative Systems in Structural and Construction Engineering, Singh (ed.) © 2001 Balkema, Rotterdam, ISBN 90 5809 161 9

Conductive concrete for bridge deck deicing and anti-icing

S. A. Yehia & C. Y. Tuan
Department of Civil Engineering, University of Nebraska, Omaha, Nebr., USA

ABSTRACT: Conductive concrete is a cementitious admixture containing electrically conductive components to attain stable and high electrical conductivity. Due to its electrical resistance and impedance, a thin conductive concrete overlay can generate enough heat to prevent ice formation on a bridge deck when connected to a power source. In 1998, Yehia and Tuan at the University of Nebraska developed a conductive concrete mix specifically for bridge deck deicing. In this application, a conductive concrete overlay is cast on top of a bridge deck for deicing and anti-icing. The mechanical and physical properties of the conductive concrete mix met the ASTM and AASHTO specifications for overlay construction. Deicing experiments were conducted during the winter of 1998 and 1999 for deicing and anti-icing. The results showed that a conductive concrete overlay has the potential to become the most cost-effective bridge deck deicing method.

1 INTRODUCTION

Bridge pavement surfaces are prone to ice accumulation, making wintry travel hazardous. Current practice is to use road salt and deicing chemicals, which cause damage to concrete and corrosion of reinforcing steel. A thin conductive concrete overlay can generate enough heat to prevent ice formation on a bridge deck. Conductive concrete is a cementitious admixture containing electrically conductive components to attain stable and high electrical conductivity. Due to the electrical resistance and impedance in conductive concrete, heat is generated when connected to a power source and can be used for deicing or anti-icing.

Conductive concrete is a relatively new material technology developed to achieve high electrical conductivity and high mechanical strength. A conductive concrete mix has been developed at the University of Nebraska specifically for bridge deck deicing. Over 150 trial batches of conductive concrete were prepared and their properties evaluated.

In this paper, mix proportioning, material evaluation, heating performance, and simplified heat transfer analysis will be briefly presented.

2 MIXTURE PROPORTIONING

In 1998, Yehia and Tuan (1999, 2000) at the University of Nebraska developed a conductive concrete mix specifically for bridge deck deicing. In this application, a conductive concrete overlay is cast on the top of a bridge deck for deicing and anti-icing. In this mix, steel shaving and fibers were added to the concrete as conductive materials. Conductive concrete test specimens from three main categories were evaluated during the optimization process:

Category 1: containing steel fibers only;
Category 2: containing steel shaving only; and
Category 3: containing both steel shaving and steel fibers.

Compressive strength, electric resistivity, workability and finishability were used as primary evaluation criteria for each trial batch. The electric resistivity test results from Category 1 and Category 2 showed that using steel fibers or steel shaving alone could not provide an electric resistivity lower than the 10 $\Omega \bullet$m necessary for deicing application.

The optimization of Category 3 specimens consisted of two stages. In the first stage, the impact of using different steel shaving and steel fiber ratios on the mechanical strength and electrical conductivity was assessed. From this stage test results, 20 percent per volume of steel shaving and 1.5 percent per volume of steel fibers were considered as an upper bound of the conductive material, beyond which poor workability and surface finishability will result. Additional experiments were conducted in the second stage to identify the optimum volumetric ratios of steel fibers and shaving to use in the mix. Steel shaving of 10, 15 and 20 percent was used with 1.5 percent of steel fibers per volume of conductive concrete, respectively. Over 25 batches of conductive concrete were prepared. The evaluation criteria in this stage were mechanical properties (compressive and flexural strength), slab heating performance, power source (DC vs. AC), size effect, electric resistivity, and electrode configuration. Detailed discussion of the optimization results is presented elsewhere (Yehia 2000).

3 SMALL SLAB HEATING TESTS

Two slabs of 305mm x 305mm x 51mm (1 ft × 1 ft × 2 in.) from each trial mix were tested using 35 volts of DC power, and the corresponding current was recorded. Two thermocouples were installed in each slab to measure the mid-depth and surface temperature, and both were located at the center of the slab. The experimental results showed that the temperature at the mid-depth of the slabs increased at a rate of approximately 0.56°C/min. (1°F/min.) with 35 volts. Average power of 516 W/m^2 (48W/ft^2) was generated by the conductive concrete to raise the slab temperature from -1.1°C (30°F) to 15.6°C (60°F) in 30 minutes. This power level is consistent with the successful deicing applications using electrical heating cited in the literature (Zenewitz 1977, Henderson 1963).

4 MATERIAL EVALUATION

The mix design containing 20% steel shaving and 1.5% steel fibers per volume from the optimization process has been tested extensively to evaluate its mechanical and physical properties. In addition, a conductive concrete patch was constructed on an interstate bridge near the Nebraska-Iowa border for durability evaluation.

4.1 Mechanical and physical properties

The mechanical and physical properties of the conductive concrete mix were evaluated in accordance with the ASTM (1990) and AASHTO (1995) specifications (Yehia 1999).

Table 1 Mechanical and physical properties of the conductive concrete

Test	Result
Compressive strength	31 MPa (4500 psi)
Flexural strength	4.6 MPa (670 psi)
Rapid freeze and thaw resistance	None of the specimen failed after 312 cycles
Shrinkage	Less than that predicted by ACI-209 equation by 20-30%
Modulus of elasticity	3634 MPa (5.27 x 10^5 psi)
Permeability	Permeability rate ranges between 0.004 to 0.007 cm^3/sec.
Thermal Conductivity	10.8 W/m-°K.

The compressive strength, flexural strength, rapid freeze and thaw resistance, permeability, and shrinkage of the conductive concrete mix after 28 days have met the AASHTO requirements for bridge deck overlay. These mechanical and physical properties are summarized in Table 1.

4.2 Durability test

A 6.4m x 3.65m x 9 cm (21ft x 12 ft x 3.5 in.) conductive concrete patch was constructed in one I-480 west bound lane over the Missouri River (nearby the Nebraska-Iowa border) to evaluate the durability of conductive concrete under traffic loads. Construction began on December 3, 1999 by grinding the old overlay and cleaning the surface from the underlying concrete. On the next day and before placing the overlay, a layer of cement grout was placed to bond between the concrete deck and the conductive concrete overlay. A total of 2.6 cubic meter (3.5 cubic yards) of conductive concrete was prepared at Ready Mixed Concrete in Omaha, Nebraska, using the mix containing 20 percent steel shaving and 1.5 percent steel fibers by volume. Figure 1 shows that placing and finishing of conductive concrete are similar to those of conventional concrete. The lane was opened to traffic on December 6, 1999 after the concrete strength had reached 18 MPa (2600 psi). The overlay was visually inspected after 4 months. As shown in Figure 2, there was no fiber exposure or cracking developed in the overlay. The concrete compressive strength was 52 Mpa (7500 psi) determined from cylinder testing.

Figure 1. Conductive concrete patch on I-480.

Figure 2. Conductive concrete patch after 4 months

5 DEICING EXPERIMENTS

Conductive concrete overlay from the same mix used for the material evaluation was also cast on the top of two slabs tested in a natural environment for deicing and power consumption evaluation.

5.1 Conductive Concrete Overlay Construction

Two 15-cm (6-in.) thick concrete slabs, one 2m by 2m (7 ft by 7 ft) and the other 1.2m by 3.6m (4 ft x 12 ft), were constructed to simulate bridge decks. A 9-cm (3.5-in.) thick conductive concrete overlay was cast on the top of each slab. Two steel plates, 64mm (2.5 in.) wide and 6mm (0.25 in.) thick, were embedded along the length of each slab for electrodes. The steel plates had perforations greater or equal to the 13-mm (0.5-in.) maximum aggregate size to allow concrete to flow through to provide good anchorage.

5.2 Instrumentation

Thermocouples were installed in the conductive overlays for temperature monitoring. An electronic

weather station was used to record the ambient air temperature, relative humidity, and wind speed/direction during each testing. The temperature, humidity, and wind sensors were mounted at 1.8m (6 ft) above the overlay surfaces. Sixteen thermocouple readings were recorded simultaneously at one sample per second during the deicing experiments. A 220V, 60 Hz, AC power was used for powering the overlays. A VARIAC was used to regulate the applied voltage. A transformer was used to elevate the applied voltage to a maximum of 420 volts. The overlays were connected to the AC power in parallel. An amp-meter was used to record the electrical current going through each overlay. The total current going through both overlays was limited to 10 Amps, the maximum capacity of the transformer. Detailed discussion of the specimen construction and instrumentations is presented elsewhere[7].

6 HEATING OPERATION

An equivalent circuit model was employed to describe the electric behavior of conductive concrete (Ferdon 1999, Hey 1978). In this model, the conductive concrete mix is represented as a resistor, in parallel with a variable resistor and a capacitor, as shown in Figure 3. The resistor represents the electrical resistance of the steel shaving in the cement paste and steel fibers that are directly connected, the variable resistor represents the electrical resistance of steel fibers and steel shaving not directly connected, and the variable capacitor represents gaps among the steel fibers and shaving with the cement paste as the dielectric.

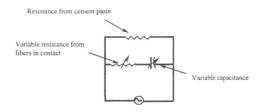

Figure 3. A model for conduction of electricity through conductive concrete

Conduction of electricity through conductive concrete may be divided into three zones: (1) linear, (2) operational, and (3) saturation zone. For efficient heating of a conductive concrete overlay, the break down point of the capacitor must be reached first, and the voltage may be reduced to the operational zone afterwards. The heating rate in the conductive concrete depends upon the current going through, which can be controlled by maintaining the applied voltage in the operational range and by limiting the current. The "break down" voltage is in the range of 450-480 volts for the 1.2m by 3.6m (4 ft by 12 ft) overlay, and 780-840 volts for the 2m by 2m (7 ft by 7 ft) overlay. For rapid heating, it is desirable to apply voltage as close to the saturation zone as possible. However, the transformer could only provide up to 420 volts with a maximum current of 10 Amps. Consequently the following operation sequence was carried out during the experiments: (1) a 420-volt voltage was applied first, and the current was recorded; and (2) once the total current going through the overlays reached the 10-Amp limit, the voltage was reduced to keep the current below 10 Amps. For an electrode spacing of 1.2m (4 ft), an applied voltage of 420 V was in the operational zone. However, for an electrode spacing of 2m (7 ft), an applied voltage of 420 V was in the linear zone and the heating rate was minimal.

7 EXPERIMENTAL RESULTS

7.1 Experiments during the winter of 1998

Deicing experiments were conducted in five snowstorms during the winter of 1998 under two scenarios: deicing and anti-icing. While the overlays were preheated 2 to 6 hours before and heated during the storms in an anti-icing scenario, they were heated only during the storms in a deicing scenario. In each experiment, the applied voltage, current going through each overlay, temperature distribution within each overlay, along with the air temperature, humidity, and wind speed/direction were recorded.

Figure 4 shows the 1.2m x 3.6m (4ft x 12ft) slab during anti-icing experiment, and Figure 5 shows the temperature distribution along the centerline of the overlay. The temperature at the centerline of the overlay was about 2 to 4°C (4 to 7°F) lower than areas close to the electrodes. Consequently, in some tests snow accumulated and formed a strip along the middle of the overlay during heavy precipitation. This behavior was also observed when the applied voltage was reduced to limit the current to 10 amps. Similar results were obtained during deicing experiment.

7.2 Experiments during the winter of 1999

One deicing experiment was conducted during the winter of 1999. The experiment started with 150 mm (6in.) snow accumulation. Figure 6 shows the 1.2m x 3.6m (4ft x 12ft) slab after the storm. The

heating performance was consistent with that of winter 1998 experiments. Table 2 summaries the essential data of the deicing experiments.

Table 2 Power consumption for deicing experiment

Date	Snow accum. mm (in.)	Air Temp. °C (°F)	Exp. Time (hours)	Power Kw-h	Cost $
Feb. 20, 1999	50 (2)	3 (37)	5	10	0.8
March 8, 1999	250 (10)	0 (32)	24	46	3.7
Feb. 18, 2000	150 (6)	-2.5 (27)	10	34	2.7

Table 3 Material Costs of Conductive Concrete vs. Conventional Concrete

Material	Cost/kg	Cost/m^3	
		Conductive Concrete	Conventional Concrete
Steel fiber	$0.88	$105	0
Steel shaving	$0.22	$115	0
Sand	$0.0053	$2.0	$3.1
½ in. limestone	$0.0053	$2.0	$6.2
Cement	$4/(sac 43 kg)	$46	$42
Total		$270.1	$51.3

8 MATERIAL AND ENERGY COSTS

The average energy cost per unit surface area is about $0.8/m^2 ($0.074/ft^2) for each storm, given that 1 kW-hr costs about 8 cents in Omaha, Nebraska. The material costs of conductive concrete are compared with those of conventional concrete in Table 3.

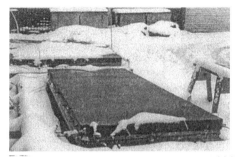

Figure 4 The 1.2m x 3.6m (4ft x 12ft) slab during anti-icing experiment – winter 1998

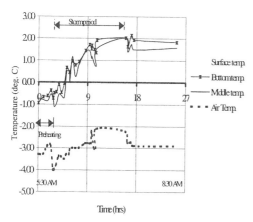

Figure 5 Temperature time history

Figure 6 The 1.2m x 3.6m (4ft x 12ft) slab during deicing experiment – winter 1999.

9 SIMPLIFIFED HEAT TRANSFER ANALYSIS

A simplified one-dimensional unsteady-state heat transfer analysis was conducted using a transient heat conduction model to determine the temperature distribution and power consumption in the conductive concrete overlay for bridge deck deicing (Incropera 1990, Karlekar 1982).

The transient model consisted of two layers, the conductive concrete overlay and the conventional concrete slab. In the deicing scenario, the conductive concrete overlay was divided into four equal layers. In the anti-icing scenario, the conductive concrete overlay was divided into three equal layers and the boundary layer was divided into two equal layers.

A 152-mm (6 in.) thick regular concrete deck was used in both scenarios. The conventional concrete layer was divided into eleven equal layers. The

boundary node was divided into two equal layers. The four sides of the model were considered to be adiabatic boundaries. The effect of radiant heat transfer was ignored in the analysis.

The conservation of energy equation, Eq. 1, was satisfied at each node.

$$E_{stored} = E_{generated} + E_{in} - E_{out} \qquad (1)$$

The implicit formulation technique was used to eliminate the restriction in the time step required to ensure stability for the explicit technique. A set of simultaneous equations was solved to determine the temperature change of each node. The temperature changes are assumed to take place during each time step, dt. The temperature at each node was updated at the end of each time step. Two algorithms were used to simulate the anti-icing and deicing scenarios. These algorithms formed the basis of a stepwise transient heat transfer analysis. In the deicing algorithm, the solution process was continued until the temperature in the ice, node 1, reached 0°C. The ice would start melting at this point and continue to absorb heat for phase change into water. The latent heat of fusion of ice is $Q_l = 333.5$ kJ/kg. During the phase change, the temperature of the ice remains at 0°C. Therefore, the stepwise solution algorithm was modified slightly to accommodate phase change and the solution was continued until the ice layer was completely melted. In the anti-icing algorithm, the solution process was continued until the surface temperature, node 1, reached 5°C. This temperature limit was determined from the experimental results; the surface temperature should be maintained between 5 to 7°C to prevent snow accumulation.

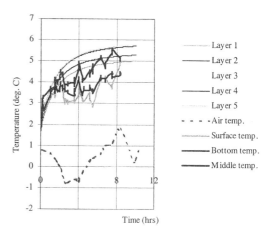

Figure 7 Temperature distribution along the thickness - 3.5-in. conductive concrete overlay (Anti-icing)

A parametric study was conducted using the anti-icing and deicing algorithms. Figure 7 shows the temperature variation with time for 3.5 in. conductive concrete overlay. Results from the heat transfer analysis were compared with the results from the experiments conducted during the winter of 1998. The variation between the analytical and the experimental results occurred because of using average ambient temperature in the analysis.

Results from the heat transfer analysis compared favorably with those obtained from the experiments. The transient heat conduction model can be used to predict the temperature distribution and power consumption in the design of conductive concrete overlay.

10 CONCLUSIONS

The optimized mix developed at the University of Nebraska for a bridge deck overlay showed excellent workability and surface finishability. The mechanical strength of the conductive concrete mix has met the ASTM and AASHTO Specifications for bridge deck overlay construction. Stable and uniform temperature distribution with gradual heating was achieved during the deicing experiments. Average power of about 590 W/m^2 (55 W/ft^2) was generated by the conductive concrete overlay to prevent snow and ice accumulation. Energy cost is in the range of $0.6 to $0.8/m^2 ($0.056 to 0.074/ft^2) per storm.

The transient heat conduction model for the simplified heat transfer analysis can be used to predict the power consumption and temperature distribution in the conductive concrete overlay

The deicing performance and the power consumption costs showed that the conductive concrete overlay has the potential to become the most cost-effective bridge deck deicing method.

REFERENCES

Yehia, S.A. & C. Tuan 1999. *Conductive Concrete Overlay for Bridge Deck Deicing*. ACI Materials Journal, 96: 382-390.
Yehia, S.A., C. Tuan, D. Ferdon, & B. Chen 2000. *Conductive Concrete Overlay for Bridge Deck Deicing: Mix Design, Optimization, and Properties*. ACI Materials Journal, 97.
Zenewitz, J. A. 1977 *Survey of Alternatives to the Use of Chlorides for Highway Deicing*. Report FHWA-RD-77-52.
Henderson, D. J. 1963 *Experimental Roadway Heating Project on a Bridge Approach*. In Highway Research Record 14, 111: 14-23.
American Society for Testing and Materials (ASTM) 1990 *Concrete and Aggregates*, Section 4, V. 04.02.
American Association of State Highway and Transportation officials (AASHTO) 1995. *Standard Specifications for Transportation Materials and Methods of Sampling and Testing*.
Yehia, S.A. & C. Tuan 2000 *Thin Conductive Concrete Overlay for Bridge Deck Deicing and anti-icing* Accepted for Publications in Transportation Research Board Record.
Ferdon, D., & B. Chen 1999. *Proposed Electrical Properties of the Conductive Concrete Mix*. Special Report, University of Nebraska, Computer Engineering Department.
Hey, J. C & W.P. Kram 1978 *Transient Voltage Suppression Manual*, 2nd Edition, General Electric Company, New York.
Incropera, F. P. & P. Dewitt 1990 *Fundamentals of Heat and Mass Transfer*, Third Edition, John Wiley & Sons, New York.
Karlekar, B. V., & M. Desmond 1982. *Heat Transfer*, Second Edition, West Publication Co., Minnesota, U.S.A.

Creative Systems in Structural and Construction Engineering, Singh (ed.) © 2001 Balkema, Rotterdam, ISBN 90 5809 161 9

Effect of loading rates on the compressive strength of high strength concrete

Y. Onggara & I. Patnaikuni
RMIT University, Melbourne, Vic., Australia

ABSTRACT: Loading rate has a significant impact on the measured compressive strength of concrete specimens and higher rate translates to higher strength, although the gain in strength is not as sensitive for high strength concrete as it is for normal strength concrete. Comparative study was carried out on 75 mm x 150 mm concrete cylinders with 56-day strength of 80 - 105 MPa using loading rates of 20, 30, 40, 60, 80 and 100 MPa/min. The ratio of strengths at all other rates to that at 20 MPa/min. (prescribed by AS1012.8) were calculated and they revealed logarithmic relationships. Loading rate as fast as 40 MPa/min. is acceptable and rates faster than this give higher strengths, but results are unpredictable and inconsistent. The ratio of average testing time are also calculated; it is deduced that loading rate does not relate linearly to loading time and apparent testing time halves as loading rate doubles.

1 INTRODUCTION

1.1 *Terminology*

Throughout this paper, the term Normal Strength Concrete (NSC) refers to concrete with 56-day compressive strength less than 50 MPa while High Strength Concrete (HSC) refers to concrete with 56-day compressive strength of 50 MPa or higher.

1.2 *Rationale & Background*

The compression test has been a standard test for years in the concrete industry which depends heavily on the results of this test. One objective is to ascertain that target strength specified in mix design is met. It has been widely recognized that the rate of loading has a significant effect on concrete strength - both normal and high strength - among other variables such as : specimen shape and size, rate of loading, mould material and design, consolidation, curing of specimens, capping materials and method, eccentricity of loading and stiffness of testing machine (Richardson 1991 and Neville 1995).

One of the most common tests on hardened concrete is compression test. It is understandably very popular because most but not all of concrete properties such as mechanical strength and durability are directly related to it. HSC has become a popular choice with concrete designers in the past two decade or so due to their superior properties to normal strength concrete such as increased strength and durability, to mention a few. They have been successfully applied in construction of high rise buildings, large span bridges, offshore structures, precast and prestressed elements, foundation piles, highway pavement and other applications where reductions in dead load, reduced creep and shrinkage, strength and durability are of paramount importance.

Numerous standards or codes contain guidelines for testing normal strength concrete and to some extent, high strength concrete. However, these guidelines are predominantly based on the experimental results obtained using NSC specimens.

Concrete specimens are tested at a prescribed loading rate in laboratory so that meaningful comparison can be made for the test results. The standard rate of loading used in each country is chosen according to standard or code of that particular country. One of the obvious advantages in performing faster testing of HSC specimens is to save time and money. More importantly, tests carried out with faster loading rates can be used to investigate the behavior of HSC under such rates and can assist engineers in simulating the effect of earthquakes and impact loading (such as pile-driving, high velocity impact, explosions, etc.) on concrete specimens or models prepared in laboratories. A limited information is available on the behavior of normal strength concrete under the effect of very high loading rates. Similarly, there are very few comprehensive investigations and well-documented tests, which have been carried out for HSC. As a general rule, the faster the loading rate, the higher the strength due to shorter loading time and reduced creep (Neville 1995).

Table 1. Mix design for experimental program

Mix No.	Bulk density (kg/m³)	Water (kg/m³)	Cement (kg/m³)	Condensed silica fume (kg/m³)	Water / binder ratio	Coarse aggregates (kg/m³)	Sand (kg/m³)	Superplasticiser (kg/m³)	Slump (mm)
LR1	2530	145.85	504	56	0.260	1288.9	552.4	10.08	15
LR6	2580	146.82	504	56	0.262	1288.9	552.4	11.76	185
LR4	2520	146.82	504	56	0.262	1288.9	552.4	11.76	40
LR7	2550	146.82	504	56	0.262	1288.9	552.4	11.76	40
LR6R	2770	146.82	504	56	0.262	1288.9	552.4	11.76	40

Binder content is taken as total of cement + condensed silica fume

This fact is reinforced by other researchers (Ahmad and Shah 1987, Carrasquillo et al. 1981, Imam et al. 1995, Ting et al. 1992, Patnaikuni et al. 1993, Reinhardt 1987, Richardson 1991, Sparks and Menzies 1973). Shorter loading time translates into increase in stiffness (modulus of elasticity), ultimate strain (due to creep) and fracture energy while Poisson's ratio remains relatively unchanged. Tests reported by a number of investigators indicated that compressive strength of concrete specimens is sensitive to rate of loading over a broad range of rates. Evans (1958) found no significant change in compressive strength for normal strength concrete in the range of 6 MPa/min. to 6,000 MPa/min.; however at higher loading rates, an increase in rate of loading resulted in an increase in strength. Neville (1995) also pointed out that increasing the rate of application from 0.042 MPa/min. to 1.2 x 10^6 MPa/min. doubles the apparent strength of NSC. Within the practical range of loading rates of specimens (which is 8.4 MPa/min. to 20.4 MPa/min.), the measured strength varies between 97% and 103% of the strength measured at 12 MPa/min.

The rate of loading for the purpose of compression test as prescribed by ASTM C39-94 are between 8.4 to 20.4 MPa/min.); for BS1881 : Part 116-1983 are between 12 to 24 MPa/min. and finally for AS1012.8-1986 is 20 MPa/min. As indicated above by Neville, the lower limits of loading prescribed by both ASTM and BS may have no significant effect on strength. In fact, test results by Patnaikuni et al. (1993) on HSC at 5, 10 and 20 MPa/min. agreed well with this fact as the differences in strengths are about 1%.

2 EXPERIMENTAL PROGRAM

2.1 Materials and Mix Design

For this program, the authors made and tested 75 mm x 150 mm HSC cylinder specimens with target strength of approximately 100 MPa at 56 days. To achieve such a strength, admixtures were utilized in concrete manufacture.

Mineral admixture used was condensed silica fume in gray, powdered form. Two types of commercially available silica fumes were used; all mixes except for LR6 were prepared using the same silica fume. Although both silica fume are byproducts from ferro-silicon industry, they were supplied by different manufacturer and test results suggest that LR6 is 12.5% lower in strength than its LR6R counterpart even though they were based on the same mix design and tested at the same age (91 days).

The chemical admixture used was sulphonated napthalene-formaldehyde superplasticizer with solid content of 41% by weight. This superplasticiser conforms to Australian Standards AS1478-1992 type WR – Water Reducing (this is equivalent to type A and F admixtures in ASTM C494).

An AS3972-Type GB (General Blended) Portland Cement which has 50% granulated blast furnace slag was incorporated in all mixes. Locally available 7 mm coarse aggregates (angular shaped and smooth textured basalt) with saturated-surface-dry (SSD) density of 2.86 x 10^3 kg/m³ and river sand with SSD density of 2.62 x 10^3 kg/m³ and fineness modulus of 2.6 were also used. All the fine and coarse aggregates were oven dried to eliminate any moisture content that may be present. The complete mix design is given in Table 1.

The specimens were cast on steel cylinder moulds (75 mm inside diameter by 175 mm height) tightened onto a vibrating table that can hold 28 cylinders at one time. They were demoulded 24 hours after casting and cured in a temperature-controlled, lime-saturated water curing tank at 22 ± 2^0 C until the day of testing. It was proposed to test at least 2 specimens for the chosen loading rates of 20, 30, 40, 60, 80 and 100 MPa/min. at any particular testing age. Before testing, all cylinders were cut on one end using a diamond saw to provide a smooth and plane testing surface. As for the endcapping, a URCS (Unbonded Restraint Capping System) which consists of two 3 mm thick rubber endcaps with accompanying restraining ring bases was used. This system was developed by Ting et al. (1992) and has proven to give consistent compressive strength test results when used in testing of both NSC and HSC concrete cylinders.

For this experiment a computerized, 1000-kN MTS hydraulic testing machine with built-in datataker for load and deflection was used. The complete experimental results are given in Table 2.

Table 2. Experimental results of HSC for and the predicted f_c values under different loading rates

Mix no & testing age	Loading rate (MPa/min.)	f_c (MPa)	s (MPa)	Coefficient of variation (%)	$f_c/f_{c(20)}$ or α	Predicted f_c Using eq. (3) (Mpa)	Error (Mpa)	Error (%)	No. of specimens
LR1	20	91.4	3.1	3.3	1.00				3
28 days	30	91.3	2.6	2.9	1.00	92.29	0.99	1.08	4
	40	93.3	6.5	7.0	1.02	93.32	0.02	0.02	3
	60	98.0	4.5	4.6	1.07	94.77	3.23	3.29	4
	80	97.1	3.4	3.5	1.06	95.80	1.30	1.34	3
	100	98.1	2.5	2.6	1.07	96.60	1.50	1.53	4
LR6	20	86.4	2.5	2.9	1.00				4
92 days	30	86.6	7.0	8.1	1.00	87.24	0.64	0.74	2
	40	86.9	5.4	6.2	1.01	88.21	1.31	1.51	4
	60	87.7	3.4	3.9	1.01	89.59	1.89	2.15	4
	80	89.6	1.7	1.8	1.05	90.56	0.96	1.07	3
	100	93.0	1.9	2.1	1.04	91.32	1.68	1.81	3
LR4	20	95.3	5.0	5.3	1.00				4
28 days	30	96.4	1.7	1.8	1.01	96.23	0.17	0.18	3
	40	97.3	1.9	1.9	1.02	97.30	0.00	0.00	3
	60	99.2	2.7	2.8	1.04	98.82	0.38	0.39	4
	80	99.3	3.8	3.9	1.04	99.89	0.59	0.60	3
	100	98.0	2.6	2.7	1.03	100.72	2.72	2.78	4
LR7	20	96.8	1.4	1.5	1.00				4
56 days	30	98.7	2.8	2.9	1.02	97.74	0.96	0.97	4
	40	99.1	3.0	3.0	1.02	98.83	0.27	0.27	4
	60	100.7	2.5	2.4	1.04	100.37	0.33	0.33	4
	80	99.3	2.0	2.0	1.03	101.46	2.16	2.17	4
	100	105.4	1.6	1.5	1.09	102.31	3.09	2.93	4
LR6R	20	97.2	5.4	5.6	1.00				4
92 days	30	97.2	3.7	3.8	1.00	98.15	0.95	0.97	4
	40	96.3	2.9	3.0	0.99	99.24	2.94	3.05	2
	60	101.2	1.2	1.2	1.04	100.79	0.41	0.41	3
	80	105.4	2.3	2.2	1.08	101.88	3.52	3.34	4
	100	102.5	7.6	7.4	1.05	102.73	0.23	0.23	2

Note : $f_{c(x)}/f_{c(20)}$ denotes the ratio of a particular f_c at various loading rates (x can be either 30, 40, 60, 80 or 100 MPa/min.) to the f_c at 20 MPa/min, which is the standard loading rate prescribed by AS1012.8-1986.

2.2 Test Results

Table 2 shows that ratio of $f_{c(x)}$ to $f_{c(20)}$ – which is also called α or correction factor - vary somewhere between 0.99 to 1.09. In general, it appears that the faster the rate applied to HSC cylinders, the higher the strength. This is consistent with findings of other researchers mentioned in this paper. The test results also indicate that for rate up to 40 MPa/min. α is very close to 1.00 (in fact, the differences in observed strengths vary only as much as 2%).

A plot of compressive strengths versus loading rates is given as Figure 1. A line of best fit is calculated for each set of data (which contains strengths for rates of 20 and up to 100 MPa/min.) for every mix. Using logarithmic regression, the equation of the line of best fit and the correlation coefficient (r) for every data set was found and tabulated in Table 3. The correlation coefficients are very close to + 1.0 suggesting a very strong, positive correlation between the two variables. It can also be seen from Figure 1 that the rate of strength gain due to an increase in loading rate is roughly similar for all mixes tested, i.e. it is independent of mix proportions and

of testing ages. The equation of best line given in Table 3 is distinct for different mixtures and as such a more general equation which gives the relationship between f_c and $f_{c(20)}$ may be postulated.

Table 3. Correlation coefficents and equations of line of best fit for experimental results.

Mix No	Correlation coefficient, r	r^2	Equation of line of best fit
LR1	0.93	0.87	$y = 75.799 + 4.938 \ln x$
LR6	0.87	0.76	$y = 78.009 + 2.589 \ln x$
LR4	0.86	0.74	$y = 88.946 + 2.230 \ln x$
LR7	0.82	0.67	$y = 84.877 + 3.916 \ln x$
LR6R	0.84	0.70	$y = 80.634 + 4.999 \ln x$

A correlation factor (β) is introduced such that :

$$f_c = \beta \, f_{c(20)} \tag{1}$$

β is found by plotting the values of $f_c/f_{c(20)}$ as given in Table 2 against loading rates. This plot is shown in Figure 2 and by applying logarithmic regression analysis, the corresponding equation of

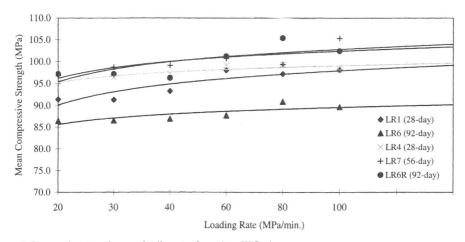

Figure 1. Compressive strength versus loading rates for various HSC mixes

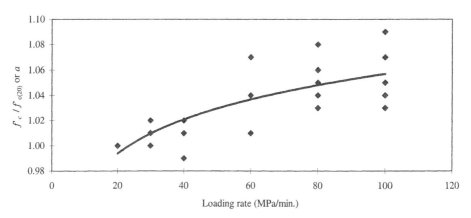

Figure 2. Plot of $f'_c / f_{c(20)}$ versus loading rate used to calculate the correlation factor β

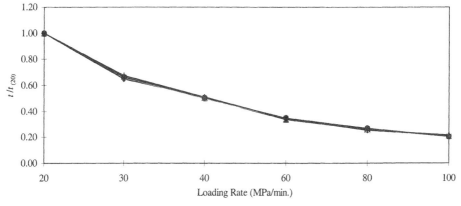

Figure 3. Time taken for crushing of specimen as a % of loading time at 20 MPa/min. versus loading rate

best line and r were calculated. Again, $r = 0.81$ means the value of β can be predicted quite accurately using the equation :

$$\beta = 0.0392 \ln x + 0.8764 \qquad (2)$$

More specifically, by combining Eqs. (1) & (2) the relationship between f'_c, $f'_{c(20)}$, β and x can be written as :

$$f'_{c(x)} = (0.0392 \ln x + 0.8764)\, f'_{c(20)} \qquad (3)$$

where : $f'_{c(x)}$ = characteristics compressive strength for loading rates limited to higher than 20 MPa/min. and up to 100 MPa/min.; $f'_{c(20)}$ = strength at 20 MPa/min. (this is obtained from testing at this particular load rate) and x = value of loading rate corresponding to $f'_{c(x)}$.

The standard estimate of error for this line is 0.0074 which means that a 0.74% error in β (correlation factor) is expected when eq. (3) is used to predict the strength of 75 x 150 mm concrete cylinders whose target strength is between 80 to 105 MPa. The range of loading rate covered is $100 \geq x > 20$ MPa/min., where x is any loading rate between this range.

In fact, eq. (3) is used to calculate the predicted strength at 30, 40, 60, 80 and 100 MPa/min. using the experimental strength at 20 MPa/min. as an input for every mix. The differences between these predicted strengths and experimental strengths define the errors arising from making this prediction; these errors are given in Table 2. Clearly, error associated with rates of up to 40 MPa/min. is very small (1.3 MPa or less) except for the huge discrepancy for Mix LR6 where error is found to be 2.94 MPa. This may be ascribed to the variations in concrete mix or to the limited number of specimens used in the test (2). Perhaps by increasing the number of samples to 4 or 5, a closer approximation to mean can be obtained and if this is so the error will be smaller.

2.3 *Loading Time*

The average loading time observed during the destructive testing of concrete specimens for each mix together with the ratios of average loading time at other rates to the average time at 20 MPa/min. - $t/t_{(20)}$ – are given in Table 4. The table indicates that loading at 30, 40, 60, 80 and 100 MPa/min. is roughly 1.5, 2, 2.9, 3.8 and 4.8 times faster, respectively, when compared to time taken for specimens crushed at 20 MPa/min. It should also be noted that the ($t/t_{(20)}$ vs. loading rate) curve for every mix coincides with one another and the relationship is non-linear as shown in Figure 3.

3 INVESTIGATION BY OTHER RESEARCHERS

Sparks and Menzies (1973) undertook an experiment with three NSC mixes made with three different types of coarse aggregates; Lytag being the least stiff and limestone being the stiffest with gravel in between. Rectangular concrete prisms with 102 x 102 x 203 mm dimensions and control cubes with 100 mm sides were made. It was reported that the 28-day compressive strength of the concrete cubes were of the order of 30 MPa for concrete made with gravel and limestone aggregates and approximately 20 MPa for concrete made with Lytag aggregates. All the prism and cube specimens were loaded to failure in the range of 0.06 to 600 MPa/min. Six or more loading rates were further chosen within the range such that the next one is always ten times higher than the previous (e.g. 0.06, 0.6, 6 MPa/min., etc.). They found that Lytag concrete showed decreasing strength with increasing loading rate below 1.8 MPa/min., while limestone and gravel concrete displayed increasing strength throughout the adopted range of loading rates. They concluded that the sensitivity of compressive strength to rate of loading was related to the stiffness of the aggregates. Concrete made with the stiffest aggregate (limestone) gave only a 4% increase in strength with a hundred-fold increase in loading rate whilst concrete made with the least stiff aggregate (Lytag) gave a 16% strength improvement (but only for rates above 0.03 MPa/s) for a similar increase in loading rate.

Test by Carrasquilo et al. (1981) have shown that the longer the duration of compressive test (ie. the lower rate of loading), the higher is the strain at failure but the lower is the resulting compressive strength. They also showed that higher strength concrete was not affected as much by different loading rates as compared to NSC. This seems to agree with experiments carried out by Reinhardt (1987) and Ahmad and Shah (1987) who concluded that the strength gain due to higher loading rate is less for higher strength concrete as compared to NSC (ie. the rate influence decreases with higher concrete strength). Tests carried out by Patnaikuni et al. (1993) confirms that testing time and loading rate is not linearly related. They made 75 x 150 HSC cylinders and tested all the specimens at 56 days. Three mixes were made; one mix was tested using sulphur capping and the other two using Unbonded Restraint Capping System. They employed loading rates of 5, 10, 20, 30, 40, 50, 80, 100, 200 and 400 MPa/min. and the study revealed that the strength measured in HSC specimen increases with increase in loading rate (using 20 MPa/min. as the standard rate), but the effect is not significant for loading rates between 5 to 100 MPa/min. as the difference in strength does

Table 4. Time taken to crush specimens at various loading rates

Loading rate (MPa/min.)	LR1 t (s)	$t / t_{(20)}$	LR6 t (s)	$t / t_{(20)}$	LR4 t (s)	$t / t_{(20)}$	LR7 t (s)	$t / t_{(20)}$	LR6R t (s)	$t / t_{(20)}$
20	280	1.00	258	1.00	284	1.00	289	1.00	291	1.00
30	182	0.65	171	0.66	187	0.66	193	0.68	193	0.66
40	142	0.51	130	0.5	141	0.50	147	0.51	147	0.50
60	97	0.35	87	0.34	98	0.35	101	0.35	101	0.35
80	72	0.26	67	0.26	72	0.25	78	0.25	78	0.27
100	58	0.21	53	0.21	58	0.21	59	0.22	59	0.20

not exceed 5%. Work by the authors suggest that the difference can be up to 9% for HSC. They also studied the effect of loading rate on different specimen sizes using 75 x 150 and 100 x 200 HSC cylinders in excess of 150 MPa. The 75 x 150 mm specimens gave an increase in strength of 2.5% and 6.1% when loaded at 30 and 40 MPa/min., respectively, as compared to the strength at 20 MPa/min. On the other hand, 100 x 200 mm gave only an increase of 1.6% and 4.5% under similar condition. However, there was no difference in strength for both sizes at 20 MPa/min. This exemplifies the fact that specimen size has a bearing on its measured strength.

4 CONCLUSIONS

Although strength measured at different loading rates is not equivalent, this study reveals that rate as high as 30 or 40 MPa/min. is acceptable to be used in testing HSC because the difference in strength compared to that obtained when loading at 20 MPa/min. is only marginal with the maximum being only 2%. It also shows that the rate of gain in strength at higher loading rates is independent of mix proportions and specimen strength.

A correlation coefficient as a function of strength obtained using 20 MPa/min. as the prescribed loading rate by AS1012.8 is suggested to assist the strength prediction of HSC specimens when faster loading rates are used. Their relationship is given as:

$f'_{c(x)} = (0.0392 \ln x + 0.8764) \, f'_{c(20)}$

Comparison between experimental and theoretical f'_c made using the above equation gives errors of up to 3% for loading rates of up to 40 MPa/min. Caution must be exercised when using the equation, because the correlation coefficient is derived using logarithmic regression analysis based on 75 x 150 HSC cylinder specimens with strength between 80 to 105 MPa and the loading rates adopted were between 20 to 100 MPa/min. Whether this equation still holds for loading rates and strengths outside the range used in this study is subject to the validation of more testing data as they become available. Specimen size also has a big influence on the strength and the above equation certainly needs modification if other cylinder sizes are used.

REFERENCES

Abrams, P.A. 1917. *Proc. ASTM.* 17:II:364.

Ahmad, S.H. & Shah, S.P. 1987. High strength concrete - A review. *Proc. of the First Sym. on Utilisation of High strength concrete.* Symposium Stavanger.

AS1012.8-1986. *Methods for testing concrete Part 8 - Method for making and curing concrete compression, indirect tensile and flexural test specimens in the laboratory or in the field.* NSW : Standards Association of Australia.

Carrasquilo, R.L., Nilson, A.H. & Slate, F.O. 1981. Properties of high strength concrete subject to short term loads. *ACI Materials journal.* 78:3:171-177.

Evans, R.H. 1958. Effect of rate of loading on some mechanical Properties of concrete. Walton, W.H. (ed). *Mechanical properties of non-metallic brittle materials. Conf. Proc.* London : Butterworths scientific publication:175-192.

FIP-CEB 1990. *High strength concrete - State of the art report.* London : Federation Internationale de la Precontrainte.

Imam, I., Vandewalle, L. & Mortelmans, F. 1995. Are current concrete strength tests suitable for high strength concrete ? *Materials and structures.* 28:384-391.

Jones, P.G. and Richart, F.E. 1936. *Proc ASTM.* 36:II:380.

Neville, A. M. 1995. *Properties of concrete (4th ed.).* London : Longman Group Limited.

Patnaikuni, I., Johansons, H.A. & Ting, E.S.K. 1993. Suitable loading rate for very high strength concrete. *Proc. of the 4th Int. Conf. on structural failure, durability and retrofitting.* Singapore 14-15 July:704-709.

Reinhardt, H.W. 1987. Simple relations for the strain rate influence of concrete. *Annual journal on concrete and concrete structures.* Darmstadt Concrete.

Richardson, D.N. 1991. Review of variables that influence measured concrete compressive strength. *Journals of materials in civil engineering ASCE.* 3:2:95-112.

Sparks, P.R. and Menzies, J.B. 1973. The effect of rate of loading on plain concrete. *CP 23/73, Conf. Proc.* 143-153.

Ting, E.S.K., Patnaikuni, I., Johansons, H.A. & Pendyala, R.S 1992. Compressive strength testing of very high strength concrete. *Proc. of the 17th conference on our world in concrete and structures.* Singapore 25-27 August:217-226.

15 Material enhancement

Creative Systems in Structural and Construction Engineering, Singh (ed.) © 2001 Balkema, Rotterdam, ISBN 90 5809 161 9

Strengthening of RC beams with CFRP sheets or fabric

P. Alagusundaramoorthy & I. E. Harik
Kentucky Transportation Center, University of Kentucky, Lexington, Ky., USA

S. Morton
Ohio Department of Transportation, Columbus, Ohio, USA

R. Thompson
Fiber Reinforced Systems, Columbus, Ohio, USA

ABSTRACT: The objective of this investigation is to study the effectiveness of carbon fiber reinforced polymer (CFRP) sheets and fabric in the flexural and shear strength of concrete beams. Four-point bending flexural tests are conducted up to failure on twelve concrete beams strengthened with different layers of externally bonded CFRP sheets and fabric. Shear tests are conducted up to failure on twelve concrete beams wrapped with four different configurations of CFRP fabric. Comparisons are made between the test results and the predictions based on analytical solutions, and the uncertainty parameters are calculated. The flexural strength is increased up to 58% on concrete beams bonded with anchored CFRP sheets. The shear strength is increased up to 33% on beams bonded with CFRP fabric at an angle of $\pm 45°$ to the longitudinal axis of the beam.

1 INTRODUCTION

The technique of bonding steel plates using epoxy adhesives is recognized as an effective and convenient method for repair and rehabilitation of existing reinforced concrete structures. However the problems associated with the steel corrosion, handling due to excessive size and weight, undesirable formation of welds, partial composite action with the surface concrete, and de-bonding lead to the need for alternative materials, and further research in this field. The high strength to weight ratio, resistance to electro-chemical corrosion, larger creep strain, good fatigue strength, potential for decreased installation costs and repairs due to lower weight in comparison with steel, and non-magnetic and non-metallic properties of carbon fiber reinforced polymer (CFRP) composites offer a viable alternative to bonding of steel plates. The emergence of high strength epoxies has also enhanced the feasibility of using CFRP sheets and fabric for repair and rehabilitation. The flexural capacity of both prestressed and non-stressed members may be increased through the external bonding of CFRP sheets and fabric. The shear strength of concrete beams can be improved by wrapping thin CFRP fabric in different configurations in conjunction with resins. Any CFRP system considered should have sufficient test data demonstrating the adequate performance of the entire system in similar applications including its method of installation.

Ross et al. (1999), Malek et al. (1998), and An et al. (1991) presented analytical procedures to calculate the flexural strength of reinforced concrete (RC) beams bonded with FRP plates. Saadatmanesh & Eshani (1991) reported test results of RC beams strengthened with GFRP plates. Spadea et al. (1998) conducted experiments on RC beams bonded with CFRP sheets. GangaRao & Vijay (1998) studied both experimentally and theoretically the behavior of RC beams wrapped with CFRP fabric. Khalifa et al. (1998) and Gendron et al. (1999) presented analytical procedures to calculate the shear strength of RC beams strengthened with externally bonded FRP sheets. Malek & Saadatmanesh (1998) studied the effect of FRP plates on the crack inclination angle and the shear capacity of concrete beams. Chaallal et al. (1998) reported the effect of epoxy-bonded unidirectional carbon fiber plastic strips on the shear strength of concrete beams. Triantafillou (1998) presented analytical and experimental results on the shear strength of concrete beams bonded with CFRP fabric. Sonobe et al. (1997) presented guidelines for the design reinforced concrete building structures using FRP.

2 OBJECTIVE

The main objective of this investigation is to study the effectiveness of CFRP sheets and fabric supplied by Fiber Reinforced Systems (FRS) on the flexural and shear strength of concrete beams. The objective is achieved by conducting the following tasks: (i) Flexural testing of concrete beams bonded with different layers of CFRP sheets and fabric; (ii) Shear testing of concrete beams wrapped with different configurations of CFRP fabric; (iii) Calculating the effect of different configurations of CFRP sheets and fabric on the flexural and shear strength of concrete beams; (iv) Evaluating the failure modes; and (v) Comparing the experimental results with analytical calculations, and predicting the uncertainty parameters. A set of conclusions is also drawn based on the experimental and analytical study carried out under this investigation.

3 FLEXURAL TESTS ON CONCRETE BEAMS BONDED WITH CFRP SHEETS/FABRIC

3.1 Test specimens

Four-point bending tests are conducted up to failure on two concrete control beams and twelve concrete beams strengthened with externally bonded CFRP sheets and fabric on the tension face. The pultruded CFRP sheet was provided by Fiber Reinforced Systems (FRS) in Columbus, Ohio. The CFRP fabric is a stitched unidirectional sheet of 0.18 mm thick. The length, breadth and depth (ℓ x b x d) of all concrete beams is kept as 4,800 mm x 230 mm x 380 mm. Each concrete beam is reinforced with two # 8 steel bars for tension and two # 3 steel bars for compression along with # 3 bars at a spacing of 150 mm center-to-center for shear reinforcement. The flexural span of all beams is kept as 4,575 mm. Concrete with compressive strength of 31 MPa and steel with yield strength of 414 MPa are used.

The concrete control beams are designated as CB1 and CB2 respectively. Five beams bonded with different layers of CFRP sheets (CB3-2S, CB4-2S, CB5-3S, CB6-3S and CB7-1S), three beams bonded with different layers of anchored CFRP sheets (CB8-1SB, CB9-1SB and CB10-2SB), and four beams bonded with different layers of CFRP fabric (CB11-1F, CB12-1F, CB13-2F and CB14-2F) are fabricated. Four numbers of bolts are used on each side of the beam to anchor the CFRP sheet in the beams CB8-1SB and CB9-1SB, and eight numbers of bolts are used on each side of the beam CB10-2SB. A two-component Magnolia epoxy primer, and Magnolia structural epoxy paste are used to bond the CFRP sheets on the concrete beams. A two-component Magnolia epoxy primer and Magnolia saturating epoxy are used to bond the CFRP fabric on the concrete beams. The details of the beams fabricated for flexural testing are presented in Table 1.

3.2 Test details

The test setup as shown in Figure 1 is used. The load is applied using two hydraulic jacks of 1,800 kN capacity. The load is transmitted through a rectangular plate of size 560 mm x 230 mm x 50 mm to the beam in order to represent the AASHTO HS25 standard truck wheel load. Hydraulic jacks having 184 mm ram and 150 mm stroke are used for testing. The top of the ram is provided with a spherical cap so that if any tilting of the plate occurs while loading, the spherical cap adjusts in such a way that only a perpendicular load is applied to the beam. Load cell is used to measure the load applied by the jacks. A rubber pad with a thickness of 13 mm is placed between the beam and the steel plate in order to minimize the abrasion between the steel plate and the beam while loading.

Electrical resistance disposable strain gages 6.35 mm long manufactured by Vishay, Measurements Group, are pasted on CFRP sheets and fabric. Reusable strain gages 76 mm long manufactured by Bridge Diagnostics are fixed on the concrete side of the beam in order to measure the compressive strains. Out-of-plane deflections are measured using Linear Variable Deflection Transducers (LVDT) manufactured by Sensotec, Ohio. The beams are loaded according to the following sequences: (i) load cycle from zero to 12 kips and back to zero. The cycle is repeated five times to study the responses of the beams under cyclic loading; and (ii) load from zero to failure. The ultimate failure load is recorded, when load shedding is initiated.

Table 1. Specimens for flexural test

Specimen[*]	Size of CFRP Sheet/Fabric (mm)		
	Length	Width	Thickness
CB1	-	-	-
CB2	-	-	-
CB3-2S	4,270	76	1.40
CB4-2S	4,270	76	1.40
CB5-3S	4,270	76	1.40
CB6-3S	4,270	76	1.40
CB7-1S	4,270	102	4.78
CB8-1SB	4,270	102	4.78
CB9-1SB	4,270	102	4.78
CB10-2SB	4,270	102	4.78
CB11-1F	4,370	203	0.18
CB12-1F	4,370	203	0.18
CB13-2F	4,370	203	0.18
CB14-2F	4,370	203	0.18

[*] # S - Number of CFRP sheets
SB - Number of anchored CFRP sheets
F - Number of layers of CFRP fabric

3.3 Results and discussion

The load/deflection and load/strain curves for cyclic loading and loading to failure are plotted for all beams tested. The failure loads are obtained from the load/deflection curves using the "top of the knee method". The failure load according to this method is essentially the load corresponding to the top of the knee of the load/deflection curve. The failure loads of the beams with their corresponding failure modes are presented in Table 2. The average value of failure loads of the concrete control beams CB1 and CB2 is calculated and used as a baseline value for comparison.

Table 2. Effect of CFRP sheets and fabric on the failure load

Specimen	Failure Load (kN)	Increase in strength[*] (%)	Mode of failure
CB1	197	-	Yielding of steel
CB2	190	-	Yielding of steel
CB3-2S	263	36	Delamination of sheets
CB4-2S	260	35	Delamination of sheets
CB5-3S	287	49	Delamination of sheets
CB6-3S	275	42	Delamination of sheets
CB7-1S	256	33	Delamination of sheet
CB8-1SB	273	41	Delamination of sheets
CB9-1SB	249	29	Delamination of sheets
CB10-2SB	306	58	Delamination of sheets
CB11-1F	219	13	Rupture of fabric
CB12-1F	223	15	Rupture of fabric
CB13-2F	263	36	Delamination of fabric
CB14-2F	270	40	Delamination of fabric

[*] Increase in strength = $\dfrac{\text{Failure load} - \text{Average of CB1 \& CB2}}{\text{Average of CB1 \& CB2}}$

The flexural strength increases from 35% to 36% on beams bonded with two CFRP sheets of 76 mm width and 1.40 mm thick ($b_s/t_s = 109$), 42% to 49% on beams bonded with three CFRP sheets of 76 mm width and 1.40 mm thick ($b_s/t_s = 163$), 29% to 41% on beams bonded with one anchored CFRP sheet of 102 mm width and 4.78 mm thick ($b_s/t_s = 21$), 58 % on beam bonded with two anchored CFRP sheets of 102 mm width and 4.78 mm thick ($b_s/t_s = 43$), 13 % to 15% on beams bonded with one CFRP fabric of 203 mm width and 0.18 mm thick ($b_f/t_f = 1128$), and 36 to 40% on beams bonded with two layers of CFRP fabric of 203 mm width and 0.18 mm thick ($b_f/t_f = 564$).

The anchoring of CFRP sheet near the supports increases the flexural strength of beams by about 8% compared to the same beam with CFRP sheet without bolting at the supports. The anchored CFRP sheet is in contact with the beams even after failure and not able to peel off easily compared to the same beam without anchoring of CFRP sheet near the supports. The failure pattern of beams bonded with two layers of CFRP fabric indicates that the CFRP fabric is behaving like sheets.

Simply adding the CFRP sheets/fabric to the beam cannot uniformly increase the strength of concrete beams. The plate slenderness ratio of CFRP sheets (b_s/t_s) and b_f/t_f of CFRP fabric are to be optimized for maximum strength.

Figure 1. Test setup for four point bending

3.4 Analytical study

An analytical procedure based on the compatibility of deformations and equilibrium of forces is developed to predict the flexural behavior of concrete beams strengthened with CFRP sheets/fabric (Alagusundaramoorthy et al. 2000). The failure load of all tested beams is calculated using the developed analytical procedure and compared with the experimental results. The uncertainty parameters such as the mean value of the modeling parameter (Experimental value/Analytical value) (x), standard deviation (σ) and co-efficient of variation (CV) are calculated. The mean value varies from 1.26 to 1.39, standard deviation varies from 0.024 to 0.062 and coefficient of variation varies from 0.019 to 0.045 for failure loads. The comparison of uncertainty parameters indicates that the developed analytical procedure underestimates the failure load.

4 SHEAR TESTS ON CONCRETE BEAMS WRAPPED WITH CFRP FABRIC

4.1 Test Specimens

Shear tests are conducted up to failure on two concrete control beams and twelve concrete beams wrapped with four different configurations of CFRP fabric. The length, breadth and depth ($\ell \times b \times d$) of all concrete beams are kept as 2,135 mm × 230 mm × 380 mm and each concrete beam is reinforced with two # 8 steel bars for tension and two # 3 steel bars for compression along with # 3 bars at a spacing of 300 mm center-to-center for shear reinforcement. The shear span of all beams is kept as 1,830 mm. Concrete with compressive strength of 31 MPa and steel with yield strength of 414 MPa are used.

Four beams wrapped with one layer of CFRP fabric inclined at an angle of 90° to the longitudinal axis of the beam (SB3-90, SB4-90, SB5-90 and SB6-90) and two beams wrapped with one layer of

451

CFRP fabric inclined at an angle of 90° with an additional layer of CFRP fabric on both sides of the web inclined at an angle of 0° (SB7-90-0 and SB8-90-0) are fabricated. Four beams wrapped with one layer of CFRP fabric inclined at an angle of ± 45° (SB9-45, SB10-45, SB11-45 and SB12-45) and two beams wrapped with one layer of CFRP fabric inclined at an angle of ± 45° with an additional layer of CFRP fabric on both sides of the web inclined at an angle of 0° (SB13-45-0 and SB14-45-0) are also fabricated for testing. A two-component Magnolia epoxy primer and Magnolia saturating epoxy are used to bond the CFRP fabric on the concrete beams The details of the beams fabricated for shear testing are given in Table 3.

Table 3. Specimens for shear test

Specimen	Strengthening details		
	Number of layers of CFRP fabric		
	Inclined at 0°	Inclined at 90°	Inclined at ±45°
SB1	-	-	-
SB2	-	-	-
SB3-90	-	1	-
SB4-90	-	1	-
SB5-90	-	1	-
SB6-90	-	1	-
SB7-90-0	1	1	-
SB8-90-0	1	1	-
SB9-45	-	-	1
SB10-45	-	-	1
SB11-45	-	-	1
SB12-45	-	-	1
SB13-45-0	1	-	1
SB14-45-0	1	-	1

4.2 Test details

The test setup for shear testing is shown in Figure 2. The details about the hydraulic jack and instrumentation used are explained in sec. 3.2. The beams are loaded according to the following sequences: (i) load cycle from zero to 12 kips and back to zero. The cycle is repeated five times to study the responses of the beams under cyclic loading; and (ii) load from zero to failure. The ultimate failure load is recorded, when load shedding is initiated.

4.3 Results and discussion

The load/centerline deflection curves for the beams SB3-90, SB4-90, SB5-90, SB6-90 and SB8-90-0, and SB9-45, SB10-45, SB11-45, SB12-45, SB13-45-0 and SB14-45-0 are plotted (Alagusundaramoorthy et al. 2000), and are used to obtain the failure loads. The failure loads are obtained from the load/deflection curves using the "top of the knee method". The failure load according to this method is essentially the load corresponding to the top of the knee of the load/deflection curve. The failure loads

of the beams with their corresponding failure modes are presented in Table 4. The average value of failure loads of the concrete control beams SB1 and SB2 is calculated and used as a baseline value for comparison.

The failure load of all beams wrapped with CFRP fabric is compared with the baseline value and the increase in strength is calculated (Table 4). The average strength is increased up to 14% on beams SB3-90, SB4-90, SB5-90 and SB6-90, 18% on beams SB7-90 and SB8-90, 33% on beams SB9-45, SB10-45, SB11-45 and SB12-45, and 33% on beams SB13-45-0 and SB14-45-0. Even though the beams SB13-45-0 and SB14-45-0 bonded with additional layers of CFRP fabric compared to beams SB9-45, SB10-45, SB11-45 and SB12-45 the average value of increase in strength is found to be same in both sets of beams.

Figure 2. Test setup for shear testing

4.4 Analytical study

An analytical procedure based on the formulations (Khalifa et al. 1998, and Malek & Saadatmanesh 1998) is developed and is used to calculate the shear strength of beams wrapped with CFRP fabric. The analytical calculations are compared with the experimental results and the uncertainty parameters are calculated (Alagusundaramoorthy et al. 2000). The mean value varies from 1.12 to 1.33, standard deviation varies from 0.04 to 0.07 and coefficient of variation varies from 0.03 to 0.06 for failure loads. The comparison of uncertainty parameters indicates that the developed analytical procedure underestimates the failure load.

5 CONCLUSIONS

The following conclusions are drawn based on the experimental and analytical calculations carried out under this investigation.

1. The flexural strength is increased up to 58% on concrete beams bonded with anchored CFRP sheets having plate slenderness ratio of 42.

2. The shear strength is increased up to 33% on concrete beams wrapped with CFRP fabric inclined at ± 45° to the longitudinal axis of the beam.

3. The uncertainty parameters indicate that the developed analytical procedures can be used to calculate the flexural and shear strength of concrete beams strengthened/retrofitted with CFRP sheets and fabric.

Table 4. Effect of CFRP fabric on the failure load

Specimen	Failure Load (kN)	Increase in strength[*] (%)	Mode of failure
SB1	378	-	Shear - compression
SB2	416	-	Shear - compression
SB3-90	463	17	Rupture of fabric followed by shear
SB4-90	434	9	Rupture of fabric followed by shear
SB5-90	468	18	Rupture of fabric followed by shear
SB6-90	436	10	Rupture of fabric followed by shear
SB7-90-0	482	22	Rupture of fabric followed by shear
SB8-90-0	453	14	Rupture of fabric followed by shear
SB9-45	528	33	Delamination of fabric followed by shear
SB10-45	546	38	Delamination of fabric followed by shear
SB11-45	509	28	Delamination of fabric followed by shear
SB12-45	531	34	Delamination of fabric followed by shear
SB13-45-0	512	29	Delamination of fabric followed by shear
SB14-45-0	542	37	Delamination of fabric followed by shear

[*] Increase in strength = $\dfrac{\text{Failure load} - \text{Average of SB1 \& SB2}}{\text{Average of SB1 \& SB2}}$

REFERENCES

Ahmed Khalifa, William Gold, J., Antonio Nanni, & Abdel Aziz. 1998. Contribution of externally bonded FRP to shear capacity of RC Flexural Members. *ASCE Journal of Composites for Construction* 2(4): 195-203.

Alagusundaramoorthy, P., Harik I.E. & Choo Ching Chiaw. 2000. Flexural testing of beams strengthened with CFRP sheets and fabric. *ASCE Journal of Composites for Construction* (in review).

Alagusundaramoorthy, P., Harik I.E. & Choo Ching Chiaw. 2000. Shear strength of concrete beams wrapped with CFRP fabric. *ACI Structural Journal* (in review).

Allen Ross, C., David Jerome, M., Joseph Tedesco, W. & Mary Hughes, L. 1999. Strengthening of reinforced concrete beams with externally bonded composite laminates. *ACI Structural Journal* 96(2): 212-220.

Amir Malek, M., Hamid Saadatmanesh & Mohammad Ehsani, R. 1998. Prediction of failure load of R/C beams strengthened with FRP plate due to stress concentration at the plate end. *ACI Structural Journal* 95(2): 142-152.

Amir Malek, M. & Hamid Saadatmanesh. 1998. Ultimate shear capacity of reinforced concrete beams strengthened with web-bonded fiber-reinforced plastic plates. *ACI Structural Journal* 95(4): 391-399.

Chaallal, O., Nollet, M.J. & Perraton, D. 1998. Shear strengthening of RC beams by externally bonded side CFRP strips. ASCE *Journal of Composites for Construction* 2(2): 111-114.

GangaRao V.H.S. & Vijay, P.V. 1998. Bending behavior of concrete beams wrapped with carbon fabric. *ASCE Journal of Structural Engineering* 124(1): 3-10.

Gendron, G., Picard, A. & Guerin, M.C. 1999. A theoretical study on shear strengthening of reinforced concrete beams using composite plates. *Composite Structures* 45: 303-309.

Hamid Saadatmanesh & Mohammad Ehsani, R. 1991. RC beams strengthened with GFRP plates: I Experimental study. *ASCE Journal of Structural Engineering.* 117(11): 3417-3433.

Spadea,G., Bencardino, F. & Swamy, R.N. 1998. Structural behavior of composite RC beams with externally bonded CFRP. *ASCE Journal of Composites for Construction* 2(3): 132-137.

Thanasis Triantafillou, C. 1998. Shear strengthening of reinforced concrete beams using epoxy-bonded FRP composites. *ACI Structural Journal* 95(2): 107-115.

Wei An, Hamid Saadatmanesh & Mohammad Ehsani, R. 1991. RC beams strengthened with FRP plates: II Analysis and parametric study. *ASCE Journal of Structural Engineering* 117(11): 3434-3455.

Yasuhisa Sonobe et al. 1997. Design guidelines of FRP reinforced concrete building structures. *ASCE Journal of Composites for Construction* 1(3): 90-115.

Creative Systems in Structural and Construction Engineering, Singh (ed.) © 2001 Balkema, Rotterdam, ISBN 90 5809 161 9

Service life of bridge decks constructed with epoxy coated rebars

Fouad S. Fanous
Department of Civil Engineering, Iowa State University, Iowa, USA

Han-Cheng Wu
Rocky Mountain Prestress, Denver, Colo., USA

ABSTRACT: The presence of cracks in bridge decks has raised some concerns among bridge and maintenance engineers in the state of Iowa. In this paper, the impact of deck cracking on the service life of a bridge deck was investigated. This was accomplished by examining several cores that were collected from cracked and uncracked areas of bridge decks. No signs of corrosion were observed on the rebars collected from uncracked locations. In addition, no delaminations or spalls were found on the decks where bars at cracked location exhibit some signs of corrosion. For a corrosion threshold ranges from 3.6 lb/yd^3 to 7.2 lb/yd^3, the predicted service life for Iowa bridge decks considering corrosion of ECR was over 50 years.

1 INTRODUCTION

To prevent the reinforcing steel from corrosion, Epoxy-Coated Rebar (ECR) was first used in the construction of a four-span bridge deck over Schuylkill River in Pennsylvania in 1973. Since then, ECR have been the most widely used corrosion protection method in bridge components in the United States.

Although the performance of ECR in corrosive environments is thought to be superior to typical black reinforcing bars, presence of cracks in bridge decks have caused some concern as to the condition of the ECR in these areas. This paper summarizes the results of an investigation the objectives of which were to determine the impact of bridge deck cracking on deck durability and to approximately estimate the remaining functional service life of a bridge deck. These objectives were accomplished by collecting core samples from 80 bridges, calculating the chloride content in these cores, develop a relationship for chloride infiltration through the deck, examining the condition of several rebar samples, and develop a rebar rating-age relationship, and estimating a bridge deck service life.

2 CORROSION PROCESS

Corrosion of reinforcing steel in concrete can be modeled in a two-stage process. The first stage is known as initiation or incubation period in which chloride ion transport to the rebar level. In this stage the reinforcing steel experiences negligible corrosion. The time, T_1, required so the chloride concentration to reach the threshold value at the rebar level can be determined by the diffusion process of chloride ion through concrete following Fick's second law (Weyers, et al, 1993). In the second stage, which is known as active and deterioration stage, corrosion of reinforcing steel occurs and propagates resulting in a noticeable change in reinforcing bar volume. This could induce cracking, delamination and spalling of the surrounding concrete. The length of the second stage, T_2, depends on how fast the corroded reinforcing bars deteriorate resulting in an observable distress. Although it is not an easy task to predict the length of the second stage, a deck will eventually reach a condition at which some types of maintenance activities must be taken.

3 ESTIMATING BRIDGE DECKS SERVICE LIFE

For a bridge deck the end of functional service life is reached when severe deterioration occurs. (Weyers et al. 1994) conducted an intensive opinion survey among 60 bridge engineers to quantify the end of functional service life. The study concluded that, it is likely that the end of functional service life for concrete bridge decks to be reached when the percentage of the worst traffic lane surface area that is spalled, delaminated, and patched ranges approximately from 9% to 14%." Also (Weyers, et al. 1994) documented that based on current local practices, it is likely that the end of functional service life for concrete decks is reached when the percentage of the whole deck surface area that is spalled, delaminated, and patched ranges from 5.8% to 10.0%.

Among the models that are available to estimate the service life of a bridge deck is the difussion-spalling model (Weyers et al. 1994). This method utilizes the concepts of a chloride diffusion period plus a deterioration period to determine when to rehabilitate a bridge deck. The length of the diffusion period can easily be calculated using Fiks law (Weyers, et al, 1993). This can be accomplished if the surface chloride exposure, corrosion threshold, the mean and the standard deviation of the cover depth and the chloride diffusion constant were known. These variables can be different for bridges from state to state or even among bridges in one state.

The diffusion-spalling model (Weyers et al. 1994) is often used to assess corrosion of uncoated reinforcing bars (Weyers et al. 1993). In this paper, this was assumed to be applicable to estimate the service life of bridge deck constructed using ECR in conjunction with a higher chloride threshold than that used for uncoated steel bars.

4 CORROSION THRESHOLD

For black bars, the corrosion threshold at the reinforcing steel level was determined to be 0.2% of weight of the cement content of concrete (Clear 1975,1976). Cady and Weyers (1983) estimated the corrosion threshold for unprotected reinforcement to be 1.2 lb/yd^3 (0.73 kg/m^3) of concrete based on 6½ sacks of cement per cubic yard of concrete. However, as it is believed that the use of ECR will delay the time required to initiate corrosion. As a result, the corrosive threshold should be higher than that for the bare steel bar. Sagues et. al, suggested the corrosive threshold for ECR of 3.6 lb/yd^3 (2.19 kg/m^3 (Sagues et al, 1995).

5 CHLORIDE IONS INGRESSION IN CONCRETE

Fick's second law (Weyers et al, 1993) to determine the length of the initiation stage, i.e., time T_1, it takes chloride ions to migrate through a bridge deck to reach the top reinforcing steel in an isotropic medium can be expressed as:

$$C_{(x,t)} = C_o \{1 - erf[\frac{x}{2\sqrt{(D_{ac}t)}}]\} \qquad (1)$$

where, $C_{(x,t)}$ is the measured chloride concentration at a desired depth x , C_o is the surface concentration measured at 0.5 in below the deck surface, lbs/yd^3 , t

is the time in years and D_{ac} is the diffusion constant, in^2/yr. The erf (y) function is the integral of the Gaussian distribution function from 0 to y.

6 CONCRETE COVER DEPTH

A sufficient cover depth can effectively provide corrosion protection for reinforcement. As reinforcing steel cover depth increases, the corrosion protection increases and hence the initiating time, T_1 increases. However, to calculate a realistic time, T_1, for chloride ion to reach the reinforcing bar level, one must make full use of the end of functional service life. Weyers et al. (1994) recommended using an average cumulative damage of 11.5%, i.e., the average of 9% to 14%, damages in the worst traffic lane for a bridge deck as the end of functional service. There is a possibility that an 11.5% of the top reinforcing steel is placed at a depth less than the mean cover depth. In this case, the depth, x, used in Equation 1 needs to be calculated as:

$$x = x_m + \alpha\sigma \qquad (2)$$

where; x_m is the mean reinforcing steel cover depth, in.; α is the a standard normal cumulative distribution of 11.5%, and σ is the standard deviation of the cover depth.

7 EPOXY-COATED REBAR CONDITION RATING

The surface condition of ECR extracted from the bridge decks can be used to indicate the effectiveness of ECR in bridge decks. Thus visual inspection on the ECR surface provides some type of an assessment of the condition and hence performance of the ECR. Similar rating scale to that used by the Pennsylvania Department of Transportation (Sohanghpurwala et al, 1997) was used in this study to rate the ECR in Iowa bridges.

Although the time required for a rebar to deteriorate from one rating to another is not explicitly stated, one can estimate the deterioration of ECR. This can be accomplished if a large population of ECRover a wide range of time is collected and rated in accordance with the rating scales listed in Table 1.

8 SELECTED BRIDGES FOR CHLORIDE ANALYSIS

Figure 1 illustrates the locations of the eighty bridge decks constructed with ECR that were selected to evaluate the surface chloride content, C$_o$, and the chloride diffusion constant, D$_{ac}$, which are needed to calculate the length of the initiation stage, T_1. Factors such as geographic location, average daily traffic, bridge structures type, number of spans were among those considered when selecting these bridges. From each deck, two core samples from cracked locations and two from uncracked locations were obtained.

9 CHLORIDE CONTENT ANALYSIS

Four concrete powder samples were collected from each core for chloride content analysis. The locations of these samples were at 1/2 in. below the surface, midway between the first sample and the rebar level, at the rebar level and, and at about 1/2 in. below the rebar level. Powder samples from cracked cores were drilled from the uncracked quadrant to avoid split the cores into half. Drilling was penetrated through the crack so that the sample contained powders collected from the cracked surface. Additional powder samples were collected from each bridge deck. The chloride concentration was tested in the material laboratory using PHILIPS PW 2404 x-ray fluorescence spectrometer which is a device used to determine and identify the concentration of element contained in a solid, powdered, and liquid sample (Schlorholtz, 1998).

10 SURFACE CHLORIDE AND DIFFUSION CONSTANTS

When reviewing the collected data, it was noticed that some data appeared to be unrealistic. For instance, the chloride analysis showed that, in some cases, higher percentage of chloride existed at deeper locations than shallower locations. These results were filtered out and the remaining results were utilized to determine C_o and D_c needed to calculate the time for the corrosion initiation stage T_1. The computational process involved the utilization of Matlab software to perform the iterative solution. Approximate ranges of C_o and D_c were selected and an iterative solution was carried

out for several combinations of these two variables. The solution was terminated when the minimum of the sum of squared errors between the predicted and measured values was reached.

Using an individual bridge data yielded the best fitting to equation 1 listed above. However, to generalize the process of calculating the service life of Iowa bridge decks, the curve fitting utilizing data from allbridges after filtering the unreliable data was employed. This process yielded a surface chloride content, C_o, and diffusion constants, D_c, for the bridges included in this study of 14 lb/yd^3 and 0.05, respectively.

11 CONDITION OF ECR VERSUS BRIDGE DECK AGE

According to the Federal Highway Administration (FHWA, 1988), bridges are inspected every two years. Thus, it was reasonable to subgroup the rebar samples from the bridge decks according to age in two-year intervals. Rebar samples were also obtained and rated utilizing the scale listed in Table 1.

All samples collected from uncracked locations appeared to have no corrosion and were given rating value of 5 or 4. In contrast, 5%, 11% and 3% of the reinforcing bar samples obtained from cracked locations were evaluated at rating of 3, 2, and 1 respectively, indicating some degree of corrosion of these rebar samples.

Since there is a range of possible values of reinforcing bar samples that can be rated at a specific rating condition, one would naturally be interested in some central value such as the average. However, there are different probabilities that different numbers of rebars in each time interval can be associated with different rating condition. Therefore, a weighted average (Ang, 1975), i.e., the expected value of the rating within each interval would be more representative rather than just using a straight average value.

Table 1 Reinforcing Steel Rating Description

Rating	Description
5	No evidence of corrosion.
4	A number of small, countable corrosion.
3	Corrosion area less than 20% of total ECR surface area.
2	Corrosion area between 20% to 60% of total ECR surface area.
1	Corrosion area greater than 60% of total ECR surface area.

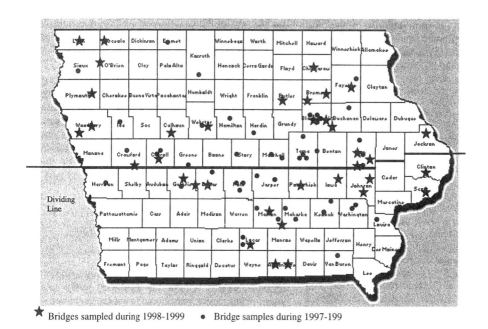

★ Bridges sampled during 1998-1999   ● Bridge samples during 1997-199

Figure 1. The State of Iowa Map Showing Locations of Cored Bridge Decks

Having calculated the expected rating value $E(r, j)$, where, r is the rating condition within an interval, j, the Matlab program was utilized in conjunction with the second order polynomial model given in the following equation to develop a rebar-condition-age relationships (Vardeman, 1994).

$$r(t) = \beta_o + \beta_1 t + \beta_2 t^2 + \varepsilon \qquad (3)$$

where, $r(t)$ = rebar rating at specified deck age t in years, β_i is a constant, and ε is an error term that represent the degree of uncertainty between predicted and measured values.

For a new bridge deck, i.e., $t = 0$, the recorded rebar rating should be always 5, i.e., β_o should equal to 5. For this reason, a second order polynomial regression analysis with forcing intercept to be five was conducted. This regression yielded the following two relationships:

(i) ECR condition - age relationship for re-bars collected from cracked locations

$$r(t) = 5.00 + .0038\, t - .0031\, t^2 \qquad (4)$$

(ii) ECR condition - age relationship for re-bars collected from uncracked location

$$r(t) = 5.00 + .0135 t - 0.00134\, t^2 \qquad (5)$$

12 ADHESION OF COATING TO THE STEEL

Dry-knife adhesion tests were performed on the collected rebar samples and rated following the approach described in section 5.3.2.1 of the NACE TM-0185. The data collected from the adhesion test revealed that the adhesion decreased as time increased. The analysis also illustrated that the age of the adhesion of the coating decreases at a faster rate for the rebars collected from cracked locations than that of rebars collected from uncracked locations. This reveals that the moisture and the high chloride concentration at cracked locations have some effects on the bond between the epoxy-coated film and the reinforcing bars.

13 SERVICE LIFE OF A BRIDGE DECK

The corrosive threshold for ECR was defined by (Sagues et al., 1995) to be about 1.2 to 3.6 lb/yd^3 (0.73 kg/m^3 to 2.19 kg/m^3); and for black steel bar is 1.2 lb/yd^3 (0.73 kg/m^3). However, the data collected in this work revealed an average chloride concentration of 7.5 lb/yd^3 (4.56 kg/m^3) existed in locations where rebar samples having rating of 3, i.e., the condition representing 0 to 20% of corrosion on ECR surface. This is the condition at which corrosion becomes noticeable on ECR. Therefore, one may selected a corrosive threshold for ECR range from 3.6 lb/yd^3 to 7.5 lb/yd^3 (2.19 kg/m^3 to 4.56 kg/m^3).

458

Utilizing Fick's Second Law one can then calculate the time in which the chloride concentration at the rebar level reached the corrosive threshold for black or epoxy coated rebars.

In general, spalling will occur a few years after enough corrosion has built up causing significant increase in the rebar volume. For spalling to take place in bridge decks with black bars between 3 to 5 years (Larson, 1969) one can then determine the service life of a bridge deck. In the case where ECR were used, spalling was assumed herein to take place a few years after approximately 60 percent or more of the rebar surface was corroded, i.e., when the rebar rating reaches condition one. In addition, a period of approximately 5 to 8 years was assumed for the time from corrosion build up to spalling. This time period is slightly longer than that associated with black bars.

The following example utilizes the diffusion – spalling model previously explained to illustrate how to incorporate the above assumptions to estimate the functional service life of a bridge deck in the state of Iowa.

14 ILLUSTRATIVE EXAMPLE

Given an Iowa Bridge deck with C_o = 14.0 lb/yd^3, and D_c = 0.05 in^2/yr. End of functional life = 11.5% which is the average of 9% to 14% damage in the worst traffic lane. For an average concrete cover depth $\bar{x}$ = 2.74 in. and a standard deviation σ = 0.444 in., the depth x in Eq. 2 can be calculated as 2.21 in.

For a threshold of 3.6 lb/yd^3 (2.19 kg/m^3) and the information summarized above, the initiation term, T_1 can be calculated using Eq. 1 to be 38 years. This time will significantly increase if one uses a higher corrosion threshold such as that listed in Section 13 above.

The corresponding rebar rating at that time following equation 6.4 is 3.6. In addition, equation 6.4 will yield an additional 22 years for the rebar to reach condition 1. Assuming five years from corrosion build up to spalling, one can then estimate the time for spalling of the above described bridge deck to = 38 + 22 + 5 = 65 years.

In comparison to bridge decks constructed with black steel bar, where the corrosive threshold is 1.2 lb/yd^3. The time, T_1, was calculated to be 17 years. The average time for spalling ranged between 2 and 5 years = 3.5 [14] years for black steel. Thus, time required to rehabilitation for unprotected steel = 17 + 3.5 = 20.5 years.

The example above illustrates the significantly increase in the service life of a bridge constructed with ECR.

REFERENCES

Ang, A. H-S and Tang, W. H. 1975. *Probability Concepts in Engineering Planning and Design, Volume I – Basic Principle*, John Wiley & Son, New York.

Clear, K.C. 1975. *Reinforcing Bar Corrosion in Concrete: Effect of Special Treatments*. Special Publication 49. American Concrete Institute, Detroit, MI, pp. 77–82.

Clear, K.C. 1976. *Time-to-Corrosion of Reinforcing Steel in Concrete Slabs*. Report FHWA-RD-76-70. FHWA, Washington, DC.

FHWA. 1988. *Recording and Coding Guide for the Structure Inventory and Appraisal of the Nation's Bridges*, U. S. Department of Transportation, Washington, D. C., December.

Sagues, A. A., Lee, J.B., Chang, X., Pickering, H., Nystrom, E., Carpenter, W. Kranc, S.C., Simmons, T., Boucher, B., and Hierholzer, S 1995. *Corrosion of Epoxy Coated Rebar on Florida Bridges. Final Report*, Florida Department of Transportation

Schlorholtz, Scott 1998. "Report of X-ray Analysis," Iowa State University, April.

Sohanghpurwala, A. A., Scannell, W.T., and Viarengo, J. 1997. *Verification of Effectivness of Epoxy-Coated Rebar*, Project No. 94-05, Pennsylvania Department of Transportation.

Weyers, R. E., Fitch, M.G., Larsen, E. P., Al-Qadi, I. L., Chamberlin, W.P., and Hoffman, P.C. 1994. *Service Life Estimate*, SHRP-S-668, Strategic Highway Research Program, National Research Council, Washington DC, pp.135-136.

Weyers, R. E., Prowell, B. D., and Springkel, M. M. 1993. *Concrete Bridge Protection, Repair, and Rehabilitation Relative to Reinforcement Corrosion: A Methods Application Manual*, SHRP-S-360, Strategic Highway Research Program, National Research Council, Washington, DC.

Creative Systems in Structural and Construction Engineering, Singh (ed.) © 2001 Balkema, Rotterdam, ISBN 90 5809 161 9

Repair of cracked steel girder using CFRP sheet

M.Tavakkolizadeh & H.Saadatmanesh
University of Arizona, Tucson, Ariz., USA

ABSTRACT: The deteriorating status of the infrastructure of the United States has received much needed attention in the last two decades. The advantages for repairing and retrofitting of existing structures rather than replacing them are obvious. The conventional technique of welding cover plates to damaged members and bridging the cracks would create a fatigue sensitive detail. The result of the researches conducted in the last few years on the use of CFRP laminate for strengthening the steel beams has been very promising. This paper examines the possibility of using CFRP laminate as a patch to repair cracked girders. A total of eight 1.30 m long S5×10 A36 steel beams were tested in four point bending. Six of those were cut and then patched over the notched section using epoxy bonded CFRP sheets. The main parameters of this study were the patch length and notch depth. Thorough width notches on the tension flange face with a depth of 3.2 and 6.4 mm were considered. The results of the experiments were compared with theoretical values. The data obtained from the experiments showed that this technique could restore the strength of the damaged girder.

1 INTRODUCTION

Since 1967, when a mandatory inspection and control on the bridges in the United States were put into place, the American Association of State Highway and Transportation Officials (AASHTO) and the Federal Highway Administration (FHWA) have developed the programs to rate the bridges through biannual inspection. As a result, it has been found that more than one third of the highway bridges in the United States are considered substandard. According to the latest National Bridge Inventory (NBI) update, the number of substandard highway bridges in this country is more than 172,600. This does not even include railroad, transient or pedestrian bridges (FHWA Bridge Program Group 2000).

More than 43% of the bridges in this country are made of steel and in any recommendation category, steel bridges were among the most recommended group for improvement. The result may be due to the sensitivity of the steel as a material to corrosion, lack of proper maintenance or age of these structures.

The cost for rehabilitation and repair in most cases is far less than the cost of replacement. In addition, repair and rehabilitation usually takes less time, reducing service interruption periods. Considering the limited resources available to mitigate the problem, the need for adopting new materials and cost-effective techniques would be clear.

This paper discussed the effectiveness of epoxy bonding of CFRP sheets to the damaged flange of steel girders. In this method, the CFRP sheets were used to patch the damage portion of the tension flange in order to carry the stresses and, furthermore, reduce the stress level in the damaged elements.

2 PREVIOUS WORK

Notable advancement in material science in the past few decades and increasing applications of these new materials for civilian purposes after the end of the Cold War created the opportunity to utilize them for repairing aging structures. Lightweight Fiber Reinforced Plastics (FRP) possess an excellent tensile strength, fatigue strength, and corrosion resistance.

A very effective technique to repair and strengthen the exciting steel girders include epoxy bonding of CFRP laminates to the tension flange of the girder in order to increase its load carrying capacity and fatigue strength. The CFRP can also be used as a patch to arrest the crack growth initiated by fatigue. Despite the higher cost of CFRP, saving in labor, machinery and application time makes this method very attractive. Meanwhile, low weight and small thickness prevent any substantial increase in weight and loss of clearance in bridges.

Several studies have concentrated on the use of epoxy-bonded steel plates for strengthening steel and concrete structures. The first reported application dated back to 1964 in Durban South Africa where the reinforcement in a concrete beam accidentally left out during construction (Dussek 1980). By 1975, more than 200 defective elevated highway concrete slabs were strengthened in Japan (Raithby 1980).

In a study conducted at the University of Maryland, adhesive bonding and end bolting of steel cover plates to steel girders provided a substantial improvement in the fatigue life of the system (Alberecht et al. 1984). They reported an increase in the fatigue life with a factor of 20, compared to the welded cover plates.

In another study directed at the University of South Florida, the possibility of using CFRP in repair of steel-concrete composite bridges was investigated (Sen and Liby 1994). They tested a total of six 6.10 m long beams made of W8×24 steel section attached to a 71.1 cm wide by 11.5 cm thick concrete slab. They used 3.65 m long, 15 cm wide, 2 mm and 5 mm thick CFRP sheets. They reported that CFRP laminates could considerably improve the ultimate capacity of composite beams.

The advantages of using advanced materials in the rehabilitation of deteriorating bridges were investigated at the University of Delaware (Mertz and Gillespie 1996). As a part of their small-scale tests, they retrofitted eight 1.52 m long W8×10 steel beams using five different retrofitting schemes. They reported an average of 60% strength increase of CFRP retrofitted systems. They also tested and repaired two corroded steel girders 6.4 m long. The girders were typical American Standard I shape with depth of 61 cm and flange width of 23 cm. Their results showed an average of 25% increase in stiffness and 100% increase in ultimate load carrying capacity.

3 EXPERIMENTAL STUDY

In order to investigate the effectiveness of CFRP sheets on repair of cracked steel girders, small-scale steel beams and pultruded carbon flat were considered. Two different notches with depth of 3.2 and 6.4 mm were taken into account. Three different laminate lengths of 10, 20, 30 cm for shallow notches and 20, 40, 60 cm for deep notches were examined.

3.1 Materials

- Epoxy: A two-component epoxy was used for bonding the laminate to steel flange surface. The mixing ratio of the epoxy was one part resin (bisphenal A based) to one part hardener (polyethylenepolyamin) by volume. The epoxy had a pot life of 30 minutes and was fully cured after 2 days at 25°C.
- CFRP: A unidirectional pultruded carbon sheet with width of 7.6 cm and thickness of 0.13 mm were used in the study. After testing a total of 16

straight strips, the average tensile strength of 2137 MPa, tensile modulus of elasticity of 144.0 GPa and Poisson ration of 0.34 were obtained. A typical failed specimen is shown in Figure 1-a.
- Steel: In order to facilitate the testing in a universal testing machine, S5×10 A36 hot rolled sections were considered for the experiments. After testing a total of 6 dogbone specimens with the width of 1.9 cm, the average yield strength of 336.4 and 330.9 MPa, modulus of elasticity of 194.4 and 199.9 GPa and Poisson's ratio of 0.308 and 0.298 were obtained for the specimens cut from the flange and web, respectively. A typical failed specimen is shown in Figure 1-b.

3.2 Specimen preparation

First, the S-sections were cut into 130 cm long pieces. Then, one of the flanges was cut in the middle by an offset band saw with a blade thickness of 0.9 mm. This created a through notch in the tension flange of each specimen. The depth of the notches were controlled with a stopper and kept as 3.2 mm and 6.4 mm with ± 0.05 mm tolerance as shown in Figure 2. Each specimen was sand blasted thoroughly by No. 50 glass bids, washed with saline solution and rinsed with fresh water just before applying the composite sheet.

CFRP sheets were cut to the approximate length with a band saw. The ends of the sheets were finished smoothly using grid 150 sand paper. After finishing, the sheets were washed with saline solution and rinsed with fresh water as well.

After the drying of the steel beam and CFRP sheet, the epoxy was mixed. The designated bond area on both pieces were covered with thin layers of epoxy and two pieces were squeezed together to force the extra epoxy and any air pocket to bleed out. Using paper clips, the edges of the CFRP sheet throughout its length were secured. After two hours, the extra epoxy around the bond area was scraped of and washed with Acetone. A typical patch is shown in Figure 3.

After 48 hours, the 5 mm long strain gages were mounted on the surface of the steel beam and CFRP sheet as shown in Figure 4.

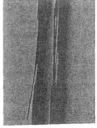

a) CFRP coupon   b)Steel strip

Figure 1. Typical failure in uniaxial tension test

Figure 2. Deep and Shallow notches in tension flange

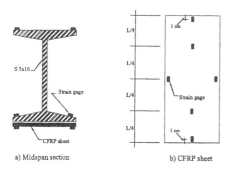

Figure 3. A typical notched steel beam patched with CFRP sheet

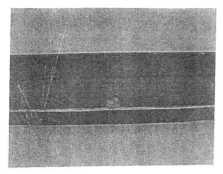

a) Midspan section   b) CFRP sheet

Figure 4. Schematic for the location of strain gages

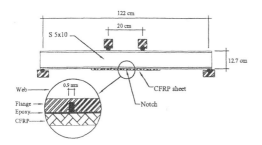

Figure 5. Schematic for 4-point bending of patched beams

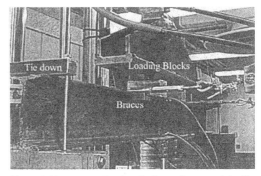

Figure 6. Test set up showing tie downs, braces and loading blocks

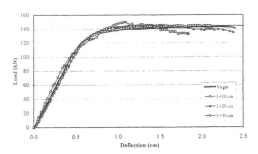

Figure 7. Load-deflection plot for 60 cm long patch on a deep notch

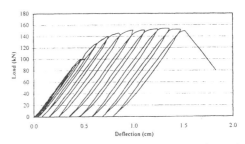

Figure 8. Load-deflection envelop for shallow notched beams

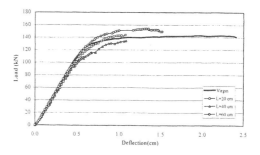

Figure 9. Load-deflection envelop for deep notched beams

463

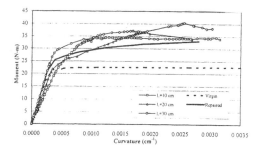

Figure 10. Moment-curvature envelop for shallow notched beams

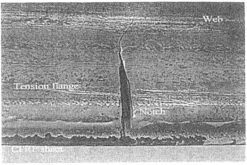

a)   Section rupture

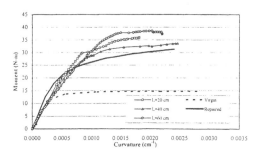

Figure 11. Moment-curvature envelop for deep notched beams

b)   Complete debonding of the sheet

3.3 *Experimental Setup*

The four point bending tests were performed using the MTS-810 testing machine. A special bending fixture was constructed. The load was measured by an MTS-661.31E-01 load cell with capacity of 2000 KN and the deflection was measured by an DUNCAN 600 series transducer with the total range of ±7.5 cm. Loading was performed under actuator displacement control with the rate of 0.025 mm/sec. Loading was applied in consecutive cycles of 2.5 mm actuator displacement increments. The values of load, midspan deflection and strains at different points were recorded with a Daytronic System 10 data acquisition system interfacing with PC through Microsoft Excel software.

The loading span was 122 cm and the loading points were 20 cm apart as shown in Figure 5. In order to prevent early lateral instability, braces were made with turnbuckles and high strength cable and used at the quarter points. Two ends of the beam were tied down to the rolling supports using two brackets. The Test setup is shown in Figure 6.

4 EVALUATION OF THE RESULTS AND ANALYSIS

4.1 *Load-deflection*

As stated earlier, the loading was applied in a progressive displacement cycles of 2.5 mm. A typical

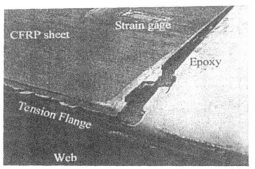

c)   End peeling of the sheet

Figure 12. Typical failure for notched beams

load-deflection plot is shown in Figure 7. The loading was continued until failure of the beam. The failure could be due to the failure of the CFRP, the delimitation of the composite sheet or the instability of the compression flange of the girder as shown in Figure 4. In the next step, the envelope curve for each specimen was generated using only the last few readings on the loading portion of each cycle.

4.1.1 *Shallow notched beam*

After drawing the load-deflection envelope curve for each specimen and comparing the behavior of the retrofitted specimens with an intact steel beam, the effectiveness of the patching technique became clear. As it is shown in Figure 8, regardless of patch length, the stiffness and ultimate load carrying capacity of the repaired beam were close to its original values. It is clear that the longer the patch the higher the total stiffness of the beam as was evident for 30 cm long patch.

The ultimate load carrying capacity of specimens showed 4.96%, 2.29% and 1.34% increase for 10 cm, 20 cm and 30 cm long patch, respectively. While the 30 cm long patch increased the stiffness of the beam by 6.23% of the intact case the 10 cm and 20 cm long patches improve the stiffness to 94.90% and 96.97% of the intact beam.

4.1.2 *Deep notched beam*
The load-deflection envelope curve for these specimens indicates the effectiveness of the method also. As it is shown in Figure 9, regardless of the patch length, the stiffnesses of the repaired beams were close to its intact values. The ultimate load carrying capacities of retrofitted beams were improved significantly for the 60 cm long patch. In case of shorter patches, the improvement in the capacity is still satisfactory. The main difference of the result for deep cuts compared to shallow cuts is the significant loss of ductility, which can be overcome by using relatively long patches.

The ultimate load carrying capacity of specimens showed 7.64%, -6.17% and 0.52% change for 20 cm, 40 cm and 60 cm long patch, respectively. The 40 cm long patch was terminated at the location with considerably high moment and shear and that was the cause of its low performance. In case of stiffness, improvements to 99.2%, 93.9% and 93.4% of the intact beam were obtained for 20 cm, 40 cm and 60 cm long patch, respectively.

4.2 *Moment-curvature*

In order to establish the validity of classical beam theory approach, the envelope plot for moment-curvature of each specimen was developed using the values of strains in the midspan. The average values of strain on top surface of the compression flange and bottom surface of CFRP sheet were used to determine the curvature.

By utilizing a simple macro procedure in Excel, the theoretical moment curvature of the section was determined. The CFRP sheet, flanges and the web were discreticized to 1, 5 and 8 layers, respectively. Changing the strain at extreme compressive fiber resulted in different values of moment and curvature. The properties of the each layer were defined separately using the result of the uniaxial testing on their coupons earlier. Steel was considered as an elastic-perfectly plastic and Composite as an elastic material.

All of the beams failed before compression flange reached the strain hardening level. Therefore, the assumption that the strain hardening part of steel would not play a role in the analysis and the material modeling was reasonable.

4.2.1 *Shallow notched beam*
The envelope moment-curvature plots for all the specimens display a significant improvement in ultimate moment capacity of the retrofitted sections. At it is shown in Figure 10, the capacity of the sections after repair increased by an average of 63% compared to its original defected condition. While the classical theory represented a conservative result for the ultimate capacity, the deviation from the experiments is not more than an average of -12%.

In case of stiffness, the theory satisfactorily predicted the result of the 30 cm long patch (+1.2%). The shorter patches tend to yield to a lower stiffness. The use of moderately soft epoxy and the inability of the theory to consider shear deformation yield to this deviation for shorter patch length (average of +28.1%).

4.2.2 *Deep notched beam*
The envelope moment-curvature plots for this set of specimens showed even more improvement in ultimate capacity compared to the unretrofitted case as shown in Figure 11. The capacity of the sections after repair increased by an average of 144% compared to its original defected condition. The deviation from the experiments is not more than an average of -15%. The overestimation of stiffness by the classical theory was more pronounced in this case. Due to a significant loss of the flange area the shear deformation at interface was high. This resulted to an average of 28.8% error.

4.3 *Failure*

There was a significant difference between the modes of failure for deep and shallow notches. Deep notches tended to fail rather suddenly by complete or partial failure of the bond between the patch and tension flange. Rupture of the notched section happened at the final stage. Figures 12-a and 12-b display typical failures for deep notched specimens. Failure in shallow notches started by debonding or peeling at the end of the sheet as shown in Figure 12-c. In all shallow cases, before debonding progress critically, the lateral instability of the beam forced the termination of the experiments.

5 CONCLUSUONS

Testing the specimens repaired by the proposed technique was very promising. In both cases of 40%

and 80% area loss of the tension flange, the method created a significant improvement in the ultimate capacity and stiffness. Based on the results of the experimental and analytical investigation, the following conclusions are made.

- Ultimate load carrying capacity of damaged steel beam can be increased significantly by the application of the CFRP patch over the damaged area. An average capacity increase of 144% and 63% was observed for 80% and 40% loss of tension flange area, respectively.
- Loss of stiffness in damaged steel members can be overcome by the patching technique to its original intact state. There was an average stiffness rebound of 99.4% and 95.5% compared to an intact case for 40% and 80% lost of tension flange, respectively.
- The classical beam bending theory can be used as a conservative but fairly accurate tool for predicting the ultimate capacity of retrofitted system. However, the theory does not seem to forecast the stiffness of the system as well due to the lack of consideration for shear deformation at the interface.
- The loss of ductility is the only evident drawback of this technique that can be incorporated into the design of the retrofitting scheme. On the other hand, the long-term durability of the bond and possibility of the galvanic corrosion are issues that should be addressed in the future.

6 REFERENCES

Albrecht, P., Sahli, A., Crute, D., Alberecht, Ph. And Evans, B. 1984. Application of adhesive to steel bridges. FHWA-RD-84-037, 106-147, the Federal Highway Administration, Washington, D.C.

Dussek, I. 1980. Strengthening of the Bridge Beams and Similar Structures by Means of Epoxy-Resin-Bonded External Reinforcement. TRB Record 785, 21-24, Transportation Research Board, Washington, D.C.

FHWA Bridge Program Group 2000. Count of Deficient Bridges by State Non Federal-Aid Highway. http://www.fhwa.dot.gov/bridges/britab.htm, The Federal Highway Administration, Washington, D.C.

Mertz, D. and Gillespie, J. 1996. Rehabilitation of Steel Bridge Girders through the Application of Advanced Composite Material. NCHRP 93-ID11,1-20, Transpiration Research Board, Washington, D.C.

Raithby, K, 1980. External Strengthening of Concrete Bridges with Bonded Steel Plates. Transport and Road Research Laboratory, Supplementary report 612, 16-18, Department of Environment, Crowthorn, England

Sen, R. and Liby, L. 1994. Repair of Steel Composite Bridge Sections Using Carbon Fiber Reinforced Plastic Laminates. FDOT-510616, Florida Department of Transportation, Tallahassee, Florida

Creative Systems in Structural and Construction Engineering, Singh (ed.) © 2001 Balkema, Rotterdam, ISBN 90 5809 161 9

Failure behavior of RC T-beams retrofitted with carbon FRP sheets

A. Saber
Department of Civil Engineering, Louisiana Tech University, Ruston, La., USA

ABSTRACT: The effects of using CFRP sheets to enhance the flexural strength of reinforced concrete members on the failure mechanisms of the retrofitted system are investigated. RC T-Beams were precracked then strengthened with CFRP sheets. The results were compared to non-precracked beams with and without CFRP material. The experimental values showed an increase in the flexural strength and stiffness of RC girders. The effect of precracking the specimens before applying CFRP materials influenced significantly the maximum deflection, but had insignificant effect on the ultimate load. The specimens were designed to have ductile behavior, however, the strengthened beams failed in brittle mode. The CFRP sheets improved the performance of the concrete in the tension zone by providing the mechanism for the tension stiffening effect, modified the crack width and crack distribution along the beam, and the failure behavior for the beams changed from ductile to brittle mode.

1 INTRODUCTION

The advantages associated with the FRP materials made the addition of fiber reinforced plastic (FRP) laminates bonded to the tension face of concrete members an attractive solution to the rehabilitation and retrofit of damaged structural systems. The most recognized advantages are the relative low cost with fast and easy installation. Composite materials as retrofit systems have been used recently in Japan, Europe and North America. Glass fiber reinforced plastic (GFRP) sheets were used to retrofit a roadway bridge to reduce the steel stresses in the tendon couplers (Rostasy et al. 1992). GFRP tow sheets were used on a bridge to increase the bending moment capacity (Nanni 1995) and on a prestressed box beam bridge (Finch et al. 1994).

The investigators have concentrated their research on strengthening reinforced concrete beams that were cast in the laboratory without subjecting these specimens to a reasonable service load. The studies have investigated a variety of retrofit systems and their responses under loading. Researchers have reported improvements in strength and stiffness of retrofitted beams. However, the reported results identified new types of failures that are often brittle, and in some cases the failure loads were significantly lower than the theoretical strength of the retrofitted system. Researchers have concluded that the failure processes and the governing criteria for reinforced concrete members retrofitted with FRP material need further study.

In this paper, experimental and analytical results are presented for the case of T-beams precracked and subsequently strengthened with CFRP sheets. The results are compared to non-precracked beams with and without CFRP sheets. The experimental phase consists of four-point bending test as well as tests to characterize material properties including the concrete-adhesive interface. The fibers of CFRP sheets are placed in the direction of the beam longitudinal axis ($0°$); one beam had two layers ($0°$ & $90°$). The T-beams represent long span beams with a shear-span to beam effective depth ratio of 10.

1.1 Research significance

The results show that the failure mode of the flexural member may be altered by strengthening and become brittle. The effect of CFRP sheets on the failure mechanism are recognized and discussed.

2 EXPERIMENTAL PROGRAM

2.1 Concrete

All the specimens are cast from one batch of concrete that is delivered by a ready mix concrete truck to the Structural and Materials Laboratory at Louisiana Tech University. The design characteristics for the concrete mix are shown in Table 1, along with fresh concrete properties. The 15-cm concrete cylinders are cured under ASTM Standard conditions.

The 28-day compressive strength is 39.5 MPa and the modulus of elasticity is 28,600 MPa.

Table 1: Concrete mix proportions

Cement Type I	1030 - kg/m^3
Crushed Granite	3151 - kg/m^3
Natural Sand	1854 - kg/m^3
Superplasticizer	6.75 - kg/m^3
Air Entrainment	3.5 - kg/m^3
Slump	18.5 - cm
Air Content	0.05
Unit Weight	232.2 - kg/m^3

2.2 CFRP Sheets

The CFRP Sheets are installed using the "wet lay-up" techniques; the fiber materials are placed on the surface dry and then impregnated with epoxy resins to form the laminate. The integrity of the system depends on the quality and strength of the concrete as well as the bond between the FRP and the concrete. All dust, dirt, oil, etc. are removed using abrasive blasting to ensure that the surface of the concrete is free of loose and unsound materials prior to applying the composite strengthening system. Table 2 shows CFRP strengthening characteristics; columns 1 and 2 identify the specimen code and description. Column 3 identifies the condition of the specimen before applying the CFRP sheet. The preloading is applied using the same configuration of the test to failure as shown in Figure 1. The applied load is 45% of collapse load (158 kN) for the control beam. Control specimen, CONT, is tested to failure without CFRP sheets.

Table2. CFRP sheets strengthening

Code	Beam Condition	CFRP Strengthening
PC1L	Precracked	One layer 0^0
PC2L	Precracked	Two layers 90^0 & 0^0
UC1L	Virgin	One layer 0^0
CONT	Virgin	NA.

2.3 TESTING

The shear strength at the concrete-adhesive interface is evaluated using simple tests. The specimens for the tension-shear and compression-shear tests are shown in Figure 2. The beam specimens have 50 x 15 x 10 cm dimensions with one # 25 bar embedded at each end to facilitate the installation of the specimen in the testing machine for the tension-shear test.

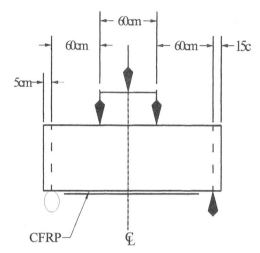

Figure 1. Test configuration.

The 15-cm cube specimens are used for the compression-shear test. After curing in ASTM Standard conditions, the specimens are cut using a wet-table saw at different angles (0°, 30°, 40°, and 60°), four groups of three specimens each. The saw-cut pieces are rejoined with a layer of the adhesive used for sheet bonding then are tested. All specimens' failure surface was in the concrete substrate. Therefore, it is concluded that the shear strength at the concrete-adhesive interface is adequate.

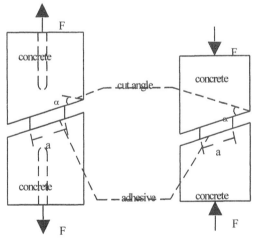

Figure 2. Tension-shear and compression-shear specimens.

2.4 T-BEAM TESTS

The T-beams have been designed to fail in flexure with adequate shear reinforcement to support the ultimate load, as would be the case for existing girders in service. The cross section details are shown in Figure 3. All specimens are tested in a four-point bending configuration, as in shown Figure 1.

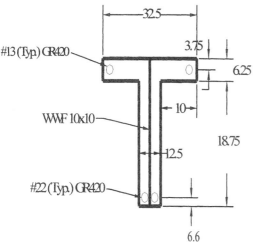

Figure 3. T-beam cross section details.

3 EXPERIMENTAL RESULTS AND ANALYSIS

3.1 Load deflection response

The mid-span deflections for each beam are measured during the test. The ultimate load, maximum deflection and deflection at yield load are presented in Table 3. The effects of the CFRP sheets on the specimens are shown as ratios of specimen to control beam for the ultimate loads and for the deflections at yield load.

Table 3. Failure loads and deflections

Beam	P_u	P_{ult} defl.	P_{yield} defl.	Effect of CFRP	
	kN	cm	cm	on delf at P_{ult}	on delf at P_{yield}
PC1L	179.3	1.71	1.03	1.14	0.92
PC2L	185.5	1.58	0.95	1.17	0.85
UC1L	192.6	2.38	1.05	1.22	0.94
CONT	158.0	1.73	1.12	NA	NA

The load deflection response of the control specimen and retrofitted beams are compared in Figures 4 and 5. The response of the CONT and UC1L specimens exhibit three regions of behavior, while the response for PC1L and PC2L specimens is approximately linear due to postcracking stiffness.

The initial stiffness before cracking is almost identical for CONT and UNC1L specimens. Before the concrete has cracked, all of the section is effec-

tive and the CFRP sheet has insignificant effect on the flexural rigidity of the section and the load that causes the concrete to crack initially.

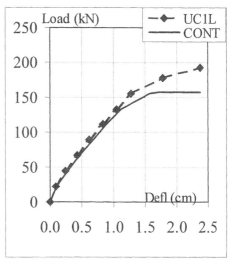

Figure 4. Load deflection response CONT and UC1L

The postcracking stiffness of the retrofitted beams is higher than that of the CONT. The CFRP sheets that are bonded to the tension face of the beam at maximum distance from the neutral axis will increase the beam stiffness. The composite material will transfer the stresses to intact concrete between cracks. Therefore, the CFRP sheet will improve the performance of the concrete in the tension zone by restraining the crack opening, a mechanism known as tension stiffening effect.

Before yielding, the tensile component of the moment couple acting on the section is shared by the tension steel reinforcement and the bonded CFRP sheet. Consequently, the yield stress of the steel is reached at a higher applied load. The strains in the reinforcing steel control the crack width in the concrete. Therefore, reducing the tensile stresses in the reinforcing steel by the CFRP sheet is structurally significant.

After yielding, the CFRP sheet continues to support the tensile stresses in the section. The flexural rigidity of the strengthened beams is higher than that for CONT, and the collapse load for the retrofitted beams is higher than that for the CONT by 22% for UC1L, 17% for PC1L and 19% for PC2L.

The slope of the load-deflection curve is higher for the strengthened beams than the CONT, and so the beam stiffness. The increase in the beam stiffness reduces the ductility of the collapse mechanism for the retrofitted beams as compared to the CONT. Similar results are reported in studies by Arduini et al. 1997, and Grace et al. 1999, using RC beams with rectangular cross sections.

The effect of precracking the specimens before applying CFRP materials influenced the maximum deflections significantly but had an insignificant effect on the ultimate load.

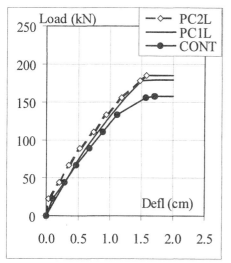

Figure 5. Load response for precracked and control specimens

4 COLLAPSE MECHANISM OF RETROFITTED BEAMS

Flexural cracks initiated in the mid-span of the beam as the tensile strength of the concrete was reached and extended vertically as loads increased. The flexural cracks remained narrow as the loads increased and were smaller than those in the control beam. The neutral axis is lower for the strengthened beam thus, the tensile cracks are reduced in height. As the applied load increased, flexural and flexural/shear cracks initiated at different locations along the shear spans and propagated towards the loading points. At this point the beam collapsed. The diagonal cracks are due to the increase in shear forces in the beam.

The cracks are narrower at the base of the beam because of the tension stiffening and confining effects of the CFRP sheets. The diagonal shear cracks that propagated towards the loading points can be seen as failure approaches. The failure mode for the beams with CFRP sheets was shear/compression failure.

The composite action between the bonded CFRP sheets and reinforced concrete member is an important structural requirement for the bonding technique to be successful as a repair method. Note that failure due to debonding of the CFRP sheets did not occur in any of the specimens tested in this study.

4.1 Uncraked vs Precracked Beams

The longitudinal strains in the CFRP sheets due to the applied loads are shown in Figures 6 and 7. The data for the shear span indicates that the longitudinal strain distribution follows the bending moment diagram. At higher loads the longitudinal strains in the shear spans increase above those of a linear variation, showing that strains are not proportional to the applied moment at these locations. At ultimate conditions, the axial strain in the CFRP sheet varies linearly between the end of the plate and the point of load. Therefore, the shear between the CFRP sheet and concrete is uniform.

Figure 6 indicates that the variation in the strain with the load at the beam center are slightly higher than those close to the load point, but the two curves are of similar form. As the applied load increased, the rate of change in the strains in the shear span is higher than that in the constant moment region. The higher rates demonstrate the initiation and progress of cracking in the region close to the support. The high level of strains in the shear span explains the shear cracking in the collapse mechanism for the beam.

Figure 7, indicates that as the applied load increases towards its maximum value, the distribution of strain in the CFRP sheet becomes unsymmetrical. The longitudinal strains attain higher values on one side of the center-line than on the other. This type of behavior is due to the defects in the bond-line, unsymmetrical distribution of cracking and the inhomogeneity of concrete.

4.2 Experimental and theoretical values

The ultimate design load for the specimens is 173.5 kN determined using the design guidelines of ACI 318, and for the strengthened beams it is 218 kN calculated based on the methodology recommended by Nanni 1993. The strengthened beams collapsed at loads lower than the predicted value, which suggest that higher safety factors should be considered for the retrofitted systems.

5 CONCLUSIONS

The data obtained from the experimental program shows that the CFRP sheet provides a substantial increase in the moment capacity and beam stiffness. The effect of precracking the specimens before applying CFRP materials influenced the maximum deflections significantly but had an insignificant effect on the ultimate load. The control beam specimen failed in flexural / tension, a ductile failure mode as designed. However, the beams strengthened with CFRP sheets failed in shear / compression mode, a sudden and brittle collapse condition which is undesirable. The widths of the concrete cracks in the strengthened beams are narrower than those in the control specimen, reflecting the effect of the CFRP sheet in confining the concrete that is subject to tension forces. Hence, the use of CFRP sheet to retrofit RC beams will improve serviceability conditions. The analytical work conducted in this research is based on ACI 318 design guidelines and conventional design approaches for concrete. The failure loads are smaller than the calculated values, therefore higher safety factors should be considered in design.

6 RECOMMENDATIONS

The research completed shows that CFRP sheets can be used to increase the flexural capacity of reinforced concrete beams. However, the failure mode can change from ductile to brittle failure. Therefore, an accurate method to predict the ductility of the retrofitted system should be developed prior to using CFRP sheets efficiently. The long-term effects of the material subjected to adverse conditions, such as environmental loads and fatigue etc., should also be tested. Further testing should be conducted to determine the effects of CFRP Sheets on long-term deflection of retrofitted beams under service loads.

ACKNOWLEDGMENTS

Support of this work was provided by Louisiana Transportation Research Center (LTRC) of the Louisiana Department of Transportation and Development (LDOTD) under research project number 00-5TIRE and state project number 736-99-0805. Master Builders, Inc. provided some of the materials. This sponsorship is gratefully acknowledged.
The contents of this study reflect the views of the author who is responsible for the facts and the accuracy of the data presented herein. The contents do not necessarily reflect the official views or policies

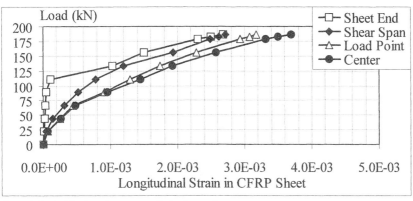

Figure 6. Typical load strain response obtained at strain gages located along CFRP sheet length on PC2L specimen.

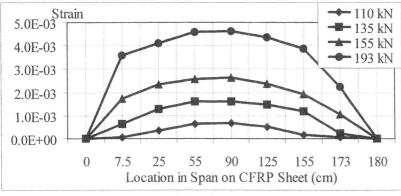

Figure 7. Typical distribution of longitudinal strain along CFRP sheet length at various applied load levels.

of the Louisiana Department of Transportation and Development or the Louisiana Transportation Research Center. This paper does not constitute a standard, specification, or regulation.

REFERNCES

Rostasy, F., Hankers, C., & Ranisch, E. 1992. Strengthening of R/C and P/C-Structures with Bonded FRP Plates. *Advanced Composite Materials in Bridges and Structures, Canadian Soc. For Civil Engrg.* 253-263.

Nanni, A. 1995. Concrete Repair with Externally Bonded FRP Reinforcement. *Concrete International*, 17(6): 22-26.

Finch, W.W., Chajes, M.J., Mertz, D.R., Kaliakin, V.N., & Faqiri,A. 1994. Bridge Rehabilitation Using Composite Materials. *Infrastructure: New Materials and Methods of Repair, Proc., Materials Eng. Conference, ASCE*: 1140-1147. NY, N.Y.

Arduini, M., & Nanni, A. 1997. Behavior of Precracked RC Beams Strengthened with Carbon FRP Sheets. *Journal of Composites for Construction,* 1(2):63-70.

Grace, N.F., Sayed, G.A., Soliman, A.K., & Saleh, K.R. 1999. Strengthening Reinforced Concrete Beams Using Fiber Reinforced Polymer (FRP) Laminates. *ACI Structural J.* 96(5):865-874.

Building Code Requirements for Structural Concrete (ACI 318-99) and Commentary. American Concrete Institute, Farmington Hills, MI.

Nanni, A. 1993. Flexural Behavior and Design of RC Members Using FRP Reinforcement. *Journal of Structural Engineering,*119(11):3344-3359.

Creative Systems in Structural and Construction Engineering, Singh (ed.) © 2001 Balkema, Rotterdam, ISBN 90 5809 161 9

Using Fibre Reinforced Plastic (FRP) reinforcing bars in RC beams and slabs

M.A. Dávila-Sänhdars
School of Civil Engineering, University of South Australia, S.A., Australia

ABSTRACT: Corrosion in RC beams and slabs can be prevented using FRP bars instead of steel bars. However, FRP bars do not yield as steel bars do. This contrast requires a different design procedure. Ductility in concrete structures relies on the yielding characteristics of the steel bars, therefore, such structures are designed as under reinforced, that is, steel bars yield before the concrete crushes. However, if the reinforcing bars do not yield, as is the case with FRP bars, then the beams have to be designed as over reinforced, that is, the concrete crushes before the FRP bars fracture. It has been found that ductility can be built by placing FRP bars in the compression zone. A mathematical model using the rectangular stress blocks is proposed herein. This model facilitates the calculation of both curvature and moment at the concrete crushing as well as at the fracture of the compression FRP bars.

1 INTRODUCTION

Concrete structures reinforced with steel bars are very vulnerable to damage in corrosive environment on the other hand concrete reinforced with FRP bars are more resistant to corrosion (Gueritze, 1992) and (Schwartz and Schwartz, 1968). In addition, the mechanical properties of FRP bars are different from steel bars, consequently, the design procedure has to be different (Pickett, 1968), (Tsai, 1968) (Ballinger, 1992) and (Benjamin, 1969). Therefore, the usual standard procedure used in concrete structures with steel bars are not applicable directly to structures reinforced with FRP bars (Daniali, 1990) and (Malvar, 1995).

Firstly, let see for instance the use of steel bars in concrete structures (Warner et al., 1989). The stress/strain relationship of the steel bars is linear between the origin and the onset of yielding, which occurs at the strength f_{ys} of 400 MPa and a Young's modulus E_s of 200, 000 MPa. Soon after of yielding, the steel become relaxed maintaining the same strength at that level, that is the steel bars posses a high ductility characteristic. Given that steel are ductile, concrete structures are designed as under-reinforced so that the steel bars yield before the concrete crushes therefore the structural elements are also ductile.

FRP bars do not behave ductile as the steel bars do (Nanni, 1993). The stress/strain relationship of the FRP bars is linear from the origin up to failure, that is, it does not show any ductility plateau

(Saadatmanesh and Ehsani, 1991). FRP bars are stronger than steel bars reaching strengths even greater than 1,000 MPa but at expenses of large strains which usually are more than 1 %. Given that FRP bars allow very large deformations before fracture, then concrete beams have to be designed as over-reinforced. However, by over-reinforcing the beams the concrete crushes prematurely without warning. Brittle failure of the beams can be prevented by placing FRP bars in the compression zone so that the FRP bars take over from the crushed concrete building a ductility plateau instead of a decaying falling branch in the moment/curvature relationship (Oehlers and Dávila-Sänhdars, 1999), (Dávila-Sänhdars, 1999) and (Dávila-Sänhdars and Oehlers, 2000).

Experimental work is first explained from the design of specimens with FRP bars in the compression and the tension zones accompanied by a description of the failure of concrete and FRP bars. Later is described a mathematical procedure for ductile design of concrete beams using the well-known rectangular stress block.

2 EXPERIMENTAL WORK

Three beams with glass FRP bars were tested up to failure to investigate ductility behavior. The cross-sections of the specimens are shown in Figure 1, which were 5.0 m long, 200mm wide and 30mm high. The specimens were made of a 50 MPa concrete strength with a Young's modulus of 34000 MPa. The FRP bars in the tension zone of the three beams were 700 MPa strength with a Young's modulus of 49000 MPa. Meanwhile the FRP bars in the compression zone were 420 MPa strength and a Young's modulus of 26000 MPa. The FRP bars used in the specimens were all 20-mm diameters. Beam 1, 2 and 3 had 3 FRP bars in the tension zone.

In the compression zone, beam 1 had two undeformed steel bar of 10-mm diameter. The aim of placing undeformed steel bars at the top of beam 1 was to prevent stress contribution by the bars to the concrete. The failure of the steel bars happened due to debonding as it was expected hence, the beam behaved as unreinforced at the top. Beam 2 had one FRP bar at the top and beam 3 had two FRP bars at the top. In addition, stirrups of R_{10} were placed uniformly at 125-mm centre to centre to prevent shear failure in the beams.

The beams were tested as simply supported with a span of 4.6 m and under 4 points loading with a constant moment region of 0.8 m at midspan.

crushing to 102 kNm at the fracture of FRP bars which in turn means that beam 2 behaved ductile.

B3 is pointing to the onset of the concrete crushing in beam 3. It can be seen in Figure 2 that the concrete crushed at an earlier curvature than it happened in beams 1 and 2.

Furthermore, the plateau in beam 3 is larger than the plateau in beam 2. The moment at the onset of the concrete crushing in beam 3 was 102 kNm and curvature $32\%10^{-6}$ mm^{-1} and the moment at the fracture of the compression FRP bars was 112 kNm and curvature $61\%10^{-6}$ mm^{-1}.

The difference in strength and curvature of the concrete crushing in the three beams was because the Young's modulus of the compression FRP bars was less than the Young's modulus of the concrete. Notice that increasing the amount of compression bars the concrete crushes earlier but the fracture of FRP bars is delayed given ductility to the beams hence, redistribution of moment is possible.

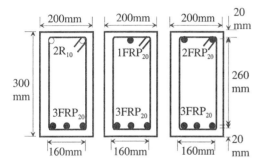

Figure 1. – Specimens cross-section

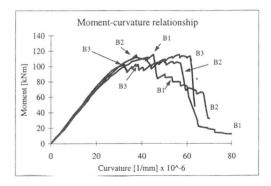

Figure 2. – Experimental moment-curvature relationship

3 ANALYSIS OF RESULTS

The three beams showed a typical failure as it can be seen in Figure 2 in which B1, B2 and B3 stand for Beam 1, Beam 2 and Beam 3 respectively. Table 1 shows the variation of moment and curvature in the specimens. Let see in Figure 2 for instance the arrows from B1, B2 and B3 at the top. B1 is pointing to the onset of the concrete crushing in beam 1 which occurred at the moment of 115.6 kNm and curvature $45\%10^{-6}$ mm^{-1}. It can be seen that the falling branch of the curve for beam 1 went down rapidly up to collapse showing a brittle failure.

B2 is pointing to the onset of the concrete crushing in beam 2. It can be seen that instead of falling rapidly as beam 1 did, beam 2 gained strength after crushing the concrete forming a plateau between the curvatures $38\%10^{-6}$mm^{-1} and $57\%10^{-6}$mm^{-1}. The moment varied from 110.9 kNm at the concrete

Table 1. Experimental results

Beam	Concrete crushing		FRP bars fracture	
	Moment [kNm]	Curvature [mm^{-1}]%10^{-6}	Moment [kNm]	Curvature [mm^{-1}]%10^{-6}
1	115.6	45	-------	-------
2	110.9	38	102	57
3	102	32	112	61

4 MATHEMATICAL MODEL

Two different sort of beams with FRP bars are discussed herein one of them is a singly reinforced beam and the another one is a doubly reinforced beam with FRP bars. The mathematical model is developed using the traditional rectangular stress block method.

4.1 Singly reinforced beam

Firstly, let us analyze a single reinforced beam as that showed in Figure 3 where b is the width of the cross-section of the beam; d is the effective depth of the section; and A_{ft} is the tension FRP reinforcement. Figure 3b shows the strains profile at the onset of the concrete crushing where ε_{cu} is the ultimate strain of the concrete; ψ is the curvature; and ε_{ft} is the strain of FRP bars in tension. The rectangular stress block is given by $n_1\gamma$ times $0.85f_c$ where γ is a reduction factor of the neutral axis depth to define the stress block (Australian Standards Concrete Structures, 1994)

The compression force of the concrete is given by C_c and T_f is the tension force in the FRP bars.

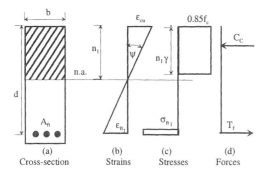

Figure 3. – Singly reinforced beam

Let be

$$C_c = T_f \tag{1}$$

for equilibrium condition, where

$$C_c = 0.85f_c b\gamma n_1 \tag{2}$$

and

$$T_f = \sigma_{ft1} A_{ft} \tag{3}$$

For strains compatibility is given

$$\frac{\varepsilon_{cu}}{n_1} = \frac{\varepsilon_{ft1}}{d - n_1} = \psi \tag{4}$$

Combining equations 1, 2, 3 and 4 and solving for ε_{ft} gives

$$\varepsilon_{ft1}^2 + \varepsilon_{cu}\varepsilon_{ft1} - \frac{0.85f_c\gamma bd\varepsilon_{cu}}{E_{ft}A_{ft}} = 0 \tag{5}$$

The strain in equation 5 corresponds to the strain of the tension FRP bars at the onset of the concrete crushing at the top the beam.

The moment is calculated using the following equation derived from 1 to 5.

$$M_n = 0.85f_c b\gamma \left(\frac{\varepsilon_{cu}}{\psi}\right)^2 \left(\frac{0.85f_c b\gamma}{A_{ft}E_{ft}\psi} - \frac{\gamma}{2} + 1\right) \tag{6}$$

where the curvature ψ can be derived from equation 4. Further values of moment in the falling branch of the moment-curvature relationship can be obtained by inserting values of curvature in Equation 6 greater than that at the onset of the concrete crushing.

4.2 Doubly reinforced beam

Let Figure 4 represents schematically the deformations of a doubly reinforced beam with FRP bars. Figure 4a depicts the cross-section of the beam with 2 FRP bars at the top and three FRP bars at the bottom. The cross-section is divided into three rectangles as follow: the blank rectangle at top represents the concrete already crushed; the hatched rectangle represents the concrete in compression; and the blank rectangle at the bottom represents the tension zone of the cross-section of the beam. d is the effective depth of the beam; d_1 is distance between the top of the beam and the centroid of the compression reinforcement; and n_1 is the neutral axis of the section of the beam. A_{fc} is the amount of compression FRP reinforcement; and A_{ft} is the amount of FRP reinforcement in tension.

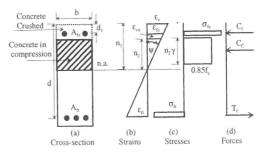

Figure 4. – Double reinforced beam

From Figure 4b and for strains compatibility it is given

$$\frac{\varepsilon_c}{n_1} = \frac{\varepsilon_{fc}}{n_1 - d_1} = \frac{\varepsilon_{cu}}{n_2} = \frac{\varepsilon_{ft}}{d - n_1} = \psi \tag{7}$$

where ε_c is the strain at the top fiber of concrete; ε_{fc} is the strain of the compression bars; ε_{cu} is the concrete ultimate strain; ψ is the curvature of the beam which at the onset of the concrete crushing is taken as ψ_1 and at the fracture of the compression bars is taken as ψ_u. ε_{ft} is the strain of the FRP bars in tension.

From Figure 4d it is given

$$c_f + c_c = T_f \tag{8}$$

where C_f is the force of the compression bars; C_c is concrete force; and T_f is the force of the tension bars. Combining equations 7 and 8 and solving for ψ_1 at the onset of concrete crushing gives

$$K_1\psi_1^2 - \varepsilon_{cu}K_3\psi_1 - K_2 = 0 \tag{9}$$

where

$$K_1 = dE_{ft}A_{ft} + d_1E_{fc}A_{fc} \tag{10}$$

$$K_2 = 0.85f_cb\gamma\varepsilon_{cu} \tag{11}$$

$$K_3 = E_{ft}A_{ft} + E_{fc}A_{fc} \tag{12}$$

E_{fc} is the Young's modulus of the compression FRP bars and E_{ft} is the Young's modulus of the bars in tension.

The curvature at the fracturing of the compression bars is given by

$$\left(\frac{K_1}{K_3}\right)\psi_u^2 - \varepsilon_{fcu}\psi_u - \frac{K_2}{K_3} = 0 \tag{13}$$

The moment at the onset of the concrete crushing as well as the moment at the fracture of the FRP bars in compression can be calculate substituting ψ by either ψ_1 or ψ_u in the following equation

$$M_n = \frac{K_2}{\psi}\left(d - \frac{1}{K_3}\left(K_1 - \frac{K_2}{\psi^2}\right) + \frac{\varepsilon_{cu}}{\psi}\left(1 - \frac{\gamma}{2}\right)\right) + \psi\left(\frac{1}{K_3}\left(K_1 - \frac{K_2}{\psi^2}\right) - d_1\right)E_{fc}A_{fc}(d - d_1) \tag{14}$$

5 NUMERICAL EXAMPLES

Calculations of moment and curvature at the onset of the concrete crushing as well as at the fracture of the compression FRP bars can be done using equations 5, 6, 9, 13 and 14. Results are presented in Table 2. In Table 2 ψ_1 and ψ_u are the curvatures at the onset of the concrete crushing and at the fracture of the compression FRP bars respectively. M_1 and M_u are the moments at concrete crushing and at the fracture of the compression FRP bars.

Table 2. - Theoretical results

Beam	A_{ft} [mm^2]	A_{fc} [mm^2]	ψ_1 [mm^{-1}]%$\times10^{-6}$	M_1 [kNm]	ψ_u [mm^{-1}]%$\times10^{-6}$	M_u [kNm]
1	942	0	52	111.8	79	39
2	942	314	54	122	85	65
3	942	628	54	125	92	90

In Figure 5 are shown the plots of the falling branch of the moment-curvature relationship of the three specimens using the rectangular stress block method. It is worthy to note in Figure 5 that the moment and curvature of the three beams are practically the same at the concrete crushing no so at the fracture of the compression FRP bars. More importantly is to note that as the amount of compression FRP bars is increased the beams tend to become more ductile.

6 ANALYSIS OF RESULTS

There is a discrepancy between the experimental results and the theoretical results using the rectangular stress block method most of all at the fracture of compression FRP bars. At the concrete crushing, the moment is almost the same except the curvature that increases with the amount of compression reinforcement. That is opposite to experimental results in which the curvature diminishes with the increase of compression bars. Furthermore, at the fracture of the compression bars the curvature is larger using the rectangular stress block method than the curvature from the laboratory test. The moment is less using

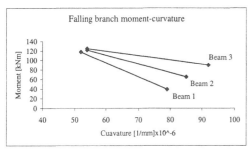

Figure 5. – Theoretical moment-curvature relationship

the rectangular stress block method than the moment plotted with the experimental results.

7 CONCLUSIONS AND RECOMMENDATIONS

It has been found that beams with FRP bars has to be designed as over reinforced beams because FRP bars do not yield as steel bars do. Furthermore, the design procedure has to be different than that used in the design of RC beams with steel bars. Theoretically and experimentally has been demonstrated in this paper that beams with FRP bars in the tension zone only, present a brittle failure. However, ductility can be built in the beams placing FRP bars in the compression zone because at the failure of concrete the stress is shed on the compression FRP bars.

There is a discrepancy between the experimental results and the results obtained using the stress block method. It is the author's believe that the stress block method needs to be reviewed in order to obtain results closer to experimental results.

REFERENCES

Australian Standards Concrete Structures AS 3600-1994. p 67.

Ballinger, C.A. (1992), Development of fibre-reinforced plastic products for construction market. *Advanced composite meterials in bridges and structures*. KW. Neale and P. Labossière, Editors. pp 3— 13.

Benjamin, B.S. (1969). *Structural design with plastics*. Van Nostrand Reinhold Company. pp 1—2.

Daniali S. (1990). Bond strength of fiber reinforced plastic bars in concrete. *Serviceability and durability of construction materials*. Proceedings of the First Materials Engineering Congress. Part (2). ASCE, pp 1182—1191.

Dávila-Sänhdars M.A. (1999). Ductility of RC beams with FRP reinforcing bars. *MESc Thesis*. University of Adelaide, Australia.

Dávila-Sänhdars M.A. and D.J. Oehlers (2000). Ductility Design of RC Beams with FRP bars. *Proceeding of the ACUN-2. International Composites Conference. Composites in the Transportation Industry*. Australia.

Gueritse, A. (1992) Durability Criteria for non-metallic tendons in an alkaline environment. *Advanced composite materials in bridges and structures*. K.W. Neale and Labossière, Editors. pp 129— 137.

Malvar L.J. (1995). Tensile and bond properties of GFRP reinforcing bars. *ACI Materials Journal* Vol 92, No. 3 pp 266—285.

Nanni A. (1993). Flexural behaviour and design of RC members using FRP reinforcement. *Journal of Structural Engineering, ASCE*. Vol 119 No. 11 pp 3344—3359.

Oehlers D.J. and M.A. Dávila-Sänhdars (1999). Ductility of R.C. beams with Compression and Tension FRP bars. *Submitted for publication*.

Pickett G. (1968). Elastic moduli of fiber reinforced plastic composites. *Fundamental aspects of fiber reinforced plastic composites* Ed. R.T Schwartz and H.S Schwartz. *Air Force US*. pp 13—27.

Saadatmanesh H. and M.R. Ehsani (1991). Fiber composite bars for reinforced concrete construction. *Journal of Composite Materials*. Vol 25.pp 188—203

Schwartz R.T. and H.S. Schwartz (1968). Introduction. *fundamental aspect of fibre reinforced plastic composites*. John wiley and sons. pp vii—xi.

Tsai E. (1968). Strength theories of filamentary structures. *Fundamental aspects of fiber reinforced plastic composites*. Ed. R.T. schwartz and H.S Schwartz. Air Force Us, p 3—11.

Warner R.F., B.V. Rangan and A.S. Hall. (1989). *Reinforced concrete*. 3rd edition, Longman Cheshire Pty Limited, editors.

Creative Systems in Structural and Construction Engineering, Singh (ed.) © 2001 Balkema, Rotterdam, ISBN 90 5809 161 9

Shear behavior of RC beam strengthened with CFRP grid

A. Yonekura
Hiroshima Institute of Technology, Japan

Y. Kawauchi, S. Okada & M. Suzuki
Kyokuto Company Limited, Hiroshima, Japan

K. Zaitsu
Sato Benec Company Limited, Oita, Japan

ABSTRACT : Recently the repairing and strengthening of the reinforced concrete structures are required due to increase of deterioration of the concrete structures. The reinforcing for shear failure is very important for the concrete structures under the seismic design.

In this study , CFRP grid and CF sheet were used as shear reinforcement of Reinforced Concrete (RC) beam. The effect of the shear reinforcing for the specimens with CFRP grid was experimentally compared with that of the non-sprayed specimens with CF sheet. CFRP grid was bonded on the sides of shear span of reinforced concrete beams by spraying with polymer cement mortar. It was confirmed that this system can be available for the application of reinforcing of the concrete structure through this examination.

Additionally this system was applied for the reinforcing of the front side of abutment in order to restrain of cracks occurring due to Alkali-Aggregate-Reaction and the material included lithium nitrite was injected in the cracks in order to control Alkali-Aggregate-Reaction.

1 INTRODUCTION

The lifespan of reinforced concrete (RC) structures has abruptly become shorter overall as a result of defects in construction methods and materials, or changes in social situations and the natural environment, that were not foreseen when the structures were erected. And for the same reasons, these structures have lost durability. In order to improve the durability, improvements are frequently made to design methods, and applied to new structures of this type when they are erected.

Thus, sufficient durability can be built into new structures of this type, but for existing structures repairing and strengthening are necessary. In addition, the great Hanshin Awaji Earthquake in Japan demonstrated that these structures may be subjected to earthquakes on a scale unanticipated when they were designed, which means that the shearing strength in the traditional design is rather insufficient; if an earthquake on the same scale should occur again, it is clear that they could suffer brittle fracturing throughout. Methods that have been developed thus far to remedy this include bonding steel plates or carbon fiber to the structures. Carbon fiber (CF) has been receiving much attention in recent years, since it has excellent performance over a range of parameters, including high strength, no corrosion and being

lightweight. In the present research the authors examined an integrated thickness-enhancing shearing reinforcement technique whereby CFRP grids are attached to the sides of RC girders, which are then mechanically sprayed with acrylic polymer- cement mortar. This technique has the features of being readily implemented highly economical, and also high-quality; if its reinforcing effect can be confirmed, it is likely to enjoy widespread application as a shear reinforcing technique. Accordingly, the authors conducted static loading experiments on RC girders that had been reinforced against shear stress using the above technique, and on RC girders treated using the CF sheet bonding technique, and compared the reinforcing effects of the two techniques.

2 OUTLINE OF THE EXPERIMENTS

2.1 *Types of test-specimens and properties of their materials*

For the experiments, twelve RC girder specimens were fabricated. Ten of these girders were given reinforcing treatment consisting either of a combination of CFRP grid plus spraying with acrylic polymer-cement mortar, or of a combination of CF sheet plus an epoxy or acrylic bonding agent. One of the two remaining girders was given thickness-enhancing treatment consisting simply of spraying with the polymer-cement mortar, while the other was fabricated as an unreinforced test-specimen for the purpose of comparison.

Static loading tests were conducted on these test-specimens, and measurements were taken of the nature of the deformation in the girders, the bending and shearing cracks appearing in them, CFRP and stirrup strain, and ultimate strength. The results of the measurements were then considered.

In the application of the CFRP grids, two factors capable of affecting the reinforcing effect were varied for different girders: the cross-sectional area of the CFRP grid, and the number of anchors used for preventing ultimate flaking of the reinforcing grid. In the application of the CF sheets also, certain factors were varied among different test-specimens, namely the type of bonding resin used and the positions in which the sheets were affixed.

Table 1 shows the types of test-specimens that were fabricated, Table 2 shows the characteristics of the CFRP grid material, and Table 3 shows the characteristics of the polymer mortar material.

2.2 Test apparatus and loading method

The reinforcement arrangement of RC beam is shown in Fig.1.

The loading apparatus and method are shown in Fig. 2; as can be seen, the load was applied at 2 points on a 1800 mm span. The load was applied continuously as monotonous loading, until it resulted in fracture.

Table 1. Type of test-specimens

Name	CFRP		Bonding material
	Form	Affixing position	
RC	None	—	—
RC-PM			
CR6-PM-8	Grid (thickness) CR6(t=4mm)	Sides of girder	Polymer-cement mortar, sprayed
CR6-PM-16			
CR6-PM-24			
CR3-PM-8	Grid (thickness) CR3(t=2mm)		
CR3-PM-16			
CR3-PM-24			
CFS-RIB-S	Sheet CFS	Sides of girder	Acrylic resin
CFS-EPO-S			Epoxy resin
CFS-RIB-SU		Sides and bottom side	Acrylic resin
CFS-EPO-SU			Epoxy resin

CR6 — PM — 8

Type of CFRP

Type of bonding resin

Number of anchor and affixing position

RC :None

CR6:CFRP grid(t=4mm)

CR3:CFRP grid(t=2mm)

CFS:CF sheet

PM: Polymer-cement mortar

RIB: Acrylic resin

EPO: Epoxy resin

S: Two sides of beam

SU: Two sides and bottom side of beam

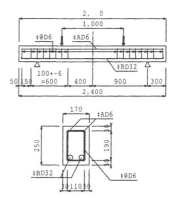

Fig. 1. Reinforcement arrangement of RC-beam

Table 2. Characteristics of CFRP grid materials

Type	CR3	CR6
Reinforcing fiber	Carbon fiber	
Impregnating plastic	Vinyl ester	
Specific gravity	1.42	
Tensile strength (N/mm^2)	1177	
Modulus of elasticity (KN/mm^2)	98.1	
Cross-sectional area (mm^2)	4.4	17.5
Maximum load (N)	5.10	20.6
Standard weight (g/m)	6.3	25

Table 3. Characteristics of spray-applied polymer cement mortar

Type	Characteristic value	Units
Age	13	Days
Compressive strength	28.6	N/mm^2
Flexural strength	7.2	N/mm^2
Tensile strength	2.3	N/mm^2
Bond strength	2.3	N/mm^2
Modulus of elasticity	23.2	KN/mm^2
Polymer type	Acrylic polymer	

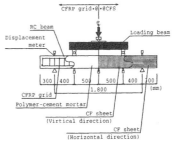

Fig. 2. Method of loading specimens

L

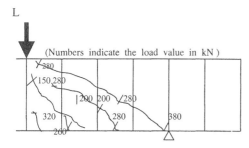

(Numbers indicate the load value in kN)

Fig.3. Shear cracking pattern of RC without repairing

L

(Numbers indicate the load value in kN)

Fig.4. Shear cracking pattern of RC-PM with polymer cement and without CFRP grid

L

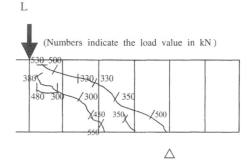

(Numbers indicate the load value in kN)

Fig.5. Shear cracking pattern of CR6-PM-16 with CFRP grid(CR6) and polymer mortar

L

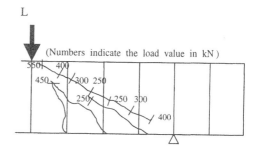

(Numbers indicate the load value in kN)

Fig.6. Shear cracking pattern of CR3-PM-16 with CFRP grid (CR3) and polymer mortar

3 RESULTS OF EXPERIMENTS

3.1 *Reinforcing effect of CFRP grids or CF sheets.*

As shown in Table 4 and Fig.9, the ultimate strength of the RC girders reinforced with CFRP grids increased by around 30%, regardless of the type of grid. Looking more closely at the portion of the figure as shown with arrows , it is fined that the increases in the load at flexural and shear cracking are much larger than the increase in the ultimate strength.

This is a major feature of the shear reinforcing effect produced by the CFRP grids in the experiments. Fig.3~6 shows the shear cracking pattern of beams strengthened with CFRP grids and polymer-cement mortar or without CFRP grids and with polymer-cement mortar as compared non-reinforced RC beam.

All specimens are broken by shear failure. In the case of the RC beams reinforced with CFRP grids and polymer-cement mortar, shear-failure occurred at the matrix of RC beam and after that , CFRP grids and polymer-cement mortar scaled off as shown in Fig7, while in the case of CF sheet , shear failure occurred by scaling off or breaking off CF sheet as shown in Fig8.

Fig. 7. CFRP grids are broken

Fig. 8. Shear failure due to breaking of CF

Table 4. Reinforced effect due to CFRP grid or CF sheet

№	Name	Flexural cracking load		Digonal cracking load		Ultimate load at shear failure	
		Con-crete	Rein-forcing effect		Rein-forcing effect		Rein-forcing effect
		$(, _l \, \overline{m})$	$(\bullet \, \rfloor$	$(, _l \, \overline{m})$	$(\bullet \, \rfloor$	$(, _l \, \overline{m})$	$(\bullet \, \rfloor$
1-1	RC-1	25	---	118	---	471	---
1-2	RC-2	25	---	127	---	431	---
2	RC-PM	39	60	206	68	510	13
3-1	CR6-PM- 8	59	140	206	68	579	28
3-2	CR6-PM-16	49	100	226	84	596	32
3-3	CR6-PM-24	54	120	245	100	608	35
3-4	CR3-PM- 8	39	60	206	68	590	31
3-5	CR3-PM-16	49	100	226	84	583	29
3-6	CR3-PM-24	54	120	216	76	530	17
4-1	CFS-RIB-S	29	20	---	---	557	23
5-1	CFS-EPO-S	29	20	---	---	538	19
4-2	CFS-RIB-SU	---	---	---	---	565	25
5-2	CFS-EPO-SU	---	---	---	---	556	23

3.2 *Reinforcing effect per unit CFRP overall volume in ultimate state*

When the overall CFRP volume is expressed as the sum of the volumes of the carbon fiber and plastic matrix, the reinforcing effect in the various grid- and sheet-reinforced test-specimens is as given in Table 5. In that table, the test-specimens serving as the standard against which the increase in ultimate strength is measured is the unreinforced RC test-specimen for the carbon sheet-reinforced test-specimens, while for the carbon grid-reinforced test-specimens it is the RC-PM test-specimen.

Consequently, the table's values for increase in ultimate strength due to CFRP grid reinforcement do not include the increase in ultimate strength produced by the spray-applied polymer-cement mortar. This permits a comparison of the different reinforcing materials.

Fig. 10 shows the reinforcing effect of the different forms (grid and sheet) of CFRP; here the reinforcing effect is taken to equal the ratio of the CFRP overall volume to the increase in tensile strength. In this figure, the larger the graph's gradient, the better the reinforcing effect, relative to the amount of CFRP used.

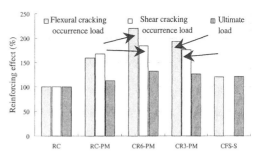

Fig.9 Yield strength and reinforcement effect

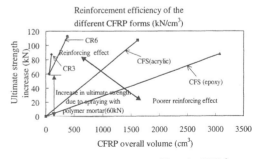

Fig. 10 Reinforcing effect produced by the different by CFRP forms

Table 5. Reinforcing effect due to unit CFRP overall volume

	CFS		CR3	CR6
	Epoxy	Acrylic		
Thickness of CF + plastic (mm)	1.28×2 2.56	0.67× 2=1.34	1.1	4
CFRP overall volume (cm³)	3072	1608	95	380
Ultimate strength of unreinforced test-specimen (kN)	450	450	510	510
Ultimate strength of reinforced test-specimen (kN)	538	557	537	563
Ultimate strength increase (kN)	88	107	27	53
Unit reinforcing effect (kN/cm³)	0.029	0.067	0.28	0.14

3.3 Increase effect of the flexural rigidity due to reinforcing.

Fig. 11 shows the curves for load deflection versus at the center span of beams, for each CFRP form. Although no large difference is observed here between the gradients of the curves for the CFRP grid-reinforced test-specimens and the RC-PM test-specimen, both these curves have a gradient about 20% greater than the unreinforced test-specimen. However, a comparison of the ultimate deflection increase effect in the CFRP grid-reinforced and RC-PM test-specimens shows figures of 10% in the CR3 type test-specimens and of 35% in the CR6 type. Additionally, the CF sheet-reinforced test-specimens have about 7% increase in ultimate deflection compared to the unreinforced test-specimens.

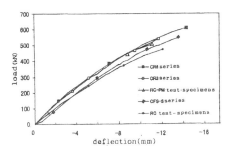

Fig. 11. Curves for load vs. deflection at span central portion

3.4 Comparison of calculated and actual shear strength of beams.

In the equations for computing shear strength that are given in the concrete standard specifications in Japan, the shear strength Vy is the strength at which the stirrup yields. Further, Vc is the strength at which diagonal cracks occur, Vc' the increase in that strength due to the polymer mortar, Vs the due to the stirrup, and Vs' the increase when the value for the CFRP grids is converted into a value for reinforcing steel. The results of calculations using these equations are given in Table 6, where the values in parentheses are the actually measured values.

Table 6 Comparison of calculated and actual values (kN)

	Vc	Vc' Increase due to sprayed mortar	Vs	Vs' Increase due to CFRP	Vy		Ultimate Strength
RC		0		0	307	(323)	451
RC-PM	156 (123)	26 (29)	151 (200)		333	(362)	510
CR3				36	368	(362)	568
CR6				141	474	(365)	594

The actual values are shown in a parenthesis

As can be seen from this table, implementing reinforcing with CFRP grids produces no large change in the actual value measured for Vy, nor any difference between the calculated and measured Vy values. This implies that the CFRP grids contribute almost nothing to the reinforcing effect up to the load at which the stirrup yields. However, the ultimate shear strength continues increasing beyond the stirrup yield point, and from this it can be inferred that the shearing reinforcing effect of the CFRP grids manifests itself from the vicinity of the stirrup yield point onwards.

4 APPLICATION IN A REAL STRUCTURE

The experiments demonstrated that the largest reinforcing effect was produced by a combination of CFRP grids and polymer mortar. Accordingly, this combination was applied to a real structure. The structure selected for this purpose was an RC abutment in which cracks had occurred due to alkali aggregate reaction ; the object of applying the reinforcing was to suppress broadening of the cracks if they reoccurred, and to prevent penetration into the abutment by moisture and carbon oxides, etc., from the exterior. After being implemented, this reinforcement was placed under follow-up observation; as of the present time, no crack reoccurrence, or rising-up of the polymer mortar or other deformation, has been observed.

Fig. 12. Cracks due to alkali aggregate reaction

Fig. 13 Arrangement of CFRP grid

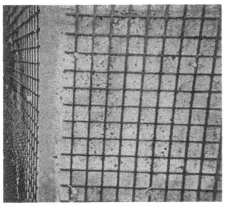

Fig. 14. Detail of CFRP grid

Fig. 15. Spraying of polymer mortar on the CFRP grid

5 CONCLUSION

The research reported in this paper experimentally examined the reinforcing effect that is produced when RC beams are given shear reinforcing consisting of CFRP grids bonded with polymer-cement mortar. The results obtained from this research may be summarized as follows:

1 A shearing reinforcing effect of 15 to 35% was observed in all of the reinforced test-specimens. However, since there were major differences among the test-specimens depending on the amount of CFRP used in them, the reinforcing effect per unit CFRP overall volume was examined. The results of such examination showed that the reinforcing effect provided by the CFRP grids was some ten times greater than that provided by the CF sheets.

2 In the CFRP grid reinforcing, it was found that the increase in the load at which flexural and diagonal cracks occur was even greater (60 to 120%) than that of non-RC beam.

3 In the CFRP grid-reinforced test-specimens, the deflection was suppressed by about 20% compared to the unreinforced test-specimen.

4 It was determined that the onset of the CFRP grids' reinforcing effect occurs in the vicinity of the stirrup yield load.

ACKNOWLEDGEMENTS

Finally, the authors would like to take this opportunity to express their gratitude to all the people who gave their assistance to the execution of this research. This study was carried out as a part of the study group RAMS (Repairing of Materials And Structures) on the repairing and strengthening of structures.

REFERENCES

Atsushi Ono and Asuo Yonekura 1999. Reinforcement Effect of CFRP Surfaces Bonded to RC Girders", in *Papers from the 51st Research Conference of the Civil Engineers' Society's Chugoku Branch*, June 1999, p. 601f

Atsuhiko Machida, ed. 1998. *University Civil Engineering and Reinforced Concrete Engineering*, Ohmsha Corporation Publishing Bureau, April 1998.

16 Concrete material behavior

Creative Systems in Structural and Construction Engineering, Singh (ed.) © *2001 Balkema, Rotterdam, ISBN 90 5809 161 9*

Experimental study on different dowel techniques for shear transfer in wood-concrete composite beams

N. Gattesco

Department of Civil Engineering, University of Udine, Italy

ABSTRACT: An experimental investigation on the behavior of four different connection techniques for wood-concrete composite beams has been carried out. In three cases dowels were anchored to the timber member without any gluing material while in the latter case were fixed with epoxy resin. Moreover, in one connection technique the concrete slab is in contact with the timber element (dowel diameter 12 mm), while in the others a 25 mm boarding is interposed between them (connector diameter 16 mm). A special test method was used in the investigation aimed to simulate as better as possible the actual behavior of dowels in the composite beam. The results evidenced that epoxy resin plays an important role in the connection leading to a significant increase in both stiffness in service and bearing capacity. The connections made without resin evidenced an almost equal stiffness in service, so that they have equal efficiency in terms of deformability. Headed dowels showed a better anchorage to the thin concrete slab than those without head.

1 INTRODUCTION

Most of the old buildings in Europe are made with masonry walls and wooden floors. Such floors are normally designed to support relatively limited live loads and frequently they present an excessive deformability, so that in most cases they need to be strengthened and stiffened in order to satisfy the actual needs. Moreover in seismic areas it is also the necessity to provide to floors a great stiffness in their plane so to restrain the walls perpendicular to the seismic action and to be able to transfer the horizontal forces to shear walls (walls parallel to the seismic action).

A good solution to such needs may be given by the realization of a thin concrete slab over the existing timber deck, adequately connected to the wooden beams, so to obtain a composite wood-concrete system. The connection may be made using different devices: nails, screwed bolts, steel profiled elements, dowels, etc.

Various researchers faced the study of the wood-concrete composite system both to determine adequate solutions for the connection (Gelfi & Giuriani 1999, Piazza & Ballerini 1998, Spinelli 1992, Tacac et al. 1999, Turrini & Piazza 1983), with concern to effectiveness and ease to use, and to model the structural behavior (Giuriani & Frangipane 1993, Gutkowski 1996, Turrini & Piazza 1983).

Most of the available structural numerical models are based on those carried out for steel-concrete composite beams (Newmark et al. 1951, Yam & Chapman 1968). Such models need, beside the mechanical characteristics of the materials of the two elements (concrete slab and timber beam), the load-slip relationship of the shear connection. Even for simplified analyses, at least the stiffness in service and the bearing capacity of the connectors is needed.

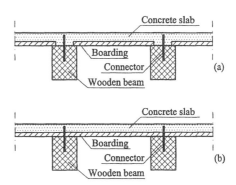

Figure 1. Slab-beam connection: without (a) and with (b) interposed boarding.

The present study concerns an experimental investigation on the behavior of four different simple techniques of dowel connections that may be used in the strengthening of existing wooden floors. One of these techniques requests the partial removal of the floor boarding so to put the slab in contact with the timber beam (Figure 1a); differently in the other techniques the concrete slab is cast over the boarding (Figure 1b). The contribution of the boarding to the stiffness and to the resistance of the connector is very low (Gelfi & Giuriani 1999).

The experimental model was set up properly so to simulate as confidently as possible the actual behavior of the connector in the composite beam. The purpose of the study was to compare the mechanical characteristics of the different dowel techniques (stiffness in service and bearing capacity) as well as to obtain reliable load-slip relationships necessary for theoretical and/or numerical models of the connection behavior.

2 INVESTIGATION PROGRAM

The experimental investigation concern shear tests on four different types of dowel connections for wood-concrete composite beams. Three connection techniques adopt dowels made with smoothed steel rods; the dowels are embedded in the timber member by forcing them, through some hammer blows, inside a drilled hole with a diameter slightly smaller then that of the rod (11.75 mm for 12 mm dowels, 15.75 mm for 16 mm dowels). The fourth connection technique adopts deformed bars as dowels: a bore hole with a diameter 2 mm larger than the connector diameter was drilled in the timber beam and the dowel was fixed with epoxy resin.
The connection techniques studied are in the following detailed.

Type A: a smoothed steel rod (12 mm diameter) is embedded 60 mm (5 times the rod diameter d) in the timber beam and 40 mm in the concrete slab; no boarding between concrete and timber beam.

Type B: a smoothed steel rod (16 mm diameter) is embedded 80 mm (5 d) in the timber beam and 40 mm in the concrete slab; a boarding 25 mm thick is interposed between concrete and timber beam.

Type C: equal to type B but using headed studs (16 mm diameter) as dowels.

Type D: a deformed steel bar (16 mm diameter) is embedded 80 mm (5 d) in the beam with epoxy resin and 40 mm in the concrete slab; a 25 mm boarding is present at the interface.

The concrete slab in all cases is 50 mm thick. The embedding depth of the dowel in the timber member (5 d) assures that the maximum bearing capacity of the dowel is reached, on the base of the theoretical model presented by Gelfi & Giuriani, 1999 (limit analysis).

Connector types B and C are interesting because they need only to drill the hole over the boarding and then to insert the dowels. Provided that the concrete slab is thin (50 mm), the connection type C

with headed studs was considered so to improve the embedment of the stud to the concrete.

3 SPECIMEN FEATURES

Shear tests to determine the connection behavior are normally carried out adopting the standard push-out test, as for steel-concrete composite beams. But this test method do not simulate correctly the actual behavior of the connector in the composite beam, as discussed in detail in Gattesco & Giuriani (1996) for steel-concrete, so that a different test method was used in the present study.

In particular the test is aimed to provoke a pure relative translation of the two members (concrete slab and timber beam), as occurs in the composite beam where the longitudinal shear force causes the slab to slip over the wooden members; the moment due to the eccentricity between the shear force Q and the axial force ΔN is equilibrated by vertical shear ΔV (Figure 2).

Similarly to steel and concrete composite beams (Gattesco & Giuriani, 1996), the specimens were designed so to perform a direct shear test. For such a goal the two opposite forces applied to the specimen (one to the concrete and the other to the timber) have to be aligned and their action line have to include the contraflexure point of the dowel (Figure 3).

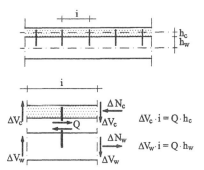

Figure 2. Forces acting on the concrete slab and on the wooden beam due to longitudinal shear.

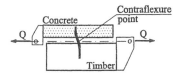

Figure 3. Schematic description of the specimen.

In detail specimens are made with a U-shaped concrete element and a rectangular timber member, connected together by means of one steel dowel

(Figure 4). For type A connection, the concrete element is in contact with the timber member, while for other connection types a 25 mm thick boarding was interposed between them. The U-shape of concrete member is necessary to perform the alignment of the forces applied to the specimen. The geometric characteristics of the specimens are illustrated in Figure 4.

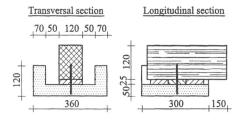

Figure 4. Geometric details of specimens.

The timber elements used for tests were cut from two beams (120x160x4500 mm) of Eastern Alps red spruce. The average relative humidity was 12.5 % and the specific gravity was 0.43. The compressive strength and the elastic modulus in the grain direction were determined from 16 prism samples (55x55x165 mm); the average values were 38.4 MPa and 8956 MPa, respectively.

The concrete of specimens tested were cast in two different times (first time connection types A and B and last time connection types C and D); the first group reached a compressive strength of 26 MPa and a secant modulus of elasticity (40% of strength) of 29600 MPa while the second group reached 36 MPa and 33400 MPa, respectively. Such values are determined from eight cylinders (100 mm diameter, 200 mm height).

The main characteristics of the dowels used for tests are summarized in Table 1.

Table 1. Characteristics of the steel of dowels.

Connection type	Type A	Type B	Type C	Type D
Dowel	φ 12	φ 16	φ 16	φ 16
	smooth	smooth	head stud	defor. bar
Steel	Fe 430	Fe 510	St 37	Fe B 44 k
Yielding stress [MPa]	328	411	480	504
Tensile strength [MPa]	424	570	509	594

4 EXPERIMENTAL APPARATUS AND TEST PROCEDURE

In order to apply forces to the specimen according to Figure 3, adequate steel devices were used (Figure

5a). At one end of the timber member a steel device (a) is fixed by means of one steel ring (b) (60 mm diameter, 20 mm height, 6 mm thick) coupled with a screwed bolt (c) (12 mm diameter) and two transversal bolts (d) M12 cl. 8.8 (Figure 5b). The tension bar (e), which is used to apply the force to the timber element, ends with a hammer head free to move vertically inside a slot; the setting screw (f) allow to register the vertical position of the tension bar.

Another steel device (g) is fixed to the concrete member at the opposite end of the specimen by means of four tension bolts (h) (Figure 5c). The tension bar (i), which is used to apply the force to the concrete member, is detailed as tension bar (e). The setting screws (f) and (j) allow to align the applied forces and to set their action line.

To perform the test, the tension bars (e) and (i) were grasped to the jaws of a hydraulic testing machine. Two inductive transducers (sensitivity 0.002 mm) were used to capture the slip between the two elements (concrete slab and timber beam).

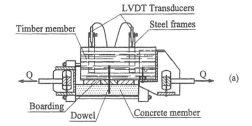

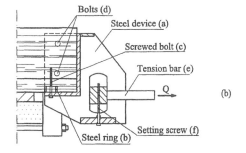

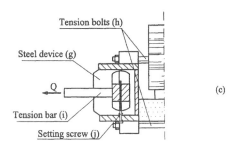

Figure 5. Experimental model: a) complete view, b) details at timber side, c) details at concrete side.

489

A displacement control procedure was adopted imposing a constant slip rate of 0.002 mm/s up to a slip value of 2 mm; then the slip rate was increased to 0.005 mm/s. Tests were stopped when the slip reached 15 mm.

Two more inductive transducers (sensitivity 0.002 mm), mounted on two steel frames fixed to concrete (Figure 5a), were used to check that during loading the relative rotation between timber and concrete members be negligible. To avoid this rotation during the plasticization of the wood beneath the dowel, which causes a shift of the contraflexure point in the dowel, a relative translation of the loading axis was necessary acting simultaneously on the setting screws (f) and (j). Friction between the hammer head of tension bars and the slot surface (Figure 5) was reduced lubricating the moving parts with stearic acid.

5 EXPERIMENTAL RESULTS

In the study 16 specimens were tested: four replications per each connection type. The results are presented in a graphic form in Figure 6, where the shear force Q is plotted against the slip s occurred between the concrete member and the timber element. In particular in Figure 6a the four curves obtained from the specimens arranged with connector type A are reported; similarly in Figures 6b, 6c, 6d the curves of the group of specimens with connection types B, C, D, respectively, are illustrated. The curves do not differ considerably one another.

The load-slip curves show a nonlinear path up to the origin. In fact, the wood and the concrete surrounding the dowel are subjected to plastic deformations rather early causing stress redistributions even for low values of the applied load. Moreover the dowel, which is subjected to shear and flexure, starts to yield for a value of the load considerably lower (~59%) than that which causes the full plasticization of the connector section. However, the curves do not differ appreciably from linearity in the first part up to approximately 50% of the maximum load; beyond this value their curvature increases appreciably. The load-slip curves become almost flat when either the wood (eventually also concrete) beneath the connector is fully plastic or two plastic hinges are formed in the shank of the dowel.

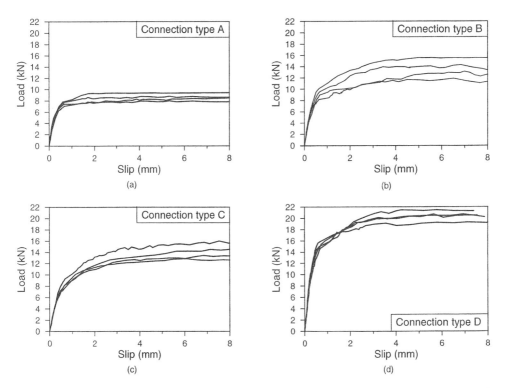

Figure 6. Shear load against slip: a) specimens with connection type A, b) specimens with connection type B, c) specimens with connection type C, d) specimens with connection type D.

Experimental curves are then characterized by three branches: a first branch almost linear up to 50% of the maximum load (linear elastic behavior of the connection), a second branch with increasing curvature (significant plastic deformations involved), a third branch almost flat (perfectly plastic behavior). All the connection types studied did show a very ductile behavior: the collapse was reached for slip values greater than 15 mm.

Characteristic values for the connection are the stiffness in service K_s, represented by the slope of the first branch of the load-slip curve (secant stiffness at 50% of the ultimate load), the conventional yielding load Q_y, which is the ordinate of the intersection point between the two straight lines approximating the first and the third branch of the load-slip curves, and the ultimate load Q_u (value of the load at 8 mm slip). The average values of the four specimens of each group are reported in Table 2.

Table 2. Stiffness in service, yielding load and ultimate load per each connection type.

Connection type	Type A	Type B	Type C	Type D
Secant stiffness [kN/mm]	22.30	20.14	17.66	38.43
Yielding load [kN]	8.35	12.50	12.28	19.60
Ultimate load [kN]	8.79	13.70	14.02	20.40

A comparison between the connection types considered put in evidence that the specimens arranged with connection types B, C, D have an appreciably greater resistance than specimens with connection type A even though the former have the boarding interposed between the concrete slab and the timber beam. The lower resistance of type A connection is, however, mainly due to the lower yielding stress of the steel of dowels with respect to that of the other connection types. So that the greater diameter of the dowels of connection types B, C, D (16 mm against 12 mm) counterbalance effectively the larger distance between dowel plastic hinges due to the boarding. Moreover connection type D shows a greater shear resistance than type B and C because from the one hand the dowels have a higher yielding stress and from the other hand the epoxy resin improves locally the embedding strength of wood.

Secant stiffnesses of connection types A, B and C are not very different one another (a slightly lower value was obtained with connection type C) which means that such connection techniques are almost equivalent in terms of deformability. Instead connection type D showed a considerably higher stiffness due to both the effect of epoxy resin, which improves the dowel-wood interaction, and to the contribution of the boarding considerably increased by epoxy resin. In fact, resin glues both the boarding to the timber beam (no slip at boarding-beam interface) and the dowel perimeter to the boarding.

Only specimens arranged with the connection technique D evidenced one or more radial splitting cracks in the concrete slab; such cracks occurred for slip values greater than 4÷5 mm. Not appreciable drops in resistance were noted after cracking.

At the end of the test the specimens with connection types A and B evidenced an appreciable local damage of the concrete beneath the dowel (crushing). On the contrary, negligible concrete damage was observed in specimens with connection type C. The reason of this different behavior is partly due to the different compressive strength of concrete but mainly is due to the better anchorage of the headed stud in the thin concrete slab.

6 CONCLUSIONS

The results of the experimental study carried out on four different connection techniques for shear transfer in wood-concrete composite beams allow to draw the following concluding remarks.

1 A special test method to study the behavior of the single connector was set up with the purpose to simulate as confidently as possible the actual behavior of the connector in the composite beam.

2 All connections studied evidenced a very high ductility (more than 15 mm slip without appreciable drops in resistance).

3 The load-slip relationships are characterized by an almost linear first branch, a curved second branch and a flat third branch. Such curves allow to determine the stiffness in service and the ultimate load of the connection as well as to analytically model the complete load-slip relationship to be used in numerical structural models.

4 Dowel connections embedded in the timber member without any gluing material (type A, B, C) may be considered almost equivalent in terms of deformability; greater bearing capacity was obtained for connection types B, C mainly because a higher yielding stress was used for the dowel.

5 Same connection efficiency may be obtained in a composite beam with the boarding interposed between the concrete slab and the timber beam by using a higher dowel diameter than that needed for a composite beam without boarding. No reduction in ductility was noted.

6 Headed studs (type C) showed a better anchorage to the thin concrete slab (limited crushing).

7 Dowels embedded in the timber member with epoxy resin (type D) evidenced considerably higher stiffness and resistance than other connection types studied. The resin improves both the dowel-wood interaction and the contribution of the boarding.

8 To extend such considerations to other connector diameters further experimental tests are needed.

ACKNOWLEDGEMENT

The financial support of the Italian Ministry of University and Scientific Research (MURST) is gratefully acknowledged. The author wish to thank Mr. Roberto Timeus for his help during the execution of the experimental tests.

REFERENCES

Gattesco, N. & Giuriani, E. 1996. Experimental study on stud shear connectors subjected to cyclic loading. *J. Construct. Steel Res.*. 38(1): 1-21.

Gelfi, P. & Giuriani, E. 1999. Stud shear connectors in wood-concrete composite beams. *Proc. 1st RILEM Symp. on Timber Engng. Stockolm, Sept. 1999*: 245-254.

Giuriani, E. & Frangipane, A. 1993. Wood-to-concrete composite section for stiffening of ancient wooden beam floors. *Dip. Mecc. Strutt., Univ. di Trento, I° Workshop Italiano sulle Strutture Composte, Trento, Giu. 1993*: 308-318.

Gutkowski, R.M. 1996. Test and analysis of mixed concrete-wood beams. *Proc. Int. Wood Engng. Conf., New Orleans, La.* (Omnipress, Madison, WI, 1996): 3.436-3.442.

Newmark, N.M., Siess, C.P. & Viest, I.M. 1951. Tests and analysis of composite beams with incomplete interaction. *Proc. Society for Experimental Stress Analysis.* 1: 75-92.

Piazza, M. & Ballerini, M. 1998. Composite wood-concrete floors: experimental comparison between different connection types. (*in Italian*). *Ist. di Scienza e Tecnica delle Costruzioni, Univ. di Ancona, III° Workshop Italiano sulle Strutture Composte, Ancona, Ott. 1998*: 349-368.

Tacac, S., Plazibat-Loncaric, S. & Bogicevic, P. 1999. Wood-concrete composite structures joined by special type dowels. *Proc. 1st RILEM Symp. on Timber Engng. Stockolm, Sept. 1999*: 255-262.

Turrini, G. & Piazza, M. 1983. A technique to restore wooden floors. (*in Italian*). *Recuperare.* Anno II, 5: 224-237.

Spinelli, P. 1992. A simplified method for the design of wood-concrete structural elements. (*in Italian*). *Bollettino degli Ingegneri.* Anno XXXIX, 10, Firenze: 7-10.

Yam, L.C.P. & Chapman, J.C. 1968. The inelastic behaviour of simply supported composite beams of steel and concrete. *Proc. Instn. Civ. Engrs.*. Part II, 41, Dec.

Creative Systems in Structural and Construction Engineering, Singh (ed.) © 2001 Balkema, Rotterdam, ISBN 90 5809 161 9

Microstructure model for estimating early-age concrete strength

S. D. Hwang & K. M. Lee
Department of Civil Engineering, University of Sungkyunkwan, Korea

J. K. Kim
Department of Civil Engineering, Korea Advanced Institute of Science and Technology, Korea

J. H. Kim
Department of Civil and Environmental Engineering, Sejong University, Korea

ABSTRACT: In this study, a microstructure model for the estimation of concrete strength considering the degree of hydration and total porosity of concrete is proposed. To establish a relationship between concrete strength and degree of hydration, theoretical hydration modeling and compressive test were performed. The test results showed that the effect of hydration on strength development changes according to curing ages, water/cement ratio, and curing temperature. The comparison of test results with estimated strength shows that the microstructure model can estimate compressive strength of early-age concrete with ages.

1 INTRODUCTION

Many researches have been attempted to predict concrete strength from which several types of strength estimation models were proposed. The most representative strength models are ACI model (ACI Committee 209, 1992), CEB-FIP model (CEB-FIP, 1990), and maturity models (Oluokun, 1990). The models can easily predict concrete strength with ages based on 28-day strength but do not consider various factors influencing on concrete strength.

Recently, the degree of hydration was employed in the strength predicting models (Bentz, 1999; Kishi, 1993). Degree of hydration is closely related to the development of concrete strength, because hydrates formed by hydration are the building block of the concrete strength.

In this study, a microstructure model estimating strength is proposed as a function of degree of hydration and total porosity in concrete. The evaluated model parameters from the prediction model agree well with the present and past test results.

2 STRENGTH ESTIMATION MODEL

2.1 Hydration model

A theoretical hydration model suggested by Byfors (1980) is adopted in this work. In this model, degree of hydration is described as

$$\alpha_c = \exp\left\{-\lambda_1\left\{\ln(1+\frac{t_{eq}^c}{t_1})\right\}^{-k_1}\right\} \tag{1}$$

where t_{eq}^c = the equivalent maturity time, λ_1, t_1, and k_1 = material constants, and t_1 decreases but λ_1 increases as water/cement ratio decrease.

The equivalent maturity time for cement t_{eq}^c is given by

$$t_{eq}^c = \int_0^t \beta_T \beta_{w/c} \beta_w \, dt \tag{2}$$

where β_T, $\beta_{w/c}$ and β_w are the factors that consider a curing temperature, water/cement ratio, and water distribution, respectively.

The factor expressing the influence of curing temperature β_T is presented as Arrhenius-type rate equation for thermal activation (Hansen and Pedersen, 1977; Byfors, 1980).

$$\beta_T = \exp\left(\frac{E}{R}\left(\frac{1}{T_{ref}+273} - \frac{1}{T+273}\right)\right) \tag{3}$$

where E/R = activation temperature (K), T_{ref} = reference temperature (20$\square$). The temperature dependency of the activation temperature is described as

$$\frac{E}{R} = \theta = \theta_{ref}\left(\frac{30}{T+10}\right)^{k_3} \tag{4}$$

where the parameters θ_{ref} and k_3 are material constants (Byfors, 1980; Jonasson, 1994; Hedlund, 1996).

The factor for water/cement ratio $\beta_{w/c}$ is described by maximum degree of hydration α_{max}.

$$\beta_{w/c} = \left(\frac{\alpha_{max} - \alpha_c}{\alpha_{max}}\right)^r \tag{5}$$

$$\alpha_{max} = \frac{1.031 \times W/C}{0.194 + W/C} \tag{6}$$

Degree of hydration was calculated by the theoretical hydration model associated with curing temperature, relative humidity, water/cement ratio, and distributions of pore and water.

2.2 *Porosity of concrete*

Strength of concrete is influenced by total volume of various types of voids in concrete (i.e., entrapped air, capillary pores, gel pores, and entrained air, et cetera). However, the most important voids influencing strength of concrete are capillary pores with diameter over 50 nm. The other types of pores have effects on shrinkage rather than strength of concrete. Thus, in this study, capillary pores and air content of fresh concrete were considered in estimating strength of concrete. Total porosity in concrete with ages can be estimated by degree of hydration calculated by hydration model presented in section 2.1 of this paper.

2.3 *Microstructure model*

Degree of hydration, porosity of concrete, and water/cement ratio were employed in a microstructure model for estimating incremental concrete strength is given by

$$df_c' = f_{c28} \times \log[100 - P(t)] \times d\alpha^{n(t)} \qquad (7)$$

$$n(t) = n_{28} \times \exp\left[s\left(1 - \frac{28}{t}\right) \right] \qquad (8)$$

where t = time (day), $P(t)$ = total porosity of concrete with ages, $n(t)$ = material property associated with effects of hydration increment on strength increment, f_{c28} = compressive strength at 28 days expressed as a function of water/cement ratio, n_{28} = n value at 28 days, and s = material parameter representing the shape of n(t) curve.

3 EXPERIMENTAL WORK

3.1 *Scope*

An experiment was conducted to measure compressive strength of concrete. Test parameters are (1) w/c ratio – 0.58 to 0.46, (2) curing temperature – 10, 20, 30 □, (3) curing age of concrete – 1, 2, 3, 7, 28 day(s).

3.2 *Materials and mix proportion*

Test specimens were made with Type□Portland cement, coarse aggregates with a nominal maximum size of 25 mm, specific gravity of 2.65 and fineness

modulus of 6.50, and fine aggregates with specific gravity of 2.70 and fineness modulus of 2.35. Table 1 lists the mix proportions of four types of concrete, where the mix type C320 represents concrete with unit cement content of 320 kg/m^3.

Table 1. Mix proportions of specimens (kg/m^3).

Mix Type	W/C	W	C	Coarse Agg.	Fine Agg.	S/a (%)	Air (%)
C320	0.58	185	320	1025.6	712.7	41	5.0
C350	0.53	185	350	1008.3	703.1	41	5.0
C370	0.50	185	370	1016.2	680.8	40	5.0
C400	0.46	185	400	1032.0	691.0	40	5.0

3.3 *Testing procedure*

All specimens were stripped after placed in controlled humidity room for 1 day and transferred to a water bath at curing temperatures of 10, 20, 30 □, respectively.

UTM with 1,000 kN load capacity was used to test compressive strength of concrete cylinders. Three specimens from each mix type were tested at ages of 1, 2, 3, 7, 28 days, respectively. Specimens were loaded at a rate of 140 to 340 kPa/sec as recommended by ASTM C39.

4 RESULTS AND DISCUSSIONS

4.1 *Compressive strength*

Table 2 tabulates the experimental compressive strength results of four types of concrete with various curing ages.

Table 2. Compressive strength of concrete (MPa).

Mix.	Temp. (°C)	Age (day)				
		1	2	3	7	28
	10	0.8	4.6	8.4	16.1	27.4
C320	20	6.0	10.0	13.8	21.3	28.2
	30	7.5	11.9	14.8	21.6	27.7
C350	20	7.2	11.8	17.7	26.6	35.6
	10	1.2	8.0	13.5	23.9	35.5
C370	20	8.6	15.1	19.2	27.5	37.1
	30	11.4	17.4	21.7	30.0	35.7
C400	20	10.0	16.2	22.4	32.1	39.6

4.2 *Determination of parameters*

Figure 1 shows a plot of parameter $n(t)$ versus ages which represents a relationship between an increment of degree of hydration and an increment of compressive strength with ages. The value of $n(t)$ is dependent on the curing temperature.

At early age, the strength increase of concrete cured at 10 °C is much smaller than that of concrete cured at 20 °C while the degree of hydration is relatively not much different from each other. Therefore, the effects of hydration on strength development of concrete cured at 10°C are less than that of concrete cured at 20°C. Thus, n value in curing temperature 10 °C at early age is higher than that in 20 °C. On the contrary, the effects of hydration in early age is very high in curing temperature of 30 °C, because the strength increase in concrete cured at 30°C in early age is higher than that of concrete cured at 20°C. However, an effect of hydration on strength development is reduced when compared to those at temperature 20 °C in later age.

In this model, factor s in Eq. (10) is used to determine the shape of $n(t)$ curve. It is found that s linearly increases with curing temperature. A constant n_{28} in Eq. (9) represents the relationship between an increment of degree of hydration and strength increment at 28 days, which varies according to curing temperature as well.

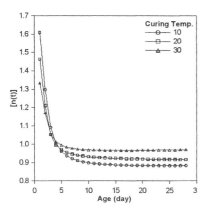

Figure 1. Effect of hydration on concrete strength.

The test results show that f_{c28} representing compressive strength at 28 days is inversely proportional to water/cement ratio. Porosity with ages is presented in Figure 2. Compressive strengths at 28 days with water/cement ratio are presented in Figure 3.

Consequently, parameters in Eqs. (7) and (8) are determined from the test results as

$$n_{28} = 0.0041T + 0.83 \qquad (9)$$

$$s = 0.00055T - 0.029 \qquad (10)$$

$$f_{c28} = \frac{1.867}{0.042 + (w/c)^{6.75}} \qquad (11)$$

where T = temperature ($\square$).

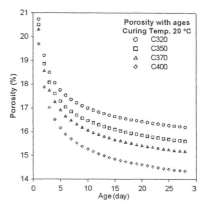

Figure 2. Porosity of concrete with ages

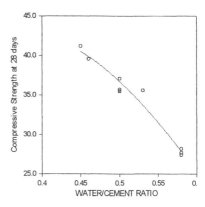

Figure 3. Water/cement ratio vs. compressive strength at 28 days.

4.3 Comparison of test results with strength estimation model

Figure 4 shows that the accuracy of strength prediction in curing temperature of 30$\square$ is a little lower than curing temperature of 10 and 20$\square$. This trend might be dependent on water temperature at the time of mixing. Moreover, it can be inferred that strength revelation ratio in early age is slightly higher in concrete with lower water/cement ratio and higher curing temperature.

Figure 4 reveals that the strength prediction model proposed in this study can estimate compressive strength of concrete with curing temperature and ages within an acceptable margin of error.

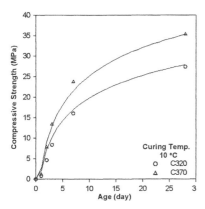

Figure 4(a). Comparison of test results with estimated strength by Eq. (7) in curing temperature 10° C

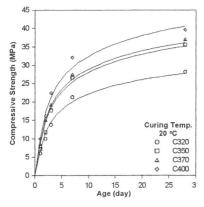

Figure 4(b). Comparison of test results with estimated strength by Eq. (7) in curing temperature 20° C

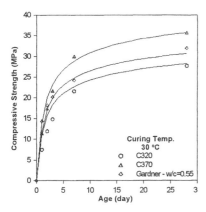

Figure 4(c). Comparison of test results with estimated strength by Eq. (7) in curing temperature 30° C

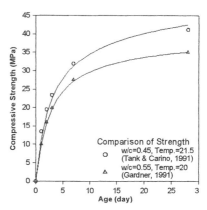

Figure 4(d). Comparison of test results with estimated strength by Eq. (7).

5 CONCLUSION

The conclusions drawn from this study are as follow:

1. A microstructure model is proposed to estimate compressive strength of early-age concrete. Degree of hydration and porosity of concrete are employed in the model. Curing temperature and water-cement ratio that strongly influence the strength development of concrete are also considered in the proposed model.

2. Test results are compared with the values estimated by the model and good agreements are obtained. Therefore, it is concluded that the strength model proposed in this study can be used to estimate compressive strength of early-age concrete with curing ages and temperature.

ACKNOWLEDGEMENTS

The authors would like to express their appreciation for the financial support provided by KISTEP.

REFERENCE

Bentz, D. P. 1999. "Modeling Cement Microstructure: Pixels, particles, and property prediction," *Materials and Structures/Materiaux et Constructions*, Vol. 32, pp. 187-195.

Byfors, J. 1980. *Plain concrete at early ages*, CBI report FO 3:8, Sweden.

Comite Euro International Du Beton 1990. CEB-FIP MODEL CODE(DESIGN CODE), pp. 33□81.

Gardner, N. J. 1990. "Effect of Temperature on the Early-Age Properties of Type□, Type □, Type /Fly Ash Concretes," *ACI Materials Journal*, Vol. 87, No. 1, pp. 68-78.

Hansen, F. P. and Pedersen, E. J. 1977. "Maturity Computer for Controlled Curing and Hardening of Concrete," *Journal of the Nordic Federation*, No 1, pp. 21-25, Stockholm, Sweden.

Jonasson, J. E. 1994. *Modeling of Temperature, Moisture, and Stresses in Young Concrete*, PhD dissertation, Lulea University of Technology, Lulea, Sweden, No. 153D.

Kishi, T. and Maekawa, K. 1993. "Multi-component Model for Hydration Heat of Concrete based on Cement Mineral Compounds," *Proceedings of the JCI,* Vol. 15, No. 1, pp. 1211-1216.

Mehta, P.K. and Monteiro, P. J.M. 1993. *Concrete,* second edition, Prentice Hall, pp. 57.

Neville, A. M. 1996. *Property of Concrete,* 4th edition, pp. 269-279.

Oluokun, F. A., Burdette, E. G., and Deatherage, J. H. 1990. "Early-Age Concrete Strength Prediction by Maturity Another Look," *ACI Materials Journal,* Vol. 87, No. 6, pp. 565-572.

Tank, R. C. and Carino, N. J. 1991. "Rate Constant Functions for Strength Development of Concrete," ACI Materials Journal, Vol. 88, No. 1, pp. 74-83.

Creative Systems in Structural and Construction Engineering, Singh (ed.) © 2001 Balkema, Rotterdam, ISBN 90 5809 161 9

Application of nonlinear fracture mechanics on concrete structures

A. S. Morgan & T. Ishikawa
Nihon Hitek Corporation, Tokyo, Japan

J. Niwa
Tokyo Institute of Technology, Japan

T. Tanaba
Nagoya University, Japan

ABSTRACT: One of the goals of this paper is to apply nonlinear fracture mechanics to design problems encountered by engineering society. The nonlinear fracture mechanics approach was utilized to solve the flexure strength problems in 2 dimensional analysis, the pullout test problem in axisymmetric analysis and torsion strength problem in 3-dimentional analysis. The size effect on ultimate strength is detected and compared with empirical equations for all investigated cases. Using Arc-length solution technique, the post peak behavior is captured and discussed.

1 INTRODUCTION

The increasing size of concrete structures in recent years is making size effect consideration of growing importance. With the introduction of fracture mechanics, however, it became clear that the tensile properties play a dominant role in the failure of concrete structures where a discrete crack is formed in continuum and it is predominant for the failure of the whole structure. The behavior of concrete under tensile loading in conjunction with strain gradients can be described by means of the so-called fictitious crack model proposed by Hillerborg (1976 & 1983) and modified by Ingraffea & Gerstle (1984).

The two orthogonal rod elements, originally proposed by Ngo & Scordelis (1967), have been commonly used for modeling the bond-slip behavior of reinforced concrete structures. The expansion of the orthogonal rod elements concept to the nonlinear fracture mechanics approach to simulate concrete structure failure, can be considered as a successful new technique for failure expression. In this approach the fracture energy of concrete G_F is associated with a stress-crack width curve, consequently, the strain can be calculated that is equivalent to the crack width of failure surface separation, and then the corresponding stress can be obtained. Moreover, the arc-length method was employed to capture the post peak behavior; therefore, the failure type whether it is ductile or brittle with or without snap-back can be recognized.

In this study, the validity of the extended fictitious crack approach in calculating the ultimate strength of a concrete structure will be discussed, and the comparison of the numerical results with empirical equations will be carried out. The comparison with experimental results can certainly be done only for the small specimens, but for large specimens, it is very difficult to perform the experiments.

2 FLEXURE FAILURE OF PLAIN CONCRETE BEAMS

The concrete beam geometry and the chosen finite element mesh are illustrated in Figure 1. Six different sizes were considered h=100,500,1000,2000, 3000 and 5000 mm. The concrete properties are identical for all six beams are: f'_c=30.0 MPa, f_t=3.0 MPa, G_F=100 N/m and E_C=30.0 GPa.

Two sets of calculations are performed. One is concentrated load at distance 2h from the support as shown in Figure 1 and the other is uniformly distributed loads along the top edge of the beam. The failure was initiated by formation of a crack process zone with a discrete crack in the region of tensile stresses along the beam centerline.

Figure 2 shows the tendency of flexure strength to decrease with the increase in beam size in both cases of loading. This behavior is known as the size effect. Also, as shown in Figure 2 the size effect disappears for large heights and the peak stress of the beams tends to become equals to the tensile strength of concrete. These results was supported by Carpinteri multifractal scaling law (1994), which noted that the scale effect should vanish in the limit of structure sized tending to infinite, where an asymptotic finite strength can be determined. On the other hand, for small specimens (i.e. small when compared to the microstructural characteristic size), the effect of disordered microstructure becomes progressively more important, and the strength increases with decreasing size, ideally tending to infinite as the size tends to zero.

The solution was obtained by the arc-length control and the post peak behavior can be achieved as shown in Figures 3 & 4, which show the load-displacement, diagrams of beam sizes 100 mm and 5000 mm, respectively. The degree of brittleness or ductility exhibited by a concrete structure in response to external loading depends on the size of the structure. From Figures 3 & 4 It can be concluded that the flexural behavior of plain concrete beams is significantly affected by the beam size and the small beams fail in ductile or plastic manner, while large beams of the same material fail in a brittle, and often, catastrophic manner. For more details, refer to Morgan et al. (1997).

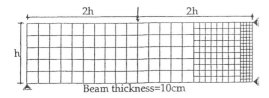

Fig. 1 The mesh for plain concrete beams

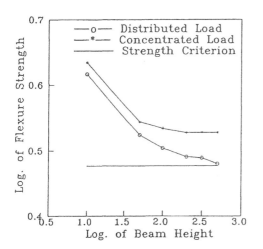

Fig. 2 Predicted size effect on flexure strength

3 THE PULL-OUT TESTS

In this investigation, three different support conditions were considered. The first support condition case is carried out by considering the reaction ring as inverted roller support on the top surface of the specimen (Fig. 5). The second case is analyzed without ring support, but by considering hinged sup-

port at the top edge surface of the specimen (Fig. 5) The third case is carried out without considering neither the reaction ring support, nor the top edge support. In this analysis the pullout failure is simulated by the dominant failure cause of circumferential cracking, and no hoop cracking is considered, refer also to Sonobe et al. (1994)

3.1 Studying the size effect on the inclination of the failure cone surface

Nine different embedded depths are considered, such as d=50, 150, 450, 600, 1000, 2000, 5000, 10000, and 12500 mm. The concrete properties are identical for all nine concrete blocks: f'_c=30.0 MPa, f_t=3.0 MPa, G_F=100 N/m, and E_C=30.0 GPa. The failure cone surface is assumed to be oriented at angle ranges between (26°-76°).

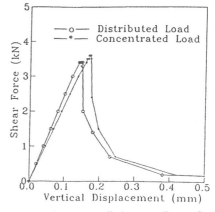

Fig. 3 Shear force versus displacement diagram for 100 mm beam height

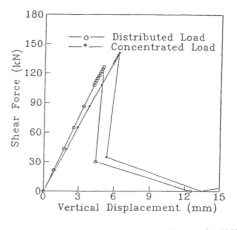

Fig. 4 Shear force versus displacement diagram for 5000 mm beam height

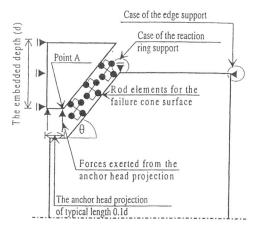

Fig. 5 Schematic diagram to illustrate failure cone surface and boundary conditions in pullout tests

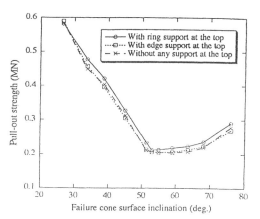

Fig. 6 Variation of pullout strength w.r.t. cone failure inclination angle θ (Embedded depth = 150 mm)

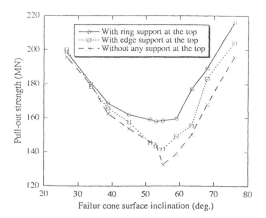

Fig. 7 Variation of pullout strength w.r.t. cone failure inclination angle θ (Embedded depth = 5000 mm)

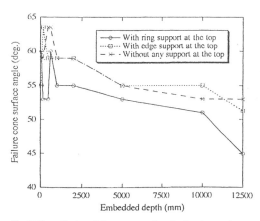

Fig. 8 Size effect on failure cone surface inclination angle

Sample of the results is shown in figures. 6 & 7. It has been found that the inclination of the failure cone surface which gives the minimum pull-out strength is about (60°) for embedded depths up to 2000 mm, for more details refer to Morgan et al. (1999). On the other hand, for huge embedded depths such as 5000 mm or more, it has been found that the inclination of the failure cone surface, which gives the minimum pullout strength, is about (50°). Figure 8 shows that the size effect on the failure cone surface inclination angle. The figure illustrates that for huge embedded depth, the cone failure surface inclination getting more flat than the case of small embedded depths. Also, it can be noticed that the inclination of the failure cone surface, which gives the minimum pullout strength in the second and third supporting condition cases is getting more steeper than the case of the reaction ring support, i.e. the produced failure cone surface area becomes smaller, consequently the ultimate pullout strength. The previous conclusion indicates that the commonly adapted method assuming 45° failure surface yields exaggerated resisting load. The pullout strength of cone failure is mainly dependent on mode I fracture.

3.2 Size effect analysis

The resulting minimum pullout strengths from the results are utilized for this study. Figure 9 shows the tendency of the nominal pullout strength to decrease with the increase in the embedded depth whatever the boundary condition is. This behavior is known as the size effect.

Figure 9 shows that, the size effect moderates for large embedded depths, and the nominal pullout strength of the concrete blocks tends to be bounded with a certain limit. Figure 9 proves that the proposed analytical model can predict the size effect of Pullout tests.

501

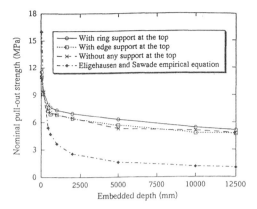

Fig. 9 Predicted size effect on pullout strength

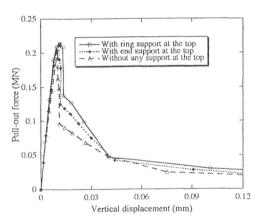

Fig. 10 Pullout force versus displacement diagram for 150 mm embedded depth

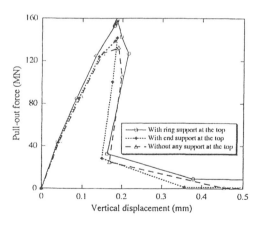

Fig. 11 Pullout force versus displacement diagram for 150 mm embedded depth

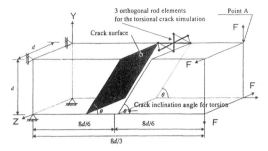

Fig. 12 Schematic diagram to illustrate the failure surface and the boundary conditions

It is better to mention that Eligehausen & Sawade (1989) empirical equation was verified up to 450 mm. Moreover, Figure 9 shows that the presented results have a rather good agreement with Eligehausen & Sawade empirical equation for embedded depths up 150 mm.

Figures 10 & 11 show the load-displacement diagrams of embedded depths 150 mm and 5000 mm, respectively for the three boundary conditions. By using the arc-length method, the full pullout force versus displacement diagram can be obtained. The convergence criterion is maintained in all load levels before and after the peak load.

In the case of small embedded depths such as 150mm, Figure 10 shows that the failure is ductile and the snap back phenomenon will not occur for such small embedded depths whatever the boundary condition is. On the other hand, for large embedded depths such as 5000 mm, Figure 11 shows a post peak snap back response, which reflects the brittle behavior of such large embedded depths.

4 THE RESULTS OF THREE DIMENSIONAL TORSION STRENGTH ANALYSIS

Figure 12 shows the schematic diagram of beam geometry, which is used, in the numerical study. The geometrical sizes of the analyzed model are taken the same as those tested by Bazant et al. (1988).

The numerical results were compared with Bazant's size effect equation, which fits the test results. The relevant fracture material parameters are determined to be the same as in the experiments by Bassinet et al. (1988). Concrete fracture properties are identical for all seven concrete beams as follows: tensile strength f_t=2.7 Map, fracture energy G_F=100.0 N/m, and Young's modulus E_C=35.0 Gap. In contrast to the experiments, in which the maximum size of the beam was d = 150 mm, the numerical analysis for seven geometrically similar beams with d = 37.5, 75, 150, 200, 300, 750, and 900 mm has been carried out.

502

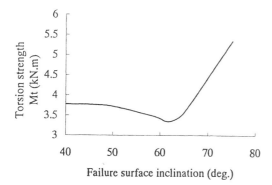

Fig. 13 Variation of torsion strength w.r.t. cone failure inclination angle θ (d = 150 mm)

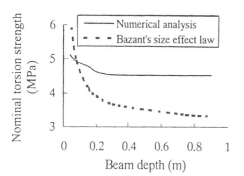

Fig. 14 Predicted size effect on torsion strength

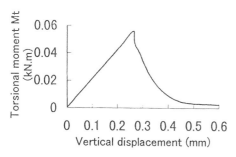

Fig. 15 Torsion moment versus displacement diagram for d = 37.5 mm

The failure surface is assumed to be oriented at angles between (40°-75°). Within this range, many inclination angles were selected to perform this study. For the sake of studying the size effect on the crack inclination that gives the minimum torsion strength, 5 beams were selected with d = 37.5, 75, 150, 200 and 750 mm. Sample of the results is shown in Figure 12 It has been found that the inclination of the failure surface to give the minimum torsion strength ($Mt = 2Fd$) is about 62° for all selected beams up to 750 mm depth. For more details

refer to Morgan et al. (1991). Also, it is noticed that the inclination of failure surface is almost the same for all concrete beams, which reflects that there is no size effect on the inclination angle of the failure surface can be detected. This result indicates that the commonly adapted method assuming 45° torsional failure surface yields exaggerated resisting torsion strength. The torsion strength is mainly dependent on mode I fracture.

For the seek of studying the size effect on torsion strength of concrete beams, the inclination angle of 62° which gives the minimum strength is adopted for the analyzed beams.

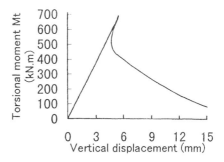

Fig. 16 Torsion moment versus displacement diagram for d = 900 mm

Figure 14 shows the tendency of the nominal torsion strength to decrease with the increase in the beam depth. On the other hand, the size effect moderates for large depths and the nominal torsion strength of the concrete beams tends to be bounded with a certain limit. Figure 14 shows that the presented results have rather good agreement with Bazant's size effect law for depths up to 150 mm.

Figures 15 & 16 show the load-displacement diagrams for the case of beam depths 37.5 mm and 900 mm, respectively. In the both figures the vertical displacements are determined at the point A (Fig. 12).

In the case of small beam depths such as 37.5 mm, Fig.15 shows that the failure mode is ductile and the snap back phenomenon does not take place. On the other hand, for large beam depths such as 900 mm, Figure 16 shows a post peak snap back response, which reflects the brittle behavior of such large beams. The snap back occurs, as a result of a bifurcation process, which leads to a sudden drop in both load and deflection. It can be concluded that the torsion failure of concrete beams is significantly affected by the beam depth. The failure mode changes from ductile to brittle as the size of concrete beam increases. In other words, fracture of concrete leads to brittle failures due to the size effect of decreasing nominal strength with increasing structural size.

5 CONCLUSIONS

The nonlinear fracture mechanics approach offers a possibility to explain the size effect in flexure strength of concrete beams, pullout tests of anchor bolts and torsion strength of concrete beams. Incorporation of the arc-length method enables detecting the post peak behavior even for snap back instability. It has been observed that for small structure size the ultimate strength capacity is profoundly affected by the size effect. On the other hand, for large structure size, the numerical predictions showed that the size effect becomes insignificant. Moreover, the snap back phenomenon occurs when the structure size increase, and the brittle behavior of concrete blocks becomes significant. In other words, the behavior of the concrete structures changes from ductile to brittle if the structure size increases. Also, it is found that the failure cone surface with inclination angle about (60°) gives the minimum pull-out strength for a wide range of embedded depths up to 2000mm, on the other hand for huge embedded depths such as 5000mm, or more the inclination of the failure cone surface with range (50°) gives the minimum pull-out strength. Moreover, in the case of considering edge support or in the case of considering top free surface, the inclination of the failure cone getting steeper, consequently the ultimate pull-out strength becomes smaller than the case of considering the reaction ring support. The previous conclusion indicates that the commonly adapted method assuming 45° failure surface yields exaggerated resisting load. The pullout strength of cone failure is mainly dependent on mode I fracture.

Also, it is found that the inclination of torsional failure surface with angle 62° gives the minimum torsion strength. This result indicates that the commonly adapted method assuming 45° torsional failure surface yields exaggerated resisting torsion strength. The torsion strength is mainly dependent on mode I fracture.

REFERENCES

Bazant, Z.P., Sener, S. and Prat, P. 1988. Size Effect Tests of Torsional Failure of Plain and Reinforced Concrete Beams. *Materials and Structures*, Vol. 21, pp. 425-430

Carpintary, A., Chiaia, B. and Ferro, G. 1994. Multifractal Scaling Law for the Nominal Strength Variation of Concrete Structures. *Size Effect in Concrete Structures*, Edited by Mihashi, H., Okamura, H. and Bazant, Z.P., JCI International Workshop, Sendai, Japan, pp. 193 - 206.

Eligehausen, R. and Sawade, G. A., 1989. Fracture Mechanics Based Description of the Pull-out Behavior of Headed Studs Embedded in Concrete. *RILEM report on Fracture Mechanics on Concrete - Theory and Application*, Chapman Hall Ltd, London.

Hillerborg, A., Modeer, M. and Peterson, P.E. 1976. Analysis of Crack Formation and Crack Growth in Concrete by Means of Fracture Mechanics and Finite Elements. *Cement and Concrete Research*, Vol. 6, pp. 773-782.

Hillerborg, A. 1983 Analysis of One Single Crack. *Developments in Civil Engineering, Fracture Mechanics of Concrete*, Elsevier, London, Vol. 7.

Ingraffea, A.R. and Gerstle, W.H. 1984. Nonlinear Fracture Model for Discrete Crack – propagation. *Proc., Research Workshop on Application of Fracture Mechanics to Cementitious Composites*, Northwestern University, Evanston.

Morgan, A. S.E., Niwa, J., and Tanabe, T. 1997. Detecting the Size Effect of Concrete Beams Based on Nonlinear Fracture Mechanics. *Journal of Engineering Structures*, Vol. 19, No. 8, pp. 605-616.

Morgan, A. S.E., Niwa, J., and Tanabe, T. 1999. Size Effect Analysis for Pull-out Strength Under Various Boundary Conditions. *ASCE Journal of Engineering Mechanics*, Vol. 125, No. 2, February, pp. 165-173.

Morgan, A. S.E., Niwa, J., Nishigaki, Y., and Tanabe, T. 1999. Predicting The Size Effect on Concrete Beams in Torsion Based on Fictitious Crack Approach. *Proc. of JCI Transaction*, Japan, Vol. 21, pp. 377-384.

Ngo, D., and Scordelies, A. C. 1967. Finite Element Analysis of Reinforced Concrete Beams. *Journal of ACI*, Vol. 64, No. 3, pp. 152-163.

Sonobe, Y., Tannabe, S., Yokozawa, K. and Mishima, T. 1994. Experimental Study on Size Effect in Pull-out Shear Using Full Size Footings. *Size Effect in Concrete Structures*, Edited by Mihashi, H., Okamura, H. and Bazant, Z.P., JCI International Workshop, Sendai, Japan, pp. 323 – 333.

Creative Systems in Structural and Construction Engineering, Singh (ed.) © 2001 Balkema, Rotterdam, ISBN 90 5809 161 9

Torsional capacity of normal and high-strength concrete deep beams – General review

T.A.Samman
Civil Engineering Department, King Abdulaziz University, Jeddah, Saudi Arabia

ABSTRACT: The review of literature indicates that any investigation on torsion of concrete members reported so far has been largely concerned with normal and high-strength concrete shallow beams with and without a transverse opening. The study of deep beams has been done mostly to study the shear and flexural behavior of normal-strength concrete beams. The works dealing with torsion in deep beams with and without a transverse opening reported so far also investigated normal-strength concrete members. One research work appears to have been carried out so far on the torsional behavior of high-strength concrete deep beams with and without a transverse opening. Two studies appear to have been carried out so far on normal and high-strength concrete deep T-beams under torsion. Accordingly a literature survey concerning the theoretical torsional capacity of normal and high-strength concrete deep beams with and without transverse openings is presented.

1 INTRODUCTION

Over the last three decades, a large number of shallow beams with concrete strength up to 41 MPa have been tested under either pure torsion or with combined loading conditions. The results of these studies have been the primary input in the development of the current torsion design recommendations (ACI Committee 318, 1995). However, these guidelines do not have any provision for the design of deep beams under torsion. Structural members with span/depth ratios less than five and loaded on the compression faces are classified by the ACI code (ACI Committee 318, 1995) as deep beams.

In many instances, structural members with low span/depth ratios are used. These include transfer girders in multistory buildings providing column offsets, rectangular tank walls, floor diaphragms and shear walls (Hassoun, 1998). Shear tests (Chow et. al. 1953, De Paiva and Seiss 1965, Kong and Robins 1971, Ramakrishnan and Ananthanarayan 1968, Desayi 1974, Rogowsky et. al. 1986, Kong and Sharp 1977, Kong et. al. 1978) on normal-strength concrete members with low span/depth ratios have revealed important differences in their behavior compared to members with usual proportions. This has led the ACI Code (ACI Committee 318, 1995) to specify special shear design provisions for flexural members with span/depth ratio less than five. A recent study (Akhtaruzzaman and Hasnat

1989) has shown that the torsional behavior of concrete deep beams differs significantly from that of shallow beams. Thus, for deep beams, it may not be appropriate to use the provisions recommended by the ACI Code (ACI Committee 318, 1995) for shallow beams.

As a result of recent developments in material technology, the application of high-strength concrete (concrete with compressive strength in excess of 41 MPa) has gained wide acceptance in the construction industry in many parts of the world. Due to the lack of sufficient research data in this area. The American Concrete Institute formed Committed 363 to look into the various aspects of production, use and application of high-strength concrete. The committee (ACI Committee 363, 1987) reported that extensive research works are necessary on many aspects of high-strength concrete behavior both at material and structural levels. Largely as a result of this report, a number of works have been carried out. Most of these are on the material properties of high-strength concrete, and several work on its structural behavior (ACI Committee 363, 1992). A limited number of studies appear to have carried out on the torsional behavior of deep beams using high-strength concrete (Samman and Al-Siyoufi 1994, Samman and Radain 1993, Samman 1995, Radain et. al. 1999, Al-Shareef 1995, Samman and Radain 1995, Ashour et. al. 1999). However, the present ACI code (ACI Committee 318, 1995) does not have design provisions for various aspects of the behavior of high-strength concrete members except a limited provision for shear and flexure.

Beams in high rise buildings are often provided with transverse openings for various service facilities. These openings affect the torsional and shear capacities of a beam section. It needs to be suitably strengthened against any premature failure (Naser, 1990). The existing design code (ACI Committee 318, 1995), however, do not contain any design provisions for beams with a transverse opening. It appears that this is due to lack of information on the torsional behavior of beams with an opening. Therefore, the objective of this study is to present a literature survey concerning the theoretical torsional capacity of normal and high-strength concrete deep beams with and without transverse openings.

2 TORSIONAL THEORIES

Torsion is a major factor to be considered in the design of many types of reinforced concrete structures, including space frame, beams that support cantilever slabs or balconies, spandrel beams, and spiral staircases. These torsional moments occasionally cause excessive shearing stresses. As a result, severe cracking can develop well beyond the allowable serviceability limits unless special torsional reinforcement is provided (Winter and Nelson 1991). Due to the lack of sufficient research data and the knowledge-gap in the area of torsion during the first half of this century, torsion received relatively scant attention and was omitted from design considerations. However, since 1960's numerous aspects of torsion in concrete have been examined in various parts of the world and advanced the understanding of the torsion problems significantly (Hsu, 1984). Consequently, several theories have been developed to predict the torsional strength of plain and reinforced concrete members. The ACI Code (ACI Committee 318, 1971) incorporated for the first time detailed design recommendations for normal-strength concrete shallow beams under torsion. However, the present ACI Code (ACI Committee 318, 1995) has only provisions for the design of normal-strength concrete shallow beams under torsion.

2.1 Plain concrete beams

Concrete structural members subjected to torsion will normally be reinforced with special torsional reinforcement. However, the contribution of concrete in resisting the applied twisting moments before any cracking can be reasonably investigated. Four theories are available in literature (Hsu 1984, Winter and Nilson 1991, Hsu and May 1985, Hassoun 1995) to predict the torsional capacity of normal-strength plain concrete shallow members with rectangular sections: elastic, plastic, skew-bending and space truss analogy theories.

2.1.1 Elastic theory
The behavior of a plain concrete torsional member can be described by St. Venant's elastic theory (Winter and Nilson 1991) as follows:

$$T_e = \alpha_e X^2 Y f_t' \tag{1}$$

where:
X, Y = Shorter and longer sides of the rectangular section, respectively.
α_e = St Venant's coefficient which depends on Y/X, and varies from 0.208 to 0.33.
f_t' = Tensile strength of concrete. A reasonable value of 0.42 MPa has been suggested by (Hsu, 1984).

2.1.2 Plastic theory
In the plastic theory (Hassan, 1998), failure torque, T_p, is expressed as:

$$T_p = \alpha_p X^2 Y f_t' \tag{2}$$

where:
α_p = (0.5-X/6Y), a plastic coefficient which depends on Y/X, and varies from 0.333 to 0.5.

2.1.3 Skew-bending theory
The skew-bending theory (Hsu, 1984) has been proposed to predict the torsional strength, T_{np}, as:

$$T_{np} = (X^2 Y/3)(0.85 f_r) \tag{3}$$

where:
f_r = modulus of rupture of concrete.

Hsu (1984) developed the following empirical equation to predict the nominal torsional strength of plain concrete beam in terms of the concrete compressive strength provided that $X \geq 4$ inches.

$$T_{np} = 6(X^2 + 10) Y^3 \sqrt{f_c'} \tag{4}$$

where :
f_c' = compressive strength of concrete.

2.1.4 Space Truss analogy
This theory (Hsu and May, 1985) does not explicitly quantify the torsional strength of plain concrete beams, it provides the cracking torque of a reinforced concrete element as:

$$T_{cr} = 2 A_{ot} \tau \tag{5}$$

where :
$$t = (3/4) (A_{cp}/P_{cp})$$
$$A_o = (2/3) A_{cp}.$$

A_{cp} and P_{cp} are the area and perimeter of the concrete enclosed by center line of stirrup, respectively. τ is considered as the tensile strength of concrete and limited to $4\sqrt{f_c}'$ for normal strength concrete.

2.2 *Reinforced concrete beams*

The general behavioral pattern of a reinforced concrete beam under torsion is different from that of a plain concrete beam. However the failure itself occurs in a skew bending mode. Torsional strength of a reinforced concrete beam is partly due to that of concrete and partly due to reinforcement. The contribution of concrete is considered to be only about 40% of the torsional capacity of an unreinforced section in torsion (Hsu 1984, Winter and Nilson 1991, Hassoun 1998).

2.2.1 *Skew-bending theory*
Hsu (1984) developed the following expression to predict the nominal torsional strength of normal-strength reinforced concrete solid shallow beams, T_{nr}:

$$T_{nr} = (X^2 Y/\sqrt{x})(0.2 \sqrt{f_c}') + \alpha_t (X_1.Y_1.A_t.f_{sy}/s) \quad (6)$$

where:
X_1 & Y_1 = center-to-center distance of shorter and longer leg of stirrups.
α_t = $\sqrt{m} (f_{ly}/f_{sy})(1+0.2(Y_1/X_1)) < 1.6$
m = ratio of volume of longitudinal steel to volume of stirrups
f_{ly} & f_{sy} = longitudinal and stirrups steel yield strength
A_t = area of one leg of a closed stirrup resisting torsion within a distance s.

2.2.2 *Space truss analogy*
Hsu and May (1985) developed the following expression to predict the nominal torsional strength of normal-strength reinforced concrete solid shallow beams, T_{nr};

$$T_{nr} = (2A_o.f_d.\,t_d) \sin\alpha.\cos\alpha \quad (7)$$

where:
$$t_d = (A_l f_l/P_o(\sigma_d-\sigma_f)) + (A_s f_s/s(\sigma_d-\sigma_f))$$
$$Cos^2\alpha = (A_l f_l / P_o(\sigma_d-\sigma_t))$$

$$Sin^2\alpha = [(A_s f_s / s\, t_d)+ \sigma_l) / \sigma_d]$$

3 ACI TORSIONAL DESIGN PROVISIONS

The ACI Code (ACI Committee 318, 1989) procedure for the torsional design of normal-strength shallow beams is based on the Skew Bending theory as follows:

$$T_{nr} = (X^2 Y/3) (0.2 \sqrt{f_c}') + \alpha_t (X_1.Y_1.A_t.f_{sy}/s) \quad (8)$$

where:
$\alpha_t = 0.6 + 0.33 (Y_1/X_1) < 1.5$

In the present ACI code (ACI Committee 318, 1995), the design for torsion is based on the space truss analogy as follows:

$$T_{nr} = (2 A_O.A_t.f_{sy}/s) \cot 45^o \quad (9)$$

where:
A_O = gross area enclosed by shear flow bath

4 RESEARCH EFFORTS

The design guidelines and theories discussed above were further modified based on additional information from recent research efforts.

4.1 Normal-strength plain concrete deep beams

Akhtaruzzaman and Hasnat (1989) demonstrated that the torsional capacity of plain normal-strength concrete deep beams with solid rectangular sections, T_{np}, remains practically constant for $L/Y \geq 3.0$ and can be adequately predicted by the skew-bending theory provided that the splitting tensile strength is considered as the tensile strength of concrete, i.e.,

$$T_{np} = 0.935 (X^2 Y/3)f'_{sp} \quad (10)$$

where :
f'_{sp} = spliting tensile strength of concrete.
The torsional strength, T_{np}, for members with $L/Y < 3.0$, varies inversely with L/Y ratio and can be predicted within reasonable accuracy as reported by:

$$T_{np} = 0.85 (X^2 Y/3) f'_{sp} (1.34 - 0.08 L/Y) \quad (11)$$

where :
L = center-to-center span
L/Y = span/depth ratio and must be < 3.0

4.2 High-strength plain concrete deep beams

Samman and Al-Siyoufi (1994) observed the same behavior as reported by Akhtaruzaman and Hasnat (1989) and proposed the following changes in Eq. (11) to predict the torsional capacity of high-strength plain concrete solid deep beams with L/Y < 3 as:

$$T_{np} = (X^2Y/3) \, f'_{sp} \, (1.43 - 0.180 \, (L/Y)) \qquad (12)$$

Samman and Radain (1993) modified Eq.(3) to predict the nominal torsional capacity of high-strength plain concrete deep beams as follows:

$$T_{np} = (X^2Y/3)(0.68 \, \sqrt{f'c}) \qquad (13)$$

Samman (1995) reported that Eqs.(10 and 11) gives the best prediction for torsional behavior of high-strength plain concrete solid deep beams.

Radain, Ashour, and Samman (1999) proposed the following equation to predict the torsional capacity of high-strength plain concrete solid deep beams as:

$$T_{np} = (0.95/6) \, f_{sp}' \, X^2Y \, [1 \times e^{-(L/Y)0.8}] \qquad (14)$$

4.3 *Normal-strength concrete deep beams with an opening*

Akhtaruzzaman and Hasnat (1989) reported that the nominal torsional strength of deep beams with circular transverse openings, T_{npo}, remains practically constant for $L/Y \geq 3.0$ and can be adequately predicted by the equation proposed by (Mansur and Hasnat 1979), with f_r replaced by f'_{sp} as follows:

$$T_{npo} = (0.85/6) \, X^2Y \, (\sec \theta - (D_O/Y)) \, \text{cosec}\theta \, f'_s \qquad (15)$$

where :
$\text{Sec}^3\theta - 2\sec\theta + D_o/Y = 0$
D_o = opening diameter.

For members with $L/Y < 3.0$, the torsional strength varies inversely with L/Y ratio and a modification to Mansur and Hasnat's equation (Mansur and Hasnat 1979) has been proposed for predicting the nominal torsional strength, T_{npo}, as follows:

$$T_{npo} = (0.85/6) \, X^2Y \, (\sec \theta - (D_O/Y)) \, \text{cosec}\theta \, f'_{sp} \\ (1.34 - 0 - 08 \, L/Y) \qquad (16)$$

4.4 *High-strength concrete deep beams with an opening*

Radain, Ashour, and Samman (1999) showed that a better modification for Eqs. (15 and 16) to predict the nominal capacity of high-strength plain concrete deep beams with transverse opening is:

$$T_{npo} = (0.95/6) \, f'_{sp} \, X^2Y(\text{Sec } \theta - (Do/Y) \\ \text{cosec } \theta \, [1 + e^{-[L/Y)(1-(Do/Y)0.8]}] \qquad (17)$$

4.5 *Reinforced concrete solid deep beams*

Al-Shareef (1995) modified the ACI Code (ACI Committee 318, 1989) equation to predict the torsional capacity of reinforced high-strength concrete deep beams as follows:

$$T_{nr} = (X^2Y/3) \, (0.2 \, \sqrt{f_c'}). \, e[1+(Y/L)]^{0.3} \\ + \alpha_t \, (x_1.y_1.At.f_{sy}/s) \qquad (18)$$

Samman and Radain (1999) proposed the following changes in the ACI Code (ACI Committee 318, 1989) equation to predict the torsional capacity of reinforced high-strength concrete deep beams as follows:

$$T_{nr} = 0.52 \, X^2Y \, \sqrt{f_c'}). \, [1+5 \, e^{-L/Y}] \\ + \alpha_t \, (x_1.y_1.At.f_{sy}/s) \qquad (19)$$

Ashour, Samman, and Radain (1999) modified the space truss equation (Eq. 7) developed by Hsu and May (1985) to predict the torsional capacity of reinforced high-strength concrete deep beams as follows: -

$$T_{nr} = (2A_o.f_d. \, t_d) \, \sin\alpha.\cos\alpha - \beta \qquad (20)$$

where:
$\beta = [1 + (1/2(L/h) \, (2.5-0.5(L/h)]$

4.6 *Reinforced concrete deep beams with an opening*

Samman and Radain (1999) modified the ACI Code (ACI Committee 318, 1989) equation to predict the torsional capacity of reinforced high-strength concrete deep beams with an opening as follows:

$$T = T_c + T_s \qquad (21)$$

where:
T_c = nominal torsional strength provided by concrete.
$T_c = 0.52 \, X^2Y \, f_c' \, (\sec\theta - (D_o/y) \, \text{cosec } \theta \\ [1+5e^{-L/Y\{1-(Do/Y)4\}}]$
T_s = nominal torsional strength provided by total torsional reinforcement,
$T_s = T_{s1} + T_{s2} + T_{s3}$
T_{s1} = nominal torsional strength provided by regular (longitudinal and stirrups) torsional reinforcement.
$T_{s1} = \alpha_t \, x_1 \, y_1 \, A_{s1} \, (F_{ys1}/S)[1-\cos\theta(D_o/y)]$
T_{s2} = nominal torsional strength provided by vertical and horizontal bars around the transverse opening only
$T_{s2} = n\alpha_t \, x_1 \, F_{ys2} \, A_{s2}$
T_{s3} = nominal torsional strength provided by inclined bars around the transverse opening only
$T_{s3} = n\alpha_1 \, A_{s3} \, F_{ys3} \, x_1 \, (\sin \alpha' + \cos \alpha')$

A_{s3} = area of an inclined bar around the opening
A_{s2} = area of horizontal bar around the opening
F_{ys2} = horizontal bar yield strength
F_{ys3} = inclined bar yield strength
A_{s1} = area of the leg of a closed stirrup resisting torsion within a distance S
S = stirruping spacing
F_{sy1} = stirrup yield strength
n = 2, number of reinforcement bars crossed by the failure plain on tension side
α' = inclination of the bars around the opening

5 CONCLUSION

The literature review above, thus, indicates that there is an urgent need for some work on torsion of high-strength concrete deep beams with or without transverse opening, such investigations will also be in line with the recommendations of ACI Committee 363 (1987).

REFERENCES

ACI Committee 318, 1995, Building Code Requirements for Reinforced Concrete, American Concrete Institute, Detroit, Michigan, 369.

Hassoun, N. M, 1998, Structural Concrete; Theory and Design, Addison-Wesley Publishing Company, California, 824.

Chow, L., Conway, H.D., and Winter, G., 1953, Stress in Deep Beams, Transaction ASCE, Vol. 118, 686-696.

De Paiva, H.A.P., and Seiss, C.P., 1965, Strength and Behavior of Deep Beams in Sehar, Journal of Structural Engineering, Structural Division, ASCE, Vol. 91, No. ST 5, October, 19-41.

Kong, F.K., Robins, P.J., and Cole, D.F., 1970 Web Reinforcement Effects in Deep Beams, ACI Journal, Proceedings Vol. 67, No. 12, December, 1010-1017.

Kong, F.K., and Robins, P.J., 1971 Web Reinforcement in Lightweight Concrete Deep Beams, ACI Journal, Proceedings Vol. 68, No.7, 514-520.

Ramakrishnan, V. and Ananthanarayan, Y., 1968 Ultimate Strength of Deep Beams in Sehar, ACI Journal, Proceedings Vol. 65, No. 1, 87-98.

Desayi, P., 1974 A Method for Determining the Shear Strength of Reinforced Concrete Beams with Small a_v/d Ratios, Magazine of Concrete Research, Vol. 26, No.86, 29-38.

Rogowsky, D.M., MacGregor, J.G., and Ong, S.Y., 1986 Tests on Reinforced Concrete Deep Beams, ACI Journal, Proceedings Vol. 83, No. 4, 1986, 614-623.

Kong, F.K., and Sharp, G.R., 1977 Structural Idealization for Deep Beams with Web Openings, Magazine of Concrete Research, Vol. 29, No. 99, 81-95.

Kong, F.K., Sharp, G.R., Apleton, S.C., Beaumont, C.J., and Kubic, L.A., 1978 Structural Idealization for Deep Beams with Web Openings: Further Evidence, Magazine of Concrete Research, Vol. 30, No. 103, 89.

Akhtaruzzaman, A.A., and Hasnat, A. 1989, Torsion in Concrete Deep Beams with an Opening, ACI Structural Journal, 86(1), 20-25.

ACI Committee 363, 1987, Research Needs for High-Strength Concrete (ACI 363-IR-87), ACI Materials Journal, 84, (6), 559-561.

ACI Committee 363, 1992, State-of-the-Art Report on High-Strength Concrete, American Concrete Institute, Detorit, Michigan, 55.

Samman, T.A., and Al-Siyoufi, M.M. 1994, Effect of Span/Depth Ratio on Torsional Capacity of High-strength Plain Concrete Deep Beam", Dirasat Hundasia Journal of United Arab Emirates University, United Arab Emirates. 7(1), 1-21.

Samman, T.A., and Radain, T.A. 1993, Effect of Aspect Ratio on Torsional Capacity of High-strength Plain Concrete Deep Beams", Engineering Journal of Qatar University, Qatar State, 6, 115-134.

Samman, T.A. 1995, High-strength Plain Concrete Deep Beams under Torsion, Structural Engineering Review Journal, 7(2), 93-105.

Radain, T.A, Ashour, S.A. and Samman, T.A. 1999, Torsion in High-strength plain concrete Deep Beams with Transverse Opening, Emirates Journal for Engineering Research, 4(1), 10-21.

Al-Shareef, A.A. 1995, Torsional behavior of high-strength reinforced concrete deep beams, M.S. Thesis, Civil Engineering Department, King Abdulaziz University, Jeddah, Saudi Arabia, 164.

Samman, T.A., and Radain, T.A. 1999, Torsion in High-strength Plain and Reinforced Concrete Deep Beams with and without Transverse Opening, Final Report, Project No. AR-12-36, King Abdulaziz City for Science and Technology (KACST), Riyadh, Saudi Arabia, 290.

Ashour, S.A., Samman, T.A. and Radain, T.A. 1999, Torsional behaviour of Reinforced High-strength Concrete Deep Beams, ACI Structural Journal, Vol. 96, No. 6, 1049-1058.

Naser, T.A.S., 1990, Concrete T-Beams with Opening Under Torsion, M.S. Thesis, Civil Engineering Department, King Abdulaziz University, Jeddah, Saudi Arabia, 150.

Winter, G., and Nilson, A.H. 1991, Design of Concrete Structures, "McGraw-Hill Book Company, New York, 11th Edition, 904.

Hsu, T.C. 1984, Torsion of Reinforced Concrete, Van Nostrant Reinhold Company, New York, First Edition, 516.

ACI Committee 318, 1971, Building Code Requirements for Reinforced Concrete, American Concrete Institute, Detroit, Michigan, 153.

Hsu, T.C., and May. L, 1985, Softening of Concrete in torsion Member: Theory and Tests, ACI Journal, Proceedings V82, No. 3, 290-303.

ACI Committee 318, 1989, Building Code Requirements for Reinforced Concrete, American Concrete Institute, Detroit, Michigan, 280.

Mansur, M.A., and Hasnat, A. 1979, Concrete Beams with Small Opening Under Torsion, Journal of Structural Engineering, Structural Division, ASCE, 105 (ST11), 2433-2447.

Creative Systems in Structural and Construction Engineering, Singh (ed.) © 2001 Balkema, Rotterdam, ISBN 90 5809 161 9

Further insight into flexural bond behavior of prestressed concrete beams

I. R. A. Weerasekera
Department of Civil Engineering, University of Moratuwa, Sri Lanka

R. E. Loov
Department of Civil Engineering, University of Calgary, Alb., Canada

ABSTRACT: In recent times large strands (13, 15 and 16 mm in diameter) have become increasingly popular in the pretensioned prestressed concrete industry and have found wide applications in varying geometries of sections. However use of such elements and their behavior in several situations have been questioned with respect to anchoring of these strands in concrete; which is accomplished by bond. The experimental results available on bond are limited and information relating to large strands is rare. Laboratory investigations have been conducted to determine the influence of the inadequately examined parameters on flexural bond. The principal variables considered were strand size, concrete strength, beam width, clear side cover and bottom cover. The experimental program consisted of fabricating and testing thirteen full-scale beam specimens. Results of these beam tests are presented.

1 INTRODUCTION

Bond in pretensioned concrete is a twofold phenomenon. The first effect is prestress transfer bond. The subject of this paper is the second effect. Here additional bond stresses develop similar to those in reinforced concrete due to bending moment gradient and as a consequence of flexural cracking under applied loading. The action of these bond stresses is known as flexural bond.

The structural performance of prestressed concrete members depends on the satisfactory transmission and anchorage of the prestressing force. This is an important design consideration as both the flexural capacity and shear resistance are affected by the bond strength. Therefore it is important to determine the distribution of bond stresses: namely magnitude and the length over which these stresses are distributed have to be found realistically. At present it is not possible to estimate bond parameters accurately. Due to the complex nature of the bond phenomenon and inadequacy of knowledge on bond, there is lack of consensus on various influences.

2 BACKGROUND

In different countries the flexural bond length in prestressed concrete beams is either estimated by empirical formulae based on previous tests or estimated by performing tests. Some codes indicate that several variables influence bond but are not explicit as to which variables should be considered.

The variables often taken into account are strand diameter and the level of prestress in the steel. There is generally no recognition of the importance of concrete properties.

The ACI code uses a flexural bond length, (in ksi and inches)

$$l_f = (f_{ps} - f_{se})d_b \tag{1}$$

(in MPa and mm)

$$l_f = 0.145(f_{ps} - f_{se})d_b \tag{2}$$

where f_{ps} = Stress in prestressing steel
f_{se} = Stress in prestressing steel at transfer
d_b = Strand diameter

These equations were based on the test results of Hanson and Karr (1959). A more complete literature review of the factors affecting flexural bond length are found in Martin and Scott (1976), Zia and Mostafa (1977), Cousin et al (1990), Deatherage and Burdette (1991) and Weerasekera (1991).

3 EXPERIMENTAL PROGRAM

3.1 Outline of Experiments

An experimental program has been carried out to investigate the bond characteristics of large sizes (13 mm and 15 mm diameter) of normal 7-wire prestressing strands. The scope of the experimental program included fabrication and testing of thirteen prestressed concrete beams. The objective of the testing, was to make measurements connected with flexural bond. The principal variables are concrete strength, strand size, side cover, bottom cover and effective beam width. These parameters are shown in Figure 1.

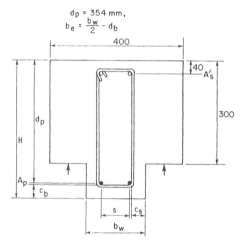

$d_p = 354$ mm,
$b_e = \dfrac{b_w}{2} - d_b$

Figure 1 : General Shape of Specimens and Definition of Symbols

3.2 Idealization of Variables

One of the difficulties encountered in bond studies has been separation of the variables. To overcome this deficiency, a general specimen shape in the form of a deep flanged T-section was selected. This general shape shown in Figure 1 offers several advantages. Primarily it provides a convenient geometry so that the cover and strand spacing can be varied easily. Also the underside of the flange provides a suitable face to locate beam supports in a zone above the anchorage region so that concrete around the tendon is not compressed.

3.3 Development of Experimental Program

The principal variables are shown in Table 1. All variables chosen comply with the CSA standard (Canadian Standards Association (1994)) To obtain as much information as possible six different arrangements (later referred to as groups A-F) were considered. Arrangement A is treated as the base and all variables are related to this case. A total of 13 beams consisting of 3 of the base arrangement (A) and 2 of every other arrangement (B, C, D, E and F) are included in the experimental program.

3.4 Testing Technique

Considering the importance of the Prestressing Steel-Concrete interface, a method that does not interfere with the bonding is preferred. For the determination of flexural bond behavior, beam tests were performed. A two point symmetrical loading arrangement producing a constant moment region was adopted. A bond failure was anticipated when the moment diagram due to load reached the bond capacity envelope as illustrated in Figure 2.

4 TEST SPECIMENS

All the specimens have the same depth to the prestressing steel, and identical flange dimensions of 400 mm x 300 mm. The beam length was planned to be at least twice as long as the expected development length for 13 mm strand. At the top ordinary reinforcing bars were provided to control cracking due to eccentric prestress. Here a nominal amount of reinforcing steel was provided in accordance with the CSA code provisions. A few stirrups were provided in the shear span to hold the beam together after failure.

The main design objective was to develop specimens which would resist flexure and shear. The intent was to obtain bond failures. The flexural capacity was determined and all beams were intentionally made to be under-reinforced. The shear strength was also determined based on equations in the CSA standards. The flexural capacity and shear strength are given in Table 2. Also included is the cracking moment.

Details of the loading positions of all thirteen beams and their theoretical and experimental strength parameters are given in Table 2. All beams were 5680 mm in overall length, with 90 mm overhang when supported to provide a 5500 mm span.

5 RESULTS

There were several types of failure modes. Flexural failure, shear failure, bond-flexural and bond-shear failure modes were encountered.

As there were two strands and two ends the end slips vary. The slip of the first strand was considered to be the initial bond slip. Bond failure refers to the case when the end slip exceeded 0.01 mm in both strands. Flexural cracking was observed during bond failure, but before a flexural or shear failure occurred, considerable strand slip was measured.

A typical flexural failure would have considerable flexural cracking, yielding of steel and finally crushing of the concrete in the compression zone at the point of maximum moment. No measurable strand slip could be seen during a flexural failure. Flexural failure occurred in the two beams (A-3 and B-2).

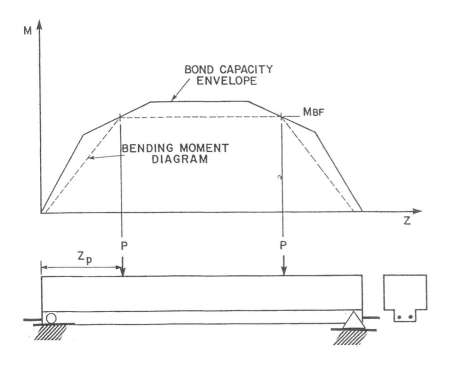

Figure 2. Loading arrangement and moment envelopes

Table 1. Beam designation & section properties

Group	Beam Designation		f'_{ci} MPa	d_b mm	b_e mm	c_s mm	c_b mm	S mm	b_w mm	H mm
A	A-1	(1)	20	12.7	80	40	40	93	185	400
	A-2	(2)	23	12.7	80	40	40	93	185	400
	A-3	(3)	20	12.7	80	40	40	93	185	400
B	B-1	(6)	38	12.7	80	40	40	93	185	400
	B-2	(7)	36	12.7	80	40	40	93	185	400
C	C-1	(8)	22	15.3	78	39	39	93	185	400
	C-2	(9)	24	15.3	78	39	39	93	185	400
D	D-1	(4)	22	12.7	80	40	20	93	185	380
	D-2	(5)	21	12.7	80	40	20	93	185	380
E	E-1	(10)	22	12.7	80	20	40	133	185	400
	E-2	(11)	22	12.7	80	20	40	133	185	400
F	F-1	(12)	22	12.7	39	20	40	53	105	400
	F-2	(13)	21	12.7	39	20	40	53	105	400

In Table 1 the numbers in parentheses indicate the data points referred to in Figure 2.

Table 2. Flexural bond test result

| Beam | Distance From Load to Support m | Theoretical | | | Experimental | | |
		Cracking Moment kN.m	Flexural Strength kN.m	Shear Strength kN	Cracking Moment kN.m	Maximum Steel Stress MPa	Flexural Bond Strength kN.m
A-1	0.910	72.7	121	143	72.8	(1557)	(100.1)
A-2	0.910	72.7	121	143	72.8	1620	104.7
A-3	1.510	72.7	121	113	60.4	(1632)	(105.7)
B-1	1.010	79.0	126	165	60.6	(1632)	(106)
B-2	1.610	79.0	126	113	64.4	(1685)	(111)
C-1	1.510	93.3	164	113	75.5	1191	90.6
C-2	1.110	93.3	164	143	66.6	1100	44.4
D-1	1.510	72.5	121	113	60.4	1428	90.6
D-2	1.010	72.5	121	143	50.5	1191	70.7
E-1	1.010	72.7	121	143	60.6	1159	60.6
E-2	1.510	72.7	121	113	75.5	1227	75.5
F-1	1.010	68.6	121	81.4	70.7	1127	50.5
F-2	1.510	68.6	121	64.4	60.8	1158	60.4

In Table 2 the values in parentheses indicate a failure condition other than bond failure.

A typical shear failure was always preceded by flexural cracking before a diagonal shear crack would develop from the support. No strand slip was found during a shear failure. Shear failure occurred in two beams (A-1 and B-1).

Shear failure was encountered at very short development lengths. For these cases it could only be said that the development length is longer than the embedment length as no other information could be derived.

Combined failure modes were encountered in most situations. But flexure or shear failure did not govern the primary mode of failure. Bond-shear failure occurred in six cases (A-2, C-2, D-2, E-1, F-1 and F-2). Bond-flexural failure occurred in three occasions (C-1, D-1 and E-2). Bond-shear failures were sometimes associated with longitudinal cover splitting. In some cases after bond failure occurred the maximum moment could no longer be supported although a significant but lower moment capacity remained as the strand slipped further. This can be attributed to a combination of friction and mechanical interlock components of bond mechanisms.

6 DISCUSSION

The test results are presented in Table 2. Test results show that except Group A and Group B other cases fall below the failure envelope specified in the ACI and CSA codes. When concrete strength is increased there is an enhancement of flexural bond stresses. But the slope of the curve can not be established as these specimens in Group B did not fail in bond. Specimens in Groups C, D, E and F show the danger of current practices. Group D results show that variation of steel stress beyond transfer is lower but almost parallel to the code curve. The curves in Groups C, E and F are lower and flatter than the code curve. These results collectively show that flexural bond stresses are influenced by reduced cover, spacing, effective beam width and large diameter strand. It can be concluded that the current code equations are not reliable for small covers and spacing. Also the current equations are not conservative for large diameter strand. The points corresponding to all the beams are shown in Figure 3.

The specimens were loaded incrementally to failure.

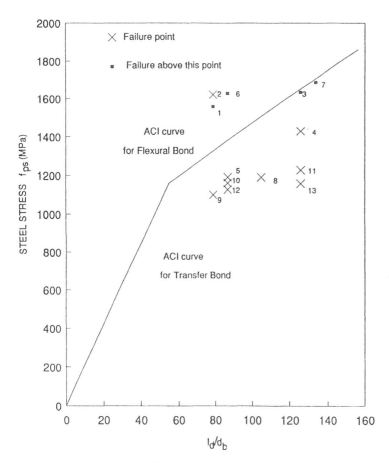

Figure 3. Strand development for all test beams

During loading the static loads, sustained loads and cyclic loads were examined.

Slippage of the tendons under different loading conditions can vary.

Slippage may also depend on the degree of cracking involved. Initial bond slip occurred in 8 specimens at loads below the service load. General bond slip or bond failure occurred in 7 specimens at loads below the service load. Small cover can cause premature bond failure.

Applying cyclic loading after general bond slip has occurred but at loads below 0.55 of ultimate failure capacity did not increase slip. After general bond slip has occurred, cyclic loading causes slip to increase in each cycle at 0.67 of ultimate flexural capacity. For example Beam E-1 failed after cycling. Bond slip failure can occur even before a beam begins to crack. Therefore there may not be any warning if that type of slip occurs.

7 CONCLUSIONS

1. Flexural bond tests reveal the influence of strand size, concrete cover, concrete strength and loading on flexural bond performance.
2. Beams that failed prematurely in bond-shear and bond-flexure show the inadequacy of currently used flexural bond lengths and the dangers associated with such bond failures.
3. Beams that failed in shear or flexure indicated the availability of reserve bond capacity when concrete strength is high or cover is adequate.
4. Strand development curves obtained from this investigation can be used to determine flexural bond length or development length, for specific conditions.
5. The current code equations are not safe and need revision. The test results of this study can be used to improve code provisions.

REFERENCES

American Concrete Institute (1989), ACI Committee 318, Building Code Requirement for Reinforced Concrete (ACI 318-89), Farmington Hills, Michigan, USA..

Canadian Standards Association (1994), CSA A23.3-94, Design of Concrete Structures, Toronto, Ontario, Canada.

Cousins, TE, Johnston, DW and Zia, P (1990), "Development Length of Epoxy-Coated Prestressing Strand", ACI Materials Journal, Vol 87, No 4, 1990, pp 309-318.

Deatherage, JH and Burdette, EG (1990), "Development Length and Lateral Spacing Requirements of Prestressing Strand for Prestressed Concrete Bridge Products", Final Report to PCI, April 1990, pp 58.

Hanson, NW and Karr, PH (1959), "Flexural Bond Tests of Pretensioned Prestressed Beams", ACI Journal, Proceedings, Vol. 55, January 1959, pp 783-803.

Martin, LD and Scott, NL (1976), "Development of Prestressing Strand in Pretensioned Members", ACI Journal, Proceedings, Vol. 73, No. 8, August 1976, pp 453-456.

Weerasekera, IRA (1991), "Transfer and Flexural Bond in Pretensioned Prestressed Concrete", PhD thesis, Department of Civil Engineering, The University of Calgary, Calgary, Canada, July 1991, pp 1-322.

Zia, P and Mostafa, T (1977), "Development Length of Prestressing Strands", PCI Journal, Vol 22, No 5, September-October 1977, pp 54-65.

Creative Systems in Structural and Construction Engineering, Singh (ed.) © 2001 Balkema, Rotterdam, ISBN 90 5809 161 9

Experimental study on deterioration of concrete structure due to acid rain

Yoshikazu Akira – *Graduate School of Science and Engineering, Kagoshima University, Japan*

Koji Takewaka – *Department of Civil Engineering, Asian Institute of Technology, Thailand*

Takayuki Sato – *Dai-Nippon Toryo Company Limited, Japan*

Toshinobu Yamaguchi – *Department of Ocean Civil Engineering, Kagoshima University, Japan*

ABSTRACT: Acid rain caused by human and volcanic activities affects not only human body and natural environment but also concrete structures. However, it is difficult to evaluate deterioration of concrete due to it throughout the long term. In this study, the influences of the volcanic acid rain on the durability of concrete structure were examined experimentally by the five years exposure test at the natural environment. And also, in order to evaluate the effects of the acid rain in the short period, authors have developed the original equipment for accelerated test. In the accelerated test, effect of acid rain on neutralization of concrete and corrosiveness of steels in concrete were investigated. As the results of both experiments, the following conclusions were obtained. The acid rain causes yellowish coloring on the surface of concrete. And the phenomenon that a part for the concrete paste began to melt was recognized in the fifth year. As for the neutralization of concrete, it was found small effects due to acid rain. However, as for the corrosiveness of steels in concrete, it is clearly accelerated due to acid rain. Therefore, for reinforced concrete structure especially with cracks, deterioration due to acid rain should be considered.

1 INTRODUCTION

Acid rain problem is one of the most serious environmental problems of the earth, in these days. In Europe and North America, it has been reported that the atrophy of forest, the acidification of soil, lakes and marshes, and extinction of flora and fauna of these area, and so on, as the influences of acid rain.

As for concrete structures, also it must be affected by acid rain, because of that concrete is a high alkali material and should be maintain its alkalinity. For example, the decomposition of C-S-H and the acceleration of corrosion of reinforcement steels in concrete, due to the pH lowering in pore solution of concrete, were indicated in the past researches. However, it has not come to the quantitative grasp of these phenomena (Kobayashi, 1997; Sato, 1999). Considering the allowable crack under the present design concept of reinforced concrete structures and causing possibility of artificial defects under construction, effects of acid rain on the neutralization of concrete and corrosion of re-bars can not be disregarded.

In this study, the experimental examination was carried out for the purpose of having the quantitative grasp of the effects of acid rain on concrete structure. Firstly, outdoor exposure tests were carried out to investigate the effect of acid rain on reinforced concrete at the two types of actual environments.

Secondly, for the purpose of evaluating the effect of the acid rain in the short period, authors developed the original accelerated machine and performed accelerated test using that equipment. In the test, effect of acid rain on neutralization of concrete and corrosiveness of steels in concrete were investigated.

Mt.Sakurajima

Kamakura city

Figure 1 Exposure area

Table 1 Mix proportions

W/C (%)	unit weight (kgf/m³)				AE entraining agent (cement×ml)	water reducing agent (cement×%)
	water	cement	fine aggregate	coarse aggregate		
50	180	360	820	946	7	1
70	180	257	897	955	7.5	1

Table 2   Factors and levels in exposure test

	Type-A	Type-B
exposure area	Mt.Sakurajima Kamakura city	Mt.Sakurajima
W/C (%)	50 , 70	50 , 70
correspondent years of pre-neutralization by acceleration	0 , 20	20
cover depth of rebar (cm)	3, 2	3
crack width (mm)	0	0.1

Table 3 The ratio of corrosion area of Sakurajima to Kamakura and cover depth 2cm to 3cm in 5 years

exporsuer area	$\dfrac{\text{Mt.Sakurajima}_{5year}}{\text{Kamakura city}_{5year}}$		cover depth	$\dfrac{2cm_{5year}}{3cm_{5year}}$	
	3cm	2cm		Mt.Sakurajima	Kamakura city
50% pre-	1.6	1.2	50% pre-	4.3	5.9
70% non-	-	5.3	70% non-	5.4	-
70% pre-	2.0	4.4	70% pre-	4.9	2.2

Table 4 Mix proportions

W/C (%)	unit weight (kgf/m³)			flow value (mm)
	water	cement	fine aggregate	
50	317	633	1267	210
70	298	426	1490	209
90	273	303	1667	210

Table 5 Factors and levels in accelerated test

	Series-1	Series-2
W/C (%)	50 , 70	70 , 90
correspondent years of pre-neutralization by acceleration	0 , 20	0 , 50
cover depth (cm)	3	
crack width (mm)	0 , 0.2	

Table 6 Setting condition of the promotion equipment which follows the natural environment

	Series-1	Series-2
correspondent years	2	1
CO_2	10%	5%
temperature	30℃	30℃
humidity	100～70%	100～70%
spraying quantity	2250mm³/mm²	2250mm³/mm²
drying time	80h	74h
spraying time	16h	22h
spraying solution	acid rain similar solution (pH3.0) distilled water	
pH control	at the begining of each cycle	continuously

2 EXPOSURE TEST

2.1 *Experimental outline*

2.1.1 *Outline of exposure area*

Exposure tests were carried out in Mt.Sakurajima (Kagoshima prefecture) and Kamakura city (Kanagawa prefecture) was shown Figure 1.

Mt.Sakurajima is one of active volcanoes located in southern part of Japan and just beside the center of Kagoshima city. Therefore, a rain in this area is apt to be an acid rain due to the volcanic activities of Mt.Sakurajima. On the other hand, a rain in Kamakura city, which is residential area near by Tokyo, is considered to have less acidity than Mt.Sakurajima.

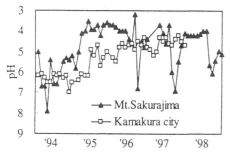

Figure 2 pH of the rainfall water in each exposure area

Figure 2 shows the changing of pH of the rainfall water for 5 years (1994-1998) in both exposure areas. In Kamakura city, acidity of the rain is getting higher year by year due to the increase of human activities, and an average acidity during exposure period is calculated as pH4.85. In Mt.Sakurajima, however, because of the volcanic activities, acidity of the rain is changing around pH 4 for all through the 5 years and an average acidity is pH 4.08. It means that the concentration of hydrogen ion of the rain in Mt.Sakurajima is approximately 6 times of that in Kamakura city. Addition to that, to comparing the concentration of sulfate ion in the rain, that in Mt.Sakurajima is also very higher than that in Kamakura city.

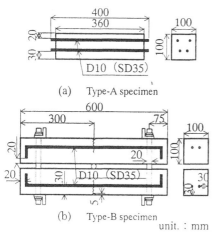

(a)   Type-A specimen

(b)   Type-B specimen

unit. : mm

Figure 3 Outline of specimens

2.1.2 *Specimens*

For every specimen, ordinary portland cement was used and mixture proportions are shown in table1. For the exposure test, two types of specimens, sound reinforced concrete beam (type-A) and reinforced concrete beam with bending crack (type-B), were prepared as shown in Figure 3. These beams had deformed reinforcing bars (SD35/D10) embedded with

a cover depth of 2 or 3 cm. For type-B, by making pre-cutting part at the center of specimen and fastening two specimens together, crack of which width was 0.1mm at the surface was induced before exposure as shown in figure 3(b).

Additionally, for both of types, in order to examine the difference of deterioration properties between newly and not newly concrete structures, some specimens were induced pre-neutralization by acceleration method before exposure. Table 2 gives factors and levels of the experiment.

2.1.3 *Experimental method*

Specimens had been exposed in Mt.Sakurajima (Type-A) and Kamakura city (Type-A and B) since November 1993. After 1, 3 and 5 years exposure, neutralized thickness of concrete and corrosion area of reinforcing steel in concrete were measured.

2.2 *Results of exposure test*

2.2.1 *Effect on concrete surface*

Appearances of specimens after 5 years exposure are shown in Figure 4. As to the specimens exposed in Kamakura city, there are not any changes of shape, but the color of concrete surface was changed to be dark due to the reproduction of fungus such as molds and etc.. On the other hand, in the case of that in Mt.Sakurajima, due to chemical reaction and the dissolving of cement paste, concrete surface was yellowed and roughened. It is considered that these differences of apparent are based on the difference between the acidity of rains in both exposure areas.

Mt. Sakurajima

Kamakura city

(W/C70% pre-neutralized)

Figure 4 Appearance photograph after the exposure of 5 years

2.2.2 *Effect on neutralization of concrete*

The results of measurement of the neutralized thickness of Type-A specimens exposed in Mt.Sakurajima and Kamakura city are shown in figure 5. As to the specimens of W/C50%, since neutralization of specimen hardly progressed, difference of exposure areas can not be recognized.

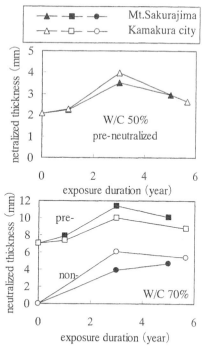

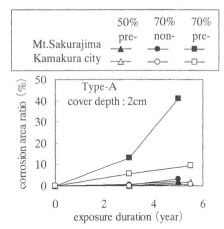

Figure 5 The neutralized thickness of the type-A specimen in exposure test

Figure 6 Corrosion area after exposure

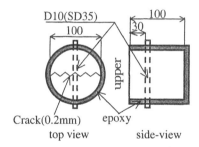

Figure 7 Outline of specimens

Figure 8 Rainfall system

Acid solution   Distilled water

5 cycle, W/C70%, non-neutralized

Figure 9 Appearance photograph by the Series-2 accelerated test

From these results, it can be considered, for the concrete which already neutralized to some extent by CO_2, the additional neutralization by acid rain will be accelerated compare to that of newly concrete.

2.2.3 *Effect on corrosion of reinforcing steel*

The corrosion areas of reinforcing steel in Type-A specimens with a cover depth of 2cm after the exposure is shown in figure 6. From the figure, in any conditions such as W/C and non- or pre-neutralization, the corrosion area of re-bars in the specimen exposed in Mt.Sakurajima is clearly larger than that in Kamakura city. Additionally, in the case of in Mt.Sakurajima, the corrosion rate is seems to be accelerated with increase of exposure time. This

However, as to the specimens of W/C70%, interested differences in exposure area were found. In the case of "W/C70%, non-neutralized specimens", the neutralization of specimen in Kamakura city progressed faster than that in Mt.Sakurajima, and this difference decreased with passing exposure time. On the contrary, in the case of "W/C70%, pre-neutralized specimens", the neutralization in Mt.Sakurajima progressed faster than that of Kamakura city, and this difference tended to increase with passing of exposure time.

phenomenon is one of the most considerable effects of acid rain on reinforced concrete structures.

Table 3 shows the results of comparing the corrosion area on exposure area and cover depth of reinforcing steel. From the ratios to compare the effect of exposure areas, with increase of W/C, the effect of acid rain on corrosion of reinforcing steel will be larger. Also, from the ratios to compare the effect of cover depths, when concrete is affected by acid rain, a cover depth is to be a main factor of corrosion rate of reinforcing steel, comparing to W/C or existence of initial neutralization.

3 ACCELERATED TEST

3.1 Experimental outline

3.1.1 Specimens
For the accelerated test, high-early-strength portland cement was used, and reinforced mortar cylinder specimens with a cover depth of 3cm were made, as shown in figure 7. Table 4 gives mixture proportions of mortars. Before the test, crack of which width was 0.2mm at the surface was induced as shown in figure 7. In order to examine the effect of the acid rain clearly, every sides and basal faces of the specimen were coated by epoxy resin. Experimental factors and levels of the specimens are shown in table 5.

3.1.2 Experimental method
In the accelerated test, dry and wet repetition was carried out under the 2 conditions (series-1,2) shown in Table 6. Setting 96 hours as a cycle, the drying time in a cycle was decided as a accelerated neutralized thickness, estimated by using Kishitani and Morinaga for each experimental condition, to be equal to the actual one in Mt,Sakurajima after 1 or 2 years (Izumi, 1994). Also, considering the actual quantity of annual rainfall in Mt.Sakurajima, the quantity of acid solution sprayed to the specimens during a cycle was decided as 2250 mm^3/mm^2 (NAO, 1998). Figure 8 shows the rainfall system originally developed by authors. Table 7 shows the chemical property of the acid solution (pH3.0) simulated the volcanic acid rain in Mt.Sakurajima.

For comparison, an experiment using distilled water (pH5.1) instead of acid solution was carried out.

3.2 Results of accelerated test

3.2.1 Effect of the acid rain on concrete surface
Appearances of "W/C70% non-neutralized specimens" after 5 cycles in accelerated test (series-1) are shown in figure 9. As same as the exposure test, surface of the specimen sprayed by acid solution was yellowed and roughened due to chemical reaction and the dissolving of cement paste.

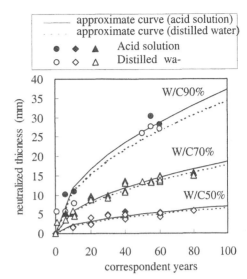

Figure 10 The aging variation of neutralized thickness in Series-1 and Series-2 accelerated test

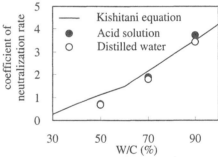

Figure 11 Neutralization velocity coefficient of the accelerated test result and Kishitani equation

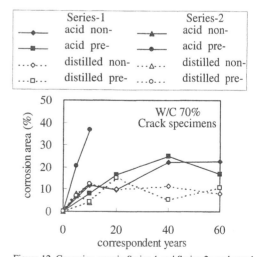

Figure 12 Corrosion area in Series-1 and Series-2 accelerated test

521

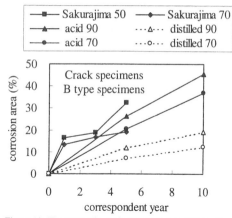

Figure 13 The comparison of corrosion area of Series-2
accelerated test and the exposure test

In the case of non-neutralized specimens with crack, the white precipitate appeared at crack portion. It is considered the elution of calcium carbonate.

3.2.2 *Effect on neutralization of concrete*
In Figure 10, averages of neutralized thickness of the specimens in each correspondent exposure years were shown. Here, in the case of "pre-neutralized specimens", experimental data was plotted in the figure considering the effect of pre-neutralization. Though the remarkable difference in neutralized thickness between the results using acid solution and distilled water was not found, the progress of neutralization tends to be a little due to the acid solution.

Figure 11 shows coefficients of neutralization rate calculated by root equation in each W/C. As in the figure, a 5 % increase of the coefficient for using acid solution approximately.

3.2.3 *Effect on corrosion of reinforcing steel*
The corrosion areas of reinforcing steel in specimens with crack after accelerated test (series-1 and 2) is shown in Figure 12. The results obtained from series-2 indicates the influence of acid solution on corrosion rate of steel in mortar obviously. However, from the results obtained from series-1, only a small effect due to acid solution can be seen, comparing to the results of series-2 especially in short correspondent yeas. It is considered that the difference in both experimental results was caused by the difference of the way to control pH of acid solution. That is to say, pH of acid solution was controlled continuously in series-2, however, in series-1, pH was controlled only at the beginning of each cycle.

Figure 13 shows the corrosion area of reinforcing steel in mortar obtained from series-2. To compare the accelerated test with the actual exposure test, corrosion areas of Type-B specimens exposed in

Mt.Sakurajima were also shown in the figure. As can be seen, the increasing rates of corrosion areas in both accelerated test by acid solution and exposure test by volcanic acid rain are similar. And it is clarified that the effect of acid rain on the corrosion of steel in concrete was remarkable, especially when concrete has cracks.

Table 7 Chemical property of acid solution

	HCl	H_2SO_4	HNO_3
pH 3.0	6.0	34.0	8.0

(unit : mg/l)

4 CONCLUSIONS

In this study, the experimental examination was carried out for the purpose of quantitatively grasping the effect of acid rain on concrete structure. As the results, it was concluded as follows.

1) Under the acidity environment, concrete surface will be yellowed and roughened due to chemical reaction and the dissolving of cement paste .
2) Only a little effect of acid rain on neutralization of concrete was found. However, for the concrete which already neutralized to some extent by CO_2, the additional neutralization by acid rain will be accelerated compare to that of newly concrete.
3) It was clarified that effect of acid rain on corrosion of steel in concrete structure was remarkable, especially under the conditions of high W/C and existence of crack in concrete.
4) On deterioration of concrete structure due to the acid rain, the effect of the acid rain can be examined by the accelerated test.

REFERENCES

Kazusuke Kobayashi, 1997. Deterioration process of concrete structures affected by acid deposition, Journal of materials, concrete structures and pavements, Japan society of civil engineers, No.564, 5-35,pp.243-251,

Takayukii Sato, 1999. Experimental study on deterioration of concrete due to acid rain and deterioration control by surface coating, journal of materials, concrete structure and pavements, Japan society of civil engineers, No.634,5-45, pp.11-25,

Itoshi Izumi, 1994. Lecture1 "neutralization", Concrete journal, Japan concrete institute, Vol.32, No.2, pp.72-83,

National Astronomical Observatory 1998. Rika nenpyo

17 Soils and pavement analysis

Creative Systems in Structural and Construction Engineering, Singh (ed.) © 2001 Balkema, Rotterdam, ISBN 90 5809 161 9

Using the AHP to prioritize pavement maintenance activities

A. El-Assaly
Civil Engineering Department, University of Alberta, Edmonton, Alb., Canada

A. Hammad
WAIWARD Steel Fabricators, Edmonton, Alb., Canada

ABSTRACT: One of the major problems facing transportation agencies is that of evaluating pavement maintenance alternatives and finding the optimum maintenance strategy for defected pavement segments. Applying the optimum maintenance strategy reduces the total maintenance costs over the service life of pavements, postpones the costs of major rehabilitation, and maximizes the benefits gained from limited budgets. Pavement maintenance activities are usually triggered based on Surface Condition Rating (*SCR*) process The main objective of this paper is to present a Decision Support System (*DSS*) for prioritizing pavement maintenance activities using the Analytical Hierarchy Process (*AHP*) for Alberta Infrastructure (*AI*) as a multi-criteria decision-making technique.

1 INTRODUCTION

Selecting the most appropriate maintenance alternative for a defective pavement segment is a problem transportation agencies face daily. The decision significantly affects the service life of the road and its overall life cycle cost. This decision has to be made by experienced engineers or technicians, who are able to identify the problem and allocate the most appropriate treatment. Most transportation agencies in North America face a lack of experienced staff due to retirement and budget constraints. There is a tremendous need to transfer the knowledge of experienced staff to the new and less experienced staff.

The current practice for selecting a maintenance alternative is mainly subjective and depends only on the decision-maker's experience. Solving this problem in a consistent manner would help transportation agencies properly manage pavement and estimate budgets for the future. After fully realizing which treatments can be applied to a defective segment, there will be more than one feasible solution to fix each type of pavement distress. It is necessary to evaluate these solutions and select the most appropriate one based on standard criteria.

The impact of computer technology on everyday aspects of life is increasing rapidly. Computer applications can now be found in management decision-making areas. Decision-making is one of the most difficult tasks in the field of management; decisions should be sound, reasonable, and acceptable to all affected people. For many years, decision-making was considered to be pure art, or a talent acquired over time through experience. However, the decision-making environment has been shown to be much more complex in recent years. The main reason for this is the implementation of new technologies that introduce more alternatives. The tremendous increase in the cost of errors makes poor decision-making unacceptable. Therefore, computer applications have been employed in the decision-making process to increase accuracy in solving complex problems (Turban, 1998). The main advantage of computer-aided decision-making is the possibility of merging the mathematical capabilities of computers with human experience.

2 THE DECISION MAKING PROCESS

The decision-making process involves three main phases, as shown in Figure 1 (Simon, 1977). The process could be illustrated in three phases

2.1 Intelligent Phase

In this phase, the reality is examined, the problem is identified, and the problem variables are defined. Any related data is collected and analyzed at this time.

2.2 Design Phase

A model will be constructed to represent the real system. The factors affecting the problem and the relationships between them and the problem variables should be determined.

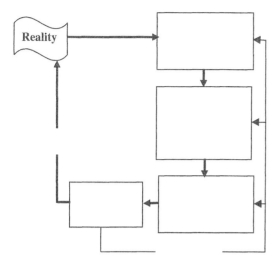

Figure 1. Decision-Making Process

To build the model, assumptions are used to simplify the actual problem, such as omitting some of the least important factors or limiting the model to a part of the problem. The model is then validated and criteria are set for the evaluation of the possible alternatives.

2.3 *The Choice Phase*

Introduces a solution, which will be tested until the results are reasonable and the solution can be implemented. Successful implementation leads to solving the original problem, while failure leads to a return to an earlier phase of the process.

3 MAIN COMPONENTS OF *DSS*

The Decision Support System (*DSS*) is composed of several main components, represented as follows:

3.1 *Data Management Subsystem*

This process requires identifying the goal or focus, defining the decision-making criteria, and then calculating priorities for feasible alternatives. The goal of the developed *DSS* is to allocate the most appropriate treatment alternative for defected pavement segments. To identify feasible treatments, human knowledge regarding technical constraints of applying treatments was captured.

3.2 *Knowledge Management Subsystem*

This component of the *DSS* deals with human preferences based on previous experience. Artificial Intelligence applications can be used to capture human

knowledge regarding the problem under investigation. There are different applications of intelligent systems, which can be used to achieve this goal. Knowledge Based Expert Systems (KBES), Artificial Neural Networks (ANN), and Fuzzy Set Theory are examples of these applications. A decision support system including this component is called intelligent an *DSS* (Turban, 1998).

3.3 *The model Management Subsystem*

This considered the main component of a *DSS*. There are different types of models, which vary in their capabilities based on the problem solving requirements. Statistical, tactical and strategic, and financial and marketing models can all be used for this component of a *DSS*. The main objective of the model component of an intelligent *DSS* is to merge the data and expert opinions together to arrive at the desired decision. The AHP is one of the most suitable models for this purpose. The concept of the AHP will be illustrated later.

3.4 *The User Interface*

Allows the end-user to communicate with the *DSS*. Power, flexibility, and ease of use are the main characteristics of a successful user interface. Some *DSS* experts argue that the user interface is the most important component of a *DSS* since it is the only part of the *DSS* the user deals with (Sprague, 1986). An inconvenient user interface is a major reason for decision-makers to prefer the traditional decision-making process to the *DSS*.

4 DEVELOPING *DSS* FOR PAVEMENT MAINTENANCE ALLOCATION

Several pavement maintenance experts within the province of Alberta attended workshops to define distress attributes as well as factors that influence the selection of treatments. It was found that the distress type and degree of severity are the most important factors for selecting maintenance alternatives; as such they formulate the technical constraints. The term "technical constraints" refers to the applicability of a specified treatment to fix a certain problem (Hammad, 1999). While density degree might affect the decision in the current practice, the researchers decided to omit it from the technical constraints because it is fiscal rather than a technical

4.1 *Technical Constraints*

The technical constraints represent the main challenge in developing the model because not all treatments can fix all problems. After several attempts, the transportation experts agreed to divide the problem into smaller problems with separate sub-models. This required identifying the applicable treatments for the three severity degrees of each distress type, identifying cases that can be fixed by the same treatments, and building separate sub-models for each case.

4.2 Pavement attributes

Pavement attributes, such as soil type and base type, might affect the actual service life of maintenance alternatives (El-Assaly, 2000). However, AI maintenance inspectors do not consider this while making their decisions. The research team and the pavement experts decided to allow the maintenance inspectors to preview these attributes through the system, which will enable them to build better understanding of the effect of these attributes on maintenance activities.

4.3 Pavement Distresses

Based on the knowledge obtained, the research team and the pavement experts decided that the *DSS* would assess nine surface distresses. The analysis of the applicable treatments revealed that four of them could be fixed using more than one treatment, and that these treatments can be prioritized using AHP. These distresses are cracks, transverse cracks, depressed transverse cracks, and loss of aggregate. For the selected distresses, different treatments are required for each degree of severity. It was found that five cases could cover all possible combinations, as shown in Table 1 (Hammad, 2000). The table illustrates five cases; each contains the distress types, their severity degrees and the three associated maintenance alternatives that can fix the problem.

Alligator cracks as well as rutting, which are structurally related distresses, require testing of the structural adequacy of the road.

4.4 The DSS Criteria

To identify the decision-making criteria, several were introduced to the pavement experts committee. All of the pavement experts on the committee agree that three main criteria govern the decision to allocate treatments to the defected pavement segments. These criteria are: disruption during application, cost per year of expected service life, and previous experience with the treatment

The term disruption during application refers to all possible hazards that might occur during the treatment process.

The time labours are required to be on highway, amount of warning signs, flagmen, traffic closure requirements, and labour safety during applying

treatment are some examples of these hazards. To compare the relative importance of each criterion, AHP was introduced to the committee members with special emphasis on the idea of pair comparison.

4.5 Pairwise Comparisons

Pair comparisons are formulated in matrices. Matrix calculations are used to prioritize the elements in one set and obtain the final priorities. For human experts, it is always easier to apply pairwise comparisons than to rank a large set of different alternatives. The AHP builds a square matrix for pairwise comparisons into each set. A scale of 1-9 was found to be suitable for applying the pairwise comparison (Saaty, 1990). The scale is illustrated in Table 2; it represents the following qualitative distinctions: equal, weak, strong, very strong, and absolute preference.

As an example of pairwise comparisons, the pavement experts agreed that previous experience is:

- three times more important than disruption during application, and
- twice as important as cost per year.

Table 1. Five Treatments Cases

Cases	Treatments	Distresses
Case 1	Hot Pour	Cracks with *Moderate* Severity.
	Cold Pour	Transverse Cracks with *Slight & Moderate* Severity.
	Rout & Seal	Depressed Transverse Cracks with *Slight* Severity.
Case 2	Hot Pour	Cracks with *Extreme* Severity.
	Cold Pour	
	Spray Patch	
Case 3	Hot Pour	Transverse cracks with *Extreme* Severity.
	Spray Patch	
	Mill & Fill	
Case 4	Mill & Fill	Depressed Transverse Cracks with *Moderate* &
	Spray Patch	*Extreme* Severity.
	Thermo Patch	
Case 5	Fog Coat	Loss of Aggregate with *Moderate* & *Extreme*
	Seal Coat	Severity.
	Chip Seal	

4.6 The Model Management Subsystem

The AHP is used as a model to build *DSS* in different areas. The process requires identifying the goal or focus, defining the decision-making criteria, and then calculating priorities for feasible alternatives.

Figure 2 illustrates the hierarchy of the model based on the previously mentioned criteria. Six submodels were built for each of the previously described cases. All the sub-models were similar, except that each has different feasible alternatives.

Table 2. Pairwise Comparison Scale.

Intensity of Importance	Definition	Explanation
1	Equal importance	The two alternatives contribute equally to the objective
3	Weak importance of one over another	The expert slightly prefers one alternative over another
5	Strong importance of one over another	The expert strongly prefers one alternative over another
7	Very strong importance of one over another	The expert very strongly prefers one alternative over another
9	Absolute importance of one over another	The expert has the highest possible preference for one alternative over another
2, 4, 6, 8	Intermediate values between the scale values	When intermediate judgements are required

To build the six sub-models using the pre-identified criteria, twelve matrices were submitted to the pavement experts committee. For each of the previously mentioned cases, two comparison matrices were prepared; one for "previous experience" and another for "disruption during application". The data regarding cost of treatment and
expected service life will be left as a user input due to the variation in their values throughout the province and under different conditions. Figure 2 illustrates the hierarchy of the decision-making problem.

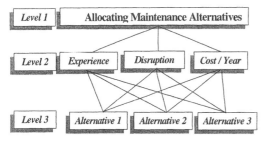

Figure 2. Hierarchy of the Decision-Making Problem.

4.7 *The Data Management Subsystem*

Figure 3 illustrates the data subsystem required for the *DSS* development. The data required for the *DSS* was found in three different databases in the Alberta Infrastructure (AI) data repository. These databases use highway numbers as well as control sections (large portions of the highway with fixed boundaries) to identify all of the records. The first database used was the Surface Condition Rating (SCR) data. The database divides the Alberta highway network into almost two thousand segments based on similar surface conditions. This method of segmenting enables *AI* maintenance inspectors to easily allocate maintenance alternatives to segments. For each segment, all information related to its surface conditions is recorded. The most important of this information was the data regarding the severity and density of different distress types located on segments.

The second database used was the inventory database, which contains all of the Alberta highway network attributes. All highway control sections are divided into smaller segments called inventory sections, and for each inventory section, many attributes are recorded. Only five of these attributes were transferred to the *DSS* database management subsystem. These attributes include base type, soil type, climatic region, Equivalent Single Axle Load (ESAL) representing traffic loads, and segment age.

The third database used was the roughness data, which is represented by the International Roughness Index (IRI). The reason for using roughness data was to enable the pavement maintenance inspectors to check the roughness of specific segments and correlate it to the surface condition and other pavement attributes.

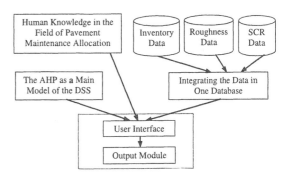

Figure 3. Data Management Subsystem.

5 THE DSS IMPLEMENTATION

The DSS is designed to be used by maintenance inspectors in each district separately. The reason for this is that the treatments' unit prices and expected service lives vary between districts across the province of Alberta. The database repository of the system contains the pre-prepared surface condition data, Roughness data, and Inventory data. The system starts by asking the user to submit the unit prices and expected service lives for the twelve assessed treatments in his/her district. After obtaining the required information, the system automatically suggests possible treatments and calculates treatment priorities for each of the segments. The priorities are calculated based on the segment surface condition as well as the unit prices and the expected service lives for treatments that have been previously submitted by the user.

The system then allows the user to select a specific segment from the stored segments in the database. For the selected segment, the user can view its surface condition, represented by the existing distresses and their ratings and the overall Surface Condition Index. The average IRI that represents the segment roughness is also calculated from the "Roughness" database and presented to the user. The user can then view segment attributes or the suggested treatments for the selected segment.

6 THE DSS VALIDATION

The system was tested using the "what-if" analysis". It is structured according to what will happen to the solution if an input value is changed (Turban, 1998). A MS Excel spreadsheet was prepared to instantly calculate the final priority vectors for the six sub-models of the problem. The experts were asked to submit their expectations if one of the parameter values is changed. Their expectations were compared to the output of the DSS and the priorities after the change were found to be acceptable and comply with the experts' expectations. The experiment was repeated several times to ensure that the obtained priorities are always satisfying. Finally, the pavement experts strongly recommended implementing the DSS as the maintenance module in the new Pavement Management System (PMS) currently under development at AI.

7 CONCLUSIONS

The Analytical Hierarchy Process (AHP) proved to be of great benefit in solving decision-making problems, which require the integration of human knowledge and mathematical calculations. The AHP was used as the main module for the DSS to prioritize the feasible alternatives based on pre-specified criteria. A hierarchy model requires identifying the goal, determining the decision-making criteria, and outlining the feasible alternatives. The human knowledge is captured using pair-wise comparison matrices. The AHP translates the pair-wise comparison matrices created by experts to priorities. Due to the technical constraints of the problem, it could not be solved using only one hierarchy model. The decision-making process was divided into six sub-models, each with the same feasible alternatives.

The DSS contains a user-friendly interface to organize the input-output process. The user can enter the treatment's unit prices and expected service lives in his/her district or accept the default ones. After that, the user can select a specified segment to plot

its characteristics, and obtain suggestions of treatment alternatives with their priorities for the distresses existing in this segment. The output module enables the user to print out all these useful reports for any specified segment in the district.

The DSS introduces consistency to the decision-making process and transfers the knowledge of some of the best pavement experts in AI to new and less experienced staff. The DSS also introduces maximum usage of surface conditions data and offers an example for integrating this data with inventory and roughness data for better pavement management.

The DSS was tested and validated by AI pavement experts who recommended implementing it as a component in the new Pavement Management System (PMS) currently under development.

8 REFERENCES

El-Assaly, A., Ariaratnam, S., and Hempsey, L., "Development of Deterioration Models for the Primary Highway Network in Alberta, Canada," Journal of Infrastructures Management Systems, ASCE, In press, 2000.

Hammad, A., "Decision Support System for Prioritizing Pavement Maintenance Alternatives," A Master of Science Thesis, University of Alberta, 1999.

Hammad, A., El-Assaly, A., and AbouRizk, S., "Decision Support System for Allocating Pavement Maintenance Alternatives," Journal of Infrastructures Management Systems, ASCE, In press, 2000.

Saaty, T. L., "Multicriteria Decision-Making: The Analytical Hierarchy Process", RWS Publications, 1990.

Simon, Herbert Alexander, "The new science of management decision." Revised edition, Englewood Cliffs, NJ, Prentice-Hall, 1977.

Sprague, R. H. and H. J. Watson, "Decision Support Systems: Putting Theory Into Practice", Prentice-Hall Inc., 1986.

Turban, E. and J. E. Aronson, "Decision Support Systems and Intelligent Systems", Fifth Edition, Prentice-Hall Inc., 1998.

Creative Systems in Structural and Construction Engineering, Singh (ed.) © 2001 Balkema, Rotterdam, ISBN 90 5809 161 9

The estimation of deflections and stresses in paved surfaces

I. Špacapan & M. Premrov
Faculty of Civil Engineering, University of Maribor, Slovenia

ABSTRACT: This paper presents the method for the analysis of deflections and stresses in paved surfaces based on simple experimental acquisition of Green's function of arbitrary heterogeneous soil. The experimental and computational procedure that yields Green's function on the soil surface in analytical form is explained. Numerical examples of the simulated experiments on the soil typical under paved surfaces are presented, and the results of identification compared with exact solutions. The results are used for the calculation of the dynamic flexibility matrix of soil. The deflections, shear forces and bending moments in a concrete road plate, as well as the contact stresses between the plate and the bearing soil due to traffic like loading are computed via method of substructure synthesis. The experimental and computational efficiency and the accuracy of the results of the presented method are briefly discussed and compared to some alternative procedures.

1 INTRODUCTION

In every engineering analysis of the stresses, strains and displacements of reinforced or paved surfaces due to dynamic or static loading it is necessary to identify the mechanical characteristics of the soil in situ. Although the soil has distinctive nonlinear and time dependent mechanical characteristics, it behaves almost as a linear material when it is firm enough and the loading is transient and relatively small in amplitude. Such circumstances are typical by paved traffic surfaces where either the natural bearing soil is sufficiently rigid, or it is improved to acquire the desired rigidity. Thus for any engineering analysis of traffic loading, or other similar kinds of surface loading, we have to identify the linear soil's characteristics. For this purpose great variety of methods is at disposal, see for instance Hoadley (1985) for traditional methods, and for more modern seismic methods Al-Hunaidi (1993) and Roesset et al. (1991). They vary from very simple methods to very complex ones, and so basically varied the quality of their results: from poor accuracy, incompleteness and unreliability to excellent results. From the engineering point of view, we cannot afford extensive soil identification by problems that have to be solved fast and must not be financially demanding. In any case, there is always a certain degree of ambiguity whether the results are sufficiently correct and complete or not, Crouse (1990). In addition, when we want to reduce the uncertainty of the results, alternative methods are used for verification.

When the mechanical constants of the soil are acquired, we have to set up a physical and a computational model of soil in order to analyze the interaction between the loading device and the paved surfaces, and the stresses in the reinforced surface. More complex soils require a considerable amount of computations. This can be accomplished in several ways: by BEM (Brebbia 1984), via computed Green's function on soil's surface by specialized computer programs (CLASSI 1983), or only approximately by FEM. So, in the analysis of traffic loading of paved surfaces there is always a three-step procedure: soil identification, computing of soil model, and numerical analysis usually via substructure synthesis. This altogether demands a considerable amount of work if reasonably reliable and accurate results are to be obtained.

We are presenting an alternative method which, we believe, has certain important advantages regarding the experimental simplicity and the accuracy of experimental results, as well as a reduced computational effort compared with alternative methods yielding an equivalent quality of results. It is based on Green's function measured on the soil's or on the paved surface.

In this paper we will first present basic formulae for the computation of displacements and stresses in paved surfaces providing that Green's function on the soil surface is given in analytical form. Then we will point out some features of Green's function pertinent to proper setting-up of the measuring system and performing analytical approximation of experimental Green's function. By numerical examples we will illustrate the acquisition of the simulated experimental Green's function, followed by the analysis of the theoretical case of a concrete reinforced road. The results are compared with the simulated ones obtained by a simplified experimental procedure, and alternatively, with the results based on simplified computations.

2 DESCRIPTION OF THE METHOD

2.1 Substructure synthesis using Green's function

Green's function in the frequency domain on the soils surface is the displacement caused by the point load. It has nine components, and depends on the frequency ω, location $\mathbf{r}_o$ of application of unit point load, and the location $\mathbf{r}$ of the displacement $\mathbf{u}$. It is symbolically presented by equation 1.

$$\mathbf{G}=\mathbf{G}(\omega,\mathbf{r}_o,\mathbf{r}) \tag{1}$$

If we are analyzing only the case of concentrated normal loading to surface, and off diagonal terms can be neglected as in case of vertical traffic loading, then only the component G_{zz} of Green's function need to be considered. This component has an important feature, which has to be considered when measuring it, and properly treated when integrating it. Namely, it is theoretically simply singular at the point of application of point load, i.e. at r = 0. The singular value is constant for all frequencies, except for small variations due to possible damping in soil. Certainly, in general it varies with the loading point location, i.e. with $\mathbf{r}_0$.

By a sufficiently dense net of surface boundary elements the simple collocation method yields accurate results. It requires simple integration, including suitable numerical treatment of singular values, when computing the elements of the flexibility matrix:

$$H_{ij}^g = \int_{A_j} G_{zz}(\omega,r_i,r)dA(r) \tag{2}$$

Inverting this matrix yields the stiffness matrix of soil $\mathbf{K}^g$. The simplest way of substructure synthesis is that the dynamic stiffness matrix of the plate on the soil's surface $\mathbf{K}^p$ is composed of elements of the same size and form as the boundary elements of soil's surface. Reducing the redundant rotational degrees of freedom yields $\mathbf{K}^p_r$, and simply adding both matrixes, yields linear equations for the interacting system:

$$\left(\mathbf{K}^g + \mathbf{K}^p_r\right)\mathbf{U} = \mathbf{P} . \tag{3}$$

Solving the equations on the displacements vector $\mathbf{U}$ for a given loading vector $\mathbf{P}$, the contact stresses between the soil and the plate are computed using $\mathbf{K}^g$. Shearing forces and bending moments in the plate are computed by $\mathbf{K}^p$, encountering also the rotations.

2.2 Measuring system and computation of Green's function

Basically we measure the impact of a hammer on the soil's surface and the displacements at various distances r_i, i=1,2,..., as is schematically shown in Figure 1. P(t) is the exciting force, and u(t) the displacements of soil's surface at the distance r_i. FT designates the force transducer which is usually integrated into the hammer, and VT is the velocity or acceleration transducer. Not going into details, we only mention that the transducers might need suitable grounding to ensure proper bonding to soil. The measuring procedure itself and the manipulation of measuring data are standard, which means the same as they are by measuring the transfer functions. This is described in standard literature (see for instance Isermann 1986). In our case, the result of measurements and data manipulation is the transfer function force-displacement, which is, apart from measuring errors, merely Green's function evaluated at $r = r_i$. For the sake of the simplicity of explanations we shall from now on suppose that the soil is axisymmetric, bearing in mind that the extension of the presented method to the general case of soil is obvious. With this supposition Green's function depend only on the distance r.

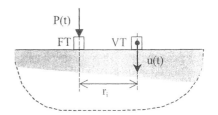

Figure 1. Measurement of Green's function.

In Figure 2 we can see the graph of the real part of theoretical Green's function. It is, as usually, multiplied by r and normalized to yield a unit singular value. This Green's function is computed, as all subsequent ones, by the computer program CLASSI (1979) for the soil profile presented in Figure 3. It is approximately a profile that might occur in road construction. The upper layer is added, or the natural one improved to get the required rigidity of the road construction. The density of soil is supposed to be uniform, $\rho=1800$ kg/m^3, and the material damping is 1%. In Figure 3 the soil is covered by a concrete

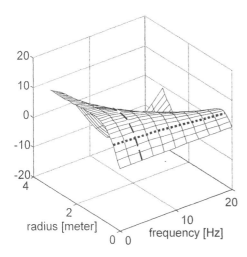

Figure 2. Real part of Green's function.

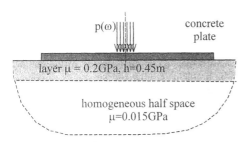

Figure 3. Soil profile for the computation of Green's function, and loaded concrete plate that is analyzed.

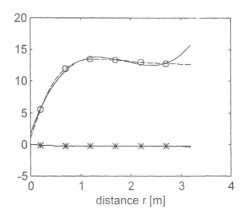

Figure 4. Static case. Real and imaginary part of Green's function at zero frequency, marked by circles and asterisks, respectively.

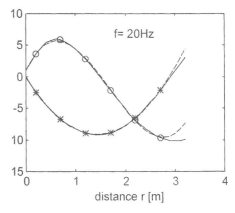

Figure 5. Dynamic case. Real and imaginary part of reduced Green's function at 20 Hz frequency, marked by circles and asterisks, respectively.

plate which we will be later analyzed for a loading that approximately corresponds in magnitude to a truck wheel load. In Figure 2 the bold dotted line represents Green's function at a constant distance r, the one we get by measurements. Indeed, due to FFT, this line is experimentally given by a considerably great number of discrete numeric values. On the other hand, the bold dashed line represents Green's function at constant frequency, the one that we need to compute the dynamic flexibility matrix of soil. It is not directly measured and has to be acquired computationally in an analytical form suitable for integration.

For an experimental procedure that is convenient for practical engineering use it is of utmost importance that the measurements be simple. Therefore we assume to have measured displacements only at few distances, and from these limited data of Green's function, we have to get it in an analytical form depending on distance r, and valid in the range of interest. For an analysis of interaction between the soil and a loaded circular plate with the diameter of 3.2 meters the necessary measurements of Green's function are only within this diameter, that is on the

contact surface. In Figures 4 and 5 is shown, for example, Green's function at zero and at 20Hz frequency, respectively. Solid lines represent exact values. Circles and asterisks show the simulated measured data at distances 0.2m and then at each consecutive half a meter. Thus, only six complex values for each frequency are given. Here, it is important to note that we cannot measure Green's function at zero distance or very near the excitation point because of its singularity. Thus, 0.2m is chosen as a typical starting distance for the first measuring distance, providing that we have transducers and an impact hammer of suitable size that will not introduce an essential systematic error. Using these six data per each frequency Green's function is approximated by best fit by a product of polynomials of lower degrees having complex coefficients b_j, and an exponential function:

$$r.G(\omega_k,r) \cong e^{-i\frac{r.\omega}{c}} \sum_{j=0}^{n} b_j.r^j . \tag{4}$$

Using a lower degree of polynomial is essential, because otherwise, by scarce data, extrapolation to singular value and interpolating values become too erroneous (Hamming 1986). An exponential term is necessary for higher frequencies and/or greater distances in order to reduce the order of approximating polynomials. This exponential term emerges from the physical fact that Green's function in the time domain represents the wave disturbance due to the impulsive load. That disturbance arrives with the time delay:

$$\Delta t = r/c, \tag{5}$$

where r is the distance, and c the group velocity of disturbance. In the frequency domain this causes a periodic variation of the function, which is difficult to approximate by a simple polynomial. Thus this variation is reduced by an exponential term. This reduction is called time shifting. In Figure 5 the time shifted Green's function is shown. It is sufficient for an effective approximation that we choose an approximate value for c. However, this value can be easily estimated from the measured signal in time. In our case we have used simply the velocity of the upper layer, which is 333m/s, and the approximation by a polynomial of the fourth degree. The approximated Green's function is shown in figures 4 in 5 by dashed lines showing good agreement to exact values presented by solid lines. For this case the graphs of exact and approximated Green's functions match perfectly when using a polynomial already of the fifth degree. Thus it is easy to acquire excellent approximation. By repeating such approximations for all the frequencies of interest, we get Green's function in analytical form presented by equation 4. More detailed discussion of measurements and approximations of experimental Green's functions of soil for various profiles is presented in the references (Špacapan 1995, Umek 1995).

3 CASE STUDIES OF PAVED STRUCTURE USING APPROXIMATED GREEN'S FUNCTION

Although the graphs of approximated and exact Green's function may look very alike there are still differences between them. The effect these differences have on engineering quantities of interest is studied on the example of a circular concrete plate. Its diameter is R=3.2 m, thickness h=0.12m, Young modulus E=30GPa and Poisson's ratio 0.25, and it is laid on top of the formerly presented soil profile. The plate is loaded by constant vertical traction p=0.6MPa within the centered circle with a radius of 0.1m (see Figure 3). The resulting force is P=18.85kN. So, we have an axially symmetric case.

The vertical displacements u and contact stresses p_s between the plate and the soil were calculated, as well as the shearing forces q and bending moments m_r in the plate, using the computing procedure presented in subsection 2.1. Some typical results are presented in Figure 6-7 by solid lines, and specifically all the values of the displacements at 20Hz in Figure 8. The accuracy of the results, solely due to the used discretisation of the plate and the soil model, is within 1%. To estimate the influence of the accuracy of Green's function approximation, we used third and fourth degree approximating polynomials, as well as the exact values. The differences in the results are very small for computed quantities and cannot be observed in the presented graphs in Figures 6-7, except for the contact stresses which are shown separately in Figure 9. The solid line belongs to the results where Green's function is approximated by a polynomial of the fourth degree, while the dashed line belongs to the approximation a polynomial of the third degree. Only the distinction of graphs near the singular point, which emerges from erroneous extrapolation to the singular value is evident. This suggests that the accuracy of the method could be problematic only for the results of contact stresses near the concentrated loading surface.

To compare the effectiveness of the presented method, the results of two alternative methods that can be considered equivalently simple, are presented. The first one is the homogeneous half space model, for which values of Green's function are published in literature, and for the static case given analytically by Boussinesq equations. Although it is disputable which material characteristics should be used for this model, and how to get them experimentally, this model is still frequently used in practice. We have used for this model the characteristics of the top layer for this model and the results are presented in Figures 6-7 by dashed lines. We can observe a considerable discrepancy between the results, but it is an engineering decision whether they are acceptable as adequate estimates for an analyzed problem.

The second model is based on the so called modulus of soil reaction. The model of soil is called a spring model and is very simple to compute. It is formed by separated elastic springs, having elastic constants that can be approximately acquired experimentally by plate loading tests. We have simulated this model simply by canceling the off-diagonal terms of the flexibility matrix of the exact soil model. The results of the analysis, using this model, are presented by dotted lines in Figures 6-7 showing intolerable deviations from the correct ones. This clearly suggests that over simplified computing models based on simple plate loading tests are unacceptable.

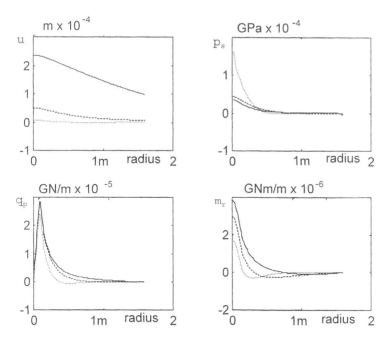

Figure 6. Static case. Absolute values of displacements u, contact stresses p_s, bending moments m_r and shearing forces q_r are shown in subfigures in clockwise direction, respectively. Solid, dashed and dotted line show the results of the computations via Green's function, of the half space and soil-spring model, respectively.

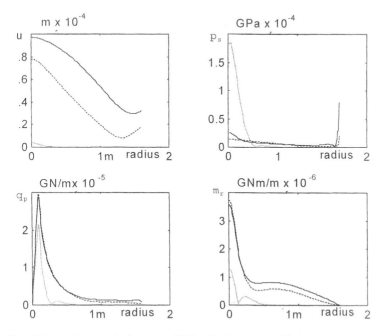

Figure 7. Dynamic case at the frequency of 20Hz. Absolute values of displacements u, contact p_s, bending moments m_r and shearing forces q_r are shown in subfigures in clockwise direction, respectively. Solid, dashed and dotted line show the results of the computations via Green's function, half space and soil-spring model, respectively

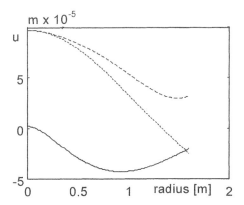

Figure 8. The real, imaginary and absolute value of the displacements at the frequency of 20Hz shown by the solid, dotted and dashed line, respectively.

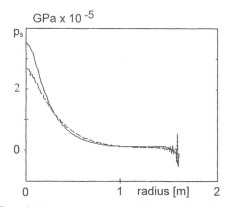

Figure 9. Contact stresses computed via approximated Green's function by the fourth and third degree polynomial shown by the solid and dashed line, respectively.

4 CONCLUSIONS

Simple experimental estimation of soil properties or simplified computing models yield intolerable inaccuracy, or at least high unreliable results of the analysis of paved surfaces. On the other hand, extensive soil testing and computing is too complex for many practical engineering cases. Yet, in the paper presented method and the numerical examples suggest that by simple measurements of Green's function along with considerably simple computations, it is feasible to analyze paved surfaces considerably accurate with a relatively small effort.

As the presented examples suggest, we can use a more dense net of measuring points, along with some polynomials of higher degree for approximating Green's function. In this way we can easily regulate the accuracy of the engineering analysis. The reliability of the method should be very high, because Green's function comprises all the soil characteristics, regardless of its complexity, for the presented kind of analysis.

It may be necessary, in case of a considerable lateral variation of soil profile, to carry out excitation and measurements at more surface points. This would yield Green's function for every point of excitation r_0, and depending on the vector r rather than solely on the distance r. Consequently, the computation of the flexibility matrix of the soil's surface would be more tedious.

We see no reason why the method could not be applied to other components of Green's function, and consequently for the analysis of loading and displacements in all directions.

The method can evidently be applied to the already built paved surfaces to estimate its characteristics for further analysis or for monitoring needed in maintenance works.

We believe that the performance of the method can be suitably automated by built in computer programs for data processing and analysis that would yield desired results on the measuring site. Last but not least, we wish to add that exactly the same measured data as for Green's function can be simultaneously used to compute the soil profile by inversion of the velocity dispersion curve.

REFERENCES

Al-Hunaidi, M.O. 1993. Insights on the SASW nondestructive testing method. *Canadian J. Civ. Engrg. 20, 940-950.*
CLASSI, Wong H.L. & Luco J.E. A linear continuum mechanics approach 1979. *Report CE 79-3.* Department of CE Univ. of South. Calif., L.A.
Hamming R.W. 1986. *Numerical methods for scientists and engineers.* New York: Dover Publ. inc.
Hoadley, J.P. 1985. Measurement of dynamic soil properties, chapter 9 in *Analysis and design of foundations for vibrations.* Edited by Moore P.J. Rotterdam: Balkema.
Isermann R. 1986. *Identifikation Dynamischer systeme.* Berlin: Springer Verlag.
Roesset, J.M. & Chang, D.W. & Stokoe, K.H. 1991. Comparison of 2-D and 3-D models for analysis of surface wave tests. *Fifth Int. Conf. On Soil Dynamics and Earthq. Engrg.* Karlsruhe Germany.
Špacapan I. & Umek A. 1995. Experimental determination of near field Green's function on the surface of layered soil. *Computational Methods and experimental measurements. Pro. Conference, Capri 1995.* C.M. Publications: Boston.
Umek A. & Špacapan I. 1995. Identification of the layered half-space Green's function. *Soil dynamics and earthquake engineering. Conf., Crete 1995.* C.M. Publications: Boston.

Creative Systems in Structural and Construction Engineering, Singh (ed.) © 2001 Balkema, Rotterdam, ISBN 90 5809 161 9

Calcrete as sustainable road building material for Botswana

C.T.Ganesan
Faculty of Engineering and Technology, University of Botswana, Botswana

ABSTRACT: Botswana had been using calcrete as sub-base material for the main paved roads and in the roads built in the Kalahari Desert regions. This paper describes the types of calcrete available in Botswana, their formation and their engineering properties. Studies have been made regarding their use as moderately good bases, untreated shoulder materials and surfacing aggregates. Results of lime/cement – calcrete stabilisation are also included. The existing guidelines for the use of calcretes in paved roads are discussed with suitable modifications towards the development of appropriate specifications and construction techniques. Use of calcrete as the road building material is the challenging special priority for Botswana enabling massive saving of cost.

1 INTRODUCTION

Botswana is a centrally land locked country in the Southern African plateau. It has a total land area of 580,000 km². In the Kalahari region which covers about 80% of the area of Botswana confining to the western and northern part of the country, the predominant road building material that is suitable for use as base and sub-base is calcrete. This is geologically pedogenic material of which laterite and ferricrete are other family members. It is generally formed on flat land or in depressions where moisture is able to accumulate. Calcrete is also found on ridges or higher lying areas; however, in this case the surrounding materials have been eroded.

2 CLASSIFICATION AND DEVELOPMENT OF CALCRETE

Calcrete can be found in two ways.
• By the in situ cementation of almost any host soil with calcium carbonate deposition in the soil void space.
• By the progressive deposition of calcium carbonate replacing the parent soil that could have been eroded away. Roads Department draft.
Hence, the name calcrete can justifiably, applied to material varying from an almost pure lime stone in which there is hardly any trace of host material (hard pan calcrete) to a material consisting large extent of host material. These are white or grey in colour and take on many forms similar to ferricrete i.e. calcrete powder, calcified sand, nodular calcrete or hard pan.

2.1 *Calcified sand*

This represent the early stage in calcrete development in which the host material has not become very strongly cemented. Nodules are not formed.

2.2 *Powder calcrete*

Loose powder of silt and carbonate particles with little or no nodular development. These are different from calcified sand because of the presence of lime in the host soil grain.

2.3 *Nodular calcrete*

Nodules of carbonate cemented material in a calcareous matrix varying in size from silt to 60 mm of gravel. As a road construction material these are considered very important for Botswana.

2.4 *Honeycomb calcrete*

Intermediate stage of nodular and hard pan type. The nodules present have much greater strength; but the voids between the nodules are still getting filled with relatively uncemented material.

2.5 Hardpan calcrete

This is the final stage of development of all calcretes. They occur as harder sheet like crust or laminated layer, rarely more than half a meter thick, which overlies less well-cemented materials of lower stage development. Hardpan calcrete usually need ripping or even blasting before utilisation.

2.6 Boulder calcrete

It is the weathered stage of hardpan calcrete. Unless crushed it is unsuitable for road construction.

3 CALCRETE ROADS IN BOTSWANA

Previous history shows that Botswana had been using calcrete as sub-base material for the main paved roads and in the rural roads built in the Kalahari Desert regions. Table 1. The roads have been conconstructed with crushed or mechanically stabilised calcrete bases with lime or cement or both and are giving good performance. The guidelines formulated by Transport Research Laboratory (TRRL), UK and South African guidelines were followed in designing the pavement thickness and materials standards. (TRH 14, Ganesan1998). Roads built as part of the rural road project have used various thicknesses as per the design guides. In the selection of pavement material trial specifications based on research were followed. Calcrete suitable for use as surfacing aggregates are also exist in some parts of Botswana although their full potential is still being investigated on experimental roads.

4 PROPERTIES OF CALCRETES

Table 2 gives brief description of engineering properties of calcrete available in Botswana. Figure 1 shows standard sieve analysis and grading for the common types of calcrete (Ganesan 1998). From these results it is evident that hardpan calcrete can be used as base materials where as nodular calcrete can be used as sub-base material and as embankment soil. As a result of research studies their usefulness can be extended particularly in low volume traffic road construction. Because of the variability of calcrete even with the recognised types it is suggested that the full range of laboratory tests be conducted before its selection. Generally the properties like grading, the strength and proportions of the hard particles and soil fines (passing through 0.425 mm sieve) define the usefulness of the calcrete as road building material, TMH1 1986.

5 STABILISATION WITH LIME / CEMENT

Some locations in Botswana have calcrete of poor strength and hence are not suitable for direct use in road construction. They are mixed with cement or lime or both to have an effect on strength improvement. In this connection were conducted. When hydrated lime is added to calcrete it reduces the plasticity by initiating flocculation with any clay minerals present. In some cases when additional lime remains a pozzolanic reaction found to take place. Cement is usually added to effect a direct gain in strength by creating cementitious bonds between soil particles. When measured in the laboratory after specified curing periods high strength have been obtained with calcrete using either lime or cement. But in field the success of stabilisation depends on a number of factors. These include the time between mixing and compaction, the efficiency of curing and effective sealing over long periods and initial consumption of lime (ICL) of the unstabilised material.

The difficulties of compacting cement stabilised materials after long delays i.e. three hours or greater are well documented by previous studies. (Ganesan 1998). The only means of determining the response of a material with lime / cement stabilisation is to perform a test programme where by the density – strength – delay time relationship is determined. No specific examples are available from Botswana or elsewhere which demonstrate a rapid lime reaction although the data contained in Table 3 illustrates the change in density achieved at constant compaction at different time delays. Botswana has medium to low humid climate and with climatic condition carbonation reaction is found to occur beneath permeable surfacing. Present research has shown that with calcrete carbonation can retard and in some extreme cases even reverse the strength development in materials stabilised with cement and lime. (AASHO 1980, Ganesan 1998).

Table 1 Existing roads in Botswana with Calcrete

Main roads	km	
Francis town - Nata	28	Lime stabilised calcrete base with double surface dressing
Palaapye - serowe	45	2 stage lime and cement stabilisation of calcrete with a double surface dressing
Nata – Kazangula	100	Lime stabilised calcrete with a double surface dressing
Rural roads Sentiwa – Tsau	50	Untreated calcified sand powder calcrete with a double surface treatment comprising a single graded aggregate followed by sand seal
Tsabong - Makopang	95	Untreated mechanically stabilised calcrete base with double surface treatment comprising a single graded aggregate followed by sand seal
Molopole-Letlhakang	50	Untreated calcrete base with a double surface dressing
Mopipi - Rakopa	50	Untreated calcrete base with a double surface dressing

Table 2. Experimental results on some Calcrete samples

Calcrete type	Consistence limits			% passing				A P V	M D D	O M C	C B R	P M	GM	Remarks
	LL	PI	SL	0.08	0.425	2	19							
Hard pan 1	-	-	-	18	38	50	90	70	1978	9.50	172	170	2.0	Roadbase. Untreated, Lime modified
Hard pan 2	28	5	3.7	6	34	48	85	23	1972	9.1	110	170	2.1	Roadbase, Experimental road
Nodular 1	-	11	5.3	20	28	43	90	70	1968	10.5	70	308	2.1	Roadbase. Untreated, lime modified
Nodular 2	44	15	8.5	11	30	40	85	53	1964	10.2	105	450	2.0	Roadbase. Untreated, lime modified
Nodular 3	50	16	8	8	22	30	88	70	1960	9.8	99	352	3.0	Sub-base, untreated, lime modified
Powder	36	7	3	20	60	70	98	15	1490	25	50	420	1.5	Sub-grade. Lime/cement modified
Powder, calcified sand	41	2	7.5	38	75	85	98	25	1836	12.2	50	128	1.0	Sub-grade. Lime/cement modified

Note: PM = Plasticity modulus i.e. PI x % passing 0.425 mm sieve
GM = Grading modulus i.e. [300 – (sum of percentage oassing through sieves 2, 0.425 and 0.075)] / 100
APV = Aggregate plier's value
MDD = Mass dry density
OMC = Optimum moisture content
CBR = 4 days soaked California bearing ratio value

Table 3. Effect of time delay on the strength of stabilised Powder Calcrete

Time delay	Unconfined compressive strength MN/m^2		
	3% hydrated lime	3% cement cured 0n	
		28 days	7 days
10 minutes	7.8	7.5	5.2
1 hour	6.8	6.0	4.8
2 hours	6.1	5.5	3.5
4 hours	5.5	5.1	3.2
6 hours	4.2	4.8	3.0
8 hours	3.5	4.6	3.0

Table 4. Revised guidelines for light traffic calcrete roads

Test	% passing the 425 μ sieve		
	10-50	50-65	65-85
Max.particle size range (mm)	75μ -10	75μ -10	75μ -10
% passing 63μ sieve	5 – 25	15 - 35	20 - 35
Max. ratio of % passing 425μ and 63μ sieves	3.5	3.5	3.5
Linear shrinkage (LS)	<10	<6	<6
Plasticity Index (PI)	<25	<15	<15
Max. LS X % passing 425μ sieve	600	500	500
Min. 4-day soaked CBR at field	40	40	40
Min. % of calcium carbonate of the material passing through 425μ	12	12	12

Table 5 Revised guidelines for unsurfaced calcrete shoulders of light traffic roads

Test	% passing 425μ	
	10 - 50	50 - 65
Max. particle size range (mm)	75μ - 10	75μ - 10
% passing 63μ sieve	5 – 25	15 - 35
Max. ratio of % passing 425μ and 63μ sieves	3.5	3.5
Linear shrinkage (LS)	<10	<6
Plasticity Index (PI)	<25	15
Max. LS X % passing 425μ sieve	650	550
Max. LS X % passing 425μ sieve	80 - 180	120 -210
Min. 4-day soaked CBR at field	50	50
Min. % CaCo$_3$ material passing through 425μ	25	25

Experiments show that poorly graded calcrete which have flocculation of different degrees failed to show a permanent strength gain and infact less stable than untreated materials. Very good curing techniques and early sealing of road surfaces are probably the best measures of using calcretes of poor grades stabilised lime or cement. Research studies in this respect are still in progress to determine the most effective technique. The choice between cement and lime stabilisation may therefore depend on the initial design objectives i.e. modification plastic components or cementing materials. In either case the same curing may be required.

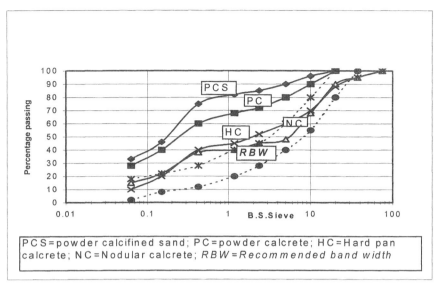

Fig. 1. Particle size distribution of common types of calcretes

Additional consideration in stabilisation design is the initial consumption of lime (ICL) required to produce and maintain sufficiently high alkaline conditions for the formation of cementitious products. This is the minimum content of lime required for long term durability. With calcrete it is likely that ICL would remain constant with time. In case of lime and cement combination it is more efficient to introduce a two stage stabilisation process; first adding lime atleast up to the determined ICL, value determined from experiments for a particular calcrete sample and for the second stage adding cement and followed by compaction. Ofcourse this process involves material costs and double handling procedures.

6   CALCIUM CARBONATE IN CALCRETES

Calcium carbonate is the main mineral component introduced in to the calcrete during its formation. The presence of this is thought to be one of the main mineralogical influences on the often-satisfactory behaviour of some of otherwise poorly graded calcrete.   In this respect it is believed that a minimum amount of carbonate is required to achieve a satisfactory response of the material. Botswana calcrete have the calcium carbonate varying from less than 10% in calcified sands of Kalahari Desert and above 50% in hard pan and powder calcretes. A simple apparatus known as 'Collina' calcimeter is used to measure the calcium carbonate (Milland 1993).

7   SOLUABLE SALTS IN CALCRETES

Highly soluable salts such as sodium chloride are found to be present in some calcretes. It is found that the salt migrates to the surface of base layers of the road and disrupt the bond with the prime coat and bituminous surfacing. Conductivity tests for the soil paste is done to determine the salt content. (Netterbarg 1981).

8   DESIGN GUIDELINES FOR CALCRETE PAVED ROADS

Until 1984 the specifications used in Europe, North America and South Africa were followed. All these specifications generally require strict control of soil plasticity, grading, CBR and aggregate hardness. As a result of studies of calcrete performance in South Africa and Namibia guidelines for paved roads under different traffic conditions have been published by NTTR (Nettarbarg 1981). The latter specifications have been incorporated into the Botswana Road Design Manual, Ministry of Communications, (1994) and have been applied in the construction of secondary and rural feeder roads receiving less than 200,000 equivalent standard axles (ESA) in a design period. A complete locally based specification for Botswana depending on its largely varied climate and local traffic conditions is required. Local expertise in the use of calcrete in paved roads has been obtained as a result of parts of rural road network and monitoring of four major full-scale road experiments over the past two decades. These projects are stated in Table 1. A revised specification which applies to a cumulative traffic loading of 160,000 ESA is shown in Table 4 and is based on a more critical examination of the performance of calcified sands, powder calcrete and gap-graded materials (Lonjanga, 1989).

As a result of the unsatisfactory performance of some calcified sands as unsurfaced shoulder materials separate guidelines for their use are required. For low volume traffic loads these are stated in Table 5 and include restrictions on the maximum allowable amount of material passing 425µ sieve and a minimum calcium carbonate. In case shoulders are provided with a protective bituminous seal the appropriate base materials for the design traffic conditions as stated in Table 4 should be used.

9 CALCRETE PROFILES OF BOTSWANA

After an elaborate reconnaissance surveys, the Transport and Road Research Laboratory (TRRL) reports the following regarding the landforms of Calcrete.

- Calcrete occurs in any one or a number of types in a particular profile.
- Calcrete is unpredictable in thickness and extent, although it is more likely to be consistent in situations where its associated surface landform is fully developed.
- The field parties were confused in digging full depth of materials due to the presence of deep overburden of hard layers of materials other than calcrete. This has necessaciated the requirement of proper mechanical plant for deep excavations and hence, the increase of cost of production.
- Due to the variability of quality in different layers and thickness sufficient care needs to be exercised during excavation to prevent contamination and mixing of materials.
- Powder calcrete layers often fade out at the base in calcareous sand leaving the thickness of unsuitable material in doubt.
- For the low traffic roads medium to poor quality calcrete is often suffice and normally occur in larger quantities.
- When interpreting calcrete bearing landforms in remote sensing images it is not always possible to positively identify the extent of a platform in aerial photographs, or the extent of a grey sand area in Landstat images.

10 RECOMMENDATIONS AND CONCLUSIONS

1. The results of the studies have shown that a wide range of calcrete can be recommended for use in light roads of Botswana.

2. The role of calcium carbonate in the performance of calcrete as a road building material is not fully known. However a minimum carbonate content is allowed.
3. For higher traffic levels it would appear that the maximum plasticity up to a plasticity index of 15 could be adopted for the base material.
4. The untreated materials used for cement and lime stabilised road bases should be well graded.
5. Efficient curing is necessary. Insufficient curing of cement / lime mixed with calcrete lead to poor hydration of cemented materials and carbonation
6. Unnecessary delay in initial mixing and compaction is to be avoided.
7. A minimum grading modulus of 1.5 and a uniformity coefficient greater than 8 is recommended to ensure mechanical stability.
8. A minimum unconfined crushing strength of 1.7Mpa at field density is recommended.
9. A minimum curing period of 7 days for cement and 28 days for lime stabilised materials is suggested.
10. For shoulder materials guidelines suggested in Table 5 can be followed.
11. Hard calcrete which satisfy normal aggregate requirement can be used successfully in bituminous surfacing up to high traffic situations. They should possess a minimum 10% fine. The aggregate crushing value should be 210 kN in the dry test and 160 kN after a period of 4 days soaking.
12. Calcrete base materials often require more bituminous prime for adequate seal of the surface materials. The correct type of bituminous application rate can be arrived at the site by trial. Coarse graded calcrete often require bituminous coating.
13. A conventional double surface treatment provides an adequate surface for all calcrete bases and can last for 5 to 7 years. Frequent application of 'fog sprays' of bitumen can refresh the surfacing and can increase the life of seal.
14. Single surface dressings with sand seals are found to be unsuitable and they may require frequent maintenance.
15. The use of calcrete aggregate seals has been successfully demonstrated on rural roads. Although quantitative design procedures are not available site trials are necessary to establish the technique of satisfactory performance.

REFERENCES

AASHTO, 1980 *Standard Specifications for Highway Materials and Methods of Sampling and Testing,* part 1. Washington.

Ganesan. C. T. 1998, Material and pavement design for sealed low traffic roads in Botswana, *Botswana Journal of Technology,* vol.7, No. 2, pages 48 – 55.

Lawrence C. J. and Toole T. 1984 The location, selection and use of calcrete for the bituminous road construction, *Laboratory report, TRRL,* report No. 1122, UK.

Lonjanga A. V, Toole T and Greening P.A.K, 1989 The use of calcrete in paved roads in Botswana, *TRRL (ODA),* Technical report.

Milland R.S 1993, Road building in tropics, *HMSO,* London.

Nettarbarg F, 1981, Salt damage to roads, *National Institute for Transport and Road Research (NITRR),* Pretoria, South Africa.

Road Design Manual, 1994, Roads Department, Ministry of Works and Communication, Gaborone, Botswana.

Road Design Manual, 1997, Road building materials, Draft compendium, Roads Department, Ministry of Works and Communication, Gaborone, Botswana.

Standard Methods of Testing of Road Construction Materials, TMH 1, 1986, CSRA, Pretoria, South Africa.

Guidelines for Road Construction Materials, TRH 14, 1986, CSRA, Pretoria, South Africa.

Creative Systems in Structural and Construction Engineering, Singh (ed.) © 2001 Balkema, Rotterdam, ISBN 90 5809 161 9

An analysis of grid structure on elastic foundation

M. H. Nam, M. O. Kang & Y. H. Lee
Department of Civil Engineering, Gyeongsang National University, Korea

D. H. Ha
Department of Civil Engineering, Changwon College, Korea

Y. K. Ryoo
Do-Myeng Engineering Limited, Korea

ABSTRACT : The aim of this study is focused on having to solve any kind of frame structures not only without solving simultaneous equations but also to get the exact solutions. For this purpose the method of distribution of the end actions, like moment distribution method, is used because unlike the slope deflection method or any of the other existing method this method does not require the solution of simultaneous equations. Instead, the results are obtained by a procedure of successive approximations, i.e, an iteration technique. But the infinite cyclic distribution of the unbalanced end actions and problem of the displacements are the disadvantages of this method. A new method by the concept of the partial leading matrix, leading matrix, main frame and connecting frame are introduced in this study to settle the disadvantages of the method of distribution of end actions. The grid structure on an elastic foundation is used to demonstrate the validity of this method. The results from this method were compared to the results of the finite element method and a good agreement has been found.

1 INTRODUCTIONS

The method of distribution of end actions which was originated by Hardy Cross is normally used to analyze all types of the statically indeterminate frames. But in the case of large space frames, plane frames, grid frames and grid frames on elastic foundations which consist of many members and joints, this method is not suitable because of the limitation of infinite cyclic distributions and the influence of the displacements which take place at each node of the structures. This new method divides the structure into two parts, the main frame and connecting frame. The main frame is used to make the leading matrix. This leading matrix is made from the partial leading matrices which are derived from the characteristics of geometric series. The connecting frame distributes the unbalanced end actions between two nodes on the main frames at a specific time also by the geometric series. As a result this method not only reduced the calculation cycles very much but also got the solutions without solving simultaneous equations.

2 THEORETICAL CONSIDERATIONS

2.1 *Distribution matrix of grid structure*

To analyse the grid structure on an elastic foundation by the proposed method the distribution and carry-over matrices are obtained by the differential equation for a beam on an elastic foundation.

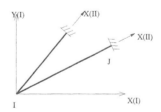

Figure 1. A part of grid frame on an elastic foundation and Coordinate system .

In Figure 1, when the unbalanced end actions $P_i(I)$ are applied at node i, the distributed end actions of member ij and carry-over end actions, namely, bending moments, axial forces, shear forces and torsions, can be expressed by distribution and carry-over matrices of the method of distribution of end actions as follows;

$$p_{ij}(I) = D_{ij}P(I) \tag{1}$$

where,

D_{ij} : distribution matrix of member ij

$P_i(I)$: unbalanced end actions at node i in global coordinate

$p_{ij}(I)$: end actions of member at node j in global coordinate.

To get the distribution matrix, one considers the force-displacement relationship and the relationship of the coordinates as follows;

$$p_{ij}(II) = k_{ij}(II)\theta_i(II) \qquad (2)$$

$$\theta_i(II) = T_{ij}\theta_i(I) \qquad (3)$$

$$p_{ij}(II) = T_{ij}p_{ij}(I) \qquad (4)$$

here,

$p_{ij}(II)$: i node end actions of member ij in local coordinates

$k_{ij}(II)$: stiffness matrix of member ij in local coordinates

$\theta_{ij}(II)$: displacements of node i in local coordinates

T_{ij} : transformation matrix of member ij.

From these equations the force-displacement in global coordinate is given by

$$p_{ij}(I) = k_{ij}(I)\theta_{ij}(I) \qquad (5)$$

$$k_{ij}(I) = T_{ij}^{T} k_{ij}(II)T_{ij}. \qquad (6)$$

When a node is connected to several frames as shown in Figure 1, the deforming node will produce some end actions at each member according to the member's stiffness. Also, the summation of end actions at each member can be given by the following equations

$$\sum p_{ij}(I) = P_i(I) = \sum k_{ij}(I)\theta_{ij}(I) \qquad (7)$$

$$\theta_i(I) = \sum k_{ij}(I)^{-1}P_j(I). \qquad (8)$$

Substituting Equation 8 into Equation 5, we obtain

$$p_{ij}(I) = k_{ij}(I)\sum k_{ij}(I)^{-1}P_i(I). \qquad (9)$$

When compared Equation 9 with Equation 1, one can find out the distribution matrix is the same as Equation 10.

$$D_{ij} = k_{ij}(I)\sum k_{ij}(I)^{-1} \qquad (10)$$

2.2 Carry-over matrix of grid structure

The end actions at node j have some relationship to the end actions at node i which is called the carry-over matrix and P_i is the end actions of node i.

$$P_j(I) = C_{ij}P_i \qquad (11)$$

C_{ij} is the carry-over matrix of member ij . The node actions in the local coordinates can be transformed into the global coordinates, and vice versa;

$$p_{ij}(II) = T_{ij}P_{ij}(I) \qquad (12)$$

$$p_{ij}(I) = T_{ij}^{T}P_{ij}(II). \qquad (13)$$

$P_i(II)$ is expressed as Equation 14 in which q_{ij} is the carry-over stiffness matrix which is derived by differential equation applied by boundary conditions.

$$P_j(II) = q_{ij}p_{ij}(II) \qquad (14)$$

Equation 15 shows the relationship between $P_j(I)$ and $P_j(II)$.

$$P_j(I) = T_{ij}^{-1}P_j(II) \qquad (15)$$

By substituting Equation 14, Equation 12 and Equation 11 into Equation 15 and in view of

$$T_{ij}^{-1} = T_{ij}^{T}$$

the end actions at node j can be obtained by the node actions on node i

$$P_j(I) = T_{ij}^{T}q_{ij}T_{ij}D_{ij}P_i \qquad (16)$$

Thus, when compared Equation 16 with Equation 11, the carry-over matrix is obtained as Equation 17.

$$C_{ij} = T_{ij}^{T}q_{ij}T_{ij}D_{ij_i} \qquad (17)$$

2.3 Concept of leading matrix

The definition of the leading matrix is the influenced end actions quantities of each node when the arbitrary node in the structure is subject to unit unbalanced end actions as shown in Figures 2a-g where,

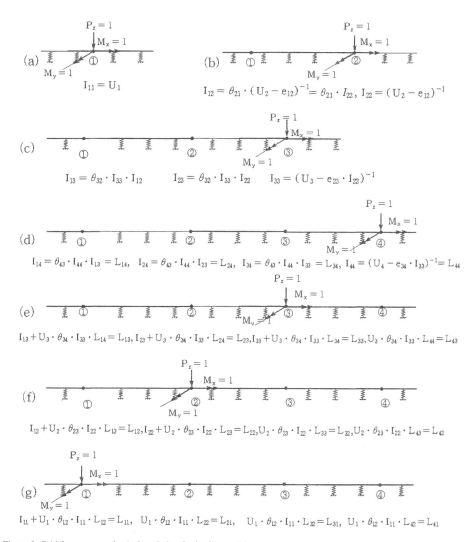

Figure 2. Grid frame on an elastic foundation for leading matrix.

U_i : matrix of unit unbalanced end actions at node i

θ_{ij} : distribution and carry-over matrix from node i to node j

I_{ij} : partial leading matrix, the end actions of node i when node j subject to unit unbalanced end actions

L_{ij} : leading matrix, the end actions of node i when node j subject to unit unbalanced end actions.

2.4 *The main frame and connecting frame*

The structures in Figure 3a,b consist of bold lines which are the main frames to make the leading matrix and ordinary lines which are the connecting frames to distribute and carry-over the unbalanced end actions of nodes on main frames to the nodes on opposite main frames rapidly by using the geometric series.

Figure 3a shows a long main frame which has so many nodes that the leading matrix along this long main frame becomes large size. The long main frame is not efficient for making the leading matrix along it and consumes much more computer memory and time. Because the main frames in Figure 3b are shorter than the main frame in Figure 3a, the size of the leading matrices in Figure 3b becomes smaller than the leading matrix in Figure 3a. In this study the authors divided the main frame in Figure 3a into many small main frames as shown in Figure 3b to reduce the size of the leading matrix. As a result, the leading matrix in Figure 3a is reduced from 75 rows and 75 columns to 15 rows and 15 columns.

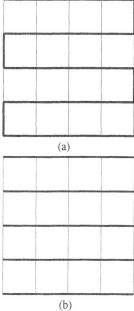

(a)

(b)

Figure 3. The main frame and
connecting frame.

2.5 Analysis procedures by the leading matrix

The first step is locking each node on the main frames and connecting frames to get the first unbalanced end actions to the structure as shown in Figure 3b. The second step is releasing the main frames and distributing and carrying-over the first unbalanced end actions of each node by the leading matrices to get the first cycle adjusted unbalanced end actions of each node. Then, locking the main frames again and then releasing the connecting frames one gets the first distributed end actions of each node from which the second unbalanced end actions are obtained by the geometrical series at a time and then lock the connecting frames again. By releasing the main frames one more time and distributing and carrying-over the second unbalanced end actions of each node by the leading matrices one gets the second cycle adjusted unbalanced end actions of each node. Then, locking the main frames again and then releasing the connecting frames to get the second distributed end actions of each node from which the third unbalanced end actions are obtained by the geometric series at a specific time and then locking the connecting frames again. The entire same procedure mentioned above is repeated continuously until the adjusted unbalanced end actions converge.

When the converged adjusted unbalanced end actions are obtained, then sum all adjusted unbalanced end actions to get the total adjusted unbalanced end actions from the first unbalanced end actions to converged adjusted end actions. Finally, get the member forces by multiplying distribution matrices of each node, which were obtained from Equation 10, to the total adjusted unbalanced end actions of each node to get the distributed end actions of each node and then carrying-over these distributed end actions to the opposite nodes by multiplying the carrying-over matrix which was obtained from Equation 17.

3 AN EXAMPLE AND RESULTS

A simple grid structure on elastic foundation is taken as an example to show the validity and simplicity of the proposed method. The results are compared with the results of the finite element method. The properties of members are as follows;

shear modulus $G = 9.83 \times 10^5 t/m^2$
elastic modulus $E = 2.30 \times 10^6 t/m^2$
cross sectional area $A = 0.16m^2$
foundation modulus $k = 10,000 t/m^2$.

Table 1. Result of torque.
(unit : kg · cm)

Element	Torque	
	Leading matrix	SAP2000
1-2	2.467e-3	2.450e-3
1-4	2.222e-4	2.097e-4
2-1	-2.467e-3	-2.450e-3
2-3	2.465e-3	2.450e-3
2-5	0	0
3-2	-2.465e-3	-2.45e-3
3-6	-2.250e-4	-2.097e-4
4-1	-2.222e-4	2.097e-4
4-5	5.346e-2	5.348e-2
4-7	-2.222e-4	-2.097e-4
5-2	0	0
5-4	-5.346e-2	-5.348e-2
5-6	5.346e-3	5.348e-2
5-8	0	0
6-3	2.250e-4	2.097e-4
6-5	-5.346e-2	-5.348e-4
6-9	2.25e-4	2.097e-4
7-4	2.222e-4	2.097e-4
7-8	2.467e-3	2.450e-3
8-5	0	0
8-7	-2.467e-3	-2.450e-3
8-9	2.465e-3	2.450e-3
9-6	-2.250e-4	-2.097e-4
9-8	-2.465e-3	-2.450e-3

Table 2. Result of bending moment .
(unit : kg · cm)

Element	Bending moment	
	Leading matrix	SAP2000
1-2	2.307e-3	2.097e-4
1-4	-2.453e-4	-2.097e-4
2-1	-6.426e-3	-6.225e-3
2-3	-6.534e-3	-6.225e-3
2-5	-4.891e-3	-4.901e-3
3-2	2.105e-4	-2.097e-4
3-6	-2.464e-3	-2.450e-3
4-1	-2.673e-2	2.674e-2
4-5	0	0
4-7	-2.673e-2	-2.674e-2
5-2	-4.500e-1	-4.465e-1
5-4	0	0
5-6	0	0
5-8	4.470e-1	4.465e-1
6-3	2.673e-2	2.674e-2
6-5	0	0
6-9	-2.673e-2	-2.674e-2
7-4	2.453e-3	2.450e-3
7-8	-2.307e-4	-2.097e-4
8-5	4.891e-3	4.901e-3
8-7	6.426e-3	6.225e-3
8-9	6.534e-3	6.225e-3
9-6	2.464e-3	2.450e-3
9-8	-2.105e-4	-2.097e-4

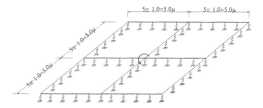

Figure 4. Grid frame on an elastic foundation.

4 CONCLUSIONS

The proposed method, dividing the grid structure on elastic foundation into the main frame and connecting frames, and applying the leading matrix method to the main frame and geometric series to the connecting frames, shows as follows;

1) The grid structure on elastic foundation was completely analysed by hand without solving any simultaneous equations.
2) The size of the leading matrix depends on the number of nodes in the main frame.
3) The computational time, which depends on the number of iterations, is short.

4) Even if the existing method of distribution of end actions is not proper to solve some complicated structures, this proposed method is easily applicable to any kind of frame structure.
5) The results of two methods, the leading matrix method and the finite element method, show that there is little difference. But the proposed method is more accurate than the existing method, because of the infinite cycle distributions and carry-overs by the characteristics of the geometric series although the difference is not important in practical engineering design.
6) If there are regular procedures to distributing and carrying-over the connecting frames, not the numerical solution but the exact solution can be obtained.

REFERENCES

Nam, M.H. 1974, A Study on the solution of Influence Line for Infinite Span Continuous Beams. *Theses Collection of Chung-Buk National Uni- versity*, Republic of Korea,8:261-267.
Nam, M.H. 1974, *A Study on the Elastic Analysis of Grid Foundation*, Master degree thesis, Busan National university, Republic of Korea.
Nam, M.H. Jang, H.D, Lee, K.H & Ha, D.H. 1997, A Grid Structure Analysis by Leading Matrix Method. *Proc. 7-th Int. Conf.*, Seoul, 19-21 August 1997, 1:215-220.

Creative Systems in Structural and Construction Engineering, Singh (ed.) © 2001 Balkema, Rotterdam, ISBN 90 5809 161 9

Real-time tracking for compaction of superpave mixes

Amr A.Oloufa

Department of Civil and Environmental Engineering, University of Central Florida, Fla., USA

ABSTRACT: Recently, the United States witnessed an explosion in the use of Superpave asphalt mixes. Florida is one of the five leading States (in addition to Indiana, Maryland, New York, Texas, and Utah) in this application. Since the quality and stability of asphalt pavements is an extremely vital requirement for roadways, attention to the quality issue of asphalt pavements is paramount. This paper reports on a research project to develop an automated data collection tool for tracking Superpave compaction using Global Positioing System (GPS) technology.

1 INTRODUCTION

The asphalt density, especially for Superpave-based mixes, is extremely dependent on the number of passes of the compactor (among other factors). If the asphalt mat is compacted with less than the appropriate number of passes, there may be a large volume of air voids and a low density. On the other hand, if the mat is overly compacted, the aggregates will be crushed further and the attributes of the mix will be changed leading again to "Tender Zone" issues and a reduction in density. However, the number of passes made by the compactors is hard to monitor in the compacting process. Figure 1 shows the results of observations made at 20 locations in a 2-mile stretch of pavement that was rolled with a 12-ton three-wheel roller (Kilpatrick and McQuate 1967). This figure shows that two areas in a pavement, the join and edge, tend to receive less compaction than the rest of the cross section.

In the most recent proceedings of "Superpave: Today and Tomorrow" that was held in St. Louis in 1998, the most talked about element of Superpave mixtures has been compaction. According to these proceedings, *"Fifty percent of the respondents indicated more effort was required for compaction. Sixty percent had to add more rollers (up to three additional) to the paving train. Sixty percent said they saw no difference in the pickup of the mix, but 30% did quit using pneumatic rollers. ..."*

A major problem facing all compaction studies is the lack of an automated data collection tool that continuously collects information related to compaction equipment positions and asphalt temperatures throughout the compaction process. This leads to a large variability in recording both the exact number of compactor passes, and asphalt temperatures. The development of such a tool will enable answering field questions such as optimal compactor distance behind pavers, number and type of compaction equipment, minimum and maximum number of passes, asphalt temperature, and so on.

2 WORK TO DATE

There have been many efforts to develop automated methods for the quality control of pavement compaction. Previous researches include Compaction Documentation System (CDS, GEODYN, 1985), MACC (Froumentin et al, 1996), and the Compactor Tracking System using GPS (CTS, Li, 1995), and later CTS-II.

The CDS system is a system of monitoring the compaction process where data such as lane change, direction change, number of passes, layer number, start and stop of the compactor are entered manually by the operator during the process. Since there is no sensor to identify the orientation and position of moving compactor, the operator must follow the moving path that is decided previously.

The MACC is a prototype operator aiding system for compactors that displays a real-time two-dimensional colored map of the compaction pattern.

The CTS system is able to provide the compaction operator with a visual display of the coverage area, color-coded with the number of coverages done by the compactor as the compaction is occurring.

The data generated by the CTS can also be stored in a permanent record as a historical document.

3 GLOBAL POSITIONING SYSTEM (GPS) AND GLOBAL NAVIGATION SATELLITE SYSTEM (GLONASS)

Global Positioning System (GPS) is a satellite based radio-navigation system. There are 24 GPS satellites orbiting the Earth in 6 planes and transmitting radio signals. Based on measurements of the amount of time that the radio signals travel from a satellite to a receiver, GPS receivers calculate the distance and determine locations in terms of longitude, latitude, and altitude, with great accuracy.

Global Navigation Satellite System (GLONASS) is developed by the Russian Federation. GLONASS is very similar to GPS in terms of the satellite constellation, orbits, and signal structure. GLONASS also uses 24 satellites in three orbital planes and these satellites transmit the same code as GPS satellites do but at different frequencies (GLONASS Group, 1997).

3.1 GPS Satellite Signals

GPS satellites transmit two carrier signals, L1 and L2. The L1 frequency carries the P-Code (Precise Code), C/A (Coarse Acquisition) Code, and navigation message. The L2 frequency carries a P-Code and navigation.

3.2 Basic Concept of GPS

Using GPS satellites as reference points, a GPS receiver determines its location based on the distances between the receiver and GPS satellites. The satellites are used as reference points since their orbital motion is constantly monitored by ground control stations, so that their instantaneous positions are always know with great precision.

The distance between a receiver and a satellite is calculated using a simple equation of *Distance = Speed x Time,* where Speed is the speed of the signal which is transmitting at the speed of light (186,000 mile/sec) and Time is the time for the signal travels from the satellite to the receiver. Time can be calculated by measuring the departure time of the signal at the satellite and the arrival time at the receiver.

3.3 GPS Receivers

GPS receivers can be categorized broadly into three types based on accuracy: C/A code, carrier phase and dual frequency receivers. Each of the three types offers different levels of accuracy, and the price of the receiver is dependent on its accuracy.

C/A code receivers typically provide 1~5 meter accuracy with differential correction, with an occupation time of 1 second. Longer occupation time (up to 3 minutes) will provide accuracy consistently within 1~3 meter and can be reduced to 30 centimeter.

Carrier phase receivers typically provide 10~30 cm accuracy with differential correction. Distance to satellites from the receiver is determined by counting the number of waves that carry the C/A code signal (referred to as ambiguity resolution). This method is much more accurate but requires a substantially higher occupation time to attain 10 ~ 30 cm accuracy.

Dual-frequency receivers are capable of providing sub-centimeter accuracy with differential correction. Dual-frequency receivers receive signals from the satellites on two frequencies simultaneously. In order to acquire and use both frequencies, the reacquisition time is longer than other receivers are.

GPS & GLONASS receivers receive the signals from both GPS and GLONASS satellites and provide the same accuracy of dual-frequency receivers, sub-centimeter. However, GPS & GLONASS receivers are not affected by the limitation that GPS-only or dual-frequency receivers have: with fewer than 5 satellites in view GPS-only receivers do not work at all, or dual-frequency receivers work so slowly to accomplish the accuracy (van Diggelen, 1997).

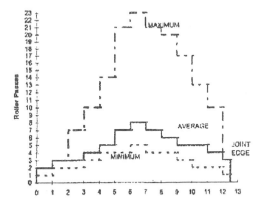

Figure 1: Observations made at locations in a stretch of pavement.

3.4 GPS Position Accuracy and Error Sources

Accuracy of GPS is the degree of conformance between the estimated or measured position, time, and/or velocity of a GPS receiver and its true time, position, and/or velocity as compared with a constant standard.

The accuracy of GPS receiver is affected by errors caused by natural phenomena, mechanical failure of elements in the system, or intentional disturbance. (Hurn, 1993).

The Selective Availability (SA) Error is intentionally introduced to the GPS signals and then is broadcast by the satellites. The U.S. has disabled SA in May 2000. The Ionospheric Error is a function of the local time of day and latitude. The Troposhperic Error is a function of the weight of the atmosphere above the GPS antenna and is modeled using the atmospheric pressure. The Orbit and Satellite Clock Error occurs because there are slight variations in the orbits of GPS satellites. Monitoring stations track this error and broadcast these corrections to the satellites. Because of the delay in sending these corrections, orbit errors exist. In addition to the satellite position, the atomic clocks drift off causing another error in time measurement and therefore in position. The Multipath Error occurs when strong signals from satellites are not along the direct line of sight between the user's antenna and the satellite. Multipath is often the dominant error source in DGPS applications on mobile applications. The Receiver Noise Error occurs as electronic devices emit electromagnetic energy, some at the GPS frequencies, which contributes a range error to the measurement.

3.5 *Differential Correction of GPS Positions*

Differential GPS (DGPS) is a means of correcting for some system errors by using the errors observed at a known location to correct the readings of another receiver (rover). A reference receiver, or base station, computes corrections for each satellite signal. Most of the errors caused by the sources discussed earlier are eliminated by differential correction. Differential corrections may be used in real time or post-processed (Dana, 1996).

4 PROPOSED SYSTEM ARCHITECTURE

In asphalt compaction, it is customary to use more than one compactor behind the paver. The first compactor named the breakdown roller, is used to create the initial compacted surface. While compacting with the breakdown roller achieves most of the final asphalt density, the roadway surface is not as smooth as it can be. To improve roadway surface smoothness, the initial breakdown roller may be followed by another breakdown roller, or by a pneumatic roller followed by a finish roller.

Therefore proposed system must not only be capable of tracking multiple compactors, it must also be able to display to all compactor operators the TOTAL number of coverages achieved over the roadway surface.

The system's architecture is composed of positioning devices, hardware, and software. For positioning, the researchers selected high accuracy, real-time kinematic (RTK) GPS devices with a RMS accuracy of less than 2 cm. To reduce the total cost of the system, and while it is conceivable that lower accuracy devices may be sufficient, it was decided to initially use higher accuracy devices to prove the concept.

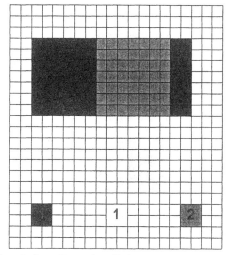

Figure 2: Raster Processing of Polygons

For the hardware architecture, the researchers selected simple inexpensive and diskless computers on each compactor. This computer interfaces with the GPS device onboard each compactor and sends the compactor's position to a remote computer (base station) on the site. The base station assembles the coverages from all compactors and broadcasts the final coverage to all compactors.

This design allows multiple compactors to be tracked while eliminating the need for expensive computers to be installed on each compactor; as the main processing chores are done on the base station. This architecture also allows the simultaneous creation of an independent record of the compaction process that can be used by the roadway's owner.

This approach also gives a detailed and complete history of areas covered by the compactor, which cannot be obtained by other tests or devices. This record serves as a detailed and complete history of areas covered by the compactor, which cannot be obtained by other tests or devices. This configuration also provides an independent data gathering mechanism that can be used by owners and inspectors to substantiate coverages done by contractors.

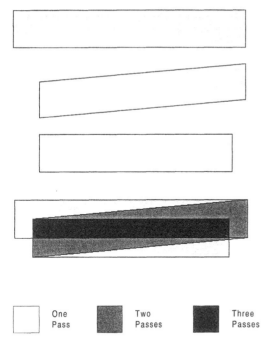

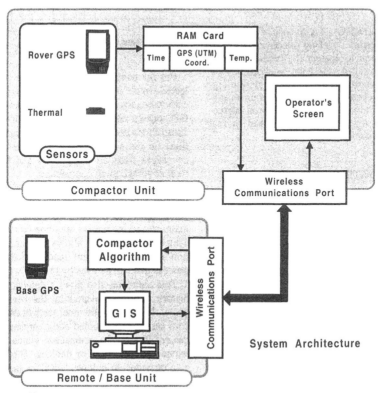

The software is used to develop the coverages developed by the compactors. There are two methods that can be used. The first is based on image processing technology where images of polygons are "rasterized" and cell values at each location on the roadway are accumulated to develop the total number of coverages (see Figure 2). The second method for processing these polygons uses vector and polygon topology to achieve this result. In this case, intersection polygons are calculated and displayed with a color code corresponding to the number of coverages (see Figure 3).

In a raster-based system, coverage files are stored as a cluster of pixels, each with a color or number corresponding to the number of coverages. In a vector-based system, coverages are stored as a set of locations corresponding to the polygons developed by the compactor footprint. In the system developed for CTS, a raster-based system was used whereas CTS-II utilized a vector- based system.

After a reasonable number of points are acquired, there is no degradation of performance from raster processing. However, in vector processing, unless data smoothing is employed, performance will continue to degrade as more points are added.

| One Pass | Two Passes | Three Passes |

Figure 3: Vector Processing of Polygons

Figure 4: CTS-III Architecture

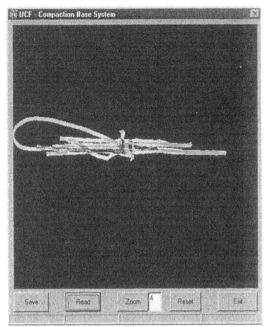

Figure 5: CTS-III User Interface (Base)

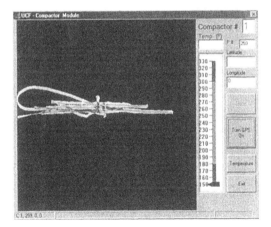

Figure 6: CTS-III User Interface (Rover)

5 COMPACTION TRACKING SYSTEM (CTS-II)

With collaboration with researchers at Penn State University, the author developed two early prototypes (CTS & CTS-II) to prove the feasibility of the proposed system along with evaluating the suitability of high-accuracy GPS devices for this application.

This system suffered from the following problems:

1. As mentioned earlier, it is customary in pavement jobs to use more than one compactor, with the second unit for finishing the asphalt surface. CTS-II cannot handle data from multiple compactors.
2. Computations are carried out using a notebook on the compactor. This requires a relatively fast processor and a hard disk on the compactor requiring extensive provisions for shock mounting that substantially increase cost.
3. Performance is too slow. This is due mainly to the large number of "sliver" polygons created after a large number of coverages is achieved. This effect can only be mitigated if a "dissolve" operation is used, where sliver polygons with the same coverage are dissolved into one larger polygon eliminating the need to store and transmit this unneeded geometry.
4. It is not known whether vector processing is truly superior to raster processing. As we mentioned earlier, raster processing is faster to compute but slower to transfer in a wireless system. Vector processing produces smaller files but requires more time to compute.

6 COMPACTION TRACKING SYSTEM (CTS-III), LATEST DEVELOPMENTS

To solve the problems listed above, the research team at the University of Central Florida developed a new version of the Compaction Tracking System (CTS-III). This system is capable of tracking up to four compactors simultaneously, with little degradation in performance as shown in simulated tests.

The system consists of a Symbol Technologies diskless color workstation. Each workstation sends its vehicle's position, via Symbol's Technology pectrum 24 (2.4 GHz) wireless network, to the base station using a TCP/IP-based protocol. The new system architecture is shown in Figure 4.

The base station receives positions from all compactors. Using the algorithms discussed earlier, the footprint of each compactor is deduced and the output coverage is developed through a raster *"overlay"* operation. All compactor tracks are overlayed in this manner and an output coverages' file is produced and broadcast back to all compactors. This provides for each compactor operator a real-time color-coded display of all passes done by all compactors.

A major challenge in the development of the new system was overcoming the performance degradation of vector-based processing as described before.

To solve this problem, the researchers used a compressed, lossless, raster-based format as opposed to the vector processing described before.

There as several image compression methods available. For the requirements of this research, the stored image had to be extremely small in size to reduce transmission bottlenecks. Also, the image stored must retain the original colors as each color represents a specific number of passes. Some image compression schemes are "lossy", which means that some of the original colors may be lost when the image is compressed. An example is the popular "*jpg*" format used extensively in the internet. While "*jpg*" files are small in size, this format was intended for displaying pictures and was deemed not suitable for this application.

This image format selected has a maximum of 16 colors representing the various passes. This allows sufficient representation of the number of passes while reducing the size of the file. Figures 5 & 6 show the software developed for the application for the base station and compactors (rovers).

ACKNOWLEDGEMENTS

The author wishes to thank Penn State collaborators Dr. H. Randolph Thomas and Mr. Won-seok Do for their contributions to this research. Also, the Ingersoll Rand Corporation, and Ben Franklin Partnership of Pennsylvania for sponsoring earlier research. Also, thanks to Symbol Technologies for hardware grants in support of this research.

REFERENCES

Compactor Operator Training Manual. Glenn O. Hawbaker Inc., Pennsylvania, USA, 1995.

Froumentin M., and F. Peyret, *An Operator Aiding System for Compactors*, Proceedings of the 13th International Symposium on Automation and Robotics in Construction (IS-ARC), Tokyo, Japan, June 1996.

Geller M., *Compaction Equipment for Asphalt Mixtures: Placement and Compaction of Asphalt Mixtures.* Philadelphia: PA, ASTM, 1984, pp. 28-47.

"*The Geodynamic Compaction Documentation System*". Sweden: GEODYN, Inc., 1985.

Hughes C. S., *Compaction of Asphalt Pavement.* Washington, D.C.: Transportation Research Board, National Research Council, 1989.

Kilpatrick K. J., and R.G. McQuate, *Bituminous Pavement Construction.* Washington, D.C., Federal Highway Administration, 1967.

Li, C., A. A. Oloufa, and H. R. Thomas, *A GIS-based System for Pavement Compaction.* Journal of Automation in Construction. (Invited Paper), May 1996.

Oloufa A. A., W. Do and H. R. Thomas, *An Automated System for Quality Control of Compaction Operations: Receiver Tests & Algorithms*". Proceedings of the 16th International Symposium on Automation and Robotics in Construction (ISARC), Madrid, Spain, September 1999.

Creative Systems in Structural and Construction Engineering, Singh (ed.) © 2001 Balkema, Rotterdam, ISBN 90 5809 161 9

Soil-structure interaction: A neural-fuzzy model to deal with uncertainties in footing settlement prediction

P. Provenzano
University of Palermo, Italy

ABSTRACT: Engineers have always been aware of uncertainties in soil – structure interaction problems, related to soil inherent variability, site conditions, construction tolerance, failure mechanisms, in particular when the structure is founded in granular soils. Geotechnical and structure design deal with these uncertainty by heuristic and expertise knowledge using input data that fall in the category of non statistical uncertainties. This paper is devoted to the description of a Neuro-Fuzzy method to predict foundation settlements in granular soils and their effects on the structure response. The learning process analyses over 200 record of settlement of building foundations, tanks and embankments on sand and gravel. Uncertainties and vagueness into parameters of in situ tests, boundary conditions and external loads have been introduced as fuzzy numbers to be later reflected in the soil-foundation-superstructure system response.

1 SOIL STRUCTURE INTERACTION PROBLEM

1.1 *Uncertainty in design*

Soil-structure interaction analysis in granular soils have motivated primarily by the need to limit differential settlements within buildings to avoid structural or architectural damage. Sand deposits are generally much more heterogeneous than clay deposits, so it is likely that differential settlements are hither. Significant and varying degrees of uncertainty are inherently involved in the design process. They can be grouped in 1.estimating loads; 2.variability of the ground condition at the site; 3.evaluation of geotechnical material properties and 4.degree to which the analytical model represents the actual behaviour of superstructure, foundation and soil (Baecker, 1996). Prediction models need to be developed to dealing with this uncertainty. Expert systems based of neural networks and fuzzy systems are considered a promising tool for reliability assessment.

1.2 *Problem formulation*

The presented soil-structural interaction analysis involves frame structures, divided into elements connected by nodes, founded in granular soils (Fig.1).A set of n nodal displacements are defined; they can be represented as a vector
$$\underline{D} = [D_1, D_2 \dots D_n]^T$$
in $\Re^n{}_D$ space. A set of n loads are also defined; they can be represented as a vector

$$Q = [Q_1, Q_2 \dots Q_n]^T$$
that includes loads and boundary conditions. A soil-structural interaction model represent the response of the structure to the applied loads with a certain level of accuracy. It defines a relationship between the forces and displacements for the nodes of the elements

$$\Re^n{}_D \to \Re^n{}_Q .$$

In linear elastic theory this relationship is represented by a stiffness operator , linear, non singular, symmetric, positive. Usually loading vector is a vector of independent variables, so the relationship is

$$\underline{\underline{K}}^{-1} : \Re^n{}_Q \to \Re^n{}_D$$
where inverse of stiffness operator is flexibility operator (Fig.2). In Fig.(1.b) the soil-structure system is separated in three elements: superstructure, foundation and soil, mechanically connected to each other, so that they can change stress satisfying equilibrium and compatibility equations.
In generally
$$\begin{aligned}
\underline{D}_s &= \underline{D}_s \left(\underline{Q}_s + \underline{R}_{sf} \right) \\
\underline{D}_f &= \underline{D}_f \left(\underline{R}_{fS} + \underline{R}_{fS'} \right) \\
\underline{D}_{S'} &= \underline{D}_{S'} \left(\underline{R}_{s'f} \right)
\end{aligned} \qquad (1)$$
where the third equation is the foundation settlement model, linear elastic in current design. Since the soil-structure interaction analysis is to be considered within serviceability limit states, soils exhibit non linear behaviour even at small strain. The aim of this research is to focus the role of an expert system, based on a fuzzy-neural network, to forecast displacement of shallow foundations in granular soils, keeping into account non linear effects.

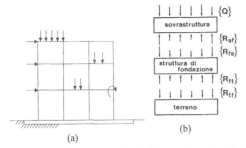

(a)

(b)

Figure1 (a) Frame structures founded in granular soils ; (b) Interaction of components (Burghignoli, 1985)

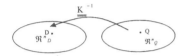

Figure2. Flexibility operator

Figure 3. Fuzzy relation

2 NEURO-FUZZY BASED SOIL STRUCTURE INTERACTION

2.1 *Expert systems in Geotechnical and structural engineering*

Expert system are knowledge-based system that contain expert knowledge. In general expert system represent and interpret heuristic in computers. Different methods can be used to achieve a solution. In this article an expert system based on neural network with fuzzy input vectors is suggested for non statistical uncertainty treatment in geotechnical and structural design. In particular neural network is used to learn from data and to predict the settlement goal function; fuzzy logic to input in soil-structure interaction problem non statistical uncertainty. The reason of this uncertainty may be because the observations made can be better categorised with linguistic variables, or because the number of factors and the contribution of each in influencing a particular property are unknown or too large, or because the relationship between this factors can be stated linguistically or can be learned from database or experience. To dealing with this uncertainty deterministic procedures adopt usually conservative design values, but for most structures, even for critical facilities, use of this values cannot be economically justified. The probabilis-

tic approach, in which all uncertainties are considered random, has a sophistication which is often in stark contrast to the crudeness of data which engineers must work with (Dong, 1991). On the other hands, expert system give a framework to dealing with heterogeneous uncertainties and heuristic knowledge. They, thought the reliability of the soil-structure system as resulting from a general and comprehensive examination of all its failure modes, can pay attention to available data, non linear aspects and synthesis of results. In this way, the work to do resembles that in an experimental study. It means that the single analysis is a simulation of the reality and the problem to be solved concerns with the exploiting of the maximum of information from these trials (Biondini, Bontempi, Malerba, 2000).

2.2 *Formulation levels*

Let $\tilde{Q}$ a loading vector with fuzzy coefficients
$$Q_i = \{\mu_Q(x)/x, x \in \Re\}$$
that includes uncertain loads and boundary conditions. In Tab.(1), for different uncertainties inputs and for different models, different levels of soil-structural interaction problem are used to dealing with uncertain information (Provenzano, Bontempi, 2000). In particular in gravel and sand foundation design, because of uncertainty affecting the model, an expert system, based on Neural Network, can be used for the analysis; for a fuzzy input the soil-structural interaction problem formulation is
$$\tilde{Q} = \tilde{Q}(\tilde{D}) \qquad (2)$$
The proposed soil-structure interaction analysis correspond to level 5, and the solution problem is given by follow neural-fuzzy system
$$\underline{\underline{\tilde{K}}}_s \tilde{D}_s = \tilde{Q} + \tilde{R}_{sf}$$
$$\underline{\underline{\tilde{K}}}_f \tilde{D}_f = \tilde{R}_{fs} + \tilde{R}_{ft} \qquad (3)$$
$$\tilde{D}_t = \tilde{D}_t(\tilde{R}_{ft})$$
The knowledge of soil-structure interaction problem maps the loading space into displacement space and approximate objective function in two-dimensional case us pictured in figure 3.

Table 1. Soil-structural interaction problem levels

Level	$\underline{Q}$	EI	$\underline{K}$	$\underline{D}$	Model	Solution
0	Crisp	Crisp	Crisp	Crisp	Mechan. (Linear)	$\underline{Q} = \underline{K}\,\underline{D}$ (crisp system)
1	Crisp, comb.	Crisp	Crisp	Crisp	Mechan. (Linear)	$\underline{Q} = \underline{K}\,\underline{D}$ (crisp system)
2	*Fuzzy*	Crisp	Crisp	*Fuzzy*	Mechan. (Linear)	$\tilde{Q} = \underline{K}\,\tilde{D}$ (fuzzy system)
3	*Fuzzy*	*Fuzzy*	Crisp	*Fuzzy*	Mechan. (Linear)	$\tilde{Q} = \tilde{EI}\,\underline{K}\,\tilde{D}$ (fuzzy system)
4	*Fuzzy*	*Fuzzy*	*Fuzzy*	*Fuzzy*	Mechan. (Linear)	$\tilde{Q} = \tilde{K}\,\tilde{D}$ (fuzzy system)
5	*Fuzzy*	*Fuzzy*	*Fuzzy*	*Fuzzy*	Neural	$\tilde{Q} = \tilde{Q}(\tilde{D})$ (Neural-Fuzzy S.)

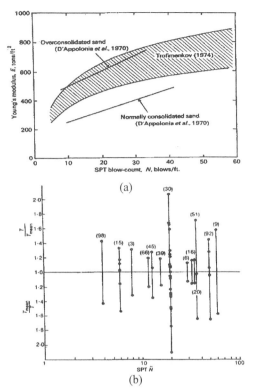

(a)

(b)

Figure 4. (a)USSR practice (Trofimenkov, 1974) (b) Inherent variability. (Burland and Burbidge, 1985)

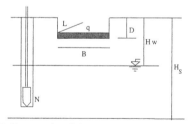

Figure 5. Geometry foundation

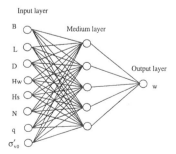

Figure 6. Network architectural

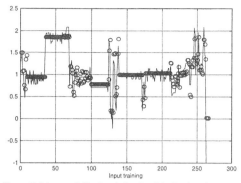

Figure 7. Measured (line) and simulated (points) settlements

2.3 Settlement prediction Neural network for shallow foundations on sand and gravel

Predictive approach currently used is either a direct empirical correlation of settlement with *in situ* testing data (Standard Penetration Test, Static Cone Penetration Test, Plate Load Test) or the development of empirical correlation of modulus for use with elastic theory for settlement prediction. Methods for predicting settlement using SPT blow-counts are used world-wide and they probably lye on the greatest body of data available for correlation purposes. Burland & Burbidge (1986) propose follow relation of settlement to the bearing pressure q', the breadth of loaded area (B) and the average SPT blow count (N) over the depth of influence:

$$w = f_s f_l f_t \left[\left(q' - \frac{2}{3} \sigma'_{v0} \right) B^{0.7} I_c \right] \qquad (4)$$

where f_s, f_l and f_t are correction factors for shape, for thickness of sand layer and for the time.

Because of limited physical similarity between driving a sampler into the ground and settlement due to a static load and because of the sensitivity of blow- counts data, the accuracy of settlement prediction from these methods is uncertain. On the other hand a number of settlement prediction techniques -
based on elastic theory are available, generally utilising either empirical or laboratory evaluated modul of elasticity.The elastic model for settlement has the general form

$$w = \frac{q_{AV} B}{E} I \qquad (5)$$

where w is the settlement, q_{AV} is the net effective pressure increase and I is a settlement influence factor which keeps into account the effects of the shape and the rigidity of foundation, the Poisson's ratio and the ratio of the depth of the compressible stratum to the foundation breadth B.

The elastic solution application requires selection of a modulus, usually E for drained conditions as occurs in the case of granular soils. Empirical correlations are generally employed for this purpose.

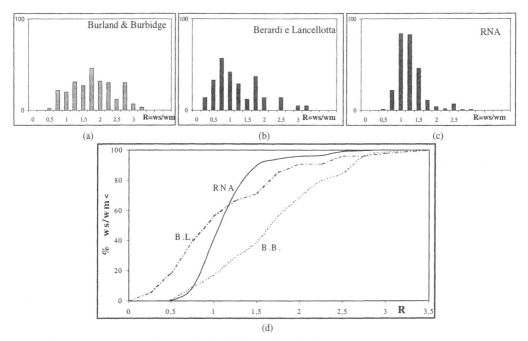

Figure 8. (a, b, c) Histograms showing the results of predictive methods; (d) Cumulative curve of the variables R=ws/wm

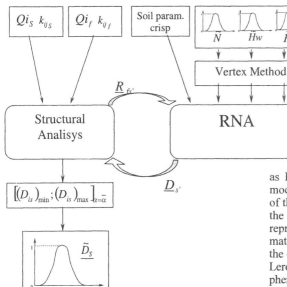

Figure 9. Soil-structure interaction analysis.

Fig.(4.a) presents correlation with SPT blow-count data in USSR practice (Trofimenkov, 1974) beside two tentative relationships presented by D'Appolonia et al.(1970). Deformation modulus can also be measured in laboratory tests, either directly as E in triaxial test samples or as the constrained modulus D, in oedometer tests. However the results of these tests require empirical calibration in view of the extreme difficulty in obtaining or preparing truly representative samples. Soil stiffness, in fact, is not a material property, but is a parameter depending on the current state of stress, specific volume and strain. Leroueil e Vaughan (1990) state that other structural phenomena, such us early diagenesis, may play an important role in determining the elastic behaviour of cohesionless soils. In particular due to inherent variability of natural deposits of granular soils, settlement can differ from expected values, generally by a factor of 1.5, but one could normally expect differences of up to a factor of about 3 in actual settlement (Fig.4.b). Correlating foundation compressibility with compressibility grade gives differences varying between factors of 4 to 8. Thus there is still room for considerable improvement in predictive methods but the limitations of inherent variability should always be borne in mind (Burland & Burbidge, 1986).

2.4 Neural network to analysis of settlement of shallow foundation on sand and gravel.

The neural network is used here as a shallow foundation settlements prediction. In figure 5 is pictured problem geometry. It is a feed-forward network based on multilayered Perceptron with a back-propagation learning algorithm. The learning process analyses over 200 record of settlement of building foundations, tanks and embankments on sand and gravel. Input signal vectors have build on values of breadth (B, [m]), length (L, [m]), depth (D, [m]), depth of water table beneath level ground (H_w, [m]), thickness of sand or gravel stratum (H_s, [m]), average SPT blow count over the depth of influence (N), gross bearing pressure at foundation level (q, [KPa]), maximum previous effective overburden pressure (σ'_{v0}, [KPa]). Output vector on foundation settlement (ρ, [mm]). In figure 6 is the network architecture. Input vectors and the corresponding output vector are used to train (fig.7) a network until it can approximate a function, associate input vectors with specific output vector.

2.5 Validation of prediction Neural network

A summary of measured settlement and of its estimate made by using the mentioned RNA is given in the figures(8), in front of Burland and Burbidge (1985) and Berardi and Lancellotta (1991) methods. Assuming as a basic variable the ratio between the simulated w_s and the measured w_m settlement, a preliminary idea about the accuracy and the reliability of RNA method, can be obtained through the inspection of the histograms in Fig.(8.a,b,c). For each method the percent of cases where the ratio $R=w_s/w_m$ is less than a given values is given. In fig.(8.d) the cumulative percents are pictured. In conclusion, Burland and Burbidge method leads to an underestimate of actual settlement of 17%, Berardi & Lancellotta method of 56%, RNA method of 41% of the examined cases; using RNA method $R= C_s/C_m <0.75$ for 8% of the examined cases, using B & L method for 40%. Moreover using RNA method yelds $R= C_s/C_m >1.25$ for 27% of the examined cases, using B & L method for 71%.

3 METHODOLOGY OF NEURO-SOIL-STRUCTURE INTERACTION ANALYSIS CONSIDERING SOIL UNCERTAINTY

3.1 Methodology

In the present study a methodology is established for conducting neuro-soil structure interaction analysis

with uncertainty in soil parameters (Fig.9). The uncertain input parameters considered are SPT blow count over the depth of influence (N), depth of water table beneath level ground (H_w, [m]) and thickness of sand or gravel stratum (H_s, [m]). These are represented as fuzzy sets and processed with Vertex method. At each selected a-level (a=0, 0.5, 1), an interval is obtained for each of the three uncertain parameter. For each combination (eight) the parameters are fixed values, just like other parameters used in the neuro-soil structure interaction analysis. According to the vertex method, the minimum and maximum of the eight settlement values define the resulting settlement "interval". By repeating this process for all selected a-levels, a set of settlement intervals corresponding to the selected a-levels is obtained. These intervals and corresponding a-levels define the final output, a settlement fuzzy set. It is the same for structural stresses.

3.2 Example

The problem consists of a frame structures, subjected to loads concentrated and uniformly distributed and founded in granular soil with 1.00x1.00 m pad foundations (Fig.10). SPT tests and site knowledge have shown that sand density is medium dense, with N_{SPT} interval base equal to [10, 30].

The depth of water table beneath level ground is subjected to seasonal excursions in interval [3, 5] m. The thickness of sand stratum is variable in interval [15, 20] m. Membership functions of N_{SPT} and Hw are shown in fig.(11). The neuro-fuzzy interaction method is used to obtain the solution for each parameters combination. Membership functions of pad foundations settlements(w) and differential settlements (Δw) are given in figure (12) and (13), and the membership functions of the bending moments at the section of maximum moment on the left and right side are given in figure (15). An important aspect of the neuro-fuzzy solution is that for a specific α-cut the solution is equivalent to that of all possible combinations of a live soil conditions.

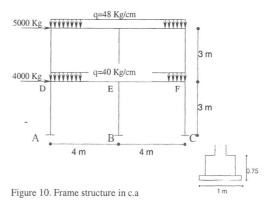

Figure 10. Frame structure in c.a

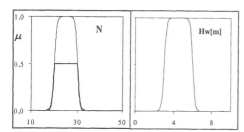

Figure 11. Membership function of N and Hw variables

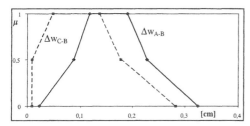

Figure 12. Membership functions of absolute settlements of pad foundations

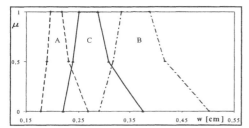

Figure 13. Membership function of differential settlements

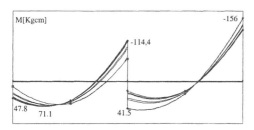

Figure 14. Bending moments at different a-levels on beam DF

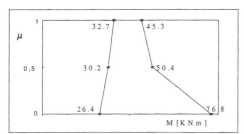

Figure 15. Membership function of bending moment on section E (on the right)

Than the range of bending moment and its distribution are available, and this information should important in making design decision. Uncertainty in loading can be treated with same methodology.

4 CONCLUSIONS

The attractiveness of proposed fuzzy-neural network lies that it is adaptive model free estimator for predict settlements of shallow foundations in granular soils. Use of fuzzy numbers is effective to dealing with soil parameter uncertainty and to calculate the extreme values of a structure's response to all possible boundary condition combinations. A procedure to analyzed soil-structure interaction, based on this neural network, is proposed to dealing also with loading uncertainty.

ACKNOWLEDGEMENTS

The writer express their gratitude to University of Catania for financing this research. The author would also like to thank University of Rome "Tor Vergata", for technical support.

REFERENCES

Berardi, R. & Lancellotta, R. 1991. Stiffness of granular soil from field performance. Geotechnique 41, No 1, 149-157

Biondini, F., Bontempi, F. & Malerba P.G. 2000. Fuzzy Theory and Genetically-Driven Simulation in the Reliability Assessment of Concrete Structures. 8th ASCE Joint Specialty Conference on Probabilistic Mechanics and Structural Reliability, July 24-26, 2000, University of Notre Dame, Indiana.

Burghignoli, A. 1985. Il metodo dei coefficienti di influenza per l'analisi di alcuni problemi di interazione tra il terreno e le strutture. Rivista Italiana di Geotecnica

Burland, J.B. & Burbidge, M.C. 1985. Settlement of foundation on sand and gravel. Proc. Instn. Civ. Engrs, Part 1 78, 1325-1381

Dong, W.M. 1987. Fuzzy information processing in seismic hazard analysis and decision making. Soil Dynamic and earthquake Eng., 6, No. 4.

Murlidharan, T.L. 1991. Fuzzy behavior of on Winkler foundation. J. of Engineering Mechanics, Vol.117, No.9, September

Provenzano, P. 2000. Neuro-Fuzzy models in Geotechnical Engineering (In Italian) Phd. Thesis. University of Palermo. Palermo. Italy

Provenzano, P. & Bontempi, F. 2000. Impostazione dell'analisi strutturale in presenza di informazioni incerte attraverso Logica Fuzzy. STUDI E RICERCHE – Vol. 21, 2000 Scuola di Specializzazione in Costruzioni in C.A. – Fratelli Pesenti. Politecnico di Milano, Italia

Provenzano, P., Bontempi, F. & Musso, A. 2000. Interazione terreno-struttura: una rete neurale fuzzy per l'elaborazione delle incertezze nel calcolo dei cedimenti in terreni granulari. 13° Congresso C.T.E. Pisa 9-10-11 Novembre

18 Structural design issues

Creative Systems in Structural and Construction Engineering, Singh (ed.) © 2001 Balkema, Rotterdam, ISBN 90 5809 161 9

Truck weight limits and bridge infrastructure demands

G. Fu, J. Feng & W. Dekelbab – *Department of Civil and Environmental Engineering, Wayne State University, Detroit, Mich., USA*

F. Moses – *Department of Civil Engineering, University of Pittsburgh, Pa., USA*

H. Cohen – *Ellicott City, Md., USA*

D. Mertz – *Department of Civil Engineering, University of Delaware, Newark, Del., USA*

ABSTRACT: This paper reports the development of a method for estimating effects of truck weight limit change on bridge network costs, funded by the US National Cooperative Highway Research Program. Four categories of cost impact are covered here as a result of changing truck weight limits: steel fatigue, reinforced concrete (RC) deck fatigue, additional inadequate exiting bridges, and higher design load for new bridges. This development has taken into account constraints of data availability at the State infrastructure system level. Results of two application examples included in the study indicate that the last two categories may be dominant contributors to the total cost impact.

1 INTRODUCTION

Trucks deliver a significant portion of the nation's product. In 1974, for example, this includes 60 percent of all inter-city shipments of manufactured products, 80 percent of all fruits and vegetables, and 100 percent of all livestock (RJHansen 1979).

While we benefit from truck transportation, highway agencies have spent a significant amount of resources to establish and maintain the highway system. Quantifying causes of these expenditures has been a focus of several studies in a number of countries (Moses 1989, Fu et al 2000). This paper presents a method for estimating the costs of truck weight limit changes for a network of bridges. (Truck weight here collectively refers to the truck's gross weight, axle weights, and spacings of axles.) This subject is important because trucking at higher gross vehicle weight (GVW) is more productive but is envisioned to be more costly to the infrastructure. Thus, transportation agencies receive constant pressure to increase truck weight limits. The suggested method is to help transportation agencies deal with such pressure quantitatively and rationally.

2 PREDICTING TRUCK LOAD SPECTRA

A direct impact of truck weight limit change is the change in truck load spectra to bridges. It includes changes in truck weight histograms (TWHs) and wheel weight histograms (WWHs). The former rep-resents the load to the entire bridge, affecting the bridge's relative strength demand. It also influences steel bridge fatigue accumulation. The latter is the load to bridge decks that transfer wheel loads to the supporting frame. A new method is developed below for predicting the TWHs and WWHs under a change in truck weight limits.

2.1 Predicting TWHs

Changes in TWHs due to truck weight limit changes may be classified into the following three types of freight shifting. 1) Load shifts without changing truck types (truck configurations), referred to as truck load shift hereafter. 2) Load shifts with changing of truck configuration, referred to as truck type shift below. 3) Exogenous shifts, such as economy growth and mode shift (e.g., from and to rail) due to competition. These shifts are individually dealt with below. Base Case used below refers to the condition before the considered change in truck weight limits. Alternative Scenario represents the condition after the change.

It is assumed that TWHs for the Base Case are available for each type of vehicles, except automobiles and 4-tire light trucks. Theses two types of vehicles are considered irrelevant to issues related to bridge strength and fatigue. These TWHs for the Base Case may be obtained using weigh-in-motion (WIM) data available with State highway agencies.

The FHWA vehicle-mile-of-travel (VMT) data for Year 2000 may also be used as the default data set (Fu et al 2000). This default data set is available for 12 functional classes and for each State.

2.1.1 Truck load shift

In truckload shifting as a result of truck weight limit increase, trucks of a given type can be loaded heavier. This is because the Alternative Scenario's practical maximum GVW (PMGVW) is higher than the Base Case PMGVW. This type of change in TWHs is expected to occur when the Alternative Scenario does not require trucks to change their configuration for carrying the new allowable loads.

In other words, only the TWH for that type of vehicle will be subject to change (shifting). This shifting should be performed only for weight-limit dependent truck traffic. For example, trucks heavier than Base Case PMGVW operating under special permits will not need to change in response to the weight limit change. The amount of traffic responding to the change is identified using a window shown in Fig.1 over the Bases Case TWH (for the impacted type of trucks). Namely, the traffic that is within this window will be subject to shifting and otherwise not. The window is defined by five parameters: a_1, a_2, b_1, b_2, and c, all in fraction or percentage. These parameters are referred to as window parameters.

Parameters b_1 and b_2 define a neighborhood of weight-limit sensitive traffic, with reference to the Base Case's PMGVW. When GVW_{BC} / $PMGVW_{BC}$ is close to 1 between $1-a_1$ and $1+a_2$, the level of weight-limit dependence is described by c. It indicates the percentage of the traffic to be changed under the Alternative Scenario. Beyond this small range to the left, the weight-limit dependence is assumed to vary linearly from c at $1-a_1$ to zero at $1-b_1$ being the lower boundary of the neighborhood. To the right from $1+a_2$ a similar behavior is assumed of weight-limit dependence up to $1+b_2$.

After the weight-limit dependent truck traffic $TT'_{GVWk,BC}$ is identified for weight interval k, the following equations will be thereto applied in modifying the Base Case TWH, as a response to the considered change in truck weight limits:

$$GVW_{AS} = GVW_{k,BC}(PMGVW_{AS}/PMGVW_{k,BC}) \quad (1)$$
$$TT_{GVW,AS} = TT'_{GVWk,BC} \times$$
$$(GVW_{k,BC}-TARE_{BC})/(GVW_{AS}-TARE_{AS}) \quad (2)$$

where subscripts $_{BC}$ and $_{AS}$ refer to the Base Case and Alternative Scenario, respectively. TARE is the empty weight of truck. $TT_{GVW,AS}$ is the truck traffic at weight GVW_{AS} under the Alternative Scenario.

Eq.2 indicates the change in operating weight or GVW. It occurs only when the operating weight is within the window in Fig.1. Eq.2 enforces the condition that the total payload travel (in kN-km or kip-mile) is conserved during load-shifting since the total amount of freight carried remains constant.

Note that when $PMGVW_{AS}$ is greater than $PMGVW_{BC}$ representing an increase in weight limit, the total amount of truck traffic will decrease since fewer trips will be required to transport the same amount of freight (payload). Note that possible payload changes are covered below addressing external factors, such as economy-growth-dependent payload increase and competition-induced payload shift from or to rail.

In applying Eqs.1 and 2, GVW_{BC} is taken at the midpoint of a weight interval falling in the window defined in Fig.1. Consequently, the value of GVW_{AS} generally will not match the midpoint of a weight interval. It is then appropriate to distribute $TT_{GVW,AS}$ between two neighboring weight intervals to achieve the desired value of GVW_{AS}, which are designated as the i th and the i+1 th intervals, respectively. The distribution ratios p_i and p_{i+1} for the i th and the i+1 th weight intervals are required to satisfy the following equations:

$$p_i + p_{i+1} = 1 \quad (3)$$
$$p_i GVW_{i,AS} + p_{i+1} GVW_{i+1,AS} = GVW_{AS} \quad (4)$$

Then the truck traffic equal to $p_iTT_{GVW,AS}$ is to be moved to the i th GVW interval and $p_{i+1} TT_{GVW,AS}$ to the i+1 th interval.

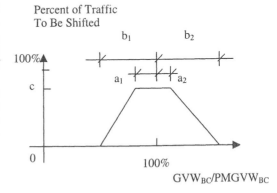

Fig.1 Window for Truck Traffic Shifting

The following assumptions have been used for the proposed method. A) For many commodities (e.g., potato chips), the cubic capacity of the truck is the limiting factor. B) Heavier trucks excessively above $PMGVW_{BC}$ and operating under special permits may not react to weight limit changes if other factors do not change. C) The total payload traveled (in kN-km or kip-miles) remains the same before and after the weight limit change, i.e., Payload x Distance of Travel = Constant. This has been expressed in Eq.2, and the distribution of this traffic over the truck-weight intervals is altered due to shifting. Selecting parameters a_1, a_2, b_1, b_2, and c for the proposed method may require measured data and appropriate engineering judgement.

2.1.2 Truck type shift

The same equations Eqs.1 and 2 for truck load shift are recommended to be used for truck type shifting. However, $TT'_{GVWk,BC}$, $PMGVW_{BC}$ and $TARE_{BC}$ now refer to the truck type from which traffic is shifted, and $TT_{GVW,AS}$, $PMGVW_{AS}$, and $TARE_{AS}$ refer to the truck type to which traffic is shifted. This reflects the fact that a new truck type of configuration will be needed in order to take advantage of the new weight limits.

2.1.3 Exogenous shift

Exogenous shifts here refer to those changes to TWHs due to external factors, other than those discussed above. The influencing factors may be, for example, economic growth, competitiveness with other transportation modes (e.g., rail), etc. (Cambridge Systematics et al. 1997) provides detailed discussions on transportation modal shifts for freight demand predictions. The guidelines presented there are of help in understanding relevant issues, and estimate the amount of truck traffic shift.

The first step of accounting for these effects is to identify the traffic in the TWHs that is subject to exogenous shift. For the case of overall economic growth as an example, all traffic should be subject to change, unless otherwise objected. This may be readily taken into account by using a growth factor to be applied to all traffic:

$$ADTT_{AS} = g\ ADTT_{BC} \qquad (5)$$

where g is the growth factor, which could be estimated based on data at the network level. ADTT stands for annual daily truck traffic.

For the case of transportation modal change due to a truck weight limit change, it would be reasonable to use the same kind of window in Fig.1 for identifying the impacted traffic. In addition, a multiplier r can be applied to the affected traffic at weight GVW:

$$TT_{GVW,AS} = r_{GVWk}TT'_{GVWk,BC}\ x$$
$$(GVW_{k,BC}-TARE_{BC})/(GVW_{AS}-TARE_{AS}) \qquad (6)$$

As indicated, r_{GVWk} can be a function of operating weight GVW at the k th interval. This multiplier is higher than 1.0 for traffic increase and less than 1.0 for decrease.

Two examples of testing the suggested prediction method have been included in the study (Fu et al 2000). Measured truck weight data are used for these examples. They include a statewide truck weight limit increase from about 73 kips to 80 kips in the 1980s for State of Arkansas and a permit weight limit change from 105 kips to 129 kips for two routes in State of Idaho. Results show that the prediction method can capture the resulting changes in TWHs, using the following window parameters: c=0.95, $a_1=a_2=0.10$ and $b_1=b_2=0.20$.

2.2 Predicting WWHs

For assessing reinforced concrete (RC) deck fatigue, truck wheel weight distributions are needed to estimate the cost effects of changes in truck weight limits. It is suggested here that predicting WWHs be based on GVW, assuming that there is a correlation between the wheel weights and the gross weight. This assumption is particularly valid for trucks loaded to the limits, which are dominant in RC deck fatigue. When a TWH is available, possibly obtained using the method recommended above or directly from truck weight data, the wheel weights can be estimated using the following empirical relations:

$$Wheel\ Weight = E + F * GVW \qquad (7)$$

where E and F are model coefficients for each axle. They can be obtained using WIM data and a regression analysis. In (Fu et al 2000), examples of E and F were obtained using data from the State of California. It is also recommended that agencies use their own WIM data to obtain those coefficients for typical truck types within the jurisdiction.

3 PRIORITIZED COST IMPACT CATEGORIES

A number of categories of cost impact are attributable to increased truck loads as a result of truck weight limit changes. Examples are increased steel bridge-joint fatigue or wearing, and increase of concrete cracking in both severity and quantity. Four cost impact categories are prioritized in this study: 1) steel bridge fatigue, 2) RC bridge deck fatigue, 3) additional inadequate load rating for existing bridges, and 4) higher load requirement for new bridges. This prioritization is based on available knowledge of the mechanism of cost impact and data availability for a reasonably reliable cost estimate. The suggested method for estimation is planned not to require data currently unavailable with State transportation agencies.

3.1 *Steel bridge fatigue*

Fatigue of steel bridge components has been extensively investigated (Moses et al. 1987). The vast majority of state agencies have experience with fatigue damage. Under an increase in truck weight limits, fatigue accumulation is expected to increase due to load (and thus tress range) increase, although the truck traffic is expected to decrease if the total payload remains constant.

The following procedure is suggested to estimate the impact cost due to additional fatigue accumulation. 1) Identify possibly vulnerable bridges. 2) Decide to analyze all or a sample of possibly vulnerable bridges. 3) For the analysis of each bridge, generate the TWH under the Base Case and predict the TWH under the Alternative Scenario. 4) Estimate remaining safe life and remaining mean life for both the Base Case and Alternative Scenario. 5) Select responding action for treating possible fatigue failure. 6) Estimate the costs for the selected action. 7) Sum the costs for all bridges. 8) Perform a sensitivity analysis to understand possible controlling effects of the input data.

For each bridge analyzed, the fatigue assessment should follow the AASHTO procedure (1994), to be consistent with current practice:

$$Y = \frac{f\,K * 10^6}{(T_a/T)\,T\,C\,(R_s\,S_r)^3} \qquad (8)$$

where Y is the total life in years. K is a constant tabulated for each type of fatigue sensitive detail in the AASHTO specifications (1994), and f equal to 1 for safe life and 2 for mean life. C is the number of cycles for a passage of the fatigue truck. R_s is a reliability factor. S_r is the stress range for a passage of the fatigue truck whose weight can be determined using WIM data. (Note that a considered weight limit change may require a change to the fatigue truck model, when a majority of trucks are to be affected in operating behavior.) For the Base Case and the Alternative Scenario, this stress range should be calculated using respective THWs. The Base Case TWH is based on site specific WIM data or the default VMT data (Fu et al 2000). The Alternative Scenario TWH is to be developed using the Base Cases TWH and the prediction method discussed above. T is the current annual daily truck volume for the outer lane. T_a is an estimated lifetime-average daily truck volume in the outer lane. The AASHTO specifications (1994) provide the values for these parameters or guidelines about determining

the values. Note that the AASHTO procedure for T_a represents an approximation, which may lead to under- or over-estimates. The following formula is recommended to improve this assessment:

$$T_a/T = \Sigma_{i=1,2,..,Y}\,(1+u)^i\,/\,[Y(1+u)^A] \qquad (9)$$

where u is the annual traffic growth rate. It may be estimated using information in the agency's bridge inventory. A is the current age of the bridge.

The expected impact costs are to be estimated as follows:

$$\text{Expected Impact Cost} = \text{Cost}\,(P_{f,AS} - P_{f,BC}) \geq 0 \quad (10)$$

where subscripts $_{AS}$ and $_{BC}$ indicate respectively the Base Case and the Alternative Scenario. In case the expected impact cost turns out to be negative, then it is set zero because no impact is expected due to the considered change in weight limits. The impact cost here depends on the action selected in response to possible fatigue failure. This action can be repair, replacement, monitoring, and/or their combinations. The default is recommended to be repair. $P_{f,AS}$ and $P_{f,BC}$ are probabilities of fatigue failure within a predetermined planning period PP and calculated as follows:

$$P_f = \Phi\,((PP - Y_{\text{Mean Remaining}})/\,\sigma_Y)$$
$$- \Phi\,(-Y_{\text{Mean Remaining}}/\,\sigma_Y) \qquad (11)$$
$$(Y_{\text{Mean Remaning}} - Y_{\text{Safe Remaining}}\,)/\,\beta = \sigma_Y \qquad (12)$$

where $Y_{\text{Mean Remaning}}$ and $Y_{\text{Safe Remaining}}$ are mean remaining life and safe remaining life according to Eq.8. β is the target reliability index equal to 2 for redundant members. Φ is the cumulative probability function for the standard normal variable and PP, a pre-selected planning period in years for which the cost effect is covered.

3.2 *Reinforced concrete deck fatigue*

Based on previous studies on RC deck fatigue under wheel load (Matsui 1991, Perdikaris et al 1993, Fu et al 1992), the following procedure is recommended for assessing fatigue accumulation using a similar format to that in Eq.8

$$Y_d = \frac{K_d\,\,K_p}{(T_a/T)\,T\,C_d\,(\,R_d\,I\,P_s\,P/P_u)^{17.95}} \qquad (13)$$

where Y_d is the service life of the deck. Y_d will be the mean service life for the reliability factor R_d set equal to 1 and the evaluation life for R_d equal to 1.35. T_a and T have been defined in Eq.9. C_d is the average number of axles per truck. P/P_u is the equivalent stress ratio caused by wheel load P

$$P/P_u = [\Sigma \ f_i(P_i/P_u) \ (P_i/P_u)^{17.95}]^{1/17.95} \qquad (14)$$

where P_u is the ultimate shear capacity of the deck. Eq.13 uses the same linear damage accumulation assumption (the Miner's Law) as for steel fatigue. K_d is a coefficient that covers the model uncertainty (with respect to the assumed Miner's Law). K_p addresses the difference between the state of deck failure recognized in the laboratory and the state of real decks when treatment applied (Fu et al 2000).

P_u is the nominal ultimate shear strength of the deck and suggested to be estimated as follows, according to the ACI code (1995):

$$P_u = (2 + 4/ \ \alpha \ ) \ (f_c')^{1/2} \ b_0 d \ \gamma < 4 \ (f_c')^{1/2} \ b_0 d \ \gamma \qquad (15)$$

where f_c' is the concrete compressive strength in psi. α is the ratio of the tire print's long side to short side, set equal to 2.5 for a nominal tire print of 0.508 m by 0.203 m (20 in. by 8 in.) for dual tires. d is the deck's effective thickness equal to the total thickness minus the bottom cover thickness. It is recommended to also subtract a 0.00635 m (0.25 in.) thick layer from the nominal thickness to account for wearing observed in bridge decks. b_0 is the perimeter of the critical section, which is defined by the straight lines parallel to and at a distance d/2 from the edges of the tire print used. γ is a model correction parameter, which is set at 1.55 based on the test data in Perdikaris et al (1993). It should be noted that the above parameters are nominal values of respective variables with uncertainty, as in many other cases for strength or fatigue assessment.

The expected impact cost can be estimated in the same way as Eqs.11 and 12 except that $Y_{d,Mean \ Remaning}$ and $Y_{d,Evaluation \ Remaining}$ are used. The cost can be for patching, overlay, or replacement, depending on the responding action selected. Portland cement concrete overlay is selected here as the default action responding to deck deterioration. It should be noted that steel rebar corrosion has been widely accepted to be a major cause of RC deck deterioration, especially in areas where a large amount of de-icing salt is used. In these areas, RC deck fatigue becomes secondary or negligible for deck deterioration or consumption.

3.3 Additional inadequate existing bridges

Currently in the US highway system, there are a number of bridges that are inadequate in load carrying capacity. This is indicated by their load rating factor lower than 1, according to the AASHTO requirement (1994, 1998). When higher truck loads are legalized or permitted, more bridges will become inadequate. Costs to correct the additional inade-

quacy are covered in this cost impact category. The new rating factor is recommended to be calculated as follows:

$$RF_{AS} = RF_{BC}/AF_{rating} (M_{BC,rating \ vehicle} / \ M_{AS,rating \ vehicle}) \qquad (16)$$

where RF_{AS} is the rating factor for the Alternative Scenario. RF_{BC} is the rating factor for the Base Case (likely the existing rating factor). $M_{BC, \ rating \ vehicle} / \ M_{AS, \ rating \ vehicle}$ is the ratio between the maximum load effects due to the rating vehicle under the Base Case and due to the new rating vehicle under the Alternative Scenario. The new rating vehicle is a model representing the practical maximum load permissible under the changed weight limits. It could be a set of vehicles. AF_{rating} is the ratio between the live load factors for the Base Case and the Alternative Scenario:

$$AF_{rating} = [2W_{AS}^* + 1.41 \ t(ADTT_{AS}) \ \sigma_{AS}^*] / \\ [2W_{BC}^* + 1.41 \ t(ADTT_{BC}) \ \sigma_{BC}^*] \qquad (17)$$

where W^* and σ^* are the mean and standard deviation of the top 20 percent of the TWH, and t is a function of ADTT (Fu et al 2000). Subscripts $_{BS}$ and $_{AS}$ respectively refer to the Base Case and the Alternative Scenario. This approach is consistent with the concept of load and resistance factor rating under development (AGLictenstein 1999) for AASHTO.

In Eq.16, $M_{BC,rating \ vehicle} / \ M_{AS,rating \ vehicle}$ reflects the adjustment to rating due to the new truck model. The adjustment factor AF_{rating} is the ratio of the live load factors to cover uncertainty changes in truck weight spectra.

For cost estimation, those bridges that are inadequate with $RF_{BC}<1$ under the Base Case should be excluded, because they do not contribute to the cost impact (additional costs). When a bridge is found to be inadequate or overstressed under the Alternative Scenario but adequate under the Base Case (i.e., $RF_{BC} \geq 1$ and $RF_{AS}<1$), an action needs to be selected as the basis for cost estimation. It can be, for example, posting, strengthening, replacing, or a combination thereof. Note that, in reality, the decision making process requires information on a number of other factors, not only the load rating.

3.4 Higher load requirement for new bridges

The bridge design load is supposed to envelope current and expected future loads for the bridge life span. When higher loads are legalized or permitted, the design load needs to be adjusted to assure relatively uniform safety of the bridges. The costs caused by this adjustment are covered in this cost impact category. The analysis will require the following steps. 1) Identify the new bridges to be constructed in the future within PP. 2) Estimate the required design load for each of these bridges under the Alternative Scenario. 3) Estimate the additional costs for each of these bridges under the new design load.

Step 1) may be approximated using the bridges constructed in recent years and averaged to an annual population of new bridges. It can be done using the agency's bridge inventory. Step 2) is to be accomplished using the following formula for the amount of design load change:

$$DLCF = (M_{AS\ design\ vehicle} / M_{BC\ design\ vehicle})\ AF_{design}$$
(18)
$$M_{AS\ design\ vehicle} / M_{BC\ design\ vehicle} \geq 1$$
(19)
$$AF_{design} = (2W_{AS}* + 6.9\ \sigma_{AS}*) / (2W_{BC}* + 6.9\ \sigma_{BC}*)$$
(20)

where DLCF stands for design load change factor indicating the ratio between the design load effects under the Base Case and the Alternative Scenario. $M_{AS,\ design\ vehicle} / M_{BC,\ design\ vehicle}$ is the ratio of the maximum load effects due to the design vehicle under the Base Case and the same under the Alternative Scenario. Practically, it should not be lower than 1. Namely, when $M_{AS\ design\ vehicle}$ is smaller than $M_{BC\ design\ vehicle}$, the design vehicle under the Base Case would be the governing load and the ratio should be taken as 1 in Eq.19. This will assure that the new design will not be lower than the current design load. AF_{design} is the ratio between the live load factors under the Base Case and the Alternative Scenario. It is an adjustment factor for the design load used to cover the change in uncertainty associated with the considered Alternative Scenario. It plays a similar role as AF_{rating} in Eq.17 for additional deficiency in existing bridges.

DLCF in Eq.18 indicates the relative increase in the design load effect. The incremental cost can be accordingly calculated as the impact cost. A set of default costs data have been prepared for this purpose, if no more specific data are available (Fu et al 2000).

4 APPLICATION EXAMPLES AND CONCLUSIONS

This study included two application examples for the proposed method of estimating costs of a bridge network resulting from truck weight limit changes. The first example investigated such effects due to an increase in permit truck weight from 105 kips to 129 kips for two routes in the State of Idaho. This weight limit increase was required to be in accordance with the federal bridge formula. The second example was for the State of Michigan for a scenario of legalizing 6-axle semi-trailers at a GVW of 97 kips. This truck type is legal in Canada and permissible in several States close to Michigan. The State

of Michigan is under pressure to permit or legalize it. Results of these examples indicate that the major contributors to the total cost impact are additional inadequate bridges and higher requirement for bridge design load.

ACKNOWLEDGEMENTS

A.G.Lichtenstein assisted in preparing part of the default costs data for applying the proposed method. The work reported here was funded by the US National Cooperative Highway Research Program. These supports are gratefully appreciated. However, the views expressed here do not necessarily reflect those of the sponsor.

REFERENCES

AASHTO (1998): LRFD Bridge Design Specifications, 2nd Ed. Washington, DC.

AASHTO (1994): Manual for Condition Evaluation of Bridges, Washington, D.C.

ACI "Building Code Requirements for Reinforced Concrete " ACI 318-95, 1995

AGLichtenstein & Associates 1999: NHCRP 12-46 Proposed Manual for Condition Evaluation and LRFR of Highway Bridges

Cambridge Systematics, Inc. et al. "A Guidebook for Forecasting Freight Transportation demand", NCHRP Report 388, TRB, 1997

Fu,G., J.Feng, W.Dekelbab, F.Moses, H.Cohen, D.Mertz, and P.Thompson "Effect of Truck Weight on Bridge Network Costs", Draft Final Report to NCHRP, Wayne State Univ., June 2000

Fu,G., S.Alampalli, and F.P.Pezze (1992) "Long-term Serviceability of Isotropically Reinforced Concrete Bridge Deck Slabs", TRB Transportation Research Record 1371, 1992, p.17

Matsui,S. 1991 "Fatigue Deterioration Mechanism and Durability of Highway Bridge RC Slabs", Report to Hanshin Expressway Public Corp.

Minnesata DOT 1991 "Truck Weight Limits and Their Impact on Minnesota Bridges", Task Force Report, Response to TRB Special Report 225

Moses,F., Schilling,C.G., and Raju,K.S,. Fatigue Evaluation Procedures for Steel Bridges. NHCRP Report 299, TRB, Washington, DC, Nov. 1987

Moses,F. (1989) "Effects on Bridges of Alternative Truck Configurations and Weights" Draft Final Report to NCHRP - TRB, 1989

Perdikaris,P.C. Petrou,M.F., and Wang,A.1993 "Fatigue Strength and Stiffness of Reinforced Concrete Bridge Decks", FHWA/OH-93/016, Civil Engineering Dept., Case Western Res. Univ., OH

RJHansen Asso. 1979 "State Laws and Regulations on Truck Size and Weight", NCHRP Report 198

Weyers,R.E., Fitch,M.G., Larsen,E.P., and Al-Qadi,I.L. "Concrete Bridge Protection and Rehabilitation: Chemical and Physical Techniques", SHRP-S-668, National Research Council, 1994

Creative Systems in Structural and Construction Engineering, Singh (ed.) © 2001 Balkema, Rotterdam, ISBN 90 5809 161 9

The quest for quality in timber roof structures in South Africa

A.C.Malherbe & I.L.Steenkamp
Department of Construction Management, University of Port Elizabeth, South Africa

ABSTRACT: Although timber engineering is very limited in the South African construction industry in general, the use of pre-fabricated "engineered" nail-plated timber trusses dominates the provision of roof structures for the home building market in S.A. The shift from site made- to rationally designed pre-fabricated trusses over the past 30 years has held out the potential for improvement in the quality of these timber structures. However, a very low level of conformance to quality standards exists. The situation has been exacerbated in recent times by structural developments in the home building industry, as well as developments in South African society.

The historical developments and their influence on the quality of timber roof structures are reviewed. The present level of non-conformance to engineering standards is assessed and the areas of major concern identified. Proposals for, and likely future developments in quality assurance in the pre-fabricated timber roof truss industry are outlined.

1 INTRODUCTION

The use of structural timber in building construction in South Africa is confined largely to the provision of roof structures for buildings of domestic scale. Timber floors and timber frame wall construction have gained only limited acceptance due to historical, social, environmental and cost factors.

Where timber is used, albeit rarely, in complex roof structures such as auditoria, exhibition centres etc., the delivery process is the same as for structural steel and concrete structures, involving design and site controls by registered professional Engineers, based on South African Bureau of Standards (SABS) Codes of Practice. These standards are constantly reviewed in the light of international experience, incorporating research findings relating to local materials and conditions. Hence the quality of such large-scale timber structures is comparable to that of similar steel or concrete structures in South Africa and equally comparable to the quality of similar structures elsewhere in the world.

However, on a domestic scale the delivery process for structural timber roofs is markedly different, not requiring the involvement of registered engineering professionals and hence not subject to the "normal" engineering controls.

This dichotomous treatment of timber as a structural material has resulted from a number of historical developments in:

– the regulatory environment,
– the roof truss industry,
– the home building industry and
– the social environment.

2 REGULATORY ENVIRONMENT

After an earlier abortive attempt at standardizing building regulations throughout South Africa, the National Building Regulations and Building Standards Act (No 103 of 1977) and the National Building Regulations (NBR) promulgated in terms of that law, finally became mandatory in 1985, placing the burden of ensuring conformance to the quality standards contained therein upon the local authorities in whose jurisdiction such building activity takes place.

These regulations spell out simply stated performance standards for, inter alia, the structural system employed and provide for two avenues of conforming to these:

– the appointment of a "responsible person" to ensure that a rational design is carried out in accordance with specified codes of practice and corresponding SABS specifications for materials, methods and workmanship or,
– construction according to "deemed-to-satisfy" regulations.

These deemed-to-satisfy regulations are so limited in their scope and restrictive with regard to maximum span (10m), spacings, roof slopes etc, as to be largely irrelevant in the modern home building scenario. The alternative is to appoint a "responsible person" who has to accept appointment by the "owner" and along with that, certain legal responsibilities, including that of taking reasonable steps to ensure that the work is carried out according to his rational design and issuing a certificate to that effect (SABS 1989). The responsible person could be a Professional Engineer or other "approved, competent person".

The term "Professional Engineer" had clear legal status in terms of the Professional Engineers Act (No. 81 of 1968), by which the statutory South African Council for Professional Engineers was tasked with maintaining a register of persons adequately qualified by virtue of education and training (normally at least an honors level University degree plus at least three years post-graduate in-service training). Since 1990 this function has been performed by a reconstituted statutory body called the Engineering Council of South Africa (ECSA) which has had the additional mandate to register Engineering Technicians, Professional Technologists (Engineering), and Certificated Engineers, all of whom could be considered "competent persons" depending on their area of specialization.

Whereas local authorities insist on the appointment of "responsible persons" for the simplest of steel or reinforced concrete structural systems, they have generally not done so for rationally designed nail-plated timber roof structures that clearly do not conform to the deemed-to-satisfy requirements. Representations by the roof truss industry at the time of introduction of the NBR, convinced the relevant authorities that pre-fabricated nailplated trusses were being designed with the aid of computer software developed by "competent persons" who also provided training and support to fabricators (Booth, pers. comm.). It was therefor presumed that the resulting rationally designed structure conformed to the relevant design codes as specified in the NBR. However nobody has been obliged to accept responsibility for the design, nor to certify that the roof has been erected according to the design - a fundamental principle of the NBR.

Mounting concern among financial institutions about the security of their mortgages, especially in large low cost housing developments in the metropolitan areas, prompted the formation of the National Home Builders Registration Council (NHBRC), initially with the support of the organized building industry, in June 1995. It was, at first, a non-statutory body with the aim of protecting consumers against poor workmanship and of raising standards in the industry. In terms of the Housing Consumer Protection Measures Act (No. 95 of 1998), the NHBRC has been reconstituted a statutory body and since July 1999 has the power to enforce the following:

– registration by the home-building contractor with the NHBRC,
– enrolment of every new home along with payment of an enrolment fee based on a percentage of the project value (presently 1.3%),
– that a 5-year warranty be issued by the builder with respect to defects affecting the structure of the building and that this warranty be honored by him, failing which the builder will be de-registered (effectively excluding him from this market), and the NHBRC will then ensure that the defects are made good,
– conformance by the builder to the NHBRC "Home Builder's Manual" - a generally more user-friendly presentation of the NBR deemed-to-satisfy rules - which is monitored by NHBRC assessors during construction. Alternatively, the builder must appoint a competent person to carry out a rational design and certify conformance with his design.

As far as roof structures are concerned, the Home Builder's Manual contains even more restrictive rules for site-made trusses and calls for the appointment of a "competent person" when nailplate connectors are used on such trusses. Alternatively the roof must be both rationally designed and certified to have been erected in accordance with the designer's specification, by "competent persons". Such persons must carry adequate professional indemnity insurance (NHBRC 1999b). However to date, this stipulation has not been universally enforced by the NHBRC for timber roof structures in the way that it has been for any other form of rationally designed structural element, such as foundations, walls, reinforced concrete floors, etc.

3 ROOF TRUSS INDUSTRY

Whereas the traditional method of roof truss construction had been site-made bolted-and-nailed trusses, the entrance by American-based timber engineering companies since the 1960's into the South African market produced a massive swing toward pre-fabricated nail-plated trusses. At present nail-plated trusses account for approximately 95% of all timber roof structures in South Africa (Blackwood-Murray, pers. comm.).

These are manufactured by independent truss fabricators licensed by four "Engineering Systems" who provide design software, training and consulting services, as well as the equipment, nailplates and related hardware for the fabrication of the timber roof components. The environment in which the fabricators operate is extremely competitive. The design and fabrication of the trusses is often viewed as a "loss leader" for access to other, more lucrative building materials supply contracts.

The Institute for Timber Construction was formed in 1972 as an industrial association with the aim of promoting and developing the design of timber structures. In order to ensure an adequate level of conformance to a mutually accepted standard of quality in the design, fabrication and delivery of timber roof trusses and other roof structure components, the ITC has been issuing Certificates of Competence to truss fabricators since 1986. Three-yearly cyclical audits have been conducted on certificated fabricators and a 70% subjective excellence grading has been required for the certificate to be renewed. The audit has evaluated such areas as:
- timber quality and -storage,
- dimensional accuracy in cutting and jointing of trusses,
- structural competence and awareness of designers
- the standard of documentation sent to site,
- handling and transportation of trusses to site.
- commercial and administrative competence.

As at June 2000, more than 80% of the approximately 150 fabricators in South Africa hold current Certificates of Competence (Blackwood-Murray, pers. comm.). Holding such a certificate, though not mandatory, has been seen as a valuable marketing asset in the very competitive roofing supply market.

Similarly, Certificates of Competence have been issued to roof erectors who have proven their competence in the following areas:
- demonstrated understanding of structural principles, especially relating to bracing,
- demonstrated ability to interpret drawings and specifications,
- correct handling and storage of roof structure components,
- conformance to correct erection procedures.

The number of roof erectors in South Africa is unknown, but is certainly many times the number of fabricators. However, as at June 2000, only 30 hold current Certificates of Competence as roof erectors. The holding of such a certificate in the roof erection market does not appear to provide the same competitive edge as it does in the equally competitive roof fabrication market.

In January 1997 the ITC commenced a program of training and certifying roof inspectors. Initially these were employed in six pilot areas throughout South Africa, in order to provide a resource of "competent persons" that could monitor and certify conformance to erection standards.

After considerable lobbying and consultation among fabricators, their engineering system suppliers and local authorities, a procedure was established whereby the ITC itself could act as the "responsible person". The process involved:

- scrutiny of fabricator designs by a professional engineer appointed by the licensor (engineering system supplier),
- inspection by an ITC inspector of the erected roof before loading,
- the issuing of a loading certificate by him,
- final approval of the rational design and its execution by an ITC-employed professional engineer who issues a completion certificate to the local authority.

This "A19" service (named after the NBR regulation requiring the appointment of a competent person) to the public and the industry was launched in Port Elizabeth in March 1998. At the same time some local authorities in the pilot areas commenced demanding that a "responsible person" accept responsibility for both design and inspection of rationally designed timber roof structures in accordance with the requirements of the NBR.

The administrative procedures proved to be too onerous for the ITC and a new A19 service was launched in February 1999. "Approved designers" were appointed in regions throughout SA, to whom the responsibilities of the ITC and systems were delegated. These have to ensure compliance with both the NBR and the NHBRC requirements when appointed by the owner or builder respectively. ITC inspectors would continue to issue certificates of compliance for the erected roof as well as the prefabricated trusses themselves, but the "approved designer" must issue the completion certificate, thereby accepting responsibility for both the design and erection of the completed roof.

4 BUILDING INDUSTRY

At the same time that the roof truss industry was undergoing a metamorphosis, the skills levels in the building construction industry in general have been declining drastically. Alman (1989) attributes the skills shortage to:
- stage of development in South Africa,
- the cyclical nature of the construction industry which results in poor job security, skilled workers finding secure employment in other industries during recessions in the construction industry,
- grossly inadequate expenditures on training and development,
- world-wide trend of 'de-skilling' i.e. craftsmen being replaced by semi-skilled process workers.

This latter factor is particularly pertinent in the prefabricated timber roof industry where, in addition, the specialization of the labor force in the two areas of fabrication and erection has been taking place in a quality vacuum.

Hindle (1990) contends that all the above factors were aggravated by the economic, educational and labor policies of the pre-democracy government.

The 1998 "Construction Green Paper" comments as follows on the increasing use of labor-only-subcontractors (LOSC), who account for up to 85% of the labor employed in the home building industry: "the distorted reliance on LOSC as adopted by the South African construction industry has resulted in a dramatic decline in health, safety, productivity and output quality" (Department of Public Works 1998).

5 SOCIAL ENVIRONMENT

The transformation of South African society to a liberal democracy is a matter of record. Active government intervention to promote greater representivity of all population groups in all spheres of economic activity, has in itself resulted in a shift in the profile of participants in the home building supply industry towards more small builders (called emerging contractors), especially from previously disadvantaged groups. At the same time the focus of Government expenditure has shifted to creating basic infrastructure in rural areas involving the provision of schools, clinics, etc. on a large scale. Together with the historical developments outlined above, this has given rise to the following quality related pressures:
– scarcity of skills in local rural communities where traditional building methods do not include the use of "Engineered Trusses",
– the prescribed need for local community involvement which tends to exclude skilled people from elsewhere in favor of local service providers, especially local labor,
– the difficulty and expense of providing normal supervision by professionals (Architects, Engineers etc.) in inaccessible or dangerous regions.

In addition, the legacy of the "struggle" is a society conditioned to non-compliance with regulations. This mindset is evident in all spheres and at all levels of South African society. Rent boycotts of the struggle years have been supplanted by "a culture of non-payment" of property taxes, service levies, traffic fines etc. Illegal occupation of property by the landless has given way to illegal use of property in direct contravention of local authority zoning regulations by citizens of all race groups and economic classes. In this regard the Chief City Estates Officer for Port Elizabeth is reported as lamenting: "The rulings are there, but the Council lacks the resources to police them.....so many precedents of wrongful using exist that following the rules has become very difficult indeed." (Randall 2000). It is thus not uncommon for building regulations to be openly disregarded. Local authorities generally do not have the physical or political capacity to enforce compliance with regulations.

Low morale and the general state of inertia of municipal officials are further aggravated by the un-certainty created by the reconstitution of local authorities by the Municipal Demarcation Board in terms of the Municipal Demarcation Act (No. 27 of 1998).

6 CURRENT LEVEL OF CONFORMANCE

On-going research into building defects conducted by the Department of Construction Management at the University of Port Elizabeth (Malherbe 1998), has revealed that, in early 1998, 93% of the newly erected timber roof structures in the Port Elizabeth/Jeffreys Bay area exhibited defects in one or more of the following areas (roofs exhibiting such defects expressed as a percentage of the total sample):
– defects in fabrication/design   20%
– defects in erection   93%
– erection defects possibly influenced by poor/ unavailable site documentation   60%

After the introduction of the "A19" system in March 1998 in the Port Elizabeth area, a further investigation was carried out into the relative importance of design and fabrication errors and the possible influence of poor or incomplete site documentation, revealing the following prevalence of defects in the design/fabrication/documentation process:
– design defects comprising: 20%
   technical defects   21%
   defects in documentation sent to site   79%
– fabrication defects comprising: 5%
   plating (size)   0%
   plating (other)   56%
   timber   31%
   jointing   13%

The high incidence of defects in site documentation was not entirely unexpected. During the period immediately preceding the above two investigations, fifteen truss fabricators in the eastern Cape region were audited by the ITC. An analysis of these audits has revealed that only 34% of fabricators achieved a better than 70% rating for both "completeness and accuracy"- and "method of presentation and issue"- of design documentation sent to site. Nevertheless 93% of the fabricators audited were awarded certificates of competence by the ITC.

Defects in fabrication of trusses were being monitored country-wide by SABS inspectors using a 16-point checklist during this time. No serious failures in fabrication were reported for the fabricators in this region. This has been taken as confirmation that, although fabrication is far from satisfactory, the major quality problem does not lie in fabrication. The present focus of the quest for quality needs to be in the areas of site documentation and erection, which are interrelated.

7 PROBLEM AREAS

The major area of non-conformance of timber roof structures is clearly erection, where fewer than 10% of completed structures are satisfactory. The aspects of erection most frequently found to be deficient, are:

- Bracing   73%
- Cleats, brackets etc.   67%
- Support conditions (up or down)   40%
- Multiple members connected   40%
- Plumb and straight   27%
- Web bracing   27%
- Truss positioning   20%

All these aspects of erection can be addressed by appropriate training of erection personnel. In addition, despite the fact that a very large percentage of roofs are not erected correctly even when the site documentation is up to standard, the poor quality of site documentation needs to be addressed, in view of its carry-over effect on the quality of erection.

Training of design personnel employed by truss fabricators would undoubtedly reduce the level of design errors. However, the far greater problem of poor/nonexistent site documentation will not improve without appropriate motivation, which can be provided by the following catalysts:

- Local authorities by enforcing NBR A19, calling on the owner to appoint a "competent person", to accept responsibility for both the design and construction supervision, which can only be assured where proper design outputs are provided,
- The NHBRC by, inter alia, blocking payment to builders when a responsible person does not certify that the roof conforms to the designer's specification, thus obliging the fabricator to issue an appropriate specification,
- Building contractors who are quality conscious for their own sake (or for any of the above reasons), insisting on a proper set of drawings from his supplier (the fabricator),
- Owners demanding a certificate of conformance by a "competent person" for the completed roof structure from the builder for any of the above reasons,
- Mortgagors demanding such a certificate for any of the above reasons, withholding final payment until it is supplied by either builder or owner,
- Home insurers demanding proof that a roof conformed to standard at the time of its original construction before honoring subsequent claims for damage to the building,
- The ITC, by withdrawing certification of errant fabricators,
- Engineering systems, by enforcing their existing licensing agreements with fabricators.

The fact that the problem of poor site documentation persists, despite technological advances in software and communications that make it possible to ensure superior documentation with minimal effort, casts serious doubt on the commitment to quality in the home construction industry in general, and the timber roof structure industry in particular.

However, even when proper site documentation exists, conformance of the completed roof structure will only result if there is adequate motivation for the erector to ensure this. Such motivation can be provided by the same parties outlined above. Ultimately the catalyst will be the need to produce a certificate that the roof conforms to standard.

Experience in the eastern Cape region has shown that, where roof inspection and certification are not anticipated there is very little likelihood that adequate erection documentation will reach the site, and no serious attempt to ensure conformance by roof erectors.

In the local authority of Jeffreys Bay the municipal officials have consistently insisted on conformance with the NBR and demanded certification by a responsible person of every timber roof structure since March 1998. Despite a high level of resistance from owners and builders, all roofs since that date have been certified to conform to the regulations by a competent person, at an average cost of approximately $50 (typically less than 0,2% of the construction cost).

By contrast, the much larger neighboring Port Elizabeth local authority has not insisted on conformance with the NBR because of political and other pressures outlined above. The resulting level of conformance of completed roof structures is consequently very low. Roofs are only inspected and certified when at least one of the catalysts identified above, is present. In these cases the major problem facing roof inspectors has been getting hold of site documentation in order to check compliance.

8 THE WAY FORWARD

Prefabricated timber roof structures will likely continue to dominate in the home building market.

In order to assure the normal engineering quality of rationally designed structures also for timber roofs, the existing legislative framework of building regulations needs to be enforced. The ITC has put in place a cost effective quality control organization in the form of regional inspectorates, making compliance with regulation A19 affordable and achievable. Existing design software enables truss fabricators to produce comprehensive drawings for erection at minimal cost, and electronic communications make these available virtually anywhere at the cost of (and speed of-) a telephone call.

The NHBRC has the mandate from a democratically elected government to ensure that new housing consumers are protected. The ITC has facilitated the establishment of a mutually accessible database to minimize the paperwork required to appoint a competent person and record his certification of the completed roof, so that the same quality can be assured for the timber roof as is demanded of other rationally designed structural systems. In any case 86% of NHBRC-registered builders already make use of their own structural engineers (NHBRC 1999b).

The major stumbling block remaining is the lack of commitment to quality at every level of the delivery process of timber roof structures. It is always "someone else's problem". Admittedly, the fragmented nature of the process, involving many independent participants, may make quality improvement more tedious.

Crosby (1984) contends that the "producing of defect-free products and services on time is mostly caused by the minds of those who hold the strings of power." Hence the lead must be set by the organs of power: central government, local authorities and the NHBRC. Until these demonstrate that they are serious about implementing quality standards, any further industry led initiatives such as training of erectors - important and necessary as these may be - will not impact significantly on the present unsatisfactory quality of timber roof structures.

REFERENCES

Alman, M. 1989. *Barriers to Quality in the S.A. Building Industry.* MBA Research Paper, The Graduate School of Business, University of Cape Town.
Blackwood-Murray, T.J. Executive Director, Institute of Timber Construction, Johannesburg, 30 June 2000.
Booth, Victor. Chairman, Joint Structural Division of S.A. Institution of Civil Engineering, Bloemfontein, 2 February 1999.
Crosby, P.B. 1984. *Quality without tears.* New York: McGraw-Hill
Department of Public Works 1998. Notice 89 of 1998: To Creating an Enabling Environment for Reconstruction, Growth and Development. *Government Gazette* No. 18615 (14 January 1998).
Hindle, R. D. 1990. Deteriorating Quality Standards of New Housing in the Western Cape. *S.A. Builder/Bouer* September 1990.
NHBRC 1999a. What do you think of the NHBRC? *Builder's Bulletin.*2(1): 7
NHBRC 1999b. *Home Builder's Manual.* National Home Builders Registration Council.
Malherbe, A.C. 1998. Aspects of Conformance to Quality Standards of Timber Roof Structures in South Africa. In Julius Natterer (ed.), *The fifth world conf. on timber eng, Proc. intern. conf., Lausanne, 17-20 August 1998.* Lausanne: Presses Polytechnique et Universitaires Romandes
Randall, Ina 2000. Illegal business ignores city rulings, but...council lacks resources to act. Algoa Sun, 22 June 2000. Port Elizabeth.
SABS 1989. The S.A. Standard Code of Practice for the Application of the National Building Regulations (SABS 0400 – 1987 as amended 1988 and 1989), The South African Bureau of Standards.

Creative Systems in Structural and Construction Engineering, Singh (ed.) © 2001 Balkema, Rotterdam, ISBN 90 5809 161 9

Development of new supporting detail for base columns which ensures superior flexibility to the columns without failure

Hisato Hotta
Department of Architecture and Building Engineering, Tokyo Institute of Technology, Japan

Hirokazu Ikebukuro & Daigo Nomura
Tokyo Institute of Technology, Japan

ABSTRACT: We propose new supporting detail of base columns that have superior flexibility without failure under high axial compression. In this detail, base column and its bed are divided by hemispheric surface. In this paper, we explain the mechanism of the detail and report fundamental properties and restoring force characterristics of the proposed supporting detail obtained through an axial compression test and several shear-bending tests with axial compression. As a result, the detail had sufficient compressive strength under axial compression and the proposed structural system consisted of the detail and the column kept the shear force equivalent to 40% of assumed dead load ($0.3 \, bD\sigma_B$) almost constant under high axial compression ($0.7bD\sigma_B$) in the range of the member angle from -0.06rad to 0.06rad. The proposed system has desirable restoring force characteristics.

1 INTRODUCTION

In a weak beam-strong column concept, ends of base columns are allowed to yield in ultimate situation. In that mechanism, failure of base columns could directly cause a whole building to collapse. Therefore the base columns are required to have sufficient ductility.

As for the base column, authors have proposed several methods to improve flexural ductility of RC columns subjected to relatively high axial compression [1] [2]. In our previous paper [3], from result of dynamic response analysis carried out for several RC frame, we showed that safety margin of columns' ends was far smaller than that of beams' in the present structure subjected to terrific ground motion. We recommended pin-supported flame as a structure in which columns of base story never fail due to bending regardless of magnitude of earthquake. From analytical investigation for its ability, it was showed that the pin-supported frame with a base dummy story could be quite effective structural system which had far more ductile property than that of the ordinary one.

In this paper, we propose new supporting detail of columns that has superior flexibility without failure under high axial compression. In this detail, construction joint with hemispheric surface is intentionally made between columns and beds, as shown in Fig.1. The joint consists of hemisphere made of reinforced concrete and central reinforcing bars which go through the hemispheric joint. Utilizing weakness of construction joint, it is expected that slipping occurs at the hemispheric joint with a certain bending moment less than the bending strength of the column. Around the reinforcing bars inside the hemisphere, cone shaped plastic clay is arranged in order that they do not prevent the joint from slipping.

The construction joint resists external forces, as shown in Fig.1, that consist of bending moment M, shear force Q and axial compression N. When those forces are loaded to the column, slipping occurs at the hemispheric joint and some friction stress and normal stress arise on the hemispheric surface. Under low axial compression or tension, the normal stress developed between the bed and the hemispheric joint mainly works as reaction force against the shear force Q. The central reinforcing bars bear

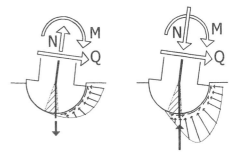

(a)under low axial compression   (b)under high axial compression
      or tension   pression

Fig.1 Mechanism of the detail

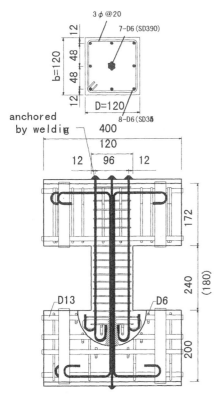

Fig. 2 Dimension and detail of specimens
(unit: mm)

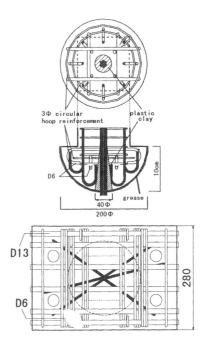

Fig. 3 Detail of the hemispheric part of the column end
and the bed (unit: mm)

the tensile force equivalent to the sum of the axial force N and vertical component of the friction and the normal stress perpendicular to the hemispheric surface. The friction developed between the surface of the bed and the hemispheric joint resists the bending moment M. Under high axial compression, the friction developed between the surface of the joint resists the bending moment M and the normal stress works as the reaction force against the shear force Q as similar to the above situation, however, the central reinforcing bars share the axial compression N with the hemispheric surface. In the following chapter, we carried out axial compression test in order to make sure that the joint has sufficient compressive strength.

2 BEHAVIOR OF SPECIMEN UNDER AXIAL COMPRESSION

2.1 Specimen and testing procedure

A specimen detailed in Fig.2 and listed in Table1 was made and tested. The specimen named C (for Compression test) had a cross section of 160 × 160mm and a length of 240mm. As longitudinal reinforcement, eight D6 deformed bars (SD345) evenly

distributed along all edges of the column section and seven D6 bars (SD390) were arranged in the center of the cross section(pg=3.33%).

As shown in Fig.3, the former eight bars were anchored inside the hemisphere. The latter central reinforcing bars went through the hemispheric joint to make the column not to come out off the bed in tension. Because the lower lateral stiffness of the central bars is more comfortable for the joint, we arranged seven thin bars as the central reinforcement instead of using one thick bar. In the seven bars, only one bar at the center was anchored to the end plate by welding and the others were anchored inside the bed. As written in the previous chapter, we put a circular cone shaped plastic clay around the central reinforcing bars inside the hemisphere. The surface of the joint was covered with thin coat of grease.

As shown in the upper figure in Fig.3, the hemispheric part of the column end was strengthened by crossed D6 bars and two circular hoop reinforcements. As for the bed, which supports the hemisphere, we arranged D13 bars and D6 bars as illustrated in the lower figurer in Fig.3, and laterally strengthened the bed to avoid splitting tensile failure of the bed due to high axial compression.

We carried out centrally loaded axial compression test in order to make sure that the joint has sufficient compressive strength. In the test, we used Amsler type universal testing machine.

Table 1 List of specimens.

Specimen		CBS	C	CBS-C1	CBS-C2	CBS-V
bθ Dθ h (mm)		120x120x180	120x120x240			
Compressive Strength of Concrete σB (MPa)	bed of column	305	375			
	column	297	349			312
Hoop Reinforcing Bars		Q-8φ @Q0 (0.59%)				

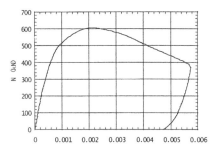

Fig.4 Axial compression - Axial displacement relationship

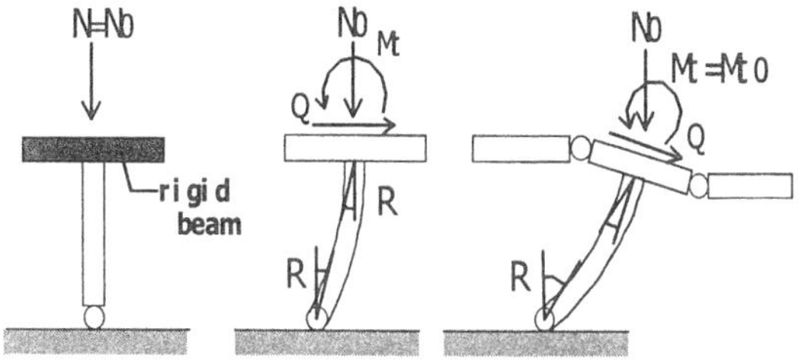

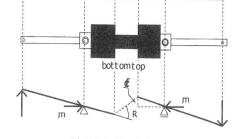

Fig.5 Ultimate state of pin supported column and rigid beam

2.2 Test results

The relationship between the axial compression and the axial displacement is shown in Fig.4. Compressive strength of the specimen was 607kN which corresponded to 95.6(%) of theoretical compressive strength and failure of the specimen was caused by the yielding of the column. As for the joint of the specimen, we could not observe the joint was damaged at all.

3 BEHAVIOR OF SPECIMEN IN SHEAR BEMDING TEST WITH AXIAL COMPRESSION

3.1 Specimen and testing procedure

A specimen with the same dimensions as the one except for the length of the column discussed in the previous chapter was made and tested. In this experiment, the specimen named CBS (for Compression, Bending-Shear test) had a length of 180mm as shown in Table 1. We carried out cyclic shear-bending test with axial compression. Loading program was decided assuming a simplified structure consisted of a pin-supported column, that is the specimen, and rigid-plastic beams as illustrated in Fig.5. The beams are assumed to yield and develop plastic hinges at the ends with a certain bending moment which is equivalent to Mto as the moment at the top end of the column. Hence, in the testing procedure, at first, we increased shear force controlling both the end rotations of the column to be equal until the moment at the top end of the column Mt reaches Mto. And then, we increased the member angle R to R=0.06rad keeping the moment at the top end of the column Mto constant. After that, we decreased the moment Mt from Mt=Mto to Mt=-Mto

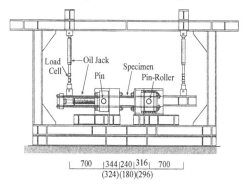

Fig.6 P-Δeffect (unit: mm)

Fig.7 Loading set up (unit: mm)

controlling decrement of both the end rotations to be equal. Then we reduced the member angle R to R= -0.06rad in the same procedure of increasing the member angle R. In those loading processes, the axial compression was kept constant. We continued two loops of cyclic loading in the same procedure, however, changed the moment Mto and the axial compression in each loop.

The outline of the loading set up is illustrated in Fig.6 and Fig.7.

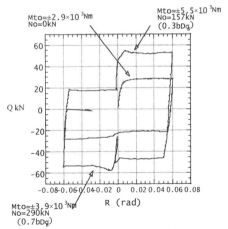

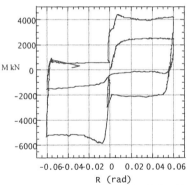

Fig.8 Shear force-member angle relationship

Fig.9 Moment -member angle relationship

When the member angle R is increased, we must give consideration to the moment added to both the ends of the column caused by the axial compression so called P-Δ effect, because, as shown in Fig.7, the length between the bottom end of the testing region and the pin support of the loading set up and the length between the top end of the testing region and the pin-roller support of the loading set up are long in comparison with the length of the testing region and central axis of the testing region deviates from axis of the axial compression according to the deflection increases. During a phase of loading cycle in which we kept the moment at the top end of the column Mto constant, we calculated the additional moment due to P-Δeffect to both the ends of the column from both the end rotations and controlled the moment at the top end of the column Mto in each loading step.

3.2 Test results

The relationships between the shear force and the member angle R is shown in Fig.8, and the moment at the bottom end of the column in Fig.9.

In each loading cycle, we controlled the axial compression No and the constant moment of the top end of the column Mto as follows.

In first loading cycle, the axial compression No=0tf and the moment Mto=$\pm2.9\times10^3$Nm, and in former part of second loading cycle, the axial compression No=130kN($0.3bD\sigma_B$) and the moment Mto=$\pm5.9\times10^3$Nm, in latter part of the second loading cycle, the axial compression No=290kN ($0.7bD\sigma_B$) and the moment Mto=$\pm3.9\times10^3$Nm. The axial compression No=0kN and No=290kN mean varying axial compression for exterior columns and No=130kN for interior columns. We determined the constant moment Mto as high as possible within the range that the top end of the column was not

yielded due to bending and the column was not failed due to shear force when the member angle R was increased. So we considered bending strength and shear strength of the column, then decided the constant moment Mto for the interior columns on Mto=$\pm5.9\times10^3$Nm and that for the exterior columns on Mto=$\pm2.9\times10^3$Nm or Mto=$\pm3.9\times10^3$Nm as about half as that for the interior columns.

As shown in Fig.8, the specimen kept almost constant shear force regardless of the member angle in each loop. Fig.9 indicates that the joint of the specimen showed constant moment while the specimen kept constant shear force.

Diagonal tension crack occurred on central surface of the testing region during second loading cycle (axial compression No=290kN), but neither cracks which imply yield of the reinforcements nor failure of the concrete were observed in the column. As for the joint, concrete was peeled off from edges of upper surface of the joint after the test. But the specimen showed superior flexibility under high axial compression, and the joint did not seem to be damaged seriously.

In this shear-bending test with axial compression, the specimen kept the shear force equivalent to 40% of assumed dead load($0.3 bD\sigma_B$) almost constant under high axial compression($0.7bD\sigma_B$) in the range of the member angle applied in the test. In the following chapter, we report the restoring force characteristics of the joint of the specimen.

4 RESTORING FORCE CHARACTERISTICS OF THE SPECIMEN

4.1 Specimens and testing procedure

Three specimens with the same dimensions as the one discussed in the chapter 2 named CBS-C1, CBS-C2 (for Bending-Shear test with Constant axial Compression) and CBS-V (for Bending-Shear test with Varying axial Compression) were made and tested. The compressive strength of the concrete ranged about from 31.2 to 34.9MPa as shown in Table 1.

To investigate the restoring force characteristics, cyclic shear-bending tests with axial compression were carried out. Loading procedures of those experiments were much about the same as the test described in the previous chapter.

In the tests of the specimens CBS-C1 and CBS-C2 assumed to be interior columns, we kept the axial compression constant and controlled the member angle R as illustrated in Fig.10. The axial compression No and the constant moment at the top end of the column Mto were controlled to be constant in all loading cycles. Applied axial compression No was equivalent to $0.3bD\sigma_B$ and the constant moment Mto decided by the same reason in the previous chapter was Mto $=\pm7.8\times10^3$Nm in the tests of each specimens.

In the test of the specimen CBS-V, we controlled the axial compression in proportion to the moment at the top end of the column Mt in order to represent the state of exterior columns. We determined the constant moment Mto on Mto=$\pm3.9\times10^3$Nm and varied the axial compression No from No=10kN to No=300kN linearly dependant on the moment at the column top Mt as shown in Fig.11 in order to represent varying axial compression. When the constant moment Mt was at 0Nm, the axial compression was 160kN equivalent to the dead load $0.3bD\sigma_B$ as shown in Fig.11.

4.2 Test results

The relationship between the shear force and the member angle of three specimens are shown in Fig.12.

As for the results of the specimens CBS-C1 and CBS-C2 subjected to the constant axial compression, during the loading procedure in which we kept the constant moment Mto and increased the member angle R, the restoring force sloped gradually up to the peak of the shear force in each loading loop of ±0.02rad, ±0.04rad and 0.06rad in a virgin loop, however, the peaks of the shear force became lower than that of the previous loop. And then, as shown in Fig.12 (b), the peak of the shear force converged on a certain shear force after three or four times repetition of the cyclic loading. When the member angle was increased bigger than once experienced, the shear force was enlarged again with the increase of the member angle.

As for the specimen CBS-V, during the regions in which the specimen was subjected to the axial compression No=0.10kN and the constant moment Mto=$\square3.9\times10^3$Nm, the specimen followed the same rule as the specimens CBS-C1 and CBS-C2. Hence, the restoring force also sloped gradually up to the peak shear force in each loading loop of

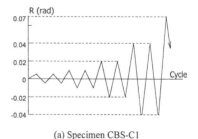

(a) Specimen CBS-C1

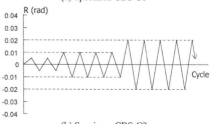

(b) Specimen CBS-C2

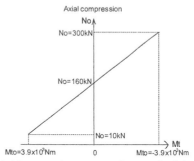

(c) Specimen CBS-V

Fig.10 Loading schedule

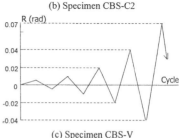

Fig.11 Axial compression-moment at the top end relationship

-0.01rad, -0.02rad and -0.04rad during the procedure, and the peak shear force of each loading loop also became lower than that of the previous loop. On the other hand, during the regions in which the specimen was subjected to the axial compression No=300kN($0.7bD\sigma_B$) and the constant moment Mto=3.9×10^3Nm, shapes of the hysteresis loops of the segment were not similar to those of the specimens CBS-C1 and CBS-C2, as shown in Fig.12 (c). In loading cycle of 0.01rad and 0.02rad, in the region that the member angle was increased bigger than once experienced, the shear force was enlarged

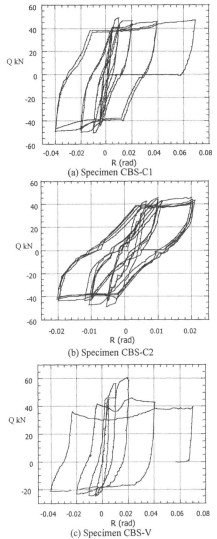

(a) Specimen CBS-C1

(b) Specimen CBS-C2

(c) Specimen CBS-V

Fig.12 Shear force-member angle relationships

with the increment of the member angle, however in loading cycle of 0.04rad and 0.06rad, the peaks of the shear force became lower than that of the previous loading loop and decrement of peak shear force was much larger than the specimens CBS-C1 and CBS-C2. Consequently, the restoring force characteristics of the proposed supporting detail under varying axial compression, although we could know that of virgin loading loop only, were similar to those of the detail under constant axial compression except that the variation of the peak shear force was much larger than the specimens under constant axial compression. From this experiment, it is expected that the restoring force characteristics of the detail depend on the axial compression rather than the

member angle. The reason is considered that the friction developed between the bed and hemispheric joint due to axial compression influences the shear force.

5 CONCLUSIONS

Through the axial compression test and shear-bending tests, the following knowledge is obtained.
1) The proposed supporting detail has sufficient compressive strength and the failure of the proposed structural system was caused by the yielding of the column under axial compression.
2) In the shear-bending test with axial compression, the specimen consisted of the joint and the column kept the shear force equivalent to 40% of assumed dead load($0.3 bD\sigma_B$) almost constant under high axial compression($0.7bD\sigma_B$) in the range of the member angle from -0.06rad to 0.06rad.
3) As for the restoring force characteristics under constant axial compression, the shear force increases gradually toward the peak with the increase of the member angle during the loading procedure in which we keep the constant moment Mto and increase the member angle. The increment of the shear force becomes smaller in the range of the member angle once experienced, and the peak shear force of each loading loop becomes lower than that of the previous loop. After three or four times repetition of the cyclic loading, the peak shear force of the hysteresis loop converges on a certain shear force. When the member angle is increased bigger than once experienced, the shear force is enlarged again with increase of the member angle.
4) The restoring force characteristics of the detail under varying axial compression are similar to those of the detail under constant axial compression except that the variation of the shear force is much larger than under constant the axial compression.

REFERENCES

Hotta, H., & Takiguchi K., 1997. "Proposal of a New Method to Improve Flexural Ductility for Reinforced Concrete Column Subject to High Axial Load□ J. Struct. Constr. Eng., AIJ, Japan, 495,pp.115-119
Hotta, H., Wakimoto K., and Takiguti, K.,1999 "Effectiveness of New Method to Improve Flexural Ductility of RC Columns Using Reinfoceing Bars Cut Near Critical Section.□ J. Struct. Constr. Eng., AIJ, Japan, 523, pp.111-116
Hotta, H., Kusunoki, J., & Kurosaka, A.,1998 "Dynamic Analysis of RC Strusture Paying Atenntion to the Hysteresis Energy of the Bottom Ends Hinges of Base Columns□ The 6th East Asia-Pcific Conferance on Struct. Eng. & Const., Taipei:Taiwan

Creative Systems in Structural and Construction Engineering, Singh (ed.) © 2001 Balkema, Rotterdam, ISBN 90 5809 161 9

Design of a PC box girder bridge with corrugated steel webs in Hanshin Expressway

A. Nanjo, H. Iguchi & H. Kobayashi – *First Kobe Construction Department, Hanshin Expressway Public Corporation, Kobe, Japan*

J. D. Zhang – *Civil Engineering Department, P.S. Corporation, Osaka, Japan*

A. Shoji – *Technical Division, Oriental Construction Company Limited, Tokyo, Japan*

K. Kobayashi & A. Kurita – *Department of Civil Engineering, Osaka Institute of Technology, Japan*

ABSTRACT: Nakano Viaduct, which is located in the North Kobe Line of Hanshin Expressway, has been planning as a prestressed concrete box girder bridge with corrugated steel webs. The structural scheme is expected to reduce the self-weight and to shorten the construction period comparing with conventional PC box girders. The following technical characteristics, which have been seldomly applied in the similar formed bridges, are presented: 1. Using field fillet welding joints between corrugated steel webs; 2. Partial prestressing on concrete slab in which the tensile stress is controlled within an allowable range and no cracks would occur; 3. Using CT shaped steel plate combining with headed studs as shear connection between concrete slab and corrugated steel webs.

Since the above mentioned points were adopted in this bridge without any precedents in Japan, three series of large-scale loading tests were carried out for the purpose of identifying its safety and structural characteristics. In this paper, the proprieties of design methods for Nakano Viaduct are discussed based on the test results.

1 INTRODUCTION

Nakano Viaduct, which is located approximately in the middle of Arimaguchi JCT and Nishinomiya Yamaguchi JCT, is being under construction at present using cantilever erection method and whole staging method. The four main-line bridges (east & west line) and two rampway bridges are all designed as the four-span continuous PC box girder bridge with corrugated steel plate webs (shortened by corrugated web bridge). Figure 1 shows the general view of the viaduct.

The corrugated web girder is a steel-concrete composite structure by using corrugated steel in stead of concrete web in PC box girder whose the concept was first developed in France in 1980's. The

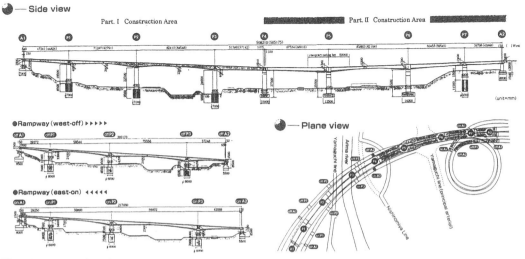

Figure 1. General View of Nakano Viaduct

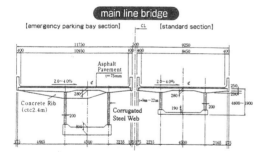

main line bridge

[emergency parking bay section] CL [standard section]

Asphalt
Pavement
t=75mm

Concrete Rib
(ctc2.4m)

Corrugated
Steel Web

Figure 2. Section of Main Girder

advantages are considered of self-weight reduction and efficient prestressing on concrete slab members and so on. So far in Japan, three same-formed bridges had been constructed and numbers of bridges are being under construction or planning by adopting the same structural type.

2 FEATURES OF NAKANO VIADUCT

Comparing with the existing corrugated web bridges, the following technical characteristics can be pointed out in Nakano Viaduct.

2.1 Curved Bridge

Nakano Viaduct is constructed as the first curved corrugated web bridge with the radius of 440m for main-line bridges, and the minimum radius of 250m for rampway bridges. Since the warping stress getting bigger due to the less torsional rigidity of corrugated web bridge comparing with conventional PC box girders (Yoda, 1993), more diaphragms were arranged to reduce warping stress based on the results of FEM analysis. Also, the field measurements are planning to verify the design stress during the construction works.

2.2 Using of Weathering Steel Plate

It is necessary to pursue the rationality of maintenance management after the construction work. Since the bridge is situated in a favorable environment, the weathering steels that have no need to the painting in the future are adopted for reducing the whole life cycle cost of the bridge.

2.3 Field Fillet Welding Joint

The friction grip joint using high strength bolts between corrugated steel webs have been adopting in Japan. In France, however, for the consideration of absorption of construction error, the field fillet welding joint have been used. For the purpose of construction rationalization, the fillet welding joint is used in the part of whole staging construction works.

Since the fillet welding joint method has never been practiced so far in Japan, large-scale loading tests were carried out in order to confirm the joint strength and the stress distribution near the joint.

2.4 New Steel-Concrete Connection

One of the typical points of Nakano Viaduct is the adopting of a new connection using CT shaped steel plates combining with headed studs(shortened by Nakano method) as shear connection between steel webs and upper concrete slab. So far in Japan, the flange-stud connection which is usually used in steel bridges, or the embedded connection was adopted in the existing bridges. Because of using weathering steel webs, flange method is adopted as the basic connection style to make form setting easier to protect from concrete mortar dropping which causes steel web stains. In this bridge, the adopted connection detail is shown in Figure 3 using CT shaped steel plates as perfobond strip to reduce the number of studs. Perfobond strip connection which has been proposed by Leonhardt et al.(1987) is the method of using pillar-shaped concrete which have been filled into the steel plate hole as the function of dowels to resist the shear flow. For the purpose of safety confirmation, push-out tests on shear connection were carried out.

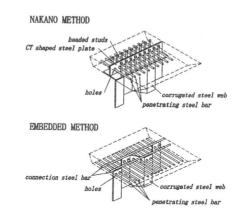

Figure 3. Connection Methods between Steel Web and Concrete Slab

2.5 Partial Prestressed Concrete Slab in Transverse Direction

It is common to design upper concrete slab perpendicular to the bridge axis as full-prestressed concrete structure in PC box girders in Hanshin Expressway Public Corporation. However, under the considering of cost-reduction, partial prestressing (shortened by PPC) design started to be adopted on concrete slab where the tensile stress is controlled within an allowable range in which no cracks occur. The quantity of transverse prestressing tendons in concrete slab is reduced by about 20% in this bridge. Due to the first time of using PPC slab, a large-scale loading test on concrete slab was carried out in order to confirm the structural characteristics.

3 EXPERIMENTAL WORKS

3.1 *Mechanical Behavior of Field Fillet Welding Joint (Series I Test)*

3.1.1 *Specimens*
Series I is the basic test for the above mentioned field fillet welding joints. The test aimed on:
1) understanding of stress distribution of corrugated steel web.
2) confirmation of loading carrying capacity of field fillet welding joints.

The specimens with four types of joints are presented in Table 1. Figure 4 shows the details of the specimen. Specimen CF with field fillet welding joint that is used for the actual bridge by different lap length of 80mm and 160mm. For specimen CN with lap splice joint only without welding at the right side of the loading point, and steel web plate continuous without joints at the left side. Due to the limits of loading equipment, 1/2 scaled specimens were used.

Table 1. Details of Specimens for Series I

	Specimen CF		Specimen CN	
	right side	left side	right side	left side
	type1	type2	type3	type4
Joint type	Fillet welding joint		lap splice	none joint
Lap length	80mm	160mm	80mm	- -

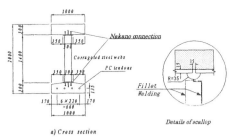

a) Cross section

Details of scallop

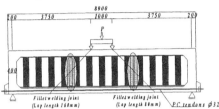

b) Side view of CF specimen

Figure 4. Specimen CF of Series I

Under the considering of welding efficiency, corrugated steel plate webs are chosen with the same shape and same thickness as those used in the actual bridge. Also, the dimension of concrete slab and the amount of prestressing tendons are adjusted to meet the actual bending stress.

3.1.2 *Outline of Measurement*
Since the stress distribution near the flange has not been clarified in the past experiments, three-axis strain gauges were used in order to analyze the stress distribution near field fillet welding joint between corrugated webs, and strain gauges were closely set out around the scallop of steel web near concrete slab. Also, vertical displacement meters were arranged at loading points, supported points and joints.

3.1.3 *Loading Method*
Figure 4 also shows the loading method. The loading hysteresis is presented following:
1) twice cyclic loading was carried out until flexural
   cracking under the loading control mode.
2) twice cyclic loading was carried out up to the yield point of tensile reinforcements.
3) then switch to displacement control mode and keep on loading to the limits (5000KN) of loading capacity.

3.1.4 *Test Results*
Even the surcharge load exceeded the calculated ultimate load, no failure could be observed on specimen CF except some flexural cracks. For specimen CN, shear failure occurred on the side of lap splice joint by the time surcharge load reached 3400KN which is 1.3 times as much as design load. Therefore, the strength of fillet welding joints on Nakano Viaduct is found to be sufficient, and there would no failure occur under design load even the welding parts are damaged in case.

Figure 5 shows the distribution of shear strain on steel webs joint. Specimen CF with field fillet welding joints behaves in the similar way as specimen CN of none joint type before reaching the design load (Pa). The stress concentration occurred near concrete slab, further studies are needed to clarify its influence on fatigue design.

The difference of changing lap length did not affect the stress behavior. Based on the test results, it was found that concrete slab can share a part of shear force (20% in this test). Therefore, the present design method based on the assumption that webs carry all the shear force is considered as a safety method.

For specimen CN with lap splice joint, the shear strain distribution is unequal along the height of lap splice joint, because steel webs without fillet welding are unable to transmit fully the shear force.

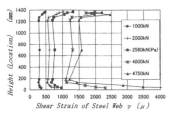
(a) Specimen CF with lap length 80mm

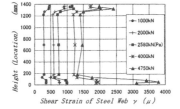
(b) Specimen CF with lap length 160mm

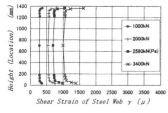
(c) Specimen CN without joint

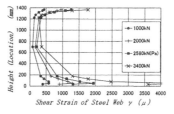
(d) Specimen CN with lap length 80mm

Figure 5. Distribution of Shear Strain on the Corrugated Steel Web Joints

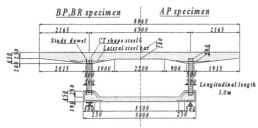

Figure 6. Specimens for Series II

3.2 Loading Tests on Concrete Slab in Transverse Direction (Series II Test)

3.2.1 Specimens

The purpose of series II is to confirm the propriety of adopting PPC slab and Nakano method connection, which are mentioned above. Special attentions were paid for comparing Nakano connection with embedded connection. Figure 6 shows the three types of specimens, and Table 2 presents the details.

For specimen AP and BP as PPC structure, tensile stress of concrete is allowable to -1.5MPa under design load. While, for specimen BR as PRC structure, -4.0MPa tensile stress is allowable by decreased prestressing force, and the tensile stress is decided based on controlling the width of crack at serviceability limit state.

3.2.2 Outline of Measurement

Since the main purpose is to confirm the stress and deformation performance, measuring items focus on:
1) vertical displacement at loading points and supported points;
2) strain in steel bar as well as prestressing tendons;
3) cracking width and strain in upper concrete slab;
4) strain distribution on corrugated steel web;
Meanwhile, the crack developing on concrete was checked out during the whole loading stage.

3.2.3 Loading Method

Test was carried out according to loading methods from A to D as shown in Figure 7 to generate the allowable tensile stress.
1) For loading method A and B, twice cyclic loading to generate tensile stress of -1.5MPa or -4.0MPa on upper side at supports or lower side in mid-span.
2) For loading method C, twice cyclic loading to generate tensile stress of 1.5 times of -1.5MPa or -4.0MPa both on upper side at supports and lower side in mid-span.
3) For loading method D, loading on supported slab edges till the slab-web connection collapsed. The center of slab in mid-span was fixed in order to avoid the premature breaking before connector collapsing.

3.2.4 Test Results

All the specimens were loaded till vertical displacements reached 150mm. Figure 8 shows the relation between load and vertical displacement at slab edges.

By studying hysteresis loops of specimen AP and BP, the strength and deformation behaviors of Nakano method is nearly same as embedded method. Moreover, as opposed to specimen AP in which cracks developed right up in the connection, cracks on specimen BP (Nakano method) dispersed well, and then CT shaped steel plate is considered effective for distributing cracks.

584

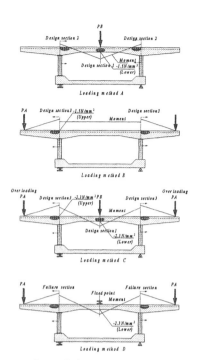

Figure 7. Loading Method

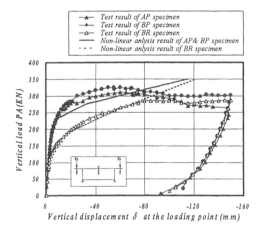

Figure 8. Relation between Load and Displacement

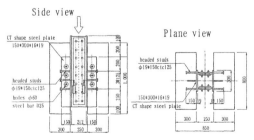

Figure 9. Specimen PBS of Series III

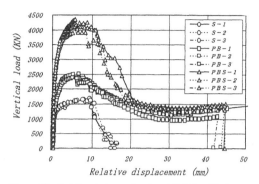

Figure 10. Relation between Load and Relative Displacement

As for two type of specimen BP and BR with Nakano connection, although the hysteresis loops showing the same maximum load, the displacement of less prestressed specimen BR was larger during initial stage, and so as the crack width.

The test results of all the specimens agree approximately with the calculated values by non-linear analysis as shown in figure 8.

3.3 Push-out Tests of Shear Connection (Series III Test)

3.3.1 Specimens

The purpose of series III test is to examine the slip-resisting effects by using perfobond strip together with headed studs as shear connection. Table 3 and Figure 9 show the specimen details. For specimen S, 12 headed studs(φ19) were welded on the flange plate. For specimen PB, CT shaped steel plate with punched holes (φ60) was used as shear connection designated as perfobond strip.

Table 2. Details of Specimens for Series II

Speci-en	Slab Structure	Allowable tensile stress	Connection
AP	PPC	-1.5MPa	Embedded method
BP	PPC	-1.5MPa	Nakano method
BR	PRC	-4.0MPa	Nakano method

For specimen S with headed studs only, the shear strength was calculated based on Japanese Specification for Highway Bridge II(1996). For specimen PB, ultimate strength was calculated by adopting the reduction factor of 0.7(Ebina,1998) for the equation which was proposed by Wayne (1994), because the diameter of holes as concrete dowel is φ60mm. The equation is shown as following. Also, the design strength is provided as 1/3 of ultimate strength.

585

Photo 1. Nakano Viaduct under Construction

$$Q_{pu} = 2 \, (\pi\varphi^2) \, / \, 4 \times 1.14\sigma ck \times 0.7 \qquad (1)$$

where φ = diameter of holes punched in steel plate; and σck = specified concrete compressive strength.

3.3.2 Outline of Measurement

Applied load was measured by load cell that was set out between surcharge jack and specimens. The slip between concrete and steel flange was measured by displacement meters. Meanwhile, the crack generating on concrete block was checked out during the loading stage.

3.3.3 Loading Method

Cyclic loading test was carried out by 1/2, 1, 2, 3 times of design shear force each time, and the surcharge load was kept on up to connection failure.

3.3.4 Test Results

Figure 10 shows the relation between load and slip-displacements of all specimens. The sufficiency of safety was confirmed considering a strength reduction factor 0.7, since the maximum strength is about 1.7 times as much as ultimate strength which was calculated based on the above Equation(1). The strength of specimen PBS is just the addition of specimen S and PB. The test results indicate explicitly the propriety of design method for actual bridge.

Table 3. Details of Specimens for Series III

Specimen	Number	Shear Connection
S	3	Headed studs only
PB	3	Perfobond strip only
PBS	3	Perfobond strip together with headed studs (Nakano method)

4 CONCLUSIONS

The following conclusions have been revealed through the three series loading tests, and the proprieties of design methods for Nakano Viaduct

were confirmed based on the test results.
1) The field fillet welding joint between corrugated steel webs can transfer shear force efficiently.
2) The stress concentration near flange and scallops of corrugated steel webs occurred. Based on these results, further studies are needed for its influence on fatigue design.
3) The efficiency of Nakano connection method is almost the same as Embedded connection between slab and steel webs, and also Nakano connection acts well for the combination with partial prestressed concrete slabs.
4) The shear transfer strength of connection using perfobond strip together with headed studs can simply be calculated by adding their individual strengths, and the design equation was verified based on the test results.

Finally, the present situation of Nakano Viaduct is shown in photo 1. The bridge is expecting to be completed in June 2001.

REFERENCES

Ebina,T., Takahashi,K., Uehira,K. & Yagishita,F. 1998. Basic study for shear capacity of perfobond strip. *Proceedings of the 8th Symposium on Developments in Prestressed Concrete 1998*: 31-36.

Japan Road Association 1996. Specification for Highway Bridge Part II, (steel bridge).

Leonhardt,F., Andra,W., Andra,H.P. & Wharre,W.1987. New improved shear connector with high fatigue strength for composite structures. *Beton-und Stahlbetonbau 1987*: 325-331.

Yoda,T. & Oura,T.1993. Torsional behavior of composite PC box girders with corrugated steel webs. *Structures Engineering Journal Vol.39A. 1993*: 1251-1258.

Wayne,S.Roberts & Robert,J.Heywood. 1994. An innovation to increase the competitiveness of short span steel concrete bridges, *Developments in Short and Medium Span Bridge Engineering'94* : 1161-1166.

Creative Systems in Structural and Construction Engineering, Singh (ed.)© 2001 Balkema, Rotterdam, ISBN 90 5809 161 9

The effect of a construction defect on the strength and behaviour of masonry structures

F.A.Santos & B.P.Sinha
Department of Civil and Environmental Engineering, University of Edinburgh, UK

H.R.Roman
Universidade Federal de Santa Catarina, Brazil

ABSTRACT: The effect of workmanship on brickwork compressive strength is well documented. It has been shown that the effect of all workmanship factors combined will reduce the strength in compression up to 50%. The effect of not filling vertical joints on the strength and behaviour of masonry structures built with hollow clay blocks is not known although the practice is common in some developing countries. This paper describes a series of experiments to study the effect of unfilled vertical mortar joints on the compressive strength, the flexural tensile strengths in two orthotropic directions, and the shear strength of masonry elements. To study the behaviour of structures under shear, one-third scale-model masonry structures with filled and unfilled vertical mortar joints were built and tested under combined compression and shear. Results show that the compressive and the flexural strengths in two orthogonal directions are substantially lower compared to filled vertical mortar joints. The deflection of the structure is up to 50% higher than structures built with filled vertical mortar joints under lateral loading. The ultimate shear strength is also lower compared to the structures built with filled vertical mortar joints.

1 INTRODUCTION

Masonry load-bearing walls have been/are being used to construct high-rise apartment blocks in Brazil. A peculiar practice of not filling the vertical mortar joints has developed in this country. This workmanship defect may have profound effect on the strength and behaviour of the structure.

In a load-bearing structure, walls primarily carry vertical loading, but frequently have to resist lateral loading due to wind. Therefore, the strength of masonry in compression, flexure and shear becomes an important consideration in the design. It was, therefore, decided to assess the effect of this defect on the strength of walls built with hollow blocks.

2 EXPERIMENTAL PROGRAMME

One-third scale extruded wire-cut hollow clay blocks were used for the tests. The average compressive strength (BS 3921 1974) was 29.2 N/mm² (net area).

A cement:lime:sand mortar (1:1:6 by volume) was used for all test structures and specimens. Three mortar cubes (100 mm x 100 mm x 100 mm) were tested on the same day as the corresponding test walls or structures. The average strength of mortar was 5.20 N/mm² with a coefficient of variation of 2%.

All materials conformed to the relevant British Standards (BS 890 1972, BS 3921 1974, BS 1200 1976, BS 5628 1992, BS 12 1995).

2.1 Compressive strength and modulus of elasticity

Five-course high prisms (Fig. 1), 3 with filled vertical mortar joints and 3 without filled vertical mortar joints were tested under axial compression.

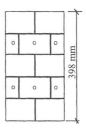

Figure 1. Prisms with *demec* points.

In addition, four-course high prisms, 3 bricks long capped with gypsum were tested parallel to the bed joints under compressive loading to determine the modulus of elasticity (Ex).

At the beginning of each test, care was taken to apply the load axially by monitoring the strains on both faces of the test specimen. The axial strains at 3 points on each face were measured with *demec* gauges. The load was applied at stages until failure. The stress strain relationship was established to determine the modulus of elasticity in two orthogonal directions.

2.2 *Flexural tensile strength*

2.2.1 *Flexural tensile strength normal to bed joints*
Wallettes, two-brick wide and nine-course high, were tested as simply supported beams (Fig. 2). The load was applied at 1/3 points simply by filling a hanging bucket with sand.

Figure 4. Flexural tensile test of a 4-course specimen with unfilled vertical mortar joints.

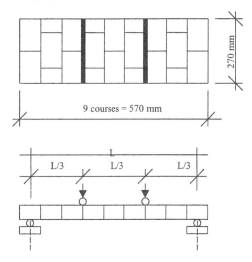

Figure 2. Test arrangement.

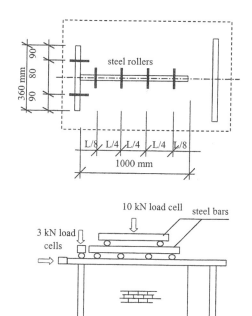

Figure 5. Test arrangement for shear.

The test apparatus is presented in Figure 3.

2.2.2 *Flexural tensile strength parallel to bed joints*
Five four-course wallette specimens were tested with filled and unfilled vertical mortar joints.

The testing apparatus is as shown in Figure 4. Load was applied in small increments by means of a hydraulic jack using a 5000 N load cell. The loading arrangement was similar to the previous case shown in Figure 2.

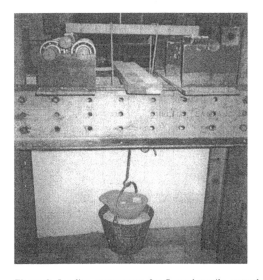

Figure 3. Loading arrangement for flexural tensile strength normal to bed joint.

2.3 Behaviour and strength of structures in shear

Six 1/3-scale masonry structures were tested under combined compression and shear (Fig. 5). To compare the behaviour, half of the structures were built with filled vertical mortar joints and half with unfilled mortar joints. Compressive stresses varied from 0.4 to 2.1 N/mm² (net area).

For all test structures with and without filled vertical mortar joints, the walls were built on a strong floor of a laboratory by an expert bricklayer. Vertical and horizontal mortar joints in the flange of each structure were filled completely.

The flange of the shear wall was grouted with 1:3 cement:sand mortar by volume. The pre-cast concrete slab was joined to walls using an epoxy mortar to avoid premature failure of the interface between wall and slab.

2.3.1 Test arrangement

Two independent frames were used to apply lateral and vertical loads. Vertical load was applied at four points along the length of the shear wall and at two points on the flange. Lateral load was applied at the slab level.

Deflection was measured along the height of the shear walls towards their loaded and unloaded sides using dial gauges reading to the accuracy of 0.002 mm. One dial gauge was also located at the level of the slab.

Strain was measured at the bottom along the length of the shear wall. Seven *demec* points were located at both faces of the wall to provide the strain distribution due to vertical and lateral loads.

2.3.2 Test procedure

Structures were tested at a minimum age of 21 days. Before each test, one mortar cube was tested to ensure that the desired mortar strength has been attained. The remaining mortar cubes were tested on the same day after testing the structure to failure.

Initial readings, prior to the application of the load, were taken on the both faces of the walls for all *demec* points. The actual strain for a given load was then obtained by subtracting the final reading from the initial reading.

Initially, the vertical load was applied up to the desired level of pre-compression both in the shear wall and the flange. It was applied in small increments, alternating between flange and shear wall.

After the desired level of pre-compression was reached and, prior to application of shear load, the readings from the *demec* gauges were taken in order to obtain the strain distribution in the bottom of flange and shear wall. Initial readings of the dial gauges were also taken after the application of the vertical load.

Shear load was applied to the structure slowly in increments and kept constant while deflection and strain readings were taken. Deflections were recorded for each increment of horizontal load until failure.

In some cases, post-cracking strains were measured. Generally, measurements of strain stopped at 65% of the ultimate shear load, which corresponded to approximately 90% of the load at which cracks were first noticed. All structures were tested to complete failure.

3 RESULTS AND DISCUSSION

3.1 Compressive strength and modulus of elasticity

The moduli of elasticity obtained from the stress-strain curves are given in Table 1.

Table 1. Modulus of elasticity of brickwork in two orthogonal directions for specimens with filled and unfilled vertical mortar joints.

Specimen	Ex parallel to bed joint		Ey normal to bed joint	
	Filled N/mm²	Unfilled N/mm²	Filled	Unfilled
1	4573	1995	4907	3564
2	3548	2180	6340	3421
3	2403	1810	6414	4039
Mean	3503	1995	5888	3675

From Table 1, it can be seen that the initial tangent modulus elasticity with unfilled vertical mortar joints is significantly lower.

In compression, initial cracks developed in the units under the vertical mortar joints propagating in the same direction of the applied load.

The average compressive strength for the prisms was 5.03 N/mm² with filled vertical joints and 3.99 N/mm² for prisms and unfilled vertical joints respectively. This represents a drop of 21% in the strength if the vertical mortar joints are not filled. The drop of strain is significant compared to works (BCRA 1972, JAMES 1993) on solid masonry walls.

3.2 Flexural tensile tests normal and parallel to the bed joints

The flexure tensile strengths in two orthogonal directions of specimens both with filled and unfilled vertical mortar joints are presented in Tables 2 and 3.

3.2.1 Normal to the bed joints

Typical failure occurred at the interface of mortar joints in the area of maximum constant bending. Cracks passed through the bed joints for both specimens with filled and unfilled vertical mortar joints.

Results show a reduction of 58% in flexural strength normal to bed joint for unfilled vertical mortar joints compared to the strength of wallettes with filled vertical mortar joints. The reduction is slightly higher than masonry built with solid bricks (Sinha 1979).

Table 2. Flexural tensile strength of specimens normal to the bed joints with filled and unfilled vertical mortar joints.

Specimen	Filled vertical mortar joints	Unfilled vertical mortar joints
	N/mm²	N/mm²
1	0.26	0.13
2	0.35	0.20
3	0.60	0.16
4	0.39	0.19
5	0.40	0.19
Mean	0.40	0.17

3.2.2 Parallel to the bed joints

Typically, two different modes of failure were observed: cracks passing through the blocks and head joints or through the bed joints only for specimens with filled vertical mortar joints. For specimens with unfilled vertical mortar joints the cracks always occurred through the mortar joints, in a zigzag manner. Failure was sudden in both cases. Table 3 gives the results.

Table 3. Flexural tensile strength of specimens parallel to the bed joints with filled and unfilled vertical mortar joints.

Specimen	Filled vertical mortar joints	Unfilled vertical mortar joints
	N/mm²	N/mm²
1	0.55	0.41
2	0.59	0.44
3	0.75	0.45
4	0.70	0.43
5	0.54	----
Mean	0.63	0.43

From Table 3, a significant reduction in strength is observed with the ultimate stress reducing by 32% if vertical joints are not filled.

3.3 Shear tests

The test results are shown in Table 4 and Figure 6.

3.3.1 Shear strength

The ultimate shear strength of structures with filled and unfilled vertical mortar joints is given in Table 4.

Table 4. Shear strength of walls with filled and unfilled vertical mortar joints.

Pre-compression	Ultimate shear stress	
	Filled vertical mortar joints	Unfilled vertical mortar joints
N/mm²	N/mm²	N/mm²
0.4	0.40	0.39
1.7	0.80	0.71
2.1	1.00	0.98

The shear strength of structures with filled and unfilled vertical mortar joints largely depends on the pre-compression. It is well established that the Coulomb's type of relationship between shear strength and pre-compression exists and is given by (Hendry 1978):

$$\tau = \tau_o + \mu\,\sigma \qquad (1)$$

where τ_o is the initial bond, μ is the coefficient of friction and σ is the pre-compression. The initial bond or adhesion is very low for both types of structures, hence shear strength is mainly derived from the frictional component. As a results, the shear strength of shear walls with unfilled vertical joints for pre-compression of 2.1 N/mm² is 5% lower than those with filled vertical mortar joints. However, for pre-compression of 0.4 N/mm² the difference in strength increases approximately to 15%.

3.3.2 Deflections

A comparison of deflections at the top of a shear wall with filled and unfilled vertical mortar joints is presented in Figure 6. The pre-compression applied to the structure was 2.1 N/mm².

The top deflection of the structures with unfilled vertical mortar is approximately 30% higher compared to structures with filled vertical mortar joints.

The load deflection relationship for both structures is linear up to approximately 75% of the ultimate shear load. However the displacements increases suddenly when the shear load is close to failure.

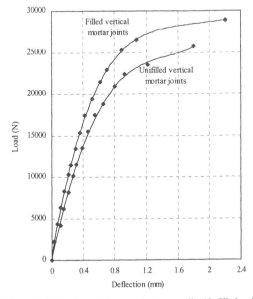

Figure 6. Deflections at the top of a shear wall with filled and unfilled vertical mortar joints.

4 CONCLUSIONS AND SUMMARY

On the basis of the investigation it appears that the workmanship defect of not filling the vertical joints with mortar results in significant drop of strength and stiffness:

(i) 38% reduction is observed in the Young's modulus of prisms compared to those with filled vertical mortar joints.

(ii) The compressive strength of prisms with unfilled vertical mortar joints is approximately 21% lower compared to prisms with filled vertical mortar joints.

(iii) The drop in flexure tensile strength normal and parallel to bed joints was approximately 32% to 58% respectively.

(iv) Shear strength is 5% to 15% lower depending on the level of pre-compression. The lower the pre-compression the higher is the reduction in the shear strength of walls with unfilled vertical mortar joints compared to structures where vertical joints are filled.

(v) The top deflection of structures when vertical mortar joints are not filled is 30% to 50% higher than structures built with filled vertical mortar joints.

In summary, the studies clearly show that the defect of not filling the vertical mortar joints has adverse effect on the mechanical properties and the strength of masonry in compression, flexural and shear. For safety, this practice must be discouraged for the load-bearing masonry construction.

ACKNOWLEDGMENTS

The authors are grateful to the CAPES/Brazil and to Faculdade de Engenharia de Sao Jose do Rio Preto (Brazil) for their financial support. The experimental work was done in the Department of Civil and Environmental Engineering of the University of Edinburgh.

REFERENCES

BS 890 1972. *Building limes*. London: British Standards Institutions.

BS 3921. 1974. *Specification for clay bricks and blocks*. London: British Standards Institutions.

BS 1200. 1976. *Building sands from natural sources*. London: British Standards Institutions.

BS 5628. 1992. *Code of Practice for structural use of masonry. Unreinforced masonry: part 1.* London: British Standards Institutions.

BS 12. 1995. *Portland cement*. London: British Standards Institutions.

B.C.R.A. 1972. *Investigation of the effect on Brickwork of not filling vertical mortar joints*. Internal Report, Stoke-on-Trent.

Hendry, A. W. 1978. A note on the strength of brickwork in combined racking shear and compression, *Proceedings of the British Ceramic Society*. Load Bearing Brickwork (6), 27, 47-52, December.

James, J. A. 1973. Investigation of the effect of workmanship and curing conditions on the strength of brickwork, *Proc. of the Third Int. Brick Masonry Conference*, Essen, Germany.

Sinha, B. P., Loftus, M. D. & Temple, R. 1979. Lateral strength of model brickwork panels. *Proc. Inst. Civ. Eng.* Part 2, 1979, Vol 67, pp 191-97

Creative Systems in Structural and Construction Engineering, Singh (ed.)© 2001 Balkema, Rotterdam, ISBN 90 5809 161 9

Seismic resistance test of beam-column connections for concrete-filled steel-box columns

Lyan-Ywan Lu
Department of Construction Engineering, National Kaohsiung First University of Science and Technology, Taiwan

Lap-Loi Chung & Chin-Hsun Yeh
National Center for Research on Earthquake Engineering, Taipei, Taiwan

Yuan-Zi Wang
Department of Civil Engineering, National Taiwan University, Taipei, Taiwan

ABSTRACT: In this study, seven full-scale connections each composed of a concrete-filled steel box column and two W-shape steel beams were subjected to cyclic loading tests, so the seismic performance of the connection detail commonly used in local steel construction can be investigated. The test result shows that most specimens suffer abrupt fracture of the welding between the beam flanges and the column. Because of this brittle failure, the full hysteretic property of the connections can not be developed. It is also shown that the stiffness and strength of the panel zones will decrease with an increase on the column width-to-thickness ratio. On the other hand, the in-filled concrete will increase the strength of the panel zone, but the increase on the stiffness of the panel zone is insignificant. A formula that can predict the yielding strength of the connection panel zones is also proposed. When compared with AIJ 1987 codes, the proposed formula predict shear strengths which show better agreement with the test data, with a maximum discrepancy less than 15%.

1 INTRODUCTION

Concrete-filled steel members can form one type of steel-concrete composite structural systems. The use of concrete-filled steel columns (with rectangular or circular section) in building construction has increased in the recent decades, because engineers are more aware of the benefits of this type of composite construction. The benefits come from utilizing simultaneously the material properties of steel and concrete in a structural system, so the strength, stiffness and ductility of the structure can be enhanced altogether.

On the other hand, the design of beam-column connections plays a critical role in the safety of steel-concrete composite structures located in strong earthquake areas (Alostaz and Schneider, 1996). Unlike the conventional pure steel or reinforced concrete structures, the composite construction is still lack of a consistent analysis and design method for beam-column connections. The diversity of composite member fabrication details, different construction procedures in different countries, incomplete knowledge about force transfer mechanism in connections and so on, are all contributed to the difficulty of proposing such a consistent method.

Hollow steel box columns are widely used in steel constructions in Taiwan, especially for highrise buildings. To increase the compression strength of these columns, some designers even fill the hollow columns with concrete; however, for conserva-

tism they do not include the stiffness and strength contributed by in-filled concrete during the analysis of lateral loading. These large-size box columns are normally fabricated in factories by welding four steel plates together. For these columns, a standard moment resisting connection is usually made by bolting the web and welding the flanges of the beams to the columns in the field. This type of connections is so common that if it can be also used in the construction involving concrete-filled box columns, the construction cost may be reduced significantly. In view of this, in this study, seven full-scale connections that connect concrete-filled steel columns and steel beams were tested, using the connection detail stated above. The seismic performance of these standard connections can be evaluated experimentally. The design detail and fabrication procedure of the test specimens all followed the local engineering practice.

The research result presented here is a part of a research program conducted at the National Center for Research on Earthquake Engineering (NCREE, Taiwan). The aim of the program is to study and promote steel-concrete composite construction in this country. The program has planed a series of tests and studies on the seismic behavior of beam-column connections with various detail designs. As a part of initiation research work, this study is expected to provide useful information and lead to a seismic-safe and cost-effective connection detail for composite structures.

2 DETAIL OF TEST SPECIMENS

There were totally seven specimens fabricated for the test. Each specimen was comprised of a moment resisting connection between an interior box steel column and two wide flange steel beams. These specimens, which simulate full-scale interior substructures of buildings, were expected to have strong column-weak beam behavior. The names and dimensions of the specimens are listed in Table 1. For the connections FS12 and S12, two specimens each were prepared. For all specimens, the beams were made of A36 H600×200×11×17 (mm) steel and the columns were made of A572 Grade 50 steel. When tested, the seven specimens are divided into two groups. The first group that includes specimens FS6, FS8 and FS10 is primary for studying the effect of column width/thickness ratio (i.e., B/t ratio), while the second group that consists of specimens FS12-1, FS12-2, S12-1 and S12-2 are for investigating the influence of in-filled concrete on the seismic resistant capacity of the connections.

Tabel 1. Name and Dimension of Specimens.

Specimen name	(1) Column Section (mm)	(2) t (mm)	(3) B/t ratio	(4) In-filled Concrete
FS6	☐ 400×400×6×6	6	66	YES
FS8	☐ 400×400×8×8	8	50	YES
FS10	☐ 400×400×10×10	10	40	YES
FS12-1	☐ 400×400×12×12	12	33	YES
FS12-2	☐ 400×400×12×12	12	33	YES
S12-1	☐ 400×400×12×12	12	33	NO
S12-2	☐ 400×400×12×12	12	33	NO

The connection detail for all seven specimens is identical. Figure 1 shows the fabrication detail of the connections. The beam web is connected to the column by a vertical shear plate that is two-side fillet welded on the column and bolted to the beam. The beam flanges are connected to the column using complete penetration groove welds. In making the flange connections, a backing bar was used on the lower side of each flange to permit all welding to be done from the top. The backing bars were not removed after the fabrication of the specimens were completed. A connection detail described above will transfer most of the beam moment through the beam flanges and most of its moment capacity is developed here. The columns of all specimens have two interior diaphragms located at the beam flange levels. The diaphragms, each with a circular opening of 140 mm in diameter for pouring concrete, are connected perpendicularly to the interior walls of the column by complete penetration square-groove welds. The purpose of installing the diaphragms is to transfer the tensile or compressive forces from the beam flanges and to stiffen the columns thus prevent the distortion of the column plates. Because the diaphragms are located inside the box column, they were welded to the column plates during the building process of the box column.

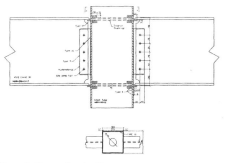

Figure 1. Connection detail.

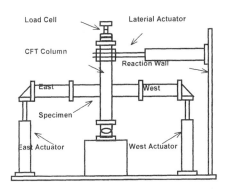

Figure 2. Test setup.

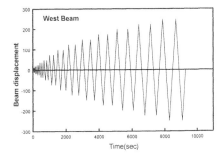

Figure 3. Cyclic Loading Pattern

3 TEST SETUP

In order to simulate the forces and moments developed in the beam-column connection vicinity of a real building subjected to a lateral load, each specimen was tested as a substructural system as shown in Figure 2. It is shown in the figure the loading facility includes two vertical and one horizontal hydraulic actuators. The two vertical actuators, which are attached to the beams, provide an angle rotation for the connection by a synchronously opposite mo-

594

tion, while the horizontal actuator is used to hold the top of the column with zero displacement. The two vertical actuators are displacement controlled. Figure 3 depicts their cyclic loading pattern. The maximum loading displacement increases from 6.25 to 150 mm (equivalent to 0.25% to 6% story drift) and each displacement was conducted for two cycles. All specimens were tested under a constant axial load of 2000 kN provided by a oil jack placed on the top of the column.

As shown in Figure 2, the ends of the two beams in each specimen are free to rotate and thus represent the points of inflection at the mid-spans of two adjacent beams. Similarly, the two ends of the column which is also free to rotate can simulate the points of inflection at the mid-height of two adjacent stories. This was to simulate a building of a story height 2.6 meter and the span length 5 meter. In order to save the specimen materials, each beam in the specimens was divided in two parts and bolted together (see Figure 2), so the part of the beam attached to the actuator (away from the column) could be repeatedly used for every specimen.

The instrumentation for measuring specimen response includes: two load cells embedded in the vertical actuators to measure the applied vertical shear in the beams (which can be converted into moments applied at the joint); two LVDTs embedded in the vertical actuators to measure the vertical displacements of the beam tips (which can be converted into equivalent story drift); π-gauges to measure the shear deformation of the connection's panel zone; tilt meters to measure the column rotation; one LVDT on the top of column to monitor the shortening of the column; strain gauges to measure local strains in the beam flanges and the panel zone. In addition, in order to observe material yielding process more clearly, prior to testing, all specimens were coated with whitewash, therefore, cracking and flaking of the applied coating would signal how seriously the yielding occurred during the test.

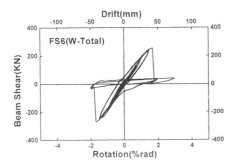

Figure 4. Hysteretic response of specimen FS6

4 YIELDING STRENGTH OF PANEL ZONE

Most of the moment transfers from beams to the column in the connections described in Section 2 is in the form of a couple that is produced by the tensile and compressive forces in the beam flanges. When these relatively large concentrated forces are applied to the connection, it may distort the skins of the box column and produce very large shear forces in the panel zone. In other words, when a structure subjected to a very large lateral load, its panel zones may yield due to shear loading. Therefore, a proper connection design must ensure that the yielding of the panel zone will not occur before the yielding of the beam flanges, under the strong column-weak beam concept. In this study, the yielding shear strengths of the panel zones obtained from the experiment data were compared with the theoretical values computed by three formulas that follows. The third formula is proposed by this study.

4.1 Theoretical values

All formulas presented below compute the total shear strength by superposing the individual strengths of steel and in-filled concrete, since the steel plates will separate from concrete core under a very small loading, if the column steel skins do not have shear studs inside (Furlong, 1968). In the formulas, the following notations are used: d_b, t_{bw}, t_{bf} represent the web depth, web thickness, flange thickness of the beams; d_c, b_c, t_{cf}, t_{cw} represent the depth, width, flange thickness, web thickness of the column; σ_y denotes the yielding strength of steel; f'_c and F_c denotes the compression and shear strengths of concrete; A_s and A_c denote the effective areas of steel and concrete.

- Formula after Krawinklar and ACI: From Krawinklar's study (1978) and the ACI provision for RC beam-column connection (ACI, 1995), the yielding shear strength of the panel zone may be computed by

$$V_{y,K} = 0.55\alpha d_c t' \sigma_y + 3.2\sqrt{f'_c} A_j \qquad (1)$$

where the unit of f'_c is kg/cm², α denotes the modification factor to account for the column axial load, t' denotes the total thickness of the column webs (in our case $t' = 2 t_{cw}$), A_j is the effective shear area of concrete, which was taken as one-half of the in-filled concrete area in this study.

- Formula given by Architectural Institute of Japan (AIJ, 1987):

$$V_{y,AIJ} = A_s \frac{\sigma_y}{\sqrt{3}} + 2.5 F_c A_c \frac{b_c}{d_b} \qquad (2)$$

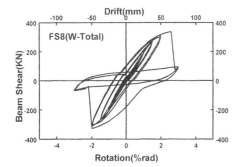

Figure 5. Hysteretic response of specimen FS8

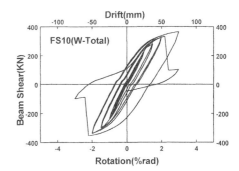

Figure 6. Hysteretic response of specimen FS10

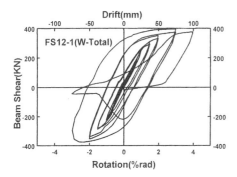

Figure 7. Hysteretic response of specimen FS12-1.

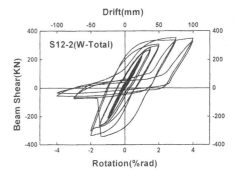

Figure 8. Hysteretic response of specimen S12-2.

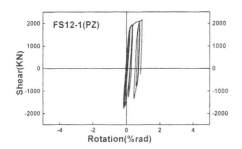

Figure 9. Panel zone hysteretic response of specimen FS12-1.

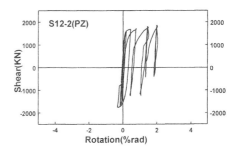

Figure 10. Panel zone hysteretic response of specimen S12-2.

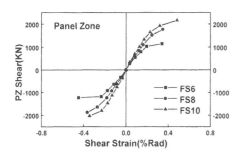

Figure 11 Shear envelope curves for panel zones.

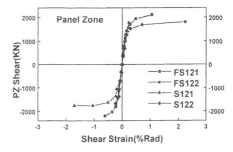

Figure 12. Shear envelope curves for panel zones.

Table 2. Yielding Shear Strength of Panel Zone

Specimen	(1) Krawinkler $V_{y,K}$ (kN)	(2) AIJ $V_{y,AIJ}$ (kN)	(3) Proposed $V_{y,pro}$ (kN)	(4) Experiment $V_{y,exp}$ (kN)	(5) Ratio $V_{y,exp}/V_{y,K}$	(6) Ratio $V_{y,exp}/V_{y,AIJ}$	(7) Ratio $V_{y,exp}/V_{y,pro}$
FS6	1512	1583	1498	1439	95%	91%	96%
FS8	1650	1723	1609	1639	99%	95%	102%
FS10	2115	2199	2058	1796	85%	82%	87%
FS12-1	2172	2252	2087	1804	83%	80%	86%
FS12-2	2227	2290	2114	1916	86%	84%	91%
S12-1	1835	1952	1810	1755	96%	90%	97%
S12-2	1835	1952	1810	1719	94%	88%	95%

In this study, A_s and A_c were taken as the halves of the cross-sectional areas of steel and concrete.

– Formula proposed by this study: by using Von Mises yielding criteria for the steel and also adopting the shear failure model of soil with axial force for the concrete that is confined in the panel zone, this study proposes the following formula

$$V_{y,pro} = \left[2(d_c - 2t_{cf})t_{cw} + \frac{5.2b_c t_{cf}^{3}}{(d_b - t_{bf})^2} \right]\sqrt{\frac{\sigma_s^2 - \sigma_p^2}{3}}$$
$$+ F_c'' A_c \frac{b_c}{d_b} \qquad (3)$$

where σ_p denotes the axial compressive stress in the steel column, F_c'' denotes the modified concrete shear strength to account for the axial compressive stress in the concrete.

4.2 Experimental values

In the following discussions, the experimental yielding strength $V_{y,exp}$ of the panel zone is defined as the shear force developed when the shear strain of the panel zone reaches the following value

$$\gamma_y = \frac{1}{G_s}\sqrt{\frac{\sigma_y^2 - \sigma_p^2}{3}} \qquad (4)$$

where G_s is the shear modulus of steel. This equation is derived by assuming the steel has reached its yielding strength defined by Von Mises yielding criteria. The corresponding shear force in the panel zone under this strain condition may be computed by

$$V_{y,exp} = \frac{M_b}{d_b - t_{bf}} - V_{col} \qquad (5)$$

where M_b is the net moment transferred from the beams and V_{col} is shear transferred from the column.

5 TEST RESULTS AND DISCUSSION

5.1 Overall Connection Behavior

Presented by the end shear (actuator load) vs. the end drift of the beam on the west side, the hysteretic response of the specimens FS6, FS8, FS10, FS12-1 and S12-2 are shown in Figures 4 ~ 8). All hysteretic loops in these figures experience one or two sudden drops in beam resistant shear. The drops are due to the abrupt fracture of the welding between the beam flanges and the box column. The rupture of welding, accompanied by a very large sound, was mostly initiated at the backing bar attached to the bottom beam flange and propagated into the steel plate of the box column. If the test continued, the beam flange would peel off the steel plate of the column and left a very large crack at the connection area. This type of brittle failure was extensively observed in 1994 Northridge earthquake that caused unprecedented damage to beam-column connections in conventional moment-resisting steel frames. Through an experimental and numerical study on connections between W-shape columns and beams, Popov et. al. (1998) have given a clearly explanation for the cause of such brittle failure. Although box columns with in-filled concrete (instead of W-shape columns) were tested in this study, the similar failure mode as described by Popov et. al. was observed in the test, except the crack propagated through column flanges but not the column webs and beam webs.

In the experiment, the test process was not stopped immediately if the fracture of beam flange welding occurred only at one side of the connection (west or east side). The test was terminated when welding fracture or sever flange yielding (or breaking) occurred at the other side of the connection. From the failure pattern, it was observed that the specimen group FS6, FS8 and FS10, which have relatively thinner plates when compared with the other four specimens, were failed by fracture of beam flange welding on the both sides of the connection. The connections of specimen FS6 completely failed even without exhibiting any plastic behavior (see Figure 4). On the other hand, for specimens FS12-1, FS12-2 and S12-1, which have relatively thicker column plates, although the welding fracture occurred in one side of the connection, the other side was failed by sever plastic deformation of the beam flange. Therefore, the specimens in second group generally have better energy dissipation capacity than that in the first group, as can be seen from the hysteretic response in Figures 7 and 8.

However, from an overall view, the connection type adopted by this study and also by local engineering practice does not possess sufficient ductility and may not be able to withstand sever seismic loads.

5.2 Panel Zone Behavior

Figures 9 and 10 show the relation of shear force and shear strain of the panel zone for specimen FS12-1 and S12-2, respectively. In the figures, the shear strain was measured by the π-gauges in the panel zone, while the shear force was computed by Equation (5). In both figures, the hysteretic curves shift to the right. This offset is primary due to un-symmetric failure of the connections that is caused by the fracture of the beam flange welding on one side of the connection, as described previously.

Figures 11 and 12 depict the envelope curves of the panel zone hysteretic loops for two different groups of the specimens, respectively. In Figure 11, where the results of the specimens with different column B/t ratios are compared, it is observed that stiffness and strength of panel zones will decrease with an increase on the column B/t ratio. The effect of column B/t ratios on the ductility of the panel zones can not be obtained from Figure 11, since the connections of the specimen suffered brittle failure. Figure 12 compares the panel zone behavior of the connections with or without in-filled concrete. It is shown that the stiffness of all four specimens are almost equal, while the strengths of FS12-1 and FS12-2 is higher than those of S12-1 and S12-2. This implies that in-filled concrete will not increase the stiffness of the panel zone, but will increase its strength.

In predicting the yielding shear strength of panel zones, Table 2 compares the experimental value with the theoretical values computed by Equations (1) to (3) (i.e., compare the fourth column with the first to third columns of the Table 2). Note that when computing the yielding strength of specimens S12-1 and S12-2 (without in-filled concrete), we have neglected the second terms of Equations (1) ~ (3), which represent the contribution of the in-filled concrete. Inspecting column (4) of Table 2, one may find that the column B/t ratio and the in-filled concrete will affect the yielding strength of the panel zone. Also, from the ratios of experimental to theoretical values shown in the fifth to seventh columns, it can be concluded that: (i) All three formulas predict the yielding strength with an error less than 20%. In general, the values given by the proposed formula are more consistent with the test data, with a maximum discrepancy less than 15%. (ii) From the data of specimens FS6, FS8, FS10, the predicted values are more accurate, when the column of the specimen has a larger width/thickness ratio (thinner steel box plates). (iii) For connections with hollow steel box columns (S12-1 and S12-2), Krawinkler's

formula is more accurate than AIJ formula, while the proposed one is even better than the former one.

6 SUMMARY AND CONCLUSIONS

- In this study, seven steel-concrete composite specimens adopting the connection detail commonly used for connecting a hollow steel box column with steel beams by local engineers were tested. The connection detail performed unsatisfactorily and mostly specimens were failed by abrupt fracture of welding between the beam flanges and the column. Because of this brittle failure mode, the advantage of using steel-concrete composite material, i.e., better ductility and hysteretic behavior, can not be obtained.
- In predicting yielding shear strength of panel zones, the proposed formula, which adopts Von Mises yielding criteria for steel and a soil failure model for concrete, is more accurate than that in AIJ design codes.
- The stiffness and strength of the panel zone in a concrete-filled steel box column will decrease with an increase on the column width/thickness ratio. On the other hand, the in-filled concrete will increase the strength of the panel zone, but the increase on the stiffness is negligible.

ACKNOWLEGEDMENT

This research work is sponsored by the National Science Council and National Center for Research on Earthquake Engineering, Taiwan, R.O.C. through the grant NSC-88-2711-3-319-200-01. This support is gratefully acknowledged.

REFERENCE

ACI 1995. *Buildings Code Requirements for Structural Concrete* (ACI 318-95). American Concrete Institute, Detroit, Michigan.

AIJ 1987, *Standard for Structural Calculation of Steel Reinforced Concrete Structures*. Architectial Institute of Japan, Tokyo.

Alostaz, Y. M. & Schneider, S. P. 1996. Analytical behavior of connections to concrete-filled steel tubes. *Journal of Constructional Steel Research*, 40(2): 95-127.

Furlong, R.W.1968. Design of steel-encased concrete beam-columns. *Journal of Structural Division, ASCE*, 94(ST1):267-281.

Krawinkler,H. 1978. Shear in Beam-Column Joints in Seismic Design of Steel Frames. *Engineering Journal*, American Institute of Steel Construction, 3rd quarter.

Popov, E. P., T.-S. Yang and S.-P. Chang 1998. Design of steel MRF connections before and after 1994 Northridge earthquake. *Engineering Structures*, 20(12):1030-1038.

19 Computer algorithms and fuzzy theory applications

Creative Systems in Structural and Construction Engineering, Singh (ed.) © 2001 Balkema, Rotterdam, ISBN 90 5809 161 9

Reliability assessment of concrete structures by using fuzzy theory and genetic algorithms

F. Biondini
Technical University of Milan, Italy

F. Bontempi
University of Rome 'La Sapienza', Italy

P.G. Malerba
University of Udine, Italy

ABSTRACT: The paper deals with the reliability assessment of reinforced and prestressed concrete framed structures. Due to the uncertainties involved in the problem, the geometrical and mechanical properties which define the structural problem cannot be considered as deterministic quantities. In this work such uncertainties are modeled by using a fuzzy criterion. The problem is formulated in terms of safety factor and the membership function over the failure interval is defined for several limit states. The work to do resembles that in an experimental study. It means that the single nonlinear analysis is a simulation of the reality and the problem to be solved concerns with the exploiting of the maximum of information from these trials. The strategic planning of the simulation is here found by a genetic optimization algorithm.

1 INTRODUCTION

The paper deals with the reliability assessment of structural systems. Specific attention is in particular devoted to concrete framed structures, but the presented methodology is thought generally applicable.

Due to the uncertainties involved in the problem, the geometrical and mechanical properties which define the structural problem cannot be considered as deterministic quantities. Such uncertainties are here modeled by using a fuzzy criterion (Bojadziev G. & Bojadziev M. 1995, Jang *et al.* 1997). The problem is formulated in terms of safety factor and the membership function over the failure interval is defined for several limit states (Biondini *et al.* 2000).

As well known, the structural reliability results from a general and comprehensive examination of all the failure modes. As a consequence, one must pay special attention to the following three aspects which define the assessment process:
1. *Available Data.*
2. *Structural Analysis.*
3. *Synthesis of the Results.*

For structural systems that show intrinsically nonlinear behavior, like the ones made of concrete, an accurate description of the response cannot be obtained without entering into the nonlinear field. Consequently, the reliability assessment of a structure belonging to such a class of systems, cannot be definitely assured without considering its actual nonlinear behavior. In the belief that nowadays nonlinear analysis poses no longer problem, one focuses on the organization of the available data and of the sequence of the analyses. In this way, the work to do resembles that in an experimental study. It means that the single analysis is a simulation of the reality and the problem to be solved concerns with the exploiting of the maximum of information from these trials. In this work, the strategic planning of the simulation is found by a genetic optimization algorithm.

2 SEARCH FOR THE RESPONSE INTERVAL

Let p a parameter belonging to the set of quantities which define the structural problem and λ a load multiplier. It is clear that to each set of parameters corresponds a set of limit load multiplier, one of them for each assigned limit state.

For sake of simplicity, we start our developments by considering the relationship between one single parameter p and one single limit state, defined by its corresponding limit load multiplier λ. At first, it is worth noting that, in general, such relationship is nonlinear even if the behavior of the system is linear. This is typical of the design processes where the structural properties which correlate loads and displacements are considered as design variables. Thus, the nonlinear law $\lambda=\lambda(p)$ can be drawn as in Fig.1.a, which shows that for each value of p, there is a corresponding value of λ. However, from Fig.1.b it is also clear that the response interval [λ_{min}; λ_{max}] corresponding to [p_{min}; p_{max}] cannot be simply obtained from $\lambda(p_{min})$ and $\lambda(p_{max})$. The problem of finding the interval response can be instead properly formulated as an optimization problem by assuming

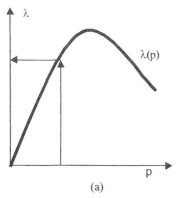
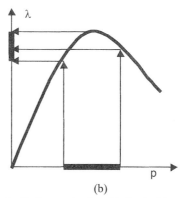

(a)   (b)

Figure 1. (a) Relationship between a structural parameter p and a limit state load multiplier λ. (b) Interval of the limit state multiplier λ corresponding to an interval of the parameter p.

the objective function to be maximized as the size of the response interval itself. In particular, for the general case of n independent parameter p, collected in a vector $\mathbf{x} = [p_1 \quad p_2 \quad ... \quad p_n]^T$, and m assigned limit states, the following objective function is introduced:

$$F(\mathbf{x}) = \sum_{i=1}^{m} \left(\lambda_{i,\max} - \lambda_{i,\min} \right) \qquad (1)$$

A solution $\mathbf{x}$ of the optimization problem which takes the side constraints $\mathbf{x}_{\min} \leq \mathbf{x} \leq \mathbf{x}_{\max}$ into account is developed by genetic algorithms. Such algorithms are heuristic search techniques which belong to the class of stochastic algorithms, since they combine elements of deterministic and probabilistic search (Michalewicz 1992).

Properly, the search strategy works on a *population* of *individuals* subjected to an evolutionary process, where individuals compete between them to survive in proportion to their *fitness* with the *environment*. In this process, population undergoes continuous reproduction by means of some *genetic operators* which, because of competition, tend to preserve best individuals. From this evolutionary mechanism, two conflicting trends appear: exploiting of the best individuals and exploring the environment. Thus, the effectiveness of the genetic search depends on a balance between them, or between two principal properties of the system, *population diversity* and *selective pressure*. These aspects are in fact strongly related, since an increase in the selective pressure decreases the diversity of the population, and vice versa (Biondini 1999).

With reference to the optimization problem previously formulated, a population of m individuals belonging to the environment $E = \{ \mathbf{x} \mid \mathbf{x}^- \leq \mathbf{x} \leq \mathbf{x}^+ \}$ represents a collection $X = \{ \mathbf{x}_1 \quad \mathbf{x}_2 \quad ... \quad \mathbf{x}_m \}$ of m possible solutions $\mathbf{x}_k^i = [x_1^k \quad x_2^k \quad ... \quad x_n^k] \in E$, each defined by a set of n design variables x_i^k ($k = 1, ..., m$). To assure an appropriate hierarchical arrangement of

the individuals, their fitness $F(\mathbf{x}) \geq 0$, which increases with the adaptability of $\mathbf{x}$ to its environment E, should be properly scaled. More details about the adopted scaling rules, internal coded representation of the population, genetic operators and termination criteria, can be found in a previous paper (Biondini 1999).

3 LIMIT STATES AND NONLINEAR ANALYSIS

Based on the general concepts of the R.C. and P.C. design, the structural performances should generally be described with reference to a specified set of limit states, as regards both serviceability and ultimate conditions, which separate desired states of the structure from undesired ones (Bontempi *et al.* 1998).

Splitting cracks and considerable creep effects may occur if the compression stresses σ_c in concrete are too high. Besides, excessive stresses either in reinforcing steel σ_s or in prestressing steel σ_p can lead to unacceptable crack patterns. Excessive displacements $\mathbf{s}$ may also involve loss of serviceability and then have to be limited within assigned bounds $\mathbf{s}^-$ and $\mathbf{s}^+$. Based on these considerations, the following limitations account for adequate durability at the serviceability stage (*Serviceability Limit States*):

1. $-\sigma_c \leq -\alpha_c f_c$ \qquad (2.a)

2. $|\sigma_s| \leq \alpha_s f_{sy}$ \qquad (2.b)

3. $|\sigma_p| \leq \alpha_p f_{py}$ \qquad (2.c)

4. $\mathbf{s}^- \leq \mathbf{s} \leq \mathbf{s}^+$ \qquad (2.d)

where α_c, α_s and α_p are suitable reduction factors of the limit strengths f_c, f_{sy} and f_{py}.

When the strain in concrete ε_c, or in the reinforcing steel ε_s, or in the prestressing steel ε_p reaches a limit value ε_{cu}, ε_{su} or ε_{pu}, respectively, the collapse of the corresponding cross-section occurs. However, the collapse of a single cross-section doesn't necessarily lead to the collapse of the whole structure. The latter is caused by the loss of equilibrium arising when the reactions **r** requested for the loads **f** can no longer be developed. So, the following ultimate conditions have to be verified (*Ultimate Limit States*):

1. $-\varepsilon_c \leq -\varepsilon_{cu}$ (3.a)

2. $|\varepsilon_s| \leq \varepsilon_{su}$ (3.b)

3. $|\varepsilon_p| \leq \varepsilon_{pu}$ (3.c)

4. $\mathbf{f} \leq \mathbf{r}$ (3.d)

Since these limit states refer to internal quantities of the system, a check of the structural performance through a nonlinear analysis needs to be carried out at the load level. To this aim, it is useful to assume $\mathbf{f} = \mathbf{g} + \lambda\mathbf{q}$, where **g** is a vector of dead loads and **q** a vector of live loads whose intensity varies proportionally to a unique multiplier $\lambda \geq 0$.

In most cases, R.C. and P.C. structures should be analyzed by taking material and, eventually, geometrical non-linearity into account if realistic results under all load levels are needed. In this work, a two-dimensional structure is modeled using a R.C./P.C. beam finite element whose formulation, based on the Bernoulli-Navier hypothesis, deals with such kinds of non-linearity (Bontempi *et al.* 1995, Malerba 1998).

In particular, both material $\mathbf{K}'_M$ and geometrical $\mathbf{K}'_G$ contributes to the element stiffness matrix $\mathbf{K}'$ and the nodal forces vector $\mathbf{f}'$, equivalent to the applied loads $\mathbf{f}'_0$ and to the prestressing $\mathbf{f}'_p$, are derived by applying the principle of the virtual displacements and then evaluated by numerical integration over the length l of the beam:

$$\mathbf{K}'_M = \int_0^l \mathbf{B}^T \mathbf{H} \mathbf{B} dx \qquad (4.a)$$

$$\mathbf{K}'_G = \int_0^l N \, \mathbf{G}^T \mathbf{G} \, dx \qquad (4.b)$$

$$\mathbf{K}' = \mathbf{K}'_M + \mathbf{K}'_G \qquad (4.c)$$

$$\mathbf{f}' = \int_0^l \mathbf{N}^T (\mathbf{f}'_0 + \mathbf{f}'_p) \, dx \qquad (4.e)$$

$$\mathbf{N} = \begin{bmatrix} \mathbf{N}_a & 0 \\ \hline 0 & \mathbf{N}_b \end{bmatrix} \qquad (4.f)$$

$$\mathbf{B} = \begin{bmatrix} \partial \mathbf{N}_a / \partial x & 0 \\ \hline 0 & \partial^2 \mathbf{N}_b / \partial x^2 \end{bmatrix} \qquad (4.g)$$

$$\mathbf{G} = \begin{bmatrix} 0 & \dfrac{\partial \mathbf{N}_b}{\partial x} \end{bmatrix} \qquad (4.h)$$

where N is the axial force and **N** is a matrix of axial $\mathbf{N}_a$ and bending $\mathbf{N}_b$ displacement functions. In the following, the shape functions of a linear elastic beam element having uniform cross-sectional stiffness **H** and loaded only at its ends are adopted. However, due to material non-linearity, the cross-sectional stiffness distribution along the beam is non uniform even for prismatic members with uniform reinforcement. Thus, the matrix **H**, as well as the

sectional load vector equivalent to the prestressing $\mathbf{f}'_p$, have to be computed for each section by integration over the area of the composite element, or by assembling the contributes of concrete and steel. In this way, after the constitutive laws of the materials are specified, the matrix **H** of each section can be computed under all load levels.

The equilibrium conditions of the beam element are derived from the already mentioned principle of the virtual work. Thus, by assembling the stiffness matrix **K** and the vectors of the nodal forces **f** with reference to a global coordinate system, the equilibrium of the whole structure can be formally expressed as follows:

$$\mathbf{Ks} = \mathbf{f} \qquad (5)$$

where **s** is the vector of the nodal displacements. It is worth noting that the vectors **f** and **s** have to be considered as total or incremental quantities depending on the nature of the stiffness matrix $\mathbf{K} = \mathbf{K}(\mathbf{s})$, or if a secant or a tangent formulation is adopted.

4 APPLICATION

The proposed procedure is applied to the reliability analysis of the two-span post-tensioned R.C. continuous beam shown in Figure 2. The span length is l=7500 mm, while the dimensions of the rectangular cross-section are b=203.2 mm and h=406.4 mm. The beam is reinforced with 2 bars $\varnothing$14 mm (A_{s1}=153.9 mm^2) placed both at the top and at the bottom edge with a cover c=32.4 mm. The prestressing steel tendon consists of 32 wires $\varnothing$5 mm (A_{p1}=19.6 mm^2), having a straight profile from the ends of the beam (e_1=0 mm) to the middle of its spans (e_2=−50 mm) and a parabolic profile from these points to the middle support (e_3=88 mm). After the time-dependent losses, at the ends the nominal prestressing force is P_{nom}=527.5 kN. This force decreases along the beam because of the losses due to friction. The curvature friction coefficient $\mu = 0.3$ and the wobble friction coefficient $K = 0.0016$ rad / m have been assumed.

The stress-strain diagram of concrete is described by the Saenz's law in compression and by an elastic perfectly-plastic model in tension. The stress-strain dia-

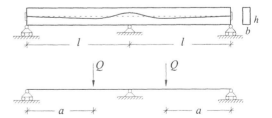

Figure 2. Prestressed continuous beam (Lin 1955). Tendon layout and load condition.

gram of reinforcing steel is assumed elastic perfectly-plastic in tension and in compression. For prestressing steel the plastic branch is instead assumed as non-linear and described by a fifth degree polynomial function (Bontempi *et al.* 1998). These constitutive laws are completely defined by the following properties:

$$f_{c,nom} = -41.3\,\text{MPa} \quad \varepsilon_{cu} = -3.4\%_0 \quad \varepsilon_{ctu} = 2\varepsilon_{ct1} \quad (6.\text{a})$$

$$f_{sy,nom} = 314\,\text{MPa} \quad E_s = 196\,\text{GPa} \quad \varepsilon_{su} = 16\% \quad (6.\text{b})$$

$$f_{py,nom} = 1480\,\text{MPa} \quad E_p = 200\,\text{GPa} \quad \varepsilon_{pu} = 1\%. \quad (6.\text{c})$$

with a weight density of the material $\gamma=25\,\text{kN/m}^3$.

Finally, two concentrated live loads having nominal value $Q_{nom}=50\,\text{kN}$ and placed at the distance $a = 4880\,\text{mm}$ from each end, as shown in Figure 2, have considered. For this load condition, the serviceability limit states are detected by assuming $\alpha_c=0.45$, $\alpha_s=0.80$, $\alpha_p=0.75$, $s^+=-s^-=l/400$.

To show the capabilities of the algorithm, the following quantities are considered to be uncertain:

- the strengths of concrete and of both reinforcing and prestressing steel for each of the ten finite elements that compose the beam (30 variables);
- the prestressing force of the strands (1 variable);
- the live loads (1 variable).

All these 32 fuzzy variables, reduced to a dimensionless form with respect to their nominal values, are assumed to have a triangular membership function with interval base [0.7 – 1.3] and mean value equal 1. Seven α-levels are considered, corresponding to the following intervals: [0.70 – 1.30], [0.75 – 1.25], [0.80 – 1.20], [0.85 – 1.15], [0.90 – 1.10], [0.95 – 1.05] and [1.00 – 1.00].

The membership functions of the limit load multiplier for the limit states previously defined are presented in Figure 3. At first we observe that, for the chosen load value (λ=1), the limit states are not violated and the structure is safe with respect to the assumed α-levels. For increasing load values (λ>1) the third service limit state (about the stress in the prestressing steel) never appears, while for the other limit states it can be appreciated the spread of the uncertainty, especially for the larger α-levels. To this regards, it is worth noting that uncertainties larger than 15% seems to appear critical, in particular for the ultimate limit states.

Figure 4 shows the histograms of the load multiplier λ resulting from about 400 simulations performed for a given α-level and by assuming alternatively: (a) a purely random choice of the data, and (b) a genetically driven simulation. A comparison of the results leads us to appreciate the higher capability of the genetic search in exploring the regions of the response interval where the limit state violations tends to occur. Finally, Figure 5 shows the load–displacement diagrams for several simulations associated to the α-levels [0.70 – 1.30] and [0.95 – 1.05], respectively.

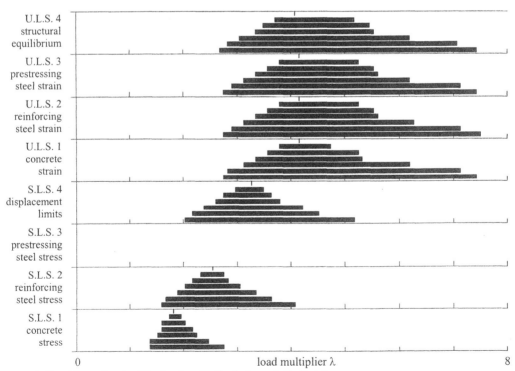

Figure 3. Resulting α-levels of the load multiplier λ for each limit state.

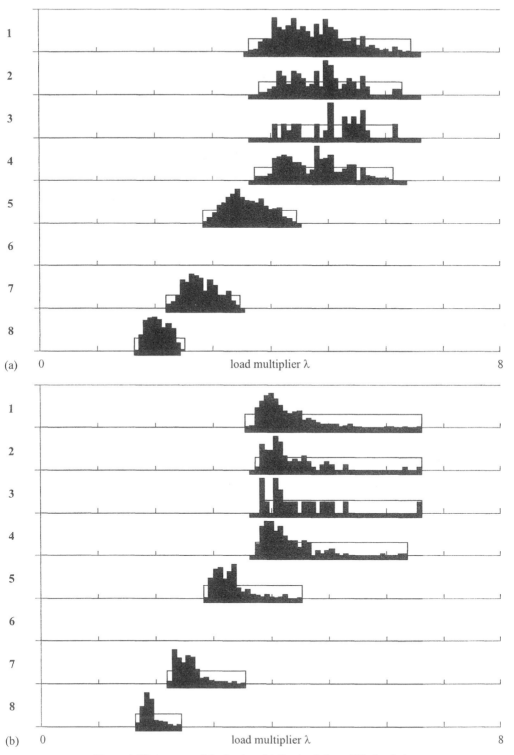

(a) 0   load multiplier λ   8

(b) 0   load multiplier λ   8

Figure 4. Histograms of the load multiplier λ of about 400 simulations.
(a) Random choice of data. (b) Genetically driven simulation.

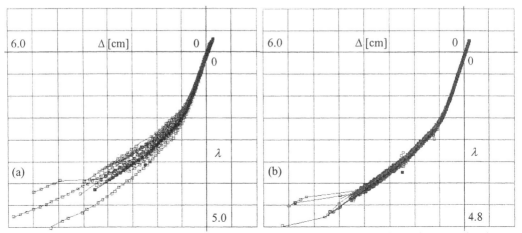

Figure 5. Load multiplier λ – maximum displacement Δ diagrams for several simulations associated to different α-levels (nominal curve in black): (a) [0.70 – 1.30]; (b) [0.95 – 1.05].

5 CONCLUSIONS

The fuzzy reliability assessment of R.C. and P.C. structures has been considered. Structural analysis techniques capable to take the non linear behaviour of the system into account have been developed. However, the work is mainly focused on the organization of the available data and of the sequence of the analyses. In particular, the work to do resembles that in an experimental study, where the single nonlinear analysis is a simulation of the reality and the problem to be solved concerns with the exploiting of the maximum of information from these trials. In particular, the strategic planning of the simulation is usefully found by a genetic optimization algorithm.

The proposed procedure has been applied to the reliability analysis of a two-span post-tensioned continuous beam. The beam is subdivided in ten finite elements and the strengths of the materials in each of them, as well as the prestressing force and the live load intensity, are considered to be uncertain. The obtained results show that, for the chosen load value (λ=1), the limit states are not violated and the structure is safe with respect to the assumed α-levels. For increasing load values (λ>1), the serviceability limit state about the stress in the prestressing steel never appears. In particular, as regard the role of the uncertainties, it is noted that a scatter larger than 15% seems to appear critical, in particular as regards the ultimate limit states. The effectiveness of the genetic algorithm in driving the simulation process is also outlined. A comparison of the results leads us to appreciate the higher capability of the genetic search in exploring the regions of the response interval where the limit state violations tend to occur.

As a concluding remark, it is worth noting that the proposed fuzzy approach should not be considered as alternative to a purely probabilistic formulation, given that both methods account for different aspects of the same problem. In fact, as known, fuzzy theory allows a treatment of uncertainties due to a lack of information, while probability theory is based on an absolute knowledge about the stochastic variability resulting from the random nature of the same quantities. However, it is also noted that an autonomous approach to the reliability structural assessment, like the probabilistic formulation proposed by the codes, should find a higher rationality in a fuzzy approach which, due either to the real nature of the involved uncertainties, or to a higher simplicity of the mathematical formulation, seems to be more suitable for the design purposes.

REFERENCES

Biondini F. 1999. Optimal Limit State Design of Concrete Structures using Genetic Algorithms. *Studi e Ricerche*, Scuola di Specializzazione in Costruzioni in Cemento Armato, Politecnico di Milano, **20**, 1-30.

Biondini F., F. Bontempi, P.G. Malerba 2000. Fuzzy Theory and Genetically-Driven Simulation in the Reliability Assessment of Concrete Structures. *Proceedings of 8th ASCE Joint Specialty Conference on Probabilistic Mechanics and Structural Reliability*, University of Notre Dame (IN), July 24-26.

Bojadziev G., M. Bojadziev 1995. *Fuzzy sets, fuzzy logic, applications*. World Scientific.

Bontempi F., P.G. Malerba and L. Romano 1995. A Direct Secant Formulation for the Reinforced and Prestressed Concrete Frames Analysis. *Studi e Ricerche*, Scuola di Specializzazione in Costruzioni in Cemento Armato, Politecnico di Milano, **16**, 351-386 (in Italian).

Bontempi F., F. Biondini, and P.G. Malerba 1998. Reliability Analysis of Reinforced Concrete Structures Based on a Monte Carlo Simulation. *Stochastic Structural Dynamics*. Spencer, B.F. Jr, Johnson E.A. (Eds.), Rotterdam: Balkema, 413-420.

Jang J.S.R., C.T. Sun, E. Mizutani 1997. *Neuro-Fuzzy and Soft Computing*. Matlab Curriculum Series, Prentice Hall.

Lin T.Y. 1955. Strength of Continuous Prestressed Concrete Beams Under Static and Repeated Loads. *ACI Journal*, **26**(10), 1037-1059.

Malerba P.G. (Ed.) 1998. *Limit and Nonlinear Analysis of Reinforced Concrete Structures*. Udine: CISM (in Italian).

Michalewicz Z.1992. *Genetic Algorithms + Data Structures = Evolution Programs*. Berlin: Springer.

Creative Systems in Structural and Construction Engineering, Singh (ed.)© 2001 Balkema, Rotterdam, ISBN 90 5809 161 9

Simplified algebraic method for computing eigenpair sensitivities of damped systems

H.K.Jo & I.W.Lee
Korea Advanced Institute of Science and Technology, Taejon, Korea

M.G.Ko
Kongju National University, Korea

ABSTRACT: A simplified method for the eigenpair sensitivities of damped systems is presented. This approach employs a reduced equation to determine the sensitivities of eigenpairs of the damped vibratory systems with distinct eigenvalues. The derivatives of eigenpairs are obtained by solving an algebraic equation with a symmetric coefficient matrix of (n+1) by (n+1) dimension where n is the number of degree of freedom. This is an improved method of the previous work of Lee and Jung. Two equations are used to find eigenvalue derivatives and eigenvector derivatives in their paper. A significant advantage of this approach over Lee and Jung is that one algebraic equation newly developed is enough to compute such eigenvalue derivatives and eigenvector derivatives. Simulation results indicate that the new method is highly efficient in determining the sensitivities of eigenpairs of the damped vibratory systems with distinct eigenvalues.

1 INTRODUCTION

Methods for computing the derivatives of natural frequencies and the corresponding mode shapes have been studied in the past 30 years. Finding the derivatives of eigenpairs are essential to determine the sensitivity of dynamic responses of the physical systems. For structural design, these eigenpair derivatives are used to optimize the natural frequencies and the mode shapes of structures by varying design parameters.

Straightforward calculation of eigenvalue derivatives is shown in Fox & Kapoor (1968), Taylor (1972) and Plaut & Huseyin (1973) etc., but the determination of eigenvector derivatives is very complicated because of the singularity problem. Hence there have been suggested a number of different methods to compute the eigenvector derivatives of singular matrix equations.

Rudisill & Chu (1975) presented an algebraic method for the eigenvector derivatives. This method is restricted to the case of non-repeated eigenvalue problem and this method has an asymmetric coefficient matrix. Nelson (1976) solved the same problem; His technique requires only the knowledge of eigenvector to be differentiated and is recommended as an efficient solver for calculating the mode shape derivatives. This method is limited to the distinct eigenvalue problem too. Because of its complicated algorithm, programming code is lengthy and clumsy. Ojalvo (1988) extended Nelson's method to the multiple eigenvalue problem. Mills-Curren (1988) and Dailey (1989) modified Ojalvo's work. Because

those methods are based on Nelson's, their algorithms are so complicated too. In addition to these techniques, modal method (Murthy & Haftka 1988, Lim & Junkins 1987) and its modified one (Wang 1985, Liu *et al.* 1987) approximate the mode shape derivatives by a linear combination of mode shapes. It takes lots of computing time when these approaches are used due to a large number of modes. Additionally, we have an iterative method (Andrew 1978, Tan 1986 and Lee & Jung 1996), whose drawback is its inaccurate solution. Lee & Jung (1997a, b) studied an algebraic method for computing the eigenvalue and eigenvector derivatives of general matrix with non-repeated and repeated eigenvalues. This approach is very efficient and simple and it needs the only corresponding eigenvalue and eigenvector. Furthermore, not only an exact solution is obtained but also numerical stability is proved in their method.

A number of the prescribed methods can be applied to the damped systems; Hallquist (1976) proposed a method for determining the effects of mass modification in viscously damped systems. Recently Zimoch (1987) presented a sensitivity analysis method, which is applied to conservative ones as well as non-conservative systems. It, however, may be restricted to mechanical systems (lumped systems) having only distinct eigenvalues. In other words, implementing this method to the systems with multiple eigenvalues is difficult. Lee & Jung's method (1997a, b) is extended to the damped systems by Lee *et al.* (1999a, b).

In this paper, an improved method over Lee *et al.*

(1999a, b) is studied with reduction of the number of equations for the eigenpair derivatives. The eigenvalue derivatives are obtained apart from the eigenvector derivatives solving two equations in Lee *et al.* (1999a, b). But the eigenpair derivatives are found by solving one modified equation in this paper. Therefore, the FLOPS are reduced for computing to get the eigenpair derivatives while maintaining the advantages of Lee *et al.* (1999a, b). The algebraic equation of the proposed method is efficiently solved by the LDL^T type decomposition method. If the derivatives of stiffness, mass and damping matrices can be obtained analytically, the proposed method can find the exact eigenpair derivatives.

The case of distinct eiegenvalue for the damped systems (Lee *et al.* 1999a) is reviewed in the second section of this paper. The proposed method and its stability proof are made in the third section. The results are illustrated with two numerical examples in the next section.

2 PREVIOUS STUDY

As in Fox & Kapoor (1968), Taylor (1972) and Plaut & Huseyin (1973) etc., the calculation of the eigenvalue derivatives is simple. But the calculation of the eigenvector derivatives is complicated because of the singularity of the corresponding equation, therefore many researchers have tried to overcome the problem with various methods.

The finite-difference method (Adelman & Haftka 1986, Lim *et al.* 1987 and Vanhonacker 1980) uses a difference formula to approximate the derivative numerically, which requires calculating the eigenvector at a nominal and at least one perturbed design point. The modal method approximates the mode shape derivatives as a linear combination of mode shapes. The modified modal method was developed to reduce the number of nodes needed to represent the derivative of mode shape. The iterative method approximates the eigenvector derivatives using iteration numerically.

By the research in sensitivity methods, Lee & Jung's method (1997a, b) and their extension to damped systems (Lee *et al.* 1999a, b) is most stable, exact and simple as mentioned in previous section. In this paper, the improved method over Lee *et al.* (1999a) for the damped systems is presented, hence Lee and Jung's method, the base of the proposed method, is reviewed.

2.1 Lee et al. (1999a)

Consider a multi-degree-of-freedom damped system described as

$$\mathbf{M}\ddot{\mathbf{y}}(t) + \mathbf{C}\dot{\mathbf{y}}(t) + \mathbf{K}\mathbf{y}(t) = \mathbf{f}(t), \qquad (1)$$

where $\mathbf{M}$, $\mathbf{C}$ and $\mathbf{K}$ are the matrices of mass , damp-

ing and stiffness respectively, and $n \times n$ symmetric matrices. $\mathbf{M}$ is positive definite and $\mathbf{K}$ is positive definite or semi-positive definite. $\mathbf{f}$ is the excitation vector and $\mathbf{y}$ is the response vector. To determine the eigenvalue and eigenvector sensitivities, first consider the free vibration system. The solution of the free vibration of equation (1) can be assumed as

$$\mathbf{y}(t) = e^{\lambda t}\phi. \qquad (2)$$

Substituting equation (2) into equation (1) gives

$$(\lambda^2 \mathbf{M} + \lambda \mathbf{C} + \mathbf{K})\phi = \mathbf{0}, \qquad (3)$$

where λ and ϕ are the eigenvalue and eigenvector and both are complex values in general. And in order to determine the eigenvalue derivatives, the differentiating of equation (3) is used. Equation (3) is differentiated with respect to a design parameter p, then

$$(\lambda_j^2 \mathbf{M} + \lambda_j \mathbf{C} + \mathbf{K})\frac{\partial \phi_j}{\partial p} =$$

$$-(2\lambda_j \mathbf{M} + \mathbf{C})\phi_j \frac{\partial \lambda_j}{\partial p} - \left(\lambda_j^2 \frac{\partial \mathbf{M}}{\partial p} + \lambda_j \frac{\partial \mathbf{C}}{\partial p} + \frac{\partial \mathbf{K}}{\partial p}\right)\phi_j$$

$$(4)$$

Premultiplying at each side of equation (4) by ϕ_j^T, the eigenvalue derivative can be obtained as

$$\frac{\partial \lambda_j}{\partial p} = -\phi_j^T \left(\lambda_j^2 \frac{\partial \mathbf{M}}{\partial p} + \lambda_j \frac{\partial \mathbf{C}}{\partial p} + \frac{\partial \mathbf{K}}{\partial p}\right)\phi_j. \qquad (5)$$

But the eigenvector derivative $\partial \phi_j / \partial p$ cannot be found directly from equation (4), since the matrix $\lambda_j^2 \mathbf{M} + \lambda_j \mathbf{C} + \mathbf{K}$ is singular. To overcome this problem, the side condition is used in Lee & Jung's method.

To obtain the side condition, one can use state-space-form described by $2n$–dimensional eigenvalue problem as

$$\begin{bmatrix} -\mathbf{K} & \mathbf{0} \\ \mathbf{0} & \mathbf{M} \end{bmatrix}\begin{Bmatrix} \phi \\ \lambda\phi \end{Bmatrix} = \lambda\begin{bmatrix} \mathbf{C} & \mathbf{M} \\ \mathbf{M} & \mathbf{0} \end{bmatrix}\begin{Bmatrix} \phi \\ \lambda\phi \end{Bmatrix}, \qquad (6)$$

which can be written conveniently as

$$\mathbf{Az} = \lambda\mathbf{Bz}, \qquad (7)$$

where

$$\mathbf{A} = \begin{bmatrix} -\mathbf{K} & \mathbf{0} \\ \mathbf{0} & \mathbf{M} \end{bmatrix}, \quad B = \begin{bmatrix} \mathbf{C} & \mathbf{M} \\ \mathbf{M} & \mathbf{0} \end{bmatrix} \text{ and } \mathbf{z} = \begin{Bmatrix} \phi \\ \lambda\phi \end{Bmatrix}. (8)$$

The solutions of the complex eigenvalue problem equation (7) can be found as shown in Caughey & O'Kelly (1960), Wilson & Penzien (1972) and Trail-Nash (1981), and they are distinct and conjugate. The eigenvectors are normalized such as

$$\mathbf{z}_j^T \mathbf{B}\mathbf{z}_j = \begin{Bmatrix} \phi_j \\ \lambda_j\phi_j \end{Bmatrix}^T \begin{bmatrix} \mathbf{C} & \mathbf{M} \\ \mathbf{M} & \mathbf{0} \end{bmatrix}\begin{Bmatrix} \phi_j \\ \lambda_j\phi_j \end{Bmatrix} = 1, \qquad (9)$$

and arranging equation (9) gives

$$\phi_j^T (2\lambda_j \mathbf{M} + \mathbf{C})\phi_j = 1. \qquad (10)$$

Suppose that all eigenpairs and matrices $\partial K/\partial p$, $\partial M/\partial p$ and $\partial C/\partial p$ are known, and all eigenvlaues are different, where p is a design parameter. To obtain the derivatives of eigenvector, the differentials of the normalization condition are used as side condition.

Differentiating equation (10) with respect to design parameter p gives

$$\phi_j^T (2\lambda_j \mathbf{M} + \mathbf{C}) \frac{\partial \phi_j}{\partial p}$$
$$+ \frac{1}{2} \phi_j^T \left[2 \left(\frac{\partial \lambda_j}{\partial p} \mathbf{M} + \lambda_j \frac{\partial \mathbf{M}}{\partial p} \right) + \frac{\partial \mathbf{C}}{\partial p} \right] \phi_j = 0 \quad .(11)$$

Equation (4) and (11) can be combined as single matrix form as

$$\begin{bmatrix} \lambda_j^2 \mathbf{M} + \lambda_j \mathbf{C} + \mathbf{K} & (2\lambda_j \mathbf{M} + \mathbf{C})\phi_j \\ \phi_j^T (2\lambda_j \mathbf{M} + \mathbf{C}) & 0 \end{bmatrix} \begin{Bmatrix} \frac{\partial \phi_j}{\partial p} \\ 0 \end{Bmatrix}$$

$$= \begin{Bmatrix} -(2\lambda_j \mathbf{M} + \mathbf{C})\phi_j \frac{\partial \lambda_j}{\partial p} - \left(\lambda_j^2 \frac{\partial \mathbf{M}}{\partial p} + \lambda_j \frac{\partial \mathbf{C}}{\partial p} + \frac{\partial \mathbf{K}}{\partial p} \right) \phi_j \\ -\frac{1}{2} \phi_j^T \left[2 \left(\frac{\partial \lambda_j}{\partial p} \mathbf{M} + \lambda_j \frac{\partial \mathbf{M}}{\partial p} \right) + \frac{\partial \mathbf{C}}{\partial p} \right] \phi_j \end{Bmatrix}$$

(12)

So the eigenvector derivative $\partial \phi_j / \partial p$ can be obtained directly by solving the algebraic equation. The coefficient matrix of equation (14) is nonsingular.

3 PROPOSED METHOD

3.1 *Derivation of the proposed method*

Equation (4) is used directly, not premultiplying ϕ_j^T to equation for the derivative of eigenvalue. Rearranging equation (4) gives

$$(\lambda_j^2 \mathbf{M} + \lambda_j \mathbf{C} + \mathbf{K}) \frac{\partial \phi_j}{\partial p} + (2\lambda_j \mathbf{M} + \mathbf{C})\phi_j \frac{\partial \lambda_j}{\partial p}$$
$$= -\left(\lambda_j^2 \frac{\partial \mathbf{M}}{\partial p} + \lambda_j \frac{\partial \mathbf{C}}{\partial p} + \frac{\partial \mathbf{K}}{\partial p} \right) \phi_j \quad . (13)$$

Rearranging equation (11) gives

$$\phi_j^T (2\lambda_j \mathbf{M} + \mathbf{C}) \frac{\partial \phi_j}{\partial p} + \phi_j^T \mathbf{M} \phi_j \frac{\partial \lambda_j}{\partial p}$$
$$= -\frac{1}{2} \phi_j^T \left(2\lambda_j \frac{\partial \mathbf{M}}{\partial p} + \frac{\partial \mathbf{C}}{\partial p} \right) \phi_j \quad . (14)$$

Because the unknown or interested values are $\partial \phi_j / \partial p$ and $\partial \lambda_j / \partial p$, equation (13) and equation (14) can be combined as single matrix form as follows

$$\begin{bmatrix} \lambda_j^2 \mathbf{M} + \lambda_j \mathbf{C} + \mathbf{K} & (2\lambda_j \mathbf{M} + \mathbf{C})\phi_j \\ \phi_j^T (2\lambda_j \mathbf{M} + \mathbf{C}) & \phi_j^T \mathbf{M} \phi_j \end{bmatrix} \begin{Bmatrix} \frac{\partial \phi_j}{\partial p} \\ \frac{\partial \lambda_j}{\partial p} \end{Bmatrix}$$

$$= \begin{Bmatrix} -\left(\lambda_j^2 \frac{\partial \mathbf{M}}{\partial p} + \lambda_j \frac{\partial \mathbf{C}}{\partial p} + \frac{\partial \mathbf{K}}{\partial p} \right) \phi_j \\ -\frac{1}{2} \phi_j^T \left(2\lambda_j \frac{\partial \mathbf{M}}{\partial p} + \frac{\partial \mathbf{C}}{\partial p} \right) \phi_j \end{Bmatrix} \quad . (15)$$

Equation (15) is the key idea of the proposed method. In Lee *et al.*, two equation is needed for the

derivatives of the eigenvalue and eigenvector, because $\partial \lambda_j / \partial p$ is calculated apart from $\partial \phi_j / \partial p$ by pre-multiplying ϕ_j^T at each side of equation (4). But the proposed method finds $\partial \phi_j / \partial p$ and $\partial \lambda_j / \partial p$ at once by solving the single matrix, one algebraic equation, composed of the differentials of the eigenvalue problem and normalization condition.

The proposed method has the characteristics of not only finding exact solutions, having the numerical stability and having a symmetric coefficient matrix which are those of Lee *et al.*, but also being more efficient for the time required to calculate the eigenpair derivatives because of being composed of more simple equation.

The numerical stability of the proposed method is proved in next section.

3.2 *Numerical stability of the proposed method*

To prove that the coefficient matrix $\mathbf{A}^\#$ of equation (15) is nonsingular, introduce the nonsingular square matrix $\mathbf{Y}$, $\det(\mathbf{Y}) \neq 0$. Matrix $\mathbf{Y}$ is used in the determinant property, $\det(\mathbf{Y}^T \mathbf{A}^\# \mathbf{Y}) = \det(\mathbf{Y}^T) \det(\mathbf{A}^\#) \det(\mathbf{Y})$.

If $\det(\mathbf{Y}^T \mathbf{A}^\# \mathbf{Y}) \neq 0$, the determinant of $\mathbf{A}^\#$ is nonzero because the determinant of $\mathbf{Y}$ is nonzero. Nonsingular matrix $\mathbf{Y}$ is assumed as

$$\mathbf{Y} = \begin{bmatrix} \Psi & \mathbf{0} \\ \mathbf{0} & 1 \end{bmatrix}, \quad (16)$$

where $\Psi = \begin{bmatrix} \psi_1 & \psi_2 & \cdots & \psi_{n-1} & \phi_j \end{bmatrix}$, ϕ_j is the j th eigenvector of the system and ψ's are arbitrary vectors to be independent of ϕ_j. Ψ is a $n \times n$ matrix, $\mathbf{Y}$ is a $(n+1) \times (n+1)$ matrix.

Pre- and post-multiplying $\mathbf{Y}^T$ and $\mathbf{Y}$ to $\mathbf{A}^\#$ yields

$$\mathbf{Y}^T \mathbf{A}^\# \mathbf{Y} = \begin{bmatrix} \Psi & \mathbf{0} \\ \mathbf{0} & 1 \end{bmatrix}^T \begin{bmatrix} \lambda_j^2 \mathbf{M} + \lambda_j \mathbf{C} + \mathbf{K} & (2\lambda_j \mathbf{M} + \mathbf{C})\phi_j \\ \phi_j^T (2\lambda_j \mathbf{M} + \mathbf{C}) & \phi_j^T \mathbf{M} \phi_j \end{bmatrix} \begin{bmatrix} \Psi & \mathbf{0} \\ \mathbf{0} & 1 \end{bmatrix}$$

$$= \begin{bmatrix} \Psi^T (\lambda_j^2 \mathbf{M} + \lambda_j \mathbf{C} + \mathbf{K})\Psi & \Psi^T (2\lambda_j \mathbf{M} + \mathbf{C})\phi_j \\ \phi_j^T (2\lambda_j \mathbf{M} + \mathbf{C})\Psi & \phi_j^T \mathbf{M} \phi_j \end{bmatrix}$$

(17)

The last column and row of the matrix $\Psi^T (\lambda_j^2 \mathbf{M} + \lambda_j \mathbf{C} + \mathbf{K})\Psi$ is zero because of ϕ_j which is the last column of Ψ. That is

$$\Psi^T (\lambda_j^2 \mathbf{M} + \lambda_j \mathbf{C} + \mathbf{K})\Psi = \begin{bmatrix} \tilde{\mathbf{A}} & \mathbf{0} \\ \mathbf{0} & 0 \end{bmatrix}, \quad (18)$$

where $\tilde{\mathbf{A}}$ is a nonzero $(n-1) \times (n-1)$ submatrix. And λ_j is a distinct eigenvalue of the system, therefore the matrices $\Psi^T (\lambda_j^2 \mathbf{M} + \lambda_j \mathbf{C} + \mathbf{K})\Psi$ of order n have a rank $n-1$, they are singular. But the submatrix $\tilde{\mathbf{A}}$ of order $n-1$ has full rank $n-1$, and it is nonsingular, $\det(\tilde{\mathbf{A}}) \neq 0$. By normalization condition the last elements of the column vector $\Psi^T (2\lambda_j \mathbf{M} + \mathbf{C})\phi_j$ and the row vector $\phi_j^T (2\lambda_j \mathbf{M} + \mathbf{C})\Psi$ are unity.

Table 1. The lowest ten eigenvalue and their derivatives

Mode Number	Eigenvalue	Eigenvalue derivative (Lee&Jung's method)	Eigenvalue derivative (Proposed method)
1	-0.0004 - 2.6248i	-0.0138 -52.4963i	-0.0138 -52.4963i
2	-0.0004 + 2.6248i	-0.0138 +52.4963i	-0.0138 +52.4963i
3	-0.0136 -16.4491i	-5.4111e-1 -3.2896e+2i	-5.4111e-1 -3.2896e+2i
4	-0.0136 +16.4491i	-5.4111e-1+3.2896e+2i	-5.4111e-1+3.2896e+2i
5	-0.0345 -26.2358i	4.7702e-7 +2.9684e-8i	4.7702e-7 +2.9684e-8i
6	-0.0345 +26.2358i	4.7212e-7 +1.5490e-7i	4.7212e-7 +1.5490e-7i
7	-0.1061 -46.0558i	-4.2416e+0 -9.2096e+2i	-4.2416e+0 -9.2096e+2i
8	-0.1061 +46.0558i	-4.2416e+0 +9.2096e+2i	-4.2416e+0 +9.2096e+2i
9	-0.4073 -90.2444i	-1.6284e+1 -1.8043e+3i	-1.6284e+1 -1.8043e+3i
10	-0.4073 +90.2444i	-1.6284e+1 +1.8043e+3i	-1.6284e+1 +1.8043e+3i

$$\Psi^T(2\lambda_j\mathbf{M}+\mathbf{C})\phi_j = \left\{\begin{matrix}\tilde{\mathbf{b}}\\1\end{matrix}\right\}, \qquad (19)$$

$$\phi_j^T(2\lambda_j\mathbf{M}+\mathbf{C})\Psi = \left\{\begin{matrix}\tilde{\mathbf{b}}\\1\end{matrix}\right\}^T \qquad (20)$$

where $\tilde{\mathbf{b}}$ is nonzero vector. Substituting equation (18), (19) and (20) into equation (17) gives

$$\mathbf{Y}^T\mathbf{A}^{\#}\mathbf{Y} = \begin{bmatrix}\tilde{\mathbf{A}} & \mathbf{0} & \tilde{\mathbf{b}} \\ \mathbf{0} & 0 & 1 \\ \tilde{\mathbf{b}}^T & 1 & \phi_j^{\ T}\mathbf{M}\phi_j\end{bmatrix}. \qquad (21)$$

Applying the determinant property of partitioned matrix gives that
$\det(\mathbf{Y}^T\mathbf{A}^{\#}\mathbf{Y}) =$

$$\det\begin{bmatrix}0 & 1 \\ 1 & \phi_j^{\ T}\mathbf{M}\phi_j\end{bmatrix}\det\left(\tilde{\mathbf{A}} - \begin{bmatrix}\mathbf{0} & \tilde{\mathbf{b}}\end{bmatrix}\begin{bmatrix}0 & 1 \\ 1 & \phi_j^{\ T}\mathbf{M}\phi_j\end{bmatrix}^{-1}\begin{bmatrix}\mathbf{0} \\ \tilde{\mathbf{b}}^T\end{bmatrix}\right) \qquad (22)$$

where

$$\begin{bmatrix}\mathbf{0} & \tilde{\mathbf{b}}\end{bmatrix}\begin{bmatrix}0 & 1 \\ 1 & \phi_j^{\ T}\mathbf{M}\phi_j\end{bmatrix}^{-1}\begin{bmatrix}\mathbf{0} \\ \tilde{\mathbf{b}}^T\end{bmatrix} = 0 \qquad (23)$$

$$\det\begin{bmatrix}0 & 1 \\ 1 & \phi_j^{\ T}\mathbf{M}\phi_j\end{bmatrix} = -1. \qquad (24)$$

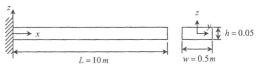

Figure 1. Cantilever beam with the thickness h as the design parameter.

Therefore rearranging equation (22) gives
$$\det(\mathbf{Y}^T\mathbf{A}^{\#}\mathbf{Y}) = -\det(\tilde{\mathbf{A}}) \neq 0. \qquad (25)$$
The determinant of $\mathbf{A}^{\#}$ thus is not equal to zero, in other words, the matrix $\mathbf{A}^{\#}$ is nonsingular.

4 NUMERICAL EXAMPLE

The efficiency and exactness of Lee *et al.* are verified in numerical examples of Lee *et al.* (1999a, b). In this section, the results by those method with cantilever beam are presented, for the case of the distinct natural frequencies. Pentium 120 having CPU capacity 120Mhz with RAM 40Mega is used for computation.

A cantilever beam with 40 elements is considered as shown in Figure 2. The number of nodes is 41 and each node has four degrees of freedom (y-translation, z-rotation, z-translation, y-rotation); the total number of degrees of freedom is 160(one node

Table 2. Some components of the first eigenvector and its derivatives

Eqn. number	Eigenvector	Eigenvector derivative (Lee&Jung's method)	Eigenvector derivative (Proposed method)
1	0	0	0
2	0	0	0
3	1.5133e-05+1.5133e-05i	-3.0267e-04-3.0267e-04i	-3.0267e-04-3.0267e-04i
4	1.2036e-04+1.2036e-04i	-0.0024-0.0024i	-0.0024-0.0024i
5	0	0	0
⋮			
157	0	0	0
158	0	0	0
159	0.0139 + 0.0139i	-0.2787 - 0.2787i	-0.2787 - 0.2787i
160	0.0019 + 0.0019i	-0.0384 - 0.0384i	-0.0384 - 0.0384i

610

is fixed). For this example, Young's modulus $(2.1 \times 10^{11} N/m^2)$, the mass density $(7.85 \times 10^3 kg/m^3)$ and the Poison's ratio 0.3 are used. The length of the beam is $10m$, width $0.5m$ and depth $0.05m$.

Rayleigh damping can be assumed as the damping matrix, which is of the form as

$$\mathbf{C} = \alpha\mathbf{K} + \beta\mathbf{M} , \qquad (26)$$

where α and β are the Rayleigh coefficients and $\alpha = \beta = 0.01$. The design parameter is the depth of beam, h .

Some sensitivity results are represented in Table 1 and Table 2.

As shown in Table 1 and Table 2, we can see that the results by Lee *et al.* and the proposed method are the same. It is exact solution. In Table 2, the first, second and fifth components and so on, are zero because first mode vibrate in z-direction(z-translation and y-rotation). The first and second component are those of y-translation and z-rotation respectively. The analysis time is 223.33 seconds in Lee *et al.* and 164.89 seconds in the proposed method for 160 eigenpair derivatives. The analysis time of the proposed method is less than that of Lee *et al.* by 26.17%.

CPU time required to calculate the 160 eigenpair derivatives is summarized for each operation in Table 3. The result of Table 3 is the average of those of ten times analysis, and standard deviations are 1.56 and 1.29 seconds for total CPU time of Lee *et al.* and the proposed method respectively.

5 CONCLUSION

To calculate the derivatives of eigenvalue and eigenvector, two equations were used in the previous work. But, by unifying the two equations by one, a simple algorithm is developed and simulated for the systems with non-repeated eigenvalues.

The proposed method is an improvement of Lee *et al.:* An exact solution is obtained and the numerical stability is proved as in Lee *et al* with an simplified algorithm. Additionally, CPU time to compute the eigenpair derivatives is remarkably reduced due to the simplification of algorithm in the proposed method as compared with Lee *et al.*

REFERENCES

Adelman, H. M. and Haftka, R. T. 1986. Sensitivity Analysis for Discrete Structural Systems, *AIAA Journal* 24: 823-832.

Andrew, A. L. 1978. Convergence of an Iterative Method for Derivatives of Eigensystems, *Journal of Comput. Phys.* 26: 107-112.

Caughey, T. K. and O'Kelly, M. E. J. 1960. Classical Normal Modes in Damped Linear Dynamic Systems, *Journal of Applied Mechanics ASME* 27: 269-271.

Dailey, R. L. 1989. Eigenvector Derivatives with Repeated Eigenvalues, *AIAA Journal* 27: 486-491.

Fox, R. L. and Kapoor, M. P. 1968. Rates of Change of Eigenvalues and Eigenvectors. *AIAA Journal* 6: 2426-2429.

Hallquist, J. O. 1976. An Efficient Method for Determining the Effects of Mass Modifications in Damped Systems, *Journal of Sound and Vibration* 44: 449-459.

Table 3. CPU time spent on the calculation of the first 160 eigenpair derivatives

Method	Operations	CPU time (sec)
Lee & Jung's Method	$\dfrac{\partial \lambda_j}{\partial p} = -\phi_j{}^T\left(\lambda_j{}^2\dfrac{\partial \mathbf{M}}{\partial p} + \lambda_j\dfrac{\partial \mathbf{C}}{\partial p} + \dfrac{\partial \mathbf{K}}{\partial p}\right)\phi_j$	33.89
	$\mathbf{A}^* = \begin{bmatrix} \lambda_j{}^2\mathbf{M} + \lambda_j\mathbf{C} + \mathbf{K} & (2\lambda_j\mathbf{M} + \mathbf{C})\phi_j \\ \phi_j{}^T(2\lambda_j\mathbf{M} + \mathbf{C}) & 0 \end{bmatrix}$	61.01
	$\mathbf{f}_j = \begin{Bmatrix} -(2\lambda_j\mathbf{M} + \mathbf{C})\phi_j\dfrac{\partial \lambda_j}{\partial p} - \left(\lambda_j{}^2\dfrac{\partial \mathbf{M}}{\partial p} + \lambda_j\dfrac{\partial \mathbf{C}}{\partial p} + \dfrac{\partial \mathbf{K}}{\partial p}\right)\phi_j \\ -\dfrac{1}{2}\phi_j{}^T\left[2\left(\dfrac{\partial \lambda_j}{\partial p}\mathbf{M} + \lambda_j\dfrac{\partial \mathbf{M}}{\partial p}\right) + \dfrac{\partial \mathbf{C}}{\partial p}\right]\phi_j \end{Bmatrix}$	47.09
	$\begin{Bmatrix} \dfrac{\partial \phi_j}{\partial p} \\ 0 \end{Bmatrix} = [\mathbf{A}^*]^{-1}\mathbf{f}_j$	81.34
	Total	**223.33**
Proposed Method	$\mathbf{A}^\# = \begin{bmatrix} \lambda_j{}^2\mathbf{M} + \lambda_j\mathbf{C} + \mathbf{K} & (2\lambda_j\mathbf{M} + \mathbf{C})\phi_j \\ \phi_j{}^T(2\lambda_j\mathbf{M} + \mathbf{C}) & \phi_j{}^T\mathbf{M}\phi_j \end{bmatrix}$	53.62
	$\bar{\mathbf{f}}_j = \begin{Bmatrix} -\left(\lambda_j{}^2\dfrac{\partial \mathbf{M}}{\partial p} + \lambda_j\dfrac{\partial \mathbf{C}}{\partial p} + \dfrac{\partial \mathbf{K}}{\partial p}\right)\phi_j \\ -\dfrac{1}{2}\phi_j{}^T\left(2\lambda_j\dfrac{\partial \mathbf{M}}{\partial p} + \dfrac{\partial \mathbf{C}}{\partial p}\right)\phi_j \end{Bmatrix}$	40.60
	$\begin{Bmatrix} \dfrac{\partial \phi_j}{\partial p} \\ \dfrac{\partial \lambda_j}{\partial p} \end{Bmatrix} = [\mathbf{A}^\#]^{-1}\bar{\mathbf{f}}_j$	70.67
	Total	**164.89**

Lee, I. W. and Jung, G. H. 1996. Numerical Method for Sensitivity Analysis of Eigensystems with Nonrepeated and Repeated Eigenvalues, *Journal of Sound and Vibration* 195(1): 17-32.

Lee, I. W. and Jung, G. H. 1997a. An Efficient Algebraic Method for Computation of Natural Frequency and Mode Shape Sensitivities: Part I, Distince Natural Frequencies, *Computers and Structures* 62(3): 429-435.

Lee, I. W. and Jung, G. H. 1997b. An Efficient Algebraic Method for Computation of Natural Frequency and Mode Shape Sensitivities: Part II, Mutilple Natural Frequencies, *Computers and Structures* 62(3): 437-443.

Lee, I. W. Kim, D. O. and Jung, G. H. 1999a, Natural Frequency and Mode Shape Sensitivities of Damped System : Part I, Distince Natural Frequencies, *Journal of Sound and Vibration* 223(3): 399-412.

Lee, I. W. Kim, D. O. and Jung, G. H. 1999b. Natural Frequency and Mode Shape Sensitivities of Damped System : Part II, Multiple Natural Frequencies, *Journal of Sound and Vibration* 223(3): 413-424.

Lim, K. B. Junkins, J. L. and Wang, B. P. 1987. Reexamination of Eigenvector Derivatives, *Journal of Guidance* 10: 581-587.

Liu, Z. S. Chen, S. H. and Zhao, Y. Q. 1994. An Accurate Method for Computing Eigenvector Derivatives for Free-Free Structures, *Computers and Structures* 52: 1135-1143.

Mills-Curran, W. C. 1988. Calculation of Derivatives for Structures with Repeated Eigenvalues, *AIAA Journal* 26: 867-881.

Murthy, D. V. and Haftka, R. T. 1988. Derivatives of Eigenvalues and Eigenvectors for a General Complex Matrix, *International Journal for Numerical Methods in Engineering* 26: 293-311.

Nelson, R. B. 1976. Simplified Calculations of Eigenvector Derivatives, *AIAA Journal* 14: 1201-1205.

Ojalvo, I. U. 1988. Efficient Computation of Modal Sensitivities for Systems with Repeated Frequencies, *AIAA Journal* 26: 361-366.

Plaut, R. H. and Huseyin, K. 1973. Derivatives of Eigenvalues and Eigenvectors in Non-Self-Adjoint Systems, *AIAA Journal* 11: 250-251.

Rudisill, C. S. and Chu, Y. 1975. Numerical Methods for Evaluating the Derivatives of Eigenvalues and Eigenvectors, *AIAA Journal* 13: 834-837.

Tan, R. C. E. 1986. Accelerating the Convergence of an Iterative Method for Derivatives of Eigensystems, *Journal of Comput. Phys.* 67: 230-235.

Taylor, D. L. and Kane, T. R. 1972. Multiparameter Quadratic Eigenvalue Problems. *Journal of Applied Mechanics* 42: 478-483.

Traill-Nash, R. W. 1981. Modal Methods in the Dynamics of Systems with Non-classical Damping, *Earthquake Engineering and Sturtural Dynamics* 9: 153-169.

Vanhonacker, P. 1980. Differential and Difference Sensitivities of Natural Frequencies and Mode Shapes of Mechanical Structures, *AIAA Journal* 18: 1511-1514.

Wang, B. P. 1985. Improved Approximate Method for Computing Eigenvector Derivatives in Structural Dynamics, *AIAA Journal* 29: 1018-1020.

Wilson, E. L. and Penzien, J. 1972. Evaluation of Orthogonal Damping Matrices, *International Journal for Numerical Methods in Engineering* 4: 5-10.

Zimoch, Z. 1987. Sensitivity Analysis of Vibrating Systems, *Journal of Sound and Vibration* 115: 447-458.

Creative Systems in Structural and Construction Engineering, Singh (ed.) © 2001 Balkema, Rotterdam, ISBN 90 5809 161 9

Minimum weight design of portal frames using a genetic algorithm

Ken Koyama
Department of Civil Engineering, Shinshu University, Nagano, Japan

Hajime Goto
Graduate School, Department of Civil Engineering, Shinshu University, Nagano, Japan

Shinichi Kikuta
Nihon Shuko Sekkei Company Limited, Tokyo, Japan

ABSTRACT: The minimum weight plastic design of portal frames is generally formulated as a linear programming problem subject to many constraints. In linear programming, the total number of active constraints is equal to the number of design or control variables. The number of control variables is relatively small compared with the number of constraints. Therefore, if possible, it would be desirable to select only the active constraints. In order to solve this problem, a genetic algorithm is applied in this paper. Among collapse mechanisms, combination mechanisms are generated by using genetic operation and adopted together with elementary mechanisms to from a set of constraints. Small sets of constraints are used to minimize a weight function and they are evolved in an optimization process. Simple two, three and four story frame examples are solved by the method proposed here.

1 INTRODUCTION

Minimum weight plastic design of portal frames usually has many constraints, depend on the frame geometry and applied load. The design is formulated as a linear programming problem to minimize the weight (Cohn 1972, Kirsch 1981). In the linear programming problem, the optimal solution will occur at an extreme point (Paul 1984). This means that if the number of design variables is n, then only n constraints among them become active. Therefore, if possible, only n constraints should be chosen as the active constraints to minimize the weight.

It is, however, difficult to choose these n constraints appropriately when the number of possible collapse mechanisms increases. The collapse mechanisms consist of elementary and linear combinations of elementary mechanisms. Combination mechanisms are called compound mechanisms in this paper.

Among elementary mechanisms, joint mechanisms are not included in the constraints, because the external work done by applied loads is zero. The combination mechanism which mutually does not eliminate the plastic hinge need not be included in the constraints, because that mechanism will not create a new feasible region. It is difficult to choose the proper mechanisms from many other possible mechanisms as the constraints. It is also difficult to create the possible compound mechanics.

2 APPLICATION OF GENETIC ALGORITHM TO MINIMUM WEIGHT PLASTIC DESIGN

In the minimum weight plastic design shown in Fig.1, for example, the number of possible plastic hinges N is 14, and the degree of statically indeterminateness R is 6, therefore the number of elementary mechanisms J is N-R=8. Elementary mechanisms consist of beam, panel and joint mechanisms and are also shown in Fig.1.Hereafter, excluding the joint mechanisms, beam and panel mechanisms are called elementary mechanisms.

The plastic conditions of mechanisms Fig.1(a) through (h) are written as:

$$4M_{p4} \geq P_1 l_1 \qquad (1.a)$$
$$4M_{p2} \geq P_2 l_2 \qquad (1.b)$$
$$4M_{p3} \geq 2H_1 h \qquad (1.c)$$
$$4M_{p1} \geq 2H_1 h + 2H_2 h \qquad (1.d)$$
$$-M_{p3} + M_{p4} \geq 0 \qquad (1.e)$$
$$-M_{p3} + M_{p4} \geq 0 \qquad (1.f)$$
$$-M_{p1} + M_{p2} + M_{p3} \geq 0 \qquad (1.g)$$
$$-M_{p1} + M_{p2} + M_{p3} \geq 0 \qquad (1.h)$$

in which $M_{pi}(i=1,2,3,4)$ are unknown design plastic moments, l is the length of a beam and h is the height of a frame column.

Compound mechanisms are generated as a linear combination of Fig.1(a) through (h). In order to generate a compound mechanism, the genetic algorithm (Goldberh 1985,1987,1989, Holland 1975, Jenkins 1991,1992, Rajeev 1992) is applied by assigning 0 or 1 to elementary mechanisms and 0, 1 or –1 to joint mechanisms. In this case, -1 implies a reverse rotation of a hinge in a joint mechanism.

0,1 and -1 are assigned randomly to a locus of strings by computer. In this example, J=8 therefore, an 8-bit string is used to code a compound mechanism. For example, if a string is coded as {0 1 1 0 0 1 1 –1}, which represents the linear combinations of Fig.1(b)+Fig.1(c)+Fig.1(f)+Fig.1(g)-Fig.1(h), then the decoded compound mechanism is given as:

$$2M_{P1}+2M_{p2}+3M_{p3}+M_{p4} \geq P_2l_2+2H_1h \qquad (2)$$

This mechanism is shown in Fig.2. In the same way, a total of 4 compound mechanisms are generated and used together with 4 elementary mechanisms Eq.(1a), (b), (c) and (d), are the constraints. Therefore the set of constraints consists of a total of 8 inequalities. In this two story example, a total of 4 constraint sets are used as the current constraint set and solved to minimize the following weight objective function by using the simplex method.

$$Z=4M_{P1}h+2M_{p2}l_1+4M_{P3}h+2M_{p4}l_1 \rightarrow min. \qquad (3)$$

Note that the elementary mechanisms are always included in each current constraint set. After solving this 4 optimal solutions are obtained from 4 current constraint sets.

The constraint sets that make the objective value greater than the average value are removed from current generation's constraint sets.

After selection,2 sets of constraints survive. Each set contains 4 elementary mechanisms and 4 compound mechanisms.

Excluding elementary mechanisms, new compound mechanisms are generated by a crossover operation using the 8 surviving compound mechanisms. In this example, a one point crossover operation is applied. The crossover point is chosen randomly. These new compound mechanisms are combined with the elementary mechanisms as the new generation's constraints. This process is repeated until the minimum weight is obtained. In order to increase the chance of finding the global optimal solution, the mutation operation is applied once for each generation. This has the effect of widening the search area in the solution space. The probability of mutation is fixed as 1/128 in the example. The optimal solution obtained after 10 generations in this example is $Z_{min}=10Pl^2$, $M_{P1}=1.0Pl$, $M_{P2}=1.5Pl$, $M_{P3}=M_{P4}=0.5Pl$, when the geometry and loads are $h=l_1=l$, $P_1=P_2=H_1=H_2=P$.

In the same manner, three and four story frames are solved. The optimal solution of plastic moments (unit=P1) is shown in Fig.3 together with the two story frame example. The span length and story height are equal to those of the two story frame.

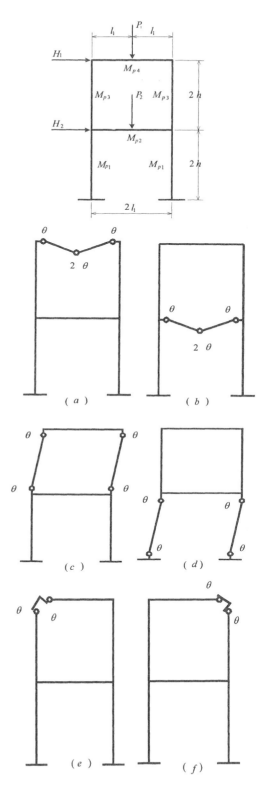

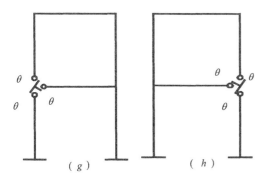

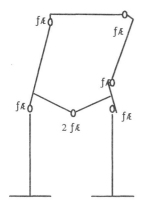

Fig.1 A story frame example and its elementary mechanisms

Fig.2 Example of a compound mechanism

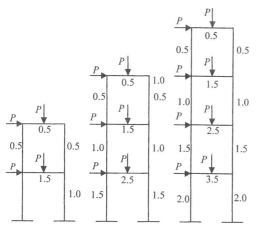

Fig.3 Optimal solutions of plastic moments for two, three and four story frame

Detailed information about the genetic algorithm are listed in Table.1.

Table 1 Details of the genetic algorithm used in frame examples

Frame stories	String length	Operations		Number of constraints per each set
		Crossover operation	Probability of mutation	
2	8	1 point	1/128	8
3	12	2 point	1/64	12
4	16	3 point	1/32	16

For the four story frame, for example, since the number of design variables is 8, at least eight inequality constraints become active. These 8 constraints are listed below.

$$4M_{P7} \geq 2Pl \quad (4.a)$$
$$4M_{P5} \geq 4Pl \quad (4.b)$$
$$4M_{P3} \geq 6Pl \quad (4.c)$$
$$4M_{P1} \geq 8Pl \quad 4.d)$$
$$3M_{P1}+M_{P2}+M_{P3}+2M_{P4}+2M_{P6}+2M_{P7} \geq 20Pl \quad (4.e)$$
$$3M_{P1}+M_{P2}+M_{P3}+2M_{P4}+M_{P5}+M_{P6}+M_{P7}+2M_{P8} \geq 20Pl \quad (4.f)$$
$$2M_{P1}+2M_{P2}+M_{P3}+M_{P4}+M_{P5}+2M_{P6}+M_{P7}+M_{P8} \geq 20Pl \quad (4.g)$$
$$2M_{P1}+2M_{P2}+M_{P3}+M_{P4}+2M_{P5}+M_{P6}+3M_{P7} \geq 20Pl \quad (4.h)$$

It is investigated if these equations satisfy the weight compatibility condition. By multiplying Eqs.4.a by $\alpha_i (I=1,2,\ldots 8)$, from top to bottom and adding to 4.h, the following equation is obtained.

$$(4\alpha_4+3\alpha_5+3\alpha_6+2\alpha_7+2\alpha_8)M_{P1}$$
$$+(\alpha_5+\alpha_6+2\alpha_7+2\alpha_8)M_{P2}$$
$$+(4\alpha_3+\alpha_5+\alpha_6+\alpha_7+\alpha_8)M_{P3}$$
$$+(2\alpha_5+2\alpha_6+\alpha_7+\alpha_8)M_{p4}$$
$$+(4\alpha_2+\alpha_6+\alpha_7+2\alpha_8)M_{P5}$$
$$+(2\alpha_5+\alpha_6+2\alpha_7+\alpha_8)M_{P5}$$
$$+(4\alpha_1+2\alpha_5+\alpha_6+\alpha_7+3\alpha 8)M_{P7}$$
$$+(2\alpha_6+\alpha_7)M_{P8}$$
$$=(2\alpha_1+4\alpha_2+6\alpha_3+8\alpha_4$$
$$+20(\alpha_5+\alpha_6+\alpha_7+\alpha_8))Pl \quad (5)$$

The objective function for the four story frame example is

$$Z=(4M_{P1}+2M_{P2}+4M_{P3}+2M_{P4}+4M_{P5}+2M_{P6}+4M_{P7}+2M_{P8}) \to \min. \quad (6)$$

Comparing the multiplier of $M_{Pi}(1,2,\ldots,8)$ in Eq.5 with Eq.6, and by solving it simultaneously, $\alpha_i(i=1,2,\ldots,8)$ are obtained as:

$$\alpha=0.16671,$$
$$\alpha_2=\alpha_3=\alpha_4=\alpha_6=\alpha_7=0.66671 \quad (7)$$
$$\alpha_5=\alpha_8=0.0$$

From this result, all $\alpha_i(i=1,2,\ldots,8)$ become greater than 0. This means that Eq.4 satisfies the weight compatible mechanism. Excluding the elementary mechanisms, four compound mechanisms are shown in Fig.4. The weight compatibility for the two and three story examples are also checked in the same way. For the four story example, the state of convergency of the weight function to the optimal solution is shown in Fig.5 as a function of generation. From this figure, it is seen that the optimal solution is obtained after about 50 generations as $Z_{min}=36Pl^2$

However, the total number of possible mechanisms is large. The formal total number of mechanisms is estimated to be $2^4 \times 3^4 = 1296$, $2^6 \times 3^6 = 46656$ and $2^8 \times 2^8 = 1679616$, respectively, since the two, three and four story portal frame examples have 8-bit, 12-bit and 16-bit string lengths. The number increases with the geometry of the frame and applied load. It is, therefore, very hard to choose proper mechanisms in constraints to minimize the weight function.

The method proposed here easily solves optimal plastic design problems involving large scale frames with many loads .

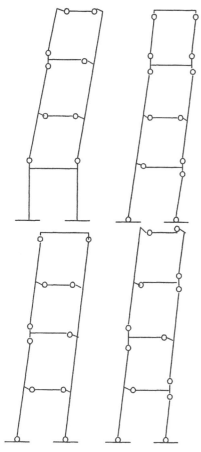

Fig.4 Weight ompatible compound mechanisms of a four story frame

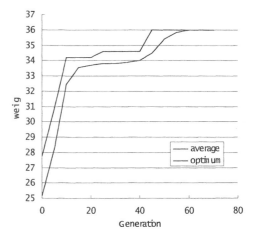

Fig.5 4 story frame optimal solution

3 CONCLUSIONS

Minimum weight plastic design problems of simple two, three and four story portal frames are solved by using a genetic algorithm. Optimal solutions are obtained using small sets of constraints. All solutions obtained here satisfy the condition of weight compatibility.

The examples solved here are relatively simple.

REFERENCES

Cohn, M. Z., Ghosh, S. K. & Parimi, S. R. (1972). "Unified approach to theory of plastic structures."J.EM5., ASCE, 1133-1158.

Gold berg, D.E. (1985)."Genetic algorithm and rule learning in dynamic system control." Proc. of The First Intl. Conference on Genetic Algorithm and Their Applications, pp.8-15, July 24-26.

Goldberg, D.E. & Samtani, M. P. (1987)."Engeneering optimization via genetic algorithm."Proc. of 9th onference on Electronic Computation, ASCE, NY.

Goldberg, D.E. (1989)."Genetic algorithm in search, optimization and machine learning." Adison Wesley, Boston, Mass.

Holland, J.H (1975)."Adaptation in Natural and Artificial Systems."University of Michigan Press.

Jenkins, W.M (1991). "Towards structural optimization via the genetic algorithm."Computers and Struct.,40(5), 1321-1327

Jenkins, W.M (1992). "Plane frame optimu design enveronment vased on genetic algorithm." J. Struct. Engrg., ASCE, 118(11), 3103-3112

Kirsch, U. (1981).OPTIMUMSTRUCTURAL DESIGN-Concepts, Methods, and Applications-" McGraw-Hill, NY

Paul J. Ossenbruggen(1984). System Analysis fpr Civil Engineers, John Wiley & Sons.

Rajeev,S..& Krishnamoothy, C.S. (1992)."Discreteoptimization of structures using genetic algorithms," J. Struct. Enorg., ASCE ASCE, 188(5), 1233-1250

Creative Systems in Structural and Construction Engineering, Singh (ed.) © 2001 Balkema, Rotterdam, ISBN 90 5809 161 9

Optimization of perforated walls for maximal fundamental frequency via genetic algorithms

C.C.Chen
Sinotech Engineering Consultants Incorporated, Taipei, Taiwan

ABSTRACT: Perforated shear walls are widely used in modern reinforced concrete buildings for better view or venting. The openings may be arranged at arbitrary location with arbitrary size, which leads to more complicated solution process. As a result, it would be of practical interest to investigate the effect of the above two factors on the natural frequencies of the perforated shear walls. The optimization is done by Genetic Algorithms which imitate the biological processes of reproduction and natural selection. These methods are ideally suited to optimization problems involve discrete, multi-modal, and non-convex design space. Through the use of Genetic algorithms, optimal design of the openings of the perforated shear walls is attempted. The design involves the selection of location and size of the openings such that the fundamental frequency of the perforated shear wall is maximal.

1 INTRODUCTION

Genetic Algorithms (GAs) (Holland, 1975) represent a class of innovative computational models based on Darwinian's principle of "survival of the fittest" to simulate mechanics of natural selection for artificial systems. GAs have gained widespread popularity in the optimization community because they impose less mathematical requirements for solving the problems and are ideally suited to problems involve discrete, multi-modal, and non-convex design space (Jenkins, 1992). The evolution process of GAs is done by firstly forming the population of strings (or individuals) which are the coding of design variables and analogous to the chromosomes of DNA. The strings in the population then go through a process of simulated evolution to evolve a new generation of population. In GAs, these strings are composed of genes which may take on some number of values called alleles. These algorithms work very well on a wide range of problems in many disciplines and their application is continuing to grow (Goldberg, 1989, Gen & Cheng, 1997). The GAs advocated by Holland and his co-workers are termed "Simple Genetic Algorithm" (SGA) which uses fixed-length binary strings, roulette wheel selection, one-point crossover and bitwise mutation (Goldberg, 1989).

Two main drawbacks are found when one applies SGA to the problems of structural optimization. The first one is so-called "premature" that genetic search is likely to be dominated by super chromosomes and converges to a local optimum. Another drawback is

the high computing cost associated with the genetic search. Since the eighties, GA researchers have investigated numerous GAs to improve the effectiveness and efficiency of SGA. These strategies deserve further attention to examine their usefulness on improving the performance of the GAs optimization.

In this study, we employ GAs for the frequency optimization of perforated shear walls. Shear walls are common structural components in modern reinforced concrete structures for resisting horizontal forces due to earthquake and wind. Perforated shear walls are walls that contain one or many openings to provide venting, lighting, viewing, and aesthetic needs. These openings result in stress concentration around the cutouts and change in the stiffness of walls. The former one has been addressed by many investigators during the past decades while no work has been located on the optimization problem despite its practical importance to engineers for structural layout and estimation of the horizontal forces in the preliminary design.

Motivated by the lack of literature, the present study aims to fill in this apparent gap in the optimization studies since natural frequencies are vital information to engineers for structural layout and estimation of horizontal forces to the structures. In perforated wall, the openings may be arranged at arbitrary location with any size that leads to more complicated solution process. Several techniques including the equivalent continuum method (Capuani, Merli & Savoia, 1994), modified beam method (Smith, 1969), Galerkin method (Coull & Mukherjee, 1973) and finite element method (Mackertich & Aswad, 1997) have been proposed for the analysis of perforated shear walls. Among them, the finite element method gains wide popularity because of its robustness and the availability of a wide range of commercial software. This approach is included in the optimization process.

In this study, we examine a number of GAs and evaluate their performance. In order to shorten the process of evaluation of the performance of various GAs, we introduce an eight dimensional Griewangk function as the test suite. The outcome of the experiment is used as the guideline for the subsequent optimization problems. Two examples are given to demonstrate the applicability of GAs.

2 SIMPLE GENETIC ALGORITHM

The design variables in GAs are encoded as binary genes in chromosomes that represent the candidate solutions for a problem. During the optimization, a population of chromosomes is generally maintained and the chromosomes are made to compete with each other for survival. The fitness of the chromosome is evaluated by a fitness function, which is closely related to the objective function. Greater chance for being selected into the parents group is given to stronger chromosomes than to weaker ones.

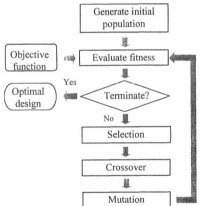

Figure 1. General procedure for optimization via GAs.

In SGA, the chromosomes are selected by the Roulette Wheel Selection (RWS) where chromosomes with higher fitness take up a larger portion for the selection. After being selected into mating pool, offspring are produced through crossover and mutating the mated parents, whereby they inherit features from each of the parents. The mutation operation allows some innovative features to be found in the offspring. In the next selection step, offspring become

candidate solutions and compete with each other. The same selection process goes through once again as shown in Figure 1. Here, the initial population is normally generated by random numbers or by perturbing user-provided chromosomes. The repeated selections of the best parents over generations are more likely to produce better offspring and eliminate those with poor fitness. These iterative procedures continue until the termination criteria are satisfied.

Crossover and mutation are controlled by the crossover probability p_c and the mutation probability p_m, respectively. These probabilities represent the expected number of chromosomes/bits to undergo crossover/mutation operations. It is generally conceived that a moderate population size, a higher probability for crossover and a lower probability for mutation yield better result. (De Jong, 1975)

3 TEST FUNCTION

As stated in first section, SGA has suffered two major setbacks which result in numerous GAs proposed over the past two decades to improve the efficiency and performance of GA. In order to evaluate the problem-solving ability of various improved GAs, we conduct simple yet efficient experiments on a test function. According to De Jong (Goldberg, 1989), the test problems are functions that possess some characteristics, including continuity, convexity, dimensions, unique or multiple modes, quadratic, and deterministic or stochastic characteristics. As pointed out by Hajela (Hajela, 1990), the structural optimization problems are found non-convex, discrete and multi-modal. Therefore, the test function must possess the above three characteristics.

Whitley (Whitley et al., 1996) listed ten common test functions for measuring the efficiency of various GAs. These functions include a unimodal function (F1), a nonlinear function over 2 variables (F2), a discontinuous function (F3), a noisy function (F4), a multi-modal function with several local optima (F5), a Rastrigin function (F6), a Schwefel function (F7), a Griewangk function (F8), a sine envelope sine wave function (F9) and a stretched V sine wave function (F10). Among them, F8 is scalable function which allows the algorithms to be tested on problems with progressively higher dimensionality. This non-convex and multi-modal function is well suited for the test function for structural optimization problems. As a result, F8 is selected for the numerical experiment of various GAs. To evaluate the efficiency of GAs that will be discussed in the following sections, we maximize a nonlinear, multi-modal, eight-dimensional F8 function. Equations 1 & 2 provide further details for maximizing these functions where A is a positive value so that $f(x)$ is always greater than 0.

$$Max_{x_i} \; f(x_i|_{i=1,n}) = 66 \times n - \sum_{i=1}^{n} \frac{x_i^2}{4000} + \prod_{i=1}^{n} \cos\left(\frac{x_i}{\sqrt{i}}\right) \quad (1)$$

subject to the following constraint,

$$-512 \le x_i \le 511 \qquad x_i \text{ is an integer} \quad (2)$$

If $n = 10$, the global optimum for Equation 1 is known to be 661 when $x_1 = ... = x_n = 0$. These design variables are encoded as binary strings with a resolution of 1024 for each variable. In other word, there are 1024^{10} possible solutions for this test problem. To avoid the probabilistic errors associated with the results, the results are taken from the average of 20 runs. Owing to the expensive computing cost of the structural analysis, it is unlikely to have large number of generations when one performs structural optimization. For this reason, an upper bound of 200 is given as the total number of generations.

4 IMPROVEMENT ON GENETIC ALGORITHMS

To assess the influence of different representation, we begin with an alternative encoding method for representing chromosomes by using Gray code (Gray, 1953). A major difference between Gray code and other representation methods, such as floating-point, list, array, etc., is that no modification is required for the genetic operations. Referring to Table 1, a Gray code takes a binary sequence and shuffles it to form some new sequence with the adjacency property.

Table 1 Comparison of three different encoding methods

Decimal code	Binary code	Gray code
0	000	000
1	001	001
2	010	011
3	011	010
4	100	110
5	101	111
6	110	101
7	111	100

Figure 2 presents the fitness of the best design at each generation while the numbers of evaluations required for yielding the best designs are illustrated in Figure 3. As GAs are also well-known as expensive tools for optimization, it is important to make a comparison on computational cost associated with the strategies. Note that only the minimal value is taken as the required number of evaluations for best-of-generation, that is, if the best design remains unchanged for several generations, the required number of evaluation for achieving it will be the same value.

In Figure 2, the use of Gray code seems to have higher convergence rate than its binary counterpart in the first 20 generations. At the last generation, the average of best fitness obtained by Gray encoding method is 652.474 which is slightly higher than 651.204, that of binary representation. From Figure 3, one observes that less evaluation is required for Gray encoding method when the generation is more than 160. It is concluded that Gray encoding method seems to be a better representation method for the test problem in terms of performance and computing cost.

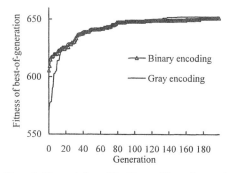

Figure 2. The evolution of the fitness of best-of-generation using two representation methods.

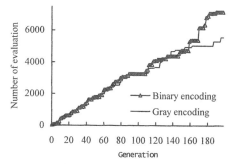

Figure 3. Number of evaluations required for the best-of-generation using two representation methods.

In the selection phase, SGA adopts generational replacement strategy where offspring replace the entire parent generation after creation. However there are strategies that retain the best designs in the enlarged space consisting of both the designs of parent and offspring generations, for example Steady-state GA (De Jong, 1975), (μ, λ) or (m, l) selection (Gen & Cheng, 1997) and elitism belong to this category of methods. Steady-state GA employs similar idea as De Jong that newly evolved offspring only replace the worst parents. On the other hand, (μ, λ) selection retains μ best unique parents and adopts RWS for the selection of $(\lambda - \mu)$ offspring.

From the fitness plots in Figure 4, it is found that generational replacement is superior to other methods in the first 50 generations. However, minor difference is seen between the above three replacement strategies in the final 20 generations. The fitnesses of the best designs are 654.271, 652.777 and 652.801 obtained from generational replacement, (μ, λ) selection and Steady-state GA, respectively. However in Figure 5, (μ, λ) selection requires only 1/5 evaluation than the rest two methods. In view of the benefit of less computing cost, we adopt (μ, λ) selection for the subsequent studies.

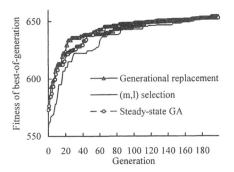

Figure 4. The evolution of the fitness of best-of-generation using various replacement strategies.

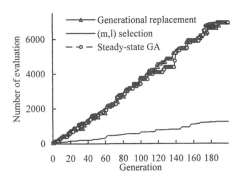

Figure 5. Number of evaluations required for the best-of-generation using various replacement strategies.

As the random process in the RWS may actually produce more or less than the expected number of copies, Brindle, (1981) proposed two famous sampling strategies: Remainder Stochastic Sampling (RSS) and Deterministic Sampling (DS). In RSS, chromosomes are allocated according to the integer part of their expected values at new generation and the rest population is filled through competing chromosomes based on the fraction part of their expected values. In DS, chromosomes receive a number of copies based on the integer part of their expected values as those do in RSS. Then, chromosomes are ranked according to the fractional

part of their expect values. Chromosomes ranked first are selected into the rest population. Clearly, if the sampling strategies only pick the best chromosome, then the population will quickly converge to that chromosome. However, this may be unfavorable in early generations because some super chromosomes may possess much higher fitness than average and dominate the search. Hence, the strategies should also pick some other chromosomes to preserve the diversity of population. Stochastic Universal Sampling (SUS) (Baker, 1985) that maintains the expected number of copies of each chromosome by creating a wheel with equally spaced markers. In this method, duplicated chromosomes are prohibited. Some randomly generated chromosomes will be included in the population if the number of offspring is less than that of parent generation. Similar effort is seen in Linear Ranking methods (LR) (Gen & Cheng, 1997) where the offsprings are allocated according to the rank of parent objective function values. Another class of selection methods named "Tournament Selection" (TS) (Gen & Cheng, 1997) combines the idea of ranking in a very efficient way. In TS, the offspring are created by repeatedly picking the best out of a tournament set of randomly generated chromosomes. Two chromosomes are normally picked in the set for competition.

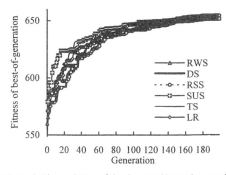

Figure 6. The evolution of the fitness of best-of-generation using various selection strategies.

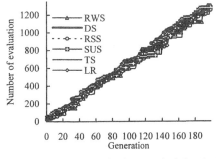

Figure 7. Number of evaluations required for the best-of-generation using various replacement strategies.

In Figures 6 and 7, the best designs yielded by RWS, DS, RSS, SUS, TS and LR are 652.207, 653.129, 651.552, 654.007, 653.444 and 654.717, respectively. From Figure 6, one knows that the curve representing SUS converges faster than other methods in the first 100 generations. However, no clear winner is seen in Figure 7. Although LR has the highest fitness of the best design, SUS seems to perform well during the whole history and it is therefore selected to be the selection strategy.

5 EXAMPLES AND DISCUSSIONS

By integrating the present improved GA with the commercial package SAP (Computers & Structures, 1997), we optimize the location and size of the openings of perforated shear walls for maximal fundamental frequency. The first example considers a rectangular shear wall with 14 feet in length, 21 feet wide, thickness of 1 inch and one rectangular opening with arbitrary size and location. To modeling the shear wall with window, this opening is assumed to maintain at least 1 foot offset from the edges of the walls. In addition, the wall is assumed to be fixed at the bottom as shown in Figure 8 and homogenous with the following properties,

$$\rho = 2.246 \times 10^{-7} \text{ lb/in}^3, E=3600\text{psi}, \nu = 0.2 \quad (3)$$

In the analysis, the shear wall without opening is modeled by 21×14 plane-stress elements. The location and shape of opening is controlled by two design variables which represent the element numbers of the lower left and upper right corners of opening. These two variables are to be optimally chosen such that the fundamental frequency λ_f of the perforated shear wall takes on the maximum value. The optimization problem may be mathematically expressed as

$$\text{Maximize } \lambda_f(e_1, e_2) \quad (4)$$

where e_1 and e_2 are arbitrary elements taking from the elements except those next to the edges.

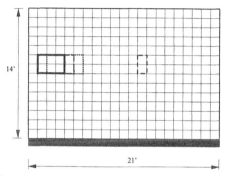

Figure 8. Geometry of the shear wall with the optimal opening.

Since the problem at hand is a maximization problem, one can use the fundamental frequency parameter as the fitness of the design, that is the fitness f_i for i-th chromosomal is given by λ_f. Based on the outcome of the test suites, we adopt SUS, (μ, λ) selection, 50%-uniform crossover in the GA. The population size, probability of crossover, and mutation rate are set to 50, 0.8 and 0.2, respectively. In addition, the maximum number of generations is assumed to be 200 and the design variables are encoded as Gray code for GA optimization.

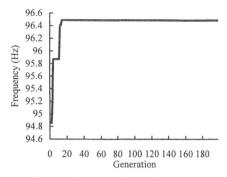

Figure 9. The evolution of best-of-generation for the perforated shear wall with an arbitrary window.

The evolution of the fundamental frequency of the best-of-generation is shown in Figure 9. It is seen that the search converges after 20 generations. The best design is found to be e_1=39 and e_2=18 as marked up with bold lines in Figure 8. The fundamental frequency of this perforated shear wall is 96.49 Hz, approximately 3% higher than that of a solid shear wall (93.78Hz). It is also observed that the stiffness of the perforated wall increases when the opening of best-of-generation (shown in dashed lines) moves from the center toward the left-hand size edge of the shear wall.

In Figure 10, we consider a six-floor shear wall with dimension 18' × 84' × 8" and door openings at each floor with dimension 4' × 8'. The material properties of the wall adopt the same values as given in Equation 3. The door is moving along the horizontal direction but maintains at least one foot with both edges of wall. Therefore, design variables are assumed to be elements at the lower left corner of openings. This optimization problem can be described by the following expression

$$\text{Maximize } \lambda_f(e_i, i = 1,6) \quad (5)$$

in which e_i represents the element number at lower left corner of the opening at i-th floor.

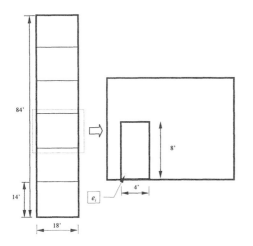

Figure 10. Geometry of the six-floor perforated shear wall.

From Figure 10, one knows that the design space consists of 13^6 possible solutions. Based on the results of the present GA, optimal design for the six-floor perforated shear wall is given in Table 2. The evolution of the present optimization is provided in Figure 11. The optimal solution is obtained before 75th generation. Maximal fundamental frequency is found to be 4.283Hz which is slightly higher than 4.27Hz, the fundamental frequency of perforated wall with all openings at the middle of the floors. Interestingly, if we look at the optimal design, its door openings at higher floors tend to be away from the middle while the door at first floor moves toward the middle.

Table 2. Optimal design of the six-floor perforated shear wall.

Design variables						Frequency
e_1	e_2	e_3	e_4	e_5	e_6	
589	603	533	799	309	1163	4.285 Hz

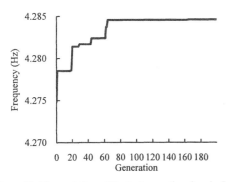

Figure 11. The evolution of best-of-generation for six-floor perforated shear wall.

6 CONCLUSIONS

This paper examines the optimization of perforated shear wall for maximal fundamental frequency by integrating the finite element method and the GA. Many GA strategies have been investigated by using a 10 dimension Griewangk function. Better strategies have been identified with their performance and efficiency in the optimization of test function and implemented into the improved versions of GAs. By selecting properly the test function, this approach provides an effective way for the GA optimization of real world problems.

Despite the increases are not significant, the results of the examples of perforated shear walls suggest that the fundamental frequencies of optimal designs are slightly higher than those of the perforated walls with traditionally arranged openings.

REFERENCES

Baker, J. 1985. Reducing bias and inefficiency in the selection algorithms. In Grefenstette, J.(ed), *Proceedings of the First International Conference on Genetic Algorithms*: 100-111. Hillsdale: Lawrence Erlbaum Associates.

Brindle, A. 1981. *Genetic Algorithms for Function optimization*, Ph.D. Dissertation, Edmonton: University of Alberta.

Capuani, D., Merli M. & Savoia, M. 1994. An equivalent continuum approach for coupled shear walls. *Engineering Structures* 16(1): 63-73.

Computers & Structures Inc. 1997. *SAP2000 Integrated Finite Element Analysis and Design of Structures Basic Analysis Reference*, version 6.1, Berkeley.

Coull A. & Mukherjee, P.R. 1973. Approximate analysis of natural vibrations of coupled shear walls. *Earthquake Engineering and Structural Dynamics* 2: 171-183.

De Jong, K.A. 1975. *An Analysis of the Behavior of a Class of Genetic Adaptive Systems*. Ph.D. Dissertation, Ann Arbor: University of Michigan.

Gen M. & Cheng, R. 1997. *Genetic Algorithms and Engineering Design*. New York: John Wiley and Sons.

Goldberg, D.E. 1989. *Genetic Algorithms in Search, Optimization and Machine Learning*. New York: Addison-Wesley.

Gray, F. 1953. *Pulse Code Communication*. U. S. Patent 2 632 058.

Hajela, P. 1990. Genetic search-an approach to the nonconvex optimization problem. *AIAA Journal* 28: 1205-1210.

Holland, J.H. 1975. *Adaptation in Natural and Artificial Systems*. Ann Arbor: The University of Michigan Press.

Jenkins, W.M. 1991. Structural optimisation with the genetic algorithm. *The Structural Engineer* 69(24): 418-422.

Mackertich S. & Aswad, A. 1997. Lateral deformations of perforated shear walls for low and mid-rise buildings. *PCI Journal* 42(1): 30-41.

Smith, B.S. 1969. Modified beam method for analyzing symmetrical shear walls. *ACI Journal* 66(12): 1005-1007.

Whitley, D. Mathias, K., Rana S. & Dzubera. J. 1996. Evaluating evolutionary algorithms. *Artificial Intelligence* 85: 245-2761.

Creative Systems in Structural and Construction Engineering, Singh (ed.) © 2001 Balkema, Rotterdam, ISBN 90 5809 161 9

Development of rate-dependent constitutive model for elastomers

A.F.M.S.Amin, M.S.Alam & Y.Okui
Department of Civil and Environmental Engineering, Saitama University, Japan

ABSTRACT: Equilibrium and instantaneous elastic responses from a solid enclose the boundary of the viscous domain where rate-dependent effect comes into play. However, due to experimental limitation, direct application of infinite slow or fast motions on the material is not practical to get such responses for determining material parameters. To this end, an experimental scheme for elastomeric materials has been proposed to identify the elastic parameters through extrapolation and using hyperelasticity model. In this course, an approach for developing finite deformation rate-dependent model incorporating modified hyperelasticity function has been introduced. The aspects of determining the viscosity parameter using the finite deformation model have been discussed. The adequacy of the proposed viscoelasticity parameter determination procedure in simulating the experimental results has been verified using the numerical model.

1 INTRODUCTION

Traditionally elastomeric materials find wide engineering applications in structural components like bridge bearings, shock absorbers etc. Recently, lead plugged natural rubber bearings (commonly known as lead rubber bearing) have been used in base isolation systems of earthquake resistant structures. Furthermore, for such specific end use, new special elastomers like high damping rubbers (HDR) with better energy absorption property have been developed. Unlike other elastomeric materials, HDR exhibits a rate-dependent elastic response. However, when subjected to cyclic loading, significant amount of hysteresis and permanent set also occur. Because of such features, the constitutive behavior of this material is not yet fully understood. Hence there exists the necessity to develop a rational constitutive model for simulating the mechanical response of HDR for using in analysis and design computations.

Generally, hyperelasticity laws are used to model the elastomeric response under monotonic loading. However, such approach can not model the rate-dependent behavior. So for developing a rate-dependent constitutive model for this range of materials, a rate-dependent hyperelasticity modeling is needed. Hence, the objective of this paper is to present an approach for developing a model to simulate the nonlinear rate-dependent behavior of the material. This paves the way for developing a complete model which will be able to simulate the cyclic response as well.

Figure 1 presents schematic representation of the viscoelastic domain. In a typical viscoelastic solid loaded in an infinitely slow rate, the stress-strain curve follows the E-E' path giving the equilibrium response. On the other hand, in case of infinitely fast motion, the response takes the I-I' path. Such response is known as instantaneous response. Both equilibrium and instantaneous response are elastic response and the domain of viscosity lies in between these two states (Huber & Tsakmakis 2000). Thus, elastic parameters determine the boundary of the domain where viscosity comes into play. These parameters need to be determined from experimental data.

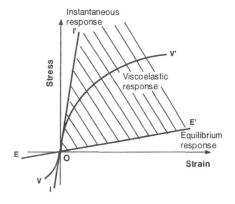

Figure 1. Typical responses from a viscoelastic solid.

However, from a practical point of view, neither infinite slow motion nor infinite fast motion is possible due to experimental limitation. In this situation, parameter determination process solely depends on numerical trials, which is not meaningful from physical point of view.

The paper presents an experimental scheme to identify the material parameters for equilibrium and instantaneous responses. In this course a finite deformation viscoelasticity model has been employed and parameter estimation process has been discussed. Finally numerical simulation results obtained from the model have been presented to discuss the adequacy of the proposed procedure.

2 EXPERIMENTAL OBSERVATION

An experimental scheme comprising of a multi-step relaxation test, monotonic compression tests and simple relaxation tests has been carried out to identify elasticity and viscosity parameters. All tests were carried out on pre-loaded specimens. The following subsections present the details of the experiments and salient features observed therein.

2.1 Experimental setup

In the present study, cubic specimens (50mm x 50mm x 50mm) were tested in a computer-controlled servohydraulic testing machine. In order to cut friction between the sample and the loading plates, polypropylene films with lubricant on top and bottom of the sample were used. The axial force and the displacement were recorded using a personal computer. The applied stretch (i.e. 1+dL/L, where L is the undeformed length) and the Cauchy stress (true stress) were calculated under the assumptions of homogenous deformation and incompressibility of the specimens, respectively.

2.2 Pre-loading

Prior to the actual experiment, all virgin specimens were subjected to specified pre-loading sequence. The objective of such preprocess was to obtain a stable state in the material by removing Mullins softening effect (Mullins 1969) from rate dependency phenomena. In the pre-process each virgin specimen was subjected to cyclic uniaxial loading for 5 cycles with a strain rate of 0.01/s. Figure 2 presents the stretch history and stress-stretch relation in a pre-loading test. Substantial softening behavior in the first loading cycle, known as Mullins effect, is evident from the figure. All specimens showed a repeatable stress-stretch response after passing 2-3 loading cycles. All tests described in the following subsections have been carried out 20 minutes after the end of respective pre-loading tests.

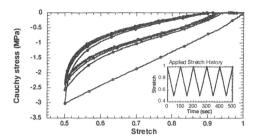

Figure 2. Applied stretch history and stretch-stress response observed in pre-loading.

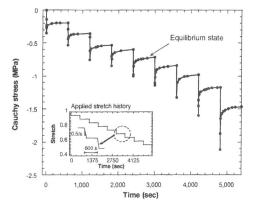

Figure 3. Applied stretch history and stretch-stress response observed in multi-step relaxation test.

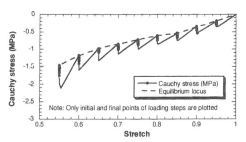

Figure 4. Stretch-stress response and approximated equilibrium locus obtained from multi-step relaxation test.

2.3 Multi-step relaxation test

Ideally, the equilibrium response is obtained when a material is loaded at infinitely slow rate. However, in case of highly viscous materials like elastomers it is quite difficult to specify a slow loading rate where the viscosity effect can be ruled out. In the current study, a multi-step relaxation test has been carried out to identify the equilibrium locus over the considered stretch range. Figure 3 presents the applied stretch and obtained stress history of the test. It is seen that, at the end of each relaxation interval of 10 min. duration, the stress relaxes to an apparent equilibrium state. Although such an equilibrium state can only be achieved in an asymptotic sense, these stages invariably indicate the neighborhood of equilibrium state. Figure 4 presents the stretch-stress response together with the approximated equilibrium locus in a multi-step relaxation test.

2.4 Monotonic compression test

The instantaneous elastic response of a solid is ideally obtained when the material is loaded at infinite fast rate. From an experimental point of view, however, there is a finite maximum value of stroke rate for any displacement controlled loading device. Although the loading rate on a specimen can be increased by using a smaller specimen dimension in the loading direction, the reduced aspect ratio of the specimen increases the boundary effects on other turn. In this context to find a method for estimating the instantaneous response of the material, a series of monotonic compression tests each at different but constant strain rate has been carried out. In the test sequence, strain rates were varied from 0.001/s to 0.96/s. For simplicity in illustration, Figure 5 shows the rate-dependent stress-strain responses observed only for the cases at strain rates of 0.001/s, 0.025/s, 0.075/s, 0.225/s, 0.47/s and 0.96/s respectively. The equilibrium locus as obtained from Figure 4 has also been compared here with the cases of different strain rates. The comparison of the curves displays the increase of stress response with increasing applied strain rate due to rate-dependency phenomena. However, at higher strain rates a diminishing trend in the increase of stress response is observed. From these data, the instantaneous response will be extrapolated which will be shown in the next section.

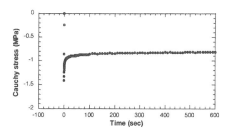

Figure 6. Stress history obtained from simple relaxation test at 0.7 stretch level.

2.5 Simple relaxation test

In order to observe the viscosity induced stress-relaxation phenomena, simple relaxation tests have been carried out at different stretch levels. Similar to the multi-step relaxation test, the strain rate in the loading phase was maintained at 0.5/s followed by a hold time of 10 min. in each test. Figure 6 presents the stress history obtained from the test at 0.7 stretch level. The figure illustrates a rapid stress relaxation feature of the material in the first 2 min. of hold time after which it approaches asymptotically towards an equilibrium state.

3 CONSTITUTIVE MODEL

The experimental observation summarized in Section 2 revealed the strain rate dependency of the material. Typically hyperelasticity laws can be used for modeling elastomeric response for a particular strain rate. However, for modeling nonlinear strain rate dependency of the material, hyperelasticity laws need to be combined with a rate-dependent model. The following subsections summarize the aspects of model configuration, the approaches for hyperelasticity, and rate-dependency modeling.

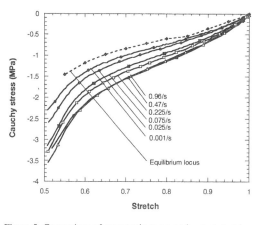

Figure 5. Comparison of monotonic compression test stretch-stress responses at different strain rates along with the equilibrium locus.

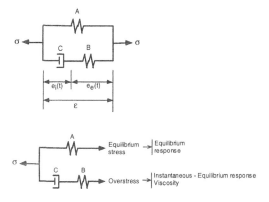

Figure 7. Three-parameter parallel model.

3.1 Model configuration

A three-parameter parallel model as illustrated in Figure 7 has been considered. In this model, the hyperelastic element A represents the equilibrium response, while the other branch consisting of hyperelastic element B and viscous dash-pot C represents the over-stress feature coming from the rate-dependent effect. The total strain ε has been decomposed into elastic strain e_e and inelastic strain e_i components.

3.2 Hyperelasticity modeling

In a phenomenological approach, under the assumption of isotropy, elastomeric materials are represented in terms of a strain energy density function W. Such functions for elastomers are expressed either in terms of the strain invarients or principal stretches. The strain-invariant-based models are easy to implement in a mathematical formulation, while the stretch based models suggest to be more flexible in fitting experimental data particularly at higher strain levels. However to follow a simpler computational approach, this paper chooses strain-invariant-based hyperelasticity models. In this approach, the three strain invarients (i.e. I, II, III) are expressed as:

$$I = \text{tr}\mathbf{B} = \lambda_1^2 + \lambda_2^2 + \lambda_3^2$$
$$II = \text{tr}\mathbf{BB} = (\lambda_1\lambda_2)^2 + (\lambda_2\lambda_3)^2 + (\lambda_3\lambda_1)^2 \quad (1)$$
$$III = \det\mathbf{B} = (\lambda_1\lambda_2\lambda_3)^2$$

where λ_1, λ_2, λ_3 are the principal stretches; left Cauchy-Green deformation tensor, $\mathbf{B=FF^T}$; $\mathbf{F}$=deformation gradient tensor.

Among the strain invariant based models, Mooney-Rivlin model is the most common one but does not perform well at higher strain levels. To solve this problem, a higher order function of I as proposed by Yamashita & Kawabata (1992) has been incorporated in this study for modeling the responses at higher strain levels. Equation 2 presents the strain energy density function of the hyperelasticity model.

$$W(I,II) = C_5(I-3) + C_2(II-3) + \frac{C_3}{N+1}(I-3)^{N+1} \quad (2)$$

where C_5, C_2, C_3, N are non-negative material constants. Here with the parameter C_3=N=0 the functions reduces to the original Mooney-Rivlin model.

In case of uniaxial loading, under the assumption of incompressibility, the third strain invariant III reduces to unity giving $\lambda_2 = \lambda_3 = (\lambda_1)^{-\frac{1}{2}} = (\lambda)^{-\frac{1}{2}}$. In such case, the function for Cauchy stress in loading direction (i.e. σ_1) for this model is expressed as:

$$\sigma_1 = 2(\lambda^2 - \frac{1}{\lambda})[C_5 + \frac{1}{\lambda}C_2 + C_3(\lambda^2 + \frac{2}{\lambda} - 3)^N] \quad (3)$$

In the stretch-stress relation, C_5 component dominates over the entire stretch regime of both tension and compression. However, the effect of C_2 component is effective only at tension zone (particularly at low stretch levels) in representing the characteristic 'S' shape stress-stretch relation. In contrast to these, the effect of C_3 component with an exponential term N is effective only at higher stretch levels.

3.3 Rate-dependency modeling

The stress and strain components of the three-parameter parallel model presented in Figure 7 has been converted into its finite deformation counterparts following the formulation of Huber & Tsakmakis (2000). The finite deformation model has been formulated under the framework of multiplicative decomposition of $\mathbf{F}$. Here the equilibrium strains e_e and e_i (Fig. 7) are related to equilibrium and intermediate equilibrium parts $\mathbf{F_e}$ and $\mathbf{F_i}$ so that $\mathbf{F = F_e F_i}$. This leads the Cauchy stress tensor $\mathbf{S}$ and rate of left Cauchy-Green deformation tensor $\mathbf{B}$ as follows:

$$\mathbf{S} = -p\mathbf{1} + \mathbf{S_E} \quad (4a)$$

$$\mathbf{S_E} = \mathbf{S_E^{(E)}} + 2\frac{\partial W'}{\partial \mathbf{I_{Be}}}\mathbf{B_e} - 2\frac{\partial W'}{\partial \mathbf{II_{Be}}}\mathbf{B_e^{-1}} \quad (4b)$$

$$\mathbf{S_E^{(E)}} = 2\frac{\partial W^{(E)}}{\partial \mathbf{I_B}}\mathbf{B} - 2\frac{\partial W^{(E)}}{\partial \mathbf{II_B}}\mathbf{B^{-1}} \quad (4c)$$

$$\dot{\mathbf{B}}_e = \mathbf{B_e L^T} + \mathbf{LB_e} - \frac{2}{\eta}\mathbf{B_e}\left(\mathbf{S_E} - \mathbf{S_E^{(E)}}\right)^D \quad 4(d)$$

where p is the hydrostatic pressure of $\mathbf{S}$ and subscript 'E' denotes the extra part of corresponding stress tensor. $\mathbf{1}$ is the identity matrix. The superscript '(E)' denotes the equilibrium stress while the subscript 'e' denotes equilibrium part of the strain tensor. $\mathbf{L}$ is the velocity gradient tensor. Super-script 'D' denotes the deviatoric part of stress. The derivative part of Equation 4b is the over-stress part (denoted by 'prime' sign) due to rate-dependent effect. η is the material parameter representing viscosity.

In deriving the explicit expressions for $\mathbf{S}$ and rate of $\mathbf{B_e}$, the hyperelasticity function presented in Equations 2 and 3 has been used together with Equation 4.

4 PARAMETER IDENTIFICATION

On the basis of experimental observation summarized in Section 2 and the constitutive model presented in Section 3, the following sub-sections present the parameter determination procedure for representing equilibrium response, instantaneous response, and viscosity effect.

4.1 *Equilibrium response*

The hyperelasticity model coefficients for the equilibrium locus obtained from the multi-step relaxation tests (Sec. 2.3) have been determined by a best-fit technique. Since the experimental observation was carried out in compression regime, there is no effect of C_2 coefficient for this zone. Hence to avoid getting negative values in curve fitting, C_2 was assigned to zero. The values of the parameters are listed in Table 1.

Table 1. Elastic material parameters

State	C_5 MJm^{-2}	C_3 MJm^{-2}	C_2 MJm^{-2}	N MJm^{-2}
Equilibrium	0.48	0.015	0.00	3.30
Instantaneous	0.85	0.120	0.00	3.30

4.2 *Instantaneous response*

The monotonic compression tests presented in Section 2.4 displayed a diminishing trend in an increase of the stress-strain response at higher strain rates indicating the approach of the instantaneous state. Interestingly, the overall stress-stretch response at each strain rate has a characteristic 'S' shaped curve, which can be described by the coefficients of the hyperelasticity model. On the basis of this feature, the hyperelasticity constants i.e. C_5 and C_3 have been determined for a constant value of N (as determined from the equilibrium locus) for different strain rate cases over the range of 0.001/s - 0.96/s. The C_5 and C_3 parameters determined this way have been plotted respectively in Figures 8 and 9 against the corresponding strain rate values.

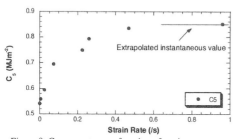

Figure 8. C_5 parameter as a function of strain rate.

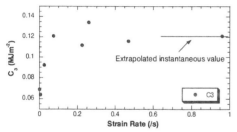

Figure 9. C_3 parameter as a function of strain rate.

It is interesting to note that the values of C_5 and C_3 parameters (Figs. 8,9) reach an asymptotic path over the strain rate of 0.25/s. This must be due to approach of the instantaneous state. The parameters for the instantaneous response are estimated from this asymptotic trend within finite strain rate region. The values of the parameters have been presented in Table 1. The subtraction of the values of C_5, C_3, C_2 from the instantaneous to equilibrium state gives the parameter values for the overstress response.

4.3 *Viscosity*

After determining the elastic parameters for the instantaneous and equilibrium response, the only remaining unknown is the viscosity parameter η representing dash-pot viscosity (Fig. 7). Here, simple relaxation test data has been used to obtain η through simulation trials of the rate-dependent hyperelasticity model (Sec. 3.3) by comparing the computed stress relaxation rate with experimental data.

For the relaxation test at 0.7 stretch level, η = 1.125 MPas represented the relaxation feature adequately (Figure 10). The parameter determined this way has been confirmed with simple relaxation data at other stretch levels.

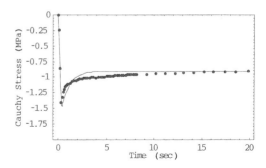

Figure 10. Optimization of viscosity parameter by simulating the simple relaxation test result (-) Numerical simulation, ($\bullet$) Experiment

5 NUMERICAL SIMULATION

Using the parameters determined in the preceding section, the model has been used to simulate the monotonic compression test at varied strain rates. Figure 11 presents the simulation results in comparison with experimental data, where a good conformity is observed. However, as expected, the representation of stress-stretch response at low stretch levels is a bit poor particularly at lower strain rates due to the limitation of hyperelastic model in that region. As a general trend, the numerical results slightly under estimated the response in all strain rate cases.

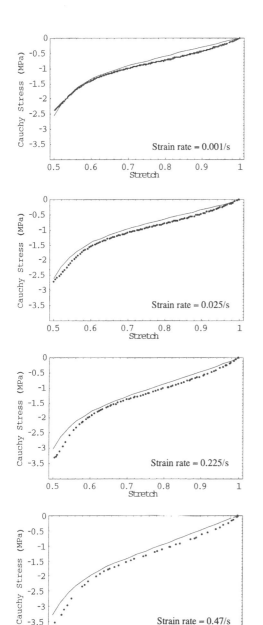

Figure 11. Numerical simulation of monotonic compression test at different strain rates. (–) Numerical simulation, (•) Experiment

tests with increasing strain rate is needed. The hyperelasticity model can then be used to find the parameters for equilibrium and instantaneous responses from these experimental data. In this connection, incorporation of an exponential term as proposed by Yamashita & Kawabata (1992) in the strain energy density formulation has been found to improve the stress-stretch representation at higher stretch level. After identifying the elastic parameters, the rate-dependent finite deformation hyperelasticity model can be used to find the viscosity parameter by comparing the simple relaxation test data. The comparison of numerical results with monotonic compression test results at varied strain rates has showed the adequacy of the proposed procedure. Although the parameter identification procedure and constitutive model presented in the paper are discussed in compression regime only, these are applicable for tension regime as well.

ACKNOWLEDGEMENT

The authors express their deep sense of gratitude to Professor H. Horii, Department of Civil Engineering, Tokyo University, Japan for extending experimental facility of his laboratory to carry out the experimental part of the work described in the paper.

REFERENCES

Huber, N. & Tsakmakis, C. 2000. Finite deformation viscoelasticity laws. *Mechanics of Materials.* 32:1-18.
Lion, A. 1996. A constitutive model for carbon black filled rubber: Experimental investigations and mathematical representation. *Continuum Mechanics and Thermodynamics.* 8:153-169.
Mullins, L. 1969. Softening of rubber by deformations. *Rubber Chemistry and Technology.* 42:339-362.
Yamashita, Y. & Kawabata, S. 1992. Approximated form of the strain energy density function of carbon-black filled rubbers for industrial applications. *Journal of the Society of Rubber Industry.* 65(9):517-528. (in Japanese)

6 CONCLUSIONS

The locus of equilibrium response of elastomers can be approximated from a multi-step relaxation experiment. However, for estimating instantaneous response, a series of monotonic uniaxial compression

Creative Systems in Structural and Construction Engineering, Singh (ed.) © 2001 Balkema, Rotterdam, ISBN 90 5809 161 9

Computer modeling of simply supported composite beam

S. Kamalanandan & I. Patnaikuni
RMIT University, Melbourne, Vic., Australia

ABSTRACT: The behaviour of composite beam is hard to predict as it depends on span, loads, geometry of sections, characters of members and stud location. Researchers have fewer clues about the behaviour of the composite beams at each point along the longitudinal direction of the beam. Most of the works carried out by researchers are limited to idealised characteristics of materials. In this study the actual characteristics of steel-concrete composite beams are analysed using actual properties of steel, concrete and stud shear connector. Push-out tests are cheaper and are easier methods to model composite beams. Once the load slip equation of studs embedded in concrete is found, modeling of composite beam becomes simpler. It is advisable to carry out push-out tests with different combinations and input load-slip equation obtained from these tests in to the model to study the behaviour of composite beams rather than testing composite beams on large scale.

1 INTRODUCTION

Full interaction of composite beams is defined as the condition of a composite beam with no slip occurring along the length of the composite beam at the concrete–steel interface. Full shear connection between the steel and concrete is defined as the least number of shear connectors required such that addition of more shear connectors will not increase the moment capacity of the beam. The term partial shear connection is used if the addition of more shear connectors increases the capacity of the beam.

2 BASIC THEORY

As explained by Johnson & Molenstra (1991), consider an element of thickness dx at a distance x from the left-hand side support as shown in figure 1. Figure 1 also shows the elements, the strain diagram and the internal forces.

Slip strain is given as $ds/dx = \varepsilon_s - \varepsilon_c - r\phi$ (1)

Where ε_s and ε_c are strains in steel and concrete at their center of gravity; ϕ is curvature; s is slip.

$r = (D_c + D_s)/2$

Where D_c is the overall depth of concrete slab and D_s is the overall depth of steel section.

To satisfy equilibrium $F = F_c = F_s$ (2)

Total moment M at the cross-section is given by

$M - M_c - M_s - rF = 0$ (2a)

The rate of shear transfer is given by the equation,

$dF/dx = C(s)/p$ (3)

Therefore $dF/dx - C(s)/p = 0$ (3a)

Where F is the force resisted by all the shear connectors in between the point of consideration and the support (equals to compressive force on concrete section); $C(s)$ is the load – slip function; p is the spacing of shear connectors.

Differentiating Equation 3a gives

$d^2F/dx^2 + (1/p)(dp/dx)(dF/dx) - (1/p)(dC/ds)(ds/dx) = 0$ (4)

The solution can be obtained by solving Equation 4 and substituting boundary conditions. A numerical solution is presented in the theoretical model.

3 THEORETICAL MODEL

The aim of this model is to calculate the parameters; moment, compressive force on concrete, force on each stud, deflection, slope, slip strain, slip and strain in steel and concrete at any cross-section for a given loading (or for a given curvature at critical section). The following model is explained for the case of a single point load at any point on the composite beam.

Assumptions and limitations for the theoretical model are as follows; (1) Steel and concrete have the same curvature at the interface; (2) Average stress in a strip of steel or concrete is assumed to be equal to the stress at the center of gravity of that strip; (3) Tensile stress of concrete is neglected.

The basic principle of composite beam is that out of three variables, namely curvature, moment and compressive force, if two variables are known then all the other parameter can be calculated.

The composite beam is divided into segments of length a. The critical point of the beam is first determined. Then each side of critical point is analysed separately. For illustration the left-hand side is considered first. Sections are numbered 0, 1, 2 etc. until the support point is reached. The notations for slip strain, slip and concrete compressive force at point 0 (critical section) are $S\_S[0]$, $S[0]$ and $F[0]$ respectively.

Critical section is analysed first. For given curvature, moment is calculated by assuming the compressive force. In other words known curvature and known compressive force can decide the position of neutral axis of steel and concrete. Computer program calculates all parameters including moment and slip strain $S\_S[0]$.

At the critical section slip is zero, thus $S[0] = 0$. Therefore $S[1]$ can be written as

$$S[1] = (a) \{S\_S[0]\} \qquad (5)$$

Using Equation 3

$$\{F[0] - F[1]\}/a = C(s)/p \qquad (6)$$

$C(s)$ is given as (Oehlers, slutter & Fisher 1971)

$$C(s) = P_u(1 - e^{-\beta s})^\alpha \qquad (7)$$

Where P_u is the ultimate shear force per connector and α, β are constants.

Equation 6 can be rewritten as

$$F[1] = F[0] - (a/p) \{P_u(1 - e^{-\beta s})^\alpha\} \qquad (8)$$

Moment at the second cross-section can be calculated using moment at first cross-section (critical cross-section). For the second cross-section concrete compressive force $F[1]$ and total moment are known. Therefore the curvature at second point can be calculated. Therefore other parameters including slip strain $S\_S[1]$ can be calculated.

This process is repeated until the last point is reached. At the last point compressive force is checked for zero value (assuming no further studs are present after support). Otherwise the process will be repeated with a new compressive force at the critical cross-section until the compressive force is closer to zero. A single program carries out all these calculations along the longitudinal direction of the composite beam.

4 COMPUTER PROGRAM

Computer program commences calculations from initial position of neutral axis depth of concrete and steel specified by the programer. Concrete above neutral axis is divided in to strips. Steel is divided in to strips above and below neutral axis.

For a given curvature compressive force is calculated based on full interaction analysis. The compressive force for a partial shear connection model will be lesser than that of full interaction model. These compressive force values give an indication of the compressive force value to be input in to the full / partial shear connection model.

Curvature and the compressive force are input in to the program and the moment is calculated. Program starts calculations from initial neutral axis depth. Firstly it calculates the strain at the center of each concrete element, and then calculates the stress on each element using stress- strain equation. Afterwards it calculates the force and moment in concrete.

If the compressive force out put from the program is lesser than that of specified compressive force (already input in to the computer program at the beginning) then the program will increase the neutral axis depth and repeat the calculations. The program keeps on adjusting the neutral axis depth of concrete until both compressive forces are closer to each other.

The program then calculates the net tensile force and moment in steel in a similar way like concrete. If the net tensile force is below the compressive force then the neutral axis depth of steel section is adjusted until the tensile force is closer to compressive force. Finally the program outputs the parameters including total moment and slip strain at the first cross-section. Moment at the other cross-sections can be calculated using the moment at the first cross-section.

The program calculates the compressive force on the next cross-section since the slip strain at the first cross-section is known. This was explained in section three. Therefore the curvature of second cross-section can be calculated since the moment and compressive force of second cross-section are known. This process is repeated for all other cross-sections.

5 MATERIAL PROPERTIES

Stress-strain relationship of concrete can be obtained from experimental results. This relationship can be compared with known stress-strain equations proposed by researchers. Equation suggested by Committee European du Beton (1970) is given by

$$\sigma = \sigma_0 \varepsilon(a - 206600\varepsilon)/(1 + b\varepsilon); \qquad (9)$$

Where $a = 39000(\sigma_0 + 7.0)^{-0.953}$;
$b = 65600(\sigma_0 + 10.0)^{-1.085} - 850.0$;

Where σ_0 is the peak stress attainable in MPa. The peak stress is taken as 0.85 of cylinder strength.

Equation suggested by Basu & Somerville (1969) can be modified without safety factor as

$$\sigma = \sigma_0\{2.41(\varepsilon/\varepsilon_u) - 1.865(\varepsilon/\varepsilon_u)^2 + 0.5(\varepsilon/\varepsilon_u)^3$$
$$- 0.045(\varepsilon/\varepsilon_u)^4\} \qquad (10)$$

Where σ_0 is peak stress; ε_u is the strain at maximum stress and it can be determined using cylinder tests. It was found that the experimental stress-strain character of concrete was closer to equation proposed by Basu & Somerville. Therefore this equation was selected to be input in to the model.

Researchers have already established stress-strain character of steel. Experimental results can be used to validate the predicted stress-strain character. In this model three straight lines were assumed to represent elastic, plastic and strain hardening regions.

Various research papers, journals have proposed the strength of shear connectors. Oehler and Johnson (1987) proposed an equation given by Equation 11

$$P_u = 5.0\,A_s f_u\,(E_c/E_s)^{0.4}\,(f_{cu}/f_u)^{0.35} \qquad (11)$$

Where P_u = Ultimate shear strength of the connector; A_s = Area of cross-section of stud shear connector; f_u = Ultimate tensile strength of stud; E_c = Elastic modulus of concrete; E_s = Elastic modulus of stud; f_{cu} = Cube strength of concrete.

Experimental results of Push-out tests show that Equation 11 gives results closer to experimental results. Equation 11 could be used to find the actual strength of stud shear connector if push-out test results are not available. Researchers (Oehlers, slutter & Fisher 1971) predicted the load –slip behaviour as

$$P = P_u(1 - e^{-\beta s})^{\alpha} \qquad (12)$$

Where P_u = Ultimate shear strength of the connector; P = Load applied; s = Slip; α, β = Constants.

Load-slip curves of different push-out tests could be analysed to find the constants α and β for different concrete strength combination.

6 EXAMPLE

As an example a simply supported composite beam is loaded at the center as shown in figure 2. The following data for the steel is taken from BHP edition (1998). Yield stress of web/flange = 320 MPa; Ultimate strength of web/flange = 460MPa; Young's modulus = 209150 MPa; Concrete strength = 51.11 MPa; P_u= 65 kN per connector; Studs were at 200mm spacing; Load-slip equation used was,

$$P = P_u\,(1 - e^{-0.7s})^{0.4} \qquad (13)$$

Computer program calculates the parameters at the center of the beam and (just) behind the studs. Three outputs are given for comparison from low curvature to a high curvature. This example shows that last the stud resists more load compare to other studs and in the plastic region load is distributed equally among the studs except the center stud.

For the first case as shown in table 1 the force on first stud closer to mid span = 200 – 179.5 = 20.5 kN; In the same way forces on second stud to last stud are 30.6 kN, 35.0 kN, 37.7 kN, 39.2 kN and 40.0 kN; The moment at cross-section 0 = 56.44 kN-m; Deflection at center of the beam = 3.93 mm.

Table 1. Outputs of computer program for a low curvature in which all the materials are in elastic range

Points	Cur x 10^6	Slip strain x 10^5	End slip	Strain x 10^4		Comp. Force
				Top of concrete	Bottom of steel	
	(per mm)		(mm)			kN
0	10.0	82.0	0	5.3	13.0	200.0
1	9.2	76.5	0.1	4.8	11.9	179.5
2	7.0	54.7	0.2	3.8	9.3	148.9
3	5.2	38.9	0.3	2.9	7.0	113.9
4	3.4	25.9	0.4	1.9	4.6	76.2
5	1.8	14.7	0.5	1.0	2.4	37.0
6	0	0	0.5	0	0	-3.0

Table 2. Outputs of computer program for a high curvature in which extreme fiber of concrete and steel are in plastic region.

Points	Cur x 10^6	Slip strain x 10^5	End slip	Strain x 10^4		Comp. Force
				Top of concrete	Bottom of steel	
	(per mm)		(mm)			kN
0	120.0	948.0	0	27.5	195.7	340.0
1	43.0	354.3	1.0	13.8	64.6	291.3
2	12.1	101.4	1.7	6.3	15.7	235.4
3	9.0	74.5	1.9	4.8	11.7	178.2
4	5.9	48.3	2.0	3.2	7.8	120.1
5	2.9	23.0	2.1	1.6	3.8	61.5
6	0	0	2.2	0	0	2.7

Table 3. Outputs of computer program when the composite beam is closer to collapse load.

Points	Cur x 10^6	Slip strain x 10^5	End slip	Strain x 10^4		Comp. Force
				Top of concrete	Bottom of steel	
	(per mm)		(mm)			kN
0	180	1391.7	0	39.6	298.3	360.0
1	70.9	574.9	1.4	18.8	111.5	306.2
2	13.6	114.2	2.5	6.9	17.8	245.9
3	9.5	79.4	2.8	4.9	12.2	184.8
4	6.3	52.8	2.9	3.3	8.1	123.3
5	3.2	27.1	3.0	1.7	4.1	61.5
6	0	0	3.1	0	0	-0.4

For the second case as shown in table 2 forces on first stud (closer to mid span) to last stud are 48.7 kN, 55.9 kN, 57.2 kN, 58.1 kN, 58.6 kN and 58.8 kN; The moment at cross-section 0 = 92.75 kN-m; Deflection at the center of the beam = 16.03 mm.

For the third case as shown in table 3 forces on first stud to last stud are 53. 8 kN, 60.3 kN, 61.1 kN,

61.5 kN, 61.8 kN and 61.9 kN; The moment at cross-section 0 = 96.88 kN-m; Deflection at the center of the beam = 23.49 mm.

7 EXPERIMENTAL PROGRAM

In stage 1 of the experimental series, push–out tests were carried out to establish the load–slip equation. The constants α and β of Equation 12 were found using the experimental results. Cylinders were tested to find the stress-strain characteristics of concrete.

In stage 2 of the experiment two composite beams B1 and B2 were cast with push-out specimens and concrete cylinders. The geometry of the beams is shown in figure 2. Beam B1 was designed for partial shear connection and beam B2 was designed for full shear connection.

Strain gauges were used to find the strain at top of top flange, bottom of bottom flange and top of concrete slab. Strain gauge location of beam B1 is shown in figure 2.

Load reached to a maximum of 190.22 kN and dropped to 183kN. At 183 kN one of the stud in the left-hand side of beam B1 has failed and the load dropped after this. Beam was loaded further and the studs in the left-hand side were sheared off one after another with a drop in load. Stud closer to the mid span did not fail.

Beam B2 reached a load of 204 kN and the load dropped to 188 kN and the concrete top surface was crushed and none of the studs failed.

8 VALIDATION OF THE MODEL

Beam B1 and beam B2 were analysed and it was found that the behaviour of these beams predicted by the model was very much closer to the experimental results. Also it was found that six composite beam results of other researchers also agreed with the proposed model. Analysis of beam B1 is presented below in detail.

It was found that for beam B1 the moment capacity of the model was six percent lesser than that of the experimental results. The deflection of the model coincided with the experimental curve in elastic region and 5-10 percent higher in the plastic moment region as shown in figure 3. The load-slip curve of the model shows 10-20 percent difference in most of the regions of the curve as shown in figure 4.

This is due to the assumptions made and due to experimental errors. The error in this experiment was probably due to friction between bottom flange of steel section and the support. Friction has some impact on the behaviour of the model, especially in load-slip character of the model. Equation 2 is not

valid since F_c & F_s are not equal. The difference could be up to 10 kN.

Concrete surrounding the stud has different stress distribution for the push-out tests compared to the beam tests. In case of push-out tests concrete near to the foot of the stud (near weld end) has more stress compare to the head end of the stud. In the case of the beam test the concrete near the foot of the stud has lesser stress compare to the head end of the stud. Slip is associated with the damage of concrete closer to the weld end in push-out tests. The damage of the test beam should be lesser and as a result the test beams could have shown lesser slip. Therefore the load-slip equation has to be modified to input in to the model. Also as per Equation 12 slip should have reached infinity at the ultimate load. Therefore the equation is not valid closer to maximum capacity of the stud shear connector. Also the post peak load – slip behaviour of studs are not defined by Equation 12.

It was found that the strain gauges placed at the bottom flange of the steel section gave readings of 10 – 20 percent less than the strains calculated from the model. This may be due to the reason that the strain gauges at the bottom of the bottom flange might have slipped. This slipping effect is more pronounced at the center of the beam compare to the end.

Also the strain gauges which were placed at the top of the top flange of the steel sections gave different readings to that of the model. Strain gauges, which were placed at the top of top flange, might have been affected by local bending of top flange due to the force on the studs.

Therefore the curvature derived from these strain gauges gave smaller curvature values compared to the value derived from the model.

Theoretical moment-curvature curve is shown in figure 5. The curve was limited up to a moment such that the stud force was 98 percent of the ultimate load (65 kN) of the stud.

9 CONCLUSION

This study shows that the ultimate moment capacity of the theoretical model is 90-95 percent of ultimate moment capacity of the experimental beams. The load-deflection behavior of the theoretical model is very close to the results of the experimental composite beams.

However the slips derived from the theoretical model show higher values than that of the experimental results.

The constants α and β have little effect on the ultimate moment capacity and on the load-deflection behaviour of the model. It suggests that the constants α derived from the push-out test need to be modified

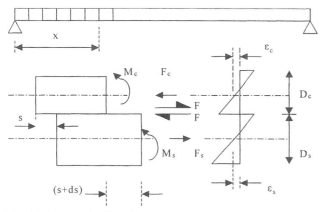

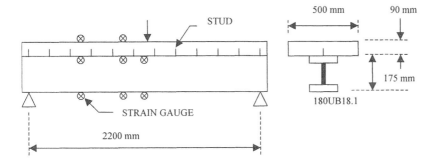

Figure 1. Element strain, slip and internal forces.

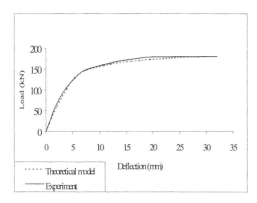

Figure 2. Geometry of beam and strain gauge location.

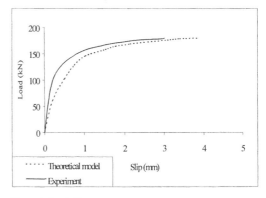

Figure 3. Load-deflection curves.

Figure 4. Load-slip curves.

to input in to the model to get the actual load-slip behaviour of the beam.

Load-slip equations derived from push-out tests for a particular type of shear connector with various concrete strengths could be modified and input in to the theoretical model to study the behaviour of composite beams.

Push-out experiments have to be done to establish more accurate load-slip equation to cover elastic, plastic and post peak regions. Also the post peak behaviour of concrete has to be researched. These accurate results could produce a perfect model. Good instrumentation could be adopted to validate the model.

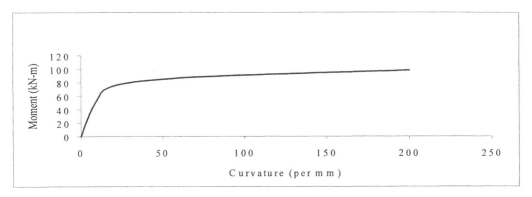

Figure 5. Moment-curvature diagram of theoretical model.

To study the behaviour of composite beams in laboratories it may be a better practice to produce accurate load-slip curves of shear connectors, stress-strain curves of materials and few composite beam tests to validate the program.

This model could be used to study the behaviour of composite beams subjected to any type of loading cases.

This model can be modified for design purpose by introducing modified stress-strain curves of concrete & steel and modified load-slip curves of shear connectors with safety factors.

Also this model could be modified to study composite beams subjected to variable repeated loading and it is currently being under preparation.

REFERENCES

Ansourian, P. & Roderick, J.W. 1978. Analysis of composite beams. Proc. AICE, Vol 104: 1631-1645.

Basu, A.K. & Sommerville, W. 1969. Derivation of formula for the design of rectangular composite columns. . Proc. Instn Civ. Engrs, Suppl.: 233-280.

BHP edition (Australia), 1998. Hot rolled and structural steel products.

Chapman, J.C. & Balakrishnan. S. 1964. Experiments on composite beams. The structural Engr Vol 42: 369-383.

Hawkins, N.M. & Roderick, J.W. 1988. The behaviour of composite beams. The Instn of Australia.

Johnson, R.P. & Molenstra, N. 1991. Partial shear connection in composite beams. Proc. Instn Civ. Engrs, Part 2: 679-704.

Menzies, J.B. 1971. Shear connectors in steel-concrete composite beams. Struct. Eng., 49, No. 3: 137-153.

Oehlers, D.J. & Johnson, R.P. 1987. The strength of stud shear connections in composite beams. Engr, 65B(2): 44-48.

Oehlers, D.J. & George, S. 1987. Composite beams with limited-slip shear capacity shear connectios. Journal of structural Engineering.

Ollgaard, J.G., Slutter, R.G. & Fisher, J.W. 1971. Shear strength of stud connectors in lightweight & normal weight concrete. AISC Eng.: 55-64.

Rotter, J.M. & Ansourian, P. 1979. behaviour and ductility in composite beams. Proc. Instn Civ. Engrs, Part 2: 453-474.

20 Structural analysis by special methods

Creative Systems in Structural and Construction Engineering, Singh (ed.) © 2001 Balkema, Rotterdam, ISBN 90 5809 161 9

A second-order fourth-moment reliability index

Yang-Gang Zhao & Tetsuro Ono
Department of Architecture, Nagoya Institute of Technology, Japan

ABSTRACT: In the second-order reliability method, the failure probability is generally estimated using simple or general paraboloids, and the accuracy of the estimation is dependent upon whether or not the general second-order limit state surface can be approximated appropriately by the paraboloids. In the present paper, for the estimation of failure probability corresponding to the general form of a second-order performance function, a second-order fourth-moment reliability index is presented. Since no parabolic approximation is performed, the proposed formulas provide better results than previous formulas of closed form. The second-order fourth-moment reliability index can also be used as an alternative reliability index that corresponds to either the simple or the general parabolic approximation, even for extremely unevenly distributed curvatures.

1 INTRODUCTION

Ultimately, structural reliability analysis is used to evaluate failure probability, the probability content in the failure set of the probability space. Difficulty in computing this probability has led to the development of various approximation methods. Of interest here is the second-order reliability method (SORM), wherein the limit state surface is approximated by a second-order surface in transformed standard normal space, and the failure probability is given as the probability content outside the second-order surface (Fiessler et al. 1979; Der Kiureghian et al. 1987, 1991; Tvedt 1988, 1990).

Since the expression for the failure probability that corresponds to a general second-order surface is quite complicated, this failure probability is usually estimated using a parabolic surface. Two such approximations have been developed. The first consists of approximating the second-order surface using a paraboloid defined by the principal curvatures of the limit state surface at the design point (Der Kiureghian et al. 1987,1991; Koyluoglu & Nielsen 1994; Cai & Elishakoff 1994), and is referred to here as general parabolic approximation. In the second, the second-order surface is approximated using a simple rotational paraboloid defined by the total principal curvature of the limit state surface at the design point (Zhao & Ono 1999a), and is referred to here as simple parabolic approximation. Several expressions have been developed for failure probability in closed form; however, the accuracy of these expressions is restricted according to whether or not the general sec-

ond-order limit state surface can be approximated appropriately by the paraboloids, which are simply special forms of second-order surfaces.

The present paper proposes a failure probability approximation that is performed directly using the general second-order performance function. Since no additional approximation, e.g. parabolic approximation is performed on the second-order performance function itself, the proposed method generally provides better results than other currently used SORM formulas of closed form.

2 REVIEW OF THE SECOND-ORDER RELIABILITY METHOD

The second-order Taylor expansion of a performance function in standardized space $G(\mathbf{U})$ at design point $\mathbf{U}^*$ can be expressed as (Fiessler, et al. 1979; Tvedt 1990):

$$G_S(\mathbf{U}) = \beta_F - \alpha^T \mathbf{U} + \frac{1}{2} (\mathbf{U} - \mathbf{U}^*)^T \mathbf{B} (\mathbf{U} - \mathbf{U}^*) \qquad (1)$$

where

$$\alpha = -\frac{\nabla G(\mathbf{U}^*)}{|\nabla G(\mathbf{U}^*)|} \qquad \mathbf{B} = \frac{\nabla^2 G(\mathbf{U}^*)}{|\nabla G(\mathbf{U}^*)|}$$

α is the directional vector at the design point in u-space, $\mathbf{B}$ is the scaled second-order derivatives of $G(\mathbf{U})$ at $\mathbf{U}^*$, known as the scaled Hessian matrix, and β_F is the first-order reliability index.

The difficult, and time consuming part of SORM is

the computation of the scaled Hessian matrix **B**. To address this problem, an efficient point-fitting parabolic approximation (Kiureghian et al. 1987, 1991) is derived, in which the major principal axis of the limit state surface and the corresponding curvature are obtained in the course of obtaining the design point, without computing the Hessian matrix; and an alternative point-fitting SORM was developed by (Zhao & Ono 1999b), in which the performance function is directly point-fitted using a second-order polynomial of standard normal random variables. The general polynomial expression of Eq. 1 is (Tvedt, 1990).

$$G_S(\mathbf{U}) = a_0 + \sum_{j=1}^{n} \left(\gamma_j \mu_j + \lambda_j \mu_j^2 \right) \quad (2)$$

where a_0, γ_j, and λ_j are $2n+1$ coefficients.

Since the expression for the probability content corresponding to Eq. 2 is quite complicated (Tvedt, 1990), the approximation is usually performed using the following parabolic surface (Fiessler, et al. 1979; Der Kiureghian et al. 1987, 1991):

$$G_S(\mathbf{U}) = \beta_F - u_n + \frac{1}{2} \sum_{j=1}^{n-1} k_j u_j^2 \quad (3)$$

where k_j, $j=1, ..., n-1$ are principal curvatures at the design point that are determined as the eigenvalues of a matrix having $n-1$ columns and $n-1$ rows approximately transformed from the scaled Hessian matrix.

For parabolas expressed by Eq. 3, a numerical integration method was developed by Tvedt (1988). Since the exact computation of the failure probability is quite complicated, numerous studies have contributed to the development of approximations of closed form. Tvedt (1988) has derived a three-term approximation. More accurate closed form formulas were derived by Koyluoglu & Nielsen (1994) and Cai & Elishakoff (1994) using McLaurin series expansion and Taylor series expansion. These formulas generally work well in the case of a large curvature radius and a small number of random variables. A simple parabolic approximation and an empirical second-order reliability index were developed (Zhao and Ono 1999a), providing a simpler and more accurate approximation method compared to the previously mentioned formulas.

The simple parabolic approximation is given as

$$G_S(\mathbf{U}) = \beta_F - u_n + \frac{1}{2R} \sum_{j=1}^{n-1} u_j^2 \quad (4)$$

where R is the average principal curvature radius:

$$R = \frac{n-1}{K_s} \qquad K_s = \sum_{j=1}^{n} b_{jj} - \alpha^T \mathbf{B} \alpha \quad (5)$$

K_s is the total principal curvature of the limit state surface at the design point, which is obtained without rotational matrix transformation or eigenvalue analysis of the scaled Hessian matrix. b_{jj}, $j=1, ..., n-1$ are the diagonal elements of **B**. α is the directional vector at the design point **U***.

Using R, K_s in Eq.5, the empirical second-order reliability index corresponding to Eq. 4 was given by Zhao & Ono (1999a).

Although the applicable ranges of the empirical reliability index are much larger than those of other SORM formulas (Zhao & Ono 1999a), an essential prerequisite of applying these formulas is that the limit state surface can be approximated appropriately by the paraboloid, a restriction that also applies to the general parabolic approximation. When this prerequisite can not be satisfied, the approximation gives significant errors, and thus requires an accurate method by which the failure probability corresponding to the performance function expressed in Eq. 2 can be obtained.

For the accurate computation of failure probability corresponding to Eq. 2, an exact integral expression for the probability content of the quadratic set is given by Tvedt (1990), and an Inverse Fast Fourier Transformation (IFFT) method is given by Zhao and Ono (1999b); however, no direct approximations of closed form for Eq. 2 have been reported. Therefore, the present paper introduces an estimation of closed form for obtaining the failure probability directly, using the performance function in Eq. 2.

3 SECOND-ORDER FOURTH-MOMENT RELIABILITY INDEX

3.1 Moment Approximation for Second-Order Reliability

In this study, the evaluation of the failure probability corresponding to Eq. 2 is performed directly utilizing the probability moment information of the second-order performance function. If the central moments of the performance function can be obtained, the failure probability, which is defined when the probability content for the performance function is less than or equal to zero, can be expressed as a function of the central moments.

If the first two moments of $G_S(\mathbf{U})$ are obtained, the second-moment reliability index is expressed as

$$\beta_{SM} = \frac{\mu_S}{\sigma_S} \quad (6)$$

where μ_S and σ_S are the mean value and standard deviation of the second-order performance function $G_S(\mathbf{U})$, respectively, and Φ is the probability distribution function of a normal random variable.

As shown in the next section, because the first two moments are generally not sufficient, high-order moments will be used.

3.2 Second-Order Fourth-Moment Reliability Index

For the second-order performance function $G_S(\mathbf{U})$, if the first four moments are obtained, using the principle of high-order moment standardization technique (HOMST) (Ono & Idota, 1986), the standardized variable

$$Z_u = \frac{Z - \mu_S}{\sigma_S} \tag{7}$$

is related to a standard normal random variable through the following equation (Zhao et al. 1999):

$$u = \frac{\alpha_{3S} + 3(\alpha_{4S} - 1) Z_u - \alpha_{3S} Z_u^2}{\sqrt{(5\alpha_{3S}^2 - 9\alpha_{4S} + 9)(1 - \alpha_{4S})}} \tag{8}$$

where α_{3S} and α_{4S} are the third and fourth dimensionless central moment ratios, i.e. the skewness and kurtosis, respectively, of $Z = G_S(\mathbf{U})$.

Since

$$Prob[Z \leq 0] = Prob\left[Z_u \leq -\frac{\mu_S}{\sigma_S}\right] = Prob[Z_u \leq -\beta_{SM}] \tag{9}$$

the fourth-moment reliability index is obtained as

$$\beta_{FM} = \frac{3(\alpha_{4S} - 1)\beta_{SM} + \alpha_{3S}(\beta_{SM}^2 - 1)}{\sqrt{(9\alpha_{4S} - 5\alpha_{3S}^2 - 9)(\alpha_{4S} - 1)}} \tag{10}$$

When $\alpha_{3S} = 0$, Eq. 10 degenerates as $\beta_{FM} = \beta_{SM}$.

3.3 Moments for the Second-Order Performance Function

The second-order performance function $G_S(\mathbf{U})$ in Eq. 2 is a second-order polynomial of independent standard normal random variables u_j, $j = 1, ..., n$. The first four central moments can be obtained utilizing the moment characteristics of the normal random variable. They are given by:

$$\mu_S = a_0 + \sum_{i=1}^{n} \lambda_i \tag{11}$$

$$\sigma_S^2 = \sum_{i=1}^{n} \left(\gamma_i^2 + 2\lambda_i^2\right) \tag{12}$$

$$\alpha_{3S} \sigma_S^3 = 2 \sum_{i=1}^{n} \lambda_i \left(3\gamma_i^2 + 4\lambda_i^2\right) \tag{13}$$

$$\alpha_{4S} \sigma_S^4 = 3 \sum_{i=1}^{n} \left(\gamma_i^4 + 20\gamma_i^2 \lambda_i^2 + 20\lambda_i^4\right) + 6 \sum_{i=1}^{n-1} \sum_{j>i}^{n} \left(\gamma_i^2 \gamma_j^2 + 4\lambda_i^2 \lambda_j^2 + 2\gamma_i^2 \lambda_j^2 + 2\gamma_j^2 \lambda_i^2\right) \tag{14}$$

In particular, for the general parabolic approximation Eq. 3, the first four central moments are given by:

$$\mu_S = \beta_F + \frac{1}{2} \sum_{i=1}^{n-1} k_i \tag{15}$$

$$\sigma_S^2 = 1 + \frac{1}{2} \sum_{i=1}^{n-1} k_i^2 \tag{16}$$

$$\alpha_{3S} \sigma_S^3 = \sum_{i=1}^{n-1} k_i^3 \tag{17}$$

$$\alpha_{4S} \sigma_S^4 = 3 + 3 \sum_{i=1}^{n-1} \left(\frac{5}{4} k_i^4 + k_i^2\right) + \frac{3}{2} \sum_{i=1}^{n-2} \sum_{j>i}^{n-1} k_i^2 k_j^2 \tag{18}$$

In particular, for the simple parabolic approximation Eq. 4, the first four central moments are given by:

$$\mu_S = \beta_F + \frac{n-1}{2R} \tag{19}$$

$$\sigma_S^2 = 1 + \frac{n-1}{2R^2} \tag{20}$$

$$\alpha_{3S} \sigma_S^3 = \frac{n-1}{R^3} \tag{21}$$

$$\alpha_{4S} \sigma_S^4 = 3 + \frac{3(n-1)}{R^2}\left(1 + \frac{(n+3)}{4R^2}\right) \tag{22}$$

Once μ_S, σ_S, α_{3S} and α_{4S} are obtained, the second-order second moment reliability index β_{SM} and the second-order fourth moment reliability index β_{FM} can be readily obtained using Eqs. 6 and 10 respectively.

4 EXAMPLES AND INVESTIGATIONS

4.1 Investigation for a Rotational Parabolic Surface

The first example considers Eq. 4 directly, the performance function used in the simple parabolic approximation, from which the empirical second-order reliability index was obtained. For this example, the empirical second-order reliability index gives better approximations than other reliability indices of closed form (Zhao & Ono, 1999a).

Variations of the computed second-order reliability indices with respect to the number of random variables n, the first-order reliability index β_F, and the curvature R are shown in (a), (b) and (c) of Fig. 1, respectively, for the exact results, for the results obtained using the empirical second-order reliability index, and for the results obtained using the proposed second-order second- and fourth-moment reliability indices. In Fig. 1, the exact results are obtained using direct integration, the curves above the lines of the first-order reliability index show the results of the limit state surface with positive curvature (convex to the origin) and those below the first-order reliability index show the results of the limit state surface with negative curvature radius (concave to the origin). Figure 1 reveals the following:

(1) In general, both the empirical second-order reliability index and the proposed second-order fourth-moment reliability index improve the first-order reliability index greatly and give good approximations of the exact results for almost the entire range of the number of random variables n (from 2 to 30), the first-

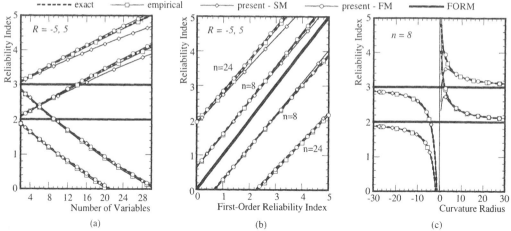

------ exact   —□— empirical   —◇— present - SM   —○— present - FM   ■■■ FORM

Figure. 1   Variation of reliability index with respect to n, β_F and R for Ex. 1

order reliability index β_F (from 0 to 5), and the curvature R (from -30 to -0.5, from 4.0 to 30), that was investigated, for not only positive curvature but also for negative curvature..

(2) In general, the second-order fourth-moment reliability index gives better approximations of the exact results than the empirical second-order reliability index. In addition, the second-order fourth-moment reliability index gives better results for the limit state surface concave to the origin than for that convex to the origin.

(3) For a small number of random variables or for limit state surfaces concave to the origin, even the simple second-order second-moment reliability index gives very good approximations of the exact results.

(4) When the absolute curvature radius is large sufficiently (implies that the limit state surface is linear enough), or the number of random variables is sufficiently small, both the exact and the proposed second-order fourth-moment reliability index indices are also close to the first-order reliability index. In other words, FORM can only provide accurate results when the curvature radius is very large.

The numerical values of the first four central moments and the second-order reliability index for various specific values of parameter β_F, n and R are listed in Table 1. Table 1, shows that for a relative large curvature radius, e.g. $R>5.0$ or a small number of variables, e.g. $n<8$, the skewness α_3 is close to 0.0 and the kurtosis α_4 is close to 3.0, which are equivalent to the values for a normal random variable. In other words, for a relative large curvature radius or a small number of variables, the second-order performance function used in the simple parabolic approximation $Z=G_S(\mathbf{U})$ obeys normal distribution approximately and the second-order reliability index can be approximately obtained from only the first two central moments of the second-order performance function. For an extremely small curvature radius, e.g. $R=0.5$ or 1.0, the skew-

ness α_3 is quite large and the second-order fourth-moment reliability index produces significant errors. This may occur because the HOMST is not accurate for large value of skewness. Note, in Table 1, for $R=0.5$ or 1.0, the curvature radius is less than half of the first-order reliability index, so the nonlinearity of the limit state surface is too strong. In these cases, the IFFT method have to be used (Zhao & Ono 1999b).

From the investigation performed in this example, the second-order fourth-moment reliability index can be used as an alternative reliability index that corresponds to the simple parabolic approximation.

4.2 Investigation for a General Paraboloid with Unevenly Distributed Curvatures

The second example considers the following performance function in standard normal space; (for this example, investigated by Zhao and Ono (1999a), none of the currently used SORM formulas, including the empirical second-order reliability index, gave satisfactory results),

$$G(\mathbf{U})= \beta_F - u_8 + \frac{1}{2}\sum_{j=1}^{7} a^j u_j^2 \qquad (23)$$

where a is a factor having a value from -0.8 to 0.8. Because a^j changes according to j, the paraboloid expressed by Eq. 23 has unevenly distributed curvatures. The smaller the absolute value of a, the more unevenly the curvature is distributed.

The variations of the second-order reliability indices with respect to a are shown in Fig. 2, for the exact results, for the results obtained using the empirical second-order reliability index, and for the results obtained using the proposed second-order second- and fourth-moment reliability indices. The first-order reliability index is assumed as 2.0, and the exact results are obtained by IFFT method (Zhao and Ono 1999b).

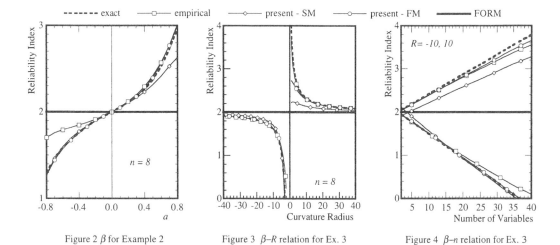

Figure 2 β for Example 2　　　　Figure 3 β–R relation for Ex. 3　　　　Figure 4 β–n relation for Ex. 3

Table 1 Numerical results for Example 1

Parameters			Moments				Reliability Index		
β_F	n	R	μ	σ	α_3	α_4	SM	FM	Exact
0	8	5.0	0.70	1.07	0.05	3.03	0.656	0.652	0.653
1	8	5.0	1.70	1.07	0.05	3.03	1.592	1.604	1.604
2	8	5.0	2.70	1.07	0.05	3.03	2.528	2.570	2.564
3	8	5.0	3.70	1.07	0.05	3.03	3.465	3.549	3.529
4	8	5.0	4.70	1.07	0.05	3.03	4.402	4.542	4.499
5	8	5.0	5.70	1.07	0.05	3.03	5.339	5.548	5.473
2	2	5.0	2.10	1.01	0.01	3.01	2.079	2.084	2.083
2	6	5.0	2.50	1.05	0.04	3.02	2.384	2.411	2.407
2	24	5.0	4.30	1.21	0.10	3.05	3.500	3.691	3.731
2	30	5.0	4.90	1.26	0.12	3.06	3.898	4.174	4.136
2	8	15.0	2.23	1.01	0.00	3.00	2.216	2.217	2.217
2	8	10.0	2.35	1.02	0.01	3.00	2.309	2.314	2.314
2	8	5.0	2.70	1.07	0.05	3.03	2.529	2.465	2.564
2	8	2.5	3.40	1.25	0.23	3.22	2.722	2.963	2.928
2	8	1.0	5.50	2.12	0.73	4.04	2.590	3.215	3.539
2	8	0.5	9.00	3.87	0.96	4.49	2.324	2.956	4.035

Figure 2 shows that for $a>0$, which implies that the curvatures have the same signs, both the empirical and the second-order fourth-moment reliability indices provide good approximations of the exact reliability index while the SOSM reliability index produces significant errors.

For $a<0$, which implies that the curvatures have different signs and that the total principal curvature is negative, both the second-order second-moment and the second-order fourth-moment reliability indices provide good approximations of the exact reliability index while the empirical reliability index produces significant errors. This implies that the proposed second-order fourth-moment reliability index can be applied to problems with unevenly distributed principal curvatures. For $a<0$, the simple second-order second

moment reliability index is sufficiently accurate to obtain the second-order estimate of the failure probability.

4.3 Investigation for a Spherical Surface

The third example considers the following performance function in standardized space, which is the general case of the practical examples used by Cai (1994) and Koyluoglu (1994).

$$G(\mathbf{U}) = \pm \left[\sum_{j=1}^{n} \left(u_j - d_j \right)^2 - R^2 \right] \quad (24)$$

The limit state surface of (24) is a hypersphere with radius R and center at point $(d_j, j=1,...,n)$, in which the sign + implies that R is positive and the surface is convex to the origin, and the sign − implies that R is negative and the surface is concave to the origin. $y=G(\mathbf{U})$ is a random variable having non-central chi-squared distribution, and the exact value of probability $P_F=Prob\{G(\mathbf{U})<0\}$ is computed directly using this distribution (Sankaran 1963). The non-central parameter, i.e. the distance from the origin to the spherical center δ, is expressed as:

$$\delta^2 = \sum_{i=1}^{n} d_i^2 = (R + \beta_F)^2 \quad (25)$$

The variations of the second-order reliability indices with respect to the curvature radius are shown in Fig. 3, in which the first-order reliability index is taken to be 2.0 and is depicted as a horizontal line, and the number of random variables is assumed to be 8. The curves above the line of the first-order reliability index show the results for positive R and those below the line of the first-order reliability index show the results for negative R. Figure 3 shows that for negative curvature radius, the second-order fourth-moment reliability index agrees very well with the exact results. For positive curvatures, the second-order fourth-moment method gives a very good approximation of the exact results when the curvature radius is relatively large. The second-order fourth-moment reliability index produces significant errors when the curvature radius is very small, perhaps due to the large skewness, as described previously.

The variations of the second-order reliability indices with respect to the number of random variables are shown in Fig. 4, in which the first-order reliability index is taken to be 2.0 and is depicted as a horizontal line, and the curvature radius is taken to be -10 or 10. The curves above the line of the first-order reliability index show the results for the limit state surface convex to the origin (R=10) and those below the line of the first-order reliability index show the results for the limit state surface concave to the origin (R=-10). Figure 4 reveals that for a negative curvature radius, the second-order fourth-moment reliability index agrees very well with the exact results. For positive curvatures, the second-order fourth-moment method also gives the very good approximation of the exact results. The second-order second-moment reliability index generally produces large errors because the first two central moments are not sufficient for approximating the failure probability.

From the investigation in this example, the second-order fourth-moment reliability index can be used as a reliability index that corresponds to a second-order polynomial without parabolic approximation.

5 CONCLUSIONS

1) For computation of the failure probability corresponding to the general form of the second-order performance function, a moment approximation of second-order reliability and a second-order reliability index with closed forms , i.e. the second-order fourth-moment reliability index, are proposed. Since no additional approximations, e.g. simple or general parabolic approximations, are performed, the proposed formulas give better results for second-order reliability than currently used formulas of closed form.

2) The second-order fourth-moment reliability index can also be used as an alternative reliability index that corresponds to the simple parabolic approximation. The reliability index gives very good approximations of the exact results for both positive and negative total principal curvatures.

3) The second-order fourth-moment reliability index can also be used as an alternative reliability index that corresponds to the general parabolic approximation. Moreover, the second-order fourth-moment reliability index can be used even for a paraboloid having extremely unevenly distributed curvatures, for which other formulas can not provide satisfactory results.

4) For both the simple parabolic approximation and general parabolic approximation, when the total principal curvature is negative, i.e. the limit state surface is concave to the origin, another very simple reliability index, i.e. the second-order second-moment reliability index, also provides very good approximations of the exact results. For the general second-order performance function, the second-order second-moment reliability is generally insufficient for the accurate computation of second-order reliability.

5) When the skewness is large, the second-order fourth-moment reliability index produces significant errors. This should be considered when using the proposed approximation.

ACKNOWLEDGMENTS

This study was in part supported by a Grant-in-Aid for Scientific Research (B) (No.10555199) from the Ministry of ESSC, Japan. The support is gratefully acknowledged.

REFERENCES

Cai, G.Q. and Elishakoff, I., (1994), "Refined Second-Order Reliability Analysis." *Structural Safety*, 14 , 267-276.

Der Kiureghian, A., Lin, H.Z. and Hwang, S.J., (1987), "Second-Order Reliability Approximations." *J. Engrg. Mech. ASCE, Vol.* 113, *No.* 8, 1208-1225.

Der Kiureghian, A. and De Stefano, M., (1991), "Efficient Algorithm for Second- Order Reliability Analysis." *J. Engrg. Mech. ASCE, Vol.* 117, *No.* 12, 2904-2923.

Fiessler, B., Neumann, H-J., and Rackwitz, R., (1979), "Quadratic Limit States in Structural Reliability." *J. Engrg. Mech. ASCE*, 105(4), 661-676.

Koyluoglu, H.U. and Nielsen, S.R.K., (1994), "New Approximations for SORM integrals." *Structural Safety*, 13 235-246.

Ono, T. and Idota, H., (1986). "Development of high-order moment standardization method into structural design and its efficiency." *J. Structural and construction engineering, AIJ*, Tokyo, 365, 40-47 (in Japanese).

Sankaran, N, (1963), "Approximations to the non-central Chisquire Distribution." *Biometrika*, 50, 199-204.

Tvedt, L. (1988), "Second-order Reliability by An Exact Integral." Proc. 2*nd IFIP Working Conference on Reliability and Optimization on structural Systems*, ed. by P. Thoft-Chistensen, Springer Verlag, 377-384.

Tvedt, L. (1990), "Distribution of Quadratic Forms in the Normal Space - Application to Structural Reliability." *Journal of Engrg. Mech., ASCE*, 116(6), 1183-1197.

Zhao, Y.G. and Ono, T., (1999a). "New approximations for SORM: Part 1." *Journal of Engrg, Mech., ASCE*, 125(1), 79-85.

Zhao, Y.G. and Ono, T., (1999b). "New approximations for SORM: Part 2." *Journal of Engrg, Mech., ASCE*, 125(1), 86-93.

Zhao, Y.G., Ono, T. and Idota H., (1999), "Response uncertainty and time-variant reliability analysis for hysteretic MDF structures." *Earthquake Engineering & Structural Dynamics*, 28,1187-1213.

Creative Systems in Structural and Construction Engineering, Singh (ed.) © 2001 Balkema, Rotterdam, ISBN 90 5809 161 9

Limit analysis of reinforced concrete columns by the yield line theory

S. Uehara
Ariake National College of Technology, Omuta, Japan

K. Sakino
Kyushu University, Fukuoka, Japan

F. Esaki
Kyushu Kyoritsu University, Kitakyushu, Japan

ABSTRACT: A new method to predict the horizontal shear capacity and the failure mechanism of reinforced concrete columns is proposed. The method is based on the yield line theory, i.e., the upper-bound theorem, considering the interaction of bending moment, shear force and axial force. By analyzing 26 column specimens, selected by the Japan Concrete Institute, it is cleared that the horizontal shear capacity and the yield line inclination angle of columns can be predicted by the proposed method.

1 INTRODUCTION

Evaluating the ultimate behavior of reinforced concrete (referred as RC, hereafter) members is quite important when the performance of RC structures are examined against severe earthquakes. The strut-and-tie models (Schlaich et al. 1987) and combined models of truss and arch (the Architectural Institute of Japan 1999), are commonly used to estimate the strength of RC members. These models are considered as lower-bound models since resisting mechanisms are assumed.

However, no upper-bound model has been proposed, considering the interaction of bending moment, shear force and axial force. Hence, a new method, based on the yield line theory, is proposed in the present paper.

Figure 1 shows a column subjected to antisymmetric loads. In the proposed method, yield lines are assumed as planes which cross the extreme compressive corner of member ends. As a result, yield lines are equivalent to slanting RC sections. The yield criterion of those RC sections is evaluated by the superposition theory.

The column fails when the yield lines reach at their yield strength. Figure 2 shows a failure mechanism. The middle element may move and rotate. Horizontal shear capacity Q can be calculated by the equilibrium condition with internal forces at yield lines. The minimum value would be the horizontal capacity of the column when the yield line inclination angle θ_f is varied,.

The features of the proposed method are (1) using

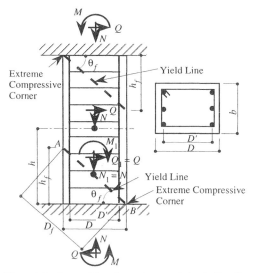

Figure 1. A column subjected to antisymmetric load of bending moment, axial force and shear force

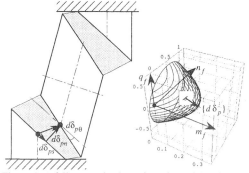

Figure 2. A failure mechanism of a column following the associated plastic flow rule

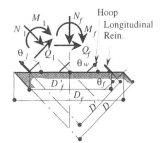

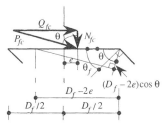

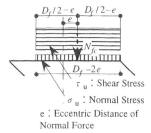

Figure 3. Forces and moment on a reinforced concrete section

Figure 4. Sectional forces on concrete at a yield line section

Figure 5. Averaged stress blocks on concrete at a yield line section

yield criterion of section forces; (2) using generalized yield line theory considering the interaction of bending moment, shear force and axial force; and (3) being possible to predict the failure mechanism of RC columns.

2 ASSUMPTIONS IN THE ANALYSIS AND YIELD CRITERION

2.1 Assumed columns and assumptions in the analysis

The objective columns of the proposed method are limited to follow the next conditions.

1. Loading of a column should be antisymmetric type, simple beam type or cantilever beam type. In other words, axial force and shear force should be uniform along the member.

2. Sections of a column should be rectangular. Longitudinal reinforcements and hoops should be arranged orthogonally and uniformly along the column.

Furthermore, following assumptions in the analysis are introduced.

1. A yield line is a plane which crosses the extreme compressive corner.

2. Increments of plastic displacements $\{d\delta_p\} = \{d\delta_{p\theta}, d\delta_{pn}, d\delta_{ps}\}^T$ are assumed to follow the associated plastic flow rule as Figure 2 shows. Hence, Separated elements move and rotate relatively according to the ratio of forces and moment at yield lines.

3. Concrete is assumed to be rigid-plastic material whose strength is $\nu\sigma_B$, when tensile strength is ignored. Reinforcement is also assumed to be rigid-plastic material and dowel action is ignored.

Here, $d\delta_{p\theta}, d\delta_{pn}, d\delta_{ps}$: increments of plastic rotation, plastic normal displacement and plastic tangential displacement respectively, σ_B : the cylinder compressive strength, ν : effectiveness factor.

2.2 Yield criterion of concrete

Yield lines are assumed to cross the extreme compressive corners of columns from existing test evidences.

Provided that an arbitrary yield line of inclination angle θ_f is a plane, the section of the yield line would be a rectangular section with breadth b and depth D_f as Figure 3 shows, where M_1, N_1, Q_1 are moment, axial force and shear force at the centroid of the section on longitudinal axis respectively; M_f, N_f, Q_f are moment, axial force and shear force at the centroid of the section on normal axis to the yield line section D' respectively, where D' : the distance from the centroid of longitudinal compressive reinforcing bars to the centroid of longitudinal tensile reinforcing bars, θ_l : crossing angle of longitudinal reinforcing bars to the yield line section, θ_w : crossing angle of hoops to the yield line section, and $D'_f = D' / \cos\theta_l$.

Concrete at a yield line section is assumed to be subjected to bending moment M_{fc}, normal force N_{fc} and shear force Q_{fc} when the yield line section of RC yields being subjected to M_f, N_f, Q_f. The rest of the forces and bending moment are to be resisted by reinforcement.

The force P_{fc} working at the eccentric point stands for M_{fc}, N_{fc} and Q_{fc}, as Figure 4 shows, where e is the eccentric distance of normal force and $M_{fc} = N_{fc} \cdot e$. When the region of failure is assumed to spread from extreme compressive corner to the symmetrical point with respect to the working point of P_{fc}, averaged stress block of concrete can be estimated as shown in Figure 5.

Assuming that concrete fails when the principal stress calculated from averaged stresses shown in Figure 5 reaches $\nu\sigma_B$, the yield criterion of concrete is expressed as

$$\frac{1}{2}\left(\frac{n_{fc}^2 + q_{fc}^2}{\nu^2} - \frac{n_{fc}}{\nu}\right) + \frac{m_{fc}}{\nu} = 0 \qquad (1)$$

where

$$m_{fc} = \frac{M_{fc}}{b\,D_f^2\,\sigma_B} \quad , \quad n_{fc} = \frac{N_{fc}}{b\,D_f\,\sigma_B} \quad , \quad q_{fc} = \frac{Q_{fc}}{b\,D_f\,\sigma_B}$$

Equation (1) is a curved surface as Figure 6 shows.

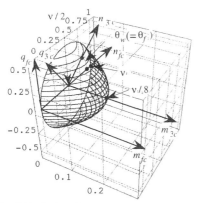

Figure 6. Yield criterion of concrete

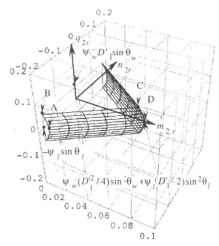

Figure 7. Yield criterion of reinforcements

The equation is a parabola when n_{fc} or q_{fc} is constant and a circle when m_{fc} is constant. As Figure 7 shows, Equation (1) may be changed to an formula on m_{3c}, n_{3c}, q_{3c} -coordinate as follows, which is the central axis of the surface:

$$\frac{1}{2}\left(\frac{n_{3c}^2}{v^2}+\frac{q_{3c}^2}{v^2}-\frac{1}{4}\right)+\frac{m_{3c}}{v}=0 \qquad (2)$$

The m_{3c}, n_{3c}, q_{3c} -coordinate is convenient when addition theory is applied in later chapter.

2.3 Yield criterion of reinforcements

Columns are supposed to be reinforced only by longitudinal reinforcing bars and hoops. As a result, the yield criterion of reinforcements is expressed by the surfaces shown in Figure 7. The surfaces are obtained by adding the yield criterion of longitudinal reinforcing bars, the triangle in Figure 7, to the yield criterion of hoops, the parabola in Figure 7, where

$$m_{2r}=\frac{M_{1r}}{bD_f^2\sigma_B} \; , n_{2r}=\frac{N_{1r}}{bD_f\sigma_B} \; , q_{2r}=\frac{Q_{1r}}{bD_f\sigma_B} \; ,$$

$$\psi_l=2\frac{a_t}{bD}\frac{\sigma_y}{\sigma_B} \; , \psi_w=p_w\frac{\sigma_{wy}}{\sigma_B}$$

and M_{1r}, N_{1r}, Q_{1r} : bending moment, axial force and shear force resisted by reinforcement respectively, a_t : area of longitudinal tension reinforcement, σ_y : yield strength of longitudinal tension reinforcement, p_w : web reinforcement ratio, σ_{wy} : yield strength of hoops, $D'_1 = D'/D$.

2.4 An approximate yield criterion of RC sections

The exact yield criterion of RC sections can be obtained by adding the criterion of reinforcements, Figure 7, to the yield criterion of concrete, Figure 6. However, the exact yield criterion is inconvenient because the criterion is consisted of many mathematical functions. Consequently, an approximate yield criterion is proposed in the present paper. Figure 8 shows the proposed yield criterion of a RC section. The curved surface is obtained by expanding the yield criterion of concrete as much as the effect of reinforcements on axes m_3, n_3 and q_3 . The yield surface in Figure 8 is expressed by

$$\frac{1}{2}\left\{\left(\frac{n_3}{v\alpha_n}\right)^2+\left(\frac{q_3}{v\alpha_q}\right)^2-\frac{1}{4}\right\}+\frac{m_3}{v\alpha_m}=0 \qquad (3)$$

where

$$\alpha_n=\frac{a}{\frac{v}{2}} \; , \quad \alpha_q=\frac{b}{\frac{v}{2}} \; ,$$

$$\alpha_m=\frac{\frac{v}{8}+\left(\frac{1}{4}D'_1\psi_w\sin^2\theta_f+\frac{1}{2}\psi_l\cos^2\theta_f\right)D'_1}{\frac{v}{8}}$$

and

$$a=\sqrt{\frac{c_n^2d_q^2-d_n^2c_q^2}{d_q^2-c_q^2}} \; , \quad b=\sqrt{\frac{d_n^2c_q^2-c_n^2d_q^2}{d_n^2-c_n^2}}$$

$$c_n=\psi_l\cos\theta_f \; , c_q=\psi_wD'_1\sin\theta_f+\frac{v}{2} \; ,$$

$$d_n=\psi_l\cos\theta_f+\frac{v}{2} \; , d_q=\psi_wD'_1\sin\theta_f$$

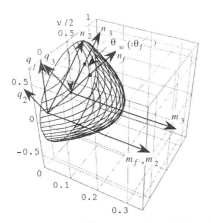

Figure 8. Approximate yield criterion surface of a RC section

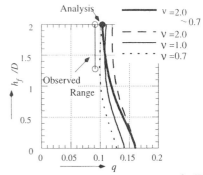

(a) Change of horizontal shear capacity q with h_f/D

$\psi_l = 0.230$, $\psi_w = 0.012$

$n = 0.137$, $M/QD = 1.0$

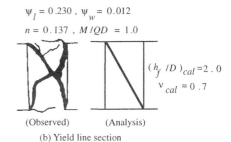

$(h_f/D)_{cal} = 2.0$

$\nu_{cal} = 0.7$

(Observed)　　　(Analysis)

(b) Yield line section

Figure 9. Analysis of specimen Mic-No.1

The variables α_n , α_q and α_m are expanding factors on axes m_3 ,n_3 and q_3 .

Exact yield criterion of RC sections will be discussed elsewhere.

3 THE HORIZONTAL SHEAR CAPACITY OF RC COLUMNS

In the proposed method, the approximate yield criterion of RC sections is used because of the convenience in the analysis. Consequently the solution by the method would be an approximate upper-bound value.

Using the equilibrium condition with the variables n_3 , q_3 and m_3 of yield criterion, and the non-dimensional axial force n , shear force q at the point of contraflexure horizontal shear capacity (shear force) q can be expressed as follows,

$$q = \frac{-B + \sqrt{B^2 - 4AC}}{2A} \qquad (4)$$

where

$$A = \frac{1}{2\nu}\frac{1}{\alpha_q^2}\cos^2\theta_f \ ,$$

$$B = \frac{1}{2\alpha_q^2}\cos\theta_f\sin\theta_f + \frac{1}{\alpha_m}\left(\frac{h}{D} - \frac{1}{2}\tan\theta_f\right)\cos^2\theta_f \ ,$$

$$C = \frac{1}{2\nu}\left\{\frac{1}{\alpha_n^2}\left(n - \frac{\nu}{2}\right)^2\cos^2\theta_f + \frac{1}{\alpha_q^2}\left(\frac{\nu}{2}\right)^2\sin^2\theta_f - \left(\frac{\nu}{2}\right)^2\right\}$$

It is difficult to obtain the smallest q by differentiating Equation (4). Hence, iteration of calculation is proposed to obtain the minimum values varying incli-

nation angle of a yield line.

On the other hand, it is not easy to estimate properly the effectiveness factor ν at yield line sections, since the strength of concrete is changeable by cracking, confinement, strain gradient, etc. It may be large at member-end sections for confinement from girders and/or strain gradient, and it may be small at diagonal sections of members for cracking. Consequently, in the present analysis, the effectiveness factor is changed lineally from $\nu = 2.0$ at a member-end section to $\nu = 0.7$ at a diagonal section of a column loaded antisymmetrically, in proportion to the section height h_f of assumed yield line . These values of ν are decided by referring the study by Takiguchi (1992,1998).

4. ANALYSES OF SPECIMENS SELECTED BY JCI

To demonstrate the proposed method, 26 specimens selected by the Japan Concrete Institute (JCI, 1983) are analyzed.

Figure 9(a) shows the change of horizontal shear capacity q with the ratio of a yield line section height h_f to the depth of column D . The thick solid line is the change of q , when varying $\nu = 2.0$ to $\nu = 0.7$. Other lines of $\nu = 0.7$, $\nu = 1.0$ and $\nu = 2.0$ are shown to indicate the effect of ν . The figure indicates that the smallest strength is at diagonal section and the horizontal shear capacity in the test is estimated fairly well by the predicted value. In addition, the Figure 9(b) indi-

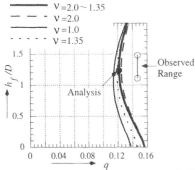

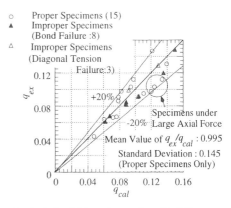

(a) Change of horizontal shear capacity q with h_f/D

$\psi_l = 0.488$, $\psi_w = 0.047$
$n = 0.217$, $M/QD = 2.0$

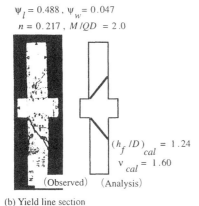

$(h_f/D)_{cal} = 1.24$
$v_{cal} = 1.60$

(Observed)　(Analysis)

(b) Yield line section

Figure 10. Analysis of specimen Mic-No.2

Figure 11. Analysis of 26 specimens selected by JCI

cates that the predicted yield line section coincides with the one of the test. In fact, the specimen failed in diagonal tension mode, which is not proper to be estimated by the proposed method because the horizontal shear capacity is effected by the tensile strength of concrete. However, the error seems to be small because the specimen is reinforced moderately and loaded with some axial force.

In Figure 10(a) $v = 1.35$ at the diagonal section is assumed since the loading of the column is simple beam type($v = (0.7+2.0)/2 = 1.35$). The Figure 10(a) indicates that the horizontal shear capacity is underestimated a little. However, the Figure 10(b) shows that the yield line section is predicted fairly well by the analysis.

Figure 11 shows the results of 26 specimens on horizontal shear capacity. The 15 specimen failed in flexural failure mode, flexural compression failure mode, shear tension failure mode or shear compression failure mode, are called "proper specimens", since these failure modes are assumed in the present method. On the other hand, the other 11 specimens failed in bond failure mode or diagonal tension failure mode, are called "improper specimens", since these failure modes

are not assumed in the present method.

The proposed method predict the horizontal shear capacity of the 15 proper specimens fairly well. The horizontal shear capacity of specimens in bond failure mode are overestimated, which does not contradict with the fact that the predicted values are from the upper-bound theorem.

Figure 12 shows the comparison of the predictions with the observed on the height incline of yield line section, h_f/D. The minimum and maximum of h_f/D in tests are decided from photos as in Figure 12(b). The predicted height inclines are relatively near to the observed.

Table 1 shows the properties of specimens analyzed in the present paper.

5 CONCLUSIONS

A method to predict the horizontal shear capacity and the failure mechanism of reinforced concrete columns is proposed. The method is based on the yield line theory, i.e., the upper-bound theorem. By analyzing 26 column specimens, selected by the Japan Concrete Institute, it is cleared that the horizontal shear capacity and the yield line inclination angle of RC columns are well predicted by the proposed method.

However, longitudinal reinforcements in most of 26 specimens yielded at failure. Hence, the examination of the proposed method on shear failure specimens, in which longitudinal reinforcements do not yield, is necessary.

REFERENCES

Architectural Institute of Japan 1999. Design *guidelines for earthquake resistant reinforced concrete buildings based on inelastic displacement concept*: Architectural Institute of Japan: 142-162 (in Japanese)

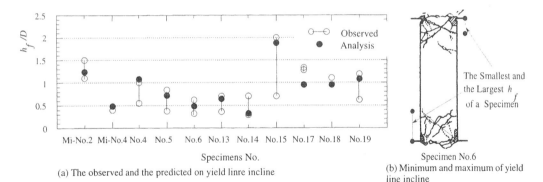

(a) The observed and the predicted on yield linre incline

(b) Minimum and maximum of yield line incline

Specimen No.6

Figure. 12 Predicted yield line incline with the observed

Table 1. Analytical results and properties of specimens selected by the Japan Concrete Institute

Specimen No.*1	b (mm)	D (mm)	M/QD	n	ψ_l	ψ_w	σ_B (MPa)	Load *2	q_{ex}	q_{ex}/q_{cal}	n_{cal}	h_f/D (Analysis)	h_f/D (test)*3	Observed Failure*4
△Mic-No.1	250	250	1	0.137	0.230	0.012	20.8	A	0.091	0.884	0.700	2.00	1.30～2.00	DT
○Mic.No.2	200	200	2	0.217	0.488	0.047	18.0	B	0.147	1.208	1.597	1.24	1.10～1.50	F,ST
▲Mic-No.3	250	250	2	0.109	0.316	0.108	23.5	A	0.082	0.931	1.766	0.72	(0.50～1.10)	F,BO
○Mic-No.4	250	250	2	0.581	0.126	0.125	23.6	A	0.098	0.811	1.844	0.48	0.4	F,C,Bu*5
△ No.1	250	250	1	0.107	0.104	0.065	24.0	A	0.099	1.080	1.610	0.60	0.70～2.00	F,DT
△ No.2	250	250	1	0.107	0.197	0.149	24.0	A	0.129	0.999	1.714	0.44	0.51～1.54	F,DT
▲ No.3	250	250	1	0.000	0.360	0.173	18.6	A	0.148	0.986	1.662	0.52	(0.25～0.81)	F,BO
○ No.4	250	250	2	0.136	0.134	0.030	18.9	A	0.066	0.923	1.532	1.08	0.55～1.0	F,ST
○ No.5	250	250	2	0.136	0.134	0.058	18.9	A	0.067	0.879	1.688	0.72	0.38～0.84	F,ST
○ No.6	250	250	2	0.272	0.134	0.120	18.9	A	0.095	0.861	1.792	0.48	0.32～0.62	F,C,Bu*5
○ No.7	250	250	2	0.137	0.230	0.012	20.8	A	0.086	1.100	1.194	1.86	not clear	F,ST
○ No.8	250	250	2	0.261	0.230	0.012	20.8	A	0.112	1.147	0.960	2.40	not clear	(F),ST
▲ No.9	250	250	2	0.136	0.230	0.154	18.9	A	0.092	0.879	1.792	0.48	(0.47～0.93)	F,ST,BO
▲No.10	250	250	2	0.136	0.354	0.192	18.9	A	0.120	0.886	1.792	0.48	(0.38～0.81)	F,ST,BO
○No.11	250	250	2	0.137	0.230	0.012	20.8	A	0.064	1.054	1.402	1.84	not clear	ST
○No.12	250	250	2	0.261	0.230	0.012	20.8	A	0.091	1.158	1.350	2.00	not clear	F,ST
○No.13	250	250	2	0.581	0.223	0.125	23.6	A	0.112	0.838	1.792	0.64	0.36～0.70	F,C,Bu*5
○No.14	250	250	2	0.436	0.223	0.232	23.6	A	0.103	0.828	1.896	0.32	0.29～0.70	F,C,Bu*5
○No.15	200	200	2	0.571	0.514	0.025	17.2	B	0.131	0.939	1.389	1.88	0.70～2.0	SC,ST*5
▲No.16	250	250	2	-0.111	0.515	0.034	18.6	A	0.068	1.008	1.532	1.44	(0.45～1.66)	BO
○No.17	200	200	3	0.100	0.421	0.080	19.6	B	0.098	1.192	1.753	0.95	1.28～1.33	F,ST
○No.18	200	200	3	0.200	0.421	0.080	19.6	B	0.103	1.094	1.753	0.95	1.11	F,ST
○No.19	250	250	3	0.136	0.230	0.040	18.9	A	0.043	0.899	1.766	1.08	0.62～1.19	F,ST
▲No.20	250	250	3	0.136	0.354	0.050	18.9	A	0.061	0.981	1.740	1.20	(0.65～1.38)	F,ST,BO
▲No.21	250	250	2	0.109	0.316	0.108	23.5	A	0.085	0.961	1.766	0.72	(0.49～1.04)	F,BO
▲No.22	250	250	2	0.109	0.316	0.108	23.5	A	0.083	0.946	1.766	0.72	(0.62～0.67)	F,BO

*1 ○: proper specimens,▲: specimens of bond failure,△: specimens of diagonal tension failure , Mic-:Specimens for micro analytical models

*2 A: antisymmetric loading, B: simple beam loading

*3 It is difficult to decide the height of yield line of bond failure specimens. Hence observed heights are shown with parentheses.

*4 They are F: Flexural Failure, DT: Diagonal Tension Failure, ST: Shear Tension Failure, SC: Shear Compression Failure, BO: Splitting Bond Failure, C: Crush of Concrete, Bu: Buckling of Reinforcing Bar.

*5 Compressive longitudinal reinforcing bars yielded but tensile longitudinal reinforcing bars did not yield.

Japan Concrete Institute 1983. *Proceedings of JCI 2nd colloquium on shear analysis of RC structures, Collected experimental data of specimens for verification of analytical models:* Japan Concrete Institute: 9-20 (in Japanese)

Schlaich, J., Schafer,K., & Jennewein, M. 1987. Toward a consistent design of structural concrete. Prestressed Concrete Institute Journal, Vol.32 No.3: 74-150

Takiguchi, K., Hotta, H., Mizobuchi, T., & Morita, S. 1992. Fundamental Experiments on compressive properties of concrete around the critical section of R/C column. Journal of Struct. Constr. Engng, AIJ No.442:123-131 (in Japanese)

Takiguchi, K.& Toyama, J.. 1998. A study on ultimate shear strength of arch mechanism in R/C column. Journal of Struct. Constr. Engng, AIJ No.503: 93-100 (in Japanese)

Creative Systems in Structural and Construction Engineering, Singh (ed.) © 2001 Balkema, Rotterdam, ISBN 90 5809 161 9

Limit analysis of conical-shaped steel water tanks based on linear programming

Z. Shi & M. Nakano
Research and Development Center, Nippon Koei Company Limited, Tokyo, Japan

ABSTRACT: The past two decades witnessed several structural failures of conical-shaped steel water tanks under hydrostatic pressure. In this paper, the limit analysis of such structures is formulated as a linear programming (LP) problem, and the accuracy of numerical solutions is verified by comparison with experimental and numerical analysis results available in the literature. The study has shown that the LP approach is quite effective for evaluating the stability of conical-shaped water tanks, due to its simplicity of problem formulation, the accuracy of numerical solutions and the very short computational time required.

1 INTRODUCTION

Regarding the structural stability of conical-shaped steel water tanks under hydrostatic pressure, the experimental investigations by Vandepitte et al. (1982) on the collapse of an elevated water tower in Belgium, and the recent numerical studies by El Damatty et al. (1997) concerning a similar accident that occurred in Canada, examined the problems from complementary perspectives, and attracted attention for their practical engineering implications in the design of similar structures.

Investigations after the collapse of the elevated water tower in Canada revealed that the shell thickness of the tank was inadequate, and that buckling of the wall of the conical shell initiated approximately 0.6 m above the bottom plate (Dawe et al. 1993). While taking into consideration the large deformation and the strain hardening behavior, El Damatty et al. carried out a finite element analysis to study the effects of initial imperfections and residual stresses on the structural stability of the tank, and reckoned that inappropriate design guidelines could partly be responsible for the collapse of water tanks.

Although linear programming (LP) has long been used for the limit analysis of truss structures, few LP applications have been available for obtaining the limit load or plastic collapse load of shells and water storage tanks. Exceptions include a limit analysis of cylindrical shells (Tin-Loi & Pulmano 1991) and another on corroded penstocks (Shi et al. 2000). An LP formulation generally involves discretization and linearization of continuous equilibrium equations on internal forces and yield conditions, and usually results in a much easier solution process than does a

finite element approach. Therefore, as an alternative to the FE approach, this paper presents an LP formulation of the limit analysis of a conical-shaped steel water tank under hydrostatic pressure, and the accuracy of the solution is verified by experimental and numerical analysis results available in the literature.

2 BASIC EQUATIONS

As shown in Figure 1, a water-filled conical-shaped tank capped with an upper cylindrical shell is assumed, which is made of a homogeneous and isotropic material with perfectly plastic behavior. In deriving the following equilibrium equations, the basic assumptions of the thin shell theory are applied with respect to thickness and displacement.

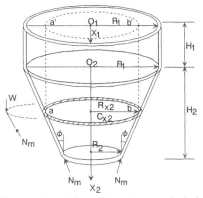

Figure 1. Internal forces due to the gravity load of water.

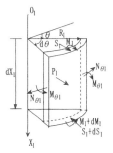

Figure 2. A cylindrical shell element.

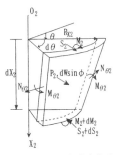

Figure 3. A conical shell element.

Figure 2 shows the internal forces acting on a cylindrical shell element under the internal water pressure P_1, and a cylindrical coordinate system (X_1, θ, R_1). In view of the symmetrical characteristic of the problem, the equilibrium conditions for internal forces are expressed as follows:

$$\frac{d^2M_1}{dX_1^2} + \frac{N_{\theta 1}}{R_1} - P_1 = 0 \tag{1}$$

Note that Equation (1) is derived by eliminating the shear force S_1 from the two individual equilibrium conditions for shear and moment.

Figure 3 shows a conical shell element with a coordinate system (X_2, θ, R_{X2}), which is subjected to not only internal water pressure, but also the self-weight of the water W from the upper part of the vessel. Assuming the tank is simply supported as shown in Figure 1, the compressive force N_m in the conical shell can be directly obtained from the equilibrium conditions in the vertical direction:

$$N_m \cos\varphi = W \tag{2}$$

Since the gravity load of water also acts in the direction normal to the conical shell, a differential component $dW\sin\phi$ must also be taken into account as an external force to obtain the following equilibrium conditions for the conical shell element:

$$\frac{d^2M_2}{dX_2^2} + \frac{N_{\theta 2}}{(R_1\cos\varphi - X_2\sin\varphi)} - (\frac{P_2}{\cos^2\varphi} + \frac{dW}{dX_2}\tan\varphi)$$
$$= \frac{d^2M_2}{dX_2^2} + \frac{N_{\theta 2}}{R_2'} - P_2' = 0 \tag{3}$$

Note that W and dW/dX_2 are based on the following integration:

$$2\pi(R_1 - X_2\tan\varphi)W =$$
$$\gamma_w\int_0^{2\pi}\int_0^{H_1}\int_{R_1-X_2\tan\varphi}^{R_1} r\,dr\,dx\,d\theta + \gamma_w\int_0^{2\pi}\int_0^{X_2}\int_{R_1-X_2\tan\varphi}^{R_1-x\tan\varphi} r\,dr\,dx\,d\theta \tag{4}$$

The above equations are then made dimensionless by adopting the non-dimensional variables $n_{\theta 1} = N_{\theta 1}/\sigma_y t$, $n_{\theta 2} = N_{\theta 2}/\sigma_y t$, $m_1 = M_1/\sigma_y t^2/4$, $m_2 = M_2/\sigma_y t^2/4$, $n_m = N_m/\sigma_y t$, $w = W/\sigma_y t$, $p_1 = P_1 R_1/\sigma_y t$, $p_2 = P'_2 R'_2/\sigma_y t$, $x_1 = X_1/H_1$, $x_2 = X_2/H_2$, as well as two shell parameters $\alpha_1^2 = 2H_1^2/R_1 t$ and $\alpha_2^2 = 2H_2^2/R'_2 t$.

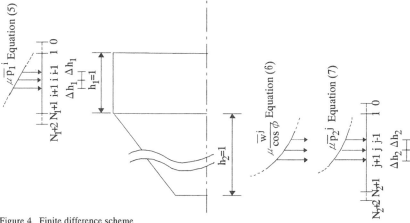

Figure 4. Finite difference scheme.

3 DISCRETIZATION OF EQUILIBRIUM EQUATIONS

The equilibrium equations are discretized using the central finite-difference approximation (Tin-Loi & Pulmano 1991). As shown in Figure 4, the upper cylindrical shell of non-dimensional height $h_1 = 1$ and the conical-shaped shell of non-dimensional height $h_2 = 1$ are divided equally into N_1 and N_2 parts at an interval of Δh_1 and Δh_2, respectively. Equations (1), (2) and (3) then become

$$\frac{m_1^{i-1} - 2m_1^i + m_1^{i+1}}{\Delta h_1^2} + 2\alpha_1^2(n_{\theta 1}^i - \mu \overline{p}_1^i) = 0; \, i = 1..N_1 + 1 \qquad (5)$$

$$n_m^j - \mu \frac{\overline{w}^j}{\cos\varphi} = 0; \, j = 1,...,N_2 + 1 \qquad (6)$$

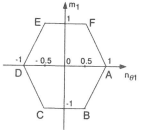

Figure 5. The hexagonal yield condition for the cylindrical shell.

$$\frac{m_2^{j-1} - 2m_2^j + m_2^{j+1}}{\Delta h_2^2} + 2\alpha_2^{j2}(n_{\theta 2}^j - \mu \overline{p}_2^j) = 0; \, j = 1.N_2 + 1 \qquad (7)$$

where the end points 0 and N_1+2 in Equation (5), and 0 and N_2+2 in Equations (7) are dummy points frequently used in finite difference schemes. Note that μ is a common load factor for the design loads $\overline{p}_1^i$, $\overline{p}_2^j$ and $\overline{w}^j$.

4 YIELD CONDITIONS

Figures 5 and 6 show Hodge's hexagonal and polyhedral approximations to von Mises's yield conditions with respect to the generalized stresses ($n_{\theta 1}$, m_1) of the cylindrical shell and (n_m, $n_{\theta 2}$, m_2) of the conical shell (Hodge 1954). For application to LP problems, these piecewise yield conditions need to be converted to linear simultaneous equations by linearly combining the vertices and intermediate variables ξ_a and ξ_b (Tin-Loi & Pulmano 1991). Therefore, for the cylindrical shell the yield conditions become

$$\mathbf{Q}_a^i = \mathbf{V}_a^i \xi_a^i; \, i = 1,...,N_1 + 1 \qquad (8a)$$

where

$$\mathbf{Q}_a^{iT} = (n_{\theta 1}^i, m_1^i) \qquad (8b)$$

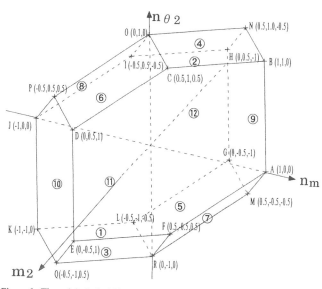

Figure 6. The polyhedral yield condition for the conical shell.

651

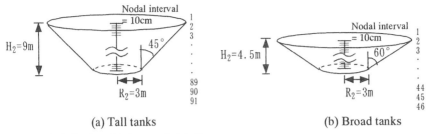

(a) Tall tanks   (b) Broad tanks

Figure 7. Conical tank models for numerical studies.

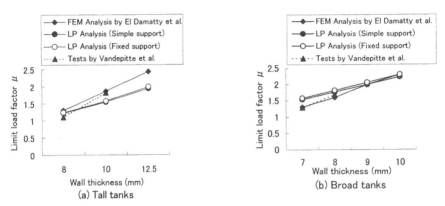

(a) Tall tanks   (b) Broad tanks

Figure 8. Relationship between the limit load and wall thickness.

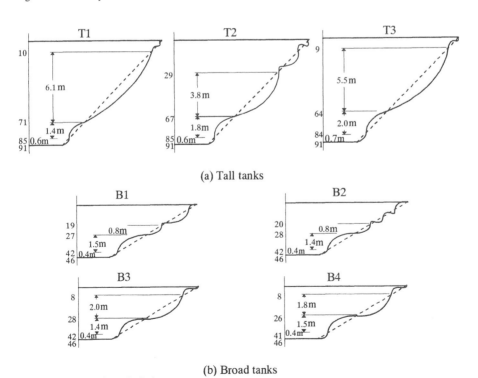

(a) Tall tanks

(b) Broad tanks

Figure 9. Failure modes at the limit state.

$$\mathbf{V}_a^i = \begin{bmatrix} 1 & 0.5 & -0.5 & -1 & -0.5 & 0.5 \\ 0 & -1 & -1 & 0 & 1 & 1 \end{bmatrix} \qquad (8c)$$

$$\xi_a^{iT} = (\xi_{a1}^i, \xi_{a2}^i, \xi_{a3}^i, \xi_{a4}^i, \xi_{a5}^i, \xi_{a6}^i) \qquad (8d)$$

$$\xi_{ak}^i \geq 0, k = 1,...,6; \sum_{k=1}^{6} \xi_{ak}^i \leq 1 \qquad (8e)$$

For the conical shell the yield conditions are

$$\mathbf{Q}_b^j = \mathbf{V}_b^j \xi_b^j; \quad j = 1,...,N_2 + 1 \qquad (9a)$$

where

$$\mathbf{Q}_b^{jT} = (n_m^j, n_{\theta 2}^j, m_2^j) \qquad (9b)$$

$$\mathbf{V}_b^j = \begin{bmatrix} 1 & 1 & 0.5 & 0 & 0 & 0.5 & 0 & 0 & -0.5 \\ 0 & 1 & 1 & 0.5 & -0.5 & -0.5 & -0.5 & 0.5 & 0.5 \\ 0 & 0 & 0.5 & 1 & 1 & 0.5 & -1 & -1 & -0.5 \end{bmatrix}$$

$$\begin{bmatrix} -1 & -1 & -0.5 & 0.5 & 0.5 & 0 & -0.5 & -0.5 & 0 \\ 0 & -1 & -1 & -0.5 & 1.0 & 1 & 0.5 & -1 & -1 \\ 0 & 0 & -0.5 & -0.5 & -0.5 & 0 & 0.5 & 0.5 & 0 \end{bmatrix} \qquad (9c)$$

$$\xi_b^{jT} = (\xi_{b1}^j, \xi_{b2}^j, \xi_{b3}^j, \cdots, \xi_{b16}^j, \xi_{b17}^j, \xi_{b18}^j) \qquad (9d)$$

$$\xi_{bl}^j \geq 0, l = 1,...,18; \sum_{l=1}^{18} \xi_{bl}^j \leq 1 \qquad (9e)$$

5 PROBLEM FORMULATION BY LINEAR PROGRAMMING

The present LP approach is based on the lower bound theorem of the plasticity theory. Application of the theorem leads to the following mathematical expressions for the plastic collapse load of a water tank:

Maximize μ;

Subject to Equations (5) to (9), and boundary conditions.

6 NUMERICAL RESULTS AND DISCUSSION

6.1 Outline of numerical studies

In total seven conical-shaped tanks of two different categories were studied by El Damatty et al., and are also adopted in the present study for the purpose of comparison; see Figure 7 and Table 1. It is worth mentioning that the dimensions of the three tall tanks (T1, T2 and T3) are close to those of the collapsed Canadian water tank.

Table 1. Dimensions of conical tank models.

Case	Wall thickness (mm)	R_2 (m)	H_2 (m)	ϕ
T1	8.0	3.0	9.0	45°
T2	10.0	3.0	9.0	45°
T3	12.5	3.0	9.0	45°
B1	7.0	3.0	4.5	60°
B2	8.0	3.0	4.5	60°
B3	9.0	3.0	4.5	60°
B4	10.0	3.0	4.5	60°

For the limit analysis using the LP method, the only material property required is the yield stress of the material, which in this case is $\sigma_y = 300$ MPa. As for the boundary conditions, the bottom supports of the tank are treated as simple supports in addition to fixed supports to study the impact on the limit load. As shown in Figure 7, a nodal interval of 10 cm is adopted for each case. To conduct the numerical analyses, the general-purpose software GAMS for mathematical programming is employed, and the computation for each case takes only a few seconds.

6.2 Limit state of conical shells under hydrostatic pressure

Comparisons of numerical results of the limit analysis using the two different approaches, namely the FE approach and the LP approach, are shown in Figure 8. The limit load factors for four perfect conical-shaped shells (T1, T2, B1 and B2) obtained from the experimental results (Vandepitte et al. 1982) are also illustrated in the same figure. A difference of up to 20% between the results of the two respective numerical approaches is found in Figure 8. Nevertheless when compared with the test results, LP-based numerical solutions are considered satisfactory.

In the numerical analyses, boundary conditions at the bottoms of tanks are found to have little impact on the limit load; the rate of increase of limit load factor for fixed supports is 3% or less either for the tall or broad tanks. Next, based on the positive or negative sign of the meridian bending moment m_2 at each nodal point, the failure mode of the conical-shaped shell is obtained for each case as shown in Figure 9. As clearly shown in the figure, buckling initiates in the wall approximately 0.6 m above the bottom plate for the tall tanks, and 0.4 m for the broad tanks. It should be emphasized that the mode of failure for the tall tanks is identical to that witnessed at the site of the accident (approximately 0.6 m above the bottom plate).

It is believed that the discrepancies between the limit load factors of this study and the study by El

Damatty et al. are caused mainly by the following factors: While the LP formulation of the problem is carried out based on the small deformation theory, El Damatty et al. incorporated the large deformation theory into their FE formulation. Other conceivable causes include the omission of strain hardening and the piecewise linear approximation of von Mises's yield conditions in the LP analysis.

7 CONCLUSIONS

The limit analysis of conical-shaped water tanks was formulated as an LP problem, and the accuracy and effectiveness of this approach were verified by comparison with experimental and numerical results available in the literature.

In the process of problem formulation by an LP method, it is the equilibrium conditions, not the constitutive equations that are directly involved. Thus, overestimation of the stiffness for thin plates or shells resulting from inadequate constitutive equations, a phenomenon often encountered during a nonlinear FE analysis for thin-walled panels or shells, does not exist. Therefore, for the limit analysis of conical-shaped tanks LP approaches provide accurate solutions, provided that the equilibrium equations, yield conditions and boundary conditions are formulated correctly.

REFERENCES

Dawe, J. L., C. K. Seah & A. K. Abdel-Zaher 1993. Collapse of a water tower. *Proceedings of the CSCE Conference*: 315-323. Canadian Society of Civil Engineering.

El Damatty, A. A., R. M. Korol & F. A. Mirza 1997. Stability of imperfect steel conical tanks under hydrostatic loading. *Struct. Eng. J.* 123(6): 703-712.

Hodge, P. G. 1954. The rigid-plastic analysis of symmetrically loaded cylindrical shells. *Appl. Mech. J.* 21: 336-442.

Shi, Z., T. Sakurai & M. Nakano 2000. Limit analysis on aging penstocks based on linear programming. To appear in *Construction and Building Materials*.

Tin-Loi, F. & V. A. Pulmano 1991. Limit loads of cylindrical shells under hydrostatic pressure. *Struct. Eng. J.* 117(3): 643-656.

Vandepitte, D., J. Rathe, B. Verhegghe, R. Paridaens, & C. Verschaeve 1982. Experimental investigation of hydrostatically loaded conical shells and practical evaluation of the buckling load. In E. Ramm (ed.), *Buckling of shells*: 375-399.Berlin: Springer-Verlag KG.

Creative Systems in Structural and Construction Engineering, Singh (ed.) © 2001 Balkema, Rotterdam, ISBN 90 5809 161 9

Dynamic optimization of a container crane structure

D. Dinevski, M. Oblak & I. Gubenšek
Faculty of Mechanical Engineering, Maribor, Slovenia

ABSTRACT: Optimization of the container crane structure is done in order to increase the values of important natural frequencies of the crane. On the basis of a full scale test and by applying the Finite Element Method a mechanical model of the container crane is made. Objective function, constraints, design and response variables are defined for an optimal design problem. Some new engineering design solutions are proposed, which results in higher natural frequencies.

1 INTRODUCTION

1.1 *General problem description*

Container crane structures (and some other hoisting appliances) are large-scale structures with very inconvenient mass distribution from the engineering point of view. A typical (see Fig. 1) container crane is up to 90 meters high with up to 600 tons dead weight, where the major part of the weight is located at the height of 40 meters. Such structure tends to have very low natural frequencies (below 1 Hz), which makes the normal operational movements hard to control. The spreader (the part that lifts containers) must be precisely positioned (tolerance is 25 millimeters) when picking up or dropping containers. Large oscillations of the structure (as a consequence of low frequencies) are very undesired characteristics of the cranes. In addition, the limits for low natural frequencies are usually set by standards for heavy-lifting appliances and\or custom-tailored orders.

It is important to emphasize that this research focuses on the structure of the crane and not on the dynamic system of the structure, the sway and the load. The sway and the load are modeled only with point mass elements.

The fact is that no serious numerical dynamic analysis can be done without validation of a mechanical model. Therefore, an experimentally verified model was used for the calculations. A short presentation of the previously conducted and reported (Dinevski, Oblak 1997) experimental verification of the container crane natural frequencies is given in Section 2.2.

The crane model and its modal analysis are briefly presented also in the reference (Dinevski & others, 1999), where a special case of emergency condition is analyzed.

1.2 *Dynamic features of the crane*

The structure is under dynamic loading at during:
- service operation; acceleration or deceleration of the hoisting motion (load on the spreader), horizontal acceleration of the sway together with the load, swinging of the load, impact effect due to collision between the load and the base, etc.
- crane travelling along the rail tracks (relatively infrequent)
- crane transportation from the manufacturer to the buyer (water-borne transport of a complete structure), where severe /but of short duration/ dynamic loading occurs
- climatic effects, seismic effects

Dynamic features of the crane structure resulting from the optimization should be:
- as high as possible natural frequencies, especially the important ones for service operation
- small amplitudes of displacements, velocities and accelerations for service operation
- minimum weight of the structure

Figure 1. The container crane with the load carrying capacity of 500/450 kN (dead weight 5500 kN, height 73.4 m), made by Metalna Maribor and located in the port of Koper in Slovenia.

2 CALCULATION AND VERIFICATION OF NATURAL FREQUENCIES

2.1 *Mechanical model and numerical analysis*

The initial structure design is shown in Figure 1.

The modeling method chosen is the Finite Element Method (FEM). For the FEM dynamic analysis a 2-node prismatic 3-D beam element with the formulation for stretching, bending, torsion and transverse shear deformation effects is applied. The element is analogous with the Timoshenko beam model. The Euler-Bernoulli model (without shear deformations and cross-sectional moment of inertia) proved less accurate on the reference tests. The second order theory is used so as to take the geometric nonlinearity into account. The FEM solvers used for the modal analysis are NISA II, NASTRAN and ABAQUS.

The numerical model of the crane structure is presented in the reference (Dinevski, Oblak 1997). The finite element model of the container crane structure has 79 line and 4 point mass elements. The model has 66 nodes and 396 degrees of freedom. The number of elements is the result of several trials, and represents a minimum to achieve satisfactory accuracy.

The results of the numerical analysis are presented in Figure 2 in the form of the first four natural frequencies and the corresponding mode shapes. The most important natural frequencies are mode No. 2 and mode No. 3.

2.2 *Experimental verification of the frequencies*

In order to validate the numerically calculated natural frequencies and vibration mode shapes, and to define the modal damping values, a full-scale forced vibration experiment was performed (Dinevski, Oblak 1997). A kind of intersection was made between forced vibration test, with the control of the input (forcing function), and step relaxation test, with the release of the structure from a statically deformed position. Instead of releasing the deformed structure, the crane was forced into a specific vibration mode with limited operational movements, by using inertial forces of the masses in motion. The ambient vibration test (wind loading forcing with crane motion), was also performed. Four accelerometers were used to register the dependence of the acceleration on the time. The agreement of results obtained by numerical analysis and those from the experiment was good; for the first three modes the relative difference between the measured and calculated frequencies was below 1.6 % for the fourth it was 4.8 %. On the basis of these results it is possible to conclude that the mechanical model of the container crane is reliable in the scope of our requirements.

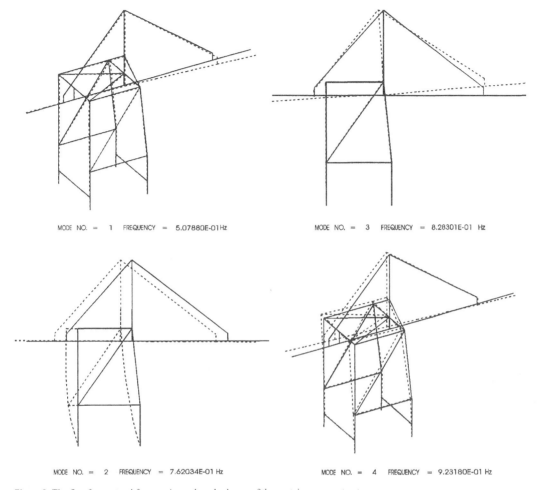

MODE NO. = 1   FREQUENCY = 5.07880E-01Hz

MODE NO. = 3   FREQUENCY = 8.28301E-01 Hz

MODE NO. = 2   FREQUENCY = 7.62034E-01 Hz

MODE NO. = 4   FREQUENCY = 9.23180E-01 Hz

Figure 2. The first four natural frequencies and mode shapes of the container crane structure

3 OPTIMIZATION

3.1 *Frequency constraints*

According to the engineers a mode shape that has the most important influence on the crane operation and its structural integrity is the mode No 2 (Fig. 2), where horizontal oscillations directly effect the positioning of the spreader during picking up or dropping the container. The decision was made that the second natural frequency should be increased to at least 0.85 Hz (from initial 0.762 Hz). The values for the other three frequencies should not be decreased. Therefore the frequency constraints are:

$$f_2 \geq 0.85 \, Hz$$
$$f_1 \geq f_1^0$$
$$f_3 \geq f_3^0 \qquad (1)$$
$$f_4 \geq f_4^0$$

where initial values are (see Fig. 2)

$$f_1^0 = 0.508 \, Hz, \, f_3^0 = 0.828 \, Hz, \, f_4^0 = 0.923 \, Hz \quad (2)$$

3.2 *Design variables*

Together with engineers dealing with the design of cranes we defined a set of design variables. According to their experience they suggested that the following structure modifications could probably increase the second natural frequency:

657

Figure 3. Original frame structure

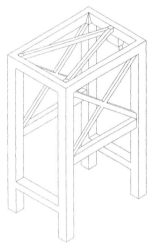

Figure 4. "The X variation"

Figure 5. "The V variation"

Figure 6. Original method of carrying the main beam

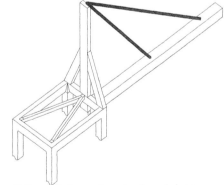

Figure 7. Proposed change of carrying the main beam

3.2.1 *Altering the frame side to the X shape*

The first proposed change of the structure is a structural change of the frame side. It is expected that due to higher stiffness of the frame the second natural frequency value will increase. The original engineering design is shown in Figure 3 and "X variation" in Figure 4.

3.2.2 *Altering the frame side to the V shape*

The second change was made with the same motivation as the first one (previous section). This time, the strengthening of the frame side is in the shape of letter "V". The "V" variation is illustrated in Figure 5.

3.2.3 *Double carrier for the main beam*

It is anticipated that changing (to the stiffer design) the method of carrying the main beam will give rise to the values of second natural frequency. The original design is shown in Figure 6 and the proposal for the change of structure in Figure 7.

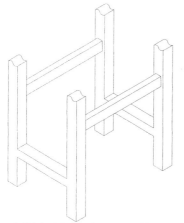

Figure 8. Original design - lower part of the crane

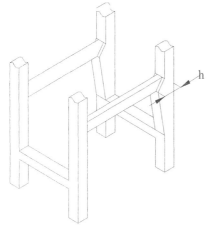

h

Figure 9. New design - lower part of the crane

3.2.4 *Strengthening the "legs"*

From the second mode shape (Fig. 2) it is clear that stiffer "legs" of the crane will give rise to the frequency value. The original design of the lower part of the crane is in Figure 8, and the fourth proposal is in Figure 9. In this case we have a continuous design variable h, which is the width of the upper part of the "leg" (Fig. 9). The initial value of h is 1.26 m.

3.3 *Design variables effect*

Due to the combination of continuous and discrete design variables it was decided that instead of defining one optimization problem, the influence of every particular design variable would be separately studied. Then we could decide easier which ones was to be considered in the optimization problem.

The qualitative results of this study are shortly listed. The quantitative results are omitted as they are not vital for the scope of this paper.

3.3.1 *Altering the frame side to the X shape*

The "X shape" of the frame side has almost no effect to 1st, 3rd, and 4th natural frequency. The second one is, however, increased to $f_2 = 0.781\ Hz$ from the initial $f_2^0 = 0.762\ Hz$. The frequency remains almost the same if the size of the X beams cross section is altered, provided that it is big enough for the structure to maintain its structural stability.

3.3.2 *Altering the frame side to the V shape*

Similar to the previous case, the "V shape" has almost no effect to 1st, 3rd, and 4th natural frequency. The second frequency is increased to $f_2 = 0.785\ Hz$ from initial $f_2^0 = 0.762\ Hz$.

3.3.3 *Double carrier for the main beam*

Double carrieer slightly decreases the values of all four natural frequencies.

3.3.4 *Strengthening the "legs"*

Strengthening the front vertical beams has an important effect on the "critical" second natural frequency, but minimal effect on other three frequencies). The cross section of the "leg" beam is therefore altered in the optimization process. The design variable is the width h of the beam (Fig. 9).

3.4 *Design variables final selection*

Let's note that all except one of the proposed design modifications turned out to be ineffective. In the cases of "X frame" and "V frame" as a matter of fact the second natural frequency increased, but the gain was relatively insignificant in comparison to the cost of the solutions proposed. As a consequence only one design variable is selected for the optimization process. This design variable is the width of the beam h as shown in Figure 9.

3.5 *Optimal design problem*

The objective function for the optimization problem is the weight of the structure (as usually is the case in structural optimization).

Response variables are first four natural frequencies, and design variable is the value h in the beam cross section.

The constraints are listed in Section 3.1.

In other words, the design problem is: "Find the optimal structure (the lowest weight) of the crane so that the second natural frequency is higher or equal to 0.85 Hz, and all other three frequencies higher or equal to frequencies from initial design .

659

The optimization problem itself is simple one (only one design variable and four constraints) therefore, it will not be discussed further. Let's just mention that the optimizer used for the solution was developed by our institute and presented in (Kegl & others 1992).

The results of the optimization:

$$h = 2.53\,m$$
$$f_1 = 0.515\,Hz$$
$$f_2 = 0.850\,Hz \qquad\qquad (3)$$
$$f_3 = 0.832\,Hz$$
$$f_4 = 0.951\,Hz$$

The weight gain as a consequence of the new value of the beam width h is 4790 kg which is still acceptable in comparison to the 550.000 kg of total weight of the structure. The solution is technically feasible without any additional modifications of the structure.

4 CONCLUSION

With the previously verified mechanical model of the crane the optimization was conducted in order to increase a critical natural frequency to an acceptable level. Several engineering solutions were proposed to achieve this goal, but only one turned out really effective. A particular solution was analyzed and concrete quantitative change of the structure proposed. With relatively small investment and acceptable weight gain the critical frequency was successfully raised to the required level.

REFERENCES

Dinevski D., Oblak M., Experimental and numerical dynamic analysis of a container crane structure, *Der Stahlbau* ISSN 0038-9145. - Jg. 66, heft Z (feb. 1997), pp. 70-77

Dinevski D., Oblak M., Gubenšek I., Novak A., Simulation of the container crane bahaviour in emergency condition, *7th East Asia-Pacific Conference on Structural Engineering and Construction*, August 27-29, 1999, Kochi, Japan Tokyo: Social System Institute, pp. 309-314

Kegl M. S., Butinar B. J.,. Oblak M. M, "Optimization of mechanical systems: On strategy of non-linear first-order approximation", *Int. j. numer. methods eng.*, 33, 223-234 (1992)

Creative Systems in Structural and Construction Engineering, Singh (ed.) © 2001 Balkema, Rotterdam, ISBN 90 5809 161 9

Stability analysis of continuous compression members by beam analogy method

Soo-Gon Lee
Chonnam National University, Kwangju, Korea

Soon-Chul Kim
Dongshin University, Naju, Korea

Kang-Il Lee
Korea Structure Safety Institute, Kwangju, Korea

ABSTRACT: In the structural design of the compression member, the elastic critical load of that member is the key factor to be considered. Contrary to the single-span member, the determination of the critical load of a continuous compression member is not an easy problem to solve, even using numerical techniques. The beam analogy method is proposed to determine the elastic critical load of multi-span compression members. In the proposed method, the given compression member is replaced by the continuous beam. To simulate the buckling mode, an imaginary lateral load at the midpoint of each span changes direction. The stress analysis of the analogous beam enables one to calculate Kinney's fixity factors. Finally, the critical load of each span of the continuous compression member is expressed by fixity factors. The proposed method is easy to apply in calculating critical load, but yields lower bound errors in some cases.

1 INTRODUCTION

In the structural design of a beam-column, the effect of the axial compressive force, P, is included by multiplying the factor $1/ (1-P/P_{cr})$ with the beam moment and deflection. In the case of a single span beam-column, the elastic critical load, P_{cr}, is easily determined, whether the sectional property of that member is constant or variable along its axis. For a multi-span beam-column, however, the conventional neutral equilibrium method, or energy principle-based Rayleigh-Ritz method cannot be efficiently applied to determine critical load. In this case, a modified slope-deflection method or a numerical method, for example, the finite difference or the finite element method, becomes a useful tool for the determination of critical load or the stress analysis of a continuous beam-column.

In this paper, the *beam-analogy method* is proposed to determine the approximate critical loads of continuous compression members. The main idea of the beam analogy method is to replace the continuous compression member with a continuous beam, to which concentrated lateral loads are applied at each mid-span of the multi-span beam. The mid-span concentrated loads are made to change their directions in order to simulate the buckling mode. The results of stress analysis of the beam are used to calculate Kinney's fixity factors(Kinney 1957). Finally, the critical load is expressed by fixity factors. The *beam analogy method* may also be applied to the determination of effective length factors of braced frame columns.

2 BEAM ANALOGY METHOD

The *beam analogy method* can be explained by the following simple example. Fig. 1(a) shows a 2-span continuous member whose critical load is to be determined by the proposed beam analogy method. *The first step* of the proposed method is to replace the given compression member with a continuous beam. Fig. 1(b) shows the analogous beam, where the direction of a concentrated lateral load, Q, at each mid-span is alternating its direction to simulate the buckling mode.

The second step is the stress analysis of the beam. In the present example, the usual slope-deflection

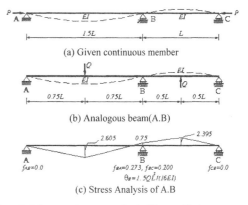

(a) Given continuous member

(b) Analogous beam(A.B)

(c) Stress Analysis of A.B

Figure 1. 2-Span continuous member(uniform section)

method may be conveniently applied to obtain end moments and rotation angles at the supports. Fig. 1(c) shows the stress analysis results.

The third step of the proposed method is to find Kinney's fixity factors(Kinney 1957) by utilizing the following relationship.

$$\left| M_{\alpha\beta} \right| = \left(\frac{4EI}{L} \right)_{\alpha\beta} \cdot \frac{f_{\alpha\beta}}{f_{\alpha\beta}'} \cdot \left| \theta_{\alpha} \right|,$$

$$f_{\alpha\beta}' = 1 - f_{\alpha\beta} \qquad (1)$$

Where | | denotes the absolute value. The above relationship was introduced by the first author(Lee 1979) to the eigenvalue problems of the tapered bars. He also applied Eq. 1 to the analysis of steel beams(Lee 1994) with partially fixed ends. In Fig.1(c), $f_{AB} = f_{CB} = 0.0$ by Kinney's definition (simply supported ends). When Eq.1 is applied to the first span AB, and to the second span BC, at the intermediate support B, f_{BA} and f_{BC} are determined in the following way.

$$\frac{0.75QL}{8} = \frac{4EI}{1.5L} \cdot \frac{f_{BA}}{1 - f_{BA}} \cdot \frac{1.5QL^2}{16EI}$$

$$\therefore f_{BA} = 0.273 \qquad (2\text{-a})$$

$$\frac{0.75QL}{8} = \frac{4EI}{L} \cdot \frac{f_{BC}}{1 - f_{BC}} \cdot \frac{1.5QL^2}{16EI}$$

$$\therefore f_{BC} = \frac{0.75}{3.75} = 0.200 \qquad (2\text{-b})$$

The final step is to determine the critical load or effective length factor (*K*-factor) using the following expressions.

$$(P_{cr})_{\alpha\beta} = (1 + f_{\alpha\beta})(1 + f_{\beta\alpha})\pi^2 \left(\frac{EI}{L^2} \right)_{\alpha\beta} \qquad (3)$$

$$= \pi^2 \cdot \left(\frac{EI}{K^2 L^2} \right)_{\alpha\beta}$$

where *K*-factor is defined by

$$K = \{ (1 + f_{\alpha\beta}) \cdot (1 + f_{\beta\alpha}) \}^{-0.5} \qquad (4)$$

The validity of Eq. 3 can be easily demonstrated ; that is, when a single prismatic member is simply supported at both ends, then $f_{\alpha\beta} = f_{\beta\alpha} = 0.0$, by definition. With $f_{\alpha\beta} = f_{\beta\alpha} = 0.0$, Eq. 3 yields $P_{cr} = \pi^2 EI/L^2$ (*K=1.0*).

In the same way, $P_{cr} = 2\pi^2 EI/L^2$ is obtained for the member with a simple($f_{\alpha\beta} = 0.0$) - fixed end($f_{\beta\alpha} = 1.0$). Finally, the critical load of the member fixed at both ends ($f_{\alpha\beta} = f_{\beta\alpha} = 1.0$) is expressed by $4\pi^2 EI/L^2$ (*K=0.5*).

With the fixity factors of Fig. 1(c), Eq. 3 and Eq. 4 give the following results:

(Span *AB*)

$$(P_{cr})_{AB} = (1.0)(1.273)\pi^2 \frac{EI}{(1.5L)^2}$$

$$= 5.584 \frac{EI}{L^2} \qquad (5\text{-a})$$

$$K_{AB} = 1 / \sqrt{1 \times 1.273} \approx 0.886 \qquad (6\text{-a})$$

(Span *BC*)

$$(P_{cr})_{BC} = (1.2)(1.00)\pi^2 \frac{EI}{(L)^2}$$

$$= 11.844 \frac{EI}{L^2} > (P_{cr})_{AB,} \qquad (5\text{-b})$$

$$K_{BC} = 1 / \sqrt{1 \times 1.273} \approx 0.913 \qquad (6\text{-b})$$

Fig. 1(a) is the very member chosen by Chen (Chen *el al.* 1987). He applied the *Neutral equilibrium method* to obtain $P_{cr} = 5.89 EI/L^2$, which is approximately 5.2% 1arger than the result given by Eq. 5(a).

Here, the *Modified slope-deflection method* (M.S.D.M)(Chajes 1974) will be briefly described. When the axial force, *P*, is considered, the end moments and rotation angles of a beam-column are related by the following :

Figure 2. Deformation of beam-column

$$M_{\alpha\beta} = (\frac{EI}{L})_{\alpha\beta} \cdot (\alpha_n \theta_{\alpha} + \alpha_f \theta_{\beta}) \qquad (7\text{-a})$$

$$M_{\beta\alpha} = (\frac{EI}{L})_{\alpha\beta} \cdot (\alpha_f \theta_{\alpha} + \alpha_n \theta_{\beta}) \qquad (7\text{-b})$$

First, these equations are applied successively to Fig. 1(a). Next, the moment equilibrium conditions at the external and intermediate supports allow one to obtain the following matrix equation:

$$\begin{vmatrix} \alpha_{n1}/1.5 & \alpha_{f1}/1.5 & 0 \\ \alpha_{f1}/1.5 & (\alpha_{n1}/1.5 + \alpha_n) & \alpha_f \\ 0 & \alpha_f & \alpha_n \end{vmatrix} \cdot \begin{Bmatrix} \theta_A \\ \theta_B \\ \theta_C \end{Bmatrix} = \begin{Bmatrix} 0 \\ 0 \\ 0 \end{Bmatrix} \qquad (8)$$

where α_n and α_f are so-called Merchant's stability functions defined by

$$\alpha_n = \frac{\phi_n}{\phi_n^2 - \phi_f^2}, \quad \alpha_f = \frac{\phi_f}{\phi_n^2 - \phi_f^2} \qquad (9\text{-a,b})$$

with

$$\phi_n = \frac{1}{(kL)^2}(1 - kL \cot kL) \qquad (10\text{-a})$$

$$\phi_f = \frac{1}{(kL)^2}(kL \csc kL - 1) \qquad (10\text{-b})$$

$$kL = \sqrt{PL^2 / EI} \qquad (10\text{-c})$$

α_{n1} and α_{f1} are Eq. 9 with $k_1 L = 1.5kL$

The characteristic equation for critical load is obtained from Eq. 8. That is :

$$\det \begin{vmatrix} \alpha_{n1}/1.5 & \alpha_{f1}/1.5 & 0 \\ \alpha_{f1}/1.5 & (\alpha_{n1}/1.5 + \alpha_n) & \alpha_f \\ 0 & \alpha_f & \alpha_n \end{vmatrix} = 0 \qquad (11\text{-a})$$

when expanded

$$\frac{\alpha_{n1}}{1.5}(\frac{\alpha_{n1}}{1.5} + \alpha_n)\alpha_n - (\frac{\alpha_{f1}}{1.5})^2 \alpha_n - \frac{\alpha_{n1}}{1.5}\alpha_f^2 = 0 \qquad (11\text{-b})$$

The least root satisfying Eq. 11(b) is found by a *trial and error procedure*, which yields $kL=2.4265$ and $P_{cr}=5.888EI/L^2$ by Eq. 10(c). This elastic critical load coincides with that of Chen's above mentioned result.

The frequently used *finite element method*(Chajes A. 1974) will also be described briefly. Fig. 3 shows an element of a beam-column subjected to constant axial force, P and a set of nodal forces $\{q\}$. In the below figure,

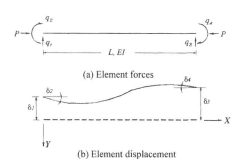

(a) Element forces

(b) Element displacement

Figure 3. Beam column element displacement

$\{\delta\}$ denotes nodal displacement vector corresponding to $\{q\}$. The element stiffness matrix, $[k]$ combines $\{q\}$ and $\{\delta\}$ in the following form.

$$\{q\} = [k]\{\delta\}, \quad [k] = [k]_b - P[k]_g \qquad (12)$$

in which

$$[k]_b = \frac{EI}{l^3}\begin{vmatrix} 12 & & & Symm. \\ -6l & 4l^2 & & \\ -12 & 6l & 12 & \\ -6l & 2l^2 & 6l & 8l^2 \end{vmatrix} \qquad (13\text{-a})$$

and

$$[k]_g = \frac{1}{15l}\begin{vmatrix} 18 & & & Symm. \\ -1.5l & 2l^2 & & \\ -18 & 1.5l & 18 & \\ -1.5l & 0.5l^2 & 1.5l & 2l^2 \end{vmatrix} \qquad (13\text{-b})$$

In the above two equations, $[k]_b$ denotes the flexural or bending stiffness matrix and $[k]_g$ is called the geometric or initial stress stiffness matrix. As one can see in Eq. 12, flexural stiffness is decreased, due to the axial compressive force, P.

The structural stiffness matrices for the entire member are obtained by transforming the individual element matrices from element to structural coordinates and then by assembling the resulting matrices. Finally, boundary conditions should be applied to the assembled matrices. The procedures are easily found

from textbooks on the finite element method or structural stability(Chajes A. 1974) hence, a detailed explanation will be omitted.

The external force vector, $\{Q\}$ and corresponding displacement vector, $\{\Delta\}$ are related by

$$\{Q\} = [K]\{\Delta\} \qquad (14)$$

where $[K]$, the structure stiffness matrix obtained after the application of boundary conditions, takes the form

$$[K] = [K]_b - P[K]_g \qquad (15)$$

With external force vector, $\{Q\}=\{0\}$, Eq.(14) and (15) constitute a typical eigenvalue problem.

$$([K]_b - P[K]_g)\{\Delta\} = \{0\} \qquad (16)$$

To obtain the least eigenvalue (here, the elastic critical load corresponding to the first mode of buckling) by a computer-aided iteration method, Eq.(16) should be transformed into the following form

$$([K]_b^{-1}[K]_g - \frac{1}{P}[I])\{\Delta\} = \{0\} \qquad (17)$$

where $[I]$ is the unit(or identity) matrix.

When the *finite element method* is applied to Fig. 1(a) one can obtain nearly the same critical load as that obtained by the modified slope-deflection method. Here, it is observed that the two representative methods, (the one, an analytical method or modified slope-deflection method, and the other a numerical or finite element method) can result in the same value of the elastic critical load.

The procedures necessary to obtain the final result using the above mentioned methods involve more complicated calculations. Furthermore, these

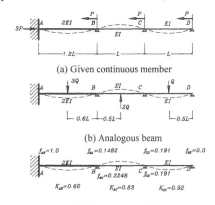

(a) Given continuous member

(b) Analogous beam

(c) Fixity factors and K-factors

Figure 4. 3-Span continuous beam

methods can neither predict the span that buckles first, nor determine the effective length factor, K, of each span.

For a clearer explanation of the proposed method, another example of a 3-span continuous compression member is used. Fig.4 (b) shows the first step of the proposed method. The second step, that is, the stress analysis of the analogous beam shown in Fig.4(b), can proceed in the following ways(the slope-deflection method is adopted).

(End moment equations)

$$M_{AB} = \frac{2EI}{1.2L}(2\theta_B) - \frac{3.6QL}{8} = -5.7792(\frac{QL}{8}) \quad (18\text{-a})$$

$$M_{BC} = \frac{2EI}{L}(2\theta_B + \theta_C) + \frac{2QL}{8} = +0.7585('') \quad (18\text{-b})$$

$$M_{CB} = \frac{2EI}{L}(\theta_B + 2\theta_C) - \frac{2QL}{8} = -0.5604('') \quad (18\text{-c})$$

$$M_{CD} = \frac{2EI}{L}(2\theta_C + \theta_D) - \frac{QL}{8} = +0.5604('') \quad (18\text{-d})$$

$$M_{DC} = \frac{2EI}{L}(\theta_C + 2\theta_D) + \frac{QL}{8} = 0.0 \quad (18\text{-e})$$

(Joint equation)

$$\sum M_B = 0, \quad \frac{6.4}{1.2}\theta_B + \theta_C = -\frac{5.6QL}{8}\cdot\frac{L}{2EI} \quad (19\text{-a})$$

$$\sum M_C = 0, \quad \theta_B + 4\theta_C + \theta_D = \frac{3QL}{8}\cdot('') \quad (19\text{-b})$$

$$\sum M_D = 0, \quad \theta_C + 2\theta_D = -\frac{QL}{8}\cdot('') \quad (19\text{-c})$$

The solution of the simultaneous equations yields the following rotation angles.

$$\theta_B = -69.3(\frac{QL^2}{848EI}) \quad (20\text{-a})$$

$$\theta_C = 72.8(''), \theta_D = -69.2('') \quad (20\text{-b, c})$$

(End moment)
Above rotation angles are substituted into slope-deflection equations to obtain the results shown at the right-hand sides of the same equation.

The third step is the determination of fixity factors by using Eq.(1). In the present problem

$$\frac{0.7585QL}{8} = \frac{8EI}{1.2L}\cdot\frac{f_{BA}}{1-f_{BA}}\cdot\frac{69.3QL^3}{848EI}$$

$$\therefore \quad f_{BA} = 0.1482 \quad (21\text{-a})$$

$$\frac{0.7585QL}{8} = \frac{4EI}{L}\cdot\frac{f_{BC}}{1-f_{BC}}\cdot\frac{69.3QL^3}{848EI}$$

$$\therefore \quad f_{BC} = 0.2248 \quad (21\text{-b})$$

$$\frac{0.5604QL}{8} = \frac{4EI}{L}\cdot\frac{f_{CB}}{1-f_{CB}}\cdot\frac{62.9QL^3}{848EI}$$

$$\therefore \quad f_{CB} = f_{CD} = 0.1910 \quad (21\text{-c})$$

The fixity factors are given in Fig. 4(c). The last step is to find the K-factor and the critical local of each span by applying Eq.(3) and (4).

Table. I. Critical load coefficient, $C(P_{cr}=CEI/L^2)$

Loading Conditions	B.C	M.S.D.M	F.E.M	B.A.M	ERROR(%)
	$f_A=0.0$ $f_E=0.0$	9.8696	9.8696	9.8689 (ALL)	0.0
	$f_A=0.0$ $f_E=0.0$	10.6221	10.6223	10.6221 (DE)	5.6
	$f_A=0.0$ $f_E=0.0$	12.7800	12.7801	12.7801 (BC,CD)	4.9
	$f_A=0.0$ $f_E=0.0$	5.6468	5.6469	5.3889 (AB)	4.6
	$f_A=0.0$ $f_E=0.0$	5.7209	5.7210	5.4226 (AB)	5.2
	$f_A=0.0$ $f_E=0.0$	7.1344	7.1344	6.6068 (CD)	7.4
	$f_A=0.0$ $f_E=0.0$	7.5874	7.5877	7.2198 (CD)	4.8
	$f_A=0.0$ $f_E=0.0$	9.8696	9.8689	9.8689 (ALL)	0.0
	$f_A=0.0$ $f_E=0.0$	10.2130	10.2131	10.0142 (AB)	1.9
	$f_A=0.0$ $f_E=0.0$	11.2438	11.2439	10.1393 (DE)	9.8
	$f_A=0.0$ $f_E=0.0$	12.6131	12.6132	11.7864 (CD)	6.6
	$f_A=0.0$ $f_E=0.0$	5.8586	5.8583	5.4897 (AB)	6.3
	$f_A=0.0$ $f_E=0.0$	5.8748	5.8746	5.5110 (AB)	6.2
	$f_A=0.0$ $f_E=0.0$	9.4846	9.4843	9.2461 (AB)	2.5
	$f_A=0.0$ $f_E=0.0$	9.7375	9.7373	9.3270 (AB)	4.2

Span AB :

$$K_{AB} = \{(1+1)(1+0.1482)\}^{-0.5} \approx 0.66$$

$$(3P_{cr})_{AB} = 2 \times 1.1482 \frac{\pi^2 (2EI)}{(1.2L)^2}$$

$$\therefore (P_{cr})_{AB} \approx 10.493 \frac{EI}{L^2} \qquad (22\text{-}a)$$

Span BC :

$$K_{BC} = \{(1.2248)(1.191)\}^{-0.5} \approx 0.83$$

$$(2P_{cr})_{BC} = 1.2248 \times 1.191 \frac{\pi^2 EI}{L^2}$$

$$\therefore (P_{cr})_{BC} \approx 7.198 \frac{EI}{L^2} \qquad (22\text{-}b)$$

Span CD :

$$K_{CD} = \{(1.191)(1)\}^{-0.5} \approx 0.92$$

$$(P_{cr})_{CD} = 1.191 \frac{\pi^2 EI}{L^2} \approx 11.754 \frac{EI}{L^2} \qquad (22\text{-}c)$$

Here is the final step, where one can easily see that the member of span BC may buckle under the magnitude of the load, $7.198EI/L^2$.

The above-mentioned modified slope-deflection method is to be applied to the member of Fig.4(a). When Eq(7-a) and (7-b) are applied to this member, one obtains

$$M_{BC} = \frac{EI}{L}(\alpha_{n2}\theta_B + \alpha_{f2}\theta_c) \qquad (23\text{-}a)$$

$$M_{BA} = \frac{EI}{L}(\frac{5}{3}\alpha_{n1}\theta_B) \qquad (23\text{-}b)$$

$$M_{CB} = \frac{EI}{L}(\alpha_{f2}\theta_B + \alpha_{n2}\theta_c) \qquad (23\text{-}c)$$

$$M_{CD} = \frac{EI}{L}(\alpha_n\theta_C + \alpha_f\theta_D) \qquad (23\text{-}d)$$

$$M_{DC} = \frac{EI}{L}(\alpha_f\theta_C + \alpha_n\theta_D) \qquad (23\text{-}e)$$

Where α_{n}, α_{f}, ... are Merchant's stability functions (see Eq.10) with

$$kL = \sqrt{PL^2/EI}, \ k_1L = \sqrt{1.5}kL, \ k_2L = \sqrt{2}kL$$

To obtain a characteristic equation, one can start with $M_{DC} = 0$ to obtain $\theta_D = -(\alpha_f/\alpha_n)\bullet\theta_C$. Then, the moment equilibrium at the support B,$(\sum M_s=0)$ and C,$(\sum M_c=0)$, give the following matrix equation

$$\begin{vmatrix} (5\alpha_{n1}/3 + \alpha_{n2}) & \alpha_{f2} \\ \alpha_{f2} & \alpha_{n2} + (\alpha_n^2 - \alpha_f^2)/\alpha_n \end{vmatrix} \begin{Bmatrix} \theta_B \\ \theta_C \end{Bmatrix} = \begin{Bmatrix} 0 \\ 0 \end{Bmatrix} \qquad (24)$$

From the above equation, one gets the characteristic equation

$$(5\alpha_{n1}/3 + \alpha_{n2}) \cdot [\alpha_{n2} + (\alpha_n^2 - \alpha_f^2)/\alpha_n]$$
$$- (\alpha_{f2})^2 = 0 \qquad (25)$$

The least root satisfying the above equation is found to be $kL=2.72787$, and gives the elastic critical load

$$P_{cr} = 7.441EI/L^2$$

The proposed beam analogy method (B.A.M.) can easily be applied to multi-span continuous compression members without regard to changes in the sectional properties or the lengths of each span. Some of the critical loads of four-span continuous compression members are shown in Table 1. In this table, the boundary conditions(B.C) are either a frictionless hinge $(f_\alpha =0.0)$ or a fixed end $(f_\alpha =1.0)$. As can be seen in Table 1, the beam analogy method (B.A.M) gives lower boundary errors for critical loads. In the column "B.A.M", "$\alpha\beta$" in the parentheses denotes that the $\alpha\beta$ span buckles first. For example, (DE) means that span (DE) will buckle first under the given conditions.

3 K-FACTORS FOR BRACED COLUMNS

Several authors (Basu & Lee 1989, Disque 1974) have proposed different methods for K-factor determination. Until now, however, no unified method is available. Now, the proposed method can be used to determine effective length factors for multi-story braced frame columns. In the braced frames of Fig. 5, columns are not permitted to move horizontally and so they can be modeled as a continuous member of Fig. 5(c).

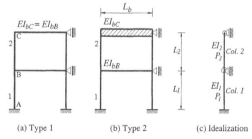

(a) Type 1   (b) Type 2   (c) Idealization

Figure 5. Braced frames and modeling

Actually, Fig. 5 is the braced frame and its modeling example was chosen by Hellesland (Hellesland et al. 1996) for effective length factors. Hellesland's K-factors for first and second story columns are $K_1=0.656$ and $K_2=0.928$, respectively, if one assumes $L_1= L_2$, $P_1= 2P_2$ in Fig. 5(c). Meanwhile, the beam analogy method applied to Fig. 5(c), gives $K_1=0.707(error, +7.8\%)$ and $K_2=1.00(error, +7.8\%)$.

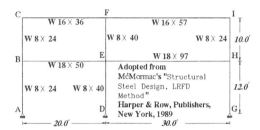

Figure 6 (a). Two story frame

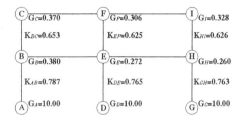

Figure 6 (b). G-Values & K-factors

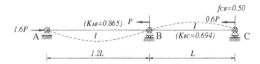

Figure 6 (c). Assumed column load

Fig. 6 shows a two-story steel frame chosen by McCormac (1989) for its effective length. factor determination. He adopted AISC's Alignment chart, which is based on the following equation

$$\frac{G_\alpha \cdot G_\beta}{4}(\frac{\pi}{K})^2 + (\frac{G_\alpha + G_\beta}{2})(1 - \frac{\pi/K}{\tan(\pi/K)})$$
$$+ \frac{2\tan(0.5\pi/K)}{\pi/K} = 1.0 \quad (26)$$

with

$$G_\alpha = \frac{\sum\limits_{\alpha}(\frac{EI}{L})_c}{\sum\limits_{\alpha}(\frac{EI}{L})_b} = \frac{sum\ of\ column\ stiffness\ at\ joint\ \alpha}{sum\ of\ beam\ stiffness\ at\ joint\ \alpha}$$

(27-a)

$$G_\beta = \frac{\sum\limits_{\beta}(\frac{EI}{L})_c}{\sum\limits_{\beta}(\frac{EI}{L})_b} = \frac{sum\ of\ column\ stiffness\ at\ joint\ \beta}{sum\ of\ beam\ stiffness\ at\ joint\ \beta}$$

(27-b)

The G-factors for each joint determined by Eq. 27 are given in the following figure (McCormac, 1989).

Finally, the K-factors satisfying Eq. 26 can be found by using a trial and error procedure. The least

roots of K for the columns are also given in Fig. 6(b).

Now, when the proposed method is applied to this frame with the assumption that the column load of each story changes in the manner of Fig. 6(c), one obtains K-factors in the parentheses.

The K-factor for the first story gives a 13% error when compared to the result obtained by the Alignment chart. But for the second-story column, the proposed method yields large upper bound errors, which can be acceptable as far as the actual structural design is concerned.

4 CONCLUSION

The elastic critical load of a multi-span compression member can be determined by using either an analytical or a numerical method. The limitation of these methods is that they permit one to obtain only the critical load. In addition to the critical load, a prediction of the span that buckles first is possible, if the proposed beam analogy method is adopted. The proposed beam analogy method can also be used to determine the effective length factors of a multi-story braced column.

The beam analogy method gives lower-bound errors of critical loads and upper-bound errors of effective length factors, which lead to a conservative design of either continuous compression members or multi-story braced columns.

REFERENCES

Basu P.K. & Lee S.L. 1989 Effective Length Factors for Type-PR Frame Members, *Proceedings of ASCE, Structural Div.* : pp185-194, May

Chajes A. 1974 *Principles of Structural Stability Theory.* Prentice-Hall Inc.

Chen W.F. and *et al.* 1987 *Structural Stability.* Elsevier Science Publishing Co. Inc.

CONSTRADO(U.K) 1983 *Steel Designer's Manual* William Collins Sons & Co., Ltd., 4th Rev.,

Disque R.O. 1974 Inelastic K-factor for Column Design, *J. Struct. Engng., AISC*, Vol.11, No.2, 2nd Qtr.: pp33-35

Hellesland Jostein and *et al.* 1996 Restraint Demand Factors and Effective Lengths of Braced Columns, *J. Struct. Engng. ASCE*, Vol.122, No.10: pp. 1216 ~ 1224

Kinney J.S. 1957 *Indeterminate Structural Analysis.* Addison Wesley Publishing Co. Inc.

Lee, S.G. 1979 *Eigenvalue Problems of Tapered Bars with Partially Fixed Ends*, Graduate School, Seoul Nat'l Univ.

Lee S.G. 1994 *Analysis of Steel Members with Partially Restrained Connections.* Singapore: The Third International Kerensky Conference,

McCormac J. C. 1989 *Structural Steel Design, LRFD Method* Harper & Row Publishers. Inc.

Creative Systems in Structural and Construction Engineering, Singh (ed.) © 2001 Balkema, Rotterdam, ISBN 90 5809 161 9

Progressive collapse of concrete frames using the Particle Flow Code

M.A. Dávila-Sänhdars & Y. Zhuge

School of Civil Engineering, University of South Australia, S.A., Australia

ABSTRACT: Concrete structures in seismic areas are very vulnerable to undergo severe damages under seismic loads. Unfortunately, the evaluation of damages so far has been done only after the earthquake ground motion. Practically, the buildings have been seen only before and after the earthquake except during the earthquake. Although numerical analyses have been developed to predict the mode of failure of the buildings, the results from such analyses have not been satisfactory at all, because the structural elements have been considered as a continuum. In this paper, the progressive collapse of a concrete frame is discussed using the Particle Flow Code. The collapse of the structure can be visualized as in a slow motion scenes keeping track of the more vulnerable areas of the structure so that the structural design can be carried out in a safer way.

1 INTRODUCTION

Concrete structures in seismic areas are vulnerable to undergo severe damage during the occurrence of an earthquake ground motion. Observing the damages in buildings after an earthquake has been found that the main damage usually occurred at the footing-column intersection as well as in the beam-column joints. Unfortunately, the buildings can been seen only before and after the occurrence of an earthquake except during the earthquake. During the occurrence of an earthquake, it is unlikely to observe the progressive failure of the buildings due to well-known reasons.

A broad variety of techniques has been developed so far to overcome the unwanted early failure in concrete buildings generated by seismic loads. A refined method for the calculation of redistributed moments in concrete frames subjected to seismic loads was developed by Cross and Morgan (1932). Later, a simplified methodology was developed for designing concrete structures subjected to dynamic loads (Norris and Wilbur 1960). Such a methodology comprises not only seismic load but also dynamic loads between the elastic range of the structure generated by other sources as for instance traffic loads and impact loads.

Finite Element Method has been broadly used in the last three decades in the analysis and design of concrete structures (Clough and Penzien 1993). The advent of fast computers facilitated the use of such a method in the analysis of structures not only under static loads but also under dynamic loads.

Using the Finite Element Method in the analysis of concrete structures, the elements are considered as a continuum. However, concrete no longer can be considered as a continuous material because it is a discrete material (Cundall et al. 1982). Concrete presents a great similarity with rocks that is to say it is a granular material. That characteristic of the concrete allows using the Particle Flow Code to simulate the progressive collapse of the structure. Assemblage of spherical particles (balls) is made up to simulate the concrete structure. The particles are linked together by parallel bonds so that when the external loads overpass the parallel bond strength then the parallel bonds break (Thornton and Barnes 1982). The breakage of parallel bonds provokes the collapse of the structure. Flow of the particles is reflected in the deformations of the concrete under external loads generating compression and shear forces between the particles (Oger et al. 1998).

Discrete Element Method has been successfully used since 1979 in the analysis of geomaterials such as rocks and soils as well as nuclear waste disposal (Cundall and Strack 1979). The analysis has been carried out considering the soil, rock and nuclear waste as a granular material formed by spherical particles. However, particles have been conceived of a particular shape depending on the researchers' point of view and what they are looking for. So for instance, the shape of particles has been assumed ellipsoidal to investigate compression stress in terms of the orientation (bedding) of the ellipsoidal particles (Ng and Elliot 1995) and (Ting and Meachum 1995).

Because concrete keeps a great similarity with rock materials then it can be analyzed using the Discrete Element Method as well (Fairhurst 1997). These similarities are manifested by the onset of plasticity as a function of the mean pressure; levels of dilation in shear; and the influence of interstitial fluids. Deformability of concrete is simulated in terms of normal and shear stiffness at contacts interparticles and the strength of the concrete is simulated by adding bond strengths not only in the normal direction but also in the shear direction. Damage in the concrete frame is simulated through the progressive breakage of the parallel bonds.

A mathematical model is discussed first to explain how the Particle Flow Code works and then a numerical example is presented using the Particle Flow Code to analyze the progressive collapse of a concrete frame.

2 MATHEMATICAL MODEL

Emphasis has been made in the displacement of particles in the sample caused by external loads i.e. particles flow throughout the solid structural element (Itasca 1999). Displacements of the particles are reflected through the deformations of the structural elements under the influence of external loads generating contact forces between particles as it can be seen in Figure 1. These contact forces can be decomposed into two component forces as follow: normal contact force F_n acting parallel to the line joining the centers of the balls A and B, and shear forces F_s acting in the transverse direction of contact normal force F_n.

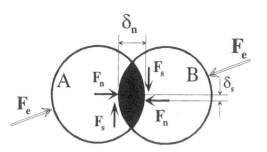

Figure 1. - Particle subjected to normal and shear forces.

The terms 'particle' and 'ball' have the same meaning herein because the particles have spherical shape, besides the term particle differs from that used in Microphysics which practically has no di-

mensions. The compression deformation δ_n of the particle is governed by the elastic spring's law as follow

$$F_n = k_n \delta_n \tag{1}$$

where k_n is the ball stiffness governed by the elasticity modulus E_c of the concrete. Meanwhile shear deformation is defined as follow

$$F_s = k_s \delta_s \tag{2}$$

where shear stiffness k_s of the ball is governed by the shear modulus G_c of the concrete.

Particles can rotate one around the others as it can be seen in Figure 2 where ball 2 has rotated clockwise around ball 1. Displacement of ball 2 is described by the dotted arc from the former position of its center C up to the current position and the net rotational displacement is defined by the vector *dis* which can be calculated as follow

$$|dis| = \sqrt{(\Delta X)^2 + (\Delta Y)^2} \tag{3}$$

and the angle of rotation as follow

$$\tan \theta = \frac{\Delta Y}{\Delta X} \tag{4}$$

In general terms, displacement of particles are governed by the law of motion regarding translation and rotation. Let us consider an ith particle in the structure therefore motions of particle i^{th} are defined by translational motion as follow

$$F_i = m(\ddot{x} - g_i) \tag{5}$$

where m is the mass of ball i; $\ddot{x}$ is the translational acceleration; and g_i is the gravity.
Rotation of paraticle i is defined as follow

$$M_i = \dot{H}_i \tag{6}$$

where M_i is the resultant moment acting on the particle; and H_i is the angular momentum of the particle. This relation is referred to a local coordinate system that is attached to the particle at its center of mass.

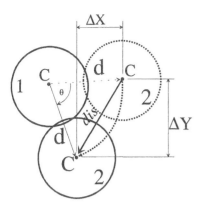

Figure 2. - Typical displacement of particles.

Rotational motion for a two dimensions analysis is defined by

$$M_z = I_z \dot{\omega}_z = (\beta m R^2) \dot{\omega}_z \qquad (7)$$

where M_z is the moment; $\beta = 2/5$ for spherical particles; R is the particle radius; and $\dot{\omega}_z$ is the angular acceleration of the particle around the z axis.

Equations of motion 5 and 7 are integrated using a centered finite-difference procedure involving a timestep of Δt and the velocities $\dot{x}_i$ and ω_z are calculated at the mid-intervals $t \pm n\Delta t/2$, and the quantities $x_i, \ddot{x}_i, \dot{\omega}_z, F_i$, and M_z are calculated at the primary intervals of $t \pm n\Delta t$ which leads to the expression of translational acceleration as follow

$$\ddot{x}_{(t)} = (\Delta t)^{-1} \left[\dot{x}_{(t+\Delta t/2)} - \dot{x}_{(t-\Delta t/2)} \right] \qquad (8)$$

and the equation of rotational acceleration as follow

$$(\dot{\omega}_z)_{(t)} = (\Delta t)^{-1} \left[(\omega_z)_{(t+\Delta t/2)} - (\omega_z)_{(t-\Delta t/2)} \right] \qquad (9)$$

Meanwhile inserting equations 8 and 9 into equations 5 and 7 and solving for the velocities at time (t + Δt / 2) gives

$$\dot{x}_{(t+\Delta t/2)} = \dot{x}_{(t-\Delta t/2)} + \left(\frac{F_{(t)}}{m} + g \right) \Delta t \qquad (10)$$

where $F_{(t)}$ is the translational force in terms of the mass m of the particle, g is the gravity and Δt is the

corresponding time-step. And the rotational volocity as follow

$$(\omega_z)_{(t+\Delta t/2)} = (\omega_z)_{(t-\Delta t/2)} + \left(\frac{M_{(t)}}{I} \right) \Delta t \qquad (11)$$

where $M_{(t)}$ is the moment in terms of the moment of inertia I of the particle during the time-step Δt.

Translational and rotational stiffnesses of an individual particle relate increments of force and moment, increments of displacement and rotation via the matrix relations as follow

$$\begin{Bmatrix} \Delta F_x \\ \Delta F_y \\ \Delta M_z \end{Bmatrix} = \begin{bmatrix} k_{xx} & k_{xy} & 0 \\ k_{yx} & k_{yy} & 0 \\ 0 & 0 & k_{zz} \end{bmatrix} \begin{Bmatrix} \delta_x \\ \delta_y \\ \theta_z \end{Bmatrix} \qquad (12)$$

where k_{xx}, k_{xy}, k_{yx}, k_{yy} are the normal and shear stiffnesses of the ball, and k_{zz} is the rotational stiffnes of the ball.

Particles in the structure are maintained together by means of parallel bonds. Parallel bonds behave as a cementatious material between two neighboring balls and they are therefore responsible for carrying the flexural strength capacity of the whole structure.

Figure 3 shows parallel bonds sticking together balls 1 and 2 and also is shown the cross-section (A—A) where it can be seen the radius of the ball and the radius of the parallel bond. Notice that the radius of the parallel bond is less than the radius of the balls. In other words, the radius of the parallel bonds may be less than of equal to the radius of the ball but never greater. Parallel bonds resemble not only concrete strength but also reinforcement strength.

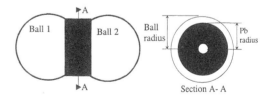

Figuer 3. Parallel bond between balls 1 and 2

Parallel bonds act in both normal and shear direction as it is shown in Figure 4. The normal parallel bond is expressed as follow

$$\Delta \overline{F}_i^n = \left(-\overline{k}^n A \Delta U_i^n \right) n_i \qquad (13)$$

where $\bar{k}^n$ is the stiffness of the normal bond; A is the area of the cross-section of the parallel bond (see Section A—A in Figure 3); ΔU_i^n is the normal deformation of the parallel bond; and n_i is a unit vector oriented from the center of one ball to the center of the neighboring ball, whereas the shear parallel bond can be given by the following expression

$$\Delta \bar{F}^s = -\bar{k}^s A \Delta U_i^s \qquad (14)$$

where $\bar{k}^s$ is the stiffness of the shear bond; A is the cross-section as in the normal bond and ΔU_i^s is the shear deformation of the parallel bond. The dash-accent at the top of the variables is only to denote parallel bond meaning. Meanwhile the elastic moment-increment is given by

$$\Delta \bar{M}_z = -\bar{k}^n I \Delta \theta_z \qquad (15)$$

where I is the moment of inertia of the bond cross-section about an axis through the contact point and in the direction of $\Delta \theta_z$. The tensile and shear stress due to external loading acting on the bond periphery are calculate using the beam theory as in equations 16 and 17 as follow

$$\sigma_{max} = \frac{-\bar{F}^n}{A} + \frac{|\bar{M}_z| \bar{R}}{I} \qquad (16)$$

$$\tau_{max} = \frac{|\bar{F}^s|}{A} \qquad (17)$$

where A and I are the area and moment of inertia of the bond cross-section. On the other hand, if the maximum tensile stress σ_{max} exceeds the normal bond strength $\bar{\sigma}_c$ $(\sigma_{max} \geq \bar{\sigma}_c)$ or the maximum shear stress τ_{max} exceeds the shear bond strength $\bar{\tau}_c$ $(\tau_{max} \geq \bar{\tau}_c)$, then the parallel bond breaks and the structure collapses.

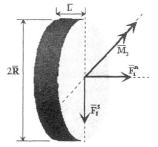

Figure 4. – Parallel bond forces (from Figure 3)

3 NUMERICAL EXAMPLE

3.1 Introduction

A small specimen was modeled to investigate the behavior of the structure during the application of seismic loads. Firstly, the structure was set up to stable equilibrium.

After reaching stable equilibrium, the frame was subjected to a simulated seismic load until collapse. The progressive collapse of the structure was followed in slow motion scenes to investigate the more vulnerable areas of the structure under seismic loads. A set of snap-shots is presented where it can be seen how and where the failure of structure happened.

Table 1. Materials properties

	Strength [MPa]	Elasticity modulus [MPa]	Normal stiffness*	Shear stiffness*	Normal stress [MPa]	Shear stress [MPa]	Density [kN/m³]
Structural balls	50	35708	$1\%10^9$	$1\%10^9$			25.0
Parallel bond			$1.7\%10^{12}$	$1.7\%10^{12}$	$6.15\%10^8$	$6.15\%10^7$	
Loading balls			$1\%10^{10}$	$1\%10^{10}$			300.0

• Normal and shear stiffness in balls are given in N/m; Normal and shear stiffness in parallel bonds are given in Pa/m.

3.2 Example

The computer model was setup using the properties outlined in Table 1. The radii of the balls in the structure were all 45 mm meanwhile the radii of the loading balls on the beams were 135, 90 and 45 mm respectively. The friction coefficient between balls was 0.0 in order to shed the stress capability of structure on the parallel bonds. Figure 5 shows the assembled structure just before the application of the earthquake. The heavy lines between the loading balls are contact forces at equilibrium.

The parallel bonds in the structure are represented by the parallel lines along the beams, columns and footing. In Figure 5 can be seen also, that parallel bonds in beams and footing are tangent to the balls meanwhile in the columns the parallel bonds are secant to the balls. That means that the width of the columns is less than the depth of the beams and footing. In other words, the sizes of structural elements are defined by the radius of the corresponding parallel bond (see Fig. 3). The structure was set up at the verge of failure so that the parallel bonds started breaking at the first cycle of the earthquake load.

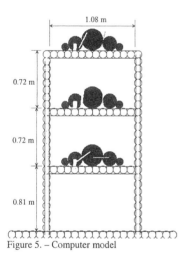

Figure 5. – Computer model

4 ANALYSIS OF RESULTS

The structure was subjected to a simulated earthquake ground motion of frequency 20.0 Hz and amplitude 10 cm during 20.0 sec. Figure 6 shows the model at $6.53\%10^{-3}$ sec of shaking in which it can be seen the breakage of the parallel bonds (see the dotted circles) in the footing-columns intersection and in the beam-column joint at the right hand side in Figure 6. The structure is off the ground because the linking bonds between the columns and the ground no longer exist.

Shortly after, at $2.44\%10^{-1}$ sec, the separation between columns and first floor was more evident and the structure started collapsing due to its self-weight as it can be seen in Figure 7. Notice in Figure 7 also that the contact forces between the loading balls look modified because those balls were moving in the direction of the settlement of the frame.

The frame was no longer vertical and it grew failing and the settlement was more evident at 1.93 sec as it can be seen in Figure 8.

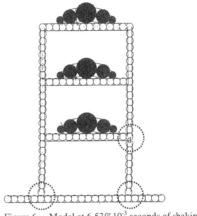

Figure 6. – Model at $6.53\%10^{-3}$ seconds of shaking.

The reader can see in Figure 8 how the structure was collapsing leaving behind the loading balls, which look like climbing up onto the structure due to inertial effect.

In Figure 9 it can be seen the first floor of the structure almost touching the ground. The loading balls on the first floor are making contact with the column meanwhile the loading balls on the top floor look like escaping over towards the left hand side. Notice that the loading balls kept verticality as a whole thing although the structure was at the last steps of collapsing resembling the motion of things and people inside the buildings during the occurrence of an earthquake ground motion.

Following the sequence of the simulated test procedure, it can be seen that the frame started failing in the footing-column connection and then at the beam-column joints at the first floor level. The remaining of the structure failed due to settlement caused by the loss of support at the base. In addition, it can be observed that the structure once it started collapsing it never recovered verticality because it was already off the ground.

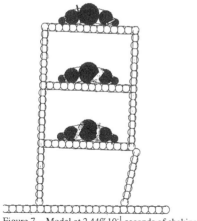

Figure 7. - Model at 2.44×10^{-1} seconds of shaking.

Figure 8. – Model at 1.93 seconds of shaking.

Figure 9. – Model at 2.93 seconds of shaking.

5 CONCLUSIONS

It is well known that is not possible to monitor the progressive collapse of a building during the occurrence of an earthquake. There are three basic reasons impeding to monitor the collapsing building such as danger, the structural elements are covered by architectural decoration and, most of all, the short duration of the earthquake.

Using the Particle Flow Code it is possible to monitor the collapse of the building through the computer simulation because the time in the computer program can be delayed in relation to the real time manipulating the size of the time-steps. Snap shots of the structure can be taken at any time to keep track of the progression in the collapse of the structure as it is shown in this paper.

More research needs to be done in order to refine the accuracy in the analysis of the progressive collapse in concrete frames.

6 REFERENCES

Clough, R. W. and J. Penzien 1993. *Dynamic of Structures*. McGraw Hill.

Cross, H & K. Morgan 1932. *Continuous Frames of Reinforced Concrete*. John Wiley & Sons, Inc. New York.

Cundall, P.A. & O.D.L. Strack 1979. A discrete numerical model for granular assemblies. *Géotechnique*, Vol 29 pp 47 — 65.

Cundall. P. A., A. Drescher & O. D. L. Strack 1982. Numerical experiments on granular assemblies; Measurements and Observations. *IUTAM Conference on Deformation and Failure of Granular Materials* / Delft / 31 Aug. — 3 Sept 1982 pp 355 — 370.

Fairhurst, C 1997. Geomaterials and Recent Developments in Micro-Mechanical Numerical Models. *International Society for Rock Mechanics*. Vol. 4 # 2. pp 11 — 14.

Itasca, 1999. *Particle Flow Code in 2 Dimensions*. Itasca Consulting Group, Inc. U.S.

Ng, T. & H. Eliot Fang 1995. Cyclic behaviour of arrays of ellipsoids with different particle shapes. AMD-Vol. 201, *Mechanics of Materials With Discontinuities and Heterogeneities*. ASME, pp 59 — 70.

Norris, C. H. & J. B. Wilbur 1960. *Elementary structural analysis*. McGraw Hill Book Company Inc. U.S.

Oger, L., S.B. Savage, D. Corriveau & M. Sayed 1998. Yield and deformation of an assembly of disks subjected to a deviatoric stress loading. *Mechanics of Materials*. ELSEVIER. pp 189 — 210.

Thornton, C. & D.J. Barnes 1982. On the mechanics of granular material. *IUTAM Conference on Deformation and Failure of Granular Materials* / Delft / 31 Aug. — 3 Sept 1982. pp 69 — 77.

Ting, J. M. & L. F. Meachum 1995. Effect of bedding plane orientation on the behavior of granular systems. AMD-Vol. 201, *Mechanics of Mateials With Discontinuities and Heterogeneities*. ASME. pp 43 —57.

21 Structural behavior of structural elements

Creative Systems in Structural and Construction Engineering, Singh (ed.) © 2001 Balkema, Rotterdam, ISBN 90 5809 161 9

Ultimate strength of reinforced concrete circular and rectangular columns

Ramon V. Jarquio
New York City Transit, Brooklyn, N.Y., USA

ABSTRACT: This paper illustrates the analytical procedures for analyzing the ultimate strength of circular and rectangular column sections using the true parabolic stress method of analysis. This methodology utilizes the standard assumptions and differs from the current procedure in that it presents the derived formulas required to analytically solve the circular and rectangular column sections. It also for the first time introduces the concept of column capacity axis for equilibrium of external and internal forces. It incorporates ACI provisions for column design but recommends minimum load eccentricity when the depth of section under compression reaches the full depth of the column section. It focuses on the capacity of the column section as being invariant, at every depth of compression and position of the column capacity axis. The external loads, on the other hand, are modified to account for multipliers due to eccentricities and other load factors. This methodology eliminates the column interaction formula for biaxial bending by using the column capacity axis for uniaxial reference of resultant loads. Finally, it will eliminate the need for finite element procedures for solving the ultimate strength capacity of reinforced concrete circular and rectangular column sections.

1 INTRODUCTION

Ultimate strength design in reinforced concrete uses the so-called parabolic-rectangular stress block method, commonly known as the Whitney equivalent rectangular stress block. This current method is crude and inefficient because the equivalent rectangle was referred to the simple parabola.

In contrast, the true parabolic stress method utilizes the basic parabola, which closely fits the stress-strain curves of concrete cylinders. This parabola is defined by the value of f_c', the depth of compression, c and the useable value of linear strain, e_c. The maximum area derived from these limiting parameters is the measure of the ultimate strength of the concrete section under consideration. Moreover, the standard method does not identify the column capacity axis for equilibrium of forces, while the true parabolic stress method utilizes this axis for equilibrium of external and external forces.

Since this methodology considers the ultimate strength capacity of a column section as invariant, slenderness effects (moment magnifications) are not included. However, standard notations are adhered to as much as possible.

2 DERIVATION

In column analysis, the basic parabola is applied in such a manner that the point of zero strain can lie outside the column section. This is so because the combined axial and bending stress/strain diagram can be a trapezoid depending on the position of the axial load. In beams, the location of zero strain is confined within the beam section. Beam rotation due to flexure assures similar triangles of the strain diagram between the compressive and tensile strains. This so-called balanced condition occurs only once as the column section is analyzed from pure axial load through axial load plus bending to a pure bending condition.

The analysis for the ultimate strength of a circular column is first developed followed by the rectangular column section. For bi-axial analysis, the equation of the basic parabola must be referred to the axes of the column section to facilitate the integration of compressive forces on the concrete section. This is accomplished by rotation of axes to convert the biaxial to a uniaxial analysis along a column capacity axis chosen along the diagonal of the rectangular section or at any axis which will satisfy the statical equilibrium conditions of forces.

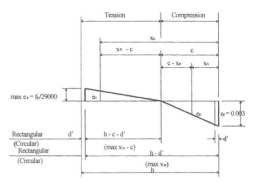

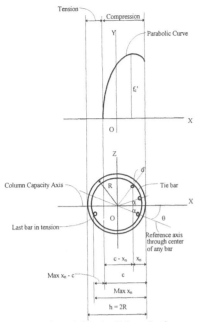

Figure 1. Circular Column Section

Bar Strain Diagram

Circular	Rectangular
When c <= 87(max x_n)/(f_y + 87)	When c <= 87(h - d')/(f_y + 87)

Compressive Strain: Compressive Strain:

$e_n = e_c(c - x_n)/c$ $e_n = e_c(c - x_n)/c$

Tensile Strain: Tensile Strain:

$e_n = (f_y/29000)(x_n - c)/(max\ x_n - c)$ $e_n = (f_y/29000)(x_n - c)/(h - c - d')$

When c > 87(max x_n)/(f_y + 87) When c > 87(h - d')/(f_y + 87)

Compressive Strain: Compressive strain:

$e_n = e_c(c - x_n)/c$ $e_n = e_c(c - x_n)/c$

Tensile Strain: Tensile Strain:

$e_n = (f_y + 87)(x_n - c)/(29000)(max\ x_n)$ $e_n = (f_y + 87)(x_n - c)/(29000)(h - d')$

Figure 2

2.1 The Circular Column

Concrete Forces - In Figure 1, the equation of the basic parabola can be easily derived as

$$y = (0.75\ f_c'/c^2)\{[(c^2 + 2Rc - 3R^2) + 2(3R - c)x - 3x^2]\} \quad (1)$$

The equation of the circle is $x^2 + z^2 = R^2$ (2)

The concrete compressive force, C is given by the expression

$$dC = 2\ yzdx \quad (3)$$

Put eq. (1) and (2) in (3) to obtain,

$$C = [1.50\ f_c'/c^2] \int [(c^2 + 2Rc - 3R^2) + 2(3R - c)x - 3x^2\ ]\ (R^2 - x^2)^{1/2}\ dx \quad (4)$$

Integrate eq. (4) using standard integration formulas and evaluate limits to obtain 15 equations not shown due to space limitations.

2.1.1 Bar Forces

Figure 2 indicates the steel strain varies linearly from zero at the neutral axis to a maximum value of $f_y/29000$ and remains at this value as a function of the concrete strain, ec = 0.003.

The position of rebars is determined from the relationship,

$$x_n = R - (R - d')\cos(n\alpha - \theta) \quad (5)$$

in which,

N = total no. of bars
n = no. of bar position from 1 to N
$\alpha = 2\pi / N$
θ = position of the column capacity axis varies from zero to α

A total of 11 equations have to written to define the strains and positions of rebars. From these, the bar forces are calculated.

2.2 The Rectangular Column with Biaxial Bending

In Figure 3 the rectangular column section is shown with width = b" and length =d. From the rotation of axes and using trigonometry,

$$h/2 = (1/2)[\ d\cos\theta + b\sin\theta\ ] \quad (6)$$

The equations of the lines representing the four sides of the rectangular section using analytic geometry are as follows:

$$z_1 = -\tan\theta\ [\ x - h/2\ ] + z_0 \quad (7)$$

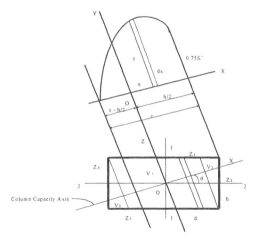

Figure 3. Concrete Forces on a Rectangle

$$z_2 = \cot\theta\,[\,x - h/2\,] + z_0 \qquad (8)$$

$$z_3 = -\tan\theta\,[x + h/2] - z_0 \qquad (9)$$

$$z_4 = \cot\theta\,[\,x + h/2\,] - z_0 \qquad (10)$$

in which, θ = inclination of the column capacity axis and

$z_0 = (1/2)[b\cos\theta - d\sin\theta]$, when $\theta < [(\pi/2) - \alpha]$ or

$z_0 = (1/2)\,[d\sin\theta - b\cos\theta]$ when $\theta > [(\pi/2) - \alpha]$

$\alpha = \arctan(b/d)$

2.2.1 *Concrete Forces*

Divide the area of the rectangle into three main sections such as V_1, V_2 and V_3 in which V_1 is the area whose limits are from $-x_2$ to x_2, V_2 is the area whose limits are from x_2 to $h/2$ and V_3 is the area whose limits are from $-h/2$ to $-x_2$. Depending upon the particular value of "c" these limits will accordingly vary within the indicated boundaries.

The equation of the basic parabola referred to the new origin of the "x" and "y" axes is written previously as

$$y = [0.75f_c'/c^2][(\,c^2 + ch - 0.75h^2\,) + (\,3h - 2c\,)\,x - 3x^2\,] \qquad (11)$$

The compressive force on the concrete section is the volume of the stress diagram and the area of the concrete section defined by "c". A total of 17 equations will define the concrete forces.

2.2.2 *Bar Forces*

Calculate bar forces as shown in Figure 2. When $c \le 87(h - d')/(f_y + 87)$, tensile strain, e_n is determined as:

$$e_n = (f_y/29000)(x_n - c)/(h - c - d') \qquad (12)$$

and when $c > 87(h - d')$ tensile strain, e_n is determined as:

$$e_n = (f_y + 87)(x_n - c)/29000(h - d') \qquad (13)$$

along width: $x_n = (n-1)2[(b-2d')/(N- 1)]\sin\theta +$

$$(\sqrt{2}\,)\,d'\cos(45° - \theta) \qquad \text{1st half}$$

$x_n = [(n - N/2) - 1]2[(b - 2d')/(N - 1)]\sin\theta +$

$$(d - 2d')\cos\theta + (\sqrt{2}\,)\,d'\cos(45° - \theta) \quad \text{2nd half}$$

in which, N = no. of bars

along length: $x_n = 2m\,[(d - 2d')/(M + 1)]\cos\theta +$

$$(\sqrt{2}\,)\,d'\cos(45° - \theta) \qquad \text{1st half}$$

$x_n = h - (\sqrt{2}\,)\,d'\cos(45° - \theta) - [m - (M/2)]\,x$

$$2[(d - 2d')/(M + 1)]\cos\theta \qquad \text{2nd half}$$

in which, M = no. of bars

Total no. of bars = N + M

3 COLUMN CAPACITY CHARTS / CURVES

The formulas derived for circular and rectangular column sections are programmed in a computer for faster calculations of column capacities at any depth "c" of the concrete under compression, including the more traditional key points of the curve such as:

1. The point where M = 0 and P = max.
2. The point where c = full depth of concrete section where the maximum allowable axial load can be specified with the minimum eccentricity.
3. The point where tension of the steel is zero.
4. The point where moment is maximum. This point can be at balanced condition or near it.
5. The point where P = 0 and M = maximum in which the section becomes a beam.

Figures 5 and 6 show the computer printouts of the ultimate strength of a given circular and rectangular column section respectively showing the values at key points and at intermediate points, when f_c' = 34.5 Mpa and f_y = 413.7 Mpa.

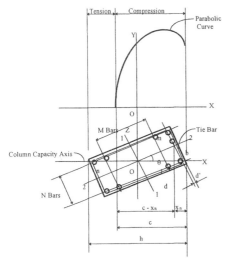

Figure 4 Rectangular Column Layout of Rebars

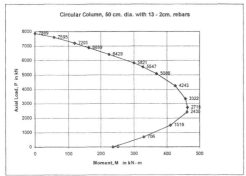

Figure 5. Circular Column Capacity Curve

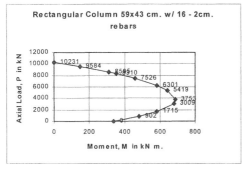

Figure 6. Rectangular Column Capacity Curve

3.1 *Rectangular Column with Uniaxial Load*

When the load axis is at either axis 1-1 or 2-2, the ultimate capacity of the column is determined when the value of "θ" in the foregoing equations is as-

sumed very nearly equal to $\pi/2$ or zero respectively. However, in practice except for retaining walls, biaxial loading always governs the analysis due to the requirements of minimum eccentricity of the applied external loads.

4 COLUMN CAPACITY AXIS

The column capacity axis is a line perpendicular to the moment axis. For equilibrium of external and internal forces the column capacity axis must be used as a reference.

4.1 *Rectangular Column*

For a rectangular column section subjected to biaxial bending, the load axis can be anywhere between zero and $\pi/2$ as determined by the values of e_1 and e_2 and the orientation of the column section. The column capacity axis can also vary from zero to $\pi/2$. For safety in design $P > P_u$ and $M > M_u$ to place the point inside the envelope of the column capacity curve. Column capacity charts can be prepared wherein the column capacity axis is assumed as the diagonal of the rectangular section.

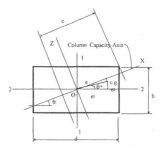

Figure 7. Column Capacity Axis

For a rectangular column section subjected to biaxial bending, the requirement that the resultant load must be collinear with the ultimate capacity of the section is satisfied when

$$e = e_u \cos(\theta - \theta_u) \tag{14}$$

θ_u = angle of eccentricity of the given loading condition ($= M_u/P_u$) with the horizontal axis

Equation (14) indicates that the resultant external load falls below the column capacity axis for a particular value of "c" in the rectangular column section. In other words, we cannot use the column capacity along the load axis since the resultant external load will not coincide with the center of gravity of internal forces of a column section defined by the value of "c".

However, the author recommends that for design purposes we can assume the diagonal as the reference axis for the column capacity. The reason for this is the fact that at other values of "θ" below the diagonal line, the column capacity is more. This assumption allows us to prepare column capacity charts or tables for any given column section. It also insures that the values being used are within the envelope of the true column capacity. This method requires that the given external load $P_u e_u$ should be resolved along the diagonal as $P_u e_u \cos(\theta - \theta_u)$ before this moment can be compared with the column capacity along the diagonal.

To find the column capacity axis of a column section subjected to an external biaxial load equal to $P_u e_u$, use the following equations:

$$\Sigma Vz = P_u e_u \sin(\theta - \theta_u) \tag{15}$$

$$\bar{z} = \Sigma Vz / \Sigma V = e_u \sin(\theta - \theta_u) \tag{16}$$

$$\Sigma V = P_u \tag{17}$$

4.1.1 Concrete Forces

$$V_1 z_1 = \tan\theta \, (V_1 x_1) \text{ when } \theta < [(\pi/2) - \alpha] \text{ and} \tag{18}$$

$$V_1 z_1 = -\cot\theta \, (V_1 x_1) \text{ when } \theta > [(\pi/2) - \alpha] \tag{19}$$

$$V_2 z_2 = -0.50(\cot\theta - \tan\theta)V_2 \bar{x}_2' - [z_0 . 0.25 \, h \, (\cot\theta - \tan\theta)]V_2 \tag{20}$$

$$V_3 z_3 = -0.50(\cot\theta - \tan\theta) \, V_3 \bar{x}_3' - [0.25 \, h(\cot\theta - \tan\theta) - z_0]V_3 \tag{21}$$

4.1.2 Bar Forces Lever Arms

M Bars: $+ z = (x_m - d_1)\tan\theta \quad < (d - 2d')\sin\theta \tag{22}$

$- z = [h - x_m - d_1]\tan\theta \quad < (d - 2d')\sin\theta \tag{23}$

N Bars: $+ z = [h - x_n - d_1]\cot\theta < (d - 2d')\sin\theta \tag{24}$

$- z = (x_n - d_1)\cot\theta \quad < (d - 2d')\sin\theta \tag{25}$

Use the initial value of θ at the diagonal and combine moments of the concrete forces and bar forces to solve for the initial value of "$\bar{z}$". Compare this value with $e_u \sin(\theta - \theta_u)$. Repeat until these two values are almost equal to obtain the correct column capacity axis for a particular external load of $P_u e_u$. This will complete the solution for the true column capacity axis to use for a particular column section subjected to a specific loading condition.

4.2 Circular Column

The load axis can fall between any two bars and hence, the column capacity axis can vary from $\theta =$ zero to $\theta = \alpha$. The load axis is determined by the value of e_1 and e_2, which are the eccentricities of the bi-axial bending moments. From the value of θ_u, the position of the column capacity axis between rebars is determined. To insure safety, $P > P_u$ and $M > Mu$ for design purposes, i.e. the point represented by P_u and M_u should be inside the envelope of the column capacity curve. Column capacity charts can be prepared wherein the column capacity axis is assumed to pass through the center of any bar.

If desired the actual column capacity axis may be determined by the centroid of bar forces. To check the very small eccentricity the lever arm of any bar is given by the expression

$$z = (R - d') \sin(n\,\alpha - \theta) \tag{26}$$

4.3 ACI Transition Zone for the Reduction Factor (Design Safety Factor), ϕ

For values of P & M from key point V to an imaginary point "W" defined by $P = 0.10 f_c' \, A_g$ of the column interaction diagram, the value of ϕ is assumed to vary linearly as a function of P, as follows:

Tied : $\phi = [0.09 f_c' \, A_g - 0.20P]/[0.10 f_c' \, A_g] \tag{27}$

Spiral: $\phi = [0.09 f_c' \, A_g - 0.15P]/[0.10 f_c' \, A_g] \tag{28}$

All values of P & M above key point 5 to imaginary point "W" are multiplied by ϕ from the above equations. In this region the interaction curve of the column capacity will be distorted. In order not to distort the interaction curve, the author recommends that ϕ be applied to the external load P_u and M_u as follows:

$$\phi = 0.45 + 0.45\{1 - 0.741[1.4\,DL + 1.7\,LL]/P_T\}^{1/2}$$

$$\phi = 0.45 + 0.45\{1 - 0.494[1.4\,DL + 1.7\,LL]/P_T\}^{1/2}$$

for tied and spiral column respectively, in which P_T = axial load capacity at Key Point 4 of the interaction curve
$P = (P_u/\phi) = (1.4\,DL + 1.7\,LL)/\phi$ and
$M = (M_u/\phi) = (1.4\,DL + 1.7\,LL)/\phi$.

5 CONCLUSIONS

1. The foregoing analyses showed that the analytical solution is feasible by using the true parabolic stress method in the analysis for ultimate strength of reinforced concrete circular and rectangular column sections.

2. This methodology will eliminate the need for finite element procedures now currently in use.

3. It also proved that the column capacity axis must be taken into consideration in order to satisfy the requirement of equilibrium of external and internal forces, which is very fundamental in any structural analysis.

NOTATIONS

A_g = gross area of concrete section

A_s = total area of steel reinforcement

A_n = area of steel reinforcement

b = width of rectangular concrete section

c = depth of concrete section in compression

d = distance from extreme compression fiber centroid of steel reinforcement

d' = distance from concrete edge to centroid of reinforcement

e = eccentricity = M/P

e_c = concrete strain

e_s = steel strain

e_n = compression (or tension) strain at nth layer of reinforcement

e_u = eccentricity of the external load ($=M_u/P_u$)

f_c = compressive stress in concrete

f_c' = specified compressive strength of concrete

f_y = specified yield strength of reinforcement

f_{yn} = steel stress $\leq f_y$

h = overall thickness of member

x_m = location of rebar from a reference axis

x_n = location of rebar from the concrete compressive edge

$\bar{z}$ = distance from x-axis of the centroid of internal forces

D = diameter of rebar , inches

L_d = development length of rebar

M_1 = moment about axis 1-1

M_2 = moment about axis 2-2

$M = M_u / \phi$ = design moment

M_u = external moment

$P = P_u / \phi$ = design axial load

P_u = external load

R = radius of circular column

ϕ = ACI capacity reduction factor (design safety factor)

θ = inclination of the resultant internal forces about the horizontal axis = arctan b/d (for design)

θ_u = arctan M_2/M_1 (inclination of the resultant external forces about the horizontal axis)

α = central angle subtended by one bar spacing

Note: All other alphabets and symbols used in mathematical derivations are defined in the context of their use in the analysis.

REFERENCES

ACI 318 R – 83. *Commentary on Building Code Requirements for Reinforced Concrete: Chapter 10.*

Ferguson, Phil, M., Breen, John, E. and Jirsa, James, O. *Reinforced Concrete Fundamentals, Edition: 45-70.*

Gerstle, Kurt, H. *Basic Structural Design.* McGraw-Hill Book Company: 25-31.

Beyer, William, H. *Handbook of MathematicalSciences, 6th edition:* 31 - 32.

Creative Systems in Structural and Construction Engineering, Singh (ed.) © 2001 Balkema, Rotterdam, ISBN 90 5809 161 9

Seismic response of asymmetric frame building designed according to eurocodes

V. Kilar
Faculty of Civil and Geodetic Engineering, IKPIR, University of Ljubljana, Slovenia

ABSTRACT: In the paper the seismic response of selected 4-stories R/C frame building designed according to Eurocodes 2 and 8 is presented. Three different structural variants are investigated: 1) symmetric variant, 2) mass eccentric variant with the same reinforcement as symmetric building and 3) redesigned mass eccentric variant considering mass eccentricity. The ultimate strength in X and Y direction is almost the same for all building variants. The seismic response is presented in terms of displacements, story drifts and ductility factors for edge frames and for selected points of the structure. The comparison of the effect of different designs is also given. It is shown that the torsional effects in the asymmetric building that was designed according to Eurocode 8 provisions for asymmetric structures, has smaller torsional rotations, smaller edge displacements and lower maximum ductility factors than the building that was not specifically designed for the mass eccentricity. Whereas the maximum damage on the flexible side is reduced due to new design, the damage is increased for the stiff side frames, due to required reduction of strength at the stiff side.

1 INTRODUCTION

It is well known that the torsional response of in plan asymmetric buildings may cause detrimental effects on the structural behavior during strong earthquake ground motions. To minimize the effect of torsion, the distance between mass centre and stiffness centre in a building (actual eccentricity) should be kept as small as possible. In practice in many cases the asymmetry can not be avoided. In these cases the torsional response can be reduced only with appropriate strength distribution among the load resisting elements. Building codes usually prescribe certain eccentricities that include actual eccentricity as well as dynamic and accidental eccentricity in the design process. These provisions result in different strength distribution that is considered as more favorable when the particular asymmetric building is subjected to strong ground shaking.

In Eurocode 8 the selection of analysis model and analysis method depend on criteria for regularity in plan and elevation. For buildings that are regular in plan and in elevation, a planar model and simplified method of analysis can be used. For the in plan asymmetric buildings that are regular in elevation and satisfy additional requirements (e.g. height less that 10 m, height/length ratio in both main directions does not exceed 0.4, load resisting elements run from foundations to the top without interruption, etc.) the analysis can be also performed using two planar models, one for each main direction. The torsional effect shall be determined separately by displacing the horizontal forces in each story for an additional eccentricity taking in account the actual eccentricity and the dynamic effects (see ENV 1998-1-2, Annex A). For each element the most unfavorable position of the load should be considered.

For general asymmetric buildings, that satisfy none of the regularity criteria, a spatial model and multi-modal response spectrum analysis method should be used.

In the last category classified also the test buildings considered in this paper. For design of buildings a spatial model and multimodal analysis were used. The masses were displaced according prescribed accidental eccentricity to the position that was the most unfavorable for each designed frame.

The accidental eccentricity should be taken into account as $e_{1i} = \pm 0,05 \cdot L_i$ where e_{1i} is accidental eccentricity of story mass i from its nominal location, applied in the same direction at all floors and L_i the floor dimension perpendicular to the direction of seismic action. Only in the case of symmetric distribution of lateral stiffness and mass, where a plannar model can be used, accidental torsional effects might be also evaluated in a very simple manner and added to the results of plannar model.

2 DESIGN OF SELECTED BUILDINGS

In order to verify the effectiveness of Eurocode 8 provisions for asymmetric structures three variants of a four-stories reinforce concrete frame building were selected as test examples (Fig. 1).

Symmetric building variant (Symm) was designed using Eurocodes 2 and 8 (soil class B, a_g=0.35g, q=3.75), considering an accidental eccentricity equal to ±5% of the relevant plan dimension of the building. The cross sections of the structural members are equal in all frames and in all stories (Fig. 2). Story masses amounted to 295 and 237 tons in bottom stories and at the roof, respectively. A spatial model and multi-modal response spectrum analysis were used to design the structure. The design base shear was equal to 23% of the total weight. Different strength levels of columns and beams, as required by analysis, were obtained by varying the amount of reinforcement. The longitudinal reinforcement in beams was determined with the program SAP2000 (1997). For columns, the Eurocode 8 rules of capacity design were applied. Reinforcement in the bottom two stories was different from that in the upper two stories. All columns of each frame in a story have equal reinforcement. The frames closer to the perimeter of the building require bigger amount of reinforcement as the frames that are closer to the centre of stiffness (S). Frames X1 and X6 are identical, as well as X2 and X5, and X3 and X4. Identical are also frames Y1 and Y4, as well as frames Y2 and Y3.

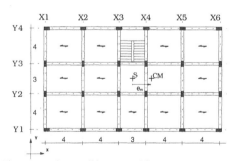

Figure 1. Analyzed building: typical floor plan.

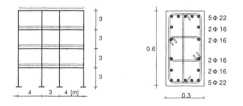

Figure 2a Elevation and reinforcement of columns in the first and second story of edge frames (symmetric structure, frames X1 and X6).

Table 1. The amount of reinforcement (cm^2) for asymmetric (AsymA) and symmetric structure (values for symmetric structure are given in brackets).

Columns			Beams	
Frame	Story	Reinforcement	Max. upper reinforcement	Max. lower reinforcement
X1	1-2	39.6 (54.0)	7.7 (15.6)	7.7 (12.5)
	3-4	25.3 (37.4)	6.0 (8.7)	6.0 (6.0)
X2	1-2	39.6 (43.8)	10.1 (12.5)	7.7 (9.4)
	3-4	28.4 (32.4)	6.0 (7.1)	6.0 (6.0)
X3	1-2	43.8 (43.8)	11.0 (11.0)	9.4 (9.4)
	3-4	32.4 (32.4)	6.0 (6.0)	6.0 (6.0)
X4	1-2	43.8 (43.8)	12.5 (11.0)	9.4 (9.4)
	3-4	32.4 (32.4	7.1 (6.0)	6.0 (6.0)
X5	1-2	54.0 (43.8)	15.6 (12.5)	12.5 (9.4)
	3-4	37.4 (32.4)	8.7 (7.1)	6.0 (6.0)
X6	1-2	65.1 (54.0)	18.8 (15.6)	17.7 (12.5)
	3-4	42.7 (37.4)	10.2 (8.7)	7.7 (6.0)

Two asymmetric building variants were obtained by shifting the centre of masses CM in the +X direction by em=0.1· L (1.9 m). In the first asymmetric variant (AsymS) the structure remained the same as for the symmetric building. The second asymmetric variant (AsymA) was redesigned considering mass eccentricity 0.1· L and accidental eccentricity equal to ±5% of the relevant plan dimension of the building. For each frame the most unfavorable position of the mass centre was considered. The cross sections of all structural elements have remained the same, as the redistribution of strength was achieved by changing the amount of reinforcement. In AsymA building variant , the reinforcement differs from frame to frame. It gradually decreases from frame X6 toward the frame X1, which has the smallest amount of reinforcement.

The amount of reinforcement for the symmetric and asymmetric variant is illustrated in Table 1. Periods for symmetric building are as follows: 1. mode 0.42s (Y), 2. mode 0.41(X), 3. mode 0.33s (T).

3 NONLINEAR STATIC AND DYNAMIC ANALYSIS AND SEISMIC INPUT

Nonlinear static (pushover) and dynamic analyses of the 3-D mathematical model were performed by using the CANNY computer program (Li, 1997). The floor diaphragms of the structure were assumed to be rigid in their own planes and to have no out-of-plane stiffness. The beams were modeled considering uniaxial bending and shear deformation. The columns were modeled considering biaxial bending, shear deformation, as well as axial and torsional deformation. The shear spring was considered elastic in all cases. Damping was taken to be a combination of mass and initial stiffness proportional damping. The damping coefficients were determined for 5% damping to 1. and 2. mode. A more detailed description of mathematical modeling of the structure can be find also in Faella and Kilar (1998).

Both horizontal components of five records were used to investigate the effects of the ground motion variation: Petrovac (Montenegro 1979), El Centro (1940), Kobe JMA (1995), and two records from the 1994 Northridge earthquake (Sylmar and Newhall). The stronger components (i.e. components with larger peak ground acceleration) were scaled to 0.7 g (2 times design ground acceleration) and applied in the building Y-direction (the ratio between X and Y components of the accelerograms remained unchanged).

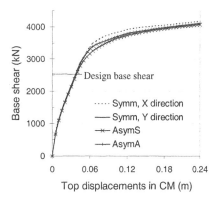

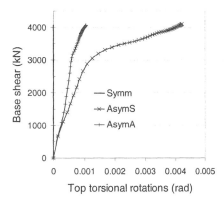

Fig. 3. Base shear - top displacement and base shear - top torsional rotation relationships for the symmetric and two asymmetric building variants.

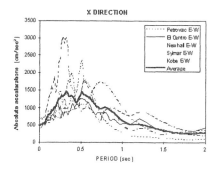

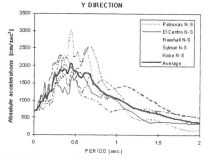

Fig. 2b Elastic acceleration response spectra for components in X and Y direction for 5% damping.

4 PUSHOVER ANALYSIS

This section describes some results obtained by pushover analysis. The load pattern was an inverted triangle. Figure 3 presents base shear - top displacement relationship for the symmetric and for two asymmetric building variants. The loads were applied in the relevant center of masses CM, independently in X and Y direction (for symmetric building) and in Y direction (for asymmetric buildings). The curves are plotted up to a top displacement equal to 2% of the building height (0.24 m).

For the symmetric structure it can be seen that the strength in X direction is slightly larger than in Y direction. The overstrength for the symmetric structure, defined as the maximum strength divided by the design base shear (2565 kN), amounts to about 1.63 for X direction and about 1.6 for Y direction. The total strength in Y direction is almost equal for all building variants. Biggest torsional rotations occur in the building AsymS. The new design with different frame strengths introduces a strength eccentricity that amounts to +1.2 m for the building AsymA. The center of strength has been moved closer to the center of masses CM and the effective eccentricity in the plastic range is reduced. As a consequence, the plastification of all frames occurs almost at the same time (similarly as in symmetric structure) what reduces torsional rotations for building AsymA. This is not the case for the building AsymS where the plastification of stiff side frames occurs much latter than the plastification of the flexible side frames.

5 NONLINEAR DYNAMIC ANALYSIS

Selected results of time-history analyses for investigated building variants are shown in Tables 2 and 3 and Figures 4, 5 and 6. In Table 2 the average maximum top displacements and maximum story drifts (average values obtained for five ground motions) of selected points of the structures are presented. Maximum top displacements are given for edge frames X1 and X6, for mass centre CM (in X and Y direction) as well as for the frame Y1. It can be seen from the Table 2 that in both asymmetric variants, as expected, eccentricity increases displacements and story drifts at the flexible side and decreases them at the stiff side (in respect to symmetric structure). The influence of torsion is smaller for the building AsymA, which was designed as asymmetric building and therefore tends to behave more like the symmetric building. The displacements of frame Y1 are very similar for all building variants.

Table 2. Maximum top displacement and maximum story drifts of different frames for different structural variants.

Variant	Maximum top displacement (cm)					Maximum story drifts (%)	
	X1	CM-Y	X6	Y1	CM-X	X1	X6
Symm	21.3	21.3	21.3	17.5	17.5	2.15	2.16
AsymS	15.0	19.4	27.0	19.6	17.2	1.59	2.73
AsymA	18.3	20.6	24.3	16.9	16.4	1.94	2.45

Figure 4 presents the time histories of top displacements of edge frames X1 and X6 for buildings AsymS and AsymA for Newhall earthquake record scaled to 0.7g. On the same figure also the top rotations for both buildings are presented. It can be seen that for the frame X6, the maximum top displacement occurs for the building AsymS (0.4 m=21% more than for AsymA), and for the frame X1, the maximum displacement occurs for the building AsymA (0.25 m=19% more than for AsymS). The difference is much bigger in terms of top torsional rotations; maximum top rotation for building AsymS amounts to 0.016 rad, which is almost 2.5 times bigger that for AsymA. It can be therefore concluded that the redistribution of strength according to Eurocode 8 successfully minimized the influence of torsion for the investigated frame building.

Figures 5 and 6 present the ductility factors for ends of beams and columns for whole both edge frames X1 and X6 (μ is defined as a ratio of response rotation to yield rotation of a flexural spring; $\mu \leq 1.0$=elastic behavior). The presented ductility factors are average values obtained for five investigated earthquake records (see chapter 3). For each frame

the average ductility factor for a whole story is also given.

The results given in Figures 5 and 6 are summarized also in the Table 3, that presents only the maximum ductility factors for frames X1 and X6 for all building variants.

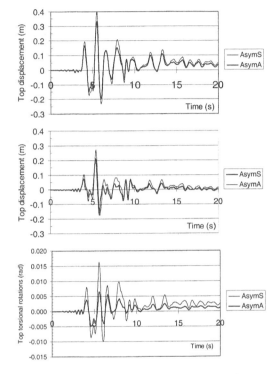

Figure 4. Time histories of top displacements of edge frames X1 and X6 and top torsional rotations for Newhall earthquake record scaled to 0.7g.

It can be seen from the Table 3 as well as from Figures 4 and 5 that, due to torsion, the ductility factors for the structure AsymS are increased for the flexible side, and decreased for the stiff side compared to the symmetric building. In the building AsymA the ductility factors are reduced for the flexible side, however, due to reduction of strength at the stiff side according to Eurocode 8, the ductility factors are increased for the frame X1. In same cases the ductility factors might be even bigger than for the symmetric structure (for example for the top of columns in the third story, see Figure 5). The new design reduces the overall maximum ductility factor from 9.1 for AsymS to 7.5 for AsymA. The pattern of ductility factors indicates that the plastic mechanism up to the third floor can be expected in most cases.

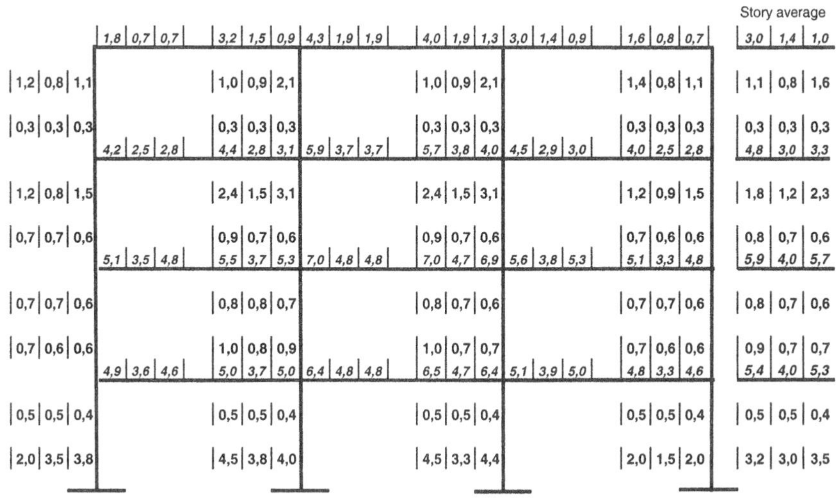

Fig. 5 (frame X1):

Row (top beams): |1,8|0,7|0,7| |3,2|1,5|0,9| |4,3|1,9|1,9| |4,0|1,9|1,3|3,0|1,4|0,9| |1,6|0,8|0,7| Story average: |3,0|1,4|1,0|

|1,2|0,8|1,1| |1,0|0,9|2,1| |1,0|0,9|2,1| |1,4|0,8|1,1| |1,1|0,8|1,6|
|0,3|0,3|0,3| |0,3|0,3|0,3| |0,3|0,3|0,3| |0,3|0,3|0,3| |0,3|0,3|0,3|
beams: 4,2|2,5|2,8 4,4|2,8|3,1 5,9|3,7|3,7 5,7|3,8|4,0 4,5|2,9|3,0 4,0|2,5|2,8 |4,8|3,0|3,3

|1,2|0,8|1,5| |2,4|1,5|3,1| |2,4|1,5|3,1| |1,2|0,9|1,5| |1,8|1,2|2,3|
|0,7|0,7|0,6| |0,9|0,7|0,6| |0,9|0,7|0,6| |0,7|0,6|0,6| |0,8|0,7|0,6|
beams: 5,1|3,5|4,8 5,5|3,7|5,3 7,0|4,8|4,8 7,0|4,7|6,9 5,6|3,8|5,3 5,1|3,3|4,8 |5,9|4,0|5,7

|0,7|0,7|0,6| |0,8|0,8|0,7| |0,8|0,7|0,6| |0,7|0,7|0,6| |0,8|0,7|0,6|
|0,7|0,6|0,6| |1,0|0,8|0,9| |1,0|0,7|0,7| |0,7|0,6|0,6| |0,9|0,7|0,7|
beams: 4,9|3,6|4,6 5,0|3,7|5,0 6,4|4,8|4,8 6,5|4,7|6,4 5,1|3,9|5,0 4,8|3,3|4,6 |5,4|4,0|5,3

|0,5|0,5|0,4| |0,5|0,5|0,4| |0,5|0,5|0,4| |0,5|0,5|0,4| |0,5|0,5|0,4|
|2,0|3,5|3,8| |4,5|3,8|4,0| |4,5|3,3|4,4| |2,0|1,5|2,0| |3,2|3,0|3,5|

Fig. 5. Maximum ductility of stiff side frame (frame X1) for different structural variants (ductility factors are given in the following order (Symm,AsymS,AsymA).

Fig. 6 (frame X6):

Story average

Row (top beams): |1,8|3,2|2,7| |3,2|4,9|3,5| |4,3|6,7|6,7| |4,0|6,4|4,7|3,0|5,1|3,6| |1,6|3,2|2,4| Story average: |3,0|4,9|3,6|

|1,2|2,0|1,3| |1,0|1,1|1,3| |1,0|1,0|1,3| |1,4|2,0|1,2| |1,1|1,5|1,3|
|0,3|0,4|0,4| |0,3|0,5|0,5| |0,3|0,5|0,5| |0,3|0,4|0,4| |0,3|0,5|0,4|
beams: 4,2|5,8|5,0 4,4|6,1|5,0 5,9|8,1|8,1 5,7|7,8|6,5 4,5|6,0|4,9 4,0|5,8|5,0 |4,8|6,6|5,5

|1,2|1,3|1,3| |2,4|2,7|2,8| |2,4|2,7|2,9| |1,2|1,3|1,3| |1,8|2,0|2,1|
|0,7|0,9|0,9| |0,9|1,5|1,4| |0,9|1,5|1,4| |0,7|0,9|0,9| |0,8|1,2|1,2|
beams: 5,1|6,6|5,5 5,5|7,2|5,9 7,0|8,9|8,9 7,0|9,1|7,4 5,6|7,0|5,9 5,1|6,9|5,5 |5,9|7,6|6,3

|0,7|0,8|0,8| |0,8|0,9|0,9| |0,8|1,5|1,4| |0,7|0,7|0,8| |0,8|1,0|1,0|
|0,7|0,9|0,8| |1,0|1,5|1,2| |1,0|1,5|1,3| |0,7|0,8|0,8| |0,9|1,2|1,0|
beams: 4,9|5,8|5,0 5,0|6,3|5,2 6,4|7,8|7,8 6,5|7,8|6,6 5,1|6,1|5,2 4,8|6,2|5,1 |5,4|6,7|5,6

|0,5|0,6|0,6| |0,5|0,6|0,7| |0,5|0,6|0,7| |0,5|0,5|0,6| |0,5|0,6|0,6|
|2,0|3,5|3,8| |4,5|3,8|4,0| |4,5|5,6|5,2| |2,0|2,5|2,2| |3,2|3,8|3,8|

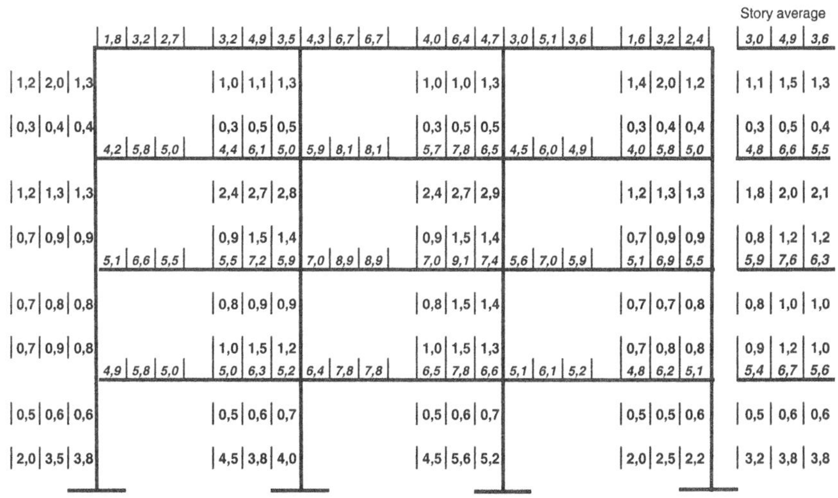

Fig. 6. Maximum ductility factors of flexible side frame (frame X6) for different structural variants (ductility factors are given in the following order (Symm,AsymS,AsymA).

Table 3 – Maximum ductility factors for different structural variants.

Variant	Frame X1 Beams	Frame X1 Columns	Frame X6 Beams	Frame X6 Columns
Symm	7	4.5	7	4.5
AsymS	4.8	3.8	9.1	5.6
AsymA	6.9	4.4	7.5	5.2

6 CONCLUSIONS

Presented results show that the torsional effects in the asymmetric building (AsymA), that was designed according to Eurocode 8 provisions for asymmetric structures, have been reduced in comparison to the asymmetric building (AsymS) where frames were identical as in the symmetric building (Symm). In the building AsymA the damage is reduced for the flexible side frames but due to reduction of strength at the stiff side, it is increased for the stiff side frames.

As a side product of the presented study, the seismic response of a building, designed according to Eurocodes 2 and 8, can be evaluated. The building was subjected to ground motions scaled to twice the design ground acceleration. The eccentricity for the building AsymS was much larger than assumed in design. Nevertheless, the seismic demand was not excessive in most of the elements. In the case of such a ground motion, the building would be severely damaged, however, most probably it would not collapse.

REFERENCES

CEC, Commission of the European Communities, (1998), Eurocode 8 – Design provisions for earthquake resistance of structures, European Committe for Standardization, ENV 1998-1-1/2/3.

Faella,G. and Kilar,V. (1998) "Asymmetric multistorey R/C frame structures: push-over versus nonlinear dynamic analysis", 11. European Conference on Earthquake Engineering, Paris, France, Proceedings, Balkema, Rotterdam.

Li,K.N. (1996), Three-dimensional nonlinear dynamic structural analysis computer program package CANNY-E, Users' manual, Canny Consultants Pte Ltd., Singapore.

SAP2000 (1997), Integrated Finite Element Analysis and Design of Structures, Computers and Structures, Inc., Berkeley, California, USA.

Creative Systems in Structural and Construction Engineering, Singh (ed.)© 2001 Balkema, Rotterdam, ISBN 90 5809 161 9

Elastic stability of half-through girder bridges

Z. Vrcelj & M. A. Bradford
School of Civil and Environmental Engineering, University of New South Wales, Sydney, N.S.W., Australia

H. R. Ronagh
Department of Civil Engineering, University of Queensland, Brisbane, Qld, Australia

ABSTRACT: This paper presents the development of a rational generic model to investigate the elastic restrained-distortional buckling of simply supported half-through girders without web stiffeners along their lengths. The distortional buckling model considers an I-section beam that is restrained completely against lateral deformations and elastically against twist rotation during buckling by the deck. A parameter is derived in the Ritz-based procedure which quantifies the various parameters affecting distortional buckling under this condition of restraint, and some design curves are presented.

1 INTRODUCTION

Half-through girder bridges provide a load path from the bridge deck to the bearing supports by means of bottom flange loading of parallel steel I-section beams. Because of this, the I-section beams of simply supported half-through girders experience compression in their top flanges and tension in their bottom flanges. At the level of the bottom (tension) flange, the deck restrains the flange against lateral and minor axis rotational deformations during buckling, and depending on the stiffness of the deck in flexure transverse to the longitudinal axis of the bridge, it provides some theoretically quantifiable degree of twist rotational restraint. At the level of the top (compression) flange of the I-section, restraint of this critical flange against buckling is effected only by the flexural stiffness of the web in the plane of its cross-section. The major consideration of the design of half-trough girders is that of instability of the steel beams, and this mode of instability must necessarily be that of restrained distortional buckling (RDB) (Bradford 1996, Ronagh & Bradford 1998), as shown in Fig. 1.

RDB is fundamentally different to the more commonly studied and familiar lateral-distortional buckling of unrestrained beams, and it can have a profound influence on the buckling behaviour of beams with continuous restraint at the level of the

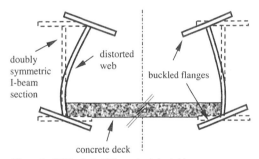

Figure 1. RDB of a half-through girder bridge

non-critical flange (Bradford 1998, Ronagh & Bradford 1998). Despite RDB being the governing buckling mode for many engineering structures that are commonly designed, such as continuous composite beams (Vrcelj et al. 1999) and half-through girder bridges (which is the subject of this paper), its accurate prediction is still a grey area in structural design. The issue of RBD in half-through girder bridges is addressed in this paper, and through the development of a generic buckling model a dimensionless parameter is identified which quantifies the variables that affect the buckling behaviour.

The most common model for considering RDB in design is the so-called U-frame method (Oehlers & Bradford 1995), in which the top compression flange

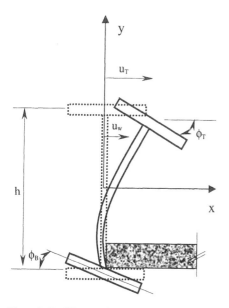

Figure 2. Buckling model

of the I-section is considered as a strut compressed uniformly along its length by the maximum bending stress that is induced in it, and which is restrained by a Winkler spring whose stiffness is that of the web in the plane of its cross-section. The U-frame approach is simplistic, and many finite element analyses have shown it to be inaccurate. A useful modification of the U-frame model was developed by Svensson (1985), in which account was taken of the variation of the bending stress in the strut model, but which retained the tensionless Winkler concept of restraint by the web. Williams & Jemah (1987) argued that the Winkler model did not account for torsional restraint, and based on finite element studies suggested that the flange-strut should be considered as a tee-section with the flange section as its table, and 15% of the web depth as its stem. This suggestion is empirical, and does not produce exact results for the elastic critical stress in the flange. Further models have been suggested by Bradford (1996), Lindner (1997) and others.

2 BUCKLING ANALYSIS

The analysis herein uses a simple Ritz-based formulation. This energy approach requires assumed deformations for the beam as it departs its trivial primary equilibrium path at the point of bifurcation into the buckled and stable secondary

path, and as the buckling deformations are assumed to be infinitesimal an eigenproblem is established that can determine the buckling load but not the postbuckling response. The deformations of the cross-section which are shown in Fig. 2 are the lateral deformation and twist rotations of the top flange u_T and ϕ_T respectively, and the twist rotation ϕ_B of the bottom flange. These buckling freedoms are consistent with the restraint conditions assumed for the half-through girder modelling. If the assumption verified elsewhere (Bradford 1994) that the flanges are stocky and deform only as rigid bodies during buckling is made, and furthermore that the web deforms in the plane of its cross-section as a cubic curve, then the vector of infinitesimal buckling deformations given by

$$\vec{u} = \left\langle u_T, \phi_T \phi_B \right\rangle^T \tag{1}$$

defines the buckled configuration of the cross-section at any position z from the origin of the beam of length L.

Herein it is assumed that the beam is simply supported and that the web is unstiffened, except at its ends where load bearing stiffeners are assumed to provide simple support with respect to out-of-plane buckling (Trahair & Bradford 1998). The deformations consistent with these kinematic boundary conditions are

$$\vec{u} = \vec{q} \sum_{i=1}^{n} \sin i\pi\xi \tag{2}$$

in which $\vec{q}$ is the vector of Ritz-coefficients which represent the maximum magnitudes of the buckling deformations, and where $\xi = z/L$. The cubic deformation of the web during buckling is further represented as

$$u_w = h\left(\alpha_o + \alpha_1\eta + \alpha_2\eta^2 + \alpha_3\eta^3\right)\sum_{i=1}^{n} \sin i\pi z \tag{3}$$

where $\eta = 2y/h$. If the condition of displacement and slope compatibility between the web and flanges is used (Bradford 1994), then the coefficients in Eqn. 3 may be found from

$$\vec{\alpha} = \overline{C}\vec{q} \tag{4}$$

The energy approach requires the strain energy stored in the top flange due to lateral deformation and twist rotation and in the bottom flange due to twist rotation U_F to be determined, as well as that stored in the web due to flexure U_W and that stored in the bottom flange by the elastic twist rotational restraint provided by the deck U_r. These may be determined from beam theory for the flanges and isotropic plate theory for the web (Bradford 1994) as

$$U_F = \frac{1}{2}\int_0^1 \left\{ \frac{EI_{yF}}{L^3} u_T{,}_{\xi\xi}^2 + \frac{GJ_F}{L}\left(\phi_T{,}_{\xi}^2 + \phi_B{,}_{\xi}^2\right)\right\} d\xi \quad (5)$$

$$U_W = \frac{1}{2}D\int_0^1 \int_{-1/2}^{1/2} \left\{ \frac{L}{h} u_w{,}_{\eta\eta} + \frac{h}{L} u_w{,}_{\xi\xi} \right.$$
$$\left. - \frac{2(1-v)}{Lh}\left(u_w{,}_{\eta\eta} u_w{,}_{\xi\xi} - u_w{,}_{\xi\eta}^2\right)\right\} d\eta d\xi \quad (6)$$

$$U_r = \frac{1}{2}L\int_0^1 k_z \phi_B^2 \, d\xi \quad (7)$$

where E = Young's modulus; v = Poisson's ratio; k_z = elastic twist rotation stiffness applied at the bottom flange; EI_{yF} = minor axis flexural stiffness of the flange, GJ = Saint Venant torsional rigidity of the flange; and the web plate rigidity is

$$D = \frac{Et_w^3}{12(1-v^2)} \quad (8)$$

The energy approach also requires the work done during buckling associated with fibre shortening under normal stress and shear. During the buckling, the loss of potential caused by the stresses $\sigma(y,z)$ and $\tau(y,z)$ associated with a given bending moment field $M(\xi)$ is

$$V_F = \frac{1}{2}\frac{1}{L}\int_0^1 \sigma \int_A \left\{u_T{,}_{\xi}^2 + u_B{,}_{\xi}^2 + v_T{,}_{\xi}^2 + v_B{,}_{\xi}^2\right\} dA \, d\xi$$
$$\quad (9)$$

$$V_W = \frac{1}{2}t\frac{1}{L}\frac{1}{h}\int_0^1 \int_{-1/2}^{1/2} \begin{Bmatrix} u_w{,}_{\xi} \\ u_w{,}_{\eta} \end{Bmatrix}^T \begin{bmatrix} \sigma & \tau \\ \tau & 0 \end{bmatrix} \begin{Bmatrix} u_w{,}_{\xi} \\ u_w{,}_{\eta} \end{Bmatrix} d\eta d\xi$$

where A = area of the cross-section, and assuming the rigidity of the flanges

$$v_{T;B} = x\phi_{T;B} \quad (10)$$

In the application herein, the moment may vary lengthwise as a cubic polynomial which may be piecewise continuous in the domain $[0,L]$. Within a particular subdomain, the moment is specified *a priori* as

$$M(\xi) = \lambda M_o\left(a_o + a_1\xi + a_2\xi^2 + a_3\xi^3\right) \quad (11)$$

with M_o being a reference value, a_i ($i = 0,1,2,3$) are predetermined coefficients defining the moment field, and λ is the buckling load factor.

In the Ritz method, the change of total potential is

$$\Pi = U_F + U_W + U_r - V \quad (12)$$

and from Eqns. 2-7 and 9 may be written as

$$\Pi = \frac{1}{2}\vec{q}^T \bar{k}\vec{q} \quad (13)$$

where $\bar{k}(\lambda)$ is a stiffness matrix that depends linearly on λ. Invoking now the variational form of neutral equilibrium that

$$\delta\Pi = 0 \; \forall \; \delta\vec{q} \quad (14)$$

where $\delta\vec{q}$ is an arbitrary variation of the Ritz coefficients, produces

$$\delta\Pi = \delta\vec{q}^T \bar{k}\vec{q} = 0 \quad (15)$$

so that

$$\bar{k}\vec{q} = \vec{0} \quad (16)$$

and since $\vec{q} = \vec{0}$ defines the primary equilibrium path, bifurcation occurs when

$$\left|\bar{k}(\lambda)\right| = 0 \quad (17)$$

Equation 17 is a standard linear eigenproblem which may be solved to produce the buckling load factor λ as well as the buckled shape defined by the eigenvector $\vec{q}$.

3 DISTORTIONAL BUCKLING PARAMETER γ

The stiffness matrix $\bar{k}$ depends on a number of parameters, including the cross-sectional and material properties and the bending moment field M. In the generic modelling of the problem in Section 2 by the use of Eqn. 2, the stiffness matrix may be expanded fully and the relative magnitudes of added terms in it can then be evaluated. The expanded form of the stiffness matrix is given in Vrcelj &

Bradford (2000), where it is identified that for a given loading regime and a particular value of the dimensionless torsional parameter

$$\alpha = \frac{k_z}{\pi^2 GJ / L^2} \qquad (18)$$

the buckling load can be specified by knowledge of only the well-known beam parameter (Trahair & Bradford 1998)

$$K = \sqrt{\frac{\pi^2 EI_w}{GJL^2}} \qquad (19)$$

and the proposed distortional buckling parameter

$$\gamma = \frac{DL^2}{GJh} \qquad (20)$$

By producing an ensemble of buckling solutions for various parameters such as L/h, h/t_w, b_f/t_f, b_f/h and α that define the problem uniquely, Vrcelj & Bradford (2000) demonstrated that indeed the buckling solutions are embodied only in a knowledge of K and γ, except for small values of K.

4 VALIDATION OF SOLUTION

The energy method has been compared with the results of the line-finite element distortional buckling program FEDBA16 developed by Bradford & Ronagh (1997), and those of the well-known software ABAQUS (1998). The results of these comparisons are reported fully in Vrcelj and Bradford (2000), and the correlation between the simplified analysis derived herein and the finite element solutions for specific cases is very good.

5 NUMERICAL RESULTS

A simply supported I-beam, as would constitute one beam in a half-through girder bridge, has been investigated. The beam is fully restrained continuously against lateral deformations and rotations during buckling, but elastically against twist rotation. In the modelling of the beam consistent with Sect. 3 of this paper, the geometry is embodied in a knowledge of the distortional restraint parameter γ in Eqn. 18 and the beam parameter K in Eqn. 19.

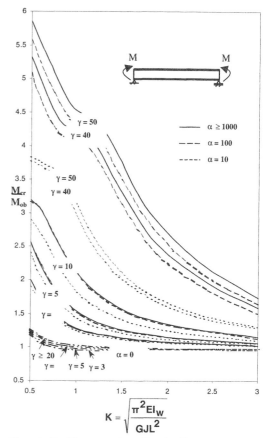

$$K = \sqrt{\frac{\pi^2 EI_w}{GJL^2}}$$

Figure 3. Buckling curves for uniform bending

Figures 3 – 5 show the variation of the dimensionless buckling moment M_{cr}/M_{ob} with the beam parameter K for uniform bending, a point load at midspan and a point load at the quarter point respectively. In these figures, M_{ob} is the elastic buckling moment for the beam restrained continuously against lateral deflection and rotation only (but which does not involve distortion of the cross-section during buckling), and α is the dimensionless torsional restraint parameter given in Eqn. 20. The trends in the three curves are very similar. With the presentation of the curves defined by the parameters K and γ, the physical significance of the results is somewhat obscured. While it might be expected that the effects of distortion would be less for high values of L (or low values of K), this is not evident intrinsically in the figures. This occurs because the high values of γ for which the effects of distortion are the most profound at low values of K are influenced by the high values of L in Eqn. 18.

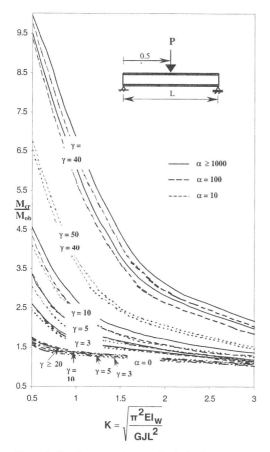

Figure 4. Buckling curves for central point load

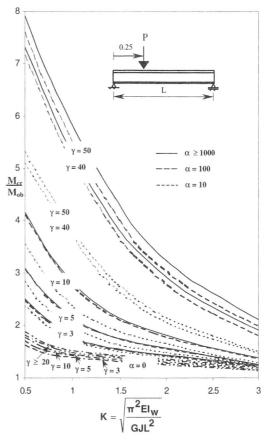

Figure 5. Buckling curves for quarter point load

Because of this, the interpretation must necessarily be in terms of other geometric parameters.

Nevertheless, the advantages in the use of Figs. 3 – 5 to determine the elastic buckling load for infinite combinations of section geometries under a given loading regime is evident.

6   CONCLUDING REMARKS

This paper has presented a Ritz-based method for determining the elastic distortional buckling loads for simply supported half-through girder bridges without transverse stiffeners along the span and which includes cross-sectional distortion during the bifurcation of equilibrium.   The generic model selected has identified a distortional buckling parameter that quantifies the effect of cross-sectional distortion, and which allows the high multiplicity of buckling curves associated with distortional

instability to be reduced to only a few.   The use of curves of the type illustrated in the paper enables hand design solutions to be obtained for restrained distortional buckling which hitherto has required complex numerical modelling.

ACKNOWLEDGEMENTS

The work reported in this paper was supported by both the 'Women in Engineering' and 'Dean's Scholarship' awards made available by the Faculty of Engineering at the University of New South Wales. The authors are grateful for the assistance of Dr Y-L Pi of that university in carrying out the work reported in this paper.

REFERENCES

ABAQUS User's Manual 1998. Pawtucket RI:Hibbitt, Karlsson & Sorensen Inc.

Bradford, M.A. 1994. Buckling of post-tensioned composite beams. *Struct. Eng. & Mechs.* 2(1):113-123.

Bradford, M.A. 1996. Stability of through-girder bridges. *Proc. Conf. on Structural Steel Developing Africa, Johannesburg,* 35-42:Johann-esburg:SAISC.

Bradford, M.A. 1998. Inelastic buckling of I-beams with continuous tension flange restraint. *J. Constr. Steel Res.* 48:63-77.

Bradford, M.A. & H.R. Ronagh 1997. Generalized elastic buckling of restrained I-beams by the FEM. *J. Struct. Eng., ASCE* 123(12):1631-1637.

Lindner, J. 1997. Lateral torsional buckling of composite beams. *J. Constr. Steel Res.* 46:1-3 (Paper 289).

Oehlers, D.J. & M.A. Bradford 1995. *Composite steel-concrete structural members: fundamental behaviour.* Oxford:Pergamon.

Ronagh, H.R. & M.A. Bradford 1998. Distortional buckling of I-shaped plate girders, a simple and efficient model. *Proc. 5th Pacific Structural Steel Conf.,* Seoul, 199-204.

Svensson, S.E. 1985. Lateral buckling of beams analysed as elastically supported columns subject to a varying axial force. *J. Constr. Steel Res.* 5:179-193.

Trahair, N.S. & M.A. Bradford 1998. *The behaviour and design of steel structures to AS4100.* 3rd Australian edn., London:E&FN Spon.

Vrcelj, Z. & M.A. Bradford 2000. Design curves for the elastic stability of half-through girder bridges. UNICIV Report, The University of New South Wales, Sydney.

Vrcelj, Z., M.A. Bradford & B. Uy 1999. Elastic buckling modes in unpropped continuous composite tee-beams. In M.A. Bradford et al. (eds), *Mechanics of Structures and Materials – Proc. ACMSM16, Sydney, 8-10 Dec. 1999*:327-333. Rotterdam:Balkema.

Williams, F.W. & A.K. Jemah 1987. Buckling curves for elastically supported columns with varying axial force, to predict lateral buckling of beams. *J. Constr. Steel Res.* 7:133-147.

Creative Systems in Structural and Construction Engineering, Singh (ed.) © 2001 Balkema, Rotterdam, ISBN 90 5809 161 9

Structural behavior of repaired cooling tower shell

T. Hara
Tokuyama College of Technology, Japan

S. Kato
Toyohashi University of Technology, Japan

M. Ohya
Matsue National College of Technology, Japan

ABSTRACT: The structural behavior of the old cooling tower shell repaired either by adding the concrete shell thickness or by placing ring stiffeners is analyzed. The models are the large cooling tower of height about 100m. The structural behaviors are computed under both repairing schemes. In the numerical analysis, the finite element scheme based on the isoparametric degenerated shell formulation is adopted. Both the geometric and material nonlinearities are taken into account. To evaluate the nonlinear behavior, the shell elements are divided into concrete layers and steel layers based on the layered approach and the new constitutive laws based on the equivalent lattice model is proposed. Applied loads are the self weight and the wind pressure. Numerical calculations are performed under the displacement controlling scheme to prevent numerical difficulties. Both the load deflection behavior and the ultimate strength are calculated for each repairing schemes under both monotonic and cyclic wind loading.

1 INTRODUCTION

Cooling towers are thin shell structures and are subjected to a cyclic loading such as wind pressure and earthquake loading. There are many old cooling towers that were subjected to an aging effects and that are required to repair. The strength of the cooling tower structures has been studied by many researchers and it is emphasized that the nonlinear behaviors of both the concrete and the reinforcement play an important role of the strength of such structures (Mang 1983, Hara 1994). However, most of these studies are based on the quasi-static analyses because of the complicated material nonlinearities of the concrete. Therefore, the characteristics of the R/C cooling tower shell under the cyclic loading have not been clarified.

In this paper, R/C cooling tower, stiffened either by adding the concrete shell thickness or by placing a ring stiffener, is analyzed under the repeated cyclic wind loading. In the numerical analysis, finite element procedure is adopted (Hinton 1984). To consider the complicated behavior of the concrete structure, the equivalent lattice constitutive model is adopted for evaluating the nonlinear behavior of the concrete element (Kato 1997).

The old cooling tower subjected to an aging effect (Golczyk 1996) is adopted as the numerical model. From this numerical model, the repeated cyclic behavior of the stiffened R/C cooling tower is presented comparing with the results under the monotonic loading.

2 ANALYTICAL MODEL

In this paper, the cooling tower at Lagisza Power Plant (Golczyk 1996), which was repaired in Poland, is adopted as the numerical model. This cooling tower was repaired by adding the shell thickness and reinforcements around the middle portion of the

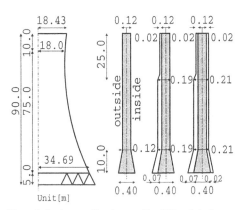

Figure 1. Lagisza cooling tower with additional shell thickness

Table 1. Material properties

Concrete		Steel	
Elastic Modulus E_C	28.3GPa	Elastic Modulus E_s	210GPa
Poisson s ratio ν	0.175	Hardening Parameter H'	0.01
Compressive Strength σ_c	25.0MPa	Yield Strength f_y	232MPa
Tensile strength σ_t	2.5MPa	Tensile Strength f_b	394MPa

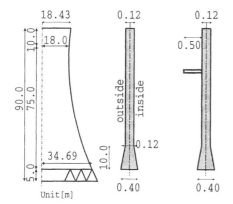

Figure 2. Lagisza cooling tower with additional stiffening ring

tower. However, in addition to such repairing scheme, it is assumed that Lagisza cooling tower is numerically repaired by placing the additional stiffening ring, instead of the additional shell thickness.

Figure 1 and 2 show the Lagisza cooling tower with additional thickness and with stiffening ring, respectively.

The details of the additional shell thickness and the cross section throughout the height are shown in references (Hara 1997, 1998b and 2000).

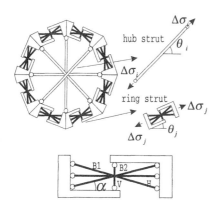

Figure 3. Lattice model

On the other hand, it is assumed that the cross section of the stiffening ring is 50cm width and 30cm thickness. The reinforcements in the stiffener are placed in both tangential and normal directions to the shell thickness and placed on both upper and lower surfaces of the ring. The position of the stiffening ring varies throughout the shell height. The material properties of the cooling tower are shown in Table 1.

3 NUMERICAL PROCEDURE

In the numerical analyses, the finite element method based on the isoparametric degenerated shell element is used and both the geometric and the material nonlinearities are taken into account (Hinton 1984). For representing the element deformation, Green Lagrange strain definitions are adopted to represent the geometric nonlinearities. Each element is subdivided into some layer (layered approach). To represent the nonlinear material properties, the equivalent lattice model (Kato 1997) is used for the concrete material and the membrane element is used for reinforcements.

3.1 Equivalent lattice model for the concrete

The lattice model adopted in the numerical analyses is shown in Figure 3. The constitutive relation of the lattice model is implemented in the FEM to represent the complicated constitutive relation of the concrete under cyclic loading. The model consists of 4 hub(radial) struts and 8 ring elements. Each ring element has 4 struts. Therefore, the model has 36 struts. They are assigned to each integration points of the FEM. Conceptually, hub struts represent the behavior under direct stresses and ring elements represent the behavior under shear stresses. Each strut is assumed to behave under the uniaxial stress strain relationship. The outline of derivation of equivalent stress strain relation is as follows:

To consider the in-plane stress strain status at any integration point, the virtual strain energy δU is evaluated as follows using the stress $\sigma_x, \sigma_y, \tau_{xy}$ and the virtual strain $\delta\varepsilon_x, \delta\varepsilon_y, \delta\gamma_{xy}$.

$$\delta U = \delta\varepsilon_x \sigma_x + \delta\varepsilon_y \sigma_y + \delta\gamma_{xy} \tau_{xy} \qquad (1)$$

Also, the virtual strain and the incremental strain in each strut arisen from the virtual strain $\delta\varepsilon_x, \delta\varepsilon_y, \delta\gamma_{xy}$ and the incremental strain $\Delta\varepsilon_x, \Delta\varepsilon_y, \Delta\gamma_{xy}$ are represented as $\delta\varepsilon_k, \Delta\varepsilon_k$, respectively. Each strain $\delta\varepsilon_k, \Delta\varepsilon_k$ is calculated by the strain transformation.

To represent the stress strain relation of each strut, following uniaxial relation is assumed.

$$\sigma_k = \eta_k \overline{E_k} \Delta\varepsilon_k + \sigma_k^0 \qquad (2)$$

where η_k, $\overline{E_k}$ and σ_k^0 are the reduction factor, the initial stiffness and initial strain, respectively. The

694

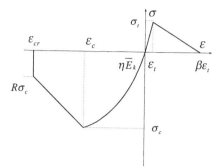

Figure 4. Nonlinear response of strut

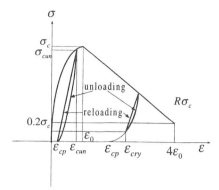

Figure 6. Stress strain in compression

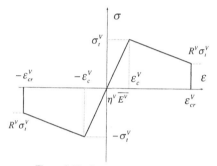

Figure 5. Nonlinear response of vertical strut

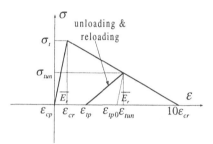

Figure 7. Stress strain in tension

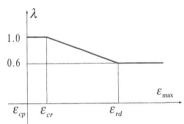

Figure 8. Nonlinear response of vertical strut

virtual strain energy of the lattice model δU_L is defined as follows.

$$\delta U_L = \sum_{k=1}^{36} \delta \varepsilon_k \sigma_k \qquad (3)$$

To equate Eq.(1) to Eq.(3), following constitutive equation of lattice model is derived.

$$\begin{Bmatrix} \sigma_x \\ \sigma_y \\ \tau_{xy} \end{Bmatrix} = \begin{bmatrix} D_{11} & D_{12} & D_{13} \\ D_{21} & D_{22} & D_{23} \\ D_{31} & D_{32} & D_{33} \end{bmatrix} \begin{Bmatrix} \Delta \varepsilon_x \\ \Delta \varepsilon_y \\ \Delta \gamma_{xy} \end{Bmatrix} + \begin{Bmatrix} \sigma_x^0 \\ \sigma_y^0 \\ \tau_{xy}^0 \end{Bmatrix} \qquad (4)$$

The initial stiffness $\overline{E_i}$ assumed in Eq.(2) is derived for each struts comparing of Eq.(4) with the constitutive relation of general in-plane constitutive relation of continuum mechanics.

To represent the nonlinear behavior of the concrete as the uniaxial stress strain status in struts, the stress strain relations of struts are assigned as shown in Figure 4 and Figure 5. For the vertical strut in the ring element, the relation defined in Figure 5 is adopted. For other struts, the relation defined in Figure 4 is adopted.

3.2 Cyclic stress strain relation in concrete

Figure 6 and 7 show the stress strain relation of struts in compression and tension under cyclic loading, respectively, The unloading stress strain

history in compression (see Figure 6) behaves on the circle that has the infinite tangent stiffness at the beginning of unloading and crosses the plastic strain ε_{cp}, epressed as Eq.(5), at zero stress (Kato 1997, Buyu kosturk 1984):

$$\varepsilon_{cp} = \left[0.334 \left(\frac{\varepsilon_{cun}}{\varepsilon_0} \right)^2 + 0.162 \left(\frac{\varepsilon_{cun}}{\varepsilon_0} \right) \right] \varepsilon_0 \qquad (5)$$

where ε_{cun} and ε_0 represent the strain at the beginning of unloading and at peak stress, respectively.

The reloading history in compression represents the straight line connecting the unloading points and the beginning of the reloading (see Figure 6).In tension, the straight line that connects the unloading point to the plastic strain ε_{tp} is defined to represent both unloading and reloading history. The plastic strain ε_{tp} is defined as Eq.(6).

$$\varepsilon_{tp} = \left(\varepsilon_{cp} + \varepsilon_{tp0}\right) \tag{6}$$

where ε_{cp} and ε_{tp0} represent the strain at the beginning of tension and at the strain of zero stress computed by unloading with initial tangent stiffness (see Figure 7).

Also, the compressive strength of the precracked concrete reduces considering the maximum tensile strength ever experienced (Kato 1997). The reduction factor changes from 1.0 to 0.6 and is defined as follows (see Figure 8).

$$\lambda = 1 - 0.4 \frac{\varepsilon_{max} - \varepsilon_{cr}}{\varepsilon_{rd} - \varepsilon_{cr}} \tag{7}$$

where $0.6 \leq \lambda \leq 1.0$ and $\varepsilon_{rd} = \varepsilon_{cp} + \kappa(\varepsilon_{cr} - \varepsilon_{cp})$. $\kappa = 2 \sim 3$. ε_{max} is the maximum tensile strain ever experienced.

3.3 Stress strain relation in steel

The reinforcing steel has a uniaxial bilinear constitutive relation both in tension and compression.

4 NUMERICAL MODEL

The numerical model is shown in Figure 9. The model is divided into 8 elements in the hoop direction and into 14 elements in the meridional direction. Numerical models are a half of the cooling tower. The cooling tower is simply supported at the base and free at the top. In this analysis, the deformation effects of the supporting V columns are not considered.

In the numerical analysis of the model by adding the shell thickness, cooling tower is analyzed for both repairing stages.

The stiffening rings are modeled not by the beam elements but by the shell elements. From this modeling, the deformation effects of the stiffening ring on the entire structural deformation are considered. Stiffening rings are divided into two elements in the normal direction to the shell thickness. Each shell and ring element is divided into 8 concrete layers and some steel layers.

As the loading condition, the self-weight is loaded gradually under the load increment control. After the applied load reaches the self-weight, the wind pressure is applied quasi statically under the displacement increment control.

In this analysis, wind pressure recommended in the IASS (1977,1979) is adopted. We assumed that the wind velocity at 10m above the ground is 44.7m/sec (100mph). In the numerical calculation, the concrete has the material properties that are considered as the deteriorated material on both inner and outer surfaces of the cooling tower shell. Therefore, we assume that the concrete possesses 50% tensile strength and the 80% compressive strength.

5 NUMERICAL RESULTS

In the numerical analysis, the old cooling tower is assumed to be stiffened either by adding the shell thickness or by placing a ring stiffener. The ultimate strength of R/C shell structures by both repairing scheme is determined as the maximum loading from the load deformation relation.

5.1 Unstiffened status

Figure 10 shows the load-deflection behavior of unstiffened cooling tower. The wind pressure is applied after loading the self weight. The solid line and the dashed line denotes the load deflection behavior under the monotonic loading and cyclic loading, respectively. The target point is the node on the windward meridian. The envelope of the maximum loads under cyclic loading shows smaller response than that under monotonic loading. Figure

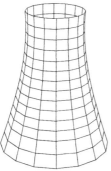

Figure 9. FE mesh

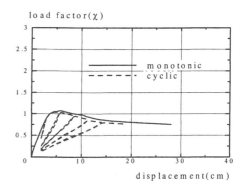

Figure 10. Load-displacement relation of unstiffened tower

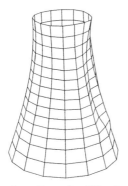

Figure 11.Deformation patterns of unstiffened cooling tower

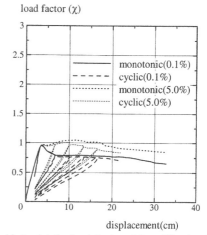

load factor (χ)

Figure 14. Load-deflection behavior of ring stiffened cooling tower

load factor(χ)

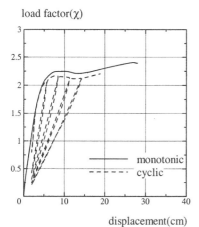

displacement(cm)

Figure 12. Load-deflection under repairing stage 1

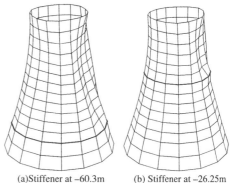

(a)Stiffener at –60.3m   (b) Stiffener at –26.25m

Figure 15. Deformation patterns of stiffened cooling tower at various stiffening position

5.2 *Cooling tower stiffened by additional thickness*

The cooling tower is stiffened under two stags of repairing scheme. At the first stage, the concrete covers of 2 cm and 7cm are added on outer surface.

In the region of the 7cm additional concrete layer, hoop and meridional reinforcements are placed. At the second stage, 2cm concrete layer is added in inner surface (see Figure 1.).

Figure 12. And 13 denote the load deflection behavior under the first and the second repairing stages, respectively. The target point is the node on the windward meridian.

The ultimate strength of the cooling tower under each repairing stages increases to 2.25 and 2.5, respectively.

load factor(χ)

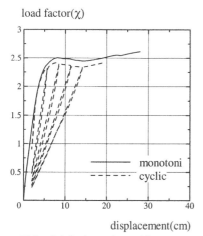

displacement(cm)

Figure 13. Load-deflection under repairing stage 2

11 shows the deformation patterns of the unstiffened cooling tower after the peak load.

5.3 *Cooling tower with stiffening ring*

A single type of ring stiffer is considered. The

effects of the stiffening position and the reinforcing ratio of the stiffening rings are analyzed for the systems.

Figure 14 shows the load deflection behavior of cooling tower with single ring stiffener at the level – 60.8m below the node. The solid line and the dashed line denote the response under monotonic and cyclic loading, respectively. The reinforcing ratio of the stiffener is 0.1%. The dotted line and the short dashed line denote the response under monotonic and cyclic loading, respectively. In this case, the reinforcing ratio of the ring stiffener is 5.0%. The ring stiffened cooling tower represents the smaller ultimate strength than that of the unstiffened cooling tower (see Figure 10). However, the stiffener with large reinforcement is more effective than that with small reinforcements. In the case of stiffener with large reinforcements, the deterioration of the cooling tower under cyclic loading is larger than that of small reinforcements, while the ultimate strength is almost the same for both loading scheme. Figure 15 shows the deformation patterns of the cooling tower with ring stiffener at different stiffening positions. The stiffening effects depend on the stiffening position.

The ring stiffener plays an important role on the deformation patterns and the ultimate strength of stiffened cooling tower. When the stiffener is placed at 26.25m below the node the strong local deformation appears between the top ring and the stiffener. Therefore, in such ring stiffening case, the ultimate strength of the stiffened cooling tower is smaller than that of the unstiffened cooling tower. In this case, the efficient stiffening effect is not obtained. Because the cooling tower adopted in the numerical calculation has thin shell thickness (12cm) and small reinforcements.

Consequently, the additional thickness and the additional reinforcements to the old reinforced concrete shell surface is effective for the cooling tower modeled in this analysis (Golczyk 1996).

6 CONCLUSIONS

In this paper, the ultimate strength of the cooling tower stiffened either by adding shell thickness or by placing a ring stiffer is analyzed. Lagisza cooling tower is adopted as the numerical model. In the numerical computation, the constitutive law derived from the equivalent lattice model is adopted and the cyclic behavior of the cooling tower is analyzed. The conclusions obtained in this analysis are as follows:
1. The cooling tower adopted in this analysis is too thin thickness to strengthen the ultimate strength by implementing the ring stiffener. Therefore, the improvement of the ultimate strength is not obtained. The cooling tower stiffened by adding

the shell thickness improves the ultimate strength.
2. The reinforcement ratio of the stiffening ring plays an important role for the stiffening of the cooling tower shell both under monotonic and cyclic behavior. In the case of stiffening ring placed on the lower portion, the ultimate strength of stiffened cooling tower with large reinforcement in the stiffener is larger than that with small reinforcement under monotonic loading.
3. To compare the load deflection behavior of ring stiffened cooling tower both under monotonic loading and cyclic loading, the ultimate strength computed by both loading conditions represent almost the same. However, the deterioration of the stiffened cooling tower is larger under cyclic loading. Especially, the cooling tower with large reinforcing ratio shows larger deterioration than that with small reinforcing ratio.
4. The load deformation behavior of the cooling tower stiffened by the additional shell thickness shows the sufficient stiffening effects both under monotonic and cyclic loading.

REFERENCES

Buyukosturk O. Ming T.T. 1984, Concrete in biaxial cyclic compression. *Journal of Structural Engineering* 110(3): 461-476

Golczyk M., Abramek W. and Centkowski J. 1996, Analysis and safety of repaired reinforced concrete cooling towers. *Proceedings of the 4th International Symposium on Natural Draught Cooling Towers* 277-281.

Hara T., Kato S.. and Nakamura H. 1994, Ultimate strength of RC cooling tower shells subjected to wind load. *Engineering Structures* 16(3):171-180

Hara T., Kato S., and Ohya M., 1997 Numerical simulation of repairing the cooling tower shell. *XIII Polish Conference on Computer Methods in Mechanics* 2:517-524

Hara T., Kato S., and Ohya M. 1998, Behavior of Repaired Cooling Tower Shell. *Proceedings of The sixth East Asia-Pacific Conference on Structural Engineering and Construction* 3:2075-2080

Hara T., Kato S., and Ohya M., 1998b Structural Behavior of Repaired R/C Cooling Tower Shell. *Proceedings of the Structural Engineering World Conference*.(on CD-ROM)

H Hara T., Kato S., and Ohya M. 2000, Ring stiffened cooling tower behavior subjected to cyclic wind load. *Proceedings of IASS Symposium*:xxx-xxx

Hinton, E. and Owen, D.J.R. 1984, *Finite element software for plates and shells*. Prineridge Press, Swansea, UK.IASS Working Group 3 1977, *Recommendation for the design of hyperbolic or other similarly shaped cooling towers*.

IASS The Working group on recommendations of IASS 1979, *Recommendations for reinforced concrete shell and folded plates*

Kato S., Ohya M. and Maeda S. 1997, A new formulation of constitutive equations for reinforced concrete element based on lattice model and application to FEM analysis of reinforced concrete shells. *Proceedings of COMPLAS-5* 1522-1527

Mang, H.A. and Floegl, H. 1983, Wind-load reinforced concrete cooling towers: buckling or Ultimate load ? *Engineering Structures*, 5:163-180.

Creative Systems in Structural and Construction Engineering, Singh (ed.) © 2001 Balkema, Rotterdam, ISBN 90 5809 161 9

Conditions for desirable structures based on a concept of load transfer courses

K.Takahashi

Department of Mechanical Engineering, Keio University, Japan

ABSTRACT: A new concept of a parameter $\hat{E}$ is introduced to express load transfer courses for a whole structure. A degree of connection between a loading point and an internal arbitrary point in the structure can be quantitatively expressed with the parameter $\hat{E}$. Based on the proposed concept, three conditions for desirable structures are introduced: (1) Continuity of $\hat{E}$, (2) Linearity of $\hat{E}$, (3) Consistency of courses. After introducing these three conditions as objective functions, structural optimization with numerical computation is carried out. Despite the fact that no concept of stresses or strains is introduced, the obtained structure has a reasonable shape. Finally, the load transfer courses for a simple structure are experimentally measured and these values demonstrate that the parameter $\hat{E}$ can effectively be used.

1 INTRODUCTION

Realization of lightweight, cost-effective structures possessing high rigidity and strength is an important aspect of structural designs, and numerical analyses utilizing the finite element method (FEM) have played a key role in this regard. Such analyses, however, do not adequately consider load transfer courses, which are inevitable in the early stages of structural planning; i.e., while conventional numerical analyses can precisely predict local stresses and strains, initial planning requires obtaining overall load transfer courses for the whole structure.

In the present study, a new concept of a parameter $\hat{E}$ is introduced to express the load transfer courses for a whole structure. A degree of connection between a loading point and an internal arbitrary point in the elastic structure can be quantitatively expressed with the parameter $\hat{E}$, and the load transfer course is defined from a contour line of the distribution diagram of $\hat{E}$.

Based on the concept of the load transfer courses, three conditions for desirable structures are introduced. These conditions are available for objective functions in numerical optimization programs. An example of structural optimization is calculated for a structure using a finite element method with a genetic algorithm (GA).

Moreover, $\hat{E}$ and the load transfer courses are experimentally measured for a thin-walled member, and it is subsequently shown that this unique parameter can be effectively used to estimate lightweight, cost-effective structures.

2 LOAD TRANSFER COURSES

2.1 Relative stiffness and a degree of connection

Figure 1 (a) shows an elastic solid with a loading point A and a supporting point B. Point C is an arbitrary point in the body. The relation between the loading and deformation is expressed as

$$
\begin{bmatrix} p_A \\ p_B \\ p_C \end{bmatrix} = \begin{bmatrix} K_{AA} & K_{AB} & K_{AC} \\ K_{BA} & K_{BB} & K_{BC} \\ K_{CA} & K_{CB} & K_{CC} \end{bmatrix} \begin{bmatrix} d_A \\ d_B \\ d_C \end{bmatrix} \tag{1}
$$

where K with suffixes is a relative stiffness(Shinobu et al. 1995) that shows an elastic constant between arbitrary two points, and p and d with suffixes are a loading vector and a displacement vector, respectively.

Considering the invariance related to the rigid translations ($p_A = 0$, $d_A = d_B = d_C$), we have the following equation.

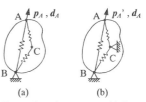

(a)   (b)

Figure 1   Expression of a structure with three springs

$$K_{AA} = -(K_{AB} + K_{AC}) \qquad (2)$$

We can express the degree of connection between A and C by using K_{AA}.

The strain energy E stored in the structure shown in Figure 1(a) is denoted by the inner product of the applied force and displacement.

$$E = \frac{1}{2} p_A \cdot d_A \qquad (3)$$

Substituting Eq. (1) into Eq.(3), we obtain the next relation.

$$E = \frac{1}{2}(K_{AA}d_A + K_{AC}d_C) \cdot d_A \qquad (4)$$

Figure 1(b) shows the same elastic body under the condition that the point C is constrained. The external force at point A for the same displacement d_A is denoted as p'_A in this case. The equation (1) is still valid, but the displacements d_B and d_C are reduced to zero. Referring Eq. (1), we have

$$p'_A = K_{AA} d_A \qquad (5)$$

and the strain energy E' stored in the structure shown in Figure 1(b) is as follows.

$$E' = \frac{1}{2} p'_A \cdot d_A = \frac{1}{2}(K_{AA}d_A) \cdot d_A \qquad (6)$$

Since the matrix K_{AA} shows the degree of connection between A and C as mentioned above, E' also expresses the degree of connection between A and C for the displacement d_A. A non-dimensional value of E' is expressed by dividing Eq. (6) by Eq. (4).

$$\frac{E'}{E} = \frac{(K_{AA}d_A) \cdot d_A}{(K_{AA}d_A + K_{AC}d_C) \cdot d_A} \qquad (7)$$

The value of E'/E also expresses the degree of connection between A and C.

For a structure modeled as a single spring, the value of E'/E decays hyperbolically. However, the value of $\hat{E}$ defined below decays linearly for a single spring model.

$$\hat{E} \equiv 1 - \frac{1}{E'/E} \qquad (8)$$

The value of $\hat{E}$ in a structure is a function of the position of C. Hereafter, we use the value of $\hat{E}$ for the estimation of the degree of connection between the points A and C.

2.2 *Load transfer courses*

The distribution of the values of $\hat{E}$ in a structure can be obtained from numerical or experimental analyses. We define the load transfer course as the line that successively connects the highest values of $\hat{E}$ in the structure; that is, the load transfer course is the curve that corresponds to the ridge line of the contour lines of $\hat{E}$. An example of the load transfer course or ridge line is shown in Figure 2.

Figure 2 Load transfer course

3 THREE CONDITIONS FOR DESIRABLE STRUCTURES

3.1 *Continuity of $\hat{E}$ distribution*

Figure 3(a) shows a course in a structure with relatively stiff domains, which are indicated by the shaded areas. Figure 3(b) denotes the distribution of $\hat{E}$ along the course, where s is the coordinate along the course and l is the length of the course. Structural discontinuities for the internal stiffness can be seen at the points D_1 and D_2 in this figure. We can more precisely recognize the discontinuities of the stiffness from a distribution curve of the curvature of $\hat{E}$ (the second derivative of $\hat{E}$ with respect to s), as shown in Figure 3(c).

Continuity of the stiffness along the course is the first condition for desirable structures. The area f_1 shown in Figure 3(c) is an example of measures to express the degree of discontinuity.

3.2 *Linearity of $\hat{E}$ distribution*

Homogeneity of the internal stiffness is also necessary for desirable structures. Because the distribution of the value of $\hat{E}$ decays linearly in a single homogeneous spring, structural homogeneity in a load transfer course is expressed with the linearity of $\hat{E}$ distribution. The linearity of $\hat{E}$ is the second condition for desirable structures. The area f_2 shown in Figure 3(b) is an example of measures used to express the degree of linearity for this condition.

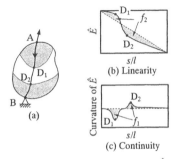

Figure 3 Continuity and linearity of $\hat{E}$

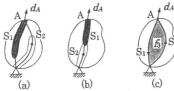

Figure 4 Consistency of courses

3.3 *Consistency of load transfer courses*

Figure 4(a) shows another model of a structure. The load transfer course from the loading point is obtained from the ridge line of the contour lines of $\hat{E}$ (S_1 in Fig. 4(a)). The load transfer course from the supporting edge is obtainable with the same procedure (S_2 in Fig. 4(a)). It is worth noting that the courses from the loading point (S_1) and from the supporting edge (S_2) are not coincident with each other in Figure 4(a).

It is desirable that the two paths S_1 and S_2 are coincident with each other in the structure, as shown in Figure 4(b). Consistency of the courses S_1 and S_2 is the third condition for desirable structures. The area f_3 shown in Figure 4(c) expresses the degree of inconsistency for this condition.

4 STRUCTURAL OPTIMIZATION

4.1 *Optimization model*

Structural optimization can be carried out satisfying the above-mentioned three conditions.

Figure 5(a) shows an initial model as an example of optimization calculations by using FEM. The objective function F is

$$\begin{cases} F = \alpha_1 f^*{}_1 + \alpha_2 f^*{}_2 + \alpha_3 f^*{}_3 \\ f^*{}_i = \dfrac{\bar{f}_i - f_i}{\bar{f}_i} \quad (i = 1,2,3) \end{cases} \quad (9)$$

where $\bar{f}_i$ is the average of f_i values in the initial generation, and α_i is the weight function.

4.2 *Desirable structure*

The most desirable structure is obtained in the final stage of the optimization process using the Genetic Algorithm (GA) as shown in Figure 5(b).

At this point, another optimization program, using conventional sensitivity analysis, is introduced to minimize stesses, and the structure obtained is shown in Figure 5(c). A comparison of the structures in Figures 5(b) and 5(c) reveals the obvious similarities between them.

Figure 6(a) shows the values of the objective function F (in Eq. 9) in the process of the optimization using FEM. Figures 6(b), 6(c) indicate the stiffness and the maximum stress in the process of the present optimization using the three conditions for desirable structures.

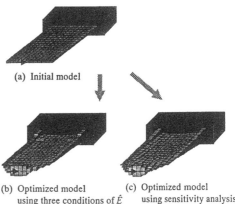

(a) Initial model

(b) Optimized model using three conditions of $\hat{E}$ (c) Optimized model using sensitivity analysis

Figure 5 Optimized structures

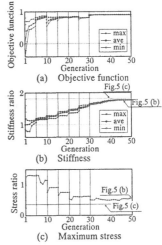

(a)   Objective function

(b)   Stiffness

(c)   Maximum stress

Figure 6 Optimization process

Note that though the optimized structure (Fig. 5(b)) using the three conditions is obtained without any aim to minimize stress or to maximize stiffness, the obtained structure has a reasonable shape for stress or stiffness.

5 MEASUREMENT OF $\hat{E}$

It is also possible to measure the values of $\hat{E}$ experimentally. Figure 7 shows the experimental models I and II, and a loading condition.

(a) Model I   (b) Model II   (c) Loading condition

Figure 7 Experiment models

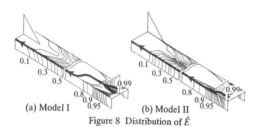

(a) Model I   (b) Model II

Figure 8   Distribution of $\hat{E}$

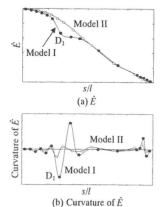

(a) $\hat{E}$

(b) Curvature of $\hat{E}$

Figure 9   Distribution curve of $\hat{E}$ and curvature

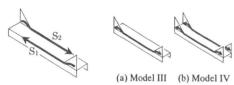

(a) Model III   (b) Model IV

Figure 10   Inconsistency of courses   Figure 11   Consistency of courses

	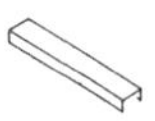 Initial model	 Model I	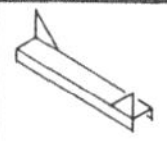 Model II	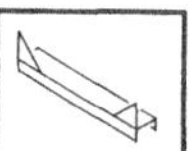 Model III
Stiffnesss (N/m)	1.91×10^4	2.19×10^4	2.19×10^4	2.23×10^4
Stiffnesss Ratio (%)	100	+5.83	+14.7	+22.1

Figure 12   Stiffness of models

The models are made of PVC (polyvinylchloride) plates. The value of energy E in Eq. (3) is obtained from the measurement of the applied force at the loading point for the unit displacement.

Under the condition that the arbitrary point is constrained, the value of E' in Eq. (6) is obtainable experimentally for the same displacement. A value of $\hat{E}$ can be derived from the values E and E' using Eq. (8).

Figures 8(a), 8(b) show the distribution of the measured values of $\hat{E}$ and a curve of the load transfer courses.

Figure 9(a) denotes the distribution curves of $\hat{E}$ along the courses. In this figure, we can observe the degree of linearity of $\hat{E}$, which is the second condition of desirable structures. The distribution curve for the model I deviates largely from the linearly decreasing line. The point D_1 indicates the existence of the structural discontinuity in the model I (cf. Figs. 3(b), 3(c)).

Figure 9(b) shows the distribution curves of the curvature of $\hat{E}$, which describe the degree of continuity of stiffness along the course, which is the first condition of desirable structures. The point D_1 indicates the existence of the structural discontinuity in model I.

Model II is more desirable than model I with regard to the first and second conditions. However, the courses from the loading point (S_1) and from the supporting point (S_2) in the model II are not coincident (Fig.10); that is, consistency, the third of the three conditions for desirable structures is unsatisfied.

For the third condition, we can conclude that model III or IV, shown in Figure 11, would transfer the applied load more effectively.

Figure 12 shows the comparison of measured values of overall stiffness (applied force / unit displacement) for the models including model III.

6 CONCLUSIONS

1. A new concept of a parameter $\hat{E}$ is introduced to express the load transfer courses in a whole structure. A load transfer course is defined from a contour line of the distribution diagram of $\hat{E}$.

2. Based on the proposed concept of $\hat{E}$ and the load transfer course, three conditions for desirable structures are introduced.

3. After introducing these three conditions as objective functions, structural optimization with numerical computation is carried out without giving consideration to the concept of stresses or strains. Further, the load transfer courses are experimentally measured and these values demonstrate that the parameter $\hat{E}$ can effectively be used.

ACKNOWLEDGEMENT

The author would like to express his gratitude to K. Soya, Y. Hatta, S. Ito, and M. Kamimura.

REFERENCES

Shinobu, M., et.al. 1995. Transferred load and its course in passenger car bodies. *JSAE Review* 16: 145-150.

22 Structural dynamics

Creative Systems in Structural and Construction Engineering, Singh (ed.) © 2001 Balkema, Rotterdam, ISBN 90 5809 161 9

Hydroelastic vibration of multiple circular plates coupled with fluid

Kyeong-Hoon Jeong, Gyu-Mahn Lee & Keun-Bae Park
Korea Atomic Energy Research Institute, Taejon, Korea

Myung-Jo Jhung
Korea Institute of Nuclear Safety, Taejon, Korea

ABSTRACT: Investigated in this paper are the modal characteristics of the multiple circular plates with fluid gaps. The model of the fluid–coupled structure simulates the bottom screen assembly of an integral reactor. Three–dimensional finite element model is constructed in order to obtain the natural frequencies and the corresponding mode shapes using a commercial finite element program when the bottom screen assembly is either in contact with coolant or in air. It is found that the fluid gaps drastically reduce the natural frequencies of the bottom screen assembly due to the hydrodynamic mass of coolant and the coupling effect between the plates.

1 INTRODUCTION

In recent literature, there has been renewed interest in the problem of plates vibrating in contact with water. This is stimulated by new technical applications and also by the availability of powerful numerical tools based on the finite element method that make numerical solutions of fluid–structure interaction problems possible. However, the use of the finite element method requires enormous amounts of time for modeling and computation.

Circular plates vibrating in contact with fluid have recently been studied. Studied were the free vibration of circular plates in contact with water on one side (Kwak, 1991 and Kwak & Kim, 1991), while the free vibration of annular plates in contact with water on one side was also investigated (Amabili et al. 1996). They considered the unbounded fluid domain and also introduced the non-dimensionalized added virtual mass incremental (NAVMI) factors in order to estimate the fluid effect on the individual natural frequency of the fluid-structure system. Chiba (1994) obtained exact solutions for the circular elastic bottom plate in a cylindrical rigid tank filled with fluid. The elastic bottom plate supported by an elastic foundation and the free surface of the fluid was considered. Montero de Espinosa et al. (1984) studied the vibration of plates submerged in water mainly to the lower modes by the approximate analytical method and experiments. Hagedorn

(1994) dealt with the theoretical free vibrations of an infinite elastic plate in the presence of water. A study on the free vibration of two circular plates coupled with fluid carried out (Jeong et al., 1997). The hydroelastic vibration of a rectangular plate in contact with fluid was theoretically investigated using Rayleigh's quotient with an assumed–mode method (Jeong et al., 1999). However, these problems are of a fundamentally different nature.

This paper is concerned with the hydrodynamic effect on the free vibration characteristics of multiple circular plates in contact with fluid. The model of the fluid–coupled structure simulates the bottom screen assembly in an integral reactor, SMART (System-integrated Modular Advanced ReacTor) that is under development at KAERI. The bottom screen assembly composed of several different circular plates is located below the reactor core and it protects the reactor pressure vessel from neutron irradiation. The hydrodynamic mass of the coolant affects the dynamic characteristics of bottom screen assembly. However, it is very difficult to theoretically estimate the hydrodynamic mass of coolant when the structure is complicated, like the bottom screen assembly, as shown in Figure 1. Therefore, the finite element analysis is still effective for a complicated fluid–coupled structure problem, although it requires enormous amounts of time to model and compute. The natural frequencies and corresponding mode shapes of the bottom screen assembly in contact with coolant are obtained using a commercial computer program, ANSYS 5.6 (1999). They will give very important information for further seismic, pump pulsation response and postulated pipe break analyses.

2 THEORETICAL BACKGROUND

2.1 Problem Formulation

As shown in Figure 1, the bottom screen assembly is composed of several circular plates that protect the pressure vessel from neutron irradiation. The circular plate assembly is always submerged in coolant during plant operation. The coolant will affect the dynamic behavior of the bottom screen assembly. The fluid gaps between the plates will mitigate the excessive increase of temperature due to the neutron flux. For a theoretical investigation, the following assumptions are made: (a) the fluid motion is small; (b) the fluid motion is incompressible, inviscid and irrotational; and (c) the plates are made of linearly elastic, homogeneous, and isotropic material.

$$\nabla^4 w_j + \frac{\rho \, h_j}{D_j} \frac{\partial^2 w_j}{\partial t^2} = \frac{p_j}{D_j}, \quad j = 1,2,\ldots,N \quad (1)$$

where $D_j = E\,h_j^3 / 12\,(1 - \mu^2)$ is the flexural rigidity of the circular plates; ρ, μ, p_j, E and N are the density, Poisson's ratio, hydrodynamic pressure on the circular plates and Young's modulus of the circular plates, and the total number of plates, respectively. ∇^4 is the biharmonic operator in polar coordinates, r and θ. The solution of equation (1) for the jth plate can be written in the form of

$$w_j(r,\theta,t) = W_j(r,\theta)\exp(i\omega t) \quad (2)$$

where ω is the coupled natural frequency of the plates.

2.2 Velocity Potential

Next, we consider the fluid region with which the hemispherical pressure vessel and the rigid upper and side walls are surrounded. The facing side of the circular plates is in contact with an ideal fluid. The three–dimensional oscillatory fluid flow in cylindrical coordinates can be described with the velocity potential. The fluid movement induced by plate vibration is described with the spatial velocity potential that satisfies Laplace's equation:

$$\nabla^2 \Phi(x,r,\theta,t) = 0 \quad (3)$$

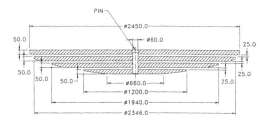

Figure 1. Bottom screen assembly of the integral reactor.

It is possible to separate function Φ with respect to r, θ and x. Thus:

$$\Phi(r,\theta,x,t) = i\omega\phi(r,\theta,x)\exp(i\omega t) \quad (4)$$

The hydrodynamic pressures on the plates can be written in terms of the velocity potential;

$$p_j = \rho_o\,\omega^2\,\phi(r,\theta,x)\big|_{\Gamma_j}\exp(i\omega t), \quad (5)$$

where ρ_o is the mass density of fluid. The compatibility conditions at the interface of the fluid domains (Γ_j) contacting along the plate surfaces are used. The compatibility conditions at the fluid interface with the plates yield

$$W_j = -\partial\phi / \partial x\big|_{\Gamma_j} \quad (6)$$

It is necessary to know the reference kinetic energies of the plates and containing fluid in order to calculate the coupled natural frequencies of the circular plates. Using the hypothesis of the irrotational movement of fluid, the reference kinetic energy of the fluid can be evaluated from its boundary motion. In fact, as a consequence of Green's theorem applied to the harmonic functions W_j, additionally function $\partial\phi / \partial r$ and $\partial\phi / \partial x$ are always zero at the boundaries of the fluid domain, except for the surfaces in contact with the plates, therefore one obtains

$$T_F = \frac{\rho_o}{2}\sum_{j=1}^{N}\int_0^{2\pi}\int_0^R \left(\frac{\partial\phi}{\partial x}\right)_{\Gamma_j}(\phi)_{\Gamma_j}\,dr\,d\theta \quad (7)$$

The reference kinetic energy of the plates in air can be written in terms of the dry mode shapes of plate, W^*_j.

$$T_a = \frac{\rho}{2}\sum_{j=1}^{N} h_j\int_0^{2\pi}\int_0^R W_j^{*2}\,r\,dr\,d\theta \quad (8)$$

On the other hand, the reference kinetic energy of the plates in contact with fluid is presented as follows:

706

$$T_d = \frac{\rho}{2} \sum_{j=l}^{N} h_j \int_0^{2\pi} \int_0^R W_j^2 \; r \; dr \; d\theta \qquad (9)$$

The natural frequency of the plates in contact with fluid can be written as

$$\omega = \sqrt{T_F / T_d} \qquad (10)$$

When the wet mode shapes of the plates are identical to the dry mode shapes of the plates, $T_a = T_d$. Hence the coupled natural frequencies of the circular plates coupled with fluid can be related to the natural frequencies in air, ω_o (Kwak,1995).

$$\omega = \omega_o / \sqrt{1 + T_F / T_d} \qquad (11)$$

However, the hydrodynamic loading on the circular plates has a significant effect on the deflection curve (Jeong et al. 1998), that is to say, the wet mode shapes are not identical to the dry mode shapes except in some special cases. Therefore equation (11) is not valid for the complicated fluid–structure interaction problem, such as the bottom screen assembly.

3 FINITE ELEMENT ANALYSIS

3.1 Finite element model

Because equation (10) can no longer be developed theoretically for the bottom screen assembly, finite

element analyses using a commercial finite element analysis computer program, ANSYS 5.6 (1999) are

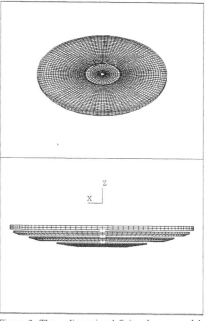

Figure 3. Three–dimensional finite element model of the bottom screen assembly only.

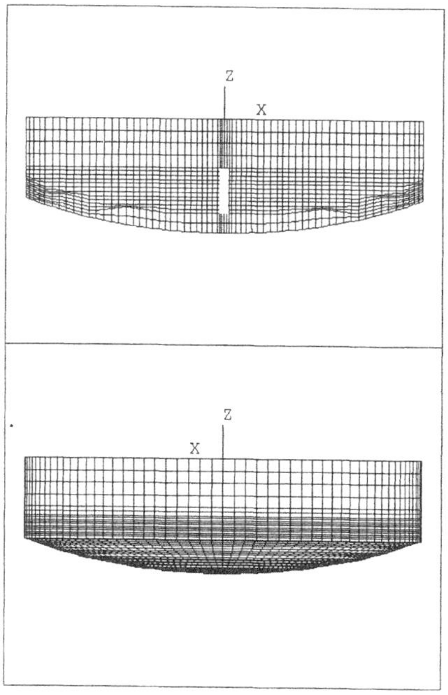

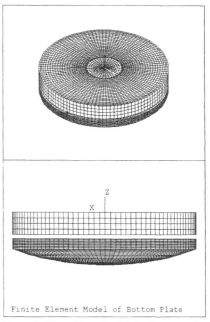

Finite Element Model of Bottom Plate

Figure 2. Three–dimensional finite element model of the bottom screen assembly with fluid.

Figure 4. Three–dimensional finite element model of the fluid gaps.

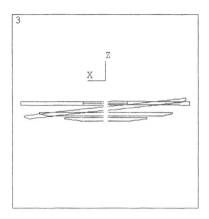

Figure 5. Fundamental mode shape of the bottom screen assembly in air (11.46 Hz).

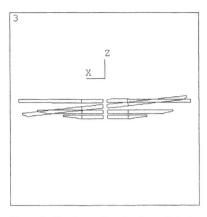

Figure 8. Fundamental mode shape of the bottom screen assembly in contact with water (4.12 Hz).

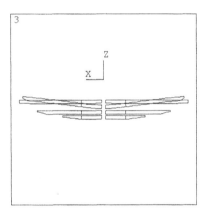

Figure 6. Second mode shape of the bottom screen assembly in air (31.88 Hz).

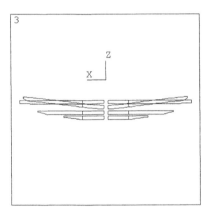

Figure 9. Second mode shape of the bottom screen assembly in contact with water (8.09 Hz).

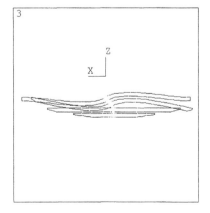

Figure 7. Fourth mode shape of the bottom screen assembly in air (96.13 Hz).

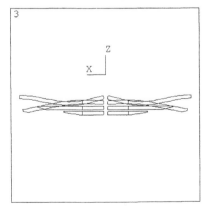

Figure 10. Fourth mode shape of the bottom screen assembly in contact with water (47.66 Hz).

Table 1. Material properties used as ANSYS input data.

Material	Property	Value
Coolant (water)	Mass density	1000 kg/m^3
Viscosity		1.131×10^{-3} N·s / m^2
	Young's modulus	2.2 GPa
Plates & pin	Mass density	7800 kg/m^3
(304SS)	Young's modulus	172.0 GPa
	Poisson's ratio	0.30

Table 2. Natural frequencies of the bottom screen assembly in air.

Mode no.	Natural freq.	Dominant mode shape
1	11.459 Hz	Top plate ($n^* = 1$, $m^* = 0$)
		2nd plate (*tilting mode*)
2	31.880 Hz	Top plate ($n = 0$, $m = 1$)
		2nd plate ($n = 0$, $m = 0$)
3	44.933 Hz	2nd plate ($n = 2$, $m = 0$)
4	96.131 Hz	Top plate ($n = 1$, $m = 1$)
		2nd plate ($n = 1$, $m = 0$)
5	103.32 Hz	2nd plate ($n = 3$, $m = 0$)
6	168.35 Hz	Top plate ($n = 1$, $m = 0$)
		2nd plate ($n = 1$, $m = 1$)
7	176.53 Hz	Top plate ($n = 0$, $m = 0$)
		2nd plate ($n = 0$, $m = 1$)
8	179.98 Hz	Top plate ($n = 0$, $m = 2$)
		2nd plate ($n = 4$, $m = 0$)
9	187.41 Hz	Top plate ($n = 1$, $m = 0$)
		2nd plate ($n = 1$, $m = 1$)
10	261.30 Hz	Top plate ($n = 2$, $m = 0$)
11	274.17 Hz	2nd plate ($n = 5$, $m = 0$)
12	287.85 Hz	Top plate ($n = 1$, $m = 1$)
		2nd plate ($n = 1$, $m = 2$)
13	293.90 Hz	2nd plate ($n = 2$, $m = 1$)
14	380.83 Hz	Top plate ($n = 3$, $m = 0$)
15	385.29 Hz	2nd plate ($n = 6$, $m = 0$)
16	438.62 Hz	2nd plate ($n = 3$, $m = 1$)
17	450.10 Hz	Top plate ($n = 0$, $m = 1$)
		2nd plate ($n = 0$, $m = 1$)
18	468.51 Hz	Top plate ($n = 1$, $m = 1$)
		2nd plate ($n = 1$, $m = 1$)

* n = number of nodal diameters, m = number of nodal circles

performed. Three–dimensional model is constructed for the finite element analysis. The fluid region is divided into a number of 3–dimensional contained fluid element (FLUID80) with eight nodes having three degrees of freedom at each node. The fluid element 'FLUID80' is generally used to model the fluid contained within the vessel having without a net flow rate and particularly well suited for calculating hydrostatic pressures and fluid–structure interactions. The bottom screens with a support pin are modeled as 3-D structural solid element (SOLID45) with eight nodes having three degrees of freedom at each node. The solid element, 'SOLID45' is generally used for the three–dimensional modeling of a solid structure. The entire fluid–structure coupled model is composed of 30024 fluid elements and 16998 solid elements, as shown in Figures 2 ~ 4. The material properties used in the analysis are summarized in Table 1. The fluid boundary conditions at the perimeter are zero displacement and rotation. The nodes connected entirely by the fluid elements are free to move arbitrarily in three–dimensional space, with the exception of those which are restricted to motion in the bottom and top surfaces of the fluid cavity. The normal velocities of the fluid nodes along the wetted plate surfaces coincide with the corresponding velocities of the plates. The clamped boundary conditions at the perimeter are considered for the top plate of the bottom screen assembly, but the other plates are free at the perimeters. The Block Lanczos method is used for the eigen-value and eigen-vector extractions of the finite element model, which is available for a large symmetric eigen-value problem. Typically, this solver is applicable to the type of problems solved using the subspace eigen-value method, however, at a faster convergence rate.

3.2 Results

The natural frequencies and corresponding mode shapes for the dry and wetted cases are obtained using the ANSYS finite element analysis code and summarized in Tables 2 and 3, respectively. These results are to be used for determining the dynamic responses due to self or external excitations such as a pump pulsation or seismic loads. The vibrational mode shapes are shown in Figures 5 ~ 7 for the dry modes and the mode shapes are also shown in Figures 8 ~ 10 for the wet modes. The fundamental natural frequency of the bottom screen assembly in air is 11.459 Hz, but the one in water is reduced to 4.116 Hz due to the hydrodynamic mass and coupling effect of the coolant. The dominant vibrational modes occur at the top and second plates up to the 18th mode, because the two plates are relatively flexible comparing with the other lower plates. The normalized natural frequency (η = natural frequency in the wetted mode / natural frequency in the dry mode) is about 0.27 for the fourth vibrational modes. This means that the relative hydrodynamic mass effect is dominant in the fourth wet mode.

Generally speaking, the normalized natural frequency gradually increases as the number of nodal diameters or nodal circles increases because of the relative hydrodynamic mass reduction. However, in this case, the tendency is very complicated due to the

Table 3. Natural frequencies of the bottom screen assembly in contact with coolant.

Mode no.	Natural freq. (Hz)	η	Dominant mode shape	Phase
1	4.116	0.3592	Top plate ($n^* = 1, m^* = 0$) 2nd plate (*tilting mode*)	N/A
2	8.092	N/A	Top plate ($n = 0, m = 2$) 2nd plate ($n = 0, m = 1$)	N/A
3	17.634	0.3924	Top plate ($n = 2, m = 0$) 2nd plate ($n = 2, m = 0$)	Out-of
4	47.662	0.2700	Top plate ($n = 0, m = 2$) 2nd plate ($n = 0, m = 1$)	Out-of
5	48.134	0.4659	Top plate ($n = 3, m = 0$) 2nd plate ($n = 3, m = 0$)	Out-of
6	58.021	0.6036	Top plate ($n = 1, m = 0$) 2nd plate ($n = 1, m = 1$)	Out-of
7	61.395	0.3478	Top plate ($n = 0, m = 2$) 2nd plate ($n = 0, m = 1$)	In
8	72.318	0.3859	Top plate ($n = 1, m = 0$) 2nd plate ($n = 1, m = 1$)	In
9	94.601	0.3620	Top plate ($n = 3, m = 0$) 2nd plate ($n = 3, m = 0$)	In
10	108.27	0.5777	Top plate ($n = 1, m = 2$) 2nd plate ($n = 1, m = 1$)	In
11	111.13	0.4253	Top plate ($n = 2, m = 0$) 2nd plate ($n = 2, m = 1$)	Out-of
12	146.96	0.5624	Top plate ($n = 2, m = 0$) 2nd plate ($n = 2, m = 1$)	In
13	157.79	0.5755	Top plate ($n = 5, m = 0$) 2nd plate ($n = 5, m = 0$)	Out-of
14	185.04	0.4859	Top plate ($n = 3, m = 0$) 2nd plate ($n = 3, m = 1$)	Out-of
15	196.76	0.4371	Top plate ($n = 0, m = 2$) 2nd plate ($n = 0, m = 2$)	Out-of
16	220.70	0.4711	Top plate ($n = 1, m = 2$) 2nd plate ($n = 1, m = 2$)	Out-of
17	238.00	0.2824	Top plate ($n = 6, m = 0$) 2nd plate ($n = 6, m = 0$)	Out-of
18	243.14	0.5543	Top plate ($n = 1, m = 2$) 2nd plate ($n = 3, m = 1$)	In

n = number of nodal diameters, m = number of nodal circles

complex structural geometry and the boundary conditions. The two different modes, that is, the in-phase and out-of-phases modes appear with respect to each dry mode when the bottom screen assembly is submerged into the coolant. It is known that the coupled natural frequency of the out-of-phase mode is always less than that of the in-phase mode. This is shown in the 6th wet mode (56.021 Hz, out-of-phase mode) and 8th wet mode (72.318 Hz, in-phase mode). Also, the fourth wetted mode natural frequency is 17.634 Hz as the out-of-phase mode, on the other hand, the seventh wetted mode natural frequency is 61.395 Hz corresponding to the in-phase mode.

4 CONCLUSIONS

The finite element model of the fluid–coupled structure is constructed to the bottom screen assembly of an integral reactor. From the three dimensional finite element model, the natural frequencies and corresponding mode shapes of the bottom screen assembly are obtained using ANSYS 5.6(1999) when it is either in contact with coolant or in air. It is found that the fluid gaps reduce the natural frequencies of the bottom screen assembly due to the hydrodynamic mass of coolant. It is also observed that the two different modes, that is, the in-phase and out-of-phase modes appear when the screen assembly is submerged into the coolant owing to the fluid coupling effect.

ACKNOWLEDGEMENT

This project has been carried out under the Nuclear R&D Program by MOST of Korea.

REFERENCES

Amabili, M. G., Frosali G. and Kwak M. K. 1996, Free vibrations of annular plates coupled with fluids, *J. of Sound & Vib.* 91, 825-846.

Chiba M. 1994, Axisymmetric free hydroelastic vibration of a flexural bottom plate in a cylindrical tank supported on an elastic foundation, *J. of Sound and Vib.* 169, 387-394.

Hagedorn, P., 1994, A note on the vibrations of infinite elastic plates in contact with water, *J. of Sound and Vib.* 175, 233-240.

Jeong, K. H., Kim, T. W., Choi, S. and Park, K. B., 1998, Free vibration analysis of two circular disks coupled with fluid, *PVP–Vol.* 366, *Proceedings, Tech. in Reactor Safety, Fluid–Stru. Interaction, Sloshing and Natural Hazards Engin.*, 157-164.

Jeong, K. H., Kim, T. W., Kim, K. S., Park, K. B. and Chang, M. H., Free vibration of a rectangular plate in contact with fluid, *Sec.11(B), Proceedings of Korea Nuclear Society Spring Meeting,* June 1999, Pohang, Korea.

Kohnke, P., 1999, *ANSYS Structural Analysis Guide*, ANSYS, Inc., Houston.

Kwak, M. K. 1991, Vibration of circular plates in contact with water. *Trans. of the ASME, J. of App. Mech.,* 58, 480-483.

Kwak, M. K. and Kim, K. C. 1991, Axisymmetric vibration of circular plates in contact with fluid, *J. of Sound & Vib.* 146, 381-389.

Montero de Espinosa, F. and Gallego-Juárez, J. A., 1984, On the resonance frequencies of water-loaded circular plate, *J. of Sound and Vib.* 94, 217-222.

Park, K. B et al., 1999, *Development of Advanced Reactor Technology*, Korea Atomic Energy Research Institute, (in Korean), KAERI/RR-1888/98.

Creative Systems in Structural and Construction Engineering, Singh (ed.)© 2001 Balkema, Rotterdam, ISBN 90 5809 161 9

Seismic behavior analysis of a bridge considering abutment-soil interaction

Sang-Hyo Kim, Sang-Woo Lee & Jeong-Hun Won
Department of Civil Engineering, Yonsei University, Korea

Ho-Seong Mha
Department of Civil Engineering, Hoseo University, Korea

Kyu-Hyuk Kyung
Hyundai Engineering and Construction Company Limited, Korea

ABSTRACT: The longitudinal dynamic behaviors of a bridge system under seismic excitations are examined with various magnitudes of peak ground accelerations. The stiffness degradation due to the abutment-soil interaction is considered in the system model, which may play the major role upon the global dynamic characteristics of the whole bridge system. The idealized mechanical model for the bridge system is proposed, which is capable of considering pounding phenomena, friction at the movable supports, rotational and translational motions of foundations, and nonlinear pier motions. The abutment-soil interaction is simulated by utilizing the one degree-of-freedom system with nonlinear spring. The stiffness degradation of the abutment-backfill system is found to increase the relative displacement under moderate seismic excitations. Soil conditions are also found to affect the relative motions of the bridge system.

1 INTRODUCTION

Earthquake related disasters are reported world widely, and the better aseismic concepts are desired correspondingly for bridge structures. Correct predictions of the dynamic behaviors of the structures under seismic excitations are more desired for the better design of the systems.

Bridge systems consisting of several simple spans can be described as the combination of multiple vibration units, and the dynamic characteristics of the most units are similar. However, the global dynamic behaviors of such a bridge system become complicated to be predicted due to many factors, which can be abutment-soil interactions, poundings between girders, frictions at movable supports, inelastic behaviors of RC piers, foundation motions (rotation and translation), and so forth. It should be noticed that the interactions between the adjacent oscillating units have drawn the interest, since it may play the major role of the span collapses. Among the interactions between vibrating units, the abutment-soil interaction may be the most important component affecting the global motions of the bridge particularly for bridge systems with similar vibrating units.

It is highly challenging to verify the effect of the abutment-soil interaction because of the uncertainties included to the abutment-backfill system in addition to the soil. The abutment-backfill system is often ignored or at most modeled as either a linear spring system. However, the abutment-soil interaction should be considered in the analysis of bridge

motions. An appropriate nonlinear model corresponding to the abutment-soil interactions should be adopted for the proper simulations.

A simple and economical way to represent the abutment-backfill system in seismic behavior analysis is to use a translational nonlinear spring (Siddharthan et al. 1997). Many important factors such as nonlinear soil behavior, abutment dimensions, superstructure loads, difference in soil conditions, and so forth, is considered to determine the nonlinear spring stiffness.

In this study, the idealized mechanical model for the multi-simple span bridge system is proposed, which can consider the abutment-soil interactions as well as the pounding and friction and other factors. The abutment-backfill system is modeled as a one-degree-of-freedom system with a nonlinear spring and a linear damper to consider the stiffness degradation. The pounding system with friction is modeled as a multi-degree-of-freedom system, composed of the individual mass-spring-damper system connected to each other by the impact and friction elements.

2 MODELING OF SYSTEMS

2.1 Bridge model

The bridge considered is a three-span simple plate girder bridge with 35m span length as shown in Fig. 1. Bent type piers, shallow foundations, and seat-type abutments are used. The pier height is 12m and the abutment height is 6.5m. In this study, only the longitudinal motions are of concern, so the total system can be divided into four individual vibrating units shown as in the Figure 1.

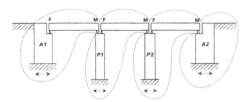

Figure 1. Bridge model.

For better efficiency, a simplified mechanical model is proposed using the lumped mass system, which is depicted in Figure 2. In the figure, m_1, m_5, m_9 are the masses of superstructures, m_2, m_6 are the masses of piers, m_3, m_7 are the masses of foundations, m_4, m_8 are the rotational mass moments of inertia of foundations, and m_{A1}, m_{A2} are masses of abutments. K_2, K_6 and C_2, C_6 are the stiffness and damping constants of the piers, K_3, K_7 and C_3, C_7 are the translational stiffness and damping constants of the foundations, and K_4, K_8 and C_4, C_8 are the rotational stiffness and damping constants of the foundations, respectively. K_{A1}, K_{A2} and C_{A1}, C_{A2} are the stiffness and damping constants of the abutments. $K_{A1,1}$, $K_{2,5}$ and $K_{6,9}$ are the stiffness of the fixed supports at individual vibration units. $F_{1,2}$, $F_{5,6}$, and $F_{9,A}$ are the friction forces at the movable supports. $S_{1,5}$, $S_{5,9}$, $S_{9,A2}$ and $C_{1,5}$, $C_{5,9}$, $C_{9,A2}$ are the stiffness and damping constants of the impact elements, and L is height of pier. $d_{1,5}$, $d_{5,9}$, $d_{9,A2}$ are the gap distances between adjacent vibration units, and $\ddot{u}_g$ is the ground acceleration.

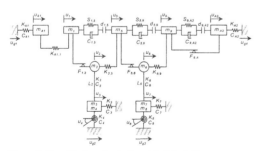

Figure 2. Simplified mechanical model of the bridge.

2.2 Abutment-backfill model

A simplest approach to represent the nonlinear behavior of the abutment due to abutment-soil interaction is to use translational nonlinear spring. The computational procedure of this methodology is re-

latively simple while the output shows great coincidence with the realistic behavior of the abutment, as been verified through a field test (Siddharthan et al. 1997; Goel & Chopra, 1997).

The abutment-backfill system is modeled in this study as one-degree-of-freedom system with nonlinear spring and linear damper to consider the abutment stiffness degradation as shown in Figure 3. The abutment is assumed to synchronize with the backfill, and the mass of the abutment-backfill system is also assumed to be the summation of the mass of the abutment and the backfill. In the Figure, m_A is the mass of the abutment-backfill system, $u_{g(A)}$ is the ground displacement, u_A is the displacement of mass m_A, and $K_A(x)$ and C_A are the translational nonlinear stiffness and linear damping constant of the soil surrounding the abutment, respectively.

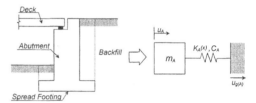

Figure 3. Abutment-backfill model.

Since the abutment undergoes rigid body movement, the abutment displacement at any point is evaluated in terms of δ_L and θ as shown in Figure 4. The nonlinear spring stiffness is obtained from estimation of the force, P_L, for a given displacement, δ_L, such that force equilibrium equation is satisfied.

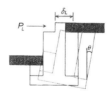

Figure 4. Abutment movement for longitudinal stiffness evaluation.

2.3 Poundings between girders

Two adjacent vibration units may produce poundings upon the applied seismic excitations with various intensities, and these poundings are governed by the relative displacements between the oscillating systems. The pounding is described in this study by placing spring-damper elements (impact elements) between the masses as shown in Figure 5. The pounding condition is defined as follows.

$$\delta_i = u_i - u_{i+4} + u_{gi} - u_{g(i+1)} - d_{i,i+4} \geq 0 \qquad (1)$$

where u_i, u_{i+4} are the displacement of mass m_i and m_{i+4}, u_{gi}, $u_{g(i+1)}$ are the ground displacement, and $d_{i,i+4}$ is the gap distance between m_i and m_{i+4}. Then the force due to pounding between m_i and m_{i+4} can be expressed as follows:

$$F_{i,i+4} = S_{i,i+4}\delta_i + C_{i,i+4}\dot{\delta}_i \quad for \ \delta_i > 0$$
$$F_{i,i+4} = 0 \quad for \ \delta_i > 0 \tag{2}$$

where $S_{i,i+4}$ and $C_{i,i+4}$ are the spring stiffness and damping constant of impact element, respectively. The stiffness of spring is typically large and highly uncertain due to the unknown geometry of the impact surfaces, uncertain material properties under impact loadings, and variable impact velocities, etc. Based on a sensitivity study, it is known that the system responses are not quite sensitive to changes in the stiffness of spring. The damping constant, which determines the amount of energy dissipated, can be obtained by following relationship (Anagnostopoulos 1988).

$$C_{i,i+1} = 2\xi_i \sqrt{S_{i,i+1} \times m_i m_{i+1}/(m_i + m_{i+1})}$$
$$\xi_i = -\ln r / \sqrt{\pi^2 + (\ln r)^2} \tag{3}$$

where r = coefficient of restitution. Value of $r = 1$ ($\xi = 0$) describes fully elastic collision, while value of $r = 0$ ($\xi = 1$) represents perfectly plastic one.

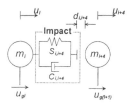

Figure 5. Idealization of pounding.

2.4 The friction of the movable supports

The frictions between the superstructures and the movable supports are usually neglected in the bridge dynamic analysis. However, this may not yield the appropriate results since it ignores the energy dissipation due to friction. In this study, a modified bilinear coulomb friction model is utilized. The simple model of the friction element between the superstructure and the support is described in Figure 6.

The relationship between the friction force and relative velocity between the adjacent oscillators can be depicted as shown in Figure 7. In stick condition, the friction force increases up to a given value, ε of the relative velocity and then sustains a constant friction force multiplying vertical force with friction coefficient, μ. The friction forces, $F_{i,i+1}$ of the stick and sliding conditions are expressed as follows.

$$F_{i,i+1} = \frac{1}{2}\mu m_i g \frac{1}{\varepsilon}\Delta_i \quad for \ |\Delta_i| < \varepsilon$$
$$F_{i,i+1} = \frac{1}{2}\mu m_i g \quad for \ |\Delta_i| \geq \varepsilon \tag{4}$$

where $\Delta_i = \dot{u}_i - \dot{u}_{i+1} + \dot{u}_{gi} - \dot{u}_{g(i+1)}$

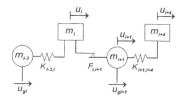

Figure 6. Friction element.

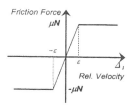

Figure 7. Friction force-relative displacement relationship.

The nonlinear pier motion is simulated by adopting the hysteresis loop function. The foundation motions are modeled as a two DOF system with translational and rotational springs and dampers (Kim et al. 2000).

3 LONGITUDINAL ABUTMENT STIFFNESS EVALUATION

The longitudinal abutment stiffness varies with types and conditions of soil surrounding the abutment. The rational evaluation of longitudinal stiffness degradation due to abutment-soil interaction is required. In this study, the backfill types such as loose sand, medium dense sand, and dense sand are considered and the longitudinal abutment stiffness for these backfill types is estimated. The soil properties for each type of backfill assumed in this study are tabulated in Table 1 (Barker et al. 1991).

The values of the earth pressure coefficient for longitudinal movements are determined by using the

Table 1. Soil properties

Type of backfill	Friction angle, ϕ	Standard penetration number, N	Relative density, D_r
Loose sand	30°	5-10	5-30%
Medium dense sand	37°	10-30	30-60%
Dense sand	45°	40-50	60-90%

Figure 8. Figure 9 shows the longitudinal abutment stiffness (passive stiffness) for all three backfill types considered. The stiffness is presented as per unit width of the abutment in the transverse direction. The longitudinal abutment stiffness decrease rapidly as the displacement increase when the displacement is small, and the stiffness degradation ratio gets larger as the relative density of soil decrease. The active stiffness of the abutment is assumed to be ten times smaller than the passive stiffness (Hambly, 1991). It is also assumed that once the degradation is started, the degraded stiffness cannot be recovered and that the stiffness after the critical displacement is reached remains constant afterwards regardless of the displacement. Table 2, obtained through experimental data and FE analyses, gives approximate displacements required to reach minimum active and maximum passive earth pressure conditions (Clough & Duncan, 1991). H is the height of the abutment.

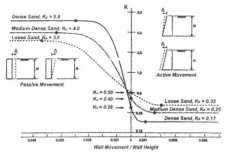

Figure 8. Relationship between movement and earth pressure.

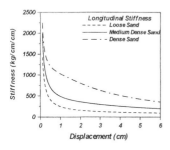

Figure 9. Longitudinal abutment stiffness (passive stiffness).

Table 2. Critical displacements to minimum active and maximum passive earth pressure conditions.

Type of backfill	Values of δ_L / H	
	Active	Passive
Loose sand	0.004	0.04
Medium dense sand	0.002	0.02
Dense sand	0.001	0.01

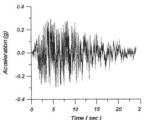

Figure 10. Simulated seismic excitation

4 RESULTS AND OBSERVATIONS

Bridge seismic responses are evaluated to see the effects of the longitudinal stiffness degradation due to abutment-soil interaction. Two types of backfill conditions are examined by introducing different abutment models; the model using the constant abutment stiffness (linear system) and the model using the nonlinear stiffness considering the abutment stiffness degradation (nonlinear system).

As the input ground motions, artificial seismic excitations are used. By using the well-known SIM-QKE code, the seismic excitations compatible to the design response spectra specified in the Korean Standard Specifications for Highway Bridges (1996) are generated. An example of the simulated seismic excitation is shown in Figure 10.

The 5cm-gap distance between adjacent vibration units is selected, and the friction coefficient for movable support is assumed to be 0.05 based on the same specification. In the case with the constant abutment stiffness, the abutment-backfill system is modeled as the two degree-of-freedom-system with translational and rotational springs and dampers. The constant abutment stiffness is determined according to Korean Standard Specifications for Highway Bridges: Seismic Design (1996).

From results, relative displacements to the ground motion and relative distances between vibration units are evaluated. Typical time histories of these displacements are depicted in Figures 11-12 for both cases with the linear and nonlinear abutment stiffness. From Figures, it can be seen that the responses become totally different in both cases, indicating that the stiffness degradation make an important effects upon bridge responses.

To see the effects of the stiffness degradation more clearly, the relative distances between the adjacent vibration units of the bridge system are evaluated under the seismic excitations with various intensities of peak ground accelerations (PGA) from 0.1g to 0.6g. The mean values of maximum relative distances are obtained. The results are tabulated in Tables 3-5 and the normalized results based on the maximum relative distances are summarized in Figure 13 for the case using constant abutment stiffness.

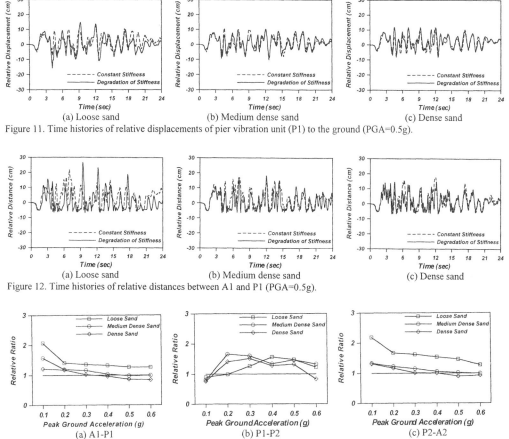

Figure 11. Time histories of relative displacements of pier vibration unit (P1) to the ground (PGA=0.5g).

(a) Loose sand   (b) Medium dense sand   (c) Dense sand

Figure 12. Time histories of relative distances between A1 and P1 (PGA=0.5g).

(a) A1-P1   (b) P1-P2   (c) P2-A2

Figure 13. Normalized results of maximum relative distances.

Table 3. Simulated results of maximum relative distances (loose sand, unit: cm)

PGA	A1-P1		P1-P2		P2-A2	
	[a]CASE I	[b]CASE II	CASE I	CASE II	CASE I	CASE II
0.1g	2.70	5.59	0.66	0.59	2.24	4.90
0.2g	8.12	11.43	1.22	1.21	6.11	10.20
0.3g	12.38	16.79	3.62	4.57	9.33	15.15
0.4g	17.01	22.56	5.12	7.98	11.73	18.02
0.5g	22.42	28.49	7.37	10.85	13.94	20.42
0.6g	26.40	33.37	11.24	13.75	17.06	21.74

a. CASE I: the case using the constant abutment stiffness
b. CASE II: the case considering the abutment stiffness degradation

Table 4. Simulated results of maximum relative distances (Medium dense sand, unit: cm)

PGA	A1-P1		P1-P2		P2-A2	
	CASE I	CASE II	CASE I	CASE II	CASE I	CASE II
0.1g	2.58	4.05	0.65	0.52	2.32	3.09
0.2g	7.75	9.25	1.37	2.27	5.81	7.13
0.3g	11.97	13.89	3.52	5.63	9.29	10.69
0.4g	16.72	17.30	5.13	6.84	11.76	12.42
0.5g	20.96	20.53	6.46	9.46	14.54	14.83
0.6g	25.05	24.91	8.80	11.64	16.47	16.35

It is observed that the maximum relative distances between the abutment and pier vibration units are commonly decreased as the relative density of backfill increases in both linear and nonlinear systems. The decrease rate of the responses is significantly higher in the nonlinear system than that in the linear system. Maximum relative distances are found to be larger in the nonlinear system than those found from the linear system. However, in some cases, the maximum relative distances with constant abutment stiffness are large under strong excitations, making exceptions.

Between pier units, the maximum relative distances of the case considering the abutment stiffness degradation is also larger than those of the case using the constant abutment stiffness in all backfill types. However, the possibility of the undesired be-

Table 5. Simulated results of maximum relative distances (dense sand, unit: cm)

PGA	A1-P1		P1-P2		P2-A2	
	CASE I	CASE II	CASE I	CASE II	CASE I	CASE II
0.1g	2.58	3.13	0.63	0.48	2.13	2.80
0.2g	7.02	8.27	0.91	1.28	5.74	6.72
0.3g	10.95	11.16	2.79	4.22	9.28	9.42
0.4g	15.52	14.95	4.49	5.72	10.67	10.69
0.5g	19.67	17.00	4.68	6.17	14.02	12.60
0.6g	23.15	19.57	7.73	6.40	15.08	14.00

havior, such as the unseating failure of superstructures, is still low since the absolute values of the relative distances are very small because of the identical dynamic characteristics between pier units.

From tables, it is found that the stiffness degradation due to the abutment-soil interaction plays the important role upon the global dynamic characteristics of the whole bridge system. The effects of stiffness degradation are found to be relatively large under moderate and weak excitations with PGA < 0.4g. The system shows much bigger relative motions than in the system with linear stiffness, which is the system without consideration of the degradation. This trend becomes clearer in the system with loser sands (Compare Tables 3-5). The seismic responses may be underestimated in the system only with the constant stiffness considered. Therefore, the stiffness degradation should be accommodated in the analysis of seismic responses of the bridges.

From results, it is suggested that the development of a relatively simple and realistic methodology to determine the abutment stiffness degradation be determined according to the soil conditions surrounding abutment to correctly estimate seismic behaviors of bridges. It is also suggested that the appropriate soil conditions and dynamic coefficients should carefully be decided for better predictions.

5 CONCLUSIONS

Responses of multi-span bridges are examined by using the simplified mathematical model, which still retains the appropriate characteristics of the corresponding bridge system. The effects of the stiffness degradation due to the abutment and soil interactions are intensively investigated. Various soil conditions are applied and results are compared with those based on the linear system, which excludes the stiffness degradation. The observed trends are as followings: First, the longitudinal abutment stiffness is found to dramatically decrease at the onset of seismic excitations. With lower relative density of the backfill, the stiffness decreases more shortly converging to the critical values. Secondly, the relative motions to both grounds and adjacent units are found to be larger in the system with nonlinear stiffness

than those found in the corresponding linear system. As PGA increases, the differences becomes insignificant, and the linear system even shows the larger relative motions in the case of the high relative density sand. Finally, under weak or moderate excitations, the relative motions are found to be larger in most cases in the nonlinear system. Hence, the response motions may be underestimated if only the linear system is considered, and care should be taken.

The stiffness degradation of the abutment and backfill system is found to take an important influence upon the global bridge motions. The soil conditions around the abutment are also found to be an important component. Hence, it can be concluded that the stiffness degradation and the soil conditions should be included in the seismic analysis of the bridge system.

ACKNOWLEDGEMENTS

This work was supported by the Brain Korea 21 Project and the Korea Science and Engineering Foundation (KOSEF) through the KEERC.

REFERENCES

Anagnopoulos, S. A. 1988. Pounding of Buildings in Series during Earthquakes. *Earthquake Engineering and Structural Dynamics*, 16. 443-456.

Barker, R. M., Duncan, J. M., Rojiani, K. B., Ooi, P. S. K., Tan, C. K., and Kim, S. G. 1991. *Manuals for the Design of Bridge Foundations*. Transportation Research Board. Washington. D. C.: NCHRP Report 343: 129-131.

Clough, G. W. and Duncan, J. M. 1991. *Foundation Engineering Handbook*. 2nd edition: New York: 223-235.

Gasparini, D. A. and Vanmarcke, E. H. 1976. *Evaluation of Seismic Safety of Buildings Simulated Earthquake Motions Compatible with Prescribed Response Spectra*. Massachusetts Ins. of Technology: Report No. 2.

Goel, R. K. and Chopra, A. K. 1997. Evaluation of Bridge Abutment Capacity and Stiffness during Earthquake. *Earthquake Spectra*, 13(1): 1-23.

Hambly, E. C. 1991. *Bridge Deck Behavior*. 2nd edition: E & FN SPON: 291-299.

Kim, S-H, Mha, H-S, Lee, S-W, and Won, J-H. 2000. Dynamic Behaviors of Bridges under Seismic Excitations with Pounding between Adjacent Girders. *12th WCEE*, PS5 1815.

Ministry of Construction and Transportation. 1996. *Korean Standard Specifications for Highway Bridges*.

Siddharthan, R. V., El-Gamal, M., and Maragakis, E. A. 1997. Stiffness of Abutments on Spread Footings with Cohesionless Backfill. *Canadian Geotechnical Journal*, 34: 686-697.

Creative Systems in Structural and Construction Engineering, Singh (ed.) © 2001 Balkema, Rotterdam, ISBN 90 5809 161 9

Structural behaviour of railway track subject to transient loading

D. M. Lilley

Department of Civil Engineering, University of Newcastle upon Tyne, UK

ABSTRACT: Measurements of train-generated vibration within steel rails have prompted an investigation of the load-deflection behaviour of sections of railway track when subject to lateral load. Modern track in the UK is mostly constructed using BS113A rails secured to sleepers (ties) using Pandrol-type spring clips. In such cases the upper sections of rails are not directly restrained and this is thought to contribute to the high levels of train-generated horizontal acceleration measured during in-situ tests. Static load tests performed in a laboratory and at locations within an operational light railway have indicated that approximately 90% of the resistance of a rail to lateral load is provided by sections of track at distances up to 2.2m away from the position of the load. A non-linear load-deflection relationship was observed during tests on a very short length of track.

1 INTRODUCTION

A study in progress at the University of Newcastle is investigating dominant modes and frequencies of vibration in steel rails, and the effects of different types of track construction. Long-term aims are to reduce environmental intrusion from sounds generated at wheel-rail interfaces and to identify defects within steel rails through dynamic analysis. Past experience (Adams et al. 1978 and Lilley 2000) has demonstrated the value of vibration testing as a form of integrity test. In 1998, approximately 800 cracked rails were reported in the UK, an increase of about 20% on those reported the previous year.

Analytical models (Mohammadi & Kabalis 1995 and Jaiswal & Iyengar 1997) of track response to dynamic loads have assumed idealized elastic beam or half-space models, and little has been published which relates to the real load-deflection behaviour of modern railway track.

The behaviour of railway track subject to moving loads from rail traffic is a complex interaction of mostly non-linear effects. The overall structural performance of track depends on the materials and form of construction, degree of wear, and effectiveness of support and restraint provided by different elements forming the track.

Vertical components of wheel loads applied to the running surfaces of a rail vary with overall vehicle weights and also as a result of the response of suspension systems to localised changes in level of the track. Transverse forces are generated by vehicles travelling over curved track and also by the behaviour of wheels seeking the line of the track.

Most modern track in the UK is formed from continuous welded steel BS113A rails supported by steel baseplates fixed to either timber or concrete sleepers (known in the USA as "ties") within stone ballast. A resilient pad is usually placed between the rail and the base-plate, and Pandrol-type clips are used to secure the base of the rail in position above the base-plate. This type of track has now almost completely replaced the older "bull-head" type rail which was held in position by substantial "chairs" fixed directly to the sleepers. Steel chocks or clips were placed within the chairs to provide lateral support to both sides of the web of the bull-head rail section. Modern rails are thus more susceptible to horizontal movement within their upper sections than the older track because of the reduction in restraint provided by the base-plate and Pandrol clips.

Both vertical and horizontal components of load from a wheel are transmitted to the rail through the area of mutual contact. Tests at the University of Newcastle have shown that the area of contact between a 740mm diameter steel wheel and a BS113A rail is approximately 100mm^2 and that this does not seem to be influenced by vertical loads up to a maximum of 120kN.

Measurements of vertical and horizontal acceleration have been made on BS113A rails supported by conventional ballasted sleepers in several different locations with the Tyne and Wear Metro system. This is a light railway system incorporating approximately

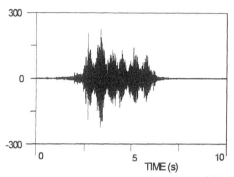

ACCELERATION (m/s^2)

Figure 1. Train-generated vertical acceleration within a steel rail

56km of twin track and forming an important part of the public transport infrastructure of the city and suburbs of Newcastle upon Tyne. Rail traffic is primarily passenger-oriented (with occasional maintenance vehicles) mostly comprising pairs of two-coach units powered by overhead electrification at speeds up to 80 km/h. Each Metro unit comprises three bogies, each with two axles.

Independent tests have shown that the minimum vertical wheel load produced by a Metro train is approximately 31.2kN. No detailed information is available about the maximum value of wheel load which is clearly related to the number and distribution of passengers, although this has been estimated at about 50kN.

Figure 1 illustrates a typical transient vertical response at the base of a rail produced by a Metro train travelling over a straight section of track at approximately 70km/h.

The effect of each passing wheel bogie can clearly be seen; maximum values of vertical acceleration of approximately 200m/s^2 were recorded. Passing trains generated similar maximum amplitudes of acceleration in the transverse horizontal direction on the rail-head.

Fourier analysis of the transient horizontal and vertical responses of the rail indicated that most of the vibration was occurring over a range of frequencies between 40Hz and 800Hz.

2 LOAD TRANSFER

Load applied to the rail-head is transmitted to the sleepers through bending, shear, and torsion of the rail. Each sleeper is supported at many points by individual stones within the ballast. Perfect contact between sleeper and ballast is very difficult to achieve in construction and to maintain during the working life of the track. Rails with attached sleepers

often "bridge" over the underlying ballast, so that passing trains produce significant vertical movement and impact as the suspended sleepers come into contact with the stone ballast. Repeated impacts create additional ground vibration and produce localised damage to the base of the sleeper. This damage is often only visible when sleepers are replaced during major track refurbishment and does not usually affect the performance of the sleeper. Regular tamping of the ballast and re-alignment of track can reduce track movement and vibration from impact between sleepers and ballast.

Rails with all sleepers in contact with ballast and supporting vertical wheel loads behave as multi-span continuous beams. Existing data (Anon., 1992, and Cope, 1993) state that the BS113A rail section has an overall height of 158.8mm and its centroid is between 83.9mm and 84.2mm from the top of the rail. Thus maximum bending stresses can be expected to occur on the top surface of the rail. Rails are usually pre-stressed at the time of construction to accommodate the effects of temperature changes and reduce the risk of buckling in very hot weather. The usual aim in the UK is to have zero axial stress in a rail when its temperature is within 21-27°C. Rail temperatures above or below this range of values produce axial forces which, in bending, cause the neutral axis to move away from the centroidal position and thus influence values of stress within the rail.

Horizontal forces applied to the top of the rail are transmitted to base-plates where there is a tendency for the rail to try to rotate about a point at the end of its bottom flange. The rail is restrained from twisting by the vertical forces provided by Pandrol clips on the side of the rail that is trying to lift off from contact with the base-plate and also by the resistance to torsion provided by the rail cross-section.

The high values of train-generated horizontal acceleration recorded within rails during track-side measurements and the apparent reduction in lateral restraint provided by modern base-plates and clips prompted an investigation to study the static horizontal load-deflection responses of sections of track constructed using "standard" components.

3 RESPONSE OF TRACK TO STATIC HORIZONTAL LOAD

Load tests were conducted on a very short section of track in a laboratory and on operational track at two different locations within Tyne and Wear Metro.

The test in the laboratory examined the load-deflection behaviour of track made from two 1m lengths of flat-bottomed BS113A rail fixed to two timber sleepers by conventional baseplates and Pandrol clips (see Figure 2). The centre-lines of the sleepers were 762mm apart. The use of a short length of rail effectively eliminated the torsional re-

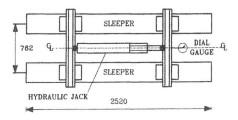

Figure 2. Schematic diagram of arrangement to test a short length of track

sistance normally provided by lengths of rail on either side of the section supporting the load, and allowed the resistance provided mainly by the Pandrol clips and the bending stiffness of the rail to be examined.

Increments of horizontal force were applied to the side of the heads of each of the two rails using a manually-operated hydraulic ram positioned midway between the two sleepers. Horizontal deflection at the top of one of the rails was measured using a dial gauge with an accuracy of ±0.01mm. The other rail was braced against the wall of the laboratory to prevent it from moving significantly due to bending or torsion.

A maximum load of 36.9kN was applied in six approximately equal increments and produced a maximum horizontal deflection of 22.4mm at the rail-head (see Figure 3). Load was subsequently removed and the deflection returned to zero. New Pandrol clips were fitted to eliminate the possible effects of plastic strains within these elements on subsequent testing. Load was again applied to the rail as before but in in-

crements of 1kN, and this time lateral deflections were measured at both the top and bottom of the rail at the mid-point position between the two sleepers.

The test was repeated using the same short lengths of rail but replacing the two timber sleepers with two concrete sleepers. Concrete sleepers are provided at the time of manufacture with cast in-situ fixings to accommodate Pandrol clips; separate baseplates are not required for concrete sleepers. Rubber pads were placed between the rail and the sleeper. In normal construction these provide a conforming layer between the rail and the sleeper to help to ensure a more even distribution of pressure in the rail seat area and to reduce the transmission of vibration and impact into the sleeper.

The load-deflection behaviour of each of the two types of track is illustrated in Figure 3. In both cases the relationship between applied horizontal load and resulting lateral deflection was non-linear.

Higher values of deflection were consistently measured in track supported by concrete sleepers than were observed with timber sleepers at the same load. In both cases, the effect of increasing lateral load was to reduce the apparent overall lateral stiffness of the rail. Maximum deflection was limited to approximately 25 mm because of the risk of the rail slipping beyond the clips. This value exceeds the limits that would normally be expected and might lead to derailment under normal operating conditions.

Measurements taken at the base of the rail are also shown in Figure 3 and indicate an overall movement at that level in the direction of applied load. This implies that the bending effect of the applied load on the rail was greater than that of torsion.

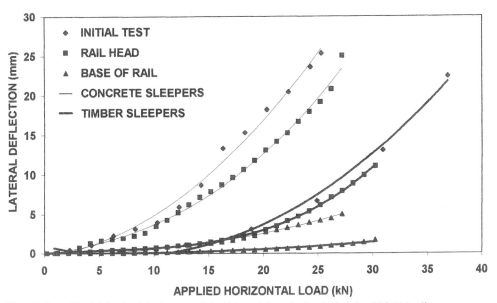

Figure 3. Lateral load-deflection behaviour recorded within track formed using 1m lengths of BS113A rail

4 IN-SITU LOAD TESTS

The tests in the laboratory were not able to include the partial torsional restraint normally provided by lengths of rail and Pandrol clips on either side of the loaded section, and further investigation was needed to evaluate this effect.

In-situ load tests were performed on two different straight sections of ballasted track within the Tyne and Wear Metro system during periods of overnight close-down. A special loading rig was constructed so that horizontal load could be simultaneously applied to the top of two rails forming the track using a manually-operated hydraulic jack. The equipment was designed so that the two rails could be either forced apart or together, depending on the position

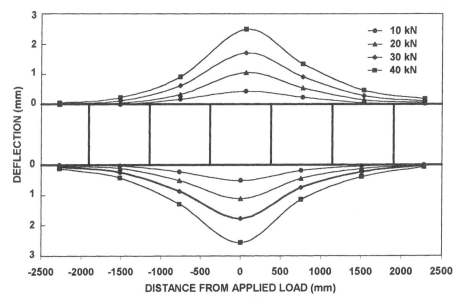

Figure 4a. Lateral load-deflection behaviour measured at rail-head (concrete sleepers)

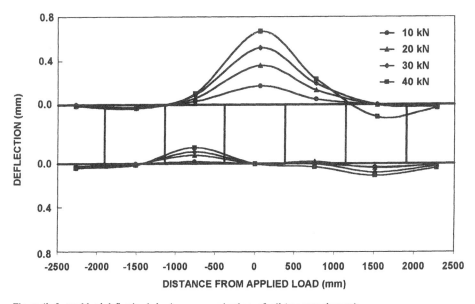

Figure 4b. Lateral load-deflection behaviour measured at base of rail (concrete sleepers)

of the jack. Measurements of lateral deflection were taken using a series of dial gauges mounted at several positions at the top and bottom of each rail. Deflections were recorded at the mid-point positions between each sleeper for a distance of approximately 2.4m on each side of the loading point. The first track (at Jesmond Station) to be tested was formed using BS113A rail supported by concrete sleepers and conventional ballast. Track with timber sleepers (at Cullercoats Station) was tested on a later occasion using an identical technique. It was essential that each test be completed within a few hours so that the track was available for normal use at the start of the next working day. This restricted the number of loading conditions that could be applied in each test. Four increments each of 10kN were applied to the track with concrete sleepers, while three similar increments were applied to that with timber sleepers.

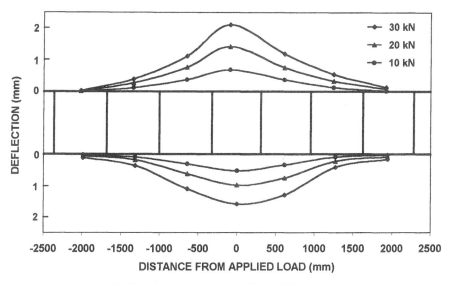

Figure 5a. Lateral load-deflection behaviour measured at rail-head (timber sleepers)

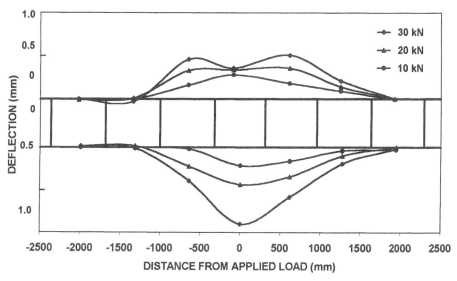

Figure 5b. Lateral load-deflection behaviour measured at base of rail (timber sleepers)

Values of lateral displacement are shown in Figures 4a, b for track with concrete sleepers; Figures 5a, b refer to that with timber sleepers. In each case the results are plotted on line diagrams that represent the track and the sleepers in their as-built positions, and represent exaggerated shapes of the deflected rails.

Figures 4a and 5a shows lateral deflections measured at the top of each set of rails. In both cases the top of each rail deflected in a single bow over a distance of about 2.2m on both sides of the rail from the point of application. Both tracks produced similar mean values of maximum deflection of 1.7-1.8mm when subjected to a load of 30kN, which is contrary to the less stiff behaviour observed in the laboratory tests on track formed with concrete sleepers. In each test the lateral deflections of the two rails forming the track were very similar, and this suggests a high degree of confidence in the results.

Lateral deflections at the loading position for both tracks were approximately 8-15% of the values determined for similar lateral loads applied to short lengths of rail in the laboratory (see Figure 3). Thus approximately 90% of the resistance of a rail to a concentrated lateral load was provided by the lateral and torsional stiffnesses of the rail and its fixings beyond the sleepers immediately adjacent to the load. Figures 4a and 5a suggest that the deflection of the two operational tracks at the loading point increased in approximately direct proportion with increasing load, but in each case there was insufficient time on site for enough load increments to confirm a linear relationship.

Values of horizontal deflection measured in each test at the base of the rail (see Figures 4b and 5b) were, as expected, significantly lower than those at the top of the rail. The behaviour of the base of the rail is not as clearly demonstrable as that at the rail-head. Maximum deflection at the base of one of the rails with timber sleepers (see the upper half of Figure 4b) was 0.74mm and occurred immediately below the applied load. The base of this rail deflected in the direction of the load for about 1.0m on either side of the load. However deflections in the rail beyond 1.0m from the load were in the opposite direction to the applied load, and suggest a more torsional response. Deflections at the base of the other rail in this section of track were generally less than 0.2mm and provide no clear evidence of its mode of response.

Figure 5b illustrates the deflected shape of the base of the track constructed with timber sleepers, and this generally resembles the same displaced shape as that of the rail-head. The maximum deflection of one rail was about 0.9mm when subjected to a load of 30kN; the other rail behaved in a more-or-less similar manner although deflections at the loading position did not increase as expected.

5 CONCLUSIONS

The structural properties of a section of railway track clearly influence its response to transient vertical and lateral loads. Evidence has been provided from static load tests to indicate that horizontal wheel loads applied to the running surfaces of a BS113A rail are resisted by the rail and its fixings over a total length of approximately 4.4m. In many cases rail vehicles may have two or more wheels within this distance, and it is very likely that their effects will superimpose upon one another to provide further complication in any analysis of train-generated vibration.

Further work is needed to investigate the behaviour of track to lateral loads applied at off-centre positions between adjacent sleepers, and to examine the modes of vibration generated within a rail when subjected to transient vibration from impact of a hand-held hammer on the side of the rail-head. It is hoped this will provide an insight into the mechanics, modes and frequencies of lateral vibration generated by passing trains, which may in turn lead to an analytical method of identifying crack development within rails using vibration measurements.

REFERENCES

Adams, R.D., Cawley, P., Pye, C.J & Stone, B.J. 1978. A vibration technique for non-destructively assessing the integrity of structures. *J. Mech. Eng. Sci.*, 20, 2, 93-100.

Anon. 1992. *The Track Handbook*, British Steel Track Products, Workington, Cumbria, UK.

Cope, G.H. 1993. *British Railway Track (6th Edition)*, The Permanent Way Institution, Barnsley, South Yorkshire, UK.

Jaiswal O. & Iyengar, R.N. 1997. Dynamic response of railway tracks to oscillatory moving masses. *Engineering Mechanics*, July, 753-757.

Lilley, D.M. (in press), Integrity testing of pile foundations using axial vibration. *J. Geo. Eng., Proc. Inst. Civ. Eng.* London, Paper No. 12092.

Mohammadi, M. & Kabalis, D.L. 1995. Dynamic 3-D soil-railway track interaction by BEM-FEM. *Int. J. Earthquake Engineering and Structural Dynamics*, London, September, 1177-1193.

Creative Systems in Structural and Construction Engineering, Singh (ed.) © 2001 Balkema, Rotterdam, ISBN 90 5809 161 9

Semi-active tuned vibration absorbers for vibration control of force and base excited structures

M. Setareh
Virginia Polytechnic Institute and State University, Blacksburg, Va., USA

ABSTRACT: This paper introduces a new class of semi-active tuned vibration absorbers "Ground Hook Tuned Vibration Absorbers" (GHTVA). This device uses a continuously variable semi-active damper (Ground-Hook) to achieve reduction in the vibration level. The ground-hook dampers have been used in the auto-industry to reduce the vibration of primary suspension systems in vehicles. This paper presents the application of ground hook damper as an element of tuned vibration absorber (TVA) to reduce vibrations of the force-excited, and base-excited single degree of freedom systems that can be representative of many civil structures such as floor vibrations, wind-excited structures, and seismically excited systems, etc. The optimum design parameters, which are evaluated in terms of non-dimensional values of the GHTVA, are obtained for different mass ratios and main mass damping ratios. Using the frequency responses of the systems with GHTVA, performance of the GHTVA is compared to that of the equivalent passive TVA, and it is found that GHTVAs can be more efficient in reducing the level of vibrations. A design guide to find the optimum tuning parameters of GHTVA is presented.

1 INTRODUCTION

The first application of tuned vibration absorber (TVA) or tuned mass damper (TMD) was reported by Frahm (1911). He used this device to reduce the rolling motion of ships. Later, Ormondroyl & Den Hartog (1928), Brock (1946), and Den Hartog (1947) studied the use of TVAs to control vibration of un-damped single degree of freedom systems. In general, TVAs work based on adding a secondary system to a main (primary) vibrating system at resonance such that the natural frequency of the secondary system is tuned to the primary system and oscillates to counteract the motion of the primary system. Typically, the secondary system has only a fraction of the primary system mass. Therefore, by adding a small amount of secondary mass, the vibration level of the primary system can be reduced significantly. Passive TVAs have been used widely in the past. They do not require any external force, and the characteristics of the spring and/or damper do not change with time.

Active TVAs have also been the subject of several studies. Based on the introduction of an external source of power to produce additional forces on the primary system to reduce its level of vibration (Udwadia & Tabaie 1981, Lund 1980, Chang & Soong 1980). Even though the active TVAs have better performance than their passive counterparts, they have several disadvantages including: need for actuators, pumps, etc., high operational costs and high power requirements.

In order to avoid the above disadvantages a new class of TVAs is introduced here using variable damping or stiffness. This class of TVAs (so called semi-active TVAs) have simple hardware requirements, low operational costs, and low power requirements. Different variations of semi-active TVAs have been introduced for transient vibration control such as wind induced vibrations in tall buildings (Hrovat et al. 1983), and for the seismic protection of civil structures (Abe 1996).

A class of semi-active dampers called "skyhook" and "groundhook" has resulted in significant control of vehicle vibrations (Miller 1988, Ahmadian 1997). However, these dampers can not be effective where the relative motion of the system is small such as vibration control of sensitive equipment, wind induced vibrations, or floor vibrations due to human movements, etc. Therefore, for these situations TVAs using semi-active dampers are being considered.

This paper studies the application of ground hook semi-active damper in a TVA configuration to reduce the vibration of single degree of freedom force and base excited systems. The resulting device is called 'Ground-Hook Tuned Vibration Absorber (GHTVA)'. The optimum design parameters of the GHTVA will be found and its effectiveness will be compared to that of equivalent passive TVA.

2 DESCRIPTION OF GHTVA

Figure 1(a) shows a single degree of freedom system consisting of mass m_1, stiffness k_1, and damping c_1 subjected to a harmonic force $F = F_0 e^{i\omega t}$. Figure 1(b) shows the same dynamic system subjected to a harmonic ground acceleration of $\ddot{x}_g = \ddot{X}_0 e^{i\omega t}$. When the exciting frequency (ω) becomes close to the system natural frequency (ω_1), the system will be in resonance and the main mass will be subjected to large amplitudes of motion (x_1). In order to reduce the vibration level, a secondary mass (m_2), is attached to the main mass through a spring with the stiffness coefficient of k_2, and a variable damper of c_2.

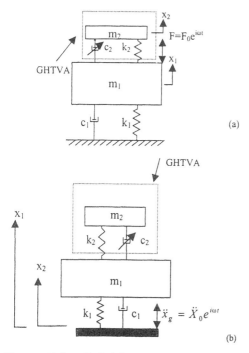

Figure 1. (a) Force-Excited System with GHTVA; (b) Base-Excited System with GHTVA.

The "ground hook control policy" assumes that the variable damping c_2 can be switched between a low and a high state.

For the continuous ground hook dampers, the damping is adjusted in the range between minimum (off) and a maximum (on) level according to:

$$\dot{x}_1(\dot{x}_1 - \dot{x}_2) \geq 0 \rightarrow c_2 = c_{on} = \min\{G|\dot{x}_1|, c_{max}\}$$

$$\dot{x}_1(\dot{x}_1 - \dot{x}_2) < 0 \rightarrow c_2 = c_{min} = c_{off} \tag{1}$$

where $\dot{x}_1$ and $\dot{x}_2$ are the velocities of masses m_1, and m_2 respectively.

The variable G is the gain that relates the damping level to the absolute velocity of the main mass, m_1.

3 RESPONSE OF GHTVA COMPARED TO TVA

3.1 Force-Excited System

The general forms of the system response can be shown as:

$$x_1 = \overline{X}_1 e^{i\omega t}$$
$$x_2 = \overline{X}_2 e^{i\omega t} \tag{2}$$

The following non-dimensional parameters as used for the passive TVAs are defined (Den Hartog 1947):

$$\mu = \frac{m_2}{m_1} = \qquad \text{mass ratio}$$

$$f = \frac{\omega_2}{\omega_1} = \frac{f_2}{f_1} = \qquad \begin{array}{l}\text{frequency ratio of the TVA or}\\ \text{GHTVA to the main system}\end{array}$$

$$\xi_1 = \frac{c_1}{2m_1\omega_1} = \qquad \text{damping ratio of the main system}$$

$$g = \frac{\omega}{\omega_1} = \qquad \begin{array}{l}\text{ratio of the excitation frequency to}\\ \text{the natural frequency of the main}\\ \text{system}\end{array}$$

In the above, ω_1, and ω_2 are in rad/sec, and f_1, and f_2 are in Hz.

Substituting (2) into system equations of motion, and simplification of the results, the equations of motion are:

$$-g^2\overline{X}_1 + 2ig\xi_1\overline{X}_1 + ig\frac{c_2}{m_1\omega_1}(\overline{X}_1 - \overline{X}_2) + \overline{X}_1 + \mu f^2(\overline{X}_1 - \overline{X}_2) = \frac{F_0}{m_1\omega_1^2}$$

$$-\mu g^2\overline{X}_2 + ig\frac{c_2}{m_1\omega_1}(\overline{X}_2 - \overline{X}_1) + \mu f^2(\overline{X}_2 - \overline{X}_1) = 0 \tag{3}$$

The displacement responses $\overline{X}_1$ and $\overline{X}_2$ in terms of non-dimensional amplification factors A_1 and A_2 are defined as:

$$\overline{X}_1 = A_1 \Delta_{st} = A_1\frac{F_0}{k_1} = A_1\frac{F_0}{m_1\omega_1^2}$$

$$\overline{X}_2 = A_2\Delta_{st} = A_2\frac{F_0}{k_1} = A_2\frac{F_0}{m_1\omega_1^2} \tag{4}$$

in which Δ_{st} is the static displacement for mass m_1.

Substitution of (4) into equation (3) will result in:

$$-g^2 A_1 + 2ig\xi_1 A_1 + ig\frac{c_2}{m_1\omega_1}(A_1 - A_2) + A_1 + \mu f^2(A_1 - A_2) = 1$$

$$-\mu g^2 A_2 + ig\frac{c_2}{m_1\omega_1}(A_2 - A_1) + \mu f^2(A_2 - A_1) = 0 \tag{5}$$

The above equations are in terms of non-dimensional parameters except for the damping term c_2.

For the passive TVAs, c_2 is a constant term which is defined as $c_2=2m_2\omega_2\xi_2$, in which ξ_2 is the TVA damping ratio. After solving equation (5), the amplification factors A_1 and A_2 can be found in terms of non-dimensional parameters.

In the case of semi-active GHTVA, c_2 varies with time based on the rule set by equation (1). When the damper is off, the damping value c_2 can be defined as $c_2=2m_2\omega_2\xi_{off}$. Substituting into equation (5) will result in a set of equations similar to ones for the passive TVA:

$$-g^2A_1 + 2ig\xi_1A_1 + 2igf\mu g_{off}(A_1 - A_2) + A_1 + \mu f^2(A_1 - A_2) = 1$$

$$-\mu g^2A_2 + 2igf\mu g_{off}(A_2 - A_1) + \mu f^2(A_2 - A_1) = 0 \quad (6)$$

According to equation (1) when c_2 is on, it is defined as $c_2 = G|\dot{x}_1|$.

In order to keep equation (5) in non-dimensional form, a non-dimensional parameter 'e' is defined as:

$$e = \frac{GF_0}{m_1^2\omega_1^2} \quad (7)$$

Substituting the parameter 'e' and c_2 into equation (5) results in the following non-dimensional set of equations of motion:

$$-g^2A_1 + 2ig\xi_1A_1 - g^2eA_1(A_1 - A_2) + A_1 + \mu f^2(A_1 - A_2) = 1$$

$$-\mu g^2A_2 - g^2eA_1(A_2 - A_1) + \mu f^2(A_2 - A_1) = 0 \quad (8)$$

3.2 Base-Excited System

For the case of based excited system, upon the substitution of (2) into the equations of motion and simplifications:

$$-g^2\bar{X}_1 + 2ig\xi_1\bar{X}_1 + ig\frac{c_2}{m_1\omega_1}(\bar{X}_1 - \bar{X}_2) + \bar{X}_1 + \mu f^2(\bar{X}_1 - \bar{X}_2) = -\frac{\ddot{X}_0}{\omega_1^2}$$

$$\quad (9)$$

$$-\mu g^2\bar{X}_2 + ig\frac{c_2}{m_1\omega_1}(\bar{X}_2 - \bar{X}_1) + \mu f^2(\bar{X}_2 - \bar{X}_1) = -\mu\frac{\ddot{X}_0}{\omega_1^2}$$

The displacement responses $\bar{X}_1$ and $\bar{X}_2$ in terms of non-dimensional amplification factors A_1 and A_2 are defined as:

$$\bar{X}_1 = A_1\frac{\ddot{X}_0}{\omega_1^2}$$

$$\quad (10)$$

$$\bar{X}_2 = A_2\frac{\ddot{X}_0}{\omega_1^2}$$

Upon substitution of (10) into equation (9), the equations of motions are:

$$-g^2A_1 + 2ig\xi_1A_1 + ig\frac{c_2}{m_1\omega_1}(A_1 - A_2) + A_1 + \mu f^2(A_1 - A_2) = -1$$

$$-\mu g^2A_2 + ig\frac{c_2}{m_1\omega_1}(A_2 - A_1) + \mu f^2(A_2 - A_1) = -\mu \quad (11)$$

As in the case of force-excited systems, c_2 is constant for passive TVAs, and defined as $c_2=2m_2\omega_2\xi_2$. Solution of equation (11) will result in the values of A_1 and A_2 in terms of non-dimensional parameters.

For the GHTVA, c_2 varies with time according to equation (1). When the damper is off, the damping coefficient c_2 is defined as $c_2=2m_2\omega_2\xi_{off}$. Substituting into equation (11) will result in a set of equations similar to ones for the passive TVA:

$$-g^2A_1 + 2ig\xi_1A_1 + 2ig\mu g\mu_{off}(A_1 - A_2) + A_1 + \mu f^2(A_1 - A_2) = -1$$

$$-\mu g^2A_2 + 2ig\mu g\mu_{off}(A_2 - A_1) + \mu f^2(A_2 - A_1) = -\mu \quad (12)$$

However, when c_2 is on, its value based on equation (1) is $c_2 = G|\dot{x}_1|$.

In order to keep equation (12) in non-dimensional form, a non-dimensional parameter 'e' is defined:

$$e = \frac{G\ddot{X}_0}{m_1^2\omega_1^2} \quad (13)$$

Substituting the parameter 'e' and c_2 into equation (11) results in the following non-dimensional format:

$$-g^2A_1 + 2ig\xi_1A_1 - g^2eA_1(A_1 - A_2) + A_1 + \mu f^2(A_1 - A_2) = -1$$

$$-\mu g^2A_2 - g^2eA_1(A_2 - A_1) + \mu f^2(A_2 - A_1) = -\mu \quad (14)$$

Here, it is shown that depending on the state of the system according to equation (1), the equations of motion switch between equations (6) and (8) for the force-excited, and equations (12) and (14) for the base-excited systems, in terms of non-dimensional parameters. Therefore, as will be shown later the parameters defined here can be used for the general design of GHTVAs.

4 PERFORMANCE OF THE GHTVA

In order to evaluate the efficiency of GHTVA in reducing the vibration of force and base excited systems, a non-dimensional parameter (p) representing the system performance is defined as follows:

$$p = \frac{(A_1)_{max} \; GHTVA}{(A_1)_{max} \; TVA} \quad (15)$$

For the GHTVA to perform better than its equivalent TVA (having, the same mass ratio as the GHTVA) the value of 'p' has to be less than 1.0.

The smaller the 'p', the more efficient GHTVA is as compared to its equivalent TVA.

5 PARAMETRIC STUDIES

In order to study and compare the behavior of GHTVA with its equivalent TVA and establishing guidelines for the design of GHTVAs, a system, with these dynamic properties is considered: $m_1 = 17.5$ N-sec^2/mm and $f_1 = 2$Hz. For the force-excited system it is assumed that $F_0 = 44,500$ N, and for the base-excited system $\ddot{X}_0 = 25.4$ mm/sec^2. It has to be noted that even though the results reported here are for a particular model, since the GHTVA design properties are defined in terms of non-dimensional parameters they can be applied to any dynamic system as will be shown later in the design guide section.   In order to study the behavior of the GHTVA with different mass ratio μ, m_2 was varied to range μ from 0.01 to 0.50. In addition the main system damping ratio,ξ_1, was assumed to be 0.0, 0.01, 0.05, and 0.10 (0.0, 0.01, 0.05 for base-excited system) to study its effect on the system behavior. In order to perform time history analyses of the system, a time step size of 0.005 second was chosen. This selection was based on the fact that it takes about 0.01 second for the valves in the semi-active dampers to respond from a fully open to a fully closed position (Miller 1988). Optimum GHTVA design parameters f, ξ_{off}, and e were computed.

Figure 2 shows the variation of the amplification factor for system with $\xi_1 = 0.01$ with various mass ratios (μ) for the force-excited and base-excited systems. The maximum main mass amplification factors (A_1) reduce with an increase in the mass ratios as can be expected in TVAs. In addition, from this figure it can be concluded that the optimality condition of TVA, which is having two equal peak amplitudes is also applicable to GHTVAs.

Figure 3 shows the variation of the main mass amplification factors with mass ratio (μ) of the GHTVA, for the force-excited and base-excited systems, respectively. Similar to the case of passive TVAs there is a much larger reduction in the amplification factor for the mass ratios of up to 10%. Beyond this point the increase in the secondary system mass (m_2) has less significant effect on the reduction of the main mass amplification factor.

Figure 4 shows the variation of the optimum frequency ratios (f) with respect to the mass ratio (μ) of the GHTVA, for the force and base excited systems, respectively. These figures demonstrate the similarity of the changes in the optimum frequency ratios for the two systems.

Figure 5 shows the changes in the optimum parameter (e) of the GHTVA with mass ratio (μ). Since the gain (G) is proportional to 'e', it can be

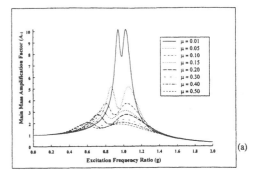

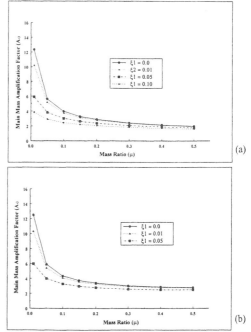

Figure 2. Variation of the GHTVA Main Mass Amplification Factor (A_1) with Excitation Frequency Ratio (g) for Different Values of μ[$\xi_1 = 0.01$] (a): Force excited system; (b): Base-excited system.

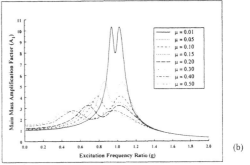

Figure 3. Variation of the GHTVA Main Mass Amplification Factor (A_1) with Mass Ratio (μ) for Different Values of ξ_1 (a): Force-excited system; (b): Base-excited system.

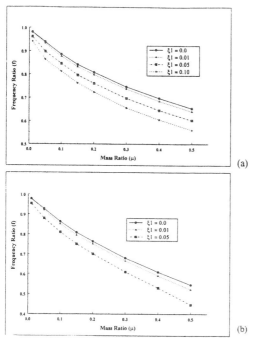

(a)

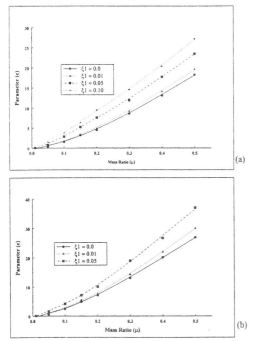

(b)

Figure 4. Variation of the Optimum Frequency Ratio (f) of the GHTVA with Mass ratio (μ) for Different Values of ξ_1 (a): Force-excited system; (b): Base-excited system.

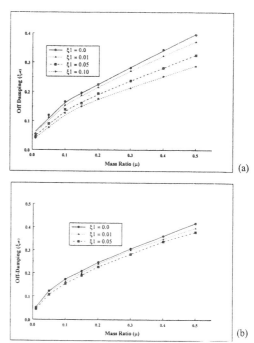

(a)

(b)

Figure 6. Variation of the Optimum Off-Damping Ratio (ξ_{off}) of the GHTVA with Mass Ratio (μ) for Different Values of ξ_1 (a): Force-excited system; (b): Base-excited system.

(a)

(b)

Figure 5. Variation of the Optimum Parameter (e) of the GHTVA with Mass Ratio (μ) for Different Values of ξ_1 (a): Force-excited system; (b): Base-excited system.

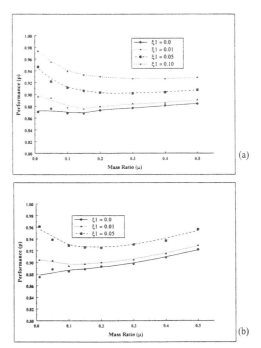

(a)

(b)

Figure 7. Variation of the Performance (p) of the GHTVA Main Mass with Mass Ratio (μ) for Different Values of ξ_1 (a): Force-excited system; (b): Base-excited system.

727

concluded that the optimum gain increases with the mass ratio (μ) and the primary system damping (ξ_1).

The variation of the optimum off-damping ratio (ξ_{off}) of the GHTVA with mass ratio (μ) is shown in Figure 6. The off damping as defined earlier increases with the mass ratio (μ) and decreases with the increase in the main system damping (ξ_1).

The variation of performance (p) as defined in equation (15) with respect to the mass ratio (μ) for different values of ξ_1 is shown in Figure 7.

As can be observed in Figure 7, the GHTVA can outperform an equivalent passive TVA. The GHTVA performs better with a decrease in primary system damping (ξ_1). In general, TVAs are effective only when the main system damping (ξ_1) is low. This fact shows that GHTVAs can be used as a substitute for passive TVAs. In the case studied here the GHTVA can outperform the equivalent TVA by about 14% when $\xi_1 = 0$, and about 12% when $\xi_1 = 0.01$ for the force-excited systems, and 12% when $\xi_1 = 0$, and 10% when $\xi_1 = 0.01$ for the base-excited systems.

6 DESIGN GUIDE FOR GHTVAs

In general, the first step in the design process is to select the GHTVA mass (m_2) or the mass ratio (μ) based on the desirable amplification factor. This can be accomplished using Figure 3. Once the mass ratio is selected, using Figures 4,5, and 6, the non-dimensional design parameters f_{opt}, e_{opt}, and $(\xi_{off})_{opt}$ can be found. Using equations (7 or 13), the optimum gain (G) can be computed, and therefore, the design parameters required in equation 1 can be obtained.

7 SUMMARY AND CONCLUSIONS

A single degree of freedom (SDOF) system subjected to harmonic force excitation was used to study the behavior of the Ground Hook Tuned Vibration Absorber (GHTVA). The optimum design parameters of the GHTVA were found for various mass ratios and system damping, and its performance was compared to the passive TVA with the same mass ratio. A procedure was proposed for the general design of GHTVAs.

From this study the following are concluded:
(1) A GHTVA can outperform an equivalent passive TVA in reducing the vibration. However, its effectiveness depends on the main system-damping ratio. The GHTVA with the control policy used here can perform up to about 14% better than its passive TVA counterpart. In addition, a GHTVA is less sensitive to off tuning due to changes in the frequency ratio (f) than an equivalent TVA.

(2) The optimum frequency ratio (f_{opt}) of the GHTVA decreases with the increase in mass ratio (μ) and main system damping ratio (ξ_1). However, the optimum value of parameter (e_{opt}) or optimum gain (G_{opt}) of GHTVA increases with the increase in the mass ratio (μ) and main system damping (ξ_1).

ACKNOWLEDGEMENTS

The research presented in this paper has been supported by the National Science Foundation under Grant No. CMS-9978610. This support is gratefully acknowledged.

REFERENCES

Abe, M. 1996. Semi-Active Tuned Mass Dampers for Seismic Protection of Civil Structures. *Earthquake Engineering and Structural Dynamics* Vol. 25: 743-749.

Ahmadian, M. 1997. Semi-Active Control of Multiple Degree of Freedom Systems. *Proc. of DETC'97, ASME Design Engineering Technical Conference*, Sept. 14-17, Sacramento, California

Brock, J.E. 1946. A note on the Damped Vibration Absorber. *Journal of the Applied Mechanics*, ASME 13, A-284.

Chang, J.C.H. & Soong, T.T. 1980. Structural Control Using Active Tuned Mass Dampers. *Journal of the Engineering Mechanics Division*, ASCE, Vol. 106, No. EM6, Proc. Paper 15882, December: 1091-1098.

Den Hartog, J.P. (3rd ed.) 1947. *Mechanical Vibrations*. New York: McGraw-Hill.

Frahm, H. 1911. *Device for Damping of Bodies*. U.S. Patent No. 989, 958.

Hrovat, D. & Barak, P. & Rabins, M. 1983. Semi-Active Versus Passive or Active Tuned Mass Dampers for Structural Control. *Journal of the Engineering Mechanics Division*, ASCE, Vol. 109, No. 3, June: 691-705.

Lund, R.A. (H.H.E. Leipholz ed.) 1980. Active Damping of Large Structures in Winds. *Structural Control*. New York: North-Holland Publishing Co.

Miller, L.R. 1988. *An Approach to Semi-Active Control of Multiple-Degree-of-Freedom Systems*. Ph.D. Thesis, Department of Mechanical and Aerospace Engineering, North Carolina State University, Raleigh, North Carolina.

Ormondroyd, J. & Den Hartog, J.P. 1928. *The Theory of the Dynamic Vibration Absorber*. Transaction of the American Society of Mechanical Engineers. Vol. 50: 9-22.

Udwadia, F.E. & Tabaie, S. 1981. Pulse Control of Single Degree of Freedom System. *Journal of Engineering Mechanics Division*, ASCE, Vol. 107, No. EM6, December: 997-1009.

Creative Systems in Structural and Construction Engineering, Singh (ed.) © 2001 Balkema, Rotterdam, ISBN 90 5809 161 9

Estimation of structural dynamic responses from partial measurements

Kyung-Taek Yang
Department of Mechatronic Engineering, Daelim College of Technology, Korea

Hak-Soo Kim
Structural Dynamics Laboratory, Honam University, Korea

ABSTRACT: In this study a method is presented for estimating dynamic responses of a structure at unmeasurable locations both in time and frequency domains. The prediction is based on measuring responses at accessible locations and transforming them into responses at desired locations using a state estimation technique. The methods are tested numerically and the feasibility for practical applications has been demonstrated through a flexible beam under moving loads, where translational and rotational dofs (degrees of freedom) of a beam at center point are estimated from partial measurements of responses at accessible points.

1 INTRODUCTION

After system identification technique has been introduced into the engineering field, performed were lots of studies to assess the structural stability by comparing the theoretical results with its experimental counterparts. Especially due to the remarkable developments in both the analyzing software and the testing equipments a lot of study results become available, however problems still remain in practical applications (P. Davis & J. K. Hammond 1984). One of the main reason for that is the incompleteness arising from linking the theoretical result with its experimental counterpart. Although a great number of dofs are required in order to establish an exact model for a structural system, the number of measurable dofs is limited. And when it comes down to rotational dofs engineers comprehend the importance of measurements, it is really difficult to measure them though. In order to overcome the difficulties, it is a pre-requisite to establish a reduced analytical model representing exactly the dynamic characteristics of the subject structure (Bruce, Irons 1965) and required is the technique estimating the dynamic responses of points whose measured data are not available because of the difficulty in measurement for rotational dofs or the limitation in the number of sensors (Daniel C. Kammer 1998).

In order to estimate the dynamic responses of dofs where sensor is not installed, suggested in this study is estimating technique in frequency domain by using measured data and reduced analytical model. State estimation equations in frequency domain are suggested in the expression of linear equations and the difficulties in calculating the repetitive matrix inversion was removed by adopting lambda-matrix of 4th order. In order to verify the feasibility of the suggested technique, for the beam under moving load rotational displacement was estimated for the center of the beam and the moment was calculated to be compared with the theoretical value.

2. MATHEMATICAL MODEL

Considering the stiffness and the damping properties of a structural system, equations of motion for the structural system having n dofs are as follows :

$$[M]\{\ddot{X}\} + [C]\{\dot{X}\} + [K]\{X\} = \{f\} \tag{1}$$

where, [M], [C], [K] are mass, damping and stiffness matrices of continuous system respectively.

For the establishment of condensed model to estimate the responses effectively, equation (1) may be transformed into equation (2) if we put Xm as master dofs and Xs as slave dofs.

$$\begin{bmatrix} [M_{mm}] & [M_{ms}] \\ [M_{sm}] & [M_{ss}] \end{bmatrix} \begin{Bmatrix} \{\ddot{X}_m\} \\ \{\ddot{X}_s\} \end{Bmatrix} + \begin{bmatrix} [C_{mm}][C_{ms}] \\ [C_{sm}][C_{ss}] \end{bmatrix} \begin{Bmatrix} \{\dot{X}_m\} \\ \{\dot{X}_s\} \end{Bmatrix}$$
$$+ \begin{bmatrix} [K_{mm}] & [K_{ms}] \\ [K_{sm}] & [Kss] \end{bmatrix} \begin{Bmatrix} \{X_m\} \\ \{X_s\} \end{Bmatrix} = \begin{Bmatrix} f \\ 0 \end{Bmatrix} \tag{2}$$

After solving the eigenvalue problem analytically using equation (2), if we transform the physical

Table. 1 Natural frequencies of full/reduced models
(Hz)

mode	1st	2nd	3rd	4th
Full model	3.6014	14.4069	32.4294	57.7107
Reduced model	3.6014	14.4069	32.4310	57.7106

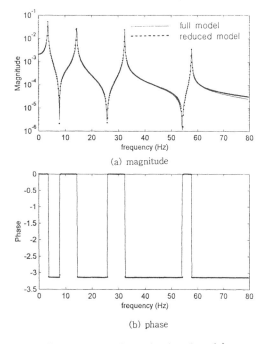

(a) magnitude

(b) phase

Fig. 3 FRFs of full and reduced models.

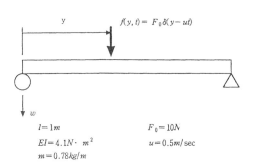

$l = 1m$ $F_0 = 10N$

$EI = 4.1N \cdot m^2$ $u = 0.5m/\sec$

$m = 0.78kg/m$

Fig. 1 Simply supported beam under moving load.

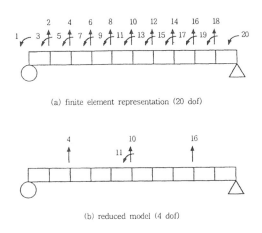

(a) finite element representation (20 dof)

(b) reduced model (4 dof)

Fig. 2 Finite element and reduced models.

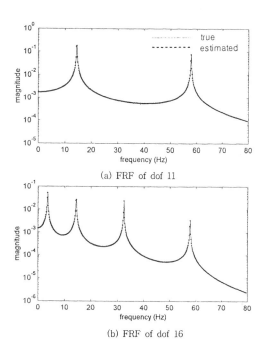

(a) FRF of dof 11

(b) FRF of dof 16

Fig. 4 Comparisons of true and estimated FRFs.

coordinate into modal coordinate in consideration of r modes within the frequency range of interest we have equation (3)

$$\{X\} = \begin{pmatrix} \{X_m\} \\ \{X_s\} \end{pmatrix} = [U]\{q\} = \begin{bmatrix} [U_m] \\ [U_s] \end{bmatrix}\{q\} \qquad (3)$$

herein, [U] is eigenvectors and [Um] is component of eigenvector for condensed coordinate and [Us] is component of eigenvector for coordinate to be removed. Condensing the coordinate by using equation (3), equation (4) is derived from equation (1) as follows (Robert J. Guyan 1965).

$$[m]\{\ddot{x}\} + [c]\{\dot{x}\} + [k]\{x\} = \{f\} \qquad (4)$$

where [m], [c], [k] are condensed mass, damping and stiffness matrices of order m respectively.

3. ESTIMATION OF DYNAMIC RESPONSES

If we express equation (4) of condensed analytical model in frequency domain through Fourier Transformation, we have equation (5) as follows

$$[D(\omega)]\{x(\omega)\} = \{F(\omega)\} \qquad (5)$$

where $[D(\omega)] = [k] - \omega^2[m] + j\omega[c]$

Equation (6) shows the relation between frequency response function and dynamic stiffness matrix.

$$[H(\omega)] = [D(\omega)]^{-1} \qquad (6)$$
$$= [[k] - \omega^2[m] + j\omega[c]]^{-1}$$

From the definition of frequency response function, we obtain equation (7) from equation (5)

$$[D(\omega)]\{h_l(\omega)\} = \{I_l\} \qquad (7)$$

where $\{h_0\}$ is the l th column of frequency response function and $\{I_0\}$ is the l th column of unit matrix [I]. By separating the measured responses(M) and those to be estimated(E), equation (7) can be rewritten as equation (8).

$$[D_M(\omega)D_E(\omega)]\begin{pmatrix} h_M(\omega) \\ h_E(\omega) \end{pmatrix} = \{I_l\} \qquad (8)$$

where $[D_M(\omega)] = [k_M(\omega)] + j\omega[c_M(\omega)] - \omega^2[m_M(\omega)]$

$$[D_E(\omega)] = [k_E(\omega)] + j\omega[c_E(\omega)] - \omega^2[m_E(\omega)]$$

In order to estimate unmeasured structural responses from the measured partial response in frequency

domain, we obtain equation (10) of loss function by defining equation error as equation (9).

$$\{\varepsilon(\omega)\} = \{I_l\} - [D_M(\omega)D_E(\omega)]\begin{pmatrix} h_M(\omega) \\ h_E(\omega) \end{pmatrix} \qquad (9)$$

$$Min\sum_k \{\varepsilon(\omega_k)\}^T\{\varepsilon(\omega_k)\} \qquad (10)$$

Equation (9) shows orthogonality between condensed analytical model and frequency response function. By minimizing the loss function, unmeasured responses for each frequency may be estimated from M measured responses and information of analytical model for the subject system as follows.

$$Min\{\varepsilon(\omega_k)\}^T\{\varepsilon(\omega_k)\} \qquad (11)$$

$$\{h_E(\omega)\} = [D_E(\omega)]^+\{\{I_l\} - [D_M(\omega)]\{h_M(\omega)\}\} \qquad (12)$$

Here []+ is a generalized inversion of a matrix and expressed as below.

$$[D_E(\omega)]^+ = [[D_E(\omega)]^T[D_E(\omega)]]^{-1}[D_E(\omega)]^T \qquad (13)$$
$$\text{(m-M) x (m-M) (m-M) x m}$$

Equation (12) is for estimating unmeasured structural dynamic responses and inverse Fourier transform gives us responses in time domain. FFT algorithm is mainly used for analysis in frequency domain and calculation is performed for 1024 frequency components by measuring equipments and software therefore if equation (12) is applied to actual problem, calculational load is high because of repetitive calculation of matrix inversion. In this study, lambda matrix of order 4 is introduced and calculation of matrix inversion is transformed into eigen value problem of higher order in order to reduce the computational burden.(Peter Lancaster 1966) The results are as below.

$$[D_E(\omega)]^+ = \left[\sum_{i=1}^{4(m-M)} \frac{\{r_i\}\{q_i\}}{j\omega - \lambda_i}\right][D_E(\omega)]^T \qquad (14)$$

where λ_i is the latent value of characteristic matrix.

r_i is the right latent vector.

q_i is the left latent vector.

4. NUMERICAL SIMULATIONS

4.1 Coordinate reduction

After solving the partial differential equation for the beam under moving load (Fo=10 N, u=0.5 m/sec)

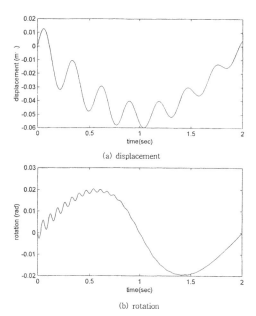

(a) displacement

(b) rotation

Fig. 5 Estimated vertical displacement and rotation at y = 0.5 m.

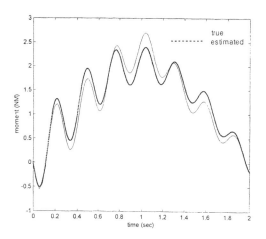

Fig. 6 Estimated moment of a beam under moving load at y = 0.5m.

in Fig. 1, we assume the vertical displacement at y=0.2 m as measured value and estimated are the vertical and the rotational displacements at center by combining the measured value and finite element model of the beam.

Fig. 2.(a) shows the finite element model of 20 degrees of freedom, which is condensed into a

model of 4 dof as shown in Fig. 2(b) by eliminating 16 dofs. Four coordinates are selected as master dofs and the frequency response functions are compared in Fig. 3.

$$\{X_M\} = \{X_4, X_{10}, X_{11}, X_{16}\} \qquad (15)$$

Modal parameters before and after coordinate reduction are shown in table 1 and the condensed model represents the dynamic characteristics of the full model.

4.2 State Estimation

By predicting the frequency response functions of dofs 10, 11, 16 from measured response of dof 4 through equations (10), we obtain Fig. 4 which shows good agreement with theoretical values. Especially in the case of dof 11 of rotational dof at center of the beam, only the 2nd and the 4th modes are reflected in dynamic responses and the 1st and the 3rd modes have no influences. It is because rotational dof at center of the beam becomes nodal point in the odd-numbered modes due to the symmetry of the beam. If we obtain the responses of dof 10 and dof 11 of the vertical and the rotational displacements at the center respectively through Inverse Fourier Transform of estimated frequency responses, then we could have Fig. 5 showing good relation with theoretical values. Also we can calculate moment of the beam from the estimated rotational displacement by using the following equation.

$$\frac{M(y,t)}{EI} = \frac{\partial \theta(y,t)}{\partial y} \qquad (16)$$

Fig. 6 shows the change of moment with time at center of the beam by differentiating the estimated rotational displacement by using equation (16) and it shows almost similar trend to the theoretical value with partial error. This is because only 4 modes are contributed to estimate the state although unlimited modes are reflected in theoretical value. In case of beam moment is an important barometer to assess the structural capability and if it is under moving load, two kinds of components, that is, the static and the dynamic components by moving load co-exist in moment as shown in Fig. 6.

5 CONCLUSIONS

In this study, reduced model is derived by condensing finite element model of subject structural system to estimate the dynamic responses of points where the measurements is not available and suggested are estimation technique of dynamic responses in frequency domain by using the measured data and the reduced model. The suggested technique is applied to the behavior of the beam under moving load and the estimated results are compared with the theoretical values.

Conclusions are as follows.

1. While estimation of dynamic responses in time domain is reduced to nonlinear state estimation problem, which becomes linear in frequency domain, λ-matrix of order 4 is adopted in this study to avoid calculational burden in repetitive generalized inverse at each frequency.

2. Whatever kind of sensors are used for measuring the rotational displacements of a structural system, it is difficult to measure them because of its characteristics however, possible is comparatively accurate estimation of the dynamic responses of rotational dofs by using the state estimation technique suggested in this study and the moment is also obtained through numerical differentiation.

REFERENCES

Irons, Bruce "Structural Eigenvalue Problems : Elimination of Unwanted Variables" AIAA Journal, Vol. 3(5), 1965, pp. 961-962.

Kammer, Daniel C. "Estimation of Structural Responses Using Remote Sensor Locations", 17th International Modal Analysis Conference, vol. 2, 1998, pp. 1379-1385.

Davies, P. and Hammond, J. K. "A Comparison of Fourier and parametric Methods for Structural System Identification", ASME Journal of Vibration, Acoustics, Stress and Reliability in Design,, Vol. 106, 1984, pp. 40-48.

Lancaster, Peter "Lambda-matrices and Vibrating Systems," Chapter 3, PERGAMON PRESS, 1966.

Guyan, Robert J. "Reduction of stiffness and Mass Matrices," AIAA Journal, vol. 3(2), 1965, pp. 380.

23 Seismic response of structural elements

Creative Systems in Structural and Construction Engineering, Singh (ed.) © 2001 Balkema, Rotterdam, ISBN 90 5809 161 9

A simplified approach to the analysis of torsional problems in seismic base isolated structures

T.Trombetti, C.Ceccoli & S.Silvestri
DISTART, Università di Bologna, Italy

ABSTRACT: Structures characterized by eccentricity in plan between the center of mass and the center of rigidity, under seismic excitations, develop torsional couples which may increase the level of dynamic loading, when compared to those developed in systems characterized by coincident centers of mass and of rigidity. This rotational behavior plays a major role in the determination of the maximum deformation developed in base seismic isolators. Two different types of isolators models have been considered in this paper: both a linear elastic and bi-linear. A new simplified approach to solve both problems is presented. In spite of the simplicity of the method and of the approximations introduced, the estimations obtained are very close to those obtained though a complete three dimensional analysis thus showing a high degree of reliability and stability of the simplified method of analysis here proposed.

1 INTRODUCTION

The dynamic behavior of eccentric structures has been the object of extensive research in the past decade both in the linear and non-linear domains. In spite of these research efforts, a number of still unresolved problems prevent the full comprehension of dynamic lateral torsional coupling in structures.

The research performed to date can be subdivided into three categories:

1. Investigation of the linear elastic response of full three-dimensional laterally-torsionally coupled systems (e.g. Hejal & Chopra 1987).
2. Investigation of the inelastic dynamic response of three-dimensional single- and multi-storey eccentric structural systems (e.g. Goel & Chopra 1990).
3. Analysis and evaluation of design code provisions for torsional effects. This body of studies mainly focuses on the development of simplified analysis and design procedures for laterally-torsionally coupled structures (e.g. Chandler & Hutchinson 1987).

The laterally-torsionally coupled behavior of seismic isolated structures can be fairly simplified due to the following reasons:

1. Most common seismic isolators are cylindrical elements whose lateral stiffness is generally well known and independent of the direction of deformation.
2. The dynamic behavior of seismic isolated structures can be fairly well captured through a simplified linear analysis in spite of the inherent non-liner force-deformation relationship of most common isolators (HDRB).
3. Under seismic excitation, the deformations of a seismic isolated structure are localized mainly in the isolators and are only marginally influenced by the interaction of the superstructure.

The new approach presented here and the resulting simplified analysis procedure enables to both understand qualitatively and estimate quantitatively the laterally-torsionally coupled response of specific eccentric systems.

2 THE ISOLATED BUILDING MODEL

Lateral-torsional coupled behaviors of base-isolated buildings can be effectively captured modeling the structure as a simple rigid deck, that can move in its own plane resting over the isolators. Three degrees of freedom are sufficient to describe this dynamic system: the displacements along two perpendicular directions and the rotations around a vertical axis.

If there is eccentricity in plan between the center of mass and the center of rigidity, torsional moments are developed under seismic excitations, which considerably increase the level of dynamic loading and deformation.

Isolators are generally characterized by non-linear behavior. In this paper two different modeling of the isolators are adopted: a linear elastic one and a bi-linear one (Fig. 1).

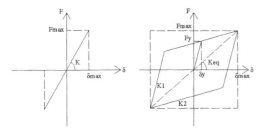

Figure 1. Linear and bilinear modeling of the isolators.

3 LINEAR ELASTIC ANALYSIS

The linear elastic response of a one-storey eccentric system in free vibrations from a given initial displacement are analyzed first.

The ratio between the maximum rotation, θ_{max} and the maximum lateral displacement, y_{max} developed in free vibration in the direction of the initial deformation by such system is termed as the α ratio:

$$\alpha = \left(\frac{\theta_{max}}{y_{max}} \right)_{f.v.} \tag{1}$$

It can be shown that, for a reference system centered at the center of the mass (Fig. 2), α depends only upon the physical characteristics of the structure, namely upon ρ_M, γ, e, ξ.
Where ρ_M is mass the radius of inertia of the superstructure with respect to the center of mass,

$$\gamma = \frac{\omega_\theta}{\omega_L} \tag{2}$$

$$\omega_L = \sqrt{\frac{k}{m}} \tag{3}$$

$$\omega_\theta = \sqrt{\frac{k_{\theta\theta}}{I_G}} \tag{4}$$

$$k = \sum_{i=1}^{N} k_i \tag{5}$$

$$k_{\theta\theta} = \sum_{i=1}^{N} k_i \left(x_i^2 + y_i^2 \right) \tag{6}$$

m is the total mass of the superstructure,
I_G is the polar mass moment of inertia with respect to the z-axis which passes through the center of mass,
N is the number of the isolators,

$$e = \sqrt{e_x^2 + e_y^2} \tag{7}$$

e_x is the transversal eccentricity (with respect to the seismic entrance direction):

$$e_x = \frac{E_x}{D_e} \tag{8}$$

E_x and E_y are the structural eccentricity along the x- and y- directions respectively,
D_e is the "equivalent diagonal" of the system defined as

$$D_e = \sqrt{12} \cdot \rho_M \tag{9}$$

ξ is the viscous damping ratio.
A number of numerical and experimental simulations have shown that the α ratio is a robust and stable parameter almost excitation independent, i.e.:

$$\alpha \cong \left(\frac{\theta_{max}}{y_{max}} \right)_{forced} \tag{10}$$

Since a number of research work have suggested that the maximum displacement, $y_{max}(e)$, developed under seismic excitation at the center of mass by laterally coupled (eccentric) systems is very similar to the maximum displacement, $y_{max}(e=0)$, developed by the corresponding non eccentric system (SDOF), the following approximation can be taken into consideration:

$$y_{max}(e) \cong y_{max}(e=0) \tag{11}$$

Therefore it is possible to effectively estimate the maximum rotation of a given eccentric system under seismic excitation as follows:

$$\theta_{max} \cong \alpha \cdot y_{max}(e=0) \tag{12}$$

Furthermore the sensitivities of the α ratio to various system characteristics provide invaluable guidance for optimum system design.
From a number of numerical and experimental simulations (Trombetti 1994, 1998), it can be concluded that the torsional factor α:
- for a given value of transversal eccentricity e_x, decreases for increasing values of the longitudinal eccentricity e_y (in the direction of the bases seismic excitation);
- increases for increasing values of the transversal eccentricity e_x (mainly for small values of eccentricities);
- is inversely proportional to the equivalent diagonal D_e (i.e. to the dimensions) of the system;
- increases as the γ parameter tends to unity;
- decreases slightly for increasing values of damping
- for undamped structures has the following closed form expression:

$$\alpha = \frac{1}{\rho_M} \frac{e_x}{e} \frac{\sqrt{48 \left(\frac{e}{\gamma^2 - 1} \right)^2}}{\sqrt{48 \left(\frac{e}{\gamma^2 - 1} \right)^2 + 1}} \tag{13}$$

– for damped structure is bounded from above by the following expression:

$$\alpha_d \leq \frac{\Lambda}{2} \alpha \cdot \max \left\{ \exp\left(-\xi \frac{\pi}{2} \sqrt{\frac{\Omega_1}{\Omega_3}} \right) + \exp\left(-\xi \frac{\pi}{2} \right) \right\} \qquad (14)$$

where, in case of $\xi_1 = \xi_2 = \xi_3$:

$$\Lambda = \sqrt{1 + \frac{\xi}{\sqrt{1 - \xi^2}}} \qquad (15)$$

Furthermore, a new parameter α'', defined as follows, can be introduced:

$$\alpha'' = \alpha \cdot \frac{D_e}{2} = \frac{\theta_{max} \cdot \frac{D_e}{2}}{y_{max}} = \frac{y_{transv.}}{y_{max}} \qquad (16)$$

The α'' factor is independent from the dimensions of the structure and allows also a precise physical meaning for the numerical value of the α'' parameter: it represents the ratio of the displacement induced by the rotational response at the corner of a rectangular structure (in the direction orthogonal to the diagonal) to the displacement of the center of the mass (which is common to all supports) (Fig. 2).

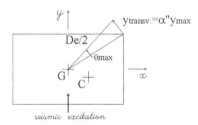

Figure 2. Analytic model. Transversal displacement and factor α''.

4 NON-LINEAR ANALYSIS

A bilinear modeling of the isolators is considered. In extending the research to the non-linear field, it is necessary to perform numerical non linear simulations and no closed form exact solutions for the α ratio can be obtained.

When a non linear response of the isolators is developed, this results in a modification of the stiffness matrix, thus resulting in a change of the position of the center of rigidity that cannot be univocally defined any more.

A non uniform, trapezoidal, distribution of the mass has been considered, in order to simulate dynamic system that are characterized by values of eccentricities between the center of mass and center of stiffness (in the initial, linear elastic conditions) in the range between 0 and 24% of the equivalent diagonal. As for structures characterized by a regular grid of isolators is simple to compute the rotational stiffness with respect to the center of the initial linear stiffness ($k_{\theta\theta}'$), it is convenient to define γ' parameter as:

$$\gamma' = \frac{\sqrt{\frac{k_{\theta\theta}'}{I_C}}}{\omega_L} \qquad (17)$$

where I_C is the polar mass moment of inertia with respect to the center of the initial linear stiffness.

The mass and the inertia of the all superstructure are condensed in the center of the mass, which is taken as origin of the reference system (Fig. 2). Systems characterized by transversal eccentricity e_x only are considered.

The new parameters, which have to be introduced in the non-linear field, are T_L, ψ, Ip, δ_y, SHR.

T_L is the "effective" longitudinal period which is assumed to be 2 sec with $Ip=5$.

ψ is the shape factor of the building's base:

$$\psi = \frac{a}{b} \qquad (18)$$

Ip is the plastic index, which indicates the level of the plastic excursion:

$$Ip = \frac{\delta_{max} - \delta_y}{\delta_y} \qquad (19)$$

δ_y is the elastic limit displacement.
SHR is the strain hardening ratio, between the plastic and the elastic rigidity, assumed to be 0.1:

$$SHR = \frac{k_2}{k_1} \qquad (20)$$

4.1 *Values of α'' under various dynamic inputs*

The non-linear analysis has started with the investigation of the values taken by the ratio α'' under a number of different dynamic excitations.

The following dynamic responses have been considered: free vibrations from given initial displacement (simulated with a first linear and then constant force), free vibration from a given initial velocity, harmonic excitation, white noises excitation, earthquake excitation. Figure 3 shows the results obtained: it is clear that the values of the ratio obtained for seismic excitation and for harmonic loading provide an upper bound. The average of the 11 considered ground motions is taken as reference.
Note that when the plastic index reaches high values, all situations here considered produce tightly bounded values of the α'' parameter.

4.2 *The α'' factor in the plastic range*

The dependence of factor α'' upon parameters e, Ip, γ' is presented in the graphics of Figure 4. α''

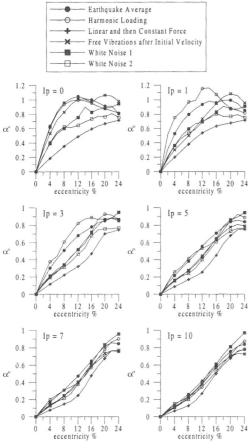

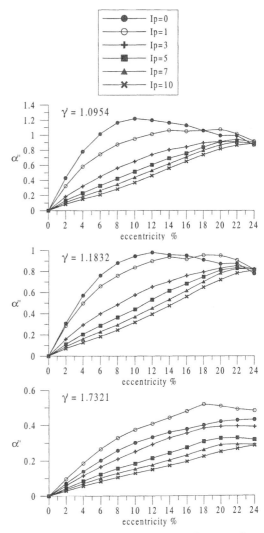

Figure 3. Responses under different seismic inputs for a square-shape structure with γ'=1.1547.

Figure 4. Plastic behavior of the α'' ratio for square-shape structures with values of γ' equal to 1.0954, 1.1832, 1.7321.

doesn't increase always proportionally to the increase in eccentricity: this is true only for values of e≤12%, while the α'' curves tend to the same value in correspondence of e=24% (γ'=1.7321 is an exception). The curves lower for increasing values of the plastic index Ip, which confirms that plasticity plays the role of hysteretic damping. It is interesting to notice that α'' decreases for increasing values of γ'. High values of γ' characterize structures with sparse isolators mesh, e.g. with few isolators located on the perimeter. In analogy to a flexional resistant section (where it is worth to centrifugate the masses in order to obtain a major inertia), in base isolated buildings is better to centrifugate isolators in order to limit the rotations of the storey (compatibly with vertical loads).

Figure 5 shows the ratio between α''$_{plastic}$ with Ip>0 and α''$_{elastic}$ with Ip=0. For common values of parameter γ', e.g. γ'<1.2 (almost uniform distribution of isolators), the elastic value of α'' is always higher than any plastic one. It is a very important result that could allow for a simple linear modeling of the isolator, which is in favor of security, instead of a more complicate bilinear one. However, there are some exceptions which concern the cases of high γ' and very low Ip. In these conditions, respectively, the isolators are too few and the hysteretic cycles are too narrow to produce a sufficient damping.

4.3 Sensitivity of the maximum longitudinal displacement to system eccentricity

The usefulness of Eq. 12 for maximum rotational response estimation relies on the fact that

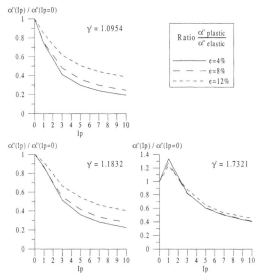

$\alpha'(Ip) / \alpha'(Ip=0)$

$\gamma = 1.0954$

Ratio $\dfrac{\alpha'\ \text{plastic}}{\alpha'\ \text{elastic}}$

——— $e=4\%$
— — — $e=8\%$
- - - - $e=12\%$

$\alpha'(Ip) / \alpha'(Ip=0)$

$\gamma = 1.1832$

$\alpha'(Ip) / \alpha'(Ip=0)$

$\gamma = 1.7321$

Figure 5. The "plastic reduction" of α'' from the elastic range to the plastic one.

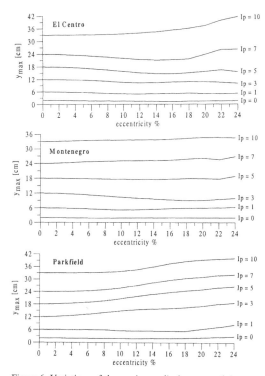

Figure 6. Variations of the maximum displacement of the center of mass with eccentricity, for a square-shape structure with $\gamma'=1.1832$, under the ground motions of El Centro 1940 NS 0.31g, Montenegro 0.36g and Parkfield 0.49g.

$y_{max}(e) \cong y_{max}(e=0)$ As this was shown to be true for linear elastic systems, investigations have been performed to verify that this approximation apply also for non–linear systems. The results obtained are shown in Figure 6 where the maximum displacement developed under seismic excitation by different systems are plotted versus the system eccentricity. Six curves are represented in every diagram, each being relative to one specific value of the plastic index. The curves of the maximum displacement remain almost horizontal thus showing a non sensitivity of the maximum displacement to a variation of the system eccentricity, this result being particularly evident for $e<12\%$.

4.4 The proposed method

A formulation of α'' for non linear systems is obtained through least squares fitting the mean values of the ratio α'' obtained numerically under a number of earthquake excitation.

For $e \leq 12\%$ two different formulations are possible. The first one gives immediately the value of α'', once the parameters e, Ip, γ' are known:

$$\alpha''(e,\gamma) = -0.75e + 11.87\frac{e}{\gamma'^2} + 12.26\frac{e}{\gamma'^4} - 125.60\frac{e^2}{\gamma'^5} \quad \text{Ip=1}$$

$$\alpha''(e,\gamma) = -1.01e + 10.84\frac{e}{\gamma'^2} + 1.79\frac{e}{\gamma'^4} - 42.63\frac{e^2}{\gamma'^5} \quad \text{Ip=3}$$

$$\alpha''(e,\gamma) = -0.93e + 8.80\frac{e}{\gamma'^2} - 0.27\frac{e}{\gamma'^4} - 16.68\frac{e^2}{\gamma'^5} \quad \text{Ip=5}$$

$$\alpha''(e,\gamma) = -0.71e + 7.24\frac{e}{\gamma'^2} - 0.99\frac{e}{\gamma'^4} - 3.93\frac{e^2}{\gamma'^5} \quad \text{Ip=7}$$

$$\alpha''(e,\gamma) = -0.75e + 6.81\frac{e}{\gamma'^2} - 2.21\frac{e}{\gamma'^4} + 5.33\frac{e^2}{\gamma'^5} \quad \text{Ip=10}$$

$$(21)$$

The second way consists in giving the ratio between the researched plastic value of α'' and its elastic value, which is given by Eq. (13):

$$\frac{\alpha''_{plastic}}{\alpha''_{elastic}} = \frac{1}{1 + q(e,\gamma') \cdot Ip} \quad (22)$$

where $q \cong 0.2 \div 0.4$, particularly:

$$q(e,\gamma) = 2.29 + 3.56e\gamma'^2 - 80.04\frac{e^2}{\sqrt{\gamma'}} + 236.10e^3 - 1.76\gamma' \quad (23)$$

For $e>12\%$, a parabolic interpolation between $\alpha''(e=0.12)$ and $\alpha''(e=0.24)$ is suggested:

$$\alpha''(e) = a \cdot e^2 + b \cdot e + c \tag{24}$$

where:

$$a = \frac{\alpha''(e=0.24) - \alpha''(e=0.12)}{(0.12)^2} - \frac{\dfrac{d\alpha''}{de}(e=0.12)}{0.12} \tag{25}$$

$$b = -\frac{\alpha''(e=0.24) - \alpha''(e=0.12)}{0.06} + 3 \cdot \frac{d\alpha''}{de}(e=0.12) \tag{26}$$

$$c = \alpha''(e=0.24) - 0.24 \cdot \frac{d\alpha''}{de}(e=0.12) \tag{27}$$

$$\alpha''(e=0.24) = -0.822 \cdot \gamma' + 1.789 \tag{28}$$

Since the α'' ratio is known, the rotation should be calculated as following:

$$\theta_{max} = \frac{\alpha'' \cdot y_{max}(e=0)}{D_e/2} \tag{29}$$

and the maximum additional displacement due to the rotational response can be easily calculated with geometric considerations.

5 SUGGESTED DESIGN PROCEDURE FOR ISOLATORS

The α'' method allows the following simplified design procedure for seismic isolators:

1 Given the planer characteristics of the structure to be seismic isolated the location of the isolators is know, hence the exact value of γ' and N (number of isolators);

2 the mass of the building allows to calculate the equivalent linear rigidity, which is to be attributed to isolators in order to obtain the desired effective longitudinal period of the isolated structure;

3 a linear analysis of the non-eccentric structure under the design ground motion is performed to estimate the maximum isolator deformation;

4 an isolator, which reaches this maximum displacement with a low shear strain (e.g. $\gamma_t = 100\%$) and which presents the previous equivalent rigidity for lower displacements (e.g. $\gamma_t \cong 60\%$ or $Ip \cong 5$), can be then selected;

5 once the specific isolator to be used is selected (using the above described criteria), the actual isolator behavior can be modeled as bi-linear,

6 the maximum displacement of the non eccentric bi-linear structure can be evaluated through non linear dynamic analysis, thus obtaining the appropriate design value of the plasticity index Ip;

7 it's now possible to apply the α'' method in order to estimate the maximum rotation of the actual eccentric structure and the maximum displacement of the most charged edge point;

8 it can be then verified that the corner isolators can sustain the obtained displacement with an acceptable level of strain. If this is not the case, it is necessary to choose a different base isolator and repeat the procedure.

6 CONCLUSIONS

1 A simplified estimation procedure for the maximum rotational response of base isolated eccentric buildings structure under seismic excitation is proposed. It involves only the determination of the torsional factor α (or α'') and of the maximum deformation of an "equivalent" SDOF oscillator, which is simple and much faster (from a computational viewpoint) than a complete three-dimensional dynamic response analysis.

2 This approach allows a general comprehension of the torsional response behavior of laterally-torsionally coupled dynamic systems both in elastic and in plastic range. Analysis of the sensitivity of torsional factor α upon system parameters provide precious guidance for the design of base isolated structures and understanding of their dynamic response.

3 The "elastic" value of the torsional factor α (here provided in closed form) represents an upper bound of the corresponding plastic torsional factor (with few exceptions for systems characterized by high γ' and very low Ip).

REFERENCES

Hejal, R. & Chopra, A.K. 1987 *Earthquake response of torsionally coupled buildings.* Report UCB/EERC-87/20, Earthquake Engineering Research Center, Univesity of California, Berkeley, CA, Dec 1987.

Goel, R.K. & Chopra, A.K. 1990. *Inelastic seismic response of one-storey, asymmetric-plan systems: effects of stiffness and strength distribution.* Earthquake Engng. Struct. Dyn., 19(3), 949-970.

Chandler, A.M. & Hutchinson, G.L. 1987. *Code design provisions for torsionally coupled buildings on elastic foundation.* Earthquake Engng. Struct. Dyn., 15, 517-536.

Trombetti, T. 1994. *Un approccio semplificato all'analisi dei problemi torsionali negli edifici isolati sismicamente alla base.* Estratto dal "Giornale del genio civile", Fascicolo 10°-11°-12° - Ottobre-Novembre-Dicembre 1994.

Trombetti, T. 1998. *Experimental/analytical approaches to modeling, calibrating and optimizing shaking table dynamics for structural dynamic applications.* Rice university, Houston, Texas.

Creative Systems in Structural and Construction Engineering, Singh (ed.) © 2001 Balkema, Rotterdam, ISBN 90 5809 161 9

Seismic evaluating of bridges using ambient vibration measurement

Y.G.Wang
Department of Basic Courses, Chang'an University, People's Republic of China

J.H.Zhang
Department of Engineering Mechanics, Xi'an Jiaotong University, People's Republic of China

ABSTRACT: The main objective of this study is to develop a seismic evaluation procedure of the existent bridge. To realize this, a bridge on the Ba River in the eastern suburb of Xi'an was tested. The scope of test covers three aspects as follows: ambient vibration measurement, estimating modal parameters and dynamic strength checking. The field measurements were conducted in three traffic states. The least inexpensive way is achieved in the busy traffic state, because it is not required to control the traffic. The seismic evaluation of the existent bridge includes the outward damage assessment and intensity evaluation. The latter is presented in this paper. The evaluating process is done by a simple model with one degree of freedom. The parameters of this model are determined with the estimated modal parameters. The maximum moment is computed for checking the maximum press stress. This simple assessment model is based on the Chinese code. The update finite element method ought to be studied further.

1 INTRODUCTION

The Ba River Bridge on National Highway 312 in the eastern suburb of Xi'an was tested and studied in the period ranging from May to October, 1999. A part of this study — the experimental ambient modal analysis is presented in this paper.

The Ba River Bridge, as shown in the picture in Fig 1, was built in 1978.The structure of the bridge with a length of 423m and 24 spans is one of the beam type. Since the sand round the piers was removed illegally, the foundation of the bridge was damaged. Although reinforced in 1998, the reliability of the bridge has to be analyzed, especially, for seismic evaluation

An ambient vibration method is employed in the evaluation procedure. The ambient vibration has been measured with the traffic controlled. The

Fig.1 BA River bridge

vibration response data of the piers has been recorded in three states: (a) free traffic on the bridge at midnight, in other words, when the excitation to the bridge is from the earth wave; (b) in heavy traffic in the afternoon, i.e. when the bridge is excited by the automobiles mainly; (c) the bridge is excited by a truck weighing 13 tons at three speeds with the traffic closed. The natural frequencies and mode shapes are estimated with the comprehensive PSD-based method developed by authors. The frequencies and /or mode shapes are used to calculate the earthquake forces and the maximum bending moment, which are employed to evaluate the seismic performance of a bridge. The procedure is worked out with an expert system ESSE (Expert System for Seismic Evaluation of structures). The ambient experimental modal analysis has studied by many authors (Bao and Ko, 1991, Zhu and Zhang, 1999, Feng, 1998). Since the ambient vibration is used without the excitation measurement, there is no mathematical formulation between the measured data and the modal parameters. The random decrement technique is discussed and applied to the ambient modal analysis (Feng, 1998). But the random decrement signal is the auto-correlation function of the response vibration. It is free vibration only when the excitation is a white noise (Zhang, 1984). In engineering, the power spectrum is applied extensively to estimate the modal parameters. In the bridge engineering, e.g. in 1985 Abdel-Ghaffar   and   Scanlan   presented   the ambient vibration of Golden Gate bridge; in 1997, Harik etc. presented the field testing of Brent-Spence Bridge and the ambient vibration properties, which are used for the seismic evaluation of the bridge. It is correct for the stationary vibration only when the FFT and the smooth technique are employed. The modern spectral analysis can be used, e.g. the maximum entropy spectrum based ARMA model and MUSIC spectrum estimation (Schmidt, 1986). In 1998, Zhu proposed a network method to estimate the transfer matrix with the responses at two positions of bridge. The transfer matrix data can be used to calculate the modal parameters, but difficulties arise from the data for training the network.

The seismic evaluation ought be done by the seismic response analysis, which can be performed by time-history method and/or response spectrum with the finite element model or one/two-DOFs model (Harik etc, 1997). In the paper, the following Chinese standards for the maximum bending moment with the first modal shape are used as the check criteria for the piers. The evaluated quantity depends on the types of the bridges. The maximum shear stress or bending moment is employed for the strength checking, and the maximum displacement is used for beam fall checking.

The primary aim of the study is to develop a seismic evaluation method for bridges and an expert system, but only the former is presented here.

2 AMBIENT VIBRATION MEASUREMENT

The ambient vibration on Piers No.3 and No.7 from the western bank was measured with a special vibration-meter, which has a high sensitivity in order to pick up the micro response. A digital magnet recorder was employed to record a great amount of data.

2.1 Measurement Positions and Recorded States

The horizontal vibration of the piers in the same di-

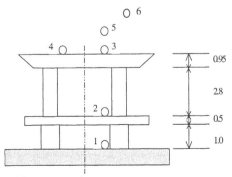

Fig.2 Measurement points

rection as the longitudinal bridge was measured with four sensors placed as shown in Fig.2. To compare with Point No.3 a sensor was set at Point No.4. The results have shown that the magnitude and phase of the vibration at Points No.4 and No.3 are just the same, and there is no torque deformation in the piers. The vibration data at Points No.1, 2 and 3 have been processed and used for seismic evaluation. Points No.5 and 6 are on the top of bridge and at two neighboring spans for measuring the vertical vibration of the super-structure.

Three states were designed as indicated in Table 1, and the ambient vibrations were recorded respec-

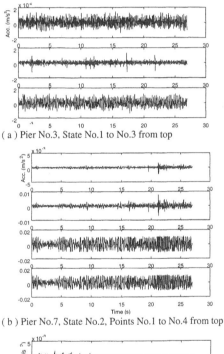

(a) Pier No.3, State No.1 to No.3 from top

(b) Pier No.7, State No.2, Points No.1 to No.4 from top

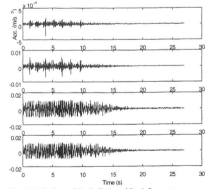

(c) Pier No.7, State No.3, Points No.4 from top

Fig.3 Ambient vibration on Pier No.7, (State No.3, Point 1-4 from top)

744

Table 1. Three states for vibration record

State No.	Excitation states	Recorded period	Note
1	Ground microseism	39 min	About at 12 PM, free traffic
2	Traffic	53 min	From 5 PM to 1 am, in several periods, busy traffic
3	Special truck	98 s	Free Traffic truck weighing 13 Tone, at speed of 30 40 50 and 80 Km/h

tively. For the different recorded states, the vibration magnitude and property are very different. Fig. 3 is a sample of the ambient response accelerations.

The data in State No.1 is of stationary random vibration. The ambient vibration in State No.2 can be considered as a stationary stochastic process if long data recording is processed with large average times. However, it is better to use short recording for saving purposes. The vibration in State No.3 is the instantaneous response of a time-variant system with a moving truck.

2.2 Estimation of Ambient Modal Parameters

For a vibration system, the modal parameters are defined with the eigen values and eigen vectors of the mass, stiffness, and damping matrices and estimated with the excitation and responses or with the frequency response function. They can not be determined only with the forced vibration response except for ideal impulse or white noise excitation. The modal parameters estimated with the ambient vibration measurement are only approximate. The modal parameters could be defined anew in order to apply the renewed concept of the mode to the ambient modal analysis. Here estimating modal parameters are also of the classical concept. Therefore, there are some key problems that have to be resolved:

1) How to judge the natural frequencies by the spectrum analysis? When is the Auto- or cross-power spectrum density (PSD or CPSD) used? The peak frequencies of PSD are only resonant frequencies, and they are polluted by the measurement noise. The modal shapes are dependent strongly on a disturbance of the natural frequencies. In this paper a comprehensive method to estimate accurately the natural frequencies is deveped, in which the phase of the CSPD and MUSIC spectrum is employed.

2) It is very important to suppress the measurement noise for estimating natural frequencies from the ambient vibration. The MUSIC method of spectrum estimation was used for getting a smooth

spectrum. The data are separated into the damped exponential components by using the MUSIC method. There fore, a short data series with more noise can be used and the natural frequencies can be estimated with higher accuracy.

3) For the different measurement states in Table 1, the properties of the ambient vibration signal are distinct from each other so that the different estimators of spectrum have to be employed. For the ambient vibration in states No.1 and No.2 there is no problem with the employment of the classical estimator of PSD with data samples long enough. But for State No.3, since the data series are non-stationary and very short, the MUSIC method of pseudo PSD can give a better result with a sharp peak of spectrum. The natural frequencies and damping ratio as well as the modal shapes can be calculated with the eigen values and eigen vectors of the data matrices in the MUSIC method. Of course, the MUSIC method is also available for States No.1 and 2.

2.2.1 Comprehensive PSD-based Method (CPM)

When the linear structures are with small damping and without mode couple, the following relationships are satisfactory

$$\hat{\Phi}_1 = \frac{S_{x_i x_j}(\omega_r)}{S_{x_j x_j}(\omega_r)} \cong \frac{\varphi_{ir}\varphi_{jr}}{\varphi_{jr}\varphi_{jr}} = \frac{\varphi_{ir}}{\varphi_{jr}} \tag{1a}$$

$$\hat{\Phi}_2 = \frac{S_{x_i x_i}(\omega_r)}{S_{x_i x_j}(\omega_r)} \cong \frac{\varphi_{ir}\varphi_{irr}}{\varphi_{ir}\varphi_{jr}} = \frac{\varphi_{ir}}{\varphi_{jr}} \tag{1b}$$

$$\hat{\Phi}_3 = \sin g(S_{x_i x_j}) \sqrt{\frac{S_{x_i x_i}}{S_{x_j x_j}}} \cong \frac{\varphi_{ir}}{\varphi_{jr}} \tag{1c}$$

where ω_r is the rth order nature frequency and φ_{ir} is the component of the rth order modal shape at the ith position, $S_{x_i x_j}$ and $S_{x_i x_i}$ are CSPD and SPD of the vibration responses respectively. The accuracy of Eq.(1a,1b and 1c) are also depended on the measurement noise and property of the excitation. When the measurement noise are correlated at ith and jth positions, and the measurement noise and the vibration response are also correlated. Hence,

$$\hat{\Phi}_1^{(N)} = \hat{\Phi}_1 \frac{1}{1 + S_{n_j n_j}(\omega_r)/S_{x_j x_j}(\omega_r)} \tag{2a}$$

$$\hat{\Phi}_2^{(N)} = \hat{\Phi}_2 [1 + S_{n_i n_i}(\omega_r)/S_{x_i x_i}(\omega_r)] \tag{2b}$$

$$\hat{\Phi}_3^{(N)} = \hat{\Phi}_3 \sqrt{\frac{1 + S_{n_i n_i}(\omega_r)/S_{x_i x_i}(\omega_r)}{1 + S_{n_j n_j}(\omega_r)/S_{x_j x_j}(\omega_r)}} \tag{2c}$$

745

Small damping makes to use the real mode and the estimators are real, so that the phase of the estimating, $\hat{\Phi}_k, k = 1, 2, 3$, is zero or 180°, which is used to identify ω_r. Because of phase $[S_{x_i x_j}(\omega \neq \omega_r)] \neq 0$ or 180°, the following formula is employed to judge ω_r.

$$\min_{\omega \in \Omega} \{ phase[(\Phi_k)_{ij}], or, (phase[(\Phi_k)_{ij}] - 180);$$
$$k = 1,2,3 \, |\Omega : (\omega_r - \Delta\omega, \omega_r + \Delta\omega)\}$$

$(\Omega : [\omega_r - \Delta\omega, \omega_r + \Delta\omega]$ to express resonant region)   (3)

Noise does not influence the phase of estimators $\hat{\Phi}_k^{(N)}, k = 1, 2, 3$, so it is better to use the information of the phase to judge ω_r.

In generally, it is true that

$$\left| \hat{\Phi}_1^{(N)} \right| \leq \left| \hat{\Phi}_1 \right| = \left| \hat{\Phi}_2 \right| \leq \left| \hat{\Phi}_2^{(N)} \right| \qquad (4a)$$

$$\left| \hat{\Phi}_1^{(N)} \right| \leq \left| \hat{\Phi}_3^{(N)} \right| \leq \left| \hat{\Phi}_2^{(N)} \right| \qquad (4b)$$

If the measurement noise is proportional to the measured vibration response, it does not influence the estimation of the modal shape.

It should be noted that the damping is the less and

$\dfrac{d \, (\, phase \, [S_{x_i x_j}(\omega)])}{d\omega}$ is the greater in the resonant region. For a structure with very small damping the other way has to be employed to determine ω_r. In this paper, the MUSIC estimator of pseudo spectrum is used.

The procedure estimating the nature frequencies and modal shapes is shown in Fig.4.

2.3 *Modal Frequencies and Shapes*

According to the Chinese seismic standards, only the first mode is employed to check the strength of the piers, so that only the first nature frequency and modal shape were estimated by the procedure in Fig.4. The estimated values are summarized in Table 2. The estimated first modal shapes are shown in Fig.5. The results in table 2 show that the modal parameters can be estimated with sufficient accuracy by using the ambient vibration in each of three states. Of course, the ideal state is No.2 since uncontrolled traffic makes the ambient vibration test least inexpensive.

Table 2. First mode at pier No 7

State No.		First freq. ω_1 (Hz)	First mode shape
1		2.8935	[0.0270 0.1672 1.0000]
2		3.0382	[0.0288 0.1270 1.0000]
3	30Km/h	3.0382	[0.0364 0.1404 1.0000]
	40Km/h	3.0382	[0.0325 0.1235 1.0000]
	50Km/h	2.9659	[0.0309 0.1350 1.0000]
	80Km/h	3.0382	[0.0336 0.1520 1.0000]

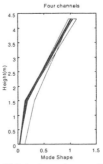

Fig.5 Mode shapes in several times

3 SEISMIC EVALUATION

The complete evaluation procedure consists of the local damage monitoring and integral behavior estimation. Both are related to one another (Wang, 1997), but only the integral behavior evaluation, which depends on the ambient vibration, is presented here.

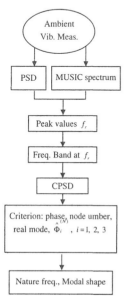

Fig 4 CPM procedure

For the measured modal parameters, the seismic evaluation procedure is denoted in Fig.6. The maximum moment calculated with the estimated modal parameters is compared with the maximum moment computed with the structure parameters.

3.1 Earthquake Force

According to "highway seismic design code", the first mode is only employed and a model with single degree of freedom is used to calculate the earthquake force with the following formula:

$$ehtp = c_i \times c_z \times k_h \times \beta \times (m_1 + m_2) \times 9.8(N) \quad (5)$$

where c_i is an significance coefficient and $c_i = 1.3$ for National Highway 312, c_z is a comprehensive coefficient, $c_z = 0.25$ when the height of a pier is less than 10m, k_h is a horizontal earthquake coefficient, $k_h = 0.1$ for 7 grade of earthquake intensity and $k_h = 0.2$ for 8 grade, β is dynamic amplifying coefficient for third class field and the measured nature frequency, m_1 is an equivalent mass of the superstructure, m is the total mass of the superstructure, and m_2 is an equivalent mass of the structure below the rubber supporter.

The calculated earthquake force is shown as the first row in table 3.

The maximum moment is

$$M_{max} = ehtp \times L_1 \quad (6)$$

where L_1 is the equivalent height of the pier. The results are given by the second row in table 3.

Table 3. Evaluation data for pier No.7

Earthquake intensity	7 grade		8 grade	
	With modal shape	Without modal shape	With modal shape	Without modal shape
ehtp (KN)	19.111	18.836	38.223	37.672
M_{max} (KN·m)	93.503	89.942	187.005	179.884
M_0 (KN·m)	140.91		140.91	

3.2 Seismic Checking

In order to check whether the maximum moment is beyond the design value, the allowable moment should be computed with the structural and material parameters. When the pressure stress of the concrete in a pier equals the limitation value, the moment is

$$M_0 = \frac{0.68 \times (R_a - \frac{G_c}{\pi \times r^2}) \times r^3 \times [w + 24 \times \pi \times n \times h_{j1} \times (\frac{r_g}{r})^2]}{96 \times k} \quad (7)$$

where R_a is compressive strength of the concrete, G_c is the calculated gravity, r is radius of pier, r_g is radius where the reinforcing steel is set, h_{j1} is the reinforcing steel ratio, n is the proportion of elasticity modulus of steel to concrete, and w is calculated by

$$w = 12\alpha - 3Sin(4\alpha) - 32Cos\alpha Sin^3\alpha \quad (8)$$

where α is the central angle of the compressive area of a pier and is gotten by solving the following equation

$$2Sin^3\alpha + 3Cos\alpha(Sin\alpha Cos\alpha - \alpha) = 3\pi h_{j1} n Cos\alpha \quad (9)$$

In Eq.(7), $k = (1 - \cos\alpha)/2$. For pier No.7, $r = 0.6$, $R_a = 11(Mpa)$, $G_c = 1124.5(KN)$, $r_g = 0.45$, $n = 8.07$, $h_{j1} = 0.0025$, $\alpha = 45°$, $w = 1.425$, $k = 0.147$. The allowing maximum moment M_0 is listed on last row in table3.

The seismic property must be satisfied on

$$M_{max} \leq M_0 \quad (10)$$

From the results in Table 3, the pier No.7 is OK for an earthquake with 7 grade earthquake intensity and not for 8 grade.

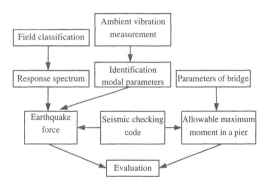

Fig.6 Evaluating procedure

4 CONCLUSION

The complete integral evaluation procedure is presented in this paper. The following data is employed:
- Ambient vibration data in any of three measurement states;
- Modal parameters estimated by the Comprehensive PSD-based Method developed in this study;
- Structural and material parameters of the bridge;
- Chinese highway seismic design code.

The procedure has been applied to the Ba River Bridge. The results show that the Comprehensive PSD-based Method has advantages in accuracy and it is most economic for ambient vibration measurement in state No.2 because of uncontrolled traffic condition. It follows that even if this bridge has been reinforced, its seismic property is also a problem.

In fact, the estimated modal parameters depend on the measurement error. It is needed to discuss further the effect of accuracy of modal parameters on the evaluation results. Of course, damage monitoring is also important.

ACKNOWLEDGMENT

The financial support for the research project was provided by Shaanxi Provincial Government and National Natural Science Fund. Authors would express their thanks to Prof. J. Q. Lei, Dr. C. C. Zhu, Dr. C. Y. He, and so on for their valuable cooperation and support.

REFERENCES

Abdel-Ghaffar, A.M., and R.H.Scanlan, 1985, "Ambient Vibration Studies of Golden Gate Bridge, I: Suspended Structure," Journal of Engineering Mechanics, ASCE, 111(4), pp463-482.

Bao,Z.W., and J.M.Ko, 1991, "Determination of modal parameters of tall buildings with ambient vibration measurement," The International Journal of Analytic and experimental Modal Analysis, Vol. 6No. 1, pp. 57-68.

Cole,H.J., 1973, "On-line failure detection and damping measurement of aerospace structures by random decrement signatures," NASACr-2205.

Feng,M.Q., J.-M.Kim, H.Xue, 1998, "Identification of a Dynamic System Using Ambient Vibration Measurements," Journal of Aplied Mechanics, Vol. 65, pp. 1000-1021.

Harik,I.E., D.L.Allen, R.L.Street, M.Guo, R.C.Graves, J.Harison and M.J.Gawry, 1997, "Free and Ambient Vibration of Brent-Spence bridge," Journal of Structural Engineering, ASCE, 123(9), pp1262-1268.

Harik,I.E., D.L.Allen, R.L.Street, M.Guo, R.C.Graves, J.Harison and M.J.Gawry, 1997, "Seismic Evaluation of Brent-Spence Bridge," Journal of Structural Engineering, ASCE, 123(9), pp1269-1275.

Schmidt, R. O., 1986, Multiple Emitter Location and Signal Parameter Estimation, IEEE trans. Antennas Propagat, Vol. AP-34, pp. 276-80, 1986.

Zhu. C. C., C. Y. He, J. H. Zhang, H. Li, 1999, Identifying the Structural Modal Parameters of Buildings by Using Ground Microseism Test, J. of Experimental Mechanics. Vol. 14. N0.2.

Zhang, J.H., Tang, C. T., 1984, The Mathematical Expression for Random Decrement Technique, Journal of Applied Mechanics Vol.1, No.2, 1984.

Wang, Y. G. 1997, "The Measurement and Assessment of Safety for Existed Bridges," Journal of Xi'an Highway University Vol.17, No.2.

Creative Systems in Structural and Construction Engineering, Singh (ed.) © 2001 Balkema, Rotterdam, ISBN 90 5809 161 9

Seismic behavior of a high-performance composite frame building made with advanced cementitious composites

V. Kilar
Faculty of Civil and Geodetic Engineering, IKPIR, University of Ljubljana, Slovenia

N. Krstulovič-Opara
Department of Civil Engineering, North Carolina State University, Raleigh, N.C., USA

ABSTRACT: The paper explores the possibilities to use a High-Performance Fiber Reinforced Concretes (HPFRCs) for design of seismic resistant cost-effective and durable infrastructural systems. Composite frame buildings are made through selective use of different HPFRCs: Slurry Infiltrated Mat Concrete (SIMCON), Slurry Infiltrated Fiber Concrete (SIFCON) and High Strength - Lightweight Aggregate Fiber Reinforced Concrete (HS-LWA FRC) which further minimizes dead load, seismic loads, and to eliminates the need for transverse reinforcement. To evaluate seismic behavior of such HPFRC-based composite frames, seismic response of a four story composite frame is compared to the response of a four-story RC building tested in the ELSA laboratory at Ispra. While the ongoing research is developing an entire composite-frame system, current paper focuses on the effect of using only composite beam members. Therefore, same (conventional) RC columns used in the RC building tested in Ispra are used in the composite building. Pushover analysis and nonlinear dynamic analysis results for the artificially generated accelerogram demonstrate substantially higher strength, approximately 20% smaller maximal top displacement and higher when composite beams are used.

1 INTRODUCTION

A promising way of cost-effectively resolving the problems of rapidly aging and deteriorating civil infrastructure is to use High-Performance Fiber Reinforced Concretes (HPFRCs) which exhibit high ductility, high tensile and compressive strengths. Such material characteristics lead to very high increases in the ability of a structure to dissipate energy — a feature particularly desirable for earthquake-resistant design. Previous use of two particular HPFRCs: Slurry Infiltrated Fiber Concrete (SIFCON) and Slurry Infiltrated Mat Concrete (SIMCON) in seismic retrofit resulted in large increases in ductility, strength and energy dissipation of the otherwise non-ductile beam-column sub-assemblages (Krstulovic-Opara et al. 2000, Dogan et-al. 2000). However, in a retrofit situation the full potential of HPFRCs is limited by geometry and properties of the existing structure. Only when HPFRCs are used in new construction could their improved features be fully exploited. This is the main objective of an ongoing National Science Foundation (NSF) - funded investigation in which beam and column members of a HPFRCC-based 12 story prototype structure are being developed and tested under reversed cyclic loading. Such a frame system, termed herein High-Performance Composite Frame (HPCF), is made by selectively using: SIMCON, SIFCON and High Strength -

Lightweight Aggregate Fiber Reinforced Concrete (HS-LWA FRC). Their material properties and effect on structural behavior are thus reviewed next.

The presented project uses steel Fiber Reinforced Concrete (FRC). In this case fibers inhibit crack growth and substantially increase energy absorption, tensile strength, fatigue, ductility and impact resistance. Based on the level of improved behavior, FRCs are divided into: conventional FRCs and High Performance FRCs (HPFRCs). Conventional FRCs are made by premixing discontinuous fibers with concrete in which case due to workability requirements the maximum fiber volume fraction of steel fibers is limited to approximately 2%. At such a low volume fraction fibers mainly contribute to post-cracking ductility and energy absorption (Shah 1988).

HPFRCs are a newer class of FRCs which exhibits behavior particularly desirable for earthquake-resistant design: significantly increased strength, ductility and energy absorption (Naaman & Harajli 1990, Naaman 1992). HPFRCs used in the presented projects include: (a) SIMCON (Krstulovic-Opara et al. 1997, Hackman et al. 1992), and (b) SIFCON (Naaman & Harajli 1990, Lankard 1984). Both are manufactured by pre-placing fibers into the form, followed by infiltration of a specially designed high strength cement-based. The key difference between the two is that SIMCON is made with continuous fiber-mats, while SIFCON is made with high fiber volume fractions of discontinuous fibers. Since fiber orientation in SIMCON fiber-mats can be controlled, very high tensile strengths and ductilities are reached for a comparatively low fiber volume which translates into high improvement in structural response (Krstulovic-Opara et al. 1997, Krstulovic-Opara & Al-Shannag 1999a). However, due to its fiber-mat configuration, SIMCON is not well suited for manufacturing of "fuses," in which case SIFCON is a better alternative (Naaman et al. 1987).

Seismic Use: Successful worldwide use of conventional FRCs has demonstrated that fibers can replace stirrups and ties, resulting in higher shear and flexural strength, higher stiffness and slower stiffness degradation, higher ductility and energy dissipation, better concrete confinement and better re-bar bond, leading to substantially better, safer and more economical seismic-resistant performance over conventional reinforced concretes. Improvement in seismic performance is so high that the joint shear reinforcement can be eliminated, thus reducing cost and congestion of the joint steel (Sood & Gupta 1987, Henager 1977). Improvements are even better if HPFRCs are used: 1 in. thick SIMCON-jacketing of non-ductile beam-column sub-assemblages (Krstulovic-Opara et al. 2000, Dogan et al. 2000) eliminated deterioration and failure of column lap splices, eliminated the pullout of the bottom beam reinforcement, eliminated the soft-story mechanism, and moved the plastic beam hinge away from the joint zone.

2 DESCRIPTION OF THE HIGH-PERFORMANCE COMPOSITE FRAME BUILDING

The ongoing NSF project is developing a novel "High-Performance Composite Frame" (HPCFs) building by selectively using SIMCON, SIFCON and HS-LWA FRC. The HPCF building that is being developed consists of three main members: (1) a composite beam with stay-in-place SIMCON formwork and cast-in-place HS-LWA FRC core, (2) a CFT column with cast-in-place HS-LWA FRC core, and (3) a precast SIFCON fuse connecting the two. To maximize cost-effectiveness and construction speed, all three members are connected on site through bolting. To achieve a good monolithic frame response, HS-LWA FRC member "cores" are cast-in-place. Since use of FRC eliminates the need for secondary reinforcement: (1) reinforcement congestion is eliminated, and (2) construction speed is further increased. Since the composite column member is still being tested, it is not analyzed in this paper. Instead the paper focuses on a HPCF building made only with high-performance beam and fuse members, while column members are assumed to be the same conventional RC members as used in the reference Ispra building, as shown in Figure 1. A more detailed description of beam and fuse members is provided next.

• Beam Member: The most economic way is to use SIMCON only in the outer layer of a beam where stresses are the highest, as shown in Figure 1. This: (1) minimizes flexural reinforcement, (2) eliminates shear reinforcement, and (3) optimizes member dimensions. The stay-in-place beam formwork was made with 5.25 % volume fraction SIMCON, as indicated in Figure 1. The formwork was easily manufactured and handled without the use of any additional reinforcing steel. The section "core" was cast-in-place, 2% fiber volume fraction, 8.0 kN/cm^2 HS-LWA FRC. Its estimated unit weight was 19.6 kN/m^3 (123 lb/ft^3). Casting-in-place of dense HS-LWA FRC required little effort since conventional reinforcement and related reinforcement congestion have been eliminated. The specimen was first tested under reversed-cyclic "cantilever" type loading. Since in the prototype structure the nonlinear response occurs only in the fuse region while the central beam region behaves (predominantly) elastic, the specimen was tested only up to the load level to be experienced in the prototype structure. Characteristical points of three-linear envelope of moment curvature response are presented in Table 1. Undamaged specimen portion was tested next in static three-point loading (shear span to beam depth ratio was 2.1). The second test demonstrated high shear resistance: even though (a) no shear reinforcement was used, and (b) the loading was shear critical, the composite beam exhibited a ductile flexural response.

• Fuse Member: Use of SIFCON fuses provides very high ductility, strength, and energy dissipation of a beam-column sub-assemblage (Vasconez et al. 1997). Work by Naaman, Wight and collaborators (Naaman et al. 1987, Abdou et al. 1988, Soubra et al. 1992, Vasconez et al. 1997) has focused on cast-in-place SIFCON fuses that act as connections between precast concrete members. While excellent seismic response of such connections was reported, a certain drawback of this approach is the increased work-effort required in manufacturing of SIFCON fuses on the site. Therefore, precast SIFCON fuses are being developed in the ongoing research, which (1) eliminates problems associated with casting-in-place of SIFCON, (2) permits the use of a more optimal reinforcement layout, and (3) since the fuse is precast, the bolted connections can be designed so that after an earthquake a damaged fuse can be replaced. Different precast fuse layouts have been tested under static reverse-cyclic loading. Resulting ductilities varied between 6.1 and 6.5 (Wood 2000). Layout of the representative SIFCON fuse modeled in this paper, and characteristical points of three-linear envelope of moment curvature response are presented in Figure 2 and Table 1, respectively.

Since the actual dimensions of the tested specimens were smaller than dimensions of the reference RC Ispra structure, tested specimen dimensions were scaled up for 19%. Experimentally determined member properties were scaled according to the "true model" scaling laws (Sabnis et al. 1983). Resulting cross-sectional dimension of both the model fuse and the model central beam region were 30 cm x 48.6 cm (11.9 in. x 19 in.). While the center-to-center column spacing of 5 m was the same as for the reference Ispra building, scaled SIFCON fuse length was 78.6 cm (30.9 in.). Since the goal of this paper is to compare effect of using only high performance composite central beam region and SIFCON fuse on the structural response, the same RC column members as used in the Ispra building and shown in Figure 4, are used in the presented HPCF.

3 DESCRIPTION OF A REFERENCE TEST BUILDINGS

The reference test structure (Negro et Al. 1994) is a four-story r/c frame building [Fig. 3], designed in accordance with the prescriptions of Eurocode 2 and 8 for High Ductility class structures. Dimensions in plan are 10 m x 10 m; interstory heights are 3.0 m, except for the ground story which is 3.5 m high. The structure was symmetric in the direction of testing (X direction), with two spans equal to 6.0 and 4.0 m. In the present study the Ispra structure was lightly modified. A doubly symmetric building was obtained by shifting the central frame in Y direction on the axis of symmetry and by changing the reinforcement in some sections of the beams. Both spans were taken to be equal to 5.0 m. All columns have square cross section with 400 mm side [Fig. 4], except for the interior column which is 450 mm x 450 mm; all beams have rectangular cross section, with total height of 450 mm and width of 300 mm. A solid slab, with thickness of 150 mm, was adopted for all stories of Ispra building [Fig. 4]. In the preliminary design of the building additional dead load to represent floor finishing and partitions equal to 2.0 kN/m2, live load equal to 2.0 kN/m2, peak ground acceleration of 0.3g, soil type B, importance factor equal to 1 and behavior factor q equal to 5 were assumed.

Table 1. Test results: characteristic points of three-linear envelope of moment curvature response of fuse and beam sections.

| Beam (bottom in tension) | | Fuse | |
Curvature (1/m)	Moment (kNm)	Curvature (1/m)	Moment (kNm)
0.0012	142.1	0.0077	255.3
0.0091	355.2	0.0119	327.1
0.0331	399.4	0.0592	392.1

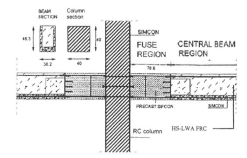

Figure 1. Beam-fuse region of HPCF building.

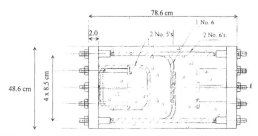

Figure 2. SIFCON fuse specimen tested in the experimental phase of the ongoing NSF research.

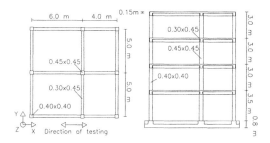

Figure 3. Reference test building [Negro et Al. 1994].

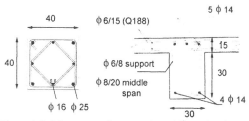

Figure 4. Reinforcement of corner column (1st story) and reinforcement of beam section (Ispra, 1st floor, middle frame).

4 DESCRIPTION OF THE SEISMIC INPUT

The seismic input is an artificial accelerogram, called S7, generated by using the waveforms derived from real signals recorded during the 1976 Friuli earthquake: its response spectrum fits the one given by EC8 for soil profile B at 5% damping. High and low level pseudodynamic tests were performed in the European Laboratory for Structural Assessment (ELSA), using such accelerogram S7 scaled by 1.5 and 0.4 respectively. [Fig. 5] compares the EC8 and S7 elastic spectra. In our study the buildings were subjected to S7 accelerogram scaled by 1.5 (high level test).

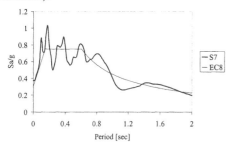

Figure 5. Eurocode 8 and S7 elastic spectra.

5 NUMERICAL MODELLING

The model was set up using all the available theoretical and experimental data. CANNY-E computer program (Li, 1997) was used to perform the static and dynamic numerical analyses.

For Ispra building the geometry of the modelled building and the masses, concentrated in the centre of mass of each floor, are taken from [Negro et Al. 1994]. The building is modelled as a planar structure consisted of three paralel frames. The damping matrix is assumed to be proportional to the instantaneous stiffness matrix, assigning a damping ratio equal to 2%. The beams arc idcaliscd by a non-linear uniaxial bending model, with elastic shear deformation; their axial and torsional deformations are not taken into account. The inelastic flexural deformation is lumped at the element ends and is given by the rotation of two nonlinear springs, which are connected to the joint by a rigid zone. The moment-rotation relationships in the two rotational springs are computed based on moment-curvature relationships, assuming an asymmetrical moment distribution along the length of the element. A tri-linear skeleton curve is assigned to represent the cross section behaviour before and after cracking and yielding. For the after yielding stiffness, when the upper part of the section is in tension, an increased value is assumed with respect to the one theoretically computed: in particular, it is 5% of the initial stiffness for the beams of the external frames and 10% for the beams of the internal one. In this way the progressive spreading of the region in which slab reinforcement yields, when the beam rotation increases, is modelled, even though the section is assumed rectangular. The hys-

teretic behaviour follows the Takeda rules, but the pinching effect is also considered. Best correlation with the experimental results are obtained assuming small values for the unloading stiffness (in each cycle it is reduced of 50% with respect to the previous one). The columns are idealised by non-linear uniaxial bending model; each column has been idealised by two coupled uniaxial bending elements to provide resistance capacity in two orthogonal directions. A detailed study of the influence of different model parameters to the response of Ispra building can be found in (Faella, Kilar and Magliulo, 2000).

Similar model assumptions were used also to model a new HPCF building. The beams were divided in to two fuse elements and one central beam element. Moment curvature relationships for fuse and beam were obtained by tests that were performed on the University of North Carolina (see Table 1). The total calculated mass is approximatelly 10% smaller than for Ispra building, due to lightweight concrete used for beams of new building. The contribution of floor slabs was not taken into account.

Table 2 presents the periods for both investigated buildings obtained before and after step-by-step analysis. It can be seen that periods for HPCF building are smaller than for Ispra building; the differences are bigger after the building is already damaged. The reason is mainly due to smaller masses and slower stiffness deterioration of HPCF building.

Table 2. Periods for Ispra and HPCF building before and after dynamic step-by-step analysis.

	Periods (s)			
	Ispra		Hpcf building	
	Elastic	After step-by-step analysis	Elastic	After step-by-step analysis
X1	0,56	1,24	0,53	0,96
X2	0,18	0,36	0,17	0,31
X3	0,10	0,20	0,09	0,18

6 RESULTS OF PUSHOVER ANALYSIS

Figure 6 shows the pushover analysis results obtained with the program CANNY for both building variants. Horizontal loads were distributed in the form of an inverted triangle and increased until the top displacement 2% (0.25 m) of building height is reached (the value suggested for live-safety of RC frames).

It can be seen from Figure 6 that the HPCF model with new type of beams gives better initial stiffness and aproximatelly 45% higher ultimate strength than Ispra building. Point "1" presents the first cracking of concrete section, point "2" the yielding of first beam (Ispra building) or fuse (HPCF) and point "3" the first yielding at the base of columns in the 1st story. It is interesting to note that the first yielding of

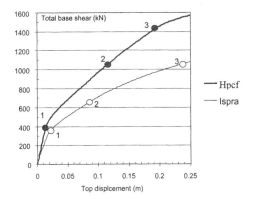

Figure 6. Base shear - top displacement relationships obtained by pushover analysis, loads are applied in X direction.

columns in NBB occurs at smaller top displacement than in Ispra building. The reason is that the stronger beams transfer stronger moments to the columns which should be for a real building separately designed according to Eurocode 8 rules of capacity design.

7 RESULTS OF NONLINEAR DYNAMIC ANALYSIS

Selected results of time-history analyses are shown in Figures 7-9 and in table 3. In Figure 7 and 8 the time histories of top displacements and base shear are presented for both investigated buildings. Figure 9 present the envelope of maximum displacements and story drifts for both building variants.

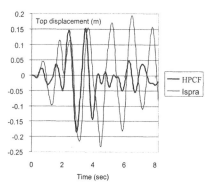

Figure 7. Top displacement for both investigated buildings.

It can be seen from the Figures 7-9 that HPCF building exhibit aproximatelly 25 % smaller maximum top displacements and approximatelly 23 % bigger base shear than Ispra building. Further more, the maximum story drift for HPCF (2.03) is 18% smaller than maximum base shear for Ispra building

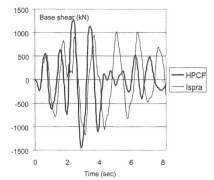

Figure 8. Base shear for both investigated buildings.

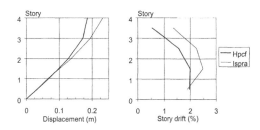

Figure 9. Displacement envelopes and maximum story drifts for accelerogram S7 scaled with factor 1.5.

(2.47). Table 3 presents the maximum ductility factors for both buildings. It can be seen, that, as expected, for the HPCF building the demage is concentrated in the fuses, while the central part of beam remains elastic. The fuse that is specially designed to sustein higher demage is supposed to be replaced after an actual earthquake.

Table 3. Ductilities of beams for Ispra and HPCF building.

	Ductility factors*	
	Ispra	Heigh performance concrete frame building (HPCF)
Floor	Maximal ductility of beams	Maximal ductility of fuse**
1	1,70	**3,94**
2	**2,66**	1,28
3	1,90	0,48
4	1,20	0,20

* Ductility factor greater than 1.0 indicates yielding
** Beams remain elastic for HPCF building

8 CONCLUSIONS

The paper presents seismic response of a novel High Performance Composite Frame (HPCF) building made through the selective use of HPFRCs, and compares it to the response of a R/C frame building tested in Ispra. While HPCF building made of composite beam and column members is being developed in the ongoing NSF research, this paper focuses only on the effect of composite beam members on the structural response and uses conventional RC columns. Effect of using high-performance composite columns will be presented in future publications.

Under same seismic excitaions the HPCF building exhibited very good seismic response. It experienced slower stiffness degradation, smaller ultimate displacements, higher ultimate strength, smaller story drifts and smaller vibration periods than the reference RC building. The damage was concentrated in highly ductile fuse zones, while central beam regions remained elastic. This is a very desireable response, because the fuse zones are precast and the bolted connections can be designed so that the damaged fuse can be relatively easier to replace after the earthquake than is the case with conventional, earthquake damaged RC members. Minimal amount of damage was observed in the column members. This can be eliminated if high-performance composite column are used, which is the part of the ongoing NSF investigation.

Finally, it could be pointed out that the two-dimensional layout of SIMCON and its unique manufacturing properties related to its fiber-mat configuration, open up novel possibilities for a cost-effective and improved structural performance that were not previously possible using construction materials. It is thus anticipated that when used in the field, the proposed approach could be less labor- and equipment-intensive and more economical than conventional methods. Finally, manufacturing of SIMCON is based on the use of widely available construction equipment and building expertise, and can thus be relatively easily introduced into the field without major re-training and changes in existing construction practices. Hence, the presented approach might provide some unique new ways of developing durable and cost-effective high-performance infrastructural systems, essential for the economic well-being of a nation in this next century.

ACKNOWLEDGEMENTS

Presented investigation was partially supported by the National Science Foundation's grants CMS-9632443 with Dr. S. C. Liu as the Program Director, and CMS-9632443B with Ms. Cassandra Dudka as the Program Director.

REFERENCES

Abdou, H., Naaman, A., Wight, J. 1988. Cyclic Response of Reinforced Concrete Connections Using Cast-in-Place SIFCON Matrix, Report No. UMCE 88-8, Department of Civil Eng., University of Michigan.

Dogan, E., Hill, H., Krstulovic-Opara, N. 2000. Suggested Design Guidelines for Seismic Retrofit With SIMCON & SIFCON, HPFRC in Infrastructural Repair and Retrofit, ACI SP-185.

Faella G., Kilar V., Magliulo G. 2000. Overstrength factors for 3D R/C buildings subjected to bidirectional ground motion, 12. World conference on Earthquake Engineering, Auckland, New Zealand.

Hackman, L. E., Farrell, M. B., Dunham, O. O. Dec. 1992. Slurry Infiltrated Mat Concrete (SIMCON), Concrete International, pp. 53-56.

Henager, C. H. 1977. Steel Fibrous, Ductile Concrete Joints for Seismic-Resistant Structures, Reinforced Concrete Structures in Seismic Zones, ACI SP-53, pp. 371-386.

Krstulovic-Opara, N., Al-Shannag, M. J. Jan.-Feb. 1999. Slurry Infiltrated Mat Concrete (SIMCON) - Based Shear Retrofit of Reinforced Concrete Members, ACI Structural Journal, pp. 105 - 114.

Krstulovic-Opara, N., Al-Shannag, M. J. May - June 1999. Compressive Behavior of Slurry Infiltrated Mat Concrete (SIMCON), ACI Materials Journal, pp. 367 - 377.

Krstulovic-Opara, N., LaFave, J., Dogan, E., Uang, C. - M. 2000. Seismic Retrofit With Discontinuous SIMCON Jackets, HPFRC in Infrastructural Repair and Retrofit, ACI Special Publication SP-185.

Lankard, D. 1984. Slurry Infiltrated Fiber Concrete (SIFCON): Properties and Applications, Potential for Very High Strength Cement Based Materials, Proc. of Material Research Society, Vol. 42, pp. 277 - 286.

Li, K.N. 1997. CANNY-E. Three-dimensional nonlinear dynamic structural analysis computer program package. Canny Consultants Pte Ltd.

Naaman, A. E. 1992. SIFCON: Tailored Properties for Structural Performance, High Performance Fiber Reinforced Cement Composites, E & FN Spon, pp. 18-38.

Naaman, A. E., Harajli, M. H. 1990. Mechanical Properties of High Performance Concretes, Report SHRP-C/WP-90-004, Strategic Highway Research Program, National Research Council.

Naaman, A., Wight, J., Abdou, H. November 1987. SIFCON Connections for Seismic Resistant Frames, Concrete International, pp. 34 - 38.

Negro, P., Verzelletti, G., Magonette, G.E. & Pinto, A.V. 1994. Tests on a Four-Storey Full-Scale R/C Frame Designed According to Eurocodes 8 and 2: Preliminary Report.

Sabnis, G. M., Harris, H. G., White, R. N., and Mirza, M. S. 1983. Structural Modeling and Experimental Techniques, Civil Engineering Series, Prentice Hall, Inc., Englewood Cliffs, NJ.

Shah, S. July 1988. Theoretical Models for Predicting the Performance of Fiber Reinforced Concrete, Journal of Ferrocement, pp. 263-284.

Sood, V., Gupta, S. 1987. Behavior of Steel Fibrous Concrete Beam-Column Connections, Fiber Reinforced Concrete - Properties and Applications, ACI SP - 105, pp. 437 - 474.

Soubra, K. S., Wight, J., Naaman, A. 1992. Fiber Reinforced Concrete Joints for Precast Construction in Seismic Areas, Report No. UMCEE 92-2, Department of Civil and Env. Eng., The University of Michigan.

Vasconez, R. M., Naaman, A. E., Wight, J. K. 1997. Behavior of Fiber Reinforced Connections for Precast Frames Under Reversed Cyclic Loading, Report Number UMCEE 97-2, The University of Michigan.

Wood, B. 2000. The Use of Slurry Infiltrated Fiber Concrete (SIFCON) in Hinge Regions for Earthquake Resistant Concrete Moment Frames, Ph.D. Thesis, North Carolina State University, Raleigh.

Creative Systems in Structural and Construction Engineering, Singh (ed.) © 2001 Balkema, Rotterdam, ISBN 90 5809 161 9

Experimental research for the seismic enhancement of circular RC bridge piers in moderate seismicity regions

Y.S.Chung, J.H.Park & C.B.Cho
Department of Civil Engineering, Chung-Ang University, Korea

Y.G.Kim
Department of Safety Engineering, Han-Kyung University, Korea

N.S.Kim
Hyundai Institute of Construction Technology, Korea

ABSTRACT: The pseudo dynamic test has been carried out so as to investigate the seismic performance of RC bridge piers strengthened with and without glass fiber sheets. Collapse or severe damage of a number of infrastructures in Kobe(1995) and Northridge(1996) earthquakes has emphasized the need to develop the retrofit measures to enhance flexural strength, ductility and shear strength of RC bridge piers nonseismically designed before 1992. Therefore, the objective of this experimental research is to investigate seismic behavior of circular reinforced concrete bridge piers by the pseudo dynamic test, and then to enhance the ductility of concrete piers strengthening with glass fiber sheets in the plastic hinge region. 7 circular RC bridge piers were made in a 1/3.4 scale. Important test parameters are confinement steel ratio, retrofitting, load pattern, etc. The seismic behavior of circular concrete piers under artificial ground motions has been evaluated through strength and stiffness degradation, energy dissipation. It can be concluded that existing bridge piers wrapped with glass fibers in the plastic hinge regions could have enough seismic performance.

1 INTRODUCTION

Even though earthquakes have several economics, social, psychological and even political effects in the areas and the countries where they take place, most Korean thought a few years ago that Korea is located rather far away from the active fault area and immuned from the earthquake hazards. Recently, it has been observed in the Korean Peninsula that the number of moderate or low earthquake motions tend to be increased year by year. Furthermore, the collapse and significant damage of RC bridge piers in 1995 Kobe earthquake and 1996 Northridge earthquake stimulate the establishment of seismic design provisions and the development of effective retrofitting method for various infrastructures which could be appropriate for geological and topographical conditions in Korea. Therefore, the objective of this pseudo dynamic test is to investigate the seismic performance of existing circular RC bridge piers which nonseismically designed before current seismic design provisions, and then to evaluate the ductility of concrete piers strengthening with glass fiber sheets in the plastic hinge region.

2 DESCRIPTION OF EXPERIMENT

2.1 Material Properties of Test Specimen

SHE-51 glass fiber sheet had been used as retrofitting material. Its yielding stress is 5,600 kgf/cm². D10 deformed steel had been used as longitudinal steel in RC piers, of which confinement steels had been laterally used with D6 deformed steel. Yielding stress is 4,700 kgf/cm² for D10 deformed steel and 4,400 kgf/cm² for D6 deformed steel. A compressive strength of concrete cylinder was $f_{ck} = 261 kgf/cm^2$.

2.2 Design of Specimen

Circular solid RC piers of Hagal bridges, being located in Kyung-Gi Do province, Korea, were adopted as a prototype of this test specimen. It had been seismically designed in accordance with the provisions of Korea Highway Design Specification. Test specimens have been nonseismically or seismically designed in accordance with the provisions of Korea Highway Design Specification. 2 test specimens nonseismically designed had been wrapped with 2 layers of glass fiber as shown in Fig. 2. Fig. 1 shows detailed dimensions of test specimen. The applied scale factor to consider dynamic similitude between the prototype and the specimen is used as 3.4. The scale factors for diameter, force, mass and time are S, S^2, S^3 and S, respectively.

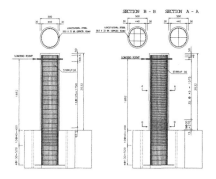

(a) None Seismic Design   (b) Seismic Design

Fig. 1 Detailed Dimension

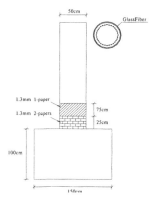

Fig. 2 Retrofit Scheme

2.3 Experimental Plan

2.3.1 Quantity of Specimen

7 test specimens have been prepared for the pseudo dynamic test to investigate their seismic performance. They are one for pilot test, three for moderate

Table 1 Test Specimen Description

Test Type	Loading Pattern	Classification (Design)	Notification
Pilot Test	I	Non-Seismic	NS-PD-LP1-A1
Pseudo Dynamic Test	I	Non-Seismic	NS-PD-LP1-A1
		Seismic	S-PD-LP1-A1
		Strengthening	T-PD-LP1-A1
	II	Non-Seismic	NS-PD-LP1-A1
		Seismic	S-PD-LP1-A1
		Strengthening	T-PD-LP1-A1
Total	7		

* NS : Non-Seismic   S : Seismic
   T: TYFO(Strengthening)   PL : Pilot Test
   PD : Pseudo Dynamic Test

artificial earthquake and three for strong artificial earthquakes, as shown Table 1. Important test parameters are input ground motion, confinement steel ratio, retrofitting, etc.

2.3.2 Input Ground Motion

As shown on Fig. 3, this pseudo dynamic test has used two input ground motions. First one is moderate artificial earthquakes with 0.2g PGA for KHC(Korea Highway Cooperation), and second one is strong artificial earthquakes with 0.36g PGA for Kaihokus, Japan. Both artificial earthquakes are based on rock soil condition and their duration is 24sec, respectively. Being different from general acceleration records, Fig. 3(B) pointed initial parts of Kaihokus ground acceleration which almost incur the failure.

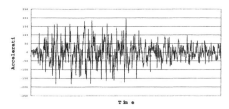

(A)   Moderate Artificial Earthquake(KHC)

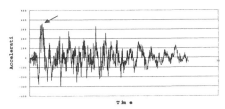

(B)   Strong Artificial Earthquake(Kaihokus)

Fig. 3 Artificial Ground Acceleration

2.3.3 Loading Pattern

Loading pattern is determined on the basis of KHBD(Korea Highway Bridge Design) code.

Table 2 Loading Patterns

Loading Pattern	Input Ground Motion	
	Ground Motion	PGA
I (Moderate Artificial Earthquake)	I –45.1(gal)	0.154g
	I +19.6(gal)	0.22g
	I +98.0(gal)	0.30g
	I +196(gal)	0.40g
	I +294(gal)	0.50g
	I +392(gal)	0.60g
	I +490(gal)	0.70g
II (Strong Artificial Earthquake)	II –201.8(gal)	0.154g
	II -137.2(gal)	0.22g
	II -98.00(gal)	0.26g
	II –58.80(gal)	0.30g

As shown on Table 2, PGA values for loading pattern I has been started from 0.154g and gradually leveled up by 0.1g from 0.3g to the failure state. Meanwhile, PGA values for loading pattern II are shown on Table 2.

3 EXPERIMENTAL PROGRAM

3.1 Loading Method

100ton MTS actuator was used for the pseudo dynamic test. Its maximum stroke is ±250mm. When lateral force will be imposed at the test specimen by the actuator, it is desirable to keep the axial force constant. So, a hydraulic axial force controller have been manufactured to keep a constant axial load. Loading rate is determined as 0.3mm/sec considering the effects of strain rate for the pseudo dynamic test.

3.2 Measurement

During the pseudo dynamic test, lateral displacements have been acquired by 2 displacement transducers, which are located at 94.2cm, 188.2cm from the loading point of test column downward, as shown on Fig. 4. Further 2 LVDTs have been placed on the footing in order to check unexpected displacements during the test. Curvatures have been also measured in the plastic hinge zone of each test column by using 8 clip gauges, as shown in Fig. 4. Strain gauges for confinement and longitudinal steels have been attached in the plastic hinge region. Plastic strain gauges have been used to measure the strain of confinement steels during elastic and plastic behavior.

3.3 Preliminary Experiment

The pseudo dynamic test is similar to standard step-by-step nonlinear dynamic analysis procedures in that the controlling computer software considers the response to be divided into a series of time step. Within each step the governing equation of motions are numerically solved for the incremental structural deformation. In the pseudo dynamic method, the ground motions as well as the structure's inertial and damping characteristic are specified numerically in a conventional dynamic analysis. However, the structure's restoring force characteristic are measured directly from the damaged specimen as the test progresses. Scheme of pseudo dynamic test is shown on Fig. 6. Explicit Newmark' β method has been used for algorithm of the pseudo dynamic test. Pilot test has been conducted before the main pseudo dynamic test. The goal of this test is to check the interface between experimental equipments and the software of pseudo dynamic test. As a result, we can have a good agreement between experimental and analytical result. Analytical result was calculated by FEM program (SARCF[4] - Nonlinear Dynamic Frame Analysis) in the inelastic range. Fig. 7 shows the results of pilot test, which should be in a good agreement.

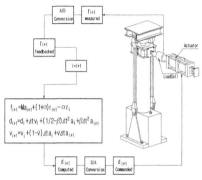

Fig. 6 Scheme of Pseudo Dynamic Test

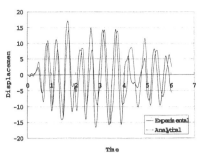

Fig. 7 Pilot Test Result

4 TEST RESULTS

4.1 Time Displacement History

Seismic and noneseismic test specimens have shown similar failure patterns. Their failure mechanism is cracking and spalling of cover concrete, breaking of

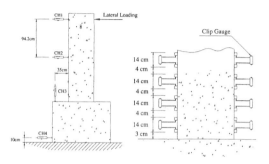

Fig. 4 LVDT Setup   Fig. 5 Clip Gauge Setup

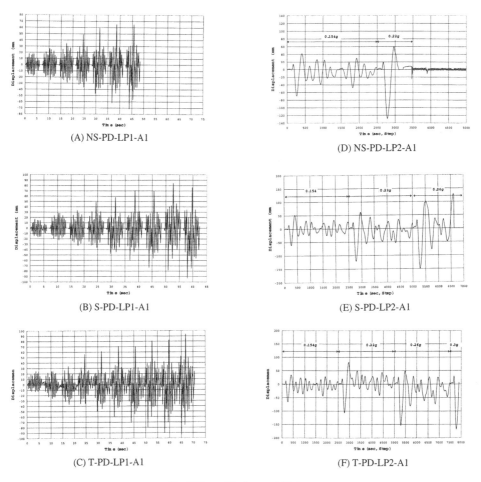

Fig. 8 Time – Displacement History

the confinement steel, crushing of the core concrete, and then buckling of the longitudinal steel in sequence. The failure specimens strengthened with glass fiber sheets is initiated by splitting glass fiber sheets in the plastic hinge region. The maximum displacement observed at the failure state is shown on Table 3. Fig. 8 shows the time-displacement history for 6 test specimens.

Table 3 Measured Maximum Displacement

Classification	Maximum Displacement
NS-PD-LP1-A1	149.41 mm
S-PD-LP1-A1	157.22 mm
T-PD-LP1-A1	104.24 mm
NS-PD-LP1-A1	125.23 mm
S-PD-LP1-A1	224.05 mm
T-PD-LP1-A1	167.72 mm

4.2 Hysteric Curve

Fig. 9 shows hysteric curves of 6 test specimens subjected to moderate and strong artificial ground accelerations. As shown on Fig. 9, it can be observed that ductility ratio for S-PD-LP1-A1(Fig. 9 (B)) and T-PD-LP1-A1(Fig. 9 (C)) is bigger than that of NS-PD-LP1-A1(Fig. 9 (A)). It is in particular noted from Fig. 9 that NS-PD-LP2-A1 specimens severely failed at 0.22g PGA, but S-PD-LP2-A1 and T-PD-LP2-A1 failed at 0.26g and 0.30g PGA, respectively. Thus, it can be said that gassfiber method is sufficient to enhance the ductility requirement for seismic design.

4.3 Strength Envelop Curve

Strength envelop curve of all test specimens under loading pattern 1 has shown a similar trend to 0.4g

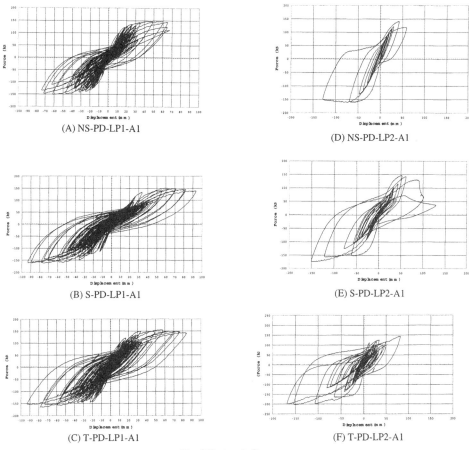

(A) NS-PD-LP1-A1

(D) NS-PD-LP2-A1

(B) S-PD-LP1-A1

(E) S-PD-LP2-A1

(C) T-PD-LP1-A1

(F) T-PD-LP2-A1

Fig. 9 Hysteresis Curve

PGA from the beginning, but have begun to be partially in upward trend from 0.5g PGA to the failure state. As shown in Fig. 10, moderate and seismic specimens show similar strength envelopes at given PGA. For strength envelope curves under loading pattern 2 in Fig. 11, meanwhile, significant strength drop was observed at 0.154g for NS-PD-LP2-A1 and 0.26g for S-PD-LP2-A1 and T-PD-LP2-A1, respectively.

4.4 Energy Dissipation

For the measurement of energy dissipation capacity at a given PGA, cumulative input energy was analyzed. As shown in Fig. 12, input energy is defined as the workdone of the actuator from the beginning of the test to the given displacement amplitude. Cumulative input energy is calculated as the sum of input energy. As shown in Fig. 13, cumulative energy dissipation capacity for loading pattern 1 enhanced by 95% for S-PD-LP1-A1 specimen and by 104%

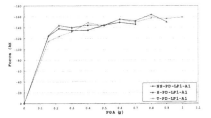

Fig. 10 Strength Envelope for LP1

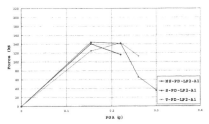

Fig. 11 Strength Envelope for LP2

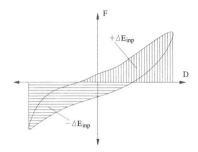

Input Energy, Einp = | +ΔEinp | + | -ΔEinp |
Cumulative Input Energy, Ecum=ΣEinp

Fig. 12 Input Energy Definition

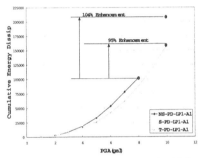

Fig. 13 Cumulative Energy Dissipation(LP1)

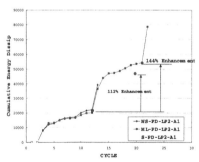

Fig. 14 Cumulative Energy Dissipation(LP2)

for T-PD-LP1-A1 specimen as against NS-PD-LP1-A1 specimen. As shown in Fig. 14, meanwhile, cumulative energy dissipation capacity for loading pattern 2 enhanced by 112% for S-PD-LP2-A1 specimen and 180% for by T-PD-LP2-A1 specimen as against NS-PD-LP2-A1 specimen.

5 CONCLUSIONS

So as to analyze seismic behavior of concrete bridge piers, pseudo dynamic test have been done 7 circular solid RC columns which are an 1 to 3.4 scaled model of Hagal bridge piers located in KyungGi-do province. It can be concluded from the test that

1) Existing RC bridge piers nonseismically designed have been proved to be by and large resistant under moderate artificial earthquakes, but exhibited a notable damage under strong artificial earthquake with 0.22g PGA.

2) Both retrofitted and seismic specimens have shown a similar seismic resistant capacity. The latter was designed in accordance with Korea Highway Bridge Design Specification.

3) It is found that the existing bridge piers wrapped with glass fibers in the plastic hinge regions should have excellent seismic performance under moderate ground acceleration.

ACKNOWLEDGEMENTS

The authors gratefully acknowledge the support from the Korea Earthquake Engineering Research Center(Contract No. 1997G0402) and research equipment support program of Chung-Ang University. The authors also thank Pung-Lim Co. for providing test specimens.

REFERENCES

Ministry of Construction and Transportation (1996), " Korea Highway Bridge Design Specification "

Y.S. Chung, G.H. Han & K.K. Lee(1999), "Research of Plastic Response by Quasi Static Test for Circular Hollow RC Bridge Pier", *Proceedings of EESK Conference Spring*

Christopher R. Thewalt & Stephen A. Mahin(1987), "Hybrid Solution Techniques for Generalized Pseudo-Dynamic Testing", *Report No. UCB/EERC-87/09*

Y.S. Chung, M. Shinozuka & C. Meyer(1988), "Sarcf User's Guide - Seismic Analysis of Reinforced Concrete Frame", *Technical Report NCEER-88-0044, November 9*

Pui-shum B. Shing and Stephan A. Mahin(1984), Pseudo-Dynamic Test Method for Seismic Performance Evaluation: Theory and Implementation, *Report No. UCB/EERC-83/12*

M.J.N. Priestley, F. Seible, G. M. Calvi, *Seismic Design and Retrofit of Bridge*, JOHN WILEY & SONS, Inc

Mehmet B. Atalay, Joseph Penzien(1975), The Seismic Behavior of critical Regions of Reinforced Concrete Components as Influenced by Moment, Shear and Axial Force, *EERC Report 75-19, Yhe University of California*

Creative Systems in Structural and Construction Engineering, Singh (ed.) © 2001 Balkema, Rotterdam, ISBN 90 5809 161 9

Seismic response analysis of an arch dam with effects of nonlinear contraction joint behavior

T. Nishiuchi
Central Research Institute of Electric Power Industry, Chiba, Japan

ABSTRACT: This paper describes a numerical study on the seismic response of a typical arch dam with the nonlinear contraction joint behavior. The seismic-induced joint opening effects, along with seasonal changes in the joint opening behavior due to temperature loads, were investigated by using a newly developed analysis procedure based on 3D FEM. The nonlinear analysis procedure specifically addresses the redistribution of forces that occur in arch dams when the contraction joints open and close in response to earthquake ground motion. Compared with the results of standard linear analyses of the monolithic model, it is found that the opening of the dam's contraction joints in winter season has a significant effect on the transient response of the membrane stress and the maximum crest displacement in stream direction at crown cantilever..

1 INTRODUCTION

Arch dams are usually constructed as cantilever monoliths separated by contraction joints. From our numerical and experimental study (Nishiuchi &c. 1998), it is found that some of the contraction joints open near the surface of an existing arch dam. Therefore, it is expected that the contraction joints may open and close through several cycles of vibration during earthquake, affecting the transfer of internal forces between the arches and cantilevers. Despite the possibility of contraction joint opening, it is impossible to simulate the non-linear joint behavior by standard seismic analysis procedures, which assume that a dam is a monolithic structure and the behavior is linear elastic. This makes it indispensable to develop a newly analysis procedure for computing the earthquake response of arch dams including the nonlinear effects of contraction joint opening. The nonlinear analysis procedure specifically addresses the redistribution of forces that occur in arch dams, when the contraction joints open and close in response to earthquake ground motion.

The objectives of this study are as follows;
(1) To validate the applicability of the newly developed analysis procedure to this specific problem.
(2) To examine the effects of contraction joint opening on the seismic response of a typical arch dam.

2 METHOD OF THE ANALYSIS PROCEDURE

2.1 *Finite element model of an arch dam*

The finite element model of the existing arch dam has a 133m high and crest length of 276m, almost perfectly symmetric doubly curved arch. The dam consists of 20 monolith cantilevers separated by contraction joints.

Fig.1 shows the finite element mesh pattern. The region of dam-foundation rock was modeled by 8 node isoparametric solid elements. The fine mesh pattern through thickness in the stream direction was used near the surface of the dam body, while coarse mesh pattern was used in the internal portion of the dam body. The contraction joints were modeled by three dimensional non-linear elements. Opening of

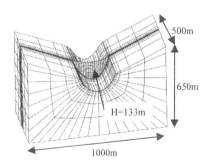

Fig.1 Calculation model of Arch Dam

the contraction joints in the model was controlled by specifying appropriate properties for the joint elements. More details about the joint elements are given in the next section. The seismic-induced hydrodynamic effects of the reservoir water were represented by an added mass concepts proposed by Westergarrd, neglecting water compressibility.

Based on the in-situ measurement results, energy dissipation in the finite element model was provided by Rayleigh damping. The damping was evaluated from the specified modal damping ratios, that is, approximately 2% of critical damping in the first and second vibrational modes (Toyoda &c.1997).

The material properties for the seismic response analyses are shown in Table 1. For analyses, concrete and rock elements were assumed to be elastic.

Table 1. Material Properties

	concrete	Joint	rock
E(MPa)	30000	9000	7000
ν	0.2	0.2	0.2

2.2 Modeling of contraction joint

Fig.2 illustrates the basic concept of contraction joint modeling in an arch dam. As shown in this figure, the joint elements develop resisting forces due to relative normal and tangential displacements between the two surfaces.

Coulomb type failure criterion, as shown in Fig. 4, was used as the failure condition of the joint ele-

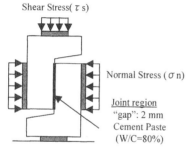

Fig.2 Characteristics of calculation method

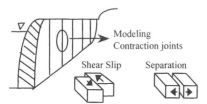

Fig.3 Static push off test

ments. The parameters in the yield function were determined from the results of "Push-off test"(Fig.3), which was conducted in the previous study (Nishiuchi &c. 1995).

The stress-strain relationship at the joint surface is shown in Fig 5 and 6.

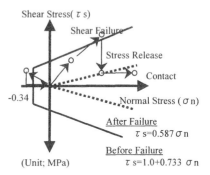

Fig.4 Failure criterion on joint elements

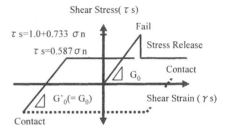

Fig.5 Shear stress - strain relationship

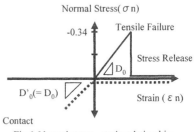

Fig.6 Normal stress - strain relationship

The normal stresses, σn, in the joint are non-linear functions of the strain, (εn). The joint has a specific tensile strength, $\sigma n=0.34$(MPa), in the normal direction, and once it is reached the joint unloads and the subsequent tensile strength is zero.

The reduced stiffness is provides for the joint when the two joint surfaces contact again. On the other hand, shear failure is assumed to begin when the shear stress of the joint exceeds a specified value defined by $\tau=1.0+0.733\sigma n$. After the threshold value is reached, the shear stress is released and the subsequent shear strength is provided by τ=0.587σn for the remainder of the analysis. The tangential stiffness coefficients are reduced when the shear strain is reversed as shown in Fig. 6

2.3 Conditions of the numerical simulation

The actual earthquake response of arch dams includes the static effects of the dam weight, the hydrostatic pressure of the impounded water and temperature loads. Several kinds of analyses were carried out to these specific problems. The separate gravity analysis was performed in order to simulate the construction sequence of the dam. For cases with a full reservoir, the hydrostatic load was applied to the entire dam.

Heat-transfer analyses were performed in order to investigate thermal stress induced by the seasonal changes in the temperature condition. The results of the above mentioned static analyses were used as the initial conditions of the seismic response analyses of the arch dam.

3 STATIC BEHAVIOR OF THE DAM

3.1 Outline of in-situ measurements of the dam

To verify the performance of joint model, we carried out several static tests and measurements (Toyoda &c. 1997), vibration test, static measurement of ordinary deformation at crown cantilever section and displacement of joint. In this paper, we showed one result of displacement of crown cantilever. The location of measurement instruments are showed in Fig.7.

3.2 Comparison between measurement results and simulation results

The comparison between the test results and the analysis results are shown in Fig.8, in which the displacement data measured at the crown cantilever were plotted.

The static deformation at crown cantilever obtained from non-linear analysis is as same as that of linear analysis.

It is confirmed that the results of the simulation are consistent with the trend found in the tests, though the slight discrepancy between the measured displacements and the computed ones is recognized

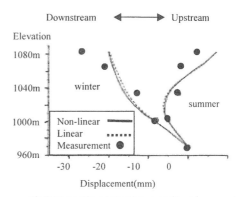

Fig.8 Comparison between test results and
static analytical results (Crown cantilever)

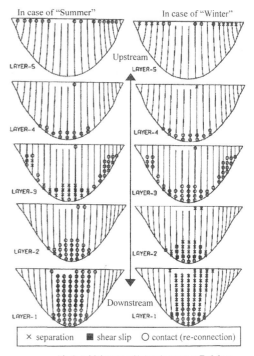

Fig.9 Initial states of joint elements at T=0.0sec

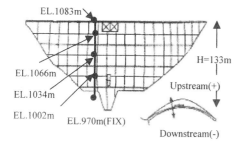

Fig.7 Measurement locations

EL CENTRO CALIF. USA NS 1940 V18

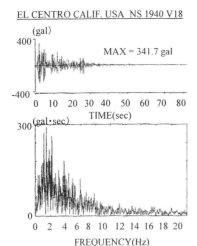

Fig.10 Seismic wave and fourier spectrum

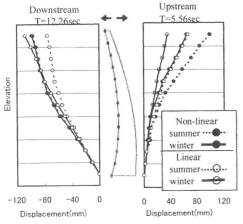

Fig.11 Deformation of Dam body at the crown cantilever

Table 2. Analytical conditions

		Contraction Joint	
		Modeling	No model
Thermal	In winter	C1	C3
Load	In summer	C2	C4

The following cases are considered:
C1. With joints, including thermal stresses
    in winter season (February)
C2: With joints, including thermal stresses
    in summer season (August)
C3: No joints, including thermal stresses
    in winter (February)
C4: With joints, including thermal stresses
    in summer season (August)

The initial states of joint elements at time zero (starting seismic load) are showed in Fig.9. Most of the joints open and close repeatedly through the seasonal changes of the deformation. In particular, the closure of the joints in winter season plays an important role to transfer the internal forces between the arches and cantilevers. Usually, the transfer region of internal forces in the dam body is called "Secondary Arch","Inner Arch" and " Effective Arch"

4 RESULTS OF EARTHQUAKE RESPONSE ANALYSES

4.1 *Analytical conditions*

In this papers, we explains the typical 4-cases which showed in Table 2.

The seismic wave and fourier spectrum which used in analysis are showed in Fig.10. The earthquake ground motion was evaluated by "SHAKE" code, and the modified ground motion was applied in the stream direction at the artificial boundary of the foundation rock

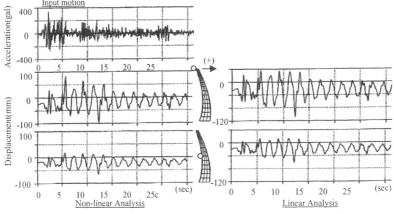

Fig.12 Numerical results (in case of "winter") - Time histories of displacement in stream direction at crown cantilever

4.2 Displacement at crown cantilever

The largest displacement in stream direction at the crown cantilever is showed in Fig.11.

The maximum displacement of the crown cantilever occurred at the time of 12.26sec as the dam displaces in the upstream direction.

When the crown cantilever moves to upstream, the difference between displacement of non-linear analysis and that of linear analysis is bigger. But when the crown cantilever moves to downstream, the difference between displacement of non-linear analysis and that of linear analysis is smaller. The reason which occurred this difference may be due to the restriction of dam-side rock. When the crown cantilever moves to upstream, there is no restriction and the crown cantilever is easy to move.

Time histories of displacement in stream direction at crown cantilever are showed in Fig.12. The maximum acceleration of seismic wave is occurred at the time of about 2.0 sec, the biggest displacement at crown cantilever occurs late for the time which occurred maximum acceleration. The difference between time history of displacement of non-linear analysis and that of linear analysis is smaller.

4.3 Non-linear behaviors of joint

Allowing the contraction joints to open reduces the maximum arch tensile stresses at the crown section.

The results of non-linear behaviors of joints in the Case C1 and C2 are shown in Fig.13.

In case of summer, at the time of 12.26 sec, crown cantilever moves to downstream and arch forces are strong at joints, therefore large area of joints transfer the arch forces and some area of joints slip caused by seismic load. In this condition, maximum vertical tensile stress occurs at concrete-rock interface at upstream heel. In case of winter, at the time of 5.56 sec, crown cantilever moves to upstream and arch forces are weak at joints, therefore large area of joints are separated and it becomes easy to move crown cantilever, because arch forces are weak at joints, therefore several cantilever blocks move independently. The maximum vertical tensile stress of concrete occurs at downstream surface.

4.4 Time history and distribution of stress

The time histories of the membrane stress "σz" in the vertical direction at point A are showed in Fig.14.

In the case of winter and in non-linear analysis, at the time of 5.56 sec, the stress level increases caused by the change of joints condition.

The membrane stress distribution at point A and B are showed in Fig.15.

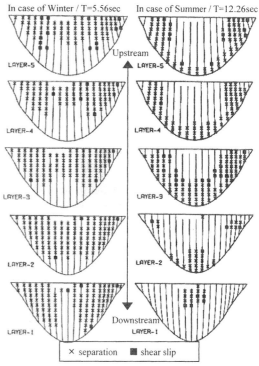

Fig.13 Numerical results - Non-linear behavior of joints -

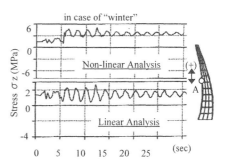

Fig.14 Time histories of the membrane stress at "point A" in the vertical direction(σz) at crown section

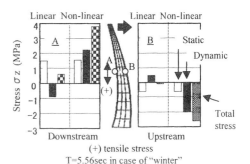

Fig.15 Membrane stress "σz" distribution at point A and B

In the case of winter, at the time of 5.56 sec, crown cantilever moves to upstream, the direction of static stress in non-linear analysis is as same as that in linear analysis, but the direction of dynamic stress in non-linear analysis is different from that in linear analysis.

In case of non-linear analysis, crown cantilever is easy to move compared to that in linear analysis, caused by this movement. So, tensile stress of concrete at downstream surface is much bigger than that in linear analysis. Oppositely, compressive stress of upstream surface is much bigger than that in linear analysis. In this results show the flexibility of crown cantilever caused by non-linear behaviors of joints.

The opening of the joints and release of arch tensile stresses has an important effect on the maximum cantilever stresses in the dam. The large cantilever tensile stresses on the downstream surface occur as the dam displaces in the stream direction, and the internal forces are redistributed from arch action to cantilever bending in the upstream direction.

In the case of winter, predicted crack patterns on the downstream surface is showed in Fig.16. In this analysis, material property of concrete is linear, therefore it can not estimate the concrete crack correctly. In this figure, crack area is the mark which occurred the tensile stress of concrete is over 2.45MPa and crack line's direction is calculating result from principle stress.

These cracks only occurred at downstream and upstream surface and there are no cracks in case of summer because of the strong arch forces.

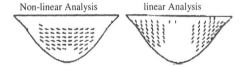

Non-linear Analysis   linear Analysis

Fig.16 Crack patterns on the downstream Surface (" in case of winter ")

5 CONCLUSION

The seismic behavior of the existing arch dam subjected to horizontal earthquake ground motion was investigated and the applicability of the newly developed analysis procedure to this specific problem was clarified. The conclusions obtained from this study are as follows.

(1) Compared with the results of standard linear analyses of the monolithic model, it is found that the opening of the dam's contraction joints in winter season has a significant effect on the transient response of the membrane stress and the maximum crest displacement in stream direction at crown cantilever.

(2) The earthquake analysis of the arch dam with contraction joint opening has shown that the arch tensile stresses predicted by a linear analysis are unrealistically large. The cantilever stresses from a linear analysis, however, are unrealistically low, and the problems associated with large downstream cantilever stresses induced by joint opening may be severe.

(3) The non-linear analysis procedure seems to be useful as a direct simulation tool for getting full understanding of the seismic behaviors of arch dams including joint opening effects.

To assess the generality of these findings, both additional numerical simulation and comparison with experiment data should be performed.

REFERENCES

T. Nishiuchi, Y. Toyoda, T. Matsuo, M. Ueda & K. Mizuno 1998 "Investigation on the Static Behavior of an Aged Arch Dam Using Non-Linear Finite Element Method", CRIEPI Report No.U97060

Y. Toyoda, T. Matsuo, T. Nishiuchi, M. Ueda & K. Mizuno 1997 "Study on the Dynamic Response Properties of an Existing Arch Dam Based on Dynamic In-situ Test Results", CRIEPI Report No.U97031

T. Nishiuchi, T. Kanazu & M. Xu 1995 "A Study on Shear-Slip Failure Criterion of Concrete Blocks Joints", CRIEPI Report No.U94051

T. Nishiuchi, T. Kanazu & M. Xu 1995 "A Study on the Effect of Shear Key Blocks on the Behavior of Vertical Joints in Arch Dams", CRIEPI Report No.U94052

Creative Systems in Structural and Construction Engineering, Singh (ed.) © 2001 Balkema, Rotterdam, ISBN 90 5809 161 9

Seismic performance of hybrid RC columns confined in aramid fiber reinforced polymer square tube

T. Yamakawa
Department of Civil Engineering and Architecture, University of the Ryukyus, Okinawa, Japan

P. Zhong
Graduate School of Engineering and Science, University of the Ryukyus, Okinawa, Japan

ABSTRACT: A new structural concept is proposed for the design of the hybrid RC columns with premanufactured Aramid Fiber Reinforced Polymer (AFRP) tube, which performs the dual function of stay-in-place formwork and transverse reinforcement. Hybrid RC columns were tested under cyclic lateral load while simultaneously subjected to constant vertical load. The theoretical flexural, shear and bond strengths were compared with experimental results. The flexural yield of hybrid RC columns can be ensured when the shear and bond strengths exceed flexural strength.

1 INTRODUCTION

The first author has published papers on the reinforced concrete columns using carbon fiber grid as transverse reinforcement instead of steel hoops (Yamakawa, Zhong & Fujisaki 1997), while conducting research on development of composite columns doubly confined by square steel tube and hoop (Yamakawa, Hao & Muranaka 1997).

In Japan, continuous fiber sheet has been applied in seismic retrofit since about 1990. Especially after the 1995 Hyogoken-Nanbu Earthquake, the seismic diagnosis and retrofit technology are taken seriously and related research results are positively utilized to retrofit construction. Therefore, the construction experience was piled up, and the cost of this expensive new material was gradually lowering with the mass consumption. In September 1999, design guidelines for seismic repair using continuous fiber materials was published in Japan (BDPAJ 1999).

In the United States of America, the carbon shell system was proposed by Seible at University of California, San Diego in 1997 (Seible et al. 1997). It is a new concept that instead of the steel tube, the carbon fiber tube filled with concrete composes a truss element and can be used for bridge. It should be noted that longitudinal and transverse reinforcements are not arranged, so the method should be seen as concrete-filled-tube.

Independent to Seible's approach, the authors devise a study to use continuous fiber tube in order to confine concrete transversely and improve ductility after strength. Impregnating the continuous fiber sheets by epoxy resin in square shape, the square tube performs the dual functions of stay-in-place formwork and transverse reinforcement. In this study, the tube can not play a role of structural member alone, but it can hybridize concrete and reinforcement to form a hybrid element, which has possibility of behaving well in ductility and seismic performance.

A pilot test on columns with the carbon fiber square tube was carried out, and experimental results showed that the hybrid column had a almost equivalent excellent strength and ductility as columns reinforced by the steel tube (Yamakawa & Zhong 1999). Thence, a research on hybrid columns using circular aramid fiber tube was carried on by authors (Yamakawa, Satoh & Zhong 1999).

In this paper, the square aramid fiber tube is utilized in hybrid columns. The purpose of this study is to validate the effectiveness and the extent prospect of the suggestion, as well as seismic performance of the hybrid columns subjected to constant axial load and lateral cyclic forces.

This study is different to seismic retrofit method using continuous fiber sheet. Seismic retrofit method is for existing buildings, but this method is for new-built structure. Moreover, this tube can protect column from corrosion such as the offshore structures. And this tube is lighter than steel, the workability also can be improved.

This research on the hybrid columns with the AFRP tube is valuable to be developed and the future studies is to establish design guidelines. Furthermore, the aramid fiber rods can be adopted as longitudinal reinforcement while fiber grid being the transverse reinforcement, replacing steel reinforcement, so a new structural element consisting of fiber material and concrete can be developed.

2 TEST PLAN

A total of eight columns is provided with 250 mm square and 750 mm height and shear span to depth ratio of 1.5. Details of columns are listed in Table 1. According to the amount of longitudinal reinforcement, all specimens are divided into two series. In H98M series, columns were reinforced vertically with 8 deformed steel bars of which nominal diameter was 19 mm. In H99M series, longitudinal reinforcement was 12 deformed steel bars with nominal diameter of 13 mm. The nominal diameter of steel hoops was 6 mm in common.

The mechanical properties of materials are shown in Table 2. The compression strengths of concrete cylinder are very close between two series, so the influence of concrete strength can be ignored when comparing their elastoplastic behavior between H98M and H99M column series.

The aramid fiber square tube is made of aramid fiber sheet, which has a width of 300 mm, fiber content of 280 g/m^2. The fiber sheet was wrapped transversely with lap of 200 mm, and impregnated by the epoxy resin. After hardening, the thickness of tube plate is about 3.0-4.9 mm for 4-ply-tube, 5.0-5.6 mm for 6-ply-tube. In order to avoid the stress concentration to fiber sheet in the column corner, curvature radius was adopted as 25 mm. Between tube and stubs, gaps of about 10 mm were provided.

All columns were tested under cyclic lateral force while simultaneously subjected to constant axial load. Loading test setup is shown in Figure 1, and loading program is depicted in Figure 2. Story drift angle was controlled at 0.5% and at each stage 3 cycles were repeated. As indicated in "Design Guidelines for Earthquake Resistant Reinforced Concrete Buildings Based on Inelastic Displacement Concept" by Architectural Institute of Japan, drift angle R =1.5% shall be ensured for columns at design limit state. In this loading program, the drift angle R = 5% is adopted as the final deformation. The drift angle R is story drift angle given by dividing story drift by the height of column.

3 EXPERIMENTAL RESULTS

Experimental results and failure modes obtained from test are summarized in Table 3. The AFRP tube reinforcing columns H98M-A44 and H98M-A44h are of bond failure, because of the large amount of longitudinal reinforcement. With too much transverse confinement, the columns of H99M series generated flexural yield except the specimen H99M-A44.

After loading test, AFRP or steel tubes were tore off and the failure conditions of concrete were observed. In H98M series, bond slip cracks occurred along longitudinal steel and shear cracks were found at the upper and lower end of columns, and the bond slip was of the side splitting failure mode. Only the

Table 1. Details of columns.

Specimen	N/(bDσ$_B$)	σ$_B$	p$_g$	p$_w$	Ply of sheet	p$_{wf}$
		MPa	%	%		%
H98M-A0	0.35	37.3	3.67	0.64	0.0	0.0
H98M-A44	0.35	37.3	3.67	0.0	4 plies	0.62
H98M-A44h	0.35	37.3	3.67	0.64	4 plies	0.62
H98M-S33h	0.35	37.3	3.67	0.64	steel tube	2.56
H99M-A0	0.35	36.8	2.44	1.28	0.0	0.0
H99M-A44	0.35	36.8	2.44	0.0	4 plies	0.62
H99M-A44h	0.35	36.8	2.44	1.28	4 plies	0.62
H99M-A66h	0.70	36.8	2.44	1.28	6 plies	0.93

* N = axial force, b = section width, D = section depth, σ_B = compression strength of concrete cylinder, p_g = longitudinal reinforcement ratio, p_w = transverse reinforcement ratio of hoop, p_{wf} = transverse reinforcement ratio of AFRP or steel tube.

Table 2. Properties of reinforcing materials.

Transverse Reinforcement	thickness or	σ$_y$	E
	section area	MPa	GPa
AFRP sheet (1 ply)	0.193 mm	2060	118
Steel tube	3.2 mm	265	223
Rebar D19 (SD345)	287 mm^2	380	189
Rebar D13 (SD295)	127 mm^2	359	202
Hoop for H98M series D6 (SD295)	32 mm^2	388	184
Hoop for H99M series D6 (SD295)	32 mm^2	466	223

* σ_y = tension yield strength of steel, or standard tension strength of AFRP sheet, E = modulus of elasticity.

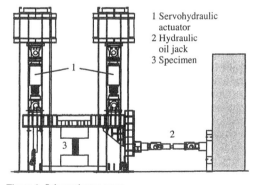

Figure 1. Schematic test setup.

1 Servohydraulic actuator
2 Hydraulic oil jack
3 Specimen

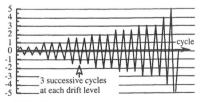

Figure 2. Loading program.

Table 3. Summary of experimental results.

Specimen	V_{exp} (kN)	R_{exp} %	R_f %	Failure mode	Mark
H98M-A0	271.0	0.41	1.00	S	•
H98M-A44	277.2	0.52	2.60	B	•
H98M-A44h	299.5	1.00	2.49	B	•
H98M-S33h	358.7	1.41	2.61	FB	*
H99M-A0	277.2	0.95	2.83	FB	*
H99M-A44	255.5	0.48	2.13	B	•
H99M-A44h	304.1	1.91	>5.00	F	•
H99M-A66h	354.2	1.96	>3.00	F	•

* V_{exp} = experimental peak shear force in push loading direction, R_{exp} = the drift angle responding to V_{exp}, R_f = the drift angle responding to the shear force being 80% of V_{exp}.
** S= shear failure, B= bond failure, FB= bond failure after flexural yielding, F= flexural failure.

steel tube reinforcing column H98M-S33h showed flexural cracks in the both ends.

Since lower amount of longitudinal reinforcement was arranged in addition to the same shear span to depth ratio, much shear and flexural cracks occurred, and seldom bond split cracks were found except the AFRP tube column H99M-A44.

Lateral shear force V versus drift angle R curves are depicted in Figure 3, in which the dotted lines represent calculating flexural yield strength V_{f-1} including N-δ effect by the simplified equation (AIJ 1999a). This equation for flexural yield strength is widely used in Japan and is introduced in the design guidelines by Architectural Institute of Japan.

In the ordinary column H98M-A0, at peak of drift angle R=0.5%, lateral shear force descended immediately when shear cracks occurred. For the sake of longitudinal steel bars of large diameter and amount, the vertical load could be sustained by confined concrete and longitudinal steel, so brittle failure did not occurred although shear failure occurred. Under cyclic load, small shear cracks linked each other and bond splitting cracks formed. After the bond cracks became dominant, responsively the V-R hysteresis curve was in the inverse shape of letter 'S'. The longitudinal steel did not yield at all during loading.

The specimen H98M-44 confined by the AFRP tube showed bond slip at R=0.5%. Trouble happened in measure instrument, so loading test was stopped when drift angle got 2.5%.

The specimen H98M-A44h doubly confined by the AFRP tube and hoops showed stable performance at R=0.5%. At R=1.0%, lateral force degraded as inverses 'S'. After loading test, bond splitting cracks was observed along longitudinal steel, and the bond failure occurred.

When confined by the steel tube and hoops, longitudinal steel of the specimen H98M-S33h yielded at R=1.0%. From 1.5%, bond slip appeared on V-R curves, and the bond deterioration became outstanding. Thus, it was thought that after flexural yield, the bond failure occurred in the end of column.

On the other hand, arranged with much transverse steel hoops and less amount of longitudinal steel, different performance was observed in H99M series.

The ordinary column H99M-A0 showed flexural cracks in the end at R=0.5%. At R=1.0%, the longitudinal steel got its tensile yield and the peak shear force was obtained. After R=2.0%, the deterioration of lateral force indicated that bond failure occurred due to falling of the cover concrete in the end region of column.

Confined by the AFRP tube, the column H99M-A44 got its peak shear force at R=0.5%, then showed degradation till the end of loading. After test, it was observed that all longitudinal bars buckled,

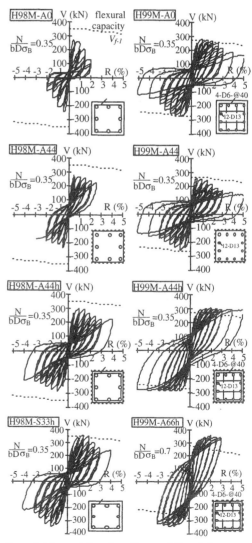

Figure 3. Measured shear force V versus drift angle R curves.

771

and the cover concrete among bond splitting cracks along the inside longitudinal bars spalled off. Comparing the confinement conditions, the inside bars are weaker than the corner bars, so the inside steel buckled prior to the corner steel. For the sake of transverse confinement, this column is of bond failure accompanying with buckling of the longitudinal bars.

The column specimen H99M-A44h of 4 plies of sheet doubly confined by the AFRP tube and hoops depicted stable flexural behavior in hysteresis loop. Although some bond splitting cracks were observed, it was thought that flexural failure governed the behavior of column.

With the AFRP tube of 6 plies of sheet and the same amount of hoops, subjected to high axial load level $N/(bD\sigma_B)=0.7$, the column specimen H99M-A66h presented similar flexural capacity. For the deformation of outplane, the loading was stopped at the end of $R=3.0\%$. Compression cracks and bond splitting cracks were observed, but the flexural failure was main collapse mode.

Average vertical strain on the centerline of column was obtained from the test. Figure 4 compares the strains of columns at the end of each drift angle stage. Depending on different failure mode, tendency of axial strains distinguishes.

In H98M series being of bond failure, because the effective compressive concrete section areas decreased for the cover concrete's spalling off, the axial strain of the column H98M-A0 was the largest. In the column H98M-S33h, because of the yielding of longitudinal bars, the axial strain was bigger than that of the columns H98M-A44 and H98M-A44h, in which longitudinal bars did not yield.

In H99M series, the axial strain decreased with increasing transverse reinforcement. When subjected

to high axial load ratio level of 0.7, the column H99M-A66h strained over 3 times as much as the others under the axial load ratio level of 0.35. This is a reason why concrete is possessed of nonlinear characteristics. The stress-strain relationship remains linear approximately when the axial load ratio is up to 0.5, but when the axial load ratio exceeds 0.5 the stress becomes nonlinear to strain.

4 THEORETICAL INVESTIGATION

When utilizing a fiber model to estimate flexural capacity of the reinforced concrete column, the stress-strain models of confined concrete are adopted. Sakino and Sun approached a relationship for confined concrete by the hoops or steel tube (Sakino & Sun 1994), and Aramid Fiber Retrofit Institute in Japan suggested a model for concrete confined by aramid fiber sheet. The calculating results are shown in Figure 5. In H98M series, the confinement effect from the AFRP tube was greater than that from the steel tube. In H99M series, the enhancement of compressive strength, due to the confinement effect of hoops with supplementary ties, was outstanding. On the other hand, for the longitudinal reinforcement, a bilinear model is used, so that the strain-hardening is not considered.

It is well known that the fiber model can be used to calculate the flexural moment-curvature relationship in the critical transverse section. In order to calculate the shear strength due to flexural capacity versus story drift angle relationships, it was assumed that flexural deformation concentrated in the flexural yield hinge region of which length was half of section depth. Simultaneously the curvature distribution along the column height was also

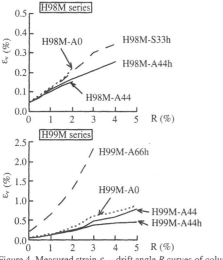

Figure 4. Measured strain ε_v – drift angle R curves of columns.

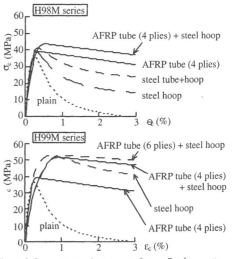

Figure 5. Stress σ_c - strain ε_c curves for confined concrete.

assumed. Finally the virtual work method is utilized to carry out the flexural capacity-drift angle relationship of columns.

Shear strength was calculated by AIJ Design Guidelines equation (AIJ 1999b). The drift angle R_y when plastic hinge occurred was calculated by Sugano's equation (AIJ 1999a), and shear strength degrades according to rotation of plastic hinge. For columns confined by AFRP tube, the stress at the strain of 0.7% could be used as yield strength of hoops according to "Design Guidelines for Retrofit Repair and Construction" by the Building Disaster Prevention Association of Japan (BDPAJ 1999).

The calculating and experimental results of flexural and shear strengths are compared in Figure 6. For flexural capacity, beside the accurate analytical results by the fiber model, results by the simplified equations are showed in the figures.

Test results of the specimen H98M-A0 is close to shear strength. The V-R skeleton curve of the column H98M-A44 does not reach flexural and shear capacity, so it agrees to bond splitting. The column H98M-A44h shows the same bond failure behavior. The column H98M-S33h obtained similar V-R skeleton curves to the analytical flexural results.

In the specimen H99M-A0, the flexural yield was ensured, but when the plastic hinge rotating, the cover concrete fell off and the bond strength descended more fast than the flexural and the shear capacity, thus finally the bond failure happened. The measured skeleton curve of the specimen H99M-A44 was lower than flexural and shear forces, because the bond slip occurred. The measured V-R skeleton curves for the specimen H99M-A44h and H99M-A66h agree to the analytical flexural results by the fiber model shown in Figure 6, although shear strength seems not sufficient as drift angle increases.

Here, the flexural capacity by the simplified equations is compared with the analytical results by the fiber model. Under low axial load level, the former can predict test results to a certain degree, but under high axial load level, it can not. This is a reason why the confined effect is not taken into account in the simplified eqation.

By the fiber model, the axial load-flexural moment interaction diagrams are depicted together with experimental results in Figure 7. When the axial load becomes higher, the enhancement of flexural moment due to confinement effect becomes greater. And it is important to consider the transverse confinement effect when designing. As shown in Figure 7, the shear failure specimen H98M-A0 and the bond failure specimens H98M-A44, H98M-A44h and H99M-A44 do not reach N-M interaction diagrams which mean the flexural strength of the standard RC column specimens.

Bond stress and bond strength are calculated (AIJ 1999b), and shear strength due to the bond slip is estimated. AFRP tube is supposed to hoop ratio as $(p_{wf} E_f / E_s)$. Above-mentioned calculated results are compared with the experimental results in Figure 8. In which, the V_{exp} shows experimental peak shear force, the V_s represents calculated shear strength before plastic hinge occurred, the V_{f-l} means

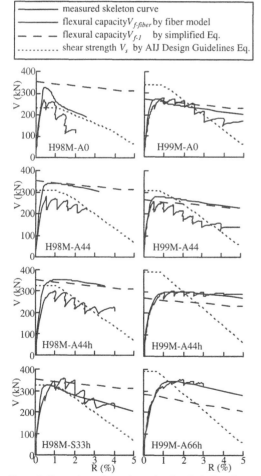

———	measured skeleton curve
———	flexural capacity $V_{f\text{-}fiber}$ by fiber model
– – –	flexural capacity $V_{f\text{-}l}$ by simplified Eq.
··········	shear strength V_s by AIJ Design Guidelines Eq.

Figure 6. Comparison of experimental and analytical results.

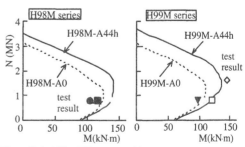

Figure 7. Axial load N - moment M curves.

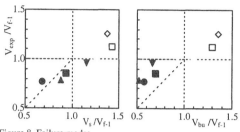

Figure 8. Failure modes.

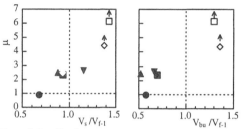

Figure 9. Ductility factors.

calculated flexural capacity by simplification equation, the V_{bu} is calculated shear strength due to the bond slip. Especially, the degradation of bond strength is dominant in comparison with shear strength in the specimens of the bond failure type, such as H99M-A44, H98M-A44 and H98M-A44h through calculation as illustrated in Figure 8. When shear and bond strengths exceed flexural capacity, the flexural yield shall occur. This also can be applied in designing.

In Figure 9, the ductility factor $\mu=R_f/R_y$ is compared with the analytical strength, where the R_f is experimental drift angle when shear force descends to 80% of peak shear force, the R_y is drift angle when columns yield, calculated by Sugano's approach (AIJ 1999a). The arrows depicted in the figures indicate that the column should have more ductility because the drift angle R_f corresponding to the ultimate state deformation could not be obtained in the loading test. Namely, shear force does not descend to 80% of the peak shear force until final drift angle during the cyclic loading test.

As shown in Figure 9, the greater the value of V_s/V_{f-1} and V_{bu}/V_{f-1} becomes, the larger the ductility factor μ is. But for the ordinary column H99M-A0, it is very difficult to prevent cover concrete from spalling off, so the ductility can not be expected. Even if too much hoops are arranged, there is a limit of ductility to the ordinary hooped columns without AFRP tube, just as the specimen H99M-A0.

5 CONCLUSIONS

This paper proposes a new hybrid column utilizing aramid fiber reinforced polymer tube. The AFRP tube is premanufactured by impregnating aramid fiber sheet by epoxy resin, and plays the roles of stay-in-place formwork when casting concrete, and transverse reinforcing material.

With longitudinal reinforcement ratios of 3.67% and 2.44%, eight columns were loaded under cyclic lateral forces while simultaneously subjected to constant axial load. Experimental observations shows that bond failure occurred in the hybrid column doubly confined by the AFRP tube and hoop when the longitudinal ratio was 3.67%. On the other hand, with the longitudinal ratio of 2.44%, columns showed significantly improved strength and ductility even when under high axial load level of 0.7.

For the hybrid columns confined by the AFRP tube and hoop, when shear and bond strengths exceed flexural capacity simultaneously, flexural failure with high ductility and high strength can be obtained.

REFERENCES:

Architectural Institute of Japan (AIJ). 1999. *AIJ Standard for Structural Calculation of Reinforced Concrete Structures-Based on Allowable Stress Concept-revised 1999*: 154-155, 52-57. (in Japanese)

Architectural Institute of Japan (AIJ). 1999. *Design guidelines for earthquake resistant reinforced concrete buildings based on inelastic displacement concept*: 142-162, 175-192. (in Japanese)

Sakino, K., Sun, Y.P. 1994. Stress-strain curve of concrete confined by rectilinear hoop. *J. of Struct. and Constn. Engrg*: No. 461, 95-104. Arch. Inst. of Japan. (in Japanese)

Seible, F., Hegemier. G., Karbhari, V., Burgueno.R. & Davol, A. 1997. The carbon shell system for modular short and medium span bridges. *Intnl. Composites EXPO'97*: Session 3-D/1-66.

The Building Disaster Prevention Association of Japan (BDPAJ). 1999. *Design Guidelines for Retrofit Repair and Construction for Reinforce Concrete and Steel Reinforced Concrete Buildings Utilizing Continuous Fiber Reinforcing Material*: 43-45.

Yamakawa, T., HAO, H.T. & Muranaka, K. 1997. Elastoplastic behavior of doubly confined R/C columns in steel tube and hoops. *J. of Struct. and Constn. Engrg*: No. 500, 83-89. Arch. Inst. of Japan.

Yamakawa T., Satoh H., & Zhong P. 1999. Seismic performance of Hybrid Reinforced concrete circular columns confined in aramid fiber reinforced polymer tube. *4th Intnl. Symp. on Fiber Reinforced Polymer Reinforcement for Reinforced Concrete Structures*: 1089-1102.

Yamakawa T.& Zhong P. 1999. Seismic performance of RC columns laterally confined by carbon fiber reinforcing plastic tube. *Specialist Techniques and Materials for Concrete Construction, International Congress "Creating with Concrete"*: 455-463.

Yamakawa, T., Zhong, P. & Fujisaki, T. 1997. Elastoplastic behavior of hybrid R/C columns with CFRP grids. *NON-Metallic (FRP) Reinforcement for Concrete Structures Proc. of the 3rd Intnl. Symp.* : Vol.2, 607-614

24 Mechanical behavior

Creative Systems in Structural and Construction Engineering, Singh (ed.) © 2001 Balkema, Rotterdam, ISBN 90 5809 161 9

True parabolic stress method of analysis in reinforced concrete

Ramon V. Jarquio

New York City Transit, Brooklyn, N.Y., USA

ABSTRACT: This paper describes the application of the true parabolic stress method of analysis for predicting the ultimate strength of reinforced concrete members. It is different from the so-called parabolic-rectangular stress block method of analysis in that it utilizes the basic parabola, which approximates the stress/strain diagram of the compressive properties of the concrete material. The methodology of the true parabolic stress method not only satisfies the basic equilibrium conditions of external and internal forces, but also opens the way for the analytical analysis for ultimate strength of reinforced concrete sections.

1 INTRODUCTION

For more than half a century, analysis of reinforced concrete structural members is being done using the so-called parabolic-rectangular stress block method first proposed by Whitney. This paper will present the true parabolic stress method as the analytical alternative analysis for the ultimate strength of reinforced concrete sections. This analytical method uses the parabolic nature of the concrete stress/strain curve and the linear stress strain property of steel reinforcement. The interaction of these two materials is related by the common deformation as they resist applied loads to their ultimate capacities.

2 DERIVATION

Figure 1 shows the overlay of the basic parabola to the concrete stress strain curve and the linear strain based on the steel stress/strain curve. From Figure 2, the equation of the basic parabola is given by the relationship

$$y/x(L - x) = m/(L/2)^2 \qquad (1)$$

The concrete force is defined by the maximum useable concrete strain allowed and the maximum area bounded by the curve OMN and the X-axis as shown in Figure 3. From equation (1)

$$y/x(2 x_m - x) = f_c' /(x_m)^2 \text{ or}$$

$$y = f_c' x (2x_m - x)/ (x_m)^2 \qquad (2)$$

and from the strain diagram,

$$x_m / e_m = c/ e_c \text{ or } x_m = c\, e_m / e_c \qquad (3)$$

Substitute eq. (3) in (2)

$$y = (e_c f_c' / c^2 e_m^2) [2 c e_m x - e_c x^2] \qquad (4)$$

The area under this curve by calculus is given by

$$A = \int (e_c f_c' / c^2 e_m^2) [2 c e_m x - e_c x^2] dx \qquad (5)$$

Integrate and evaluate limits from 0 to c to obtain

$$A = (c f_c' e_c / 3 e_m^2) [3 e_m - e_c] \qquad (6)$$

Differentiate equation (6) with respect to e_m, equate to zero and solve for the value of e_m to yield the maximum area under the curve.

$$A' = (c f_c' e_c / 3)[e_m^2 (3) - (3 e_m - e_c)(2 e_m)]/e_m^4$$

$$= 0$$

$$e_m = (2/3) e_c \qquad (7)$$

Equation (7) indicates that the maximum compressive stress f_c' passes through the centroid of the triangular strain diagram. From equation (3)

$$x_m = (2/3) c \qquad (8)$$

Substitute equation (8) in equation (2) to obtain the equation of the basic parabola to use in the analysis, i.e.,

$$y = (3 f_c' / 4 c^2) [4 c x - 3 x^2] \qquad (9)$$

Equation (9) is the basic parabola to use in the analysis for the ultimate strength of any reinforced concrete section. At the assumed point of failure, "N"

$$f_c = c(2x_m - c) f_c' / (x_m)^2 \qquad (10)$$

Put equation (8) in equation (10) and simplify

$$f_c = 0.75 f_c' \qquad (11)$$

In Figure 1 using ACI value of $e_c = 0.003$ and the depth of concrete in compression, denoted by "c", the total compressive force on this section per unit width is given by the area bounded by the parabola, the X-axis and the equation of the line x = c, i.e.

$$A = \int (3 f_c' / 4 c^2) [4 c x - 3x^2] dx \qquad (12)$$

Integrate and evaluate limits from 0 to c to obtain

$$A = 0.75 c f_c' \qquad (13)$$

Equation (13) indicates that the equivalent rectangle to the parabolic area must contain the product of the multipliers to the sides of the rectangle equal to the constant 0.75. Current ACI method indicates this product varies according to the value of f_c' and is less than 0.75 for all values of f_c'.

It is also obvious from equation (13) that the compressive force is equal to the product of the compressive depth, "c" and the compressive stress, "f_c" at the assumed point of failure "N".

The centroid or point of application of this force can be calculated from

$$A\bar{x} = \int (3 f_c' / 4 c^2) [4 c x^2 - 3 x^3] dx \qquad (14)$$

Integrate and evaluate limits from 0 to c to obtain

$$A\bar{x} = 0.4375 c^2 f_c' \qquad (15)$$

Therefore,

$$\bar{x} = 0.583 c \qquad (16)$$

$$c - \bar{x} = 0.417 c \qquad (17)$$

It is apparent from the above equations that the value of the concrete force is not affected by change in the value of the concrete strain, e_c. On the other hand, bar force varies in accordance with the value of e_c. Canadian uses a value of $e_c = 0.0035$, which will increase the bar force from $e_c = 0.003$ allowed under the ACI methodology.

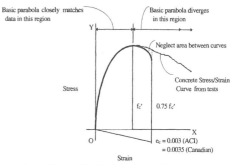

Figure 1. Overlay of Basic Parabola to Concrete Stress/Strain Data

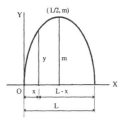

Figure 2. Parabola

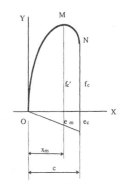

Figure 3. Basic Parabola for Ultimate Strength Analysis

2.1 Singly-Reinforced Concrete Beam

In reinforced concrete beams, the maximum compressive force is developed from the edge to a depth equal to "c". In figure 4, the compressive and tensile forces are as follows:

$$C = 0.75 c f_c' b \qquad (18)$$

$$T = A_s f_y = pbdf_y \qquad (19)$$

$$\sum F = 0: \quad C = T$$

$$0.75 c f_c' b = pbdf_y \quad \text{or} \quad p = 0.75c f_c' / df_y \qquad (20)$$

Using linear distribution of strains with respect to the neutral axis,

$$c = 87d/(87 + f_y) \qquad (21)$$

Substitute equation (21) in equation (20) to solve for "p" at balanced condition, i.e.

$$p_b = 65 f_c' / f_y (87 + f_y) \qquad (22)$$

To insure ductile failure, $p_{max} = 0.75 p_b$ is mandated in the ACI code. Therefore, for design use

$$p_{max} = 49 f_c' / f_y(87 + f_y) \qquad (23)$$

in which f_c' and f_y are expressed in ksi.

2.1.1 *Minimum Reinforcement*

In Figure 4, assume plain concrete so that $c = d/2$ and for f_c' substitute the value of the modulus of rupture equal to $7.5 (f_c')^{1/2}$. Hence, from equation (20)

$$p_{min} = 0.75(d/2)\{237(f_c')^{1/2} \}/1000f_y \quad \text{or}$$

$$p_{min} = 0.09(f_c')^{1/2} / f_y \qquad (24)$$

2.1.2 *Moment Capacity*

Summing moments about the compressive force, C

$$M = A_s f_y(d-0.417c) = pbdf_y(d-0.417c) \qquad (25)$$

Substitute equation (19), (20) and (23) in equation (25) to obtain

$$M = Kbd^2 \qquad (26)$$

in which,

$$K = 49(51 + f_y) f_c' /(87 + f_y)^2 \qquad (27)$$

2.2 *Doubly-Reinforced Concrete Beam*

In Figure 5, the compressive force on the steel rein-forcement and the concrete force are obtained from the following expressions, i. e.

$$C_1 = A_s' f y' \qquad (28)$$

$$C_2 = 0.75c f_c' b \qquad (29)$$

The tensile force is determined from

$$T = A_s f_y \qquad (30)$$

From the strain diagram,

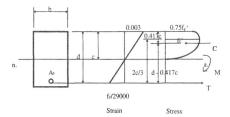

Figure 4. Singly-Reinforced Beam (balanced condition)

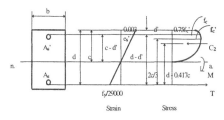

Figure 5. Doubly - Reinforced Beam

$$e_s' = 0.003(c-d')/c \qquad (31)$$

From the property of the parabola,

$$f_c = (c-d')\{(c/3) + d'\} f_c' /(2c/3)^2 \qquad (32)$$

$$f_y' = (0.003/c)\{(c-d')(29000)\} - f_c \qquad (33)$$

Put equation (32) in equation (33) and simplify

$$f_y' = (1/c^2)\{(87-0.75f_c')c^2-(87d' +1.50d'f_c')c + 2.25d'^2f_c' \leq f_y \text{ max} \qquad (34)$$

Put equation (34) in equation (28) to obtain

$$C_1 = (A_s'/c^2)\{(87-0.75f_c')c^2 - (87d' + 1.50d'f_c')c + 2.25d'^2f_c') \qquad (35)$$

$$\Sigma F = 0: \quad C_1 + C_2 = T \qquad (36)$$

The resulting equation for " c" is a cubical equa-tion, which can be solved analytically by Cardan's and Trigonometric formulas, depending on the value of the determinants.

2.2.1 *Moment Capacity*

Moment capacity is derived from $\Sigma M = 0$, to obtain

$$M = C_1 (d-d') + C_2(d-0.417c) \qquad (37)$$

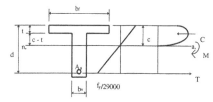

Figure 6. T - Beam Diagram

2.3 *T-Beam Analysis*

Two cases of T- beam are analyzed below.

2.3.1 *Concrete Forces*

Case 1: When $c \geq t$ use beam formulas derived above.

Case 2: When $c > t$, i.e., $d > [(87 + f_y/87]t$, the concrete forces are determined from Figure 6 as follows:

$$C_F = [0.75f_c'/c^2] \int [4cx - 3x^2] \, dx \quad (38)$$

Integrate between limits (c- t) to t to yield

$$C_F = [0.75f_c'/c^2][ct^2 + c^2t - t^3] \quad (39)$$

$$C_F \bar{x}_F' = [0.75f_c' \, b_F/c^2] \int [4cx^2 - 3x^3] dx \quad (40)$$

Integrate and evaluate limits from (c-t) to t to yield

$$C_F \bar{x}_F' = [0.75f_c' \, b_F/12c^2][12c^3t + 6c^2t^2 - 20ct^3 + 9t^4] \quad (41)$$

$$C_W = [0.75f_c' \, b_W/c^2] \int [4cx - 3x^2] \, dx \quad (42)$$

Integrate and evaluate limits from 0 to (c-t) to yield:

$$C_W = [0.75f_c' \, b_W/c^2][c^3 - c^2t - ct^2 + t^3] \quad (43)$$

$$C_W \bar{x}_W' = [0.75f_c' \, b_w / c^2] \int [4cx^2 - 3x^3] \, dx \quad (44)$$

Integrate and evaluate limits from 0 to $(c - t)$ to yield:

$$C_W \bar{x}_W' = [0.75f_c' \, b_W /12c^2][7c^4 - 12c^3t - 6c^2t^2 + 20ct^3 - 9t^4] \quad (45)$$

2.3.2 *Bar forces*

The total bar force is calculated as $T = A_s f_y$. Applying the equilibrium condition of forces,

$\Sigma F = 0$: $T = C_F + C_W$, from which

$$A_s = [0.75f_c'/c^2f_y][(b_F - b_W)(ct^2 + c^2t - t^3) + b_Wc^3](46)$$

$\Sigma M = 0$: $M = T(d - c) + C_F \bar{x}_F' + C_W \bar{x}_W'$, from which

$$M = A_s f_y(d - c) + [0.75f_c'/12c^2][(b_F - b_W)(12c^3t + 6c^2t^2 - 20ct^3 + 9t^4 + 7c^4b_W] \quad (47)$$

3 CONCLUSIONS

1. The true parabolic stress method of analysis is a more rational and clearer methodology as compared to the current ACI parabolic-rectangular stress block method.
2. The true parabolic stress method proved that it is not necessary to use the equivalent rectangular stress block to solve for the ultimate strength of reinforced concrete sections.
3. It is now possible to analytically calculate the ultimate strength of reinforced concrete circular and rectangular column sections without resorting to finite-element procedures.

NOTATION

A_g = gross area of concrete section
A_s = total area of steel reinforcement (rebars)
A_s' = area of compression reinforcement
B = width of rectangular concrete section
C = total concrete compressive force for singly-reinforced concrete beam
C_f = compressive force on T-Beam Flange
C_w = compressive force on T-Beam Web
C_1 = compressive force on steel reinforcement
C_2 = total concrete compressive force for doubly-reinforced beam
d = distance from extreme compression fiber to centroid of steel reinforcement
d' = distance from concrete edge to centroid of steel reinforcement
e_c = concrete strain
e_s' = steel strain in the compression steel reinforcement
f_c = compressive stress in concrete
f_c' = specified compressive strength of concrete
f_y = specified yield strength of reinforcement
$p = A_s / A_g$ = ratio of steel reinforcement
T = total tensile force in rebars
x_F = lever arm from neutral axis of concrete force on T-Beam Flange
x_w = lever arm from neutral axis of concrete force on T-Beam Web

Note: All other alphabets and symbols used in the mathematical derivations are defined in the context of their use in the analysis.

REFERENCES

ACI 318 R – 83. *Commentary on Building Code Requirements for Reinforced Concrete: Chapter 10.*

Ferguson, Phil, M., Breen, John, E. and Jirsa, James, O. *Reinforced Concrete Fundamentals, Edition: 45-70.*

Gerstle, Kurt, H. *Basic Structural Design.* McGraw-Hill Book Company: 25-31.

Beyer, William, H. *Handbook of Mathematical Sciences, 6th edition*: 31 - 32.

Creative Systems in Structural and Construction Engineering, Singh (ed.) © 2001 Balkema, Rotterdam, ISBN 90 5809 161 9

Numerical equations for member design of concrete filled steel tubular columns

M. Chao
Cheng Shiu Institute of Technology, Taiwan

J. Q. Zhang
University of Western Sydney, N.S.W., Australia

ABSTRACT: In this paper a set of empirical equations were proposed for the prediction of member strength of concrete filled tubular columns for both normal and high strength concrete. Firstly, accurate numerical analyses of hundreds of composite columns were performed using a general analysis procedure developed by the authors. Secondly, the coefficients of the proposed numerical equations were derived by curve fitting the large amount of results obtained through the numerical analyses. Finally, the proposed equations were verified by comparison with the published test data from various sources.

1 INTRODUCTION

The calculation of axial force and bending moment strength diagrams plays an important part in the design of steel-concrete composite columns (Bridge et al, 1997). Although various codes of practice attempt to simplify the complex calculation procedures, most of them still remain tedious and difficult to use (Eurocode 4, 1992).

Design interaction equations for beam-column members have been proposed by several researchers for steel and reinforced concrete members. Chen and Duan (1989) published interaction equations for steel members (short and slender) under the combined bending moments and axial force. Griffis (1992) examined and compared the interaction diagrams for composite columns plotted according to the ACI and the AISC-LRFD design approach.

In this paper a set of empirical equations were proposed for the prediction of member strength of concrete filled tubular columns for both normal and high strength concrete. Firstly, accurate numerical analyses of hundreds of composite columns were performed using a general analysis procedure developed by the authors (Chao, 2000). Secondly, the coefficients of the proposed numerical equations were derived by curve fitting the large amount of results obtained through the numerical analyses. Finally, the proposed equations were verified by comparing the results with the published test data from various sources.

2 THE EQUATION OF MEMBER STRENGTH LINE

2.1 Equation of CFST member strength line

For a CFST beam-column, the load-carrying capacity of the member with the influence of slenderness and imperfection can be represented by an *N-M* interaction diagram, which is referred to in Fig. 1 as the member strength line. Generally the member strength line is derived from the section strength line. There has been some research effort to successfully represent the section strength line by using a set of empirical equations (Hajjar and Gouley, 1996; Chao and Zhang, 1999). A similar approach is attempted here with the member strength line.

A general equation representing the member strength line subjected to combined bending moment and axial load can be written as follows:

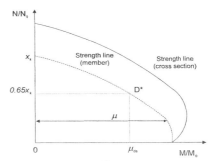

Fig. 1 Member strength diagram

$$N_i \geq N_{ds}$$

$$(\frac{N_i - N_{ds}}{N_k - N_{ds}})^\alpha + \frac{M_i}{M_{ds}} = 1.0 \qquad (1a)$$

$$N_i < N_{ds}$$

$$\frac{N_i}{N_{ds}} + (\frac{M_i}{M_0 - M_{ds}})^\beta = 1.0 \qquad (1b)$$

where N_k, M_0 are reduced axial compression and pure bending moment strengths of the column, respectively. Their computation is described in details in Chao (2000). N_{ds}, M_{ds} are axial compression and bending moment strengths at a special point D*, respectively. N_i, M_i are axial compression and bending moment strengths of a particular point on the member strength line.

$$\alpha = 1 + 0.01\frac{L}{D} \qquad (2a)$$

$$\beta = 2.95 - 0.095\frac{L}{D} + 0.0011(\frac{L}{D})^2 \qquad (2b)$$

2.2 Development for the coefficient of the force points

The member strength interaction diagram for a CFST column is constructed based on the three basic points as indicated in Fig. 1. These are, (1) the reduced maximum axial compressive load, N_k; (2) the reduced axial load and moment strength at point D*, and (3) the pure bending moment, M_0.

The concentric member strength N_k is given as in Eurocode 4 (1992)

$$N_k = \chi_k N_0 \qquad (3)$$

The force point D* is defined as:

$$N_{ds} = 0.65\chi_k N_0 \qquad (4)$$

$$M_{ds} = \mu_{ds} M_0 \qquad (5)$$

where μ_{ds} is a coefficient, and is determined empirically in this paper.

The first step in the development of an expression for the CFST member strength line entails selecting a wide range of cross-sections and length of columns representative of CFST members. There are three properties that most directly affect the behaviour of the CFST member strength. These may be identified by three dimensionless ratios: the ratio of the column length to section depth (or diameter for circular section) (L/D), the ratio of the section depth to wall thickness (D/t) and the ratio of the steel yield design strength to the concrete compression design strength (f_{yd}/f_{cd}).

Six series of length to diameter were selected, with L/D ratios being 5, 10, 20, 30, 40 and 50. Within each series, five series of sections were selected, with D/t ratios being 20, 40, 50, 80 and 100. Within each of these different f_{yd}/f_{cd} ratios were chosen. The design concrete strength f_{cd}

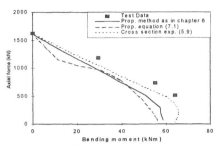

Fig. 2(a) Comparisons with Matsui's result ($L/D = 4$, $f_{cd} = 32$ MPa, $f_{yd} = 410$ MPa)

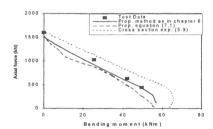

Fig. 2(b) Comparisons with Matsui's result ($L/D = 12$, $f_{cd} = 32$ MPa, $f_{yd} = 410$ MPa)

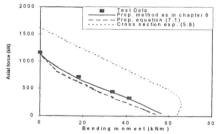

Fig. 2(c) Comparisons with Matsui's result ($L/D = 24$, $f_{cd} = 32$ MPa, $f_{yd} = 410$ MPa)

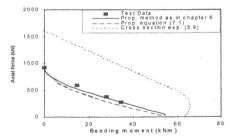

Fig. 2(d) Comparisons with Matsui's result ($L/D = 30$, $f_{cd} = 32$ MPa, $f_{yd} = 410$ MPa)

ranged from 20 MPa to 120 MPa, a 10 MPa uniform increment in the value. The design steel yield strength of tubes f_{yd} ranged from 200 MPa to 600

MPa, a 100 MPa uniform increment in the value. Therefore f_{yd}/f_{cd} ranged from 2.5 to 30.

Rigorous numerical analyses were performed using a general analysis procedure developed by the authors (Zhang and Chao, 1999). The results of these accurate analyses form the basis of the empirical expression proposed in this paper.

The coefficient μ_{ds} is determined by a least-squares procedure using the axial force and moment data from the analysis of each member. The coefficient is expressed in terms of the L/D, D/t and f_{yd}/f_{cd} ratio of the member:

$$\frac{M_{ds}}{M_0} = \mu_{ds} = [a_2(\frac{D}{t})^2 + a_1(\frac{D}{t}) + a_0](\frac{L}{D})^2$$

$$+[b_2(\frac{D}{t})^2 + b_1(\frac{D}{t}) + b_0](\frac{L}{D})$$

$$+[c_2(\frac{D}{t})^2 + c_1(\frac{D}{t}) + c_0] \leq 1.0 \qquad (6)$$

where a_i, b_i, and c_i are the coefficients. They are given in the appendix for rectangular and circular section for normal strength concrete. The coefficients are also derived for high strength concrete (Chao 2000).

3 RESULTS AND DISCUSSION

In this work the proposed empirical expression was used to predict the member strength of CFST beam-columns. The results were compared with the test results in published reports.

3.1 Normal strength concrete

To provide confirmation of the validity of the proposed design equation, comparisons have been made with results of full-scale test of rectangular and circular sections on 40 uniaxially bent beam-columns as reported by Matsui, et al (1995).

The specimen sectional dimensions are 150×150 *mm* with a wall thickness of 4.5 *mm* for rectangular CFST sections. Various lengths were tested with an L/D ratio of 4, 8, 12, 18, 24 and 30. The concrete strength is 32 MPa and the yield stress of the steel is 410 MPa. The test data for the 24 tests reported by Matsui, et al (1995) are selectively shown in Fig. 2 (*a*)-(d).

For circular CFST sections, the specimen sectional dimensions are 165 *mm* in diameter with a wall thickness of 4.5 *mm*. Various lengths were tested with an L/D ratio of 4, 8, 12, 18, 24 and 30. The concrete strength is 42 MPa and the yield stress of the steel is 358 MPa. The data for the 24 tests reported by Matsui (1995) are selectively shown in Fig. 3 (*a*)-(d).

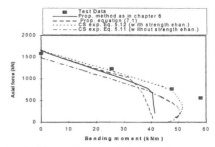

Fig. 3(*a*) Comparisons with Matsui's result ($L/D = 4$, $f_{cd} = 42$ MPa, $f_{yd} = 358$ MPa)

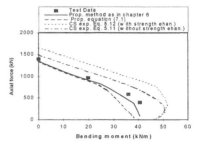

Fig. 3(b) Comparisons with Matsui's result ($L/D = 12$, $f_{cd} = 42$ MPa, $f_{yd} = 358$ MPa)

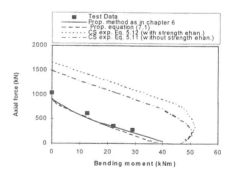

Fig. 3(c) Comparisons with Matsui's result ($L/D = 24$, $f_{cd} = 42$ MPa, $f_{yd} = 358$ MPa)

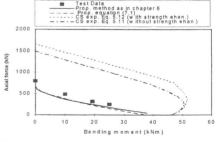

Fig. 3(d) Comparisons with Matsui's result ($L/D = 30$, $f_{cd} = 42$ MPa, $f_{yd} = 358$ MPa)

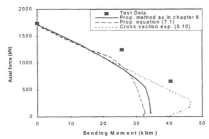

Fig. 4(a) Comparisons with Chung's result
(L/D = 4, f_{cd} = 86 MPa, f_{yd} = 450 MPa)

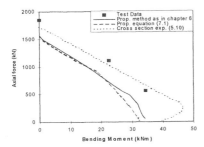

Fig. 4(b) Comparisons with Chung's result
(L/D = 12, f_{cd} = 86 MPa, f_{yd} = 450 MPa)

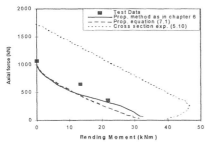

Fig. 4(c) Comparisons with Chung's result
(L/D = 24, f_{cd} = 86 MPa, f_{yd} = 450 MPa)

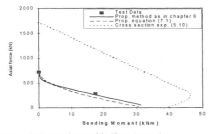

Fig. 4(d) Comparisons with Chung's result
(L/D = 30, f_{cd} = 86 MPa, f_{yd} = 450 MPa)

3.2 High strength concrete

Tests for high strength concrete filled steel tube columns were reported by Chung [Chung *et. al.*, 1999]. Seventeen columns with square sections

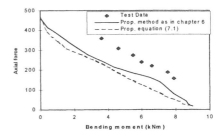

Fig. 5 (a) Comparisons with Kilpatrick's result
(L/D = 19, f_{cd} = 57 MPa, f_{yd} = 450 MPa)

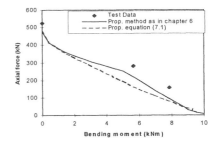

Fig. 5(b) Comparisons with Kilpatrick's result
(L/D = 21.5, f_{cd} = 96 MPa, f_{yd} = 450 MPa)

were constructed using a 165.2 × 165.2 *mm* with a 4.5 *mm* wall thickness. Various lengths were tested with an *L/D* ratio of 4, 8, 12, 18, 24 and 30. The concrete strength is 86 MPa and the yield stress of the steel is 450 MPa. The data for the 17 tests reported by Chung, et al (1999) are shown in Fig. 4 (*a*)-(*d*).

Analyses were also carried out on specimens tested by Kilpatrick and Rangan (1997). The specimen was constructed using a 101 *mm* diameter circular steel tube (CHS) with 2.4 *mm* wall thickness. The first group has a column length of 1947 *mm*. The concrete strength is 57 MPa and the yield stress of the steel is 410 MPa. The second group has a column length of 2175 *mm* and the concrete strength is 96 MPa and the yield stress of the steel is 450 MPa. The results of CFST columns are shown in Figs. 4 (*a*)-(*b*).

3.3 Experimental verification

In Figs. 2-5, solid squares represent experimental results, solid lines represent the results from modified Eurocode 4 method (Chao, 2000) and the dashed lines represent the proposed equation described in this paper. In addition to these, the cross-section strength presented in Zhang and Chao (1999) is also shown by dotted line. For circular sections, it also included the strength curve with or without confinement effect. It is shown that good agreement is observed between the strength of columns from the tests and the proposed equation for slender columns with L/D ratio greater than 12. However, it appears that the prediction of column strength may become increasingly conservative as L/D ratio becomes less than 8. This could be due to the proposed method is based Eurocode 4 recommendations with regard to the second order effect and imperfections, and it is overly conservative for short columns.

4 CONCLUSIONS

In this paper a set of empirical equation were presented for the prediction of member strength of CFST columns for both normal and high strength concrete. The proposed equations were based on accurate numerical analyses, performed using a general analysis procedure developed by the authors. The numerical equations have been verified extensively by comparing with 75 CFST tests.

From comparison between the proposed method based on Eurocode 4 recommendation, the proposed equations here and the test results, it can conclude that both the proposed method and the proposed equations can provide accurate predictions of the strengths of concrete filled slender tubes with a wide range of end eccentricities and dimension. Short columns with L/D ratio less than 12 can be designed according to the proposed equation for section strength only given elsewhere without considering the effect of secondary effect.

It is found that it is feasible as in the section design to provide a set of numerical equations to cover all the possible practical design parameters for member design with acceptable accuracy. The equations and the coefficients are numerically lengthy, however they can be readily coded into a programmable calculator or spreadsheet for design purposes.

REFERENCES

Bridge, R.Q., O'Shea, M.D. and Zhang, J.Q. (1997), "Load moment interaction curves for concrete filled tubes", Proceeding of 15th Australasian Conference on the Mechanics of Structures and Materials, Melbourne, pp141-146.

Chao, M. (2000), The Design and Behaviour of Concrete Filled Steel Tubular Beam-Columns, PhD Thesis, University of Western Sydney, Australia.

Chao, M. and Zhang, J.Q. (1999), "Accuracy of Numerical Expressions for the Section Analysis of CFST Beam-columns", Proceedings of EASEC-7, August, Japan, pp 967-972.

Chen, W. F., and Duan, L. (1989), "Design Interaction Equation for Steel Beam-Columns", Journal of Structural Division, ASCE, Vol. 115, No.5, pp1225-1243.

Chung, J., Tsuda, K. and Matsui, C. (1999), "High-Strength Concrete Filled Square Tube Columns Subjected to Axial Loading", The Seven East Asia-Pacific Conference on Structural Engineering & Construction, August, Kochi, Japan, pp955-960.

Eurocode 4 (1992) ENV 1994-1-1 Eurocode 4, Design of Composite Steel and Concrete Structures, Brussels, European Committee for Standardisation.

Griffis, L. G. (1986), "Some Design Considerations for Composite-Frame Structures", Engineering Journal of the AISC Second Quarter, pp59-64.

Hajjar, J. F. and Gouley, B. C. (1996), "Representation of Concrete-Filled Steel Tube Cross-Section Strength", Journal of Structural Engineering, Vol. 122, No. 11, Nov. pp1327-1336.

Kilpatrick, A. E. and Rangan, B. V. (1997), Tests on High-Strength Composite Concrete Columns, Research Report No. 1/97, School of Civil Engineering, Curtin University of Technology, Perth, March, 202pp.

Matsui, C., Tsada, K. and Ishibashi, Y. (1995), "Slender Concrete Filled Steel Tubular Columns under Combined Compression and Bending", 4th Pacific Structural Steel Conference, Vol. 3, Singapore, pp29-36.

Zhang, J.Q. and Chao, M. (1999), "A Numerical Procedure for the Analysis of Steel Concrete Composite Columns", Proceedings of EASEC-7, August, Japan, pp961-966.

APPENDIX 1

A1.1 *Rectangular CFST Columns*

<u>Normal Strength Concrete ($f_{cd} \le 50$ MPa)</u>

$$a_0 = [-0.47(\frac{f_{yd}}{f_{cd}})^2 + 12.89(\frac{f_{yd}}{f_{cd}}) - 73.82] \times 10^{-5}$$

$$a_1 = [0.23(\frac{f_{yd}}{f_{cd}})^2 - 5.72(\frac{f_{yd}}{f_{cd}}) + 24.94] \times 10^{-6}$$

$$a_2 = [-0.21(\frac{f_{yd}}{f_{cd}})^2 + 5.20(\frac{f_{yd}}{f_{cd}}) - 22.68] \times 10^{-8}$$

$$b_0 = [0.16(\frac{f_{yd}}{f_{cd}})^2 - 5.15(\frac{f_{yd}}{f_{cd}}) + 38.30] \times 10^{-3}$$

$$b_1 = [-0.12(\frac{f_{yd}}{f_{cd}})^2 + 3.34(\frac{f_{yd}}{f_{cd}}) - 19.35] \times 10^{-4}$$

$$b_2 = [0.13(\frac{f_{yd}}{f_{cd}})^2 - 3.22(\frac{f_{yd}}{f_{cd}}) + 15.68] \times 10^{-6}$$

$$c_0 = [-0.22(\frac{f_{yd}}{f_{cd}})^2 + 6.28(\frac{f_{yd}}{f_{cd}}) + 10.14] \times 10^{-2}$$

$$c_1 = [0.2(\frac{f_{yd}}{f_{cd}})^2 - 6.4(\frac{f_{yd}}{f_{cd}}) + 45.6] \times 10^{-3}$$

$$c_2 = [-0.23(\frac{f_{yd}}{f_{cd}})^2 + 5.71(\frac{f_{yd}}{f_{cd}}) - 32.6] \times 10^{-5}$$

A1.2 *Circular CFST Columns*

<u>Normal Strength Concrete ($f_{cd} \leq 50$ MPa)</u>

$$a_0 = [0.68(\frac{f_{yd}}{f_{cd}})^2 + 17.26(\frac{f_{yd}}{f_{cd}}) - 422.19] \times 10^{-6}$$

$$a_1 = [-0.38(\frac{f_{yd}}{f_{cd}})^2 - 9.6(\frac{f_{yd}}{f_{cd}}) + 209.64] \times 10^{-7}$$

$$a_2 = [0.3(\frac{f_{yd}}{f_{cd}})^2 + 12(\frac{f_{yd}}{f_{cd}}) - 205] \times 10^{-9}$$

$$b_0 = [-0.16(\frac{f_{yd}}{f_{cd}})^2 + 1.63(\frac{f_{yd}}{f_{cd}}) + 16.0] \times 10^{-3}$$

$$b_1 = [0.5(\frac{f_{yd}}{f_{cd}})^2 + 2.6(\frac{f_{yd}}{f_{cd}}) - 156.9] \times 10^{-5}$$

$$b_2 = [-0.5(\frac{f_{yd}}{f_{cd}})^2 - 3.6(\frac{f_{yd}}{f_{cd}}) + 133.2] \times 10^{-7}$$

$$c_0 = [0.5(\frac{f_{yd}}{f_{cd}})^2 - 8.72(\frac{f_{yd}}{f_{cd}}) + 54.64] \times 10^{-2}$$

$$c_1 = [-0.5(\frac{f_{yd}}{f_{cd}})^2 - 11(\frac{f_{yd}}{f_{cd}}) + 322] \times 10^{-4}$$

$$c_2 = [-0.8(\frac{f_{yd}}{f_{cd}})^2 + 38.1(\frac{f_{yd}}{f_{cd}}) - 365.6] \times 10^{-6}$$

Creative Systems in Structural and Construction Engineering, Singh (ed.) © 2001 Balkema, Rotterdam, ISBN 90 5809 161 9

An experimental study on the strength of grouted conduit connections under cyclic axial loading

F. N. Rad
Department of Civil Engineering, Portland State University, Oreg., USA

R. R. Imper
Morse Brothers, Prestressed Concrete, Harrisburg, Oreg., USA

ABSTRACT: The purpose of this research project was to devise a confinement scheme to enhance the bond capacity of imbedded reinforcement. Several specimens were manufactured in which segments of reinforcement were cast in concrete, with reinforcement confined by a steel conduit as well as steel box and spiral wire. In the first series of specimens, two samples were tested as preliminary experiments. In a second series, a dozen new specimens were cast and six were tested. Two test loading types were used: Cyclic Tension and Static Tension. The static pull and cyclic test results showed that the bond capacity of reinforcements is improved significantly with the inclusion of steel conduit to provide confinement. The confinement scheme of providing conduit plus spiral reinforcement shows great potential to be an innovative means to enhance the cyclic load capacity of reinforcement cast in concrete.

1 INTRODUCTION

In a reinforced concrete beam, slab, column, or wall, tensile forces are provided by reinforcement and compressive forces by concrete and/or steel. A force transfer or bond between concrete and steel is essential to maintain strength. Absence of bond strength would result in the bar pulling out of the concrete, and the tensile force decreasing to zero causing member failure.

Bond strength is dependent on several factors, including:
- Shear interlock between bar deformations and concrete
- Adhesion between steel and concrete
- Friction against sliding as the steel is stressed
- Drying shrinkage of concrete causing gripping effects
- The quality of concrete
- Development length, splicing, hooks, transverse reinforcement
- Diameter, spacing, and cover of reinforcement, and epoxy coating

In this investigation, the primary focus was on the enhancement of bond capacity by addition of transverse confinement for bars in tension.

2 PURPOSE OF THIS RESEARCH AND PHYSICAL TEST SPECIMENS

In this research project, we set out to devise a confinement scheme to enhance the bond capacity of imbedded reinforcement. Several specimens were manufactured in which segments of reinforcement were cast in concrete, with reinforcement confined by a variety of means. Specimens 1-12 were cast, and two were tested as preliminary experiments. As a result of the first experiments, it was decided to change the confinement scheme. New specimens, denoted by specimen numbers 13 through 24 were cast. Table 1 shows a summary of the specimens manufactured in the second series. All specimens were 10x10-in. cross-section, and other important features are given.

3 PHYSICAL TESTS PERFORMED

Two test loading types were used:
1. Cyclic Tension Test
2. Static Tension Test

Several sources of testing procedures were consulted, including: ASTM E 488-90, "Standard Test Methods for Strength of Anchors in Concrete and Masonry Elements," and ICBO AC 58, "Acceptance Criteria for Adhesive Anchors in Concrete and Masonry Elements." The actual testing procedure was selected based on the recommendations of an ad-hoc

Table 1. Specimens manufactured in series 2.

Specimen Number	Length of Specimen (in)	Rebar Size	Offset (in)	Grout	Test Date	Grout Strength (psi)
16*	12	#6	1.0,1.0	5-Star	15-Dec-99	9318
17*	18	#10	None	Con-bextra	15-Dec-99	14441
19*	18	#10	None	5-Star	14-Dec-99	9318
20*	18	#10	0.8,0.8	5-Star	15-Dec-99	9318
22**	24	#10	0.8,0.8	Con-bextra	18-Apr-00	14441
24**	24	#10	0.8,0.8	5-Star	12-Apr-00	9318

*Specimens No. 16, 17, 19 and 20 were tested in static tension mode after cyclic loading.
**Specimens No. 22 and 24 were cut in half (12-in nominal length) and tested in static tension.

Figure1. Test Specimen #16 under cyclic loading

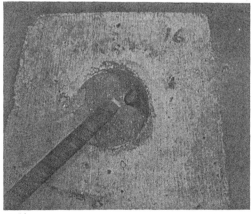

Figure 2. Test Specimen #16 after testing.

technical committee consisting of the authors and a consulting structural engineer.

For Specimens #16, 17, 19 and 20, the loading was "tension-zero" cyclic loading for a large number of cycles. The load values were increased to go well into the strain hardening range, and close to the ultimate bar capacity. The cyclic loading was then stopped, and the specimens were tested using a monotonically increasing tension load. Specimens #22 and 24 were tested under a monotonically increasing tension load only, without cyclic loading.

4 THEORETICAL CONSIDERATIONS

ACI 318-99 specifies the basic development length of deformed bars as:

$$\lambda_d / d_b = (3/40) \times (f_y / \sqrt{f_c'}) \times (\alpha\beta\gamma\lambda) / \left[(c + K_{tr}) / d_b \right]$$

The 1989 ACI Code specified a minimum
$$\lambda_d / d_b = .03(f_y / \sqrt{f_c'})$$
which is the same as the 1999 Code when the maximum "cover plus transverse reinforcement" bracket term equal to 2.5 is substituted in the general equation.

Development of deformed bars in tension enclosed in rigid corrugated tube and transverse reinforcement cannot be easily determined without significant experimental research. However, for the specific details of specimens used in this study, if one assumes that the 24 gage corrugated tube provides confinement similar to closed stirrups, with area = 0.0276x3in = 0.0883 square inches at 3-in pitch, and that a combined effect can be estimated to be 0.11 (for #3) plus 0.0883 (for the tube), for a total of 0.198 square inch, then K_{tr} = transverse reinforcement index = $A_{tr}f_{yt}/1500sn$ would be around 2.6 in. The actual concrete cover in the specimens was 5 inches, thus making the total "cover plus transverse reinforcement" term $[(c + K_{tr}/d_b)]$ increase to a high value close to 7.6/1.27 = 6. For such a high confinement value, the required λ_d / d_b decreases to a number close to 12, for 60 grade reinforcement and 4000 psi concrete. For #10 bar, this would be about 15 inches. Also, as concrete capacity grows, the development length decreases. This would be true for the specimens tested, since both the grout and the concrete capacities were in excess of 4000 psi.

Following the aforementioned approximate analysis, it seems plausible to estimate the development length of 60-grade #10 bars under the confinement scheme provided in this experimental research (including the high strength grout) as "approximately 15 inches or less."

5 SUMMARY OF RESULTS

Table 2 shows the number of cycles to which each specimen was subjected. After the cyclic loading, the specimens were subjected to static tension tests.

Table 2. Summary of results.

Specimen Number	Type of Loading	Load Increments (kips)	Hz	No. of Cycles
16	Cyclic	0.5	0.2	171
(specimen was also tested in static tension after cyclic loading)				
17	Cyclic	0.5	0.2	179
(specimen was also tested in static tension after cyclic loading)				
19	Cyclic	0.5	0.1	353
(specimen was also tested in static tension after cyclic loading)				
20	Cyclic	0.5	0.2	162
(specimen was also tested in static tension after cyclic loading)				
22&24	Static Tension Test without Cyclic Loading			One

In all specimens, the maximum static load approached the ultimate strength of the bar, except for Specimen #16 (#6 bar), in which the bar failed. In other cases, the ram extension or the test set-up load capacity was reached.

6 OBSERVATIONS

The 12-inch bar length in Specimens #22 and 24 (tested under monotonically increasing tension load) appeared adequate for static bond development length. For Specimens #16, 17, 19 and 20 (tested under tension-zero cyclic loading) the specimens did not fail after a large number of cycles, ranging from 162 to 353.

The confinement scheme of providing conduit plus spiral reinforcement shows great potential to be an innovative means to enhance the cyclic load capacity of reinforcement cast in concrete.

REFERENCES

American Concrete Institute 1999. *Building Code Requirements for Structural Concrete (318-99) and Commentary (318R-99)*. Farmington Hills: ACI.
American Society for Testing and Materials 1990. Standard Test Methods for Strength of Anchors in Concrete and Masonry Elements, E 488-90. West Conshohocken: ASTM.
International Conference of Building Officials 1995. *Acceptance Criteria for Adhesive Anchors in Concrete and Masonry Elements*. Whittier: ICBO.

Creative Systems in Structural and Construction Engineering, Singh (ed.) © 2001 Balkema, Rotterdam, ISBN 90 5809 161 9

Nondestructive strength evaluation of early-age concrete using longitudinal wave velocity

H. K. Lee & K. M. Lee
Department of Civil Engineering, Sungkyunkwan University, Korea

D. S. Kim
Department of Civil Engineering, Korea Advanced Institute of Science and Technology, Korea

ABSTRACT: In this study, a reliable nondestructive strength evaluation method for early-age concrete using the longitudinal wave velocity is proposed. Compression tests were performed to examine the factors influencing the velocity-strength relationship of concrete, such as water-cement (w/c) ratio, fine aggregate ratio, curing temperature, and curing condition. The test results show that a change in the w/c ratio and curing temperature have minor effect on the velocity-strength. However, curing condition significantly influences the velocity-strength relationship of early-age concrete. Moreover, the longitudinal wave velocity increases with decreasing fine aggregate ratio. We can conclude from the study that the strength evaluation of early-age concrete can be achieved by a nonlinear equation which considers the effects of curing condition and fine aggregate ratio.

1 INTRODUCTION

The usage of nondestructive testing leads to enhanced safety and allows effective scheduling of construction, thus making it possible to maximize the time and cost efficiencies. Furthermore, if the quality assurance of new concrete was achieved, the repair and maintenance costs can be significantly reduced.

Several methods for measuring in-situ concrete strength have been used in the past (i.e., rebound hammer, probe penetration, ultrasonic pulse velocity, maturity, and cast-in-place cylinder (ACI Committee 208 1988)). However, the results obtained by these methods do not agree satisfactorily with actual concrete strength. Therefore, a more reliable and easier method for evaluating concrete compressive strength than previously established methods is needed.

In this paper, the nondestructive evaluation method of early-age concrete strength using the longitudinal wave velocity measured by impact-resonance method is presented.

2 IMPACT RESONANCE TEST

The longitudinal wave velocity can be easily obtained using the impact-resonance method on concrete cores or standard cylinders (Pessiki and Johnson 1996). It is assumed that the specimen length L is greater than twice the diameter D. The

unconstrained longitudinal wave velocity (in this case, it is called rod-wave velocity, V_c) of the specimen is given by

$$V_c = f_1 \cdot \lambda \qquad (1)$$

where f_1 = the first mode resonance frequency; λ = the wave length of first mode ($2 \times L$).

For Poisson's ratio of 0.18, the longitudinal wave velocity measured by impact-resonance method in infinite solid is about 4 percent greater than that of unconstrained one.

$$V_c = 0.96 \, V_p \qquad (2)$$

where V_c = rod-wave velocity; V_p = constrained longitudinal wave velocity.

Ultrasonic pulse velocity is approximately 5 percent greater than longitudinal wave velocity measured by impact-resonance method in infinite solid (Sansalone and Streett 1997).

3 EXPERIMENTAL WORK

3.1 *Scope*

The relationship between velocity and strength is influenced by variables such as moisture content, aggregate content, type of aggregate, curing condition of concrete, aging, et cetera (Metha and Monteiro 1993). Thus, it is essential to investigate the factors influencing the velocity-strength relationship.

An experiment was performed to investigate the following factors influencing the velocity - strength relationship : (1) w/c ratio - 0.46 to 0.58, (2) fine aggregate ratio - 36 to 44%, (3) curing temperature - 10, 20, and 30□, and (4) curing condition – moisture and drying curing. The curing conditions used in this study are listed in Table 1.

Table 1. Curing conditions of specimen.

Curing Condition	Details
Moisture (A)	Specimen was placed in a water bath at 10, 20, and 30□ until the testing date
Drying (B)	After casting, specimen was covered with wet-burlap for 3 days followed by exposure to ambient temperature of 23 ±5□ and humidity of 50±10 % until the testing date

3.2 Materials and mix proportion

The concrete specimens were made with the following materials: Type□Portland cement, crushed granite with nominal maximum size of 25 mm and with specific gravity of 2.65, and river sand with specific gravity of 2.70. Table 2 tabulates the mix proportions of the tested concrete specimens. As shown in Table 2, the mixtures of C1 to C4 have the same water content and similar aggregate content.

Table 2. Mix proportions for four types of concrete (kg/m³).

Mix Type	W/C	W	C	Coarse Agg.	Fine Agg.	S/a (%)	Air (%)
C1	0.58	185	320	1025.6	712.7	41	5.0
C2	0.53	185	350	1008.3	703.1	41	5.0
C3	0.50	185	370	1016.2	680.8	40	5.0
C4	0.46	185	400	1032.0	691.0	40	5.0

3.3 Specimen preparation

Cylinders with φ100 x 200 mm were made for compression testing. The fresh concrete was produced by normal mixing procedures. After fresh concrete was placed into cylinder molds, they were compacted using a vibrator to ensure proper placing of the concrete and to reduce the entrapped air. All cylinders were covered with plastic sheets for 24 hours after the fresh concrete was placed in the molds. The cylinders were then placed in controlled humidity room at temperature of 20□ and relative humidity of 50%. The cylinders were stripped after placed in controlled humidity room for 1 day. Then, the cylinders were cured until tested according to curing conditions listed in Table 1.

3.4 Test procedure

To measure the longitudinal wave velocity of concrete, impact-resonance test was performed. In the impact-resonance testing, a small steel sphere with a diameter of 7.6 mm was used as an impact source. Piezoelectric accelerometer (PCB Model 353B15) which operates reliably over a wide range of amplitude and frequency was used to monitor the response of the specimen. The output signals from the accelerometer was conditioned and recorded with a waveform analyzer. The waveform analyzer had the capacity to perform for data acquisition and signal processing. The number of data points in displacement waveform was 4096 and sampling frequency was 125 Hz.

For concrete cylinders cured in a water bath, measurement of the rod-wave velocity was conducted 3 hours after cylinders were removed from water bath to maintain moisture content in cylinder.

A typical waveform of acceleration obtained from one day old cylinder after casting is shown in Fig. 1(a). Transformation of this waveform results in the amplitude spectrum shown in Fig. 1(b). The dominant peak in the spectrum at 7.538 kHz is the resonance frequency. Other peaks are attributed to other modes of vibration of the cylinder excited by the impact.

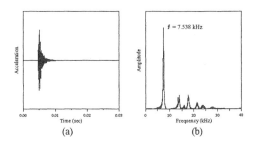

(a)   (b)

Figure 1. Typical measurements of (a) waveform of acceleration and (b) amplitude spectrum obtained from φ100 x 200 mm cylinder.

4 TEST RESULTS AND DISCUSSIONS

4.1 Factors influencing velocity-strength relationship at early age

4.1.1 Water-cement ratio and curing condition
Figure 2 is a plot of the velocity vs. strength data for 72 concrete cylinders cured in 20□ water bath. From the test results, the w/c ranging from 0.46 to 0.58 minutely alter the velocity-strength relationship under similar aggregate content in concrete and same curing condition.

Figure 3 shows the velocity vs. strength data for 36 cylinders cured under curing condition B. Moreover, there is a considerable difference in velocity-strength relationships when cured under different curing conditions even if concrete was manufactured with same mix proportion. For example, when the rod-wave velocity was 4.0 km/s, the corresponding compressive strength was approximately 40 MPa in curing condition A and 30 MPa in curing condition B.

The moisture content of concrete affects the velocity rather than the in-situ strength. In other words, the velocity is faster through a water-filled void than through an air-filled one (Sturrup et al 1984).

4.1.2 Curing temperature

Figure 4 is a plot of the velocity vs. strength data for cylinders cured at different curing temperature in curing condition A. Figure 4 shows that the fitted curve agrees well with the data of 10, 20 and 30□-cured cylinders implying that curing temperature of 10 to 30□ does not influence the velocity-strength relationship. There is a one-to-one correlation at early age regardless of w/c ratio and curing temperature.

4.1.3 Fine aggregate ratio

Figure 5 is a plot of velocity vs. strength data with varying fine aggregate ratio from 36 to 44%. Figure 5 shows that the velocity increases with decreasing fine aggregate ratio. On the other hand, the compressive strength is not significantly affected by a minute change in fine aggregate ratio.

By increasing the aggregate content, there is an increase in the proportion of higher elastic modulus materials resulting in the increase of concrete velocity (Pessiki and Carino 1988, Neville 1995). The compressive strength, on the other hand, is not significantly affected nither by the aggregate content nor the elastic modulus of the aggregates.

4.1.4 Determination of nondestructive estimation equation

The test results show that the velocity-strength relationship is affected by curing condition and fine aggregate ratio variation. Finally, nondestructive estimation equation is proposed as

$$f_c(t) = a \exp[\, b k_c k_{s/a} V_c(t)\,] \qquad (3)$$

where f_c = compressive strength (MPa); V_c = rod-wave velocity (km/s); t = aging of concrete (day); a and b = constants; k_c= correction factor for curing condition; $k_{s/a}$= correction factor for fine aggregate ratio. There may be other empirical functions that fit the data better than Equation 3. However, the fit is reasonable and adequate for the purpose of this study.

Coefficients in Equation 3 are tabulated in Table 3. A basic equation for evaluating early-age concrete strength is obtained from 20□-cured cylinders with fine aggregate ratio of 40~41%. By means of nonlinear regression, constants of a and b in the

basic equation were fixed at 0.024 and 1.79, respectively, which make k_c and $k_{s/a}$ become 1.0. In other cases, k_c and $k_{s/a}$ were evaluated according to curing condition and fine aggregate ratio. As shown in Table 3, k_c in curing condition B is 1.04 and $k_{s/a}$ varies from 0.98 to 1.02 with fine aggregate ratio.

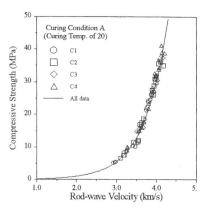

Figure 2. Velocity vs. strength data for concrete cured under curing condition A.

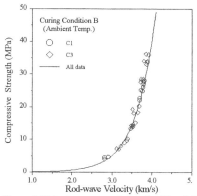

Figure 3. Velocity vs. strength data for C1 and C3 cured under curing condition B.

5 CONCLUSIONS

The following conclusions are drawn from this study.

1. Water-cement (w/c) ratio and curing temperature had little effects on the velocity-strength relationship of early-age concrete.

2. The relationship between velocity and strength was considerably influenced by the curing condition and fine aggregate ratio.

3. For nondestructive strength evaluation of early-age concrete, correction factors of curing condition and fine aggregate ratio are introduced in a strength estimation equation.

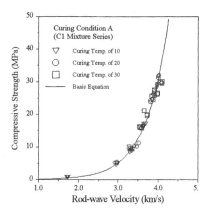

Figure 4(a). Velocity vs. strength data for C1 cured under condition A with varying curing temperatures.

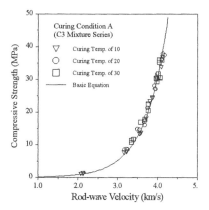

Figure 4(b). Velocity vs. strength data for C3 cured under condition A with varying curing temperatures.

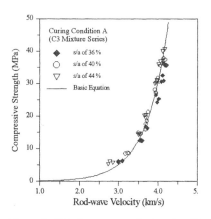

Figure 5. Velocity vs. strength data for C3 cured under condition A with varying fine aggregate ratios.

Table 3. Coefficient in the equation (3)

Curing Condition	Curing Temp. ($\square$)	S/a (%)	Constant		Coefficient	
			a	b	k_c	$k_{s/a}$
A	10	40~41	0.024*	1.79*	1.0*	1.0
	20	36~37				0.98
		40~41				1.0*
		44~45				1.02
	30	40~41				1.0
B	-	40~41			1.04	1.0

* Coefficients in basic equation.

ACKNOWLEDGEMENT

The authors gratefully acknowledge the funding of the research by SAFE Research Center in Sungkyunkwan University.

REFERENCES

ACI Committee 208. 1988. "In-Place Methods for Determination of Strength of Concrete," *ACI Material Journal*, Vol. 85, No. 5, pp. 446-471.

Mehta, P. K. and Monteiro, J. M. 1993. *Concrete*, 2nd edition, Prentice Hall, pp. 348.

Neville, A. M. 1995. *Properties of Concrete*, 4th edition, Longman, England, pp. 631-632.

Pessiki, S. P. and Carino, N. J. 1988. "Setting Time and Strength of Concrete Using the Impact-Echo Method," *ACI Materials Journal*, Vol. 85, No. 4, pp. 389-399.

Pessiki, S. and Johnson, M. R. 1996. "Nondestructive Evaluation of Early-Age Concrete Strength in Plate Structures by the Impact-Echo Method," *ACI Materials Journal*, Vol. 93, No. 3, pp. 260-271.

Sansalone, M. and Street, W. B. 1997. *Impact-Echo*, Bull brier Press, pp. 51.

Sturrup, V. R., Vecchio, F. J. and Caratin, H. 1984. "Pulse Velocity as a Measure of Concrete Compressive Strength," In-Situ/Nondestructive Testing of Concrete, SP-82, *Edited* by Malhotra, V. M., ACI, Detroit, pp. 201-227.

Creative Systems in Structural and Construction Engineering, Singh (ed.) © 2001 Balkema, Rotterdam, ISBN 90 5809 161 9

Elastic analysis of isotropic infinite plane with two circular holes by using constraint-release technique

Takashi Tsutsumi
Department of Civil Engineering, Fukushima National College of Technology, Japan

Ken-ichi Hirashima
Department of Civil and Environmental Engineering, Yamanashi University, Japan

ABSTRACT: This paper presents the analytical solution of isotropic infinite plane containing two circular holes of different size. The method of solution is constraint-release technique which repeats superposing two kinds solution of single-connected problem until converging to both boundary conditions. Some numerical results are shown by graphical representation.

1 INRODUCTION

The general solution of an infinite plane containing two circular holes was obtained by Jeffery (1921), and Ling (1948). Among these methods, stress and displacement were discussed in bipolar coordinate. However, the only case that the radii of two circular holes are same size was dealt in these papers.

On the other hand, authors already induced the solution for anisotropic elastic elliptic ring subjected to arbitrary loads on inner or outer boundary (1997, 2000a), isotropic elastic semi-infinite plane with a circular hole subjected to arbitrary loads on hole boundary (1999a, 2000b), and isotropic elastic semi-infinite plane with an elliptic hole subjected to arbitrary loads on hole boundary (1999b). These studies were carried out by using constraint-release technique which induces the solution for double connected problem in which two kinds of solutions for simply-connected problems are repeatedly superposed to converge to the boundary conditions respectively.

It is the purpose of this paper to show the elastic solution for the isotropic infinite plane contained two circular hole in different size by using constraint-release technique.

2 FUNDAMENTAL EQUATION

Consider the two-dimensional infinite plane containing two circular holes, as presented in Figure 1. In the case of isotropy, the stress components σ_x, σ_y, τ_{xy} and displacement components u_x, u_y are given by

$$\sigma_x = 2\operatorname{Re}[\varphi'(z)] - \operatorname{Re}[\bar{z}\varphi''(z) + \psi''(z)] \qquad (1)$$

$$\sigma_y = 2\operatorname{Re}[\varphi'(z)] + \operatorname{Re}[\bar{z}\varphi''(z) + \psi''(z)] \qquad (2)$$

$$\tau_{xy} = \operatorname{Im}[\bar{z}\varphi''(z) + \psi''(z)] \qquad (3)$$

$$u_x - iu_y = \frac{1}{2G}[\frac{3-\nu}{1+\nu}\varphi(z) - \{\bar{z}\varphi'(z) + \overline{\psi'(z)}\}] \qquad (4)$$

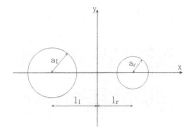

Figure 1. Infinite plane containing two circular holes

Where $\varphi(z)$ and $\psi(z)$ are Airy's stress functions, Re and Im represent the real part and imaginary part of complex functions in brackets, respectively. And G, υ in Eq.(4) are shear elastic modulus and poisson's ratio. Additionally, the upper bars of the terms represent complex conjugates.

The formulas which map stress and displacement components into curvilinear coordinates (ξ, η) are shown:

$$\sigma_\xi + \sigma_\eta = \sigma_x + \sigma_y \qquad (5)$$

$$\sigma_\eta - \sigma_\xi + 2i\tau_{\xi\eta} = e^{2i\theta}(\sigma_y - \sigma_x + 2i\tau_{xy}) \qquad (6)$$

$$u_\xi - iu_\eta = e^{i\theta}(u_x - iu_y) \qquad (7)$$

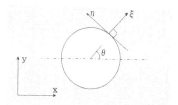

Figure 2. Curvilinear coordinate

3 FORMULATION OF THE PROBLEM

The purpose of this paper is to obtain the isotropic elastic solution for a problem in which arbitrary load to an infinite plane containing two circular holes is applied as shown in Figure 3. In this section, the solution for isotropic elastic infinite plane containing two circular hole is shown, superposing solutions for infinite planes containing a circular hole. First, the coordinate corresponding to right circular hole boundary Γ_r are

$$z_r = a_r e^{i\theta} + l_r \qquad (8)$$

Normal stress $\sigma_{\xi,0}^r$ and shear stress $\tau_{\xi\eta,0}^r$ on the boundary of right circular hole only, as shown in Figure 3., are expanded into finite Fourier expansions,

$$\sigma_{\xi,0}^r - i\tau_{\xi\eta,0}^r = \bar{c}_{0,0}^r$$
$$+ \sum_{m=1}^{M} (\bar{c}_{0,m}^r \cos m\theta + \bar{d}_{0,m}^r \sin m\theta) \qquad (9)$$

Airy's stress functions for the infinite plane containing a circular hole are expanded as follows

$$\varphi_{r,0}(z) = M_0^r \log(z - l_r) + \sum_{m=1}^{M} A_{0,-m}^r (z - l_r)^{-m} \qquad (10)$$

$$\psi_{r,0}(z) = N_0^r (z - l_r) \log(z - l_r) + K_0^r \log(z - l_r)$$
$$+ \sum_{m=1}^{M} B_{0,-m}^r (z - l_r)^{-m} \qquad (11)$$

The complex coefficients $M_0^r, N_0^r, K_0^r, A_{0,-m}^r, B_{0,-m}^r$ are obtained as following form:

$$M_0^r = \frac{1-\upsilon}{8}(c_{0,1}^r + id_{0,1}^r) \qquad (12)$$

$$N_0^r = -\frac{3-\upsilon}{8}(\bar{c}_{0,1}^r - i\bar{d}_{0,1}^r) \qquad (13)$$

$$K_0^r = \bar{c}_{0,0}^r a_r^2 \qquad (14)$$

$$A_{0,-m}^r = -\frac{a_r^{m+1}}{m}\frac{c_{0,m+1}^r + id_{m+1}^r}{2} \quad (m \geq 1) \qquad (15)$$

$$B_{0,-1}^r = \frac{a_r^3}{2}(\frac{1+\upsilon}{2}\frac{c_{0,1}^r + id_{0,1}^r}{2} - \frac{\bar{c}_{0,1}^r + i\bar{d}_{0,1}^r}{2}) \qquad (16)$$

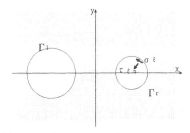

Figure 3. Arbitrary load on circular hole

$$B_{0,-m}^r = \frac{a_r^{m+2}}{m}\{\frac{c_{0,m}^r + id_{0,m}^r}{2} - \frac{\bar{c}_{0,m}^r + i\bar{d}_{0,m}^r}{2(m+1)}\} \ (m \geq 2) \ (17)$$

The coordinates corresponding to virtual boundary Γ_l, shown by the broken line in Figure 4., are represented by following equations:

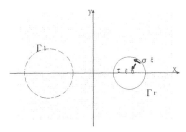

Figure 4. Infinite plane containing only right hole

$$z_l = a_l e^{i\theta} - l_l \qquad (18)$$

Therefore, the normal stress and shear stress on boundary Γ_l are represented as

$$\sigma_{\xi,1}^l = 2\,\mathrm{Re}[\varphi'_{r,0}(z_l)]$$
$$- \mathrm{Re}[\{\bar{z}_l \varphi''_{r,0}(z_l) + \psi''_{r,0}(z_l)\}e^{2i\theta}] \qquad (19)$$

$$\tau_{\xi\eta,1}^l = \mathrm{Re}[\{z_l \varphi''_{r,0}(z_l) + \psi''_{r,0}(z_l)\}e^{2i\theta}] \qquad (20)$$

The same holds for the right circular hole, with the right side of above equations expanded into Fourier expansions:

$$\sigma_{\xi,1}^l - i\tau_{\xi\eta,1}^l = \bar{c}_{1,0}^l$$
$$+ \sum_{m=1}^{M} (\bar{c}_{1,m}^l \cos m\theta + \bar{d}_{1,m}^l \sin m\theta) \qquad (21)$$

To cancel out the normal stress and shear stress on the left virtual hole boundary Γ_l in the infinite plane containing a right circular hole, the negative values of above normal stress and shear stress are loaded on the boundary of the infinite plane containing a left circular hole as shown in Figure 5.

798

In this case, the stress functions are represented by following equations:

$$\varphi_{l,1}(z) = M_1^l \log(z + l_l) + \sum_{m=1}^{M} A_{1,-m}^l (z + l_l)^{-m} \qquad (22)$$

$$\psi_{l,1}(z) = N_1^l (z + l_l)\log(z + l_l) + K_1^l \log(z + l_l)$$
$$+ \sum_{m=1}^{M} B_{1,-m}^l (z + l_l)^{-m} \qquad (23)$$

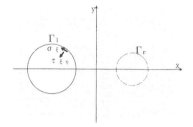

Figure 5. Infinite Plane containing only left hole

The complex coefficients in Eq.(22) and (23) are obtained as following form:

$$M_1^l = \frac{1-\upsilon}{8}(c_{1,1}^l + id_{1,1}^l) \qquad (24)$$

$$N_1^l = -\frac{3-\upsilon}{8}(\overline{c}_{1,1}^l - i\overline{d}_{1,1}^l) \qquad (25)$$

$$K_1^l = \overline{c}_{1,0}^l a_l^2 \qquad (26)$$

$$A_{1,-m}^l = -\frac{a_l^{m+1}}{m}\frac{c_{1,m+1}^l + id_{1,m+1}^l}{2} \quad (m \geq 1) \qquad (27)$$

$$B_{1,-1}^l = \frac{a_l^3}{2}(\frac{1+\upsilon}{2}\frac{c_{1,1}^l + id_{1,1}^l}{2} - \frac{\overline{c}_{1,1}^l + i\overline{d}_{1,1}^l}{2}) \qquad (28)$$

$$B_{1,-m}^l = \frac{a_l^{m+2}}{m}\{\frac{c_{1,m}^l + id_{1,m}^l}{2} - \frac{\overline{c}_{1,m}^l + i\overline{d}_{1,m}^l}{2(m+1)}\} \quad (m \geq 2) \qquad (29)$$

Therefore, the normal stress $\sigma_{\xi,1}^r$ and shear stress $\tau_{\xi\eta,1}^r$ on Γ_r are represented as

$$\sigma_{\xi,1}^r = 2\text{Re}[\varphi'_{l,1}(z_r)]$$
$$- \text{Re}[\{\overline{z}_r\varphi''_{l,1}(z_r) + \psi''_{l,1}(z_r)\}e^{2i\theta}] \qquad (30)$$

$$\tau_{\xi\eta,1}^r = \text{Re}[\{z_r\varphi''_{l,1}(z_r) + \psi''_{l,1}(z_r)\}e^{2i\theta}] \qquad (31)$$

The same holds for the left circular hole, with the right side of above equations expanded into Fourier expansions:

$$\sigma_{\xi,1}^r - i\tau_{\xi\eta,1}^r = \overline{c}_{1,0}^r$$
$$+ \sum_{m=1}^{M}(\overline{c}_{1,m}^r \cos m\theta + \overline{d}_{1,m}^r \sin m\theta) \qquad (32)$$

To cancel out the normal stress and shear stress on the right virtual hole boundary Γ_r in the infinite plane containing a left circular hole, the negative values of above normal stress and shear stress are loaded on the boundary of the infinite plane containing a right circular holeas shown in Figure 4. In following procedure the above operation is repeated N times to satisfy both boundary conditions. As a result, the stress funcyions for the isotropic infinite plane containing two circular holes are obtained by

$$\varphi(z) = \sum_{n=0}^{N} \varphi_{r,n}(z) + \sum_{n=1}^{N} \varphi_{l,n}(z) \qquad (33)$$

$$\psi(z) = \sum_{n=0}^{N} \psi_{r,n}(z) + \sum_{n=1}^{N} \psi_{l,n}(z) \qquad (34)$$

4 RESULTS AND DISCUSSION

Figure 7. shows the dependence of the normalized normal stress $\sigma_{\xi,N}^r / p$ arising at the left edge of right circular hole through calculation as a function of the number of repeated calculation N when hydrostatic pressure p is loaded on only right circular hole, as shown in Figure 6. The number of repeated calculations N increases as arising normal stress $\sigma_{\xi,N}^r / p$ decreases. And, the distance between two holes increases as arising normal stress $\sigma_{\xi,N}^r / p$ decreases.

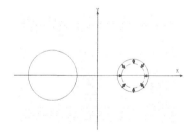

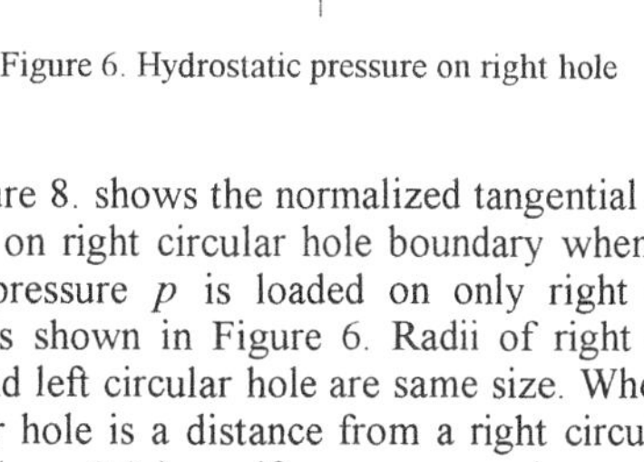

Figure 6. Hydrostatic pressure on right hole

Figure 8. shows the normalized tangential stresses σ_{η}^r / p on right circular hole boundary when hydrostatic pressure p is loaded on only right circular hole, as shown in Figure 6. Radii of right circular hole and left circular hole are same size. When a left circular hole is a distance from a right circular hole (i.e. $l_l / a_l = 3.0$), uniform compression arises all around right circular hole. Nevertheless, when a left circular hole is close to a right circular hole, compression decreases at the left edge of right circular hole and increases at neighborhood of the left edge of the right circular hole.

Figure 9. shows the normalized tangential stresses σ_{η}^r / p when hydrostatic pressure p is loaded on

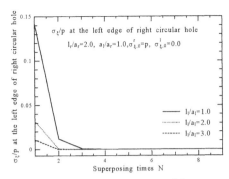

Figure 7. Convergence of σ_ξ^r / p

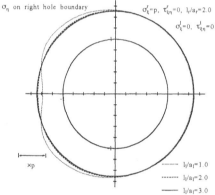

Figure 8. σ_η^r when the parameter is l_l / a_l.

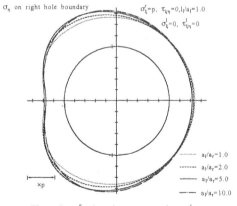

Figure 9. σ_η^r when the parameter is a_l / a_r

5 CONCLUSION

In this paper, the solution is proposed for isotropic elastic infinite plane containing two circular holes under arbitrary loaded on the edge of circular hole by using constraint-release technique. Furthermore, some numerical results are shown by graphical representation.

REFERENCES

Jeffery, G. B. 1921. Plane stress and plane strain in bipolar co-ordinate, *Phil. Trans.* A.221:265.

Hetenyi, M. 1960. A method of solution for the elastic quarter-plane, *Jour. Appl. Mech.*, 27, 289-296.

Ling, C. B. 1948. On the stresses in a plate containing Two Circular holes, *Jour. Appl. Phy.*, 19:77-82.

Moriguchi, S. 1957. *Two Dimensional Elasticity*, (in Japanese): Kyouritsu Pub.

Noda, N. & Matsuo, T. 1997. Numerical solution of singular integral equations in stress concentration problems, *Int. J. Solids Structures*, 34:2429-2444.

Tsutsumi, T. & Hirashima, K. 1997. Analysis of Anisotropic Elliptic Ring By Using Constraint Release Technique (in Japanese), *Trans. of JSME, Series A*, 63:2411-2416.

Tsutsumi, T., Arai, H. & Hirashima, K. 1999a. Elastic Analysis of Ground Displacements Due to Shallow Circular Tunnel by Using Constraint-Release Technique (in Japanese), *Proc. 29th Sym. Rock Mech. JSCE*:50-56.

Tsutsumi, T., Arai, H. & Hirashima, K. 1999b. Elastic analysis for the ground displacement due to elliptic tunnel by using constraint-release technique, *Proc. 1st Int. Conf. on Advances in Structural Eng. and Mech.* Vol.2:1381-1386.

Tsutsumi, T. & Hirashima, K. 2000a. Analysis of orthotropic circular disks and rings under diametrical loadings, *Structural Eng. and Mech.* 9:37-50.

Tsutsumi, T., Arai, H. & Hirashima, K. 2000b. Displacement Fields for an Isotropic Elastic Semi-infinite Plane with Circular Hole Subjected to Arbitrary Loads at the Circular Hole (in Japanese), *J. Soc. Mat. Sci. Jpn.* 49:774-778.

only right circular hole, as shown in Figure 6. Where a_l / a_r is the parameter representing the ratio of the radius of left circular hole to that of right circular hole. Radius of left circular hole a_l increases as compression increases at left upper area and left lower area of right circular hole.

Creative Systems in Structural and Construction Engineering, Singh (ed.) © 2001 Balkema, Rotterdam, ISBN 90 5809 161 9

Effects of partial shear connection on the dynamic response of semi-continuous composite beams

B.Uy
University of New South Wales, Sydney, N.S.W., Australia

A.Chan
Rooney and Bye Consulting Engineers, Sydney, N.S.W., Australia

ABSTRACT: Semi-continuous composite beams are becoming a popular method of design in modern steel construction as they enable the span to depth ratios to be increased with very little increased cost in terms of labour. This method of construction relies on steel reinforcement placed over the column lines to increase the rotational stiffness at a traditional simply supported end. Partial shear connection is also becoming of increasing interest as it also causes a reduction in construction labour cost, without a great decrease in strength or stiffness. To properly identify the effects of both semi-continuity and partial shear connection on the vibration response of composite beams, a detailed dynamic analysis has been undertaken herein. This study uses an existing mathematical model to consider the vibration of composite beams, which is one of the most important serviceability limit states in the design of floor systems in steel framed buildings.

1 INTRODUCTION

Semi-continuous composite beams are developed by the use of longitudinal steel reinforcement being placed over the column lines in conventional simply supported composite beam construction as illustrated in Figure 1. These beam systems rely on the use of semi-rigid composite connections, which have been subjected to extensive in depth studies throughout the world, (Nethercot 1997 and Leon 1998).

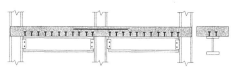

Figure 1. Semi-continuous composite beam

Serviceability of semi-continuous composite beams has been considered in a detailed manner by Uy and Belcour (1999 and 2000). These studies looked at the effects of semi-rigid joints when employed in simply supported beam construction, however the effects of partial shear connection and the reduced stiffness were not included in this study. Eurocode 4 (British Standards Institution 1994) and AS 2327.1 (Standards Australia 1996) does include provisions for vibration, however it does not include a detailed method for the determination of natural frequencies of beams with semi-rigid joints which

are germane to semi-continuous composite beam construction.

Girhammar and Pan (1993) developed a mathematical model to consider the effects of partial shear connection on the vibration behaviour of simply supported composite beams. This paper will augment this model by incorporating the effects of semi-rigid joints using the approach of Crisinel and Carretero (1996) and thus will allow the development of solutions for natural frequencies of semi-continuous composite beams with varying levels of shear connection.

2 MATHEMATICAL MODEL

2.1 *General*

A mathematical model developed by Girhammar and Pan (1993) which considered the dynamic behaviour of continuous composite beams with partial shear connection was augmented in this paper to include the effects of semi-rigidity.

The mathematical model is able to determine both the natural frequency and the slip in a composite beam. These are the two major issues, which were of concern in the analysis. The mathematical formulation involves using the two ordinary differential equations to solve the free vibration problem.

$$\frac{d^6\phi_n}{dx^6} - \alpha^2 \frac{d^4\phi_n}{dx^4} - \omega_n^2 \frac{m}{EI_o} \frac{d^2\phi_n}{dx^2} + \omega_n^2\alpha^2 \frac{m}{EI_\infty} = 0 \quad (1)$$

$$\frac{d^2 f_n}{dt^2} + \omega_n^2 f_n = 0 \qquad (2)$$

Equation 1 expresses the eigenmodes, ϕ_n in terms of the eigenvalues ω_n, where m is the mass per unit length of the steel and concrete subelements, EI_0 =bending stiffness of the non-composite section and EI_∞ =bending stiffness of the fully composite section.

The general solution given by Girhammar and Pan (1993) can be expressed as

$$\phi_n = \sum_{i=1}^{6} c_i e^{k_{i,n} x} \qquad (3)$$

where c_i are arbitrary coefficients which are to be determined and e^{kx} is the set of orthogonal polynomials and the roots of the equation can be determined from

$$\left(k_{i,n}^2\right)^3 - \alpha^2 \left(k_{i,n}^2\right)^2 - \omega_n^2 \frac{m}{EI_0} k_{i,n}^2 \qquad (4)$$

$$+ \omega_n^2 \alpha^2 \frac{m}{EI_\infty} = 0$$

solving the cubic equation it can be shown that

$$k_{1,n}^2 < 0, k_{2,n}^2 > 0, k_{3,n}^2 > 0$$

The general solution can then be written as

$$\phi_n = c_1 \sin\left(k_{1,n} x\right) + c_2 \cos\left(k_{1,n} x\right) +$$
$$c_3 \sinh\left(k_{2,n} x\right) + c_4 \cosh\left(k_{2,n} x\right) + \qquad (5)$$
$$c_5 \sinh\left(k_{3,n} x\right) + c_6 \cosh\left(k_{3,n} x\right)$$

where the term α^2 is given by the equation

$$\alpha^2 = \frac{f_y z^2}{EI_0 \left(1 - \frac{EI_0}{EI_\infty}\right)} \qquad (6)$$

where f_y is the yield stress of the steel section in MPa and z is the lever arm between the compressive and tensile forces.

In the general case of arbitrary boundary conditions the eigenvalues are obtained from large and complicated transcendental equations, which require numerical solutions. The iterations were conveniently carried out in the following way using the MATLAB program

1. Assume a natural frequency ω_n
2. Calculate the coefficients k_1, k_2 and k_3;
3. Satisfy the transcendental equation.

Once an eigenvalue, ω_n for a particular set of boundary conditions is known, its associated eigenmode, ϕ_n can be determined. This is done by removing one of the equations for the boundary conditions. This leaves five equations in six unknowns, which may be solved by assuming the values of the integration constants.

2.2 Boundary conditions

The boundary conditions for the standard beam cases of either pinned, clamped or free ends require one to identify the deflection, slope, moment or shear at the left or right hand supports. The following table lists these boundary conditions for various cases

Table 1. Boundary conditions for beams.

B.C	Deflection		Slope		Moment		Shear	
L/R	L	R	L	R	L	R	L	R
PP	0	0			0	0		
CF	0		0			0		0
CP	0	0	0			0		
CC	0	0	0	0				

B.C-Boundary Condition
L-Left
R-Right
P-Pinned
C-Clamped
F-Free

2.3 Semi-rigid joints

The mathematical model presented can be modified to include the boundary condition, whereby the moment is expressed as a function of the end rotation. The augmentation of the model to include the effects of semi-rigidity uses the approach of Crisnel and Carretero (1996) to determine the rotational stiffness of the joint. In the modelling of the semi-continuous composite beam, a linear spring is used to represent the stiffness of the end of the beam to the column. This spring is assumed to have a linear stiffness for the serviceability behaviour of the beam.

This linear stiffness is determined using the approach of Crisnel and Carretero (1996) where the stiffness of the joint S_j is calculated using Equation 7

$$S_j = \frac{z^2}{\frac{1}{k_s} + \frac{1}{k_v} + \frac{1}{k_c}} \qquad (7)$$

where z = the lever arm between the compressive and tensile forces; k_s = the stiffness of the reinforced concrete slab; k_v = the stiffness of the shear connection; and k_c = the stiffness of the compression zone.

3 CALIBRATION AND PARAMETRIC STUDY

3.1 *General*

The model presented is required to be calibrated against existing results. Once calibration has taken place a parametric study is required to be undertaken. A parametric study is undertaken here to consider the effects of level of shear connection η, and percentage of reinforcement p. To consider the dynamic behaviour of semi-continuous composite beams a range of parameters must be considered which reflect real design situations in practice. A parametric study is undertaken here to consider the effects of cracking, reinforcement percentage and span length on the dynamic response of semi-continuous composite beams. The parametric study has been conducted on beams which have been designed for typical office floor loading according to Australian Standard AS1170.1-1989 (Standards Australia 1989) in a gravity loaded multi-storey braced frame. The beams thus satisfy both the serviceability and strength limit states of the Australian Standard AS2327.1-1996 (Standards Australia 1996) with appropriate modifications to account for semi-rigid and partial strength joints.

3.2 *Calibration of Model*

The model developed here was compared with the results of Uy and Belcour (1999 and 2000) as well as the results, which are suggested by the Australian Standard AS 2327.1 (Standards Association of Australia 1980), which is outlined in Equation 8.

$$f = K\sqrt{\frac{E_s I'_{cs}}{wL^4}} \qquad (8)$$

where $K = 155$ for simply supported beams; E_s = the modulus of elasticity of the steel joist (N/mm^2); I'_{cs} = the second moment of area of the composite section (mm^4); w = the dead load on the beam (kN/m); and L = the beam span length (mm).

The comparisons are shown in Figure 2, which shows the relationship between natural frequency and spacing of the shear connectors. This shows that the natural frequency reduces as the level of shear connection reduces, and is to be expected as a reduction in the level of shear connection reduces the beam flexural rigidity.

3.3 *Effects of level of shear connection, η and percentage of reinforcement,* p

The effects of the level of shear connection, η on the beam fundamental natural frequency of semi-continuous composite beams is illustrated in Figures 3, 4, 5 and 6. One can see that there is a reduction in

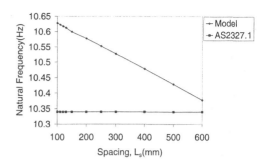

Figure 2. Calibration of model with AS2327.1

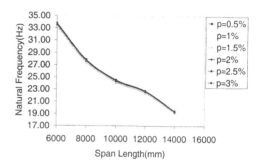

Figure 3. Natural frequencies (η=1.0)

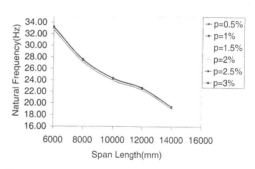

Figure 4. Natural frequencies (η=0.75)

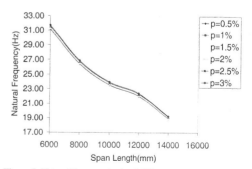

Figure 5. Natural frequencies (η=0.50)

the beam natural frequency as the level of shear connection is reduced. The figures show that there is approximately a 2.5 % decrease in beam natural frequency with a 25 % reduction in level of shear connection. This relationship is in fact non-linear and an expression for the reduction in beam natural frequency could be derived to incorporate these effects. Whilst increasing the level of reinforcement increases the stiffness and subsequently the natural frequency, the effects are not as great as those for increasing the level of shear connection.

All the analyses undertaken for typical spans and loading in buildings were found to provide natural frequencies in the acceptable range of structural response. All natural frequencies were found to be greater than the maximum accepted limit of 15 Hz. By extrapolation, it can be clearly seen that for spans greater than 14 metres, natural frequencies could fall below 15 Hz and thus resonance effects would need to be checked more closely according to the imposed loading conditions. This may include considering the effects of damping provided by partitions and other parts of the structure.

4 SUMMARY AND DISCUSSION

4.1 Effects of degree of shear connection

The results of the previous section showed that the beam fundamental natural frequency is influenced by both the level of reinforcement and the degree of shear connection between the steel beam and concrete slab. Figure 7 shows the relationship between each of these three parameters. Whilst the percentage of reinforcement can have an effect on the beam natural frequency, it is shown that the results are much more sensitive to the degree of shear connection. The degree of shear connection can have the effect of increasing the beam natural frequency by as much as 10 %, whereas this is not able too be achieved by an increase in reinforcement percentage.

4.2 Longitudinal slip behaviour

In addition to the beam natural frequency, the model is able to predict the longitudinal slip between the steel beam and the concrete slab. Figure 8 shows the longitudinal end slip when compared with an applied uniform load. Again the longitudinal slip is extremely sensitive to the degree of shear connection. In particular, Figure 8 highlights the flexible nature of the shear connectors when low levels of shear connection are used. Generally most stud headed shear connectors have ultimate slip capacities of between 4 and 6 mm and thus a beam with a level of shear connection η of 0.25 with a service load of 40 kN/m would be very close to ultimate slip failure under service loads.

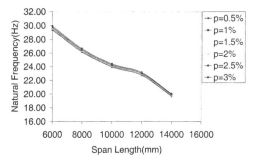

Figure 6. Natural frequencies (η=0.25)

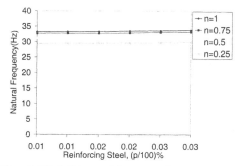

Figure 7. Effect of degree of shear connection

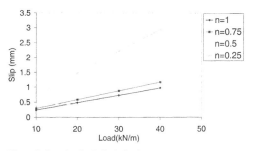

Figure 8. Longitudinal slip behaviour

5 CONCLUSIONS

This paper has described an augmented mathematical model, which can deal with the serviceability effects of composite beams with semi-rigid joints and partial shear connection. This has been calibrated with other results and a parametric study has been carried out to identify the effects of both degree of shear connection and percentage of reinforcement.

The results within the paper have shown that the level of shear connection is an extremely important parameter in the vibration performance of a semi-continuous composite beam. The fundamental natural frequency of a semi-continuous composite beam can vary by as much as 10% by increasing the level of shear connection from 0 through to 1.0. The effects of level of shear connection on the dynamic response of semi-continuous composite beams are therefore much more influential than the percentage of reinforcement.

6 FURTHER RESEARCH

The analysis conducted herein has been on the fundamental natural frequency of semi-continuous composite beams considering free vibrations only. The effects of forced vibrations should be considered in future studies, as the effects of human reaction to vibrations are of extreme importance to the serviceability behaviour of composite floors in steel frames buildings.

This study has also constrained the analysis to consider only the fundamental mode and mode superposition has not been considered. Further research is necessary to be conducted on the effects of mode superposition, which can be extremely important in influencing the behaviour of very long span floor systems.

To properly calibrate the model presented in this paper it is necessary to undertake some full-scale tests on semi-continuous composite beams, which display partial shear connection. Such tests are currently being planned to be undertaken at the Randwick Heavy Structures Laboratory at The University of New South Wales.

7 ACKNOWLEDGEMENTS

The authors would like to thank the Australian Research Council, which provided financial support for this project. This paper is part of an ongoing research programme on the behaviour and design of composite structural steel-concrete members being undertaken by the Structural Engineering Group at The University of New South Wales.

Furthermore the authors would also like to thank Messrs Phillip Butcher and Charles Blunt of Rooney and Bye Consulting Engineers, Sydney for their support of the second author during the conduct of the work in this paper, which constituted the majority of his dissertation project.

8 REFERENCES

British Standards Institution 1994. Eurocode 4, ENV 1994-1-1 1994. Design of composite steel and concrete structures, Part 1.1, General Rules and Rules for Buildings.

Crisinel, M. & A. Carretero 1996. Simple prediction method for moment-rotation properties of composite beam-to-column joints, *Proceedings of the Engineering Foundation Conference on Composite Construction:Composite Construction III, Irsee, 9-14 June 1996:* 823-835.

Girhammar, U.A. & Pan, D.H. 1993. Dynamic analysis of composite members with interlayer slip. *International Journal of Solids and Structures* 30 (6): 797-823.

Leon, R.T. 1998. Composite connections, *Progress in Structural Engineering and Materials, Construction Research Communications,* 1: 159-169.

Nethercot, D.A. 1997. Behaviour and design of composite connections, *Composite Construction, Conventional and Innovative, Innsbruck, 16-18 September 1997:* 657-662. Zurich: IABSE.

Standards Association of Australia 1980. *AS2327.1-1980, SAA Composite Construction Code, Part 1: Simply supported beams.*

Standards Association of Australia 1989. *AS1170.1-1989, SAA Loading Code, Part 1: Dead and live loads and load combinations.*

Standards Australia 1996. *AS2327.1-1996, Australian Standard, Composite Structures, Part 1: Simply supported beams.*

Uy, B. & Belcour, P.F.G. 1999. Dynamic response of semi-continuous composite beams, *16th Australasian Conference on the Mechanics of Structures and Materials, Sydney, 8-10 December:* 613-618.

Uy, B. & Belcour, P.F.G. 2000. Serviceability of semi-continuous composite beams in buildings, *6th ASCCS Conference, Los Angeles, 22-24 March:* 623-630.

25 Concrete shear

Creative Systems in Structural and Construction Engineering, Singh (ed.)© 2001 Balkema, Rotterdam, ISBN 90 5809 161 9

Research of shear-lag effect in curved box girder with variation calculus

Dawen Peng & Hai Yan
College of Civil Engineering and Architecture, Fuzhou University, People's Republic of China

ABSTRACT: In this paper, the curved box girder, esp. its shear-lag effect, has been systematically studied on the basis of conclusions on linear-box girder drawn by many researchers. In accordance with the coupled bending-torsion as well as the variable curvature in the cross-section of a curved-box girder, the strain function of the flange plate is proposed. The calculus equations and boundary conditions in a curved-box girder have been derived by the principle of variation, and then a numerical solution has been obtained. Finally, a three-dimensional model aided with the SAP has adopted to confirm the method proposed in this paper, whose results showed that this method provided a theoretical basis to calculate the shear-lag effect in a curved-box girder, and thus was applicable in practice.

1 INTRODUCTION

In the past two decades, the structure with a box-shaped cross-section has been adopted widely in bridge projects in China due to its lightweight, large anti-torsion rigidity and its convenience to advanced construction techniques. The longitudinal force applied to a box-shaped girder is usually transferred from its belly plate to its flange plate when it is flexed. Because the shear strain reaches its maximum at the intersection of the belly plate and the flange plate, that is: the farther away the belly plate, the less the shear strain, the longitudinal displacement difference within the flange plate is resulted and the bending positive stress distributes in certain curved shape along the flange plate. That is called "the shear-lag effect". The study on the shear-lag dates back to 1924 when T.V.Karman firstly discussed on the effective width of T-shaped girder with a broad flange. In 1964, E.Reissner also proposed a numerical solution to a rectangle-shaped girder without a cantilever by the principle of variation, and suggested that the larger spacing between the rib plates, the more noticeable the shear-lag effect in a box-shaped girder. From November 1969 to November 1971, four accidents, structure unstability or failure, took place in Austria, England, Australia and German, respectively. Among the factors that should be responsible for these accidents, one is that the designers did not take account into the shear-lag effect seriously. However, the study on the theory of shear-lag effect in the

box-shaped girder lags far behind that on its structure. There is much less researches or papers on the shear-lag effect in the curved box-shaped girder, esp. one with large curvature.

In this paper, the displacement model of the shear-lag effect and the formula of the total potential energy in the curved box-shaped girder have been proposed incorporating with project experience. Based upon the principle of variation, the calculus equations and boundary conditions have been derived, and furthermore a numerical solution to the calculus equations has been presented. Using the method of FEM (finite element method), some helpful conclusions have been reached and that the method above is applicable in practice is confirmed.

2 ANALYZING THE SHEAR-LAG EFFECT IN THE CURVED BOX-SHAPED GIRDER USING THE PRINCIPLE OF VARIATION

2.1 *The curvature and the longitudinal strain in a curved box-shaped girder*

In the process of analyzing a bridge with a curved girder subject to torsion and bending, it is convenient to take the axis curvature radius as the curvature of the whole cross-section because the axis curvature radius, R_0, is usually much larger than the dimension of curved plane by an order, and then the Vlasov equation is derived. However, as for the broad curved box-shaped girder, such an assumption, that is: the curvature radius in whole cross-section is constant, often results in the calculated σ_{max}(the maximal stress) less than the exact one. As shown in figure 1:

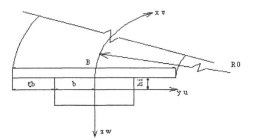

Figure 1. Sketch map of a curved box girder.

The curvature usually varies along the girder width in the shape of a parabola, and it is represented as equation 1:

$$\lambda(y) = \frac{1}{R} = \frac{1}{R_0 - y} \qquad (1)$$

Aided with FEM, Prof. Li Guohao, a famous bridge expert in China, studied the box-shaped girder with large curvature which is subject to plane torsion, and concluded that the error could be as much as 3.2% or 10.9%, respectively, when the ratio of R_0 to R (R_0/B) is 10 or 3, if taking R_0 as the value of R. Unfortunately, the calculation is virtually too complex to be applicable when the exact variation of curvature radius is considered. Thus, it is advisable to express the curvature as equation 2:

$$\lambda(y) = \frac{1}{R} \approx \frac{1}{R_0}(1 + \frac{1}{R_0} \times y) \qquad (2)$$

It can be seen from the equation 2 that the curvature is assumed as the sum of constant R_0 and a liner amendment, or the sum of first two components of Taylor extension.

The longitudinal strain had already been derived by S. Timoshenko, and given as equation 3:

$$\varepsilon_x = \frac{dv}{dx} - \frac{u}{R} \qquad (3)$$

where: ε_x is the axis strain; v, u is the displacement in the x, y axis, respectively;

Taking into the variation of curvature along width, equation 4,5 is given:

$$\varepsilon_x = \frac{dv}{Rd\varphi} - \frac{u}{R} = (1+\frac{y}{R_0})\frac{dv}{R_0 d\varphi} - (1+\frac{y}{R_0})\frac{u}{R_0}$$

$$= (1+\frac{y}{R_0})(\frac{dv}{dx} - \frac{u}{R_0}) \qquad (4)$$

$$v = h_i \times \frac{dw}{dx}, \qquad u = h_i \times \phi \qquad (5)$$

where: w is the displacement in z-axis, hi is the z coordinate of flange plate; φ is the rotation angle of the curvature center when the cross-section rotates around the girder axis; ϕ is the rotation angle when the cross-section rotates around x-axis.

Thus, equation 5 simplified as:

$$\varepsilon_x = h_i \times (1+\frac{y}{R_0})(\frac{d^2 w}{dx^2} - \frac{\phi}{R_0}) \qquad (6)$$

When the shear-lag effect is considered, ε_x in this paper is expressed as

$$\varepsilon_x = h_i \times (1+\frac{y}{R_0})[(\frac{d^2 w}{dx^2} - \frac{\phi}{R_0}) + (1-\overline{y}^3) \times u'(x)] \quad (7)$$

where: $u'(x)$ is the first order derivative when the difference of shear distortion in the flange plate reaches its peak●

The first component in the right side of equation (7) is the amendment to curved box-shaped girder on the basis of the assumption that the shear-lag effect in linear box-shaped girder distributes transversely in the shape of a third order parabola.
Defining:

$$\overline{y} = \begin{cases} \dfrac{y}{b} & 0 \le y \le b \\[2mm] \dfrac{(b+\xi b - y)}{\xi b} & b \le y \le b+\xi b \end{cases} \qquad (8)$$

where: $\overline{y}$ is usually positive; b is the half of the spacing between of belly plates; ξb is the width of cantilever;

2.2 Fundamental assumption of the principle of variation

Such assumptions as following have been adopted in this paper:

1 The torsion warp and irregular warp are neglectable, that is: the cross-section remains a plane and boundary shapes are unchanged. As the matter of fact, if the condition as equation 9 is available. Then the warped distortion in curved box-shaped girder is not very large.

$$k = L \times \sqrt{\frac{G \times I_d}{E \times I_w}} \ge 30 \qquad (9)$$

where: K is the warped distortion coefficient; L is the span; G is the shear modulus; E is the elastic modulus; I_d is the anti-torsion inertial moment; I_w is the anti-bending inertial moment; As for the concrete bridge with a closed box-shaped cross-section, the assumptions above have always been met if its cross-section is close to square.

2 The longitudinal or shear strain is represented respectively as equation 10,11 when the bending shear-lag effect, the coupled bending and torsion,

and the variation of curvature along the girder width are incorporated at the same time.

$$\varepsilon_x = h_i \times (1+\frac{y}{R_0})[(\frac{d^2w}{dx^2}-\frac{\phi}{R_0})+(1-\bar{y}^3) \times u'(x)] \quad (10)$$

$$\gamma_{xy} = -\frac{3h_i}{b^3} \times y^2 \times u(x) \quad (11)$$

3 As for the rib part, only its bending stress energy is considered because the plane assumption is roughly satisfied.

4 As for the up and down flange plates, the vertical stress $\sigma_z = 0$, and the shear strain γ_{xz}, γ_{yz} and transverse strain ε_y are all neglectable.

2.3 *Defining the coefficient of shear-lag*

The bending stress distributes diagonally in stead of asymmetrically when the shear-lag effect is not incorporated because the curvature varies along the width of girder. Traditionally, to define the coefficient of shear-lag is to express the transversely asymmetrical distribution of bending stress. However, it is virtually impossible to take bending or torsion into account separately in the box-shaped girder. A generalized coefficient of shear-lag, λ equal to the ratio of the real bending stress within the cross-section to the calculated bending stress by the common curved girder theory, is proposed as so to incorporate the coupled bending-torsion. To define such a generalized coefficient not only features the asymmetrical distribution of the longitudinal stress in a cross-section but makes calculating stresses at certain locations available as well. In addition, it should be noticed that all the conclusions drawn here are only associated with the coefficient of shear-lag defined in this paper because researchers did not yet agree with each other about the definition of shear-lag coefficient.

In the process of analyzing the shear-lag effect, analysis mainly focuses on both the longitudinal shear-lag effect, that is: the shear-lag effect at the intersections of the flange plate and the inside or outside rib plate, and the transverse shear-lag effect, that is: the shear-lag effect within a cross-section of the flange plate. It can also see that the shear-lag is positive when $\lambda > 1$ while negative when $\lambda < 1$.

2.4 *Deriving the fundamental variation equations*

According to the principle of minimum potential energy, a structure is usually in a balanced stage when an external force is applied on it. Therefore, the variation of the total potential energy is equal to zero when any imaginary displacement happens, that is:

$$\delta\Pi = \delta(\bar{V}-\bar{W}) = 0 \quad (12)$$

where: $\bar{V}$ is the system strain energy; $\bar{W}$ is the potential energy of the external force;

The potential energy of the external force in a simple-supported girder can be expressed as:

$$\bar{W} = \bar{W}_q + \bar{W}_m + \bar{W}_M + \bar{W}_Q + \bar{W}_T \quad (13)$$

where: transverse force is:

$$\bar{W}_q = -\int_0^l q \times w dx \quad (14)$$

distributing torque is:

$$\bar{W}_m = -\int_0^l m \times \phi dx \quad (15)$$

end bending moment is:

$$\bar{W}_M = -[M \times w']\,|_0^l \quad (16)$$

end shear is:

$$\bar{W}_Q = -[Q \times w]\,|_0^l \quad (17)$$

end torque is:

$$\bar{W}_T = -[T \times \phi]\,|_0^l \quad (18)$$

The stress energy of a curved girder can be expressed as equation 19:

$$\bar{V} = \bar{V}_w + \bar{V}_{so} + \bar{V}_{su} + \bar{V}_T \quad (19)$$

Because the plane assumption is roughly satisfied in the belly plate, only the bending stress energy is taken into account. And the bending stress energy of the belly plate is given as equation 20:

$$\bar{V}_w = \frac{1}{2}EI_w \int_0^l (w''-\frac{\phi}{R_0})^2 dx \quad (20)$$

where: E is the material elastic modulus; I_w is the anti-bending inertial moment in the belly plate;

Assuming the thickness of up or low plate as t_0, t_u respectively, then the strain energy in up or low flange plate is respectively given as equation 21,22:

$$\bar{V}so = \frac{1}{2}\iint t_0(E\varepsilon_{x0}^2 + G\gamma_0^2)dydx \quad (21)$$

$$\bar{V}su = \frac{1}{2}\iint t_u(E\varepsilon_{xu}^2 + G\gamma_u^2)dydx \quad (22)$$

where: G is the shear modulus;
Torsion strain energy is:

$$\bar{V}_T = \frac{1}{2}GI_d \int(\phi'^2 + \frac{2}{R_0}\phi'w' + \frac{w'^2}{R_0^2})dx \quad (23)$$

where: I_d is the anti-torsion inertial moment in the cross section• `Thus, total potential is:

$$\Pi = \frac{1}{2}EI\int_0^l[w''-\frac{\phi}{R_0}]^2 dx + \frac{1}{2}EI_s\int_0^l(\frac{3}{2}w''u' + \frac{9}{14}u'^2$$

$$-\frac{3\phi u'}{2R_0})dx + \frac{1}{2}EI_s\int_0^l \frac{1}{3} \times (\frac{b}{R_0})^2 \times (w''^2 - \frac{2\phi w''}{R_0} + \frac{\phi^2}{R_0^2}$$

$$+w''u'+\frac{1}{3}u'^2-\frac{\phi u'}{R_0})dx + \frac{1}{2}GI_s\int_0^l \frac{9}{5b^2}u^2 dx$$

$$+\frac{1}{2}GI_d\int_0^l(\phi'^2+\frac{2}{R_0}\phi'w'+\frac{w'^2}{R_0^2})dx - \int_0^l q_y w dx - \int_0^l m\phi dx$$

$$+[Mw']\,|_0^l - [Qw]\,|_0^l - [T\phi]\,|_0^l \quad (24)$$

where: I_s is the anti-bending inertial moment of up or low flange plate.

The fundamental equations and boundary conditions are also derived when the principle of variation is applied to the equation above.

Eula equation:

$$\begin{cases} F_w - \dfrac{d}{dx}F_{w'} + \dfrac{d^2}{dx^2}F_{w''} = 0 \\[2mm] F_u - \dfrac{d}{dx}F_{u'} = 0 \\[2mm] F_\phi - \dfrac{d}{dx}F_{\phi'} = 0 \end{cases} \quad (25)$$

Boundary conditions:

$$\begin{cases} [F_{w'} - \dfrac{d}{dx}F_{w''}]_{x=0*\delta x=l} = 0 \\[2mm] F_{u'} \mid_{x=0*\delta x=l} = 0 \\[2mm] F_{\phi'} \mid_{x=0*\delta x=l} = 0 \\[2mm] F_{w''} \mid_{x=0*\delta x=l} = 0 \end{cases} \quad (26)$$

In combination of the equation 25 and the equation 26, the fundamental equations and boundary conditions are represented as following, when both the shear-lag effect and the variation of curvature along the width in a simple-supported curved box-shaped girder are considered.

1 the fundamental equations:

$$[EI + \frac{1}{3}(\frac{b}{R_0})^2 EI_s] \times w^{IV} - \frac{EI + GI_d + \frac{1}{3}(\frac{b}{R_0})^2 EI_s}{R_0} \times \phi''$$

$$-\frac{GI_d}{R_0^2} \times w'' + [\frac{3}{4} + \frac{1}{6}(\frac{b}{R_0})^2] \times EI_s u''' - q = 0 \quad (27)$$

$$-\frac{EI + GI_d + \frac{1}{3}(\frac{b}{R_0})^2 EI_s}{R_0} \times w'' + \frac{EI + \frac{1}{3}(\frac{b}{R_0})^2 EI_s}{R^2_0} \times \phi$$

$$-GI_d \phi'' - \frac{[\frac{3}{4} + \frac{1}{6}(\frac{b}{R_0})^2]EI_s}{R_0} \times u' - m_z = 0 \quad (28)$$

$$[\frac{3}{4} + \frac{1}{6}(\frac{b}{R_0})^2]EI_s w'' - \frac{[\frac{3}{4} + \frac{1}{6}(\frac{b}{R_0})^2]EI_s}{R_0} \times \phi'$$

$$+ [\frac{9}{14} + \frac{1}{9}(\frac{b}{R_0})^2]EI_s u'' - \frac{9GI_s}{5b^2} \times u = 0 \quad (29)$$

2 boundary conditions:

$$-\{[EI + \frac{1}{3}(\frac{b}{R_0})^2 EI_s] \times w''' - \frac{EI + GI_d + \frac{1}{3}(\frac{b}{R_0})^2 EI_s}{R_0} \times \phi'$$

$$-\frac{GI_d}{R_0^2} \times w' + [\frac{3}{4} + \frac{1}{6}(\frac{b}{R_0})^2] \times EI_s u'' - Q\}\delta w \mid_0^l = 0 \quad (30)$$

$$\{[EI + \frac{1}{3}(\frac{b}{R_0})^2 EI_s] \times w'' - \frac{[EI + \frac{1}{3}(\frac{b}{R_0})^2 EI_s]}{R_0^2} \times \phi$$

$$-[\frac{3}{4} + \frac{1}{6}(\frac{b}{R_0})^2] \times EI_s u' + M\}\delta w' \mid_0^l = 0 \quad (31)$$

$$\{GI_d \phi' + \frac{GI_d}{R_0} \times v' - T\}\delta\phi \mid_0^l = 0 \quad (32)$$

$$\{[\frac{3}{4} + \frac{1}{6}(\frac{b}{R_0})^2]EI_s w'' - \frac{[\frac{3}{4} + \frac{1}{6}(\frac{b}{R_0})^2]EI_s}{R_0} \times \phi$$

$$+ [\frac{9}{14} + \frac{1}{9}(\frac{b}{R_0})^2]EI_s u'\}\delta u \mid_0^l = 0 \quad (33)$$

The equation 27~29 and the equation 30~33 are simplified as following without incorporating the shear-lag effect or the variation of curvature.

$$EI \times w^{IV} - \frac{GI_d}{R_0^2} \times w'' - \frac{EI + GI_d}{R_0} \times \phi'''' u - q = 0 \quad (34)$$

$$-\frac{EI + GI_d}{R_0} \times w'' - GI_d \phi'' + \frac{EI}{R^2_0} \times \phi - m_z = 0 \quad (35)$$

The equations above are as same as the Vlasov equation in which the distortion is not incorporated. When the radius is very close to infinity, this equation is changed to be:

$$EI \times w^{IV} + \frac{3}{4}EI_s u' - q = 0 \quad (36)$$

$$EI_s[\frac{9G \times u}{5E \times b^2} - \frac{9}{14} \times u'' - \frac{3}{4}w''']= 0 \quad (37)$$

The equations above are the calculus equation of the shear-lag in the linear girder. Therefore, it can be seen from the analysis above that the derived fundamental equations here are of universality, and that the equations 34~35 and 36~37 can be taken just as two special examples.

As to the general boundary conditions:
1 End gemel without end bending moment: w=0 ϕ=0 w''=0 u'=0
2 Fixed end: w=0 ϕ=0 w'=0 u=0
3 Free end: the boundary conditions can be obtained with zeroing the contents in the large parentheses of equations 30~33.

Referred to the Hanson's analytical solution to calculus equations in curved girder, an analytical solution to the displacement of curved box-shaped girder is available, and then the stress as well as the strain is obtained, in combination with equations 27~29. However, the analytical solution, virtually, is too complex to be applicable in practice. Here, a more convenient solution based on the method of Gliaojing is proposed.

2.5 Numerical solution using the method of Garliaojing

Assuming the displacement curves as:

$$w = \sum_{l}^{k} C_k \times \sin(\frac{k\pi}{l}) \qquad (38)$$

where: l is the axis span of a curved box-shaped girder;

$$\phi = \sum_{l}^{i} C_i \times \sin(\frac{i\pi}{l}) \qquad (39)$$

where: C_k, C_i, C_j are the uncertain constants;

$$u = \sum_{l}^{j} C_j \times \sin(\frac{j\pi}{l}) \qquad (40)$$

The calculation above is aided with a Matlab-based program, and the schematic diagram of program is given as following:

Assuming the displacement models: (including unknown coefficients)

⇩

Calculating all order derivatives of displacements

⇩

Calculating the integral of all components of total potential energy respectively

⇩

Calculating the differentiate calculus of total potential energy to each coefficient

⇩

Assuming the differentiate calculus above as zero, then linear equations with (I+j+k) order are obtained

⇩

Solving these linear equations and calculating the coefficients:

⇩

Replacing the coefficients into the displacement models, then displacement is derived

⇩

Calculating the strains

⇩

Calculating the stress

⇩

Distribution of shear-lag coefficient in space is available

2.6 Confirming the method

To confirm the method proposed in this paper, a simple-supported curved box girder with over-restriction, as shown in figure 2 is adopted, and the calculated results are compared with those by Sap. The dimensions of the bridge are specified as following: the curvature radius of axis: $R_0 = 100m$, the span: $L=42m$, the width of box: 15.4m, the bottom width: 7.2m, the thickness of flange plate: tu=30cm, the thickness of belly plate: tw=50cm, the height of box: h=2.5m, distributing force: q=0.1MN/m. Others necessary parameters: the material elastic modulus, E=20000Mpa; the Possion ratio: $\gamma = 0.16$. It can be seen from figure 3 that there is a good agreement between calculated results by the method proposed here and those by Sap, and that this method has enough precision and thus are applicable in practice.

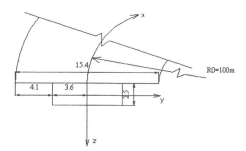

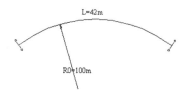

Figure 2. Sketch map of simple-supported curved box girder.

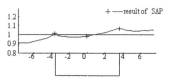

Figure 3. Sketch map of the calculated curved box girder.

3 CONCLUSIONS

From the analysis above, some suggestive conclusions as following have been reached.

1. It can been seen that the method proposed here to analyze the shear-lag effect of the curved box

girder, which is based on the principle of variation, is rational and applicable.

2. It can been also seen that the combination of the coupled bending-torsion, the variation of curvature along the width of curved girder, and the superposition of shear factor, have been not only proposed firstly in this paper, but considered as well in the process of calculation. Subsequently, the calculated results here that have a good agreement with the real values have been reached.

3. The calculus equations derived here are not only applicable to a numerical solution based on the Garliaojing method but to other aspects as well, specifically those so as to obtain an analytical solution and to derive the stiffness equations, which would be adopted in the FEM of curved box-shaped girder in the near further.

4. The calculus equations derived in this paper are virtually of universality, and both the calculation of common curved box girder and of the shear-lag effect in the linear box girder can be taken just as particular examples.

REFERENCES

German Standards 1981. Betonbr Ucken Bemessung Und Ausfuhrung. DIN 1075.

Guo Jinqiong, Fan Zhenzheng & Luo Xiaodeng 1983. Study on the Shear-lag Effect in the Box-shaped Girder Bridge. Journal of Chinese Civil Engineering, Vol. 16(1). 1-13.

K.R.Moffatt & P.J.Dowling, 1978. British shear lag rules for Composite Girder. ASCE. Journal of the Structural Division, Vol. 104.

Li Guohao 1986. The torsion and the bending in the sheet box girder with a large curvature. Journal of Chinese Civil Engineering, Vol. 20(1). 65-75.

Shidou Chang 1992. Prestress influence on the shear-lag effect in continuous Box-Girder. ASCE. Journal of structural Engineering, Vol. 118. No. 11.

Zhang Shiduo 1998. The Shear-lag Effect in the Box Sheet Girder. The People Transportation Press.

Creative Systems in Structural and Construction Engineering, Singh (ed.) © 2001 Balkema, Rotterdam, ISBN 90 5809 161 9

Failure mechanism in I-beam subjected to transverse bending

S. Nakano
Department of Civil Engineering, Kure National College of Technology, Hiroshima, Japan

E. Kishimoto
Civil Department, City of Hiroshima, Japan

ABSTRACT: The design of a web with both shear force and transverse bending moment acting is based after concrete cracks. In this paper, the failure mechanism of the web of I-Beams is investigated by the superposition of the reinforcement forces and the concrete forces that are obtained using the in-plane force theory, and the superposition of forces caused by the transverse bending moment. Using the equations obtained from the above-mentioned method, the stirrup stress in the compressive zone of the web due to the transverse bending moment and the ultimate load are obtained. To investigate the stirrup stress and the ultimate load obtained using the said analytical method, experiments were performed on 15 specimens. The variables in the experiment are the area of the stirrup and the transverse bending moment. The theoretical values of the stirrup stress and the ultimate load nearly coincide with experimental values.

1 INTRODUCTION

The effect of the transverse bending moment appears not only in the plate but also in the web subjected to strong shear force. The effect of transverse bending on the shear strength of T-beams was investigated (Kupfer & Ewald 1973, Kaufmann & Menn 1976). On the other hand, the stress condition of I-beams subjected to shear force and transverse bending moment was investigated by superposing the in-plane stresses and the plate stresses (Ewald 1977, 1982).

The design of the web with both shear force and transverse bending moment acting is based after concrete cracks. Because the inner forces spread after concrete cracks, the equilibrium of shear force and transverse bending moment should at the same time be considered.

In this paper, the failure mechanism of the web of I-Beams is investigated by superposition of the reinforcement forces and the concrete forces obtained using in-plane stresses, and stresses caused by bending and shear force. There are two known methods (Baumann 1972, Zararis 1988) using the in-plane force theory. In this paper, the method of the latter is used because the investigation of the crushing of concrete is possible.

2 EXPERIMENTS

To investigate the shear force when stirrup of the transverse bending compressive zone yields or when concrete crushes, experiments were performed on 15 specimens.

In the following, the specimens, the loading process and the experimental results are described.

2.1 Specimens

Table 1 shows the dimensions of the cross-section, reinforcing bar and concrete characteristic values of the specimens.

The primary steel bar of the tension flange was arranged largely to prevent bending failure. The transverse steel bar was made up of the stirrup that was extended up to the flange edge and arranged on the outside of both upper and lower flanges.

To reinforce the supporting point, the lower flange was designed with a thickness of 9.5 cm from the edge of specimen up to 25 cm of its length.

2.2 Loading condition

Figure 1 describes the loading condition. The primary load P was placed on the web at the center of span using the Universal Testing Machine. The transverse load $\bar{P}$ was produced by two upper and lower channels pressed by four hydraulic jacks arranged at intervals of 40 cm. With this type of loading, the edge of flange was loaded at intervals of 20 cm. This loading condition, except for the transverse bending moment, produced the normal force in the web. By lengthening the channel length, the effect of the normal force on the stirrup stress was 2~3 % of the yield stress of the stirrup.

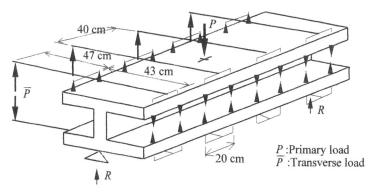

Figure 1. Loading condition

P :Primary load
$\bar{P}$:Transverse load

2.3 Experimental results

Figure 2 describes the cracking diagram for specimen Z9. The first shear crack appeared on both the tension and compressive zone at the same time by increasing only the primary load. An increase of the transverse load involves the appearance of more shear cracks along the already shear cracked compressive zone, and the increase of crack opening in the tension zone. Specimen Z9 collapsed by shear failure.

Figure 3 describes the cracking diagram for specimen Z13. The first shear crack appeared on both the tension and compressive zone at the same time by loading only the primary load. Just after applying the transverse load, the transverse bending compressive zone cracked diagonally and the tension zone cracked horizontally. By increasing more the transverse load, the horizontal cracks on the tension zone extended. By increasing only the primary load, more

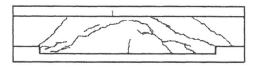

a. Tension zone

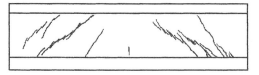

b. Compressive zone
Figure 2. Cracking diagram for Z9

cracks appeared on the compressive zone and specimen Z13 collapsed by the crushing of concrete.

Table 1. Dimension of cross-section and characteristic values

Specimen	Span (cm)	Flange width (cm)	Height (cm)	Web width (cm)	Flange thickness Upper (cm)	Lower (cm)	Stirrup diameter (cm)	Stirrup spacing Tension side (cm)	Compression side (cm)	Yielding stress of stirrup (kNcm⁻²)
Z1	160	51	35.0	7.0	5.0	5.0	0.55	8	8	50.67
Z2	160	51	35.0	7.0	5.2	5.0	0.55	8	12	50.67
Z3	160	51	35.0	7.0	5.0	5.0	0.55	8	16	50.67
Z4	160	51	35.5	6.8	6.3	6.0	0.55	10	10	31.36
Z5	160	51	35.5	6.8	6.3	5.6	0.55	10	15	31.36
Z6	160	51	35.7	6.8	6.3	6.4	0.55	10	20	31.36
Z7	160	51	35.5	7.0	5.5	7.0	0.55	10	10	31.36
Z8	160	51	35.5	6.8	6.1	5.9	0.55	10	15	31.36
Z9	160	51	35.5	6.6	6.2	5.4	0.55	10	20	31.36
Z10	160	51	30.4	6.6	6.1	5.8	0.55	10	10	35.28
Z11	160	51	30.7	6.9	6.3	5.9	0.55	10	15	35.28
Z12	160	51	30.6	6.9	6.1	6.1	0.55	10	20	35.28
Z13	160	51	30.7	6.8	6.2	6.1	0.55	10	10	35.28
Z14	160	51	30.5	7.0	6.1	6.1	0.55	10	15	35.28
Z15	160	51	30.5	7.0	6.1	6.0	0.55	10	20	35.28

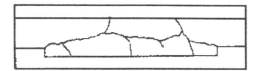

a. Tension zone

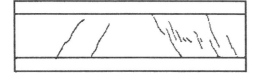

b. Compressive zone

Figure 3.Cracking diagram for Z13

3 THEORETICAL ANALYSIS

It is possible that a plane structure element subjected to bending be divided into two layers, namely, a tension plate in the tension zone, and a compressive plate in the compressive zone.

In this paper, the stress condition of the transverse bending compressive zone of the web is investigated using the principal tensile force (Baumann 1972), and the reinforced concrete force obtained from the in-plane force theory (Zararis 1988). By superposition of these forces and the forces due to the transverse bending moment, shear force when the rein-

forcing bar yields and shear force when the concrete crushes were obtained.

3.1 Stresses in plates carrying in-plane forces

Figure 4 shows the forces acting on a unit element of the plate in the direction of the crack. N_1 and N_2 are the internal principal forces of the plate, σ_{sy} the stress of reinforcing bar in the y direction, A_{sx} and A_{sy} the area of reinforcing bar per unit length in the x and y direction, respectively (Zararis 1988). The forces acting on the crack are the forces of reinforcing bar, while the forces acting orthogonal to the crack are the compressive force C of concrete and shear force V.

Equating the internal forces (N_1, N_2) and reinforced concrete forces (σ_{sy}, V, C), the following equations are obtained:

$$\sigma_{sy}[A_{sx}\cot^2\phi_1\cos^2\phi_1 + 0.4(A_{sx}+A_{sy})\cos^2\phi_1$$
$$+ A_{sy}\sin^2\phi_1] = N_1\sin^2(\phi_1+\theta) + N_2\cos^2(\phi_1+\theta) \quad (1)$$

$$\sigma_{sy}[(A_{sy}-A_{sx}\cot^2\phi_1)\sin\phi_1\cos\phi_1$$
$$+ 0.4(A_{sx}\cos^2\phi_1 - A_{sy}\sin^2\phi_1)\cot\phi_1]$$
$$= (N_1-N_2)\sin(\phi_1+\theta)\cos(\phi_1+\theta) \quad (2)$$

$$C = N_1\cos^2(\phi_1+\theta) + N_2\sin^2(\phi_1+\theta) \quad (3)$$

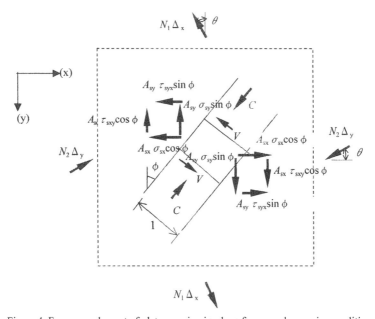

Figure 4. Forces on element of plate carrying in-plane forces under service conditions

3.2 Stress condition in webs subjected to transverse bending

The cross-sectional forces and inner forces in webs subjected to the transverse bending moment and shear force are described in Figure 5. $A_{sy} \cdot f_y$ (f_y is the yielding stress of the stirrup) and $A_{sy} \cdot \sigma_{sy}$ are the stirrup forces per unit length in the transverse bending compressive and tension area respectively, M_e the transverse bending moment per unit length, and $\bar{x}$ the width of the transverse bending compressive area.

In the following, the equilibrium conditions are described. To simplify, it is assumed that the stress condition over $\bar{x}$ is uniform. From Figure 5, the following equation is obtained from the equilibrium of moment around A:

$$C \cdot \cos\phi_1 \cdot (h - \frac{\bar{x}}{2}) - A_{sy}\sigma_{sy} \cdot (h - h') - M_e = 0 \qquad (4)$$

From the equilibrium of forces in the vertical direction, the following equation is obtained:

$$A_{sy}\sigma_{sy} + A'_{sy}f_y - C \cdot \cos\phi_1 = 0 \qquad (5)$$

3.3 Superposition of in-plane forces and transverse bending stresses

In the following, it is assumed that the stirrup for the transverse bending tensile zone yields. Using Equa-

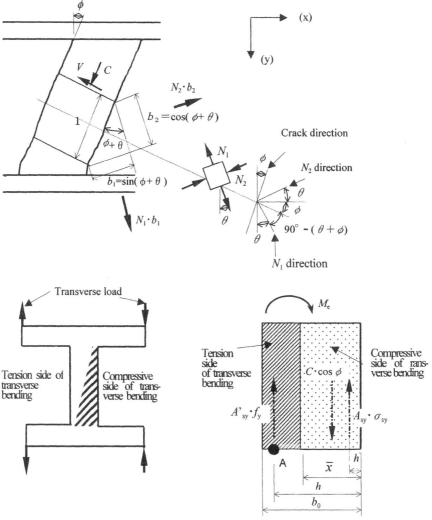

Figure 5. Stress condition of web

tions (1)~(5), the following items are studied:
1. Stirrup stress before the yielding of stirrup in the transverse bending compressive zone
2. Failure due to yielding of the stirrup in the transverse bending compressive zone
3. Failure due to crushing of concrete in the transverse bending compressive zone

3.3.1 *Stirrup stress before yielding of stirrup in transverse bending compressive zone*

Using above-mentioned equations, initial crack direction and stirrup stress of the transverse bending compressive zone are obtained. The unknown values are ϕ_1, C, $\bar{x}$, θ and σ_{sy}. On the other hand, the reinforcement area in the x direction $A_{sx} = 0$, stirrup stress of the transverse tensile area $\sigma_{sy} = f_y$, M_e, shear force $T = Q/z$ (Q is shear force, and z is the distance between the tensile and compressive force), including the principal stress $N_1 = T \cdot \tan\theta$ are known values. The relationship between ϕ_1 and θ is obtained as follows:

$$\{a_1 - a_2 \cdot \tan(\phi_1 + \theta)\} \cdot \{a_1 \cdot \cot(\phi_1 + \theta) - a_2\}$$
$$= \cot(\phi_1 + \theta) \qquad (6)$$

where:

$$a_1 = 0.4 A_{sy} \cdot \cos^2\phi_1 + A_{sy} \cdot \sin^2\phi_1$$
$$a_2 = A_{sy} \sin\phi_1 \cdot \cos\phi_1 - 0.4 A_{sy} \cdot \sin^2\phi_1$$

The stirrup stress in the transverse bending compressive zone is obtained as follows:

$$\sigma_{sy} = \frac{-b_2 + \sqrt{b_2^2 + 4b_1 \cdot b_3}}{2b_1} \qquad (7)$$

where:

$$b_1 = \left(h - h'\right) \cdot A_{sy}^2$$
$$b_2 = A_{sy} \cdot [-T \cdot c_1 \cdot \cos\phi_1 \cdot h + T \cdot c_1 \cos\phi_1 \cdot h'$$

$$-\left(h - h'\right) \cdot A_{sy}' \cdot f_y + M_e]$$
$$b_3 = M_e \cdot A_{sy}' \cdot f_y$$
$$c_1 = \tan\theta \cdot \cos^2(\phi_1 + \theta) - \cot\theta \cdot \sin^2(\phi_1 + \theta)$$

3.3.2 *Failure due to yielding of stirrup in transverse bending compressive zone*

After the first cracking of the web, an increase in loading involves an increase of crack opening corresponding to an increase in the strain orthogonal to the cracks. After the yielding of the stirrup, the width of cracks increases, resulting to the loss of aggregate interlock. With this, reinforcement shear force is lost and the stirrups, considered to be built-in between the crack surfaces, are no longer able to carry shear forces for a transverse displacement.

Ignoring the reinforcement shear force, and using Equations 1 and 2, the relationship between ϕ_2 and θ can be obtained:

$$[\sin^2\phi_2 - \sin\phi_2 \cdot \cos\phi_2 \cdot \tan(\phi_2 + \theta)] \cdot [\cos(\phi_2 + \theta)$$
$$\cdot \cot(\phi_2 + \theta) + \sin(\phi_2 + \theta) \cdot \cos(\phi_2 + \theta)]$$
$$-[\sin^2\phi_2 \cdot \cot(\phi_2 + \theta) - \sin\phi_2 \cdot \cos\phi_2] = 0 \qquad (8)$$

Shear force at failure can be obtained as follows:

$$Q = -\frac{1}{c_1 \cdot \cos\phi_2} \cdot \left(A_{sy} + A_{sy}'\right) \cdot f_y \cdot z \qquad (9)$$

3.3.3 *Failure due to crushing of concrete in transverse bending compressive zone*

Before the formation of the ultimate crack and the failure of I-beams due to the yielding of stirrups, failure due to the crushing of the concrete in the transverse bending compressive zone between the first cracks may occur.

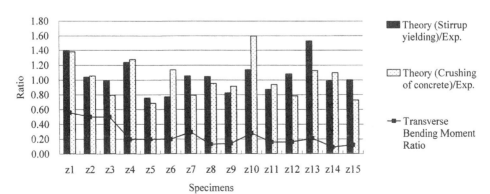

Figure 6. Comparison between theoretical and experimental results

The tensor of reinforced concrete forces in the direction of cracks may be analyzed in three tensors, namely a tensor of axial forces of steel bars, a tensor of shear forces of steel bars and a tensor of the additional concrete forces (Zararis 1988). Knowing the internal forces of the web and the axial forces of the stirrup, stresses of concrete between cracks can be obtained. And from these stresses, the principal compressive stress and the principal tensile stress of concrete are obtained.

The shear force Q at failure is obtained as follows:

$$d_1 \cdot \left(\frac{Q}{z}\right)^2 + (d_2 + d_3 \cdot \sigma_{sy}) \cdot \frac{Q}{z}$$
$$+ (d_4 \cdot \sigma_{sy} + d_5 + d_6 \cdot \sigma_{sy}^2) = 0 \quad (10)$$

where:

$$d_1 = -f_c' \{\tan\theta\cos^2(\phi_1 + \theta) - \cot\theta\sin^2(\phi_1 + \theta)\}^2$$

$$d_2 = -f_{ct}^2 \{\tan\theta\sin^2(\phi_1 + \theta) - \cot\theta\cos^2(\phi_1 + \theta)\}$$

$$d_3 = 2 f_c' \{\tan\theta\cos^2(\phi_1 + \theta) - \cot\theta\sin^2(\phi_1 + \theta)\}A_{sy} \cdot \cos^2\phi_1$$

$$d_4 = f_{ct}^2 \cdot A_{sy} \cdot \sin^2\phi_1$$

$$d_5 = f_c' \cdot f_{ct}^2$$

$$d_6 = -f_c' \cdot A_{sy}^2 \cdot \cos^4\phi_1$$

where f_c is the compressive strength of concrete, f_{ct} the tensile strength of concrete and σ_{sy} the stirrup stress in the transverse bending compressive zone obtained from Equation 7.

4 DISCUSSION

In the following, experimental results are described. Cracking pattern can be separated into two types: one of which is due to shear force, and another of which is due to the crushing of concrete. Shear failure was caused by the yielding of stirrup in the transverse bending compressive zone. In the case of the crushing of concrete, cracks in the vertical as well as in the diagonal direction appear in the compressive zone, though in most cases stirrups don't yield.

In specimens Z3, Z6, Z9, Z12 and Z15, the cross-sectional area of stirrup in the transverse bending compressive zone is half of that in the tensile zone. Based from experimental results, except for specimens Z3 and Z6, concrete in the transverse bending compressive zone of these specimens was under compressive stress.

Figure 6 shows a comparison between theoretical and experimental results. From the figure, it is evident that the crushing of concrete obtained using this theory was the cause of failure in specimens Z3, Z7, Z12, Z13 and Z15.

5 CONCLUSION

Failure due to shear force is of two types: the yielding of the stirrup and the crushing of the diagonal compressive strut. In this paper, in the case of I-beams, the forces of stirrup and concrete, which are brought about by primal tensile forces, are obtained using the in-plane force theory. Using the superposition of these forces and the transverse bending moment, the failure mechanism of the web is investigated. Furthermore, comparisons between theoretical and experimental results were made with regards to ultimate load and stirrup stress. The comparisons were made using crack diagrams and the relationship between shear force – stirrup strain for 15 specimens.

Comparison between the theoretical and experimental results leads to the following conclusions:

1. Theoretical and experimental results of the stresses of stirrup are in good agreement.
2. Stress condition of concrete in the transverse bending compressive zone is not affected by the amount of stirrup.

REFERENCES

Baumann, T. 1972. Tragwirkung orthogonaler Bewehrungsnetze beliebieger Richtung in Flächentragwirken aus Stahlbeton, *Deutscher Ausschuss für Stahlbeton*, Heft 217: W. Ernst & Sohn.

Ewald, G. 1977. Uberlagerung von Scheiben- und Plattentragwirkung am Beispiel star profilierter Stahlbeton- und Spannbetonträger bei hoher Schub- und begrenzter Querbiegungspruchung, *Technische Universität München Universitat*.

Ewald, G. 1982. Zur Tragwirkung und Bemessung von Kastenträger unter Berücksichtigung wirklichkeitsnahen Werkstoffverhaltens, *BETON- UND STAHLBETONBAU*, 12:301-305

Kaufmann, J. & C. Menn 1976. Versuche über Schub bei Querbiegung, *Institut für Baustatik und Konstruktion, ETH Zürich*, Bericht Nr. 7201-1.

Kupfer, H. & G. Ewald 1973. Bericht zu dem Forschungsprogramm Uberlagerung von Scheibenschub und Plattenbiegung im Spannbeton - brückenbau , *Versuchsergebnisse an dem Träger V I/68 dem Bundesverkehrsminisrerium*.

Zararis, P. D. 1988. Failure Mechanisms in R/C Plates Carrying In-Plane Forces, *Journal of Structural Engineering, ASCE*, 114(3): 553-574.

Creative Systems in Structural and Construction Engineering, Singh (ed.) © 2001 Balkema, Rotterdam, ISBN 90 5809 161 9

Ductility versus shear strength of partially prestressed concrete bridge piers

W.A. Zatar & H. Mutsuyoshi

Department of Civil and Environmental Engineering, Saitama University, Japan

ABSTRACT: In order to clarify the interaction between ductility and shear strength of partially prestressed concrete (hereafter known as PPC) bridge piers, results of specimens with low strength ratios tested under statically reversed cyclic loading were analyzed. The strength ratio is defined herein as the ratio of shear capacity to flexural capacity. It was clarified how concrete shear strength decreases as a result of increasing ductility. The applicability of the model by Priestley et al. for RC columns was examined for PPC bridge piers. To account for the differences between the results obtained experimentally and by Priestley's et al. model, some modifications were proposed.

1 INTRODUCTION

Satisfactory seismic response of concrete structures requires that brittle failure modes should be inhibited. Since it is common practice to rely on ductile inelastic flexural response of plastic hinges to reduce the strength requirements for structures responding to strong seismic attack, it is necessary to inhibit shear failure by ensuring that the shear strength exceeds the shear corresponding to maximum feasible flexural strength (Ang et al. 1989). In other words, in order to prevent occurrence of such failure of piers, the strength ratios should be more than unity. Nevertheless, It is generally known that for RC members, although the strength ratios may exceed unity, the members can still fail in shear because of the effect of cyclic load in decreasing the shear strength.

On the other hand, prestressed concrete is known for its ability to increase the shear strength. The use of PPC piers was previously proposed to obtain low residual displacements after an earthquake (Zatar et al. 1998, JPCEA 1998, Ikeda 1998, Zatar & Mutsuyoshi 1998, Zatar and Mutsuyoshi 2000). Yet it is not understood what is the optimum economical strength ratio that can ensure that shear failure will not occur for PPC piers when it is subjected to earthquake loads. Therefore, the objectives of this study were to quantitevely clarify how much increase in shear strength can be attained as well as to clarify how the strength ratio can influence the inelastic response and failure mode of PPC piers.

Additionally, concrete shear strengths given in most codes are almost assumed constant. Conversely, when a bridge pier is exposed to an earthquake excitation, it is shown that concrete shear strength contribution is generally variable as it usually degrades with increasing displacement ductility until it becomes negligible at high ductility levels. Therefore, it was proposed herein to study the interaction of shear strength and ductility for the previously studied PPC piers. Applicability of methods previously developed for RC bridge piers (Ang et al. 1989, Wong et al 1993, Priestley et al. 1994a, b) were examined. Finally, modifications based on experimental data were proposed to predict the interaction of shear and displacement ductility of PPC piers with low strength ratios.

2 EXPERIMENTAL PROGRAM

2.1 *Specimen variables and testing setup*

Since the objectives of this study were to identify the inelastic response behavior of PPC bridge piers and to study the effect of having low strength ratios on the behavior of the specimens, four specimens with low strength ratios were tested. The shear capacities are calculated based on ACI Code 318 requirements. Strength ratios of the specimens ranged from 1.07 up to 1.18 to study their effect on the resulting failure and to ensure that shear failure will not occur before yielding of the longitudinal reinforcing bars. The strength ratios were obtained through employing different shear resistance values resulted from chan-

Table 1. Variables of test specimens

Spec. No.	Variables of PC Pier Specimens											
	Strength Ratio	Mech. Prestressing Ratio	Cross Sec.		Reinforcing Bars & PC Tendons in Cross Section				Shear Reinforcement		Normal Stress (MPa)	
			Dim.	a/d	Rein.	%	PC	%	Ties	A_{sh}/b.s	Axial	PC
E-1	1.18	0.00	30*30	4.20	12D16	2.68	x	x	D6@7cm	0.27	1.0	0.0
E-2	1.07	0.30	30*30	4.20	10D16	2.22	2D13	0.30	D6@7cm	0.27	1.0	1.5
E-3	1.14	0.21	30*30	4.20	12D16	2.68	2D13	0.30	D6@5.5cm	0.34	1.0	1.5
E-4	1.11	0.37	30*30	4.20	10D16	2.22	2D17	0.50	D6@5.5cm	0.34	1.0	3.0

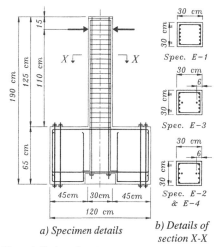

a) Specimen details   b) Details of section X-X

Figure 1. Test specimens.

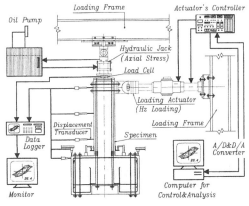

Figure 2. Experimental loading setup.

ging the shear reinforcement and the prestressing level of the specimens. The 1st specimen (named E-1) was a RC control specimen while the other three specimens were PPC specimens. Nominal compressive strength of concrete was 35 MPa. The mechanical prestressing ratio (λ), which is defined as the contribution of the prestressing tendons to the overall capacity of the cross section (Zatar & Mutsuyoshi 1999) ranged from 0.0 for the RC control specimen to 0.37 for the PPC specimens. Specimens were tested under statically reversed cyclic loading. Variables of test specimens are shown in Table 1 while details of the specimens are shown in Figure 1.

An axial stress level of 1 MPa was imposed to the specimens. Specimens were mounted vertically and were loaded horizontally through a loading actuator. The experimental loading setup is shown in Figure 2 while full description and instrumentation can be found elsewhere (Zatar et al. 1998, JPCEA 1998, Zatar & Mutsuyoshi 1998). Predetermined increasing displacement amplitudes were fed to the specimens during testing. The first displacement amplitude was the cracking displacement followed by the first yielding displacement. Then, consecutive multiple integers of the yielding displacements were applied. The used testing system consisted of the specimen, loading actuator, loading jack, loading cell, displacement transducers, data logger, personal computer that controls the input data and another personal computer that controls the output data.

3 TEST RESULTS

The inelastic behavior of each specimen in terms of load-displacement relationship was obtained. As expected, the load carrying capacities of specimens increased by increasing the mechanical prestressing ratios through addition of prestressing tendons to the control specimen.

Figure 3 shows the load-displacement curve of the control specimen E-1 with a strength ratio of 1.18. It was found that pinching of the hysteretic behavior was noticeable due to shear. Also, strength degradation was pronounced at a displacement of 52

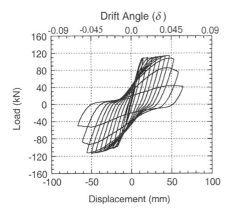

Figure 3. Load-displacement curve of specimen E-1.

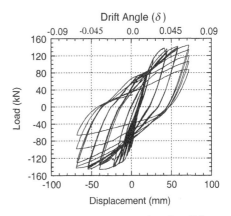

Figure 5. Load-displacement curve of specimen E-3.

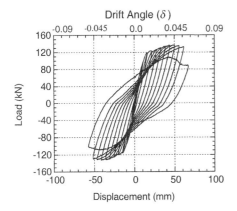

Figure 4. Load-displacement curve of specimen E-2.

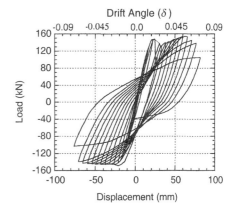

Figure 6. Load-displacement curve of specimen E-4.

mm (drift angle $\delta = 0.047$) due to a final shear failure mode after flexural yielding of the reinforcing bars. For ductility considerations of PPC specimens and since there are no clear yielding points, the yielding point was assumed to be the intersecting point of extrapolating the stiffness after cracking and stiffness after yielding. Figure 4 shows the load-displacement curve of the PPC specimen E-2 with a mechanical prestressing ratio of 0.3 and a strength ratio of 1.07. It was found that although the strength ratio was less than that of specimen E-1, no pinching was observed. Strength degradation commences at higher displacement amplitude of that of specimen E-1 followed by a final shear failure mode after flexural yielding of the prestressing tendons. Figure 5 shows the load-displacement curve of the PPC specimen E-3 with a mechanical prestressing ratio of 0.21 and a strength ratio of 1.14. No pinching was encountered. Strength degradation leading to a shear failure after flexural yielding

commenced at a displacement of 70 mm ($\delta = 0.064$). Figure 6 shows the load-displacement curve of the PPC specimen E-4 with the highest mechanical prestressing ratio of 0.37 and with a strength ratio of 1.11. Almost, no pinching was observed. Strength degradation after flexural yielding of the prestressing tendons was observed to start at displacement amplitude of 70 mm ($\delta = 0.064$) followed by a final shear failure.

It can be observed that slight decreases in the strength ratios of the PPC specimens than that of the control RC specimen did not necessarily lead to shear failures at displacement amplitudes lower than that of the RC specimen since the overall shear capacity was enhanced by prestressing.

Then, it was intended to have an overview about the degradation of shear strength as a result of the cycling effect that usually result from any earthquake excitation as can be seen in the following part.

4 DUCTILITY AND SHEAR STRENGTH INTERACTION

Examination of bridge columns in earthquakes enables a clear distinction to be made between brittle modes of failure and ductile shear failure, where a degree of ductility develops hinges before shear failure occurs. Therefore, it is necessary to inhibit shear modes of failure by increasing the strength ratios (Ang et al. 1989). This is acknowledged in the conceptual model for shear strength proposed by the *Applied Technology Council (ATC-Seismic Design 1981)*. In this model the shear strength is assumed to decrease in a linear fashion as the member displacement ductility increases. If the shear force corresponding to flexural strength is less than the residual shear strength, ductile flexural response is ensured. If it is greater than the initial shear strength, a brittle shear failure results. If the shear force is between the initial and the residual shear strength, the shear failure occurs at ductility corresponding to the intersection of the strength and force-deformation characteristics. Although this behavior is reasonably well accepted, it has not found its way into concrete design codes, except in very few codes. Codes such as the *ACI-318-89* and the *New Zealand Concrete Code* simply assume that the concrete shear strength can be ignored when $P < 0.05 f'_c A_g$ and $P < 0.1 f'_c A_g$, respectively. Neither of them includes an explicit relationship between ductility and shear strength (Priestley et al. 1994c). The AIJ *recommendations* proposed that for a ductile member, the permissible diagonal compressive stress is progressively reduced as the plastic rotation increases (Yoshikawa and Miyagi 1999) which is somehow similar to the *ATC model*. Nevertheless a deficiency in the *AIJ recommendations* (Design 1990) exists where its recommendations were proposed only for rectangular sections.

4.1 Model by Priestley et al.

Recently, considerable experimental research for RC columns, particularly by Ang et al. (1989) and Wong et al. (1993) has been directed towards a better definition of the shear strength/ductility relationship. Additional results from Priestley et al. (1994a, b) have supplemented this data. Furthermore, Priestley and Benzoni (1996) refined the models by Ang et al. and Wong et al. His model was shown to be accurate enough to a full range of experimental database. Priestley et al implemented further experimental data by Mattock and Wang, Jirsa and Woodward, Xiao et al. and an extensive Japanese database by Watanabe and Ichinose to verify the superior accuracy of his

proposed model over the others (Priestley et al. 1994c). Additionally, Priestley and Benzoni (1996) and Xiao and Marirossyan (1998) presented modifications of the original model to account for special cases of RC columns.

In the approach by Priestley et al. for RC columns, the components of the overall shear resistance were separated in the form of steel truss component (V_s), axial force component (V_p) and concrete component (V_c). The magnitude of V_s depends on the transverse reinforcement content. The V_p component depends on the column aspect ratio. The concrete component (V_c) is equal to $k . (f'_c . A_e)^{0.5}$ where k depends on the level of ductility (Fig. 7) and A_e is the effective shear area $= 0.8 A_g$.

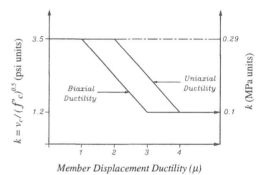

Figure 7. Model of concrete shear strength factor versus ductility by Priestley et al.

4.2 A proposed model for PPC piers with low strength ratios

Since the model by Priestley et al. was proposed for RC columns and not for PPC piers, the effect of the prestressing tendons in increasing the shear strength was not accounted for. Therefore, there was a necessity to have a model for such contribution as

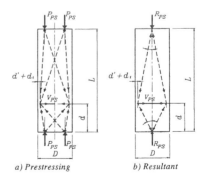

a) Prestressing   b) Resultant

Figure 8. A truss model for prestressing component of shear strength.

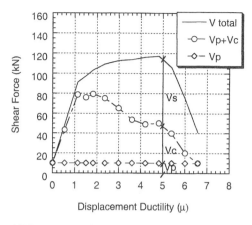

a) RC specimen E-1.

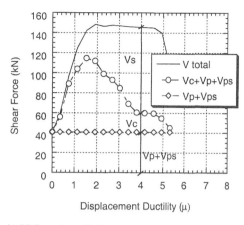

b) PPC specimen E-4.

Figure 9. Displacement ductility versus shear components of specimens E-1 & E-4.

proposed herein. It was considered that the column axial stress due to prestressing (ΣP_{PS}) could enhance the shear strength by an arch action that forms inclined struts (Fig. 8). The critical section was assumed at a distance $= d$ from the footing surface. Individual prestressing tendons could participate in the overall shear strength through formation of individual compression struts. A simple calculation method was suggested through identification of the resultant (R_{PS}) magnitude and point of application of the external applied prestressing forces calculated from static and solving the equilibrium truss mechanism in Figure 8(b).

4.3 *Justification of methodology*

During testing of specimens, strains in the reinforcing ties were measured and the associated stresses were obtained. Using these measured stresses and the determined inclination angle of the major shear crack, the V_s component could be experimentally obtained at different loading stages. Calculating V_p and V_s and then subtracting from the experimental lateral loads at each loading stage, the V_c component could be calculated. Because of space limitations, Figure 9 shows only a comparison between the experimental components of shear resistance of both the RC specimen E-1 and the PPC specimen E-4. From the V_c component, the k factors could be calculated and could be plotted versus ductility. A comparison between these results and those of Priestley's et al. approach for RC columns (Fig. 10) shows some differences at almost all ductility levels. Similar findings were observed for Priestley and Benzoni (1996) and Xiao and Mariros-

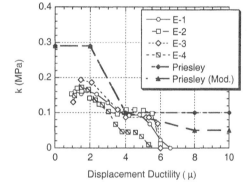

Figure 10. Experimental and Priestley's et al. k factors versus displacement ductility.

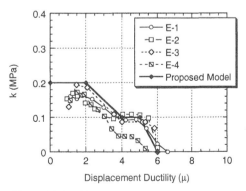

Figure 11. justification of the proposed k factors versus displacement ductility.

825

syan (1998) where it was found that the upper limits, which identify Priestley's model, are conservative especially at low displacement ductility factors.

In order to have a better simulation of the obtained experimental k values than those obtained by Priestley's et al. approach, a modification in Priestley's et al. model was proposed herein to account for cases of PPC piers with low strength ratios. The proposed model which is shown in Figure 11 assumes that an initial value of 0.2 is to be used as k value, if MPa units are used, up to a ductility of 2.0 followed by a gradual decrease in k when increasing the ductility up to 4. Then, a constant k of 0.1 can be used up to a ductility of 5.0 followed by a gradual decrease in k till it vanishes at a ductility of 6.0. It should be noted that, opposite to Priestley's model, k value vanishes in the proposed model because this model was intended for PPC piers with low strength ratios. It is expected that if the PPC piers have higher strength ratios, the k values might have values rather than zero for ductility more than 6. Nevertheless such expectation requires further experimental verification.

5 CONCLUSIONS

In order to clarify the inelastic response behavior and the interaction between ductility and shear strength of partially prestressed concrete bridge piers, specimens with low strength ratios were tested under statically reversed cyclic loading. It was found that the addition of prestressing tendons could increase the shear resistance of the bridge piers.

It was also clarified how concrete shear strength degrades as a result of increasing ductility. The interaction between the concrete component of shear resistance and displacement ductility was clarified. The applicability of the interaction model by Priestley et al. for RC columns was examined for PPC bridge piers.

To have a better simulation of the experimental results, some modifications to Priestley's et al. model were proposed. A separate axial prestressing component was added to Priestley's et al. model. Such component has the advantage of being constant when increasing the ductility. Furthermore, the model of concrete component was changed to account for PPC bridge piers with low strength ratios.

REFERENCES

Ang, B.G., Priestley, M.J.N. & Paulay, T. 1989. Seismic Shear Strength of Circular Reinforced Concrete Columns. ACI Structural Journal: 86(1): 45-59.

Architectural Institute of Japan 1990. Design Guidelines for Earthquake Resistant Reinforced Concrete Buildings Based on Ultimate Strength Concept, Japan.

Ikeda, S. 1998. Seismic Behavior of Reinforced Concrete Columns and Improvement by Vertical Prestressing. Proceedings of the 13th FIP Congress on Challenges for Concrete in the Next Millennium. Vol. 2.

JPCEA report 1998. Seismic Behavior of Prestressed Concrete Pier. Japan Prestressed Concrete Engineering Association. March.

Priestley, M.J.N., Seible, F., Xiao, Y. & Verma, R. 1994a. Steel Jacket Retrofitting of Reinforced Concrete Bridge Columns for Enhanced Shear Strength-Part 1. ACI Structural Journal.

Priestley, M.J.N., Seible, F., Xiao, Y. & Verma, R. 1994b. Steel Jacket Retrofitting of Reinforced Concrete Bridge Columns for Enhanced Shear Strength-Part 2. ACI Structural Journal.

Priestley, M.J.N., Verma, R. & Xiao, Y. 1994c. Seismic Shear Strength of Reinforced Concrete Columns. ASCE, Journal of Structural Engineering, 120(8):2310-2329.

Priestley, M.J.N. and Benzoni, G. 1996. Seismic Performance of Circular Columns with Low Longitudinal Reinforcement Ratios. ACI Structural Journal, 93(4):474-485.

Wong, Y.L., Paulay, T. & Priestley, M.J.N. 1993. Response of Circular Reinforced Concrete Columns to Muli-Directional Seismic Attack. ACI Structural Journal, 90(2):180-191.

Xiao, Y. and Martirossyan A. 1998. Seismic Performance of High-Strength Concrete Columns. ASCE, J. of Structural Engineering, 124(3):241-251.

Yoshikawa, H. & Miyagi, T. 1999. Ductility and Failure Modes of Single Reinforced Concrete Columns. Seminar on Post-peak Behavior of RC Structures Subjected to Seismic Loads: Vol. 2:229-244.

Zatar, W., Mutsuyoshi, H. & Inada, H. 1998. Dynamic Response Behavior of Prestressed Concrete Piers under Severe Earthquake. Proc. of Japan Concrete Institute, Vol. 20:1003-1008.

Zatar, W. and Mutsuyoshi, H. 1998. Seismic Behavior of Partially Prestressed Concrete Piers. Proc. of 2nd Symposium on Ductility Design Method for Bridges, Japan Society of Civil Engineers:189-192.

Zatar, W. and Mutsuyoshi, H. 1999. A Restoring Force Model for Partially Prestressed Concrete Piers. Proc. of Japan Concrete Institute, Vol. 21.

Zatar, W. and Mutsuyoshi, H. 2000. Reduced Residual Displacements of Partially Prestressed Concrete Bridge Piers. 12th Conference on Earthquake Engineering. Paper No. 1111.

Creative Systems in Structural and Construction Engineering, Singh (ed.)© 2001 Balkema, Rotterdam, ISBN 90 5809 161 9

Cyclic loading test of wall type piers with several configuration types of shear reinforcement

T.Watanabe, T.Fujiwara, T.Tsuyoshi & T.Ishibashi
East Japan Railway Company, Tokyo, Japan

ABSTRACT: This paper reports the experimental results of ten model wall type reinforced concrete members. Experimental parameter is configuration of cross ties and hoops in order to find configuration of cross ties which is easier to construct and have sufficient deformational capacity compared with cross hoops used widely in real structures like wall type piers and box culvert nowadays. From the test results, for example, specimen with cross ties which have lap splices with splice length of twenty times of cross bar diameter and also have standard hooks specified in current code for railway structures showed sufficient deformational capacity and also show a little decrease in load-carrying capacity even after the specimen had undergone large displacement ductility levels.

1 INTRODUCTION

In railway structure, shear reinforcements of vertical members like column which needs large deformational capacity against seismic loading are specified to be arranged as hoops with 135° hooks in its end which are developed to inner concrete core[RTRI(1991)]. In this way, transverse reinforcements must be pull down from top of the longitudinal bars to arrange ties in case of structures such as wall type piers and box culvert. And sometimes it is difficult to construct because the cross ties are arranged in zigzag position in vertical direction.

While, in seismic code of Japan Society of Civil Engineers (1996) which were established after the South Hyogo Prefecture Earthquake (1995), lap splice is allowed if the ties have standard hooks in its end and lap splices are placed inside the core concrete. But such experimental results of deformational capacity of members, which have lap splices inside the core concrete, are still lacking.

Therefore, we conducted cyclic loading tests with wall type specimens changing configuration of cross ties aiming to simplify arranging works of the cross ties in site. In this paper, we made experimental study on the influence of configuration of cross ties to the failure mode and deformational capacity of specimens.

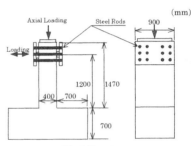

Fig 1 Specimen

2 EXPERIMENTAL PROGRAM

2.1 *Specimen Design*

Ten model columns with 900mm width scaling wall members have been designed for cyclic loading tests. As shown in Fig.1, the columns had a clear height of 1,470mm with a cross section of 900mm ×400mm. The bottom of the model columns were constructed into 1,800×900×700mm load stub. The stubs were heavily reinforced to assure a rigid behavior during the tests.

Fig.2 shows cross section of each specimen with different configuration of cross ties. As mentioned before, we paid attention on configuration of cross ties and all the arrangements of bars except of cross ties are same in all specimens. All the specimens were reinforced with 22 D19 longitudinal bars and transversely reinforced with one D6 peripheral hoop per section besides cross ties. Four cross ties are arranged in each section. The transverse bars were placed at 100mm center-to-center spacing in all model columns. Shear span to effective depth ratio is 3.3 in all specimens.

Model column No.1 was transversely reinforced with a peripheral hoop and without any cross ties in each section.

Model column No.2 ∼ 10 were transversely reinforced with peripheral hoop and four pairs of cross ties and only changing configuration of cross ties. The cross ties were arranged in zigzag position between adjacent section 1 and 2 as shown Fig.2 (b) in these 9 model columns.

Model column No.2 has four cross hoops in each section, which is widely used in current railway structures.

Model column No.3 has cross ties consist of one pair of cross ties which have lap splices with 20 ϕ in length(ϕ :cross tie diameter) and standard hooks. Model column No.4 has the same bar arrangement as model column No.3 but without standard hooks.

Model column No.5 has the cross ties arranged inside the longitudinal bars and don't surround any longitudinal bars. Each pair of cross ties has lap splices in the core concrete as same as model column No.3.

Model column No.6 has the cross hoops which surround longitudinal bars at one side and arranged inside the longitudinal bars at another side of the column.

Model column No.7 has cross ties with lap splices at the middle part of core concrete as same as model column No.3 but each ties have large width and thus longitudinal bars weren't arranged at the corner of each cross tie. Model column No.8 has similar cross ties arrangements as No.7, but lap splices are placed at the shallow part of the core concrete.

Model column No.9 and 10 have straight shape of cross ties with standard hooks. Therefore, eight cross ties are arranged in each section in order to make transverse reinforcement ratio as same as other seven model columns. Each cross ties hooked around the longitudinal bars in model column No.9 and didn't hook around the longitudinal bars in model column No.10, respectively. Meanwhile, Each peripheral hoop is divided into four parts.

Shear strength to flexural strength ratio V_y/V_{mu} (where V_y:shear strength and V_{mu}:shear force corresponding to the first yielding of longitudinal bars), which is the main parameter of displacement ductility factor are 1.2 in model column No.1 and 2.0 in model column No.2∼10. All reinforcing bars used in the columns conformed to SD345 of JIS standard G3112. The D19 and D6 bars had yield strength of 343MPa in the standard and concrete compressive strength used in specimens were 29.2 ∼38.3MPa.

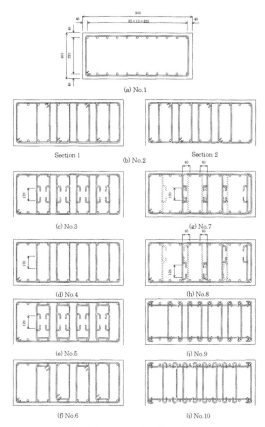

Fig 2 Cross section details

2.2 Test Setup and Loading Program

As shown in Fig.1, the lateral load are applied by 980kN hydraulic jack and the tests were conducted under constant axial load using 980kN hydraulic jack moving parallel in lateral loading direction in order to maintain vertical loading. The axial stress is 0.98MPa for ten model columns.

The cyclic loading tests were conducted under following program. Load reversals in push and pull directions were symmetric. First, lateral force was applied until the first yielding of longitudinal reinforcements appeared in both push and pull directions. Subsequent loading was carried out under displacement control, attempting one cycle for each of the peak displacement ductility levels of μ =1,2,3, ⋯ until the specimen lose their load carrying capacity significantly.

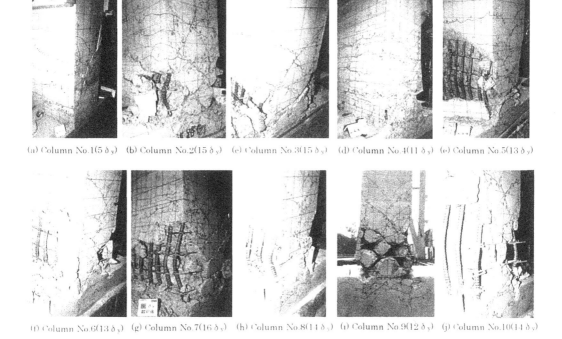

(a) Column No.1(5 δy)　(b) Column No.2(15 δy)　(c) Column No.3(15 δy)　(d) Column No.4(11 δy)　(e) Column No.5(13 δy)

(f) Column No.6(13 δy)　(g) Column No.7(16 δy)　(h) Column No.8(14 δy)　(i) Column No.9(12 δy)　(j) Column No.10(14 δy)

Fig 3 Failure pattern of each model column

3 EXPERIMENTAL RESULTS AND DISCUSSIONS

3.1 *General Observations*

Although all ten specimens developed approximately the same maximum flexural strength, their ultimate performances and the ductility levels achieved were different for different test conditions. As shown in Fig.3, the ultimate failure modes for all model columns were developed according to following scenarios.

Column No.1. Model column No.1 was failed during the loading to achieve the peak at ductility level μ =5 in the push direction. The failure was initiated by the penetration of shear cracks through the core concrete, as shown in Fig.3 (a).

Column No.2. Plastic hinge was fully formed at bottom end of model column No.2 and it developed a displacement ductility factor of over μ =10; however, it lost the capacity upon the loading reversal at μ =14 due to the opening of 135° hooks of D6 hoops near column ends caused by the buckling of compression longitudinal bars within the column's plastic hinge.

Column No.3. The ultimate failure modes for model column No.3 were developed as same scenario as model column No.2 before losing the

load-carrying capacity. No significant slipping of cross ties which have lap splices at the middle part of the core concrete were observed until the end of the test but some rupture of the D6 cross ties were observed as the column lose its load-carrying capacity.

Column No.4. Model column No.4 successfully underwent until the ductility level μ =9, but it failed in shear during the ductility level of μ =10. Compression cover concrete scaling was observed at 600mm range of bottom end of the model column during the ductility level of μ =10 in the push direction, and the model column lost its load-carrying capacity upon the loading at μ =10 in the pull direction, due to the penetration of shear cracks accompanied by bond slipping of the D6 cross ties from the core concrete.

Column No.5. The compression concrete scaling were observed at μ =7 in the pull direction and μ =8 in the push direction at 500mm range of bottom end of the column No.6 because the D6 cross ties were all placed inside the longitudinal bars. After the compression cover concrete scaling, the column lost its load-carrying capacity gradually until the end of the test.

Column No.6. The compression concrete scaling was observed at μ =10 in the pull direction, which

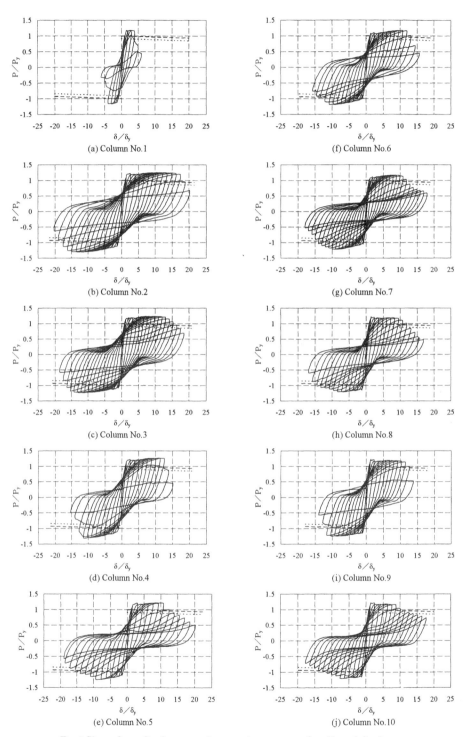

Fig 4 Shear force-displacement hysteretic responses for all model columns

was more late than model column No.5 because its cross hoops were placed outside the longitudinal bars at one side of the column. After the compression cover concrete scaling, the specimen suddenly lost its load-carrying capacity due to the column member slipping along the diagonal cracks at bottom end of the column ranged 0～200mm from the surface of the footing during the loading in the push direction. Meanwhile, compression concrete was severely damaged during the loading in the pull direction accompanied by the member slipping. The column lost its load-carrying capacity rapidly.

Column No.7. Compression cover concrete scaling were observed at μ =10 and 11 which were earlier than model column No.2 and 3 because the model column was transversely reinforced with D6 peripheral hoops and D6 large width cross ties which corner weren't placed at position of longitudinal bars and didn't restrict the longitudinal bars significantly.

Compression concrete was damaged largely at μ =13 in the push direction and the specimen lost its load-carrying capacity accompanied by the column member slipping along the diagonal cracks at bottom end of the column. However, the cross ties resist against the degradation of load-carrying capacity even after the load had decreased smaller than yield force, changing its shapes from rectangular to octagon due to the buckling of compression longitudinal bars. The column was able to maintain stable resistance against the loading at the end of the test.

Column No.8. Model column No.8 failed in similar mode with model column No.7.

Column No.9. The compression cover concrete scaling were observed upon the loading reversals at μ =9. At μ =11, the compression concrete was severely damaged, and column member's slipping was occurred during the loading to achieve the peak at μ =11 in the push direction. Meanwhile, the model column significantly lost its capacity at μ =11 and 12, accompanied by the openings of many of 180° hooks of peripheral hoops and cross ties near column ends.

Column No.10. The compression cover concrete scaling were observed at μ =7 in the pull direction and μ =8 in the push direction, as same as column No.5. The damage seemed to be distributed at wide range from the end of the column and it was progressed very slowly until the end of the test.

3.2 Shear Force – Lateral Displacement Hysteretic Responses

Fig.4 shows shear force-lateral displacement hysteretic relationships for the model columns. The lateral axis represents lateral displacement divided by predicted value of displacement corresponding to the first yielding of the longitudinal bars and actual material strengths [Umihara et al.(1999)]. The vertical axis represents shear force divided by predicted value of shear force corresponding to the first yielding of the longitudinal bars and actual material strengths. The predicted capacity lines are shown by dashed lines and also the capacity lines estimated on the assumption that the cover concrete had crushed are shown by dotted lines. The slopes of the predicted capacity lines express the effects of extra bending due to axial load.

As shown in Fig.4, although all ten specimens developed almost same ratio of shear force to predicted yield force which are approximately 1.2, their ductility levels achieved were different for different configuration of transverse reinforcements.

Column No.1. As shown in Fig.4(a), sudden degradation of load-carrying capacity occurred at an experimental ultimate displacement ductility factor μ =5 due to shear failure.

Column No.2. As shown in Fig.4(b), column No.2 exhibited excellent hysteretic behavior and developed an ultimate displacement ductility factor of over μ =10. The significant peak lateral force degradation was very small until the column reached μ =13～15, where the column lost its load-carrying capacities due to the compression cover concrete scaling and opening of 135° hooks of peripheral hoops and cross hoops.

Column No.3. A stable response of model column No.3 up to μ =12～13 can be seen from Fig.4(c). After the displacement ductility level, gradual lateral force degradation occurred until the end of the test. The lateral force degradation of model column No.3 was smaller than column No.2, due to no significant slipping of cross ties which have lap splices in the middle part of the core concrete, until the end of the test.

Column No.4. Satisfactory response up to μ =10 in the pull direction, where shear failure occurred for model column No.4 is shown in Fig.4(d). The shear failure was due to the slipping of cross ties from the core concrete.

Column No.5. As shown in Fig.4(e), model column No.5 developed a stable hysteretic response up to μ =7 ～ 8, where compression cover concrete were crushed due to the longitudinal bars' buckling and lateral force degradation which deserves to no cover concrete occurred. The following gradual lateral force degradation was observed until the end of the test and the load-carrying capacity seemed to be satisfactory.

Column No.6. As shown in Fig.4(f), the degradation of lateral force due to compression cover concrete scaling were occurred at μ =10～11, a little later than column No.5. This is because one side of each cross hoop was placed outside the longitudinal bars and thus resist stronger than column No.5 against the buckling of the longitudinal bars. However, the degradation of lateral force was larger and thus failed earlier than column No.5.

Column No.7. Response for model column No.7, as shown in Fig.4(g), indicates stable force-displacement characteristics up to $\mu = 11 \sim 12$, where the degradation of lateral force first occurred. However, the following degradation of force became smaller as the displacement ductility factor became larger, due to the resistance of deformed cross hoops.

Column No.8. Although the similar load-displacement behavior to column No.7 can be seen from Fig.4 (h), its degradation of lateral force was a little larger than column No.7.

Column No.9. A stable response of model column No.9 up to $\mu = 10$ can be seen from Fig.4 (i). However, significant degradation of force occurred at $\mu = 11$ due to the openings of 180° hooks of peripheral hoops and cross ties, which led to the rapid failure of the column.

Column No.10. Response for model column No.10, as shown in Fig.4 (j), indicates similar behavior as column No.5, which cross ties were placed inside the longitudinal bars and compression cover concrete scaling occurred relatively earlier than the other specimens due to the buckling of the longitudinal bars, and thus the degradation of lateral force also occurred. However, the following degradation of lateral force was gradual and the column maintained a stable behavior even after the end of the test.

4 CONCLUSIONS

Ten wall type columns with different configuration of transverse reinforcements have been experimentally studied. Based on the comparisons of ten model columns, the following conclusions can be reached:

1. Stable response up to large displacement ductility level can be achieved by model column No.3 in which cross ties with lap splices in the middle part of the core concrete with 20ϕ in length (ϕ: cross tie diameter) and 180° hooks in its end, as well as model column No.2 which is transversely reinforced with cross hoops which were currently used.

2. Model column No.4 in which cross ties with lap splices as same as column No.3 but without any hooks in its end failed in shear after the yielding of the longitudinal bars due to the slipping of cross ties accompanied by the buckling of the longitudinal bars, which proved to be less ductile than column No.2 and 3. Hooks are needed to

place in the end of cross ties in this case.

3. Model column No.7 in which cross ties have large width and the longitudinal bars aren't placed at the corner of cross ties, exh?bited stable hysteretic behavior until the end of the test. The degradation of force are small particularly corresponding to the large displacement ductility levels. The configuration of the cross ties are easy to construct and it seems to be effective to use it in real structures.

4. Model column No.8 in which cross ties have lap splices at the shallow part of the core concrete, exhibited relatively stable behavior compared with column No.7. This can also be used in real structures too.

5. Model column No.5 and 10 in which cross ties were all placed inside the longitudinal bars suffered the degradation of lateral force due to the compression cover concrete scaling at $\mu = 7 \sim 8$. However, the following degradation of force was gradual and the columns exhibited stable behavior until the end of the test. Further study will be needed on the ratio of inside-placed to outside-placed cross ties avoiding the bond failure of the longitudinal bars.

6. The model column No.6 in which cross hoops were placed inside and outside of the longitudinal bars at its each side of each cross hoops suffered large degradation of lateral force in the large displacement ductility level. It is considered that this type of configuration cannot be used in real structures.

7. Model column No.9 in which cross ties have straight figure and hooked around the longitudinal bars in its both end significantly lost its load-carrying capacity due to the opening of the 180° hooks caused by the buckling of the longitudinal bars. Therefore, this type of configuration of transverse reinforcements cannot be used in real structures too.

REFFERENCES

Railway Technical Research Institute 1992. *Design standards for railway structures (concrete structures)*. Maruzen: Tokyo.

JSCE Concrete Committee 1996. *Concrete standard codes (seismic design)*. Japan Society of Civil Engineers: Tokyo.

Umihara,T.,Kobayashi,K, & Ishibashi,T 1999. Study on evaluation of yield displacement of reinforced concrete column with large transverse reinforcement ratio subject to earthquake loading. *Proceedings of the Japan Concrete Institute Vol.22*. Tokyo.

Creative Systems in Structural and Construction Engineering, Singh (ed.) © 2001 Balkema, Rotterdam, ISBN 90 5809 161 9

Anchor effectiveness of inclined meshes as concrete web reinforcement

F.J.Orozco
Monterrey Institute of Technology, Mexico

W.C.McCarthy
New Mexico State University, Las Cruces, N.Mex., USA

ABSTRACT: A series of prestressed concrete beams were tested to extract their shear carrying capacity for differing web reinforcement types. Explored were meshes of intertwined elements, welded wire fabric (WWF), and glass fiber reinforced plastic (GFRP). Of particular importance was a determination of anchor effectiveness and optimization of the material by inclined mesh configuration as compared to standard stirrups. American Concrete Institute (ACI) and modified compression field theory (MCFT) methods were applied to predict the web shear capacity of the beams and the veracity of both methods was evaluated. WWF and GFRP inclined meshes were found to be effective alternatives to conventional web reinforcement; however, failure of the WWF welds before full anchor effectiveness was reached provoked the conclusion that a need exists for strengthening of the connection points. No such restrictions existed for GFRP material with the recommendation that full investigation of fiberglass reinforced plastic, in an inclined mesh configuration as web reinforcement, be undertaken to firmly establish its design limits.

1 INTRODUCTION

Shear stresses by themselves or jointly with normal stresses act to produce diagonal tension. If the diagonal tension reaches the cracking tensile strength of the concrete, shear cracks appear. Web reinforcement is provided and, almost without exception, stirrups are the reinforcement of choice to prevent shear crack propagation and extension in width leading to premature failure.

Although it is obvious that inclined rebars optimize the material as shear reinforcement, traditionally vertical rebars are preferred for economic and practical reasons. The savings in material with inclined rebars do not compensate for higher labor costs. Additionally, the use of inclined rebars requires special supervision to prevent an incorrect inclination. Inclined elements in a prefabricated mesh configuration would overcome these disadvantages. The prefabrication process, in addition to the ease of installation, will reduce the labor costs while a mesh made of two directional inclined elements solves the incorrect placement problem.

Because of a high tensile strength, and thermal compatibility with concrete, steel has long been used as web reinforcement. However, the emerging use of glass fiber reinforced plastic (GFRP) as an alternative reinforcement also suggests its potential use for web shear as indicated by tests conducted by Nawy et al. (1971) and Nawy & Neuwerth (1977). In this instance, the GFRP rods were placed in the standard vertical stirrup configuration. Characteristics such as high corrosion resistance, absence of magnetic interference, and light weight greatly increase the attractiveness of GFRP as a reinforcing material.

Whichever material is used, meshes are normally made by smooth elements that cannot develop effective anchorage due to weak bonding. This leaves mechanical anchorage which is provided by the welding or intertwining of the elements as the primary mechanism to prevent them from slipping. Xuan et al. (1988) reported acceptable shear behavior for a vertical steel mesh placement in prestressed concrete T-beams. Acceptable anchorage was attained by welding the vertical wires to their horizontal counterparts. The same mechanism would work for meshes formed by inclined elements made by either steel or GFRP material.

Intertwined wire, similar to chain link fencing, is another anchor mechanism that might be acceptable in spite of possible high residual stresses at the twisting points. Because chain link is an easy-to-obtain and a relatively inexpensive commercial product, it is worth investigating.

Welded wire fabric (WWF), GFRP, and chain link mesh types were investigated as to their suitability for web shear reinforcement. The mesh elements had an inclination of 45° with respect to the longitudinal axis of the beam. The soundness of the mesh types was judged on anchor effectiveness and optimization of the material with standard stirrup capacity as the base line.

Twelve prestressed beams were tested to shear failure in this study, three for each mesh type and three with standard stirrups. These model beams were 152.5 mm wide and 305 mm deep and held a single line of web reinforcement midline to the beam. Although an insufficient number of beams were tested to obtain definitive results suitable for code modification, the number was sufficient to identify trends and to point to additional avenues of further testing that would ultimately lead to code changes or eliminate the need for further testing if the reinforcement proves to be totally unsuitable.

ACI 318-99 provisions for shear that reflect two directional inclined elements, and the Modified Compression-Field Theory (MCFT) developed by Vecchio and Collins (1986) were applied to predict shear capacity both as a measure of their relative accuracy and to gain greater insight into the shear failure mechanism. Dowel action is normally not considered in the development of shear capacity. However, the authors felt that its contribution was significant and would skew the results if it were disregarded.

2 PREDICTED SHEAR CAPACITY

2.1 ACI Approach

To consider inclined elements in two directions, it was established that the forward inclined elements of the mesh had an inclination of α while the backward inclined elements had an inclination of $180°-\alpha$ with respect to the longitudinal axis of the beam, and that both elements had the same cross-sectional area (A_v). Since it is assumed that the contribution of concrete to the shear capacity (V_c) is independent of the form of web reinforcement, only the contribution of reinforcement to the web shear capacity (V_s) needs to be modified. Thus, a condition of equilibrium was applied to the free-body diagrams in Figure 1 to obtain the following equation:

$$V_s = \frac{A_v f_{v_1} jd}{s}\left(\frac{\sin\alpha}{\tan\theta} + \cos\alpha\right) +$$
$$\frac{A_v f_{v_2} jd}{s}\left(\frac{\sin\alpha}{\tan\theta} - \cos\alpha\right) \qquad (1)$$

The ACI method determines the ultimate shear force provided by the web reinforcement by assuming yielding of the reinforcement. Therefore, for meshes, $f_{v_1} = f_{v_2} = f_{vy}$, where f_{vy} is the yield stress for steel and the ultimate tensile strength (f_u) for GFRP material. Also the ACI method considers the angle of the crack inclination to be 45° and replaces jd by d. With these assumptions, equation (1) becomes:

$$V_s = 2\frac{A_v f_{vy} d}{s}\sin\alpha \qquad (2)$$

2.2 MFT Approach

The MCFT approach was based on work conducted by Mitchell & Collins (1974), and treats the crack inclination as an unknown. Together with f_{v_1}, f_{v_2}, f_{sx}, f_{px}, and f_2, θ can be determined using equilibrium and compatibility conditions. The quantities f_{sx}, f_{px}, and f_2 are the actual stresses, due to shear, in the non-prestressed longitudinal reinforcement, the prestressed longitudinal reinforcement, and the concrete struts, respectively. Then, the web reinforcement's contribution to the shear capacity is found using equation (1).

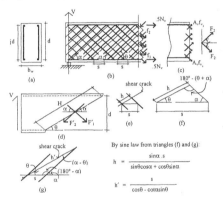

Figure 1. Truss model for meshes.

Unlike the ACI method which adopts semiempirical equations to evaluate the contribution of concrete to the shear capacity, the MCFT method considers the tensile strength of the web reinforced cracked concrete, f_l, in a more rational approach. Vecchio and Collins established this tensile strength by using an empirical constitutive law with an upper limit that is a function of the capacity of both web reinforcement and longitudinal reinforcement to transmit forces across cracks. Since the present research involves two directional inclined elements, the relationships which limit f_l were re-derived as presented below:

$$f_l \leq v_{ci}\tan\theta + \frac{2A_v}{b_w s}\left\{ \begin{array}{l} \sin\alpha\left[f_{vy} - \left(\frac{f_{v_1} + f_{v_2}}{2}\right)\right] + \\ \frac{\cos\alpha}{\cot\theta}\left(\frac{f_{v_1} - f_{v_2}}{2}\right) \end{array} \right\} \qquad (3)$$

$$A_{sx}f_y + A_{px}f_{pu} + \frac{2A_v f_{vy} jd\cos\alpha}{s\tan\alpha} \geq \qquad (4)$$

$$A_{sx}f_{sx} + A_{px}f_{px} + \frac{A_v jd}{s\tan\alpha}\left(f_{v_1} + f_{v_2}\right)\cos\alpha$$

$+f_1 b_w jd +$

$$\left[\left\{ f_1 - \frac{2A_v}{b_w s} \left[\frac{\sin\alpha \left[f_{vy} - \left(\frac{f_{v_1}+f_{v_2}}{2} \right) \right] +}{\frac{\cos\alpha}{\cot\theta} \left(\frac{f_{v_1}-f_{v_2}}{2} \right)} \right] \right\} b_w jd \cot^2\theta \right] \quad (4)$$

2.3 Contribution of Dowel Action to Shear Capacity

Despite the significance of the dowel action to shear capacity as published by investigators such as Johnston & Zia (1971), and Taylor (1972); it is commonly neglected as a component of shear design. However, in investigating anchor effectiveness for web reinforcement, dowel action must be considered. Otherwise, the experimental shear capacity to predicted shear capacity ratio will be amplified reflecting a false overestimate of anchor effectiveness. Various equations for evaluating the contribution of dowel action to the shear capacity were reviewed. Reineck's equation (1991), presented below, was selected for this research because the results were consistent with information gathered by other researchers, e.g., Taylor (1972).

$$V_d = \frac{6}{f_c^{1/3}} b_n d_b f_{ct} \quad (5)$$

where:

$$b_n = b_w - \sum d_b \quad \text{(m)} \quad (6)$$

$$f_c = 0.95 f_c' \quad \text{(Mpa)} \quad (7)$$

$$f_{ct} = 0.246 f_c^{2/3} \quad \text{(MPa)} \quad (8)$$

Equation (5) is applied to each layer of longitudinal reinforcement.

3 EXPERIMENTAL SHEAR CAPACITY

3.1 Materials

The twelve pretensioned concrete beams used in this study were constructed in the Materials Laboratory at New Mexico State University. The 152.5 mm by 305 mm rectangular beams required a special form to provide easy access and maneuverability and to facilitate pretensioning.

The beams were 2362 mm long and spanned 2210 mm. The longitudinal reinforcement was designed as shown in Figure 2 to ensure that shear failure would occur before flexural capacity was reached. The deformed bars had a yield strength of 414 MPa, The 7-wire strand had an ultimate tensile strength (f_{pu}) of 1862 MPa and was prestressed at 0.5 f_{pu}. Three #3

compression rebars supported both the web and tensile longitudinal reinforcements (see Fig. 2).

Four series of differing web reinforcement were tested with three beams in each series. Web reinforcement investigated included #3 (f_{vy} = 276 MPa) single-legged vertical stirrups, a 9- gauge chain link mesh, a WWF6X6-W3.0X3.0 welded wire mesh, and a GFRP mesh (cut from a commercially molded GFRP grating) with each type designated as the A, CH, W, and F beam series, respectively. The CH, W, and F meshes were configured in square patterns of 51 mm, 152 mm, and 102 mm, respectively, so that their elements were inclined at 45°. An average yield stress was experimentally determined to be 310 MPa for the chain link steel and 352 MPa for the welded wire fabric. Since the yield point does not exist for GFRP material, the ultimate tensile strength, f_u = 103 MPa, was used in this research. The GFRP modulus of elasticity was 14.8 GPa.

Light gage wire was used to attach stirrups and meshes to longitudinal reinforcement for holding them in place while concrete was poured. Only for stirrups standard hooks were used as required by ACI code.

To assess levels of strain and stress, two strain gages were attached to adjacent web reinforcing elements prior to casting the concrete. The position of the gages, between web reinforcement anchors, will be a general indicator of the effectiveness of strain transferal from the concrete to the web reinforcement. They also provide a strain history that can be compared to shear crack development.

The compressive strength of the concrete was targeted to be 34.5 MPa at 28 days after casting with a slump between 152 and 203 mm. The actual compressive strength for the test beams are given in Table 1.

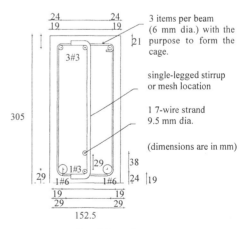

Figure 2. Cross-section of the test beam.

Table 1. Effectiveness of the web reinforcemnet based on resisted shear.

Beam	f'_c (MPa)	A'_{v_2} (mm²)	a/d	Shear failure location*	(Vu) Ultimate shear (kN)	(Vp) Predicted shear (kN)		(Vu/Vp) x 100 Anchor Effectiveness			
						Vaci	Vmcft	ACI(%)	Ave.	MCFT(%)	Ave.
A-1	44	71	3.03	533	91.10	82.66	91.54	110		100	
A-2	40	71	3.03	457	95.54	85.32	90.21	112	111	106	102
A-3	43	71	3.03	533	91.10	82.21	91.10	111		100	
CH-1	50	13	2.21	305	68.88	97.90	98.34	070		070	
CH-2	41	13	3.03	457	72.75	79.37	84.26	092	080	086	076
CH-3	41	13	3.03	457	61.37	79.81	84.52	077		073	
W-1	40	19	2.85	457	76.70	73.15	73.86	105		104	
W-2	38	19	2.71	381	65.90	78.84	80.88	084	093	081	091
W-3	35	19	2.67	381	68.88	77.10	77.81	089		089	
F-1	33	39	2.62	381	71.50	75.19	77.77	095		092	
F-2	41	32	2.67	457	101.90	71.77	69.41	142	114	147	115
F-3	42	52	2.67	381	87.37	84.12	81.41	104		107	

*   Distance from the center of end support to the shear failure section in mm.

3.2 Procedure

A concentrated load was applied by an hydraulic actuator mounted on a steel frame. This load was transformed into two concentrated loads by means of a small steel beam supported by the test beams to ensure that the shear failure would occur in the outer third of the beam close to the position of the strain gages. This was accomplished in most of the tests by increasing one of the shear spans and positioning the hydraulic load slightly off center of the small steel beam.

The beams were loaded in increments of 4.4 kN until shear failure was reached. At each increment, the magnitude of the load and strain gage readings were recorded. The shear cracks were marked as they appeared and when failure occurred they were inspected along with the condition of the web reinforcement. Crack inclination and crack width were measured.

3.3 Test Results

Evidence of anchor effectiveness was revealed by the ratio of ultimate to predicted shear capacity presented in Table 1. It is very clear from Table 1 that the chain link series exhibits substantially less effectiveness as compared to the other series. A lack of tautness was the apparent reason in this case even though residual stress problems were detected as several intertwining wires broke at the turning point. Since achieving a satisfactory level of tautness would require an excessive amount of time and labor, it is obvious that chain link would be totally without merit as web reinforcement.

Next in line in terms of lack of anchor effectiveness was the welded wire fabric. Full effectiveness was not realized due to weakness of the welds, most of which fractured prior to failure. If the welds could be strengthened, then welded wire fabric would serve as an ideal web reinforcement. Whether or not the added cost of strengthening the welds would be justified by the savings in material is not known.

Both the standard stirrups and GFRP reinforcement exhibited full anchor effectiveness. This is particularly important with reference to GFRP material since bond between the concrete and reinforcement is minimal. As such, the mechanical anchors alone were sufficient to transfer the stresses. The GFRP material (Polyester/Glass Fiber) utilized in this research is a commercially available product (IKG Fiberglass Systems, Inc.) and not specifically designed for use as web reinforcement. Yet the material still produced excellent results in this role. Improved performance may be expected if the GFRP material were tailored to the job. The ultimate strength, modulus of elasticity, surface roughness, and other parameters can be controlled in the manufacturing process. With mass production, the cost of GFRP should be more than competitive with its steel counterparts.

Excessive crack widths would destroy the aggregate interlock and the concrete's contribution to shear capacity. If this occurs early in the load cycle, it would adversely affect the predictability of the ACI and MCFT methods and raise questions on the viability of the mesh. This did not happen in the test series. Basically, crack width development in the four web reinforcements were essentially similar. The welded wire fabric was best at inhibiting crack growth. Tests with the stirrups and GPRP meshes produced comparable crack results, although neither were inordinately larger than crack development with the welded wire fabric. Only near failure did crack widths reach a significant magnitude (about 1") to nullify aggregate interlock. Figure 3 shows one load versus strain plot for the stirrup, WWF, and GFRP series.

Table 1 demonstrates the shear capacity predictability of both the ACI and MCFT methods. Although the MCFT is slightly more accurate, the added complexity works against its adoption for code use. What is perhaps not surprising, the ACI

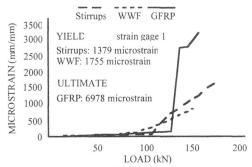

Figure 3. Strain curves for different web reinforcement series (strain gage 1)

approach works very well as a predictor. This should lend confidence to practicing engineers as they experiment with new materials and configurations when they design a better product.

3.4 Optimization

Optimization of the material accomplished by mesh configuration was difficult to quantify since the many parameters that dictate shear strength varied from beam to beam. A valid comparison, therefore, is possible only if the variability of those parameters is factored into the comparison. With conventional stirrups setting the standard, the shear capacity relationship between the series was taken to be proportional to the cross sectional tributary area (A'_v), the concrete compressive strength, the yield strength of the web reinforcement material, and inversely proportional to the shear span to depth ratio as shown in equation (9) below:

$$V'_u = V_u \left(\frac{f''_c}{f'_c} \right) \left(\frac{f_{vy}^*}{f_{vy}} \right) \left(\frac{A_v'^*}{A'_v} \right) \left[\frac{a/d}{(a/d)^*} \right] \qquad (9)$$

V'_u/V^*_u is defined as the shear capacity ratio or optimization factor, where V^*_u is the experimentally derived shear capacity for conventional stirrups. With equation (9), V'_u becomes an equivalent shear strength for comparative purposes. Assume, for instance, that the concrete strength f'_c for a beam with a nonconventional mesh is greater than f'^*_c for a beam with standard stirrups, all other factors being equal. Then, the test shear capacity would also be greater as the concrete contributes more to the shear capacity. Thus, the test shear capacity should be reduced accordingly if a valid comparison with the stirrup capacity is to be made.

A question arose as to whether the variability of the concrete ought to be reflected by the square root of the strength. However, Elzanaty et al. (1986) demonstrated that the concrete's impact on web shear is directly related to the ultimate strength and not to the square root. Also Kani (1966) identified

the shear span to depth ratio to be inversely related to the web shear strength. A'_v is the web reinforcement area corresponding to the tributary spacing (229 mm) for the A-beam series which was calculated using the relationship given below where s is the horizontal spacing and A_v is the cross sectional area of a given mesh doubled to take into account elements in two directions.

$$A'_v = 2A_v \left(\frac{229}{s} \right) \qquad (10)$$

The optimization ratios in Table 2 once again verify the ineffectiveness of chain link as web reinforcement. At the same time, the welded wire fabric and GFRP beams achieved full optimization of the material bettering standard stirrups by 13% and 33%, respectively. This is particularly impressive for welded wire fabric which produced full optimization without full anchor effectiveness. With the F-beam series, it could be argued that the F-2 beam greatly exceeded its predicted capacity somewhat distorting the capabilities of GFRP reinforcement. Yet, the F-1 beam went in the opposite direction counterbalancing the argument. On average, the F-2 and F-3 beams closely match the optimization ratio produced by the F-1 beam. Thus, in the authors' opinion, the shear capacity ratios in Table 2 suggest not only GFRP as a viable web reinforcing material in an inclined mesh configuration, but also the necessity to increase the number of tests in order to remove any uncertainties.

Table 1 demonstrates the relative merits of the ACI versus MCFT methods. The ACI procedure is conservative across the board and, except for standard stirrups, closely parallels the MCFT predictions. The MCFT method is a better predictor for standard stirrups and its use could be advantageous in this case particularly as advancements in computer technology circumvents its complexity. Definite savings are possible with MCFT, as stirrups are the most commonly used web reinforcement. Another conclusion is that the ACI code should be modified to include equation (1) incorporating the effect of bidirectional inclined web reinforcement and the variability of the shear crack inclination on shear capacity. Such a code change would jointly require some form of verification of full anchor effectiveness. For instance, the welds in beams W-2 and W-3 fractured prematurely producing experimental shear capacities well below predictions. Neither the ACI or MCFT methods are configured to account for the lack of anchor effectiveness and both greatly overpredicted the beams ability to carry shear in these cases. Additional testing is needed to validate anchor effectiveness for specific materials or to devise a reduction factor that would decrease the predicted capacity in response to partial anchor effectiveness.

Table 2. Optimization of the web material by mesh configuration

Beam	V_u (kN)	f'_c (MPa)	f_{vy} (MPa)	A'_{v2} (mm²)	a/d	$f'*/f'_c$	$f_{vy}*/f_{vy}$	A'_v*/A'_v	a/d/(a/d)*	V'_u** (kN)	V'_u/V_u*	Ave.
A-1	Ave.	Ave.	276	71	3.03	1.00	1.00	1.000	1.00		1.00	
A-2	92.57	42	276	71	3.03	1.00	1.00	1.000	1.00	92.57	1.00	1.00
A-3			276	71	3.03	1.00	1.00	1.000	1.00		1.00	
CH-1	68.88	50	310	65	2.21	0.85	0.90	1.076	0.73	41.39	0.45	
Ch-2	72.75	41	310	65	3.03	1.05	0.90	1.076	1.00	73.97	0.80	0.64
Ch-3	61.37	41	310	65	3.03	1.02	0.90	1.076	1.00	60.62	0.66	
W-1	76.70	40	352	39	2.85	1.06	0.79	1.852	0.94	111.81	1.21	
W-2	65.90	38	352	39	2.71	1.10	0.79	1.852	0.89	94.39	1.02	1.13
W-3	68.88	35	352	39	2.67	1.20	0.79	1.852	0.88	106.42	1.15	13%
F-1***	71.50	33	103	123	2.62	1.29	2.67	0.567	0.86	120.08	1.30	
F-2***	101.90	41	103	110	2.67	1.04	2.67	0.651	0.88	162.10	1.75	1.33
F-3***	87.37	42	103	168	2.67	1.00	2.67	0.426	0.88	87.45	0.95	33%

*** For GFRP the streee of rupture is taken instead of the yielding stress
** Ultimate shear capacity factored by the parameters influencing the shear stength.
* Parameters corresponding to the A-beam series.

Without consideration of the effect of dowel action in Table 1, V_u/V_p would increase by about 20%. The implication is that dowel action plays a major role in shear stress resistance. A study of the beam failure patterns (Orozco 1994) indicate that dowel action was present and effective essentially throughout the load history, losing its viability only when bond failure of the longitudinal reinforcement occurred.

4 CONCLUSIONS

A summary of pertinent data gathered in this research is as follows:
1. Chain link or intertwined meshes are wholly unsuitable as web reinforcement.
2. The potential of welded wire fabric as effective web reinforcement in an inclined mesh configuration is high, with premature weld failure as the only drawback. Strengthening the welds or making hooks at the ends of the wires to achieve full anchor effectiveness would address this problem. Whether or not this would be economically viable has not been determined.
3. The ACI code formula for shear capacity provided by the web reinforcement, modified to account for sloped reinforcement in two directions, may be used to accurately predict shear capacity. The code should adopt this modification with the provision that full anchor effectiveness must be verified prior to its use.
4. The MCFT method may also be used to accurately predict shear capacity; however, it is much more complex in its application. Therefore, adoption of MCFT for primary code use is not recommended, although provisions should be made that would permit engineers to apply MCFT as an alternative approach. There are definite circumstances where substantial savings could be realized by using MCFT.
5. With mass production and fiberglass reinforced plastic manufactured to meet the specific demands of web reinforcement, GFRP sloped meshes offer the best alternative to standard stirrups as effective web reinforcement. Every indicator points to GFRP web reinforcement, in an inclined mesh configuration, as giving engineers an attractive material both in terms of cost and shear carrying capacity.

5 REFERENCES

American Concrete Institute. 1999. *Building Code Requirments for Reinforced Concrete (ACI 318-89) and Commentary (ACI 318R-89)*. Farminton Hills, MI: ACI.

Elzanaty, A.H., Nilson, A.H., and Slate, F.O. 1986. Shear capacity of prestressed concrete beams using high-strength concrete. *ACI Journal, Proceedings*, Vol. 83, No. 3, May-June: 359-368. Farmington Hills, MI: ACI

Johnston, D.W. and Zia, P. 1971. Analysis of dowel action. *J. Struct. Div., Proceedings ASCE* 97(ST5): 1611-1630. Reston, VA: ASCE

Kani, G.N.J. 1966. Basic facts concerning shear failure. *ACI Journal* 63(6): 675-692

Mitchell, D. and Collins, M.P. 1974. Diagonal compression-field theory: A rational model for structural concrete in pure torsion. *ACI Journal* 71(8): 396-408.

Nawy, E.G. and Neuwerth, G.E. 1977. Fiberglass reinforced concrete slabs and beams. *J. Struct. Div., Proceedings ASCE* 103(ST2): 421-440. Reston, VA: ASCE.

Nawy, E.G., Neuwerth, G.E., and Phillips, C.J. 1971. Behavior of fiberglass reinforced concrete beams. *J. Struct. Div., Proceedings ASCE* 97(ST9): 2203-2215. Reston, VA: ASCE.

Orozco, F.J., 1994. *Non-Conventional Shear Reinforcement Effectiveness for Partial Prestressed Concrete Beams of Rectangular Cross Section*. Ph.D. thesis, New Mexico State University, Las Cruces, New Mexico.

Reineck, K. 1991. Ultimate shear force of structural concrete members without transverse reinforcement derived from a mechanical model. *ACI Structural Journal* 88(5): 592-602.

Taylor, H.P.J. 1972. Shear strength of large beams. *J. Struct. Div., Proceedings ASCE* 98(ST11): 2473-2490. Reston, VA: ASCE

Vecchio, F. and Collins, M.P. 1986. The modified compression field theory for reinforced concrete 'elements subjected to shear. *ACI Journal* 83(2): 219-231.

Xuan, X., Rizkalla, S., and Maruyama, K. 1988. Effectiveness of welded wire fabric as shear reinforcement in pretensioned prestressed concrete t-beams. *ACI Structural Journal* 85(4): 429-436.

26 Steel structures

Creative Systems in Structural and Construction Engineering, Singh (ed.) © *2001 Balkema, Rotterdam, ISBN 90 5809 161 9*

Learning from structural failures

D.A.Cuoco
LZA Technology, Thornton-Tomasetti Group, New York, N.Y., USA

ABSTRACT: Structural failures can range from deflection or cracking of floor slabs to total building collapses. Major failures typically receive wide publicity, and questions are immediately raised with regard to the cause of the failure, the liability of involved parties, and whether the failure could have been prevented. This paper discusses some of the causes of structural failures, and provides some suggestions to minimize the risk of structural failures in the future.

1 INTRODUCTION

Structural failures can take various forms. They can range from excessive deflections to unsightly cracks to annoying vibrations to total collapse.

When a major structural failure occurs, several questions immediately come to mind. Why did it happen? Who is responsible? What are the consequences, not only from the standpoint of deaths or injuries, but from the standpoint of liability on the part of the various parties involved? What could have been done to prevent it? How can we learn from our mistakes?

Through the use of case studies of actual structural failures, this paper will discuss various issues, such as constructability being part of the design process, minimizing risk on "performance" projects, consistency between design details and analysis assumptions, over-reliance on the computer, and the need for more involvement by the design engineer during construction.

2 WHY FAILURES OCCUR

The reason for a structural failure usually falls into one or several of the following categories:

- Forces of Nature
- Deficient Design
- Improper Construction
- Material Deficiency
- Improper Use/Maintenance

Extreme forces of nature, often considered to be *forces majeures*, include hurricanes, tornadoes, and earthquakes. Building codes provide design requirements for wind and seismic loads, which are based upon previous known events and long-term recurrence intervals, e.g. 100 years, that are not expected to be exceeded during the life of the structure. Occasionally, however, structures are subjected to extreme events, which induce forces that significantly exceed the code-specified design requirements, thereby resulting in failures. Attempting to design structures for these extreme events (even if they could be accurately predicted) would not be economically feasible.

Deficiencies in design account for many structural failures. Sometimes these deficiencies result from an improper analysis of the structure. Examples include lack of consideration of certain loading cases (e.g. thermal loads, snow drifting, unbalanced loading) or failure to consider critical combinations of various load cases. Notwithstanding the fact that many excellent computer analysis programs are readily available, the computer only calculates the load cases and load combinations that are input by the designer.

Design deficiencies can also result if the structure is properly analyzed, but is incorrectly detailed on the design drawings. One example would be a truss structure for which all of the member forces have been correctly determined, but for which the connections do not account for any eccentricities resulting from members not intersecting at a common working point. Too often designers only worry about the sizing of the main members and assume that the member forces will somehow find the correct paths by "osmosis."

Improper construction is another culprit frequently encountered. This can initiate during the shop drawing stage if details are not correctly interpreted from the design drawings. (Of course, the review of shop drawings by the designer is intended to mitigate this problem.) Other problems can result when the actual construction does not comply with the design documents or the shop drawings. Examples include misfabrication of members, improper welding, understrength concrete, insufficient temporary bracing or shoring, introduction of excessive erection stresses, and other factors related to the means and methods of construction.

Though not as common, a material deficiency can cause a structural failure. In this case, both the design and construction can be adequate, but the material supplied does not meet the specified mechanical or chemical properties. This is not likely to be apparent to the designer or constructor, but could be determined by subsequent testing after the failure. The submittal of proper mill reports and other certifications of material properties prior to construction is intended to ensure compliance with the project specifications.

Finally, structural failures can result from improper use or lack of maintenance by the user of the structure. These failures can occur immediately upon occupancy of the structure or many years afterward. An example of improper use of the structure would be the installation of heavy equipment or file storage in a floor area that was originally designed for typical office loading, thereby creating an overload condition. Structures can also fail due to a lack of maintenance, such as failure to paint exposed steel, keep roof drains clean, wash down parking decks, etc.

2.1 Constructability is Part of Design

Designers typically focus on the design of a structure, and feel that it is up to the contractor to figure out *how* to build it. This is not necessarily problematic. In fact, it allows for innovative thinking on the part of each contractor bidding a project with regard to the means and methods he will use to come up with the most competitive bid.

A problem arises, however, if the designer feels that it is up to the contractor to figure out *if* the structure can be built. There are too many situations in which a detail on a design drawing fits together perfectly on paper, but is virtually impossible to construct, e.g. welding in inaccessible areas, bolts that cannot be reached, etc. Even if caught during the shop drawing stage, these problems can result in claims for added costs and delays. And if these problems are not discovered until during

construction, they will most certainly result in claims. It is thus incumbent upon the designer to think about the details he draws from the standpoint that there is some reasonable way to build them.

The two-span canopy extension of a new roof structure (Fig. 1) had to have its metal deck lower than the balance of the roof framing in order to accommodate insulation and additional concrete topping. The designer specified the sizes of the shelf angles attached to the beam webs along lines A and B, but unfortunately the outstanding leg of the shelf angle was the same length as the half-flange width of the beam to which it was attached. Thus, if the metal deck was cut short enough to fit between the beam flanges, it would have no bearing on the shelf angles. Instead, the deck was fabricated slightly longer so that it could be "angled in" between the beams and still have some bearing on the shelf angles. When the deck was erected, however, it was pushed closer to line A, which left it barely in contact with the shelf angle at line B. The deck was welded into place and, while pouring the additional concrete topping, it slipped off the shelf angle at line B and five workers fell 40 feet (12 m) to the ground.

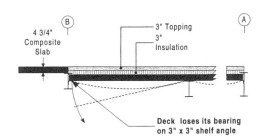

Figure 1. Roof canopy collapse.

For some projects, in order to achieve the proper design behavior, the sequence of construction is critical. When doing renovation work, for example, the removal and addition of structural members in conjunction with providing temporary shoring sometimes must be done in a particular sequence in order to avoid collapse. In these cases, the designer should include a *"suggested"* construction procedure" in the design documents, and require that the contractor develop the final procedure and submit it for review. By doing this, the contractor is still responsible for means and methods, even if he adopts the designer's suggested procedure, but at least the designer has conveyed any constraints that need to be considered in order to achieve an end result that will perform as designed.

2.2 *Minimizing Risk on "Performance" Projects*

Performance-type projects are those for which a portion of the design responsibility is delegated from the engineer-of-record to another party, usually the contractor. The contractor then provides the required design services, either through an engineer on staff or by retaining an independent engineering consultant. Some contractors complain that this is simply a method by which the designer shuns his responsibility by shifting it to the contractor. In reality, there are many situations, such as those where proprietary products are involved, where the performance approach is essential.

There are several key elements to making the performance approach successful. First and foremost, the contract documents must be absolutely clear in establishing the scope of design responsibility that is being delegated to the contractor. Failure to do this can result in the design of critical components "slipping through the cracks." The landmark case that highlights this issue, of course, is the collapse of the Kansas City Hyatt walkways in 1981, in which no one claimed the design responsibility for the hangar detail that failed (Fig. 2), and which resulted in 114 deaths.

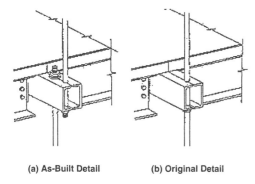

| (a) As-Built Detail | (b) Original Detail |

Figure 2. Kansas City Hyatt walkway hanger detail.

Second, the contract documents must clearly define the design criteria that must be used by the contractor's engineer. In some cases, this can simply be a reference to the applicable building code. In other cases, where additional or more stringent design criteria are required, the contract documents must clearly specify these requirements.

Third, the contract documents should require submittals by the contractor, e.g. drawings, calculations, etc., for the work he is responsible to design, signed and sealed by the contractor's engineer licensed in the State of the project. This is intended to ensure adequate quality of the submittals.

Fourth, these submittals should actually be reviewed by the engineer-of-record. Sometimes designers do not properly review these submittals because of a false sense of security, thinking that if there's a problem, it's someone else's, namely that of the contractor's engineer. Unfortunately, however, if a problem arises it usually becomes everyone's. Therefore, the engineer-of-record should diligently review the contractor's submittals as he would for shop drawings in general, perhaps more so since the contractor's submittals essentially become the design documents for that component of the project.

2.3 *Design Details v. Analysis Assumptions*

When designing a structure, the designer makes certain assumptions in order to develop the structural model and perform the analysis, such as whether to consider connection joints as being hinged or fixed. Once the analysis is complete, the designer sizes the members and develops connection details. Sometimes it is not possible to develop a connection detail that is consistent with the design assumption, due to too many members framing into a connection or some other constraint. In these cases, the connection detail needs to take priority, and the design should be revised and/or re-analyzed so that the design details and analysis assumptions ultimately match.

The collapse of the Hartford Civic Center roof structure in January 1978 represents a classic case in which the design details vastly differed from the analysis assumptions. This custom-designed space frame structure utilized a joint comprised of bent steel plates at intermediate brace points, which greatly facilitated the connection of intersecting chord and web members (Fig. 3). In the analysis, this joint was assumed to be a rigid brace point. However, because the members did not intersect at a common working point and the bent plates were relatively flexible, the bracing effect of the web members was essentially that of a spring as opposed

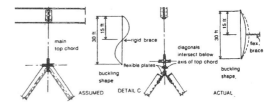

Figure 3. Hartford Civic Center connection detail.

to a rigid brace, thereby significantly reducing the compression capacity of the chord member.

2.4 *Over-Reliance on the Computer*

With today's powerful computers and analysis software, designers have the ability to analyze a structure for hundreds of load cases and loading combinations. For complex structures, the end result is a structure that is designed much more accurately and much less conservatively than in the past. Greater economies in structures have thus been achieved, and innovative structural systems have become viable. The key to all this is that the designer must properly model the structure, input the correct loads, generate the correct loading combinations, and, most importantly, review the results to make sure that they make sense. Hand calculations, even if crude and approximate, can usually detect a blunder in the analysis of even the most complicated structure. Too often designers automatically assume that the computer results are correct without checking the basic laws of statics, whereas a simple hand calculation could reveal that the results are meaningless!

The airplane hangar roof shown in Figure 4 is comprised of a post-tensioned concrete folded plate structure, with the end plate spanning 250 feet (76 m) over a hangar door opening. Over a period of

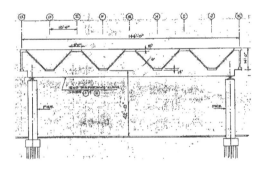

Figure 4. Post-tensioned concrete roof structure.

several months after the roof was constructed and the hangar door was installed, the end plate deflected excessively and the hangar door had to be removed, shortened, and re-installed, at substantial cost. Over the next year, the end plate continued to deflect until the door once again had to be removed, shortened, and re-installed, at substantial cost. An investigation ensued, and hand calculations indicated that there should have been twice as many tendons in the bottom of the end plate as compared to what was designed and built. Upon review of the

original computer analysis, it was discovered that an input error caused the dead load to be acting horizontally instead of vertically and, as a result, the computer analysis determined that only half the correct number of tendons would be sufficient.

2.5 *More Involvement during Construction*

There was a time when the designer's scope of services was frequently expanded to include full-time observation of the construction work. Nowadays, the norm is for the owner to have these services performed by a testing agency, at much reduced cost. Some designers feel that this is beneficial and reduces their liability because, if a construction deficiency subsequently manifests itself, they can claim that they weren't at the jobsite to rectify it. The fallacy of this concept, of course, is that if there is a major problem, the designer will be drawn into a lawsuit, regardless of whether or not he was at the jobsite during construction. Therefore, it is a much better approach to be at the jobsite during the construction work (assuming the owner is willing to pay for this service) so that there is a greater likelihood that deficiencies will be discovered and corrected, thus minimizing or eliminating the occurrence of problems later on.

Another advantage of the engineer being on-site during construction is that he is in the best position to provide an overview of structural integrity as the structure is being built. For example, he may realize that there is a lack of temporary bracing and notify the contractor, even though it is ultimately the contractor's responsibility to provide such bracing. Also, on-site observation by the engineer represents the last chance to find things that may have been overlooked in the design.

If the owner is unwilling to pay the designer to provide on-site observation services, the designer should require the testing agency to provide a resume of its proposed inspector. The qualifications of this inspector should be carefully reviewed, and in some cases an interview process should be undertaken to ensure that the proposed inspector is properly qualified. Likewise, if the owner is willing to pay the designer to provide these services, it is incumbent upon the designer to provide an on-site representative with adequate qualifications.

3 CONCLUSION

The one positive thing that emerges from a structural failure is the lesson that can be learned. By discussing and disseminating information on failures, designers can learn from past mistakes and minimize the risk of structural failures in the future.

Creative Systems in Structural and Construction Engineering, Singh (ed.) © 2001 Balkema, Rotterdam, ISBN 90 5809 161 9

Simple analytical method of super steel frame

X. H. Ju & T. K. Mitsukura – *Geo Center M. Company Limited, Miyazaki, Japan*

H. Yokota – *Miyazaki University, Japan*

E. I. Goto – *TTK Corporation, Tokyo, Japan*

Y. C. Zhang – *Harbin Institute of Technology, People's Republic of China*

ABSTRACT: In order to get a simple analytical model for practical design, the equivalent theory of super steel frame was studied thoroughly in this paper. We found that the super column with inverted V braces don't obey the hypothesis of plane, when bearing pure moment. According to the new discovery, the former equivalent theory was studied again, and on the basis of it, a new equivalent method, which has become easier and more integrated, is presented in this paper.

1 INTRODUCTION

The super steel frame is a new kind of high-rise structure, which is accompanied by the development of modern high-rise buildings occurring in recent years. It usually consists of two parts: the main structure comprising super columns and super beams; the secondary structure including common columns and beams. The super component is a kind of latticed and battened system, which is made of many members. Compared with normal steel frame, the super structure has lots of advantages such as very large lateral rigidity and good mechanical characteristic as a whole, furthermore, a big flexibility can be brought on architectural layout. As a typical example, the plan and elevation drawing of Kobe TC Building is shown in Fig.1.

To anti-seismic high-rise building, it's necessary not only for structural elastic analysis but also for dynamic elasto-plastic analysis. The structure of super frame with braces is very complicated, though it's possible to analyze elasto-plastic seismic response of original structure by accurate model on the basis of members' restoring force behavior, it's very hard to be used in practical design. By the equivalent method in this paper, the job for analyzing the super frame in earthquake can be simplified greatly and the structural characteristics can be grasped easily. For the reasons, it is important and necessary to establish a simplified equivalent model instead of the accurate model.

In this paper, by the analytical result about the mechanical behavior of super components, a special characteristic has been found. That is the hypothesis of plane doesn't work, while super column with inverted V braces, which has been usually adopted on

practical super frame, suffers pure bending moment. On the opposite, other kinds of super components such as the super beam with K braces can obey the hypothesis of plane well. According to the new discovery, the former equivalent theory of Fukada and Uchiyama was studied again, and on the basis of it, a new equivalent method is presented in this article. Compared with the former equivalent theory, the main different points in this paper may be summarized as follows.

a) We found that the super column with inverted V braces do not obey the hypothesis of plane. On the opposite, other kinds of super components such as the super beam with K braces can obey the hypothesis of plane well. Furthermore, the reasons and its influence to equivalent stiffness are discussed in this paper.

b) The former formula to solve the equivalent stiffness of super column is modified and presented in detail.

c) The former complicated methods to solve the equivalent stiffness of super beam are abandoned and a new unified method is presented in this paper.

d) The compensatory ability of bending and shearing of equivalent model is found in this paper, so the reason why the new equivalent method can be fit for any kinds of super components is explained.

2 THE SOLUTION OF EQUIVALENT STIFFNESS

2.1 Equivalent stiffness of super column

2.1.1 Resolution of deformations
To any one layer of super column, the deformations can be divided into several types shown in Fig. 2. According to the horizontal and vertical displace-

ment {U, V} of A, B, C, D nodes, the deformations can be solved as follows:

Vertical displacement of AB section:
$$V_{AB} = (V_A+V_B)/2 \tag{1}$$
Vertical displacement of CD section:
$$V_{CD} = (V_C+V_D)/2 \tag{2}$$
Axial deformation of one layer:
$$\Delta V = V_{AB} - V_{CD} \tag{3}$$
Angle of rotation of AB section:
$$\theta_{AB} = (V_A - V_B)/L \tag{4}$$
Angle of rotation of CD section:
$$\theta_{CD} = (V_C - V_D)/L \tag{5}$$
Angle of rotation of one layer:
$$\Delta\theta = \theta_{AB} - \theta_{CD} \tag{6}$$
Whole horizontal displacement of one layer:
$$\delta = (U_A+U_B)/2 - (U_C+U_D)/2 \tag{7}$$
Inclined displacement:
$$\delta_1 = \theta_{CD} \times h \tag{8}$$
Local bending displacement:
$$\delta_M = \Delta\theta \times h/2 + Qh^3/12EI \tag{9}$$
EI: bending equivalent stiffness in Eq.(20)
Shear deformation:
$$\delta_S = \delta - \delta_1 - \delta_M \tag{10}$$

2.1.2 *Resolution of internal force*
According to the end force {N, Q, M} of 1, 2, 3, 4 elements (Fig.2), provided that the positive directions are shown in Fig.2(a), the internal forces can be solved as follows:

Horizontal force of C point:
$$FH_C = Q_1 + N_3 \times Cos\alpha + Q_3 \times Sin\alpha \tag{11}$$
Vertical force of C point:
$$FV_C = N_1 + N_3 \times Sin\alpha - Q_3 \times Cos\alpha \tag{12}$$
Horizontal force of D point:
$$FH_D = Q_2 - N_4 \times Cos\alpha + Q_4 \times Sin\alpha \tag{13}$$
Vertical force of D point:
$$FV_D = N_2 + N_4 \times Sin\alpha + Q_4 \times Cos\alpha \tag{14}$$
Axial force of one layer:
$$N = FV_C + FV_D \tag{15}$$
Shearing force of one layer:
$$Q = FH_C + FH_D \tag{16}$$
Bending moment of CD section:
$$M_{CD} = (FV_C - FV_D) \times L/2 + \sum_{i=1}^{4} M_i \tag{17}$$
Average bending moment of one layer:
$$M = M_{CD} - Q \times h/2 \tag{18}$$

By former equivalent method presented by Fukada, while calculating bending moment of one layer, the end moments of elements are neglected. It means that $\sum_{i=1}^{4} M_i$ was missed from the formula. But according to the studying result of this paper, the equivalent model becomes more accurate when it is considered. In elastic calculation, the precision of equivalent model can be improved about 1%.

2.1.3 *Equivalent stiffness*
Using the deformations and internal forces above, the equivalent stiffness can be solved by the elastic

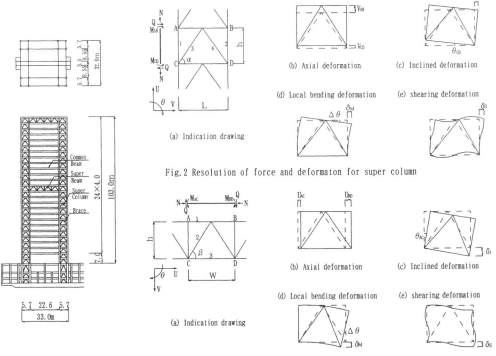

Fig.1 Kobe TC building

Fig.2 Resolution of force and deformaton for super column

Fig.3 Resolution of force and deformaton for super Beam

mechanical theory.

Axial equivalent stiffness:

$$EA = N \times h / \Delta V \qquad (19)$$

Bending equivalent stiffness:

$$EI = M \times h / \Delta \theta \qquad (20)$$

Shearing equivalent stiffness:

$$GS = Q \times h / \delta_s \qquad (21)$$

2.1.4 *Explanation about equivalent stiffness*

By the study of this paper, the bending equivalent stiffness of super column with inverted V braces is decided not only by the material characteristics of the members and boundary conditions, but also by the types of external loads. Under different load case, the bending equivalent stiffness is not same and the change is irregular. The reason is that this kind of super column just doesn't obey the hypothesis of plane.

In order to solve the problem, shearing displacement is defined as the whole one minus inclined and local bending displacement. Though the shearing and bending deformation of the equivalent model can't correspond with the original model one by one, the sum of this two kind of displacement is equal between equivalent and accurate models. It means that we use the shearing deformation to compensate the error of bending one. This feature of equivalent model may be called compensatory ability of bending and shearing.

Therefore, the equivalent method can adapt to any kinds of super components. Furthermore, the equivalent model, which is built in a simple load case, can get good accuracy in any kind of load cases including dynamic load. To the super component, which obeys the hypothesis of plane, its equivalent model can show the real deformation, both of the shearing deformation and the bending deformation are equal between the equivalent and accurate models. To the super component, which does not obey the hypothesis of plane, the total deformation of the equivalent model is equal to that of the accurate model, even though the shearing or the bending deformation can not correspond one by one.

2.2 *The equivalent stiffness of super beam*

By the former research, the methods to solve the equivalent stiffness of super beam were completely different from that of super column. In the former method of Fukada, it was supposed that the reverse bending point would be in the middle of the super beam. In order to solve the equivalent stiffness, a complicated method to resolve the deformation is used. In the former method of Uchiyama, half of the super beam, which was taken from the super frame, was considered as cantilever beam of half-span. Besides the same assumption with the first former method, the support of super beam was supposed fixed end as well. For the assumptions above were

used when calculating the equivalent stiffness of super beam, it is natural to be different from the fact and cause some error.

However, according to the studying result of this paper, the super beam with K braces can obey the hypothesis of plane section perfectly when bearing pure bending moment, so its equivalent stiffness is not effected by the load cases. Compared with the super column of inverted V braces, its equivalent stiffness become more regular and its mechanical characteristics can be grasped more easily.

Therefore, the equivalent methods of super column and super beam are unified and the equivalent formulae of super beam are derived newly in this paper. Compared with the former methods, the former assumption is not used and the model is able to be near to the real condition as soon as possible, so the equivalent stiffness of super beam can be solved more fully and accurately, furthermore it can be understood more easily.

2.2.1 *Resolution of deformations*

By the indication in Fig.3, the deformations can be divided as follows:

Horizontal displacement of AC section:

$$U_{AC} = (U_A + U_C)/2 \qquad (22)$$

Horizontal displacement of BD section:

$$U_{BD} = (U_B + U_D)/2 \qquad (23)$$

Axial deformation of one segment:

$$\Delta U = U_{AC} - U_{BD} \qquad (24)$$

Angle of rotation of AC section:

$$\theta_{AC} = (U_A - U_C)/h \qquad (25)$$

Angle of rotation of BD section:

$$\theta_{BD} = (U_B - U_D)/h \qquad (26)$$

Angle of rotation of one segment:

$$\Delta \theta = \theta_{BD} - \theta_{AC} \qquad (27)$$

Whole vertical displacement of one segment:

$$\delta = (V_B + V_D)/2 - (V_A + V_C)/2 \qquad (28)$$

Inclined displacement:

$$\delta_1 = \theta_{AC} \times W \qquad (29)$$

Local bending displacement:

$$\delta_M = \Delta \theta \times W/2 + QW^3/12EI \qquad (30)$$

    EI: bending equivalent stiffness in Eq.(39)

Shear deformation:

$$\delta_s = \delta - \delta_1 - \delta_M \qquad (31)$$

2.2.2 *Resolution of internal force*

According to the end force {N, Q, M} of 1, 2, 3 elements (Fig.3), provided that the positive directions are indicated in Fig.3(a), the internal forces can be solved as follows:

Horizontal force of C point:

$$FH_C = N_3 + N_2 \cos\beta - Q_2 \sin\beta \qquad (32)$$

Vertical force of C point:

$$FV_C = Q_3 + N_2 \sin\beta + Q_2 \cos\beta \qquad (33)$$

Axial force of one segment:

$$N = N_1 + FH_C \qquad (34)$$

Shearing force of one segment:

$$Q = Q_1 + FV_C \qquad (35)$$

Bending moment of AC section:

$$M_{AC} = (FH_C - N_1) \times h/2 + \sum_{i=1}^{3} M_i \qquad (36)$$

Average bending moment of one segment:

$$M = M_{AC} - Q \times W/2 \qquad (37)$$

2.2.3 *Equivalent stiffness*

Axial equivalent stiffness:

$$EA = N \times W / \Delta U \qquad (38)$$

Bending equivalent stiffness:

$$EI = M \times W / \Delta \theta \qquad (39)$$

Shearing equivalent stiffness:

$$GS = Q \times W / \delta_s \qquad (40)$$

3 THE MECHANICAL BEHAVIOR OF SUPER COMPONENTS

Compared with a common member, the super component is made of many framed elements and its deformation become more complicated. When bearing loads, even if just a pure moment, the axial, shearing and bending deformation will emerge in every inner members and the total deformation of super component is their synthetical reaction under the control of deformed harmonious conditions.

Because of the difference between super component and common member, it is necessary to study if the super component can also obey the hypothesis of plane. In order to prove it, three kinds of super components (Fig.4) are tested in this paper. When being given a couple of opposite concentrated force P on the top, the super component bears a pure moment $P \cdot L$. According to the equivalent methods above, the equivalent model can be set up and be solved. Compared with the accurate model, A typical result of lateral displacement is shown in Table 1.

By this study, if the super components can obey the hypothesis of plane, two conditions below should be satisfied.

a) The lateral displacement in the position of height h should equals $PLh^2/2EI$, compared with the displacement U_1 of first layer, the displacement U_i of i layer is $U_i = i^2 U_1$.

b) The lateral displacements between accurate and equivalent models are equal.

From the result of Table 1, it is easy to find that the super component with inverted V braces can not satisfy the two conditions above, so it has been proved that when suffering pure moment, the super component with inverted V braces does not obey the hypothesis of plane. On the opposite, other kinds of super component with K or X braces etc. can satisfy the conditions above and obey the hypothesis of plane well.

After changing the section parameters of members and analyzing many kinds of super components, the following conclusions have been found.

a) When super component with inverted V braces

Table 1. The lateral displacement of super components

Layer	Model with inverted V braces		Model with K braces		Model with X braces	
	A	E	A	E	A	E
1	0.32	0.69	0.57	0.57	0.70	0.70
2	1.88	2.77	2.29	2.29	2.78	2.78
3	4.83	6.23	5.15	5.15	6.25	6.26
4	9.16	11.07	9.17	9.16	11.12	11.12
5	14.88	17.30	14.32	14.31	17.37	17.38
6	21.93	24.91	20.62	20.61	24.95	25.03

* A: accurate, E: equivalent

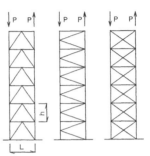

Fig.4 Models of super components

bears pure moment, big axial force will emerge in the brace and it makes the side column bear shearing force. These bring about shearing deformation of the super component and cause the total deformation is not pure bending, so the hypothesis of plane can not be satisfied, even if the super component with inverted V braces just bears a pure moment.

b) It has been discovered that the axial stiffness of inverted V braces affects the characteristic of deformation greatly. When the section area of brace becomes bigger, the total deformation wills far from the pure bending; when the section area tends to zero, the total deformation will become pure bending. That is to say the hypothesis of plane can not be obeyed, unless the axial stiffness of inverted V brace becomes zero.

c) Except the super component with inverted V braces, the other kinds of super components such as the super beam with K braces and the others with X braces; single inclined braces or without braces etc. can obey the hypothesis of plane well. By the analytical result, less axial force emerges in brace and shearing force of side column tends to zero, so axial deformation of brace and local bending deformation of side column happen scarcely and the total deformation becomes pure bending.

4 STRUCTURAL ANALYSIS

In order to test and verify the proposed equivalent method, two kinds of super frames with inverted V braces or X braces (Fig.5 and Fig.6) are selected and their corresponding equivalent stiffness is solved when the super frame just bears a lateral concentrated force on the top. By lots of analysis, some typical analytical results to compare with its own equivalent model (Fig.7) are presented here.

When bearing a random complicated load case, which includes gravity, lateral concentrated and uniformed forces etc., the elastic static analytical results with American SAP program are shown in Fig.8 and Fig.9. Compared with the results of accurate model, the average equivalent error of deformation is less than 1%.

Furthermore, using American DRAIN-2D program and inputting EL-CENTRO earthquake wave, the elastic dynamic analytical results are shown from Fig.10 to fig.13. By statistical results, the average error about the maximum of dynamic response displacement is less than 2%.

Now, it has been verified that the equivalent model, which is built in a simple load case, can get very good accuracy in any kind of load cases including dynamic load. It means that the affection of load cases has been overcome by the compensatory ability of bending and shearing discussed above. That is why the equivalent method can adapt to any kind of super frames and the equivalent model built by simple static analysis can be used to complicated dynamic analysis.

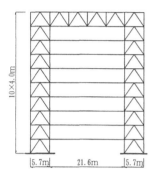

Fig. 5 Accurate model A
with inverted V braces

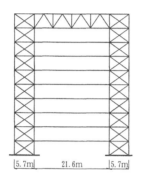

Fig. 6 Accurate model B
with X braces

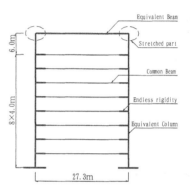

Fig. 7 Equivalent model
for A or B

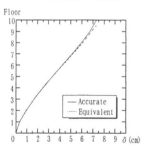

Fig. 8 Lateral static deformation
with inverted V braces

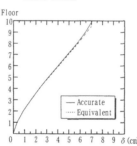

Fig. 9 Lateral static deformation
with X braces

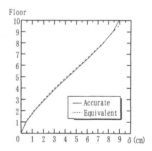

Fig. 10 Lateral dynamic deformation
with inverted V braces

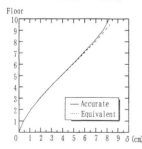

Fig. 11 Lateral dynamic deformation
with X braces

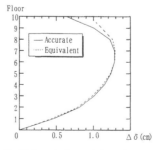

Fig. 12 Dynamic layer deformation
with inverted V braces

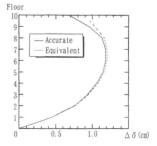

Fig. 13 Dynamic layer deformation
with X braces

5 CONCLUSION AND NEXT WORK

5.1 *The Conclusions*

a) When super column with inverted V braces bears pure bending moment, the characteristic of its bending deformation does not obey the hypothesis of plane section. On the opposite, other super components can obey the hypothesis well.

b) According to the proposed equivalent method, the equivalent model has compensatory ability of bending and shearing, so the problem of item a) has been solved. The equivalent method can adapt to any kind of super components and the equivalent model built by a simple static analysis can be used to any complicated static and dynamic analysis.

c) The equivalent methods to super column and super beam have been unified, so it becomes easier and more integrated.

d) By structural analysis, the good accuracy of equivalent models has been tested and the equivalent method has been verified preliminarily.

5.2 *The next work*

a) Though the equivalent method is derived on the basis of elastic mechanical theory, according to the static elasto-plastic analytical results of the accurate model, it can also be used to establish the elasto-plastic equivalent model. It's very important to continue studying the validity and adaptability of the equivalent method in static and dynamic elasto-plastic analysis.

b) It is also valuable to study and spread the equivalent method in other practical structures.

REFERENCES

Fukada, Y., Hirao, Y., Ikezaki, M., Okano, M., et al: A Study on Non-linear Response Analysis of Super-structure, Part 1- Part 4, Summaries of Technical Papers of Annual Meeting, Architectural Institute of Japan, pp. 229-236, 1988.10 (in Japanese)

Uchiyama, F., Kato, Y., et al: The Analytical Study for Structural Characteristics of Super Frame, Part 1- Part 2, Summaries of Technical Papers of Annual Meeting, Architectural Institute of Japan, pp. 1399-1402, 1990.10 (in Japanese)

Inoue, R., et al: The Analytical Study for Structural Characteristics of Super Frame, Part 3, Summaries of Technical Papers of Annual Meeting, Architectural Institute of Japan, pp. 1543-1544, 1991.9 (in Japanese)

Richard, L.: Current Evolution of Tall Building Structures, Australian Structural Engineering Conference, Sydney 1994. 9.

Ju, X.H.: The Equivalent Models of Super Steel Frames and Structural Analysis, Thesis for Master Degree of Harbin University of Architecture and Civil Engineering, Harbin China, pp. 25-94, 1996. 12 (in Chinese)

Adachi, M.: Structural Design and Analysis of New Tokyo City Hall Tower, Tall Building: 2000 and Beyond, Hong Kong, 1990.

Creative Systems in Structural and Construction Engineering, Singh (ed.) © 2001 Balkema, Rotterdam, ISBN 90 5809 161 9

A sensitivity study in advanced analysis of steel frames comprising non-compact sections

Q. Xue
Civil and Hydraulic Engineering Research Center, Sinotech Engineering Consultants Incorporated, Taipei, Taiwan

M. Mahendran
Physical Infrastructure Centre, Queensland University of Technology, Brisbane, Qld, Australia

ABSTRACT: During the past decades, extensive research has been conducted to develop, implement and verify the advanced analysis methods to be used as a reliable simplified design tool. This study presents the refined plastic hinge method of steel frames comprising sections, which are non-compact and subject to the effects of local buckling. The effects of different modeling of the distributed loads are illustrated based on the results verified by comparison with the ABAQUS finite element solutions.

1 INTRODUCTION

The Advanced analysis methods have been being more and more attractive due to their transparency in representing the strength and stability of structures, and their feasibility for design use such that separate specification member capacity checks are not required. In such kind of analysis, all the significant nonlinear effects in the analysis are expected to model.

The advanced analysis methods have been subjected to extensive development and application during the past decade. The distributed plasticity analysis (Clarke & Hancock 1991, Toma & chen 1992) and the concentrated plasticity analysis (Liew 1992, Attalla et al. 1994, Liew et al. 1994, King & Chen 1994, Chen & Chan 1995) are the two groups of such methods. The former is regarded as impractical for general design use. The application of such methods is mainly restricted to fully laterally restrained two-dimensional steel frames comprising compact sections. Avery (1998) has recently incorporated the effects of local buckling into a second-order inelastic plane frame analysis with the distributed loads approximated to lumped nodal loads. In this paper, a concentrated plasticity method named the refined plastic hinge method implicitly accounts for the effects of gradual cross-sectional yielding, longitudinal spread of plasticity, initial geometric imperfections, residual stresses, and local buckling is presented. The effects of accurate and approximate modeling of the distributed loads are illustrated based on the solutions verified by comparison with that obtained from ABAQUS finite element analyses.

2 NUMERICAL FORMULATION AND SOLUTION TECHNIQUE

The refined plastic hinge method is the best to account for the effects of local buckling (Avery 1998) among five of the most significant concentrated plasticity advanced analysis formulations: the refined plastic hinge method (Liew 1992), the notional load plastic hinge method (Liew et al. 1994), the hardening plastic hinge method (King & Chen 1994), the quasi plastic hinge approach (Attalla et al. 1994), and the springs in series method (Yau & Chan 1994).

In this study, the formulation of a frame element force-displacement relationship is based on the refined plastic hinge method. Local buckling effects such as the reduction in section capacity, gradual stiffness reduction and hinge softening is implicitly accounted for by the application of simple equations

2.1 *Element force-displacement relationship*

The conventional beam-column element is employed to formulate the force-displacement relationship through the co-rotational approach (Xue 1997). The basic member force-deformation relations are derived in the local co-ordinates (Fig. 1). In the presence of nonlinear geometric and/or nonlinear material behavior, incremental and/or iteration strategies should be used to solve the problem. For the solutions up to the limit point, when employing a pure incremental method or the (Modified) Newton-Raphson Method, the tangent stiffness matrix must be formulated. For classification as advanced analysis, the tangent stiffness matrix must account for all factors that may significantly influence the behavior of a structure, including:

- second-order effects,
- material properties,
- residual stresses,
- geometric imperfections, and
- local buckling.

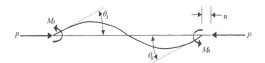

Figure 1. Beam-column element in local coordinates

Chen and Lui (1987) proposed the *second-order elastic* force-displacement incremental relationship for an elastic beam-column element in the local co-rotational co-ordinate system without plastic hinges could be expressed as:

$$
\left\{\begin{array}{c} \dot{M}_A \\ \dot{M}_B \\ \dot{P} \end{array}\right\} = \frac{EI}{L} \begin{bmatrix} s_1 & s_2 & 0 \\ s_2 & s_1 & 0 \\ 0 & 0 & A/I \end{bmatrix} \left\{\begin{array}{c} \dot{\theta}_A \\ \dot{\theta}_B \\ \dot{u} \end{array}\right\} \tag{1}
$$

in which (s_1, s_2) are the stability functions:

$$
s_1 = 4 + \frac{2\pi^2 \rho}{15} - \frac{(0.01\rho + 0.543)\rho^2}{4 + \rho} + \frac{(0.004\rho + 0.285)\rho^2}{8.183 + \rho}
$$

$$
s_2 = 2 - \frac{\pi^2 \rho}{30} + \frac{(0.01\rho + 0.543)\rho^2}{4 + \rho} - \frac{(0.004\rho + 0.285)\rho^2}{8.183 + \rho}
$$

$$\tag{2}$$

where:

$$
\rho = \frac{P}{P_e} = \frac{PL^2}{\pi^2 EI} \tag{3}
$$

In the refined plastic-hinge method, zero-length plastic hinges are inserted at the element ends to account for material yielding at the stage when forces at any cross-section equals or exceeds its section capacity. The modified incremental force-displacement relationship of a beam-column element with a plastic hinge at end A can be expressed as:

$$
\left\{\begin{array}{c} \dot{M}_A \\ \dot{M}_B \\ \dot{P} \end{array}\right\} = \frac{EI}{L} \begin{bmatrix} 0 & 0 & 0 \\ 0 & (s_1 - s_2^2)/s_1 & 0 \\ 0 & 0 & A/I \end{bmatrix} \left\{\begin{array}{c} \dot{\theta}_A \\ \dot{\theta}_B \\ \dot{u} \end{array}\right\}
$$
$$
+ \left\{\begin{array}{c} 1 \\ s_2/s_1 \\ 0 \end{array}\right\} \dot{M}_{scA} \tag{4}
$$

A similar relationship is obtained for an element with a plastic hinge at end B:

$$
\left\{\begin{array}{c} \dot{M}_A \\ \dot{M}_B \\ \dot{P} \end{array}\right\} = \frac{EI}{L} \begin{bmatrix} (s_1 - s_2^2)/s_1 & 0 & 0 \\ 0 & 0 & 0 \\ 0 & 0 & A/I \end{bmatrix} \left\{\begin{array}{c} \dot{\theta}_A \\ \dot{\theta}_B \\ \dot{u} \end{array}\right\}
$$
$$
+ \left\{\begin{array}{c} s_2/s_1 \\ 1 \\ 0 \end{array}\right\} \dot{M}_{scB} \tag{5}
$$

If plastic hinges form at both ends of the element, the relationship is modified as:

$$
\left\{\begin{array}{c} \dot{M}_A \\ \dot{M}_B \\ \dot{P} \end{array}\right\} = \frac{EI}{L} \begin{bmatrix} 0 & 0 & 0 \\ 0 & 0 & 0 \\ 0 & 0 & A/I \end{bmatrix} \left\{\begin{array}{c} \dot{\theta}_A \\ \dot{\theta}_B \\ \dot{u} \end{array}\right\} + \left\{\begin{array}{c} \dot{M}_{scA} \\ \dot{M}_{scB} \\ 0 \end{array}\right\} \tag{6}
$$

where $\dot{M}_{sc}$ stands for the change in moment at the plastic hinge represented by subscription A or B.

The effects of local buckling can be accounted for by using the section capacity equations and provisions for local buckling provided in either the AISC LRFD (1995) or the AS4100 (SAA 1990) specifications. Only the latter is considered in this paper. The effects of local buckling are accounted for by the use of the form factor (k_f) representing the reduction in pure axial compression section capacity, normalized effective section modulus (Z_e/S) representing the reduction in pure bending moment capacity, and web slenderness ratio (λ_w).

$$
\frac{1}{k_f} p_{sc} + \frac{1}{c_1 Z_e/S} m_{sc} = 1 \quad \text{for } p/m \geq c_3
$$

$$
\frac{c_2}{k_f} p_{sc} + \frac{1}{Z_e/S} m_{sc} = 1 \quad \text{for } p/m < c_3
$$

$$\tag{7}$$

where:

$$c_1 = 1 + 0.18\left(\frac{82 - \lambda_w}{82 - \lambda_{wy}}\right) \le 1.18 \ ;$$

$$c_2 = 0 \ ; \tag{8}$$

$$c_3 = \frac{k_f}{Z_e/S}\left(1 - \frac{1}{c_1}\right)$$

$$k_f = \frac{A_e}{A_g} \tag{9}$$

and

$$\lambda_s \le \lambda_{sp} : \quad Z_e/S = 1$$

$$\lambda_{sp} < \lambda_s \le \lambda_{sy} : \quad Z_e/S = Z/S + \left(1 - Z/S\right)\left(\frac{\lambda_{sy} - \lambda_s}{\lambda_{sy} - \lambda_{sp}}\right) \tag{10}$$

$$\lambda_s > \lambda_{sy} : \quad Z_e/S = Z/S\left(\frac{\lambda_{sy}}{\lambda_s}\right)$$

Two functions are used to account for gradual yielding, distributed plasticity and the associated instability effects: the tangent modulus (E_t) instead of the elastic modulus and the flexural stiffness reduction factor (ϕ). These functions represent the distributed plasticity along the length of the member due to axial force effects and the distributed plasticity effects associated with flexure, respectively.

The tangent modulus functions recommended by Liew (1992) are appropriate for compact hot-rolled I-sections but are not appropriate for non-compact sections subject to local buckling effects. The additional stresses associated with the local buckling deformations that occur in non-compact sections cause a reduction in stiffness which must be included in the tangent modulus function. The effects of local buckling on the tangent modulus can be accounted for by using the compression member capacity equations and provisions for local buckling. Such a procedure can be referred to Avery (1998).

To account for the effects of local buckling on the flexural stiffness reduction, Avery (1998) proposed a new stiffness reduction factor (ϕ) based on that described by Liew (1992) to represent the gradual transition for the formation of a plastic hinge at each end of an initially elastic beam-column element.

$$\phi = 1 \quad \text{for } \alpha \le \alpha_{iy}$$

$$\phi = \frac{\alpha\left(\alpha_{sc} - \alpha\right)}{\alpha_{iy}\left(\alpha_{sc} - \alpha_{iy}\right)} \quad \text{for } \alpha_{sc} \ge \alpha > \alpha_{iy} \tag{11}$$

where, the force state parameter α and that corresponding to the section capacity (denoted by α_{sc}) can be obtained from code provisions and α_{iy} is taken as 0.5 or can be determined from the initial yield interaction equation described by Avery (1998).

Non-compact sections subject to local buckling exhibit hinge softening behavior due to the increasing stresses caused by increasing local buckling deformations, resulting in a reduction in the effective section core available to resist the applied axial force and bending moment. It is modeled by using negative values of the tangent modulus and the flexural stiffness reduction factor at the location of a plastic hinge in a non-compact section.

2.2 System force-displacement relationship

Standard co-ordinate transformation and assemblage can be performed to obtain the force-displacement relationship of the system.

2.3 Solution technique

A pure incremental method is employed in this study.

2.4 Applied loads

Member loads are either modeled exactly as distributed loads or approximated as nodal loads to both the member ends. For members subjected to transverse distributed loads, the *slope-deflection equations* must be modified by adding an extra term for the fixed-end moment of the member as

$$\begin{Bmatrix} M_A \\ M_B \\ P \end{Bmatrix} = \frac{EI}{L}\begin{bmatrix} s_1 & s_2 & 0 \\ s_2 & s_1 & 0 \\ 0 & 0 & A/I \end{bmatrix}\begin{Bmatrix} \theta_A \\ \theta_B \\ u \end{Bmatrix} + \begin{Bmatrix} M_{FA} \\ M_{FB} \\ 0 \end{Bmatrix} \tag{12}$$

The fixed-end moment for a member subjected to uniformly distributed loads may be expressed as

$$M_{FA} = -M_{FB} = -\frac{WL^2}{12}\left[\frac{3(\tan u - u)}{u^2 \tan u}\right] \tag{13}$$

The terms in the brackets represent the effect of axial force on the fixed-end moment of the member.

The fixed-end moment for a member subjected to non-uniformly distributed loads may vary with the distributed load pattern. Thus in such a case, the externally applied distributed loads can be replaced by the equivalent internal nodal forces.

3 NUMERICAL EXAMPLES

The example comprised 8 analyses of single bay, single story, pinned base non-sway (braced) frames (Figure 1) with non-compact I-sections subject to major axis bending under distributed beam loading. Each member is modeled by either 2 (with nodal loads only) or 4 (with distributed loads) elements. Two values of column slenderness ratio L_c/r, one beam-column stiffness ratio γ ($\gamma = \dfrac{I_c/L_c}{I_b/L_b}$, s/h=1.5), two load cases ($P + w$, and $P = 0$) and two types of section slenderness (Figure 2) were considered. The identification system used to describe the parametric variables for the analysis is illustrated in Figure 2, where 'bc' stands for beam-column model.

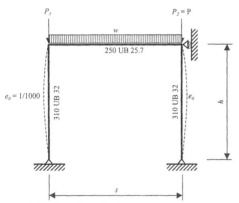

Figure 1. Configuration of the frame

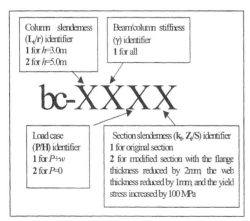

Figure 2: identifier system

A summary of the ultimate vertical column loads (P_u) and beam distributed loads (w_u) obtained from

ABAQUS finite element package is presented in Table 1 (Avery 1998). A comparison between the ultimate loads obtained from the refined plastic hinge (RPH) analyses and the finite element analysis (FEA) is presented in Table 2.

Table 1: ABAQUS solutions

Frame	$P = 0$	$P + w$	
	w_u (kN/m)	w_u (kN/m)	P_u (kN)
bc-11X1	100	31.7	815
bc-11X3	87.9	26.7	685
bc-21X1	35.1	12.1	862
bc-21X3	31.0	9.85	704

Table 2: Comparison of the solutions with different model

Frame	RPH (nodal loads) / FEA	RPH (distributed loads) / FEA
bc-1111	1.04	1.03
bc-1121	1.035	0.998
bc-2111	0.98	0.97
bc-2121	1.009	0.965
bc-1113	1.09	1.07
bc-1123	1.006	0.97
bc-2113	1.06	1.05
bc-2123	0.997	0.98

The model with approximate nodal loads is, on average, 2.7 percent unconservative compared to the finite element model. The maximum unconservative error is 6 percent, slightly higher than the recommended 5 percent limit. The model with distributed loads is, on average, 0.4 percent unconservative compared to the finite element results. The maximum unconservative error is 5 percent, within the recommended 5 percent limit.

The horizontal load-deflection (mid-height) curves for each analysis, by using the approximate nodal loads and the distributed loads are illustrated through Figure 3 to Figure 10. The horizontal load-deflection curves of bc-1121 with 2 elements to model the member subjected to the distributed load are also presented here in Figure 11. In general, the distributed load model gives better results and is preferable when fewer elements are adopted to model each member.

4 CONCLUSIONS

The advanced analysis of frames comprising non-compact sections using the refined plastic hinge method has been presented. Member transverse loads can be modeled either accurately by the distributed loads or approximately by the nodal loads.

bc-1111

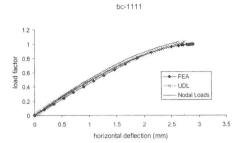

Figure 3: Horizontal load-deflection for bc-1111

bc-1113

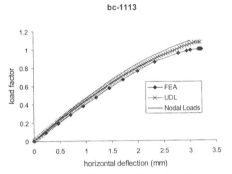

Figure 4: Horizontal load-deflection for bc-1113

bc-2111

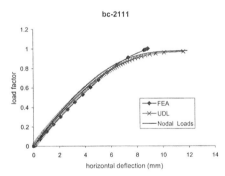

Figure 5: Horizontal load-deflection for bc-2111

The distributed load model is more precise to represent the ultimate load and deflection behavior of frames subjected to member transverse loads. However, the difference becomes smaller when more elements are used to model the member. When fewer elements are used to save the computational effort, as for a structure with a large number of members, the distributed load model is preferable.

bc-2113

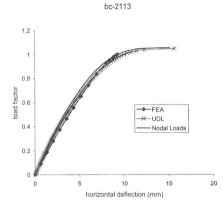

Figure 6: Horizontal load-deflection for bc-2113

bc-1121

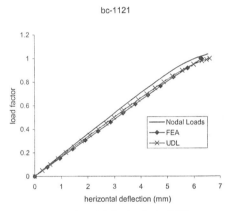

Figure 7: Horizontal load-deflection for bc-1121

bc-1123

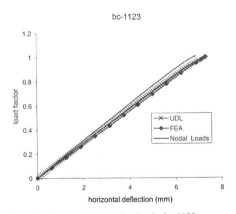

Figure 8: Horizontal load-deflection for bc-1123

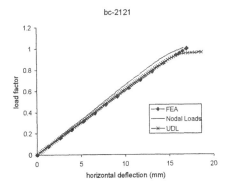

Figure 9: Horizontal load-deflection for bc-2121

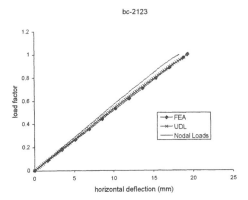

Figure 10: Horizontal load-deflection for bc-2123

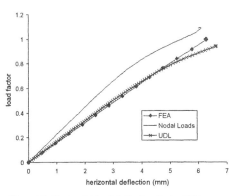

Figure 11: Horizontal load-deflection for bc-1121 modeled by 2 elements per member.

ACKNOWLEDGEMENT

Many thanks are given to Dr. P. Avery for his enthusiastic technique support.

REFERENCES

AISC, 1995. *Manual of Steel Construction, Load and Resistance Factor Design, 2nd Edition.* American Institute of Steel Construction, Chicago, IL, USA.

Attalla, M. R., Deierlein, G. G., & McGuire, W. 1994. Spread of plasticity: quasi-plastic-hinge approach. *Journal of Structural Engineering, ASCE* 120(8): 2451-2473.

Avery, P. 1998. *Advanced analysis of steel frame structures comprising non-compact sections.* PhD dissertation. School of Civil Engineering, Queensland University of Technology, Brisbane, Australia.

Chen, W. F. & Chan, S. L. 1995. Second-order inelastic analysis of steel frames using element with midspan and end springs. *Journal of Structural Engineering, ASCE* 121(3): 530-541.

Chen, W. F. & Lui, E. M. 1987. *Structural stability - theory and implementation.* Elsevier Applied Science, New York, NY, USA.

Clarke, M. J. & Hancock, G. J. 1991. Finite-element nonlinear analysis of stressed-arch frames. *Journal of Structural Engineering, ASCE* 117(10): 2819-2837.

King, W. S., & Chen, W. F. 1994. Practical second-order inelastic analysis of semi-rigid frames. *Journal of Structural Engineering, ASCE* 120(7): 2156-2175.

Liew, J. Y. R. 1992. *Advanced analysis for frame design.* PhD dissertation. School of Civil Engineering, Purdue University, West Lafayette, IN, USA.

Liew, J. Y. R., White, D. W., & Chen, W. F. 1994. Notional-load plastic-hinge method for frame design. *Journal of Structural Engineering, ASCE* 120(5): 1434-1454.

SAA 1990. *AS4100-1990 Steel Structures.* Standards Association of Australia, Sydney, Australia.

Toma, S. & Chen, W. F. 1992. European calibration frames for second-order inelastic analysis. *Journal of Engineering Structures* 14(1): 7-14.

Xue, Q. 1997. *Nonlinear dynamic analysis for the instability problem of plane and spatial frames.* PhD dissertation. Civil Engineering Department, The University of Queensland, Brisbane, Australia.

Yau, C. Y. & Chan, S. L. 1994. Inelastic and stability analysis of flexibly connected steel frames by springs-in-series model. *Journal of Structural Engineering, ASCE* 120(10): 2803-2819.

Creative Systems in Structural and Construction Engineering, Singh (ed.) © 2001 Balkema, Rotterdam, ISBN 90 5809 161 9

Effects of initial imperfections on local and global buckling behaviors of thin-walled members

Mitao Ohga & Hiroshi Izawa
Department of Civil and Environmental Engineering, Ehime University, Matsuyama, Japan

Akira Takaue
Chodai Limited, Takamatsu, Japan

Takashi Hara
Tokuyama Technical College, Japan

ABSTRACT: The combined behaviors of the local and overall buckling deformations of the U-section and Box-section members subjected to the axial loads are analyzed by the finite element method. In the finite element method, the isoparametric degenerated shell element is adopted and only the geometrical nonliniarity is considered based on the Green Lagrange strain definition, To solve the nonlinear equation, the displacement incremental scheme employed, in which the incremental axial displacements are added to the both ends of the members. As the initial imperfections, four types of initial imperfections are introduced, 1) the bulge type of imperfection, 2) the local type of imperfection, 3) the global type of imperfection, and 4) the combined imperfection of local and global ones.

1 INTRODUCTION

Thin-walled members are widely used in a broad range of structural applications to reduce the material cost as well as the dead weight of a structure. The cross sections of the members are composed of thin plate panels, therefore, the local deformation (local buckling), overall deformation (overall buckling) and combination of these two types of deformations (combined buckling) are occurred when the member is subjected to the axial load. The examination about these deformations itself and the effects of these deformations to the behavior and strength of the members do not seem to be sufficient as far as the authors know (Ohga, M et al. 1999).

In this paper, the effects of the initial imperfections on the behavior with the local, overall and combined deformations of the thin-walled members subjected to the axial loads, obtained by the finite element stability analysis are examined. As the initial imperfections, four types of initial imperfections: 1) the bulge type of imperfection, 2) the local type of imperfection, 3) the global type of imperfection, 4) the combined imperfection of local and global ones, are introduced.

For the local type of imperfection, the local buckling mode corresponding to the local buckling load is introduced. The local buckling loads and mode shapes of the members subjected to the axial loads are obtained by the transfer matrix method (TMM). The transfer matrix is derived form the differential equations for the plate panels composing the cross section of the members, and the point matrix relating the state vectors between consecutive panels is used

to allow the transfer procedure over the cross section of the thin-walled members. Therefore, the buckling loads and mode shapes gauged the interaction between the panels of the cross section can be obtained (Ohga, M. et al. 1995 Tesar, A. & L. Fillo 1988 Uhring, R. 1973).

For the finite element stability analysis, the isoparametric degenerated shell element is adopted, and only the geometrical nonlinearity is considered based on the Green Lagrange strain definition (Hinton, E. & D.R.J. Owen 1984, Hinton, E. 1988). To solve the nonlinear equation, the displacement incremental scheme is employed, in which the incremental axial displacements are added to the both ends of the members.

2 FINITE ELEMENT STABILITY ANALYSIS

In the finite element stability analysis, the isoparametric degenerated shell element with 9 nodes is adopted, and only the geometrical nonlinearity is considered based on the Green Lagrange strain definition. To solve the nonlinear equation, the displacement incremental scheme employed, in which the incremental axial displacements are added to the ends of the members. Two types of displacements are adopted, 1) the equal displacements are added to all over the cross section of the member as shown in Figure 2a (constant displacement condition), 2) the displacement is added to only one point through the loaded plate as shown in Figure 2b.

As the initial imperfections, four types of initial imperfections are introduced, 1)the bulge type of

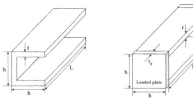

(a) Constant displacement   (b) One point displacement

Figure 1. Analytical models (U-section).

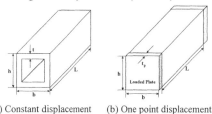

(a) Constant displacement   (b) One point displacement

Figure 2. Analytical models (Box-section).

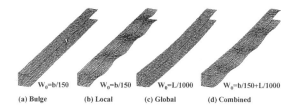

$W_0=b/150$ $W_0=b/150$ $W_0=L/1000$ $W_0=b/150+L/1000$

(a) Bulge   (b) Local   (c) Global   (d) Combined

Figure 3. Initial imperfection (U-section).

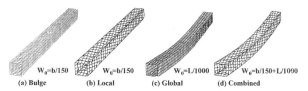

$W_0=b/150$ $W_0=b/150$ $W_0=L/1000$ $W_0=b/150+L/1000$

(a) Bulge   (b) Local   (c) Global   (d) Combined

Figure 4. Initial imperfection (Box-section).

imperfection (Bulge: Fig. 3a $w_0=b/150$), 2) the local type of imperfection (Local: Fig 3b $w_0 = b/150$), 3) the global type of imperfection (Global: Fig. 3c $w_0= L/1000$), and 4) the combined imperfection of local and global ones (Combined: Fig 3d). In the bulge type of imperfection a displacement is applied to the point on the flange. The global type of imperfection is the beam type imperfection where there is no distortion of the cross section. In the analysis, adding to above four imperfections, the analysis without initial imperfection (Non-Imperfection) is performed.

3 LINEAR BUCKLING STRENGTH AND MODE SHAPES

In this paper, the linear buckling strength and buckling mode using for the local initial imperfection (Figs 3b, 4b) in the finite element stability analysis of the members subjected to the axial loads are obtained by transfer matrix method (TMM).

In the method, introducing the trigonometric series into the governing equations for the plate panels composing the cross section of the thin-walled members, the transfer matrix, $\mathbf{F}$, relating the state vectors at both ends of the cross section are obtained, and considering the relation between the state vectors of the consecutive two panels the point matrix, $\mathbf{P}$, is obtained.

Applying the transfer and point matrices, $\mathbf{F}$, and $\mathbf{P}$, to the U-section members, the relation between the state vectors at both ends of the cross section is obtained (Fig. 1a):

$$\mathbf{Z}_3 = \mathbf{F}_3\mathbf{P}_2\mathbf{F}_2\mathbf{P}_1\mathbf{F}_1\mathbf{Z}_0 = \mathbf{U}\mathbf{Z}_0 \qquad (1)$$

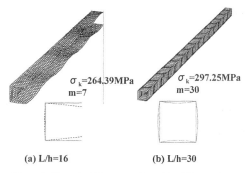

$\sigma_k=264.39\text{MPa}$
$m=7$

$\sigma_k=297.25\text{MPa}$
$m=30$

(a) L/h=16   (b) L/h=30

Figure 5. Linear buckling strengths and buckling modes.

Considering the boundary conditions at both ends of the cross section, the stability equations for the U-section members are obtained.

$$\mathbf{U}'\mathbf{Z}'_0 = \mathbf{0} \qquad (2)$$

In the case of box-section members, applying the transfer and point matrices, the stability equation can be obtained as follows:

$$[\mathbf{U} - \mathbf{I}]\mathbf{Z}_0 = \mathbf{0}, \qquad \mathbf{U} = \mathbf{P}_4\mathbf{F}_4\mathbf{P}_3\mathbf{F}_3\mathbf{P}_2\mathbf{F}_2\mathbf{P}_1\mathbf{F}_1 \qquad (3)$$

Solving the stability equations (2), (3), the linear buckling strength for U and box section members are obtained.

Using the linear buckling strength obtained above and proceeding the transfer procedure again, the mode shape corresponding the linear buckling strength can be obtained.

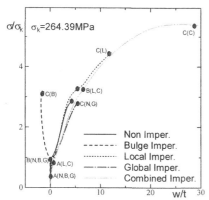

Figure 6. Load-deformation curves (U-section, constant displacement).

Figures 5a and 5b show the linear buckling strength and their mode shapes of U-section (L/h=16) and box-section members (L/h=30). As shown in Figure 5a, the buckling mode shape of U-section member indicates the local buckling mode, and the mode number in axial direction is m=7. The linear buckling strength is σ$_k$=264.39MPa.

The buckling mode shape of box-section member shows the local buckling mode as the case of U-section member, and the mode number in axial direction is m=30. The linear buckling strength is $\sigma_k = 297.25MPa$.

4 NONLINEAR BEHAVIOURS OF U-SECTION MEMBERS UNDER AXIAL LOADS

4.1 *Constant displacement condition*

Figure 6 shows the comparison of the load-deflection curves of the U-section members (L/h=16) under the constant displacement condition (Fig. 1a) for the various initial imperfections (Fig. 3: Bulge, Local, Global and Combined). Figure 7 shows the deformation shapes at the various load levels (point A, B and C in Fig. 6). In Figures 6 and 7, the results for without initial imperfection (Non-Imperfection) are also shown. The vertical line of Figure 6 indicates the axial load divided by the linear buckling strength, and the horizontal line the vertical displacement of the flange divided by the thickness of the plate panel. In the analysis, the member is divided into 480 (12x40) finite elements as shown in Figure 3.

In the case of Non-Imperfection, as shown in Figures 6 and 7a, in the early load stage (Fig. 6: A(N)), the tiny local deformation occurs at the both ends of the member (Fig. 7a). At around the linear buckling strength ($\sigma/\sigma_k = 1.0$), the deformation increases suddenly and the deformations at the both ends of the member expand throughout the member. After the local buckling happens, the geometrical

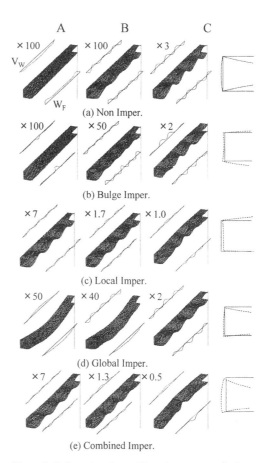

(a) Non Imper.

(b) Bulge Imper.

(c) Local Imper.

(d) Global Imper.

(e) Combined Imper.

Figure 7. Deformation shapes (U-section, constant displacement).

nonlinearity appears, and just after the overall deformation happens at around $\sigma/\sigma_k = 2.8$, the solution is divergent. As shown in Figure 7a, the deformation shape at the last load stage (Fig. 6: C(N)) is similar to that for the linear buckling strength obtained by TMM (Fig. 5a), and although very small overall deformation appears, the local deformation excels. (Fig. 7a)

In the case of the bulge type imperfection (Fig. 3a), as shown in Figures 6 and 7b, in the early load stage (Fig. 6: A(B)), the tiny local deformation occurs around the bulge imperfection. At around the linear buckling strength ($\sigma/\sigma_k = 1.0$), the deformation increases suddenly as the case of Non-Imperfection, and the local deformation around the imperfection expands throughout the member. Then just after the overall deformation happens ($\sigma/\sigma_k = 3.2$), the solution is divergent. As shown in Figure 7b, the deformation shape at the last load stage (Fig. 6: C(B)) shows the distortional shape (combined shape of local and overall deformation).

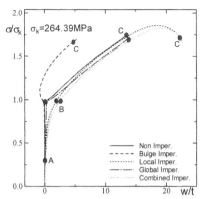

Figure 8. Load-deformation curves (U-section, one point displacement).

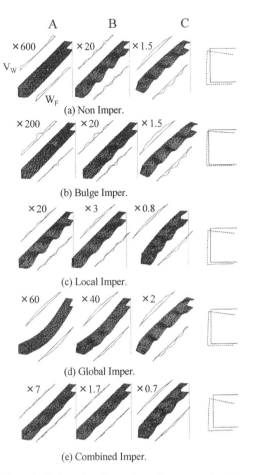

(a) Non Imper.

(b) Bulge Imper.

(c) Local Imper.

(d) Global Imper.

(e) Combined Imper.

Figure 9. Deformation shapes (U-section, one point displacement).

The mode number in the axial direction coincides with that of the linear buckling strength (Fig. 5a).

In the case of the local type imperfection (Fig. 3b), as shown in Figure 6, the deformation increases gradually from the early load stage. This is the different tendency from previous two cases (Non-Imperfection, Bulge). The shape of the load-displacement curve, in this case, changes at around $\sigma/\sigma_k = 3.2$. This originated from the fact that at this load stage, the overall deformation appears adding to the local deformation as shown in Figure 7c.

In the case of the global type imperfection (Fig. 3c), the load-displacement curves are similar to that of Non-Imperfection. About the mode shape, although the difference caused by the initial imperfections (Non-Imperfection and Global) is shown in the early load stage, the deformation shapes for both imperfection cases are similar to each other (Fig. 7c).

In the case of the combined type imperfection (Fig. 3d), although the load-displacement curve is similar to that for the local type imperfection in the early load stage, the stable solution can be obtained even after the ultimate strength as shown in Figure 6.

These results indicate that the combined type imperfection can cope with the various deformations (local, global and combined ones).

4.2 One point displacement condition

Figure 8 shows the load-deformation curves of U-section members (L/h=16) under the one point displacement condition (Fig. 1b), and Figure 9 shows the deformation shapes at the various load levels (point A, B and C in Fig. 8).

In the case of three types of the initial imperfections, which do not include the components of the local deformation (Non-Imperfection, Bulge and Global), the deformations increase suddenly at around the linear buckling strength as shown in Figure 8. Although in the case of the constant displacement condition only the local deformation appears at this load stage (Fig. 6), in the case of one displacement condition the overall deformation appears simultaneously adding to the local deformation, and the geometrical nonliniarity does not appear (Fig. 9).

In the case of other two types of initial imperfections, which include the components of the local deformation (Local and Combined), the deformations increase gradually from the early load stage as shown in Figure 8. As described above in the case of the constant displacement condition only the result with combined initial imperfection is stable beyond the ultimate strength, the results with local and combined types of imperfection are stable through out the analysis in the one point displacement condition.

As shown in Figure 9, the deformation shape for every initial imperfection indicates the distortional shape, and the lateral deformation at the web is distinguished.

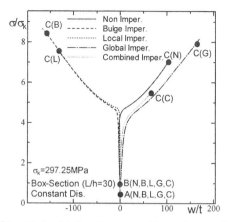

Figure 10. Load-deformation curves (Box-section, constant displacement).

5 NONLINEAR BEHAVIOURS OF BOX-SECTION MEMBERS UNDER AXIAL LOADS

5.1 Constant displacement condition

Figure 10 shows the comparison of the load-deflection curves of the box-section members (L/h=30) under the constant displacement condition (Fig. 2a) for the various initial imperfections (Fig. 4: Bulge, Local, Global and Combined). Figure 11 shows the deformation shapes at the various load levels (point A, B and C in Fig. 10).). In Figures 10 and 11, the results for without initial imperfection (Non-Imperfection) are also shown. In the analysis, the box-section member is divided into 480 (8x60) finite elements as shown in Figure 4.

In the case of three types of initial imperfections, which do not include the components of the overall deformation (Non-Imperfection, Bulge and Local, Fig. 4a, b), as shown in Figure 11, at around the linear buckling strength (Fig. 10, B(N)), the tiny local buckling appears. Then the displacements increase suddenly at around $\sigma / \sigma_k = 4.5$. The deformation at this load stage is mainly depended on the overall deformation as shown in Figure 11a. After this load level the geometrical nonlinearity appears. At the last load stage, the web indicates the local deformation, on the other hand the flange the combined one of local and overall deformations. The mode in the axial direction is closes to that for the linear buckling mode (Fig. 5b, m=30).

deformation only, on the other hand the flange shows the combined deformations.

5.2 One point displacement condition.

Figure 12 shows the load-deformation curves of the box-section members (L/h=30) under the one point displacement condition (Fig. 2b), and Figure 13

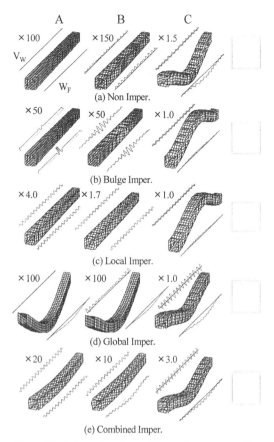

(a) Non Imper.

(b) Bulge Imper.

(c) Local Imper.

(d) Global Imper.

(e) Combined Imper.

Figure 11. Deformation shapes (Box-section, constant displacement).

shows the deformation shapes at the various load levels (point A, B and C in Fig. 12).

In the case of three types of initial imperfections, which do not include the component of the overall deformation (Non-Imperfection, Bulge and Local, Fig. 4a, b), the displacements increase suddenly at around the $\sigma / \sigma_k = 2.0$, as shown in Figure 12, and the overall deformations appear (Fig. 13). The load level at which the overall deformation appears is less than that of the constant displace condition (Fig. 10).

In the cases of global and combined type imperfections that include the component of the overall deformation (Fig. 4c,d), as shown in Figure 12 the vertical displacement at the web increases gradually from early load stage. At around the linear buckling strength (Fig. 12, B(G, C)), the local buckling happens in the web as shown in Figures 13d, e. At the last load step, the web shows the local deformation, and the flange shows the combined one as the case of the constant displacement condition.

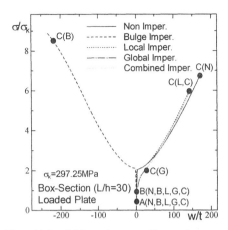

Figure 12. Load-deformation curves (Box-section, one point displacement).

6 CONCLUSION

In this paper, the effects of the initial imperfections on the behavior with the local, overall and combined deformations of the thin-walled members subjected to the axial loads, obtained by the finite element stability analysis are examined. The following conclusions are obtained.

1) In the case of the U-section members, the deformation increases suddenly at around the linear buckling strength for both displacement incremental schemes. Although in the case of the constant displacement condition only the local deformation appears at this load stage, in the case of one displacement condition the overall deformation happens simultaneously adding to the local deformation.

2) In the case of the U-section members, the stable solution can be obtained even after the ultimate strength for the combined imperfection, this shows that the combined imperfection can cope with the various deformations.

3) In the case of the box-section members, the overall deformation appears suddenly at certain load level in the cases of initial imperfection that include the component of the overall deformation. This load level for one point displacement scheme is less than that for the constant displacement scheme.

Although only the geometrical nonliniarity is considered in this paper, the material nonliniarity should be considered as well as the geometrical one.

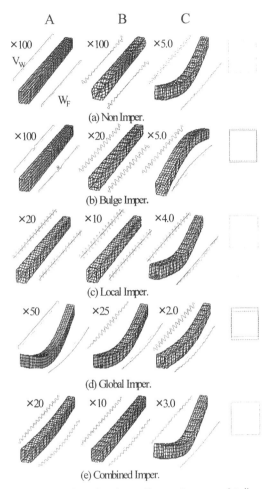

Figure 13. Deformation shapes (Box-section, one point displacement).

REFERENCES

Hinton, E. & D.R.J. Owen 1984. *Finite Element Software for Plates and Shells*. Pineridge Press: Swansea.

Hinton, E. 1988. *Numerical Methods and Software for Dynamic Analysis of Plates and Shells*. Pineridge Press: Swansea.

Ohga, M. et al. 1995. Buckling mode shapes of thin-walled members. *Computers & Structures* 54: 767-773.

Ohga, M et al. 1999. Effects of initial imperfections on nonlinear behaviors of thin-walled members. *Advances in Structural Engineering and Mechanics, Seoul, Korea*: 585-590.

Tesar, A. & L. Fillo 1988 *Transfer Matrix Method*. Dordrecht: Kluwer Academic Publishers.

Uhrig, R. 1973. *Elastostatik und Elastokinetik in Matrizen Schreibweise*. Belrin: Springer-Verlag.

Creative Systems in Structural and Construction Engineering, Singh (ed.) © 2001 Balkema, Rotterdam, ISBN 90 5809 161 9

Numerical analysis on fatigue failure modes in load-carrying fillet welded cruciform joints

S. Kainuma
Department of Civil Engineering, Nagoya University, Japan

T. Mori
Department of Civil Engineering, Hosei University, Tokyo, Japan

Y. Itoh
Center for Integrated Research in Science and Engineering, Nagoya University, Japan

ABSTRACT: Load-carrying fillet welded joints in welded steel structures are extensively used for connecting the secondary structural members. Many fatigue cracks have been reported in the fillet welded joints of plate girder bridges in Japan. The failure modes of the welded joint mainly depend on three factors including the plate thickness, the weld size and the weld penetration. Hence, the fatigue life evaluation of load-carrying fillet welded joint is more difficult than that of other types of joints. In this research, parametric fatigue crack propagation analyses were carried out on the fillet welded cruciform joints in order to quantitatively clarify the influence of the plate thickness, the weld size and the weld penetration on the failure mode. According to the numerical analysis results, a quantitative evaluation method was proposed for the failure modes in consideration of these three parameters.

1 INTRODUCTION

Many fillet welded joints are used in the connections between secondary members of steel bridges. Recently in Japan, fatigue cracks have been observed with the increase of traffic weight and volume beyond expectation. Most cracks initiate at the fillet welded joints such as the upper end of the web gap plates and the vertical stiffeners with sway bracings of plate girder bridges, as shown in Fig.1. The joint is considered as the load-carrying fillet welded cruciform joint. The cracks initiate and propagate from weld toes and/or weld roots. In case that the cracks initiate from weld toe, it is comparatively easy to detect. On the other hand, it is difficult for the cracks from the weld roots to detect since the cracks propagate inside of the welded joints. Hence, the fatigue failure mode has been investigated until now.

The weld leg length of transition point between root failure and toe failure is called as the critical leg length. The origin of fatigue failure has generally been evaluated at the ratio of critical leg length for the main plate thickness. For instance, fatigue tests were carried out on load-carrying fillet welded cruciform joints with plate thickness of 13 mm and 15 mm and hardly with the weld penetration (Soete & Crombrugge 1952). According to the experimental results, it was found that the ratio of critical leg length is around 0.85. The fatigue tests were also performed on the cruciform joints with 16 mm and 32 mm of the plate thickness (Ouchida & Nishioka

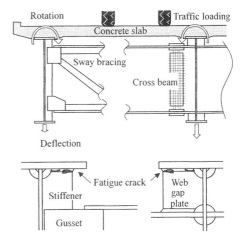

Figure 1. Example of fatigue crack

1964). As a result, it was clarified that the ratio is approximately 1.0 and it decreases a little with the increase in the plate thickness. The effect of groove depth on the ratio in the fatigue failure mode was examined. It has been confirmed that the ratio decreases at approximately 85% when the joints have weld penetration, which is considered as additional pat of the weld size.

In order to investigate the influence of penetration depth and welding leg length on the ratio of critical leg length, the parametric fatigue crack

863

propagation analyses were carried out on several cruciform joint models (Mori & Kainuma 1999). The ratio of the joint with plate thickness of 20 mm is about 1.2. When the penetration depth is large, the ratio decreases to around 1.0. However, the effect of the leg length, the penetration depth and the plate thickness on the fatigue failure mode has not quantitatively been clarified. Hence, in order to establish the evaluation method of the fatigue failure mode, it is necessary to clarify the behavior of fatigue crack propagation from the weld root. Therefore, fatigue crack propagation tests, thermal conduction analyses and elasto-plastic stress analyses were performed in order to examine the behavior of the fatigue crack propagation (Kainuma et al. 1997). From these results, it was found that high compressive residual stress existed at the vicinity of root tip and the residual stress delays the propagation rate when the crack length is comparatively short. Furthermore, it was clarified that the fatigue life under the small amplitude stress such as actual steel bridges is prolonged by the compressive residual stress. However, since restraining stress is generated at the joints connected with other members in actual structures, the compressive residual stress cannot be considered. This is because that the evaluation becomes the dangerous side when the compressive residual stress is expected. In addition, it is also impossible to measure the residual stress exactly by non-destructive testing. Therefore, the fatigue failure mode should be examined not considering the residual stress at the weld root.

In this study, the effect of weld leg length and penetration depth on the ratio of critical leg length, which has already been investigated (Mori & Kainuma 1999), is reexamined and quantitatively clarified by further parametric fatigue crack propagation analyses. The fatigue failure mode considering the effect of plate thickness is also investigated on the basis of the numerical analysis results. The parametric fatigue crack propagation analyses were carried out for load-carrying fillet welded cruciform joints without considering the compressive residual stress at the weld root.

2 PROCEDURE OF FATIGUE CRCAK PROPAGTION ANALYSIS

Analytical models were load-carrying fillet welded cruciform joints of which plate thickness t is 9, 20, 40 and 75 mm. There are seventy kinds of models with various weld leg lengths and weld penetrations. Shape and dimensions of the model are shown in Fig.2 and Table 1. In these analyses, the leg length on the main plate side is the same as that on the cross plate side considering the general fillet welds. The angle and the radius of the weld toe are 45 de-

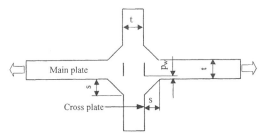

Figure 2. Numerical analysis model

gree and 0.5 mm, respectively. In particular, the radius was decided on the basis of the measurement of actual welded joint fabricated by semi-automatic arc welding.

The fatigue crack propagation analyses were performed in cases of both the toe failure and the root failure. In these analyses, the fatigue life was calculated by substituting stress intensity factor range of each crack length for da/dN-ΔK relationship of JSSC (1995). It is repeated from the initial crack length to the critical crack length. Due to not expecting compressive residual stress at the weld root in the actual welded structures, the stress was neglected in the analyses. For the analyses failing from the weld toe, the initial crack was defined as elliptical shape, of which depth and the width are 0.1 mm and 0.2 mm, respectively. The critical crack length was defined as 80 % of the main plate thickness. On the other hand, for the analyses failing from the root, the non-penetrating length was defined as initial crack since the weld root tip is sharp like fatigue crack where the stress concentration is high. The critical crack length was defined as the sum of a half of the main plate thickness and 80 % of the leg length of the fillet weld. The mean design curve (Japanese Society of Steel Construction 1995) was used as the fatigue crack propagation rate. The stress intensity factor range was calculated for the toe failure analyses by using the expression presented in Albrecht & Yamada (1977). For the root failure analyses, the expression of Frank & Fisher (1979) was adapted for the calculation of the stress intensity factor range. The validity of these analytical conditions has been confirmed by comparing the analytical results with the fatigue test results.

In the toe failure analyses, the stress distribution of weld toe along crack growth cross section is required. Hence, the finite element stress analysis was carried out on all cruciform joint models, which are shown in Table 1. The models were decided to be 1/4 model considering the symmetry of the joint. The smallest element size at the vicinity of weld toe is 0.025×0.025 mm.

Table 1. Dimensions of analytical model

Plate thickness t (mm)	Leg length s (mm)	Weld penetration p_w (mm)	s/t	Plate thickness t (mm)	Leg length s (mm)	Weld penetration p_w (mm)	s/t
9	4	0	0.444	20	24	1	1.2
		3				1.5	
		3.5				2.5	
		4			25	0	1.25
	4.5	0	0.5		30	0	1.5
	6	0	0.667	40	18	0	0.45
	7	2	0.778			14	
		3				16	
	9	0	1			18	
		1			20	0	0.5
		2			31	11	0.775
		2.5				13	
	11	0	1.222			15	
		1			40	0	1
		2				4	
	13	0	1.444			8	
	13.5	0	1.5		48	0	1.2
20	9	0	0.45		52	0	1.3
		8			60	0	1.5
		9		75	34	0	0.453
	10	0	0.5		35	20	0.467
	13	0	0.65			28	
		1				32	
		2.5			45	12	0.6
		5				24	
		7.5				30	
		9			53	22	0.707
		10				24	
	17	0	0.85			26	
		4			75	0	1
		5				8	
		6				16	
	20	0	1		98	0	1.307
		3			112.5	0	1.5
		5			120	0	1.6

3 INFLUENCES OF VARIOUS FACTORS ON CRITICAL LEG LENGTH

The effects of welding leg length and penetration depth have already been examined (Mori & Kainuma 1999). In this chapter, the number of analytical model is increased to seventy and the critical leg length is quantitatively clarified considering the effect of the plate thickness.

3.1 Influence of weld leg length

The relationship between fatigue strength of 2×10^6 cycles in cross section of the main plate and s/t is shown in Fig. 3. In this figure, the analytical results are shown when the plate thickness t and weld penetration p_w are 20 mm and 0 mm, respectively. Weld leg length is varied from 9 mm to 30 mm. The fatigue strength for root failure increases with the increase of s/t. This tendency is roughly the same as the weld toe failure. The reason may be

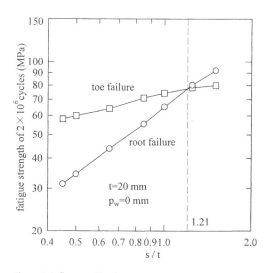

Figure 3. Influence of leg length on critical leg length

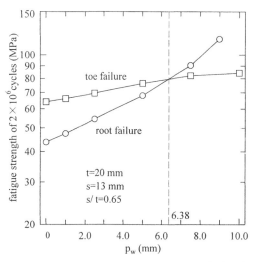

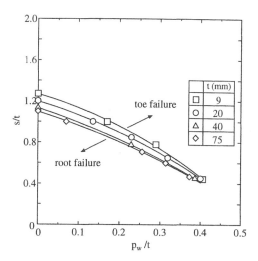

Figure 4. Influence of weld penetration on critical leg length

Figure 6. Relationship between s/t and p_w/t

that the welded joints transmit all external force through the fillet welds. This means that the stress concentration at the weld toes becomes larger with the increase of s/t. In the region where s/t is smaller than 1.21, the fatigue strength of toe failure is higher than that of root failure. On the other hand, in the region where s/t is larger than 1.21, the fatigue strength for toe failure becomes smaller than that for root failure. According to these results, the ratio of critical weld leg length is 1.21 when the plate thickness and weld penetration are 20mm and 0 mm, respectively.

3.2 *Influence of weld penetration (groove depth)*

The relationship between the fatigue strength of 2×10^6 cycles and p_w/t is shown in Fig. 4. Here, the plate thickness t and the leg length s are 20 mm and 13 mm, respectively. p_w increases with the increase of fatigue strength for root failure. For the toe failure, the fatigue strength also increases with the increase in p_w. The reason may be that the stress transmits directly to the main plate to the cross plate when the penetration depth increases. The stress concentration at the weld toe is decrease by

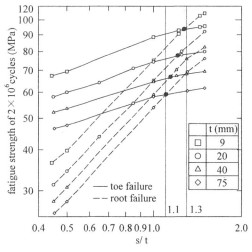

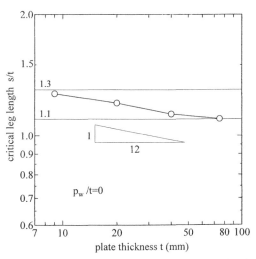

(a) Influence of plate thickness

(b) Relationship between plate thickness and critical leg length

Figure 5. Influence of main plate thickness on critical leg length

stress transmitted to main plate directly. By increasing the p_w from 0 mm to 10 mm (full penetration), the stress concentration factor at the weld toe decreases to approximately 65 % (from 5.46 to 3.59). The fatigue strength for toe failure and root failure is crossed at the point where p_w is 6.38 mm. When p_w is smaller than 6.38 mm, the fatigue strength for root failure is smaller than that for toe failure. On the other hand, when p_w is larger than 6.38 mm, the fatigue strength for root failure is higher than that for the toe failure. In the JSSC Fatigue Design Recommendation, the weld throat of the cruciform joint failing from weld root with the weld penetration is defined as $(s+ p_w)/\sqrt{2}$. This means that only the penetration depth increases the leg length. When p_w is 6.38 mm, the ratio of critical leg length s/t becomes 0.97. It corresponds the s/t decreases to 80 % when there is no penetration (s/t is 1.21).

3.3 Influence of main plate thickness

The influence of main plate thickness t on the ratio of critical leg length to plate thickness s/t is shown in Fig.5. In this figure, the analytical results of the joint models with non-penetration are shown in Fig. 5(a). The ratio of critical leg length s/t decreases with the increase of the plate thickness from 9 mm to 75 mm. This tendency is also similar in the case of the joint with the weld penetration. The relationship between t and the ratio of critical leg length s/t is shown in Fig. 5(b). By increasing the plate thickness t from 9 mm to 75 mm, the ratio of critical leg length is decreased with the inverse proportion to power of 1/12 of plate thickness (from 1.3 to 1.1). However, the effect of the plate thickness on the fatigue failure mode is comparatively small.

4 PROPOSAL OF EVALUATION METHOD FOR FATIGUE FAILURE MODE

The relationship between the ratio of critical leg length s/t and p_w/t is demonstrated in Fig. 6 when s, p_w and t are varied. The regression curve were calculated using the analytical results of the joints with variance from 9 mm to 75 mm. The ratio of critical leg length s/t becomes smaller with the increase in t, as shown in Fig. 5. From the relationship between s/t and p_w/t, the joint seems to fail from the weld toe on the upper side of the regression curve. On the opposite side, which is on the lower side of the curve, the joint seems to fail from the weld root.

5 SUMMARY OF FINDINGS

In this research, parametric fatigue crack propagation analyses were carried out on the seventy kind of load-carrying cruciform joint models of which the leg length, the weld penetration and plate thickness were varied. From these analytical results, the influences of three fundamental factors on the fatigue failure mode were quantitatively clarified. Moreover, the evaluation method of the fatigue failure mode was proposed.

The followings summarize the findings.

(1) The ratio of the critical leg length to the main plate thickness of load-carrying fillet welded cruciform joint without the weld penetration is varied from 1.1 (plate thickness is 75 mm) to 1.3 (plate thickness is 9 mm).

(2) The ratio of the leg length to the main plate thickness decreases with the decrease of the penetration.

(3) Although the ratio of the critical leg length decreases with the inverse proportion to power of 1/12 of plate thickness, the effect of the main plate thickness on the ratio is comparatively small. This reduction effect by plate thickness is approximately 20 %.

(4) The method for evaluating the fatigue failure mode was proposed.

REFERENCES

Albrecht, P. & Yamada, K. 1977. Rapid calculation of stress intensity factors. *Journal of the Structural Division, Proceedings of ASCE*. Vol.103, No.2: 377-389.

Frank, K.H. & Fisher, J.W. 1979. Fatigue strength of fillet welded cruciform joints. *Journal of the Structural Division, Proceedings of ASCE*. Vol.105, No.9: 1727-1740.

Japanese Society of Steel Construction (JSSC). 1995. Fatigue Design Recommendations for Steel Structures. *JSSC Technical Report*.

Kainuma, S., Mori, T. & Ichimiya, M. 1997. Propagation behavior of fatigue crack originated from the weld root of fillet welded cruciform joints. *Steel Construction Engineering. JSSC*. Vol.4, No.14: 1-8 (in Japanese).

Mori, T. & Kainuma, S. 1999. A study on fatigue crack initiation points in load-carrying type of cruciform fillet welded joints. *The Seventh East Asia-Pacific Conference on Structural Engineering & Construction*. Vol.1, 219-224.

Ouchida, A. & Nishioka, A. 1964. Study of fatigue strength of fillet welded joints. *Hitachi Review*, Vol.13, No.2: 3-14.

Soete, W. & Van Crombrugge, R. 1952. A study of the fatigue strength of welded joints. *British Welding Journal*. Vol.31, No.2: 100-103.

27 Structural behavior of bridges

Creative Systems in Structural and Construction Engineering, Singh (ed.) © 2001 Balkema, Rotterdam, ISBN 90 5809 161 9

Improved capacity determination of historic steel truss bridges

C. M. Bowen & M. D. Engelhardt
University of Texas, Austin, Tex., USA

ABSTRACT: There are a significant number of historic steel truss bridges in the United States, and many have inadequate structural capacity according to current load standards. Improving the load rating while maintaining the historic integrity of these bridges poses a major challenge. One portion of a historic steel truss bridge that is frequently problematic from a load rating point of view is the floor system. A number of techniques to strengthen floor systems have been developed and implemented for historic bridge preservation projects. However, as an alternative to strengthening the floor system, it may sometimes be possible to improve the rating through refined structural analysis and field load testing. The objective of this approach is to demonstrate that the in-situ floor system actually has adequate load capacity. This may, in turn, preclude the need for costly strengthening measures that potentially compromise the historic fabric of the bridge. This paper summarizes the results of this approach on a historic steel truss bridge in Llano, Texas.

1 INTRODUCTION

Currently, there are several hundred older metal truss bridges that remain in vehicular service in the state of Texas, many in the range of 70 to over 100 years in age. A number of these bridges are of historical interest due to their age or other unique features. Maintaining a historic truss bridge in service poses significant challenges due to structural or functional deficiencies frequently found in these bridges. Structural deficiencies can often be addressed by structural modifications or strengthening measures. However, such modifications can sometimes be costly and can affect the aesthetics and historical integrity of the bridge.

An alternative approach for addressing inadequate structural capacity of a historic bridge is to apply more refined methods of structural analysis combined with field load testing to develop an improved load rating for the bridge. The use of improved analysis techniques and field load testing can sometimes be effectively used to demonstrate a significantly higher load rating than determined by current standard load rating techniques. This higher load rating may preclude the need for major structural modifications, resulting in cost savings and avoiding disturbance to the historical fabric of the bridge.

This paper presents a brief summary of a case study currently underway for a historic steel truss bridge located in Llano, Texas. As is typical of many older truss bridges, the floor system controlled the load rating for the bridge, and was well below HS 20 based on standard load rating techniques. However, these load-rating techniques include a number of conservative simplifying assumptions. These include, for example, neglecting composite action between the concrete slab and the steel floor beams, neglecting the load carrying contribution of the concrete slab, neglecting rotational restraint at the ends of beams, and the use of conservative load distribution factors. As part of this case study, the bridge floor system was evaluated in greater detail through the use of more refined structural analysis techniques and through the use of field load testing,

Figure 1. Historic bridge in Llano, Texas

in order to determine if an improved load rating could be justified. This paper presents some the key preliminary findings from this study.

2 CASE STUDY

The subject of this study was the Roy Inks Bridge, a historic steel truss bridge located in Llano, Texas. The bridge consists of 4 trusses in line spanning some 800 feet over the Llano River. Built in the 1930's, it is a Parker truss with a slab on steel girder floor system. (See Figure 1).

Analysis using AASHTO load rating methods indicated that the truss was more than adequate for an HS 20 loading with many members having reserve capacities of well over 200% to 300%. Similar analysis also indicated that the floor system was not up to current AASHTO HS 20 standards. As this deficiency is very common in older steel truss bridges, the floor system was the focus of the investigation.

The floor system of this bridge consists of a series of both transverse and longitudinal steel members. The transverse beams are located at the lower panel points of the truss, connected to the truss verticals. The connection between the beam and truss is a pair of riveted clip angles. This connection, coupled with the lack of flexural stiffness of the vertical member, led to the assumption of a simply supported end condition for the beams. Longitudinally there are 6 stringers connected to the webs of the transverse members. Again, because of the angle connection detail between the stringer and beam, no end restraint was assumed. Resting on top of the steel framework is a reinforced concrete slab. There are no mechanical shear connectors between the slab and steel support frame.

In the Llano bridge, the transverse members are WF 33x132's, the longitudinal members are WF 18x50's, and the concrete slab is 6.5" thick with both longitudinal and transverse 5/8" reinforcement bars. The Carnegie Steel Company manufactured both steel members, and dimensions were obtained from the manual, *AISC Iron and Steel Beams 1873 To 1952*. Figure 2 shows one of the four trusses spanning the Llano River, and Figure 3 shows the plan view of the floor system between two of the truss bottom panel points.

3 INITIAL ANALYSIS

The initial structural analysis of the floor system was done using both AASHTO allowable stress design (ASD) and load factor design (LFD) methods. Since the yield stress was not specified on any of the available bridge plans, AASHTO requires assuming

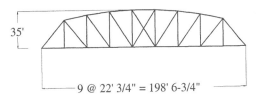

Figure 2. Single span of the Llano Parker truss bridge

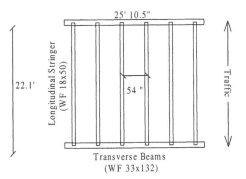

Figure 3. Plan view of a single bay of bridge deck

Figure 4. Section of beam flange removed

a yield stress based on the year the bridge was built. For the Llano bridge, the yield stress was assumed to be 30 ksi. Subsequently, the original mill certificates were located on microfilm. The certificates indicated the average yield strength to be between 37 and 38 ksi.

To confirm the data obtained from the mill certificates, a small section of the first transverse beam flange was removed (see Figure 4) and a standard tensile test performed. These tests confirmed a 37 ksi yield strength. AASHTO load ratings were revised, conservatively using a yield strength of 36 ksi, and the results are summarized in tables 1 and 2.

Results indicate that the longitudinal members are

satisfactory for an HS 20 load rating, but the transverse members are still inadequate. The inventory level is the maximum load allowable for expected normal traffic, the operating level is maximum load allowable for infrequent traffic, and the overload is an LFD criteria that ensures no yielding of the steel members in service loadings.

4 IMPROVED ANALYSIS

To obtain a better assessment of the actual bridge deck capacity, finite element analysis (FEA) was done, using the commercial structural analysis program SAP 2000.

Standard isotropic shell elements were used for the slab. As with the yield stress of the steel, the original drawings did not indicate the concrete compressive strength (f'c). Again, according to AASHTO code, an assumed strength based on the date built must be used. For the Llano bridge, the assumed strength was 2500 psi. Shell elements were 6"x6" for elements between the stringers, and 10"x6" for elements outside the two outermost stringers. For the purpose of having a node at a loading point for a wheel spacing of 6', the center of the model had two 3"x6" rows of shell elements.

Standard beam elements (simply supported at all ends) were used for the transverse and longitudinal members. Figure 5 shows the FEM model used.

Construction details where such that the longitudinal reinforcing steel was not continuous over the transverse members. This characteristic prevented moment and shear from being transferred by the slab across the transverse members. Hence the behavior

Figure 5. Finite element model of 2 bays of the bridge deck

of the entire truss deck could be captured accurately with only a single bay. Two bays were used however, since the eventual field test loading would span two bays. Three and four bay models were constructed, but gave results virtually identical to the two bay model. Therefore, the two bay model was used to simplify the analysis.

To insure no moment or shear would be transferred by the shell elements across the transverse member, a 0.25" gap was inserted in the mesh over the middle transverse beam. Displacement compatibility between the transverse member and the edge of the shell elements was achieved by attaching the shell nodes along the gap to the middle transverse member with axially stiff beam elements. The ends of these small beam elements had moment releases at each end to assure no moment would be transferred from one bay to the next. Alternately, the gap was not inserted, but the shell elements adjacent to the center transverse member where given an infinitesimal modulus of elasticity (E). This approach gave essentially the same results as the gap model.

The model was loaded with the equivalent of HS 20 loading in order to compare to previous hand calculations using AASHTO load ratings.

5 FEA RESULTS

5.1 Longitudinal Members

An AASHTO analysis requires a series of point loads placed in line on each member positioned in a manner to achieve the maximum possible moment. This moment is subsequently multiplied by a distribution factor to take into account the sharing of the moment between all the longitudinal members.

The value and spacing of the point loads depends on the type of load rating one is trying to achieve. For example, for a HS 20 load rating, there are three point loads; an 8 kip point load followed by a 16 kip point load spaced 14 feet away, and another 16 kip load spaced from 14-30 feet. The spacing of the last

Table 1. Load Rating results for longitudinal members

Stringers – $F_y = 30$ ksi			
Level →	Inventory	Operating	Overload
ASD	HS 18.9	HS 20.53	N/A
LFD	HS 20.97	HS 35.01	HS 18.93
Stringers – $F_y = 36$ ksi			
Level →	Inventory	Operating	Overload
ASD	HS 24.91	HS 36.29	N/A
LFD	HS 25.93	HS 43.29	23.48

Table 2. Load Rating results for transverse members

Beams– $F_y = 30$ ksi			
Level →	Inventory	Operating	Overload
ASD	HS 13.13	HS 21.84	N/A
LFD	HS 15.94	HS 26.60	HS 14.31
Beams– $F_y = 36$ ksi			
Level →	Inventory	Operating	Overload
ASD	HS 18.22	HS 27.87	N/A
LFD	HS 20.12	HS 33.58	HS 18.17

load is done in a manner to produce the maximum moment of the member. Since the longitudinal members in this bridge are relatively short (approx. 22'), only a single point load of 16 kips placed at midspan produces the largest moment in the stringers. The distribution factor for this bridge is 0.82, resulting in an unfactored design moment for a load of 16 kips and length of member 22' equal to 72.16 k-ft.

Note the loading is only one wheel line; in other words only the right (or left) side wheels. Essentially it is half the total load of the HS 20 truck.

Since, as mentioned previously, AASHTO does not allow any slab contribution in resisting loads, this moment estimate is conservative. FEA results show the slab taking up to 15% of the total moment. Table 3 summarizes the FEA results and compares them to an AASHTO analysis. (The deck is symmetric; hence only three members are shown. Member one is the outer stringer, with members 2 and 3 the adjacent ones towards the center).

5.2 *Transverse Members*

A similar analysis was done with the transverse members. The important difference in the AASHTO approach is that there is no allowance for lateral distribution of the load in transverse members. Consequently, the wheel loads not directly on top of the member are distributed back to the member proportionately to the distance to the next member as point loads. This conservative method does not reflect the spreading of the load out over the length of the beam by the slab. For example, if a concentrated load is placed at the midpoint between two transverse members, the code approach requires you to assign half of the load on each member as a concentrated load at midspan. In actuality, the slab would spread the load over each member as some type of distributed load along the length. Since a concentrated load will produce a higher response than an equal magnitude distributed load, the AASHTO approach results in a higher moment. For the example given above with a point load centrally located midway between the two transverse members, FEA returns a moment of approximately 68% of the simplified AASHTO approach. Table 4 summarizes the HS 20 loading comparisons for the transverse members.

6 FIELD LOAD TESTING

To confirm the improved capacity obtained by FEA, a load test was conducted on the Llano Bridge. The test consisted of strain gauging the steel floor members, and using a sand loaded dump truck approximately the size and shape of an H 20 truck as a load vehicle. Figure 6 shows the configuration of the load truck.

Table 3. Analysis results for longitudinal members

Member Loaded	Moment (k-ft)	% of AASHTO Moment	Slab Moment (k-ft)	Slab as % of Total Moment
1	31.57	44.08%	13.63	15.49%
2	33.79	47.11%	12.97	14.74%
3	52.95	73.93%	10.31	11.71%

Table 4. Transverse member results

AASHTO Moment	FEA Moment	FEA/AASHTO
216.8 k-ft	192.05 k-ft	88.6%

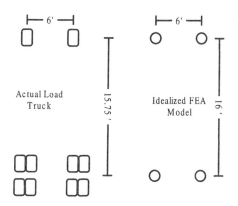

Figure 6. Llano load truck

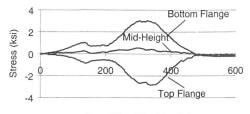

Figure 7. Typical data from field load test for longitudinal member

The front axle of the load truck weighed 10.32 kips (5.33 kips per wheel), and the rear tandem axle weighed 34.22 kips (17.11 kips per side). The trucks were driven over the bridge at approximately 5 mph, slow enough to minimize any dynamic effects. Multiple runs were taken, varying the position of the truck from side to side of the deck. Data was obtained using a portable data acquisition system.

The field test yielded several expected results, as well as some unexpected ones. A typical result is shown in Figure 7. The abscissa indicates the position of the front wheel as it traverses the deck, and the ordinate shows the stress at a specific point on the beam. For the stringers, the plots essentially show back-to-back influence lines. Each rise represents the influence line for one of the axles. The graph for the transverse member is similar, but the distinctive rises are not present since the longitudinal members and the slab gradually transfer the load. As can be seen in the graphs, the top and bottom stresses are essentially equal in the transverse member. Slightly lower stress is seen in the top flange in the stringers. This indicates virtually no composite action in the transverse members and very little in the longitudinal members. Other researchers have found evidence of composite action despite the lack of mechanical connections. Composite action was not observed this field test.

Strain gages near the ends of both the transverse and longitudinal members showed very little stress, indicating the accuracy of the original assumption of no end restraint.

Converting the stresses to moments for the above data, the field load test shows a maximum moment of approximately 109 k-ft in the transverse members and 24 k-ft in the longitudinal members.

The maximum moments occurred as expected, with the rear tandem at midspan for the longitudinal members and when directly on top of the transverse members. Due to the narrowness of the bridge and the large mirrors on the load trucks, the outermost stringers could not be loaded with the wheel line directly on top.

7 COMPARISON OF RESULTS

Tables 5-7 summarizes the maximum moment produced by the field load test truck based on the a standard AASHTO analysis, a finite element analysis, and the field load test. These tables are for the maximum response loading positions, but they reflect the general trend found overall. As mentioned previously, the outermost stringers could not be directly loaded during the field load test, and are not included in the comparisons.

The results indicate that a finite element analysis and field load testing can establish a significant lowering of expected flexural response, as expected. However, the magnitude of the difference was much larger than anticipated.

Figure 9 graphically illustrates the difference in responses of the three methods for the transverse beam.

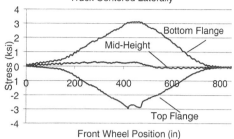

Figure 8. Typical data from field load test for transverse members.

Table 5. Comparison of maximum response for a transverse member

Method	Maximum Moment (k-ft)	Moment as % of AASHTO Moment
AASHTO	184.5	100%
FEA	166.2	90.1%
Load Test	109.1	59.1%

Table 6. Comparison of maximum response for 2nd longitudinal member

Method	Maximum Moment (k-ft)	Moment as % of AASHTO Moment
AASHTO	77.2	100%
FEA	47.8	61.9%
Load Test	23.8	30.8%

Table 7. Comparison of maximum response for 3rd longitudinal member

Method	Maximum Moment (k-ft)	Moment as % of AASHTO Moment
AASHTO	77.2	100%
FEA	45.3	58.7%
Load Test	23.8	30.8%

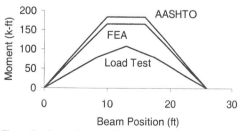

Figure 9. Comparison of the three methods for a transverse beam

The results above clearly show significantly lower predicted moments for the members of the bridge floor system based both on the finite element analysis and on the field load test. However, before these results can be used to justify an increased load rating, a rational explanation is needed for the lower moments. Although this case study is still in progress, a preliminary explanation for the observed results is that the slab (acting non-compositely) is resisting substantial moment and that the slab is providing a more favorable distribution of load to the supporting steel members. As noted earlier, neither composite action between the concrete slab and steel beams, nor rotational restraint at the ends of the steel members appears to be present in this bridge.

Interestingly, the results of this study also show substantial differences between the FEA predictions and the field load test (see Figure 9). This discrepancy is still under investigation. An increase in the transverse stiffness of the concrete slab in the FEA model results in a closer correlation with the field test data. Such an increase in transverse stiffness can be partially justified by the layout of reinforcement in the slab. Longitudinally, the slab is very lightly reinforced, with No. 5 bars spaced at about 18-inches. In the transverse direction, No. 5 bars are placed every 5-inches. As noted earlier, isotropic shell elements were used for the initial finite element analysis. It appears that orthotropic elements provide a better model for the concrete slab.

8 CONCLUSIONS

This paper has described preliminary findings from a structural evaluation of the floor system for a historic steel truss bridge located in Llano, Texas. The objective of this investigation was to determine if the use of advanced structural analysis techniques and field load testing could be used to justify an increased load rating. An improved load rating, in turn, can minimize or preclude the need for strengthening measures, thereby saving cost and avoiding modifications that may affect the historical integrity of the bridge.

Preliminary results of this study indicate that both finite element analysis and field load test data show lower stresses in the bridge floor members than obtained from standard load rating techniques. Thus, it appears that a significantly improved load rating is indeed possible for this bridge without the need for any strengthening measures. Work is still in progress on this investigation to further refine the finite element model for the bridge floor system to achieve closer correlation with the field test data. Laboratory tests are also currently underway on a full-scale replicate of one bay of the floor system of the case study bridge, in order to obtain a better understanding of its behavior. The results of these experiments will be reported in subsequent publications

ACKNOWLEDGEMENTS

The writers gratefully acknowledge funding for this project provided by the Texas Department of Transportation (TxDOT). The assistance of Charles Walker and Steve Landers of TxDOT, and Barbara Stocklin and Steve Sadowsky, formerly of TxDOT, is greatly appreciated. The assistance of Dilip Maniar, Matt Haberling and Prof. Dan Leary of the University of Texas at Austin is also acknowledged. The writers also thank Mr. Abba Lichtenstein for advice provided on this project. Additionally, the writers wish to thank project members Dr. Karl Frank and Dr. Joe Yura for their aid.

REFERENCES

Chajes, M.J., Mertz, D.R. & Commander, B. 1997. Experimental Load Rating of A Posted Bridge. *Journal of Bridge Engineering,* ASCE Vol. 2, No. 1, 1-4.

Saraf,, S. & Nowak, A.S. 1998. Proof Load Testing of Deteriorated Steel Girder Bridges. *Journal of Bridge Engineering,* ASCE Vol. 3, No. 2, 82-89.

AASHTO Manual for Condition Evaluation of Bridges, 1992. American Association of State Highway and Transportation Officials, Washington, D.C.

AISC Iron and Steel Beams, 1873 to 1952, 1953. American Institute of Steel Construction, Chicago, Il.

Creative Systems in Structural and Construction Engineering, Singh (ed.) © 2001 Balkema, Rotterdam, ISBN 90 5809 161 9

Bridges emerging from the structural cooperation between straight and curved superstructures

F. Biondini
Technical University of Milan, Italy

F. Bontempi
University of Rome 'La Sapienza', Italy

P.G. Malerba
University of Udine, Italy

ABSTRACT: The paper deals with the optimization of the structural morphology of plane framed structures subjected to a given set of loading conditions. The optimal structural morphology is found through a two-level approach by using optimality criteria and soft-computing techniques. In particular, the *external* structural morphology, i.e. the geometrical dimensions and the topology of the structural type, is optimized at the first level (macro-level) by means of genetic algorithms, while the internal morphology, like the geometry and the shape of the cross-sections, is selected at the second level (micro-level) through a fully stressed design. In the application, specific attention is paid to a framed bridge and the structural morphology which leads to the optimal cooperation between the straight deck and the curved supporting superstructure is searched for.

1 INTRODUCTION

Among the catalyst elements of the actual global growth of economy, is the creation of new communication thoroughfares and a radical upgrading of the actual infrastructural network. In the next decade we can foresee not only the steadiness of this development, but also its acceleration, especially in those countries which are presenting the highest increase in their economy.

In this case there will be a strong demand to renew the concepts at basis of the bridge design. As an objective, such an innovation must be able not only to reach a suitable carrying capacity, but also to have optimal characteristics in terms of stiffness, of ductility and of shape. Therefore it must be understood how important it is to give a unitary formulation to the research of shapes and proportions efficient as regards the static and dynamic (seismic) behaviour, durable and recyclable as regards the economy and environment and in harmony with aesthetics.

These results can be reached through a fruitful collaboration among structural and environmental engineers and architects. The forerunners attempted conceptual bridge design in their own ways and drew their ideas from natural principles, intuition, logic, comparative studies, productivity, practicality and so on.

At present, with the progress of computer sciences, of structural analysis and of optimization techniques, the structural engineer possesses extensive and powerful tools to contribute to a systematic and rational approach to conceptual design and, by combining high-tech with creativity, he is to be responsible for the overall quality of his work.

The developments presented in this paper are devoted to the problem of finding the optimal morphology of plane framed structures, but can be usefully applied also to other kind of structures. A set of loading conditions, the boundary constraints and the feasible region containing the structure are assumed to be given, while the geometry and the topology of the structure, i.e. the resistant scheme, as well as the size of its elements must be determined in an optimal way.

The optimal structural design can be considered as an iterative process where each iteration consists of two steps: (1) An analysis step, which verifies the structure associated to the current design. (2) A redesign step, where the design variables are updated to improve some target requirement and to satisfy the design constraints.

In this work, such problem is solved through a two-level approach by using optimality criteria and soft-computing techniques (Biondini *et al.* 2000). In particular, the *external* structural morphology, i.e. the geometrical dimensions and the topology of the structural type, is optimized at the first level (macro-level) by means of genetic algorithms (Michalewicz 1992), while the internal morphology, regarding the geometry and the shape of the cross-sections, is selected at the second level (micro-level) through the well known fully stressed design (Gallagher & Zienkiewicz 1973).

A final application is devoted to the optimal design of a framed bridge, where the structural morphology which leads to the optimal cooperation between the deck and the superstructure is searched for.

2 FORMULATION OF THE OPTIMIZATION PROBLEM

Let $\mathbf{x}$ and $\mathbf{a}$ two vectors of design variables which lead to define, respectively, the *external* morphology of the structure, like the geometrical dimensions and the topology of the structural type, and its *internal* morphology, like the geometry and the shape of the cross-sections. By assuming linear elastic behavior, the equilibrium equations between the vectors of the applied loads $\mathbf{f}$ and of the nodal displacements $\mathbf{q}$ can be written as follows:

$$\mathbf{f} = \mathbf{K}(\mathbf{x},\mathbf{a})\mathbf{q} \quad \Rightarrow \quad \mathbf{q} = \mathbf{K}(\mathbf{x},\mathbf{a})^{-1}\mathbf{f} \qquad (1)$$

where the stiffness matrix of the structure $\mathbf{K}=\mathbf{K}(\mathbf{x},\mathbf{a})$ is clearly depending on the unknown morphology. Based on the displacement field $\mathbf{q}=\mathbf{q}(\mathbf{x},\mathbf{a})$, the internal stress state can be also derived. In particular, let $\mathbf{s}_{top}=\mathbf{s}_{top}(\mathbf{x},\mathbf{a})$ and $\mathbf{s}_{bot}=\mathbf{s}_{bot}(\mathbf{x},\mathbf{a})$ the vectors of the normal stresses at the top and at the bottom of the more critical sections over the whole structure. By assuming for sake of simplicity that shear failures and buckling phenomena as well as excessive displacements are avoided, safe conditions can be assured simply by limiting such stresses within assigned bounds $\mathbf{s}^-$ and $\mathbf{s}^+$ under the considered load conditions.

On the basis of the previous considerations, the objective of the design process is to find the vectors $\mathbf{x}$ and $\mathbf{a}$ in such a way that the structural volume $V=V(\mathbf{x},\mathbf{a})$ is minimum, according to either side constraints with bounds $\mathbf{x}^-$, $\mathbf{x}^+$, and $\mathbf{a}^-$, $\mathbf{a}^+$, or inequality behavioral constraints on the stress state $\mathbf{s}_{top}(\mathbf{x},\mathbf{a})$ and $\mathbf{s}_{bot}(\mathbf{x},\mathbf{a})$:

$$\min \left\{ V(\mathbf{x},\mathbf{a}) \left| \begin{array}{c} \mathbf{s}^- \leq \mathbf{s}_{top}(\mathbf{x},\mathbf{a}) \leq \mathbf{s}^+, \ \mathbf{s}^- \leq \mathbf{s}_{bot}(\mathbf{x},\mathbf{a}) \leq \mathbf{s}^+ \\ \mathbf{x}^- \leq \mathbf{x} \leq \mathbf{x}^+, \ \mathbf{a}^- \leq \mathbf{a} \leq \mathbf{a}^+ \end{array} \right. \right\} \quad (2)$$

From the mathematical point of view, such formulation leads to a nonlinear programming problem (Rao 1996) which, due especially to the ill conditioning introduced by the different nature of the design variables $\mathbf{x}$ and $\mathbf{a}$, usually results hard to solve, at least for the absolute optima. In this work, as already mentioned, instead of the mathematical programming techniques, a two-level approach which accounts for the actual physical meaning of the structural problem is considered. The external morphology $\mathbf{x}$ is optimized at the first level (macro-level) in a direct way by using genetic algorithms, while the internal morphology $\mathbf{a}$ is selected at the second level (micro-level) in an indirect way by means of a fully stressed design.

3 MICRO-OPTIMIZATION OF THE INTERNAL MORPHOLOGY (FULLY STRESSED DESIGN)

In the following, the vector $\mathbf{a}$ is selected at the micro-level through a fully stressed approach by assuming the vectors $\mathbf{s}_{top}(\mathbf{x},\mathbf{a})$ and $\mathbf{s}_{bot}(\mathbf{x},\mathbf{a})$ as known (Rozvany 1996). Since stress criteria are based on the assumption that

each element of the optimal structure is subjected to its allowable stresses under at least one load condition, an optimal vector $\mathbf{a}$ can be found on the frontier of the following feasible domain D, namely $\mathbf{a} \in \mathrm{Fr}(D)$:

$$D = \left\{ \mathbf{s}^- \leq \mathbf{s}_{top}(\mathbf{a}) \leq \mathbf{s}^+, \mathbf{s}^- \leq \mathbf{s}_{bot}(\mathbf{a}) \leq \mathbf{s}^+, \mathbf{a}^- \leq \mathbf{a} \leq \mathbf{a}^+ \right\} \ (3)$$

This concept is shown with reference to a single cross-section which is assumed to expand or contract homotetically according to a factor $\alpha \in [\alpha_{min}; \alpha_{max}]$. Thus, both the area $A=A(\alpha)$ and the inertia modulus $W=W(\alpha)$ of the actual section are related to those of the original one A_0 and W_0 ($\alpha=1$) by the following scaling laws, here verified for the simple case of the rectangular shape shown in Fig. 1:

$$A(\alpha) = BH = \alpha^2 B_0 H_0 = \alpha^2 A_0 \qquad (4)$$

$$W(\alpha) = \tfrac{1}{6} BH^2 = \alpha^3 \tfrac{1}{6} B_0 H_0^2 = \alpha^3 W_0 \qquad (5)$$

By denoting with N and M the axial force and the bending moment acting on the section, respectively, the normal stresses at the top and at the bottom are deduced through the plane section hypothesis (Fig. 1):

$$\sigma_{top}(\alpha) = \frac{N}{A(\alpha)} - \frac{M}{W(\alpha)} = \frac{N}{\alpha^2 A_0} - \frac{M}{\alpha^3 W_0} = \frac{n_0}{\alpha^2} - \frac{m_0}{\alpha^3} \ (6)$$

$$\sigma_{bot}(\alpha) = \frac{N}{A(\alpha)} + \frac{M}{W(\alpha)} = \frac{N}{\alpha^2 A_0} + \frac{M}{\alpha^3 W_0} = \frac{n_0}{\alpha^2} + \frac{m_0}{\alpha^3} \ (7)$$

where $n_0=N/A_0$ and $m_0=M/W_0$. Such stresses must verify simultaneously the following conditions:

$$\sigma^- \leq \sigma_{top}(\alpha) = \frac{n_0}{\alpha^2} - \frac{m_0}{\alpha^3} \leq \sigma^+ \qquad (8)$$

$$\sigma^- \leq \sigma_{bot}(\alpha) = \frac{n_0}{\alpha^2} + \frac{m_0}{\alpha^3} \leq \sigma^+ \qquad (9)$$

which, at the limit, lead to four equation of third degree in the unknown α:

$$\sigma^-\alpha^3 - n_0\alpha + m_0 = 0 \qquad \sigma^-\alpha^3 - n_0\alpha - m_0 = 0 \quad (10)$$

$$\sigma^+\alpha^3 - n_0\alpha + m_0 = 0 \qquad \sigma^+\alpha^3 - n_0\alpha - m_0 = 0 \quad (11)$$

each of them having only one real solution, say α_i. A feasible solution is then obtained by choosing the larger section, namely $\alpha = \max \alpha_i$ ($i=1,\dots,4$).

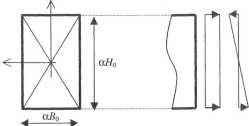

Figure 1. α-Homotetic variation of a rectangular-shaped section and typical normal stress diagrams.

4 MACRO-OPTIMIZATION OF THE EXTERNAL MORPHOLOGY (GENETIC ALGORITHMS)

At the macro-level, the vector **a** is assumed as already defined by using the previous procedure. At this stage, the optimization problem depends on the design variables **x** alone and can be restated as follows:

$$\min \left\{ V(\mathbf{x}) \mid \mathbf{x}^- \leq \mathbf{x} \leq \mathbf{x}^+ \right\} \qquad (12)$$

The problem so formulated is usefully processed by using genetic algorithms, which are heuristic search techniques which belong to the class of stochastic algorithms, since they combine elements of deterministic and probabilistic search. Despite their heuristic nature, genetic algorithms operate according to strong theoretical foundations which assure a high level of performance (Holland 1975).

The search strategy works on a *population* of *individuals* subjected to an evolutionary process where individuals compete between them to survive in proportion to their *fitness* with the *environment*. In this process, population undergoes continuous reproduction by means of some *genetic operators* (basically *selection*, *crossover* and *mutation*) which, because of competition, tend to preserve best individuals. From this evolutionary mechanism, two conflicting trends appear: exploiting of the best individuals and exploring the environment. Thus, the effectiveness of the search depends on a balance between them, or between two properties of the system, *population diversity* and *selective pressure*. These aspects are in fact strongly related, since an increase in the selective pressure decreases the diversity of the population, and vice versa.

With reference to the optimization problem previously formulated, a population of m individuals represents a collection $X = \{ \mathbf{x}_1 \ \mathbf{x}_2 \ ... \ \mathbf{x}_m \}$ of m possible feasible solutions $\mathbf{x}_k^T = [x_1^k \ x_2^k \ ... \ x_n^k]$, each defined by a set of n design variables x_i^k ($k = 1, ..., m$). The fitness of each individual **x** is measured by a non negative scalar function $F(\mathbf{x})$ which increases with the adaptability of **x** to its environment. In this work, the fitness function to be maximized is related to the volume of the structure as follows:

$$F(\mathbf{x}) = C - V(\mathbf{x}) \qquad (13)$$

being C a suitable constant able to assure $F(\mathbf{x}) \geq 0$. However, for an appropriate hierarchical arrangement of the individuals, such function should be properly scaled. Details about scaling rules, internal coding of the population, genetic operators and termination criteria, can be found in a previous paper (Biondini 1999).

5 APPLICATION TO BRIDGES

A bridge having an overall length $L = 150$ m is considered. As shown in Figure 2, the straight deck is carried by a framed structure supported at both the ends and at two internal points. The material is linear elastic with Young modulus $E = 28$ GPa, weight density $\gamma = 25$ kN/m^3 and stress limits $\sigma^+ = \sigma^- = 150$ MPa. The bridge is subjected to five load conditions defined by the self weight and by a uniform load $q = 2$ kN/m distributed in each load case as shown in Figure 2.

The optimal structural scheme of the bridge is searched for. To this aim, the free edge of the structure is described by a natural cubic spline over six differently spaced intervals. Moreover, the eight different structural typology shown in Figure 3 are considered. The design variables **x** which define the external morphology are then represented by the 6 independent parameters which define either the symmetric profile of the structure or the location of the internal supports, and by an integer parameter which identifies the typology, leading to a total of 7 external design variables. Design constraints are introduced in order to limit the profile within two vertical bounds located at ±30 m from the deck. As regards the internal morphology, the cross-sections vary homotetically according to a factor α from the initial conditions $A_0 = 3.5$ m^2 and $W_0 = A_0/48$ m^3. The design variables **a** which define the internal morphology are then identified by such α–factors and their number, according to the structural symmetry, is varying between 37 and 73 depending on the structural typology.

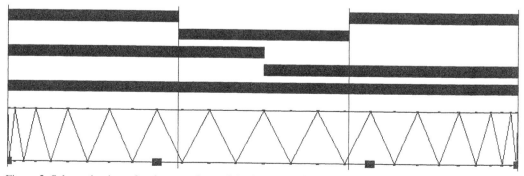

Figure 2. Schematic view of a given typology of the framed bridge and of the load conditions.

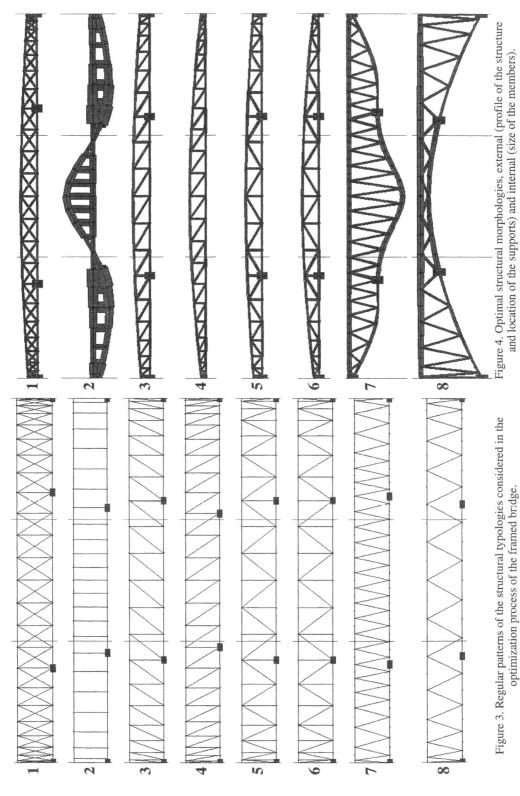

Figure 4. Optimal structural morphologies, external (profile of the structure and location of the supports) and internal (size of the members).

Figure 3. Regular patterns of the structural typologies considered in the optimization process of the framed bridge.

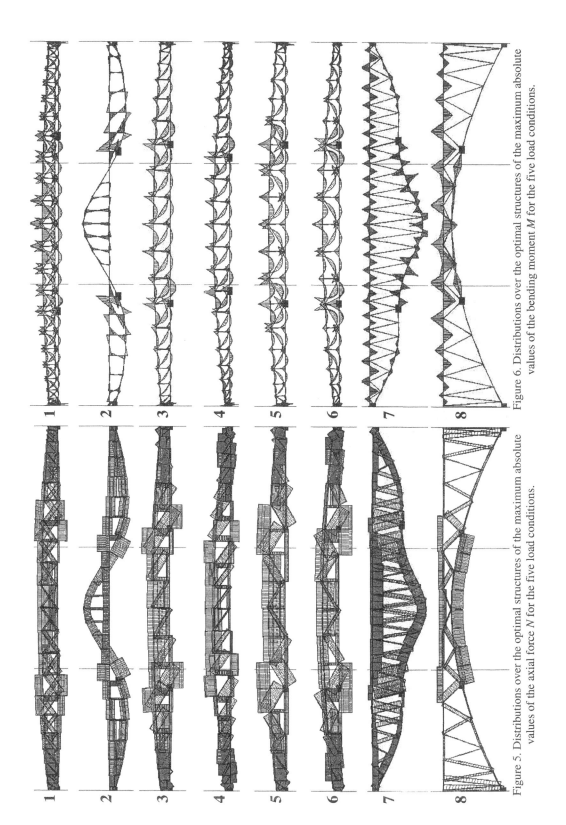

Figure 6. Distributions over the optimal structures of the maximum absolute values of the bending moment M for the five load conditions.

Figure 5. Distributions over the optimal structures of the maximum absolute values of the axial force N for the five load conditions.

Table 1. Comparison between total volumes, maximum displacements and maximum internal forces for the optimal solutions (*italic* style for the worst solutions and **bold** style for the best ones).

Structural Typology	1	2	3	4	5	6	7	8
Max Displacement [cm]	6.8	*15.8*	6.2	7.9	5.7	6.6	**4.8**	14.7
Length/Max Displacement Ratio	2221	*954*	2442	1896	2642	2313	**3188**	1036
Max Axial Force [kN]	**12**	*16480*	28	23	20	24	986	2621
Max Shear Force [kN]	25	*7859*	23	50	25	**13**	582	713
Max Bending Moment [kNm]	8642	13380	7573	10330	7295	8332	7471	*14430*
Volume [m³]	31	*293*	**26**	27	**26**	**26**	77	114
Number of Realizations N	660	**387**	576	600	*1070*	892	474	393
Mean Volume $\Sigma V_k/N$ ($k=1\div N$) [m³]	142	*1282*	189	**136**	158	152	667	784

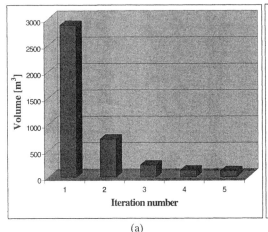 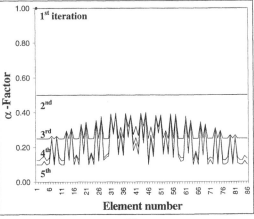

(a)   (b)

Figure 7. Fully stressed design (micro-optimization) of a bridge genetically selected (macro-optimization). Evolution of (a) the total volume of the structure and of (b) the sectional α-factors ($|\Delta\alpha| \leq 0.50\alpha$).

Figure 4 shows the morphology of the optimal solution, external (profile of the framed structure and location of the supports) and internal (size of the members), for each structural typology. A comparison between such solutions in terms of the total volumes, the maximum displacements and the maximum internal forces, is made in Table 1. Figures 5 and 6 make a further comparison between the optimal distributions of the maximum absolute values over the five load conditions of both the normal force and the bending moment, respectively.

Finally, with reference to the basic steps of the fully stressed design (micro-optimization), performed for each bridge selected by the genetic algorithm (macro-optimization), Figure 7 shows a typical evolution of the total volume of the structure and of the cross-sectional α–factors of the members. To improve the numerical process, a maximum variation $\Delta\alpha_{max}=\pm0.50\alpha$ has been allowed at each iteration.

REFERENCES

Biondini F. 1999. Optimal Limit State Design of Concrete Structures using Genetic Algorithms. *Studi e Ricerche*, Scuola di Specializzazione in Costruzioni in Cemento Armato, Politecnico di Milano, **20**, 1-30.

Biondini F., F. Bontempi, P.G. Malerba 2000. The Search for Structural Schemes by Optimality Criteria and Soft-Computing Techniques. *Proceedings of Structural Morphology Conference*, Delft, The Netherlands, August 17–19.

Gallagher R. H., Zienkiewicz O. C. (Eds.) 1973. *Optimum Structural Design*. John Wiley & Sons.

Holland J.H. 1975. *Adaptation in Natural and Artificial Systems*. University of Michigan Press.

Michalewicz Z. 1992. *Genetic Algorithms + Data Structures = Evolution Programs*. Springer.

Rao S.S. 1996. *Engineering Optimization. Theory and Practice*. John Wiley & Sons.

Rozvany G.I.N. 1989. *Structural Optimization via Optimality Criteria*. Kluwer Academic Press.

Creative Systems in Structural and Construction Engineering, Singh (ed.) © 2001 Balkema, Rotterdam, ISBN 90 5809 161 9

Stability of composite bridge girders

H. R. Ronagh

Department of Civil Engineering, University of Queensland, Brisbane, Qld, Australia

M. A. Bradford

School of Civil and Environmental Engineering, University of New South Wales, Sydney, N.S.W., Australia

ABSTRACT: A simple method for the distortional buckling analysis of composite bridge girders is revisited here in a new form. The approach is based on the analogy of beam on elastic foundation. In this analogy, bottom flange of the steel joist is simulated as a beam that rests on the elastic foundation of the web and is subjected to an axial force with a distribution that can be determined from the bending moment diagram of the girder. Here in order to model the problem mathematically, a simple finite element with two nodes and three degrees of freedom at each node is developed using the Energy approach. The method is then used in order to provide design curves for a range of loading patterns. Graphs comparing the method to other techniques are also provided. Because of simplicity and clarity of the model, it may be found useful in real practice.

1 INTRODUCTION

Usual composite steel/concrete bridge girders are composed of a concrete slab and a steel I-beam (or joist) that are connected together by means of stud shear connectors. The shear connectors may be shot-fired or welded on the top flange of the steel joist and are embedded in the concrete. Trying to utilise the maximum section capacity, while being mindful of economy, often results in a web which is very slender, and the resulting composite girder is classified as "slender" (Oehlers & Bradford, 1995). Web stiffeners may then be used along the length in order to stiffen the section against local buckling or vertically to eliminate buckling in shear. Slenderness of the web of the joist, however, may make the girder susceptible to another type of instability, which can be referred to as "Restricted Distortional Buckling (RDB)".

The RDB phenomenon is opposed to the usual lateral-torsional buckling in which the web maintains its original un-bent shape during buckling, while both flanges twist by the same angle. It is also different from the usual distortional instabilities in which both top and bottom flanges are free to deflect laterally and twist while the web bends out-of-plane as a plate. The I-shaped steel joist has a top flange that is fully embedded in, or restricted from movement, by the concrete deck. The bottom flange, on the other hand, is unrestricted and can deflect laterally out-of-plane and twist. The only restriction to this movement is provided by the flexural stiffness of the web. In the negative bending moment region, the bottom flange is subjected to significant compressive stresses, and thus it may become unstable, twist and deflect out-of-plane while pulling the web as it buckles. The result is a web with a distorted shape that is bent about its minor axis. A cross-section that undergoes RDB is shown in Figure 1.

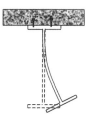

Figure 1. Buckled Cross-Section

Unlike the lateral torsional buckling which has been scrutinized prolifically for most of the last century, or the usual distortional buckling which has been well-researched internationally, the RDB phenomenon has been studied more recently by only few researchers.

Most of the studies in this area are based on the similarity of the problem to that of a beam resting on an elastic foundation. Svensson (1985) was the first who proposed this. He presented a simple method for the instability analysis of simply supported, one and two span composite girders. Svensson modeled the free bottom flange of the steel joist as a column that is subjected to a varying axial force $N(x)$, and is elastically supported by a Winkler foundation having a constant modulus k as shown in Figure 2.

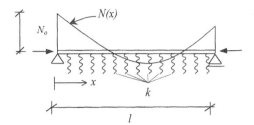

Figure 2. Svensson's Model

This idealised situation occurs in the bottom flange when the composite girder is in negative bending, where the effect of compressive stresses on the bottom flange can be represented by a compressive axial force. The Winkler foundation, then, simulates the distributed resistance provided by the web against the out-of-plane deflection of the bottom flange.

In order to calculate the stiffness of the Winkler foundation, Svensson chose to simplify the problem. He considered a unit longitudinal portion of the web as a cantilever fully restrained at the top flange, and ignored the torsional resistance of the web. He then calculated k as

$$k = \frac{E t_w^3}{4(1-v^2)h^3} \tag{1}$$

in which E is the Young's modulus, t_w and h are the thickness and height of the web respectively, and v is the Poisson's ratio introduced to include plate behavior. The Winkler strut problem can then be represented mathematically by the following differential equation.

$$EI\frac{d^4w}{dx^4} + \frac{d}{dx}\left(N(x)\frac{dw}{dx}\right) + kw = 0 \tag{2}$$

in which w is the lateral deflection, and I is the corresponding second moment of area of the column (or in reality, the bottom flange of the joist).

Introducing the non-dimensional quantities

$$\xi = x/l \tag{3}$$

$$\beta l = \left(\frac{k}{EI}\right)^{1/4} l = \left(\frac{kl^4}{EI}\right)^{1/4} \tag{4}$$

$$n(\xi) = N/N_o \tag{5}$$

$$\lambda = N_o/N_E, \tag{6}$$

where N_E is the Euler buckling load $(= \pi^2 EI/l^2)$, Equation (2) can be re-written as

$$EI\frac{d^4w}{d\xi^4} + \pi^2\lambda \frac{d}{d\xi}\left(n(\xi)\frac{dw}{d\xi}\right) + (\beta l)^4 w = 0 \tag{7}$$

This clearly shows that the critical load multiplier, λ, depends only on the axial force shape function $n(\xi)$, and on βl. Non-dimensional design curves for different shape functions could therefore be obtained.

Svensson (1985) considered nine axial load patterns as shown in Figure 3. He applied these load patterns to Equation (7), and solved the problem considering zero lateral deflection at the boundaries (i.e. simply supported ends). In order to solve the problem, Svensson made use of a general numerical procedure based on the Galerkin technique. He then plotted the buckling loads in a convenient non-dimensional format.

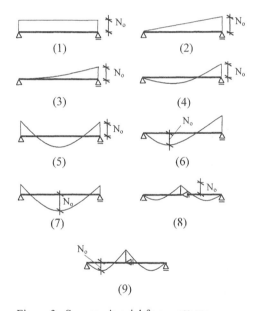

Figure 3. Svensson's axial force patterns

The advantage of Svensson's model is simplicity, as it is both easy to understand and to implement. The method, however ignores some parameters that may affect the results. For example, it ignores the Saint Venant torsional resistance of the web and the bottom flange, and also ignores the participation of the web as part of the buckling column. In order to correct the latter, Williams and Jemah (1987), suggest that one should include 15% of the web in the buckling column in addition to the bottom flange.

Several other publications can be found in the literature which are improved versions of this work.

These are the works of Goltermann & Svensson (1990), and Williams et al (1993). One study which views the problem from a different angle is the study of Bradford & Gao (1992).

As mentioned previously, Svensson's concept has the significant advantage of simplicity. Svensson's solution of the concept, however, is not as simple and one generally has to rely on the provided curves in order to obtain the solution or to set up a new system as soon as the boundary conditions are not simply supported or the pattern for the axial forces are different from those provided. For this reason, here an alternative solution method is presented. This method uses the same assumptions as Svensson but is based on Finite elements. It also includes the torsional rigidity of the bottom flange. The method is described in the following.

2 CURRENT METHOD

The current method is based on finite elements and Svensson's analogy. Finite elements here are special two noded elements with three degrees of freedom at each node, as shown in Figure 4. These freedoms are, transverse displacement v and it's derivative with respect to x, ϕ_z, and the twist of the element ϕ_x. The element models the bottom flange of the steel joist that is resting on the elastic foundation provided by the web.

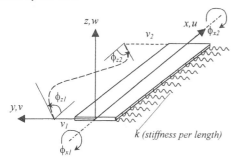

Figure 4. Current Finite Element

Transverse deflection is assumed to vary cubically while the twist shape function is assumed to be linear. In addition, variation of the axial force along the length is chosen to be linear, so that with N_1 and N_2 being the end values for the axial force, axial force along the length is written as:

$$N(x) = N_1 (1 - \frac{x}{l}) + N_2 (\frac{x}{l}) \tag{8}$$

in which l is the length of the element.

In order to calculate the required stiffness and stability matrices, an energy approach is taken. The energy stored in the system during a buckling deformation can be written as:

$$U = U_{Cf} + U_{Ct} + U_S \tag{9}$$

where U_{Cf}, is the energy stored due to the column's flexural deformation, U_{Ct}, is the energy associated with the twist of the column, and U_S, is the energy stored in the springs.

These energy terms can be calculated as:

$$U_{Cf} = \frac{1}{2} \int_0^l E I_z \left(\frac{d^2 v}{dx^2} \right)^2 dx \tag{10}$$

$$U_{Ct} = \frac{1}{2} \int_0^l G J \left(\frac{d\phi_x}{dx} \right)^2 dx \tag{11}$$

$$U_S = \frac{1}{2} \int_0^l k \, v^2 \, dx \tag{12}$$

where I_z and J are the second moment of area of the bottom flange about the z axis and it's torsional rigidity respectively and G is the shear modulus.

Assuming a cubic variation for v, and a linear variation for ϕ_x in Equations (10-12), as

$$v = a_1 + a_2 x + a_3 x^2 + a_4 x^3 \tag{13}$$

$$\phi_x = a_5 + a_6 x \tag{14}$$

and substituting in Equation (9), gives U as:

$$U = \frac{1}{2} \underline{a}^T \underline{K}_a \, \underline{a} \tag{15}$$

where $\underline{a}$ is the column vector containing the coefficients a_1 to a_6, and can be written as:

$$\underline{a} = \{ a_1 \ a_2 \ a_3 \ a_4 \ a_5 \ a_6 \}^T \tag{16}$$

The stiffness matrix $\underline{K}_a$ is not of much use, as it is not assemblable. In order to make it useable, a relation matrix between these coordinates and the co-ordinate system which contains the end freedoms, $\underline{q}$, must be established. By looking at the boundaries of the column, this relation matrix, can be easily written as:

$$\underline{q} = \begin{Bmatrix} v_1 \\ \phi_{z1} \\ \phi_{x1} \\ v_2 \\ \phi_{z2} \\ \phi_{x2} \end{Bmatrix} = \begin{bmatrix} 1 & 0 & 0 & 0 & 0 & 0 \\ 0 & 1 & 0 & 0 & 0 & 0 \\ 0 & 0 & 0 & 0 & 1 & 0 \\ 1 & l & l^2 & l^3 & 0 & 0 \\ 0 & 1 & 2l & 3l^2 & 0 & 0 \\ 0 & 0 & 0 & 0 & 1 & l \end{bmatrix} \underline{a} \tag{17}$$

or

$$\underline{q} = \underline{C} \, \underline{a} . \tag{18}$$

Substituting this, the energy Equation can be rewritten in the form

$$U = \frac{1}{2} \underline{q}^T \left(\underline{C}^{-T} \underline{K}_a \underline{C}^{-1} \right) \underline{q} \tag{19}$$

or as

$$U = \frac{1}{2} \underline{q}^T \underline{K}_q \underline{q} \tag{20}$$

where $\underline{K}_q$ is the stiffness matrix in terms of the assemblable coordinates $\underline{q}$. Performing the calculations, $\underline{K}_q$ would be found as:

$$\underline{K}_q = \begin{bmatrix} \underline{K}_{11} & \underline{K}_{12} \\ \underline{K}_{21} & \underline{K}_{22} \end{bmatrix} \tag{21}$$

where

$$\underline{K}_{11} = \begin{bmatrix} \dfrac{13kl}{35} + \dfrac{12EI_z}{l^3} & \dfrac{11kl^2}{210} + \dfrac{6EI_z}{l^2} & 0 \\ \dfrac{11kl^2}{210} + \dfrac{6EI_z}{l^2} & \dfrac{kl^3}{105} + \dfrac{4EI_z}{l} & 0 \\ 0 & 0 & \dfrac{GJ}{l} \end{bmatrix} \tag{22}$$

$$\underline{K}_{22} = \begin{bmatrix} \dfrac{13kl}{35} + \dfrac{12EI_z}{l^3} & -\dfrac{11kl^2}{210} - \dfrac{6EI_z}{l^2} & 0 \\ -\dfrac{11kl^2}{210} - \dfrac{6EI_z}{l^2} & \dfrac{kl^3}{105} + \dfrac{4EI_z}{l} & 0 \\ 0 & 0 & \dfrac{GJ}{l} \end{bmatrix} \tag{23}$$

$$\underline{K}_{12} = \begin{bmatrix} \dfrac{9kl}{70} - \dfrac{12EI_z}{l^3} & \dfrac{-13kl^2}{420} + \dfrac{6EI_z}{l^2} & 0 \\ \dfrac{13kl^2}{420} - \dfrac{6EI_z}{l^2} & \dfrac{-kl^3}{140} + \dfrac{2EI_z}{l} & 0 \\ 0 & 0 & -\dfrac{GJ}{l} \end{bmatrix} \tag{24}$$

$$\underline{K}_{21} = \underline{K}_{12} \tag{25}$$

Similarly, the stability matrix can be calculated from the loss of the potential energy of the stresses. This loss can be written as

$$W = \int_0^l \int_A \sigma(x)\, dA (ds - dx) \tag{26}$$

where $\sigma(x)$ is the stress at x, A refers to the cross-sectional area of the column (bottom flange) and ds is the differential length of the deformed column.

In the usual manner, $(ds - dx)$ can be approximated by:

$$ds = \frac{w'^2 + v'^2}{2}\, dx \tag{27}$$

where primes denote first order derivatives with respect to x, and

$$\sigma(x) = \frac{N(x)}{A} \tag{28}$$

Replacing these in Equation (26), and noticing that

$$w = -\phi_x\, y \tag{29}$$

gives

$$W = \frac{1}{2} \int_0^l N(x) v'^2\, dx + \frac{I_z}{2A} \int_0^l N(x)(\phi'_x)^2 dx \tag{30}$$

Using the shape functions described before in Equations (13) and (14), and substituting in Equation (30), then transforming the result into the assemblable coordinate system as described previously gives:

$$W = \frac{1}{2} \underline{q}^T \underline{S}_q \underline{q} \tag{31}$$

in which $\underline{S}_q$ is the stability matrix calculated as:

$$\underline{S}_q = \begin{bmatrix} \underline{S}_{11} & \underline{S}_{12} \\ \underline{S}_{21} & \underline{S}_{22} \end{bmatrix} \tag{32}$$

where

$$\underline{S}_{11} = \begin{bmatrix} \dfrac{3(N_1 + N_2)}{5l} & \dfrac{N_2}{10} & 0 \\ \dfrac{N_2}{10} & \dfrac{(3N_1 + N_2)l}{30} & 0 \\ 0 & 0 & \dfrac{(N_1 + N_2)I_z}{4Al} \end{bmatrix} \tag{33}$$

$$\underline{S}_{12} = \begin{bmatrix} \dfrac{-3(N_1 + N_2)}{5l} & \dfrac{N_1}{10} & 0 \\ \dfrac{-N_2}{10} & \dfrac{-(N_1 + N_2)l}{60} & 0 \\ 0 & 0 & \dfrac{-(N_1 + N_2)I_z}{4Al} \end{bmatrix} \tag{34}$$

$$\underline{S}_{22} = \begin{bmatrix} \dfrac{3(N_1 + N_2)}{5l} & \dfrac{-N_1}{10} & 0 \\ \dfrac{-N_1}{10} & \dfrac{(N_1 + 3N_2)l}{30} & 0 \\ 0 & 0 & \dfrac{(N_1 + N_2)I_z}{4Al} \end{bmatrix} \tag{35}$$

These matrices can be used in order to study a variety of beams and boundary conditions. The advantage of the current method over Svensson's method is that it allows study of different boundary conditions in an easy way. Svensson's results are limited to simply supported boundary conditions and a new set-up is required for other boundaries. Here, changing a boundary condition, is a matter of restraining or not restraining a freedom only. The new method includes the twist of the bottom flange as well. It can therefore pick-up torsional buckling modes, if any, and stiffens the system a little bit closer to reality. Williams and Jemah's suggestion regarding the inclusion of 15% of the web area in the buckling column can also be accommodated very easily. Some of the results of the study are presented in the following.

3 RESULTS

In order to find the rate of convergence of the method, an analysis was performed for a simply supported case with loading types 1 and 5 of Figure 3. Number of elements was increased from one to ten and the relative difference between the results for a particular number of elements and that of ten elements were recorded. These are shown in Figure 5. As is seen and was expected, loading type 1 has a faster convergence rate, but in both cases, eight elements are enough to ensure a result with 99% accuracy. In all results presented in the following, ten elements are used in order to increase the accuracy even further.

Percent Error

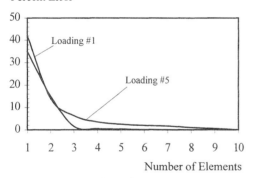

Figure 5. Finite element errors

In order to compare the results with previous studies, a simply supported composite girder similar to Svensson (1985) is considered. All nine loading cases are applied using ten elements along the length while in order to allow a direct comparison, terms associated with the twist of the bottom flange are set to zero. In Figure 6, current results are shown together with the results of Svensson. Vertical axis represents the effective length factor or the ratio (l_e/l). This factor is also equal to $\lambda^{-1/2}$ (As given in Equation 6). Horizontal axis represents the non-dimensional parameter βl of Equation (4). The continuous graph is the original Svensson's graph

$l_e/l \; (= \lambda^{-1/2})$

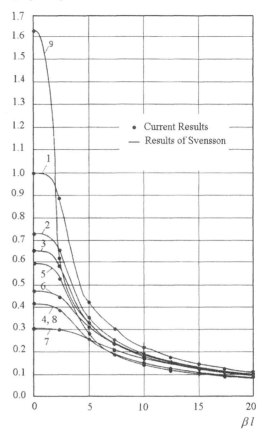

Figure 6. Comparing the two methods

while the black dots represent the current results. A very good agreement is seen in all cases.

One of the facilities that the method provides, is the ease of including boundary conditions other than simply supported. In Figures 7, 8 and 9, some of the results for load cases 1, 2 and 3 of Figure 3 are presented. Vertical and horizontal axes are similar to Figure 6, (l_e/l, represents the vertical, and βl the horizontal). Boundary conditions in Figures 8 and 9 include, Clamped-Clamped (C-C), Clamped-Simple (C-S), Simple-Clamped (S-C), Clamped-Free (C-F) and Free-Clamped (F-C). The cases (F-C) and (S-C) are not presented in Figure 7, as they produce similar results to (C-F) and (C-S) cases, respectively, for that particular load case.

It can be seen that (C-F) condition provides the highest effective length factor for all values of βl. On the other hand, the relative magnitude of the effective length factor for other cases varies depending on βl.

$l_e/l \ (\lambda^{-1/2})$

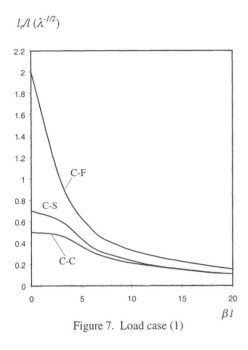

Figure 7. Load case (1)

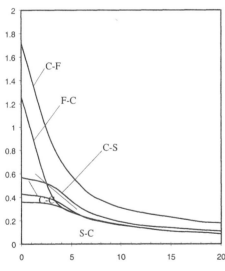

Figure 8. Load case (2)

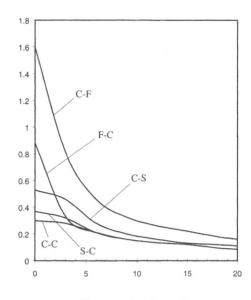

Figure 9. Load case (3)

the analysis are accurate and found to be in agreement with the results of other researchers. Due to simplicity and clarity of the model, it may be found useful in real practice.

REFERENCES

Bradford, M.A. & Gao, Z. 1992. Distortional Buckling Solutions For Continuous Composite Beam, *Journal of Structural Engineering, ASCE*, 118(1):73-89.

Goltermann, P. and Svensson, S.E. 1990. Lateral Distortional Buckling: Predicting Elastic Critical Stress, *Journal of Structural Engineering, ASCE*, 116(5):1465-67.

Oehers, D.J. and Bradford, M.A. 1995, *Composite Steel and Concrete Structural Members: Fundamental Behavior*, Pergamon Press, Oxford.

Svensson, S. E. 1985. Lateral Buckling of Beams Analysed as Elastically Supported Columns Subject to a Varying Axial Force, *Journal of Constructional Steel Research*, 5:179-93

Williams, F.W. and Jemah, A.K. 1987. Buckling Curves for Elastically Supported Columns with Varying Axial Force, to Predict Lateral Buckling of Beams, *Journal of Constructional Steel Research*, 7(2):133-147.

Williams, F.W., Jemah, A.K. and Lam, D.H. 1993. Distortional Buckling Curves For Composite Beam, *Journal of Structural Engineering, ASCE*, 119(7):2134-49.

4 CONCLUSION

A finite element with 2 nodes and three degrees of freedom per node is developed for the restricted distortional buckling (RDB) analysis of composite steel/concrete bridge girders. The method is based on simulating the bottom flange of the girder to a beam resting on an elastic foundation. The results of

Creative Systems in Structural and Construction Engineering, Singh (ed.) © *2001 Balkema, Rotterdam, ISBN 90 5809 161 9*

Micromechanics of fatigue failure in steel bridges

K. M. Mahmoud

URS Corporation, New York, N.Y., USA

ABSTRACT: A study of the effects of microscopic cracks and microcavities on fatigue life of steel bridge structures is presented in this paper. Under fatigue conditions, migration of dislocation results in localized plastic deformation at the microscopic level. Consequently, microscopic cracks form, grow and eventually coalesce together to produce major cracks. This paper is interested in the deterioration of the material structure at early stages (e.g., during the initiation or nucleation phase of fatigue cracking). This is achieved by introducing a variable that reflects the damage in this early stage of deterioration of the material. A demonstration of the ramifications of microvoids formation on a bridge fatigue life is presented.

1 INTRODUCTION:

The phenomenon of fatigue takes place in components and structures subjected to cyclic repetitive stressing, resulting in fracture at a load much smaller than that required under a static load. For many years, fatigue has been a significant and difficult problem for bridge engineers. One of the reasons for that is the fact that cracking and failure due to fatigue provide little evidence of plastic deformation. Therefore, fatigue cracking is difficult to detect before serious deterioration develops in the bridge component. It is believed that 50 to 90% of all failures in metallic structures are fatigue related. Taking the lower bound of this figure, simply means that every other bridge failure is attributed to fatigue.

The first reports of fatigue failure date back to the early days of railroads. After locomotives and railroad cars have been running satisfactorily for several years, thin cracks appeared in their axles at points of sudden change in cross-sectional dimensions, particularly at sharp, re-entrant corners. The close relationship between the failure of axles and the repetitiveness of loading was amply recognized by (McConnell 1849). The first systematic and quantitative investigation of fatigue damage was provided by (Wohler 1871), who performed many experiments that resulted in the widely known S-N curve (i.e., stress, S, versus number of cycles to failure, N). However, the breakthrough was using fracture mechanics in the assessment of fatigue damage in bridge components containing well-developed cracks (Fisher et al. 1989).

Microscopic cracks, in certain bridge details, are practically unavoidable due to fabrication process, construction practices or even misunderstanding of a detail orientation at the time of design. This may lead to unaccounted for fatigue prone stresses that were never anticipated and crack propagation may be considered to begin with the first load application. Nucleation and crack growth are commonly regarded as basic causes of fatigue damage accumulation and ultimate fatigue failure. Nucleation and crack growth also constitutes two principal phases in the fatigue damage process, as shown in Figure 1.

The fatigue crack nucleation process is not yet fully understood. However, for ductile bridge steels, as indicated by experimental observations, the crack initiation period may consume significant percentage of the usable fatigue life. This suffices to indicate how prudent it is to comprehend the behavior of the material structure in this nucleation, or initiation period.

The effect of microvoids in a crack-free continuum is described within the framework of continuum damage mechanics. In reality, a macroscopic crack and microvoids may exist simultaneously in a bridge component. To develop a sound fracture criterion, it is reasonable to analytically recognize both deficiencies.

This paper presents a damage model for the interaction between a macroscopic crack and the microvoids ahead of its tips. This is achieved by accounting for microvoid accumulation in the vicinity of the crack-tip through the introduction of a damage variable that represents the fraction of original cross-section area occupied by voids. A power-law relates the stress to the strain of the material and the deformation theory of plasticity is used to develop the stress, strain and displacement fields ahead of the crack-tip. A bridge component loaded in Mode I, with a remote cyclic stress, is studied. The effects of initial crack size, final crack size, type of detail and cyclic stress level on the service life of the bridge are assessed and found to be in good agreement with Paris power-law for fatigue crack growth.

2 PROPOSED MODEL

The first characterization of damage, as an internal variable, was suggested by (Kachanov 1958). On the basis of the hypothesis of isotropic damage, he defined D as follows:

$$D = A_{void} / A = 1 - (A_{net} /A) \quad , 0 \leq D \leq 1 \quad (1)$$

where A and A_{net} are the macroscopically observable original and net cross-sectional areas, respectively, while A_{void} is the cross-sectional area of the damaged material. It follows that the nominal stress, σ, and the actual stress, s, are related as:

$$\sigma = s (1 - D) = \Omega s \quad (2)$$

where Ω is the continuity function. It is assumed that the damage accumulation is concentrated in a small zone centered at the crack-tip, and it is called damage-zone. For a strain hardening material, obeying a linear relation to the yield point (σ_0 , ε_0) and a power-hardening law thereafter, the strain is related to the nominal stress, σ, according to:

$$\varepsilon = (\varepsilon_0 / \sigma_0) \sigma \quad , \quad \sigma < \sigma_0 \quad (3a)$$

$$\varepsilon = \varepsilon_0 (\sigma / \sigma_0)^n \quad , \quad \sigma > \sigma_0 \quad (3b)$$

where σ_0 and ε_0 are reference stress and strain, respectively, and n is the power hardening exponent. The reference stress is a material parameter and may be considered as the yield stress. Inside the damage-zone, the measure of damage is assumed to vary linearly with the crack opening displacement, i.e. D = η u_y, where η is a material damage parameter and u_y is the displacement normal to the crack plane. A

linear relationship is considered here as a first approximation. However, selecting a nonlinear relationship, while it may enhance the model, would increase the numerical work. Within the zone, the net stresses are assumed equal to the net yield stress, s_0. The stress, strain and displacement fields ahead of the crack-tip are expressed in terms of the HRR asymptotic solution (Hutchinson 1968, Rice & Rosengren 1968). The damage field should alter the HRR field singularity, however, for mathematical simplicity, it is assumed, in this paper, that the continuity function Ω = constant, within the damage-zone. This assumption reads:

$$\Omega = (1 - D) = \text{constant.} \quad (4)$$

throughout the damage-zone. With this approximation, the HRR asymptotic solution near the crack-tip retains its classical singularity. The HRR singular field is modified to account for the microscopic damage ahead of the crack-tip. A detailed analysis of this damage model is presented in (Mahmoud & Kassir 1999).

3 FATIGUE CRACK PROPAGATION

The conditions of fatigue crack propagation under cyclic tensile loading (i.e., Mode-I), varying from 0 to σ_∞, is investigated in this section. An instantaneous damage increment is accumulated in the damage-zone in each cycle. No damage is caused upon unloading from σ_∞ to 0. Within the damage-zone, the damage increment, ΔD, per each cycle of loading is proportional to the normal displacement. Performing the numerical solution of crack growth, the new crack length, a, which is found from the condition:

$$D(a_N , N) = 1 \quad (5)$$

is incrementally computed from:

$$a_N = a_0 + da_N \quad (6)$$

where a_0 is the initial crack size and da_N is given by:

$$da_N = [1 - \{1/ D(r, \pi/2;N)\}] \cdot [(1/12\pi) (K_I /\Omega(N) s_0)^2] \quad (7)$$

where D(r, $\pi/2$;N), denotes damage at a point (r, $\pi/2$), in terms of polar coordinates, from the crack-tip and K_I is the elastic stress-intensity factor for Mode-I. Using the new crack length given by Equations (6) and (7), the iterative process is repeated until instability occurs, i.e., the critical crack length to cause crack instability is reached. In the following, the effects of initial crack size, final crack size, type of detail and cyclic stress level are presented and discussed.

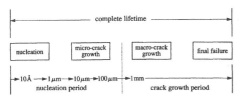

Figure 1 — Fatigue life of a structural element.

3.1 Effect of initial crack size

To illustrate the effect of initial crack size, a_0, the remaining fatigue life, $N = N_f - N_i$, as predicted from the damage model is plotted versus initial crack size in Figure 2, where N_i is the number of cycles elapsed to the initiation of crack extension and N_f is the number of cycles elapsed to failure. As shown in the Figure, if the initial crack size is increased from 0.05 inch (1.27 mm) to 0.1 inch (2.54 mm), the remaining life reduces by about 50% of the life for a 0.05 inch (1.27 mm) initial crack. Then, for instance, if the weld quality is such that the maximum crack, which cannot be detected, is 0.1 inch (2.54 mm) instead of 0.05 inch (1.27 mm), half the life of the bridge has already been used up, by allowing larger initial cracks to remain in the structure. If the final crack size is taken to be the same and equal to 1 inch (25.4 mm), one can see that a threefold increase in initial crack size from 0.1 inch (2.54 mm) to 0.3 inch (7.62 mm) reduces the remaining life to about 25%. This demonstrates the strong effect of initial crack size on bridge life and emphasizes the importance of taking into account the quality of initial crack size in the prediction of fatigue life of welded bridges subjected to cyclic loading.

3.2 Effect of final crack size

The influence of final crack size, a_f, on fatigue life is demonstrated by plotting the remaining fatigue life versus final crack size using an initial crack size of 0.15 inch (3.81 mm), as shown in Figure 3. In this Figure, a final crack size ranging from one inch (25.4 mm), which is used for Figure 2, to 5 inches (127 mm) has been used. Only marginal increases in remaining life are obtained for quite substantial changes in final crack size. For instance, an increase in final crack size from one inch (25.4 mm) to two inches (50.8 mm) results in a mere 7% increase in remaining life and an increase from one inch (25.4 mm) to 5 inches (127 mm) yields only about 12% increase in remaining fatigue life. This illustrates that the final crack size has little effect on bridge life when the controlling failure mechanism is fatigue.

3.3 Effect of type of detail and cyclic stress level

Using Paris power-law, the rate of fatigue crack growth, da/dN, is given by:

$$da/dN = C_p (\Delta K)^p \qquad (8)$$

where ΔK is the stress intensity factor range, and C_p and p are material parameters. A growth exponent, p, of four is appropriate for most welded steel structures (Maddox 1969). The stress intensity factor range, ΔK, in this case is: $\Delta K = \sigma_\infty \sqrt{\pi a}$. Then,

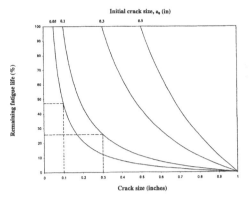

Figure 2 - Effect of initial crack size, a_0, on remaining fatigue life

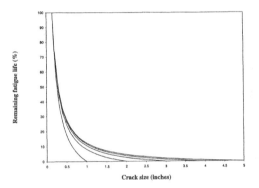

Figure 3 - Effect of final crack size, a_f, on remaining fatigue life

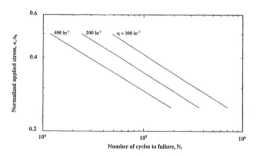

Figure 4 - Number of cycles to failure, N_f, versus external stress ratio, σ_∞ / S_0

the number of cycles required to cause failure, N_f, may, therefore, be expressed as:

$$N_f = [a_0]^{1 - p/2} \sigma_\infty^{-p} / [(p/2) - 1] [C_p \pi^{p/2}] \qquad (9)$$

The above equation reveals that $N_f \alpha \sigma_\infty^{-4}$ with $p = 4$. Figure 4 predicts that $N_f \alpha \sigma_\infty^{-3.9}$, which is in very good agreement with the prediction of Equation (9).

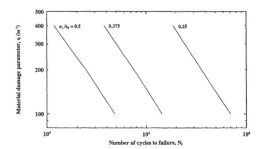

Figure 5 - Number of cycles to failure, N_f, versus material damage parameter, η

One more noteworthy feature of Figure 4, is that different values of the material damage parameter, η, could be looked at as indicators of the category of the weld detail, i.e. Category D, E, or E' as defined by the AASHTO Specifications, with the worst case corresponding to the higher value of η reflecting severer state of damage.

It is also noted that equation (9) predicts a relation $N_f \ \alpha \ a_0^{-1}$, which is in good agreement with the results shown in Figure 5, where the number of cycles to failure, N_f, is depicted versus the damage parameter, η, which is normalized with respect to s_0 a_0 / E.

Using equation (9), and assuming that the cyclic stress is reduced by a factor of two, the initial crack size that would give the same predicted service life is given by:

$$a_{0\,(0.5\,\sigma\infty)} = a_{0\,(\sigma\infty)}\ [16\,a_f / (\,a_f + 15\,a_0\,)] \qquad (10)$$

Since $0 < a_0 < a_f$, numerical values of the above formula range between one and 16. For high cycle fatigue in welded bridges, $a_f >> a_0$. Thus, a 16-fold increase in initial crack size can result if stresses are reduced by a factor of two. The same 16-fold increase in number of cycles required to cause failure is also predicted by equation (10), when the stresses are reduced by a factor of two. This is in a very good agreement with the 15.5-fold increase predicted by the damage model as shown in Figure 5.

4 CONCLUSIONS

1. Damage mechanics correctly predicts the premature fatigue cracking of bridge components.

2. It is shown that microvoids have detrimental ramifications on bridge fatigue life and they should be considered in assessing safe fatigue life of bridges.

3. Initial crack size has a strong influence on bridge life as most of the fatigue life is consumed in the initiation phase.

4. Final crack size has very little influence on total bridge life.

REFERENCES

Fisher, J.W., B.T.Yen, & W. Dayi 1989. *Fatigue of* bridge *structures- A commentary and guide for design, evaluation and investigation of cracking.* ATLSS Report No. 89-02.
Hutchinson, J.W. 1968. Singular behavior at the end of a tensile crack in a hardening material. *Journal of the Mechanics and Physics of Solids*: 16: 13-31.
Kachanov, L.M. 1958. Time of the rupture process under creep conditions.@ *Izv. Akad. Nauk SSSR. Otd. Tekh. Nauk*, 8, 26-31.
Maddox,S. J. 1969. Fatigue crack propagation weld metal and heat affected zone material.@ *Welding Institute Report* E/29/69, 1969.
Mahmoud, K. M. & K.M. Kassir 1998. Damage field ahead of a tensile crack in an elastic-plastic and viscoplastic material. *International Journal of Fracture:* 96: 149-165.
McConnell, J.E. 1849. On railway axles. *Proc. Inst. Mech. Engrs.*, October: 1847-1849, London.
Rice, J.R. and G.F. Rosengren 1968. Plane strain deformation near a crack tip in a power-law hardening material. *Journal of the Mechanics and Physics of Solids*: 16: 1-12.
Wohler, A. 1871. Tests to determine the forces acting on railway carriage axles and the capacity of resistance of the axles. *Engineering*: 11.

Creative Systems in Structural and Construction Engineering, Singh (ed.) © 2001 Balkema, Rotterdam, ISBN 90 5809 161 9

The effects of fabrication and construction on curved bridge behavior

D.G. Linzell

Department of Civil and Environmental Engineering, Pennsylvania State University, Pa., USA

ABSTRACT: Designers can often misinterpret the effects of fabrication and construction on the behavior of bridge structures. These effects are magnified for curved bridges, whose curvature leads to forces and deformations that can deviate considerably from straight bridges of similar design. The behavior of a curved steel bridge during construction was one item examined for the Curved Steel Bridge Research Project (CSBRP) funded by the Federal Highway Administration (FHWA). A series of elastic tests of an experimental curved bridge system were performed during erection. Analytical models were created and their predictions were compared against experimental results. Two types of analytical models were developed: (1) sophisticated finite element models using ABAQUS and (2) a simplified model using the V-Load method.

1 INTRODUCTION

Curved steel bridges have seen increased implementation as new and replacement structures in the United States and abroad over the past three decades. For interchanges and river crossings that contain severe site restrictions and complicated bridge geometries, steel girders can offer advantages over other types of materials with regards to fabrication, erection, and serviceability.

Increase in the design and construction of curved steel bridges in the U.S. helped to initiate a program in the 1990's to reexamine and revise the AASHTO Guide Specifications for Horizontally Curved Highway Bridges (AASHTO 1980, 1993). The Curved Steel Bridge Research Project (CSBRP), funded by the Federal Highway Administration (FHWA), was one phase of this program. The goal of the CSBRP was to conduct fundamental research into the behavior of curved steel members and to develop rational design procedures for curved steel bridge structures though examination of a large-scale experimental system (Zureick et al. 1994).

Initial CSBRP testing focused on behavior of the experimental bridge system during construction. Following the construction tests, a number of investigations of the behavior of curved I-girder sections under predominantly flexural loads were performed. The construction investigations, which are the focus of this paper, consisted of a total of nine elastic tests that examined six partial framing plans of the overall structure.

Data produced from the tests was used to: (1) study the behavior of curved steel plate girder systems during construction with differing levels of lateral and shoring support, (2) evaluate the accuracy of detailed and simplified analytical models used to predict the behavior of curved steel bridge systems and (3) develop data reduction schemes for future investigations of the experimental curved bridge system. Detailed analytical models were assembled and analyzed using ABAQUS/Aqua Versions 5.4 to 5.8 (Hibbitt, Karlsson & Sorenson, Inc. 1998a, 1998b, 1998c). The simplified model was analyzed using the V-Load method (AISC 1986). This paper contains brief summaries of the construction tests that were completed and presents sample comparisons between experimental and analytical data and some conclusions drawn from those comparisons.

2 PAST RESEARCH

In the late 1960's, the Consortium of University Research Teams (CURT) was formed to help develop a specification for curved steel bridge systems designed and constructed in the United States. This constituted the first organized effort to investigate the behavior of curved steel bridge systems in the US. The research team involved five universities and the results of their efforts led to eventual publication of the first edition of the AASHTO Guide Specifications in 1980. During approximately the same time period, another concerted research effort into curved steel bridge behavior was underway in Japan. A large portion of that research led to development and publication of a draft copy of the Hanshin Expressway Public Corporation's Guidelines for the Design of Horizontally Curved Girder Bridges (Hanshin Expressway Public Co. 1988). The AASHTO Specifications and Hanshin Guidelines are currently the only specifications in the world that deal with curved steel bridge design and construction.

Both bodies of research incorporated extensive laboratory testing, which examined either scale models of curved bridge systems or studied individual medium-scale components. Scale model tests of curved I-girders examined twin and multi-girder structures (Brennan 1970, 1971, Brennan et al. 1970, Brennan 1974, Brennan and Mandel 1979, Culver and Christiano 1969, Mozer et al. 1973, Nakai and Kotoguchi 1983). Component tests typically studied single I-girders with imposed support and loading conditions (Heins and Spates 1970, Mozer and Culver 1970, Mozer et al. 1971, Fukumoto and Nishida 1981, Nakai et al. 1983, 1984a, 1984b).

Since the first editions of the AASHTO Specifications and Hanshin Guidelines have been published, isolated independent experimental investigations of curved I-girder behavior have occurred. These projects have typically involved testing single girders to study specific aspects of their behavior (Yoo and Carbine 1985, Shanmugam et al. 1995, Thevendran et al. 1998).

None of the aforementioned laboratory studies have explicitly examined the behavior of curved steel bridges during construction. There have also been limited field studies of curved steel bridge construction behavior. A recent study in the US, which focused on a two-span continuous horizontally curved and superelevated I-girder structure being constructed in Minnesota, has been the most comprehensive to date (Galambos et al. 1996, Huang 1996, Pulver 1996, Hajjar and Boyer 1997).

Therefore it was hoped that the research described herein would help expand the knowledge base of curved steel I-girder bridge behavior during construction.

3 TESTING PROGRAM

The curved bridge superstructure tested for the CSBRP was a simply-supported system that consisted of three radially braced plate girders (Figure 1). It was designed so that the mid-span portion of the exterior girder (G3) failed under predominantly flexural loads while the rest of the system remained elastic. The remainder of the

Figure 1. Bridge Plan (Courtesy FHWA/HDR).

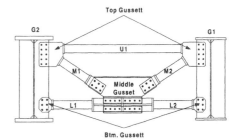

Figure 2. Typical Cross Frame Elevation.

system acted as a reusable testing frame that provided realistic load and boundary conditions to specimens spliced at mid-span of G3.

Girder spans along the arc were 26.2 m (86'-0^3/$_4$") for G1, 27.4 m (90'-0") for G2 and 28.6 m (93'-11^1/$_4$") for G3 with radii of curvature of 58.3 m (191'-3"), 61.0 m (200'-0") and 63.6 m (208'-9"), respectively. Girder dimensions were from 121.9 cm x 0.8 cm (48"x^7/$_{16}$") to 121.9 cm x 1.3 cm (48"x^1/$_2$") for the webs and from 40.6 cm x 1.7 cm (16"x^{11}/$_{16}$") to 61.0 cm x 5.7 cm (24"x2^1/$_4$") for the flanges. Girders were radially braced using "K" type cross frames shown in Figure 2. Cross frame design and placement was a direct result of the stipulation that failure would originate at mid-span of G3.

Girders were supported with radially oriented steel abutments. Both spherical bearings and Teflon pads were used to minimize frictional forces and allow translation and rotation in any direction except downward at the ends of G1 and G3. Guided bearings and a restraining frame were used to restrict G2's movement and stabilize the system. Lower lateral bracing was placed in the exterior bays to limit differential displacements between adjacent girder lines.

Extensive instrumentation was placed onto the bridge to ensure that all aspects of its behavior were tracked and recorded. Load cells were positioned at girder support points at the abutments and at intermediate shoring locations. Strain gages were affixed to the girders, cross frames and lower lateral bracing. Figure 3 indicates where girder strain gages were positioned for the final construction test, ES3-1. Standard transducers were used to measure displacements and rotations at select cross sections while laser and total station measurement systems tracked overall deformations.

As more structural components were added to the bridge the number of instruments increased. The final test, ES3-1, recorded over 1050 individual data points.

Nine tests, involving six framing plans, were performed during construction. Systems that were tested were variations of the final structure shown in Figure 1 and they involved G1 and G2 or all three girders (Figure 4).

Each construction test began with the system shored to its theoretical dead load camber position, commonly referred to as the "no-load" position. Once the "no-load" position was established, testing was initiated. ES1 series tests involved removing and replacing shoring from beneath G1 while G2 remained shored. ES2 tests examined the behavior of twin-girder systems as shoring was incrementally removed and replaced from beneath both G1 and G2. As Figure 5 indicates, ES3-1 studied the behavior of the three-girder system with differing levels of shoring support.

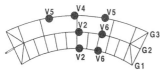

Figure 3. Girder Strain Gage Locations.

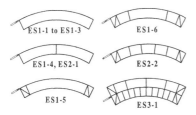

Figure 4. Construction Testing Framing Plans.

Figure 5. Test ES3-1.

4 ANALYTICAL MODELS

4.1 *Finite Element Models*

Detailed finite element models were constructed and analyzed using ABAQUS/Aqua Versions 5.4 to 5.8 (Hibbitt, Karlsson & Sorenson, Inc. 1998a, 1998b, 1998c). These models contained over 8400 elements and 50000 degrees of freedom and they were used extensively during the design phase to predict behavior during flexural testing (Zureick et al. 1997). Comparisons were made between finite element predictions and experimental results for five of the construction tests: ES1-4, ES1-6, ES2-1, ES2-2, and ES3-1. The comparisons helped to:

1 validate the accuracy of the finite element models,
2 determine if instrument locations and densities were adequate, and
3 examine the effects of fabrication and construction on behavior.

Finite element analyses of the construction tests mimicked shoring placement and removal sequences used in the laboratory. To study the effect of fabrication tolerances on system behavior, analyses were performed using both nominal geometric dimensions and geometric dimensions from an extensive set of measurements recorded during construction. Component self-weights were applied to the models by using a steel density of 77 $^{kN}/_{m^3}$ (490 $^{lb}/_{ft^3}$) and through application of additional dead loads that accounted for connection details not explicitly reproduced in the models.

Three types of comparisons between finite element predictions and experimental results were typically made:

1 plots of girder support reactions verses girder mid-span displacements,
2 plots of girder flange and web strain variations at select points during the tests, and
3 plots of cross frame member axial forces verses girder mid-span displacements.

Representative girder support reaction and displacement comparisons are shown in Figure 6 for ES1-4 and Figure 7 for ES3-1. Figure 6 examines G1 and Figure 7 examines G3. Both figures study the variation in support reactions at the abutments (1L and 1R) and at the mid-span shoring point (7) for select steps during each test. Each testing step involved a change in girder shoring conditions. Steps shown for ES1-4 in Figure 6 were during the shoring removal portion of the test while the three steps shown for ES3-1 in Figure 7 were for: (1) the initial "no-load" configuration, (2) the removal of all interior shores and (3) the G3 mid-span shore at Cross Frame 7 supporting a load of approximately 71.2 kN (16 kips).

Both figures indicate good agreement between predicted and actual girder support reactions and displacements. While agreement was generally acceptable, some discrepancies are apparent. These discrepancies were attributed to:

1 imperfections introduced during fabrication that led to fit-up problems during construction, and
2 zero shifts in the data caused by heating of the instrumentation and data acquisition system circuitry prior to testing.

During fabrication of the girders, G2 was incorrectly cambered. To correct this error, the girder was heated and forced back to its design camber using a series of "V" heats of segments of the girder web. Correcting the camber resulted in out-of-plumb transverse stiffeners that subsequently caused fit-up problems during construction. Forces that were introduced into the system during construction were not measured and, as is shown in the figures, these forces resulted in a redistribution of loads in the experimental structure that could not reproduced in the finite element models. While these discrepancies are clearly evident in Figures 5 and 6, generally their magnitudes were small when compared against overall forces and deformations developed under self-weight. For example, support reaction discrepancies shown in Figure 6 for ES3-1 approached 18 kN (4 kips), which is less than 5% of the total self-weight of the system, which approached 521 kN (117 kips). Figures 7 and 8 also show that minimal benefit was realized when nominal geometric properties were replaced with measured properties in the finite element models.

Figures 8 and 9 contain sample comparisons between experimental and analytical girder strains and cross frame axial forces. G3 mid-span top and bottom flange strains are examined for a single ES3-1 testing step in Figure 8 while axial forces in the lower chord members of mid-span Cross Frame 7 between G2 and G3 are examined in Figure 9.

Generally good agreement between analytical and experimental girder strains and cross frame forces exists in both Figures, but discrepancies similar to those discussed for Figures 6 and 7 do exist.

4.2 V-Load Method

It was of interest to examine the accuracy with which a common curved girder approximate analysis method, the V-Load method (AISC 1986), predicted the experimental curved bridge system's behavior during construction. This was accomplished by comparing predicted girder mid-span moments to

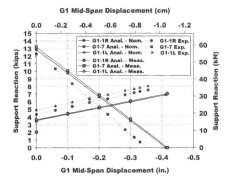

Figure 6. Analytical and Experimental G1 Support Reactions vs. Mid-Span Displacement, ES1-4.

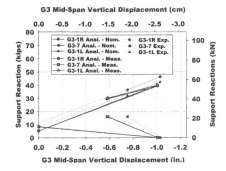

Figure 7. Analytical and Experimental G3 Support Reactions vs. Mid-Span Displacement, ES3-1.

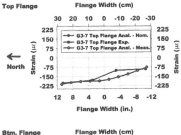

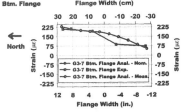

Figure 8. Analytical and Experimental G3 Flange Strains at Mid-Span, G3 Mid-Span Displacement = 2.5 cm (1"), ES3-1.

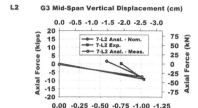

L2

G3 Mid-Span Vertical Displacement (cm)

Legend: 7-L2 Anal. - Nom. / 7-L2 Exp. / 7-L2 Anal. - Meas.

G3 Mid-Span Vertical Displacement (in.)

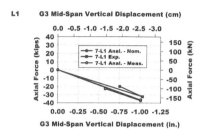

L1

G3 Mid-Span Vertical Displacement (cm)

Legend: 7-L1 Anal. - Nom. / 7-L1 Exp. / 7-L1 Anal. - Meas.

G3 Mid-Span Vertical Displacement (in.)

Figure 9. Analytical and Experimental Axial Forces, Cross Frame 7 Between G2 and G3, ES3-1.

Table 1. G1 and G3 Mid-Span Moments, Unshored Condition					
Girder	Mid-Span Moment (kN-m)			Difference from Experiment (%)	
	Exp.	FEM	V-Load	FEM	V-Load
G1	41	39	-50	-3	-220
G3	1306	1395	1333	7	2

bending at mid-span of G1 showed that interior girder moment predictions using this approach could not be considered as accurate as those for the exterior girder. These results correspond with previously published material from an extensive study of the V-Load method (Fiechtl et al. 1987). This research showed that V-Load estimations of dead load flexural behavior in an exterior girder were accurate but nonconservative results could be obtained for the interior girder in a multi-girder system.

values obtained from an ABAQUS model and from a construction test. Comparisons were made for ES3-1.

The V-Load method was derived for curved girder systems in which the cross frames were continuous between all girder lines. Since the experimental curved bridge contained some discontinuous cross frame lines (Figure 1), it was decided that V-Loads would be calculated assuming that these discontinuous cross frame lines had been removed. Primary moments would be determined using measured dead loads, which included the discontinuous cross frames. Additional ABAQUS analyses were performed to examine the effect of removing cross frames and they indicated that mid-span dead-load moments would not change appreciably when discontinuous cross frame lines were removed.

Comparisons between experimental mid-span moments and those predicted from the ABAQUS analyses and the V-Load method are shown for ES3-1 in Table 1. The table indicates that both analytical and V-Load results estimated experimental G3 mid-span moments quite well, with differences being 10% or less. V-Load estimations for G1 mid-span moments were not nearly as accurate and were nonconservative, with a moment of -50 kN-m (-37 k-ft) being predicted at mid-span of G1 while the experimental magnitude was +41 kN-m (+30 k-ft). The ABAQUS model predicted the experimental mid-span moment for G1 quite well. Even though mid-span moment magnitudes for G1 were small, the fact that the V-Load method predicted negative

5 CONCLUSIONS

Summaries of a series of elastic tests of a full-scale horizontally curved steel bridge system are presented. Descriptions of the bridge are given and levels of instrumentation and documentation are discussed.

Comparisons between experimental results and predictions from ABAQUS finite element analyses and a V-Load analysis are presented. It was demonstrated that the finite element models provided acceptable predictions of construction behavior with discrepancies that existed being predominantly attributed to load redistribution in the experimental structure during construction. The V-Load analysis was shown to accurately predict mid-span dead-load moments in the exterior girder but to inaccurately predict interior girder dead load moments.

ACKNOWLEDGEMENTS

HDR Engineering through FHWA Contract No. DTFH61-92-C-00136 supports this research for the Georgia Institute of Technology. Sheila Duwadi serves as the Contracting Officer's Technical Representative for FHWA. Advisement provided by Dr. Abdul-Hamid Zureick and Dr. Roberto Leon of The Georgia Institute of Technology during the author's involvement with the project is gratefully acknowledged. Technical input from Dann Hall and Mike Grubb of BSDI and John Yadlosky of HDR have been invaluable to this investigation. The author would also like to thank Bill Wright of FHWA, Joey Hartmann of Professional Service Industries, Inc., and James Burrell of Qualcomm, Inc. for their assistance.

897

REFERENCES

American Institute of Steel Construction 1986. *Highway Structures Design Handbook, Volume I, Ch.12 - V-Load Analysis.*

American Association of State Highway and Transportation Officials 1980, 1993. *Guide Specifications for Horizontally Curved Bridges.* USA.

Brennan, P.J. 1970. Horizontally Curved Bridges First Annual Report: Analysis of Horizontally Curved Bridges through Three-Dimensional Mathematical Model and Small Scale Structural Testing. *Syracuse University, First Annual Report, Research Project HPR-2(111).* USA.

Brennan, P.J. 1971. Horizontally Curved Bridges Second Annual Report: Analysis of Seekonk River Bridge Small Scale Structure through Three-Dimensional Mathematical Model and Small Scale Structural Testing. *Syracuse University, Second Annual Report, Research Project HPR-2(111).* USA.

Brennan, P.J. 1974. Analysis and Structural Testing of a Multiple Configuration Small Scale Horizontally Curved Highway Bridge. *Syracuse University, Research Project HPR-2(111).* USA.

Brennan, P.J., Antoni, C.M., Leininger, R. and Mandel, J.A. 1970. Analysis for Stress and Deformation of a Horizontally Curved Bridge Through a Geometric Structural Model. *Syracuse University Dept. of Civil Eng., Special Research Report No. SR-3.* USA.

Brennan, P.J. and Mandel, J.A. 1979. Multiple Configuration Curved Bridge Model Studies. *Journal of the Structural Division, ASCE,* 105(ST5): 875-890, USA.

Culver, C.G. and Christiano, P.P. 1969. Static Model Tests of Curved Girder Bridge. *Journal of the Structural Division, ASCE,* 95(ST8): 1599-1614, USA.

Fiechtl, A.L., Fenves, G.L. and Frank, K.H. 1987. Approximate Analysis of Horizontally Curved Girder Bridges. *Texas University, Austin, Center for Transportation Research, Final Report No. FHWA-TX-91-360-2F.* USA.

Fukumoto, Y. and Nishida, S. 1981. Ultimate Load Behavior of Curved I-Beams. *Journal of the Engineering Mechanics Division, ASCE,* 107(EM2): 367-384, USA.

Galambos, T.V., Hajjar, J.F., Leon, R.T., Huang, W., Pulver, B.E., and Rudie, B.J. 1996. Stresses in Steel Curved Girder Bridges. *Minnesota Dept. of Trans. Report No. MN/RC-96/28.* USA.

Hanshin Expressway Public Corporation 1988. *Guidelines for the Design of Horizontally Curved Girder Bridges (Draft).* Japan.

Hajjar, J.F. and Boyer, T.A. 1997. Live Load Stresses in Steel Curved Girder Bridges. *Progress Report on Task 1 for Minnesota Dept. of Trans. Project 74708.* USA.

Heins, C.P. and Spates, K.R. 1970. Behavior of Single Horizontally Curved Girder. *Journal of the Structural Division, ASCE,* 96(ST7): 1511-1529, USA.

Hibbitt, Karlsson & Sorenson, Inc. 1998a. *Introduction to ABAQUS/Pre, Version 5.8.* USA.

Hibbitt, Karlsson & Sorenson, Inc. 1998b. *ABAQUS/Post Manual, Version 5.8.* USA.

Hibbitt, Karlsson & Sorenson, Inc. 1998c. *ABAQUS/Standard User's Manual, Version 5.8.* USA.

Huang, W.H. 1996. Curved I-Girder Systems. *Ph.D. Dissertation, Department of Civil Engineering, University of Minnesota.* USA. •

Mozer, J., Culver, C. 1970. *Horizontally Curved Highway Bridges – Stability of Curved Plate Girders. Carnegie Mellon University, Report No. P1, Research Project HPR – 2(111).* USA.

Mozer, J., Ohlson, R. and Culver, C. 1971. Horizontally Curved Highway Bridges – Stability of Curved Plate Girders. *Carnegie Mellon University, Report No. P2, Research Project HPR – 2(111).* USA.

Mozer, J., Cook, J., and Culver, C. 1973. Horizontally Curved Highway Bridges – Stability of Curved Plate Girders. *Carnegie Mellon University Report No. P3, Research Project HPR-2(111).* USA.

Nakai, H. and Kotoguchi, H. 1983. A Study on Lateral Buckling Strength and Design Aid for Horizontally Curved I-Girder Bridges. *Transactions of the Japanese Society of Civil Engineers,* 339: 195-205, Japan (in Japanese).

Nakai, H., Kitada, T., and Ohminami, R. 1983. Experimental Study on Bending Strength of Web Plate of Horizontally Curved Girder Bridges. *Proceedings of the Japanese Society of Civil Engineers.* 340: 19-28, Japan (in Japanese).

Nakai, H., Kitada, T., and Ohminami, R. 1984a. Experimental Study on Ultimate Strength of Web Panels in Horizontally Curved Girder Bridges Subjected to Bending, Shear, and Their Combinations. *Proceedings, Annual Technical Session – Structural Stability Research Council: Stability Under Seismic Loading:* 91-102, USA.

Nakai, H., Kitada, T., Ohminami, R. and Fukumoto, K. 1984b. Experimental Study on Shear Strength of Horizontally Curved Plate Girders. *Transactions of the Japanese Society of Civil Engineers,* 350: 291-290, Japan (in Japanese).

Pulver, B.E. 1996. Measured Stresses in a Steel Curved Bridge System, *M.C.E. Report, Department of Civil Engineering, University of Minnesota.* USA.

Shanmugam, N.E., Thevendran, J.Y., Liew, R., Tan, L.O. 1995. Experimental Study on Steel Beams Curved in Plan. *Journal of Structural Engineering, ASCE,* 121(2): 249-259, USA.

Thevendran, J.Y., Shanmugam, N.E., Liew, R. 1998. Flexural Torsional Behavior of Steel I-Beams Curved in Plan. *Journal of Constructional Steel Research,* 46(1-3): 79-80, UK.

Yoo, C.H. and Carbine, R.L. 1985. Experimental Investigation of Horizontally Curved Steel Wide Flange Beams. *Proceedings Annual Technical Session: Stability Aspects of Industrial Buildings, SSRC:* 183-191, Cleveland, OH, USA.

Zureick A., Leon, R.T., Burrell, J., and Linzell, D. 1997. Curved Steel Bridges: Experimental and Analytical Studies. *Proceedings - Innovations in Structural Design: Strength, Stability, Reliability. A Symposium Honoring Theodore V. Galambos, SSRC:* 179-190, Minneapolis, MN, USA.

Zureick, A., Naqib, R., and Yadlosky, J.M. 1994. Curved Steel Bridge Research Project, Interim Report I: Synthesis. *HDR Engineering, Inc., Publication Number FHWA-RD-93-129.* USA.

Creative Systems in Structural and Construction Engineering, Singh (ed.) © 2001 Balkema, Rotterdam, ISBN 90 5809 161 9

Search for fundamental patterns in bridge structural design

A. Baseggio & M. Gambini – *Milan, Italy*

F. Biondini – *Technical University of Milan, Italy*

F. Bontempi – *University of Rome 'La Sapienza', Italy*

P.G. Malerba – *University of Udine, Italy*

ABSTRACT: The paper deals with the identification of optimal structural morphologies through evolutionary procedures. Two main approaches are considered. The first one simulates the Biological Growth (BG) of natural structures like the bones and the trees. The second one, called Evolutionary Structural Optimization (ESO), removes material at low stress level. Optimal configurations are addressed by proper optimality indexes and by a monitoring of the structural response. Design graphs suitable to this purpose are introduced and employed in the optimization of a pylon carrying a suspended roof and of a bridge under multiple loads.

1 INTRODUCTION

One of the most promising research field which has been recently applied to the identification of optimal structural morphology deals with evolutionary procedures which operate on the basis of some analogies with the growing and the evolutionary processes of natural systems. Such methods are based on the simple concept that by slowly removing and/or reshaping regions of inefficient material, belonging to a given over-designed structure, its shape and topology evolve toward an optimum configuration.

Two main approaches among those proposed in literature are here considered. In the first one, the structural morphology is modified by simulating the Biological Growth (BG) of natural structures like the bones and the trees (Mattheck & Burkhardt 1990, Mattheck & Moldenhauer 1990). In the second one, called Evolutionary Structural Optimization (ESO), material at low stress level is removed by degrading its constitutive properties (Xie and Steven 1993, 1994). The basic steps of these procedures should be repeated until optimal configurations appear. However, to this regards, no well established convergence criteria exist.

In this work, the better structural solutions emerging from the evolutionary process are identified on the basis of proper optimality indexes and by monitoring the actual structural response. In the first part, *design graphs* suitable to this purpose are presented (Baseggio *et al.* 2000). In the subsequent part, both BG and ESO methods are briefly recalled and these graphs are usefully employed in the selection of the optimal morphology of a pylon carrying a suspended roof and of a bridge type structure under a multiple load condition. These structures are considered to be in plane stress and made of linear elastic material having symmetric or non symmetric behavior in tension and in compression. The structural analyses needed during the evolutionary process are carried out by a LST-based (Linear Strain Triangle) finite element technique.

2 OPTIMALITY INDEXES AND DESIGN GRAPHS

As mention before, at each step of the evolutionary process the present structure is modified in such a way that a better configuration with respect to given evolutionary criteria is hopefully achieved. However, the optimality of such solutions needs often to be judged with respect to design criteria which are not necessarily coincident with those which regulate the evolution. In this work, design criteria are synthesized by one or more optimality indexes able to measure the quality of the present solution with respect to the initial one.

It is generally recognized that Nature tends to build structures in such a way that the internal strain energy, or the external work done by the applied loads, is minimum. Based on such a consideration, a proper optimality index may be represented by the following Performance Structural Index (Zhao et al. 1997):

$$PSI = \frac{V_0 \cdot W_0}{V \cdot W} \tag{1}$$

being W the external work per unit of volume V and where 0 denotes the initial configuration. This formulation implicitly refers to a single load condition, but it can be easily extended to account for multiple loads, for example by a weighted average of the contributions PSI^i of each load condition $i=1,...,NC$:

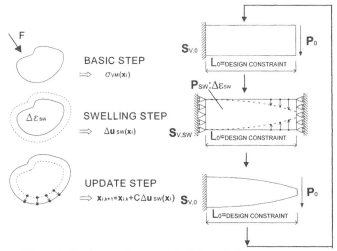

Figure 1. Fundamental steps of the BG evolutionary process.

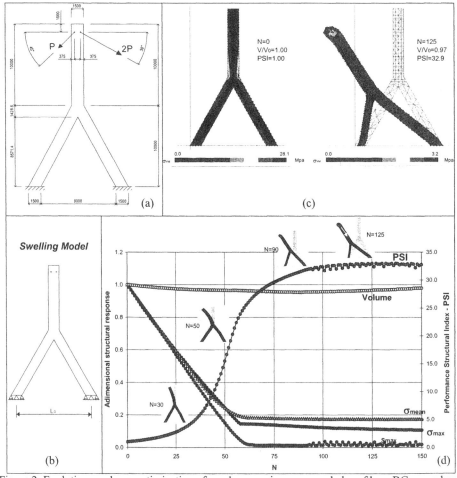

Figure 2. Evolutionary shape optimization of a pylon carrying a suspended roof by a BG procedure.

$$PSI = \frac{\sum_{i=1}^{NC} w_i \cdot PSI^i}{\sum_{i=1}^{NC} w_i} = \sum_{i=1}^{NC} \omega_i \cdot PSI^i \qquad (2)$$

Of course, depending on the specific problem to be examined, additional indexes may be introduced. For example, structures made of material having low tensile strength, like stone or concrete, should be designed by limiting the amount of tensioned material. Thus, by denoting by V_c the portion of the volume V which is compression dominated (see Fig. 3), the following percentage of Compressed Material Volume:

$$CMV = \frac{V_c}{V} \qquad (3)$$

appears to be as well a meaningful optimality index. Moreover, sometimes may be useful to optimize not only the mechanical behavior, but also some geometrical properties of the structure. A measure of the present free Perimeter Γ of the structural boundary with respect to the initial one Γ_0:

$$2P = \frac{\Gamma}{\Gamma_0} \qquad (4)$$

gives for instance an idea about the advantages in terms of cost of formworks and structural durability. In addition, being related to the weight of the structure, the percentage of Removed Material Volume:

$$RMV = \frac{V_0 - V}{V_0} \qquad (5)$$

may be itself an important indicator about the total cost.

After some optimality indexes are selected and eventually grouped in a single averaged index, hierarchical arrangements of the solutions explored during the evolutionary process become possible. However, some additional design constraints on the structural response, for example in terms of maximum displacement and maximum stress level, are usually needed to assure the feasibility of the solution which seems to appear optimal. Thus, the best morphology requires to be identified by a monitoring of both the optimality indexes and the structural response. To this aim, *design graphs* which contemporarily describe the evolution of all such quantities, for example versus the *RMV* index, are introduced.

3 BG PROCEDURES (BIOLOGICAL GROWTH)

The evolutionary procedures considered in the following, work by simulating the Biological Growth (BG) of natural structures (Mattheck & Burkhardt 1990, Mattheck & Moldenhauer 1990). Such structures are known to evolve by adapting themselves to the applied loads according to the *axiom of uniform stress*, which states that in the optimal configuration the stress field distribution tends to be fairly regular over the structure (Mattheck 1998). Thus, structural shape and topology are gradually modified in such a way that material is added in zones with high stress concentrations and removed from under-loaded zones (*swelling*).

The simplest form of a *swelling law* able to regulate such modifications is assumed as follows:

$$\dot{\varepsilon}_{SW} = \frac{1}{V} \frac{dV}{dt} = K \cdot (\bar{\sigma}^n - \bar{\sigma}_{REF}^n) = DF \qquad (6)$$

being: $\dot{\varepsilon}_{SW}$ the swelling strain rate; V the evolutionary time-dependent volume $V = V(t)$; K an artificial constant; $\bar{\sigma}$ the actual von Mises stress and $\bar{\sigma}_{REF}$ its reference, or far-field, value; n a suitable exponent ($n=1$ for a stress-based and $n=2$ for a energy-based criterion); DF the Driving Force of the evolutionary process. Based on this law, the numerical simulation of the growth mechanism is obtained through the following three steps (Fig. 1).

(1) Basic Step. A finite element analysis is performed to obtain the stress distribution $\bar{\sigma}$ over the structure.

(2) Swelling Step. The Driving Forces are firstly computed. In particular, for structure in plane state the following isotropic swelling strain increment vector $\Delta \mathbf{e}_{SW} = \frac{1}{2} \Delta \varepsilon_{SW} [1 \ 1 \ 0]^T$ is considered for the time increment Δt. Based on such strain distribution, the load vector $\Delta \mathbf{f}_{SW}$ equivalent to swelling is derived and the corresponding incremental displacement vector $\Delta \mathbf{u}_{SW}$ is evaluated as follows:

$$\mathbf{f}_{SW} = \int_V \mathbf{B}^T \mathbf{D} \Delta \mathbf{e}_{SW} dV \quad \Rightarrow$$
$$\Rightarrow \quad \mathbf{K} \Delta \mathbf{u}_{sw} = \Delta \mathbf{f}_{SW} \quad \Rightarrow \quad \Delta \mathbf{u}_{sw} = \mathbf{K}^{-1} \Delta \mathbf{f}_{SW} \qquad (7)$$

being $\mathbf{B}$ the compatibility matrix of the finite element, $\mathbf{D}$ the constitutive matrix of the material and $\mathbf{K}$ the stiffness matrix of the structure. It is worth noting that, in this work, additional geometrical design constraints are accounted for directly by replacing the actual boundary conditions of the swelling model in such a way that swelling displacements which violate the constraints are not allowed. This concept is shown in Fig. 1, where the cantilever beam is forced to maintain its initial length during the evolution.

(3) Update Step. The location $\mathbf{x}_{i,k}$ of each node $i = 1, \ldots, N$ of the finite element model at the current generation k is updated according to the swelling displacements $\Delta \mathbf{u}_{SW}$ just obtained as follows:

$$\mathbf{x}_{i,k+1} = \mathbf{x}_{i,k} + C \Delta \mathbf{u}_{SW} \qquad (8)$$

being C a suitable extrapolation factor which implicitly contains the constant K. Such factor may be either fixed at the first generation and then considered time-independent, or varied during the evolution. In any case, its value should be chosen to assure noticeable shape variations and progressively decreasing driving forces (Mattheck & Moldenhauer 1990).

The BG procedure is applied to the shape optimization of a pylon carrying a suspended roof. The geometry of the initial structure and the load condition are shown in Fig. 2.a. Since the distance between the supports is retained, the swelling model in Fig. 2.b is adopted. The design graph in Fig. 2.d shows the progressive convergence of the evolutionary process towards higher levels of the optimality index *PSI* and lower levels of the structural response, while the structural volume remains practically the same. Noteworthy the end of the pylon tends to lie along the line of action of the resultant of the applied loads. Finally, Fig. 2.c allows us to compare the maps of the von Mises stress corresponding, respectively, to the initial structure and the optimal one, and to appreciate how the latter presents a stress distribution having lower maximum intensity nearly uniform.

4 ESO PROCEDURES (EVOLUTIONARY STRUCTURAL OPTIMIZATION)

The Evolutionary Structural Optimization (ESO) procedures modify the topology of a given overdesigned structure by slowly removing regions of inefficient material (Xie and Steven 1993, 1994). The initial domain is subdivided in finite elements and a structural analysis is carried out. A representative quantity of the structural response, say the von Mises stress, is then evaluated at the element level and compared with a portion *RR* (*Rejection Ratio*) of a reference value, for instance the maximum stress over the whole structure. If such a lower limit is not reached (Criterion 1 in Table 1), the material inside the corresponding elements is considered to be inefficient and it is removed by degrading its constitutive properties, typically the Young modulus. The parameter *RR*, who determines the portion of material which is removed at each step of the evolutionary process, is usually assumed as follows:

$$RR(SS) = A_0 + A_1 \cdot SS \qquad (9)$$

being *SS* an integer counter which is added by a unity whenever a *Steady State* is reached, while A_0 and A_1 are numerical constants able to assure a grad-

ual evolution. Proper values seems to be $A_0 \cong 0.0$ and $A_1 \cong 0.005$. In this work, however, the rate of the process is controlled also by introducing an upper limit on the percentage of removed material volume V_{REM} at each step (*Rate of Removed Material*):

$$RRM = \frac{V_{REM}}{V} \le RRM_{\max} \qquad (10)$$

Criterion 1 is appropriate for materials having good strength in both tension and compression. However, many structures exhibit low strength in tension, like those made of stone or concrete, or in compression, like those subjected to buckling phenomena. To account for the cases in which the optimal structural morphology should be defined by limiting the amount of material subjected to critical stress states, the concept of *tension* and *compression dominated material* has been introduced (Guan *et al.* 1999). As shown in Fig. 4, material is considered tension (compression) dominated if the maximum (minimum) principal stress is of tension (compression) type. Based on this concept, the actual domain Ω is subdivided at each step in two parts, Ω_T and Ω_C, and in each of them the efficiency of the material is verified by using the absolute values of the principal stresses instead of the von Mises stresses (Criteria 2 and 3 in Tab. 1).

The criteria just introduced implicitly refer to a single load condition, but can they be easily applied to the case of multiple loads by removing material only if the rejection criterion is verified for every load condition.

In the basic formulation the minimum portion of removable material is identified with a single finite element. However, it is worth noting that a more general formulation can be achieved if the control of efficiency is performed on a *minimum elimination unit* formed by a *group* of elements. Several grouping criteria are clearly possible. By joining for example two adjacent triangular elements, structural solutions characterized by more regular boundaries are usually obtained. Moreover, the discrete nature of some structural types like masonry can be also better modeled, f.i. by building blocks representing one or more bricks.

Table 1. Efficiency and rejection criteria for symmetric and asymmetric material.

CRITERION AND MATERIAL TYPE	FORMULATION
(1) symmetric	$\sigma^{VM} \le RR(SS) \cdot \sigma_{\max}^{VM}$
(2) asymmetric with low tensile strength	$\sigma_{11} \ge 0.0$ and $\lvert \sigma_{22} \rvert \le RR(SS) \lvert \sigma_{22,\max} \rvert$
(3) asymmetric with low compressive strength	$\sigma_{22} \le 0.0$ and $\lvert \sigma_{11} \rvert \le RR(SS) \lvert \sigma_{11,\max} \rvert$

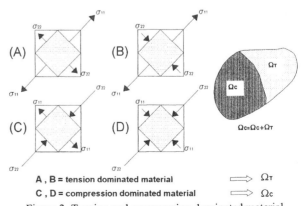

A , B = tension dominated material $\quad \Longrightarrow \Omega_T$
C , D = compression dominated material $\quad \Longrightarrow \Omega_C$

Figure 3. Tension and compression dominated material.

Of course, since in each group a different rejection criteria can be considered, the previous approach also allows us to take the case of non homogeneous structures into account. Finally, by introducing a no-rejection criterion, is possible to *freeze* a sub-region of the initial domain (*Non Design sub-domain*) which cannot be never removed. This is useful for instance dealing with bridge type structures, where the deck level is usually fixed.

Fig. 4 shows the ESO procedure applied to the optimization of the structural morphology of a bridge (Ito 1996) subjected to six load conditions.

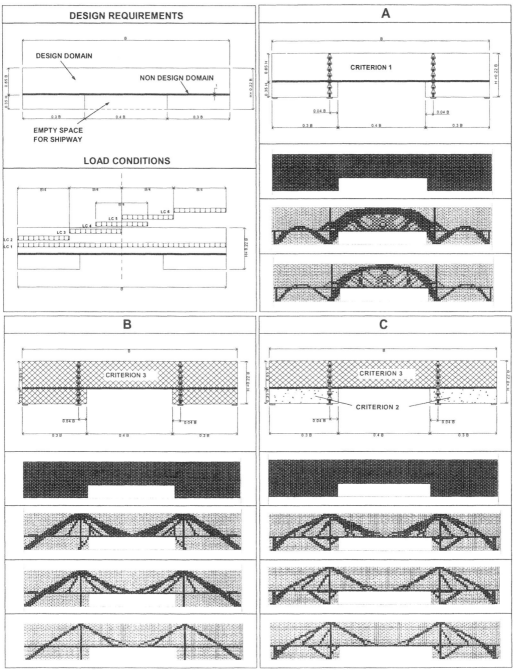

Figure 4. Some optimal structural morphologies of a bridge type structure.

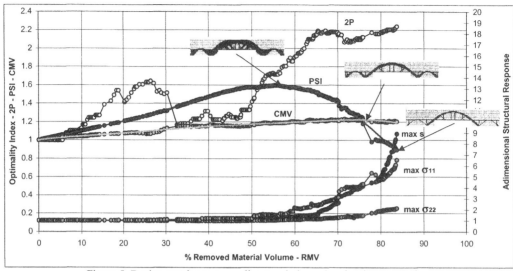

Figure 5. Design graph corresponding to a balanced arch scheme solution.

The position of both the deck and the supports is assumed to be fixed and a free space for navigation is provided under the deck. The first window of Fig. 4 shows the geometric proportions, the design requirements, the load conditions and the initial domain chosen for the procedure. At first, a cable-stayed scheme is searched for by adding two axially rigid pylons to the Non Design domain. Windows A, B and C of Fig. 4 show some layouts obtained during the evolutionary process for different material types. By assuming symmetric material (Criterion 1), a balanced arch scheme emerges instead of the expected one (Fig. 4A). To be winning, the cable-stayed scheme should favor tensioned fields and then work on asymmetric material having low compression strength (Criterion 3, Fig. 4B). However, such a solution tends to anchor some tensioned elements directly on the lateral supports. A more rational scheme can be achieved if the rejection criteria are differentiated over the structure, for instance by assuming the material under the deck to be asymmetric with low tensile strength (Criteria 2-3, Fig. 4C).

Despite of the found solutions, the balanced arch scheme initially obtained should be preferred if material having low tension strength is used. Fig. 5 shows the design graph and some of the corresponding optimal configurations resulting from the evolutionary process for the case of symmetric material without pylons. Such diagram allows us either to appreciate the optimality level of found schemes with reference to several optimality indexes, or to control the corresponding feasibility of the structural response.

5 CONCLUSIONS

The Biological Growth (BG) and the Evolutionary Structural Optimization (ESO) have been applied to morphology optimization problems. The swelling step of the BG procedures has been extended to take geometrical constraints into account. A formulation of ESO suitable to deal with asymmetric (tension and compression dominated) materials, fixed geometrical boundaries and alignments (non design domains) and multiple load conditions has been presented. Such processes may lead to many final optimal choices, as has been shown by an application searching for the optimal structural layout of a bridge having clearance limitations. Among these choices, the final actual optimum may be judged by using suitable design graphs, with reference to design criteria not necessarily coincident with those which control the evolutionary process.

REFERENCES

Baseggio A., F. Biondini, F. Bontempi, M. Gambini, P.G. Malerba 2000. Structural Morphology Optimization by Evolutionary Procedures. *Proceedings of Structural Morphology Conference*, Delft, The Netherlands, August 17–19.

Guan H., Steven G.P., Querin O.M., Xie Y.M. 1999. Optimisation of Bridge Deck Positioning by the Evolutionary Procedure. Structural Engineering and Mechanics 7(6), 551-559.

Ito M. 1996. Selection of Bridge Types from a Japanese Experiences. Proc. of IASS Int. Symposium on Conceptual Design of Structures, University of Stuttgart, Stuttgart, **1**, 65-72.

Mattheck C., Burkhardt S. 1990. A New Method of Structural Shape Optimization based on Biological Growth. Int. J. of Fatigue. 12(3), 185-190.

Mattheck C., Moldenhauer H. 1990. An Intelligent CAD-Method based on Biological Growth. Fatigue Fract. Engng. Mat. Struct. 13(1), 41-51.

Mattheck C. 1998. Design in Nature. Learning from Trees. Springer Verlag.

Xie Y.M, Steven G.P. 1994. Optimal Design of Multiple Load Case Structures using an Evolutionary Procedure. Engineering Computation, 11, 295-302.

Xie Y.M., Steven G.P. 1993. A Simple Evolutionary Procedure for Structural Optimization. Computers & Structures, 49(5), 885-896.

Zaho C., Hornby P., Steven G.P., Xie Y.M. 1998. A Generalized Evolutionary Method for Numerical Topology Optimization of Structures under Static Loading Conditions. Structural Optimization, 15, 251-260.

28 Structural behavior of plates and slabs

Creative Systems in Structural and Construction Engineering, Singh (ed.) © 2001 Balkema, Rotterdam, ISBN 90 5809 161 9

Lightweight perforated infill panels for retrofit of flat-plate buildings

F. K. Humay & A. J. Durrani
Rice University, Houston, Tex., USA

ABSTRACT: A retrofit solution involving lightweight perforated infill panels was developed and experimentally tested at Rice University. Four-tenth scale slab-column subassemblies were retrofitted with prefabricated infill panels and subjected to quasi-static loading conforming to FEMA 273. Effects of uniformly distributed perforations (circular holes and rectangular openings) on the behavior of the infill wall system were compared with results of solid panels. Preliminary results demonstrated a substantial increase in the initial stiffness and overall strength capacity of retrofitted specimens as compared to that of the bare flat-plate frame. Furthermore, the energy dissipation, calculated as the area under the load-displacement curves, was also greatly enhanced.

1 INTRODUCTION

A large portion of existing reinforced concrete (RC) flat-plate buildings are seismically deficient and pose a risk of catastrophic failure if subjected to earthquakes of low to moderate intensity. Many of these buildings, particularly in the central and eastern parts of the United States, were only designed for gravity and wind loading and therefore do not have the lateral stiffness and proper connection detailing needed to withstand a seismic event. Experience from past earthquakes including the 1971 San Fernando (Mahin et al. 1983), 1985 Mexico City (Meli & Rodriguez 1988) and others, clearly elucidates the potential for overwhelming damage and loss of life. The vulnerability of flat-plate buildings is mainly attributed to excessive lateral drift and punching shear failure at the slab-column connection. In many instances, the lack of continuous reinforcement (particularly bottom reinforcement) through the joint has resulted in the progressive collapse of floors during moderate earthquakes. Laboratory tests have demonstrated that slab-column connections in existing non-ductile flat-plate buildings exhibit insufficient lateral stiffness and have a limited moment-transfer capacity (Durrani et al. 1995).

Because punching shear failure usually results in local collapse, and often triggers progressive collapse, non-ductile flat-plate buildings pose a substantial threat to life safety. It is, therefore, imperative to develop strategies for the retrofit of these potentially hazardous structures. Results from past research reveal two possible retrofit solutions: strengthening of the connections and control of lateral drift. Luo and Durrani previously investigated schemes that involved increasing the connection capacity to protect against progressive collapse once punching failure has occurred (Luo & Durrani 1994). An external steel capital attached to the column directly underneath the slab was proposed. Tests conducted on interior slab-column connections retrofitted in this manner demonstrated the feasibility of this technique. The second option can be achieved by the addition of stiffening elements such as shear walls, steel bracing or infill walls. Since the addition of shear walls or steel bracing may be architecturally intrusive and relatively expensive, the use of infill panels emerges as a more attractive solution. Previous experimental findings indicate the ability of infill panels to significantly improve the seismic performance of bare beam-column frames (Bertero & Brokken 1983, Frosch et al. 1996, Kahn 1976), but no test data exists for slab-column frames containing infills.

2 INFILLED FRAMES

Since seismic retrofit has become an increasingly critical issue, infill panels are being used to repair damaged buildings and upgrade existing ones. In particular, the rehabilitation of RC buildings has been researched using various infill materials and panel configurations. Experimental testing has been conducted with cast-in–place concrete, precast concrete (using both single and multiple panels), masonry, steel panels and shotcrete. In general, the strength and stiffness of an infilled frame was found to be much higher than that of a bare frame. Most researchers also agree that infills can dissipate a large amount of energy when they are subjected to seismic reversals. Because of these qualities, infill panels are an effective solution for the seismic retrofit of non-ductile frames.

Although infill panels can greatly enhance a structure's seismic performance, if not designed properly, however, they may have a detrimental impact on the existing surrounding frame. The addition of infill panels increases a building's mass, and as a result, the inertial forces imparted to the structure are also increased. If the extra dead load is substantial, problems may arise with the existing capacity of the foundation. Furthermore, infill panels will stiffen a structure and ultimately alter the load path and failure mechanisms of the infilled frame. Because of large stirrup spacing and lap splices designed for compression, columns in "weak" (non-ductile) frames are not designed to resist high concentrated shear forces or large overturning moments. In addition, brittle failures of unreinforced infills, such as masonry, have resulted in shedding of debris into streets or stairwells, with great hazard to life (Priestley 1980). Connectivity between the infill wall and the bounding frame is also an extremely important issue that will determine the new load path and resultant behavior.

3 EXPERIMENTAL PROGRAM

3.1 *LWPSC Infill Panels*

The infill panels developed at Rice University were composed of a lightweight pumice stone concrete (LWPSC) with a unit weight of approximately 1234 kg/m^3 (77 lbs/ft^3) and an average compressive strength of 8.97 MPa (1300 psi). Dimensions of the individual panels were sized to allow laborers to construct the infill wall without the need for any additional mechanical equipment. The reduced-scale standard unit was 519 mm (1'-8 1/4") long, 231 mm (9") high and 78 mm (3") thick. A total of sixteen panels were required to infill each subassembly. All of the panel-to-panel and panel-to-frame joints contained mild reinforcing and were tightly grouted. In addition, dowel connections were provided between both the top and bottom slabs and the infill wall.

3.2 *Test Configuration*

The specimens were tested in a self-stressing steel reaction frame. Top and bottom column stub connections were designed and fabricated to behave in a pinned manner. A 489-kN (110-kip) hydraulic actuator, attached to a steel transfer beam at the top

of the specimen, applied the load. The overall testing set-up is illustrated in Figure 1. Positive loading was associated with retraction of the actuator and negative loading with extension.

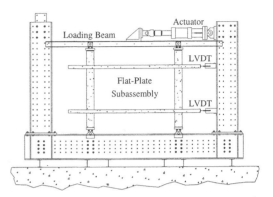

Figure 1. Bare-Frame Test Configuration

3.3 *Loading History*

All of the tests were performed under displacement control and consisted of fully reversed cyclic loading at increasing levels of deformation. Due to the configuration of the testing equipment, the system's control and feedback signals were both located at the elevation of the actuator. Peak displacements for each level of deformation were, therefore, determined in terms of the overall drift (the actuator displacement divided by the total height of the specimen) of the subassembly. LVDT's located at each floor level were used to monitor the interstory drift, Δ_{id}, throughout the loading history. The interstory drift was calculated as the relative displacement of the top and bottom slab divided by the story height as shown in Equation 1 below:

$$\Delta_{id} = \frac{\delta_{rel}}{h_{story}} = \frac{\delta_{ts} - \delta_{bs}}{1.22 \text{ m} (48")} \tag{1}$$

where δ_{ts} = displacement of LVDT at the top slab; δ_{bs} = displacement of LVDT at the bottom slab.

The loading protocol conformed to the provisions of FEMA 273 (1997) and consisted of increasing the overall drift in increments of 0.25% up to 1.0%. Three full cycles were performed at every drift level to investigate the degradation in stiffness, strength and energy dissipation under repeated motions. After completing the series of cycles at 1.0% drift, one cycle each at 0.75%, 0.50% and 0.25%, respectively, were executed. This sequence of cycles was included to quantify the residual stiffness after damage had occurred. The remaining portion of the load history was designed to develop the ultimate failure

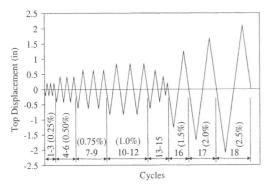

Note: Values in () represent overall drift of specimen.
1 in = 25.4 mm

Figure 2. Cyclic Loading History

mechanism of the subassembly and is shown in Figure 2.

3.4 *Retrofitted Subassemblies*

All of the specimens were retrofitted with 78 mm (3") thick LWPSC infill panels with identical dimensions. Dowel connections between the slabs and infill wall were provided in all of the specimens with perforations. Two different types of perforations were investigated. Specimen S2-HCA consisted of uniformly distributed 64 mm (2 1/2") diameter holes (three per panel) located throughout the wall. Specimen S4-OCX, on the other hand, had two 128 mm (5") long x 103 mm (4") high rectangular openings per unit also located over the entire wall area. For comparison, a bare slab-column subassembly (Fig. 1), specimen S1-BF, was also tested.

4 EXPERIMENTAL RESULTS

4.1 *Observed Behavior*

The bare frame specimen was tested to a maximum overall drift of 1.5% at which time the test was stopped and the subassembly retrofitted. 1.5% drift was chosen as the termination point for the bare frame for two reasons. First, this drift level was considered a reasonable upper bound for flat-plate buildings prior to experiencing punching shear failure. Secondly, since the specimen would ultimately be retrofitted and retested, an attempt was made to limit any excessive damage.

The first visible cracking occurred in the specimen at an overall drift of 0.25%. Each slab-column joint experienced a similar crack situated at the interior face of the column and spanning the entire width of the slab. These cracks were the result of flexure. No other visible damage to the subassembly was apparent at this time. As the loading increased,

additional flexural cracks developed spreading out toward the center of the slabs.

Cracking in specimen S2-HCA was first observed in the infill wall during the cycles to 0.50% overall drift. All of the cracks originated from the circular openings and spread outward at approximately 45° angles. As the subassembly was loaded in the negative direction cracks formed parallel to the diagonal spanning from the lower west corner of the wall to the upper east corner. As the direction of loading was reversed the cracking pattern followed the opposite diagonal. The damage was uniformly distributed throughout the entire area of the infill panel. The greatest amounts of distress were initially observed within the two middle stacks of standard panels. Of these eight panels, the four directly adjacent to the top and bottom slabs had the largest concentration of cracking. This indicated the transfer of shear forces between the slab and infill wall through the dowel connections. Similarly, the outer stacks of standard panels experienced the same type of damage but to a lesser degree. Figure 3 is a photograph of the infill wall at the end of the testing. Because of the strength of the infill wall, the top column stubs were significantly damaged due to the transmission of high shear forces and bending moments. The test was ultimately stopped due to excessive shear damage in the columns.

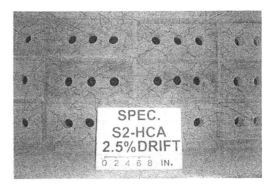

Figure 3. Close-up of Infill Wall for Specimen S2-HCA @ 2.5% Overall Drift (End of Test)

Similarly, cracking in specimen S4-OCX was distributed throughout the entire wall area. At the early stages of loading small cracks emanated from the corners of the rectangular openings. Cracking became more significant during the cycle to 0.75% overall drift. Piers between the openings experienced X-shaped shear cracking that started in the central portion of the wall and moved outward. The ultimate capacity of the wall was controlled by the shear strength of the LWPSC piers. Damage to the surrounding slab-column frame was minimal compared to specimen S2-HCA. Distress within the slabs was similar, but the columns of specimen S4-OCX only experienced a small number of flexural cracks. The final state of specimen S4-OCX is illustrated in Figure 4.

Figure 4. Specimen S4-OCX @ 2.5% Overall Drift (End of Test)

4.2 Load – Displacement Relationship

Plots of lateral load vs. interstory drift and lateral load vs. overall drift were generated for the entire loading history of each specimen. Because of the test set-up, the true difference in performance due to the addition of the infill walls was more accurately reflected in the interstory drift data. To better compare the results of different tests, maximum lateral load – interstory drift envelopes were developed and plotted on the same graph (Fig. 5). Connecting the maximum load points from the first cycle at each drift level created these envelopes.

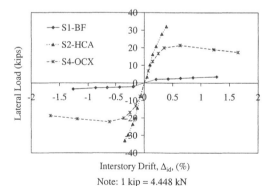

Interstory Drift, Δ_{id}, (%)

Note: 1 kip = 4.448 kN

Figure 5. Maximum Load – Interstory Drift Envelopes

4.3 Strength

The increase in strength observed in the retrofitted specimens ranged from 9.5 to 6.5 times that of the bare slab-column subassembly. Strength enhancement was more pronounced in specimen S2-HCA. The envelope curve was fairly linear up to the peak load. The maximum load – displacement envelope for specimen S4-OCX, on the other hand, was approximately bi-linear. The maximum load was reached after the apparent yield point.

4.4 Stiffness

The actual stiffness of a specimen for any given cycle was difficult to determine because it continuously changed as the subassembly was damaged and the loading progressed. For this reason, the secant, or peak-to-peak, stiffness was determined to be a practical substitution of the true stiffness. The secant stiffness for each cycle was found by computing the slope of the line connecting the peak point in the positive direction with that in the negative. Plots of the interstory secant stiffness, K_{rel}, vs. the interstory drift are given in Figure 6.

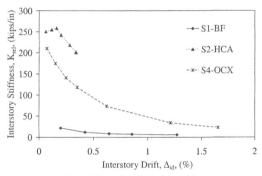

Interstory Drift, Δ_{id}, (%)

Note: 1 kip = 4.448 kN; 1 in = 25.4 mm

Figure 6. Interstory Peak-to-Peak Stiffness vs. Interstory Drift

Initial stiffness values were calculated as the peak-to-peak stiffness for the first cycle to 0.25% overall drift. Specimen S1-BF had an initial stiffness of 3783 kN/m (21.6 kips/in). In comparison, specimens S2-HCA and S4-OCX had initial stiffness values of 43,832 kN/m (250.3 kips/in) (11.6 times greater) and 36,810 kN/m (210.2 kips/in) (9.7 times greater) respectively.

4.5 Energy Dissipation

When evaluating the inelastic response of a structure under reversed loading, energy dissipation capacity is an important parameter to consider. The energy dissipated during a complete cycle is defined as the area within the corresponding load-displacement curve. To maximize the amount of energy released, it is most desirable to have load-displacement loops that are as open and full as possible.

Energy dissipation capacity was found for each specimen from the cyclic plots of both lateral load vs. top displacement and lateral load vs. relative displacement. The amount of energy dissipated was determined by using a linear variation between the recorded data points. Plots of the cumulative dissipated energy based on the lateral load – interstory drift plots are presented in Figure 7. The curves represent the sum of the dissipated energy for all cycles of loading.

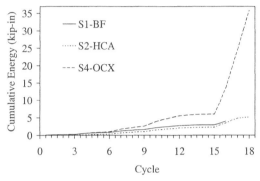

Note: 1 kip = 4.448 kN; 1 in = 25.4 mm

Figure 7. Cumulative Energy Dissipation (Calculated from Lateral Load – Relative Floor Displacement Plots)

The bare frame specimen had very low energy dissipation as evidenced by severely pinched hysteresis loops. Energy dissipation occurred at the slab-column connections, predominately due to cracking within the slab. Although a substantial amount of cracking was observed within the infill panels of specimen S2-HCA, the wall was too strong to experience complete failure. Uniformly distributed cracking did dissipate some energy, but the overall behavior of the infill was fairly linear. Conversely, the infill panels used to retrofit specimen S4-OCX dissipated the most energy of all the subassemblies tested. Diagonal shear cracking, especially between openings, dissipated the majority of the energy. Because of the configuration of the perforations, the cracking was distributed throughout the entire wall area.

5 CONCLUSIONS

An infill panel system using lightweight pumice stone concrete (LWPSC) was developed at Rice University. In order to assess the promise of this system as a retrofit option for flat-plate buildings, reduced-scale quasi-static experiments were conducted. A number of variables were investigated including the addition of uniform perforations throughout the area of the infill wall.

From preliminary test results, LWPSC perforated infill panels appear to be a very promising retrofit solution. As previously shown, the strength and stiffness characteristics were greatly enhanced in both specimens S2-HCA and S4-OCX. In addition, energy dissipation was significantly improved using panels with rectangular openings. S2-HCA did not see an increase in energy dissipation because the wall was too strong and behaved in a very linear manner. Of the two systems, the infill panels containing rectangular openings are a better alternative. This configuration provides better control over the location of cracking, maximum wall load and ultimate failure mechanism.

REFERENCES

Bertero, V. & Brokken, S. 1983. Infills in seismic resistant buildings. *Journal of Structural Engineering, ASCE* 109(6): 1337-1361.

Durrani, A. J. & Du, Y. 1995. Seismic resistance of nonductile slab-column connections in existing flat-slab buildings. *ACI Structural Journal* 92(4): 479-487.

FEMA 273 1997. NEHRP guidelines for the seismic rehabilitation of buildings. Federal Emergency Management Agency.

Frosch, R. J., Li, W., Jirsa, J. O. & Kreger, M. E. 1996. Retrofit of non-ductile moment-resisting frames using precast wall panels. *Earthquake Spectra* 12(4): 741-760.

Kahn, L. F. 1976. Reinforced concrete infilled shear walls for aseismic strengthening. *Doctor of Philosophy Thesis*: University of Michigan.

Luo, Y. H. & Durrani, A. J. 1994. Seismic retrofit strategy for non-ductile flat-plate connections. *Proceedings of the Fifth U.S. National Conference on Earthquake Engineering*: 627-636.

Mahin, S. A., Bertero, V. V., Chopra, A. K. & Collins, R. G. 1976. Response of the Olive View hospital main building during the San Fernando earthquake. *Report No. EERC 76-22*: University of California at Berkeley.

Meli, R. & Rodriguez, M. 1988. Seismic behavior of waffle-flat plate buildings. *Concrete International* 10(7): 33-41.

Priestley, M. J. N. 1980. Design of Earthquake Resistant Structures (Chapter 6). New York: John Wiley & Sons.

Creative Systems in Structural and Construction Engineering, Singh (ed.) © 2001 Balkema, Rotterdam, ISBN 90 5809 161 9

Time-dependent behavior of reinforced concrete flat slabs at service loads – An experimental study

R. I. Gilbert

School of Civil and Environmental Engineering, University of New South Wales, Sydney, N.S.W., Australia

ABSTRACT: An experimental program of long-term testing of large-scale reinforced concrete flat slab structures is described and the results from the first series of tests on five continuous flat slab specimens are presented. Each specimen was subjected to sustained service loads for periods up to 500 days and the deflection, extent of cracking and column loads were monitored throughout. The measured long-term deflection is many times the initial short-term deflection, due primarily to the loss of stiffness associated with time-dependent cracking under the combined influences of transverse load and drying shrinkage. This effect is not accounted for in the current code approaches for deflection calculation and control, and the test results will be used to suggest improvements to the current design procedures for the serviceability of flat slabs.

1 INTRODUCTION

The design of a reinforced concrete structure is complicated by the difficulties involved in estimating the service load behavior. The deflection and extent of cracking in reinforced concrete flexural members depend primarily on the non-linear and inelastic properties of concrete and, as such, are difficult to predict with confidence. The problem is particularly difficult in the case of slabs, which are typically thin in relation to the their spans and are therefore deflection sensitive. It is stiffness rather than strength that usually governs the design of slabs, particularly in the cases of flat slabs and flat plates.

In most concrete codes (including ACI 318-95 and AS3600-1994), the procedures specified for calculating the final deflection of a slab are necessarily design-oriented and simple to use, involving crude approximations of the complex effects of cracking, tension stiffening, concrete creep and shrinkage and the load history. Unfortunately, in most code procedures, the effects of cracking in a slab are not adequately taken into account, particularly those resulting from the time-dependent cracking caused by restraint to shrinkage and temperature deformations.

The final deflection of a slab depends on the extent of initial cracking which, in turn, depends on the construction procedure (shoring and re-shoring), the amount of early shrinkage, the temperature gradients in the first few weeks after casting, the degree of curing and so on. It also depends on the degree of restraint and the quality of the concrete, in particular the magnitude and rate of development of shrinkage.

Many of these parameters are, to a large extent, out of the control of the designer. In field measurements of the deflection of many identical flat slab panels (Sbarounis, 1984, Jokinen and Scanlon, 1985), large variability was reported. Final deflections of identical panels differed by over 100% in some cases.

Over the past several years, numerous cases have been reported, in Australia and elsewhere, of flat slabs for which the calculated deflection is far less than the actual deflection. Many of these slabs complied with the code's serviceability requirements, but still deflected excessively. Evidently, the deflection calculation procedures embodied in the code do not adequately model the in-service behavior of slabs.

Surprisingly, no laboratory controlled long-term measurements of deflection and crack propagation in large-scale reinforced concrete slabs have been reported in the literature to date. This lack of reliable experimental data has hindered both analytical research and the development of reliable design procedures. In this paper, an extensive experimental program of long-term testing of large-scale flat slab structures currently underway at the University of New South Wales is described and the results are presented for the first series of tests. These benchmark results include the measured time-varying material properties, slab deflection, extent of cracking and column loads and will be used to develop and evaluate alternative design procedures.

With the trend towards higher strength reinforcing steels, the design for serviceability will increasingly assume a more prominent role in the design of slabs. Designers will need to pay more attention to the specification of both the concrete mix, particularly the creep and shrinkage characteristics, and a suitable construction procedure, involving acceptably long stripping times, adequate propping, effective curing and rigorous on-site supervision.

2 EXPERIMENTAL PROGRAM

A three-year experimental program to measure the time-dependent in-service behavior of reinforced concrete flat slabs commenced at the University of New South Wales in 1998. The work is funded by the Australian Research Council and will eventually involve testing eight large-scale flat plate structures. At the time of writing this paper, five slab specimens (designated S1 to S5) have been tested under sustained service loads for periods up to 500 days. S3 to S5 remain under load. The tests are described and some significant initial results are presented here.

2.1 *Slab specimens and test set-up*

Each slab is continuous over two spans in two orthogonal directions and is supported on nine 200 mm x 200 mm x 1250 mm long columns below the slab. The plan dimensions of each slab and the top and bottom reinforcement layouts are identical and are shown in Figure 1. Also shown are the 16 points (#1 - #16) on the slab soffits at which deflections were measured. Slabs S1 and S2 are 100 mm thick, while S3, S4 and S5 are 90 mm thick. The slab reinforcement consists of 10 mm diameter deformed bars (Y10) and the clear concrete cover to the outer layer of bars is 15 mm. The concrete for each specimen has significantly different properties (except for S4 and S5 where the same batch of concrete was used).

The base of each column for specimen S1 is pinned with all exterior columns mounted on roller supports to minimize restraint to drying shrinkage in each direction (see Figure 2). The base of each exterior column in specimens S2 to S5 is poured monolithically with a 700 x 700 x 300 mm pad footing fixed to the laboratory floor. These supports prevent translation at the column bases and thereby provide restraint to shrinkage in the slab. This is more typical of the support conditions in practical slabs.

All slabs were cast and initially moist cured for nine days (14 days for slab S3), at which time the formwork was removed and the slab back-propped to the laboratory floor. At age 14 days (age 28 days for slab S3), each slab was subjected to a uniformly distributed load applied via concrete blocks carefully constructed and arranged to ensure uniform loading and uninhibited air flow over both the top and bottom surfaces of the slab. The props were removed and testing commenced. The loading arrangement for three of the slabs (S1, S2 and S3) is shown in Figure 3.

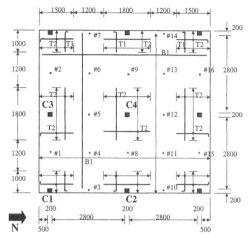

Bottom reinforcement: B1 = Y10 bars @ 220 mm centers
Top reinforcement: T1 = Y10 bars @ 250 mm centers
                         T2 = Y10 bars @ 140 mm centers
E-W bars placed 1st and last.   N-S bars placed 2nd and 3rd.

Figure 1. Plan of slab specimens and reinforcement layout.

Figure 2. Slab S1 just prior to first loading.

Figure 3. Slabs S1 to S3 under load.

2.2 *Load histories of slab specimens*

The load history of each slab is given in Figure 4. For example, in Figure 4a, S1 was initially loaded at age 14 days with a uniformly distributed load consisting of self-weight (2.40 kPa) and a single layer of concrete blocks (3.15 kPa). The load was sustained until age 169 days, when a second layer of blocks was added (3.10 kPa). The load was then held constant until age 301 days when the second layer of blocks was removed. At age 433 days, the remaining layer of blocks was removed and, until the test ended at age 512 days, the slab carried just its self-weight.

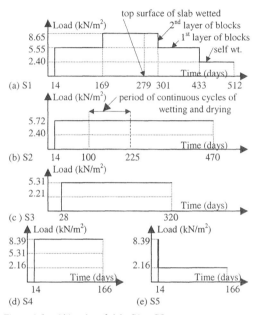

Figure 4. Load histories of slabs S1 to S5.

Between ages 100 and 225 days, the top surface of S2 was exposed to wetting and drying. Initially the exposure to water was accidental, but it unexpectedly and significantly influenced slab behavior (see Figure 5b). To explore this effect further, the top surface of S1 was intentionally wetted at age 279 days.

2.3 Test measurements

For the duration of the tests, deflections were measured at the center of each bay and at the mid-span on each column line. Strains on slab edges at various locations were recorded, as were the loads in columns C1 to C4 (Figure 1). The time-dependent development of cracking on both the top and bottom slab surfaces was also monitored.

Concurrent with the slab tests, material properties were measured on companion specimens. Compressive strength and elastic modulus of concrete were measured at various ages on cylinders, and the flexural tensile strength was measured on prisms. Creep was measured on cylinders loaded at 14 days in standard creep rigs. Shrinkage strains were recorded from 600 x 600 x 100 mm thick slab specimens (with edges sealed to ensure that drying only took place at the top and bottom surfaces).

2.4 Test results

The measured compressive strength, f_c, elastic modulus, E_c, and flexural tensile strength, f_t, for the concrete at first loading (averaged over four test specimens for each batch) are given in Table 1. The average yield stress for the reinforcement is $f_y = 650$ MPa and its elastic modulus is $E_s = 219,000$ MPa. The measured creep coefficient, ϕ, and shrinkage strain, ε_{sh} (x10^{-6}), are given in Table 2.

Table 1. Material properties at age of first loading.

Slab Specimen	E_c (MPa)	f_c (MPa)	f_t (MPa)
S1	30,020	34.5	4.39
S2	29,100	29.0	2.72
S3	22,620	18.0	2.48
S4	22,010	19.9	2.76
S5	22,010	19.9	2.76

Table 2. Creep coefficient and shrinkage strain.

Age (days)	S1 ϕ	S1 ε_{sh}	S2 ϕ	S2 ε_{sh}	S3 ϕ	S3 ε_{sh}	S4 & S5 ϕ	S4 & S5 ε_{sh}
14	0	124	0	302			0	20
20	0.73	153	0.84	400	0	150	0.65	80
28	0.89	228	1.29	520	0.58	220	1.20	140
40	1.26	293	1.58	569	0.81	280	1.41	195
80	1.64	386	2.41	752	1.30	400	2.20	318
120	1.84	438	2.59	770	1.55	500	2.41	392
200	2.11	573	2.74	830	1.89	630	2.55	550
300	2.29	585	2.81	890	2.05	700		
450	2.34	597	2.89	960				

The mid-panel deflection versus time curves at points #4, #6, #11 and #13 on each slab are plotted in Figure 5 and the average deflections at these four points, Δ_1, for each slab are given in Tables 3 and 4. Also given in Table 3 for S1 are the average deflections at the symmetric points #8 and #9 (designated Δ_2), #1, #2, #15 and #16 (Δ_3), #5 and #12 (Δ_4) and #3, #7, #10 and #14 (Δ_5).

Table 3. Average deflections for slab S1.

Age (days)	Average deflections (mm) Δ_1	Δ_2	Δ_3	Δ_4	Δ_5
14	2.62	2.06	1.53	2.04	1.56
20	3.85	3.10	2.25	3.04	2.18
28	4.63	3.73	2.72	3.68	2.59
40	5.18	4.18	3.05	4.17	2.89
80	6.39	5.11	3.78	5.17	3.48
120	7.07	5.64	4.21	5.74	3.84
169*	7.60	6.03	4.53	6.12	4.09
169**	8.67	6.93	5.24	6.99	4.77
200	9.06	7.34	5.48	7.40	5.01
279	9.45	7.68	5.69	7.62	5.19
280	8.35	6.80	5.30	7.05	5.25
301*	10.52	8.72	6.73	8.61	6.64
301**	9.53	7.90	6.09	7.83	6.02
350	9.92	8.23	6.35	8.15	6.27
433*	10.27	8.52	6.58	8.43	6.49
.433**	9.21	7.63	5.88	7.57	5.81
512	8.83	7.31	5.63	7.26	5.57

* before and ** after load applied or removed

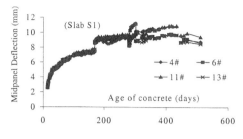

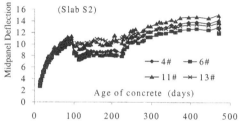

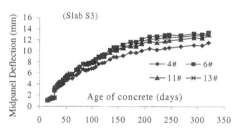

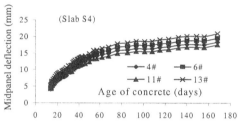

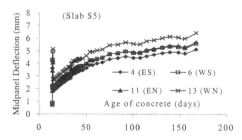

Figure 5. Midpanel deflection versus time (Slabs S1 to S5).

The vertical reactions at the bases of the columns C1 to C4 in S1 were measured throughout the test using load cells and are presented in Table 5. Also presented are the measured loads in column C4 in specimens S2, S3, S4 and S5.

Table 4. Average deflection Δ_1 for slab S2, S3, S4 and S5.

S2		S3		S4		S5	
Age (days)	Δ_1 (mm)	Age (days)	Δ_1 (mm)	Age (days)	Δ_1 (mm)	Age (days)	Δ_1 (mm)
14	2.84	14	1.08	14	4.50	14	4.59
20	4.58	20	1.38	20	7.55	15	2.27
28	6.15	28	1.56	28	9.19	20	2.52
40	7.70	28	2.87	40	12.41	28	3.01
80	10.17	40	4.22	100	16.93	40	3.73
100*	10.66	60	5.57	150	18.30	100	4.94
101	9.15	100	7.64	166	19.17	154	5.72
120	8.80	120	8.80				
200	9.46	160	10.39				
225	9.49	200	11.49				
230	10.46	300	12.40				
250	11.47	320	12.75				
320	12.88						
400	13.67						

* top surface first wetted

Table 5. Column loads.

Age	S1				S2	S3	S4	S5
	C1 (kN)	C2 (kN)	C3 (kN)	C4 (kN)	C4 (kN)	C4 (kN)	C4 (kN)	C4 (kN)
(days)								
14	15.9	29.7	40.7	63.7	72.4	26.3	86.9	23.5
30	14.3	29.0	40.8	64.8	75.1	52.7	87.5	25.5
100	12.5	28.4	41.7	64.6	75.6	56.0	88.9	29.4
169	12.8	28.8	42.0	65.0				
169	23.2	43.7	64.5	97.1				
190	23.6	44.0	64.0	94.7				

At first loading, no cracking was observed in S1, although very fine flexural cracks occurred within the first two weeks under load on the top surface of the slab, radiating from the interior column C4. These cracks increased in width with time and extended, but remained quite serviceable with a maximum recorded crack width of less than 0.15mm at age 169 days, prior to increasing the load. When the superimposed load was doubled at age 169 days, the existing cracks extended and widened but no new cracks developed. The maximum crack width at age 279 days, prior to wetting the top surface of the slab was 0.375 mm. No cracking occurred on the soffit of the slab up to this time. The crack patterns on the top surface of S1 are shown in Figures 6a, 6b, and 6c.

At age 279 days, the top surface of slab S1 was wetted for 48 hours. Over the next week, fine cracks occurred in the midspan regions on the soffit of the slab as shown in Figure 6d, with a maximum crack width of 0.125mm. The extent and width of the top cracks during this period did not change appreciably.

In specimen S2, flexural cracks formed at first loading on the top surface of the slab over columns C2, C3 and C4 and additional cracking occurred on the top surface at all columns with time and the crack widths gradually increased. The soffit of S2 remained uncracked throughout the test. The maximum crack width on the top surface was 0.35mm. Crack patterns on the top surface of S2 at various ages are shown in Figures 6e and 6f.

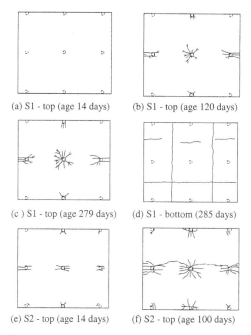

(a) S1 - top (age 14 days)	(b) S1 - top (age 120 days)
(c) S1 - top (age 279 days)	(d) S1 - bottom (285 days)
(e) S2 - top (age 14 days)	(f) S2 - top (age 100 days)

Figure 6. Crack patterns for S1 and S2.

The top surfaces of both slabs S4 and S5, initially loaded with two layers of concrete blocks (8.39 kPa), cracked at first loading. The initial cracks, radiating from the interior columns on the top surface, had a maximum width in the range 0.15 - 0.18mm.

Under the sustained load of 8.39 kPa, the maximum width of the cracks in the top surface of S4 increased with time to 0.25 - 0.375mm at age 25 days, 0.35 - 0.625mm at 40 days and 0.65 - 1.13mm at 150 days. Clearly, the cracks on the top surface of S4 became unserviceable with time. By age 25 days, fine cracks (of maximum width 0.1mm) had developed on the soffit of S4 in the positive moment regions, increasing in width to 0.175 - 0.225 mm at age 40 days and 0.2 - 0.275mm at age 150 days.

For S5, initially heavily loaded and thereafter subjected only to self-weight, no cracks appeared at any stage on the slab soffit. The cracks on the top surface almost closed on removal of the superimposed load. However, the width of the top cracks gradually increased with time, reaching 0.25 - 0.375 mm at 150 days.

2.5 Discussion of results

For the slab reinforcement shown in Figure 1, taking f_y = 550 MPa and f'_c = 20 MPa, the calculated collapse load in flexure for a 90mm thick slab is 14.34 kPa. The test loads applied to all slab specimens, therefore, were typical service loads, with the maximum applied load being 58.5% of the collapse load in S4 and S5, and significantly less in S1, S2 and S3.

Of the slabs tested, S1 and S5 behaved in a serviceable manner, with maximum final deflections less than Span/250 and reasonably fine crack widths. The concrete in the 100 mm thick specimen S1 had relatively high tensile strength and elastic modulus at first loading and relatively low shrinkage. S5 also had low shrinkage and zero superimposed sustained load. S2 (100 mm thick) and S3 (90mm thick), despite being relatively lightly loaded, suffered final deflections in excess of Span/250 and significant time-dependent cracking. The concrete in S2 had high shrinkage. S4 was subjected to a sustained load of 58.5% of the collapse load and suffered excessive deflection and excessively wide cracks (increasing in width by a factor of about 5 by age 150 days).

The results confirm that time-dependent cracking greatly affects the serviceability of flat slabs. In all specimens, new cracking occurred with time and existing cracks extended and widened (usually on the top surface). The loss of stiffness resulting from time-dependent cracking, caused the final deflection of each specimen to be significantly larger than that predicted by the code (ACI 318-95). For the slabs under constant sustained load (S2, S3, S4 and S5), the ratio of the measured incremental or time-dependent deflection, $(\Delta_1)_{time}$, to the initial deflection due to the sustained load, $(\Delta_1)_{init}$, is shown in Table 6, together with that specified in ACI 318. Clearly, the code greatly underestimates the time-dependent deflection in all cases (even for the unloaded S5).

Table 6. Long-term to instantaneous deflection ratios.

Specimen	Deflection ratio $(\Delta_1)_{time}/(\Delta_1)_{init}$		Total days under load
	Measured	ACI 318	
S2	3.81	1.5	386
S3	3.44	1.4	292
S4	3.26	1.15	152
S5	1.52	1.1	140

The significantly larger long-term deflections and the more extensive distribution of cracking in S2 (compared to S1) are due to many factors including lower concrete tensile strength, higher shrinkage, higher creep and increased restraint to shrinkage provided by the footing pads under each exterior column. Further research is underway to ascertain the relative significance of each of these factors.

In all specimens the column reactions in C1 and C2 decreased with time and the reactions at C3 and C4 increased. Therefore, the peak negative moment over the interior column (in the E-W direction) increased with time and the positive span moments decreased. This moment redistribution is primarily due to shrinkage and depends to a large extent on the reinforcement layout (Gilbert, 1988). The increase in negative moments with time is consistent with the gradual increase in flexural cracking observed on the top surface of all slab specimens and the relatively small amount of cracking in the positive moment regions. It is also typical of observations on flat slabs with serviceability problems (Gilbert, 1999), where the slab soffit is often free from cracking while the top surface is extensively and excessively cracked.

The average mid-panel deflection of S2 increased from 2.84mm at first loading to 10.66mm at age 100 days (Table 4). The relatively high shrinkage in the first 100 days (in excess of $750\mu\varepsilon$, see Table 2), was responsible for significant time-dependent cracking and the resulting rapid increase in deflection. At age 100 days, a severe hailstorm damaged the roof of the laboratory over S2 and the top surface of the slab was thoroughly wetted. Unexpectedly, within 24 hours, the average mid-panel deflection decreased by 1.51 mm (more than the instantaneous deflection due to the superimposed load). For the next 120 days, the slab was subjected to repeated periods of wetting and drying and deflection did not increase substantially (see Figure 5). No further wetting occurred after age 220 days and the average mid-panel deflection increased from 9.49 mm at age 225 days to 13.67 mm at 400 days. The immediate recovery of deflection on first wetting the slab was due to the rapid absorption of water on the top surface, thus reducing the positive curvature in the mid-span regions. Apparently, the degree of exposure is of paramount importance to the serviceability of slabs.

To explore this effect further, the top surface of S1 was wetted intentionally at age 279 days. During the previous 110 days under a heavy superimposed load, the average mid-panel deflection had increased by just 0.78mm (see Table 3) and the slab appeared to be approaching its final deflection. Within 24 hours of wetting, the average mid-panel deflection decreased by 1.10mm (which is greater than the instantaneous increase in deflection when the superimposed load was doubled at age 169 days). Within two days of wetting flexural cracking occurred in soffit of the slab (see Figure 6d) and deflection increased significantly (from 8.35mm at 280 days to 10.52mm at 301 days, see Table 3). The rapid absorption of water and the resulting reduction of positive curvature on wetting caused a redistribution of slab moments, with the positive span moments increasing and the negative support moments decreasing. The increase in positive bending in the heavily loaded slab caused cracking on the slab soffit and a sudden loss of stiffness. These significant effects on deflection and cracking need further research, but clearly should be considered in design.

3 COMMENTS ON DEFLECTION CALCULATION PROCEDURES

Current design approaches for the calculation of the deflection of flat slabs have recently been shown to be inadequate, Gilbert (1999). They fail to adequately account for the loss of stiffness due to cracking, in particular time-dependent cracking resulting from shrinkage. The tests described here confirm the significance of time-dependent cracking and its influence on long-term deflection.

In general, slabs are lightly reinforced and lose a large percentage of their stiffness when cracking occurs. Shrinkage (and, in practical slabs, temperature changes) usually results in a continuing, gradual expansion of the cracked region in flat slabs, with a resultant gradual reduction in slab stiffness with time. Consequently, the ratio of final deflection to short-term deflection in flat slabs is usually greater than 4.

Recently, several alternative methods were proposed to include the influence of shrinkage induced cracking in deflection calculations, Gilbert (1999). The experimental data generated in the present study will be used to evaluate and calibrate these design proposals.

4 CONCLUSIONS

The time-dependent in-service behaviour of five large-scale flat slab structures has been presented. The deflection, the extent of cracking and the column loads were monitored with time, together with the strength and deformation characteristics of the concrete. Significant time-dependent cracking was observed and the long-term to short-term deflection ratios were significantly greater than those predicted using current codes. The tests described here are part of an on-going experimental study in which eight flat slabs will eventually be tested.

ACKNOWLEDGEMENTS

The study is funded by the Australian Research Council. The assistance of Mr Patrick Zou, Mr X. Guo and the staff of the Heavy Structures Laboratory at UNSW is gratefully acknowledged.

REFERENCES

ACI 318-95. *Building Code Requirements for Reinforced Concrete.* American Concrete Institute. Michigan. USA.
AS3600-1994. *Australian Standard for Concrete Structures.* Standards Australia. Sydney. Australia.
Gilbert, R.I. 1988, *Time effects in concrete structures.* Amsterdam: Elsevier Science Publishers.
Gilbert, R.I. 1999. Deflection calculation for reinforced concrete structures - why we sometimes get it wrong. *ACI Structural Journal* 96 (6): 1027-1032.
Jokinen, E.P. & Scanlon, A. 1985. Field measured two-way slab deflections. *CSCE Annual Conference.* Saskatoon. Canada. 16pp.
Sbarounis, J.A. 1984. Multi-storey flat plate buildings-measured and computed one-year deflections. *Concrete International* 6 (8): 31-35.

Creative Systems in Structural and Construction Engineering, Singh (ed.) © 2001 Balkema, Rotterdam, ISBN 90 5809 161 9

Estimation for influence of initial imperfections to ultimate strength of high tensile steel plates subjected to uniform compression

T. Kaita, K. Fujii & I. Ario
Department of Civil and Environmental Engineering, Hiroshima University, Japan

ABSTRACT: It is well-known that the initial imperfection makes the ultimate strength decrease sensitively. Therefore, the buckling strength curve in lot of specifications is formulated by taking its influence into account, where initial imperfections are usually treated by initial deflection-width ratio (w_0/b) and residual stress-yielding stress ratio (σ_0/σ_y). High tensile steel is widely used in bridges recently. It is clarified that the buckling strength curve for ordinary mild steel plate underestimates that of high tensile steel plate because of different yielding stress. Therefore, the ultimate strength curve that also can predict accurately the ultimate strength of the comprehensive steel plates is required. This paper presents the buckling strength curve for comprehensive material such as high tensile steel, as well as mild steel. The influence of initial imperfection to the buckling strength is investigated by the finite element analysis for the square steel plate subjected to uniform compression. From the analytical results, the ultimate strength of all types of steel can be obtained from the buckling strength curve which used the parameter of initial deflection-thickness ratio (w_0/t) instead of w_0/b, and the initial imperfection affects sensitively to the ultimate strength when the non-dimensional parameter of width-thickness ratio (λ) is in the range of about $\lambda < 1.2$.

1 INTRODUCTION

Recently, many kinds of high performance steel have been developed, such as high tensile steel, low-yielding steel, stainless clad steel etc. It would be feasible and economical to utilize their high abilities for steel bridges. However, the present specifications or design codes are not sufficient for bridge designing of high performance steel. Therefore, there is a necessity to reconsider them in order to utilize the abilities of high performance steels.

(Kitada & Tanaka 1999) indicated that the ultimate strength of high tensile steel plates is underestimated by the usual specifications comparing with its practical strength, because the ratio of residual stresses to yielding stress is relatively lower than that of ordinary mild steel. This fact shows that the magnitude of residual stresses by welding is almost same in both steels. Therefore, another ultimate strength curve, which is different from that for mild steel, has been proposed for high tensile steel (Usami, Fukumoto and Aoki 1981). On the other hand, initial out-of-flatness of a plate has been ordinarily considered by the parameter of w_0/b (w_0: the maximum value of initial deflection, b: the width of plate) in most specifications (JSHB 1996). Though the initial deflection is mainly limited by the accepted tolerance of the applicable fabrication technology, the ultimate

strength curve derived by using the parameter w_0/b varies according its yielding stress. A strength curve, which can express the plate strength without concerning the difference of yielding stresses, would be more useful for bridge design.

In order to utilize comprehensively the performance of high tensile steel, this paper proposes a treatment of initial imperfections in the stability design through the finite element analyses considered with geometrical and material non-linearity. By showing the parameters for evaluating the ultimate buckling strength of simply supported square plates under uniform compression, a feasible parameter for initial deflections is proposed in order to get a single strength curve without concerning its yielding stress. Then, the influence of initial deflection is investigated in the case of with and without residual stresses.

2 NON-LINEAR ANALYSIS WITH INITIAL IMPERFECTIONS

Finite element non-linear analysis is performed for simply supported square plates under uniform compression, by considering initial imperfections. The analytical model is shown in Figure 1. A quarter region of the plate is analyzed with its symmetry, and uniform compressive displacements are given along

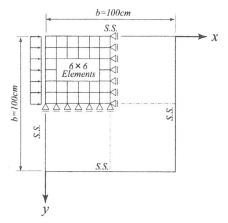

Figure 1. Analytical model

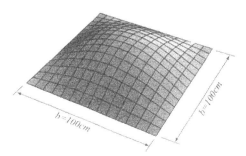

Figure 2. Shape of initial deflection

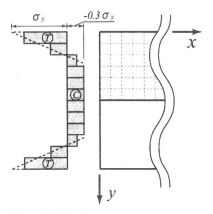

Figure 3. Distribution of residual stresses

an edge, and in-plane boundary conditions are given as in Figure 1.

For initial imperfections, the initial deflection is assumed as in Figure 2, and given as the function:

$$w = w_o \sin\frac{\pi x}{b}\sin\frac{\pi y}{b}$$

Residual stress induced by welding is also assumed as in Figure 3, The iso-parametric shell element (Nukuchal 1979) with four nodal points is used and the up-dated Lagrangean method based on incremental theory (Kawai 1974) is adopted in the analysis. The average compressive stress is calculated by dividing the sum of reaction forces on the loading edge with the cross sectional area. In the analysis, the material non-linearity of both mild and high tensile steel is assumed as perfect elasto-plasticity based on Von-Mises's yielding criteria and Plandtl-Reuss's flow rule (Zienkiewicz 1977).

3 PARAMETERS

The buckling strength curve, which predicts the ultimate compressive strength of square steel plates is usually expressed by the ultimate strength-yielding stress ratio (σ_u/σ_y) and the non-dimensional width-thickness ratio (λ). This curve considers the influence of initial imperfection, which is the initial deflections and the residual stresses occurred by welding. Where,

$$\lambda = \frac{b}{t}\sqrt{\frac{\sigma_y}{E}}\sqrt{\frac{12(1-\nu^2)}{\pi^2 k}} \qquad (1)$$

b: width of plate
k: buckling coefficient
ν: Poisson's ratio
E: modulus of elasticity
t: thickness of plate

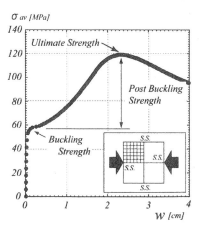

Figure 4. Typical load vs. deflection curve
(σ_y=280MPa, λ=2.0, w_0=0.0096cm, t=0.96cm)

920

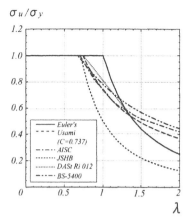

Figure 5. Buckling strength curves

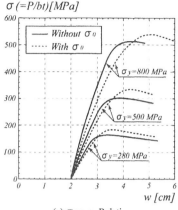

(a) σ vs. w Relation

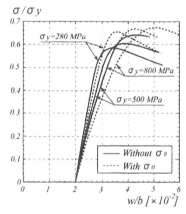

(b) σ/σy vs. w/b Relation

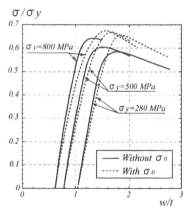

(c) σ/σy vs. w/t Relation

Figure 6. Load vs. deflection curves under uniform compression ($\lambda=1$, $w_0=2$cm)

In the case of Euler's buckling curve, the following equation is derived.

$$\frac{\sigma_u}{\sigma_y} = \frac{1}{\lambda^2} \qquad (2)$$

However, it is well-known that post buckling strength is reliable in thin plate where the practical compressive strength is larger than equation (2), and that initial imperfections make the strength decrease in thick plates. A typical load-deflection curve is shown in Figure 4.

A lot of strength curves have been proposed considering experimental data and analytical results, as shown in Figure 5.

In order to take initial deflection into account, the buckling strength curve is expressed by non-dimensional parameters, which are maximum initial deflection and the thickness or width of plate.

4 RESULT AND DISCUSSION

Based on the finite element analyses, the treatment and the influence of initial imperfection are discussed in the following.

4.1 Treatment of initial deflections

The relation of average compressive stress and deflection at the center of plate is shown in Figure 6, where the average compressive stress is calculated by dividing reaction force along loading edge with the cross-sectional area. The compressive load is given as uniformly prescribed displacements along the edge. In Figure 6, the existence of residual stress (σ_0) is considered and not considered in solid and

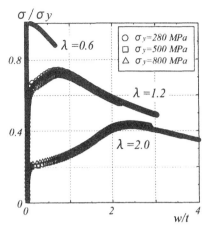

Figure 7. Load vs. deflection curves ($w_0/t=0.01$, without residual stress)

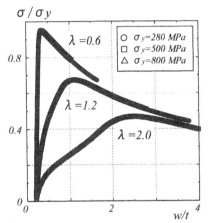

Figure 8. Load vs. deflection curves ($w_0/t=0.2$, without residual stress)

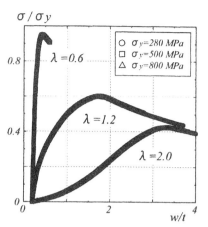

Figure 9. Load vs. deflection curves ($w_0/t=0.2$, with residual stress)

broken line, respectively. Although each figure is expressed by different parameter, Figure 6(a)-(c) show the exactly same load-deflection relationship for each yielding stress. All curves have the same maximum initial deflection $w_0=2$cm and non-dimensional parameter of width-thickness ratio $\lambda=1$. However, yielding stresses of each curve are 280, 500 and 800MPa, respectively. When λ is constant, as the yielding stress becomes larger, the plate becomes thicker as 1.92, 2.57 and 3.25cm corresponding to each yielding stress.

As shown in Figure 6(a), various ultimate strengths are obtained according to the magnitude of each yielding stress of plate. Figure 6(b), which represents the present design method, also gives various ultimate strengths even by taking w/b (deflection divid-

ed by width of plate) and σ/σ_y for abscissa and ordinate, respectively. It means that even under the same initial deflection, the buckling curve becomes totally different for different value of yielding stress. In other words, typical buckling strength curves (as Figure 5) for each type of steel is necessary for buckling design.

The thickness of plate obtained from equation (1) varies according to the magnitude of each yielding stress when non-dimensional parameter w/t is used for abscissa under the condition of constant width-thickness ratio (λ). Therefore, by using parameter w/t as abscissa, the starting point of each curve (w_0/t) varies according to its yielding stress in Figure 6(c). In other words, even under the same initial deflection (w_0), the relative initial imperfection (w_0/t) decreased as yielding stress of steel plates get higher.

Considering the results above, it is decided to observe the influence of initial deflection from a different point of view. The load-deflection curve for each yielding stress of square plate with constant $w_0/t=0.01$, 0.2 are shown in Figure 7 and 8, respectively. They show clearly that for constant λ, all load-deflection curves for various yielding stresses are exactly the same with no concerns of initial deflection (w_0/t) when residual stress does not exist. From these facts mentioned above, it can be concluded that buckling curves for various type of steel can be simplified in a single curve by using parameter (w/t) as abscissa and σ/σ_y as ordinate. This fact is very important for the bridge design using high tensile steel plates since the obtained buckling strength curve can give a more accurate estimation of its ultimate strength.

4.2 Influence of residual stress

Each broken line in Figure 6(a)-(c) shows the load-deflection curve in the case where residual stress is considered as shown in Figure 3. It is noticed that the deflection in general and the ultimate strength becomes larger when residual stress is also considered. The fact is caused by delayed yielding phenomenon, which is occurred by existing residual tensile stresses near the edges of plate and post buckling behavior which concentrates the compressive stresses near the edges. Thus, the residual stresses do not always make the ultimate strength of plate decrease. It seems to be enlarged by the residual stresses in the case of large initial deflections such as $w_0/t=1.0$.

The σ/σ_y vs. w/t relationship in the case of considering residual stress is shown in Figure 9, where three types of yielding stress are used under the condition of constant σ_0/σ_y and $w_0/t=0.2$. As a result, it is noticed that all load vs. deflection curves for various yielding stresses are exactly the same as well as the case without residual stress. Therefore, it could be concluded that when parameters w_0/t and σ_0/σ_y are constant, σ_u/σ_y can be obtained as a unique value with no concerns of the magnitude of yielding stress. The degradation of ultimate strength for $\lambda=1.2$ is the largest among three types of width-thickness ratio $\lambda=0.6, 1.2$ and 2.0.

On the other hand, Figure 9 shows a cautious behavior in $\lambda=2.0$ that the stiffness is extremely lower than that of without residual stress in the region of $\sigma/\sigma_y<0.2$ in Figure 8 and 9. The fact shows the plate has already been buckled by only the residual compressive stresses. Therefore, the extremely lower stiffness is dominated by the post buckling behavior.

4.3 Ultimate compressive strength

As shown in Figure 10, the buckling strength curve can be obtained independently of the magnitude of yielding stress when initial deflection is expressed by w_0/t. It is also noticed that the ultimate strength becomes higher than Euler's buckling curve because of the influence of post buckling when λ is relatively large. As shown in Figure 7 and 8, larger w_0/t makes σ_u/σ_y lower, especially it decrease significantly around $\lambda=1$.

So far, the behavior mentioned above is without residual stress. On the other hand, the broken line shown by black color square is the buckling strength curve considered with residual stresses under $w_0/t=0.2$ and the corresponding curve without residual stress is shown by white color square in the Figure 10. Comparing both curves, the ultimate strength considered with residual stress is lower than that without residual stress when $w_0/t=0.2$. On the other hand, as shown in Figure 6, in the case of large initial deflection (for example, $w_0/t>1$), the residual

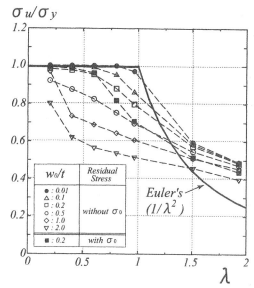

Figure 10. Buckling strength curve with w_0/t

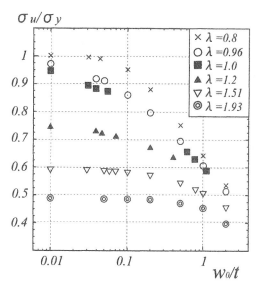

Figure 11. Influence of initial deflection (without residual stress)

stresses make the ultimate strength increase by the reason mentioned in the section 4.2.

It is noticed from Figure 10 that the decrease of ultimate strength due to residual stress is especially influenced around $\lambda=1$, when $w_0/t=0.2$. The residual stress does not affect the ultimate strength of thick plates ($\lambda<0.6$) because the buckling does not occur before the plate yields. It also does not affect the ultimate strength thin plates ($\lambda > 1.5$) since the post buckling behavior dominates the ultimate strength.

Figure 11 shows the influence of initial deflection to ultimate strength for various width-thickness ratio λ, where the residual stress is not considered. It is noticed that the ultimate strength decreases as the initial deflection increases. For example, the initial deflection makes the ultimate strength decrease remarkably at $\lambda = 0.8$ when $w_0/t = 0.1$ and at $\lambda = 1.5$ when $w_0/t = 0.2$.

5 CONCLUSIONS

A treatment of initial imperfections for simply supported square plates under uniform compression is considered in order to utilize high tensile steel for steel bridges. As a result, the followings are concluded:

1. If parameters w_0/t and σ_0/σ_y are used in buckling strength curve, it can express the ultimate compressive strength without concerning the yielding stress of plates. Therefore, the treatment of initial imperfections mentioned above is applicable used for high tensile steel as well as mild steel.

2. In case of remarkably large deflection such as $w_0/t > 1$, the residual stresses make the ultimate compressive strength increase. The fact is caused by the residual tensile stresses near the edges and concentration of compressive stress due to post buckling of the plate.

3. Residual stresses decrease the ultimate strength in the region of width-thickness parameter $0.6 < \lambda < 1.5$.

REFERENCES

Fujii, K., Miki, C. & Fujii, T. 1999. Compressive Strength of Strainless Cladding Steel Plates and Residual Stress Distribution in the Thickness. *Journal of the Japan Society of Civil Engineers*. No.633. I-49. :181-192. (In Japanese)

Fukumoto, Y. (eds) 1997. Structural Stability Design. *Pergamon press.*

Japanese Specifications for Highway Bridges (JSHB). 1996. *Japan Road Association.*

Kawai, T. 1974. The Buckling Problem Analysis. *Baifukan.* (In Japanese)

Kitada, T. & Tanaka, K. 1999. Elasto-Plastic Finite Displacement Analysis and Ultimate Strength of Columns with Box Cross Section. *Journal of Constructional Steel.* Vol.7 November 1999:443-450. (In Japanese)

Nukuchal, W. K. 1979. A Simple and Effective Finite Element for General Shell Analysis. *int. J. for Numerical Method in Engineering.* Vol.4: 179-200.

Usami, T., Fukumoto, Y. & Aoki, T. 1981. TEST ON THE INTERACTION STRENGTH BETWEEN LOCAL AND OVERALL BUCKLING OF WELDED BOX COLUMNS. *Journal of the Japan Society of Civil Engineers.* No.308. :47-58. (In Japanese)

Washizu, K. (eds) 1983. Finite Element Method Handbook. I Basic Edition. (In Japanese)

Washizu, K. (eds) 1983. Finite Element Method Handbook. II Applicable Edition. (In Japanese)

Zienkiewicz, O.C. 1977. THE FINITE ELEMENT METHOD THIRD EDITION. *Mc Grow-Hill.*

Creative Systems in Structural and Construction Engineering, Singh (ed.) © 2001 Balkema, Rotterdam, ISBN 90 5809 161 9

On reinforcement of square steel plates with a hole under cyclic shearing force

K. Fujii & Ronny
Department of Civil and Environmental Engineering, Hiroshima University, Japan

M. Nakamura & M. Uenoya
Department of Civil Engineering, Fukuyama University, Japan

ABSTRACT A manhole set for maintenance in steel bridge pier decreases its ductility and ultimate strength. This paper investigated analytically the relationship between the ductility of stiffened plates with a hole and several factors such as flexural rigidity of longitudinal stiffener and size of doubling plate. Elasto-plastic behavior of perforated square plates subjected to cyclic shearing forces is analyzed by using the finite element analysis considering with geometrical and material non-linearity. The ultimate strength and the ductility of perforated panel are investigated for the reinforcement around a hole by using longitudinal stiffener or doubling plates. From the results, the followings are concluded. 1) The longitudinal stiffener with large relative stiffness is effective to improve the ductility of stiffened plate. 2) The longitudinal stiffener, which still remains after manhole setting, is not effective at all to improve the ductility and the ultimate strength of stiffened plate. 3) Doubling plate can improve the ductility as well as the ultimate strength more effectively when it is thicker.

1 INTRODUCTION

Box bridge pier with stiffened steel plate is widely used in Japan, and it usually has a hole for manufacturing and maintenance. However, there is no confirmed standard for buckling durability design of stiffened plate with a hole. Therefore, steel piers have been designed usually based on the past actual data or the designers' own experiences.

The Great Hanshin-Awaji Earthquake brought a lot of steel pier damages caused by local buckling, which occurred in stiffened plate with a hole at the bottom of pier. As long as the structure has capability to perform enough and stable plastic displacement under excessive earthquake, the structure would not totally collapse even when it is partially destroyed. Based on this concept, the design and strengthening method of stiffened-plate panel with a hole should be established from the ductility's point of view.

In order to investigate the ultimate behavior of stiffened-plate with a hole under cyclic shearing forces and its effective strengthening method in a view of seismic ductility, parametric analyses are carried out by using non-linear finite element method based on incremental theory.

2 FEM ANALYSIS OF STIFFENED PLATE

2.1 *Analyzed Model*

Stiffened plate model is determined by regarding the size and shape of experiments whose scales are corresponded to actual steel piers. As shown in Figure 2, it is a 400mm square stiffened plate with 4.5mm thickness. The width of hole is 80mm transversally and 120mm longitudinally, and it is placed in the center of the panel. Based on this stiffened plate, the models are made according to the relevant parameters. The initial deflection of the panel is given as a half-wave of sinusoidal curve in the direction of plate's width and length.

2.2 *Material Properties and Analytical Condition*

Two-surface model, which based on kinematic hardening rule, is used for being capable to appraise cyclic elasto-plastic behavior. Table 1 shows the material properties based on the result of tensile coupon tests.

Table 1. Material properties.

Yielding stress σ_y (MPa)	343
Modulus of elasticity E(GPa)	206
Poisson ratio ν	0.28

Table 2 shows the list of used parameters, name of analytical models and parameter values, while Figure 5 shows the outline of each analyzed model, where

γ : relative stiffness of stiffener
γ^* : minimum required relative stiffness of stiffener under uniform axial compression
a_c : detachment length of center longitudinal stiffener
B : width of stiffened plate with a hole
A_h : sectional area of hole ($=B^2$)
A_p : sectional area of panel
A_d : reinforced area by doubling plate

Table 2. Models and used parameters.

Used parameters	Model names	Value of parameter
Relative stiffness of stiffener (γ/γ^*)	PLI & HLI	0.65, 1.21, 3.03
Detachment ratio (a_c/B)	PLIC & HLIC	0,0.25,0.625,1.00
Ratio of hole area (A_h/A_p)	HSL & HLL	0.013, .029, 0.101, 0.204
Ratio of reinforcement area (A_d/A_h)	DOL & DSL	0,0.67,1.0, 1.33

In Table 2, the notation P and H in model names show the condition without and with hole respectively. In model PLI and HLI, the plate thickness of longitudinal stiffener is changed and relative stiffness of longitudinal stiffener is taken as a parameter. Model PLIC and HLIC are used to investigate the detachment length of center longitudinal stiffener. The size of hole area is considered in model HSL and HLL, while the size and shape of doubling plate are investigated in model DOL (oval) and DSL (rectangular).

2.3 Loading and Boundary Condition of Cyclic Shear

Figure 6 shows the boundary conditions of analytical model. Four edges surrounding are taken as simply supported for out-of-plane. In-plane displacement along one edge is restrained, while uniformly distributed load is given as pure shearing stress along the other three edges. The displacement controlling method is applied at the center on the opposite side. Controlled displacement in x-direction is corresponded to the yielding displacement δ_y ($=\gamma_y l$). The cyclic shearing loads are given while gradually increasing the displacement amplitude as shown in Figure 7.

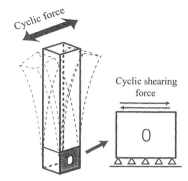

Figure 1. Steel pier under cyclic shearing force.

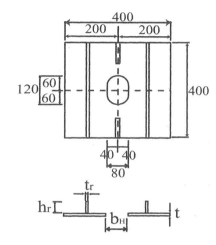

Figure 2. Basic dimensions for analysis.

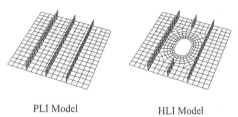

PLI Model   HLI Model

Figure 3. Element mesh for analysis.

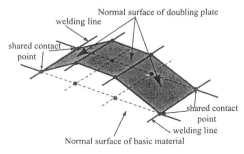

Figure 4. Joint condition between doubling plate and panel.

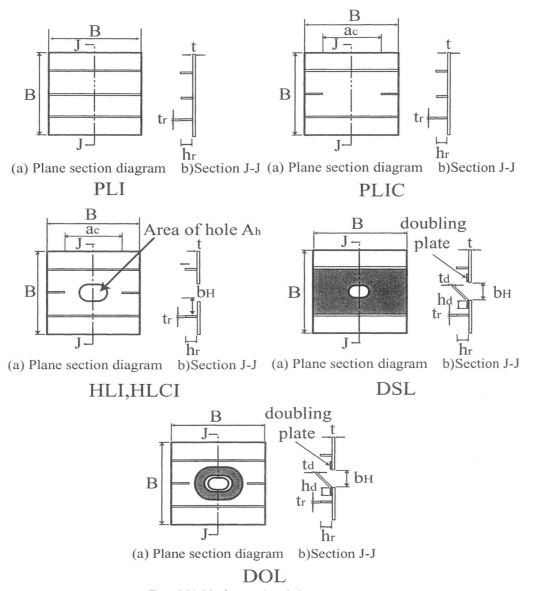

(a) Plane section diagram   b)Section J-J

PLI

(a) Plane section diagram   b)Section J-J

PLIC

(a) Plane section diagram   b)Section J-J

HLI,HLCI

(a) Plane section diagram   b)Section J-J

DSL

(a) Plane section diagram   b)Section J-J

DOL

Figure 5. Models of parametric analysis.

3 RESULTS AND CONSIDERATIONS

3.1 Relative Stiffness of Longitudinal Stiffener

Figure 8 shows the envelope curve of models that took the relative stiffness of longitudinal stiffener as a parameter. From this figure, because all stiffened panels in both cases of with or without a hole have enough ductility, each longitudinal stiffener in this study has an enough relative stiffness against the cyclic shearing load. However, when $\gamma/\gamma* < 0.65$, the buckling of longitudinal stiffener occurs and its ductility decreased remarkably after $6\delta_y$. Furthermore, it is noticed that the ultimate strength of panel with a hole can not rise up to that of without a hole, even though the relative stiffness becomes larger.

Figure 9 shows the displacement state at $8\delta_y$ of models with hole. The colored area shows the expansion of plastic region. In models without hole, while being independent to relative stiffness, the stiffener restrains the deflection effectively and as a result, the stiffener performs enough ductility. On

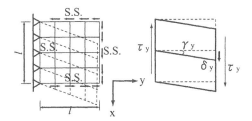

Figure 6. Boundary conditions and shearing load.

displacement

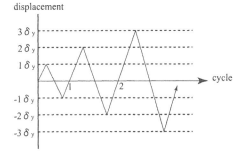

Figure 7. Loading condition.

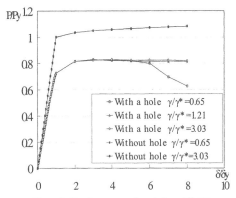

Figure 8. Envelope curve for relative stiffness.

the contrary, in models with hole, the sectional area loss increases the deflection of center panel.

3.2 *Detachment of Longitudinal Stiffener along the Center Line*

Figure 10 shows the envelope curve of models for the detachment ratio of longitudinal stiffener. Since the detachment length is found to decrease the ultimate strength, it is preferable to make the detachment as short as possible. However, the detachment ratio in actual steel pier might be usually greater than 0.625. Therefore, it can be concluded that the detached stiffener is not effective for the ultimate strength improvement.

By comparing models at $8\delta_y$ as shown in Figure 11, it is noticed that there is almost no difference in deflection state between models with and without center longitudinal stiffener. This fact also emphasizes that the detached longitudinal stiffener is not effective at all.

3.3 *Ratio of hole area*

Figure 12 shows the envelope curve for ratio of hole area. Model without longitudinal stiffener is also shown in Figure 12 in order to clarify the influence of area ratio. As easily expected, the initial stiffness and maximum load become smaller than model without hole. For small ratio, the decrease of ultimate strength after maximum load, is almost the same with that of without hole.

3.4 *Reinforcement by Doubling Plate*

Figure 13 shows the envelope curves for reinforcement by doubling plates. Since the displacement around hole is restrained in oval doubling plate model, the ultimate strength and ductility of reinforced plate are successfully improved to the level of models without hole. On the other hand, maximum load greatly increases the ultimate strength and duc-

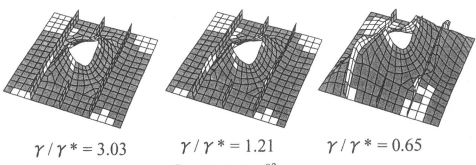

$$\gamma / \gamma * = 3.03 \qquad \gamma / \gamma * = 1.21 \qquad \gamma / \gamma * = 0.65$$

Figure 9. Deflection at $8\delta_y$.

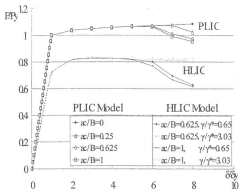

Figure 10. Envelope curve for detachment ratio

tility of plate in rectangular doubling plate model. It can be concluded from the result above that rectangular doubling reinforcement is more effective to improve the ductility and ultimate strength.

As easily expected, the ultimate strength shows better improvement for thicker doubling plates. However, it should be noticed that the ductility of a global steel pier might not increase when it is partially over reinforced.

4 CONCLUSION

Non-linear finite element analysis is performed for stiffened-plate panels with a hole at the center, and their ductility and ultimate behavior are investigated under cyclic shearing force. From the results, the followings are concluded:

1. When the relative stiffness of longitudinal stiffener is small, overall buckling occurs in stiffened panel. Since the global buckling of stiffened plate is restrained in the model with relative stiffness greater than 3.03, it would be preferable to make it greater than 3.03.

2. The longitudinal stiffener, which still remains after manhole setting, is not effective to improve the ductility of stiffened-plate panel.

3. Implementing oval or rectangular doubling plate can restrain the deflection of panel around hole. Therefore, the maximum load and ductility can be improved effectively. However, when the center sub-panel is over reinforced, the ultimate strength might not increase so much because local buckling of sub panels would occur.

4. Comparing the shapes of doubling plates, the ductility and ultimate strength greatly increase in rectangular doubling plates. Therefore, a new index that also considers the shape of doubling plate should be attached to the present design concept in which cross sectional area of doubling plate is considered only for reinforcement around hole.

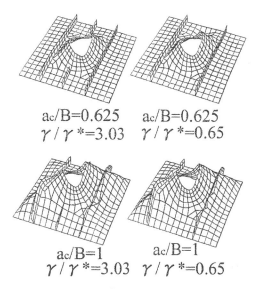

$a_c/B=0.625$ $a_c/B=0.625$
$\gamma/\gamma^*=3.03$ $\gamma/\gamma^*=0.65$

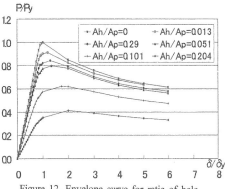

$a_c/B=1$ $a_c/B=1$
$\gamma/\gamma^*=3.03$ $\gamma/\gamma^*=0.65$

Figure 11. Deflection at $8\delta_y$

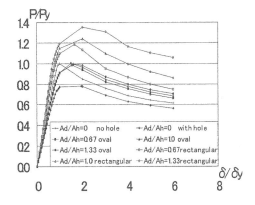

Figure 12. Envelope curve for ratio of hole

Figure 13. Envelope curve for doubling plate

REFERENCES

Dafalias, Y.F. & Popov, E.P. 1975. A Model of Nonlinear Hardening Materials for Complex Loading. pp.173-192.

Japanese Specifications for Highway Bridges (JSHB) 1996 Japan Road Association.

Nakamura, M., Fujii, K., Uenoya, M. & Kotaniguchi, Y. 1998. Influence of Opening on Strength and Ductility of Steel Box Column Under Gradually Increasing Cyclic Loading. *Journal of Structural Engineering.* Vol. 44A pp.159-168. (In Japanese)

Nukuchal, W.K. 1979. A Simple and Effective Finite Element for General Shell Analysis, *int.J. for Numerical Method in Engineering*, Vol.4, pp.179-200.

Shen, C., Mizuno, E., Tanaka, Y. & Usami, T. 1992. A Two-Surface Model for Steel with Yield Plateau. *Structural Eng./Earthquake Eng.*, Vol.8, No.4, pp.179-188.

Shen, C., Mizuno, E. & Usami, T. 1993. A Generalized Two-Surface Model for Structural Steels under Cyclic Loading. *Structural Eng./Earthquake Eng.*, Vol.8, No.4, pp.59-69.

Usami, T. & Ge, H.B. 1998. Cyclic Behavior of Thin-Walled Steel Structures-Numerical analysis, *Thin-Walled Structures*, Vol. 32, No. 1-3, 9-11.

Washizu, K. 1983. Finite Element Method Handbook I Basic Edition. In Washizu, K.(eds).(In Japanese)

Washizu, K. 1983. Finite Element Method Handbook II Application Edition. In Washizu, K.(eds). (In Japanese)

Zienkiewicz, O.C. 1977. The Finite Element Method Third Edition, Mc Grow-Hill.

Creative Systems in Structural and Construction Engineering, Singh (ed.)© 2001 Balkema, Rotterdam, ISBN 90 5809 161 9

Stress analysis of precast slabs subjected to bending and torsion

D.A.Uy
Department of Structural Engineering, Hiroshima University, Japan

S.Nakano
Department of Civil Engineering, Kure National College of Technology, Hiroshima, Japan

ABSTRACT: When joints connecting precast slabs together carry not only shear force and bending moment but also large torsion, diagonal reinforcements to resist crushing and failure at these points are laid out. So far, the analysis of slabs has been limited to those reinforced in 2 directions which gives ultimate strengths 20~30% less than its experimental value.

Analysis of the combined ultimate torsion and bending moment in slabs reinforced in the transverse, longitudinal, and diagonal directions and the effect of joints are presented. Two failure mechanisms due to the yielding of steel in slabs, one where the applied bending moment is greater than torsion, and another where torsion is greater than bending moment are presented. Using derived equations, crack directions and ultimate loads are obtained. Theoretical values and experimental results show good agreement.

1 INTRODUCTION

Composite precast slabs composed of in-situ concrete and precast concrete has often been used. In the case where joints connecting precast slabs in two-way slabs exist in areas subjected to torsion, torsional stiffness and torsional moment decrease (Ackermann 1995 & Zararis 1986, 1988).

Reinforced concrete members subjected to torsion can be analyzed using the skewed bending method or space truss model. In this paper, extending the failure mechanism (Zararis 1986) of an orthogonal two-way reinforced plate subjected to bending and torsion, the failure mechanism of slabs reinforced in three directions is investigated. In this case, based on the fracture process obtained from experiments performed, considering that only the section of the pre-

cast slab is valid against torsion, an effective depth for the composite precast slab is determined.

2 EXPERIMENT

To investigate the effect of torsion and bending on reinforced concrete slabs, two series of experiments were performed.

The first series, named SK series, consisted of 5 square slabs, one of which was homogeneous and four of which were composite, a combination of two precast slabs joined together by in-situ concrete. The homogeneous slab had side dimensions, b and l, of 53 cm × 53 cm and a thickness h of 7.5 cm (Fig.1a), while the composite slabs had dimensions of 50 cm

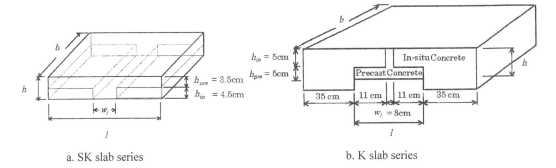

a. SK slab series   b. K slab series

Figure 1. Slab series used in the experiment

Table 1. Details of test slabs

Slab no.	Concrete cylindrical compressive strength (kN/cm²)	Steel yield strength f_y (kN/cm²)	Cross section dimensions Width b (cm)	Thickness h (cm)	b/h	Reinforcement in each face Longitudinal / Transverse Diameter/spacing	A_{sx}、A_{sy} (cm²/cm)	Diagonal Diameter/spacing	A_{sxy} (cm²/cm)	Arms of applied load e_m (cm)	e_t (cm)
I. SK Series											
SK0	3.31	32.37	53	7.5	7.07	5.8mm@5.3cm	0.0498	5.8mm@3.75cm	0.0705	0.0	44.6
SK1	3.13 (2.74)	32.37	50	8.0	6.25	5.8mm@6.2cm	0.0426	5.8mm@4.38cm	0.0603	0.0	43.0
SK2	3.13 (2.74)	32.37	50	8.0	6.25	5.8mm@6.2cm	0.0426	5.8mm@4.38cm	0.0603	0.0	43.0
SK3	3.13 (2.74)	32.37	50	8.0	6.25	5.8mm@6.2cm	0.0426	5.8mm@4.38cm	0.0603	0.0	43.0
SK4	3.13 (2.74)	32.37	50	8.0	6.25	5.8mm@6.2cm	0.0426	5.8mm@4.38cm	0.0603	0.0	43.0
II. K Series											
K1	3.04 (3.16)	34.34	40	10	4	5.8mm@6.0cm	0.0440	5.8mm@4.243cm	0.0623	0.0	35.0
K2	2.57 (3.16)	34.34	40	10	4	5.8mm@6.0cm	0.0440	5.8mm@4.243cm	0.0623	7.5	15.0
K3	3.20 (3.16)	34.34	40	10	4	5.8mm@6.0cm	0.0440	5.8mm@4.243cm	0.0623	15 0	15.0
K4	2.62 (3.16)	34.34	40	10	4	5.8mm@6.0cm	0.0440	5.8mm@4.243cm	0.0623	30.0	15.0

×50 cm and a thickness of 8.0 cm (Fig.1b). Widths of web w_j were varied (12, 12, 8 and 5 cm) to investigate its effect on slabs subjected to pure torsion.

The second series, K series, consisted of 4 slabs measuring 40 cm in width and 100 cm in total length. All 4 slabs were made up of two parts of in-situ slabs thickened 3 cm at its edges to prevent failure due to the crushing of concrete at support and load points when both torsion and bending are applied. The 4 slabs were the same in composition and steel ratio, and all had the same 8cm-web width. The only difference was the applied bending moment to torsion load ratio M/T (Fig. 2).

Reinforcements on both series consisted of six layers of steel bars inclined in three directions, two of which were inclined to the direction of the bending moments and a third inclined to resist the widening of cracks due to torsion. It was set in a way so that there were three layers each on its upper and lower half. These bars have a yield strength f_y of 330 MPa, a modulus of elasticity E_s of 200000 MPa, and a concrete ultimate strain ε_u of 0.0035. Details of the

test slabs and loading conditions are given in Table 1.

3 FAILURE MECHANISMS

3.1 *Strains and stresses on cracks*

When slabs crack, the direction of crack opening is usually assumed to be perpendicular to the direction of crack. As reinforcing bars are subjected to the forces acting on, above and between the cracks, they are forced to take the direction perpendicular to the crack.

Considering that strains of steel bars at different levels are approximately proportional to effective depths, strains (Fig.3) and stresses can be obtained. Letting the angle of the initial cracks to be ϕ_1 and the lever arm of steel forces at crack to be z, the tensor of reinforcement is shown in Figure 4. When this tensor is equated with the moment tensor of a reinforced concrete plate carrying moments that was transformed into the direction of cracks, the boundary conditions at crack are obtained. Making

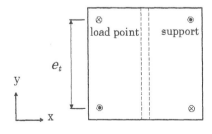

a. SK slab series

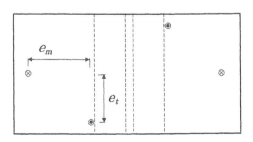

b. K slab series

Figure 2. Loading conditions

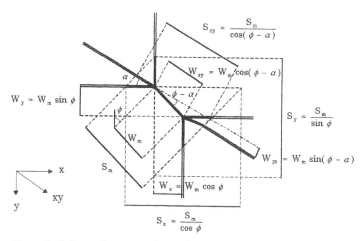

Figure 3. Deformations at crack opening

where W_m=crack width; S_m=crack spacing; α =angle of the diagonal reinforcement from the y-axis; and ϕ =angle from the y axis to the crack direction.

the appropriate transformations, the equation for the initial crack direction, which is independent of steel stresses, can be derived.

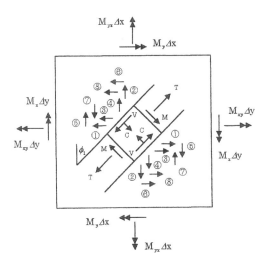

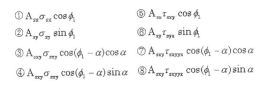

① $A_{sx}\sigma_{sx}\cos\phi_1$

② $A_{sy}\sigma_{sy}\sin\phi_1$

③ $A_{sxy}\sigma_{sxy}\cos(\phi_1-\alpha)\cos\alpha$

④ $A_{sxy}\sigma_{sxy}\cos(\phi_1-\alpha)\sin\alpha$

⑤ $A_{sx}\tau_{sxy}\cos\phi_1$

⑥ $A_{sy}\tau_{syx}\sin\phi_1$

⑦ $A_{sxy}\tau_{sxyyx}\cos(\phi_1-\alpha)\cos\alpha$

⑧ $A_{sxy}\tau_{sxyyx}\cos(\phi_1-\alpha)\sin\alpha$

Figure 4. Forces on slab element before yielding

3.2 Ultimate crack direction

After the slab's initial cracking, increasing the applied load leads to an increase in ε_{cr}, strain at crack, until its yield point is reached. Steel reinforcements then yield in one or two directions, lose its ability to carry shear force and approach a modulus of elasticity of zero. These and the fact that a significant increase in shear stresses between cracks is observed result in the formation of new cracks in other directions, which are the ultimate ones and are denoted as the angle ϕ_2.

3.3 First failure mechanism

This mechanism, known as the bending mechanism, occurs when the ratio between moment and torsion M/T is greater than 1. Due to a smaller torsional load, slip between surfaces doesn't occur. And because one of the principal moments is too small to cause cracks, slabs with this failure mechanism usually aren't fully cracked and the yielding of steel occurs on one face of the slab.

For slabs reinforced in three directions, three cases are presented to account for the yielding of steel in whichever direction, that is, be the reinforcing bars in the x direction yield first, or that in the y or in the third xy direction.

However, as can be seen from the loading conditions, it is usually steel in either the x or xy direction that will yield first. In order to limit the cases that has to be presented, consideration is given only for these two cases and for the case where reinforcement in both the x and xy directions yield simultaneously.

After the formation of ultimate cracks, an increase in loading causes an increase in the widening of the crack opening until steel in the other direction yields as well. Depending on which reinforcement yields first, five cases of steel stresses and its corresponding equilibrium conditions are determined for different regions of ϕ_2 at failure.

3.4 Second failure mechanism

This mechanism occurs when M/T is less than 1. Due to greater torsion, slip between surfaces occurs and complete cracking is observed on both faces of the slab. Thus, the lever arm z of the internal forces is taken to be equal to the vertical distance between the steel layers reinforcing both faces of the slab.

For both mechanisms, to solve for the internal moments of the slab, ϕ_2 is substituted to the equilibrium condition in the appropriate case where it failed. Converting this tensor to the direction of steel bars, torsion and bending moment at failure can be obtained.

4 DISCUSSION OF EXPERIMENTAL AND THEORETICAL RESULTS

Data from the experiment were used to check and determine the difference between experimental and theoretical values for ultimate torsion and bending

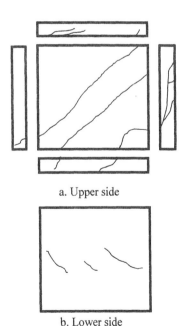

a. Upper side

b. Lower side

Figure 5. Crack diagram of SK4 slab

moment of each slab.

A program was made to simplify calculations. In this program, equations derived were used and different cases were considered.

The slabs tested, except for SK0, were all a combination of reinforced precast and in-situ slabs. In this case, the effective depths used were decreased in proportion to the decrease in strength of the concerned slabs brought about by the horizontal and vertical sliding surfaces between precast and in-situ concrete, where the slabs ultimately collapsed (Fig. 5). Because slabs more or less cannot carry additional load once sliding between precast and in-situ slab occurs, the effective cross-sectional area of the whole slab is considered to be only the precast slab's cross-sectional area. This effective cross-sectional area is assumed to be distributed equally along the original length at uniform thickness (Fig. 6).

a. SK slab series

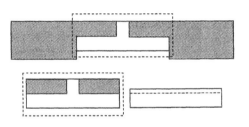

b. K slab series

Figure 6. Equivalent theoretical cross-section

Expressing the reduced effective depth in equation form,

$$d_{reduced} = \frac{\dfrac{w_j}{l} h_{in} + h_{pre}}{h} \times d_{original} \qquad (1)$$

Comparing experimental and Finite Element Method values of steel stresses for SK0 slab which is homogeneous, FEM results show good agreement with experimental ones proving that FEM is applicable for the analysis of slabs subjected to torsion. FEM results for composite slabs were obtained after decreasing the effective areas in proportion to the decrease in strength of the concerned slabs brought about by the horizontal and vertical sliding surfaces between precast and in-situ concrete as discussed above. This method of reduction gives values for ultimate torsion close to experimental ones.

Table 2. Comparison between experimental and theoretical ultimate moments

Slab No.	Experimental Results		Theoretical Results		$T_{u,exp}/T_{u,theo}$
	M_u (kN cm)	T_u (kN cm)	M_u (kN cm)	T_u (kN cm)	
I. SK Series					
SK0	0	939.626	0	941.995	0.997
SK1	0	970.209	0	835.714	1.161
SK2	0	759.294	0	736.074	1.032
SK3	0	715.002	0	908.818	0.787
SK4	0	674.928	0	743.098	0.908
II. K Series					
K1		978.548	0	913.537	1.071
K2	312.694	625.388	321.13	498.839	1.254
K3	338.445	338.445	353.2	353.199	0.958
K4	500.310	250.155	338.38	169.183	1.479

Comparison of experimental and theoretical values obtained for ultimate torsion and bending moment is given in Table 2.

It must be noted, however, that because the known values needed to solve for ultimate torsion and bending moment are particularly defined in the derivation of equations, results vary according to how exact these values are taken.

4.1 Width of web–ultimate torsion relationship for SK slab series

Although there were several differences in the composition of the slabs tested, focusing only on the magnitude of the width of web and the ultimate torsion, experimental results show that the narrower the width of web, the lesser is the ultimate torsional capacity. After varying the width of web of the slabs and applying the theory discussed, width of web was plotted against the calculated ultimate torsion, the

result of which showed a linear relationship. From Figure 7, the following assumptions can be said,

1. The greater the width of web, the greater is the ultimate torsional capacity the slab can carry; therefore, the stronger is the slab.
2. The greater the steel ratio, the steeper the curve; therefore, the greater is the ultimate torsional capacity.
3. The lesser the effective depth, the gentler is the slope of the line. (Note that SK1, SK2, SK3 and SK4 had the same steel ratio. The only difference was the slabs' effective depths.)

4.2 M/T–ultimate torsional capacity relationship for K slab series

Experimental results show that slabs of the same steel ratio and design but with different bending moment to torsion ratio show higher torsional capacity when M/T ratio is smaller. Using data from the experiment and varying only the said ratio, M/T ratio was plotted against theoretical ultimate torsional capacity, the result of which was a non-linear curve line. From Figure 8, the following can be said,

1. The greater the M/T ratio, the lesser the ultimate load the slab can carry; therefore, the lower is its ultimate torsional capacity.
2. The rate of decrease in ultimate torsional capacity slackens when M/T ratio is greater than the unit. (Except for a small difference in effective depths of the reinforcing bars, K series slabs were all the same in composition and dimension, resulting in curves that almost overlap one another when M/T is greater than 1).
3. The slope of the curves when M/T is less than the unit is steep, that is, the rate of increase in ultimate torsion increases as M/T approaches 0. (If all data, except those of effective depths, are the same, as in the K series slabs, the greater the effective depth, the greater is the ultimate torsional capacity as M/T approaches 0).

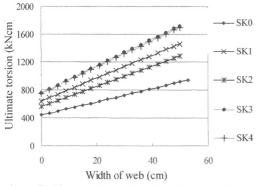

Figure 7. Theoretical relationship between width of web and ultimate torsion

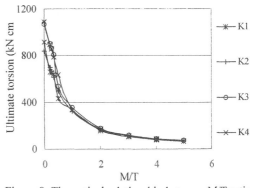

Figure 8. Theoretical relationship between M/T ratio and ultimate torsion

5 SUMMARY AND CONCLUSIONS

Failure mechanisms due to the yielding of steel in slabs reinforced with bars perpendicular to each other and a third diagonal direction are presented. The first mechanism occurs when the applied moment to torsion M/T ratio is greater than the unit, and usually doesn't lead to full cracking of slabs. The second failure mechanism occurs when the applied moment to torsion ratio is less than the unit and is accompanied by a slip between crack surfaces, which leads to full cracking. Theoretical results for homogeneous and composite slabs with varying width of web and M/T ratio show good agreement with those of experimental ones. Analysis of theoretical and experimental results leads to the following conclusions,

1. Initial and ultimate crack directions of slabs are dictated by the compatibility and equilibrium conditions considering that reinforcing bars possess shear strengths.
2. Effective depths of steel reinforcements for composite slabs need to be reduced in proportion to the effective cross-sectional area to account for the weakening in ultimate strength brought about by sliding surfaces where the slabs ultimately failed.
3. Composite slabs subjected to pure torsion showed that width of web is directly proportional to ultimate torsional capacity. Slopes of the width of web-ultimate torsion linear curve for different slabs vary according to other factors such as the reinforcing bars' effective depth, steel ratio, etc.
4. Theoretical calculation for slabs subjected to various M/T load ratio shows that the said ratio possesses a non-linear relationship with ultimate torsion. Slope of the resulting curve at $M/T > 1$ is gentle while at $M/T < 1$ is steep. Magnitude of ultimate torsional capacity for different slabs vary according to slab composition and design, i.e. effective depth, steel ratio, etc.

REFERENCES

Ackermann, G. 1995. Die Drillweiche orthotrope, elastische Platte als Berechnungsmodell fur zweiachsig gespannte Filigrandecken. *Beton- und Stahlbetonbau* 90 H.3, S.57-63.

Peter. S. 1996. Drillstefigkeit von Fertigplatten mit Statisch mitwirkender Ortbeton-schnicht. *Beton- und Stahlbetonbau* 91, Heft 3.

Schiessl, P. 1995. Zuschrift zu [3]. *Beton- und Stahlbetonbau* 90, H.10, S.269-270.

Zararis, P.D. & Penelis G.Gr 1986. Reinforced Concrete T-Beams in Torsion and Bending. *ACI Journal, Technical Paper Title no.* 83-17.

Zararis, P.D. 1988. Failure Mechanisms in RC Slabs. *J. Struct. Engrg., ASCE* 114(5): 997-1016.

Creative Systems in Structural and Construction Engineering, Singh (ed.) © 2001 Balkema, Rotterdam, ISBN 90 5809 161 9

Instabilities in continuous composite beams induced by quasi-viscoelastic slab behaviour

M.A. Bradford & Z. Vrcelj
School of Civil and Environmental Engineering, University of New South Wales, Sydney, N.S.W., Australia

ABSTRACT: This paper describes models for the rational in-plane analysis of continuous composite beams subjected to creep and shrinkage of the slab, and the out-of-plane analysis of the buckling of the steel joist. The results of the in-plane analysis are used in the finite-element based out-of-plane analysis to determine the load factor against out-of-plane (restrained-distortional) buckling. This problem is a generalisation of a situation where quasi-viscoelastic rheology in one component is coupled with instability in the other component of a bi-material composite. Although the quasi-elastic shrinkage and creep behaviour in concrete are conventionally associated with serviceability limit states, it is shown that this rheology can in theory reduce the load factor against buckling in the steel, which is a strength limit state. The ramifications of this erosion of the local buckling factor are illustrated quantitatively and discussed.

1 INTRODUCTION

Conventionally, the quasi-viscoelastic rheology of reinforced concrete that induces shrinkage and creep deformations is associated with the serviceability limit state in engineering structures. The associated service responses are usually those of time-dependent deflections and cracking, and provisions to control these are included in most national design codes of practice. However, it has been shown that quasi-viscoelastic deformations in concrete and composite steel-concrete structures can lead to geometric instability, which is usually considered to be a strength limit state.

The so-called creep buckling behaviour of slender, eccentrically-loaded concrete columns is fairly well-known and documented (Gilbert 1988, Gilbert & Bradford 1990, Bradford 1997, 1998a,b). Less well-known is the instability which may occur in thin steel sheeting that is juxtaposed with concrete that undergoes quasi-viscoelastic deformation, and which acts compositely with the concrete. This behaviour has been observed and quantified in composite profiled beams (Uy & Bradford 1995) and quantified theoretically in thin-walled concrete-filled steel tubes (Bradford

1998c) and composite profiled walls (Bradford et al. 1998). The purpose of this paper is to report the instability that may arise in the steel joist of a continuous composite T-beam due to quasi-viscoelastic deformations which occur in the concrete slab.

Gilbert & Bradford (1995) presented a flexibility-based approach for determining the response of a shored composite propped-cantilever T-beam which undergoes deformations due to creep and shrinkage, and showed that the bending moment redistribution that takes place predominantly due to shrinkage is substantial. Of particular significance is the increase in both the magnitude and extent of negative or hogging bending that occurs near the fixed support, and this was examined in the light of the serviceability limit states of deflection and concrete cracking. Bradford et al. (1999) extended the flexibility-based approach to consider two-span beams with point loads placed arbitrarily in within the spans, and which could model both shored and unshored construction. While the time-dependent increase in the negative bending region was again quantified, the ramifications that this may have on instability of the joist was were not alluded to.

The buckling which may occur in the negative moment region of a composite T-beam is well-known, but quantifying this accurately is still a grey area in the research of engineering structures. This difficulty arises because the buckling mode is associated with distortion of the cross-section (herein denoted a restrained distortional buckle or RDB), and in a composite beam both the bending moment and axial force in the steel joist vary along the length of the member. An elastic rational finite element method of analysis of this buckling has been consolidated and presented (Vrcelj et al. 1999). Because the in-plane analysis including quasi-viscoelastic deformations developed by Bradford et al. (1999) is able to determine the varying stress resultants in the steel joist in the time domain, these stress resultants may be used as input for the (uncoupled) out-of-plane method of Vrcelj et al. (1999) that uses a line-type finite element developed by Bradford & Ronagh (1997). This paper therefore makes recourse to the numerical uncoupled in-plane (quasi-viscoelastic) and out-of-plane (RBD) methods of analysis to investigate erosion of the elastic buckling load factor due mainly to shrinkage in unshored composite T-beams that would be typical of bridge girders. It is shown that the elastic RDB load factor is indeed eroded quite significantly in the time domain in this theoretical treatment. With the fairly-well accepted knowledge that contemporary design of composite T-beams against buckling is very conservative, such an erosion of the elastic buckling load factor would not be considered to be of concern in existing beams. However, since more rational and accurate methods of predicting RDB and which remove the conservatism of existing techniques are evolving, the consideration of quasi-viscoelastically induced buckling must be borne in mind in these more accurate buckling models.

2 THEORETICAL MODELS

2.1 *General*

The present idealised model that of an unshored propped cantilever as shown in Fig. 1. The in-plane method of analysis summarised very briefly in Sect. 2.2 is used firstly to determine the stress resultants, and these are used subsequently in the finite element RDB model described very briefly in Sect. 2.2. The use of the latter finite element

model enables the elastic buckling load factor to be determined in the time domain.

2.2 *In-plane model*

The in-plane model has been described in detail by Bradford et al. (1999). It is based on the flexibility method of analysis, with the quasi-viscoelastic rheology being included by recourse to Bazant's Age-Adjusted Effective Modulus Method for concrete (Gilbert 1988). This structural idealisation is one-fold indeterminate, and the redundant reaction (such as the vertical reaction at the propped end) varies in time. This reaction is calculated by iteration in the model of Bradford et al. (1999), and following the iterative scheme the bending moment and axial force distributions at a number of points along the steel joist are established. These stress resultants are then input into the out-of-plane RDB model described in the following sub-section.

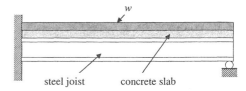

Figure 1. Propped cantilever model

2.3 *Out-of-plane model*

The out-of-plane model is a line-type bifurcative elastic finite element method developed by Bradford & Ronagh (1997). It accounts for cross-sectional distortion experienced by the web by treating this plate as deforming as a cubic curve, as illustrated in Fig. 2. As noted previously, the overall buckling mode that occurs in a composite beam in negative bending is necessarily one of RDB, since the concrete slab provides significant restraint to the top (tension) flange of the steel joist. In this paper, this restraint is assumed to be that of complete rigidity against lateral, rotational and twist deformations.

When input from the in-plane analysis, the vector of stress resultants $\bar{r}$ are used to generate the geometric stiffness matrix $\bar{s}$. Because the geometry of the cross-section is fixed, the elastic stiffness matrix $\bar{k}$ remains unchanged. The standard buckling eigenproblem represented by

$$\left(\bar{k} - \lambda \bar{s}\right)\bar{Q} = \vec{0} \tag{1}$$

is then established, where λ is the elastic buckling load factor, and $\bar{Q}$ is the vector of buckling degrees of freedom. Invoking a standard eigensolver enables the eigenvalue λ and eigenvector $\bar{Q}$ to be determined readily.

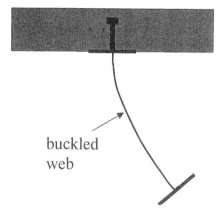

Figure 2. Distorted web during RDB

3 NUMERICAL RESULTS

The in-plane quasi-viscoelastic analysis has been applied to an unshored composite propped cantilever subjected to a sustained uniformly distributed load of 5 kN/m. The slab a width of 1500 mm and depth of 130 mm, and full interaction between concrete slab and steel joist at the interface was assumed. Figures 3 to 7 show some results for different geometries of the steel joist, in which the ratio of the long-term buckling load factor to its short-term counterpart, λ_L/λ_S is plotted as a function of time. In modelling the creep and shrinkage, the aging coefficient was taken as 0.8, and the creep coefficient ϕ and shrinkage strain ε_{sh} were assumed to be given by the same compliance function $F(t)$ as

$$\begin{pmatrix} \phi \\ \varepsilon_{sh} \end{pmatrix} = F(t) \begin{pmatrix} \phi^* \\ \varepsilon_{sh}^* \end{pmatrix} \tag{2}$$

where (Terrey et al. 1994)

$$F(t) = \frac{t^{0.7}}{t^{0.7} + 32} \tag{3}$$

in which t is in days, and the values at $t \to \infty$ are $\phi^* = 3.5$ and $\varepsilon_{sh}^* = 1000 \times 10^{-6}$.

It can be seen from the figures that the elastic buckling load factor can be eroded due to the effects of creep and, particularly, shrinkage. The figures in numerical sequence from Figs. 3 to Fig. 7 have steel joist geometries with simultaneous increasing web slenderness (h/t_w) and flange stockiness (t_f/b_f) ratios. Apart from the cross-section depicted in Fig. 3, the remaining curves are reasonably consistent numerically. With the combination of the final values ϕ^* and ε_{sh}^* and the compliance function $F(t)$, the short-term elastic buckling load factor is eroded in the long-term, up to about 70% in some cases.

The magnitude of the erosion of the load factor up to times of around 100 days is exaggerated, since significant shrinkage takes place during curing when the composite action has not mobilised. Nevertheless, the effect is shown in the figures to be quite severe. It is also worth noting that the final buckling factor that determines the buckling strength of the beam is derived from both the elastic buckling load and the plastic moment of the cross-section, as noted in Oehlers & Bradford (1995). This effect also reduces the severity of the time-dependent erosion of the load factor.

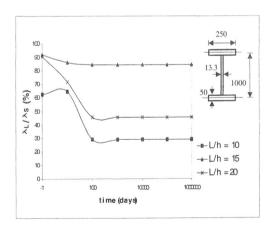

Figure 3. Buckling results for beam B1

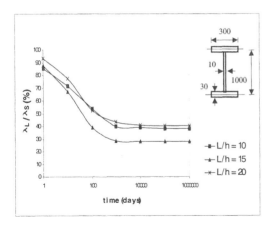

Figure 4. Buckling results for beam B2

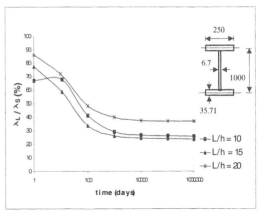

Figure 6. Buckling results for beam B4

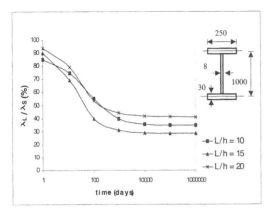

Figure 5. Buckling results for beam B3

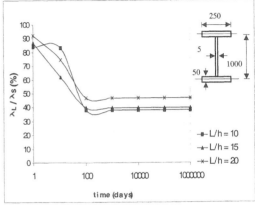

Figure 7. Buckling results for beam B5

4 CONCLUDING REMARKS

This paper has made recourse to two methods of analysis to address the issue of the dependence of the elastic buckling load factor of a continuous composite beam caused by shrinkage and creep. Firstly, a quasi-viscoelastic model has been used to determine the redistribution of bending moment and axial force within the steel joist in the time domain. The results of this in-plane analysis were then used as input data for a finite element method for analysing elastic distortional buckling. It was shown that the buckling load factor in the long term decreased somewhat from its short-term value owing to the quasi-viscoelastic rheology of the concrete slab.

This issue has hitherto been ignored in design, but the perceived conservatism of past and contemporary design methods, combined with the dependence of the strength load factor on both the elastic buckling load factor and on the plastic moment in the method of 'design by buckling analysis', would account for the propensity of existing bridge girders to be remote. However, with the impetus of evolving advanced and rational design procedure, the possibility of quasi-viscoelastic induced instabilities, as have been observed in other steel/concrete composite applications, is at least a potential issue that requires further investigation.

ACKNOWLEDGEMENT

The work reported in this paper was supported by both the 'Women in Engineering' and 'Dean's Scholarship' awards made available by the Faculty of Engineering at The University of New South Wales.

REFERENCES

Bradford, M.A. 1997. Service load analysis of slender R-C columns. *ACI Struct J.* 94(6):675-685.

Bradford, M.A. 1998a. Service-load behaviour of braced reinforced concrete columns with rotational springs. *Adv. in Struct. Eng.* 1(3):193-201.

Bradford, M.A. 1998b. Analysis of slender R-C columns at service loads. *Proc. EASEC6, Taipei, 14-16 January 1998:*925-931 Tapiei:NTU.

Bradford, M.A. 1998c. Service-load deformations of concrete-filled RHS columns. In Y.S. Choo & G.J. van der Vegte (eds.), *Tubular Structures VIII – Proc. 8th Int. Symp. on Tubular Structs., Singapore 26-28 August 1998:*425-424-433. Rotterdam:Balkema.

Bradford, M.A. & H.R. Ronagh 1997. Generalized elastic buckling of restrained I-beams by the FEM. *J. Struct. Eng., ASCE* 123(12):1631-1637.

Bradford, M.A., H. Vu Manh & R.I. Gilbert 1999. Numerical analysis of continuous composite beams under service loading. UNICIV Report R-389, The University of NSW, Sydney.

Bradford M.A., H.D. Wright & B. Uy 1998. Short and long-term behaviour of axially loaded composite profiled walls. *Proc. ICE, Structs. & Bldgs.* 128(1):26-37.

Gilbert, R.I. 1988. *Time effects in concrete structures.* Amsterdam:Elsevier.

Gilbert R.I. & M.A. Bradford 1990. Design of slender reinforced concrete columns for creep and shrinakge. *Proc. 2nd Int. Conf. on Comp. Aided Anal. & Design of Conc. Structs. held at Zell am See, Austria:*739-748. Swansea:Pineridge Press.

Gilbert, R.I. & M.A. Bradford 1995. Time-dependent behavior of continuous composite beams at service loads. *J. Struct. Eng. ASCE* 121(2):319-327.

Oehlers, D.J. & M.A. Bradford 1995. *Composite steel-concrete structural members: fundamental behaviour.* Oxford:Pergamon.

Terrey P.J., M.A. Bradford & R.I. Gilbert 1994. Creep and shrinkage of concrete in concrete-filled circular steel tubes. In P. Grundy et al. (eds), *Tubular Structures VI – Proc. 6th Int. Symp. on Tubular Structs., Melbourne, 14-16 Dec. 1995:*293-298. Rotterdam:Balkema.

Uy B. & M.A. Bradford 1995. Local buckling of cold formed steel sheeting in profiled composite beams at service loads. *Struct. Eng. Review* 7(4):289-300.

Vrcelj, Z., M.A. Bradford & B. Uy 1999. Elastic buckling modes in unpropped continuous composite tee-beams. In M.A. Bradford et al. (eds), *Mechanics of Structures and Materials – Proc. ACMSM16, Sydney, 8-10 Dec. 1999:*327-333. Rotterdam:Balkema.

Vrcelj, Z., M.A. Bradford & H.R. Ronagh 2000. Elastic stability of half-through girder bridges. *Proc. ISEC-01, Hawaii, 24-26 January 2001.* Rotterdam:Balkema.

29 Inelastic behavior of structures

Creative Systems in Structural and Construction Engineering, Singh (ed.) © 2001 Balkema, Rotterdam, ISBN 90 5809 161 9

Numerical modeling for large displacement analysis of inelastic structures

Y.J.Chiou, P.A.Hsiao & Y.L.Chen
Department of Civil Engineering, National Cheng Kung University, Taiwan

ABSTRACT: A numerical model is proposed to study the nonlinear behavior of inelastic structures. The convected material frame approach is adopted. This approach is a modification of the co-rotational approximation by incorporating an adaptive convected material frame in the basic definition of the displacement vector and strain tensor. The rigid body motion and deformation displacements are decoupled for each increment. The nonlinearities associated with the large geometrical changes are incorporated in the analysis through the continuous updating of the material frame geometry. By assuming a lumped mass matrix of diagonal form, the explicit finite element analysis involves only vector assemblage and vector storage. The algorithm is verified by comparing the numerical solutions with the results obtained by the ANSYS code. The convected material frame approach is shown to be accurate and capable of investigating large deflection of inelastic structures.

1 INTRODUCTION

Large displacement analysis of elastic structures has been extensively studied. Many researchers presented explicit algorithms for finite element analysis during the last two decades. Belytschko and Hsieh (1973), Belytschko et al. (1977) formulated the frame elements by using the traditional co-rotational approach. Belytschko and Marchertas (1974) formulated the three-dimensional plate element following a similar approach. Alternatively, Rice and Ting (1993) proposed an approach of updated geometry to develop a plane frame analysis procedure. Recently, Wang et al. (1998) formulated a general curved elastic frame element based on a convected material frame approach, and developed a general explicit algorithm for the analysis of flexible structures subjected to large geometry changes. This approach is a modification of the co-rotational approximation by incorporating an adaptive convected material frame in the basic definition of the displacement vector and strain tensor.

Researchers also have extensively studied the structures with nonlinear materials. Tang et al. (1980) adopted the traditional co-rotational approach and bi-linear constitutive relation to analyze the large deflection of the plane frame. Yang and Saigal (1984) studied the static and dynamic response of

beam with bi-linear material and large deflection. Wang et al. (1995) proposed a numerical model to study the large deflection of an elastoplastic cantilever. Recently, Mamaghani et al. (1996) developed an elastoplastic finite element formulation for beam-columns to analyze the structural steel members under cyclic loading.

The convected material frame approach (Wang et al., 1998) is applied in this study to investigate the geometrical nonlinearity of inelastic structures. The inelastic constitutive relation is derived by using a generalized nonlinear function and the kinematic hardening is adopted to account for the Bauschinger effect. In the following, the formulation of the explicit finite element method based on a convected material frame approach is presented first. The algorithm is then verified by comparing the numerical solutions with the results obtained by the ANSYS code (1998), and the nonlinear behavior of inelastic frame is fully studied.

2 PROBLEM FORMULATION

Referring to Fig. 1, the displacement history of a body subjected to large displacement is decomposed into four stages: (1) the initial geometry X at the reference time $t = 0$, (2) the convected material reference frame x at time $t = n\Delta t$, (3) the deformed body geometry x' at time $t' = (n+1)\Delta t$, and (4) the convected geometry $\hat{x}'$ at time t'. The convected geometry $\hat{x}'$ is related to the deformed body geometry x' by a pure rotation Q.

Fig. 2 shows a segment of a curved frame element under loading. The section $EFGH$, the radius of curvature R, and the central angle $d\theta$ describe the element material geometry x at time t, while section $E'F'G'H'$, R', and $d\theta'$ describe the element convected geometry x' at time t'.

The frame element is assumed to follow the Bernoulli-Euler beam theory. The thickness and width of the frame element are small compared to the element length. The effect of Poisson's ratio is negligible, and the dimensions of the cross section remain unchanged. When the geometry changes from x to the x', a plane cross section of the element remains to be a plane. Referring to Fig. 2, the axial strain in the fiber $C'D'$ can be written as

$$\hat{\varepsilon} = \frac{1}{2}\left[\left(\frac{C'D'}{CD}\right)^2 - 1\right] \tag{1}$$

The frame thickness is assumed to be small compared to the radius of curvature, $\hat{\varepsilon}$ can then be simplified as

$$\hat{\varepsilon} \cong \hat{\varepsilon}_m + \overline{y}\Delta k \tag{2}$$

where ε_m is a uniform normal strain due to the change of the element length, and Δk is the change of the element curvature. Specific forms of ε_m and Δk are obtained by neglecting the higher order terms.

$$\hat{\varepsilon}_m \cong \frac{\hat{v}^d}{R} + \frac{\partial \hat{u}^d}{\partial s} \tag{3}$$

$$\Delta k \cong (\frac{\hat{v}^d}{R^2} + \frac{1}{R}\frac{\partial \hat{u}^d}{\partial s}) - \frac{\partial^2 \hat{v}^d}{\partial s^2} \tag{4}$$

where $\hat{u}^d$ and $\hat{v}^d$ are the longitudinal and transverse displacements in the convected coordinates.

Fig. 3 shows a typical frame element, where s is the length measured, α_1 and α_2 the nodal slopes of the reference geometry, θ_1 and θ_2 the slope changes, l_0 and l the lengths of the element, and l_m and l_n the distances between nodes. The change of distance between two end nodes along $\hat{x}$-axis Δ, and the changes of end slopes of the element θ_1 and θ_2, are the independent nodal displacements. The displacements are expressed in terms of the nodal displacements and written as

$$\hat{u}^d = L(s)\Delta \tag{5}$$

$$\hat{v}^d = N_1(s)\Delta + N_2(s)\theta_1 + N_3(s)\theta_2 \tag{6}$$

where

$$L(s) = \frac{s}{l}\cos(\alpha_2 + \theta_2)$$

$$N_1(s) = (-3\frac{s^2}{l^2} + 2\frac{s^3}{l^3})\sin(\alpha_2 + \theta_2)$$

$$N_2(s) = s - 2\frac{s^2}{l} + \frac{s^3}{l^2}$$

$$N_3(s) = -\frac{s^2}{l} + \frac{s^3}{l^2}$$

The normal strain ε_m and the change of curvature Δk are then expressed as

$$\hat{\varepsilon}_m = \mathbf{B}_0^T\hat{\mathbf{d}}_e^* \tag{7}$$

$$\Delta k = \mathbf{B}_r^T\hat{\mathbf{d}}_e^* \tag{8}$$

A structural system in the deformed geometry x', at time $t' = (n+1)\Delta t$, is said to be in equilibrium if it satisfies the virtual work formulation

$$\delta U = \delta W \tag{9}$$

where U is the internal work, W the external work.

$$\delta U = \sum_e \delta U_e^d = \sum_e \int_{Ve} \delta\hat{\varepsilon}^T \hat{\sigma} \ dV \tag{10}$$

where $\hat{\varepsilon}$ is the normal strain along the longitudinal direction, and $\hat{\sigma}$ the total longitudinal stress. The internal work for each element is

$$\delta U_e^d = \delta(\hat{\mathbf{d}}_e^*)^T\hat{\mathbf{f}}_e^* = \begin{bmatrix}\delta\Delta & \delta\theta_1 & \delta\theta_2\end{bmatrix}\begin{bmatrix}\hat{f}_{2x} \\ \hat{m}_{1z} \\ \hat{m}_{2z}\end{bmatrix} \tag{11}$$

where

$$\hat{\mathbf{f}}_e^* = \hat{\mathbf{f}}_e^m + \Delta\hat{\mathbf{f}}_e^* + \hat{\mathbf{f}}_e^n \tag{12a}$$

$$\hat{\mathbf{f}}_e^m = \int\left(\mathbf{B}_0 P^m + \mathbf{B}_r M_z^m\right)ds \tag{12b}$$

$$\Delta\hat{\mathbf{f}}_e^* = \hat{\mathbf{k}}_e\hat{\mathbf{d}}_e^*$$
$$= \left\{\int_s\left[\left(\int_A EdA\right)\mathbf{B}_0\mathbf{B}_0^T + \left(\int_A Ey^2dA\right)\mathbf{B}_r\mathbf{B}_r^T\right]ds\right\}\hat{\mathbf{d}}_e^* \tag{12c}$$

$$\hat{\mathbf{f}}_e^n = \left\{\int_s\left[\left(\int_A E\overline{y}dA\right)\left(\mathbf{B}_0\mathbf{B}_r^T + \mathbf{B}_r\mathbf{B}_0^T\right)\right]ds\right\}\hat{\mathbf{d}}_e^* \tag{12d}$$

$$P^m = \int_A \hat{\sigma}^m dA \tag{12e}$$

$$M_z^m = \int_A \hat{\sigma}^m \overline{y}dA \tag{12f}$$

where $E = E(\varepsilon_j) = d\sigma/d\varepsilon|_{\varepsilon_j} = f'(\varepsilon_j)$ is the tangent modulus and it is a function of transverse coordinates $\overline{y}$, $f(\varepsilon)$ is a generalized nonlinear function used to derive the inelastic constitutive relation.

The element masses are lumped at the nodes. Each node is assumed to be in dynamic equilibrium while each element is in static equilibrium. The static equilibrium yields

$$\hat{f}_{1x} = -\hat{f}_{2x}$$
$$\hat{f}_{1y} = (\hat{m}_{1z} + \hat{m}_{2z})/l_n \tag{13}$$
$$\hat{f}_{2y} = -\hat{f}_{1y}$$

The full internal force and deformation displacement vectors for each element are

$$\left(\hat{\mathbf{f}}_e^{\text{int}}\right)^T = \begin{bmatrix} \hat{f}_{1x} & \hat{f}_{1y} & \hat{m}_{1z} & \hat{f}_{2x} & \hat{f}_{2y} & \hat{m}_{2z} \end{bmatrix} \quad (14a)$$

$$(\hat{\mathbf{d}}_e^d)^T = \begin{bmatrix} 0 & 0 & \theta_1 & \Delta & 0 & \theta_2 \end{bmatrix} \quad (14b)$$

The internal virtual work is thus given as

$$\delta U = \sum_e \delta U_e^d = \sum_e (\delta \hat{\mathbf{d}}_e^*)^T \hat{\mathbf{f}}_e^* = \sum_e (\delta \hat{\mathbf{d}}_e^d)^T \hat{\mathbf{f}}_e^{\text{int}} \quad (15)$$

Using the transformation matrix $\mathbf{T}$ between x and $\hat{x}'$, one obtains

$$\hat{\mathbf{d}}_e = \mathbf{T}\mathbf{d}_e$$

Then

$$\delta U = \sum_e \delta \mathbf{d}_e^T \mathbf{f}_e^{\text{int}} \quad (16)$$

where

$$\mathbf{f}_e^{\text{int}} = \mathbf{T}^T \hat{\mathbf{f}}_e^{\text{int}}$$

is the internal nodal force vector for the element written in the global coordinates. The external virtual work is

$$\delta W = \sum_e \delta \mathbf{d}_e^T \left(\mathbf{f}_e^{\text{ext}} - \mathbf{M}_e \ddot{\mathbf{d}}_e \right) \quad (17)$$

where,

$$\mathbf{f}_e^{\text{ext}} = \begin{bmatrix} f_{1x} \\ f_{1y} \\ m_{1z} \\ f_{2x} \\ f_{2y} \\ m_{2z} \end{bmatrix}, \mathbf{M}_e = \begin{bmatrix} \hat{M}_{1x} & 0 & 0 & 0 & 0 & 0 \\ 0 & \hat{M}_{1y} & 0 & 0 & 0 & 0 \\ 0 & 0 & \hat{I}_{1z} & 0 & 0 & 0 \\ 0 & 0 & 0 & \hat{M}_{2x} & 0 & 0 \\ 0 & 0 & 0 & 0 & \hat{M}_{2y} & 0 \\ 0 & 0 & 0 & 0 & 0 & \hat{I}_{2z} \end{bmatrix}$$

with

$$\hat{M}_{1x} = \hat{M}_{1y} = \hat{M}_{2x} = \hat{M}_{2y} = \frac{1}{2}\rho Al$$

$$\hat{I}_{1z} = \hat{I}_{2z} = \frac{1}{24}\rho l \left(Al^2 + 12 I_z \right)$$

ρ = mass density

The principle of virtual work yields

$$\sum_e \delta \mathbf{d}_e^T \mathbf{f}_e^{\text{int}} = \sum_e \delta \mathbf{d}_e^T \left(\mathbf{f}_e^{\text{ext}} - \mathbf{M}_e \ddot{\mathbf{d}}_e \right) \quad (18)$$

If one introduces a global assembled nodal displacement vector $\mathbf{d}$, the above equation becomes

$$\delta \mathbf{d}^T \left(\sum_e \mathbf{f}_e^{\text{int}} \right) = \delta \mathbf{d}^T \left(\sum_e \mathbf{f}_e^{\text{ext}} \right) - \delta \mathbf{d}^T \left(\sum_e \mathbf{M}_e \right) \ddot{\mathbf{d}} \quad (19)$$

Since $\delta \mathbf{d}$ is arbitrary, the equation of motion in global coordinates yields

$$\mathbf{M}\ddot{\mathbf{d}} = \mathbf{F}^{\text{ext}} - \mathbf{F}^{\text{int}} \quad (20)$$

where

$$\mathbf{M} = \sum_e \mathbf{M}_e, \quad \mathbf{F}^{\text{ext}} = \sum_e \mathbf{f}_e^{\text{ext}}, \text{ and } \mathbf{F}^{\text{int}} = \sum_e \mathbf{f}_e^{\text{int}}$$

To find the quasi-static solution through a dynamic relaxation procedure, a damping force may be added.

$$\mathbf{M}\ddot{\mathbf{d}} = \mathbf{F}^{\text{ext}} - \mathbf{F}^{\text{int}} - \mathbf{F}^{dmp} \quad (21)$$

The damping force may be written by assuming a standard Rayleigh damping

$$\mathbf{F}^{dmp} = C\dot{\mathbf{d}} = (\alpha \mathbf{M} + \beta \mathbf{K})\dot{\mathbf{d}}$$

in which α and β are constants, and $\mathbf{K}$ is the global stiffness matrix. In this study, since the global stiffness matrix is not available in the formulation, β is assumed to be zero. The explicit time integration method is adopted to solve the equation and a second order central difference formulation is used as the time integration technique.

$$\begin{aligned} \mathbf{d}_{i+1} = {}& (\frac{2}{2 + \alpha \Delta t}) \Delta t^2 \mathbf{M}^{-1}(\mathbf{F}_i^{\text{ext}} - \mathbf{F}_i^{\text{int}}) \\ & + (\frac{4}{2 + \alpha \Delta t})\mathbf{d}_i - (\frac{2 - \alpha \Delta t}{2 + \alpha \Delta t})\mathbf{d}_{i-1} \end{aligned} \quad (22)$$

Note that using Eq. (22), the displacements at $t + \Delta t$ are calculated by using the mass values and the external and internal forces of the previous time step. An important simplification can be introduced by assuming a diagonal mass matrix. All the calculation in Eq. (22) involves vector operation only.

3 RESULTS AND DISCUSSION

A rigid frame subjected to a static horizontal load is studied first. The geometry, constitutive relation, and load history of the rigid frame are shown in Fig. 4. The Young's modulus is $E_1 = 10.6 \times 10^3 ksi$ ($7.3034 \times 10^4 MPa$), the slope of the linear work hardening is $E_2 = 2.53 \times 10^3 ksi$ ($1.7431 \times 10^4 MPa$), the yield stress is $\sigma_y = 52.0 ksi$ ($358.28 MPa$), and the mass density ρ is $2.589 \times 10^{-3} lb \cdot \sec^2/in^4$ ($2.76794 \times 10^4 kg/m^3$). The relation of horizontal displacement versus load is presented in Fig. 5. It is found that the numerical results agree satisfactorily with the results obtained by the ANSYS code. The deformed shape of this frame is shown in Fig. 6. One can observe the characteristic of large displacement and its agreement with the result obtained by ANSYS code.

A rigid frame subjected to a dynamic horizontal load is further investigated. The geometry, constitutive relation, and load history of the rigid frame are shown in Fig. 7. The kinematic hardening is adopted to account for the Bauschinger effect. The Young's modulus, the slope of the linear work hardening, the yield stress, and the mass density are the same as the previous one. The relation of horizontal displacement versus time is presented in Fig. 8. It is also found that the numerical results agree satisfactorily with the results obtained by the ANSYS code.

4 CONCLUSIONS

A convected material frame approach for large deformation combined with the explicit finite formulation is proposed to analyze the inelastic structures.

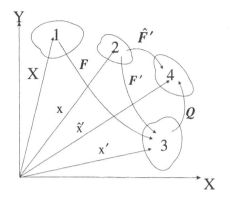

Fig.1 Schematic description of the convected material reference frame

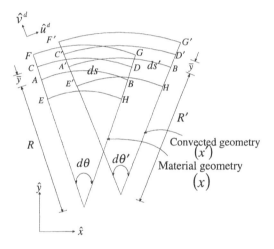

Fig.2 Kinematics for normal strain

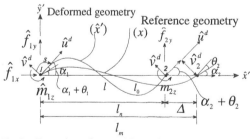

Fig. 3 Comparison of a curved frame element

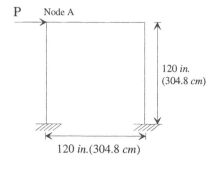

(a) Geometry

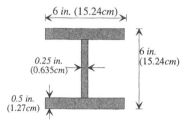

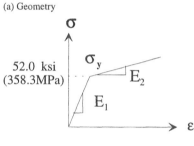

(b) Schematic diagram of constitutive relation

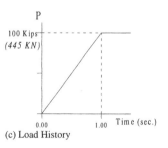

(c) Load History

Fig. 4 A rigid frame subjected to a static horizontal load

This approach is a modification of the co-rotational formulation that has been implemented in many widely used computer codes. By assuming a lumped mass matrix of diagonal form, the explicit finite element analysis involves only vector assemblage and vector storage. Through the numerical verification with the solutions obtained by the ANSYS code (1998) and the investigation of nonlinear inelastic frame, the convected material frame approach is shown to be accurate and capable of investigating large deflection of inelastic structures. This numerical model is proposed to be an alternative efficient

948

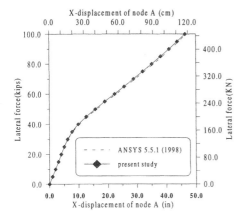

Fig. 5 Load –deflection relation of a rigid frame subjected to a static horizontal load

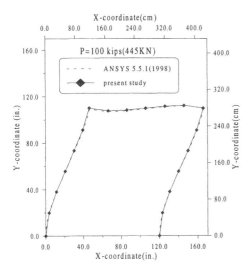

Fig. 6 Deformed shape of a rigid frame subjected to a static horizontal load

approach for nonlinear analysis of inelastic structures.

5 ACKNOWLEDGMENTS

This study is supported by the National Science Council of Republic of China under grant NSC88-2211-E-006-030.

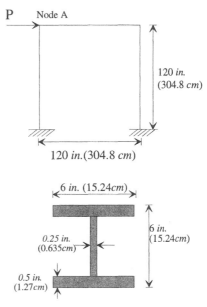

(a) Geometry

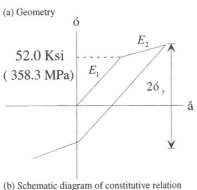

(b) Schematic diagram of constitutive relation

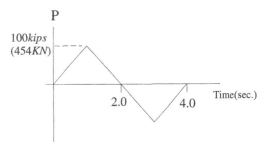

(c) Load History

Fig. 7 A rigid frame subjected to a dynamic horizontal load

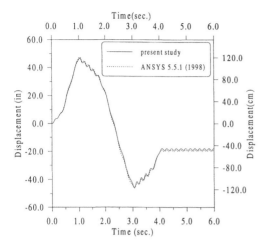

Fig. 8 Displacement-time relation of a rigid frame subjected to a dynamic horizontal load

Yang, T. Y. & S. Saigal 1984. A simple element for static and dynamic response of beams with material and geometric nonlinearities. *Int. J. Num. Meth. Eng.*, 20, 851-867.

REFERENCES

ANSYS Release 5.5.1 1998. ANSYS, Inc., Canonsburg, P. A., U. S. A.
Belytschko, T. and B. J. Hsieh 1973. Nonlinear transient finite element analysis with convected coordinates. *Int. J. Numer. Meth. Eng.*, 7, 255-271.
Belytschko, T. & A. H. Marchertas 1974. Nonlinear finite element method for plates and its application to dynamics response of reactor fuel subassemblies. Trans. ASME, *J. Pressure Vessel Tech.*, 96(4), 251-257.
Belytschko, T., L. Schwer & M. J. Klein 1977. Large displacement, transient analysis of space frames. *Int. J. Numer. Meth. Eng.*, 11, 65-84.
Mamaghani, I. H. P., T. Usami & E. Mizuno 1996. Inelastic large deflection analysis of structural steel members under cyclic loading. *Engng. Struct.*, 18(9), 659-668.
Rice, D. L. & E. C. Ting 1993. Large displacement transient analysis of flexible structures. *Int. J. Numer. Meth. Eng*, 36, 1541-1562.
Tang, S. C., K. S. Yeung & C. T. Chon 1980. On the tangent stiffness matrix in a convected coordinate system. *Comput. Struct.*, 12, 849-856.
Wang, B., G. Lu & T. X. Yu. 1995. A numerical analysis of the large deflection of an elastoplastic cantilever. *Struct. Engng. Mech.*, 3(2), 163-172.
Wang, Y. K., C. Shih & E. C. Ting 1998. *A general curved element for very flexible frames*, CE-ST-1998-001, Department of Civil Engineering, National Central University, Taiwan, R. O. C.

Creative Systems in Structural and Construction Engineering, Singh (ed.) © 2001 Balkema, Rotterdam, ISBN 90 5809 161 9

Plasticity model for reinforced concrete elements subjected to overloads

A. I. Karabinis
Democritus University of Thrace, Xanthi, Greece

P. D. Kiousis
Colorado School of Mines, Golden, Colo., USA

ABSTRACT: A theoretical and computational framework is presented for the analysis of structures that are subjected to inelastic overloads. Current practice in structural engineering applies elastic solutions to design methods that assume elastoplastic or perfectly plastic material response. This approach generates inaccuracies, which depending on the amount of overload, can be excessive due to the fact that force distributions within statically indeterminate structures depend on the relative stiffnesses of the individual structural elements (i.e. beams and columns). The relative element stiffnesses within a structure change continuously under inelastic loading and can be significantly different from their initial elastic values. For this purpose, a new plasticity model, that combines the nonlinear material response and the geometric characteristics of a structural element is developed. The model provides the computational efficiency and simplicity of matrix structural analysis and avoids modeling at the material level.

1 INTRODUCTION

Accurate analysis of concrete framed structures under static and dynamic overloads is an important and difficult task in structural engineering. Current practice follows a rather simple, but unorthodox approach, whereby the analysis of internal forces is elastic, but the subsequent design is based on inelastic material behavior. Internal forces within statically indeterminate structures are distributed based on the relative stiffness of the individual structural elements (i.e. beams and columns). Relative element stiffnesses within a structure change continuously under inelastic loading and can be significantly different from their assumed elastic values. Thus, an approach that evaluates forces elastically, and sizes members based on plastic behavior may result in large inaccuracies in cases of significant overloads. Inelastic analyses of structures are therefore desirable. However, current nonlinear approaches are based on solid block finite elements, which are computationally very expensive, require a high level of expertise, and are mainly used for research purposes.

This paper presents a novel approach in the development of constitutive equations, which is intended to represent the combined effects of material behavior and geometric properties of a structural element. The typical infinitesimal element of the classical plasticity theory is replaced by a beam or column element and the stress fields are replaced by generalized force fields. The intent is to encapsulate the nonlinear response of a concrete structural element, as this is affected by the combinations of the applied loads and the loading history. This is achieved by incremental elastoplastic relations. The resulting equations are similar to those of the elastic matrix structural analysis and can be incorporated to existing structural analysis software with minor changes. It will be shown that the present model can capture efficiently and accurately the inelastic response of structures subjected to overloads.

2 PLASTICITY EQUATIONS

2.1 *Yield Criterion*

The concepts developed in this study combine classical concepts of concrete design with fundamental approaches of the theory of plasticity. The aim is to produce a working model that can be easily adapted to existing computational tools of structural engineering. For this purpose, a yield criterion is established as a function of the axial force and bending moment of the element based on typical concrete column interaction diagrams:

$$M_r - \alpha P_r + (1 + \alpha) P_r^\gamma - 1 = 0 \qquad (1)$$

where, α, and γ are material parameters to be evaluated, M is the bending moment applied on the element, P is the axial load applied on the element, $M_r = M / M_{max}$, $P_r = P / P_{max}$, and P_{max} and M_{max} are "hardening and softening" functions and are equal to the yield values of P and M respectively during elastic deformations. After the initial yield, P_{max} and M_{max} are the largest values of axial force and bending moment that can be developed under the current state of deformations. The geometry of the yield function is demonstrated in Figure 1.

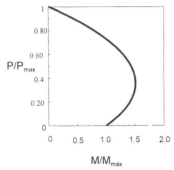

Figure 1: Illustration of the yield criterion for a concrete beam-column.

The material parameters α, and γ can be evaluated easily if a) two (P_r, M_r) yield combinations are known, or b) the balanced yield point (B.Y.P.) (P_{ro}, M_{ro}) is known.

The first case requires the iterative solution for α and γ of the following systems of equations:

$$M_{r1} - \alpha P_{r1} + (1 + \alpha) P_{r1}^{\gamma} - 1 = 0 \qquad (2)$$

$$M_{r2} - \alpha P_{r2} + (1 + \alpha) P_{r2}^{\gamma} - 1 = 0 \qquad (3)$$

where (P_{r1}, M_{r1}), and (P_{r2}, M_2) are the two known yield combinations.

The second case (more common approach) requires the iterative solution for α and γ of the equations:

$$\alpha - \gamma (1 + \alpha) P_{r_o}^{\gamma} = 0 \quad (\frac{dM_r}{dP_r}\Big|_{at\ BFP} = 0) \qquad (4)$$

$$M_{r_o} - \alpha P_{r_o} + (1 + \alpha) P_{r_o}^{\gamma} - 1 = 0 \quad (Y.F. @ B.F.P.) \qquad (5)$$

2.2 Flow rule

The "conjugate strains" of the generalized forces P and M are the axial strain ϵ and the curvature κ. The plastic increments of these strains can be evaluated from the flow rule:

$$de^p = \begin{Bmatrix} d\varepsilon^p \\ d\kappa^p \end{Bmatrix} = d\lambda \frac{\partial G}{\partial s} = d\lambda \begin{Bmatrix} \dfrac{\partial G}{\partial P} \\ \dfrac{\partial G}{\partial M} \end{Bmatrix} \qquad (6)$$

where G is a potential function, $d\lambda$ is a non-negative infinitesimal number and $s^T = [P, M]$. The gradient $(\partial G/\partial s)$ equals to the gradient $(\partial F/\partial s)$ except as M approaches 0, in which case $(\partial G/\partial s)$ approaches $[1\ 0]^T$.

Equation (6) is the "normality" rule to evaluate the plastic strain increment as a quantity that is proportional to the gradient of a potential function G. Of course, the quantity $d\lambda$ is yet to be determined.

2.3 Hardening / Softening Development

P_{max}, and M_{max} provide the pre-peak hardening and post-peak softening characteristics of the model and are expressed as functions of the plastic trajectory $\hat{e}$, which expresses the history of loading and is defined as:

$$\hat{e} = \int \sqrt{d\varepsilon^{p^2}} + \frac{1}{\eta} \int \sqrt{d\kappa^{p^2}} = \hat{\varepsilon} + \frac{1}{\eta} \hat{\kappa} \qquad (7)$$

where η is a convenient constant in units of length with the dual purpose of nondimensionalizing the plastic curvature κ^p, and making the two plastic trajectory contributors ϵ^p and κ^p of the same magnitude.

The mathematical expression of P_{max} describes the relation $P_{max} - \epsilon^p$ at zero M, while the expression of M_{max} describes the relation $M_{max} - \kappa^p$ at zero P. The models that are used here have been chosen from earlier work of the authors [Karabinis and Kiousis, 1994, 1996], and are expressed as follows:

$$P_{MAX} = P_Y + \frac{\hat{e}}{\dfrac{1}{K_\varepsilon} + \dfrac{\hat{e}}{P_U - P_Y}} - A_\varepsilon \hat{e} \qquad (8)$$

$$M_{MAX} = M_Y + \frac{\hat{e}}{\dfrac{1}{K_\kappa} + \dfrac{\hat{e}}{M_U - M_Y}} - A_\kappa \hat{e} \qquad (9)$$

In the above expressions P_Y and M_Y represent initial yield values, K_ϵ and K_κ are the initial hardening slopes, A_ϵ and A_κ are the ultimate softening slopes, and P_U and M_U are expressions of the peak (or ultimate) values of P and M respectively. These expressions are illustrated in Figure 2.

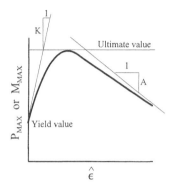

Figure 2 : Schematic of P_{max} and M_{max} as functions of the plastic trajectory $\hat{e}$.

2.4 *Development of Incremental Elastoplastic Equations:*

To develop the complete incremental elastoplastic relations, the following expressions are needed: 1) incremental elastic; 2) flow rule; 3) hardening rule; 4) consistency equation.

The incremental elastic relations are as follows:

$$dS = E\,de \quad or \quad \begin{Bmatrix} dP \\ dM \end{Bmatrix} = \begin{bmatrix} EA & 0 \\ 0 & EI \end{bmatrix} \begin{Bmatrix} d\varepsilon - d\varepsilon^p \\ d\kappa - d\kappa^p \end{Bmatrix} \quad (10)$$

The flow and hardening rules are presented in Equations (6-9).

The consistency equation is an expression that forces the load point (P, M) to stay on the yield function during elastoplastic loading. This is mathematically expressed as:

$$dF = 0 \quad \therefore \quad \frac{\partial F}{\partial N} dN + \frac{\partial F}{\partial M} dM + \frac{\partial F}{\partial \hat{e}} d\hat{e} = 0 \quad (11)$$

Following standard practices of the theory of plastic flow, the incremental load deformation relations are generated:

$$dS = D^{ep} de \quad (12)$$

where, the elastoplastic matrix D^{ep}:

$$D^{ep} = E - \frac{E \dfrac{\partial G}{\partial S} \dfrac{\partial F^T}{\partial S} E}{\dfrac{\partial F^T}{\partial S} E \dfrac{\partial G}{\partial S} - \dfrac{\partial F}{\partial \hat{e}} \sqrt{\dfrac{\partial G^T}{\partial S} \dfrac{\partial G}{\partial S}}} \quad (13)$$

Equations (12) and (13) can be easily transformed to one that relates loads to deformations:

$$dS = D^s\,du \quad or$$

$$\begin{Bmatrix} dP \\ dM \end{Bmatrix} = \begin{bmatrix} D_{11}^{ep} & D_{12}^{ep} \\ D_{21}^{ep} & D_{22}^{ep} \end{bmatrix} \begin{bmatrix} \dfrac{1}{L} & 0 \\ 0 & \dfrac{1}{L} \end{bmatrix} \begin{Bmatrix} \Delta u \\ \Delta \varphi \end{Bmatrix} =$$

$$\begin{bmatrix} \dfrac{D_{11}^{ep}}{L} & \dfrac{D_{12}^{ep}}{L} \\ \dfrac{D_{21}^{ep}}{L} & \dfrac{D_{22}^{ep}}{L} \end{bmatrix} \begin{Bmatrix} \Delta u \\ \Delta \varphi \end{Bmatrix} \quad (14)$$

Finally, Equation (14) can be used in common expressions of matrix structural analysis.

An example of an implementation for a plane frame element is presented in Equation (15) below:

$$\begin{Bmatrix} dN_1 \\ dV_1 \\ dM_1 \\ dN_2 \\ dV_2 \\ dM_2 \end{Bmatrix} = \begin{bmatrix} D_{11}^s & 0 & -D_{12}^s & -D_{11}^s & 0 & D_{12}^s \\ 0 & \dfrac{12EI}{L^3} & \dfrac{6EI}{L^2} & 0 & -\dfrac{12EI}{L^3} & \dfrac{6EI}{L^2} \\ -D_{21}^s & \dfrac{6EI}{L^2} & D_{22}^s & D_{21}^s & -\dfrac{6EI}{L^2} & D_{22}^s \\ -D_{11}^s & 0 & D_{12}^s & D_{11}^s & 0 & -D_{12}^s \\ 0 & -\dfrac{12EI}{L^3} & -\dfrac{6EI}{L^2} & 0 & \dfrac{12EI}{L^3} & \dfrac{6EI}{L^2} \\ D_{21}^s & \dfrac{6EI}{L^2} & D_{22}^s & -D_{21}^s & -\dfrac{6EI}{L^2} & D_{22}^s \end{bmatrix} \begin{Bmatrix} du_1 \\ dv_1 \\ d\varphi_1 \\ du_2 \\ dv_2 \\ d\varphi_2 \end{Bmatrix}$$

Currently, the shear contribution is considered linearly elastic due to lack of adequate experimental data to support alternative theories.

3 MODEL PROCESSES

The generalized load-deformation relation of a concrete beam element can be easily produced using the present model, if $P-\epsilon$ or $M-\kappa$ relations are known for two separate loading cases. These are typically the uniaxial compression $P-\epsilon$ relation, and the pure bending $M-k$ relation. Such relations can be experimental or theoretical, based on accepted models [e.g. Sheikh and Uzumeri, 1982; Mander et al. 1988; Sheikh and Yeh, 1992]. An example corresponding to the column shown in Figure 3 is illustrated in Figure 4.

The steps to produce the necessary parameters for the model are:

1. The P-ε and M-κ relations must be re-plotted as "P-ε^p" and "M-κ^p" (Figure 4 - Unmarked curves). Producing these curves is simple. For example $\varepsilon^p = \varepsilon - \varepsilon^e = \varepsilon - P/EA$.

2. The resulting graphs are equivalent to the P_{max} - $\hat{\varepsilon}$ and M_{max} - $\hat{\kappa}$ relations. Mapping these graphs to Equations (8) and (9) enables the evaluation of the hardening parameters of these equations.

3. Mapping an experimental or theoretical interaction diagram to Equation (1) (Figure 1) produces the parameters α and γ.

$f_c' = 32$ MPa
Longitudinal reinforcement
8 rebars Ø 19 mm - ρ = 2.44%
$f_y = 436$ Mpa; $f_u = 823$ MPa
Transverse reinforcement
Ø 10 mm at 108 mm spacing
$f_y = 483$ Mpa; $f_u = 727$ MPa

Figure 3: Column Example

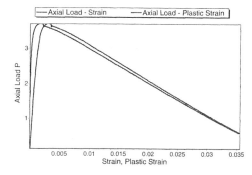

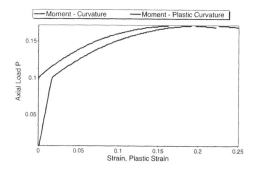

Figure 4: Axial and bending response

The model can now be implemented using Equation (12) to produce load-deformation relations or Equation (15) to solve incrementally a specific structural problem.

4 MODEL IMPORTANCE

There exist numerous models of various levels of success to justify the concern over the usefulness of "yet another" model. The following discussion is necessary to explain the importance of this study. Typical models [e.g. Sheikh and Uzumeri, 1982; Mander et al. 1988; Sheikh and Yeh, 1992] produce moment curvature relations of a concrete column for a specific axial loading. However, consider a framed structure subjected to some overload (e.g. earthquake, excessive wind, etc). Such loading may cause *continuous variation of the moment and the axial load*. This reduces the predictive capabilities of the existing models except for the cases where the variations of axial loads are not significant. The present model is based on the integration of incremental elastoplastic relations of M, P, κ, and ε without such restrictions. It is thus uniquely qualified to simulate the general case of load deformation of a beam-column, and can be integrated in existing structural programs to analyze nonlinear response.

5 MODEL APPLICATION

Consider the beam-column example shown in Figure 3, and the corresponding curves for axial compression and pure bending characteristics shown in Figure 4. These can be experimental or theoretical relations, or both.

We assume that the P-M interaction diagram is defined by Equation (1) with material parameters α = 0.5 and γ = 4.16. These have been produced from Equations (4) and (5) with the assumptions that balanced loading occurs at P/P_{max} = 0.45 with M/M_{max} = 1.17. A common source of such knowledge includes national and international design codes [ACI Committee 318[a,b]; CEN 1991]

The hardening/softening parameters that are needed in equations (8) and (9) are produced from Figure 4 as:

K_ε = 17300 MN; P_u = 3.7 MN; P_y = 1.25 MN; A_ε = 150 MN;

K_κ = 0.6 MN-m; M_u = 0.17 MN-m; M_y = 0.1 MN-m; A_κ = 1.5 MN-m

The parameters above are calculated from the iterative solution of equations (8) and (9) using the following descriptive data: 1) The peak point (P_o, ε_o^p) or (M_o, κ_o^p) and 2) the ultimate softening slope A_ε or A_κ.

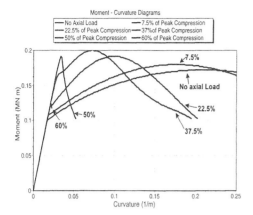

Figure 5: Moment vs Curvature of example column for different initial axial loads.

The resulting M-κ relations for different initial axial loads are shown in Figure 5. The effect of the initial axial load to the bending strength and ductility become obvious. Note also the gradual loss of ductility for axial loads less $P_{balance} = 0.45\ P_{max}$, followed by more severe loss of ductility for axial loads larger than $P_{balance}$.

The uniqueness of the predictive capabilities of this model is demonstrated in Figure 6, where the same column is subjected to the following two different paths: 1) Axial load of 1.37 MN is applied first, followed by increase of curvature to failure (at curvature 0.195 m⁻¹). 2) The Axial load of 1.37 MN and the curvature 0.195 m⁻¹ are applied in a linear incremental fashion so that their ultimate values are reached simultaneously. The difference in response between the two loading cases is significant even though the final axial load and applied curvature are

identical. The softening nature of the bending response is the reason why curvature rather than moment is applied.

The second case of the example presented in Figure 6 illustrates the difficulty in using currently existing models to simulate the behavior of a frame column subjected to overloads where both the axial compression and curvature change to comply with the overall structural equilibrium. The continuous change of axial load causes the moment vs curvature behavior to be unlike any predicted by a constant axial load model. However, to the incremental elastoplastic approach of this study, such continuous variations of axial load present no special problem.

6 SUMMARY AND CONCLUSIONS

A new model is presented to simulate the nonlinear response of concrete structural elements subjected to overloads. The model combines the geometric characteristics of a structural element with the nonlinear material behavior to produce constitutive relations between the internal forces of the element (M, N) and their conjugate "strains" (ε, κ). The model lends itself to direct implementation into classical incremental matrix structural analysis.

This study provides a "blueprint" to developing elastoplastic constitutive equations for nonlinear structural elements. As such, the specific equations used as yield criteria or load - deformation hardening/softening rules should be viewed only as examples that can be modified to fit better the behavior of specific classes of structural elements or materials. In this respect the model is equally applicable to steel structures as long as the interaction diagram and the hardening/softening criteria are modified appropriately.

REFERENCES

ACI Committee 318[a], Building Code Requirements for Structural Concrete (ACI 318 – 99) and Commentary (ACI 318R-99), American Concrete Institute, Detroit, 1999.

ACI Committee 318[b], Design Handbook In Accordance with the Strength Design Method of ACI 318-83 Vol 2 - Columns. Publication SP-17A(85)

CEN Techn. Comm. 250 SG2 (1991) Eurocode 2 : Design of Concrete Structures – Part 1: General Rules and Rules for Buildings (ENV 1992-1-1), CEN Berlin

Karabinis, I. A., and Kiousis, P. D., (1994) "Effects of Confinement on Concrete Columns: Plasticity Approach," Journal of Structural Engineering, Vol. 120, No. 9, pp. 2747-2767.

Karabinis, I. A., and Kiousis, P. D. (1996) "Strength and Ductility of Rectangular Concrete Columns: A Plasticity Approach," Journal of Structural Engineering, Vol. 122, No. 3, pp. 267-274.

Mander,J.B., Priestley,M. J. and Park, R. : ''Theoretical Stress-Strain Model for Confined Concrete'', Journal of

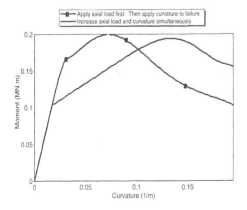

Figure 6: Moment vs. Curvature for two different paths with the same final axial load and curvature.

Structural Engineering, ASCE, V.114, No 8, August 1988, p.1804-1826

Sheikh, S. A. and Uzumeri, S. M. : ''Analytical Model For Concrete Confinement in Tied Columns'', Journal of Structural Engineering, ASCE, V.108, No ST12, December 1982, p.2703-2722

Sheikh, S.A. and Yeh, C.C. : ''Analytical Moment-Curvature Relations for Tied Concrete Columns'', Journal of Structural Engineering, ASCE, V.118, No 2, February 1992, p.529-545

Creative Systems in Structural and Construction Engineering, Singh (ed.) © 2001 Balkema, Rotterdam, ISBN 90 5809 161 9

Elastoplastic analysis of cable-strut structures as a mathematical programming problem

F. Tin-Loi & S. H. Xia
School of Civil and Environmental Engineering, University of New South Wales, Sydney, N.S.W., Australia

ABSTRACT: This paper proposes a computation-oriented approach for determining the structural response of cable-strut systems under prescribed, not necessarily monotonic, loading histories. In such analyses, it is of primary importance to deal properly and efficiently with the possible slackening and retensioning of the cables. We outline how this can be mathematically modeled for inclusion in a robust and direct solution strategy embedded within a mathematical programming framework. An example is given to illustrate application of the approach proposed.

1 INTRODUCTION

Cable-strut systems, such as pretensioned nets for large-span hanging roofs, exhibit a manifold appeal: to the architect, for their economic, functional and aesthetic potentials; to the technologist, for the many complex structural details and components to be manufactured; and finally, to the structural engineer and analyst, for the challenging nonlinear behavior to be considered in the analysis and design phases.

The statical analysis of cable systems has been the subject of a fairly abundant literature – too large to survey in this paper. Two early monographs (Krishna 1978, Buchholdt 1985) still represent a reasonably up-to-date, valuable and comprehensive conspectus of the subject. The early emphasis in this area has been on nonlinear elastic behavior and large geometry.

There is, however, currently a need to perform accurately the inelastic analyses of cable-strut systems. Such analyses must be able to account effectively for such phenomena as (reversible) cable slackening – quite a common occurrence (e.g. Pellegrino 1992) – and (irreversible) strut softening. Moreover, as indicated in some recent work (Wang et al. 98), classical Newton-Raphson type iterative algorithms encounter severe difficulties in the presence of slackening. Simply setting the cable stiffness to zero, as is typical, often leads to severe ill-conditioning and divergence of the iterative process. Further, under some loading conditions, all cables connected to the unloaded ends of some contiguous struts may slacken leading to finite mechanisms even though the system can still carry load.

This paper deals with prestressed cable-strut structures from the structural mechanics point of view only. Specifically, it addresses the aforementioned key issues of cable slackening and strut softening. Without undue loss of generality, we restrict ourselves to problems in the quasistatic domain characterized by physical (material) nonlinearities and a small displacement regime.

We focus on an elegant and computationally advantageous constitutive description of cable behavior as a special mathematical structure, referred to in the mathematical programming literature, as "complementarity", describing the orthogonality of two nonnegative vectors. Softening models for struts can be described within the same complementarity format. The approach we adopt is a finite incremental formulation involving direct solution of a special and important type of mathematical programming problem known as a "Mixed Complementarity Problem" (MCP).

A note concerning notation is appropriate. Bold faced characters indicate vectors and matrices; a null vector is indicated by $\mathbf{0}$, a transpose by the superscript T. The complementarity condition is indicated by the symbol $\perp$; for nonnegative vectors $\mathbf{w}$ and $\mathbf{z}$, this is equivalent to the condition $\mathbf{w}^T\mathbf{z} = \mathbf{0}$ which also obviously applies componentwise.

2 CONSTITUTIVE LAWS

Typical diagrammatic relationships between force (Q) and deformation (q) for a cable of negligible self-weight and for a strut are shown in Figure 1; tension (elongation) is assumed to be positive.

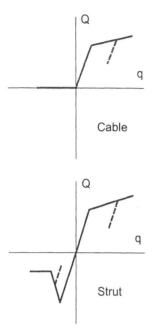

Figure 1. Cable and strut constitutive models.

Piecewise linear representations offer both an elegant formalism as well as a better representation of actual behavior (Ramberg-Osgood type laws, so often used, are smooth and fail to describe observed sudden changes of slopes).

Two important, and often also computationally difficult features, need to be mathematically modeled: the reversible slackening of a cable, and the irreversible softening of a strut under compression. We do so in the same manner as the elegant and powerful complementarity formalism introduced by Maier in the early 70s (Maier 1970).

Noting that the reversible (holonomic) behavior can be carried out in total quantities, but the irreversible (nonholonomic) behavior must be cast in rates in the yield condition, we can compactly describe the element constitutive laws for a generic cable m (assuming for simplicity, as shown in Figure 1, a single yield mode in tension) as follows:

$$q^m = e^m + p^m \tag{1}$$

$$Q^m = S^m e^m \tag{2}$$

$$p^m = z_1^m - z_2^m \tag{3}$$

$$f_1^m = h_1^m z_1^m + r_1^m - Q^m \geq 0, \quad \dot{z}_1^m \geq 0, \quad \perp \tag{4}$$

$$f_2^m = Q^m \geq 0, \quad z_2^m \geq 0, \quad \perp \tag{5}$$

where a subscript 1 refers to quantities dealing with nonholonomic cable yielding and a subscript 2 to

holonomic cable slackening. For struts, the holonomic component obviously disappears.

The total strain q is given in (1) as the sum of elastic e and "plastic" p components (the effect of imposed distortions can also be accommodated if required, as when turnbuckling operations are carried out). Elasticity is described in (2) through the assumed constant elastic stiffness S. The plastic strain in (3) is defined by an associated flow rule and expressed in terms of the "plastic multipliers". Two sets of yield functions (f_1, f_2) are introduced to describe nonholonomic (yielding) behavior and holonomic (slackening) behavior, respectively. Yielding in tension is described through the hardening (or possibly also softening) parameter h_1 and yield limit r_1.

Mechanically, a slack cable ($Q = 0$), for instance, can have nonzero shortening ($z \geq 0$), whereas when the cable has yielded in tension ($f_1 = 0$) it can either unload elastically ($\dot{z} = 0$) or yield further ($\dot{z} > 0$).

Paucity of space precludes us from describing in detail the yield conditions for struts which typically exhibit hardening in tension and hardening and/or softening in compression. Suffice it to mention that such conditions are of the type given by (4). For further clarification, the interested reader is referred to Contro et al. (1975) for cable modeling and to Tin-Loi & Xia (2000) for softening of struts.

Finally, as is conventional in finite element methodology, extension to the whole structure can be carried out easily by collecting (in fully equivalent but indexless symbols) the contributions of elemental components as concatenated vectors and block-diagonal matrices, e.g. $\mathbf{q}$ collects in a single vector all element q, and $\mathbf{S}$ is the unassembled diagonal matrix containing the elemental stiffnesses S on its diagonal. For compactness, we also gather all yield functions (pertaining to cables and struts) in a single vector $\mathbf{f}$ and all plastic multipliers in vector $\mathbf{z}$. Hence at the structure level, the corresponding constitutive relations become

$$\mathbf{q} = \mathbf{e} + \mathbf{p} \tag{6}$$

$$\mathbf{Q} = \mathbf{Se} \tag{7}$$

$$\mathbf{p} = \mathbf{Nz} \tag{8}$$

$$\mathbf{f} = \begin{bmatrix} \mathbf{f}_1 \\ \mathbf{f}_2 \end{bmatrix} = \mathbf{Hz} + \mathbf{r} - \mathbf{N}^T \mathbf{Q} \tag{9}$$

$$\mathbf{f}_1 \geq \mathbf{0}, \quad \dot{\mathbf{z}}_1 \geq \mathbf{0}, \quad \perp \tag{10}$$

$$\mathbf{f}_2 \geq \mathbf{0}, \quad \mathbf{z}_2 \geq \mathbf{0}, \quad \perp \tag{11}$$

where a matrix $\mathbf{N}$ (of outward unit normals to the yield functions) has been introduced, as is common in conventional computational plasticity. It is also interesting to note that matrix $\mathbf{H}$ for the rate formulation is a symmetric, block-diagonal matrix.

Roughly speaking, for rate formulations, even when the stress point is on a particular nonholonomic branch (whether perfectly plastic, hardening or softening) it "sees" only that particular branch.

3 FINITE INCREMENTAL FORMULATION

The constitutive relations presented in the preceding section must first be supplemented by the linear equilibrium and compatibility conditions expressed, respectively, as

$$\mathbf{C}^T \mathbf{Q} = \mathbf{P} \tag{12}$$

$$\mathbf{q} = \mathbf{C}\mathbf{u} \tag{13}$$

where $\mathbf{C}$ is a constant compatibility matrix and $\mathbf{P}$ is the vector of nodal applied forces, conjugate with the nodal displacements $\mathbf{u}$.

To avoid the computational burden of solving exactly the rate problem, we resort to a finite incremental formulation, which we are now in a position to write. This requires the subdivision of the loading history into a finite number of steps. Within each step, holonomy is assumed and any updating is carried out at the beginning of a step. The stepwise holonomic assumption is reasonable when stresses, as in the present case, increase nearly monotonically. Integration of the rate equations can be achieved by a stable scheme based on the familiar and well-researched backward difference algorithm (e.g. Comi at al. 1992).

The entire evolution of the structural response can thus be analyzed as a sequence of finite problems, each involving a configuration change from a previously known state due to a finite increment of load step $\Delta \mathbf{P}$. Any finite increment is thus represented by Δ, e.g. $\mathbf{z} = \hat{\mathbf{z}} + \Delta \mathbf{z}$, where hatted and nonhatted symbols denote known and unknown values of the variables respectively.

The stepwise holonomic counterparts of relations (6)-(13) then clearly become

$$\mathbf{C}^T \Delta \mathbf{Q} = \Delta \mathbf{P} \tag{14}$$

$$\Delta \mathbf{q} = \mathbf{C}\Delta \mathbf{u} \tag{15}$$

$$\Delta \mathbf{q} = \Delta \mathbf{e} + \Delta \mathbf{p} \tag{16}$$

$$\Delta \mathbf{Q} = \mathbf{S}\Delta \mathbf{e} \tag{17}$$

$$\Delta \mathbf{p} = \mathbf{N}\Delta \mathbf{z} \tag{18}$$

$$\mathbf{f} = \mathbf{H}(\hat{\mathbf{z}} + \Delta \mathbf{z}) + \mathbf{r} - \mathbf{N}^T(\hat{\mathbf{Q}} + \Delta \mathbf{Q}) \tag{19}$$

$$\mathbf{f}_1 \geq \mathbf{0}, \quad \Delta \mathbf{z}_1 \geq \mathbf{0}, \quad \perp \tag{20}$$

$$\mathbf{f}_2 \geq \mathbf{0}, \quad \hat{\mathbf{z}}_2 + \Delta \mathbf{z}_2 \geq \mathbf{0}, \quad \perp \tag{21}$$

It is important to note that the conversion of the exact rate formulation into a finite incremental one,

enforcing holonomic behavior for each step, does not necessarily retain the same $\mathbf{N}$, $\mathbf{H}$ and $\mathbf{r}$ as in the original rate formulation (6)-(11). These quantities are altered in the presence of softening laws. In essence, a stress point on a particular branch now needs to see other branches as well. This point is fully elucidated in Bolzon et al. (1994) and Tin-Loi & Xia (2000), in which detailed mathematical descriptions of holonomic softening laws are provided. For the sake of simplicity, we use the same symbols for holonomic and nonholonomic processes in this paper.

Relation set (14)-(21) can be simplified in three possible ways (through various straightforward substitutions) to give corresponding formulations in $(\Delta \mathbf{Q}, \Delta \mathbf{u}, \Delta \mathbf{z})$, $(\Delta \mathbf{u}, \Delta \mathbf{z})$, or $(\Delta \mathbf{z})$.

In the present work, we prefer to retain the (kinematic) $(\Delta \mathbf{u}, \Delta \mathbf{z})$ form. This is in spite of the fact that the $(\Delta \mathbf{z})$ formulation is seemingly more compact. The reason for our choice is that we intend to use a modeling framework to efficiently and automatically manage the algebraic structures present, and the $(\Delta \mathbf{u}, \Delta \mathbf{z})$ system is preferable for this processing. This form can be written as

$$\begin{bmatrix} \Delta \mathbf{P} \\ \hat{\mathbf{f}} \end{bmatrix} = \begin{bmatrix} \mathbf{C}^T \mathbf{SC} & -\mathbf{C}^T \mathbf{SN} \\ -\mathbf{N}^T \mathbf{SC} & \mathbf{H} + \mathbf{N}^T \mathbf{SN} \end{bmatrix} \begin{bmatrix} \Delta \mathbf{u} \\ \Delta \mathbf{z} \end{bmatrix} - \begin{bmatrix} \mathbf{0} \\ \mathbf{f} \end{bmatrix},$$

$$\mathbf{f}_1 \geq \mathbf{0}, \quad \Delta \mathbf{z}_1 \geq \mathbf{0}, \quad \perp, \tag{22}$$

$$\mathbf{f}_2 \geq \mathbf{0}, \quad \hat{\mathbf{z}}_2 + \Delta \mathbf{z}_2 \geq \mathbf{0}, \quad \perp$$

where we have defined

$$\hat{\mathbf{f}} = \mathbf{H}\hat{\mathbf{z}} + \mathbf{r} - \mathbf{N}^T \hat{\mathbf{Q}}$$

Problem (22) is an MCP (Dirkse & Ferris 1995) and is characterized (loosely) by the presence of free variables $\Delta \mathbf{u}$ and, of course, by complementarity constraints. The existence and uniqueness of solutions to the above MCP to any finite loading step is dependent on the nature of the hardening or softening matrix $\mathbf{H}$. Assuming the stiffness matrix $\mathbf{C}^T \mathbf{SC}$ is symmetric and positive definite (as in the case of small displacements), a positive definite $\mathbf{H}$ (pure hardening) implies a unique solution; a positive semidefinite $\mathbf{H}$ (perfect plasticity and/or cable slackening) implies that the solution is not necessarily unique; and a sign indefinite $\mathbf{H}$ (softening) implies a possible multiplicity of solutions (bifurcation), if the problem is feasible.

4 FORM-FINDING

It will be useful to mention briefly how, using the developed relations, we can tackle the initial form-finding problem. As it is not possible to deal with the many facets of this often complex exercise, we only provide a generic idea of how this can be carried out. We assume that inelastic yielding will not occur (this can be checked a posteriori) and hence the system, which involves slackening, is fully reversible or holonomic. Denoting by $\mathbf{P}_o$ the initial self-weight loads and, if necessary, forces applied at the boundary nodes, and by $\mathbf{d}$ the imposed deformations (to simulate some turnbuckle operations), we can write the holonomic system as follows:

$$\mathbf{C}^T\mathbf{Q} = \mathbf{P}_o \qquad (23)$$

$$\mathbf{q} = \mathbf{C}\mathbf{u} \qquad (24)$$

$$\mathbf{q} = \mathbf{e} + \mathbf{p} + \mathbf{d} \qquad (25)$$

$$\mathbf{Q} = \mathbf{S}\mathbf{e} \qquad (26)$$

$$\mathbf{p} = \mathbf{N}\mathbf{z} \qquad (27)$$

$$\mathbf{f} = -\mathbf{N}^T\mathbf{Q} \qquad (28)$$

$$\mathbf{f} \geq 0, \quad \mathbf{z} \geq 0, \quad \perp \qquad (29)$$

As before, we proceed to simplify the relation set (23)-(29) to furnish the following MCP in ($\mathbf{u}$, $\mathbf{z}$) variables:

$$\begin{bmatrix} \mathbf{P}_o + \mathbf{C}^T\mathbf{S}\mathbf{d} \\ -\mathbf{N}^T\mathbf{S}\mathbf{d} \end{bmatrix} = \begin{bmatrix} \mathbf{C}^T\mathbf{S}\mathbf{C} & -\mathbf{C}^T\mathbf{S}\mathbf{N} \\ -\mathbf{N}^T\mathbf{S}\mathbf{C} & \mathbf{N}^T\mathbf{S}\mathbf{N} \end{bmatrix} \begin{bmatrix} \mathbf{u} \\ \mathbf{z} \end{bmatrix} - \begin{bmatrix} \mathbf{0} \\ \mathbf{f} \end{bmatrix}, \qquad (30)$$

$$\mathbf{f} \geq 0, \quad \mathbf{z} \geq 0, \quad \perp$$

The form-finding stage can be regarded as being simply an iterative process involving repeated solves of MCP (30). In essence, starting from some stressless and weightless configuration, we can solve MCP (30) for self-weight and prestressing forces ($\mathbf{P}_o$ and/or $\mathbf{d}$) to give the displacements $\mathbf{u}$. The new configuration can then be evaluated and the process iteratively repeated to attain the desired configuration. It may be possible to automatically select the prestress subject to various constraints by setting up an optimization problem.

5 NUMERICAL SOLUTION ENVIRONMENT

The numerical solution of the finite step problem (22) is conceptually straightforward and involves the following key steps:

1. Carry out an initial form-finding analysis as described in Section 4.

2. If the maximum specified load has been reached then exit, else update MCP (22).

3. For a prescribed load increment $\Delta\mathbf{P}$, solve MCP (22) to obtain all possible solutions and go to step 2; if no solution go to the next step 4.

4. Solve MCP (22) for $-\Delta\mathbf{P}$ and go to step 2.

The computational scheme outlined above should easily work with perfect plasticity and hardening (even in the presence of cable slackening) as the underlying so-called "monotone" MCP (Dirkse & Ferris 1995) can be solved by known standard methods. The case when softening is involved is more difficult, albeit possible (see e.g. Tin-Loi & Xia 2000) to deal with, as multiple solutions may exist. Following our successful computational experiences in dealing with such MCPs, we have adopted the industry standard MCP solver PATH for the present work.

PATH, developed by Dirkse & Ferris (1995), is a nonsmooth Newton method for finding a zero of a problem (so-called "normal map") associated with the MCP. Similar to the classical Newton method for solving equations, the PATH algorithm uses a globalization scheme (i.e. one that attempts to find a solution independent of how good or bad the starting point is) that extends damped Newton methods for smooth equations to this complementarity situation.

PATH is available within the powerful modeling system GAMS (an acronym for General Algebraic Modeling System) (Brooke at al. 1998). It is within this modeling environment that we have implemented and tested, using PATH as the solver, the cable-strut models for the present work. In addition to providing an automatic facility for solving our problems in the MCP form given by (22), the GAMS framework provides several other advantages. Of these, we need only mention: the facility it provides in constructing, maintaining and solving large and complex mathematical programming models; its powerful, easy to understand and write language; simplicity and compactness of model construction; an internal efficient sparse data representation; and automatic differentiation capabilities.

The GAMS file is written using a standard text editor and solved via a simple "gams foo" command where "foo.gms" is the name of the file. For a simple and complete example of an optimization GAMS model (related to minimum weight design), see Ferris & Tin-Loi (1999).

6 NUMERICAL EXAMPLE

As an illustration of the capability of the approach described, we present an example concerning the familiar cable net hyperbolic paraboloid structure shown in Figure 2.

The cables are oriented in the principal x-y directions and form a net with 9 x 9 m meshes in plan view. There is a 5.4 m level difference between the center point and both the highest and lowest points. The ends of all cables are assumed to be anchored to a stiff supporting ring.

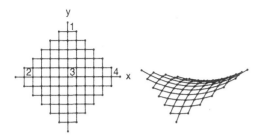

y

1

2   3   4   x

Figure 2. Cable net.

The following data have been assumed: Young's modulus for all cables = 21,000 kN/cm^2; the cross-sectional areas of hogging and sagging cables are 23 cm^2 and 35 cm^2, respectively; the tensile yielding behavior of all cables is assumed to be hardening at 5% of their elastic stiffnesses, with a yield stress of 23.5 kN/cm^2.

In the first instance, we introduced a prestress state by applying 380 kN to the ends of all cables (i.e. in both x and y directions). This did not alter the original configuration by very much.

Then, with respect to the configuration so obtained, we analyzed the cable net for two load cases (with all loads applied as vertical point loads and governed by a load factor α):

Case (a): 20α kN at all nodes.

Case (b): 20α kN at nodes on one side (right side in Figure 2) of the longest hogging cable, 8α kN loads at nodes on the other side of the longest hogging cable, and 14α kN loads at all nodes along the longest hogging cable.

The results for both load cases are shown in Figures 3-4.

Figure 3 shows the responses for the symmetric load case (a): load factor versus deflections at nodes 1, 2 and 3, and the deflection profiles of the longest hogging and sagging cables for three load levels. Similarly, Figure 4 shows the corresponding responses for the unsymmetric load case (b).

For load case (a), the longest sagging cable first yields at its ends at $\alpha = 1.188$. With an increase in α, yielding spreads rapidly in the sagging cables which have all yielded when $\alpha = 1.284$ is reached. At $\alpha = 1.377$, the ends of the longest hogging cable start to slacken and the structure rapidly loses its stiffness, with the ends of the longest sagging cable reaching rupture levels.

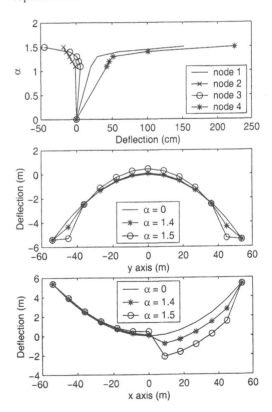

Figure 3. Results for load case (a).

Figure 4. Results for load case (b).

For load case (b), slackening of the hogging cables (on the more heavily loaded side) starts at $\alpha = 1.0928$ and at $\alpha = 1.1349$ all hogging cables in that heavily loaded region have slackened. At $\alpha = 1.349$, the ends of the longest sagging cable yield, and at $\alpha = 1.375$ all sagging cables have yielded. The structure stiffness rapidly deteriorates when the hogging cables on the more lightly loaded side start to yield at $\alpha = 1.485$.

It should be noted that the deflections are significant in the latter stages of loading and the assumption of small displacements is not tenable for these load levels.

7 CONCLUDING REMARKS

The nonlinear analysis of large-scale cable-strut systems can be formulated and efficiently solved as an MCP, particularly with the use of standard solvers such as PATH and from within a modeling environment.

The constitutive model used accounts for both cable slackening and possible softening of struts in compression in a single, unified and elegant complementarity representation.

For simplicity, only physical nonlinearities have been considered in this paper; extension to account for the often mandatory effects of geometrical nonlinearities should not pose any difficulties and can be achieved via standard updating procedures. In fact, a second-order geometric assumption, which is often sufficient for practical structures, leads to essentially the same MCP form for the governing equations.

The form-finding problem is briefly described and can in effect be viewed a special (holonomic) analysis problem.

Current work is aimed (a) at implementing a large displacement capability, (b) at finding out if the same successful methodology can be still used for very large size structures, as would often be encountered in practice, and (c) at developing procedures for automatic form-finding.

ACKNOWLEDGEMENTS

This research was supported by the Australian Research Council. We would also like to thank Professor Michael Ferris, Computer Sciences Department, University of Wisconsin – Madison for help regarding the use of PATH and GAMS.

REFERENCES

Bolzon, G., Maier, G. & Novati, G. 1994. Some aspects of quasi-brittle fracture analysis as a linear complementarity problem . In Z.P. Bazant, Z. Bittnar, M. Jirasek & J. Mazars (eds), *Fracture and damage in quasibrittle structures*: 159-174, London: E&FN Spon.

Brooke, A., Kendrick, D., Meeraus, A. & Raman, R. 1998. *GAMS: A user's guide*. Gams Development Corporation, Washington, DC 20007.

Buchholdt, H.A. 1985. *An introduction to cable roof structures*. Cambridge: Cambridge University Press.

Comi, C., Maier, G. & Perego, U. 1992. Generalized variable finite element modeling and extremum theorems in stepwise holonomic plasticity with internal variables. *Computer Methods in Applied Mechanics and Engineering* 96: 213-237.

Contro, R., Maier, G. & Zavelani, A. 1975. Inelastic analysis of suspension structures by nonlinear programming. *Computer Methods in Applied Mechanics and Engineering* 5: 127-143.

Dirkse, S.P. & Ferris, M.C. 1995. The PATH solver: a non-monotone stabilization scheme for mixed complementarity problems. *Optimization Methods and Software* 5: 123-156.

Ferris, M.C. & Tin-Loi, F. 1999. On the solution of a minimum weight elastoplastic problem involving displacement and complementarity constraints. *Computer Methods in Applied Mechanics and Engineering* 174: 107-120.

Krishna, P. 1978. *Cable suspended roofs*. New York: McGraw-Hill.

Maier, G. 1970. A matrix theory of piecewise linear elastoplasticity with interacting yield planes. *Meccanica* 5: 54-66.

Pellegrino, S. 1992. A class of tensegrity domes. *International Journal of Space Structures* 7: 127-141.

Tin-Loi, F. & Xia, S.H. 2000. Nonholonomic analysis involving frictionless contact as a mixed complementarity problem. *Computer Methods in Applied Mechanics and Engineering* (to appear).

Wang, B.B., Xu, S.Z. & Liu, X.L. 1998. Linear complementary equation method applied in the load response of cable-strut systems. *International Journal of Space Structures* 13: 35-40.

Creative Systems in Structural and Construction Engineering, Singh (ed.) © 2001 Balkema, Rotterdam, ISBN 90 5809 161 9

Nonlinear analysis of load-carrying capacity for composite beams with partial interaction

Y. Okui & K. Seki
Department of Civil and Environmental Engineering, Saitama University, Japan

D. C. Peckley
Department of Engineering Science, University of Philippines, Los Banos, Philippines

ABSTRACT: A finite element model that considers partial interaction between a concrete slab and a steel girder in composite beams is developed. The model takes into account material non-linearity in concrete and steel, and shear-slip behavior in shear connectors as well as geometric non-linearity due to large displacements. The proposed model is validated through comparison with reported experimental data for load-carrying capacity of simply supported composite beams. It is shown that the model can simulate reasonably well the load-deflection and interfacial slip in composite beams. After this validation, parametric studies are carried out on a continuous composite girder bridge to investigate the effect of the shear-slip characteristics of shear connectors. Installing flexible shear connectors near an interior support is effective for reduction of extension strain in a concrete slab, but reduces the load-carrying capacity.

1 INTRODUCTION

Recent design codes for continuous composite girders allow tensile cracking in a concrete slab near internal supports due to negative bending. Of course, the crack width must be limited within an allowable level to ensure durability of the concrete slab. An amount of reinforcement in the concrete slab is commonly increased to reduce the crack width. In structural analysis, concrete within the crack region is neglected, and only steel girder and reinforcement in the concrete slab are considered as an effective member. One of the issues in designing continuous composite girder is the functionality as well as a rational design method for shear connectors embedded in the cracked concrete slab. To clarify the function and to establish the design method, it is necessary to consider the effect of relative slip between the concrete slab and steel girder on mechanical behavior of composite girders.

Other examples where the slip effects become important are flexible shear connectors and precast concrete slabs. The flexible shear connectors are installed at internal supports in order to reduce tension and accordingly cracking in concrete slab. In composite girder bridges with precast concrete slabs, due to the limit of the spacing for in-situ concrete casting, it is not always possible to accommodate enough studs for full interaction.

In this paper, a two-dimensional nonlinear finite element program for load-carrying capacity of steel-concrete composite beams with partial interaction

has been developed. The program considers geometrical non-linearity due to large displacement and material non-linearity for steel and concrete. In order to take into account partial interaction effects, additional degree of freedoms representing the slip at an interface between the concrete deck and steel girder is introduced. Although Oven et al. (1997) already presented a formulation that could take into account the shear-slip effect, we show here a bit different modeling.

Numerical results are given for a two-span continuous bridge with conventional stud-type connectors and flexible shear connectors to discuss the effectiveness of the flexible shear connector.

2 FINITE ELEMENT FOR COMPOSITE GIRDER WITH PARTIAL INTERACTION

In this section, the proposed finite element model is briefly introduced; see Peckley (1998) and Peckley & Okui (2000) for a detailed formulation. A composite beam is modeled as two beam elements and an interfacial spring, which connects the beam elements. Figure 1 illustrates the modeling a composite beam by means of these beam elements and the interfacial spring as well as definitions of coordinates and symbols for displacements. The upper beam represents the concrete slab, and the lower beam the steel girder. The effect of local buckling in a steel section is neglected in this modeling.

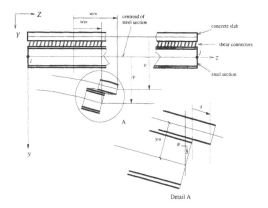

Figure 1. Composite beam element and definition of coordinates and displacements.

Since the vertical displacements, rotations, and curvatures of these two beams are assumed to be identical, the displacements of both concrete and steel section can be expressed in terms of the axial and transverse displacements at the steel section centroid w_{sn}, v, and slip at the interface s. These displacements in an element are interpolated in terms of the shape function $\mathbf{H}$ and the nodal displacements:

$$w_{sn} = \mathbf{Hw}, \quad v = \mathbf{Hv}, \quad s = \mathbf{Hs} \tag{1}$$

where

$\mathbf{w} = \{w_i, w_i', w_j, w_j'\}^T, \mathbf{v} = \{v_i, \theta_i, v_j, \theta_j\}^T, \mathbf{s} = \{s_i, s_i', s_j, s_j'\}^T$
and the prime stands for differentiation with respect to z; the Hermite function is employed as a shape function:

$$\mathbf{H} = [1 - 3(z/l)^2 + 2(z/l)^3, 1 - 2z^2/l + z^3/l^2,$$
$$3(z/l)^2 - 2(z/l)^3, z^2/l + z^3/l^2] \tag{2}$$

where l = the length of an element.

The element stiffness matrix is evaluated by using the finite element method and the nonlinear strain-displacement equation including the finite displacement effect. An updated Lagrangian formulation is employed. The tangent stiffness matrix is obtained by integrating over the volume of an element including nonlinear stress-strain relations for steel and concrete, and slip-shear force relation at the interface. In the current formulation, effects of shear stress on the nonlinear stress-strain relations are neglected, and a simple fiber model with a uniaxial stress-strain relation is employed.

Finally, we have an incremental equilibrium equation for the nodal displacement and applied force:

$$[\mathbf{K} + \mathbf{K}_G]\Delta\mathbf{u} = \Delta\mathbf{P} \tag{3}$$

where

$$\Delta\mathbf{u} = \{\Delta\mathbf{w}, \Delta\mathbf{v}, \Delta\mathbf{s}\}^T \tag{4}$$

and $\mathbf{K}$ = the tangential stiffness matrix, $\mathbf{K}_G$ = the geometric stiffness matrix. Equation 3 is solved for displacement increment $\Delta\mathbf{u}$ in an iteration manner until the unbalance forces are within allowable tolerance.

3 VALIDATION THROUGH COMPARISON WITH EXPERIMENTS

To check the proposed model and program, comparison has been made with the test data reported by Nakajima & Ikegawa (1996). Figure 2 illustrates the experimental set up, and Figure 3 shows the cross section of a specimen. This test specimen is designed to behave as a girder with partial interaction. The relative slip between the concrete slab and the steel girder on the shorter shear span is measured with clip-type gages. In addition to this experiment, Nakajima & Ikegawa (1996) carried out the push out tests of the same studs as the load-carrying test shown in Figure 2. The reported slip-shear curve is used in the numerical analysis.

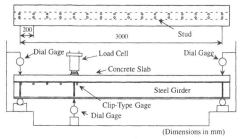

Figure 2. Test set-up from Nakajima et al. (1996).

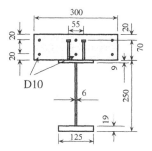

(Dimensions in mm)

Figure 3 Cross section of specimen from Nakajima et al. (1996).

Figure 4 shows the comparison of the load-deflection curve, and Figure 5 is that of the load-slip curve. In both load-displacement and -slip behavior, the numerical results are in good agreement with experiment ones.

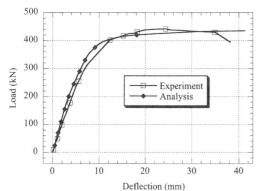

Figure 4. Comparison with experiment; Load-deflection curve.

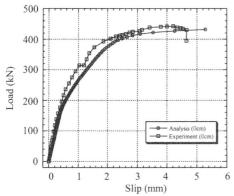

Figure 5. Comparison with experiment; Load-slip curve.

4 CONTINUOUS COMPOSITE BRIDGE MODEL

4.1 *Structural model*

In this section, we apply the proposed model to the two-span continuous composite bridge shown in Figure 6 (Japan Association of Steel Bridge Construction, 1995). The cross section of the model is shown in Figure 7. The dimensions and yield stresses of the flange and web plates of the steel section are listed in Table 1 and 2, respectively. In designing this model bridge, cracking of the concrete slab near the internal support is assumed in accordance with EUROCODE 4 (1996).

Figure 6. Two-span continuous composite girder model (53+53 m) and dimensions of the steel girder (unit: mm).

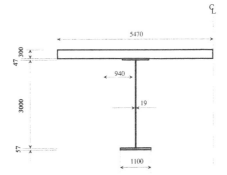

Figure 7. Cross section near the interior support.

Table 1. Dimensions and yield stress of upper and lower flange plates

Position	U. Flg. Size	L. Flg. Size	σ_y
node-node	mm	mm	MPa
1-2	430x22	640x40	215
2-3	430x22	650x43	325
3-6	430x28	760x45	325
6-7	350x18	580x40	325
7-8	370x18	880x46	325
8-9	640x33	910x47	420
9-10	940x47	1100x57	420

Table 2. Thickness and yield stress of web plates

Position	thickness	σ_y
node-node	mm	MPa
1-2	13	215
2-3	13	325
3-7	12	325
7-8	13	325
8-9	17	420
9-10	19	420

For the stress-strain relation of steel, the simple elastic-perfectly-plastic model is used, while for concrete a parabolic-linear model (Figure 8) is implemented in the program. The parabolic-linear model is given as

$$\sigma = \begin{cases} \sigma_{cm}\left[\dfrac{2\varepsilon}{\varepsilon_{cm}} - \left(\dfrac{\varepsilon}{\varepsilon_{cm}}\right)^2\right] & \text{for } (0 < \varepsilon < \varepsilon_{cm}) \\ \sigma_{cm}\left[1 - \dfrac{\varepsilon - \varepsilon_{cm}}{\varepsilon_{cu} - \varepsilon_{cm}}\right] & \text{for } (\varepsilon_{cm} < \varepsilon < \varepsilon_{cu}) \end{cases} \quad (5)$$

where the concrete strength σ_{cm}=35 MPa is used in the following analysis.

Note that Equation 5 is only valid in compression. The concrete in tension is neglected, but reinforcement steel in the RC slab is accounted as effective structural members. The tension stiffening effect in cracked RC members is also neglected in this treatment of concrete. The cross-sectional area ratio of reinforcement to the RC section is assigned to 1.5% in a negative bending region.

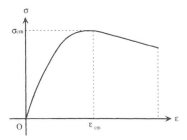

Figure 8. Stress-Strain relation of concrete in compression.

4.2 *Shear connector*

Two types of shear connectors are considered, namely conventional stud type connectors and a flexible shear connector proposed by Abe et al. (1989). The flexible shear connector is made of W shapes (called H-shapes in Japan), whose web plate is covered with expanded polystyrene to enhance flexibility when it is embedded in a concrete slab. This flexible shear connector is specially designed to reduce tensile stress in the concrete slab near interior supports in continuous composite bridges. The flexible shear connectors have been installed in a railway bridge (Okuda et al., 1990).

The shear-slip relationships of these shear connectors are shown in Figure 9, which is based on the push-out test results reported by Hosaka et al. (1998). Since it can be seen that the tangential stiffness of the flexible shear connector after yielding is smaller than that of the shear studs, the flexible shear connector is more effective after the first yielding.

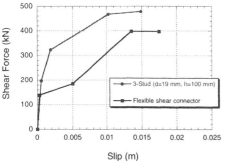

Figure 9. Shear force-slip relationships of studs and flexible shear connector.

Two cases for arrangement of shear connector are considered in the numerical analysis. Figure 10 shows the distribution and types of shear connectors for both cases. In Case A, stud type connectors are used, and the pitch of shear studs is determined based on an elastic analysis following Japanese Specification for Highway Bridges (1992). On the other hand, in Case B, the flexible shear connectors

are installed in the negative bending moment region near the interior support, and their pitch is determined according to that of slab anchors in non-composite bridges.

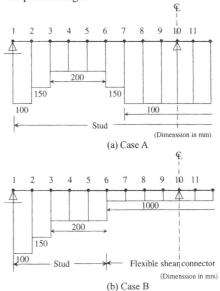

Figure 10. Distribution of stud pitches and shear connector type.

4.3 *Load cases*

In the following numerical analysis, assuming unshored construction, both dead and live loads are applied to a composite section. First, 130 % of dead load is applied to the whole bridge, and then live load is increased by increasing the live load factor α. Hence, the total load TL is given as

$$TL = 1.3D + \alpha L \qquad (6)$$

where D = Dead load, and L = Live load. Figure 11 shows considered loading cases in the numerical analysis. The intensity of the live and dead loads are 18.0 and 41.5 kN/m, respectively.

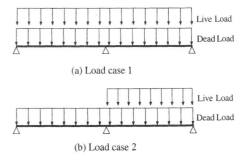

Figure 11. Load cases; Dead and live load combination.

5 EFFECT OF FLEXIBLE SHEAR CONNECTOR

Figures 12 and 13 show the load-displacement curves for Load Case 1 and 2 for both shear-connector Case A and B. In spite of changing shear connectors, almost the same load-displacement curve is obtained, and only the difference is the maximum loading factor. By introducing the flexible shear connector in Case B, the maximum loading factors of Load Case 1 and 2 are reduced to 67 % and 65 % of Case A, respectively. From a load-carrying capacity point of view, installation of the flexible shear connector near the interior support is not profitable.

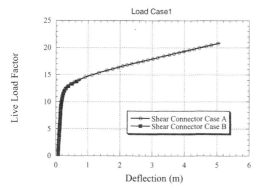

Figure 12. Load-deflection curve for Load Case 1.

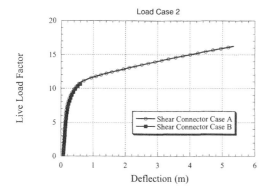

Figure 13. Load-deflection curve for Load Case 2.

The distributions of the relative slip along the length of the bridge at the maximum loading state are shown in Figures 14 and 15 for Load Case 1 and 2, respectively. In shear-connector Case B, the relative slip attains its ultimate value at the boundary between the stud-type shear connector and the flexible shear connector. This can be interpreted as concentration of shear force at the boundary due to abrupt change in stiffness of shear connectors, which governs the maximum load-carrying capacity. Gradual change in shear-connector stiffness at the boundary

seems to be necessary to enhance the load-carrying capacity.

On the other hand, for shear-connector Case A, the slips for both loading cases are small enough to be considered as full-interaction behavior. The maximum load-carrying capacity of shear-connector Case A is limited by concrete crashing near the span center.

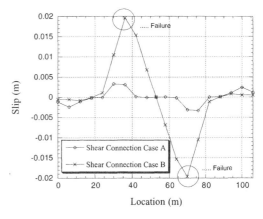

Figure 14. Slip distribution along bridge at maximum loading for Load Case 1.

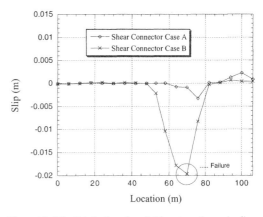

Figure 15. Slip distribution along bridge at maximum loading for Load Case 2.

Let us leave load-carrying capacity at a strength limit state to behavior under a service load. Figure 16 shows the bending moment diagram of both shear-connector cases at a service load of $D+L$. For these two cases, the difference of the maximum negative bending moments at the interior support is only 4%. Under service level loading, the reduction of negative bending moment at interior supports is not expected in spite of installing the flexible shear connector. However, it can be said that increase in the maximum positive bending moment by installing the flexible shear connector is negligibly small as well.

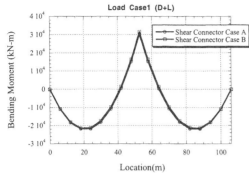

Figure 16. Bending moment diagram for Load Case 1 at a service load level: Effect of shear connector cases on bending moment.

To ensure the durability of continuous composite bridges, it is important to control tensile cracks in a concrete slab owing to the negative bending moment. One objective for installing the flexible shear connector is to reduce crack width in a concrete slab. To check this aspect, the normal strain variations in the composite section at the interior support are plotted in Figure 17, where the vertical axis stands for the vertical distance from the top of a concrete slab. In the shear-connector Case A, there is a slight strain jump at the interface between the concrete slab and steel girder. However, the behavior in Case A is practically full-interaction behavior. On the other hand, in Case B (flexible shear connector case) a considerable strain jump due to the slip occurs at the interface. Furthermore, the maximum strain in the concrete slab is reduced to 40 % of the strain in Case A. It is shown that the flexible shear connector is effective to reduce tensile strain, and accordingly tensile crack width in concrete slab.

6 CLOSURE

In this paper, a finite element model for load-carrying capacity of composite beams with partial interaction was introduced. The proposed model was applied to a two-span continuous composite bridge with two types of shear connectors. The numerical analysis shows that:
(1) By installing flexible shear connectors, the load carrying capacity is decreased.
(2) However, tensile strain at the internal support decreases, which is preferable from a crack-width control point of view.
(3) The bending moment distribution at the service load level is not significantly affected by the flexible shear connector.

Since these results are concluded from the numerical results from a continuous composite model, further parametric studies is necessary to draw gen-

eral conclusions. The present numerical results are based on the shored construction. Numerical study assuming the unshored construction is of future interest.

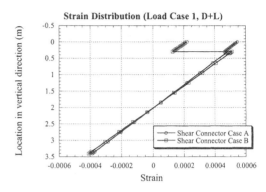

Figure 17. Normal strain distribution in a plane at the interior support (profile view) for Load Case 1 at the service load.

REFERENCES

Abe, H., Nakajima, A. & Horiuchi H., 1989. Effect of division of slab in composite girder and development of flexible connectors, *J. Structural Eng., JSCE.* 35A:1205-1214. (in Japanese)
CEN, 1996. EUROCODE4: Design of composite steel and concrete structures, Part 2: Bridge, 2nd draft.
Hosaka, T., Hiragi, H., Korda, Y., Tachibana, Y. & Watanabe, H. 1998. An experimental study on characteristics of shear connectors in composite continuous girder for railway bridges, *J. Structural Eng., JSCE.* 44A:1497-1504. (in Japanese)
Japan Road Association (JRA) 1992. Specification for high way bridges. Tokyo: Maruzen. (in Japanese)
Japan Association of Steel Bridge Construction (JASBC) 1995, Report on design of continuous composite bridges. (in Japanese)
Nakajima, A & Ikegawa, M. 1996. Elasto-plastic analysis of composite girder with inelastic behavior of shear connectors, *J. of Structural Mechanics & Earthquake Engineering, JSCE.* 43(591):97-106. (in Japanese)
Okuda, M., Satou, T., Yamaki, Y., Iseki, J. & Sasaki, H. 1990. Design and construction of Tokeidou3.4.20 viaduct in Hokusou line, *Bridge & Foundation.* 12: 13-22. (in Japanese)
Oven, V.A., Burgess, I.W., Plank, R.J. & Adbul Wali, A.A. 1997. An analytical model for the analysis of composite beams with partial interaction. *Computers & Structures,* 62(3): 493-504.
Peckley Jr., D.C. 1998. Nonlinear analysis and ultimate strength of composite beams with partial interaction, Master thesis, Saitama University.
Peckley Jr., D.C & Okui, Y. 2000. A finite element model of steel-concrete composite beams and girders for near-collapse analysis, *Proc. Int. Conf. Concrete Art, Science and Technology, CAST2000, Manila*: Theme E: 1-17.

Creative Systems in Structural and Construction Engineering, Singh (ed.) © 2001 Balkema, Rotterdam, ISBN 90 5809 161 9

Strength and behavior of slender SRC beam-columns

K. Tsuda
Faculty of Human-Environment Studies, Graduate School of Kyushu University, Fukuoka, Japan

T. Fujinaga
Department of Architecture and Civil Engineering, Kobe University, Japan

D. Fukuma
Toda Corporation, Japan

ABSTRACT: Steel reinforced concrete columns are tested. Columns are subjected to concentric and eccentric axial compressive force at both ends. As the experimental parameters, the buckling length-section depth ratio L_k/D and magnitude of eccentricity e are selected. End moment (M_u) -axial force (N_u) relations are obtained by the test. A total of 36 specimens are tested. Strength and behavior are examined, and design methods for slender SRC columns are investigated.

1 INTRODUCTION

Steel reinforced concrete (SRC) structures have been used widely for building structures in Japan. The strength of steel and concrete for building structures is getting higher with the development of new materials. The cross section with high strength materials becomes smaller, and consequently a column becomes more slender. The design of a column considering buckling and $P\delta$ effect becomes more important in such situation.

There are many researches on SRC columns in Japan. However, these researches are mainly on short columns subjected to earthquake loading. There are no systematic and fundamental studies on SRC columns under concentric and eccentric axial force in wide range of slenderness ratios.

AIJ (Architectural Institute of Japan) Standards for Structural Calculation of Steel Reinforced Concrete Structures was published first in 1958 and the latest fourth edition was revised in 1987 (AIJ 1987). The superposed strength method has been used for calculating the strength of SRC members since the outset of the Standards. In the latest revision, modified method of the superposed strength of a column section is adopted for slender steel-concrete composite columns considering the effect of additional bending moment ($P\delta$ moment).

The accuracy of the formula, however, was not examined in detail. Authors have proposed an equation to evaluate the strength of slender SRC columns on the basis of numerical analysis (Tsuda et al. 1999, Fujinaga et al.2000). In this paper, the method is examined on the basis of experimental work.

Objectives of this paper are to obtain the maximum strength and behavior by performing an experimental work, and to examine the proposed design formulas for the slender SRC beam-columns comparing with the experimental strength.

2 TEST PROGRAM

2.1 *General*

The test specimen is a SRC beam-columns of which encased steel is a wide flange section. Figure 1 shows the cross section of a specimen. The material of steel portion is mild steel (SS400, Japanese Industrial Standards). The design strength of concrete is 300kg/cm^2, and the concrete casting is carried out in a horizontal position. Specimens are subjected to concentric and eccentric axial force at both ends as shown in Figure 2.

2.2 *Experimental parameters*

As the experimental parameters, buckling length (L_k)-the section depth (D) ratio L_k/D and magnitude of eccentricity e, and a direction of the loading point are selected, and they vary as follows; $L_k/D = 4, 8, 12, 18, 24, 30$ and $e = 0, \kappa, 3\kappa, 5\kappa$ (κ: core of a section). Thirty-six specimens are tested in total. Test conditions of specimens are shown in Table 1.

2.3 *Mechanical Properties*

In addition to the tensile test of steel coupon, stub columns of shaped steel and concrete cylinders are tested under compression to examine the stress-strain relations.

Average yield stress of shaped steel is equal to 3.34 t/cm^2, and average ultimate strength is equal to 4.50 t/

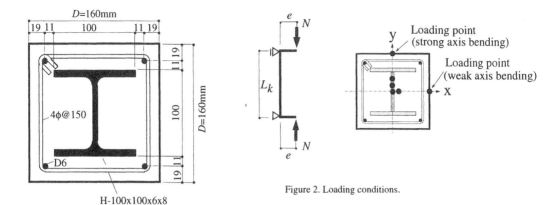

Figure 1. Cross section.

Figure 2. Loading conditions.

cm², both obtained from tensile tests as shown in Table 2. As the results of stub column test of shaped steel, yield stress σ_y of a steel is $3.6t/cm^2$ and ultimate strength σ_u is equal to $4.10t/cm^2$.

Design standard strength of concrete F_c is $300kg/cm^2$. The average compressive strength $_c\sigma_B$ of concrete obtained from the cylinder compressive test ranges from 340 to $386kg/cm^2$ as shown in Table 1. Measured dimensions of steel shape are shown in Table 3.

2.4 Loading Apparatus

Loading apparatus is shown in Figure 3. Exact pin-ended conditions are obtained because the specimens are loaded through hemispherical oil film bearing at each end. The assigned eccentricity e is given to the specimen by moving the bearing plates. Axial load in one direction is applied to a specimen.

3 Design Formula

3.1 Fundamental equation of superposed strength method

The strength of a SRC slender column is calculated by Equations 1-2, where conventional equations of simple superposition is modified based on the Wakabayashi's study (Wakabayashi 1977).

i) preferable for bending in strong axis

When $N_u < {_{rc}N_{cu}}$ or $M_u > {_sM_{u0}}(1 - {_{rc}N_{cu}}/{_{src}N_k})$

$$N_u = {_{rc}N_u}$$

$$M_u = {_{rc}M_u} + {_sM_{u0}}(1 - {_{rc}N_u}/{_{src}N_k}) \qquad (1.1)$$

Table 1. Test conditions.

Name	L_k/D	L_k (cm)	e/κ	e (cm)	$_c\sigma_B$ (kg/cm²)	N_e (ton)
C30-00			0.0	0.0		55.0
S30-05			0.5	1.3		39.8
W30-05	30	480	0.5	1.3	340	31.4
S30-10			1.0	2.7		29.7
S30-30			3.0	8.0		21.2
W30-30			3.0	8.0		15.2
C24-00			0.0	0.0		85.6
S24-05			0.5	1.3		-
W24-05	24	384	0.5	1.3	368	47.3
S24-10			1.0	2.7		43.0
S24-30			3.0	8.0		30.2
W24-30			3.0	8.0		17.2
C18-00			0.0	0.0		106.8
S18-05			0.5	1.3		89.0
W18-05	18	288	0.5	1.3	349	69.2
S18-10			1.0	2.7		59.9
S18-30			3.0	8.0		36.8
W18-30			3.0	8.0		25.6
C12-00			0.0	0.0		146.5
S12-05			0.5	1.3		111.7
W12-05	12	192	0.5	1.3	386	85.5
S12-10			1.0	2.7		106.2
S12-30			3.0	8.0		44.0
W12-30			3.0	8.0		31.3
C8-00			0.0	0.0		160.0
S8-05			0.5	1.3		116.0
W8-05	8	128	0.5	1.3	369	91.1
S8-10			1.0	2.7		115.8
S8-30			3.0	8.0		48.6
W8-30			3.0	8.0		35.0
C4-00			0.0	0.0		162.0
S4-05			0.5	1.3		138.1
W4-05	4	64	0.5	1.3	366	100.6
S4-10			1.0	2.7		124.1
S4-30			3.0	8.0		51.4
W4-30			3.0	8.0		38.6

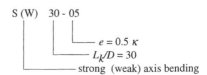

S (W)   30 - 05

— $e = 0.5\,\kappa$
— $L_k/D = 30$
— strong (weak) axis bending

970

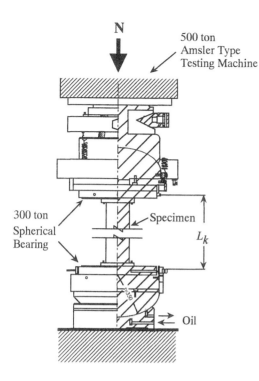

Figure 3. Loading apparatus.

Table 2. Mechanical properties.

steel	Yield stress σ_y (kg/cm²)	Ultimate stress σ_u (kg/cm²)	Yield ratio σ_y/σ_u
flange	3.34	4.50	0.74
web	3.35	4.56	0.73
main re-bar	3.48	5.69	0.61
hoop	4.14	5.00	0.83

Table 3. Measured dimensions of shaped steel.

height H (cm)	width B (cm)	flange thickness t_f (cm)	web thickness t_w (cm)
10.1	9.88	0.72	0.59

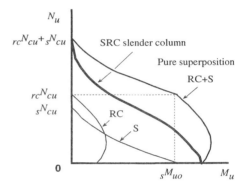

Figure 4. Superposed strength.

When $N_u > {}_{rc}N_{cu}$ or $M_u < {}_sM_{u0}(1 - {}_{rc}N_{cu}/{}_{src}N_k)$

$$N_u = {}_{rc}N_{cu} + {}_sN_u$$

$$M_u = {}_sM_u(1 - {}_{rc}N_{cu}/{}_{src}N_k) \tag{1.2}$$

i) preferable for bending in weak axis

When $N_u < {}_sN_{cu}$ or $M_u > {}_{rc}M_{u0}(1 - {}_sN_{cu}/{}_{src}N_k)$

$$N_u = {}_sN_u$$

$$M_u = {}_{rc}M_{u0}(1 - {}_sN_u/{}_{src}N_k) + {}_sM_u \tag{2.1}$$

When $N_u > {}_sN_{cu}$ or $M_u < {}_{rc}M_{u0}(1 - {}_sN_{cu}/{}_{src}N_k)$

$$N_u = {}_sN_{cu} + {}_{rc}N_u$$

$$M_u = {}_{rc}M_u(1 - {}_sN_{cu}/{}_{src}N_k) \tag{2.2}$$

where N_u =ultimate compressive strength of member; M_u =ultimate flexural strength of member; ${}_{rc}N_{cu}$ =ultimate compressive strength of RC portion subjected to compression alone; ${}_{rc}N_u$ =ultimate compressive strength of RC portion; ${}_{rc}M_u$ =ultimate flexural strength of RC portion; ${}_{rc}M_{u0}$ =ultimate flexural strength of RC portion subjected to bending alone; ${}_sN_{cu}$ =ultimate compressive strength of steel portion subjected to compressive alone; ${}_sN_u$ =ultimate compressive strength of steel portion; ${}_sM_u$ =ultimate flexural strength of steel portion; ${}_sM_{u0}$ =ultimate flexural strength of steel portion subjected to bending alone; ${}_{src}N_k$ =buckling strength of column. Subscripts s and rc indicate forces carried by the steel and RC portions of a SRC columns.

The concept of Equation 1 is shown in Figure 4.

3.2 Strength of slender steel column

As an interaction between ${}_sN_u$ and ${}_sM_u$ appearing in Equations 1-2, a conventional formula used in the plastic design of steel structures is adopted in the form of

$$\frac{{}_sN_u}{{}_sN_{cr}} + \frac{{}_sM_u}{{}_sM_{u0}(1 - \frac{{}_sN_u}{{}_sN_k})} = 1 \tag{3}$$

where ${}_sN_u$ = axial load; ${}_sN_{cr}$ ($={}_sN_{cu}$)=critical load; ${}_sN_k$ = Euler buckling load; ${}_sM_u$ = the applied end moment; ${}_sM_{u0}$= full plastic moment (AIJ 1975).

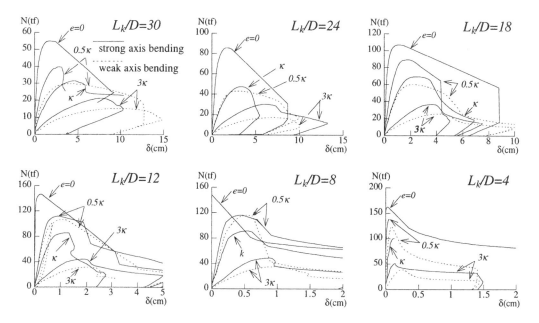

Figure 5. Axial load-lateral deflection relation.

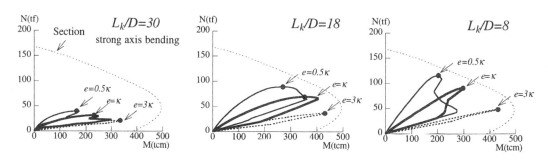

Figure 6. Moment-axial load relation (strong axis bending).

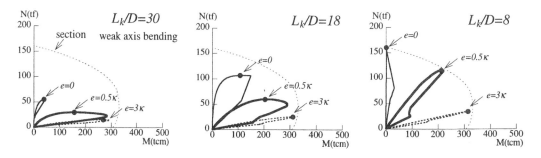

Figure 7. Moment-axial load relation (weak axis bending).

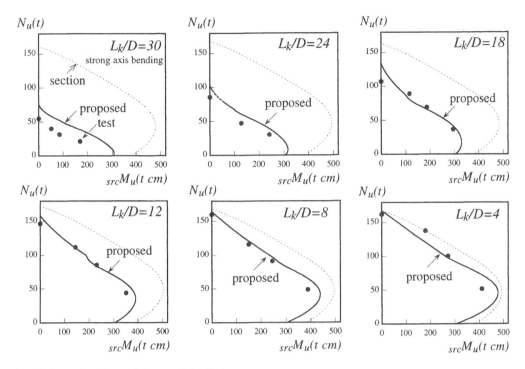

Figure 8. Comparison of strength (strong axis bending).

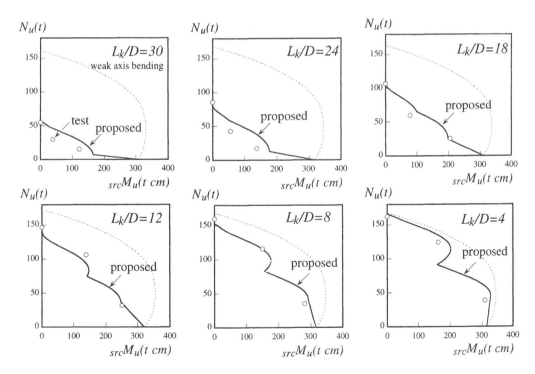

Figure 9. Comparison of strength (weak axis bending).

3.3 *Strength of slender RC column*

Strength of slender RC columns is calculated by using the superposed strength of concrete column and steel column, in which reinforcing bars are considered as slender steel column (Tsuda et al. 1999, Fujinaga et al.2000). Strength of slender concrete column is proposed before for calculating the strengths of CFT column (Tsuda 1995).

3.4 *Strength of slender SRC column*

Equation for calculating the strength of slender SRC column is obtained by substituting the strength of steel column and that of RC column for Equation 1-2.

4 RESULTS AND DISCUSSIONS

4.1 *Elasto-plastic behavior*

Figure 5 shows the relations between the axial load and the deflection at the mid-span section. In each figure, the buckling length -section depth ratio is kept constant, changing the value of eccentricity. Solid line indicates the result of strong axis bending and dashed line weak axis bending.

It is observed that as the eccentricity becomes large the maximum load decreases, while the deflection at the maximum load increases. Effect of the magnitude of eccentricity on the strength becomes small as the L_k/D ratio becomes large.

Figures 6-7 show examples of orbit of the moment at the mid-span M ($=N$ ($e+\delta$)) and axial load N. In the figure, black circle indicates a maximum load point, and dashed line full plastic moment (strength of a section). As to the strength of cross section, full plastic moment is computed by assuming the rectangular stress distribution with yield stress σ_y of steel and compressive strength $_c\sigma_B$.

It is observed that most of the specimens cannot attain the full plastic moment at the maximum load due to instability phenomenon.

4.2 *Comparison between experimental maximum load and design strength*

The experimental maximum strengths N_e are shown in Table 1. Comparison of strength is shown in Figures 8-9 in the end moment M_u (=N_u e) -axial load N_u relations. Experimental maximum loads are shown by circle, and proposed strength are shown by solid line. In addition to these strength, strength of a section is shown by dotted line.

In case of strong axial bending, when the value of L_k/D ratio less equal than 24, proposed strength can predict the experimental maximum load fairly well, while when the value of L_k/D equal to 30 it becomes unsafe. This is due to out of plane behaviors effect on the strength and behavior, which was not taken into

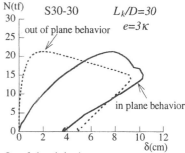

Figure 10. Out of plane behavior

consideration in the proposed method (see Fig.10). Hereafter, the design formula of beam-columns taking the biaxial flexural behavior into consideration should be required.

In case of weak axial bending, when the value of L_k/D ratio less equal than 18, proposed strength can predict the experimental maximum load well, while when the value of L_k/D equal to 24, 30 it becomes unsafe.

5 CONCLUSIONS

1. In case of strong axial bending, proposed design method agree with the experimental results well when L_k/D ratio ranges from 4 to 24. In case of L_k/D=30, it gives a unsafe strength.

2. The design formula of beam-columns taking the biaxial flexural behavior into consideration should be required.

3. In case of weak axial bending, when the value of L_k/D ratio less equal than 18, proposed strength can predict the experimental maximum load well.

REFERENCES

Architectural Institute of Japan (AIJ) 1987. *Standard for Structural Calculation of Steel Reinforced Concrete Structures.* (in Japanese).

Architectural Institute of Japan (AIJ) 1975. *Guide to the Plastic Design of Steel Structures.*(in Japanese)

Fujinaga, T.,K. Tsuda & C. Matsui. 2000. Superposed strength of slender SRC beam-columns with various steel sections. *Proceedings of the Sixth ASCCS Conference. Vol.1* : 331-338.

Tsuda, K., C. Matsui & Fujinaga. T 1999. Strength estimation of slender SRC beam-columns by using the superposed method. *Proceedings of the Seventh East Asia-Pacific Conference on Structural Engineering & Construction*: 985-990.

Tsuda, K., C. Matsui & Y. Ishibashi 1995. Stability design of slender concrete filled steel tubular columns", *Proceedings of the Fifth East Asia-Pacific Conference on Structural Engineering & Construction*:439 - 444.

Wakabayashi, M. 1977. A New design method of long composite beam-columns. *Proceedings of International Colloquium on Stability of Structures under Static and Dynamic Loads. SSRC/ASCE*: 742-756.

Creative Systems in Structural and Construction Engineering, Singh (ed.) © 2001 Balkema, Rotterdam, ISBN 90 5809 161 9

RC column retrofit technique by PC rod prestressing as external hoops

T. Yamakawa, T. Tagawa & W. Li
University of the Ryukyus, Okinawa, Japan

M. Kurashige
Neturen Company Limited, Tokyo, Japan

S. Kamogawa
Nagasaki Prefectural Government, Japan

ABSTRACT: By paying attention to the fact that the transverse confinement of concrete is very useful in order to improve ductility for RC members, a new seismic retrofit technique utilizing PC rod prestressing was proposed as one of techniques in order to prevent shear failure and improve ductility of RC columns. The technique is an active confinement by PC rod prestressing as external hoops. These extreme short column specimens with shear span to depth ratio $M/(VD)=1.0$ are modeled based upon the school buildings designed by the old seismic design code before 1971 in Japan. These extreme short RC column specimens were tested under the combination of cyclic lateral forces and constant axial load. As a result, the retrofit of column specimens illustrated an excellent seismic performance through the experimental tests, even if the shear failure for RC column specimens were likely to happen.

1 INTRODUCTION

In Japan, after the Hanshin Awaji Earthquake (the 1995 Hyogoken-Nanbu Earthquake) disaster occurred in 1995, the diagnosis and retrofit for existing buildings are being carried out gradually. It is roughly classified into three retrofit methods at present. The first is the retrofit method in which the ductility of the building increases. The second method is to increase the lateral capacity of the building. The third method is to reduce the seismic input motion. The method, that is newly proposed in this study, is one in which the ductility of the RC column drastically increases. Also its lateral capacity increases.

Authors have carried out the experimental investigation on hybrid RC column confined with square steel tube and PC rod prestressing, (Yamakawa, Muranaka & Kurashige 1997), (Yamakawa 1998). As the result, it was proven to be very effective in the column subjected to high axial compression especially. This is because shear strength, bond strength, compression strength and ductility are greatly improved by high confinement of concrete. Paying attention to this fact, a new seismic retrofit utilizing PC rod prestressing and corner blocks was proposed by authous (Yamakawa, Kamogawa & Kurashige 1999). In order to validate of this seismic retrofit, the extreme short RC column specimens whose shear failure was likely to happen were tested under the combination of cyclic lateral forces and constant axial load in this paper. The purpose of this paper is to verify that this seismic retrofit is one of the effective methods to improve the ductility of the existing poor extreme short RC columns.

Unlike steel plate or continuous fiber polymer sheet jacket using the passive confinement, this method is unique for the positive utilization of the active confinement due to PC rod prestressing. In addition, the method can be applied easily to the RC column with sleeve or spandrel walls and window frames. This is a reason why it is easy that PC rods penetrate into them. As an emergency reinforcement of the damaged structures immediately after earthquake hazard, this method seems to be suitable, because the retrofit can be conveniently carried out without the heavy machine.

2 EXPERIMENT

The experiment is planned on the case in which concrete strength is low and the case with high in this paper in order to verify whether this retrofit is effective for ductility improvement of the extreme brittle column of shear span to depth ratio of 1.0. The ductility improvement becomes easily possible on the column of shear span to depth ratio which is larger than this, if this retrofit is useful for ductility improvement of the extreme brittle column. Seismic loading test of the 6 extreme short columns with shear span to depth ratio of 1.0 with 250mm square section·is planned. Any column is also the 1/2.4

scale model of the column (on the assumption of the about 600x600 mm square section) of the school building designed according to previous old code 1971 in Japan (in which the shear reinforcement is the 250mm interval).

The mechanical characteristic values of the employed material in column test specimens are listed in Table 1. The details of this retrofit are illustrated in Fig. 1. The list of column test specimens is shown in Table 2. The column test specimens before retrofit are extreme brittle columns in which shear failure will be likely to happen. The retrofit places the PC rods of 3.8ϕ or 5.4ϕ on the identical plane like the circumference tie-hoop through the corner blocks placed in four column corners. Besides, the prestress is introduced into these PC rods. The level of introducing prestress (490MPa) adopted about 2450μ which was equal to about 1/3 of the yield point strain (about 6000μ). The prestressing could be manually easily introduced measuring the strain of the PC rod by strain gauges.

Table 1. Properties of steel bars and PC rods.

Type		a(cm^2)	f_y(MPa)	ε_y(%)	E_s(GPa)
	D10	0.71	360	0.19	189.6
Steel bar	ϕ3.7	0.11	333	0.17	200.0
	D10'	0.71	371	0.20	185.6
	ϕ3.7'	0.11	333	0.17	195.9
	ϕ3.8	0.11	1245	0.60	201.0
PC rod	ϕ5.4	0.23	1245	0.60	201.0
	ϕ3.8'	0.11	1202	0.61	200.0
	ϕ5.4	0.23	1202	0.61	200.0

* a = cross section area, f_y = yield strength of steel, ε_y =yield strain of steel, E_s = modulus of elasticity, ' = R99S series.

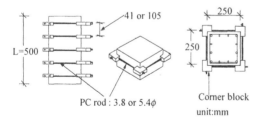

Figure 1. Details of seismic retrofit for RC column.

Table 2. Column specimens.

Specimen	σ_B (MPa)	PC rod (p_p(%))
R99S-P0	20.7	———
R99S-P105	20.7	3.8ϕ-@105 (0.08)
R99S-P41'	20.7	5.4ϕ-@41 (0.45)
R98S-P0	31.6	———
R98S-P41	31.6	3.8ϕ-@41 (0.21)
R99S-P41'H	30.9	5.4ϕ-@41 (0.45)

* Common details
  Rebar:12-D10(P_g=1.36%), Hoop:3.7ϕ-@105(P_w=0.08%), PC rods:3.8ϕ, 5.4ϕ(Prestress 490MPa).

The PC rods 3.8ϕ and 5.4ϕ are respectively correspondent to the PC rods 9.2ϕ and 13.0ϕ of the smallest diameter which prevails in actual construction at the present. Because it is intended in order to ensure the penetration hole (the diameter is 15-20ϕ) for the PC reinforcement rod, even if window frames are adjacent to the column.

The loading apparatus with the parallel supporting mechanism developed by the Building Research Institute of Japan, which was called as Ken-Ken type, was used in the experiment as illustrated in Fig. 2. The parallel supporting mechanism always maintained the horizontal loading beam to be parallel with the strong floor and constrained the rotations of both top and bottom stubs of the RC column specimens during cyclic loading test.

The instrumentation methods were designed to obtain the following data: (1) applied forces; (2) horizontal and vertical displacements; (3) strains of the PC rods and the reinforcement bars.

Cyclic lateral forces were applied to the column test specimen simultaneously subjected to constant axial load $0.2f_c$'Ag, where f_c' is the concrete cylinder strength and Ag is the gross area of section of the column. Cyclic lateral forces were referred in the

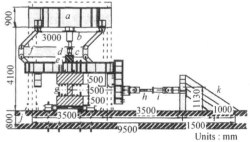

a. Vertical loading reaction frame, *b.* Servohydraulic actuator,
c. Load cell, *d.* Spherical ball bearing, *e.* Roller bearing plate,
f. Parallel supporting mechanism, *g.* Specimen, *h.* Load cell,
i. Double acting hydraulic jack, *j.* Strong floor,
k. Horizontal loading reaction frame.

Figure 2. Detail of test setup.

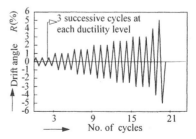

Figure 3. Loading program.

976

range of drift angle $R=\pm 0.5$, ± 1.0, ± 1.5, ± 2.0, ± 2.5, $\pm 3.0\%$ at three cycles respectively, and $R=\pm 4.0\%$, $\pm 5.0\%$ at one cycle if the ductility of the specimen was expected. The $R=\delta/h$ is a story drift angle of the column where δ is the story drift and h is the height of the column. The loading program is illustrated in Fig. 3.

3 TEST RESULTS AND DISCUSSION

Observed cracking patterns are illustrated in Fig. 4. As the shear span to depth ratio of the column specimens is small, the shear cracks on the depth side are dominant. However the shear crack width is not enlarged as the confinement by the PC rod prestressing increases.

The experimental results on the relationship between the shear force V and the story drift angle R are shown in Fig. 5. Also the measured hysteresis loops on the mean axial strain ε_v along the member axis versus the story drift angle R are illustrated in Fig. 5. The mean axial strain ε_v is given by dividing relative vertical displacement between top and bottom stubs by the column height h. As seen in the V-R and the ε_v-R hysteresis loops, these loops are improved as the PC rod reinforcement becomes larger, namely the confinement effect by the PC rod prestressing increases. The V-R hysteresis loops correspond with the ε_v-R ones. If the V-R hysteresis loops are large and do not descend, the ε_v-R ones present desirable configuration which means good seismic performance and high ductility.

The location of measured strains of the PC rods is illustrated in Fig. 6. And also the location of yielding

PC rods is presented in Fig. 6. Some representative measured strains of the PC rods are shown in Fig. 7. When the diameter of the PC rod is small or the space of the PC rod reinforcement is large, namely the PC rod reinforcement is not sufficient, the PC rods are likely to yield as shown in Fig. 7. The lateral

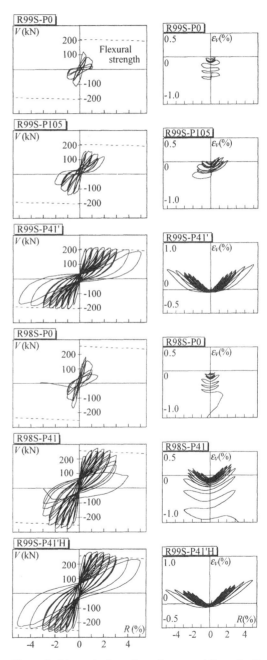

Figure 5. Measured shear force V versus drift angle R relationships and mean axial strain ε_v-R versus R relationships.

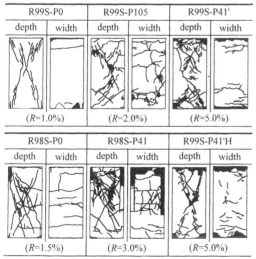

R99S-P0		R99S-P105		R99S-P41'	
depth	width	depth	width	depth	width
(R=1.0%)		(R=2.0%)		(R=5.0%)	

R98S-P0		R98S-P41		R99S-P41'H	
depth	width	depth	width	depth	width
(R=1.5%)		(R=3.0%)		(R=5.0%)	

* R =final story drift angle (see Figure 5.)

Figure 4. Observed cracking patterns after loading test.

capacity rapidly decreased on the standard column test specimens R99S-P0 and R98S-P0 which corresponded to the existing poor column, while the shear cracks happened in virgin loading. This failure type was shear failure before longitudinal reinforcement did not reach tensile yield.

In the test specimen R99S-P105 retrofitted using the PC rod 3.8ϕ in the 105mm interval, the shear cracks were generated at $R=-0.5\%$, and the maximum lateral capacity was recorded at $R=1.0\%$. The lateral capacity greatly lowered with the increase in the crack width, and the experiment ended in the shear failure at $R=-2.0\%$. It is not possible that the PC rods prevent shear failure of the column, and yield

phenomenon has been partially generated in the PC rods (see Fig. 7). This is a reason why the core concrete of the RC column expands by shear and compression forces owing to poor PC rod prestressing reinforcement.

In the test specimen R98S-P41 retrofitted using the PC rod 3.8ϕ in the 41mm interval, the largest shear strength occurred at $R=1.5\%$ without observing the lateral capacity decreasing in spite of generating the shear cracks at $R=0.5\%$. Desirable flexural behavior continued till $R=2.0\%$. However, the expansion began at $R=2.5\%$ by shear and compressive stresses in the upper part of the column due to insufficient PC rod prestressing reinforcement, and the lateral capacity lowered by the cyclic loading. This fact corresponds to the strain measurement results of the PC rods (see Figs. 5, 6 and 7).

On the other hand, flexural cracks were dominantly observed near the top and bottom of the column specimens R99S-P41' and R99S-P41'H retrofitted by the PC rod 5.4ϕ in the 41mm interval. However, shear cracks at the section depth side happened (see Fig. 4). They are small, if the shear crack width is compared with standard test specimens R99S-P0 and R98S-P0 without PC rod retrofitting. These two test specimens do not lower lateral capacity, even if drift angle of the column is made to increase to $R=3.0\%$ (see Fig. 5). Though the longitudinal reinforcement perfectly yields through experiment, the PC rods completely do not yield, as they are shown in Figs. 6 and 7. These PC rods perfectly hold the expansion of the core concrete of the column by shear force and axial compression. The reverse S-shape is slightly observed here at R99S-P41' in the hysteresis loops (see Fig. 5), when it is compared with R99S-P41'H, and the bond slip of the longitudinal reinforcement seems to have been generated a little. It is proven that the seismic performance has been improved on R99S-P41'H of

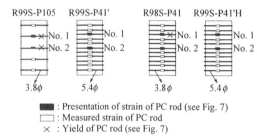

R99S-P105 R99S-P41'   R98S-P41 R99S-P41'H

3.8ϕ 5.4ϕ 3.8ϕ 5.4ϕ

■ : Presentation of strain of PC rod (see Fig. 7)
☐ : Measured strain of PC rod
✕ : Yield of PC rod (see Fig. 7)

Figure 6. Measurement and yielding of PC rods at depth side.

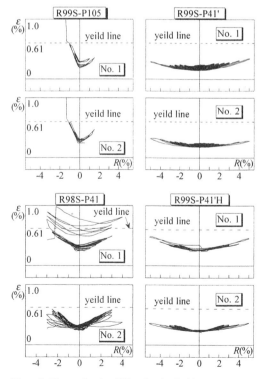

Figure 7. Measured stain of PC rod at depth side.

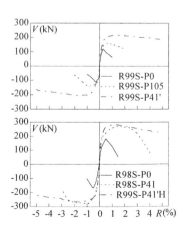

Figure 8. Measured skeleton curves.

978

which the concrete strength is high. Because it is a reason why the bond slip of longitudinal reinforcement can be hardly observed.

Measured skeleton curves are illustrated in Fig. 8. The retrofitting effect by the PC rod prestressing is sufficiently recognized, and the specimen R99S-P41'H of which concrete strength is high is excellent. And the specimen R99S-P41' is also excellent on the seismic performance. It was not possible for lack of the PC rod reinforcement that the specimen R99S-P105 improves the seismic performance.

4 ANALYTICAL INVESTIGATION

Experimental results are shown in Table 3. Passive confinement by the PC rod prestressing is considered as well as hoop. In addition, the active confined effect by the PC rod prestressing as $4.1\sigma_r$ (σ_r =lateral pressure) is added in the concrete cylinder strength σ_B according to Richart's proposed formula (Richart 1928). In the calculation, this added value is used. The constitutive law of confined concrete calculated based on Sakino and Sun's proposed equation (Sakino & Sun 1994), whose result is almost the same as Mander's formula (Mander, Priestley & Park 1988) in case of low or ordinary concrete strength, is shown in Fig. 9. With the increase of transverse reinforcement by the PC rod, it is proven to gradually improve the stress-strain curve as illustrated in Fig. 9. This constitutive law is applied to the fiber model, and the V-R curve is calculated by the assumption of the curvature distribution along the member axis of the column. In addition, the comparison of calculated shear strength with the measured skeleton curve is presented in Fig. 10. With regard to the flexural failure type specimen, it is proven to mostly follow the skeleton curve in the fiber model through Fig. 10. By comparing calculated flexural strength with calculated shear strength, failure mode of specimen can be almost estimated as shown in Fig. 10.

The comparison of experimental results and calculated N-M interaction curves is shown in Fig. 11. Though standard test specimens which fracture

by shearing stress are not reaching the flexural strength, the retrofit test specimens governed by the flexural failure are existing on the curve of the flexural strength. In this experiment, since the axial stress to concrete cylinder strength ratio $N/(bD\sigma_B)$=0.2 is comparatively small, the increment of flexural strength by the PC reinforcement rod is small. It is proven that the flexural strength also considerably increases by the retrofit with the PC rods, when the axial force rises. The rate of increase seems to be also bigger a little in case of low concrete strength.

The relationship between measured lateral capacity V_{exp} and the transverse reinforcement ratio p_w, in which the PC reinforcement rod is also contained as well as hoops, is shown in Fig. 12(a). In addition, the relationship between V_{exp}/V_f and V_s/V_f is shown in Fig. 12(b). The shear strength V_s of V_s/V_f ,

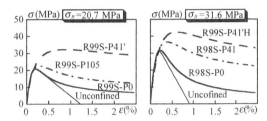

Figure 9. Stress-stain curves for concrete.

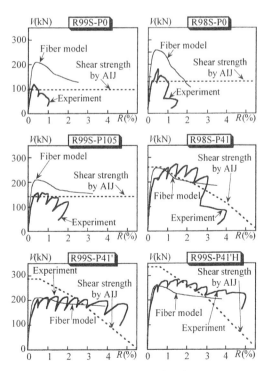

Figure 10. Calculated and experimental results.

Table 3. Experimental results.

| Specimen | σ_B (MPa) | Experiment | | Failure mode | PC rod yield or not | Mark |
		V_{exp}(kN)	R_f(%)			
R99S-P0	20.7	119.2	0.48	S	——	■
R99S-P105	20.7	158.1	1.81	S-B	Yield	▨
R99S-P41'	20.7	215.8	>5.0	F	Non	□
R98S-P0	31.6	181.1	0.61	S	——	●
R98S-P41	31.6	272.6	3.15	FC	Yield	⊙
R99S-P41'H	30.9	282.3	4.61	F	Non	○

* V_{exp} = measured lateral capacity, R_f = the drift angle in 80% of measured lateral capacity, S = shear failure, FC = flexural-compression failure, F = flexural failure, S-B = bond slip after shear failure.

is calculated according to the design guidelines by AIJ, and the flexural strength V_f is respectively calculated by the simplified formula. With the increase of p_w and V_s/V_f, it is proven that seismic performance has been improved from shear failure to flexural failure. Especially, it is necessary that the V_s/V_f exceeds 1.0. The relationship between V_{exp}/V_f and V_{bu}/V_f is shown in Fig. 12(c). The shear strength V_{bu} due to bond failure is calculated according to the design guidelines by AIJ, and the flexural strength V_f is calculated by above-mentioned formula. Two kinds of shear strength V_{bu} and V_s are compared as shown in Fig. 12(d). The V_s due to shear failure is in general larger than the V_{bu} due to bond failure. This fact suggests that it is a little difficult to improve the bond behavior by the PC rod retrofitting in comparison with the shear strength enhancement.

The ductility factor $\mu = R_f/R_y$ is also presented in Fig. 13. The R_f is measured drift angle when shear force descends to 80% of peak shear force and the R_y is drift angle when columns yield. The arrow in Fig. 13 shows that it can be expected at the ductility factor over it. This is because the cyclic loading test has finished at the largest drift angle $R=5\%$. Namely, shear force does not descend to 80% of the peak shear force until final drift angle during the cyclic loading test. As the p_w and V_s/V_f increase, the μ is improved through Fig. 13. This fact suggests that this retrofit is effective for improvement of ductility.

5 CONCLUSIONS

(1) As a result of the cyclic loading test of the extreme short RC column retrofitted by the PC rod prestressing as external hoops, it is proven that the seismic performance can be improved as a ductile column even in the case of brittle columns.
(2) It is proven that this retrofit is effective in the extreme short RC column of shear span to depth ratio of 1.0 from the analytical investigation such as the application of calculation formula and design equations.

REFERENCES

Yamakawa, T., Muranaka, K. & Kurashige, M. 1997. An Experimental Study on Seismic Behavior of High Confined RC Columns by Steel Tubes and Prestressing. *Proceedings of the Japan Concrete Institute.* 19. 1431-1442 (in Japanese)

Yamakawa, T. 1998. Seismic Performance of Hybrid RC Short Columns Highly Confined in Square Steel Tube and Prestressing. *23rd Conference on Our World in Concrete & Structures.* 17: 205-212

Yamakawa, T., Kamogawa, S. & Kurashige, M. 1999. An Experimental Study on the Seismic Retrofit Technique Confined with PC Bar Prestressing as External Hoops for RC columns. *Journal of Structural and Construction Engineering.* 526: 141-145 (in Japanese)

Richart, F. E. et al.1928. A Study of the Failure of Concrete under Combined Compressive Stresses. *University of Illinois, Engineering Experimental Station, Bulletin* 185

Sakino, K. & Sun, Y. 1994. Stress-strain curve of Concrete confined by rectilinear hoop. *Journal of Structural and Construction Engineering of Architectural Institute of Japan (AIJ).* 461: 95-104 (in Japanese)

Mander, J. B., Priestley, M. J. N. & Park, R. 1988. Theoretical stress-strain model for concrete. *ASCE. Journal of Structural Engineering.* 144: 1804-1826

Architectural Institute of Japan (AIJ). 1999. *Design guidelines for earthquake resistant reinforced concrete buildings based on inelastic displacement concept.* (in Japanese)

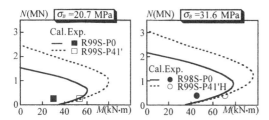

Figure 11. *N-M* interaction curves.

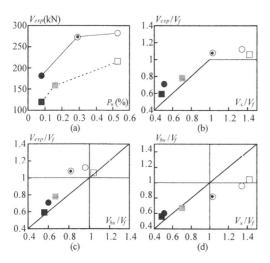

Figure 12. Lateral capacity.

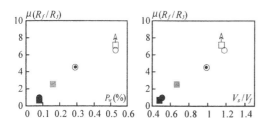

Figure 13. Ductility factor.

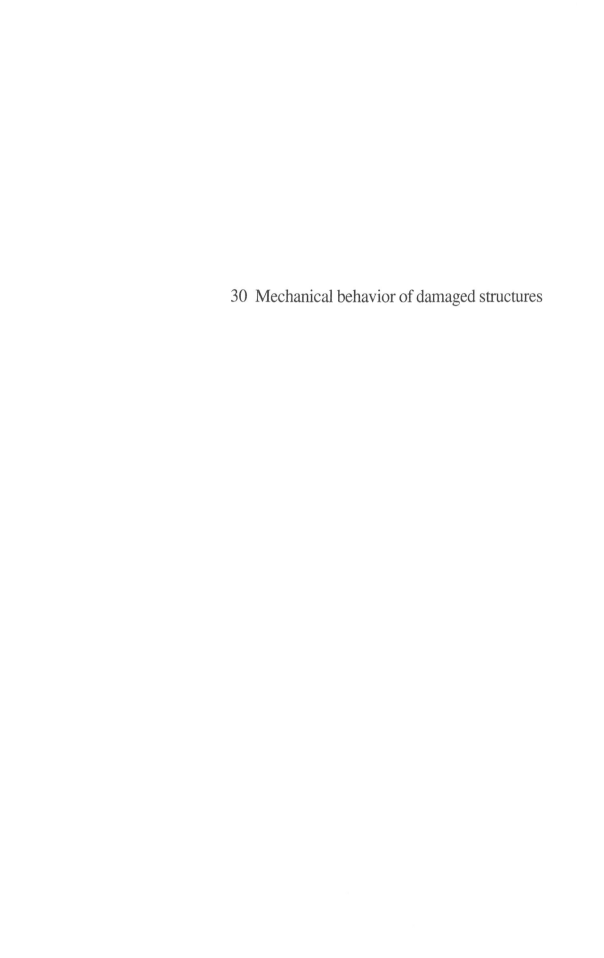

30 Mechanical behavior of damaged structures

Creative Systems in Structural and Construction Engineering, Singh (ed.) © 2001 Balkema, Rotterdam, ISBN 90 5809 161 9

Weak-link analysis for seismic retrofit of bridge structures

K. M. Mahmoud

URS Corporation, New York, N.Y., USA

ABSTRACT: This paper investigates the seismic retrofit strategy of a "critical" bridge in New York City. Site response spectra for 500-year and 2500-year earthquakes were developed in accordance with the New York City Department of Transportation (NYCDOT) new seismic design criteria. The analysis is based on softening the structure, by introducing a weak-link concept, in which guide angles yield under the 2500-year earthquake, shifting the fundamental frequency in the lower region of the response spectrum. As a part of this softening approach, six elastomeric bearings were incorporated into the structural system. This results in reducing the seismic forces demand on the substructure, thus eliminating the need for footings and piles retrofit.

1 INTRODUCTION

The vast majority of existing bridges in the United States were designed without adequate consideration for earthquakes. Several bridges have collapsed at relatively low levels of ground motions. When rehabilitating existing bridges, the common approach is strengthening. However, it is more effective to soften the structure by shifting the fundamental frequency in the lower region of the response spectrum of the seismic event.

Seismic isolation provides a practical alternative for the seismic retrofit of bridges. The primary difference in response that, in a conventional design, the deck displacement occurs in the column/pier, whereas in the isolated structure, the majority of the deck displacement occurs across the isolator. A cylindrical or rectangular block of rubber constitutes the simplest isolator for a bridge superstructure but presents a number of inconveniences, essentially related to its high deformability under vertical loads. The insertion of a number of horizontal steel plates, as with elastomeric bearing pads, solves most problems by increasing the vertical stiffness and improving the stability of the behavior under horizontal load, (Priestley et al. 1996). In this paper, the seismic retrofit strategy of a "critical" bridge is based on incorporating six elastomeric bearing reinforced with steel plates into the structural system. As per the New York City Department of

Transportation new seismic design criteria, "critical" bridges should be analyzed under a 500-year and 2500-year return earthquakes, as defined below. The seismic motion input is based on site specific response spectra developed for the two return period earthquakes.

The bridge in this case study was originally built in 1940 and is a four span structure. Due to its deteriorated condition, its superstructure was scheduled for replacement. The pier columns, the abutments and all foundations will remain. The new superstructure has four continuous composite deck spans over three piers. Each pier consists of a transverse steel plate girder cap beam supported on steel laminated elastomeric bearings installed on top of two columns. The floor system consists of steel stringers seated over the transverse cap beams at each pier. The bearing movements are restrained, longitudinally, at the center pier only by guide angles with zero gaps against the sole plate of the bearing assembly. At the abutments, the bearing movements are restrained, transversely, by guide angles with zero gaps at each stringer similar to above.

2 NYCDOT NEW SEISMIC DESIGN CRITERIA

In December 1998, NYCDOT issued a new Seismic Hazard Criterion (NYCDOT 1998), based on the new soil classes A thru F adopted in (NEHRP 1997).

Two level approach is required for evaluation of "critical" bridges, namely;

1. lower level-functional evaluation: (500-year return period or 10% probability of exceedance in 50 years), the bridge sustains no collapse, no damage to primary structural elements, minimal repairable damage and full access to normal traffic conditions.

2. upper level-safety evaluation: (2500-year return period or 2% probability of exceedance in 50 years), the bridge shall provide limited access for emergency traffic within 48 hours and full access within months, with repairable damage and no collapse.

3 SITE RESPONSE ANALYSIS

The energy released by earthquakes is propagated by different types of waves. Body waves, originating at the rupture zone, include P-waves (primary waves) which displace objects parallel to the direction of propagation and S-waves (shear waves) which displace objects at right angles to their path.

Velocities of P-waves, v_p, and S-waves, v_s, in elastic medium are frequency independent, and are given, respectively, by:

$$v_p = \{(\lambda + 2\,G)\,g_c\,/\,\rho\}^{1/2} \qquad (1)$$

$$v_s = \{G.\,g_c\,/\,\rho\}^{1/2} \qquad (2)$$

where λ is Lame's coefficient of elasticity and G is the shear modulus, given by:

$$\lambda = \{v\,E\,/\,[(1+v)\,(1-2v\,)]\} \qquad (3)$$

$$G = \{E\,/\,[2\,(1+v)]\} \qquad (4)$$

E is the modulus of elasticity of the soil, v is Poisson's ratio, ρ the density and g_c is the gravitational acceleration. While S-waves travel slower than P-waves, they transmit more energy and cause the majority of damage to structures. It is also note worthy that the S-waves velocity has a significant influence on the soils and the response spectra at the bridge foundation. For this project, the site response analysis developed the S-waves velocity, shear modulus, damping curves and idealized soil profile. This response analysis was produced using the one-dimensional program SHAKE, which solves the problem of S-waves and P-waves motion propagating vertically through idealized horizontal soil layers. In SHAKE, the total

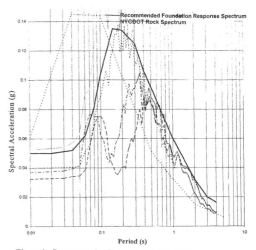

Figure 1 - Recommended Response Spectra, 500-year Return Period

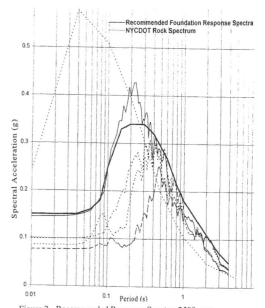

Figure 2 - Recommended Response Spectra, 2500-year Return Period

stress method is used, where the non-linearity of the soil is treated by an equivalent linear procedure.

The foundation response spectra used for the structural analysis of the bridge are shown in Figures 1 and 2 for the 500-year return period and 2500-year return period, respectively.

4 LIQUEFACTION INVESTIGATION

Liquefaction is characterized by the formation of boils and mud spouts at the ground surface, by seepage of water through ground cracks and by development of quicksand like conditions over large areas. Saturated sand deposits subjected to ground vibration, tend to compact and decrease in volume. If drainage is unable to occur, an increase in pore-water pressure builds up to the point at which it is equal to the overburden pressure, the effective stress becomes zero, the sand loses its strength completely and it develops a liquified state (Seed & Idriss 1982). Liquefaction has been reported to occur essentially in saturated deposits of loose sands or silts, according to the gradation curves given in Figure 3 (Caltrans 1994).

An assessment of the liquefaction potential can be performed evaluating the average cyclic shear stress, τ_l, required to cause liquefaction in N cycles and the average cyclic stress, τ, induced by the expected earthquake in N cycles. The factor of safety, FS, against liquefaction can then be defined as:

$$FS = \tau_l / \tau_d \tag{5}$$

τ_d can be estimated approximately from the following expression (Seed & Idriss 1982):

$$\tau_d = 0.65 \, \rho \, H \, A \, r_d \tag{6}$$

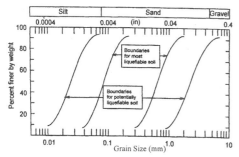

Figure 3 - Limits in the Gradation Curves Separating Liquefiable and Unliquefiable.

where H is the depth of the location under consideration, A the seismic coefficient, and r_d is a stress reduction coefficient which depends on the deformability of the soil column above the location under consideration and therefore depends on the soil profile. The range of possible values for r_d for various soil profiles is shown in Figure 4. In the

upper 40ft (12 m) the average values can be used, with errors generally limited to less than 5%. The number of cycles to be considered depends on the duration of ground shaking and therefore on the magnitude of the design earthquake. The following number of cycles have been suggested (Seed & Idriss 1982) magnitude M6, 5 cycles; M6.75, 10 cycles; M7.5, 15 cycles; and M8.5, 26 cycles.

The average shear stress required to induce liquefaction, τ_l, can be obtained from cyclic simple shear tests or from cyclic triaxial compression tests. In both cases, extreme care should be exercised to obtain good-quality "undisturbed samples". A convenient parameter to express the effective equivalent seismic action on a sand element to evaluate its liquefaction potential is presented by the ratio of average cyclic shear stress to the initial vertical effective stress, σ_0, acting on the sand before the cyclic stresses were applied. Dividing both terms by σ_0, Equation (6) becomes:

$$\tau_d / \sigma_0' = 0.65 \, A \, r_d \, (\sigma_0 / \sigma_0') \tag{7}$$

where σ_0 is the total vertical stress at the depth considered, equal to ρH, while to compute σ_0, the effect of the presence of the water table is added. This parameter (τ_d / σ_0) therefore takes into account the earthquake intensity and reflects the influence of the depth of the soil element, the soil relative density and the depth of the water table.

According to (Seed & Harder 1990), excess pore pressures begin to develop when factors of safety are less than 1.4. For this project, the average factors of safety are greater than 1.5 for both return periods.

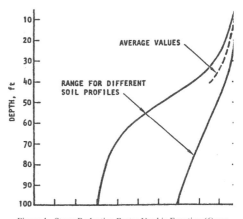

Figure 4 - Stress Reduction Factor Used in Equation (6) as a Function of Soil Profile (Seed & Idriss 1982)

Therefore, liquefaction hazard is not significant at this site.

5 SEISMIC ANALYSIS AND MODELING

Computer modeling is an essential part of seismic analysis. However, modeling techniques have not been standardized and substantial engineering judgement is required to develop appropriate models. In the following, a brief summary of the different methods of seismic analysis is presented, the method used for the example bridge is outlined and the weak-link concept is explained.

5.1 Methods of seismic analysis

There are four main methods for seismic analysis of bridge structures, single mode response spectrum analysis, multi mode response spectrum analysis, push-over (non-linear analysis) or non-linear time-history analysis. Single mode analysis is used for "regular" bridges. According to AASHTO, regular bridges do not have abrupt changes in mass, stiffness or geometry along its span and have no large differences in these parameters between adjacent spans. Multi mode analysis is recommended by AASHTO for "irregular" bridges. According to AASHTO, bridges that cannot be classified as regular are considered irregular. This method should be employed when several vibration modes are required to characterize the dynamic behavior of the bridge. Computations can be easily performed on a personal computer using any structural analysis program with capabilities to perform multi mode spectral analysis. This method serves as a starting point for a very critical bridge that requires push-over analysis or non-linear time-history analysis. The push-over analysis is a static non-linear analysis that can be used to estimate the dynamic demands imposed on a structure by earthquake ground motions. A predetermined lateral load pattern that approximately represents the seismic forces generated during an earthquake is applied to the structure. The structure is then "pushed over", by applying displacement, to the level of deformation expected during the earthquake while maintaining the applied load pattern. Non-linearities in form of member yielding, buckling, plastic hinge rotations are introduced in the analysis to account for the possibility that members will be stressed well beyond the elastic limit during an earthquake. Usually, push-over analysis is performed before (or

in lieu of) non-linear time-history analysis. Push-over analysis clearly identifies redundant load paths in the structure and actual lateral load capacity of the structural system. It is more realistic as compared to response spectrum analysis when the structure is stressed beyond its elastic limit. For structures with short periods of vibrations (stiff structures) whose response is governed by first mode, the push-over deflection profiles correlate well with deflected profiles of the structure during an earthquake. However, special analysis software with non-linear capabilities is required to perform this analysis. Also, it is often difficult to determine the appropriate lateral load pattern required for push-over analysis. Particularly, for irregular bridges when response cannot be characterized by the first mode of vibration, push-over analysis loses its accuracy when applied to an entire bridge. Usually push-over analyses in such cases are performed on a component basis and results are related to the global model of the bridge. Thus, either response spectrum or time-history analysis may be required in addition to push-over analysis in case of irregular bridges. Time-history analysis is by far the most comprehensive method for seismic analysis. The earthquake record in the form of time versus acceleration is input at the base of the structure. Th response of the structure is computed at every second, or even less, for the entire duration of an earthquake. This method differs from response spectrum analysis because the effect of "time" is considered as an initial boundary condition for computation of stresses in the next step. Furthermore, non-linearities that commonly occur during an earthquake can be included in the time-history analysis. Such non-linearities cannot be easily incorporated in response spectrum analysis. Unlike the response spectrum method, non-linear time-history analysis does not assume a specific method for mode combination. Hence, results are realistic and not conservative. In addition, this method is equivalent to getting 100% mass participation. Full mass participation is necessary to generate correct earthquake forces. Usually, only 90-95% participation is obtained in response spectrum analysis. All types of non-linearities can be accounted for in this analysis. This could be very important when seismic retrofit involves energy dissipation using yielding of members or plastic hinge rotation. However, this method is very expensive and time consuming to perform. Large amount of information are generated and special analysis software with non-linear material models and hysteresis models is required to perform non-linear time-history analysis.

5.2 Multi Mode Spectral analysis

Due to the expenses associated with the time-history analysis and based on the recommendations of (FHWA 1995), multi mode spectral analysis was deemed sufficient for the analysis of the example bridge. The dynamic analysis was performed using LARSA 98 software. The response spectra of the 500-year and 2500-year return period earthquakes were used as the input ground motions. AASHTO requires the response, as a minimum, to include the effects of a number of modes equivalent to three times the number of spans up to a maximum of 25 modes. As required by AASHTO, the response spectra were applied 100% along one direction and 30% along the mutually perpendicular direction. The member forces and displacement were computed by combining the respective response quantities from individual modes by the Complete Quadratic Combination (CQC) method.

5.3 Weak-link analysis

At the center pier, the bearings are fixed, for service loads only, through guide angles with zero gaps against the sole plate of the bearing assembly. These guide angles provide a weak-link that is designed to yield under the 2500-year return earthquake. When the guide angles yield, the structure softens, thus, elongating the period and reducing the demand on the substructure at the center pier, eliminating the need to retrofit its piled foundation. Moreover, stiffened restrainer assemblies have been provided at both sides of the bearings to prevent the steel cap beams from getting unseated off the columns in case of excessive displacements. To ensure that the displacement of the superstructure would not engage the stiffened assemblies and accidentally increase the stiffness of the bearings, these stiffened assemblies have been located at a specified distance away from the guide angles, as determined from the seismic analysis. In fact, this specified distance also takes into account the thickness of the yielded guide angle. Under the 500-year return earthquake, only guide angles at the abutments sustain damage and will need to be replaced.

This weak-link analysis provides a cost-effective and an easily repairable detail as compared to the standard New York State Department of Transportation (NYSDOT) detail, wherein a pin is used inside the bearing to make it fixed. In the latter case, it would be necessary to replace the entire bearing that would involve jacking of the girders which is costlier and prevents access to the bridge for a longer time. On the other hand, with the proposed weak-link concept, the guide angles can be replaced in a timely manner without the need to jack the bridge.

6 CONCLUSIONS

1. When rehabilitating an existing bridge for seismic action, one should think in terms of dynamics, because bridges respond dynamically to earthquakes.
2. Softening instead of strengthening may often be a better strategy for seismic retrofit.
3. The simplest approach of providing softening is by incorporating elastomeric laminated bearings.
4. The resulting shift in the frequency to the lower region of the earthquake spectrum reduces the demand seismic forces.

REFERENCES

California Department of Transportation (Caltrans) 1994. Seismic design of highway bridge foundations: Training course manual.

Federal Highway Administration (FHWA) 1995. Seismic retrofitting manual for highway bridges.

New York City Department of Transportation (NYCDOT) 1998. New York City seismic hazard study and its engineering application. Final report.

National Earthquake Hazard Reduction Program (NHRP) 1997.

Priestley, M.J.N. et al. 1996. Seismic design and retrofit of bridges. New York: John Wiley & Sons.

Seed, H.B. & L.F., Jr. Harder 1990. SPT based analysis of cyclic pore pressure generation and undrained residual strength. Proc. of H.B. Seed Memorial Symposium, May 1990: 351-376. Bi Tech publishers, Ltd.

Seed, H.B. & I.M. Idriss 1982. Ground motion and soil liquefaction during earthquakes. Earthquake Engineering Research Institute Monograph.

Creative Systems in Structural and Construction Engineering, Singh (ed.)© 2001 Balkema, Rotterdam, ISBN 90 5809 161 9

Shear strength of cracked R/C beam

T.Tamura, T.Shigematsu & M.Imori
Tokuyama College of Technology, Japan

K.Nakashiki
Kyuken Sekkei Corporation, Fukuoka, Japan

ABSTRACT: This paper reports the shear strength of R/C beams cracked due to the axial tensile force experimentally. In the case of the non-cracked beam the axial tensile force decreased the shear strength of it. However, the experimental results show that the beam increased the ultimate shear strength when the axial force is released after some cracks appear the beam face. We discuss the mechanism of the shear fracture of the pre-cracked reinforced concrete member. Also, the tendencies of the numerical predictions by finite element method are in good agreement with the experimental results.

1 INTRODUCTION

Nominal shear strength of R/C beam is generally determined by concrete strength, reinforcement ratio, effective depth of cross section, shear span to depth ratio and applied axial force. For example, in the JSCE design equation the shear strength of a R/C beam without shear reinforcement (V_{cd}) is calculated as follows:

$$V_{cd} = \beta_d \cdot \beta_p \cdot \beta_n \cdot f_{vcd} \cdot b_w \cdot d / \gamma_b \qquad (1)$$

in which β_d, β_p and β_n is the parameter of effective depth, of reinforcement ratio and of applied axial force respectively. f_{vcd} is a function of compressive strength of concrete and b_w is web width of beam and d is effective depth. γ_b is shape factor which is generally 1.3.

The authors have studied the shear strength of reinforced concrete beams subjected to axial tensile force experimentally and numerically (Tamura, T. 1994, etc.). From the present investigation, it becomes evident that the influence of axial force on the shear strength of reinforced concrete members is dependent upon the reinforcement ratio, shear span to depth ratio and so on. For example, the smaller the shear span to depth ratio is the larger the influence of the axial tension is, also the smaller the reinforcement ratio is the less the maximum loading capacity of the beam is.

However, in the experimental results, we found the different behavior in shear between the pre-cracked reinforced concrete member and non-cracked one. That was the pre-cracked member increased the ultimate shear strength when the axial stress is released. As a matter of fact, in statically indeterminate reinforced concrete structures there are many cracks caused by longitudinal tensile stress such as shrinkage of concrete, temperature change and other severe conditions.

In this paper, we discuss the effects of the initial cracks for the fracture mechanism of reinforced concrete members.

2 EXPERIMENTAL PROCEDURE

2.1 *Specimen*

Nine specimens were provided and tested. The parameters adopted here in shear span depth ratios and axial force levels. All the specimens are same dimension in width, height and length. The arrangement of the rebar is shown in Fig.1 (The reinforcing ratio p=0.022). Also, no stirrups are placed except the supporting points to clarify the structural behaviors. To introduce the axial tensile force holes are placed at the both ends. Table 1 represents the information on material properties of the specimens.

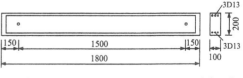

Fig.1. Specimens   (unit: mm)

Table 1. Material properties

Reinforcement D10(SD295)		
Yield Stress	Tensile Strength	Elastic Modulus
331 Mpa	477 Mpa	194 Gpa
Concrete		
Compressive Strength	Tensile Strength	Elastic Modulus
33.1 Mpa	2.79 Mpa	30.6 Gpa

Table 2. Parameters and experimental results

Spacimen		Condition		Exper. results	
Series	No.	a/d	fn	P b	Type
A	a-0	2.0	0	57.09	S
	a-1		1970	76.20	S
	a-2		2970	104.43	S
B	b-0	2.5	0	48.50	S
	b-1		1951	52.86	S
	b-2		3030	92.99	S
C	c-0	3.0	0	44.92	S
	c-1		1951	56.51	S
	c-2		2955	66.68	S

a/d: Shear span effective depth ratio
fn: Axial tensile stress (kPa)
Pb: Ultimate Load (kN)
Type: Status of failure   S-Shear failure

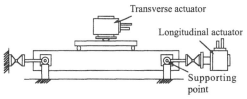

Fig.2. Test apparatus

2.2 Test apparatus

Fig.2 shows the test apparatus. It is composed of two oil pressure actuators (transverse and longitudinal direction) controlled by the electro-oil servomechanism. The axial force is induced by longitudinal actuator from both of ends of the beam and at the center of gravity of the beam height. Both supporting points are hinges of bearing style.

2.3 Bending test of cracked beam

Before the bending test the axial tensile force is introduced into both ends of the beam via longitudinal actuator to make the initial cracks. Then the volume of the axial tensile force controls the quantity of the cracks. Transverse load is introduced into the specimen through a loading beam from the two loading points as shown in Fig.2. The transverse load increases continuously until the beam fails by displacement controlled system. Dial gauges are placed at the center of the span to measure the deflection of the beam. Also the bending strains are measured at the center of tensile reinforcement and the top of the beam. During the test, new cracks are marked on the surface of the beam at each loading step.

3 EXPERIMENTAL RESULTS

3.1 Load deflection relationships

Fig.3(a),(b),(c) show the load deflection relationships of the beams of each series. All of the beams failed in shear. Figures show that the greater the initial axial stress is the smaller the deflection is. Also all the pre-cracked beams improve the ultimate strength compared with the non-cracked beams' one.

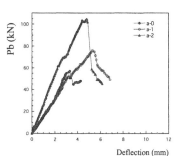

Fig.3(a). Load deflection Relationsips of Type-A series

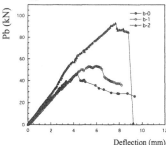

Fig.3(b). Load deflection Relationsips of Type-B series

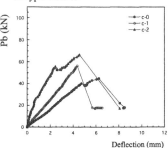

Fig.3(c). Load deflection Relationsips of Type-C series

990

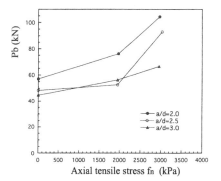

Fig.4. Relationsips between Pb and fn

3.2 *Ultimate strength*

Ultimate strength of each beam is shown in Table 2. Fig.4 shows the relationship between the ultimate load Pb and initial axial tensile stress. From this figure, it is quite obvious that the larger the initial axial stress is the greater the shear strength of the beam is.

3.3 *Ultimate state of the beams*

In case of the beam subjected axial tensile stress approximately fn=3000kpa the pre-cracks appear in all the beams. However, in case of fn=2000kpa, the cracks disappear on the face of the beam. Fig.5 compare the ultimate state of non axial stressed beams and the beams pre-cracked by initial axial stress fn=2.8MPa. In the figures, the broken lines show the pre-crack by initial axial force. Every the beam made the diagonal shear cracks. However, in case of the pre-cracked beam, almost the shear cracks occurred between two pre-cracks. Also, when a shear crack across the pre-crack, the crack does not spread straightly.

4 FRACTURE MECHANISM

Fig.6 shows a model of the non-cracked beam and of the pre-cracked one. It is evident that the fracture mechanisms of pre-cracked beam are different from non-cracked one. As shown in this figure, in case of pre-cracked beam, the shearing force cannot be propagated through concrete continuously, therefore the shear stress of the non-cracked beam is greater than the pre-cracked beam. Consequently, pre-cracked beam improves the ultimate shear strength than the non-cracked beam. In case of the beam is subjected to the initial axial stress fn=2000Mpa, it is considered to be some micro cracks in the beam. Also, the shear force is propagated through only re-bars at the pre-cracked point because of the occurring of the pre-cracks. So, when the reinforcement ratio is small, the beam will be failed in bending because the stress of the main re-

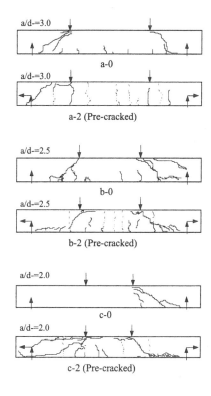

Fig.5. Ultimate state of beams

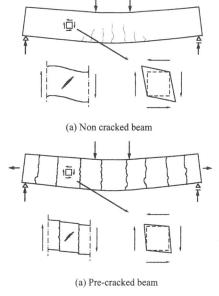

(a) Non cracked beam

(a) Pre-cracked beam

Fig.6. Shear stress in non-cracked or cracked beam

991

bar grows to yield stress easily. Thus both of the shear problems of the pre-cracked or non-cracked member are considered as the problem of the strain energy. Then the finite element analysis will be one of the effectiveness methods for the response analysis of such as problem.

5 NUMERICAL INVESTIGATION

5.1 FEM procedure

Recently, many researchers have investigated the development of computational models for nonlinear analysis of reinforced concrete structures based on finite element method (Hinton et al.1984). For the numerical analysis, the finite element approach is adopted using the degenerated isoparametoric Heterosis shell elements. The geometrical nonlinearlity is treated by using Green-Lagrange strain definition. A layered approach is employed to represent the steel reinforcement and to discretize the concrete behaviour through the thickness. The progress of cracked zones, as well as the compressive behaviour of the concrete is analysed and monitored for each layer and each element Gauss point. The load incremental procedure is adopted for the inducement of the initial axial force. Then the the displacement incremental procedure, which is useful for the instability analysis, is adopted for the lateral loading instead of the usual load incremental procedure.

5.2 Analyzed model

In the numerical investigation of the rectangular beam, only half of the beam is analyzed because of the symmetry of supporting and loading condition. Therefore the boundary conditions at the center of the span are provided that there are fixed on the deflection in x-direction and the rotation at the a-a axis. The finite element mesh and dimensions of the beam for this problem are shown in Fig.7. The model consists of 8 times18 elements and 629 nodes. All elements are divided into two concrete layers through the thickness and some elements include the steel layers. (So called 'layered approach' is employed here.) Table 3 represents the information on material properties of analyzed models. In the analysis for the member cracked by impact loads (pre-cracked member), the finite element model includes the cracked element on the assumption that the tensile strength st=0kN/cm2 and the compressive strength sc=28kN/cm2, as shown in Fig.7(b). That is the compressive strength of concrete is reduced 30% for the non-cracked element's one in the material properties of the cracked elements.

5.3 Fracture processes and ultimate state

Fig.8 indicates calculated cracking patterns of the pier

Table3. Material properties of analyzed models.

Concrete	
non-cracked element	
Elastic modulus (Ec)	2.91 (MN/cm²)
Poisson's ratio (υ)	0.2
Density (ρ)	2.45
Compressive strength (σc)	4.0 (kN/cm²)
Tensile strength (σt)	0.27 (kN/cm²)
cracked element	
Compressive strength (σc)	2.8 (kN/cm²)
Tensile strength (σt)	0 (kN/cm²)

Steel	
Elastic modulus (E₁)	20.7 (MN/cm²)
Hardening modulus (E₂)	0.207 (MN/cm²)
Yield stress (σy)	37.71 (kN/cm²)

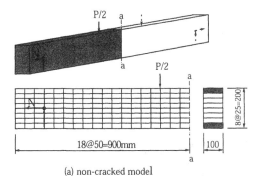

(a) non-cracked model

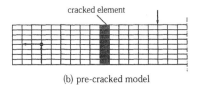

(b) pre-cracked model

Fig.7 FEM mesh of the rectangular beam

under the same transverse loading stage. At the analytical prediction in Fig.8, the short line denotes the cracking direction at the integration point. From these figures, it can be seen that the bending crack initiates in the middle span and progressively spreads towards the supporting point with the progress of the loading stage. Here it is observed that the state of the cracks of the non-stressed beam and pre-cracked beam are equal at the 50 step or 100 step, but as for the beam subjected to the axial tension N=60kN, the bending cracks grow faster than on the other beams. However, the bending cracks of the pre-cracked beam also grow faster than on the non-stressed one at the 200 step. In the 300 step of both of the beams, it is denoted that the diagonal shear cracks branch within the area of the shear span between the loading point and the support-

(Non-stressed beam)   (Pre-cracked beam)

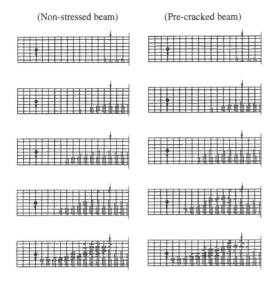

Fig.8 Fracture processes (Numerical results)

ing point. However, it is observed that there are some differential states among those diagonal shear cracks.

When further transverse loads are applied, the cracks grow in the area of the shear span. Then the concrete of the mid span in compress is split and consequently the beam collapses.

In the final step, these beams are failed in shear mode. Here it is certain that the main reinforcements don't yield in any beam. The gradient of the diagonal cracks of pre-cracked beam is in a middle state between non-stressed beam and the beam subjected to axial tension N=60kN. These failure states and crack direction obtained by the numerical analysis agree with those by the experimental results.

6 CONCLUSION

Experimental results show the pre-cracked beams improve the ultimate strength than the non-cracked beams it. Also the larger the initial axial stress is the greater the shear strength of the beam is. For the results, it is considered that the shearing force can be not propagated through concrete continuously by the initial cracks. Also, if the beam subjected axial tension the member improves the ultimate strength, even if the cracks disappear the face of the member.

The comparison between the numerical investigation and the experimental results shows that the fracture process by the numerical analysis agree with those obtained by the experimental results.

REFERENCES

Japan Society of Civil Engineers (1986). Specification for Concrete, pp. 50-55.
Tamura, T. (1994). Analysis of Shear Strength of Reinforced Concrete Beam Subjected to Axial Tension and Bending, The 3rd International Conference on the Concrete Future, 229-236.
Tamura, T., Shigematsu, T. and Hara, T. (1994). Schubtragberhalten durch auf Längszug und Biegung beanspruchte Stahlbetonbalken, Benton- und Stahlbetonbau, Heft 11, 289-294.
Tamura, T., Shigematsu, T., Hara, T., and Nakano, S. (1991). Experimental Analysis of Shear Strength of Reinforced Concrete Beams Subjected to Axial Force, Proc. Of JCI, Vol. 2, 153-160
Tamura, T., Shigematsu, T., Hara, T., and Manruyama, K. (1995). A Study of Proposed Design Equation for the Shear Strength of R/C Beams Subjected to Axial Tension, Proc, of JSCE, No.520/V-28, 225-234.
Hinton, E. and Owen, D.R.J., "Finite Element Software for Plates and Shells," Pineridge Press Swansea U.K., 1984.
Hinton, E. and Owen, D.R.J., "Computational Modeling of Reinforced Concrete Structures," Pineridge Press Swansea U.K., 1986

Creative Systems in Structural and Construction Engineering, Singh (ed.)© 2001 Balkema, Rotterdam, ISBN 90 5809 161 9

Damage and preloading in concrete

S. Yazdani

Department of Civil Engineering and Construction, North Dakota State University, Fargo, N.Dak., USA

ABSTRACT: This paper presents a general damage theory within the framework of continuum thermodynamics with its application to structural concrete. To capture the effects of preloading on concrete, an effective damage parameter is identified and used. In addition, by utilizing an effective compliance tensor the sudden stiffening behavior in concrete due to crack closure in load reversals is addressed.

1 INTRODUCTION

The overall response of structural concrete to external loads is generally governed by the kinetics of microdefects in the microstructures of the material. The shape, size, density, nucleation, and growth of defects are some of the reasons for the observed nonlinearity in the stress-strain response of concrete. It is experimentally shown that when concrete is subjected to low confining stresses, no significant amount of residual strain is accumulated. Under these circumstances, the loss of stiffness becomes one acceptable measure to describe the irreversible dissipative processes.

Also influencing the phenomenological behavior of concrete is closing of cracks under load reversals, that is when the sign of the applied stress is reversed. This aspect of material is of considerable importance in many engineering applications and can rise in a number of different forms, such as seismic excitation, and redistribution of moments and forces in highly indeterminate bodies.

Two studies dealing with nonproportional loading paths have surfaced in the literature; one is the experimental study by van Mier (1984) and the other is failure simulation by Stankowski (1990). The experimental study conducted by van Mier focused on a series of 90-degree rotations of loading paths in the triaxial compression test. His experiments involved two basic loading patterns, namely: (1) Planar mode rotation test; and (2) Cylindrical mode rotation test. In both tests, a cubical concrete specimen was subjected to triaxial compression until a certain point was reached on the ascending portion of the stress-strain curve, then the samples were unloaded and reloaded in an orthogonal direction by simply changing the major and minor compressive directions. Due to high levels of confining pressure, the stress-strain curves displayed a significant amount of residual strain, which is consistent with other investigations. Physically, the permanent deformations described above are related to the tangential motion of slip planes, particle relocation, and/or void closure, and not to the development and propagation of microcracks. In fact, the propagation of microcracks is suppressed due to added confinement.

In the other study, a numerical setup was reported by Stankowski (1990) in which a mortar-aggregate specimen was first subjected to uniaxial compression up to the prescribed level of loading, followed by an unloading to the zero stress level. The specimen was subsequently given a 90-degree rotation and reloaded in compression up to failure. A similar sequence of loading was conducted in tension.

Both studies mentioned above reported that the preloading reduced the load carrying capacity of the material when it was subjected to a new set of loading path. Although the stress-strain response differed significantly for the material preloaded in uniaxial tension rather than in triaxial compression, both showed significant reduction in strength. This important feature of concrete response needs to be addressed experimentally as well as theoretically. This paper focuses on Stankowski's results and considers the induced damage as a sole contributor in the energy dissipation mechanism.

2 GENERAL THEORY

The general formulations derived in this section come from the widely accepted approach of thermodynamics with internal variables (Lubliner 1972; Coleman and Gurtin 1967). With the assumptions of a purely mechanical theory and isothermal, rate-independent, and infinitesimal deformations in the stress space, one can start with the Gibbs free energy per unit volume, G, as

$$G(\sigma, k) = \frac{1}{2}\sigma : \mathbf{C}(k) : \sigma + \sigma : \varepsilon^i(k) - A^i(k) \qquad (1)$$

where σ represents the total stress tensor and k is the scalar damage variable. The colon (:) represents the tensor contraction operation and ε^i represents the total inelastic strain tensor. The inelastic strains may arise due to a mismatch of the crack surfaces, referred to as an inelastic damage processes (Yazdani and Schreyer, 1988). The form of Equation (1) also implies the dependence of the material compliance on the state of microcracking. This is concurrent with the arguments made by several researchers that during the damage process the fourth-order elastic compliance tensor is no longer constant but evolves with damage (Ju 1989; Ortiz 1985; Horii and Nemat-Nasser 1983). The dependence of $\mathbf{C}$ on damage parameter allows induced anisotropy to be captured through components of the material compliance. The scalar function, A^i, represents an inelastic component of the Helmholtz free energy.

In this work, we assume the following additive decomposition of the material compliance:

$$\mathbf{C}(k) = \mathbf{C}^0 + \mathbf{C}_I^c(k) + \mathbf{C}_{II}^c(k) \qquad (2)$$

in which $\mathbf{C}^0$ is the compliance tensor of an uncracked solid, and $\mathbf{C}_I^c$ and $\mathbf{C}_{II}^c$ denote the added flexibility tensor due to active microcracks in modes I and II, respectively. Similarly, an inelastic damage strain tensor can be separated into two parts as

$$\varepsilon^i = \varepsilon_I^i + \varepsilon_{II}^i \qquad (3)$$

where ε_I^i and ε_{II}^i denote the inelastic strain tensors due to inelastic damage processes in modes I and II, respectively. Note that for infinitesimal deformations and uncoupled theory, such additive decompositions of terms of any order is permissible. Following the Clausius-Duhem inequality and utilizing the standard thermodynamic arguments (Coleman and Gurtin 1967) the dissipation inequality is expressed as:

$$d_s = \frac{\partial G}{\partial k}\dot{k} \geq 0 \qquad (4)$$

where d_s is the dissipation rate and $\partial G / \partial k$ represent the thermodynamic force associated with the conjugate effective flux, $\dot{k}$. For all admissible processes, the inequality presented by Equation (4) must be satisfied. The substitution of Equations (1) to (3) and A^i into Equation (4) yields

$$d_s = \frac{1}{2}\sigma : \dot{\mathbf{C}}_I^c : \sigma + \frac{1}{2}\sigma : \dot{\mathbf{C}}_{II}^c : \sigma + \dot{\varepsilon}_I^i : \sigma + \dot{\varepsilon}_{II}^i : \sigma - \dot{A}^i \quad (5)$$

To proceed further, let the rates of added flexibility and inelastic strain tensors in Equation (5) be represented by the following rate-independent linear damage evolution laws:

$$\dot{\mathbf{C}}_I^c = \dot{k}\,\mathbf{R}_I(\sigma), \quad \dot{\mathbf{C}}_{II}^c = \dot{k}\,\mathbf{R}_{II}(\sigma) \qquad (6)$$

and

$$\dot{\varepsilon}_I^i = \dot{k}\,\mathbf{M}_I(\sigma), \quad \dot{\varepsilon}_{II}^i = \dot{k}\mathbf{M}_{II}(\sigma) \qquad (7)$$

where $\mathbf{R}_I(\sigma)$ and $\mathbf{R}_{II}(\sigma)$ are the fourth-order response tensors, which determine the direction of damage in modes I and II, respectively. Similarly, inelastic deformations in modes I and II are represented by the second-order response tensors, $\mathbf{M}_I(\sigma)$ and $\mathbf{M}_{II}(\sigma)$, respectively. Substitution of Equations (6) and (7) into Equation (5) yield the dissipation as

$$d_s = \dot{k}\left(\begin{array}{c} \dfrac{1}{2}\sigma : \mathbf{R}_I : \sigma + \mathbf{M}_I : \sigma + \dfrac{1}{2}\sigma : \mathbf{R}_{II} : \sigma \\ + \mathbf{M}_{II} : \sigma - A^i,_k \end{array}\right) \geq 0 \qquad (8)$$

where, subscript comma (,) is used to denote the partial differentiation with respect to the variable that follows. Since k is the measure of energy dissipation, by definition, irreversible character of damage implies $\dot{k} \geq 0$. Further, in the absence of any internal constraints, as indicated by Ortiz (1985), the coefficient of $\dot{k}$ must be nonnegative. Therefore, Equation. (8) can be expressed as

$$\left(\begin{array}{c} \dfrac{1}{2}\sigma : \mathbf{R}_I : \sigma + \mathbf{M}_I : \sigma + \dfrac{1}{2}\sigma : \mathbf{R}_{II} : \sigma \\ + \mathbf{M}_{II} : \sigma - A^i,_k \end{array}\right) \geq 0. \qquad (9)$$

At this point, one can define the damage potential or damage surface by simply introducing a positive valued function to force the right hand side of Equation (9) to zero. Assuming that positive function is $H(\sigma, k)$, the damage surface Ψ takes on the following form:

$$\Psi(\sigma, k) = \left(\begin{array}{c} \dfrac{1}{2}\sigma : \mathbf{R}_I : \sigma + \mathbf{M}_I : \sigma + \dfrac{1}{2}\sigma : \mathbf{R}_{II} : \sigma \\ + \mathbf{M}_{II} : \sigma - \dfrac{1}{2}t^2(\sigma, k) \end{array}\right) = 0 (10)$$

where

$$t^2(\sigma, k) = 2\left(A^i,_k + H(\sigma, k)\right) \qquad (11)$$

is identified as the damage function. After employing the standard Kuhn-Tucker loading-unloading criteria (i.e., $k \geq 0$, $\Psi \leq 0$, and $k\Psi = 0$), Equation (10) essentially completes the general formulation of the proposed model; however, the details must be worked out to fully describe a particular solid. Specific expressions for response tensors, $\mathbf{R}_I$, $\mathbf{R}_{II}$, $\mathbf{M}_I$, $\mathbf{M}_{II}$, and particular form of damage function, t, must be specified.

3 LOAD REVERSAL

When concrete is loaded in tension, unloaded, and subsequently reloaded in compression, crack closure becomes an important phenomenon, which exhibits recovery of stiffness in the reversed loading direction. The experimental works of Mazars et al. (1990) clearly show the stiffening effect of crack closure. These experimental investigations also reveal the presence of a reference pressure where the closing pressure is first noticed. This indicates that the recovery of the original stiffness of the material does not commence at the point where the sign of stress is changed, nor does it begin where the strain becomes compressive. The reference stress should be interpreted as a pressure where the majority of microcracks, in the direction of the applied stress, have become effectively closed. It is not clear at this point what this reference pressure would be under different stress and strain paths and how it would change accordingly. It should be pointed out that the complexities of the testing procedures in load reversals make the determination of the reference pressure a rather formidable task as various loading conditions must be considered. To this end, postulates on the form of the reference pressure must be based on engineering judgment until further results are available.

A number of researchers have made attempts to address these issues (see, e.g., Hansen and Schreyer 1995; Chaboche 1993; Ju 1989; Ortiz 1985). Some have used the stress-based approach while other preferred strain-based approach. Chaboche (1992) has produced a rather comprehensive review on the concepts of crack closure effects, which he refers to as unilateral effects.

In this paper, similar to Ortiz (1985), the crack closure effects are addressed by employing the stress-based projection operator in the definition of an effective compliance tensor in mode I, $\mathbf{C}_I^{ce}$, as

$$\mathbf{C}_I^{ce} = \mathbf{P}^+ : \mathbf{C}_I^{ca} : \mathbf{P}^+ + \left(\mathbf{I} - \mathrm{H}(-\underline{\lambda})\,\mathbf{I}\right) : \mathbf{C}_I^{ch}, \quad (12)$$

where $\underline{\lambda}$ is the minimum eigenvalue of σ^-. Note that when all the eigenvalues of σ are positive (tensile), the Projection Operator, $\mathbf{P}^+$, will be the fourth order identity tensor and $\mathrm{H}(-\underline{\lambda}) = 0$; therefore, $\mathbf{C}_I^{ce} = \mathbf{C}_I^{ca} + \mathbf{C}_I^{ch}$. Conversely, if all the eigenvalues

of σ are negative (compressive), $\mathbf{P}^+$ becomes the null tensor and $\mathrm{H}(-\underline{\lambda}) = 1$, which forces $\mathbf{C}_I^{ce}$ to disappear from contributing to the overall material compliance (that does not mean k is necessarily zero). It is assumed that during the crack closing process, the damage in mode I does not decrease or heal but rather is just not active due to the compressive loading condition. The damage in tensile loading direction picks up from where it had left off after the new tensile load exceeds the previously attained load-level. Note that coefficient in the second term of Equation. (12) is necessary to deactivate any changes in the apparent Poisson's effect (during mode-I type deformations) under the reversed loading paths, which guarantees the recovery of the original stiffness.

4 CONCLUSIONS

Continuum damage mechanics approach was utilized to develop a general continuum theory for structural concrete for some complex loading paths. It was shown that with some plausible assumption on the distributed nature of damage, sudden stiffening effects of crack closure under load reversal and the softening effect of preloading stress paths on the strength properties of concrete can be addressed within the framework of continuum thermodynamics.

REFERENCES

Chaboche, J. L. 1993. Development of continuum damage mechanics for elastic solids sustaining anisotropic and unilateral Damage. *Int. J. Damage Mech.*, 2, 311-329.

Chaboche, J. L. 1992. Damage induced anisotropy: on the difficulties associated with the active/passive unilateral conditions. *Int. J. Damage Mech.*, 1, 148-171.

Coleman, B. D., & M. E Gurtin. 1967. Thermodynamics with internal state variables, *J. Chem. Phys.*, 47(2), 597-613.

Hansen, N. R. & H. L Schreyer,. (1995). Damage deactivation. *J. App. Mech.*, 62, 450-458.

Horii, H. & S. Nemat-Nasser, 1983. Overall moduli of solids with microcracks: load-induced anisotropy. *J. Mech. Phys. Solids*, 31(2), 155-171.

Ju, J. W. 1989. On energy-based coupled elastoplastic damage theories: constitutive modeling and computational aspects, *Int. J. Solids Struct.*, 25(7), 803-833.

Lubliner, J. 1972. On the thermodynamic foundations of nonlinear solid mechanics. *Int. J. Non-Linear Mech.*, 7(3), 237-254.

Mazars, J., Berthaud, Y., & S. Ramtani, 1990. The Unilateral Behavior of Damaged Concrete. *Engng. Fracture Mech.*, 35(4/5), 629-635.

Ortiz, M. 1985. A constitutive theory for the inelastic behavior of concrete. *Mech. Mater.*, 4(1), 67-93.

Stankowski, T. 1990. Numerical simulation of progressive failure in particle composites. *Ph. D. Dissertation*, University of Colorado, Boulder.

van Mier, J. G. M. 1984. Strain-softening of concrete under multiaxial loading conditions. *Ph. D. Dissertation*, University of Eindehoven, The Netharlands.

Yazdani, S. & H. L. Schreyer (1988). "An anisotropic damage model with dilatation for concrete." *Mech. of Mat.*, 7(3), 231-244.

Creative Systems in Structural and Construction Engineering, Singh (ed.) © *2001 Balkema, Rotterdam, ISBN 90 5809 161 9*

Evaluation of damage level and ductility of reinforced concrete members

H. Tanaka, Y. Tanimura & T. Sato
Railway Technical Research Institute, Tokyo, Japan

M. Takiguchi
Kyushu Railway Company, Kitakyushu, Japan

T. Watanabe
Hokubu Consultant Corporation, Sapporo, Japan

ABSTRACT: A new seismic design code for railway structures in Japan has been established to adopt a performance based design. In the flow of this design, it is necessary to evaluate the damage level and ductility for assessing the seismic performance of structural members. In this study, damage levels of reinforced concrete (RC) member are estimated by considering its characteristics and difficulty of repair. The damage to verify the ductile performance is divided into four levels by considering the load-displacement characteristics and the damage of RC member obtained through a horizontal cyclic loading test of full-scale RC columns of rigid frame viaduct under axial load. Based on the result of the cyclic loading test of members, the authors propose quantitative evaluation of bearing capacity and ductility correspondent damage degree of RC members.

1 INTRODUCTION

The new seismic design standard for railway structures in Japan is a system to define the seismic performance of structures and check the response calculated through dynamic analysis by inputting the design seismic motion. The performance of structures is evaluated in relation to the damage level of their component members. The damage level of reinforced concrete members is divided into four stages from the viewpoint of damage conditions and repair work (Tanaka 1999).

In order to estimate the damage level from the response calculated through dynamic analysis, it is necessary to evaluate the limit value corresponding to each damage level in the load-displacement relationship of the members. Therefore, in this study, a calculation method of load and displacement corresponding to the damage level is discussed for RC members with a flexural failure mode based on the data of full-scale cyclic loading.

The full-scale test specimens model the columns of rigid frame railway viaduct. The dimensions of test specimens are shown in Table 1, and the result of tests and material strength are shown in Table 2.

Table 1 Dimensions of test columns

	Width b mm	Depth h mm	Effective depth d mm	Shear span L_a mm	Shear span ratio L_a/d	Longitudinal reinforcement	Hoop bar	Axial compressive stress σ_0 N/mm²	Tensional rebar ratio P_t (%)	Hoop ratio P_w (%)	Number of loading cycles
H95-1							SD345 D13@100×1			0.28	
H95-2						SD345 30×D32	SD345 D13@100×1.5		1.07	0.42	
H95-3							SD345 D16@100×1.5			0.66	
H96-11						SD390 16×D32+12×D29	SD390 D13@100×1.5	387	0.95	0.42	
H96-12	900	900	821	330	4.0					0.42	Three cycles for 1δy
H97-1							SD345 D16@100×1.5			0.66	
H97-2							SD345 D16@100×2			0.88	
H97-3						SD345 30×D32	SD345 D19@100×2		1.07	1.27	
H97-4							SD345 D16@100×2			0.88	
H97-5							SD345 D16@100×1.5	97		0.66	
H97-6								387			One cycle for 1δy
T97-1							SD345 D13@80×2	368		0.78	Three cycles for 1δy
T97-2	800	800	728	300	4.1	SD345 32×D25		0	0.78		Three cycles for 1δy
T97-3							SD345 D10@80×2	368		0.45	Three cycles for 2δy

Table 2 Summary of test results and material strengths

	Limit of damage level 2 (Yielding)				Limit of damage level 2				Limit of damage level 3		Concrete f'_c	Longitudinal reinforcement f_{ry}	Hoop bar f_{wy}
	Load kN		Displacement mm		Load kN		Displacement mm		Displacement mm		kN/mm²	kN/mm²	kN/mm²
	+	−	+	−	+	−	+	−	+	−			
H95-1	1160	-1200	26.4	-27.5	1310	-1320	79.4	-83.4	86.7	-87.4	30.0	380	376
H95-2	1200	-1180	27.8	-26.7	1320	-1310	84.3	-80.3	122.8	-116.4	31.0	380	376
H95-3	1170	-1180	27.2	-27.4	1290	-1320	109.6	-111.0	141.0	-141.5	31.4	380	387
H96-11	1180	-1220	29.3	-29.3	1460	-1470	110.1	-113.1	144.3	-125.9	29.2	425	469
H96-12	1180	-1180	29.3	-27.0	1500	-1530	115.5	-109.7	152.5	-149.8	30.3	425	469
H97-1	1230	-1190	27.1	-25.7	1480	-1520	108.6	-103.2	150.1	-136.7	26.9	368	409
H97-2	1230	-1200	26.0	-24.9	1560	-1570	129.7	-124.5	170.7	-164.9	28.2	368	409
H97-3	1240	-1200	26.3	-24.6	1570	-1580	158.1	-148.3	196.9	-185.7	29.2	368	366
H97-4	1230	-1200	25.5	-24.2	1550	-1590	126.7	-123.4	179.4	-173.1	30.9	368	409
H97-5	989	-966	24.0	-23.0	1230	-1230	120.3	-115.2	149.2	-141.2	30.7	368	409
H97-6	1250	-1210	25.9	-24.8	1600	-1540	153.9	-150.4	197.2	-192.5	31.8	368	409
T97-1	762	-773	24.3	-27.6	976	-918	147.3	-148.4	188.3	-186.5	30.0	371	373
T97-2	585	-569	19.9	-18.8	762	-695	134.4	-134.3	185.9	-172.0	32.6	387	373
T97-3	805	-807	25.0	-24.6	940	-943	110.6	-105.2	200.1	-148.4	24.3	377	402

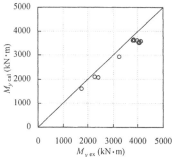

Figure 1. Comparison of experimental and calculated values of yield load.

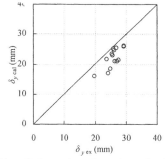

Figure 2. Comparison of experimental and calculated values of yield displacement.

2 THE LIMIT POINT OF DAMAGE LEVEL 1

The limit point of damage level 1 is defined as the point where the strain of longitudinal reinforcement reaches the tensional yield strength (Tanaka 1999).

2.1 Load

The yield load is calculated based on the Bernoulli principle. The stress-strain relationship in the following discussion is modeled as shown in the design standard for railway structures (RTRI 1992). Comparison of calculation values and test results of yield load is shown in Figure 1. In this Figure, the calculation values agree approximately with the test results.

2.2 Displacement

The displacement at the limit of damage level 1, δ_y is calculated as the summation of the displacement by the flexural deformation δ_{y0} and the displacement by ejection of longitudinal bar from footing δ_{y1}:

$$\delta_y = \delta_{y0} + \delta_{y1} \qquad (1)$$

where δ_y = displacement when the tensional bar yields; δ_{y0} = displacement due to the body deformation at yield; calculated by dividing the member longitudinally and integrating the curvature of each section; and δ_{y1} = rotational displacement due to ejection of longitudinal reinforcement, calculated by:

$$\delta_{y1} = L_a \cdot \Delta L_y / (d - X_y) \qquad (2)$$

where L_a = Shear span; d = Effective depth; and X_y = Neutral axis depth yield; ΔL_y = Ejection of tensional bar from footing:

$$\Delta L_y = 7.4\alpha \cdot \varepsilon_y (6 + 3500 \, \varepsilon_y)\phi/f'^{2/3}_{cf} \qquad (3)$$

where, ε_y = yield strain of tensional bar; ϕ = diameter of tensional bar; f'_{cf} = compressive strength of footing concrete; α = factor expressing the effect of bar spacing calculated by Equation (4) in the case of single layer bar arrangement (JSCE 1996):

$$\alpha = 1 + 0.9 \, e^{0.45 (1 - D/\phi)} \qquad (4)$$

where D = Center to center spacing of tensional bar.

Figure 2 shows the relationship between calculation values of yield displacement and experimental values. The calculation value estimates somehow

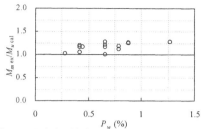

Figure 3. Relationship between experiment/calculation ratio of load at the limit of damage level 2 and hoop ratio.

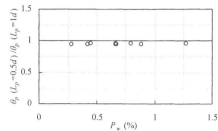

Figure 4. Effect of difference between the equivalent plastic hinge region and the plastic hinge rotational angle (limit of damage level 2).

tend to be smaller than experimental values. However, on load displacement curves, experimental values of yield stiffness of calculation values are almost equal, because the yield load is estimated a little smaller than experimental values. Therefore, calculation values of yield displacement give approximately correct estimates of experimental value.

3 LIMIT POINT OF DAMAGE LEVEL 2

The limit point of damage level 2 is defined as the maximum displacement to keep an approximate maximum horizontal resistant load (Tanaka 1999). In the test data, the experimental value of the limit point of damage level 2 is defined as the maximum displacement, where the capacity does not drop after the second cycle of load repetition to the same displacement. This is the reason why a little reduction of horizontal resistance load due to spalling of cover concrete is not considered to affect the extent of repair of the members. As the positive side and negative side of the limit point of damage level 2 were not much different, the positive side is taken into consideration in the following discussion.

3.1 Load

The load at the limit point of damage level 2 is defined as the load when compressive strain of concrete edge reaches 0.0035 (corresponding to the flexural capacity M_u). Figure. 3 shows the relationship between the ratio of the experimental calculation load and the hoop ratio P_w. The calculation values tend to be estimated lower than experimental values, but almost all data are estimated correctly.

3.2 Displacement

Displacement at the limit of damage level 2 is calculated as the sum of displacement δ_{m0} due to the flexural deformation of the body and rotational displacement δ_{m1} due to the ejection of longitudinal reinforcement from footing, where the displacement δ_{m0} due to the flexural deformation of the body is calculated after being divided into the displacement δ_{mp} due to flexural deformation in the plastic hinge region and the displacement δ_{mb} due to flexural deformation outside the plastic hinge region.

$$\delta_m = \delta_{m0} + \delta_{m1} = \delta_{mb} + \delta_{mp} + \delta_{m1} \quad (5)$$

$$\delta_{mp} = \theta_{pm} \cdot (L_a - L_p / 2) \quad (6)$$

where δ_m = displacement at the limit of damage level 2, δ_{m0} = displacement due to body deformation at the limit of damage level 2 (= $\delta_{mb} + \delta_{mp}$); δ_{mb} = displacement due to flexural deformation outside the plastic hinge region in the body deformation at damage level 2, calculated by dividing the member longitudinally and integrating curvatures of each section; δ_{mp} = displacement due to flexural deformation in the plastic hinge region in the body deformation at damage level 2; δ_{m1} = rotational displacement due to the ejection of longitudinal reinforcement from footing at the limit of damage level 2.

In this model, the flexural deformation in the plastic hinge, δ_{mp} is calculated by the following equation for the rotational angle of plastic hinge, θ_{pm} and equivalent plastic hinge length L_p, with the center of rotation assumed as the center of plastic hinge.

$$\delta_{mp} = \theta_{pm} \cdot (L_a - L_p / 2) \quad (6)$$

where θ_{pm} = rotational angle at the limit of damage level 2; L_p = Equivalent plastic hinge length.

3.2.1 Equivalent plastic hinge length

To model the flexural deformation in the plastic hinge region, it is necessary to note that the calculated deformation tends to vary due to the method for setting the plastic hinge length (JSCE 1997). Before calculating the rotational angle of plastic hinge from the experimental value of displacement in order to estimate the effect of the equivalent plastic hinge length, we first compare the equivalent plastic hinge lengths when they are $0.5d$ and $1.0d$. Figure 4 shows the relationship between the plastic hinge rotational angle due to the different plastic hinge length assumed $0.5d$ to $1.0d$ and hoop ratio. This figure clarifies that the difference of plastic hinge

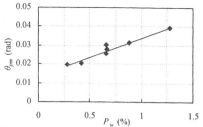

Figure 5. Relationship between the plastic hinge rotational angle and the hoop ratio (SD345).

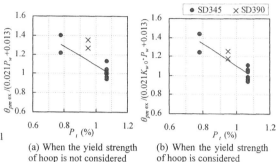

(a) When the yield strength of hoop is not considered

(b) When the yield strength of hoop is considered

Figure 6. Relationship between the plastic hinge rotational ratio and the tensional reinforcement ratio.

length affects the modeling of plastic hinge rotational angle very slightly.

Moreover, when judged from the measurement of strain gages set discretely, yield lengths of longitudinal reinforcements at the limit of damage level 2 varied approximately from $0.8d$ to $1.6d$ in the full-scale specimens.

In this study, we use Mattoc's equation (Mattoc 1967) to calculate the equivalent plastic hinge length.

$$L_p = 0.5d + 0.05L_a \qquad (7)$$

where, L_p = Equivalent plastic hinge length; d = Effective depth of the section; and L_a = Shear span.

3.2.2 *Plastic hinge rotational angle*

Figure. 5 shows the relationship between the hoop ratio P_w and plastic hinge rotational angle θ_{pm} for the specimens with SD 345 for the hoop bars. The following equation is introduced from the data of tensional reinforcement ratio $P_t = 1.07\%$.

$$\theta_{pm} = 0.021P_w + 0.013 \qquad (8)$$

From the comparison of the data of $P_t = 1.07 \%$ and that of $P_t = 0.78 \%$ in Figure. 5, the plastic hinge rotational angle tends to become large when the tensional reinforcement ratio is small.

Figure. 6 shows the relationship between the ratio of experimental value and the calculated value (Equation (8)) of plastic hinge rotational angle and tensional reinforcement ratio P_t. Figures 6 (a) and (b) show the cases where the difference of hoop strength is not and is considered, respectively. This figure suggests that it is necessary to consider the yield strength of hoop f_{wy} appropriately. Therefore, by assuming that these effects are proportional to the yield strength when it is equal to or less than corresponding to SD 390, we defined the factor for considering the hoop strength, K_{w0}:

$$\theta_{pm} = (0.021K_{w0} \cdot P_w + 0.013) / (0.79P_t + 0.153) \qquad (9)$$

where:

$$0.021K_{w0} \cdot P_w + 0.013 \leq 0.04 \qquad (10)$$

$$0.79P_t + 0.153 \leq 0.78 \qquad (11)$$

where P_w = hoop ratio (%); P_t = tensional reinforcement ratio (%); and K_{w0} = factor for considering the hoop strength, calculated by the following equation in this study:

$$K_{w0} = f_{wy} / 390 \qquad (12)$$

where f_{wy} = yield strength of hoop.

3.2.3 *Flexural deformation outside the plastic hinge region*

The displacement due to the flexural deformation outside the plastic hinge region δ_{mb} is calculated by integrating the curvature in the same way as that for at the yield.

3.2.4 *Rotational displacement due to bar ejection*

The ejection of longitudinal reinforcement is affected by various parameters, e.g. strength of concrete, spacing of tensional reinforcement, diameter of bar and so on. Here, the effect of these parameters is assumed to be the same as at the limit of damage level 2. From the effect of balanced axial force ratio, the following equation is introduced.

$$\delta_{m1} = \theta_{m1} \cdot L_a \qquad (13)$$

$$\theta_{m1} = [(2.7K_{w1} \cdot P_w + 0.22)(1 - N_0/N_b) + 1] \, \theta_{y1} \qquad (14)$$

$$\text{where } 2.7K_{w1} \cdot P_w + 0.22 \leq 3.7 \qquad (15)$$

where δ_{m1} = rotational displacement due to the ejection of longitudinal reinforcement from footing at the limit of damage level 2; θ_{m1} = rotational angle due to the ejection of longitudinal reinforcement from footing at the limit of damage level 2; θ_{y1} = rotational angle due to the ejection of longitudinal reinforcement from footing at the yield of member; K_{w1} = factor for considering the hoop strength calculated by the following equation in this study, in the same way as for K_{w0} of Equation (12); and N_0/N_b = Axial force ratio for balanced failure.

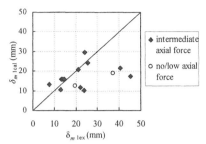

Figure 7. Comparison of experimental and calculated values of rotational displacement due to bar ejection.

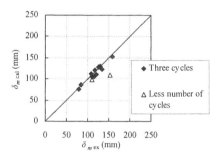

Figure 8. Copparison of ezperiment and calculation of limit displacement of damage level 2

Figure 7 compares the calculation value and experimental value of rotational displacement due to the ejection of reinforcement. The proposed equation approximately evaluates the rotational angles of bar ejection for a safety margin.

3.2.5 *Comparison of total displacement δ_m between the calculation and experimental values*
Figure 8 compares the calculation value and experimental value of displacement δ_m at the limit of damage level 2. The proposed equation evaluates the experimental value appropriately.

4 THE LIMIT POINT OF DAMAGE LEVEL 3

The limit point of damage level 3 is defined as the maximum displacement to keep the horizontal resistance corresponding to the yield load (Tanaka 1999). Because the effect of loading cycle is important for the load-displacement relationship after the limit point of damage level 3, the data, at the three loading cycles is used for processing experimental data.

4.1 *Load*

The definition of the load at this point is the yield load.

4.2 *Displacement*
Calculation of the displacement δ_n at the limit of damage level 3 is similar to that for the limit of damage level 2:

$$\delta_n = \delta_{n0} + \delta_{n1} = \delta_{nb} + \delta_{np} + \delta_{n1} \qquad (16)$$

where δ_n = displacement at the limit of damage level 3; δ_{n0} = displacement due to the body deformation at the limit of damage level 3 ($\delta_{nb} + \delta_{np}$); δ_{nb} = displacement due to flexural deformation outside the plastic hinge region in the body deformation at damage level 3, calculated by dividing the member longitudinally and integrating curvatures of each section; δ_{np} = displacement due to flexural deformation in the plastic hinge region in the body deformation at damage level 3; and δ_{n1} = rotational displacement due to the ejection of longitudinal reinforcement from footing at limit of damage level 3.

The flexural displacement of plastic hinge region δ_{np} is calculated by the equation (17) by using the plastic hinge rotational angle θ_{pn} and equivalent

$$\delta_{np} = \theta_{pn} \cdot (L_a - L_p / 2) \qquad (17)$$

where θ_{pn} = plastic hinge rotational angle at the limit of damage level 3, L_p = equivalent plastic hinge length calculated by the equation (7).

4.2.1 *Plastic hinge rotational angle*
We calculated the plastic hinge rotational angle at the limit of damage level 3, θ_{pn}, by using the increment of rotational angle from damage level 2 to damage level 3, $\Delta\theta_{pmn}$:

$$\theta_{pn} = \theta_{pm} + \Delta\theta_{pmn} \qquad (18)$$

$$\Delta\theta_{pmn} = 0.10 \, (M_u - M_y) / M_u \qquad (19)$$

where $\Delta\theta_{pmn}$ = increment of plastic hinge rotational angle from damage level 2 to damage level 3.

Figure 9 shows the relationship between the negative inclination of $M/M_u - \theta_p$ relationship and the hoop ratio P_w. The Figure shows that the number of loading cycles tends to affect the negative inclination. However, the data with 3 cycles of loading are

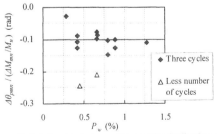

Figure 9. Relationship between the negative inclination of non-dimensional bending moment-plastic hinge rotational angle relationship and hoop ratio.

approximately constant for the variable hoop ratio. Further, with the current loading method to use the displacement at the yield load, δ_y, the value of δ_y varies according to the axial load and also the cumulative loading cycle up to the same displacement vary according to axial load. Therefore, the effect of axial load on the negative inclination is left for further studies.

4.2.2 Flexural deformation outside the plastic hinge region

The displacement due to flexural deformation outside the plastic hinge region, δ_{nb} is calculated by curvature integration in the same way as for the yield displacement. Further, it is necessary to consider the effect of unloading on the negative inclination of the load-displacement relationship. However, since the proportion of flexural deformation out side the plastic hinge region in the total deformation is minor, the effect of unloading is neglected.

4.2.3 Rotational displacement due to bar ejection

Our past study (Tanaka 1998) describes the tendency that the rotational angle due to bar ejection is constant or descends after the limit of damage level 2. Then, for making a model, the rotational angle due to bar ejection at the limit of damage level 3 is regarded as the same as that at the limit of damage level 2.

$$\delta_{n1} = \theta_{n1} \cdot L_a \qquad (20)$$

$$\theta_{n1} = \theta_{m1} \qquad (21)$$

where θ_{n1} = Rotational angle due to the ejection of longitudinal reinforcement from footing at the limit of damage level 3.

Figure 10 compares the calculated and experimental values of the displacement at the limit of damage level 3, δ_n. Proposed equations appropriately evaluate the experimental values at the loading cycle of 3 for δ_y.

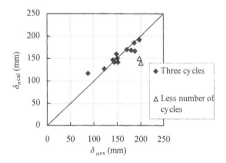

Figure 10. Comparison of experiment and calculation of limit displacement of damage level 3

5 CONCLUSIONS

A method to calculate the load and displacement for the limit of damage level 1 - 3 is studied based on the experimental data with a full-scale model simulating a column of RC rigid frame viaduct. The following conclusions are obtained.

1. The displacement of the limit of damage level 1 can be appropriately evaluated as the summation of the body deformation calculated through curvature integration and the rotational displacement due to bar ejection.

2. The displacement of the limit of damage level 2 can be appropriately evaluated as the summation of the flexural deformation off the plastic hinge region, the flexural deformation outside the plastic hinge region, and the rotational displacement of bar ejection. The flexural deformation of the plastic hinge region can be estimated by an experimental equation with parameters, such as the hoop ratio, tensional reinforcement ratio and so on.

3. The displacement of the limit of damage level 3 can be calculated as the same idea as that for the limit of damage level 2. The flexural displacement of the plastic hinge region can be estimated by considering the increment of the plastic hinge rotational angle from the limit of damage level 2.

REFERENCES

Japan Society of Civil Engineering (JSCE) 1996. Damage Analysis on Great Hanshin Awaji Earthquake Disaster and Equation to Evaluate Ductility Ratio. Concrete Engineering Series (in Japanese). 12: 52-53.

Japan Society of Civil Engineering (JSCE) 1997. Seismic Technology for Concrete Structures. Concrete Engineering Series (in Japanese). 20: 92-95.

Mattock, A. H. 1967. Discussion of 'Rotation Capacity of Reinforced Concrete Beams' by W. G. Corley. Journal of Structural Division, ASCE. 93(ST2): 519-522.

Railway Technical Research Institute (RTRI) 1992. Design Standard for Railway Structures (Concrete Structures) (in Japanese). Tokyo: Maruzen

Tanaka, H., M. Okamoto, M. Takiguchi & T. Sato 1998. An Experimental Study on the Ductility and Damage Level of RC Columns. Proceeding of the Japan Concrete Institute (in Japanese). 20(3): 1045-1050

Tanaka, H. M. Takiguchi & T. Sato 1999. Estimation of the Damage Level of RC Members. RTRI Report (in Japanese). 13(4): 5-8

Creative Systems in Structural and Construction Engineering, Singh (ed.)© 2001 Balkema, Rotterdam, ISBN 90 5809 161 9

On behavior of rivet joint repaired by replacing corroded rivet with high-tensile bolt

K. Hirai & K. Fujii
Department of Civil and Environmental Engineering, Hiroshima University, Japan

O. Minata
Department of Civil Engineering, Hiroshima Institute of Technology, Japan

K. Kajimoto
Ryomei Engineering Company Limited, Japan

ABSTRACT: The corrosion damages occur recently at the rivet joints of steel bridges. The corroded rivet is usually replaced with high-tensile bolt (HT-bolt) for corrosion repairment. However, the stress states of remained healthy rivets and HT-bolt after replacement have not been evaluated yet. The purpose of this paper is to investigate the stress behavior after replacing with HT-bolt and to clarify the strength of rivet joint . By using four rivet joint specimens (width of 8cm, length of 40cm) cut out from the riveted railroad bridge which has been used about 70 years, shearing tensile tests are performed in order to clarify the strength of joint before and after replacing with HT-bolt. The residual axial force of rivet is also measured from rivets which are taken out from the four specimens. The stress condition of splice plate, rivet and HT-bolt are investigated experimentally. It is concluded that the residual axial force is caused by the thermal shrinkage at rivet placing, and that the sliding load of a HT-bolt joint is almost the same as the shearing strength of a rivet joint. Therefore HT-bolt replacing is one of the suitable methods for repairing.

1 INTRODUCTION

Fatigue and corrosion are recognized as important factors in maintenance of steel bridge. Since there are a lot of researches which investigate fatigue measurement and rating method. Several guide-lines or specifications have been published. On the other hand, since there are not so many investigations for corrosion, feasible treating or rating method of corrosion damage should be established as soon as possible, considering the appearance of corrosion damage in bridges recently. Especially, it sometimes has appeared at rivet joints, which are usually repaired by replacing the rusted rivets with HT-bolts. However, the stress of rivets or HT-bolts and the frictional coefficient between splice plate and main plate have not been clarified yet.

Considering the facts above, the purpose of this study is to obtain i) the joint strength and the rivet stress in rivet joint utilized for a long time, and ii) the ultimate behavior and strength of the joint after replacing rivet with HT-bolt.

2 TEST SPECIMENS

Four segments are made from the removed railroad bridge which has been used for about 70 years, by cutting out rivet and L-shape angle along the flange-web connection line of it, as shown in Figure1. It also shows double shear plates with about 80 mm width and 450~600 mm length.

These segments are utilized for making shearing strength test specimen whose rivets are taken out except the shadowed rivet in Figure 1. The rivets taken out from the segments are used for measuring residual axial force of rivets.

Table1 shows the material properties of rivets obtained from tensile coupon test based on Japanese Industrial Standard.

Table 1. Material properties of rivets

Yielding stress (N/mm^2)	Tensile strength (N/mm^2)	Elongation (%)	Reduction of area (%)
326	443	38.7	74.0

3 RESIDUAL AXIAL FORCE OF RIVET

The residual axial force of rivet is measured with two ways, strain gauge method and direct measurement using axial strain gauge. The former is a method by the following procedure: (i) flattening the rivet head on one side and (ii) pasting up a 2-axial

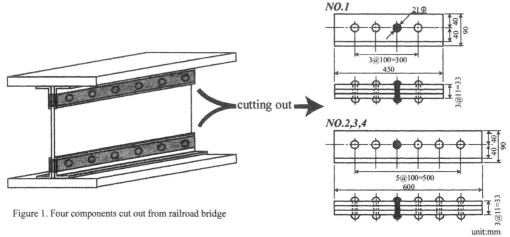

NO.1

NO.2,3,4

unit:mm

Figure 1. Four components cut out from railroad bridge

strain gauge on the flattened rivet head. (iii) after zero-adjustment of the strain gauge, releasing the axial force of the rivet by scraping another head according to grindstone and pushing it out from the plate, then (iv) measuring the strains on the head by the 2-axial strain gauge. (v) using the rivet (Photograph 1), which has been released free from residual stress, tensile test is carried out by the instrument as shown in Photograph 2. (vi) The residual axial force of the rivet can be obtained as the axial load in tensile test if the strains of 2-axial strain gauge are equal to the previous value of them when being released from residual stress.

On the other hand, the latter method can measure the residual axial force of the rivet directly by measuring the strains of axial strain gauge without removing the rivet from the plate.

Figure 2 shows the obtained value of residual axial force by both methods. It is veriflied that both residual axial force measured by the two methods are not so different. As shown in this figure, residual axial forces are distributed in 15~65 kN which would be caused by thermal cooling process after placing rivets. If the thermal change of rivet occurs 800 °C after placing, the residual axial force is calculated 50 kN, assuming the coefficient of thermal expansion is 12×10^{-6} $1/°C$. Therefore, it can be concluded that the residual axial force of a rivet is caused by thermal change of it.

4 SHEARING STRENGTH OF RIVET JOINT

Shearing strength test is carried out by using two specimens which have a remaining rivet from the segments No1,No2 in Figure1. The dimensions of specimens are shown in Figure3 and are designed to occur shearing failure of a rivet before bearing failure of main plate. Photograph3 is a set up of shearing strength test.

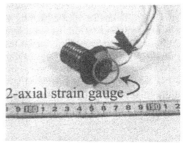

Photograph 1. Rivet for strain gauge method

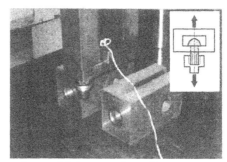

Photograph 2. Instrument for tensile test

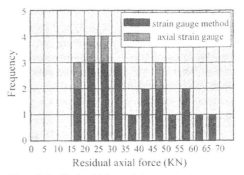

Figure 2. Residual axial force of rivets

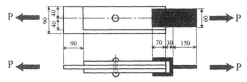

Figure 3. A specimen for shearing strength test

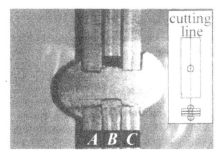

Photograph 3. Set up of shearing strength test

Photograph 4. Deformation of a rivet

The strength of a joint is determined by the smaller value between shearing strength (ρ_s) of a rivet and bearing strength (ρ_b) of plates, where the shearing strength and the bearing strength are calculated from the following equation.

where

$$\rho_s = 2\tau \frac{\pi d^2}{4} \tag{1}$$

$$\rho_b = \sigma dt \tag{2}$$

ρ_s: shearing strength (ρ_s=130 kN)
ρ_b: bearing strength (ρ_b=226 kN)
d: diameter of a rivet (d=21mm)
τ: yielding shear stress of a rivet (τ=188 N/mm^2)
σ: yielding stress of a plate (σ=326 N/mm^2)
t : thickness of a plate (t=11 mm)

Yielding stress is determined by tensile test. From the tensile test result each strength are calculated as ρ_s=130 kN and ρ_b =226 kN.

The relationship between load and relative gap is shown in Figure 4. As shown in Figure 4 relative gap increase rapidly about 130kN which is good agreement with the value calculated by equation (1) and (2). Relative gap shows the gap displacement between main plate and splice plates. As a result rapid increase of relative gap is caused by shearing failure of a rivet. The maximum loads for both specimens rise up about 200 kN after the shearing failure.

In order to clarify the deformation of a rivet, main plate and splice plate, the specimen is cut transversally after shearing strength test. The deformation of a rivet and plates is shown in Photograph 4 and for plates shown in Photograph 5,6. Black circles in

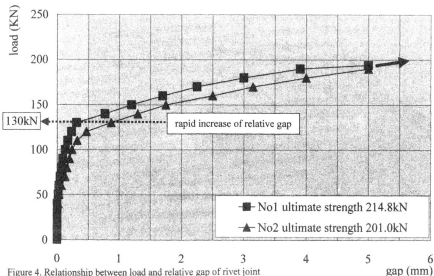

Figure 4. Relationship between load and relative gap of rivet joint

Photograph 5,6 show the location of rivet before the shearing strength test. Photograph 4 shows clearly the shearing failure of a rivet. It is clear from Photograph 5,6 that the bearing deformation is also observed in main plate (B), but it does not appear in both splice plates (A) and (C). Therefore, the bearing deformation of main plate would occur after shearing failure of a rivet.

By the way, each boundary of main plate and splice plate is found to be not so corroded from the observation of specimens in Photograph 5,6.

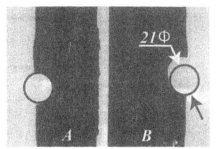

Photograph 5. Bearing deformation of a main plate and splice plates

5 SHEARING STRENGTH OF HT-BOLT JOINT

From the shearing strength test of another specimen (No3 and No4) whose rivet has been replaced by HT-bolt (F10T, M=20), the load vs. relative gap curves are obtained. The frictional load is also measured when relative gap rapidly increases and the frictional coefficient is calculated by following equation.

$$\mu = \frac{P}{N \cdot n} \tag{3}$$

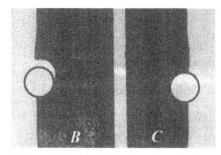

Photograph 6. Bearing deformation of a main plate and splice plates

where
μ: frictional coefficient
P: sliding load when relative gap increase rapidly
N: axial force of HT-bolt (N=178 kN)
n: number of sliding surfaces (n=2)

The relationship between load and relative gap is shown in Figure 5 for HT-bolt joint. In order to compare the relationship for rivet joint No1 is also indicated in this figure. It is clear that relative gap

increase rapidly after 130kN and 110kN for specimen No3 and No4, respectively. By substituting N=178 kN and n=2 to equation (3), derived out the frictional coefficient μ are 0.31 and 0.36, respectively. These frictional coefficients are slightly smaller than 0.40 which is used in JSHB. The sliding

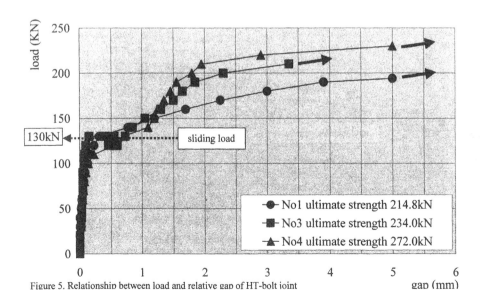

Figure 5. Relationship between load and relative gap of HT-bolt joint

Photograph 7. Deformation of a HT-bolt

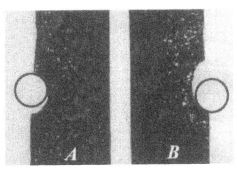

Photograph 8. Bearing deformation of a main plate and splice plates

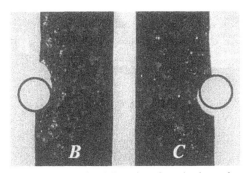

Photograph 9. Bearing deformation of a main plate and splice plates

load of bolt joint is almost the same as shearing strength of rivet joint as shown in Figure 5. The maximum loads are 234 kN and 272 kN for specimen No3 and No4 respectively, which are larger than those of rivet joint No1. It is confirmed from this test that HT-bolt joint is almost same as shearing strength of rivet joint in sliding load and superior to rivet joint in ultimate strength. Based on this result, it might not be necessary to improve the frictional coefficient of HT-bolt greater than 0.4.

Deformation of a HT-bolt and plates is shown in Photograph 7 and for plates in Photograph 8 and 9. As shown in Photograph 7, the shearing deformation of HT-bolt is never observed. This phenomenon is

quite different from that of rivet joint. In Photograph 8 and 9, the failure of HT-bolt joint is due to bearing failure of plates. Being different from rivet joint, the bearing failure is observed not only in main plate (B) but also in splice plates (A) and (C).

6 CONCLUSION

The axial force of the rivets is measured from the rivet joint which has been utilized for about 70 years. The ultimate behavior of the rivet joint and HT-bolt which replaced are investigated experimentally.
Followings are concluded ;

1. The residual axial forces of rivets, whose value spread around 15~65 kN, are caused by thermal cooling process after placing rivets. The strain gauge method is verified to be useful for measuring axial force as well as the axial strain gauge method.

2. In this test, the failure of rivet joint is caused by shearing failure of a rivet, and the maximum load exceed the value of shearing strength.

3. It is also confirmed from the shearing test that sliding load of a HT-bolt joint is almost the same as shearing strength of a rivet joint and the maximum loads are larger than those of rivet joint.

The frictional coefficients are obtained 0.31 and 0.36. Although the values are smaller than the value required for Ht-bolt joint in JSHB, the strength after replacing with HT-bolt is almost the same as that of rivet. Therefore, it can be concluded that replacing the rusted rivet with HT-bolt is one of the suitable methods in repairing the corrosion damage.

REFERENCES

Abe, M & Koshiba, A & Sugimoto, I & Sugidate, M. 1991. Test of Joints for Aged Steel used in Steel Bridges. *RTRI REPORT* Vol.5, No 12. 65-78. (In Japanese)

Japanese Specifications for Highway Bridges (JSHB). 1996. *Japan Road Association.*

Nishihara,T & Kosaka, H & Oono, A. 1988. A test for alteration of axial force and its the sliding strength for HT-bolt. *PROCEEDINGS OF THE 41TH ANNUAL CONFERENCE OF THE JAPAN SOCIETY OF CIVIL ENGINEERS.* 561-562(In Japanese)

Nishimura, H & Tahara, H & Nishimura, A. 1980. Measurement of axial forces of HT-bolts utilized for 15 years. *PROCEEDINGS OF THE 35TH ANNUAL CONFERENCE OF THE JAPAN SOCIETY OF CIVIL ENGINEERS,* 1-113. 223-224 (In Japanese)

Nishioka, T & Otoguro, Y & Yahata, T & Naganuma, T & Yoshioka, O. 1997. Various Problems Incurred to Prevent Maintenance for Steel Bridge High Strength Bolt Joint Being in Service for 10 Years Since Its Construction. *Journal of Structural Engineering* Vol.43A. 961-966 (In Japanese)

Tanihara, T & Kamei, M & Ishihara, Y & Taido, Y. 1990. Carrying Capacity Test for Friction Joint of High-Strength Bolt from a Removed Foot-way Bridge Used under 17 Years. *Journal of Structural Engineering* Vol. 36A. 1087-1096 (In Japanese)

Creative Systems in Structural and Construction Engineering, Singh (ed.)© 2001 Balkema, Rotterdam, ISBN 90 5809 161 9

Contribution study on the behavior of damaged R.C. frames

E. H. Disoky
Military Technical College, Cairo, Egypt

M. K. Zidan
Faculty of Engineering, Ain Shams University, Cairo, Egypt

A. N. Sidky
Housing and Building Research Center, Giza, Egypt

ABSTRACT: The behavior of R.C. frames is examined for different cases of supporting elements damage giving details about the changes in vertical deflections and the distribution of internal forces (bending moments and normal forces) induced in the beams and columns of the frame members duo to each case of damage. Also, the effect of the existence of infill panels in different places in the frame on this behavior is studied by assuming an equivalent strut to replace the infill panel. The best locations of infill panels in each case of damage are investigated to withstand the induced load due to the damage of a supporting element.

1 INTRODUCTION

Building structures that designed to resist the effects of ordinary loads may, however, be exposed to additional local effects arising from various accidents. In completed structures, pressure loads induced by the explosion of natural gas or detonation of bombs and impact loads caused by vehicle collisions, are examples of loads that usually not considered in structural design. Previous studies, show that the probability of structural failure due to these abnormal loads is of order of 10^{-4}, which may exceed probabilities associated with unfavorable combinations of ordinary design loads which is of order of 10^{-5} or less.

If the structure is not designed to dissipate the energy of the accidental load or to absorb the damage, a catastrophic chain reaction of failures that propagates horizontally and vertically throughout major portion of the structure may ensure. Although absolute safety cannot be achieved, occupants of multistory buildings should enjoy reasonable protection against this type of failure. So, it is useful to understand the limits and capabilities of ordinary structures to resist propagation of damage due to abnormal loads.

2 OBJECTIVE OF THIS WORK

The main aim of this work is to study the behavior of reinforced concrete multi-story frames when failure of columns in different locations of these frames occurs. Also, to study the effect of brickwork infill panels on this behavior. Another objective of this work is to make an assessment of the performance of structural components as well as the architectural features of reinforced concrete buildings to withstand these effects.

3 BEHAVIOR OF INFILLED FRAMES

Fig. 1.a. shows the typical behavior an infilled frame subjected to horizontal loads. It is assumed for this structure that the infill and frame are not constructed integrally, nor are they deliberately bonded together, which is the common case in real life.

When the load is applied, the infill and frame are separated over a large part of the length of each side and contact remains only adjacent to the corners at the ends of the compression diagonal as shown in fig. 1a. In effect, the infill behaves as diagonal strut and the analogous structure shown in fig. 1b. may be

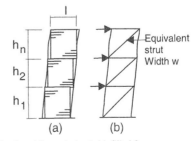

Fig. 1   a) Lateraly loaded infilled frame.
b) Equivalent frame.

postulated, with equivalent struts replacing the infill panels.

4   STUDIED MODELS

The study is carried out mainly on a 4-bay, 6-story R.C. frame (M10). All dimensions and cross sections of this frame are shown in fig.2. Applied loads are vertical uniformly distributed loads with an intensity of 4 t/m` on all stories. The modules of elasticity for steel, concrete and infill are taken 2100, 210& 60 t/cm², respectively.

Damage cases are shown in fig. 3: Central column in the ground story (M11), Exterior column in the ground story (M12).

To study effect of the existence of infill panels, different locations for these panels shown in fig. 3. are investigated. Equivalent strut is used in place of each infill panel. The effective width of these struts is taken equal to its length divided by 3, while its breadth is the same as the thickness of the infill panel (25 cm).

5   ANALYSIS OF RESULTS

5.1   *Behavior of bare frames under damage of structural elements*

The bare frame (M10) was analyzed at first without any damage and the obtained vertical deflection, N.F. & B.M. in the structural elements are used as reference lines to clarify the changes which may occur in each case of damage. The hatched areas in these figures represent the increase in the values of the concerned function. This analysis was made using a standard finite element package (SAP90), using the well known frame element with 3-d.o.f. at each node (2-translation and 1-rotation) for the frame members and a truss element with 2-d.o.f. at each node (2-translation) for the equivalent strut elements.

5.1.1   *Damage of the central column in the ground story (M11)*

1. Change in vertical deflection

Points over the damage columns (at column row c) have the max. vertical deflection with an average value of 1.75 cm which is about 6.6 times the average of the original values at that column row.
In column rows b, d, the vertical deflection increases to be 1.51, 1.47 times the original values in the 1ˢᵗ, 6ᵗʰ stories respectively.
There is no increase in vertical deflection at column rows (a, e).

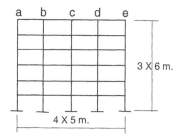

Fig. 2   4-bay, 6-story R.C. frame (M10)

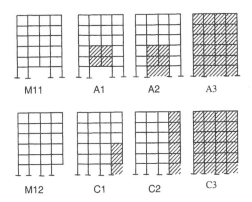

Fig. 3   Studied models

2. Change in B.M.

In sections over the damaged column, B.M. inverses its direction relative to the original case to be +ve. B.M. with an increase in value of 2.9, 1.6 times the original values in the 1ˢᵗ, 6ᵗʰ stories respectively.
The –ve. B.Ms. in beams over the damaged zone (span b, c) increase in value to be 4.6, 2.9 times the original values in the 1ˢᵗ, 6ᵗʰ stories respectively.
The –ve. B.Ms. in spans a, d increase by an average of 1.33 times the original values. There is an increase in B.Ms. which appears in column rows (b, d) to be within an average value of 11.4 m.t. and with max. of 14.6 m.t. at the 2ⁿᵈ story.

3. change in N.F.

In columns, the only increase in N.F. relative to the original case is in rows (b, d). The new values are averagely about 1.45 times the average of the original values, with a max. of 1.51 and min. of 1.4 times the original values in the 1ˢᵗ, 6ᵗʰ stories respectively.
The rest of columns have a decrease in N.F. values.

There is an increase in N.Fs. which appears in beams over the damaged zone, the max. compressive N.F. appears at the 6th story beam with a value of 8.26 t and the max. tension appears in the 1st story beams with a value of 4.95 t.

5.1.2 *Damage of the exterior column in the ground story (M12)*

1. change in vertical deflection .

Points over the damaged column, (column row e) have the max. vertical deflection with an average value of 2.8 cm which is about 16 times the average of the original values.

In column row (d), the vertical deflection has increasing values with an average of 1.6 times the average of the original values at the column row.

Approximately there are negligible changes in the vertical deflection along the column row (a, b, c).

2. Change in B.M.

In sections over the damaged column, the B.M. inverses its direction relative to the
original frame to be +ve. B.M. with increased values to be 2.8, 2.1 times the original
values in the 1st, 6th stories respectively. The max. increase is at the 2nd Story with 3.3 times the original value.

The –ve. B.Ms. in beams right to the column line d, (span d) increase in value to be
4.8, 3.0 times the original values in the 1st, 6th stories respectively.

In spans (a, b& c), the-ve. B.Ms. in beams increase in value to be with averages of 1.7, 1.8& 2.4 times the original values, respectively. In the farthest end of beams, B.Ms. inverse their direction relative to the original frame in the 2nd Story in spans (a, c) and in the 2nd &3rd stories in span (b), and have values with an average of 15% of the original values.

There is an increase in B.Ms. which appears in column rows (b, c, d& e), to have values with an average of 8.3, 8.6, 9.7& 8.94 m.t. respectively and, with max. values of 12.3, 13.76, 15.9& 18.08 m.t. respectively. Note that all the max. values appear in the 2nd story except in column row (d) where it appears in the 1st story.

3. Change in N.F.

The only increase in N.F. appears in column row d. The increase has values of 1.63, 1.4 times the original values in the 1st, 6th stories respectively.

The rest of columns have a decrease in N.F. values.

There is an increase in N.Fs. which appears in beams on spans (b, c & d), the max. compressive N.F. appears in beams is in the 1st story with a value of 10.8 t, and the max. tension appears in beams in

the 6th story with a value of 8.78 t. Note that, the last two max. values are in beams of span (c).

5.2 *Effect of the existence of infill panels*

The 4-bay, 6-story frame of fig. 2. is investigated with infill panels provided in different places as shown in fig. 3. under only two cases of damage (central& exterior column in the ground story). Also, the original vertical deflection, B.M.& N.F. in the frame (M10), will be used as a reference lines.

5.2.1 *Damage of the central column in the ground story*

1. Change in vertical deflection

For cases of infill panels only over the damaged column zone (cases A1, A2) the average of vertical deflection over the damaged column decreases with an average of 70.2, 79.3% respectively of the average of vertical deflection in case of bare frame (M11).

The existence of infill panels all over the frame (case A3) has the most effective influence on reducing the vertical deflection over the damaged column, as it reduced with an average of 87.0% of the average of the vertical deflection in case of bare frame (M11).

2. Change in B.M.

In beam sections over the damaged column, B.Ms. are still in their original direction with a decrease in their values.

Case of infill panels around the damaged column zone except in the ground story, case (A1) has a good effect in reducing the –ve B.M. relative to the bare frame (M11), in beams over the damaged zone where it has an average of about 1.87 times the original value in the 1st story beam (spans b, c).

The existence of infill panels in the ground story around the damaged column and all over the frame (cases A2, A3) represents the most significant effect on reducing the –ve. B.M. in beams all over the frame relative to the bare frame (M11). Where the –ve. B.M. in the 1st story beam over the damaged column--which has the max. value--has values of 1.34, 1.26 times the original values respectively in each case and there is a slight increase in spans (a, e). Also has a good effect in reducing B.M. in columns relative to the case of bare frame (M11).

3. change in N.F.

For case of infill panels around the damaged column except in the ground story, case (A1) the N.Fs. have values which not differs approximately than the values of the bare frame (M11), moreover, there is a high increase in N.Fs. in beams specially beams in the 1st story over the damaged column with a max. tension of 45.16 tons.

The existence of infill panels in the ground story around the damaged column and all over the frame (cases A2, A3), has the most effective influence on reducing the N.Fs. in columns relative to the bare frame (M11) N.F. in the 1st story column (rows b, d) increases to have a value of 1.22 times the original values in case (A2) while, there is a decrease in these rows in case (A3). In column rows (a, e), there is a slight increase in N.F. in case (A2), while there is an increase in value of about 1.43 times the original value in case (A3). Also, there is an increase in N.Fs. in beams, specially in the 1st story with max. tension of 22.1, 12.79 t, in the two cases (A2, A3), respectively.

5.2.2 *Damaged of the exterior column in the ground story*

1. change in vertical deflection

The existence of infill panels in the ground story and all over span (d), cases (C1&C2) has a good effect on reducing the vertical deflection over the damaged column with an average of 66.12% of the value in case of bare frame (M12).

The existence of infill panels beside the damaged zone and every where in the frame, case (C3) has the most effective influence on reducing the vertical deflection either over the damaged column or elsewhere. The vertical deflection over the damaged column is reduced with an average of 85.9% of the average value in case of bare frame (M12).

2. Change in B.M.

In all the studied infill cases, B.Ms. in beam sections just over the damaged column still in its original direction with a decrease in values.

The existence of infill panels only over the damaged zone (cases C1, C2) has a small effect in reducing B.Ms. all over the frame beams. Note that in beams, B.Ms. in the farthest ends of the damaged column, specially in the 1st 2-story beams, inverse their direction relative to the original one with values more thane that in case of bare frame (M12). Also, we can notice that, there is an increase in B.Ms. in columns, specially column rows (b, c & d).

The existence of infill panels beside the damaged zone and every where in the frame, case (C3) has the most effective influence on reducing the B.Ms all over the frame members (beams and columns), compared with the case of infill around and over the damaged column (case C2).

3. Change in N.F.

For cases C1, C2 & C3, the N.Fs. in column row (d) decrease relative to the in case of bare frame (M12) and the N.F. in the ground story (column row d)

have values of 1.4, 1.41 times the original value in that column, respectively in each case.

There is an increase in N.F. in beams and we notice that tension appear in all cases (C1, C2 & C3) with max. values of 41.01, 36.09 & 39.47 t, respectively. Note that max. tension appears in the 1st story beams span(d) in all cases.

6 CONCLUSIONS AND RECOMMENDATIONS

The following conclusions can be drawn out from the present study.

In sections just over the damaged column- where the max. vertical deflection exists- the vertical deflection increase to be about 7 times the average of the original values in case of central column damage, and about 16 times the average of the original values in case of exterior column damage.

In beam sections just over the damaged column, B.Ms. inverse their direction relative to the original one with values.

The-ve. B.Ms. in beams over the damaged zone have values with about 4.1 times the average of the original values in these sections. The-ve. B.Ms. in the beams beside the damaged zone have values with about 1.75 times the average of the original values in these sections.

There is an increase in B.Ms. in columns, specially in the neighbor column rows to the damaged one in case of central column damages, and approximately in all column rows in cases of exterior column damages.

The two neighbor column rows to the damaged have the max. increase in N.Fs.

Damage of the central column affects a limited zone of the structure which is just over the damaged zone, while the damage of an exterior column has a wide effect which extends approximately all over the frame.

In case of central column damage, the existence of infill panels symmetrically over the damaged column has a significant effect in reducing vertical deflections as it reduces them with about 74.8% of the vertical deflection in the similar bare frame case. In case of exterior column damage, the existence of infill panels only over the damaged column reduces vertical deflections with about 66.12% of the vertical deflection in the similar bare frame case.

In sections just over the damaged column, B.Ms. keep their original direction with a high decrease in their values.

The existence of infill panels in the ground story around the column has a good effect on reducing B.Ms. all over the frame in case of central column damage. In case of exterior column damage, the infill panels have a limited effect on reducing-ve. B.Ms. on beams.

The existence of infill panels beside and everywhere in the frame have a limited effect on reducing both vertical deflection or B.Ms. in case of central column damage compared with the case of infill panels only over the damaged column, while, it has a very good effect in case of exterior column damage both in reducing vertical deflection and redistribution of B.Ms.

7 REFERENCES

McGuire, W., "Prevention of progressive Collapse", Proceeding of the Regional Conference on tall buildings, Institute of technology, Bankok, Thailand, Jun. 1974.

Leyendecker, E.V., and, Burnett, E.F.P., "The incidence of abnormal loading in residential buildings", Building Science, Series 89, National Bureau of standards, U.S. Government Printing office, Washington, D.C. Sept. 1976.

Bruce Ellingwood, and, E.V. Leyendecker, " Approaches for design against progressive collapse", ASCE, Vol. 104,No. ST3, March 1978.

Bruce Ellingwood, E.V. Leyndecker and James T.P. You, "Probability of failure from abnormal load", ASCE, Vol. 109, No.4, April 1983.

M. Holmes, " Steel frames with brickwork and concrete infilling", Proc. Inst. Civ. Eng., Vol. 19, P.P. 473-478, 1961.

B. Stafford Smith, C. Carter, "a method analysis for infilled frames", Proc. Inst. Civ. Eng., Vol. 44, P.P. 31-48,1969.

Shaker El-Behairy, Aly Abdel-Rahman, " Stability of multi-story infilled frames against failure of columns", Bulletin of faculty of Eng., Ain shams Univ., 1984.

Creative Systems in Structural and Construction Engineering, Singh (ed.) © 2001 Balkema, Rotterdam, ISBN 90 5809 161 9

Author index